ENGINEERING ELECTROMAGNETICS AND WAVES

Umran S. Inan

Koç University
Stanford University

Aziz S. Inan

University of Portland

Ryan K. Said

Vaisala Inc.

Global Edition

Second Edition

Boston   Columbus   Indianapolis   New York   San Francisco   Hoboken
Amsterdam   Cape Town   Dubai   London   Madrid   Milan   Munich   Paris   Montréal   Toronto
Delhi   Mexico City   São Paulo   Sydney   Hong Kong   Seoul   Singapore   Taipei   Tokyo

Vice President and Editorial Director:
  Marcia J. Horton
Acquisitions Editor: Julie Bai
Director of Product Management: Erin Gregg
Team Lead Product Management: Scott Disanno
Program Manager: Joanne Manning
Assistant Acquisitions Editor, Global Editions:
  Murchana Borthakur
Assistant Project Editor, Global Editions:
  Amrita Kar
Manager, Media Production, Global Editions:
  Vikram Kumar

Senior Manufacturing Control, Production,
  Global Editions: Trudy Kimber
VP of Marketing: Christy Lesko
Director of Field Marketing: Tim Galligan
Product Marketing Manager: Bram van Kempen
Manager, Rights Management: Rachel Youdelman
Procurement Manager: Mary Fischer
Cover Designer: Lumina Datamatics
Cover Photo: Eternalfeelings/Shutterstock

Pearson Education Limited
Edinburgh Gate
Harlow
Essex CM20 2JE
England

and Associated Companies throughout the world

Visit us on the World Wide Web at: www.pearsonglobaleditions.com

© Pearson Education Limited 2016

ISBN 10: 0273793233
ISBN 13: 9780273793236

British Library Cataloguing-in-Publication Data
A catalogue record for this book is available from the British Library

10 9 8 7 6 5 4 3 2 1

Typeset by Laserwords Pvt. Ltd.

Printed and bound by Courier Westford

To our families

Contents

Preface

This book provides engineering students with a solid grasp of electromagnetic fundamentals and electromagnetic waves by emphasizing physical understanding and practical applications. The topical organization of the text starts with an initial exposure to transmission lines and transients on high-speed distributed circuits, naturally bridging electrical circuits and electromagnetics.

Engineering Electromagnetics and Waves is designed for upper-division (3rd and 4th year) college and university engineering students, for those who wish to learn the subject through self-study, and for practicing engineers who need an up-to-date reference text. The student using this text is assumed to have completed typical lower-division courses in physics and mathematics as well as a first course on electrical engineering circuits.

Key Features

The key features of this textbook are

- Modern chapter organization, covering transmission lines before the development of fundamental laws
- Emphasis on physical understanding
- Detailed examples, selected application examples, and abundant illustrations
- Numerous end-of-chapter problems, emphasizing selected practical applications
- Historical notes on the great scientific pioneers
- Emphasis on clarity without sacrificing rigor and completeness
- Hundreds of footnotes providing physical insight, leads for further reading, and discussion of subtle and interesting concepts and applications

Modern Chapter Organization

We use a physical and intuitive approach so that this engineering textbook can be read by students with enthusiasm and interest. We provide continuity with electric circuit theory by first covering transmission lines—an appropriate step, in view of the importance of transmission line concepts, not only in microwave and millimeter-wave applications but also in high-speed digital electronics, microelectronics, integrated circuits, packaging, and interconnect applications. We then cover the fundamental subject material in a logical order, following the historical development of human understanding of electromagnetic phenomena. We base the fundamental laws on experimental observations and on physical grounds, including brief discussions of the precision of the fundamental experiments, so that the physical laws are easily understood and accepted.

Once the complete set of fundamental laws is established, we then discuss their most profound implications: the propagation of electromagnetic waves. We begin this discussion with the propagation of waves through empty space or unbounded simple media. We then discuss the reflection and refraction of electromagnetic waves from simple planar boundaries, followed by the guiding of electromagnetic waves within planar metallic or dielectric structures. We conclude with an introduction to field-matter interactions and electromagnetic wave propagation in metamaterials.

Emphasis on Physical Understanding

Future engineers and scientists need a clear understanding and a firm grasp of the basic principles so that they can interpret, formulate, and analyze the results of complex practical problems. Engineers and scientists nowadays do not and should not spend time obtaining numerical results by hand. Most of the number crunching and formula manipulations are left to computers and packaged application and design programs, so a solid grasp of fundamentals is now more essential than ever before. In this text we maintain a constant link with established as well as new and emerging applications (so that the reader's interest remains perked up), while at the same time emphasizing fundamental physical insight and solid understanding of basic principles. We strive to empower the reader with more than just a working knowledge of a dry set of vector relations and formulas stated axiomatically. We supplement rigorous analyses with extensive discussions of the experimental bases of the laws, of the microscopic versus macroscopic concepts of electromagnetic fields and their behavior in material media, and of the physical nature of the electromagnetic fields and waves, often from alternative points of view. Description of the electrical and magnetic properties of material media at a sufficiently simple, yet accurate manner at the introductory electromagnetics level has always been a challenge, yet a solid understanding of this subject is now more essential than ever, especially in view of many applications that exploit these properties of materials. To this end we attempt to distill the essentials of physically-based treatments available in physics texts, providing quantitative physical insight into microscopic behavior of materials and the representation of this behavior in terms of macroscopic parameters. Difficult three-dimensional vector differential and integral concepts are discussed when they are encountered—again, with the emphasis being on physical insight.

Detailed Examples and Abundant Illustrations

We present the material in a clear and simple yet precise and accurate manner, with interesting examples illustrating each new concept. Many examples emphasize selected applications of electromagnetics. Over 190 illustrative examples are detailed over eleven chapters, with five of the chapters having at least 20 examples each. Each example is presented with an abbreviated topical title, a clear problem statement, and a detailed solution. In recognition of the importance of visualization in the reader's understanding, especially in view of the three-dimensional nature of electromagnetic fields, over 500 diagrams, graphs, and illustrations appear throughout the book.

Numerous End-of-Chapter Problems

Each chapter is concluded with a variety of homework problems to allow the students to test their understanding of the material covered in the chapter, with a total of over 400 exercise problems spread over eleven chapters. The topical content of each problem is clearly identified in an abbreviated title (e.g., "Digital IC interconnects" or "Inductance of a toroid"). Many problems explore interesting applications, and most chapters include several practical "real-life" problems to motivate students.

Historical Notes

The history of the development of electromagnetics is laden with outstanding examples of pioneering scientists and development of scientific thought. Throughout our text, we maintain a constant link with the pioneering giants and their work, to bring about a better appreciation of the complex physical concepts as well as to keep the reader interested.

Emphasis on Clarity without Sacrificing Rigor and Completeness

This textbook presents the material at a simple enough level to be readable by undergraduate students, but it is also rigorous in providing references and footnotes for in-depth analyses of selected concepts and applications. We provide the students with a taste of rigor and completeness at the level of classical reference texts—combined with a level of physical insight that was so well exemplified in some very old texts—while still maintaining the necessary level of organization and presentation clarity required for a modern textbook. We also provide not just a superficial but a rigorous and in-depth exposure to a diverse range of applications of electromagnetics, in the body of the text, in examples, and in end-of-chapter problems.

Hundreds of Footnotes

In view of its fundamental physical nature and its broad generality, electromagnetics lends itself particularly well to alternative ways of thinking about physical and xengineering problems and also is particularly rich in terms of available scientific literature and many outstanding textbooks. Almost every new concept encountered can be thought of in different ways, and the interested reader can explore its implications further. We encourage such scholarly pursuit of enhanced knowledge and understanding by providing many footnotes in each chapter that provide further comments, qualifications of statements made in the text, and references for in-depth analyses of selected concepts and applications. Over 550 footnotes are spread over eleven chapters. These footnotes do not interrupt the flow of ideas and the development of the main topics, but they provide an unusual degree of completeness for a textbook at this level, with interesting and sometimes thought-provoking content to make the subject more appealing and satisfying.

Electromagnetics and Waves in Engineering

The particular organization of this textbook, as well as its experimentally and physically based philosophy, are motivated by our view of the current status of electromagnetics in engineering curricula. Understanding electromagnetics and appreciating its applications require a generally higher level of abstraction than most other topics encountered by electrical engineering students. Beginning electrical engineers learn to deal with voltages and currents, which appear across or flow through circuit elements or paths. The relationships between these voltages and currents are determined by the characteristics of the circuit elements and by Kirchhoff's current and voltage laws. Voltages and currents in lumped electrical circuits are scalar quantities that vary only as a function of time, and are readily measurable, and the students can relate to them via their previous experiences. The relationships between these quantities (i.e., Kirchhoff's laws) are relatively simple algebraic or ordinary differential equations. On the contrary, electric and magnetic fields are *three-dimensional* and *vector* quantities that in general *vary in both space and time* and are related to one another through relatively complicated vector *partial differential* or vector *integral equations*. Even if the physical nature of electric and magnetic fields were understood, visualization of the fields and their effects on one another and on matter requires a generally high level of abstract thinking.

Most students are exposed to electromagnetics first at the freshman physics level, where electricity and magnetism are discussed in terms of their experimental bases by citing physical laws (e.g., Coulomb's law) and applying them to relatively simple and symmetrical configurations where the field quantities behave as scalars, and the governing equations are reduced to either algebraic equations of first-order integral or differential relationships. Freshman physics provides the students with their first experiences with fields and waves as well as some of their measurable manifestations, such as electric and magnetic forces, electromagnetic induction (Faraday's law), and refraction of light by prisms.

The first course in electromagnetics, which most students take after having had vector calculus, aims at the development and understanding of Maxwell's equations, requiring the utilization of the full three-dimensional vector form of the fields and their relationships. It is this very step that makes the subject of electromagnetics appear insurmountable to many students and turns off their interest, especially when coupled with a lack of presentation and discussion of important applications and the physical (and experimental) bases of the fundamental laws of physics. Many authors and teachers have attempted to overcome this difficulty by a variety of topical organizations, ranging from those that start with Maxwell's equations as axioms to those that first develop them from their experimental basis.

Since electromagnetics is a mature basic science, and the topics covered in introductory texts are well established, the various texts primarily differ in their organization as well as range and depth of coverage. Teaching electromagnetics was the subject of a special issue of *IEEE Transactions on Education* [vol. 33, February, 1990]. Many of the challenges and opportunities that lie ahead in this connection were summarized well in an invited article by J. R. Whinnery.[1] Challenges include (1) the need to return to fundamentals (rather than relying on derived concepts), especially in view of the many emerging new applications that exploit unusual properties of materials and that rely on unconventional device concepts, [2] submillimeter transmission lines,[3] and optoelectronic waveguides,[4] and (2) the need to maintain student interest in spite of the decreasing popularity of the subject of electromagnetics and its reputation as a difficult and abstract subject.[5] Opportunities are abundant, especially as engineers working in the electronics industry discover that as devices get smaller and faster, circuit theory is insufficient in describing system performance or facilitating design. Transmission line concepts are not only important in microwave and millimeter-wave applications; due to modern GHz clock rates and nano-scale fabrication technology, they are also necessary in high-speed digital electronics, microelectronics, integrated circuits, interconnects, [6] and packaging applications.[7] In addition, issues of electromagnetic interference (EMI) and electromagnetic compatibility (EMC) limit the performance of system-, board-, and chip-level designs, and electrostatic discharge phenomena have significant impacts on the design and performance of integrated circuits.[8] The need for a basic understanding of electromagnetic waves and their guided propagation is underscored by the explosive expansion of the use of optical fibers, which enables extremely high data rates, ranging to 100 Gbits/s[9]. Fundamental bandwidth and power constraints in traditional copper-based transmission lines is also driving the development of optical interconnects with per-channel bandwidths

[1]J. R. Whinnery, The teaching of electromagnetics, *IEEE Trans. on Education,* 33(1), pp. 3–7, February 1990.

[2]D. Goldhaber-Gordon, M. S. Montemerlo, J. C. Love, G. J. Opiteck, and J. C. Ellenbogen, Overview of nanoelectronic devices, *Proc. IEEE,* 85(4), pp. 521–540, April 1997.

[3]L. P. B. Katehi, Novel transmission lines for the submillimeter region, *Proc. IEEE,* 80(11), pp. 1771–1787, November 1992.

[4]R. A. Soref, Silicon-based optoelectronics, *Proc. IEEE,* 81(12), December, 1993.

[5]M. N. O. Sadiku, Problems faced by undergraduates studying electromagnetics, *IEEE Trans. Education,* 29(1), pp. 31–32, February, 1986.

[6]S. H. Hall and L. H. Heck, *Advanced signal integrity for high-speed digital designs,* John Wiley & Sons, 2011.

[7]H. B. Bakoglu, *Circuits, Interconnections, and Packaging for VLSI,* Addison Wesley, 1990.

[8]J. E. Vinson and J. J. Liou, Electrostatic discharge in semiconductor devices: an overview, *Proc. IEEE,* 86(2), pp. 399–418, February 1998.

[9]N. Cvijetic, D. Qian, and J. Hu, 100 Gb/s optical access based on optical orthogonal frequency-division multiplexing, *Communications Magazine, IEEE* 48(7), pp. 70-77, 2010.

in excess of 10 Gbits/s for high-performance computing applications.[10] Other important applications that require better understanding of electromagnetic fields are emerging in biology[11] and medicine.[12]

In organizing the material for our text, we benefited greatly from a review of the electromagnetic curriculum at Stanford University that one of us conducted during the spring quarter of 1990. A detailed analysis was made of both undergraduate and graduate offerings, both at Stanford and selected other schools. Inquiries were also made with selected industry, especially in the aerospace sector. Based on the responses we received from many of our colleagues, and based on our experience with the teaching of the two-quarter sequence at Stanford, it was decided that an emphasis on fundamentals and physical insight and a traditional order of topics would be most appropriate. It was also determined that transmission line theory and applications can naturally be studied before fields and waves, so as to provide a smooth transition from the previous circuits and systems experiences of the typical electrical engineering students and also to emphasize the importance of these concepts in high-speed electronics and computer applications.

New to this Edition

This book represents an effort to merge the most important concepts from our two previous textbooks: Engineering Electromagnetics[13] and Electromagnetic Waves[14]. Some of the advanced topics from these two books, such as using transmission lines as resonant circuits and cylindrical waveguides, were moved to a web addendum (see Online Addendum section below). By moving some of these sections to the web, we are better able to focus the reader on the core concepts central to transmission lines, electromagnetics, and electromagnetic waves. We also introduce two new sections on increasingly relevant modern topics: Microelectromechanical Systems (MEMS) and Metamaterials. While these are relatively advanced topics, some of the fundamental physics underpinning these two areas of active research and development connect directly to the core ideas presented in this book, and so they give concrete examples of how a solid foundation in electromagnetics and waves is still very relevant to modern technology.

The list below summarizes the changes and additions we introduced in the second edition:

- We merged topics from the two first edition textbooks into a single volume covering both engineering electromagnetics and electromagnetic waves.
- We added two new sections: Microelectromechanical Systems (Section 4.14) and Metamaterials (Section 11.3).

[10]L. Chrostowski and K. Iniewski (Eds.), *High-speed Photonics Interconnects (Vol. 13)*, CRC Press., 2013.

[11]R. H., Funk, T. Monsees, and N. Ozkucur, Electromagnetic effects—From cell biology to medicine. *Progress in histochemistry and cytochemistry*, 43(4), 177-264, 2009.

[12]E. J. Bond, et al., Microwave imaging via space-time beamforming for early detection of breast cancer, *Antennas and Propagation, IEEE Transactions on*, 51(8), pp. 1690-1705, 2003.

[13]U.S. Inan and A. S. Inan, *Engineering Electromagnetics*, Addison Wesley Longman, 1999.

[14]U.S. Inan and A. S. Inan, *Electromagnetic Waves*, Prentice Hall, 2000.

- We added an appendix with a proof of the uniqueness theorem for Poisson's equation (Appendix B).
- We moved several advanced topics in the first two editions to standalone addendum chapters that are available on the web.
- We introduced numerous edits throughout the text to add clarity and improve the presentation of some of the more challenging topics.
- We updated, modified, or added over 100 new end-of-chapter problems.
- We corrected errata that have been reported since the publication of the first two editions.

Recommended Course Content

The wider breadth of topics covered by this single volume allows the instructor to tailor the content based on the duration of the course. Tables 1 and 2 list the suggested course content for a single course and two course sequence, respectively. Each table details suggested content based on the quarter system (32 contact hours per course per quarter) and semester system (42 contact hours per course per semester). The sections marked under "Cover" are recommended for complete coverage, including illustrative examples, whereas those marked "Skim" are recommended to be covered lightly, although the material provided is more complete in case individual students want to have more in-depth coverage. The one-course sequences provide the students with (1) a working knowledge of transmission lines, (2) a solid, physically based background and a firm understanding of Maxwell's equations and their experimental bases, and (3) an introduction to electromagnetic waves. In addition to a more in-depth coverage of the transmission lines chapters and the development of Maxwell's equations, the two-course sequences give the student a working knowledge of electromagnetic wave phenomena and their applications.

TABLE 1   SUGGESTED COURSE CONTENT: SINGLE QUARTER OR SEMESTER

Chapter	Quarter Course (32 Hours)		Semester Course (42 Hours)	
	Cover	Skim	Cover	Skim
1	All		All	
2	2.1–2.4	2.7	2.1 – 2.5	2.7
3	3.1–3.3		3.1–3.6	
4	4.1–4.9	4.10	4.1–4.10	4.12
5	5.1–5.5		5.1–5.5	5.7
6	6.1–6.7	6.8	6.1–6.8	6.10
7	7.1, 7.2, 7.4	7.3, 7.5	7.1, 7.2, 7.4	7.3, 7.5
8	8.1, 8.2		8.1–8.4	
9				
10				
11				

TABLE 2   SUGGESTED COURSE CONTENT: TWO QUARTERS OR SEMESTERS

| Chapter | Two Quarter Course (64 Hours Total) | | Two Semester Courses (84 Hours Total) | |
	Cover	Skim	Cover	Skim
1	All		All	
2	All		All	
3	3.1–3.6		All	
4	4.1–4.12		4.1–4.13	4.14
5	5.1–5.7	5.8	All	
6	6.1–6.9		All	
	Quarter Break		Semester Break	
7	All		All	
8	8.1–8.6		All	
9	9.1–9.3, 9.5–9.7		All	
10	All		All	
11	11.1, 11.2		11.1, 11.2,	11.3

Instructor's Manual

We firmly believe that practice is the key to learning and that homework and exams are all instruments of teaching—although they may not be regarded as such by the students at the time. In our own courses, we take pride in providing the students with detailed solutions of homework and exam problems, rather than cryptic and abbreviated answers. To aid the instructors who choose to use this text, we have thus taken it upon ourselves to prepare a well-laid-out solutions manual, describing the solution of *every* end-of-chapter problem, in the same step-by-step detailed manner as our illustrative examples within the chapters. This instructor's manual is available to instructors upon request at www.pearsonglobaleditions.com. Supplemental information about the book and errata will be available at www.pearsonglobaleditions.com/Inan.

As authors of this book, we are looking forward to interacting with its users, both students and instructors, to collect and respond to their comments, questions, and corrections. We can most easily be reached by electronic mail at inan@stanford.edu (http://vlf.stanford.edu/), uinan@ku.edu.tr (http://www.ku.edu.tr/en/about-ku/president), ainan@up.edu (faculty.up.edu/ainan/), and ryan.said@vaisala.com.

Online Addendum

The topics in this book were carefully selected to give the student a solid foundation of transmission lines, Maxwell's equations, and the propagation and guiding of electromagnetic waves. The goal of this book is to develop an intuitive understanding of these fundamental concepts, so that the student is well equipped to apply these principles to

TABLE 3   ADDENDUM: ADVANCED TOPICS (AVAILABLE ONLINE)

Addendum A	Transmission Lines: Advanced Topics
Addendum B	Miscellaneous Wave Topics
Addendum C	Cylindrical Waveguides
Addendum D	Cavity Resonators
Addendum E	Field-Matter Interactions: Advanced Topics
Addendum F	Electromagnetic Radiation and Elementary Antennas

new challenges and to expand his/her study to more advanced topics. Due to the breadth of discussion given to each topic, in order to maintain a manageable page count, a few of the more advanced topics from the first edition are moved to an online addendum. This addendum, which is available for free at www.pearsonglobaleditions.com/Inan, contains the chapters listed in Table 3.

The Addendum chapters cover advanced topics that expand on the core material of this text. Addendum A includes selected advanced transmission line topics, including transients on lossy transmission lines and the use of transmission lines as resonant circuits. The application examples of Addendum A present further reading that can be covered after Chapters 2 and 3. The remaining addendum chapters cover advanced material related to the subject matter from the final four chapters of the book. Addendum B expands on two advanced topics introduced in Chapters 8 and 9: examples of nonplanar waves and oblique-incidence reflection from a good conductor. Addendum C extends the treatment of planar waveguides from Chapter 10 to those that are bounded in two dimensions. Addendum D introduces cavity resonators, whose treatment follows naturally from the two-dimensional waveguides covered in Addendum C. Addendum E extends the field-matter interaction topics encountered in Sections 11.1 and 11.2. Addendum F introduces electromagnetic radiation and elementary antennas, which connect the electromagnetic fields studied in Chapters 8–11 to their sources.

Acknowledgments

We gratefully acknowledge those who have made significant contributions to the successful completion of this text. We thank Professor J. W. Goodman of Stanford, for his generous support of textbook writing by faculty throughout his term as department chair. We thank Professor Gordon Kino and Dr. Timothy F. Bell of Stanford, for course-testing a preliminary version of the manuscript for our first edition. We thank numerous colleagues and former students who have identified errors and suggested clarifications. We thank Mrs. Jun-Hua Wang for typing parts of the first edition manuscript and drawing some of the illustrations. We thank Dr. Robert Marshall of Stanford for preparing the two first edition manuscripts for use in a single unified book, and Güneş Aydindoğan for typesetting large portions of the solution manual. We owe special thanks to our reviewers on the first edition for their valuable comments and suggestions, including J. Bredow of University of Texas—Arlington; S. Castillo of New Mexico State University; R. J. Coleman of University of North Carolina—Charlotte; A. Dienes of University of California—Davis; J. Dunn of University of Colorado; D. S. Elliott of Purdue University; R. A. Kinney of Louisiana State University; L. Rosenthal of Fairleigh Dickinson University; E. Schamiloglu of University of New Mexico; T. Shumpert of Auburn University; D. Stephenson of Iowa State University; E. Thomson of University of Florida; J. Volakis of University of Michigan; and A. Weisshaar of Oregon State University.

We greatly appreciate the efforts of our managing editor at Pearson, Scott Disanno, and his staff, including Michelle Bayman, William Opaluch, Joanne Manning, and Julie Bai. We also thank Haseen Khan and her staff at Laserwords for their dedication and attention to detail in the layout and production of the book.

We dedicate this book to our parents, Mustafa and Hayriye Inan, for their dedication to our education; to our wives, Elif and Belgin, for their persistent support and understanding as this project expanded well beyond our initial expectations and consumed most of our available time for too many years; and to our children, Ayse, Ali, Baris, and Cem, and grandchildren, Ayla and Nisa, for the joy they bring to our lives.

Umran S. Inan
Aziz S. Inan

I would like to thank Professors Umran Inan and Aziz Inan for giving me the opportunity to contribute to this manuscript. When I was a new graduate student at Stanford University, I had the pleasure of learning electromagnetics from the two first edition textbooks that form the basis for this second edition. It is an honor to help create a unified book that spans the content of these courses that I so enjoyed taking over a decade ago. I dedicate this book to my parents, James and Connie, for cultivating my passion for science, and to my sister, Amirah, for her unwavering moral support.

Ryan K. Said

Pearson would like to thank and acknowledge Nikhil Marriwala (Kurukshetra University) for his contribution to the Global Edition, and Duli Chand Meena (Delhi Technological University), Yaduvir Singh (Harcourt Butler Technological Institute) and Sakthivel Murugan (SSN College of Engineering) for reviewing the Global Edition.

Biography

Umran S. Inan is President of Koç University in Istanbul, Turkey where he is Professor of Electrical Engineering and Physics. He is also Professor (Emeritus) of Electrical Engineering at Stanford University where he supervised 60 PhD dissertations. Inan is a Fellow of IEEE, American Geophysical Union and American Physical Society. He is the recipient of Appleton Prize of URSI and Royal Society, Allan Cox Medal of Stanford, Special Science Award of the Scientific & Technological Research Council of Turkey, Stanford Tau Beta Pi Award for Excellence in Teaching, and numerous NASA and ESA Group Achievement Awards.

Aziz S. Inan is Professor of Electrical Engineering at the University of Portland, where he has also served as Department Chairman. A winner of the University's faculty teaching award, he conducts research in electromagnetic wave propagation in conducting and inhomogeneous media. He is a member of Tau Beta Pi and IEEE.

Ryan K. Said received his Ph.D. in Electrical Engineering from Stanford University. As a research scientist and systems engineer at Vaisala Inc., his work focuses on the development of lightning detection technologies and their applications. He is a recipient of the HMEI Award for Young Engineers and a member of the American Geophysical Union.

1

Introduction

This book is an introduction to the fundamental principles and applications of *electromagnetics.* The subject of electromagnetics encompasses *electricity, magnetism,* and *electrodynamics,* including *all* electric and magnetic phenomena and their practical applications. A branch of electromagnetics, dealing with electric charges at rest (static electricity) named *electrostatics,* provides a framework within which we can understand the simple fact that a piece of amber, when rubbed, attracts itself to other small objects.[1] Another branch dealing with static magnetism, namely *magnetostatics,* is based on the facts that some mineral ores (e.g., lodestone) attract iron[2] and that current-carrying wires produce magnetic fields.[3] The branch of electromagnetics known as *electrodynamics* deals with the time variations of electricity and magnetism and is based on the fact that magnetic fields that change with time produce electric fields.[4]

Electromagnetic phenomena are governed by a compact set of principles known as Maxwell's equations,[5] the most fundamental consequence of which is that electromagnetic energy can propagate, or travel from one point to another, as *waves.* The propagation of electromagnetic waves results in the phenomenon of *delayed action at a distance;* in other words, electromagnetic fields can exert forces, and hence can do work, at distances far away from the places where they are generated and at later times. Electromagnetic radiation is thus a means of transporting energy and momentum from one set of electric charges and currents (at the source end) to another (those at the receiving end).

---

[1]First discovered by the Greek mathematician, astronomer, and philosopher Thales of Miletus [640–548 B.C.].

[2]First noted by the Roman poet and philosopher Lucretius [99?–55? B.C.], in his philosophical and scientific poem titled *De rerum natura (On the Nature of Things).*

[3]First noted by Danish physicist H. C. Oersted in 1819.

[4]First noted by British scientist M. Faraday in 1831.

[5]J. C. Maxwell, *A Treatise in Electricity and Magnetism,* Clarendon Press, Oxford, 1892, Vol. 2, pp. 247–262.

Since whatever can carry energy can also convey information, *electromagnetic waves* thus provide the means of transmitting energy and information at a distance. The latter half of this book provides an introduction to electromagnetic waves, their propagation in empty space or material media, their reflection from boundaries, and their guiding within planar boundaries.

The concept of waves is one of the great unifying ideas of physics and engineering.[6] Our physical environment is full of waves of all kinds: seismic waves, waves on oceans and ponds, sound waves, heat waves, and even traffic waves. The idea of delayed action as manifested in wave phenomena is familiar to us when we hear a sound and its echo or when we create a disturbance[7] in a pool of water and observe that waves reach the edge of the pool after a noticeable time. We also appreciate that it might take minutes or hours for heat to penetrate into objects; that the thunderclap is delayed with respect to the lightning flash by many seconds; and that when we are lined up in front of a traffic light, it often takes a long time for us to be able to move after the light turns green. Light, or electromagnetic waves, travel so fast that their delayed action is not perceptible to our senses in our everyday experiences. On the other hand, in astronomy and astrophysics we deal with vast distances; light waves from a supernova explosion may arrive at earth millions of years after the brightness that created them has been extinguished.

Example 1.1: Time delay between Mars and Earth. The distance between Earth and Mars varies from 54.6×10^6 to 401×10^6 km. How long does it take for a message sent from Earth to reach NASA's Curiosity Rover on Mars?

Solution: The travel time t is determined by the distance l and speed of propagation v:

$$t = \frac{l}{v}$$

All electromagnetic waves, including radio waves used in broadcast telecommunications, travel in free space at the speed of $c \simeq 3 \times 10^8$ m-s^{-1}. Using $v = c$, the travel time from Earth to Mars ranges from

$$t_{\min} = \frac{54.6 \times 10^6 \text{ km}}{3 \times 10^5 \text{ km-s}^{-1}} \cdot \frac{1 \text{ minute}}{60 \text{ s}} \simeq 3.03 \text{ minutes}$$

to

$$t_{\max} = \frac{401 \times 10^6 \text{ km}}{3 \times 10^5 \text{ km-s}^{-1}} \cdot \frac{1 \text{ minute}}{60 \text{ s}} \simeq 22.3 \text{ minutes}$$

---

[6]For an excellent qualitative discussion, see J. R. Pierce, *Almost All About Waves,* MIT Press, Cambridge, Massachusetts, 1974. For more extensive treatment of waves of all kinds, see K. U. Ingard, *Fundamentals of Waves and Oscillations,* Cambridge University Press, Cambridge, England, 1990.

[7]On the scale of a pond, we can simply think of dropping a stone; on a larger scale, earthquakes in oceans produce giant *tsunami* waves. A 9-meter-high tsunami produced by the 1964 Alaskan earthquake hit the Hawaiian islands (at a distance of 2000 km) about 5 hours later, causing more than 25 million dollars of damage.

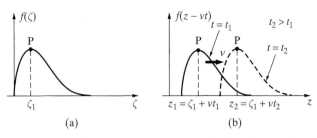

Figure 1.1 Example of a wave. (a) An arbitrary function $f(\zeta)$. (b) The function $f(z - vt)$, where v is a positive constant, plotted versus z at $t = t_1$ and $t = t_2$. The wave nature is evident as the pattern in space at time t_1 is shifted to other values of z at a later time t_2.

At a qualitative level, we recognize a wave as some pattern in space that appears to move in time. Wave motion does not necessarily involve repetitive undulations of a physical quantity (e.g., the height of the water surface for water waves in a lake). If a disturbance that occurs at a particular point in space at a particular time is related in a definite manner to what occurs at distant points in later times, then there is said to be wave motion. To express this mathematically, let z be distance, t be time, and v be a fixed positive parameter. Consider any arbitrary function $f(\zeta)$ of the argument $\zeta = (z - vt)$. Figure 1.1a shows a sketch of $f(\zeta)$, identifying a point P on the curve that corresponds to $f(\zeta_1)$. Note that this peak value of the function (i.e., point P) is obtained whenever its argument equals ζ_1. Shown in Figure 1.1b is a sketch of $f(z - vt_1)$ with respect to z at a fixed time t_1, where the point P is at $z_1 = \zeta_1 + vt_1$. A plot of the function at $t = t_2 > t_1$ (i.e., $f(z - vt_2)$) is also shown in Figure 1.1b, where we see that the point P has now moved to the right, to the new location $z_2 = \zeta_1 + vt_2$. It is clear that the entire curve $f(\zeta)$, which comprises the function $f(z - vt)$, moves in the z direction as time elapses. The velocity of this motion can be determined by observing a fixed point on the curve—for example, point P. Since this point is defined by the argument of the function being equal to ζ_1, we can set $(z - vt) = \zeta_1$, which upon differentiation yields $dz/dt = v$, since ζ_1 is a constant. It thus appears that the speed with which point P moves to the right is v, which is identified as the velocity of the wave motion. Note that the function $f(\cdot)$ could represent any physically observable entity; it may be a scalar,[8] such as voltage, or it may be a vector, such as the velocity of an object in motion. If $f(\cdot)$ is a vector, each of its components must be a function of $(z - vt)$ for it to be a propagating wave. Quantities varying as functions of $(z - vt)$ constitute natural solutions of the fundamental equations of electromagnetics and distributed electrical circuits. Chapters 2, 3, and 8–11 of this textbook are devoted to the study of voltage, current, and electromagnetic waves that vary in space and time as functions of $(z - vt)$.

---

[8]A scalar is a quantity that is completely specified by its value, such as the number of coins in your pocket, the number of people or the density of air in a room, pressure, or temperature. Other physical quantities have direction; for example, velocity, momentum, force, or displacement. Specification of a vector quantity requires both a magnitude and direction. A brief review of basic principles of vector analysis is provided in Appendix A.

Most waves travel through substances, whether they be earth, water, air, steel, or quartz, without actually carrying the substance bodily with them.[9] Like moving objects, traveling waves carry energy, albeit by different amounts depending on the nature of the waves and the medium they propagate in. Electromagnetic waves have the special property that they can also propagate in vacuum, without any matter present. However, the propagation of electromagnetic waves is nevertheless affected by the presence of matter, and this property often allows us to confine or guide waves and in doing so utilize them more efficiently. Electromagnetic engineering problems generally involve the design and use of materials that can generate, transmit, guide, store, distribute, scatter, absorb, and detect electromagnetic waves and energy.

The 20th century has witnessed rapid advances in electrical engineering, which have largely come about by our ability to predict the performance of sophisticated electrical circuits accurately. Central to this tremendous progress is our ability to utilize the simple but powerful tool called electric *circuit theory*. Classical circuit theory considers a voltage or current source applied to an electrical circuit consisting of series and/or parallel connection of simple *lumped* (see Section 1.1) circuit elements, such as resistances, capacitances, inductances and dependent sources, which may be idealized models of more complex physical devices. The behavior of circuits is described by ordinary differential equations that are derived on the basis of Kirchhoff's voltage and current laws. Circuit theory is a simplified approximation to the more exact electromagnetic theory.[10] The classical theory of electricity and magnetism relies on a set of physical laws known as Maxwell's equations, which are based on experimental facts and which govern all electromagnetic phenomena. Electromagnetic theory is inherently more complicated than circuit theory, primarily because of the larger number of variables involved. In general electromagnetic problems, most of the physical quantities that we deal with are vectors, whose values may depend on all three coordinates of space (i.e., x, y, and z in rectangular coordinates) and time (t). In classical circuit theory, on the other hand, voltages and currents are scalar quantities and are typically functions of only one variable, namely time. The theory of *distributed* circuits (see Section 1.1), or transmission lines, represents an intermediate level of complexity where, in many cases, we can continue to deal with scalar quantities, such as voltages and currents, that are now functions of two variables, namely a single spatial dimension and time. In this regime, we can continue to benefit from the relative simplicity of circuit theory, while treating problems for which the lumped circuit theory is not applicable.

In this text, and in view of the preceding discussion, we choose to study distributed circuits or transmission lines using a natural extension of circuit theory before we formally introduce the physical laws of electricity and magnetism. This approach presents the general fundamental concepts of waves and oscillations at the outset, which

---

[9]Leonardo da Vinci [1452–1519] wrote of waves, "The impetus is much quicker than the water, for it often happens that the wave flees the place of its creation, while the water does not; like the waves made in a field of grain by the wind, where we see the waves running across the field while the grains remain in place" [J. R. Pierce, *Almost All About Waves,* MIT Press, Cambridge, Massachusetts, 1974].

[10]Kirchhoff's voltage and current laws, which provide the basis for classical circuit theory, can be derived from the more general electromagnetics equations; see Chapter 4 of S. Ramo, J. R. Whinnery, and T. Van Duzer, *Fields and Waves in Communication Electronics,* 3rd ed., John Wiley & Sons, Inc., New York, 1994.

are expanded upon later when we study propagation of electromagnetic waves. In this way, the reader is provided an unhindered initial exposure to properties of waves such as frequency, phase velocity, wavelength, and characteristic impedance; to energy relations in oscillating systems and waves; and to concepts such as reflection and bandwidth, as well as to fundamental mathematical techniques necessary to describe waves and oscillations. All of these concepts subsequently extend to more complicated problems and applications where a full electromagnetic treatment becomes necessary. An initial exposure to transmission line analysis also enables us to address a wide range of increasingly important engineering applications that require the use of wave techniques, but for which a full vector electromagnetic analysis may not be necessary.[11] In our coverage of transmission lines or distributed circuits, we assume that the reader is familiar with the elementary physics of electricity and magnetism (at the level of freshman physics) and with electrical circuits at the level of understanding Kirchhoff's voltage and current laws and terminal behavior (i.e., voltage–current relationships) of circuit elements such as inductors, capacitors, and resistors. A more complete discussion of the concepts of inductance, capacitance, and resistance is provided in later chapters using the concepts and principles of electromagnetic fields as we introduce the fundamental laws of electromagnetics.

1.1 LUMPED VERSUS DISTRIBUTED ELECTRICAL CIRCUITS

A typical electrical engineering student is familiar with circuits, which are described as *lumped, linear,* and *time-invariant* systems and which can be modeled by *ordinary, linear,* and *constant-coefficient* differential equations. The concepts of linearity and time invariance refer to the relationships between the inputs and outputs of the system. The concept of a lumped circuit refers to the assumption that the entire circuit (or system) is at a single point (or in one "lump"), so that the dimensions of the system components (e.g., individual resistors or capacitors) are negligible. In other words, current and voltage do not vary with space across or between circuit elements, so that when a voltage or current is applied at one point in the circuit, currents and voltages of all other points in the circuit react instantaneously. Lumped circuits consist of interconnections of lumped elements. A circuit element is said to be lumped if the instantaneous current entering one of its terminals is equal to the instantaneous current throughout the element and leaving the other terminal. Typical lumped circuit elements are resistors, capacitors, and inductors. In a lumped circuit, the individual lumped circuit elements are connected to each other and to sources and loads within or outside the circuit by conducting paths of negligible electrical length.[12] Figure 1.2a illustrates a lumped electrical circuit to which an input voltage of $V_{in} = V_0$ is applied at $t = 0$. Since the entire circuit is considered in one lump, the effect of the input excitation is instantaneously felt at all points in the

---

[11]Examples are on-chip and chip-to-chip interconnections in digital integrated circuits and many other computer engineering applications. See A. Deutsch, et al., When are transmission-line effects important for on-chip interconnections, *IEEE Trans. Mic. Th. MTT,* 45(10), pp. 1836–1846, October 1997.

[12]The electrical length equals the physical length of the circuit element divided by the wavelength. As we discuss in Section 1.1.3, this ratio determines whether a given circuit element should be treated as a lumped or distributed element.

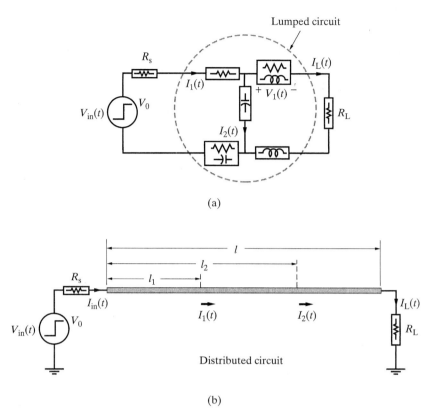

(a)

(b)

Figure 1.2 Lumped versus distributed electrical circuits. (a) When a step voltage $V_{in}(t)$ is applied to a lumped circuit, we assume that all currents and voltages start to change at $t = 0$, implicitly assuming that it takes zero travel time for the effect of the input to move from any point to any other point in the circuit. (b) In a distributed circuit, the nonzero travel time of the signal from one point to another cannot be neglected. For example, when a step voltage $V_{in}(t)$ is applied at one end of the circuit at $t = 0$, the load current at the other end does not start to change until $t = l/v$.

circuit, and all currents and voltages (such as I_1, I_2, V_1, and the load current I_L) either attain new values or respond by starting to change at $t = 0$, in accordance with the natural response of the circuit to a step excitation as determined by the solution of its corresponding differential equation. Many powerful techniques of analysis, design, and computer-aided optimization of lumped circuits are available and widely used.

The behavior of lumped circuits is analogous to rigid-body dynamics. In mechanics, a rigid body is postulated to have a definite shape and mass, and it is assumed that the distance between any two points on the body does not change, so that its shape is not deformed by applied forces. Thus, an external force applied to a rigid body is assumed to be felt by all parts of the body simultaneously, without accounting for the finite time it would take for the effect of the force to travel elastically from one end of the body to another.

With the "lumped" assumption, one does not have to consider the travel time of the signal from one point to another. In reality, however, disturbances or signals caused by any applied energy travel from one point to another in a nonzero time. For electromagnetic signals, this travel time is determined by the speed of light,[13] $c \simeq 3 \times 10^8$ m-s^{-1} = 30 cm-(ns)$^{-1}$. In practical transmission systems, the speed of signal propagation is determined by the electrical and magnetic properties of the surrounding media and the geometrical configuration of the conductors and may in general be different from c, but it is nevertheless of the same order of magnitude as c. Circuits for which this nonzero travel time cannot be neglected are known as *distributed* circuits. An example of a distributed circuit is a long wire, as shown in Figure 1.2b. When an input voltage V_0 is applied at the input terminals of such a distributed circuit (i.e., between the input end of the wire and the electrical ground) at $t = 0$, the voltages and currents at all points of the wire cannot respond simultaneously to the applied excitation because the energy corresponding to the applied voltage propagates down the wire with a finite velocity v. Thus, while the input current $I_{in}(t)$ may change from zero to I_0 at $t = 0$, the current $I_1(t)$ does not flow until after $t = l_1/v$, $I_2(t)$ does not flow until after $t = l_2/v$, and no load current $I_L(t)$ can flow until after $t = l/v$. Similarly, when a harmonically (i.e., sinusoidally) time-varying voltage is connected to such a line, the successive rises and falls of the source voltage propagate along the line with a finite velocity so that the currents and voltages at other points on the line do not reach their maxima and minima at the same time as the input voltage.

In view of the fundamentally different behavior of lumped and distributed circuits as illustrated in Figure 1.2, it is important, in practice, to determine correctly whether a lumped treatment is sufficiently accurate or whether the circuit in hand has to be treated as a distributed circuit. In the following, we quantify on a heuristic basis the circumstances under which the travel time and/or the physical size of the circuit components or the length of the interconnects between them can be neglected. Note that in problems related to heat, diffusion, sound waves, water waves, traffic waves, and so on, the travel time is readily observable and almost always has to be accounted for. In the context of electromagnetics, on the other hand, we find a wide range of applications where lumped analysis is sufficiently accurate, is substantially simpler, and provides entirely satisfactory results. However, in an equally wide range of other applications we find that the lumped treatment is not sufficiently accurate and that one has to resort to field and wave techniques, which are generally more involved, both mathematically and conceptually. It is thus important, particularly in the context of electromagnetic applications, that we develop criteria by which we can determine the applicability of lumped circuit formulations. We provide below a heuristic discussion from three different but related points of view.

1.1.1 Rise Time versus Travel Time

It is apparent from the preceding discussion that we can consider the delay time (travel time) over the signal path as long or short, important or negligible only relative to

---

[13]The more accurate empirical value of the speed of light is $c = 299{,}792{,}458$ m-s^{-1} [*CRC Handbook of Chemistry and Physics,* 76th ed., CRC Press, Inc., Boca Raton, Florida, 1995].

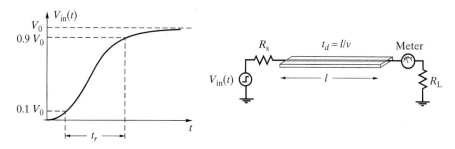

Figure 1.3   Rise time versus one-way travel time.

or in comparison with some other quantity. In terms of positive step changes in an applied signal, we note that the input signal typically exhibits a nonzero *rise time* (usually measured as the time required for the signal to change from 10% to 90% of its final value), which may be denoted as t_r (Figure 1.3). We can then compare t_r to the one-way propagation time delay through the signal path (also called the one-way transit time or time of flight), $t_d = l/v$, where v is the velocity of propagation and l is the length of the signal path. For example, in practical design of interconnects between integrated circuit chips, one rule of thumb is that the signal path can be treated as a lumped element if $(t_r/t_d) > 6$, whereas lumped analysis is not appropriate for $(t_r/t_d) < 2.5$. Whether lumped analysis is appropriate for the in-between range of $2.5 < (t_r/t_d) < 6$ depends on the application in hand and the required accuracy. To summarize:

$$\boxed{\begin{array}{ll} (t_r/t_d) > 6 & \text{(lumped)} \\ (t_r/t_d) < 2.5 & \text{(distributed)} \end{array}} \tag{1.1}$$

In terms of the particular application of high-speed integrated circuits, the on-chip rise times range from 0.5–2 ns for CMOS to 0.02–0.1 ns for GaAs technologies. The speed of signal propagation within the chips depends on the material properties; for example, it is $\sim$0.51c for SiO_2. Thus, for on-chip interconnections (typical $l \simeq 1$ cm), lumped circuit analysis breaks down (i.e., $t_r/t_d < 2.5$) for rise times less than $\sim$0.165 ns. For printed circuit boards made of a commonly used glass epoxy material, the speed of propagation is $\sim$0.47c, so that for a $\sim$10 cm interconnect, lumped analysis is not appropriate for rise times less than $\sim$1.8 ns. As clock speeds increase and rise times become accordingly shorter, distributed analyses will be required in a wider range of digital integrated circuit applications.

The importance of considering rise time versus travel time is underscored by advances in the generation of picosecond pulses.[14] For such extremely short pulse durations, with subpicosecond rise times, distributed circuit treatment becomes necessary for one-way propagation time delay of $t_d > 10^{-13}$ seconds. Assuming propagation at the

---

[14]See D. W. Van der Weide, All-electronic generation of 0.88 picosecond, 3.5 V shockwaves and their application to a 3 Terahertz free-space signal generation system, *Appl. Phys. Lett.,* 62(1), pp. 22–24, January 1993.

speed of light, the corresponding distances are > 0.03 mm! In other words, for picosecond rise times, lumped analysis is not appropriate for circuits with physical dimensions longer than a few tens of microns (1 micron $= 1$ μm).

Example 1.2: Lumped- or distributed-circuit element. Consider a 10-cm long microstrip transmission line having a propagation speed of 15 cm-ns^{-1}. For digital signal transmission with a 100 ps rise time, should this microstrip line be treated as a lumped- or distributed-circuit element?

Solution: The one-way time delay of the microstrip line can be found as

$$t_d = l/v = 10 \text{ cm}/15 \text{ cm-ns}^{-1} \simeq 0.67 \text{ ns}$$

Based on rule-of-thumb criteria (1.1), since $t_r/t_d \simeq 0.1 \text{ ns}/0.67 \text{ ns} = 0.15 < 2.5$, the microstrip line should be modeled as a distributed-circuit element.

1.1.2 Period versus Travel Time

For sinusoidal steady-state applications, the suitability of a lumped treatment can be assessed by comparing the one-way propagation delay t_d with the period T of the propagating sine wave involved. As an illustration, consider the telephone line shown in Figure 1.4. If a sinusoidal voltage at frequency f is applied at the input of the line so that $V_{AA'}(t) = V_0 \cos(2\pi f t)$, the voltage at a distance of l from the input is delayed by the travel time, $t_d = l/v$. In other words,

$$V_{BB'}(t) = V_0 \cos[2\pi f (t - t_d)] = V_0 \cos\left[2\pi f t - 2\pi \frac{t_d}{T}\right] \tag{1.2}$$

where $T = 1/f$ is the period of the sinusoidal signal. It is apparent from (1.2) that if $t_d \ll T$, then the voltage $V_{BB'}(t)$ is very nearly the same as the input voltage $V_{AA'}(t)$ at all times, and the telephone line can be treated as a lumped system. If, on the other

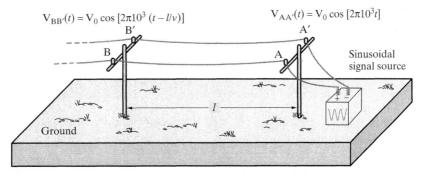

Figure 1.4   Period versus one-way travel time.

hand, t_d is comparable to T, the instantaneous voltage at a certain time t at points further down the line can be quite different from that at the input; for example, for $t_d = 0.5T$, $V_{BB'}(t) = -V_{AA'}(t)$ at any time t!

In practice, a useful rule of thumb is that lumped analyses can be used when $t_d < 0.1T$ or, under more stringent applications, when $t_d < 0.01T$:

$$
\boxed{
\begin{aligned}
t_d &< 0.01T \quad \text{(lumped)} \\
t_d &> 0.01T \quad \text{(distributed)}
\end{aligned}
}
\qquad (1.3)
$$

In the telephone line application, the signal frequency is of order ~ 1 kHz, so that $T \simeq 1$ ms, and a lumped treatment would be appropriate for $t_d < 0.01$ ms. Assuming propagation at the speed of light,[15] $t_d = 0.01$ ms corresponds to a line length of $l = 3$ km, which is the maximum length for which the lumped approximation is valid.

Example 1.3: Power transmission line. If the line shown in Figure 1.4 were instead a power transmission line operating at 60 Hz, calculate the maximum length for which lumped analysis can be used.

> **Solution:** The propagation delay t_d is related to the propagation distance l by $t_d = l/v$, where v is the propagation speed. Noting that the signal period $T = f^{-1}$, we may write the range of lengths where a lumped analysis is appropriate as
>
> $$
> l_{\text{lumped}} < 0.01 \frac{v}{f} \qquad (1.4)
> $$
>
> Once again assuming propagation at the speed of light, a 60 Hz signal corresponds to a maximum line length of
>
> $$
> l_{\text{max}} = 0.01 \cdot \frac{3 \times 10^5 \text{ km-s}^{-1}}{60 \text{ s}^{-1}} = 50 \text{ km}
> $$

Table 1.1 provides examples of other kinds of waves. We note that the speeds of propagation of different types of waves are in general different, as shown in Table 1.1.

It thus appears that for electric circuits excited by sinusoidal signals, whether a lumped circuit assumption is applicable depends on the relationship between period (or frequency $f = 1/T$), speed of travel, and length of the signal path. Note that in the frequency range $f < 10$ MHz, the lumped assumption is indeed very good for most of the lumped electrical circuit applications and models, as long as the lengths involved, or the sizes of the components and interconnecting wires or traces, are less than 30 cm!

---

[15]We shall see in later chapters that speed of propagation of voltage and current waves on a transmission line such as that shown in Figure 1.4 is indeed c.

TABLE 1.1   MAXIMUM LENGTHS FOR LUMPED ANALYSIS IN DIFFERENT APPLICATIONS

Application	Frequency	Propagation Speed $m \cdot s^{-1}$	Maximum Length (Based on $t_d = 0.01T$)
Power transmission	60 Hz	$c \simeq 3 \times 10^8$	50 km
Telephone line	1 kHz	c	3 km
AM radio broadcast	~1 MHz	c	3 m
TV broadcast	~150 MHz	c	2 cm
Radar/microwave	~10 GHz	c	0.3 mm
Sound waves in air	60 Hz	~340	5.7 cm
Sound waves in water	60 Hz	~1500	25 cm
Seismic surface waves	0.1 Hz	~5000	500 m
Heat in a concrete dam	1 cycle/hr	~1.64	59 m
Visible light	~5×10^{14} Hz	c	6×10^{-9} m
X-rays	10^{18} Hz	c	3×10^{-12} m

1.1.3 Component Size versus Wavelength

The applicability of a lumped versus distributed treatment can also be determined in terms of *wavelength* λ. For a periodic wave, wavelength is the length of one complete wave pattern at any given instant of time. In other words, wavelength is the distance between any two points at corresponding positions on successive repetitions of the wave shape. In many applications, electromagnetic quantities (electric fields, magnetic fields, currents, voltages, charges, etc.) can be assumed to vary as

$$I(z,t) = A \cos \left[(2\pi f)t - \left(\frac{2\pi}{\lambda} \right) z \right] \qquad (1.5)$$

where the amplitude A, the frequency f, and the wavelength λ are constants. In later chapters, we shall see that for sinusoidal steady-state applications, quantities with variations as described by (1.5) are natural solutions of the fundamental equations describing wave motion, which are in turn derived from the fundamental physical laws of electromagnetics known as Maxwell's equations. Note also that the form of (1.5) is similar to the general form of a wave function, or $f(z - vt)$, discussed in connection with Figure 1.1.

Now consider the behavior of this quantity, or in the case of (1.5), the current $I(z,t)$, in both space and time. At fixed points in space (i.e., $z = $ const.; for example, $z = 0$ and $z = \lambda/4$), we have

$$I(z = 0, t) = A \cos(2\pi ft - 0) \qquad \text{and} \qquad I(z = \lambda/4, t) = A \cos(2\pi ft - \pi/2)$$

The variation of current at these two fixed locations is shown in Figure 1.5. We see that at every fixed point the current varies sinusoidally with time, as expected for sinusoidal steady state. However, as is apparent from Figure 1.5, at different points along z, the current reaches its maxima and minima at different times.

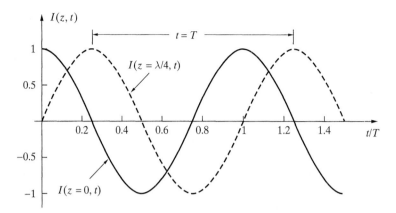

Figure 1.5 Oscillations in time. Plots versus time of $A\cos[(2\pi f)t - (2\pi/\lambda)z]$ at $z = 0$ and $z = \lambda/4$. Note that the abscissa is t/T, where $T = 1/f$, and the plot shown is for $A = 1$.

At fixed points in time (i.e., $t = $ const.; for example, $t = 0$ and $t = T/4$), we have

$$I(z, t = 0) = A\cos\left[0 - \left(\frac{2\pi}{\lambda}\right)z\right] \quad \text{and} \quad I(z, t = T/4) = A\cos\left[(\pi/2) - \left(\frac{2\pi}{\lambda}\right)z\right]$$

The variation of current in space at these two different times is shown in Figure 1.6. We see that the behavior of $I(z, t)$ at a given time is also sinusoidal, with its peak value (i.e., the point where the argument of the cosine is zero) occurring at different points in space at different times. The speed $v = dz/dt$ with which the crests (the maxima) or the valleys (the minima) of the cosinusoid propagate (move from one point in space to another) can be determined by setting its argument equal to a constant (e.g., zero, which

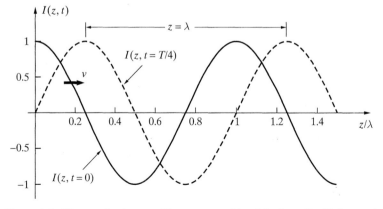

Figure 1.6 Wave motion in space. Plots versus z of $A\cos[(2\pi f)t - (2\pi/\lambda)z]$ at $t = 0$ and $t = T/4$. Note that the abscissa is z/λ and the plot shown is for $A = 1$.

corresponds to the maximum) and differentiating:

$$\left[(2\pi f)t - \left(\frac{2\pi}{\lambda}\right)z\right] = \text{const.} \qquad \longrightarrow \qquad v = \frac{dz}{dt} = \lambda f$$

For electromagnetic waves in free space, and for voltage and current waves on some simple transmission lines (e.g., those consisting of two wires separated by air or air-filled coaxial lines), the speed of propagation v is equal to the speed of light c.

Since the current at any given instant of time varies sinusoidally in space, whether we can assume a given circuit to be lumped (i.e., that the instantaneous current at a given time is approximately the same at different points across the component carrying the current) depends on the size of the element compared with wavelength, λ. This concept is illustrated in Figure 1.7, which shows a 2-cm long resistor (including leads) in comparison with wavelength at three frequencies: 100 MHz, 1 GHz, and 10 GHz. For 10 GHz, it is clear that the notion of a current through the resistor is not useful, since at any given time, one lead of the resistor has a current of different polarity than the other,[16] since according to the current expression (1.5) and Figure 1.6 the current at any fixed time switches polarity every $\lambda/2$ change in position, where $\lambda = c/f = 3$ cm for 10 GHz. Even for the 1 GHz case, the dimensions of the resistor are a significant fraction of a wavelength so that the current at any time is significantly different at different points across the circuit component. Thus, of the cases shown, the lumped assumption

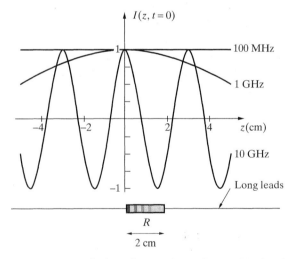

Figure 1.7   Current through 2-cm long resistor. Current $I(z,t) = A\cos[2\pi ft - (2\pi/\lambda)z]$ is shown as a function of z at $t = 0$ for frequencies of 100 MHz, 1 GHz, and 10 GHz. For the purposes of this plot, we have taken $A = 1$.

---

[16]Note that at 10 GHz, not only would we not be able to define the value of current but there would be substantial radiation losses because the resistor leads, being comparable in size to λ, would serve as efficient antennas.

is appropriate only for 100 MHz. Using the rule of thumb that the feature size L should be smaller than 1% of the wavelength for a lumped analysis to be valid, we have

$$\boxed{\begin{array}{ll} L < 0.01\lambda & \text{(lumped)} \\ L > 0.01\lambda & \text{(distributed)} \end{array}} \tag{1.6}$$

Example 1.4: Component size versus wavelength. A wireless unit operating at 4 GHz is connected to an antenna via a 2.5-cm long microstrip transmission line. Determine whether this microstrip transmission line can be approximated as a lumped-circuit element. Assume a propagation speed of 15 cm-ns^{-1} for the microstrip line.

Solution: The wavelength in the microstrip transmission line can be calculated as

$$\lambda = \frac{v}{f} = \frac{(15 \text{ cm-ns}^{-1})}{4 \times 10^9 \text{ Hz}} \left(\frac{10^9 \text{ ns}}{1 \text{ s}} \right) = 3.75 \text{ cm}$$

The ratio between the length L of the microstrip transmission line and the wavelength is

$$L/\lambda = 2.5 \text{ cm}/3.75 \text{ cm} \simeq 0.67$$

Using the rule-of-thumb criteria (1.6), a lumped-circuit model is not appropriate to use in this case.

1.2 ELECTROMAGNETIC COMPONENTS

The electromagnetic circuit components used at high frequencies can differ conspicuously in appearance from the often more familiar lumped-element circuits used at low frequencies. The connecting wires of conventional circuits provide conductors for the electric currents to flow, and the resistors, capacitors, and inductors possess simple relationships between their terminal currents and voltages. Often overlooked is the fact that the wires and circuit components merely provide a framework over which charges move and disperse. These charges set up electric and magnetic fields that permeate the circuit, often having almost indescribably complicated configurations. It would, in principle, be possible to treat the behavior of circuits entirely in terms of these electromagnetic fields instead of the usual practice of working in terms of circuit voltages and currents. It can be argued, however, that much of the progress in modern electrical and electronic applications would not have come about if it were not for the simple but powerful circuit theory.

As the operating frequency of circuits increases, however, the effects of stray capacitances and inductances alter the effective circuit behavior radically compared with its low-frequency characteristics. Radiation from the circuit also increases rapidly with frequency, which can cause significant power loss.[17] This radiative power loss may be prevented by confining the fields to the interior of metallic enclosures. For example,

---

[17]In antenna applications, we take advantage of such radiation "losses" when we design antennas to maximize the radiated power in selected directions.

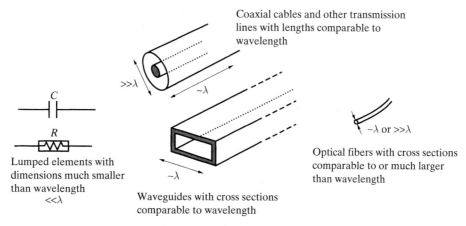

Figure 1.8   Different categories of electromagnetic circuit components.

in hollow metallic tubes (referred to as waveguides) the charges move exclusively on the interior surfaces of conductors, and because of the simple geometry of the enclosures, the electromagnetic fields can have simple analytical forms. Since it is often not possible to define voltages and currents uniquely within waveguides, analysis of waveguides is usually carried out on the basis of the full electromagnetic theory. At even higher frequencies, metallic enclosures become too lossy and impractical. Efficient guiding of electromagnetic energy at optical frequencies occurs in optical fibers, consisting of hair-thin glass strands. The light wave in an optical fiber is guided along the fiber by means of multiple reflections from its walls. The principles underlying the guiding of electromagnetic energy in metallic enclosures is discussed in Chapter 10.

In summary, component sizes in electromagnetic applications can be categorized as shown in Figure 1.8:

- Component sizes much smaller than λ (lumped elements). Examples: Resistors, capacitors, inductors, ICs, transistors, and interconnects used for $< \sim 30$ MHz.
- Component sizes comparable to λ. Examples: Hollow waveguides, long coaxial cables, some optical fibers, and some antennas.
- Component sizes much greater than λ. Examples: Graded index optical fibers and some antennas.

1.3 MAXWELL'S EQUATIONS AND ELECTROMAGNETIC WAVES

The principles of guiding and propagation of electromagnetic energy in these widely different regimes differ in detail, but are all governed by one set of equations, known as *Maxwell's equations*,[18] which are based on experimental observations and which provide

[18]J. C. Maxwell, *A Treatise in Electricity and Magnetism,* Clarendon Press, Oxford, Vol. 2, pp. 247–262, 1892.

the foundations of *all* electromagnetic phenomena and their applications. The sequence of events in the late 19th century that led to the development of these fundamental physical laws is quite interesting in its own right. Many of the underlying concepts were developed by earlier scientists, especially Michael Faraday, who was a visual and physical thinker but not enough of a mathematician to express his ideas in a complete and self-consistent form to provide a theoretical framework. James Clerk Maxwell translated Faraday's ideas into strict mathematical form and thus established a theory that predicted the existence of electromagnetic waves. The experimental proof that electromagnetic waves do actually propagate through empty space was supplied by the experiments of Heinrich Hertz, carried out many years after Maxwell's brilliant theoretical work.[19]

Maxwell's equations are based on experimentally established facts, namely Coulomb's law,[20] Ampère's law,[21] Faraday's law,[22] and the principle of conservation of electric charge. When most of classical physics was fundamentally revised as a result of Einstein's introduction[23] of the special theory of relativity, Maxwell's equations remained intact.[24] To this day, they stand as the most general mathematical statements of the fundamental natural laws that govern all of classical electrodynamics. The basic justification and validity of Maxwell's equations lies in their consistency with physical experiments over the entire range of the experimentally observed electromagnetic spectrum (see Table 8.1), extending from cosmic rays at frequencies greater than 10^{22} Hz to the so-called micropulsations at frequencies of 10^{-3} Hz. The associated practical applications cover an equally wide range, from the use of gamma rays (10^{18}–10^{22} Hz) for cancer therapy to the use of waves at frequencies of a few hertz and below for geophysical prospecting. Electromagnetic wave theory as embodied in Maxwell's equations has provided the underpinning for the development of many vital practical

---

[19]H. Hertz, On the finite velocity of propagation of electromagnetic actions, *Sitzb. d. Berl. Akad. d. Wiss.*, Feb. 2, 1888; for a collected English translation of this and other papers by H. Hertz, see H. Hertz, *Electric Waves,* MacMillan, London, 1893.

[20]Coulomb's law states that electric charges attract or repel one another in a manner inversely proportional to square of the distance between them; C. A. de Coulomb, *Première mémoire sur l'électricité et magnétisme (First Memoir on Electricity and Magnetism), Histoire de l'Académie Royale des Sciences,* p. 569, 1785.

[21]Ampère's law states that current-carrying wires create magnetic fields and exert forces on one another, with the amplitude of the magnetic field (and thus force) depending on the inverse square of the distance; A. M. Ampère, *Recueil d'observations électrodynamiques,* Crochard, Paris, 1820–1833.

[22]Faraday's law states that magnetic fields that change with time induce electromotive force or electric field; M. Faraday, *Experimental Researches in Electricity,* Taylor, London, Vol. I, pp. 1–109, 1839.

[23]Einstein, A., *Annalen der Physik,* 1905. The English translation of this paper is remarkably readable and is available in a collection of original papers titled *The Principle of Relativity,* Dover Publications, New York.

[24]Maxwell's formulation was in fact one of the major motivating factors that led to the development of the theory of special relativity. The fact that Galilean relativity was consistent with classical mechanics but inconsistent with electromagnetic theory suggested that either Maxwell's equations were incorrect or that the laws of mechanics needed to be modified. For discussions of the relationship between electromagnetism and the special theory of relativity, see Section 15 of D. M. Cook, *The Theory of the Electromagnetic Field,* Prentice-Hall, Inc., New Jersey, 1975; Chapter 10 of D. J. Griffiths, *Introduction to Electrodynamics,* 2nd ed., Prentice-Hall, Inc., New Jersey, 1989; Chapter 2 of R. S. Elliott, *Electromagnetics,* IEEE Press, Piscataway, New Jersey, 1993; Chapter 11 of J. D. Jackson, *Classical Electrodynamics,* 2nd ed., Wiley, New York, 1975.

tools of our technological society, including broadcast radio, radar, television, cellular phones, optical communications, the Global Positioning Systems (GPS), microwave heating and processing, and X-ray imaging.

Maxwell's equations embody all of the essential features of electromagnetics, including the ideas that light is electromagnetic in nature, that electric fields that change in time create magnetic fields in the same way as time-varying voltages induce electric currents in wires, and that the source of electric and magnetic energy resides not only on the body that is electrified or magnetized but also, and to a far greater extent, in the surrounding medium. However, arguably the most important and far-reaching implication of Maxwell's equations is the idea that electric and magnetic effects can be transmitted from one point to another through the intervening space whether that be empty or filled with matter.

To appreciate the concept of propagation of electromagnetic waves in empty space, it is useful to think of other wave phenomena that we may observe in nature. When a pebble is dropped into a body of water, the water particles in the vicinity of the pebble are immediately displaced from their equilibrium positions. The motion of these particles disturbs adjacent particles, causing them to move, and the process continues, creating a wave. Because of the finite velocity of the wave, a finite time elapses before the disturbance causes the water particles at distant points to move. Thus the initial disturbance produces, at distant points, effects that are *retarded* in time. The water wave consists of ripples that move along the surface away from the initial disturbance. Although the motion of any particular water particle is essentially a small up-and-down movement, the cumulative effects of all the particles produce a wave that moves radially outward from the point at which the pebble is dropped. Another excellent example of wave propagation is the motion of sound through a medium. In air, this motion occurs through the to-and-fro movement of the air molecules, but these molecules do not actually move along with the wave.

Electromagnetic waves consist of time-varying electric and magnetic fields. Suppose an electrical disturbance, such as a change in the current through a conductor, occurs at a point in a region. The time-changing electric field resulting from the disturbance generates a time-changing magnetic field. The time-changing magnetic field, in turn, produces an electric field. These time-varying fields continue to generate one another in an ever-expanding region, and the resulting wave propagates away from the location of the initial disturbance. When electromagnetic waves propagate in vacuum, there is no vibration of physical particles as in the case of water and sound waves. Nevertheless, the velocity of wave propagation is limited by the speed of light, so that the fields produced at distant points are *retarded* in time with respect to those near the source.

1.4 SUMMARY

There are two kinds of electrical circuits: *lumped* and *distributed*. In classical lumped circuit theory, the entire circuit is assumed to be at a single point such that all parts of the circuit respond to an excitation at the same time. Distributed circuits or transmission lines can be treated using a natural extension of classical circuit theory by taking account of the nonzero travel time of signals from one point to another in a distributed circuit, the speed of travel being bounded by the speed of light. In any given application, a number of criteria can be used to determine the applicability of a lumped analysis. A given system must be treated as a distributed one if:

- The rise time t_r of the applied signal is less than 2.5 times the one-way travel time t_d across the circuit, that is, $t_r < 2.5 t_d$.
- The one-way travel time t_d across the circuit is greater than one-hundredth of the period T of the applied sinusoidal signal, that is, $t_d > 0.01T$.
- The physical dimensions of the circuit are a significant fraction of the wavelength at the frequency of operation.

In Chapter 2, we will assume that all step sources are ideal and have zero rise time. We will also assume that all time delays are due to transmission lines, and consider lumped elements such as resistors, capacitors, and inductors to be ideal with zero time delay.

Electromagnetic components can be categorized in three basic groups with sizes much smaller than, comparable to, or much larger than the wavelength. Electromagnetic applications involving any one of these three classes of components are governed by a set of physical laws known as Maxwell's equations, which are based on experimental facts and describe the propagation of electromagnetic waves through empty space or material media and, in general, *all* other electrical and magnetic phenomena.

In the next two chapters, we will study a special class of electromagnetic waves called *transverse electromagnetic* (TEM) waves, which propagate on electrical transmission lines consisting of two metallic conductors. In the case of TEM waves, both the electric and magnetic fields are everywhere perpendicular to each other and to the direction of propagation. The behavior of TEM waves on two-conductor lines can be understood in the context of a voltage–current description by modeling the transmission line as a distributed circuit and utilizing many of our electrical circuits concepts. The fact that the voltage and current waves that propagate on such two-conductor transmission lines are indeed transverse electromagnetic waves will become clear in Chapter 10.

After we study TEM waves in the context of electrical transmission lines in Chapters 2 and 3, we will introduce and discuss the fundamental components of Maxwell's equations in Chapters 4 through 7. In Chapter 4 we begin by studying the branch of electromagnetics that deals with electric charges at rest, called *electrostatics*. In this chapter, we investigate the consequences of Coulomb's law, an experimentally derived statement on the forces between charged particles. In Chapter 5 we consider the steady motion of charged particles through a material, which is sustained by electrostatic forces. We then note another experimental fact: wires containing steady-state current exert a force on each other. This observation, which is captured by Ampère's law, underpins the branch of electromagnetics called *magnetostatics*, the topic of Chapter 6. Finally, Chapter 7 introduces a third experimental fact that involves currents that change with time: Faraday's law of electromagnetic induction. This final observation, coupled with the fundamental fact of charge conservation, completes the formulation of Maxwell's equations, and ultimately leads to the phenomenon of electromagnetic waves.

In the remainder of the book, we take a deeper look at the propagation of electromagnetic waves. In Chapter 8 we restrict our attention to electromagnetic waves in an *unbounded, simple,* and *source-free* medium. In Chapter 9 we remove the first condition and consider the reflection and transmission of electromagnetic waves at planar boundaries between different material media. Many practical problems encountered in electromagnetics involve reflection of waves from interfaces between dielectrics and perfect conductors (e.g., air and copper) or between two different dielectrics (e.g., glass and air), and the treatment of such problems requires that we take into account the complicating effects of the boundary surfaces. In Chapter 10 we consider electromagnetic waves guided by planar metallic and dielectric boundaries. Finally, in Chapter 11, we return to considering an unbounded medium but now allow it not to be simple, but rather have material properties which significantly affect the nature of electromagnetic waves.

PROBLEMS

1.1 **Travel time around the Earth.** Calculate the approximate time it takes for an electromagnetic signal to travel around the circumference of the Earth. Assume the Earth to be a perfect sphere with an average diameter of 1.274×10^4 km.

1.2 **Travel time between one celestial body and another.** The average distance from one celestial body to another is 454,000 km. How long will it take an electromagnetic wave to travel this distance?

1.3 **Earth–Moon communication.** During the Moon landing of one of the Apollo spaceflights, a spoken message from Earth is heard returning to Earth approximately 2.7 seconds later. The message signal traveled from Earth to the Moon, from the earphone to the microphone attached to the astronaut's space helmet, and back to Earth.[25] Use this time delay to estimate a rough value for the speed of light.

1.4 **Time delay of a radar signal.** Radars use radio waves to determine the distance to a target. Radar systems consist of a transmitting antenna that emits a radio pulse, and a receiving antenna that measures the (typically small) reflected component of the emitted waves. If the transmitter and receiver are co-located, and if the receiver detects the pulses reflected from a target 40 μs after the radar pulses leave the transmitter, what is the distance of the target from the radar?

1.5 **Echo from a cliff.** A man shouts in the direction of a cliff, which causes an echo that he hears 5 seconds later. If sound travels in air at 340 m-s^{-1}, how far away is the cliff?

1.6 **Sonar.** Sonar stands for SOund NAvigation and Ranging and is a device that uses reflected sound waves to measure underwater depths. A ship sends a sonar signal downward into the ocean and receives a return signal 6 seconds later. Assuming the speed of sound waves to be the same at all depths given by 2.5×10^3 m-s^{-1}, how deep is the ocean at that location?

1.7 **Sonar.** It takes a sonar pulse 3.7 seconds to return to its ship after being transmitted vertically towards the ocean floor. Assuming a constant speed of 3 km-s^{-1}, what is the depth of the ocean at this point?

1.8 **Auto-focus camera.** An auto-focus camera can focus on objects by use of an ultrasonic sound wave. Just like radar, the camera sends out sound waves which reflect off the distant objects and return back to the camera. A sensor detects the time it takes for the waves to

---

[25]David Keeports, "Estimating the speed of light from Earth–Moon communication," *The Physics Teacher*, Vol. 44, pp. 414–415, October 2006.

return and then determines the distance of the object from the camera. If an ultrasonic sound wave returns to the camera 0.1 second after leaving the camera, how far is the object from the camera? Assume speed of the ultrasonic sound wave to be 340 m-s^{-1}. (Polaroid developed an ultrasonic autofocus system that used 50 kHz waves, well outside the human hearing range of 20 Hz–20 kHz. Modern autofocusing cameras typically use reflected infrared light to triangulate objects in the field of view.)

1.9   **Lightning and thunder.** If a person heard the sound of a thunder approximately 5 seconds after observing the lightning flash at a distance, how far away from this person did the lightning strike? Assume the speed of sound to be 340 m-s^{-1}.

1.10   **A light-year.** A light-year (ly) is a unit of distance and is defined as the distance that light can travel in 1 year. Find the equivalent of 1.5 light-year in kilometers.

1.11   **A light-nanosecond.** What is 1 light-microsecond in meters?

1.12   **1 Astronomical Unit.** The average distance between the Earth and the Sun is defined as 1 Astronomical Unit (AU) is about 1.5×10^8 km. What is 2 AU in light-seconds?

1.13   **Distance between Proxima Centauri and Earth.** The distance between Proxima Centauri, the nearest known star to Earth (besides our Sun), is about 4×10^{13} km. Find this distance in light years.

1.14   **Seismic waves.** A geological disturbance in California is detected at a seismograph station in Washington approximately 900 km away from the epicenter of the earthquake. If the seismic wave traveled at an average speed of 5 km-s^{-1}, determine the time delay between the epicenter and seismograph station.

1.15   **Tsunami waves.** *Tsunami* is a Japanese word. It means harbor ("*tsu*") waves ("*nami*"). Tsunamis are fairly common in Japan and caused many thousands of deaths in recent centuries. A massive earthquake hit the Pacific Ocean near Honshu, Japan in 2011 producing a tsunami wave which reached Crescent City, California about 10 hours later. If the distance between the epicenter of the earthquake and Crescent City is about 8020 km, determine the average speed of the tsunami wave.

1.16   **The Indian Ocean tsunami.** The Indian Ocean tsunami triggered by a massive underwater earthquake that occurred in 2004 devastated the shores of Indonesia, Sri Lanka, India, Thailand, and other countries with waves up to 15 m high, even reaching as far as the east coast of Africa. The epicenter of the earthquake was about 160 km west of Sumatra, Indonesia and 4500 km from Africa. Assuming the average speed of the tsunami to be 800 km per hour, approximately how long did it take for the tsunami to travel from the epicenter to Sumatra? How about to Africa?

1.17   **Micro-bats.** Micro-bats use a form of radar called echolocation to navigate and find their prey such as flying insects. They locate the surrounding objects by bouncing sound wave pulses off these objects and detecting the time delay between the emitted pulses and the reflected pulses. Determine the time delay between the pulse emitted by the micro-bat and the detected pulse reflected from an insect located 10 m away from the micro-bat. Assume the approximate speed of sound waves to be 340 m-s^{-1}.

1.18   **Overhead power lines.** The alternating current (ac) high-voltages overhead power lines in India operate at a standard frequency of 50 Hz. Assuming the propagation at the speed of light determine the maximum length for lumped analysis.

1.19   **Maximum path length.** Consider an electrical signal path in a high-speed digital circuit application to carry signals with rise times as low as 250 ps. If the speed of propagation is $v = c/3$, determine the maximum path length for lumped-circuit analysis.

1.20 **Maximum coax length.** Consider a commercial coaxial cable with $v = 2 \times 10^8$ m-s^{-1}. If this cable is to be used at frequencies around 200 MHz, what is the maximum cable length for lumped-circuit analysis?

1.21 **Travel time of a microstrip transmission line.** What is the one-way travel time of a 9 cm long microstrip transmission line with $v = 1.5 \times 10^8$ m-s^{-1}?

1.22 **Maximum cable length.** A ham radio operator connects his 30 MHz transmitter to an antenna using a coaxial cable. Assuming the speed of propagation along the cable to be 2×10^8 m-s^{-1}, determine the maximum coaxial cable length beyond which it can't be modeled as a lumped-circuit element.

1.23 **Microstrip transmission line.** A 6 cm long microstrip signal trace with speed of propagation $v = c/3$ is to be used to carry 1 ns rise time signals. Is it appropriate to neglect transmission line effects and use lumped circuit analysis?

1.24 **Stripline transmission line.** A 12 cm long stripline transmission line having a speed of propagation $v = c/2$ is used to carry digital signals with 0.5 ns rise time. Is it appropriate to consider transmission line effects in this case?

1.25 **On-chip GaAs interconnect.** Consider an on-chip GaAs interconnect with speed of propagation $v = 8 \times 10^7$ m-s^{-1} to be used to carry a digital signal with rise time $t_r = 50$ ps. What is the maximum interconnect length for lumped-circuit analysis?

1.26 **A coaxial cable-lumped or distributed analysis?** A coaxial cable of length 10 m having a speed of propagation $v = 0.75c$ is used to connect the antenna and the receiver of a microwave communication system. Assuming sinusoidal steady state, what is the highest frequency for lumped-circuit analysis?

2

Transient Response
of Transmission Lines

Wave motion is said to occur when a disturbance of a physical quantity at a particular point in space at a particular time is related in a definite manner to what occurs at distant points in later times. *Transient* waves occur in response to sudden, usually brief disturbances at a source point, leading to temporary disturbances at distant points at later times. They are thus different from *steady-state* waves, which are sustained by disturbances involving periodic oscillations at the source point.

Transient waves are of importance in many different contexts. Consider, for example, a line of cars waiting at a red traffic light. When the light turns green, the cars do not all start moving at the same time; instead, the first car starts to move first, followed by the car behind it, and so on, as the act of starting to move travels backwards through the line. This wave travels at a speed that depends on the response properties of the cars and the reaction times of the drivers. As another example, when the end of a stretched rope is suddenly moved sideways, the action of moving sideways travels along the rope as a wave whose speed depends on the tension in the rope and its mass. If the rope is infinitely long, the disturbance simply continues to propagate away from its source. If, on the other hand, the distant end of the rope is held fixed, the wave can be reflected back toward the source.

Other examples of transient waves include the thunderclap, the sound wave emitted from an explosion, and seismic waves launched by an earthquake. Transient waves are often used as tools to study the disturbances that create them. The sound wave from a blast can be used for detecting the source of the blast from a long distance; a seismograph measures the strength of a distant earthquake based on tiny transient motions of the earth.

The purpose of this chapter is to study voltage and current transients on electrical transmission lines. A transmission line may consist of two parallel wires (as it will often be illustrated in this book), of coaxial conductors, or of any two conductors separated by an insulating material or vacuum. Some types of two-conductor transmission lines are shown in Figure 2.1.

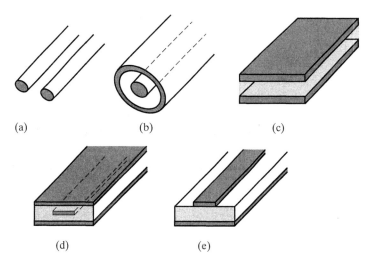

Figure 2.1 Different types of uniform transmission lines: (a) parallel two-wire; (b) coaxial; (c) parallel-plate; (d) stripline; (e) microstrip.

Many important electrical engineering applications involve *transients:* temporary variations of current and voltage that propagate down a transmission line. Transients are produced by step-like changes (e.g., sudden on or off) in input voltage or current. Digital signals consist of a sequence of pulses, which represent superposition of successive step-like changes; accordingly, the transient response of transmission lines is of interest in most digital integrated circuit and computer communication applications. Transient transmission line problems arise in many other contexts. The transient response of lines can be used to generate rectangular pulses; the earliest applications of transmission lines involved the use of rectangular pulses for telegraphy. When lightning strikes a power transmission line, a large surge voltage is locally induced and propagates to other parts of the line as a transient.

This chapter is unique; the following chapters are mostly concerned with applications that either involve static quantities, which do not vary in time or involve steady signals that are sinusoids or modified sinusoids. However, with the rapid advent of digital integrated circuits, digital communications, and computer communication applications, transient responses of transmission lines are becoming increasingly more important. Increasing clock speeds make signal integrity[1] analysis a must for the design of high-speed and high-performance boards. Managing signal integrity in today's high-speed printed circuit boards and multichip modules involves features such as interconnect lengths, vias, bends, terminations, and stubs and often requires close attention to transmission line or distributed

---

[1]The term *signal integrity* refers to the issues of timing and quality of the signal. The timing analysis is performed to ensure that the signal arrives at its destination within a specified time interval and that the signal causes correct logic switching without false switching or transient oscillations, which can result in excessive time delays. See R. Kollipara and V. Tripathi, Printed wiring boards, Chapter 71 in J. C. Whitaker (Ed.), *The Electronics Handbook,* CRC Press, Boca Raton, Florida, pp. 1069–1083, 1996.

circuit effects.[2] It is thus fitting that we start our discussion of engineering electromagnetics by studying the transient response of transmission lines. Also, analysis of transients on transmission lines requires relatively little mathematical complexity and brings about an intuitive understanding of concepts such as wave propagation and reflection, which facilitates a better understanding of the following chapters.

2.1 HEURISTIC DISCUSSION OF TRANSMISSION LINE BEHAVIOR AND CIRCUIT MODELS

Typically, we explain the electrical behavior of a two-conductor transmission line in terms of an equal and opposite current flowing in the two conductors, as measured at any given transverse plane. The flow of this current is accompanied by magnetic fields set up around the conductors (Ampère's law), and when these fields change with time, a voltage (electromotive force) is induced in the conductors (Faraday's law), which affects the current flow.[3] This behavior can be represented by a small inductance associated with each short-length segment of the conductors. Also, any two conductors separated by a distance (such as the short sections of two conductors facing one another) have nonzero capacitance, so that when equal and opposite charges appear on them, there exists a potential drop across them. Hence, each short section of a two-conductor line exhibits some series inductance and parallel capacitance.[4] The values of the inductance and capacitance depend on the physical configuration and material properties of the two-conductor line, including the surface areas, cross-sectional shape, spacing,[5] and layout of the two conductors, as well as the electrical and magnetic properties[6] of the substance filling the space between and around the conductors.

2.1.1 Heuristic Discussion of Transmission Line Behavior

We can qualitatively understand the behavior of a two-conductor transmission line by considering a lossless circuit model of the line, consisting of a large number of series inductors and parallel capacitors connected together, representing the short sections Δz of the line, as illustrated in Figure 2.2.

---

[2] See R. Goyal, Managing signal integrity, *IEEE Spectrum*, pp. 54–58, March 1994.

[3] Detailed discussion of these experimentally based physical laws will be undertaken later; here we simply rely on their qualitative manifestations, drawing on the reader's exposure to these concepts at the freshman physics level.

[4] Neglecting losses for now.

[5] This can be seen at a qualitative level, from the reader's understanding of capacitance and inductance at the freshman physics level. For example, the closer the conductors are to each other, the larger the capacitance is. On the other hand, the inductance of a two-conductor line is smaller if the conductors are closer together, since the magnetic field produced by the current flow is linked by a smaller area, thus inducing less voltage.

[6] At the simplest level, the magnetic properties of a material represent the ability of a material medium to store magnetic energy. Similarly, by electrical properties we refer to the ability of a material to store electric energy or its response to an applied electric field. The microscopic behavior of the materials that determines these properties will be discussed in Chapters 4 and 6.

Figure 2.2   Circuit model of a two-conductor lossless line.

To illustrate the behavior of a lossless transmission line, we now consider the simplest possible transient response: the step response, which occurs upon the sudden application of a constant voltage. At $t = 0$, a battery of voltage V_0 and source resistance R_{s1} is connected to the infinitely long two-conductor transmission line represented by the lossless circuit shown in Figure 2.2, where each pair of inductances corresponds to the inductance of a short section of the line of length Δz, and each capacitor corresponds to the capacitance of the same section of length Δz. Initially, the transmission line is completely discharged, so all the capacitances have zero charge (and thus zero voltage) and the inductances have zero current flowing through them. The switch in Figure 2.2 is moved to position 1 at $t = 0$, so that, starting at $t = 0^+$, the source voltage V_0 appears across the source resistance R_{s1} and the terminals of the capacitance C_1, which takes time to charge;[7] and until it charges, there is no voltage across it to drive currents through L_{1a} and L_{1b}. As the voltage across C_1 builds up, the currents in L_{1a} and L_{1b} also take time to increase.[8] When these currents increase enough to cause appreciable flow through C_2, this capacitance now takes some time to charge. As it charges, current starts to flow in inductors L_{2a} and L_{2b}, but this takes time. This same process continues all the way down the line, with the capacitor C_k not starting to charge until the preceding capacitors are charged, just as if it did not know yet that the voltage step had been applied at the input. In this way, the information about the change in the position of the switch travels down the line.

When the switch moves back to position 2 at $t = t_1$, the reverse happens: C_1 has to discharge through R_{s2}, which pulls current (not suddenly) from L_{1a} and L_{1b}, which in turn allows C_2 to discharge, and so on. All of this takes time, so C_k is not affected by the removal of the input signal until the preceding capacitors are discharged first. The rate of charging and discharging depends only on the circuit element values, so the charging and discharging disturbances both continue down the line at the same speed, since $L_{ia} = L_{jb}$ for all i, j and all C_i values are equal, assuming a uniform transmission line.

Note from the above discussion that if the inductance of the line segments is negligible, the line can be approximated as a lumped capacitor (equal to the parallel combination of all of its distributed shunt capacitances); all the points on the line are then at the same potential, and traveling-wave effects are not important. The line inductance

---

[7]Voltage across a capacitance cannot change instantaneously.
[8]Current through an inductor cannot change instantaneously.

becomes important if the line is relatively long or if the rise time of the applied signal (as defined in Figure 1.3) is so fast that the current through the inductor increases very rapidly, producing appreciable voltage drop ($\mathcal{V} = L\, d\mathcal{I}/dt$) across the inductor even if the value of L is small. By the same token, it is intuitively clear that, even if the line is long (and thus L is large), transmission line effects will be negligible for slow enough rise times, as was discussed in Chapter 1.

2.1.2 Circuit Models of Transmission Lines

It is often more useful to describe transmission line behavior in terms of inductance and capacitance *per unit length* rather than viewing the line as an infinite number of discrete inductances and capacitances, as implied in Figure 2.2. We must also note that, in general, the conductors of a transmission line exhibit both inductance and resistance and that there can be leakage losses through the insulating material surrounding the conductors. The inductance per unit length (L) of the line, in units of henrys per meter, depends on the physical configuration of the conductors (e.g., the separation between conductors and their cross-sectional shape and dimensions) and on the magnetic properties of the material surrounding the conductors. The series resistance per unit length R, in units of ohms per meter, depends on the cross-sectional shape, dimensions, and electrical conductivity of the conductors[9] and the frequency of operation. Between the conductors there is a capacitance (C), expressed as farads per meter; there is also a leakage conductance (G) of the insulating material surrounding the conductors, in units of siemens per meter. The capacitance depends on the shape, surface area, and proximity of the conductors as well as the electrical properties of the insulating material surrounding the conductors. The conductance depends on the shape and dimensions of the outside surface of the conductors[10] and on the degree to which the insulating material is lossy. A *uniform* transmission line consists of two conductors of uniform cross-section and spacing throughout their length, surrounded by a material that is also uniform throughout the length of the line. An equivalent circuit of a uniform transmission line can be drawn in terms of the per-unit-length parameters, which are the same throughout the line. One such circuit is shown in Figure 2.3, where each short section of length Δz of the line is modeled as a lumped circuit whose element values are given in terms of the per-unit parameters of the line. The electrical behavior of a uniform transmission line can be studied in terms of such a circuit model if the length of the line (Δz) represented by a single L-R-C-G section is very small compared with, for example, the wavelength of electromagnetic waves in the surrounding material at the frequency of operation. Four different circuit models are shown in Figure 2.4.

Expressions for L, R, C, and G for some of the commonly used uniform transmission lines shown in Figure 2.1 are provided in Section 2.7. The values of these quantities depend on the geometric shapes and the cross-sectional dimensions of the

---

[9]The resistance simply represents the ohmic losses due to the current flowing through the conductors; hence, it depends on the cross-sectional area and the conductivity (see Chapter 5) of the conductors.

[10]The leakage current flows from one conductor to the other, through the surrounding material, and in the direction transverse to the main current flowing along the conductors of the line; hence, it depends on the outer surface area of the conductors.

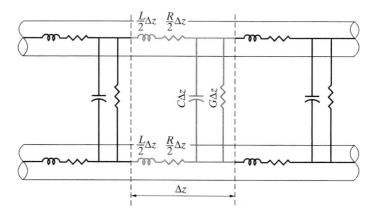

Figure 2.3 Distributed circuit of a uniform transmission line.

Figure 2.4 Lumped-circuit models for a short segment of a uniform transmission line.

line, the electrical conductivity of the metallic conductors used, the electrical and magnetic properties of the insulating surrounding medium, and the frequency of operation. The expressions for L, R, C, and G for the various transmission lines can be obtained by means of electromagnetic field analysis of the particular geometries involved. For some cases (such as the parallel two-wire, coaxial line, and parallel plate lines shown in Figure 2.1), compact analytical expressions for R, L, C, and G can be found. For other, more complicated structures (e.g., the stripline and microstrip lines in Figure 2.1), calculation of the R, L, C, and G parameters usually requires numerical computation. Methods for the determination of transmission line parameters are discussed in Chapters 4 through 6 as we introduce the governing electromagnetic equations, so that we can formally derive the expressions for the transmission line parameters. For our distributed circuit analyses of transmission lines in this and the following chapter, it suffices to know that the values of L, R, C, and G are directly calculable for any uniform transmission line configuration. We can thus proceed by using their values as specified by the expressions in Section 2.7, as given in handbooks, or as measured in specific cases.

2.2 TRANSMISSION LINE EQUATIONS AND WAVE SOLUTIONS

In this section we develop the fundamental equations that govern wave propagation along general two-conductor transmission lines. Various lumped-circuit models of a single short segment of a transmission line are shown in Figure 2.4. In the limit of $\Delta z \to 0$, any one of the circuit models of Figure 2.4 can be used to derive the fundamental transmission line equations. In the following, we use the simplest of these models (Figure 2.4b), shown in further detail in Figure 2.5.

2.2.1 Transmission Line Equations

The section of line of length Δz in Figure 2.5 is assumed to be located at a distance z from a selected point of reference along the transmission line. We consider the total

Figure 2.5 **Equivalent circuit of a short length of Δz of a two-conductor transmission line.**

voltage and current at the input and output terminals of this line section: that is, at points z and $(z + \Delta z)$. The input and output voltages and currents are denoted as $\mathcal{V}(z,t)$, $\mathcal{I}(z,t)$ and $\mathcal{V}(z + \Delta z, t)$, $\mathcal{I}(z + \Delta z, t)$, respectively. Using Kirchhoff's voltage law, we can see that the difference in voltage between the input and output terminals is due to the voltage drop across the series elements $R\Delta z$ and $L\Delta z$, so we have

$$\mathcal{V}(z + \Delta z, t) - \mathcal{V}(z,t) = -R\Delta z\, \mathcal{I}(z,t) - L\Delta z \frac{\partial \mathcal{I}(z,t)}{\partial t}$$

Note that we shall consider Δz to be as small as needed so that the lumped circuit model of the segment can accurately represent the actual distributed line. Upon dividing by Δz and expanding $\mathcal{V}(z + \Delta z, t)$ in a Taylor series,[11] in the limit $\Delta z \to 0$ we have

$$\lim_{\Delta z \to 0} \frac{\mathcal{V}(z + \Delta z, t) - \mathcal{V}(z,t)}{\Delta z} = \frac{\partial \mathcal{V}(z,t)}{\partial z} = -R\mathcal{I}(z,t) - L\frac{\partial \mathcal{I}(z,t)}{\partial t}$$

or

$$\boxed{\frac{\partial \mathcal{V}(z,t)}{\partial z} = -\left(R + L\frac{\partial}{\partial t}\right)\mathcal{I}(z,t)} \tag{2.1}$$

Similarly, using Kirchhoff's current law, the difference between the current at the input and output terminals is equal to the total current through the parallel elements $G\Delta z$ and $C\Delta z$, so we have

$$\mathcal{I}(z + \Delta z, t) - \mathcal{I}(z,t) = -G\Delta z\, \mathcal{V}(z + \Delta z, t) - C\Delta z \frac{\partial \mathcal{V}(z + \Delta z, t)}{\partial t}$$

Upon dividing by Δz and expanding $\mathcal{V}(z + \Delta z, t)$ in a Taylor series, and taking $\Delta z \to 0$, we have:

$$\lim_{\Delta z \to 0}\left\{ \frac{\mathcal{I}(z + \Delta z, t) - \mathcal{I}(z,t)}{\Delta z} \right\} = -G\mathcal{V}(z,t) - C\frac{\partial \mathcal{V}(z,t)}{\partial t} - \lim_{\Delta z \to 0}\{\Delta z(\cdots)\}$$

or

$$\boxed{\frac{\partial \mathcal{I}(z,t)}{\partial z} = -\left(G + C\frac{\partial}{\partial t}\right)\mathcal{V}(z,t)} \tag{2.2}$$

Equations (2.1) and (2.2) are known as the *transmission line equations* or *telegrapher's equations*. We shall see in Chapter 8 that uniform plane electromagnetic wave propagation is based on very similar equations, written in terms of the components of electric and magnetic fields instead of voltages and currents. Most other types of wave phenomena are governed by similar equations; for acoustic waves in fluids, for example, one replaces voltage with pressure and current with velocity.

---

[11]$\mathcal{V}(z + \Delta z, t) = \mathcal{V}(z,t) + \dfrac{\partial \mathcal{V}(z,t)}{\partial z}\Delta z + \dfrac{1}{2!}\dfrac{\partial^2 \mathcal{V}(z,t)}{\partial z^2}\Delta z^2 + \cdots$

2.2.2 Traveling-Wave Solutions for Lossless Lines

Solutions of (2.1) and (2.2) are in general quite difficult and require numerical treatments for the general transient case when all of the transmission line parameters are nonzero. However, in a wide range of transmission line applications the series and parallel loss terms (R and G) can be neglected, in which case analytical solutions of (2.1) and (2.2) become possible. In fact, practical applications in which transmission lines can be treated as lossless lines are at least as common as those in which losses are important. Accordingly, our transmission line analyses in this chapter deal exclusively with lossless transmission lines.

We now apply (2.1) and (2.2) to the analysis of the transient response of lossless transmission lines. For a lossless line we have $R = 0$ and $G = 0$, so that (2.1) and (2.2) reduce to

$$\frac{\partial \mathcal{V}}{\partial z} = -L \frac{\partial \mathcal{I}}{\partial t} \tag{2.3}$$

$$\frac{\partial \mathcal{I}}{\partial z} = -C \frac{\partial \mathcal{V}}{\partial t} \tag{2.4}$$

By combining (2.3) and (2.4) we obtain the *wave equations* for voltage and current,

$$\boxed{\frac{\partial^2 \mathcal{V}}{\partial z^2} = LC \frac{\partial^2 \mathcal{V}}{\partial t^2}} \tag{2.5}$$

or

$$\boxed{\frac{\partial^2 \mathcal{I}}{\partial z^2} = LC \frac{\partial^2 \mathcal{I}}{\partial t^2}} \tag{2.6}$$

Either one of (2.5) or (2.6) can be solved for the voltage or current. We follow the usual practice and consider the solution of the voltage equation (2.5), which can be rewritten as

$$\frac{\partial^2 \mathcal{V}}{\partial z^2} = \frac{1}{v_p^2} \frac{\partial^2 \mathcal{V}}{\partial t^2}; \qquad v_p = \frac{1}{\sqrt{LC}} \tag{2.7}$$

Note that once $\mathcal{V}(z,t)$ is determined, we can use (2.3) or (2.4) to find $\mathcal{I}(z,t)$. The quantity v_p represents the speed of propagation of a disturbance, as will be evident in the following discussion. For reasons that will become clear in Chapter 3, v_p is also referred to as the *phase velocity;* hence the subscript p.

Any function[12] $f(\cdot)$ of the variable $\xi = (t - z/v_p)$ is a solution of (2.7). To see that

$$\mathcal{V}(z,t) = f\left(t - \frac{z}{v_p}\right) = f(\xi)$$

---

[12]An important function of $(t - z/v_p)$ that is encountered often and that we shall study in later chapters is the sinusoidal traveling-wave function, $A \cos[\omega(t - z/v_p)]$. Depending on the location of the observation point z along the z axis, this function replicates the sinusoidal variation $A \cos(\omega t)$ observed at $z = 0$, except delayed by (z/v_p) seconds at the new z. Thus, (z/v_p) represents a time shift, or delay, which is a characteristic of the class of wave functions of the variable $(t - z/v_p)$.

is a solution of (2.7), we can express the time and space derivatives of $\mathcal{V}(z,t)$ in terms of the derivatives of $f(\xi)$ with respect to ξ:

$$\frac{\partial \mathcal{V}}{\partial t} = \frac{\partial f}{\partial \xi}\frac{\partial \xi}{\partial t} = \frac{\partial f}{\partial \xi} \quad \text{and} \quad \frac{\partial^2 \mathcal{V}}{\partial t^2} = \frac{\partial^2 f}{\partial \xi^2}\frac{\partial \xi}{\partial t} = \frac{\partial^2 f}{\partial \xi^2}$$

since $\partial \xi/\partial t = 1$. Similarly, noting that $\partial \xi/\partial z = -(1/v_p)$, we have

$$\frac{\partial \mathcal{V}}{\partial z} = -\frac{1}{v_p}\frac{\partial f}{\partial \xi} \quad \text{and} \quad \frac{\partial^2 \mathcal{V}}{\partial z^2} = \frac{1}{v_p^2}\frac{\partial^2 f}{\partial \xi^2}$$

Substituting in (2.7) we find that the wave equation is indeed satisfied by any function $f(\cdot)$ of the variable $\xi = (t - z/v_p)$.

That an arbitrary function $f(t - z/v_p)$ represents a wave traveling in the $+z$ direction is illustrated in Figure 2.6 for $v_p = 1$ m-s^{-1}. By comparing $f(t - z/v_p)$ at two different times $t = 2$ and $t = 3$ s, we note that the function maintains its shape and moves in the $+z$ direction as time t advances, as seen in Figure 2.6a. Similarly, by comparing $f(t - z/v_p)$ at two different positions $z = 0$ and $z = 1$ m, we note that the function maintains its shape but appears at $z = 1$ m exactly 1 s after its appearance at $z = 0$, as seen in Figure 2.6b. Figure 2.6c shows a three-dimensional display of $f(t - z/v_p)$ as a function of time at different points z_1, z_2, and z_3. To determine the speed with which the function moves in the $+z$ direction, we can simply keep track of any point on the

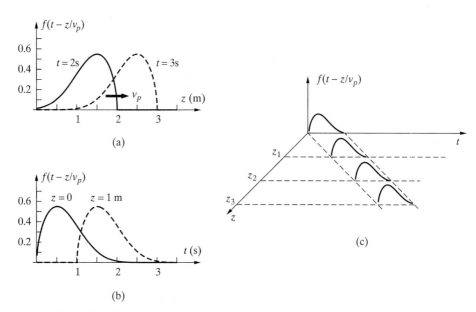

Figure 2.6 Variation in space and time of an arbitrary function $f(t - z/v_p)$**.** The phase velocity is taken to be equal to unity, that is, $v_p = 1$ m-s^{-1}. (a) $f(t - z/v_p)$ versus z at two different times. (b) $f(t - z/v_p)$ versus t at two different locations. (c) Three-dimensional display of $f(t - z/v_p)$ as a function of time at different points z_1, z_2, and z_3.

function by setting its argument to a constant. In other words, we have

$$t - \frac{z}{v_p} = \text{const} \quad \rightarrow \quad \frac{dz}{dt} = v_p$$

The speed of propagation of waves on a transmission line is one of its most important characteristics. It is evident from (2.7) that v_p depends on the line inductance L and capacitance C. In the case of most (all except the microstrip line) of the commonly used two-conductor transmission lines shown in Figure 2.1, the phase velocity v_p in the absence of losses is not a function of the particular geometry of the metallic conductors but is solely determined by the electrical and magnetic properties of the surrounding medium.[13] When the medium surrounding the metallic conductors is air, the phase velocity is equal to the speed of light in free space, namely $v_p = c$. The propagation speeds for some other insulating materials are tabulated in Table 2.1.

It is clear from the above analysis that any function of the argument $(t + z/v_p)$ is an equally valid solution of (2.7). Thus, the general solution for the voltage $\mathcal{V}(z,t)$ is

$$\mathcal{V}(z,t) = f^+\left(t - \frac{z}{v_p}\right) + f^-\left(t + \frac{z}{v_p}\right) \tag{2.8}$$

TABLE 2.1   PROPAGATION SPEEDS IN SOME MATERIALS

Material	Propagation Speed at $\sim 20°C$ (cm-(ns)$^{-1}$ at 3 GHz)
Air	30
Glass	(3–15)*
Mica (ruby)	12.9
Porcelain	(10–13)*
Fused quartz (SiO$_2$)	15.4
Alumina (Al$_2$O$_3$)	10.1
Polystyrene	18.8
Polyethylene	20.0
Teflon	20.7
Vaseline	20.4
Amber (fossil resin)	18.6
Wood (balsa)	27.2
Water (distilled)	3.43
Ice (pure distilled)	16.8**
Soil (sandy, dry)	18.8

*Approximate range valid for most types of this material.
**At $-12°C$.

[13]This result will become evident in Chapters 4 and 6 as we determine the capacitance and inductance of selected transmission lines from first principles. That $v_p = (LC)^{-1/2}$ does not depend on the geometrical arrangement of the conductors can also be seen by considering the inductance and capacitance expressions given in Table 2.2, Section 2.7. For transmission lines that do not exhibit symmetry in the cross-sectional plane, such as the microstrip line of Figure 2.1e, the phase velocity depends in a complicated manner on the properties of the surrounding material, the shape and dimensions of the conductors, and the operating frequency. See Section 8.6 of S. Ramo, J. R. Whinnery, and T. Van Duzer, *Fields and Waves in Communication Electronics,* 3rd ed., John Wiley & Sons, New York, 1994.

where $f^+(t - z/v_p)$ and $f^-(t + z/v_p)$ represent waves traveling in the $+z$ and $-z$ directions, respectively. Note that $f^+(\cdot)$ and $f^-(\cdot)$ can in general be completely different functions.

To find the general solution for the current $\mathcal{I}(z,t)$ associated with the voltage $\mathcal{V}(z,t)$, we can substitute (2.8) in (2.3) (or (2.4)), integrate with respect to time (or position), and then take the derivative with respect to z (or time) to find

$$\mathcal{I}(z,t) = \frac{1}{Z_0}\left[f^+\left(t - \frac{z}{v_p}\right) - f^-\left(t + \frac{z}{v_p}\right)\right]; \qquad Z_0 \equiv \sqrt{\frac{L}{C}} \qquad (2.9)$$

where Z_0 is known as the *characteristic impedance* of the transmission line.[14] The characteristic impedance is the ratio of voltage to current for a single wave propagating in the $+z$ direction, as is evident from (2.8) and (2.9). Note that the current associated with the wave traveling in the $-z$ direction (i.e., toward the left) has a negative sign, as expected, since the direction of positive current as defined in Figure 2.5 is to the right. In other words, since the polarity of voltage and the direction of current are defined so that the voltage and current have the same signs for forward (to the right)-traveling waves, the voltage and current for waves traveling to the left have opposite signs.

The characteristic impedance of a line is one of the most important parameters in the equations describing its behavior as a distributed circuit. For lossless lines, as considered here, Z_0 is a real number having units of ohms. Since Z_0 for a lossless line depends only on L and C, and since these quantities can be calculated from the geometric shape and physical dimensions of the line and the properties of the surrounding material, Z_0 can be expressed in terms of these physical dimensions for a given type of line. Formulas for Z_0 for some common lines are provided in Section 2.7. Characteristic impedances for other types of transmission lines are available in other publications.[15] The following example illustrates the meaning of the characteristic impedance of a line. An infinitely long and initially uncharged line (i.e., all capacitors and inductors in the distributed circuit have zero initial conditions) is considered, so there is no need for the $f(t + z/v_p)$ term in either (2.8) or (2.9), which would be produced only as a result of the reflections of the voltage disturbance at the end of the line.

Example 2.1: Step response of an infinitely long lossless line. As a simple example of the excitation of a transmission line by a source, consider an infinitely long lossless transmission line characterized by L and C and connected to an ideal step voltage source of amplitude V_0 and source

---

[14]To find $\mathcal{I}(z,t)$, we can also note that the wave equation (2.6) for the current is identical to equation (2.5) for voltage, so its solution should have the same general form. Thus, the general solution for the current should be

$$\mathcal{I}(z,t) = g^+\left(t - \frac{z}{v_p}\right) + g^-\left(t + \frac{z}{v_p}\right)$$

Substituting this expression for $\mathcal{I}(z,t)$ and (2.8) into (2.3) or (2.4) yields $g^+ = Z_0^{-1}f^+$ and $g^- = -Z_0^{-1}f^-$.

[15]*Reference Data for Engineers*, 8th ed., Sams Prentice Hall Computer Publishing, Carmel, Indiana, 1993.

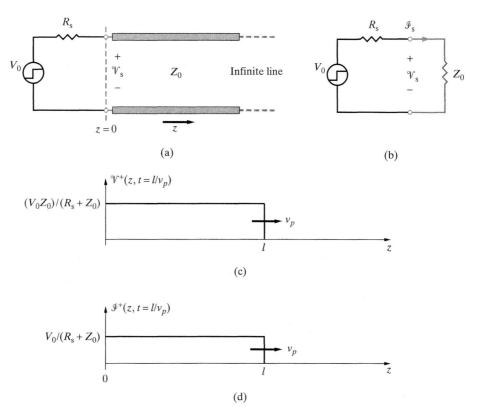

Figure 2.7   Step excitation of a lossless line. (a) Step voltage applied to an infinite lossless line. (b) Initial equivalent circuit seen from the source. (c) The voltage disturbance $\mathcal{V}^+(z,t)$. (d) The current disturbance $\mathcal{I}^+(z,t)$.

resistance R_s, as shown in Figure 2.7a. Find the voltage, the current, and power propagating down the transmission line.

Solution: Before $t = 0$, the voltage and current on the line are identically zero, since the line is assumed to be initially uncharged. At $t = 0$, the step voltage source changes from 0 to V_0, launching a voltage $\mathcal{V}^+(z,t)$ at $z = 0$ propagating toward the right. In the absence of a reflected wave (infinitely long line), the accompanying current is $\mathcal{I}^+(z,t) = (Z_0)^{-1}\mathcal{V}^+(z,t)$, as can be seen from equations (2.8) and (2.9). In other words, the characteristic impedance $Z_0 = \sqrt{L/C}$ is the "resistance" that the transmission line *initially* presents to the source, as shown in the equivalent circuit of Figure 2.7b. Accordingly, the initial voltage and current established at the source end ($z = 0$) of the line are

$$\mathcal{V}_s(t) = \mathcal{V}^+(z = 0, t) = \frac{V_0 Z_0}{Z_0 + R_s}$$

and

$$\mathcal{I}_s(t) = \mathcal{I}^+(z = 0, t) = \frac{V_0}{Z_0 + R_s}$$

The propagation of the voltage $\mathcal{V}^+(z,t)$ and the current $\mathcal{I}^+(z,t)$ down the line are illustrated in Figures 2.7c and d at $t = l/v_p$ as a function of z.

Note that the flow of a current $\mathcal{I}_s$ outward from a source producing a voltage V_0 represents a total power of $P_0 = \mathcal{I}_s V_0$ supplied by the source. Part of this power, given by $\mathcal{I}_s^2 R_s$, is dissipated in the source resistance. The remainder, given by

$$P_{\text{line}}^+ = \mathcal{I}^+(0,t)\mathcal{V}^+(0,t) = \mathcal{I}_s \mathcal{V}_s = \frac{V_0^2 Z_0}{(Z_0 + R_s)^2}$$

is supplied to the line. Because the line is lossless, there is no power dissipation on the line. Therefore, the power P_{line}^+ goes into charging the capacitances and the inductances[16] of the line, as discussed in connection with Figure 2.2. Note that as P_{line}^+ travels down the line the amounts of energy stored respectively in the capacitance and inductance of a fully charged portion of the line of length dl are given by $\frac{1}{2}(C\,dl)\mathcal{V}_s^2$ and $\frac{1}{2}(L\,dl)\mathcal{I}_s^2$.

2.3 REFLECTION AT DISCONTINUITIES

In most transmission line applications, lines are connected to resistive loads, other lines (with different characteristic impedances), reactive loads, or combinations of resistive and reactive loads. Such discontinuities impose boundary conditions, which cause reflection of the incident voltages and currents from the discontinuities, resulting in new voltages and currents that are launched in the opposite direction. In this section we consider the reflection process and also provide examples of step responses of transmission lines terminated with short- and open-circuited terminations.

Consider a transmission line terminated in a load resistance R_L located at $z = l$, as shown in Figure 2.8, on which a voltage of $\mathcal{V}_1^+(z,t)$ is initially ($t = 0$) launched by the source at $z = 0$. Note that for an ideal step voltage source of amplitude V_0 and a source resistance R_s, as shown in Figure 2.8, the amplitude of $\mathcal{V}_1^+(z,t)$ is given by

$$\mathcal{V}_1^+(0,0) = \frac{Z_0 V_0}{R_s + Z_0}$$

In general, a new reflected voltage $\mathcal{V}_1^-(z,t)$ is generated when the disturbance $\mathcal{V}_1^+(z,t)$ reaches the load position at time $t = t_d$, where t_d is the one-way travel time on the line, or $t_d = l/v_p$. In order to determine the amplitude of the reflected wave, we write the total voltage and current at the load position (i.e., $z = l$) at $t = t_d^+$ (i.e., immediately after the arrival of the incident wave) as

$$\mathcal{V}_L(t) = \mathcal{V}_1^+(l,t) + \mathcal{V}_1^-(l,t) \tag{2.10a}$$

$$\mathcal{I}_L(t) = \frac{1}{Z_0}\mathcal{V}_1^+(l,t) - \frac{1}{Z_0}\mathcal{V}_1^-(l,t) \tag{2.10b}$$

---

[16]"Charging" an inductance can be thought of as establishing a current in it.

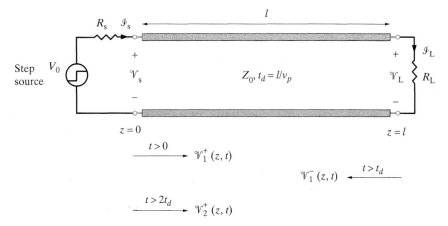

Figure 2.8   A terminated transmission line. The load R_L is located at $z = l$, while the source end is at $z = 0$.

Using (2.10a) and (2.10b) and the *boundary condition* $\mathcal{V}_L(t) = \mathcal{I}_L(t)R_L$ imposed by the purely resistive termination R_L, we can write

$$\mathcal{I}_L(t) = \frac{\mathcal{V}_L(t)}{R_L} \quad \rightarrow \quad \frac{\mathcal{V}_1^+(l,t)}{Z_0} - \frac{\mathcal{V}_1^-(l,t)}{Z_0} = \frac{\mathcal{V}_1^+(l,t) + \mathcal{V}_1^-(l,t)}{R_L} \qquad (2.11)$$

From (2.11) we can find the ratio of the reflected voltage $\mathcal{V}_1^-(l,t)$ to the incident one $\mathcal{V}_1^+(l,t)$. This ratio is defined as the *load voltage reflection coefficient*, Γ_L,

$$\boxed{\Gamma_L \equiv \frac{\mathcal{V}_1^-(l,t)}{\mathcal{V}_1^+(l,t)} = \frac{R_L - Z_0}{R_L + Z_0}} \qquad (2.12)$$

and it follows that

$$\frac{R_L}{Z_0} = \frac{1 + \Gamma_L}{1 - \Gamma_L} \qquad (2.13)$$

The reflection coefficient is one of the most important parameters in transmission line analysis. Accordingly, the simple expression (2.12) for Γ_L should be memorized. For lines terminated in resistive loads, Γ_L can have values in the range $-1 \leq \Gamma_L \leq +1$, where the extreme values of -1 and $+1$ occur when the load is, respectively, a short circuit (i.e., $R_L = 0$) and an open circuit (i.e., $R_L = \infty$). The special case of $\Gamma_L = 0$ occurs when $R_L = Z_0$; meaning that the load resistance is the same as the characteristic impedance,[17] that there is no reflected voltage, and that all the power carried by the incident voltage is absorbed by the load. It is important to note that expression (2.12) for the reflection coefficient was arrived at in a completely general fashion. In other words, whenever a voltage $\mathcal{V}_k^+(z,t)$ is incident on a load R_L from a transmission line with characteristic

---

[17]This condition is referred to as a *matched* load and is highly preferred in most applications.

impedance Z_0, the amplitude of the reflected voltage $\mathcal{V}_k^-(l,t)$ is $\Gamma_L \mathcal{V}_k^+(l,t)$, with Γ_L given by (2.12).

The generality of (2.12) can also be used to determine the reflection of the new voltage $\mathcal{V}_1^-(z,t)$ when it arrives at the source end of the line. Having originated at the load end at $t = t_d$, the reflected voltage $\mathcal{V}_1^-(z,t)$ arrives at the source end (terminated with the source resistance R_s) at $t = 2t_d$. At that point, it can be viewed as a new voltage disturbance propagating on a line with characteristic impedance Z_0 that is incident on a resistance of R_s. Thus, its arrival at the source end results in the generation of yet another reflected voltage traveling toward the load. We denote this new voltage $\mathcal{V}_2^+(z,t)$, where the subscript distinguishes it from the original voltage $\mathcal{V}_1^+(z,t)$ and the superscript underscores the fact that it is propagating in the $+z$ direction. The amplitude of the new reflected voltage $\mathcal{V}_2^+(z,t)$ is determined by the *source reflection coefficient* Γ_s, which applies at the source end of the line and is defined as

$$\Gamma_s \equiv \frac{\mathcal{V}_2^+(0,t)}{\mathcal{V}_1^-(0,t)} = \frac{R_s - Z_0}{R_s + Z_0} \tag{2.14}$$

Thus we have $\mathcal{V}_2^+(z,t) = \mathcal{V}_2^+(0,t) = \Gamma_s \mathcal{V}_1^-(0,t)$.

Note that the voltage $\mathcal{V}_1^+(z,t)$ was created at $t = 0$ and still continues to exist. Thus the source-end voltage and current at $t = 2t_d^+$ are

$$\mathcal{V}_s(t) = \mathcal{V}_1^+(0,t) + \mathcal{V}_1^-(0,t) + \mathcal{V}_2^+(0,t)$$

$$= \mathcal{V}_1^+(0,t)(1 + \Gamma_L + \Gamma_s\Gamma_L) \tag{2.15a}$$

$$\mathcal{I}_s(t) = \frac{1}{Z_0}\mathcal{V}_1^+(0,t) - \frac{1}{Z_0}\mathcal{V}_1^-(0,t) + \frac{1}{Z_0}\mathcal{V}_2^+(0,t)$$

$$= \frac{1}{Z_0}\mathcal{V}_1^+(0,t)(1 - \Gamma_L + \Gamma_s\Gamma_L) \tag{2.15b}$$

The newly generated voltage $\mathcal{V}_2^+(z,t)$ will now arrive at the load end at $t = 3t_d$ and lead to the creation of another new reflected voltage $\mathcal{V}_2^-(z,t)$, and this process will continue indefinitely. In practice, the step-by-step calculation of the successively generated voltages becomes tedious, especially for arbitrary resistive terminations. In such cases, the graphical construction of a *bounce diagram* is very helpful. We introduce the bounce diagram in the following subsection.

2.3.1 Bounce Diagrams

A bounce diagram, illustrated in Figure 2.9, also called a reflection diagram or lattice diagram, is a distance–time plot used to illustrate successive reflections along a transmission line. The distance along the line is shown on the horizontal axis and time is shown on the vertical axis. The bounce diagram is a plot of the time elapsed versus distance z from the source end, showing the voltages traveling in the $+z$ and $-z$ directions as straight

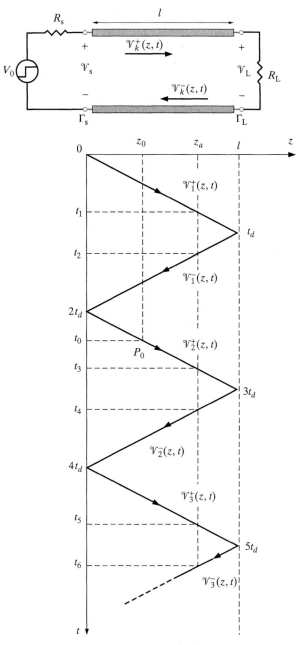

Figure 2.9   Bounce diagram.

lines sloping[18] downward from left to right and right to left, respectively. Each sloping line corresponds to an individual traveling voltage and is labeled with its amplitude. The amplitude of each reflected voltage is obtained by multiplying the amplitude of the preceding voltage by the reflection coefficient at the position where the reflection occurs.

The time sequence of events starting with the first application of the step voltage at the source end can be easily visualized from the bounce diagram. The application of the step voltage launches $\mathcal{V}_1^+(z,t)$ traveling toward the load. This voltage arrives at the load end at $t = t_d$, and its arrival leads to the generation of $\mathcal{V}_1^-(z,t) = \Gamma_L \mathcal{V}_1^+(z,t)$ propagating toward the source. This new voltage $\mathcal{V}_1^-(z,t)$ arrives at the source end and leads to the generation of $\mathcal{V}_2^+(z,t)$, and this process continues back and forth indefinitely.

Once constructed, a bounce diagram can be used conveniently to determine the voltage distribution along the transmission line at any given time, as well as the variation of voltage with time at any given position. Suppose we wish to find the voltage distribution $\mathcal{V}(z,t_0)$ along the line at $t = t_0$, chosen arbitrarily to be $2t_d < t_0 < 3t_d$. To determine $\mathcal{V}(z,t_0)$, we mark t_0 on the time axis of the bounce diagram and draw a horizontal line from t_0, intersecting the sloping line marked $\mathcal{V}_2^+(z,t)$ at point P_0, as shown in Figure 2.9. Note that all sloping lines below the point P_0 are irrelevant for $\mathcal{V}(z,t_0)$, since they correspond to later times. If we now draw a vertical dashed line through P_0, we find that it intersects the z axis at z_0. At time $t = t_0$, all points along the line have received voltages $\mathcal{V}_1^+(z,t)$ and $\mathcal{V}_1^-(z,t)$. However, only points to the left of z_0 have yet received the voltage $\mathcal{V}_2^+(z,t)$, so a discontinuity exists in the voltage distribution at $z = z_0$. In other words, we have

$$\mathcal{V}(z,t_0) = \begin{cases} \mathcal{V}_1^+(z,t)(1 + \Gamma_L + \Gamma_s\Gamma_L) & z < z_0 \\ \mathcal{V}_1^+(z,t)(1 + \Gamma_L) & z > z_0 \end{cases}$$

Alternatively, we may wish to determine the variation of voltage as a function of time at a fixed position, say z_a. To determine $\mathcal{V}(z_a,t)$, we simply look at the intersection points with the sloping lines of the vertical line passing through z_a, as shown in Figure 2.9. Horizontal lines drawn from these intersection points, crossing the time axis at $t_1, t_2, t_3, t_4,\ldots$, are the time instants at which each of the new voltages $\mathcal{V}_1^+(z,t)$, $\mathcal{V}_1^-(z,t)$, $\mathcal{V}_2^+(z,t)$, $\mathcal{V}_2^-(z,t),\ldots$, arrive at $z = z_a$ and abruptly change the total voltage at that point. Thus, the time variation of the voltage at $z = z_a$, namely $\mathcal{V}(z_a,t)$ is given as

$$\mathcal{V}(z_a,t) = \begin{cases} 0 & 0 < t < t_1 \\ \mathcal{V}_1^+(z_a,t) & t_1 < t < t_2 \\ \mathcal{V}_1^+(z_a,t)(1 + \Gamma_L) & t_2 < t < t_3 \\ \mathcal{V}_1^+(z_a,t)(1 + \Gamma_L + \Gamma_s\Gamma_L) & t_3 < t < t_4 \\ \mathcal{V}_1^+(z_a,t)(1 + \Gamma_L + \Gamma_s\Gamma_L + \Gamma_L\Gamma_s\Gamma_L) & t_4 < t < t_5 \\ \cdots & \cdots \end{cases}$$

where $\mathcal{V}_1^+(z_a,t) = \mathcal{V}_1^+(z,t)$.

---

[18]The slope of the lines on the bounce diagram (i.e., dt/dz) can be thought of as corresponding to $(\pm v_p)^{-1}$.

A bounce diagram can also be constructed to keep track of the component current waves. However, it is also possible and less cumbersome to derive the component line current, $\mathcal{I}_k^{\pm}(z,t)$, simply from the corresponding voltage $\mathcal{V}_k^{\pm}(z,t)$. In this connection, all we need to remember is that the current associated with any voltage disturbance $\mathcal{V}_k^{+}(z,t)$ propagating in the $+z$ direction is simply $\mathcal{I}_k^{+}(z,t) = (Z_0^{-1})\mathcal{V}_k^{+}(z,t)$, whereas that associated with a voltage disturbance $\mathcal{V}_k^{-}(z,t)$ propagating in the $-z$ direction is $\mathcal{I}_k^{-}(z,t) = -(Z_0^{-1})\mathcal{V}_k^{-}(z,t)$.

2.3.2 The Reflection Process

Before we proceed with specific examples, we now provide a heuristic discussion of the reflection process for the case of a transmission line terminated in an open circuit.[19] This discussion involves the same considerations as the heuristic discussion in connection with Figure 2.2 of the propagation of disturbances along a transmission line in terms of successive charging of capacitors and inductors. When the voltage disturbance reaches the open-circuited end of the line, its orderly progress of successively charging the distributed circuit elements is interrupted. Consider the approach of a disturbance to the end of an open-circuited transmission line, shown in Figure 2.10a. Figure 2.10b shows the voltage and current progressing together; L_y carries current but L_z does not, and C_y is charged to the source voltage V_0 but C_z is not. The voltage on C_y, however, causes current to flow through L_z, and thus through C_z, charging it to a voltage V_0 (Figure 2.10c).

At the time C_z is charged, all of the inductances (including L_z) carry the full current I_0. The progress of the disturbance along the line cannot continue any more, since there is no inductance beyond C_z to serve as an outlet for current as C_z is charged. As a result, C_z becomes overcharged, since the current through L_z cannot stop until the stored magnetic energy is exhausted. Thus, current continues to flow through C_z until it is charged[20] to twice its normal value ($2V_0$), at which time the current through L_z drops to zero (i.e., L_z now acts like an open circuit) (Figure 2.10d).

When L_z stops carrying current, the current carried by L_y is now driven solely into C_y, doubling its voltage and forcing the current in L_y to stop. At the same time, the voltages at the two ends of L_z are both $2V_0$, so that the current through L_z remains zero and the doubly charged capacitor C_z stays at $2V_0$. We now have the condition depicted in Figure 2.10e, where C_y and C_z are both at $2V_0$ and L_y and L_z both have zero current. This procedure now continues along the line from right to left as the voltage on the line is doubled and the current drops to zero.

---

[19]The qualitative discussion presented herein is based on a similar discussion in Chapter 14 of H. H. Skilling, *Electric Transmission Lines,* McGraw-Hill, New York, 1951.

[20]It is not obvious why the capacitor would charge to precisely twice its normal value. The circuit model of Figure 2.10a, consisting of discrete elements, is not adequate for the determination of the precise value of the reflected voltage, which is unambiguously determined by the governing wave equations (2.5), (2.6), and their solutions (2.10) as applied to an open-circuited termination, as shown in the next section. Nevertheless, consider the fact that the amount that the capacitor voltage is charged to is determined by $\int I\,dt$; thus, with no other outlet for the inductor current, twice the normal current goes through the capacitance, charging it to twice its normal value.

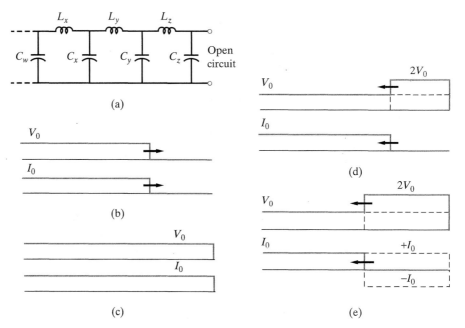

Figure 2.10 The reflection process. The orderly progress of the disturbances of current and voltage, propagating initially from left to right, is interrupted when they reach the open end of the line, leading to the reflection of the disturbance.

The above described phenomenon can be viewed as a reflection, since the original disturbance, traveling from left to right, appears to be reflected from the end of the line and to begin to progress from right to left. In Figure 2.10e, the reflected voltage disturbance of amplitude V_0 travels toward the left, and adds to the previously existing line voltage, making the total voltage $2V_0$. It is accompanied by the reflected current of amplitude $-I_0$, which is added on top of the existing line current I_0, making the total current on the line zero. Although the front of the current disturbance is progressing toward the left, it should be noted that this does not imply any reversal of current flow. The current originally flowing from the source toward the end of the line (i.e., $\mathscr{I}_1^+(z,t)$) continues to do so even after reflection. This current flows from left to right to charge the capacitances when the disturbance progresses toward the end of the line, and it continues to flow from left to right as the reflected disturbance returns, doubling the charge on the line as the voltage is raised to $2V_0$.

All of these physical effects of charging of capacitors and establishing current through inductors are simply accounted for by the general solutions for voltage and current as given respectively by (2.8) and (2.9) and by the application of the boundary condition at the termination—namely, that there must always be zero current at the end of an open-circuited line. The purpose of this heuristic discussion is simply to provide a qualitative understanding of the reflection process in terms of the equivalent circuit of the line.

2.3.3 Open- and Short-Circuited Transmission Lines

We now consider examples of step responses of transmission lines with the simplest type of terminations, namely an open or a short circuit. The circumstances treated in Examples 2.2 and 2.3 are commonly encountered in practice, especially in computer-communication problems; for example, when the voltage at one end of an interconnect switches to the HIGH state due to a change in the status of a logic gate. The resultant response is then similar to the short-circuited line case (Example 2.2) if the interconnect is a short-circuited matching stub or drives (i.e., is terminated in) a subsequent gate (or another interconnect) with low input impedance ($R_L \ll Z_0$). Alternatively, and more commonly in practice, the interconnect would be driving a gate with a very high input impedance ($R_L \gg Z_0$), corresponding to an open-circuited termination (Example 2.3).

Example 2.2: Step voltage applied to a short-circuited lossless line. Consider the transmission line of length l terminated with a short circuit at the end (i.e., $R_L = 0$), as shown in Figure 2.11a. Sketch the voltages $\mathcal{V}_s$ and $\mathcal{V}_{l/2}$ as a function of time.

> **Solution:**   Initially, the applied voltage is divided between the source resistance R_s and the line impedance Z_0 in the same manner as for the infinite line in Example 2.1, and at $t = 0^+$, a voltage $\mathcal{V}_1^+(z,t)$ of amplitude $\mathcal{V}_1^+(0,0) = (V_0 Z_0)/(Z_0 + R_s)$ is launched at the source end of the line. The corresponding current is $\mathcal{I}_1^+(z,t) = \mathcal{V}_1^+(z,t)/Z_0$ and has an amplitude of $\mathcal{I}_1^+(0,0) = \mathcal{V}_1^+(0,0)/Z_0 = V_0/(Z_0 + R_s)$. The equivalent circuit initially presented to the source by the line is thus simply a resistance of Z_0, as shown in Figure 2.11b. Eventually, when all transients die out, the equivalent circuit of the line is a short circuit (Figure 2.11c); thus, the voltage everywhere on the line (e.g., $\mathcal{V}_s$, $\mathcal{V}_{l/2}$, and $\mathcal{V}_L$) must eventually approach zero.

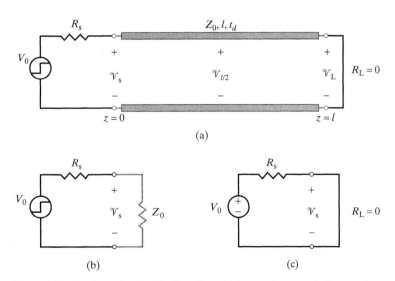

Figure 2.11   A short-circuited lossless line. (a) Step voltage applied to a short-circuited lossless line. (b) The initial equivalent circuit. (c) Steady-state (final) equivalent circuit.

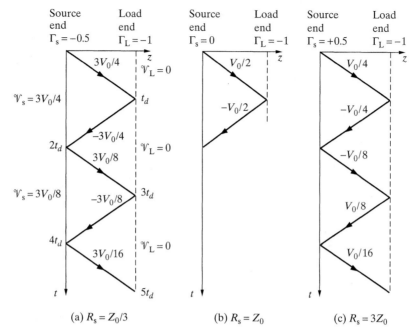

Figure 2.12   Bounce diagram for Example 2.2 (Figure 2.11). (a) $R_s = Z_0/3$, (b) $R_s = Z_0$, and (c) $R_s = 3Z_0$.

To analyze the behavior of the line voltage, we use a bounce diagram as shown in Figure 2.12. When the voltage disturbance $\mathcal{V}^+(z,t)$ reaches the end of the line (at $t = t_d = l/v_p$), a reflected voltage of $\mathcal{V}^-(z,t) = -\mathcal{V}^+(z,t)$ is generated,[21] since the total voltage at the short circuit ($R_L = 0$) has to be $\mathcal{V}_L(t) = \mathcal{V}^+(l,t) + \mathcal{V}^-(l,t) = 0$. In other words, the reflection coefficient at the load end is

$$\Gamma_L = \frac{\mathcal{V}_1^-(l,t)}{\mathcal{V}_1^+(l,t)} = \frac{0 - Z_0}{0 + Z_0} = -1$$

However, the current $\mathcal{I}_1^-(z,t)$ associated with the voltage traveling in the $-z$ direction is $-\mathcal{V}_1^-(z,t)/Z_0$, resulting in a reflected current of $V_0/(Z_0 + R_s)$, which adds directly to the incident current $\mathcal{I}_1^+(z,t) = V_0/(Z_0 + R_s)$ traveling in the $+z$ direction, doubling the total current on the line.

As it travels toward the source during $t_d < t < 2t_d$, the reflected voltage makes the total voltage everywhere on the line zero and the total current on the line equal to $2V_0/(Z_0 + R_s)$. When this disturbance reaches the source end at $t = 2t_d$, the source presents

---

[21]The reflection process at the short circuit occurs rather differently than that from an open circuit as discussed in connection with Figure 2.10. As the next-to-the-last capacitance (C_y) is charged to V_0, L_z begins to carry current; however, C_z cannot take any charge, since it is short-circuited. Current flows freely from L_z through the short circuit and into the return conductor (just like the flow of water from the open end of a pipe), until the charge on C_y is exhausted. As a result, the current through L_z becomes twice as much as its normal value ($2I_0$), and the voltage across C_y drops to zero. In this way, "reflection" reduces the voltage from V_0 to zero and increases current from I_0 to $2I_0$.

an impedance of R_s, and the reflection coefficient at the source end (i.e., Γ_s) depends on the value of R_s relative to Z_0. For $R_s = Z_0$, we have $\Gamma_s = 0$, and no further reflections occur, so the voltage on the line remains zero; in other words, a steady state is reached. However, for $R_s \neq Z_0$, it takes further reflections to eventually reach steady state, as illustrated in Figure 2.13, where the time evolution of the voltages at the source end $\left(\mathcal{V}_s(t)\right)$ and at the center $\left(\mathcal{V}_{l/2}(t)\right)$ of the transmission line are shown. Note that the load voltage $\mathcal{V}_L(t)$ is

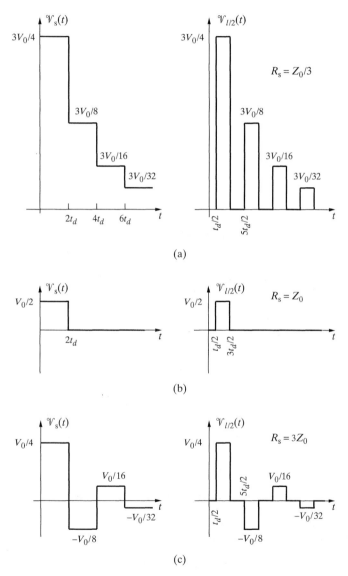

Figure 2.13 Step voltage applied to a short-circuited lossless line. Voltages at the source end $\left(\mathcal{V}_s\right)$ and at the center $\left(\mathcal{V}_{l/2}\right)$ of a short-circuited transmission line for source impedances of (a) $R_s = Z_0/3$, (b) $R_s = Z_0$, and (c) $R_s = 3Z_0$.

identically zero at all times, as dictated by the short-circuit termination. Note also that the voltage everywhere on the line, including at the center and at the source end, eventually approaches zero; however, we see from Figure 2.13 that the particular way in which $\mathcal{V}_s(t)$ and $\mathcal{V}_{1/2}(t)$ approach their final value of zero depends critically on the ratio R_s/Z_0.

Example 2.3: Overshoot and ringing effects. The distributed nature of a high-speed digital logic board commonly leads to *ringing,* the fluctuations of the voltage and current about an asymptotic value. Ringing results from multiple reflections, especially when an unterminated

(a)

(b)

Figure 2.14 Step response of an open-circuited line. (a) Circuit for Example 2.3. (b) Voltage and current distributions along the line at different time intervals.

(i.e., open-circuited)[22] transmission line is driven by a low-impedance buffer. Consider the circuit shown in Figure 2.14a, where a step voltage source of amplitude V_0 and a source resistance $R_s = Z_0/4$ drives a lossless transmission line of characteristic impedance Z_0 and a one-way propagation delay of t_d seconds. Sketch $\mathcal{V}_s$, $\mathcal{V}_L$, and $\mathcal{I}_s$ as a function of t.

> **Solution:** At $t = 0$, the source voltage rises from 0 to V_0, and a voltage $\mathcal{V}_1^+(z,t)$ of amplitude $\mathcal{V}_1^+(0,0) = Z_0 V_0/(R_s + Z_0) = 0.8 V_0$ is applied from the source end of the line. During $0 < t < t_d$, the line charges as the front of this voltage disturbance travels toward the load. At $t = t_d$, the disturbance front reaches the open end of the line, and a reflected voltage of amplitude $\mathcal{V}_1^-(l,t_d) = \Gamma_L \mathcal{V}_1^+(l,t_d) = 0.8 V_0$ is launched back toward the source, since $\Gamma_L = 1$. (In other words, the total current at the open end of the line remains zero, so we have $\mathcal{I}_1^+(l,t) + \mathcal{I}_1^-(l,t) = 0$, and thus $\mathcal{V}_1^-(l,t) = \mathcal{V}_1^+(l,t) = 0.8 V_0$.) Note that as long as the source voltage does not change, $\mathcal{V}_1^+(z,t)$ remains constant in time and also constant with z once it reaches the end of the line ($z = l$) at $t = t_d$. Once $\mathcal{V}_1^-(z,t)$ is launched (at $t = t_d^+$), it travels toward the source, reaching the source end at $t = 2t_d$, after which time it also remains constant and coexists along the line with $\mathcal{V}_1^+(z,t)$. The arrival of the reflected voltage $\mathcal{V}_1^-(z,t)$ at the source end of the line at $t = 2t_d$ leads to the generation of a new reflected voltage disturbance of amplitude $\mathcal{V}_2^+(0,2t_d) = \Gamma_s \mathcal{V}_1^-(0,2t_d) = -0.48 V_0$ [since $\Gamma_s = (R_s - Z_0)/(R_s + Z_0) = -0.6$] is launched toward the load. Note that $\mathcal{V}_2^+(z,t)$ is now superimposed on top of $\mathcal{V}_1^+(z,t)$ and $\mathcal{V}_1^-(z,t)$.
>
> This process continues indefinitely, with the total voltage and current on the line gradually approaching their steady-state values. The voltage and current distribution along the line is shown in Figure 2.14b for various time intervals. The variation of the voltage (and thus the current) with time can be quantitatively determined by means of a bounce diagram, as shown in Figure 2.15b, which specifies the values of the source- and load-end voltages at any given time. The temporal variations of the source- and load-end voltages and the source-end current, as derived from the bounce diagram, are also shown in Figure 2.15c. Both voltage waveforms oscillate about and asymptotically approach their final value V_0 (shown as dashed lines)—the process referred to earlier as ringing. The source-end current waveform eventually approaches zero, as expected for an open-circuited termination. For the case shown, the percentage maximum overshoot, defined as the percentage difference between the maximum value and the asymptotic value, for $\mathcal{V}_L$ is $[(1.6 V_0 - V_0)/V_0] \times 100 = 60\%$.

2.4 TRANSIENT RESPONSE OF TRANSMISSION LINES WITH RESISTIVE TERMINATIONS

Our discussions in the preceding section served to introduce the concepts of reflection at discontinuities in the context of the relatively simple open- and short-circuited terminations. In this section, we study the response of transmission lines terminated with an arbitrary resistance R_L to excitations in the form of a step change in voltage (e.g., an applied voltage changing from 0 to V_0 at $t = 0$) or a short voltage pulse of a given duration. Step excitation represents such cases as the output voltage of a driver gate changing from LOW to HIGH or HIGH to LOW state at a specific time, while pulse excitations are relevant to a broad class of computer communication problems. We consider two

---

[22]Note that if the input impedance of a load device is very high compared with Z_0 (i.e., $R_L \gg Z_0$), R_L can be approximated as an open circuit.

Figure 2.15 Step response of an open-circuited lossless line. (a) Circuit for Example 2.3. (b) Bounce diagram. (c) Normalized source- and load-end voltages and source-end current as a function of t/t_d.

cases of resistive terminations: (i) single lossless transmission lines terminated in resistive loads and (ii) lossless transmission lines terminated in other lossless transmission lines. Resistively terminated lines and transmission lines terminated in other lines are encountered very often in practice. In digital communication applications, for example, logic gates are often connected via an interconnect to other gates with specific input resistances, and interconnects often drive combinations of other interconnects.

2.4.1 Single Transmission Lines with Resistive Terminations

We start with a general discussion of the step response of transmission lines with resistive terminations. Consider the circuit shown in Figure 2.16a where a step voltage source (0 to V_0 at $t = 0$) with a source resistance R_s drives a lossless transmission line of

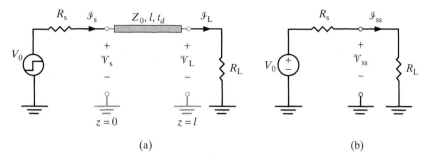

Figure 2.16   Resistively terminated line. (a) Step excitation of a lossless transmission line terminated with a resistive load. (b) Steady-state equivalent circuit seen by the source.

characteristic impedance Z_0 and one-way time delay t_d, terminated in a load resistance R_L. At $t = 0$, when the source voltage jumps to V_0, a voltage disturbance of amplitude $\mathcal{V}_1^+(0,0) = Z_0 V_0/(R_s + Z_0)$ is launched at the source end of the line; it travels down the line (during $0 < t < t_d$) and arrives at the load end at $t = t_d$, when a reflected voltage of amplitude $\mathcal{V}_1^-(l, t_d) = \Gamma_L \mathcal{V}_1^+(l, t_d)$, where $\Gamma_L = (R_L - Z_0)/(R_L + Z_0)$, is launched back toward the source. Note that $\mathcal{V}_1^+(z,t)$ remains constant in time (and also with z once it reaches $z = l$ at $t = t_d$), with its value given by $Z_0 V_0/(R_s + Z_0)$. The reflected voltage travels along the line (during $t_d < t < 2t_d$) and reaches the source end at $t = 2t_d$, when a new voltage is reflected toward the load. The amplitude of the new reflected voltage is $\mathcal{V}_2^+(0, 2t_d) = \Gamma_s \mathcal{V}_1^-(0, 2t_d)$, where $\Gamma_s = (R_s - Z_0)/(R_s + Z_0)$. As this process continues with successive reflections at both ends, the total voltage at any time and at any particular position along the line is given as an algebraic sum of all the voltages at that particular location and at that time. For example, at $t = 4t_d$, the total voltage at the center of the line is given by

$$\mathcal{V}_{l/2}(4t_d) = \mathcal{V}_1^+(l/2, 4t_d) + \mathcal{V}_1^-(l/2, 4t_d) + \mathcal{V}_2^+(l/2, 4t_d) + \mathcal{V}_2^-(l/2, 4t_d)$$

$$= \mathcal{V}_1^+(l/2, 4t_d)[1 + \Gamma_L + \Gamma_s \Gamma_L + \Gamma_s \Gamma_L^2]$$

$$= \frac{Z_0 V_0}{R_s + Z_0}[1 + \Gamma_L + \Gamma_s \Gamma_L + \Gamma_s \Gamma_L^2]$$

where we have used the fact that $\mathcal{V}_1^+(l/2, 4t_d) = \mathcal{V}_1^+(0,0) = Z_0 V_0/(R_s + Z_0)$; in other words, as long as the source voltage does not change again, $\mathcal{V}_1^+(z,t)$ remains constant in time and also is the same everywhere (i.e., at all z) once it reaches the end of the line $z = l$ at $t = t_d$. Similarly, the total current at the center of the line at $t = 4t_d$ is

$$\mathcal{I}_{l/2}(4t_d) = \mathcal{I}_1^+(l/2, 4t_d) + \mathcal{I}_1^-(l/2, 4t_d) + \mathcal{I}_2^+(l/2, 4t_d) + \mathcal{I}_2^-(l/2, 4t_d)$$

$$= \frac{V_0}{R_s + Z_0}[1 - \Gamma_L + \Gamma_s \Gamma_L - \Gamma_s \Gamma_L^2]$$

In general, since $|\Gamma_L| \le 1$ and $|\Gamma_s| \le 1$, we have $|\mathcal{V}_{i+1}^\pm| \le |\mathcal{V}_i^\pm|$, so the contribution of new individual reflected components to the total voltage or current at any position along the line diminishes as $t \to \infty$. The sum of the contributions of the voltage components traveling in both directions converges[23] to a finite steady-state value for the voltage at any position z, given as

$$\mathcal{V}(z,\infty) = \mathcal{V}_1^+(z,\infty) + \mathcal{V}_1^-(z,\infty) + \mathcal{V}_2^+(z,\infty) + \mathcal{V}_2^-(z,\infty) + \mathcal{V}_3^+(z,\infty) + \mathcal{V}_3^-(z,\infty) + \cdots$$

$$= \mathcal{V}_1^+(z,\infty)[1 + \Gamma_L + \Gamma_s\Gamma_L + \Gamma_s\Gamma_L^2 + \Gamma_s^2\Gamma_L^2 + \Gamma_s^2\Gamma_L^3 + \Gamma_s^3\Gamma_L^3 + \cdots]$$

$$= \mathcal{V}_1^+(z,\infty)[1 + (\Gamma_s\Gamma_L) + (\Gamma_s^2\Gamma_L^2) + \cdots] + \Gamma_L[1 + (\Gamma_s\Gamma_L) + (\Gamma_s^2\Gamma_L^2) + \cdots]$$

$$= \mathcal{V}_1^+(z,\infty)\left[\left(\frac{1}{1-\Gamma_s\Gamma_L}\right) + \left(\frac{\Gamma_L}{1-\Gamma_s\Gamma_L}\right)\right] = \mathcal{V}_1^+(z,\infty)\left(\frac{1+\Gamma_L}{1-\Gamma_s\Gamma_L}\right)$$

This expression can be further simplified by substituting for $\mathcal{V}_1^+(z,\infty) = Z_0V_0/(R_s + Z_0)$, $\Gamma_L = (R_L - Z_0)/(R_L + Z_0)$, and $\Gamma_s = (R_s - Z_0)/(R_s + Z_0)$, yielding

$$\mathcal{V}(z,\infty) = V_{ss} = \left(\frac{R_L}{R_s + R_L}\right)V_0$$

a result that is expected, on the basis of the steady-state equivalent circuit shown in Figure 2.16b.

At steady state it appears from Figure 2.16b as if the transmission line is simply not there and that the source is directly connected to the load. While this is essentially true, the transmission line is of course still present and is in fact fully charged, with all of its distributed capacitors charged to a voltage V_{ss} and all of its distributed inductors carrying a current V_{ss}/R_L. If the source were to be suddenly disconnected, the energy stored on the line would eventually be discharged through the load resistance, but only after a sequence of voltages propagating back and forth, reflecting at both ends and becoming smaller in time (see Example 2.5).

Example 2.4: Step response of a resistively terminated lossless line. Consider the circuit shown in Figure 2.17a for the specific case of $R_s = 3Z_0$ and $R_L = 9Z_0$. Sketch $\mathcal{V}_s, \mathcal{V}_L, \mathcal{I}_s$, and $\mathcal{I}_L$ as a function of t.

Solution: Based on the above discussion, an incident voltage $\mathcal{V}_1^+(z,t)$ of amplitude $\mathcal{V}_1^+(0,0) = Z_0V_0/(3Z_0 + Z_0) = V_0/4$ is launched on the line at $t = 0$. When this disturbance reaches the load at $t = t_d$, a reflected voltage $\mathcal{V}_1^-(z,t)$ of amplitude $\mathcal{V}_1^-(l,t_d) = \Gamma_L\mathcal{V}_1^+(l,t_d) = V_0/5$, where $\Gamma_L = (9Z_0 - Z_0)/(9Z_0 + Z_0) = 4/5$, is launched toward the source. The reflected voltage arrives at the source end at $t = 2t_d$, and a new voltage $\mathcal{V}_2^+(z,t)$ of amplitude $\mathcal{V}_2^+(0,2t_d) = \Gamma_s\mathcal{V}_1^-(0,2t_d) = V_0/10$, where $\Gamma_s = (3Z_0 - Z_0)/(3Z_0 + Z_0) = 1/2$, is produced, traveling toward the load. At $t = 3t_d$, a voltage $\mathcal{V}_2^-(z,t)$ of amplitude $\mathcal{V}_2^-(l,3t_d) = \Gamma_L\mathcal{V}_2^+(l,3t_d) = 2V_0/25$ is launched from the load end, traveling toward the source, and so on. The bounce diagram is shown in Figure 2.17b. The source- and load-end voltages and the source- and load-end currents are shown in Figure 2.17d. The steady-state circuit seen by the source is also shown in Figure 2.17c.

---

[23]Noting that $|\Gamma_s\Gamma_L| < 1$ and using the fact that for $|x| < 1$ we have

$$1 + x + x^2 + x^3 + \cdots = \frac{1}{1-x}$$

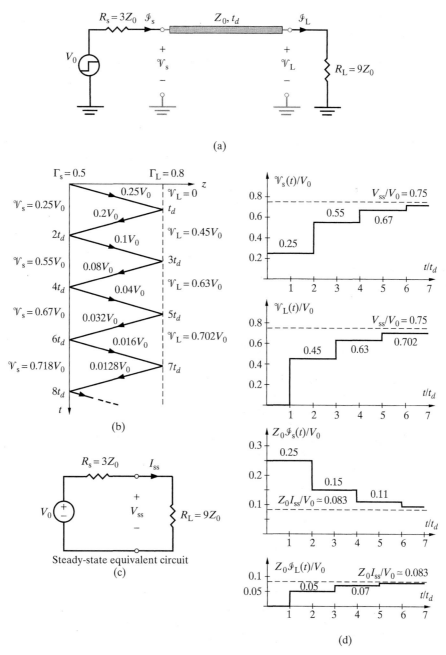

Figure 2.17   Step excitation of a resistively terminated lossless line. (a) The circuit configuration of Example 2.4. (b) Bounce diagram. (c) Steady-state equivalent circuit seen by the source. (d) Normalized source- and load-end voltages and source- and load-end currents as a function of t/t_d.

Example 2.5: A charged line connected to a resistor. Consider a transmission line that is initially charged to a constant voltage $\mathcal{V}(z,t) = V_0$ (such as the steady-state condition of the circuit in Example 2.3), as shown in Figure 2.18a. At $t = 0$, the switch is moved from position 1 to position 2. Analyze and sketch the variation of the source- and load-end voltages $\mathcal{V}_s$ and $\mathcal{V}_L$ as a function of t for three different cases: (a) $R_{s2} = Z_0/3$, (b) $R_{s2} = Z_0$, and (c) $R_{s2} = 3Z_0$.

> **Solution:** Before $t = 0$, the steady-state condition holds, and $\mathcal{V}_s(0^-) = \mathcal{V}_L(0^-) = V_{ss} = V_0$ and $\mathcal{I}_s(0^-) = \mathcal{I}_L(0^-) = I_{ss} = 0$. At $t = 0$, the switch moves to position 2, which causes both $\mathcal{V}_s$ and $\mathcal{I}_s$ to change immediately. The change in the source-end voltage $\mathcal{V}_s$ (and the source-end current $\mathcal{I}_s$) can be interpreted as a new voltage disturbance $\mathcal{V}_1^+(z,t)$ (and a new current disturbance $\mathcal{I}_1^+(z,t)$) launched on the line from the source end. The amplitude of this new voltage $\mathcal{V}_1^+(z,t)$ (and the new current $\mathcal{I}_1^+(z,t)$) is determined by the change in $\mathcal{V}_s$ (or in $\mathcal{I}_s$) between $t = 0^-$ and $t = 0^+$, namely,
>
> $$\mathcal{V}_1^+(0,0) = \mathcal{V}_s(0^+) - \mathcal{V}_s(0^-) = \mathcal{V}_s(0^+) - V_0$$
>
> and
>
> $$\mathcal{I}_1^+(0,0) = \mathcal{I}_s(0^+) - \mathcal{I}_s(0^-) = \mathcal{I}_s(0^+)$$
>
> Using the new boundary condition at the source end imposed by R_{s2}, namely,
>
> $$\mathcal{V}_s(0^+) = -R_{s2}\mathcal{I}_s(0^+) = -R_{s2}\mathcal{I}_1^+(0,0) = -R_{s2}\mathcal{V}_1^+(0,0)/Z_0$$
>
> we can write
>
> $$\mathcal{V}_1^+(0,0) = -R_{s2}\frac{\mathcal{V}_1^+(0,0)}{Z_0} - V_0 \rightarrow \mathcal{V}_1^+(0,0) = -\frac{Z_0 V_0}{R_{s2} + Z_0}$$
>
> Note that the negative sign in the source-end boundary condition $\mathcal{V}_s = -R_{s2}\mathcal{I}_s$ is due to the defined direction of $\mathcal{I}_s$, with positive current coming out of the terminal of positive voltage. At $t = t_d$, the new voltage disturbance reaches the open end of the line, where a reflected voltage $\mathcal{V}_1^-(z,t)$ with amplitude $\mathcal{V}_1^-(l,t_d) = \mathcal{V}_1^+(l,t_d)$ is produced, traveling toward the source end. At $t = 2t_d$, $\mathcal{V}_1^-(z,t)$ arrives at the source end, and a reflected voltage $\mathcal{V}_2^+(z,t)$ with amplitude $\mathcal{V}_2^+(0,2t_d) = \Gamma_s\mathcal{V}_1^-(0,2t_d) = (R_{s2} - Z_0)\mathcal{V}_1^-(0,2t_d)/(R_{s2} + Z_0)$ is launched back toward the load. This process continues until a new steady-state condition is reached, when the line voltage eventually becomes zero. Figures 2.18b, c, and d show the bounce diagrams for three different values of R_{s2}, namely $Z_0/3$, Z_0, and $3Z_0$. Figures 2.18e, f, and g show the variation of the source- and load-end voltages for all three cases as a function of time t.

The Transmission Line as a Linear Time-Invariant System.

In some cases it is useful to think of the transmission line as a linear time-invariant system, with a defined input and output. For this purpose, the input $\mathcal{V}_{in}(t)$ can be defined as the voltage or current at the input of the line, while the output $\mathcal{V}_{out}(t)$ can be a voltage or current somewhere else on the line—for example, the load voltage $\mathcal{V}_L(t)$ as indicated in Figure 2.19.

Note that since the fundamental differential equations ((2.1) and (2.2)) that govern the transmission line voltage and current are linear, the relationship between $\mathcal{V}_{in}(t)$ and

Figure 2.18   Discharging of a charged line. (a) A charged line connected to a resistor R_{s2} for Example 2.5. (b), (c), and (d) Bounce diagrams for $R_{s2} = Z_0/3$, $R_{s2} = Z_0$, and $R_{s2} = 3Z_0$. (e), (f), and (g) Normalized source- and load-end voltages as a function of t for the three different cases.

(b)

Figure 2.19   The transmission line as a linear time-invariant system. The input $\mathcal{V}_{in}(t)$ to the system can be defined as the line input voltage $\mathcal{V}_s(t)$, whereas the output $\mathcal{V}_{out}(t)$ could be any voltage or current of interest anywhere on the line, such as the load voltage $\mathcal{V}_L(t)$.

$\mathcal{V}_{out}(t)$ is linear. In other words, for two different input signals $\mathcal{V}_{in_1}$ and $\mathcal{V}_{in_2}$, which individually produce two different output signals $\mathcal{V}_{out_1}$ and $\mathcal{V}_{out_2}$, the response due to a linear superposition of the two inputs, $\mathcal{V}_{in_{1+2}} = a_1 \mathcal{V}_{in_1} + a_2 \mathcal{V}_{in_2}$, is

$$\mathcal{V}_{out_{1+2}} = a_1 \mathcal{V}_{out_1} + a_2 \mathcal{V}_{out_2}$$

Since the physical properties of the transmission line (L, C, t_d, Z_0) do not change with time, the relationship between $\mathcal{V}_{in}(t)$ and $\mathcal{V}_{out}(t)$ is also time-invariant. In other words, if the output due to an input $\mathcal{V}_{in_1}(t)$ is $\mathcal{V}_{out_1}(t)$, then the output due to a time-shifted version of the input, namely $\mathcal{V}_{in_1}(t - \tau)$, is simply a similarly shifted version of the output, namely $\mathcal{V}_{out_1}(t - \tau)$.

As with any linear time-invariant system, the response of a transmission line to any arbitrary excitation signal can be determined from its response to an impulse excitation. In the transmission line context, an input pulse can be considered to be an impulse if its duration is much shorter than any other time constant in the system or the one-way travel time t_d in the case of lossless lines with resistive terminations. In most applications, however, it is necessary to determine the response of the line to step inputs, as were illustrated in Examples 2.3 through 2.5. For this purpose, it is certainly easier to determine the step response directly rather than to determine the pulse (or impulse) response first and then use it to determine the step response.

Treatment of a transmission line as a linear time-invariant system can sometimes be useful in determining its response to pulse inputs, as illustrated in Example 2.6.

Example 2.6:   The transmission line as a linear time-invariant system. Consider the transmission line system of Figure 2.20a, the step response of which was determined in Example 2.4. Determine the load voltage $\mathcal{V}_L(t)$ for an input excitation in the form of a single pulse of amplitude V_0 and duration $0.5t_d$ (i.e., $\mathcal{V}_{in}(t) = V_0[u(t) - u(t - t_d/2)]$).

Figure 2.20   Pulse response of a transmission line. (a) The line configuration and source and load impedances. (b) The input pulse of amplitude V_0 and duration $t_d/2$ is represented as a superposition of a $\mathcal{V}_{in_1}$ and $\mathcal{V}_{in_2}$. (c) The response is computed as a superposition of the responses due to the individual step inputs. Note that $\mathcal{V}_{out_1}(t)$ was already computed in Example 2.4.

Solution: For the circuit of Figure 2.20a, it is convenient to define the input signal to be the excitation (source) voltage and the output as the load voltage, as indicated. As shown in Figure 2.20b, the input pulse of amplitude V_0 and duration $t_d/2$ can be viewed as a superposition of two different input signals: a step input starting at $t = 0$, namely $\mathcal{V}_{in_1}(t) = V_0 u(t)$, and a shifted negative step input, namely, $\mathcal{V}_{in_2}(t) = -V_0 u(t - t_d/2)$, where $u(\zeta)$ is the unit step function, $u(\zeta) = 1$ for $\zeta > 0$, and $u(\zeta) = 0$ for $\zeta < 0$. The output $\mathcal{V}_{out_1}(t)$ due to the input $\mathcal{V}_{in_1}(t)$ was determined in Example 2.4 and is plotted in the top panel of Figure 2.20c. Since the transmission line is a linear time-invariant system, the response $\mathcal{V}_{out_2}(t)$ due to the input $\mathcal{V}_{in_2}(t)$ is simply a flipped-over and shifted version of $\mathcal{V}_{out_1}(t)$, as shown in the middle panel of Figure 2.20c. The bottom panel of Figure 2.20c shows the superposition of the two responses, which is the desired pulse response.

Note that our solution of this problem using the linear time-invariant system treatment was simplified by the fact that the step response of the system was already in hand from Example 2.4.

2.4.2 Junctions between Transmission Lines

We have seen that reflections from terminations at the source and load ends of transmission lines lead to ringing and other effects. Reflections also occur at discontinuities at the interfaces between transmission lines, connected either in cascade or in parallel, as shown, for example, in Figures 2.21a and 2.22a and as often encountered in practice. For example, consider the case of the two lossless transmission lines A and B (with characteristic impedances Z_{0A} and Z_{0B}) connected in tandem (i.e., in series) as shown in Figure 2.21a. Assume a voltage disturbance of amplitude $\mathcal{V}_{1A}^+(z,t)$ (with an associated current of $\mathcal{I}_{1A}^+(z,t) = \mathcal{V}_{1A}^+/Z_{0A}$) to arrive at the junction between lines A and B (located at $z = l_j$) from line A at $t = t_0$. A voltage $\mathcal{V}_{1A}^-(z,t)$ of amplitude $\mathcal{V}_{1A}^-(l_j, t_0) = \Gamma_{AB}\mathcal{V}_{1A}^+(l_j, t_0)$ reflects back to line A, where the reflection coefficient Γ_{AB} is given by $\Gamma_{AB} = (Z_{0B} - Z_{0A})/(Z_{0B} + Z_{0A})$, since line B presents a load impedance of Z_{0B} to line A. In addition, a voltage $\mathcal{V}_{1B}^+(z,t)$ of amplitude $\mathcal{V}_{1B}^+(l_j, t_0) = \mathcal{T}_{AB}\mathcal{V}_{1A}^+(l_j, t_0)$ is transmitted into line B, where $\mathcal{T}_{AB}$ is called the *transmission coefficient,* defined as the ratio of the transmitted voltage to the incident voltage, that is, $\mathcal{T}_{AB} \equiv \mathcal{V}_{1B}^+(l_j, t_0)/\mathcal{V}_{1A}^+(l_j, t_0)$. To find $\mathcal{T}_{AB}$, we apply the boundary condition at the junction, which states that the total voltages on the left and right sides of the junction must be equal:

$$\mathcal{V}_{1A}^+(l_j, t_0) + \mathcal{V}_{1A}^-(l_j, t_0) = \mathcal{V}_{1B}^+(l_j, t_0)$$

yielding $\mathcal{T}_{AB} = 1 + \Gamma_{AB} = 2Z_{0B}/(Z_{0B} + Z_{0A})$. The transmission coefficient $\mathcal{T}_{AB}$ represents the fraction of the incident voltage that is transferred from line A to line B. Note that depending on the value of the reflection coefficient, the transmitted voltage can actually be larger in amplitude than the incident voltage, so that we may have $\mathcal{T}_{AB} > 1$ (in those cases when $\Gamma_{AB} > 0$).

Similarly, if a voltage disturbance $\mathcal{V}_{1B}^-(z,t)$ of amplitude $\mathcal{V}_{1B}^-(l_j, t_1)$ (produced by reflection when $\mathcal{V}_{1B}^+(z,t)$ reaches the end of line B) arrives at the same junction between A and B from line B at $t = t_1$, a voltage $\mathcal{V}_{2B}^+(z,t)$ of amplitude

$\mathcal{V}_{2B}^{+}(l_j, t_1) = \Gamma_{BA}\mathcal{V}_{1B}^{-}(l_j, t_1)$ reflects back to line B and a voltage $\mathcal{V}_{2A}^{-}(l_j, t_1) = \mathcal{T}_{BA}\mathcal{V}_{1B}^{-}(l_j, t_1)$ is transmitted into line A, where Γ_{BA} and $\mathcal{T}_{BA}$ are given by

$$\Gamma_{BA} = \frac{Z_{0A} - Z_{0B}}{Z_{0A} + Z_{0B}} = -\Gamma_{AB}$$

$$\mathcal{T}_{BA} = 1 + \Gamma_{BA} = \frac{2Z_{0A}}{Z_{0A} + Z_{0B}} = \frac{Z_{0A}}{Z_{0B}}\mathcal{T}_{AB}$$

Example 2.7: Cascaded transmission lines. Consider the transmission line system shown in Figure 2.21a, where a step voltage source of amplitude 1.5 V and source resistance 50Ω excites two cascaded lossless transmission lines (A and B) of characteristic impedances 50Ω and 25Ω and lengths 5 cm and 2 cm, respectively. The speed of propagation in each line is 10 cm-ns^{-1}.[24] The second line (B) is terminated with a load impedance of 100Ω at the other end. Draw the bounce diagram and sketch the voltages $\mathcal{V}_s(t)$ and $\mathcal{V}_L(t)$ as functions of time.

> **Solution:** With respect to Figure 2.21a, we note that $l_j = 5$ cm and $l = 7$ cm. At $t = 0^+$, voltage $\mathcal{V}_{1A}^{+}(z, t)$ of amplitude $\mathcal{V}_{1A}^{+}(0, 0) = 0.75$ V is launched on line A. This voltage reaches the junction between the two lines at $t = t_{d1} = 5$ cm/(10 cm-(ns)$^{-1}$) $= 500$ ps, when a reflected voltage $\mathcal{V}_{1A}^{-}(z, t)$ of amplitude $\mathcal{V}_{1A}^{-}(l_j, t_{d1}) = \Gamma_{AB}\mathcal{V}_{1A}^{+}(l_j, t_{d1}) = (-1/3)(0.75) = -0.25$ V, and a transmitted voltage $\mathcal{V}_{1B}^{+}(z, t)$ of amplitude $\mathcal{V}_{1B}^{+}(l_j, t_{d1}) = \mathcal{T}_{AB}\mathcal{V}_{1A}^{+}(l_j, t_{d1}) = (2/3)(0.75) = 0.5$ V are created. The reflected disturbance arrives at the source end at $t = 2t_{d1} = 1$ ns and is absorbed completely since $\Gamma_s = 0$. The transmitted wave reaches the load at $t = t_{d1} + t_{d2} = 700$ ps, and a reflected voltage $\mathcal{V}_{1B}^{-}(z, t)$ of amplitude $\mathcal{V}_{1B}^{-}(l, 700 \text{ ps}) = \Gamma_L\mathcal{V}_{1B}^{+}(l, 700 \text{ ps}) = (0.6)(0.5)$ V $= 0.3$ V is launched toward the source. This reflected disturbance arrives at the junction from line B at $t = t_{d1} + 2t_{d2} = 900$ ps, and reflected and transmitted voltages $\mathcal{V}_{2B}^{+}(z, t)$ and $\mathcal{V}_{2A}^{-}(z, t)$ of amplitudes respectively $\mathcal{V}_{2B}^{+}(l_j, 900 \text{ ps}) = \Gamma_{BA}\mathcal{V}_{1B}^{-}(l_j, 900 \text{ ps}) = (1/3)(0.3) = 0.1$V and $\mathcal{V}_{2A}^{-}(l_j, 900 \text{ ps}) = \mathcal{T}_{BA}\mathcal{V}_{1B}^{-}(l_j, 900 \text{ ps}) = (4/3)(0.3) = 0.4$ V are launched respectively toward the load and the source. The continuation of this process can be followed by means of a bounce diagram, as shown in Figure 2.21b. The source- and load-end voltages are plotted as a function of t in Figure 2.21c.

Example 2.8: Three parallel transmission lines. Consider three identical lossless transmission lines, each with characteristic impedance Z_0 and one-way time delay t_d, connected in parallel at a common junction as shown in Figure 2.22a. The main line is excited at $t = 0$ by a step voltage source of amplitude V_0 and a source resistance of $R_s = Z_0$. Find and sketch the voltages $\mathcal{V}_s(t)$, $\mathcal{V}_{L1}(t)$, and $\mathcal{V}_{L2}(t)$ for the following two cases: (a) $R_{L1} = R_{L2} = Z_0$ and (b) $R_{L1} = Z_0$ and $R_{L2} = \infty$.

> **Solution:** For any voltage disturbance $\mathcal{V}_i(z, t)$ of amplitude V_i arriving at the junction from any one of the three lines, the parallel combination of the characteristic impedances of

---

[24]Note from Table 2.1 that 10 cm-(ns)$^{-1}$ is approximately the speed of propagation in alumina (Al$_2$O$_3$), a ceramic commonly used for electronics packaging.

(a)

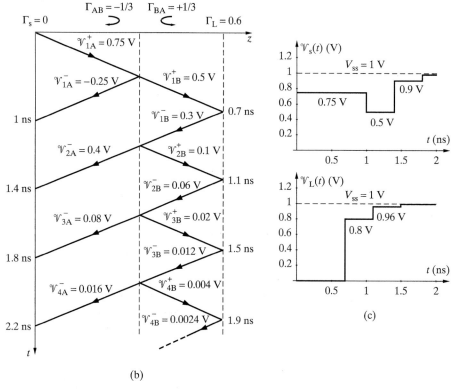

(b)

Figure 2.21 Cascaded transmission lines. (a) Circuit diagram for Example 2.7. (b) Bounce diagram. (c) Source-end voltage $\mathcal{V}_s$ and load-end voltage $\mathcal{V}_L$ as a function of t.

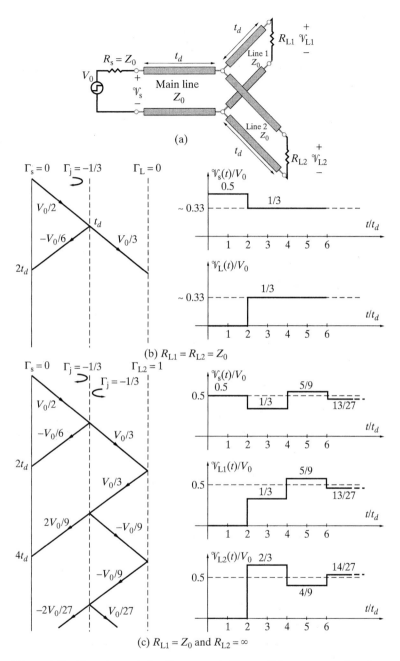

Figure 2.22   Three transmission lines connected in parallel at a common junction. (a) Circuit diagram for Example 2.8. (b) Bounce diagram and the variation of the $\mathcal{V}_s$ and $\mathcal{V}_{L1} = \mathcal{V}_{L2} = \mathcal{V}_L$ voltages as a function of t for $R_{L1} = R_{L2} = Z_0$. (c) Bounce diagram and the variation of the $\mathcal{V}_s$, $\mathcal{V}_{L1}$, and $\mathcal{V}_{L2}$ voltages as a function of t for $R_{L1} = Z_0$ and $R_{L2} = \infty$.

the other two lines acts as an equivalent load impedance at the junction. Once a voltage is incident at the junction, a voltage of amplitude $V_r = \Gamma_j V_i = (Z_0/2 - Z_0)V_i/(Z_0/2 + Z_0) = -V_i/3$ is reflected and a voltage of amplitude $V_t = \mathcal{T}_j V_i = (1 + \Gamma_j)V_i = 2V_i/3$ is transmitted to both of the other lines. Note that the reflection coefficient at the junction is denoted simply as Γ_j, because all of the transmission lines are identical and the reflection coefficient at the junction is thus the same regardless of which line the wave is incident from. For the circuit shown in Figure 2.22a, the bounce diagram and the sketches of voltages $\mathcal{V}_s$, $\mathcal{V}_{L1}$, and $\mathcal{V}_{L2}$ for both cases are shown in Figures 2.22b and c. When $R_{L1} = R_{L2} = Z_0$, only one of lines 1 and 2 is shown in the bounce diagram, since what happens on the other is identical, as seen in Figure 2.22b. When $R_{L1} = Z_0$ and $R_{L2} = \infty$, only line 2 is shown in the bounce diagram in Figure 2.22c, since no reflection occurs on line 1.

2.5 TRANSIENT RESPONSE OF TRANSMISSION LINES WITH REACTIVE TERMINATIONS

Up to now, we have studied only transmission lines with resistive terminations. In this section, we consider reactive loads. Reactive loads are encountered quite often in practice; in high-speed bus designs, for example, capacitive loading by backplanes (consisting of plug-in cards having printed circuit board traces and connectors) often becomes the bottleneck when high-speed CPUs communicate with shared resources on the bus. Inductive loading due to bonding wire inductances is also important in many integrated-circuit packaging technologies. Packaging pins, vias between two wiring levels, and variations in line width can often be modeled as capacitive and inductive discontinuities. The capacitances and inductances of these various packaging components can range between 0.5 and 4 pF and between 0.1 and 35 nH, respectively.[25]

For transmission lines with resistive loads, the reflected and transmitted voltages and currents have the same temporal shape as the incident ones and do not change their shape as a function of time. For a step excitation, for example, the reflected voltage produced by a resistive termination remains constant in time, as discussed in preceding sections. However, in the case of capacitive or inductive terminations, the reflected and transmitted voltages and currents do not have the same temporal shape as the incident ones. The terminal boundary condition at the reactive termination must now be expressed as a differential equation whose general solution can be exceedingly complicated, whether the solutions are carried out in the time domain or by the use of Laplace transformation. We illustrate the basic principles by considering a line terminated at an inductance, as shown in Figure 2.23.

When the traveling disturbance $\mathcal{V}_1^+(z,t)$ (taken in Figures 2.23b, c as a constant voltage V_0) and its associated current $\mathcal{I}_1^+(z,t)$ (taken in Figures 2.23b, c as a constant current $I_0 = V_0/Z_0$) first reach the end of the line, the inductive load acts as an open circuit, since its current cannot change instantaneously. Thus the disturbance is initially reflected in the same manner as an open circuit, the terminal voltage jumping to $2V_0$

[25]See Chapter 6 of H. B. Bakoglu, *Circuits, Interconnections, and Packaging for VLSI*, Addison-Wesley, Reading, Massachusetts, 1990.

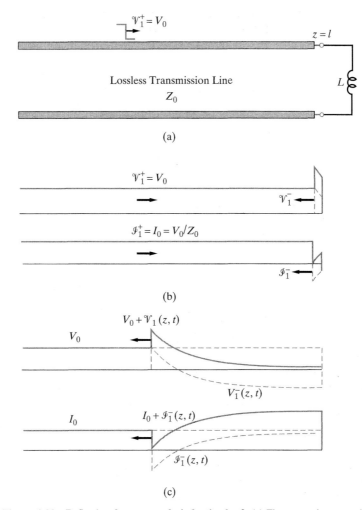

Figure 2.23   Reflection from a purely inductive load. (a) The source is assumed to be matched (i.e., no reflections back from the source) and to supply a constant voltage V_0. The distributions of the voltage and current are shown at two different times: (b) immediately after reflection when the inductor behaves like an open circuit and (c) later in time when the inductor behaves as a short circuit.

and the terminal current being zero, for that instant (Figure 2.23b). However, since a voltage of $2V_0$ now exits across the inductor, its current builds up, until it is practically equivalent to a short circuit. At steady state, when the voltage across the inductance reduces to zero, the current through it becomes $2I_0$ (Figure 2.23c), similar to the case of a short-circuit termination (see Example 2.2).

Between an initial open circuit and an eventual short circuit, the voltage across the inductive load goes through all intermediate values, and the reflected voltage changes accordingly. To determine the analytical expression describing the variation of the voltage across the inductance, we need to simultaneously solve the transmission line equations (or the general solutions dictated by them, namely (2.8) and (2.9)) along with the differential equation describing the boundary condition imposed by the inductive load as

$$\mathscr{V}_L(t) = L\frac{d\mathscr{I}_L(t)}{dt}$$

For a general incident voltage $\mathscr{V}_1^+(z,t)$, the load voltage $\mathscr{V}_L(t)$ and current $\mathscr{I}_L(t)$ are given by

$$\mathscr{V}_L(t) = \mathscr{V}_1^+(l,t) + \mathscr{V}_1^-(l,t)$$

$$\mathscr{I}_L(t) = \mathscr{I}_1^+(l,t) + \mathscr{I}_1^-(l,t) = \frac{\mathscr{V}_1^+(l,t)}{Z_0} - \frac{\mathscr{V}_1^-(l,t)}{Z_0}$$

where the location of the load (i.e., the inductive termination) is assumed to be at $z = l$. We thus have

$$\mathscr{V}_1^+(l,t) + \mathscr{V}_1^-(l,t) = L\frac{d}{dt}\left(\frac{\mathscr{V}_1^+(l,t)}{Z_0} - \frac{\mathscr{V}_1^-(l,t)}{Z_0}\right)$$

or

$$\frac{d\mathscr{V}_1^-(l,t)}{dt} + \frac{Z_0}{L}\mathscr{V}_1^-(l,t) = \frac{d\mathscr{V}_1^+(l,t)}{dt} - \frac{Z_0}{L}\mathscr{V}_1^+(l,t)$$

which is a differential equation with $\mathscr{V}_1^-(l,t)$ as the dependent variable, since $\mathscr{V}_1^+(l,t)$ is presumably known, it being the incident voltage disturbance arriving from the source end of the line. The right-hand side is therefore a known function of time, and the equation is simply a first-order differential equation with constant coefficients. Note that we assume the source to be either far enough away or matched, so that the voltage $\mathscr{V}_1^-(z,t)$ does not reach the source end and generate a reflected voltage $\mathscr{V}_2^+(z,t)$ before $\mathscr{V}_1^-(l,t)$ reaches its "steady-state" value ($-V_0$ in the case when $\mathscr{V}_1^+(z,t) = V_0$ as shown in Figure 2.23c).

To study the simplest case, let us consider an incident voltage with a constant amplitude V_0 (i.e., $\mathscr{V}_1^+(l,t) = V_0$) launched by a step source reaches the inductor at $t = 0$. The above differential equation then simplifies to

$$\frac{d\mathscr{V}_1^-(l,t)}{dt} + \frac{Z_0}{L}\mathscr{V}_1^-(l,t) = -\frac{Z_0V_0}{L}$$

The solution of this first-order differential equation is[26]

$$\mathscr{V}_1^-(l,t) = -V_0 + Ke^{-(Z_0/L)t}$$

---

[26]The solution can be found via Laplace transformation or as a superposition of the homogeneous solution and the particular solution; the validity of the solution can be shown by simply substituting it into the differential equation.

where the coefficient K needs to be determined by the known initial conditions as they relate to $\mathcal{V}_1^-(l, t)$. At the instant of the arrival of the voltage disturbance ($t = 0$), the current through the inductance $\mathcal{I}_L(t = 0) = 0$, and we thus have the incident voltage fully reflected, or $\mathcal{V}_1^-(l, t = 0) = V_0$. Thus, we must have $K = 2V_0$. The solution for the reflected voltage is then

$$\mathcal{V}_1^-(l, t) = -V_0 + 2V_0 e^{-(Z_0/L)t}$$

which varies from its initial value of V_0 to an eventual value of $-V_0$, as shown in Figure 2.23c.

Example 2.9: Lossy capacitive load. Consider the transmission line system shown in Figure 2.24a where a step voltage source of amplitude V_0 and source resistance $R_s = Z_0$ excites a lossless transmission line of characteristic impedance Z_0 and one-way time delay t_d connected to a load consisting of a parallel combination of R_L and C_L. Find and sketch the source- and load-end voltages as a function of time.

Solution: At $t = 0$, an incident voltage $\mathcal{V}_1^+(z, t)$ of amplitude $\mathcal{V}_1^+(0, 0) = V_0/2$ is launched at the source end of the line. This disturbance reaches the capacitive load at $t = t_d$, when a voltage $\mathcal{V}_1^-(z, t)$ of initial amplitude $\mathcal{V}_1^-(l, t_d)$ reflects toward the source. For $t \geq t_d$, the total load voltage $\mathcal{V}_L(t)$ and current $\mathcal{I}_L(t)$ are given by

$$\mathcal{V}_L(t) = \mathcal{V}_1^+(l, t) + \mathcal{V}_1^-(l, t)$$

$$\mathcal{I}_L(t) = \mathcal{I}_1^+(l, t) + \mathcal{I}_1^-(l, t) = \frac{\mathcal{V}_1^+(l, t)}{Z_0} - \frac{\mathcal{V}_1^-(l, t)}{Z_0}$$

where $\mathcal{V}_1^+(l, t) = \mathcal{V}_1^+(0, 0) = V_0/2$. These two equations are related by the boundary condition imposed by the load; that is,

$$\mathcal{I}_L(t) = \frac{\mathcal{V}_L(t)}{R_L} + C_L \frac{d\mathcal{V}_L(t)}{dt}$$

Substituting the first two equations into the third equation yields

$$\frac{d\mathcal{V}_1^-(l, t)}{dt} + \left[\frac{R_L + Z_0}{R_L Z_0 C_L}\right]\mathcal{V}_1^-(l, t) = \left[\frac{R_L - Z_0}{R_L Z_0 C_L}\right]\mathcal{V}_1^+(l, t)$$

which is a first-order differential equation for $\mathcal{V}_1^-(t)$. Note that in deriving it, we have used the fact that the incident voltage is constant in time, so that $d\mathcal{V}_1^+(l, t)/dt = 0$. The solution of this first-order differential equation can be found by noting that $\mathcal{V}_1^+(l, t) = V_0/2$ and by writing the general solution as

$$\mathcal{V}_1^-(l, t) = K_1 + K_2 e^{-[(R_L + Z_0)/(R_L Z_0 C_L)](t - t_d)}$$

(a)

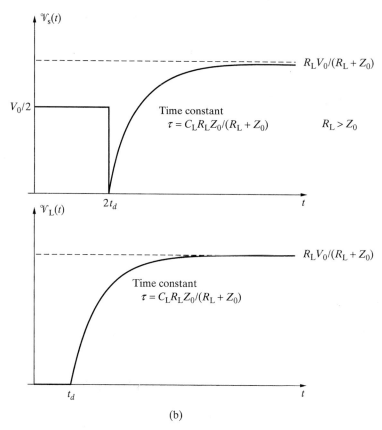

(b)

Figure 2.24 Step response of a capacitively loaded line. (a) Circuit diagram.
(b) Time variation of the source- and load-end voltages.

in which case the constants K_1 and K_2 can be determined by using the initial and final conditions. Note that the reflected voltage must vary exponentially from $-\mathcal{V}_1^+(l,t)$ at $t = t_d$ to $(R_L - Z_0)\mathcal{V}_1^+(l,t)/(R_L + Z_0)$ for $t \to \infty$. We thus have

$$\mathcal{V}_1^-(l,t) = \mathcal{V}_1^+(l,t)\left[\frac{R_L - Z_0}{R_L + Z_0} - \frac{2R_L}{R_L + Z_0}e^{-[(R_L+Z_0)/(R_L Z_0 C_L)](t-t_d)}\right]$$

which is valid for $t \geq t_d$. This behavior can be understood as follows: When the incident voltage reaches the capacitive load at $t = t_d$, the capacitor C_L is initially uncharged and acts like a short circuit, resulting in $\mathcal{V}_1^-(l, t_d) = -\mathcal{V}_1^+(l, t_d) = -V_0/2$. However, at steady state the capacitor is fully charged and acts like an open circuit, resulting in $\mathcal{V}_1^-(l, \infty) = (R_L - Z_0)\mathcal{V}_1^+(l, \infty)/(R_L + Z_0) = (R_L - Z_0)V_0/[2(R_L + Z_0)]$, as expected. Note also that the time constant of the exponential variation is $\tau = R_L Z_0 C_L/(R_L + Z_0) = R_{Th}C_L$, where R_{Th} is the Thévenin equivalent resistance,[27] as seen from the terminals of the capacitor. Substituting $\mathcal{V}_1^-(l, t)$ into $\mathcal{V}_L(t)$ yields

$$\mathcal{V}_L(t) = \frac{R_L V_0}{R_L + Z_0}[1 - e^{-[(R_L+Z_0)/(R_L Z_0 C_L)](t-t_d)}]$$

valid for $t \geq t_d$. When the reflected voltage reaches the source end at $t = 2t_d$, it is completely absorbed, since the source end of the line is matched (i.e., $R_s = Z_0$). The voltage at the source end is given by $\mathcal{V}_s(t) = \mathcal{V}_1^+(0, t) = V_0/2$ for $t < 2t_d$, and

$$\mathcal{V}_s(t) = \frac{R_L V_0}{R_L + Z_0}[1 - e^{-[(R_L+Z_0)/(R_L Z_0 C_L)](t-2t_d)}]$$

is valid for $t \geq 2t_d$. Sketches of $\mathcal{V}_s(t)$ and $\mathcal{V}_L(t)$ are shown in Figure 2.24b.

Example 2.10: A lumped series inductor between two transmission lines. Microstrip transmission lines on printed circuit boards are often connected together with bonding wires, which are inherently inductive. In this example, we consider a typical model for such a connection, namely a lumped series inductor between two different transmission lines. The measurement of the bonding-wire inductance is considered later in Example 2.13. Two microstrip transmission lines having equal line parameters of $L = 4$ nH-(cm)$^{-1}$ and $C = 1.6$ pF-(cm)$^{-1}$ and lengths 15 cm and 10 cm are connected by a wire represented by a series lumped inductance of $L_w = 5$ nH, shown in Figure 2.25a. The end of the shorter line is matched with a 50Ω load, and the circuit is excited at $t = 0$ by a unit step voltage source of $R_s = 50\Omega$. Find and sketch the variations with time of the source- and load-end voltages. Assume lossless lines.

Solution: Using the given line parameters L and C, the characteristic impedance and the phase velocity of the microstrip lines can be calculated as

$$Z_0 = \sqrt{\frac{L}{C}} = \sqrt{\frac{4 \times 10^{-9}}{1.6 \times 10^{-12}}} = 50\Omega$$

and

$$v_p = \frac{1}{\sqrt{LC}} = \frac{1}{\sqrt{4 \times 10^{-9} \times 1.6 \times 10^{-12}}} = 12.5 \times 10^9 \text{ cm-s}^{-1}$$

Therefore, the one-way delay times of the two lines are $t_{d1} = 15$ cm/(12.5 cm-(ns)$^{-1}$) = 1.2 ns and $t_{d2} = 10$ cm/(12.5 cm-(ns)$^{-1}$) = 0.8 ns, respectively, as indicated in Figure 2.25a.

---

[27]In this case being simply equal to the parallel combination of the load resistance R_L and the characteristic impedance Z_0 of the line.

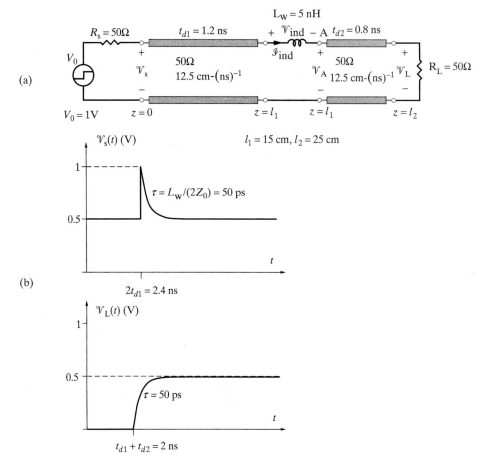

Figure 2.25 A lumped series inductor between two different transmission lines. (a) Circuit diagram. (b) Time variation of source- and load-end voltages for Example 2.10.

As in the previous example, at $t = 0$, an incident voltage $\mathcal{V}_1^+(z,t)$ of amplitude $\mathcal{V}_1^+(0,0) = 0.5$ V is launched from the source end of the line. Note that the amplitude of this incident voltage remains constant in time, as long as it is supplied by the source. When this disturbance arrives at the junction at $t = t_{d1} = 1.2$ ns, the uncharged inductor initially acts like an open circuit (i.e., it resists the flow of current), producing a reflected voltage $\mathcal{V}_1^-(z,t)$ of initial amplitude $\mathcal{V}_1^-(l_1, t_{d1}) = \mathcal{V}_1^+(l_1, t_{d1}) = 0.5$ V, the same as one would have if the two lines were not connected. Current flows into the second line through the inductor as the inductor charges exponentially and eventually behaves like a short circuit at steady state. The following equations apply for $t \geq t_{d1}$:

$$\mathcal{V}_1^+(l_1, t) + \mathcal{V}_1^-(l_1, t) = \mathcal{V}_{ind}(t) + \mathcal{V}_A(t) = L_w \frac{d\mathcal{I}_{ind}(t)}{dt} + Z_0 \mathcal{I}_{ind}(t)$$

where $\mathcal{V}_1^+(l_1, t) = 0.5$ V and $\mathcal{V}_A(t)$ is the voltage at position A, at which the impedance seen looking toward the load is $Z_0 = 50\Omega$. We also have

$$\mathcal{I}_{\mathrm{ind}}(t) = \mathcal{I}_1^+(l_1, t) + \mathcal{I}_1^-(l_1, t) = \frac{\mathcal{V}_1^+(l_1, t)}{Z_0} - \frac{\mathcal{V}_1^-(l_1, t)}{Z_0}$$

where $\mathcal{I}_{\mathrm{ind}}(t)$ and $\mathcal{V}_{\mathrm{ind}}(t)$ are, respectively, the current through and the voltage across the inductor, as defined in Figure 2.25a. Substituting the second equation into the first and using $d\mathcal{V}_1^+(l_1, t)/dt = 0$ yield

$$\frac{d\mathcal{V}_1^-(l_1, t)}{dt} + \frac{2Z_0}{L_{\mathrm{w}}}\mathcal{V}_1^-(l_1, t) = 0$$

which is a first-order differential equation for $\mathcal{V}_1^-(l_1, t)$. The solution of this equation is

$$\mathcal{V}_1^-(l_1, t) = \mathcal{V}_1^+(l_1, t_{d1})e^{-(2Z_0/L_{\mathrm{w}})(t - t_{d1})} = 0.5e^{-2 \times 10^{10}(t - 1.2 \times 10^{-9})} \text{ V}$$

which is valid for $t \geq t_{d1} = 1.2$ ns. Note that we have used the fact that $\mathcal{V}_1^-(l_1, t_{d1}) = 0.5$ and that the reflected voltage varies exponentially from $\mathcal{V}_1^-(l_1, t_{d1}) = \mathcal{V}_1^+(l_1, t_{d1})$ at $t = t_{d1}$ (when the inductor initially behaves like an open circuit) to zero at $t \to \infty$ (when the fully energized inductor eventually behaves like a short circuit). The time constant of the exponential variation is $\tau = L_{\mathrm{w}}/R_{\mathrm{Th}} = 50$ ps, where $R_{\mathrm{Th}} = 2Z_0 = 100\Omega$ is the Thévenin equivalent resistance as seen from the terminals of the inductor.

To satisfy the boundary condition at the junction, and noting that the inductor L_{w} is a lumped element, the current on both sides of the inductor must be the same. Thus, the inductor current $\mathcal{I}_{\mathrm{ind}}(t)$ can be written as

$$\mathcal{I}_{\mathrm{ind}}(t) = \mathcal{I}_1^+(l_1, t_{d1}) + \mathcal{I}_1^-(l_1, t) = \frac{\mathcal{V}_1^+(l_1, t_{d1})}{Z_0}[1 - e^{-(2Z_0/L_{\mathrm{w}})(t - t_{d1})}]$$

which is valid for $t \geq t_{d1}$. At $t = t_{d1}$, a transmitted voltage $\mathcal{V}_A^+(z, t) = Z_0\mathcal{I}_A^+(z, t)$ is launched at position A on the second line, where $\mathcal{I}_A^+(z, t) = \mathcal{I}_{\mathrm{ind}}(t)$. Therefore, the transmitted voltage $\mathcal{V}_A^+(z, t)$ can be written as

$$\mathcal{V}_A^+(z, t) = Z_0\mathcal{I}_{\mathrm{ind}}(t) = \mathcal{V}_1^+(l_1, t_{d1})[1 - e^{-(2Z_0/L_{\mathrm{w}})(t - t_{d1})}]$$

valid for $t \geq t_{d1}$.

The voltage $\mathcal{V}_A^+(z, t)$ arrives at the load end at $t = t_{d1} + t_{d2} = 2$ ns, where it is completely absorbed, since $R_{\mathrm{L}} = Z_0 = 50\Omega$. The load voltage is given by

$$\mathcal{V}_{\mathrm{L}}(t) = \mathcal{V}_A^+(l_2, t) = \mathcal{V}_1^+(l_1, t_{d1})[1 - e^{-(2Z_0/L_{\mathrm{w}})(t - (t_{d1} + t_{d2}))}]$$

$$= 0.5[1 - e^{-2 \times 10^{10}(t - 2 \times 10^{-9})}] \text{ V}$$

valid for $t > (t_{d1} + t_{d2}) = 2$ ns. Note that $\mathcal{V}_1^+(l_1, t_{d1}) = 0.5$ V for $t \geq t_{d1}$, since the incident voltage remains constant unless the source changes. Similarly, the source-end voltage is given by $\mathcal{V}_{\mathrm{s}}(t) = \mathcal{V}_1^+(0, t) = 0.5$ V for $t < 2t_{d1} = 2.4$ ns and

$$\mathcal{V}_{\mathrm{s}}(t) = \mathcal{V}_1^+(0, t)[1 + e^{-(2Z_0/L_{\mathrm{w}})(t - 2t_{d1})}] = 0.5[1 + e^{-2 \times 10^{10}(t - 2.4 \times 10^{-9})}] \text{ V}$$

for $t > 2t_{d1} = 2.4$ ns. Sketches of the time variations of source- and load-end voltages are shown in Figure 2.25b.

In the preceding two examples we used time-domain methods to determine the step response of transmission lines with terminations or discontinuities involving reactive elements. The basis for our analyses was the simultaneous solution of (2.10), describing the voltage and current of the reactive element in terms of the transmission line incident and reflected voltage and current waves, together with the differential equations that describe the terminal voltage–current relationship of the reactive load. These time-domain solutions were tractable partly because the excitation voltage was a simple step function and also because the reactive discontinuities that we analyzed involved only one energy storage element (i.e., a single capacitor or inductor). When the input voltage function is more complicated, or when the reactive discontinuity involves more than one energy storage element, it is often easier to use Laplace transform methods to determine the response of the line. We demonstrate the use of the Laplace transform method in Example 2.11.

Example 2.11: Reflections due to inductance of resistor leads. Consider the transmission line system shown in Figure 2.26, where the source voltage amplitude increases linearly from zero to V_0 over a time[28] of t_r. The output resistance of the source is $R_s = Z_0$, while the transmission line having a characteristic impedance Z_0 and a one-way time delay t_d is terminated in a reactive load consisting of a series combination of R_L and L_L. Find an expression for the reflected voltage at the load [i.e., $\mathcal{V}_1^-(l,t)$] and determine its maximum value for $R_L = Z_0$.

Solution: From Figure 2.26 we note that, starting at $t = 0$, an incident voltage is launched at the source end of the line, given by

$$\mathcal{V}_1^+(0,t) = \frac{V_0}{2t_r}[tu(t) - (t - t_r)u(t - t_r)]$$

(a)   (b)

Figure 2.26   Reflections due to inductance of resistor leads. (a) Circuit diagram of Example 2.11. (b) Time variation of the reflected voltage $\mathcal{V}_1^-(l,t)$ for the case $R_L = Z_0$.

[28]Note that t_r is not exactly the rise time discussed in Section 1.1, which was defined as the time required for the signal to change from 10% to 90% of its final value.

where $u(\cdot)$ is the unit step function. The Laplace transform of this voltage waveform is

$$\tilde{\mathcal{V}}_1^+(s) = \frac{V_0}{2t_r} \frac{1 - e^{-t_r s}}{s^2}$$

This incident voltage propagates to the load end, and no reflected voltage exists until it arrives there at $t = t_d$. For $t \geq t_d$, the total load voltage and current $\mathcal{V}_L(t)$ and $\mathcal{I}_L(t)$ are given by

$$\mathcal{V}_L(t) = \mathcal{V}_1^+(l,t) + \mathcal{V}_1^-(l,t)$$

$$\mathcal{I}_L(t) = \mathcal{I}_1^+(l,t) + \mathcal{I}_1^-(l,t) = \frac{\mathcal{V}_1^+(l,t)}{Z_0} - \frac{\mathcal{V}_1^-(l,t)}{Z_0}$$

and are related by the boundary condition imposed by the load,

$$\mathcal{V}_L(t) = R_L \mathcal{I}_L(t) + L_L \frac{d\mathcal{I}_L(t)}{dt}$$

Substituting the first two equations into the third equation yields

$$L_L \frac{d\mathcal{V}_1^-(l,t)}{dt} + (R_L + Z_0)\mathcal{V}_1^-(l,t) = L_L \frac{d\mathcal{V}_1^+(l,t)}{dt} + (R_L - Z_0)\mathcal{V}_1^+(l,t)$$

Note that unlike the case of Examples 2.9 and 2.10 the derivative of the incident voltage is not zero, since $\mathcal{V}_1^+(l,t)$ is not constant in time during the time interval $t_d \leq t \leq (t_d + t_r)$. To solve this differential equation for the reflected voltage $\mathcal{V}_1^-(l,t)$ using the given functional form of $\mathcal{V}_1^+(l,t)$, we can take its Laplace transform,

$$(sL_L + R_L + Z_0)\tilde{\mathcal{V}}_1^-(s) = (sL_L + R_L - Z_0)\tilde{\mathcal{V}}_1^+(s)$$

where $\tilde{\mathcal{V}}_1^\pm(s)$ is the Laplace transform of $\mathcal{V}_1^\pm(l,t)$. Using the previously noted form of $\tilde{\mathcal{V}}_1^+(s)$, we then have

$$\tilde{\mathcal{V}}_1^-(s) = \left[\frac{sL_L + R_L - Z_0}{sL_L + R_L + Z_0} \right]\left[\frac{V_0}{2t_r} \frac{1 - e^{-t_r s}}{s^2} \right] = \left[\frac{s + (R_L - Z_0)/L_L}{s + (R_L + Z_0)/L_L} \right]\left[\frac{V_0}{2t_r} \frac{1 - e^{-t_r s}}{s^2} \right]$$

which can be expanded into its partial fractions as

$$\tilde{\mathcal{V}}_1^-(s) = \frac{K_1}{s + (R_L + Z_0)/L_L} + \frac{K_2}{s} + \frac{K_3}{s^2}$$

where the coefficients are

$$K_2 = -K_1 = \frac{V_0 Z_0 L_L}{t_r(R_L + Z_0)^2}\left(1 - e^{-t_r s}\right) \quad \text{and} \quad K_3 = \frac{V_0}{2t_r} \frac{R_L - Z_0}{R_L + Z_0}\left(1 - e^{-t_r s}\right)$$

Taking the inverse Laplace transform[29] yields

$$\mathcal{V}_1^-(l,t) = \frac{V_0}{t_r}\frac{Z_0 L_L}{(R_L + Z_0)^2}\{[1 - e^{-(R_L+Z_0)t'/L_L}]u(t') - [1 - e^{-(R_L+Z_0)(t'-t_r)/L_L}]u(t'-t_r)\}$$

$$+ \frac{V_0}{2t_r}\frac{R_L - Z_0}{R_L + Z_0}[t'u(t') - (t'-t_r)u(t'-t_r)]$$

where $t' = t - t_d$. Note that the solution for $\mathcal{V}_1^-(l,t)$ is valid only for $t \geq t_d$, or $t' \geq 0$.

A practical case of interest is that in which $R_L = Z_0$. When a microstrip is terminated at a matched load resistance to avoid reflections, the nonzero inductance of the resistor leads may nevertheless produce reflections. To determine the maximum reflection voltage due to the inductance of the resistor leads, we substitute $R_L = Z_0$ in the solution for $\mathcal{V}_1^-(l,t)$ to find

$$\mathcal{V}_1^-(l,t) = \frac{V_0}{t_r}\frac{L_L}{4Z_0}\{[1 - e^{-(2Z_0/L_L)(t-t_d)}]u(t-t_d) - [1 - e^{-(2Z_0/L_L)(t-t_d-t_r)}]u(t-t_d-t_r)\}$$

The time variation of $\mathcal{V}_1^-(l,t)$ is plotted in Figure 2.26b, showing that the reflected voltage rises and falls exponentially with a time constant of $L/(2Z_0)$. The maximum reflected voltage occurs at $t = t_d + t_r$ and is given by

$$[\mathcal{V}_1^-(l,t)]_{max} = \frac{V_0}{t_r}\frac{L_L}{4Z_0}[1 - e^{-(2Z_0/L_L)t_r}]$$

Note that in practice the maximum reflected voltage can easily be measured—for example, by using a time-domain reflectometer (see Section 2.6), from which the value of the inductance L_L of the resistor leads can be calculated, since Z_0 and t_r are known in most cases.

2.6 TIME-DOMAIN REFLECTOMETRY

In practice, it is often necessary to make a number of measurements on a given transmission line system to characterize its transient response. The quantities that need to be measured include the nature (capacitive, inductive, or resistive) of the load termination, the characteristic impedance of the line, the maximum voltage level at which the line can be used, and others of a more specialized character. A time-domain reflectometer[30] (commonly abbreviated as TDR) is an instrument which is used to test, characterize, and

---

[29] We use the following Laplace transform pairs:

$$e^{-a(t-b)}u(t-b) \Longleftrightarrow \frac{e^{-bs}}{s+a}$$

$$(t-b)u(t-b) \Longleftrightarrow \frac{e^{-bs}}{s^2}$$

where a and b are constants.

[30] B. M. Oliver, Time domain reflectometry, *Hewlett-Packard J.*, 15(6), February 1964.

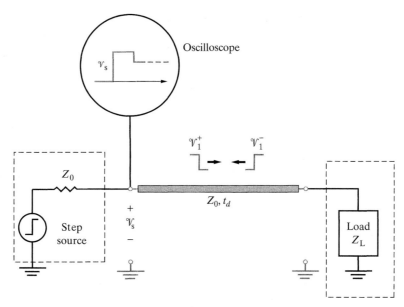

Figure 2.27   Time-domain reflectometry. Essential components of a typical TDR system.

model a system involving transmission lines and their accessories. In general, it consists of a very-fast-rise-time (typically less than 50 ps) step pulse source and a display oscilloscope in a system that operates like a closed-loop radar, as shown in Figure 2.27. The source produces an incident step voltage, which travels down the transmission line under investigation, and the incident and the reflected voltages at a particular point (typically the source end) on the line are monitored by the display oscilloscope using a high-impedance probe. The output impedance of the step source is typically well matched to the nominal characteristic impedance of the line to eliminate reflections from the source end.

The most common use of time-domain reflectometry involves the measurement of the characteristics of an unknown load termination or a discontinuity on the line. The former application is illustrated in Example 2.12 for resistive loads. A discontinuity on a transmission line could, for example, be a point of breakage on a buried coaxial line, an unwanted parasitic capacitance on an interconnect, or the inductance of a bonding wire between two interconnects. The latter case is illustrated in Example 2.13.

Example 2.12: TDR displays for resistive loads. A TDR system (represented by a step pulse source of amplitude V_0 and output impedance $R_s = Z_0$) is connected to a transmission line of characteristic impedance Z_0 terminated with a resistive load R_L, as shown in Figure 2.28. Three TDR waveforms monitored at the source end are shown for three different values of R_L. Find the load resistance R_L for each case.

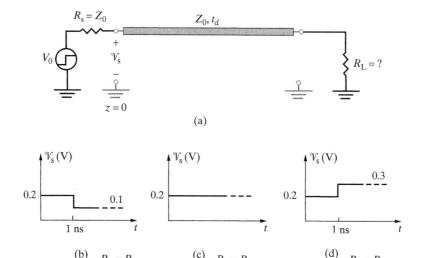

Figure 2.28 TDR displays for resistive loads. (a) A TDR system connected to a transmission line terminated with an unknown load resistor R_L. (b) $R_L = R_{L1}$. (c) $R_L = R_{L2}$. (d) $R_L = R_{L3}$.

Solution: The initial value (immediately after the application of the step input) of the source-end voltage $\mathcal{V}_s(0)$ is equal to

$$\mathcal{V}_s(0) = \mathcal{V}_1^+(z=0,0) = \frac{V_0}{2} = 0.2 \text{ V}$$

from which the amplitude of the step voltage is found to be $V_0 = 0.4$ V. At $t = 1$ ns, the reflected voltage arrives at the source end and is completely absorbed. So

$$\mathcal{V}_s(1\text{ns}) = \mathcal{V}_1^+(0,1 \text{ ns}) + \mathcal{V}_1^-(0,1 \text{ ns}) = \mathcal{V}_1^+(0,1 \text{ ns})(1 + \Gamma_L)$$

where $\Gamma_L = (R_L - Z_0)/(R_L + Z_0)$. For $R_L = R_{L1}$, we have

$$0.2(1 + \Gamma_{L1}) = 0.1$$

from which $\Gamma_{L1} = -0.5$, yielding $R_{L1} = Z_0/3$. Similarly, we find $R_{L2} = Z_0$ and $R_{L3} = 3Z_0$.

A simple summary of the TDR waveforms observed at the source end for purely resistive, capacitive, and inductive terminations is provided in Figure 2.29. Note that the case of a resistive termination was discussed in the preceding example, while a simple inductive termination was discussed in Section 2.5 and in connection with Figure 2.23. The result for the capacitive termination case corresponds to that of Example 2.9 for $R_L = \infty$.

Termination

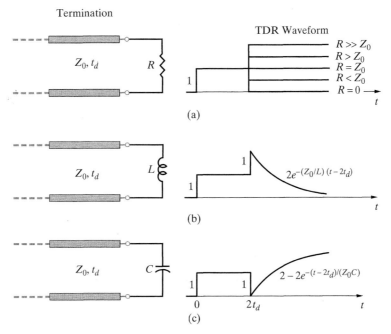

Figure 2.29   TDR signatures produced by simple terminations. Source-end TDR voltage signatures for purely (a) resistive, (b) inductive, and (c) capacitive terminations. In terms of excitation by a step voltage source of amplitude V_0 and source resistance R_s, the TDR traces shown are drawn for $V_0 = 2$ V and $R_s = Z_0$. (This figure was adapted from Figure 5 of B. M. Oliver, Time domain reflectometry, *Hewlett-Packard J.*, 15(6), pp. 14–9 to 14–16, February 1964. ©Hewlett-Packard Company 1964. Reproduced with permission.)

We now illustrate (Example 2.13) the use of the TDR technique for the measurement of the value of a reactive element connected between two transmission lines.

Example 2.13: TDR measurement of the inductance of a bonding wire connecting two transmission lines. Consider a bonding wire between two microstrip interconnects (each with characteristic impedance Z_0) on an integrated circuit board, as shown in Figure 2.30a. To measure the value of the bonding-wire inductance L_w, the circuit is terminated with a matched load (Z_0) at the load end and is excited by a matched TDR system (i.e., $R_s = Z_0$) at the input side, as shown in Figure 2.30b. The TDR waveform $\mathcal{V}_s(t)$ measured is shown in Figure 2.30c, which is similar to Figure 2.25b (Example 2.10). Determine the value of the bonding-wire inductance in terms of the area under the "glitch" seen in the TDR waveform.

Solution: In principle, the bonding-wire inductance L_w can be determined from the curvature of the glitch by accurately fitting an exponential function. However, a more accurate method is to determine L_w from the area under the curve, which can be measured more

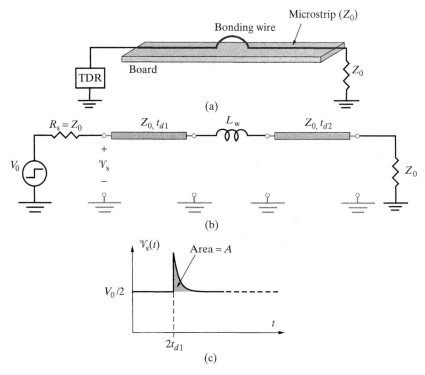

Figure 2.30 Measurement of bonding-wire inductance using a TDR system.
(a) Actual circuit configuration. (b) Equivalent transmission line circuit model.
(c) Measured TDR waveform.

accurately. Following an approach similar to that used in Example 2.10, it can be shown that the source-end voltage is

$$
\mathcal{V}_s(t) = \begin{cases} \frac{V_0}{2} & 0 < t < 2t_{d1} \\ \frac{V_0}{2}[1 + e^{-(2Z_0/L_w)(t-2t_{d1})}] & t > 2t_{d1} \end{cases}
$$

To find the area A under the glitch, we integrate $(\mathcal{V}_s(t) - V_0/2)$ from $t = 2t_{d1}$ to $t = \infty$:

$$
A = \int_{2t_{d1}}^{\infty} \frac{V_0}{2} e^{-(2Z_0/L_w)(t-2t_{d1})} \, dt = \frac{V_0}{2} \int_0^{\infty} e^{-(2Z_0/L_w)t'} \, dt' = -\frac{L_w V_0}{4Z_0} e^{-(2Z_0/L_w)t'} \Big|_0^{\infty} = \frac{L_w V_0}{4Z_0}
$$

where $t' = t - 2t_{d1}$. Therefore, the bonding-wire inductance L_w is given in terms of the area A as

$$
L_w = \frac{4Z_0 A}{V_0}
$$

A simple summary of TDR signatures of purely resistive or purely reactive discontinuities on a transmission line is provided in Figure 2.31. The signatures of discontinuities involving combinations of reactive and resistive elements are dealt with in several problems at the end of this chapter.

Figure 2.31   TDR signatures produced by simple discontinuities. Source-end TDR voltage signatures for shunt or series purely resistive, inductive, and capacitive discontinuities. In terms of excitation by a step voltage source of amplitude V_0 and source resistance R_s, the TDR voltage waveforms shown are drawn for $V_0 = 2$ V and $R_s = Z_0$, and one-way travel time t_d from the source to the disconuity. (This figure was adapted from Figure 6 of B. M. Oliver, Time domain reflectometry, *Hewlett-Packard J.*, 15(6), pp. 14–9 to 14–16, February 1964. ©Hewlett-Packard Company 1964. Reprinted with permission.)

2.7 TRANSMISSION LINE PARAMETERS

We have seen in previous sections that the response of a lossless transmission line to a given excitation depends on its characteristic impedance Z_0 and the propagation speed v_p (or the one-way travel time $t_d = l/v_p$), which in turn depends on the line inductance L and capacitance C per unit line length. The response of lossy lines is additionally influenced by the values per unit line length of the series resistance R and shunt conductance G. In general, the values of these transmission line parameters depend on (1) the geometric shapes, physical dimensions, and proximity of the two conductors that form the line; (2) the electromagnetic properties of the material surrounding the conductors; and (3) the electrical conductivity of the conductors and the frequency of operation.

In later chapters, after we have introduced the governing electromagnetic equations, we will discuss methods by which the line capacitance, inductance, resistance, and conductance per unit length can be defined and determined from basic principles. In the case of the common transmission lines shown in Figure 2.32, we will be able to find convenient analytical expressions for the line parameters. For other, more complicated, structures R, L, C, and G can be either evaluated using numerical techniques or measured. Parameters for many different transmission lines are also extensively available in handbooks.[31]

Approximate expressions for L, R, C and G parameters and for Z_0 for the common uniform transmission lines shown in Figure 2.32 are given in Table 2.2. The characteristic impedances (Z_0) provided are for lossless lines (i.e., $Z_0 = \sqrt{L/C}$). In Table 2.2, we have assumed the transmission line conductors to be made of copper and the surrounding medium to be air. Note that the parameters depend on the geometric shapes and the physical dimensions of the lines (d, a, and b). The line capacitance C and characteristic impedance Z_0 for the case when the surrounding medium is a nonmagnetic[32] material other than air can be derived from those given in Table 2.2 by using the propagation speed v_p for these media as given in Table 2.1. Specifically we have

$$[C]_{\text{material}} = \frac{c^2}{v_p^2}[C]_{\text{air}} \quad \text{and} \quad [Z_0]_{\text{material}} = \frac{v_p}{c}[Z_0]_{\text{air}}$$

where c is the speed of light in free space, or $c \simeq 3 \times 10^8$ m-s^{-1}. The line inductance L remains the same, since it is governed by the magnetic properties of the surrounding material.

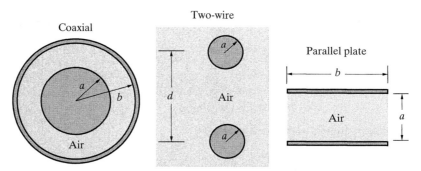

Figure 2.32   Cross-sectional view of three common uniform transmission lines. Expressions for the circuit parameters L, R, C, and G for these coaxial, two-wire, and parallel-plate lines are provided in Table 2.2.

---

[31]*Reference Data for Engineers*, 8th ed., Sams Prentice Hall Computer Publishing, Carmel, Indiana, 1993.

[32]Magnetic properties of materials are discussed in Section 6.8. In the transmission line context, all materials can be considered nonmagnetic except iron, nickel, cobalt, a few of their alloys, and some special compounds involving mixtures of magnetic materials with barium titanate.

TABLE 2.2 TRANSMISSION LINE PARAMETERS FOR SOME UNIFORM TWO-CONDUCTOR TRANSMISSION LINES SURROUNDED BY AIR

	Coaxial	Two-wire	Parallel-plate*
L (μH-m^{-1})	$0.2\ln(b/a)$	$0.4\ln\left[\dfrac{d}{2a}+\sqrt{\left(\dfrac{d}{2a}\right)^2-1}\right]$	$\dfrac{1.26a}{b}$
C (pF-m^{-1})	$\dfrac{55.6}{\ln(b/a)}$	$\dfrac{27.8}{\ln\left[\dfrac{d}{2a}+\sqrt{\left(\dfrac{d}{2a}\right)^2-1}\right]}$	$\dfrac{8.85b}{a}$
R (Ω-m^{-1})	$\dfrac{4.15\times10^{-8}(a+b)\sqrt{f}}{ab}$	$\dfrac{8.3\times10^{-8}\sqrt{f}}{a}$	$\dfrac{5.22\times10^{-7}\sqrt{f}}{b}$
G^{**} (S-m^{-1})	$\dfrac{7.35\times10^{-4}}{\ln(b/a)}$	$\dfrac{3.67\times10^{-4}}{\ln\left[\dfrac{d}{2a}+\sqrt{\left(\dfrac{d}{2a}\right)^2-1}\right]}$	$\dfrac{1.17\times10^{-4}b}{a}$
Z_0 (Ω)	$60\ln(b/a)$	$120\ln\left[\dfrac{d}{2a}+\sqrt{\left(\dfrac{d}{2a}\right)^2-1}\right]$	$\dfrac{377a}{b}$

*Valid for $b \gg a$.
**For polyethylene at 3 GHz.

The series resistance (R) is inversely proportional to the electrical *conductivity* of the particular metal that the conductors are made of, with the values given in Table 2.2 being relative to that of copper. The physical underpinnings of electrical conductivity are discussed in Chapter 5. For now, it suffices to know that it is a quantitative measure of the ability of a material to conduct electrical current and that the values of conductivity for different materials are tabulated extensively in various handbooks (see Table 1). A brief list of conductivities of some common metals relative to that of copper is provided in Table 2.3. The series resistance R is proportional to the square root of the frequency

TABLE 2.3 RELATIVE CONDUCTIVITIES OF METALS VERSUS COPPER

Material	Relative Conductivity	Material	Relative Conductivity
Aluminum	0.658	Silver	1.06
Brass	0.442	Sodium	0.375
Copper	1.00	Stainless steel	0.0192
Gold	0.707	Tin	0.151
Lead	0.0787	Titanium	0.0361
Magnesium	0.387	Tungsten	0.315
Nickel	0.250	Zinc	0.287

because of the so-called *skin effect,* which results from the nonuniform distribution of electrical current in a metal at higher frequencies, and which is discussed in Chapter 8.

With air as the surrounding medium, the shunt conductance $G = 0$, since air is an excellent insulator and leakage losses through it are generally negligible. In the case of other surrounding media for which leakage losses may not be negligible, the value of G depends on the geometrical layout of the conductors (as do the values of C, L, and R) but is more strongly determined by the loss properties of the insulating medium surrounding the conductors and is, in general, a rather complicated function of the frequency of operation. Table 2.2 provides a representative expression for G for polyethylene as the surrounding medium at an operating frequency of 3 GHz. High-frequency losses in insulating materials are discussed in Sections 8.3 and 11.2.

Example 2.14: RG58/U coaxial line. RG58/U is a commonly used coaxial line with an inner conductor of diameter 0.45 mm and an outer conductor of inside diameter 1.47 mm constructed using copper conductors and filled with polyethylene as its insulator. The line is to be used at 3 GHz. Find the line parameters (i.e., R, L, C, G, and Z_0). Note from Table 2.1 that the propagation speed for polyethylene at 3 GHz is $v_p \simeq 20$ cm-(ns)$^{-1} = 2 \times 10^8$ m-s^{-1}.

> **Solution:** We use the expressions provided in Table 2.2, except for the multipliers needed for C and Z_0 to correct for the fact that the filling is polyethylene rather than air. Note that we can also use the expression from Table 2.2 for G, since it was also given for polyethylene and for 3 GHz. We have
>
> $$R \simeq \frac{4.15 \times 10^{-8}(0.45 + 1.47) \times 10^{-3}\sqrt{3 \times 10^9}}{(0.45 \times 10^{-3})(1.47 \times 10^{-3})} \simeq 6.6\,\Omega\text{-m}^{-1}$$
>
> $$L \simeq 0.2 \ln\left(\frac{1.47}{0.45}\right) \simeq 0.237\,\mu\text{H-m}^{-1}$$
>
> $$C \simeq \left(\frac{c^2}{v_p^2}\right)\frac{55.6}{\ln\left(\frac{b}{a}\right)} = \left(\frac{3}{2}\right)^2 \frac{55.6}{\ln\left(\frac{1.47}{0.45}\right)} \simeq 106\,\text{pF-m}^{-1}$$
>
> $$G \simeq \frac{7.35 \times 10^{-4}}{\ln\left(\frac{1.47}{0.45}\right)} \simeq 6.21 \times 10^{-4}\,\text{S-m}^{-1}$$
>
> $$Z_0 \simeq \left(\frac{v_p}{c}\right)60\ln\left(\frac{b}{a}\right) \simeq 47.4\,\Omega$$

Note that the value of Z_0 is close to the nominal 50-ohm characteristic impedance of this coaxial line. Note also that the difference between 47.4Ω and 50Ω is within the range of accuracy (i.e., two digits after the decimal point) by which the physical quantities were specified (e.g., the radii of conductors, the value of v_p, etc.).

2.8 SUMMARY

This chapter discussed the following topics:

- **Transmission line parameters.** A transmission line is commonly characterized by its distributed parameters R (in Ω-m^{-1}), L (in H-m^{-1}), G (in S-m^{-1}), and C

(in F-m^{-1}), whose values are determined by the line geometry, the conductivity of the metallic conductors, the electrical and magnetic properties of the surrounding insulating material, and the frequency of excitation. Formulas for these parameters for three common transmission line geometries are provided in Table 2.2.

- **Transmission line equations.** The differential equations governing the behavior of voltage and current on a lossless ($R = 0$ and $G = 0$) transmission line, and the wave equation for voltage derived from them, are

$$\left.\begin{array}{l} \dfrac{\partial \mathcal{V}}{\partial z} = -L\dfrac{\partial \mathcal{I}}{\partial t} \\[2mm] \dfrac{\partial \mathcal{I}}{\partial z} = -C\dfrac{\partial \mathcal{V}}{\partial t} \end{array}\right\} \quad \longrightarrow \quad \dfrac{\partial^2 \mathcal{V}}{\partial z^2} = LC\dfrac{\partial^2 \mathcal{V}}{\partial t^2}$$

- **Propagating-wave solutions, characteristic impedance, and phase velocity.** The general solution of the transmission line equations leads to mathematical expressions for voltage and current along the line that are wave equations in nature, depending on both distance and time. These are

$$\mathcal{V}(z,t) = f^+\left(t - \frac{z}{v_p}\right) + f^-\left(t + \frac{z}{v_p}\right)$$

$$\mathcal{I}(z,t) = \frac{1}{Z_0}\left[f^+\left(t - \frac{z}{v_p}\right) - f^-\left(t + \frac{z}{v_p}\right)\right]$$

where Z_0 is the characteristic impedance of a lossless line, which is defined as the voltage-to-current ratio of a single disturbance propagating in the $+z$ direction and is given by $Z_0 = \sqrt{L/C}$. The characteristic impedance is one of the most important quantities that determine the response of a transmission line. Formulas for calculating Z_0 are provided in Table 2.2. The velocity with which waves on a transmission line propagate is called the phase velocity and is given by $v_p = (LC)^{-1/2}$. For a lossless transmission line, the phase velocity is determined by the properties of the material surrounding the transmission line conductors. For uniform transmission lines, v_p is a constant. Values of v_p for selected materials are tabulated in Table 2.1.

- **Transmission lines terminated in resistive loads, reflection coefficient.** Transient response of a lossless transmission line to step or pulse excitation involves reflections from discontinuities along the line or from loads at its termination. Reflection effects are described in terms of the reflection coefficient Γ, defined as the ratio of the reflected to the incident voltage at a given point. The reflection coefficients at the load and source ends of a transmission line are given by

$$\Gamma_L = \frac{R_L - Z_0}{R_L + Z_0} \quad \text{Load end}$$

$$\Gamma_s = \frac{R_s - Z_0}{R_s + Z_0} \quad \text{Source end}$$

where R_L is the resistance terminating the load end of the line, the line's characteristic impedance is Z_0, and R_s is the source resistance. In the special case of a matched termination, we have $R_L = Z_0$, so $\Gamma_L = 0$, and thus no reflection occurs from the termination. Similarly, when $R_s = Z_0$, $\Gamma_s = 0$, and no reflection occurs from the source end. In general, when a voltage disturbance is launched from the source end of a transmission line (e.g., due to a step change in input voltage), a sequence of reflections from both the load and source ends of the line occurs. The process of multiple reflections from the load and source ends of a transmission line can be described using a bounce diagram.

- **Transmission lines terminated in reactive loads.** To determine the transient behavior of lossless lines terminated in reactive elements, it is necessary to solve the differential equations that determine the voltage–current relationships of the terminations subject to the appropriate initial conditions. The reflected voltage due to a step excitation is no longer a simple step function but, in general, varies continuously at a fixed position with respect to time depending on the nature of the reactive termination.

PROBLEMS

2.1   **Lumped or distributed circuit element?** A uniform lossless transmission line with the L and C parameters and the length l given as shown in Figure 2.33 is excited by a unit step source with rise-time equal to $t_r = 0.1$ ns. Determine the following: (a) The characteristic impedance Z_0 and the one-way time delay t_d of the transmission line. (b) Whether it is appropriate or not to model the transmission line in this circuit as a lumped element.

Figure 2.33   **Transmission line.** Problem 2.1.

2.2   **Lumped- or distributed-circuit element?** A uniform lossless transmission line is used to connect a periodic pulse waveform to a load as shown in Figure 2.34. Assuming the

Figure 2.34   **Transmission line.** Problem 2.2.

characteristic impedance and phase velocity of the transmission line to be, respectively, $Z_0 = 50\Omega$ and $v_p = 2c/3$, determine the following: (a) The per-unit-length parameters L and C of the transmission line. (b) The one-way propagation time delay t_d of the transmission line. (c) Whether this transmission line can be treated as a lumped element or not.

2.3   **Open-circuited line.** Consider the circuit shown in Figure 2.35, with an ideal unit step source connected to a lossless line of characteristic impedance $Z_0 = 100\Omega$ having an open-circuited termination at the other end. Assuming a one-way propagation delay across the line of t_d, use a bounce diagram to sketch the load voltage $\mathcal{V}_L(t)$ versus time for $0 \leq t \leq 10t_d$.

Figure 2.35   Open-circuited line. Problem 2.3.

2.4   **Step and pulse excitation of a lossless line.** A uniform, lossless transmission line is excited with a step source as shown in Figure 2.36. (a) Provide an appropriate bounce diagram and use it to sketch both the source-end voltage $\mathcal{V}_s$ and the load-end voltage $\mathcal{V}_L$ as a function of time between 0 and 10 ns. Provide all the pertinent values on your sketches. (b) Repeat part (a) if the step voltage source was a pulse source given by $10\,[u(t) - u(t - 0.3\text{ ns})]$ V.

Figure 2.36   Step and pulse excitation of a lossless line. Problem 2.4.

2.5   **Resistive loads.** The circuit shown in Figure 2.37 consists of an uncharged transmission line connected to a load resistance R_L. Assuming that the switch closes at $t = 0$, sketch the load voltage $\mathcal{V}_L(t)$ over the time interval $0 \leq t \leq 3$ ns for the following load resistances: (a) $R_L = 25\Omega$, (b) $R_L = 50\Omega$, (c) $R_L = 100\Omega$.

Figure 2.37   Resistive loads.
Problems 2.5 and 2.7.

2.6   **Ringing.** The transmission line system shown in Figure 2.38 is excited by a step-voltage
source of amplitude 3.6 V and source impedance 10Ω at one end and is terminated with an
open circuit at the other end. The line is characterized by the line parameters $L = 0.5$ μH-
m^{-1}, $C = 0.2$ nF-m^{-1}, $R = 0$, $G = 0$, and has a length of $l = 30$ cm. Sketch the load voltage
$\mathcal{V}_L(t)$ over $0 \le t \le 10$ ns with the steady-state value indicated.

Figure 2.38   Ringing. Problem 2.6.

2.7   **Discharging of a charged line.** For the circuit of Problem 2.5, assume that the switch has
been closed for a long time before it opens at $t = 0$. Sketch the load voltage $\mathcal{V}_L(t)$ over
$0 \le t \le 3$ ns for the same three cases.

2.8   **Two step voltage sources.** Consider the lossless transmission-line circuit which is excited
by two step voltage sources, as shown in Figure 2.39. Use a bounce diagram to sketch the
voltages $\mathcal{V}_1$ and $\mathcal{V}_2$ as a function of time. Note that the step-voltage source on the left side
turns off at $t = 0$.

Figure 2.39   Pulse excitation. Problem 2.8.

2.9   **Pulse excitation.** The circuit shown in Figure 2.40 is excited by an ideal voltage pulse
of 1 V amplitude starting at $t = 0$. Given the length of the line to be $l = 10$ cm and the
propagation speed to be 20 cm-(ns)$^{-1}$; (a) sketch the voltage at the source end of the line,
$\mathcal{V}_s(t)$, for an input pulse duration of 10 ns; (b) repeat part (a) for an input pulse duration
of 1 ns.

Figure 2.40   Pulse excitation. Problem 2.9.

2.10 Pulse excitation. The circuit shown in Figure 2.41 is excited with a voltage pulse of amplitude A and pulse width t_w. Assuming the propagation delay of the line to be t_d, sketch the load voltage $\mathcal{V}_L(t)$ versus t for $0 \leq t \leq 10t_d$ for (a) $t_w = 2t_d$, (b) $t_w = t_d$, and (c) $t_w = t_d/2$.

Figure 2.41   Pulse excitation. Problem 2.10.

2.11 Observer on the line. A transmission line with an unknown characteristic impedance Z_0 terminated in an unknown load resistance R_L, as shown in Figure 2.42, is excited by a pulse source of amplitude 1 V and duration $t_w = 3t_d/4$, where t_d is the one-way flight time of the transmission line. An observer at the center of the line observes the voltage variation shown. (a) Determine Z_0 and R_L. (b) Using the values found in (a), sketch the voltage variation (up to $t = 4t_d$) that would be observed by the same observer if the pulse duration were $t_w = 1.5t_d$.

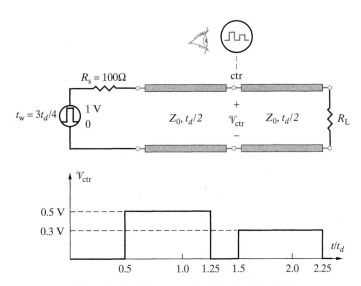

Figure 2.42   Observer on the line. Problem 2.11.

2.12 Cascaded transmission lines. For a transmission line circuit as shown in Figure 2.43, sketch $\mathcal{V}_s(t)$ and $\mathcal{V}_L(t)$ over $0 \leq t \leq 5$ ns.

Figure 2.43 Cascaded lines. Problem 2.12.

2.13 **Time-domain reflectometry (TDR).** A TDR is used to test the transmission line system shown in Figure 2.44. Using the sketch of $\mathcal{V}_s(t)$ observed on the TDR scope as shown, determine the values of Z_{01}, l_1, and R_1. Assume the phase velocity of the waves to be 20 cm-(ns)$^{-1}$ on each line. Plot $\mathcal{V}_{R1}(t)$ versus t for $0 \le t \le 4$ ns.

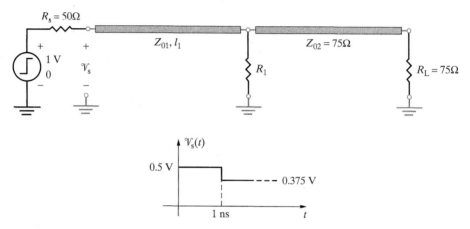

Figure 2.44 Time-domain reflectometry. Problem 2.13.

2.14 **Time-domain reflectometry (TDR).** TDR measurements can also be used in cases with more than one discontinuity. Two transmission lines of different characteristic impedances and time delays terminated by a resistive load are being tested by a TDR, as shown in Figure 2.45. (a) Given the TDR display of the source-end voltage due to a 3-V, 100Ω step excitation starting at $t = 0$, find the characteristic impedances (Z_{01} and Z_{02}) and the time delays (t_{d1} and t_{d2}) of both lines, and the unknown load R_L. (b) Using the values found in part (a), find the time and magnitude of the next change in the source-end voltage $\mathcal{V}_s(t)$, and sketch it on the display.

2.15 **Time-domain reflectometry (TDR).** Consider the circuit shown in Figure 2.46. The two line segments are of equal length l. Assuming the propagation speeds on the two lines are equal to 15 cm-(ns)$^{-1}$ each calculation Z_{01}, Z_{02}, R_L, and l using TDR display of voltage source $\mathcal{V}_s(t)$ as shown.

2.16 **Multiple lines.** In the three lossless transmission-line circuit shown in Figure 2.47, the switch closes at $t = 0$. Assuming all the lines to be uncharged before $t = 0$, draw a bounce diagram and sketch voltages $\mathcal{V}_s$, $\mathcal{V}_{L1}$, and $\mathcal{V}_{L2}$ between $t = 0$ and $t = 10$ ns using the impedance and time delay values indicated on Figure 2.47.

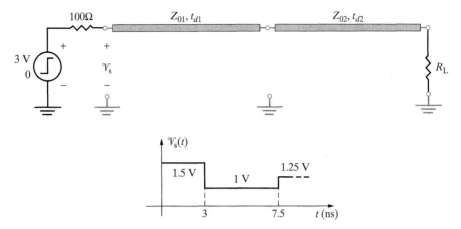

Figure 2.45   Time-domain reflectometry. Problem 2.14.

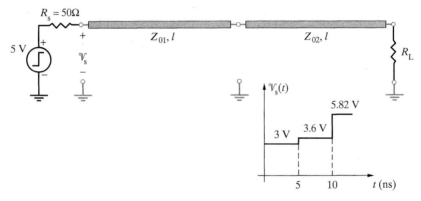

Figure 2.46   Time-domain reflectometry. Problem 2.15.

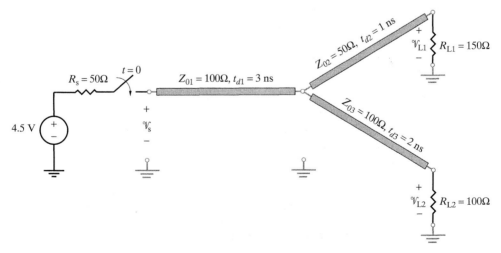

Figure 2.47   Multiple lines. Problem 2.16.

2.17 **Digital IC chips.** Two impedance-matched, in-package-terminated Integrated Circuit (IC) chips are driven from an impedance-matched IC chip, as shown in Figure 2.48. Assuming the lengths of the interconnects to be 15 cm each and the propagation velocity on each to be 10 cm-(ns)$^{-1}$, do the following: (a) Sketch the voltages $\mathcal{V}_{L1}$ and $\mathcal{V}_{L2}$ for a time interval of 10 ns. Indicate the steady-state values on your sketch. (b) Repeat part (a) if one of the load IC chips is removed from the end of the interconnect connected to it (i.e., the load point A is left open-circuited).

Figure 2.48   Digital IC chips. Problem 2.17.

2.18 **Multiple lines.** For the distributed interconnect system shown in Figure 2.49, and for $Z_{01} = Z_{02} = 50\Omega$, find and sketch the three load voltages $\mathcal{V}_1(t)$, $\mathcal{V}_2(t)$, and $\mathcal{V}_3(t)$ for a time interval of 5 ns. Assume each interconnect to have a one-way time delay of 1 ns. (b) Repeat part (a) for $Z_{01} = Z_{02} = 25\Omega$.

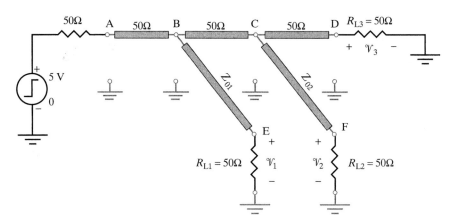

Figure 2.49   Multiple lines. Problem 2.18.

2.19 **Reflections due to parasitic effects.** The circuit shown in Figure 2.50 consists of a low-impedance driver driving a distributed interconnect system that was intended to be impedance-matched, with $Z_0 = 120\Omega$. An engineer performs some tests and measurements and observes reflections due to parasitic effects associated with the two interconnects terminated at the 120Ω loads. Assuming that the effective characteristic impedances of these interconnects (i.e., taking parasitic effects into account) is such that we have $Z_0 = 100\Omega$ instead of 120Ω, find and sketch the voltages $\mathcal{V}_1(t)$ and $\mathcal{V}_2(t)$ for $0 \leq t \leq 12$ ns, assuming the one-way time delay on each interconnect to be 2 ns. Comment on the effects of the mismatch caused by parasitic effects. Assume the initial incident wave launched at the driver end of the 60Ω line to be $\mathcal{V}_1^+ = 4$ V.

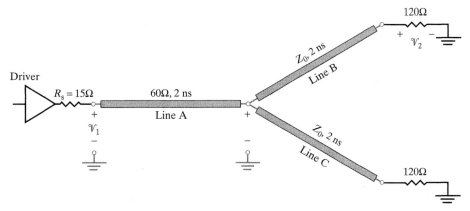

Figure 2.50   **Reflections due to mismatches.** Problem 2.19.

2.20 **Parallel multiple lines.** The transmission line system shown in Figure 2.51 consists of three lines, each having $Z_0 = 50\Omega$ and a one-way propagation delay of 1 ns. (a) Find and sketch the voltages $\mathcal{V}_s(t)$ and $\mathcal{V}_L(t)$ versus t for $0 \leq t \leq 10$ ns. (b) Repeat part (a) when the open-circuited ends are terminated with a load resistance of 50Ω each.

2.21 **Optimized multiple lines.** A multisection transmission line consists of three lossless transmission lines used to connect an ideal step source of 5V amplitude and 6Ω output impedance

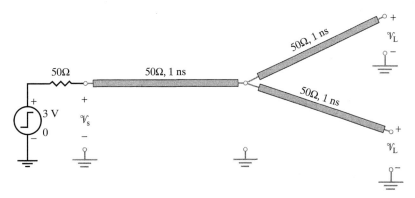

Figure 2.51   **Parallel multiple lines.** Problem 2.20.

to two separate load resistances of 66Ω each, as shown in Figure 2.52. All three lines are characterized by line parameters $L = 364.5$ nH-m^{-1}, and $C = 125$ pF-m^{-1}. To minimize ringing effects, the line lengths are optimized to be of equal length. If each line length is $l = 40$ cm, sketch the voltages $\mathcal{V}_s$ and $\mathcal{V}_L$ versus t for $0 \le t \le 20$ ns, and comment on the performance of the circuit in minimizing ringing.

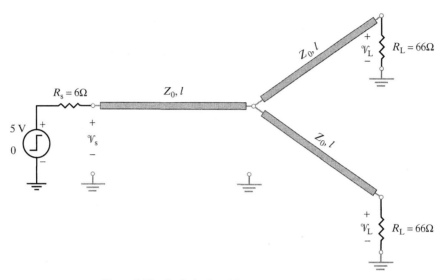

Figure 2.52   Optimized multiple lines. Problem 2.21.

2.22   Transient response of a cascaded transmission-line circuit. Consider a circuit consisting of two cascaded lossless transmission lines with characteristic impedances and one-way time delays as shown in Figure 2.53. Assuming both lines to be fully discharged before $t = 0$, sketch the source-end, junction and load-end voltages $\mathcal{V}_s$, $\mathcal{V}_J$, and $\mathcal{V}_L$ as a function of time for the time interval $0 \le t \le 10t_d$ for the following three cases: (a) $R_s = Z_0$ and $R_L = 3Z_0$, (b) $R_s = Z_0$ and $R_L = Z_0$, (c) $R_s = Z_0/3$ and $R_L = 3Z_0$.

Figure 2.53   Cascaded transmission line. Problem 2.22.

2.23   Charging and discharging of a line. For a transmission line system shown in Figure 2.54 the switch S_2 is closed at $t = 2t_d$ (where t_d is the propagation delay of each line) after switch S_1 is closed at $t = 0$. Find and sketch the voltage $\mathcal{V}_L$ versus $t = 0 \le t \le 6t_d$.

Figure 2.54   Charging and discharging of a line. Problem 2.23.

2.24   Digital IC interconnect. A circuit consists of one logic gate driving an identical gate via a 1-ft-long, 50Ω interconnect. Before $t = 0$, the output of the driver gate is at LOW voltage state and can simply be approximated as a 14Ω resistor. At $t = 0$, the output of the driver gate goes from LOW to HIGH state and can be approximated with a 5-V voltage source in series with a 14Ω resistor. The input of the load gate can be approximated to be an open circuit (i.e., $R_L = \infty$). Assuming that a minimum load voltage of 3.75 V is required to turn and keep the load gate on, (a) find the time at which the load gate will turn on for the first time. (b) Find the time at which the load gate will turn on permanently. (Assume a signal time delay of 1.5 ns-ft^{-1} along the interconnect for both parts.)

2.25   Terminated IC interconnects. The logic circuit of Problem 2.24 needs to be modified to eliminate ringing. Two possible solutions are to terminate the line in its characteristic impedance at either the source (series termination) or receiver (parallel termination) end. Both of these circuits are shown in Figure 2.55. (a) Select the value of the termination resistance R_T in both circuits to eliminate ringing. (b) Compare the performance of these two circuits in terms of their speed and dc power dissipation. Which technique is the natural choice for a design to achieve low-power dissipation at steady state?

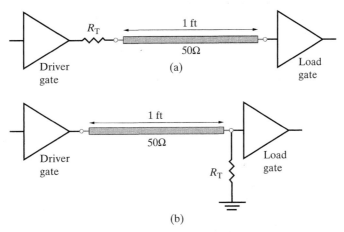

Figure 2.55   Terminated IC interconnects. (a) Series termination. (b) Parallel termination. Problem 2.25.

2.26   Digital IC gate interconnects. A disadvantage of the series termination scheme in Problem 2.25 is that the receiver gate or gates must be near the end of the line to avoid receiving a two-step signal. This scheme is not recommended for terminating distributed loads. The two circuits shown in Figure 2.56 have three distributed loads equally positioned along a 3-ft-long 50Ω interconnect on a pc board constructed of FR4 material (take $v_p \simeq 14.3$ cm-(ns)$^{-1}$). Each circuit uses a different termination scheme. Assuming the driver and all the loads to be the same gates as in Problem 2.24, find the times at which each load gate changes its logic state after the output voltage of the driver gate switches to HIGH state at $t = 0$. Comment on the performance of both circuits and indicate which termination scheme provides faster speed. (Use some of the data provided in Problem 2.24.)

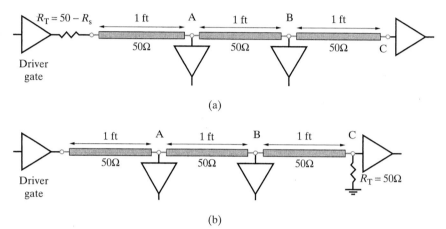

Figure 2.56   **IC gate interconnects.** (a) Series termination. (b) Parallel termination. Problem 2.26.

2.27 **Digital IC circuit.** For the digital IC circuit shown in Figure 2.57, the driver gate goes from LOW to HIGH state at $t = 0$, and its Thévenin equivalent circuit (including the series termination resistance) can be approximated as a 10-V voltage source in series with a 50Ω resistor. If the time delay for all interconnects is given to be 2 ns-ft^{-1}, find the time at which each receiver gate changes its state permanently. Assume each load gate to change state when its input voltage exceeds 4 V. Also assume each load gate to appear as an open circuit at its input. Support your solution with sketches of the two load voltages $\mathcal{V}_1$ and $\mathcal{V}_2$ as functions of time for a reasonable time interval.

2.28 **Two driver gates.** Two identical logic gates drive a third identical logic gate (load gate), as shown in Figure 2.58. All interconnects have the same one-way time delay t_d and characteristic impedance Z_0. When any one of these driver gates is at HIGH state, its Thévenin equivalent as seen from its output terminals consists of a voltage source with voltage V_0 in series with a resistance of value $R_s = Z_0$. At LOW state, its Thévenin equivalent is just a resistance of value $R_s = Z_0$. The input impedance of the load gate is very high compared with the characteristic impedance of the line (i.e., $Z_{in} \gg Z_0$). (a) Assuming steady-state conditions before both driver gates change to HIGH state at $t = 0$, sketch the load voltage $\mathcal{V}_L$ as a function of time for $0 \le t \le 7t_d$. What is the eventual steady-state value of the load voltage? (b) Assume steady-state conditions before $t = 0$ to be such that driver gate 1 is at HIGH state and gate 2 is at LOW state. At $t = 0$, gate 1 and gate 2 switch states. Repeat part (a).

2.29 **Capacitive load.** For the transmission line system shown in Figure 2.59, the switch is closed at $t = 0$. Each of the two transmission lines has a one-way time delay of 2 ns.

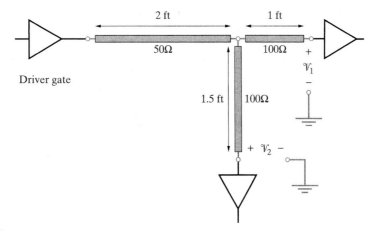

Figure 2.57   Digital IC circuit. Problem 2.27.

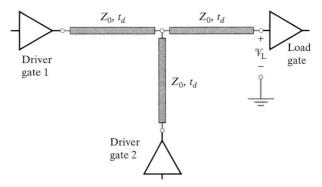

Figure 2.58   Two driver gates. Problem 2.28.

Figure 2.59   Capacitive load. Problem 2.29.

Assuming both transmission lines and the 5 pF capacitor to be initially uncharged, find and sketch the voltage $\mathcal{V}_1(t)$ across the resistor R_1.

2.30   Unknown lumped element. Using the source-end voltage waveform observed on the TDR display due to an ideal step voltage source exciting the double lossless transmission line circuit connected as shown in Figure 2.60, determine the type of the unknown lumped element at the junction and its value.

Figure 2.60   Unknown lumped element. Problem 2.30.

2.31   Unknown lumped element. The transmission line circuit has an unknown lumped element, as shown in Figure 2.61. With the source-end voltage due to step excitation measured to be as plotted, determine the type of the unknown element, and find its value in terms of the shaded area A.

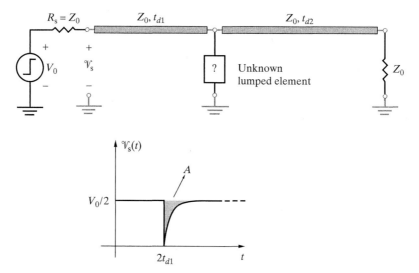

Figure 2.61   Unknown lumped element. Problem 2.31.

2.32   Source-end TDR waveform. The source-end TDR voltage waveform of a transmission-line circuit terminated with three unknown lumped elements is as shown in Figure 2.62. Using this waveform, (a) determine the characteristic impedance and the one-way time delay of the line and (b) the types and the values of the three unknown elements.

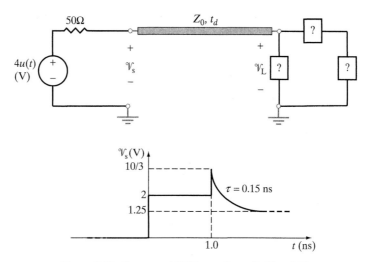

Figure 2.62   Source-end TDR waveform. Problem 2.32.

2.33   Unknown load. A 5 V dc voltage source is used to excite a lossless transmission-line circuit terminated with two unknown lumped elements, as shown in Figure 2.63a. The switch at the source end is closed at $t = 0$. The transmission line parameters Z_0, t_d, and length L are unknown. Based on the source-end TDR voltage waveform shown in Figure 2.63b, determine the type (resistor, capacitor, or inductor) and values of the two unknown lumped elements. Assume the transmission line to be at steady-state condition at $t = 0^-$.

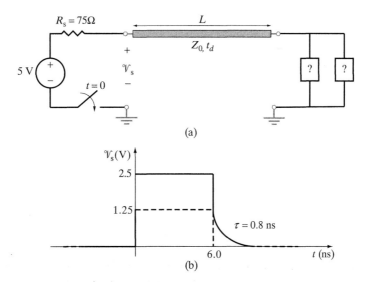

Figure 2.63   Unknown load. Problem 2.33.

2.34 **Capacitive load.** Two transmission lines of characteristic impedances 75Ω and 50Ω are joined by a connector that introduces a shunt resistance of 150Ω between the lines, as shown in Figure 2.64. The load end of the 50Ω line is terminated with a capacitive load with a 30 pF capacitor initially uncharged. The source end of the 75Ω line is excited by a step function of amplitude 3.6 V and a series resistance of 75Ω, starting at $t = 0$. Assuming the total time delay of each line to be $t_{d1} = 6$ ns and $t_{d2} = 2$ ns, respectively, find and sketch (a) the voltage $\mathcal{V}_L(t)$ at the load end of the 50Ω line and (b) the voltage $\mathcal{V}_s(t)$ at the source end of the 75Ω line.

Figure 2.64 **Capacitive load.** Problem 2.34.

2.35 **Reactive element at the junction.** In the lossless transmission-line circuit shown in Figure 2.65, find and sketch both the source end voltage $\mathcal{V}_s$ and the load end voltage $\mathcal{V}_L$ as function of time.

Figure 2.65 **Reactive element at the junction.** Problem 2.35.

2.36 **Transmission line terminated with a capacitive load.** Consider the lossless transmission-line circuit shown in Figure 2.66 where the switch at the load end closes at $t = 0$, after

Figure 2.66 **Transmission line terminated with a capacitive load.** Problem 2.36.

being open for a long time. Find and sketch the source-end and the load-end voltages $\mathcal{V}_s(t)$ and $\mathcal{V}_L(t)$.

2.37 **Transients on a transmission line.** Consider the lossless transmission-line circuit as shown in Figure 2.67. The switch is opened at $t=0$, after being closed for a long time. Find and sketch the source-end voltage $\mathcal{V}_s(t)$ as a function of time.

Figure 2.67 **Transients on a transmission line.** Problem 2.37.

2.38 **Transient response.** Consider the lossless transmission line system shown in Figure 2.68. The 20 pF capacitor is initially uncharged. The switch S is initially open but is closed at $t=7$ ns. Determine mathematical expressions for and sketch the capacitor voltage $\mathcal{V}_C(t)$ and the source-end voltage $\mathcal{V}_s(t)$ for $0 \le t \le 12$ ns. Make assumptions as needed but clearly state and justify them.

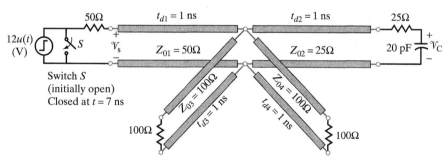

Figure 2.68 **Transients on a transmission line.** Problem 2.38.

2.39 **Inductive load.** Two transmission lines of characteristic impedances 50Ω and 75Ω are joined by a connector that introduces a series resistance of 25Ω between the lines, as shown in Figure 2.69. The load end of the 75Ω line is terminated with an inductive load. The

Figure 2.69 **Inductive load.** Problem 2.39.

inductor is initially uncharged, and a step source with amplitude s v turns on at $t = 0$. Find and sketch the voltage $\mathcal{V}_L$. First determine the initial and final values and accurately mark all points of your sketch.

2.40 Capacitive load. A step-type incident voltage wave of 1-V peak value arrives at a capacitive load at $t = 0$, as shown in Figure 2.70. After $t = 0$, find the approximate time at which the reflected voltage wave $\mathcal{V}^-$ at the load position $z = 0$ is equal to zero. Assume the load capacitor to be initially uncharged.

Figure 2.70   Capacitive load. Problem 2.40.

2.41 Step excitation. The circuit shown in Figure 2.71 is excited by a step-voltage source of amplitude 5 V and source resistance $R_s = 2Z_0$, starting at $t = 0$. Note that the characteristic impedance of the shorted stub is half that of the main line and that the second segment of the main line is twice as long, so its one-way time delay is $2t_d$. (a) Assuming the load to be an open circuit (i.e., a very large resistance), sketch the load voltage $\mathcal{V}_L(t)$ versus t for $0 \le t \le 11t_d$. (b) Repeat part (a) for the case when the input is a pulse of duration $t_w = 4t_d$.

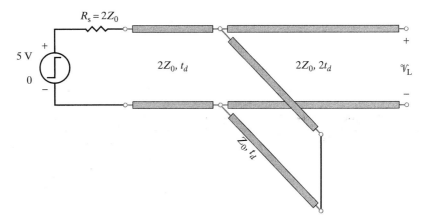

Figure 2.71   Step excitation. Problem 2.41.

2.42 Capacitive load excited by two sources. For the transmission line system shown in Figure 2.72, find the mathematical expression for the capacitor voltage $\mathcal{V}_c(t)$ and sketch it for $t \ge 0$. Assume the capacitor to be initially uncharged.

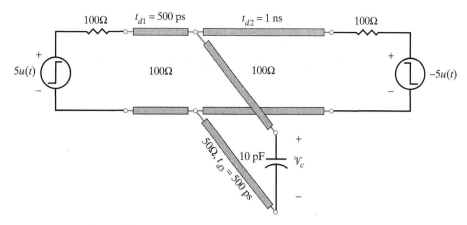

Figure 2.72   Capacitive load excited by two sources. Problem 2.42.

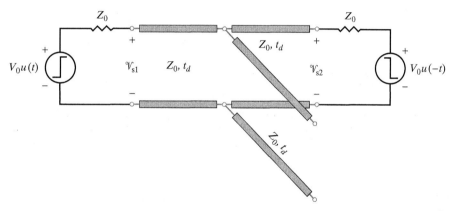

Figure 2.73   Two sources. Problem 2.43.

2.43   Two sources. For the circuit shown in Figure 2.73, sketch the voltages $\mathcal{V}_{s1}(t)$ and $\mathcal{V}_{s2}(t)$ for $0 \leq t \leq 7t_d$.

2.44   RG 8 coaxial line. A student buys an RG 8 low-loss coaxial cable for a VHF antenna project. He looks up the specifications of the RG 8 coax in a product catalog and finds out that its characteristic impedance is $Z_0 = 50$ Ω, its velocity factor is $v_p/c = 0.66$, and its line capacitance is $C = 26.4$ pF-ft^{-1}. He then cuts a portion of this coax and measures the diameter of the inner conductor and the outer diameter of the dielectric to be approximately 4 mm and 14 mm, respectively. Using these values, find or verify the values of the unit length line parameters L, C, R, and G and the characteristic impedance Z_0 of this cable at 100 MHz. Note that the dielectric inside RG 8 coax is polyethylene and that the leakage conductance per unit length of a polyethylene-filled coaxial line at 100 MHz is approximately given by $G \simeq 1.58 \times 10^{-5}/\ln(b/a)$ S-m^{-1}.

2.45   Two-wire line. Calculate the per-unit-length line parameters L, C, R, and G and the characteristic impedance Z_0 of an air-insulated two-wire line made of copper wires with wire separation of 4.2 cm and wire diameter of 2.4 mm at a frequency of 200 MHz.

3

Steady-State Waves on Transmission Lines

In digital-integrated electronics, computer communication, and many other applications, it is important to understand the response of transmission lines to steplike changes in their inputs, as we studied in detail in Chapter 2. We have seen that waves travel down a transmission line by successively charging the distributed capacitors of the line and establishing current in the distributed inductors. In this context, a wave is a function of both time and space but does not necessarily involve periodic oscillations of a physical quantity (e.g., the height of water in ocean waves). If a disturbance that occurs at a certain point and time causes disturbances at other points in the surrounding region at later times, then wave motion is said to exist. Because of the nonzero travel time along a transmission line, disturbances initiated at one location induce effects that are retarded in time at other locations. The natural response of transmission lines to sudden changes in their inputs typically involves wave motion, with disturbances traveling down the line, producing reflections at terminations, or discontinuities, which in turn propagate back to the source end, and so on. However, except in special cases, the natural response eventually decays after some time interval and is primarily described by the intrinsic properties of the transmission line (characteristic impedance and one-way travel time or length) and its termination (i.e., the load), rather than the input excitation.

In many engineering applications, however, it is also important to understand the steady-state response of transmission lines to sinusoidal excitations. Electrical power and communication signals are often transmitted as sinusoids or modified sinusoids. Other nonsinusoidal signals, such as pulses utilized in a digitally coded system, may be considered as a superposition (i.e., Fourier series) of sinusoids of different frequencies. In sinusoidal signal applications, the initial onset of a sinusoidal input produces a natural response. However, this initial transient typically decays rapidly in time, while the forced response supported by the sinusoidal excitation continues indefinitely. Once the steady state is reached, voltages and currents on the transmission line vary sinusoidally in time at each point along the line while also traveling down the line. The finite travel time of waves leads to phase differences between the voltages (or currents) at different points along the line. The steady-state solutions for voltage and current waves when reflections are present lead to *standing waves,* which also vary sinusoidally in time at each point on the line but do not travel along the line. The differences between traveling and standing waves will become clear in the following sections.

In discussing the steady-state response of transmission lines to sinusoidal excitation, we take full advantage of powerful tools[1] commonly used for analysis of alternating-current phenomena in lumped electrical circuits, including *phasors* and complex *impedance.* The phasor notation eliminates the need to keep track of the known sinusoidal time dependence of the various quantities and allows us to transform the transmission line equations from partial differential equations to ordinary differential equations. The magnitude of the complex impedance of a device or a load is a measure of the degree to which it opposes the excitation by a sinusoidal voltage source; a load with a higher impedance (in magnitude) at any given frequency requires a higher voltage in order to allow a given amount of current through it *at the given frequency.* The phase of the impedance represents the phase difference between the voltage across it and the current through it. In lumped circuits, the impedance of a load is determined only by its internal dynamics (i.e., its physical makeup as represented by its voltage–current characteristics, e.g., $V = L\, dI/dt$ for an inductor). In transmission line applications, the impedance of a load presented to a source via a transmission line depends on the characteristic impedance and the *electrical length*[2] (physical length per unit of wavelength) of the line connecting them. This additional length dependence makes the performance of transmission line systems dependent on frequency—to a greater degree than is typical in lumped circuit applications.

In this chapter, we exclusively consider excitations that are pure sinusoids. In most applications, transmission line systems have to accommodate signals made up of modified sinusoids, with the energy or information spread over a small bandwidth around a central frequency. The case of excitations involving waveforms that are other than sinusoidal can usually be handled by suitably decomposing the signal into its Fourier components, each of which can be analyzed as described in this chapter. In the first six sections we consider lossless (i.e., $R = 0$ and $G = 0$) transmission lines. Lossy transmission lines (i.e., $R \neq 0$ and $G \neq 0$) are discussed in Section 3.7. We shall see that the effects of small but nonzero losses can be accounted for by suitably modifying the lossless analysis. Note also that the loss terms are truly negligible in many applications, so that the lossless cases considered in detail are of practical interest in their own right.

---

[1]These tools were first introduced by a famous electrical engineer, C. P. Steinmetz, in his first book, *Theory and Calculation of AC Phenomena,* McGraw-Hill, New York, 1897.

[2]The electrical length of a transmission line is its physical length divided by the wavelength at the frequency of operation. For example, the electrical length of a 1.5-m-long air-filled coaxial line operating at 100 MHz (wavelength $\lambda = 3$ m) is 0.5λ, while the electrical length of the same line operating at 1 MHz (wavelength $\lambda = 300$ m) is 0.005λ.

3.1 WAVE SOLUTIONS USING PHASORS

The fundamental transmission line equations for the lossless case were developed in Chapter 2 (see equations (2.3) and (2.4)):

$$\frac{\partial \mathcal{V}}{\partial z} = -L\frac{\partial \mathcal{I}}{\partial t} \tag{3.1}$$

$$\frac{\partial \mathcal{I}}{\partial z} = -C\frac{\partial \mathcal{V}}{\partial t} \tag{3.2}$$

Equations (3.1) and (3.2) are written in terms of the space–time functions describing the instantaneous values of voltage and current at any point z on the line, denoted respectively as $\mathcal{V}(z,t)$ and $\mathcal{I}(z,t)$. When the excitation is sinusoidal, and under steady-state conditions, we can use the phasor concept to reduce the transmission line equations (3.1) and (3.2) to ordinary differential equations (instead of partial differential equations, as they are now) so that we can more easily obtain general solutions. As in circuit analysis, the relations between phasors and actual space–time functions are as follows:

$$\mathcal{V}(z,t) = \mathcal{R}e\{V(z)e^{j\omega t}\} \tag{3.3a}$$

$$\mathcal{I}(z,t) = \mathcal{R}e\{I(z)e^{j\omega t}\} \tag{3.3b}$$

Here, both phasors $V(z)$ and $I(z)$ are functions of z only and are in general complex.

We can now rewrite[3] equations (3.1) and (3.2) in terms of the phasor quantities by replacing $\partial/\partial t$ with $j\omega$. We have

$$\boxed{\frac{dV(z)}{dz} = -j\omega L I(z)} \tag{3.4}$$

and

$$\boxed{\frac{dI(z)}{dz} = -j\omega C V(z)} \tag{3.5}$$

Combining (3.4) and (3.5), we can write a single equation in terms of $V(z)$,

$$\frac{d^2V(z)}{dz^2} - (j^2\omega^2 LC)V(z) = 0 \quad \text{or} \quad \frac{d^2V(z)}{dz^2} - (j\beta)^2V(z) = 0 \tag{3.6}$$

---

[3]The actual derivation of (3.4) from (3.1) is as follows:

$$\frac{\partial \mathcal{V}}{\partial z} = -L\frac{\partial \mathcal{I}}{\partial t} \longrightarrow \frac{\partial}{\partial z}\underbrace{\left[\mathcal{R}e\{V(z)e^{j\omega t}\}\right]}_{\mathcal{V}(z,t)} = -L\frac{\partial}{\partial t}\underbrace{\left[\mathcal{R}e\{I(z)e^{j\omega t}\}\right]}_{\mathcal{I}(z,t)}$$

$$\longrightarrow \mathcal{R}e\left\{e^{j\omega t}\frac{dV(z)}{dz}\right\} = \mathcal{R}e\{-L(j\omega)e^{j\omega t}I(z)\} \longrightarrow \frac{dV(z)}{dz} = -j\omega L I(z)$$

where $\beta = \omega\sqrt{LC}$ is called the *phase constant*. Equation (3.6) is referred to as the complex *wave equation* and is a second-order ordinary differential equation commonly encountered in analysis of physical systems. The general solution of (3.6) is of the form

$$V(z) = V^+ e^{-j\beta z} + V^- e^{+j\beta z} \tag{3.7}$$

where, as we shall see below, $e^{-j\beta z}$ and $e^{j\beta z}$ represent wave propagation in the $+z$ and $-z$ directions, respectively, and where V^+ and V^- are complex constants to be determined by the boundary conditions. The corresponding expression for the current $I(z)$ can be found by substituting (3.7) in (3.4). We find

$$I(z) = I^+ e^{-j\beta z} + I^- e^{+j\beta z} = \frac{1}{Z_0}\left[V^+ e^{-j\beta z} - V^- e^{+j\beta z}\right] \tag{3.8}$$

where $Z_0 = V^+/I^+ = -V^-/I^- = \sqrt{L/C}$ is the characteristic impedance of the transmission line.

Using (3.3), and the expressions (3.7) and (3.8) for the voltage and current phasors, we can find the corresponding space–time expressions for the instantaneous voltage and current. We have

$$\begin{aligned}\mathcal{V}(z,t) &= \mathfrak{Re}\{V(z)e^{j\omega t}\} = \mathfrak{Re}\{V^+ e^{-j\beta z}e^{j\omega t} + V^- e^{+j\beta z}e^{j\omega t}\} \\ &= V^+ \cos(\omega t - \beta z) + V^- \cos(\omega t + \beta z)\end{aligned} \tag{3.9}$$

where we have assumed[4] V^+ and V^- to be real. Similarly, we have

$$\mathfrak{I}(z,t) = \frac{1}{Z_0}[V^+ \cos(\omega t - \beta z) - V^- \cos(\omega t + \beta z)] \tag{3.10}$$

The voltage and current solutions consist of a superposition of two waves, one propagating in the $+z$ direction (i.e., toward the load) and the other in the $-z$ direction (i.e., a reflected wave moving away from the load). To see the wave behavior, consider the case of an infinitely long transmission line; in this case no reflected wave is present and thus $V^- = 0$. The voltage and current for an infinitely long line are

$$V(z) = V^+ e^{-j\beta z}; \qquad \mathcal{V}(z,t) = V^+ \cos(\omega t - \beta z) \tag{3.11}$$

---

[4]If instead V^+ and V^- were complex, with $V^+ = |V^+|e^{j\phi^+}$ and $V^- = |V^-|e^{j\phi^-}$, we would have

$$\mathcal{V}(z,t) = |V^+|\cos(\omega t - \beta z + \phi^+) + |V^-|\cos(\omega t + \beta z + \phi^-)$$

and

$$I(z) = \frac{V^+}{Z_0} e^{-j\beta z}; \qquad \mathscr{I}(z,t) = \frac{V^+}{Z_0} \cos(\omega t - \beta z) \qquad (3.12)$$

Note that, everywhere along the line, the ratio of the voltage to current phasors is Z_0; hence, Z_0 is called the characteristic impedance of the line. Note, however, that this is true not only for an infinitely long line but also for a line of finite length that is terminated at a load impedance $Z_L = Z_0$, as we discuss later.

The solutions (3.11) and (3.12) are in the form of (1.5), which was introduced in Chapter 1 by stating that electromagnetic quantities with such space–time dependencies are often encountered. Here we see that this form of solution is indeed a natural solution of the fundamental transmission line equations.[5] The space–time behavior of the voltage wave given by (3.11) is illustrated in Figure 3.1. We note from Figure 3.1a that the period of the sinusoidal oscillations (as observed at fixed points in space) is $T_p = 2\pi/\omega$. The voltage varies sinusoidally at all points in space, but it reaches its maxima at different times at different positions. Figure 3.1b indicates that the voltage distribution as a function of distance (observed at a fixed instant of time) is also sinusoidal. The distance between the crests of the voltage at a fixed instant of time is the *wavelength* $\lambda = 2\pi/\beta$. As time progresses, the wave propagates to the right ($+z$ direction), as can be seen by observing a particular point on the waveform (e.g., the crest or the minimum) at different instants of time. The speed of this motion is the *phase velocity,* defined as the velocity at which an observer must travel to observe a stationary (i.e., not varying with time) voltage. The voltage observed would be the same as long as the argument of the cosine in (3.11) is the same; thus we have

$$\omega t - \beta z = \text{const.} \quad \longrightarrow \quad v_p = \frac{dz}{dt} = \frac{\omega}{\beta} = \frac{\omega}{\omega\sqrt{LC}} = \frac{1}{\sqrt{LC}} \qquad (3.13)$$

where v_p is the phase velocity, which was also introduced in Sections 1.1.3 and 2.2 (see equation (2.7)).

As discussed in Section 2.2, for most of the commonly used two-conductor transmission lines (Figure 2.1), the phase velocity v_p is not a function of the particular geometry of the metallic conductors but is instead solely determined by the electrical and magnetic properties of the surrounding insulating medium. When the surrounding medium is air, the phase velocity is the speed of light in free space, namely, $v_p = c \simeq 3 \times 10^8$ m-s^{-1} = 30 cm-(ns)$^{-1}$. The phase velocity v_p for some other insulating materials was tabulated in Table 2.1. Phase velocities for some additional materials are given in Table 3.1, together with the corresponding values of wavelength at a frequency of 300 MHz. Note that since $\lambda = 2\pi/\beta$, we have $v_p = \omega/\beta = \lambda f$, so that the phase constant β and wavelength λ depend on the electrical and magnetic properties of the material surrounding the transmission line conductors as well as on the frequency of operation.

---

[5]We will encounter the same type of space–time variation once again in Chapter 8, as the natural solution of Maxwell's equations for time-harmonic (or sinusoidal steady-state) electric and magnetic fields.

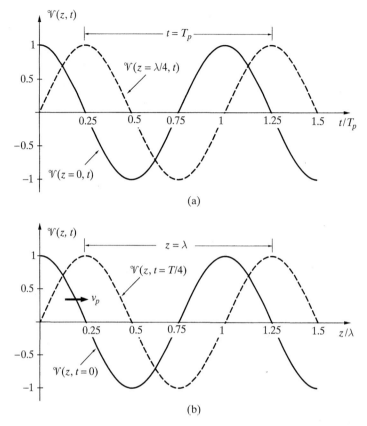

Figure 3.1   Wave behavior in space and time. (a) $\mathcal{V}(z,t) = V^+ \cos\left[2\pi(t/T_p) - 2\pi(z/\lambda)\right]$ versus t/T_p for $z = 0$ and $z = \lambda/4$. (b) $\mathcal{V}(z,t) = V^+ \cos\left[2\pi(t/T_p) - 2\pi(z/\lambda)\right]$ versus z/λ for $t = 0$ and $t = T_p/4$. In both panels we have taken $V^+ = 1$.

TABLE 3.1   PHASE VELOCITY AND WAVELENGTH IN DIFFERENT MATERIALS

Material	Wavelength (m at 300 MHz)	Phase Velocity (cm-(ns)$^{-1}$ at 300 MHz)
Air	1	30
Silicon	0.29	8.7
Polyethylene	0.67	20.0
Epoxy glass (PC board)	0.45	13.5
GaAs	0.30	9.1
Silicon carbide (SiC)	0.15	4.6
Glycerin	0.14	4.2

3.2 VOLTAGE AND CURRENT ON LINES WITH SHORT- OR OPEN-CIRCUIT TERMINATIONS

Most sinusoidal steady-state applications involve transmission lines terminated at a load impedance Z_L. Often, voltages and currents near the load end are of greatest interest since they determine the degree of matching between the line and the load and the amount of power delivered to (versus that reflected from) the load. A portion of a lossless transmission line terminated in an arbitrary load impedance Z_L is shown in Figure 3.2. We can use this setup to explore the concept of reflected waves on transmission lines, a fundamental feature of distributed circuits in general.

Assume that a forward-propagating ($+z$ direction) wave of the form $V^+e^{-j\beta z}$ produced by a source located at some position z ($z < 0$) is incident on load Z_L located at $z = 0$. Contrary to the case of an infinitely long transmission line, here the ratio of the total voltage $V(z)$ to the total current $I(z)$ at any position z along the line is not equal to Z_0. For example, at the load position ($z = 0$), we must satisfy the boundary condition $[V(z)/I(z)]_{z=0} = Z_L$, where Z_L is in general not equal to Z_0. Thus, since $V^+/I^+ = Z_0$, and in general $Z_0 \neq Z_L$, a reverse-propagating ($-z$ direction) reflected wave of the form $V^-e^{+j\beta z}$ with the appropriate value for V^- must be present so that the load boundary condition is satisfied. The total voltage and current phasors, $V(z)$ and $I(z)$, at any position on the line consist of the sum of the forward and reverse waves as specified by (3.7) and (3.8), namely

$$V(z) = V^+e^{-j\beta z} + V^-e^{+j\beta z} \tag{3.14}$$

$$I(z) = \frac{1}{Z_0}[V^+e^{-j\beta z} - V^-e^{+j\beta z}] \tag{3.15}$$

where V^+ and V^- are in general complex constants to be determined by the boundary conditions.

When a transmission line has only a forward-traveling wave with no reflected wave (e.g., in the case of an infinitely long line), the ratio of the total voltage to the current is the characteristic impedance Z_0, as was discussed in Section 3.1. When the line is terminated

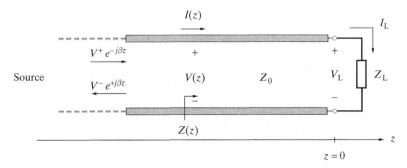

Figure 3.2   A terminated lossless transmission line. For convenience, the position of the load is taken to be $z = 0$.

so that, in general, a reflected wave exists, the ratio of the total line voltage $V(z)$ to the total line current $I(z)$ at any position z—the *line impedance*—is not equal to Z_0. The line impedance is of considerable practical interest; for example, the impedance that the line presents to the source at the source end of the line (called the input impedance of the line, denoted by Z_{in}) is the line impedance evaluated at that position. The source, or the generator, does not know anything about the characteristic impedance of the line or whether a reflected wave exists on the line; it merely sees that when it applies a voltage of V_s to the input terminals of the line, a certain current I_s flows, and thus the source interprets the ratio of V_s/I_s as an impedance of a particular magnitude and phase.

The line impedance as seen by looking toward the load Z_L at any position z along the line (see Figure 3.2) is defined as

$$Z(z) \equiv \frac{V(z)}{I(z)} = Z_0 \frac{V^+e^{-j\beta z} + V^-e^{j\beta z}}{V^+e^{-j\beta z} - V^-e^{j\beta z}} \tag{3.16}$$

In general, the line impedance $Z(z)$ is complex and is a function of position z along the line. From electrical circuit analysis, we know that a complex impedance $Z(z)$ can be written as

$$Z(z) = R(z) + jX(z)$$

where the real quantities $R(z)$ and $X(z)$ are the resistive and the reactive parts of the line impedance, respectively.

The following example considers the case of a *matched load,* defined as a load impedance equal to the characteristic impedance of the line, or $Z_L = Z_0$.

Example 3.1: Matched load. A lossless transmission line is terminated with a load $Z_L = Z_0$, as shown in Figure 3.3a. Find the magnitude of the reflected wave V^- and the line impedance $Z(z)$.

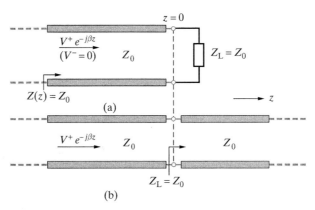

Figure 3.3   Matched load. (a) Circuit diagram. (b) Z_L is replaced by an infinitely long extension of the same line.

Solution: If $Z_L = Z_0$, the load boundary condition $V(z = 0)/I(z = 0) = Z_L = Z_0$ is satisfied without any reflected wave, so that $V^- = 0$. The line impedance at any position z along the line is $Z(z) = V(z)/I(z) = Z_0 = Z_L$ independent of z. In this case, the load impedance Z_L can be viewed as an infinite extension of the same transmission line, as shown in Figure 3.3b.

Transmission line segments terminated in short or open circuits are commonly used as tuning elements for impedance matching networks (see Sections 3.2.3 and 3.5). In the next two subsections, we study these two special cases in detail before considering (in Section 3.3) the more general case of lines terminated in an arbitrary impedance Z_L.

3.2.1 Short-Circuited Line

Figure 3.4 shows a transmission line of length l terminated in a short circuit ($Z_L = 0$). Short-circuited termination forces the load voltage V_L to be zero, so from (3.14), we have

$$V_L = [V(z)]_{z=0} = [V^+ e^{-j\beta z} + V^- e^{j\beta z}]_{z=0} = V^+ + V^- = 0$$

leading to

$$V^- = -V^+ \quad \text{or} \quad \frac{V^-}{V^+} = -1$$

Note that the load current flowing through the short circuit can be found from (3.15) using $V^- = -V^+$:

$$I_L = [I(z)]_{z=0} = \frac{1}{Z_0}(V^+ - V^-) = \frac{2V^+}{Z_0}$$

Anywhere else along the line we have

$$V(z) = V^+(e^{-j\beta z} - e^{j\beta z}) = -2V^+ j \ \sin(\beta z)$$

$$I(z) = \frac{V^+}{Z_0}(e^{-j\beta z} + e^{j\beta z}) = \frac{2V^+}{Z_0} \ \cos(\beta z)$$

Figure 3.4   Short-circuited line.   The input impedance Z_{sc} of a short-circuited line can be capacitive or inductive depending on the length l of the line.

The instantaneous space–time function for the voltage can be obtained from $V(z)$ using (3.3a). We have

$$\mathscr{V}(z,t) \equiv \mathcal{R}e\{V(z)e^{j\omega t}\} = \mathcal{R}e\{V^+(e^{-j\beta z} - e^{j\beta z})e^{j\omega t}\}$$

$$= \mathcal{R}e\left\{2|V^+|e^{j\phi^+}\sin(\beta z)e^{-j\pi/2}e^{j\omega t}\right\}$$

$$= 2|V^+|\sin(\beta z)\cos\left(\omega t - \frac{\pi}{2} + \phi^+\right)$$

where $V^+ = |V^+|e^{j\phi^+}$ is in general a complex constant, to be determined from the boundary condition at the source end of the transmission line. However, note that we can assume $\phi^+ = 0$ (i.e., V^+ is a real constant) without any loss of generality because V^+ is a constant multiplier that appears in front of all voltages and currents everywhere along the line. Accounting for a possible finite phase ϕ^+ simply amounts to shifting the time reference, with no effect on the relationships among the various quantities. Accordingly, we assume $\phi^+ = 0$ throughout the following discussion.

Note that the voltage $\mathscr{V}(z,t)$ along a short-circuited line is a cosinusoid (in time) whose amplitude varies as $2|V^+|\sin(\beta z)$ with position z along the line. It is not a traveling wave, since the peaks (or minima) of the $\cos(\omega t - \pi/2)$ always stay (i.e., stand) at the same positions (i.e., z/λ) along the line as the voltage varies in time (see Figure 3.5b). The voltage $\mathscr{V}(z,t)$ is thus said to represent a pure *standing wave*. Similarly, for current we have

$$\mathscr{I}(z,t) = \frac{2|V^+|}{Z_0}\cos(\beta z)\cos(\omega t)$$

which is also a standing wave (see Figure 3.5c). The absolute amplitudes of both the voltage and current phasors, namely $|V(z)|$ and $Z_0|I(z)|$, are shown as functions of z/λ in Figure 3.5d.

Note that the current and voltage are in time quadrature (i.e., 90° out of phase), such that when $\mathscr{I}(z,t)$ at a given point z is zero, the absolute amplitude of $\mathscr{V}(z,t)$ is a maximum, and vice versa. Standing waves stand on the line but do not travel or carry any time-average power to the load;[6] they represent reactive power in a manner analogous to the voltage and current, also in time quadrature, of a capacitor or inductor. The power relationships on a transmission line are discussed in more detail in Section 3.4.

[6]To see this, consider that the instantaneous power carried by the wave is given by

$$\mathscr{P}(z,t) = \mathscr{V}(z,t)\mathscr{I}(z,t) = \left[2|V^+|\sin(\beta z)\cos\left(\omega t - \frac{\pi}{2}\right)\right]\left[\frac{2|V^+|}{Z_0}\cos(\beta z)\cos(\omega t)\right]$$

$$= \frac{2|V^+|^2}{Z_0}\sin(2\beta z)\cos\left(\omega t - \frac{\pi}{2}\right)\cos(\omega t) = \frac{|V^+|^2}{Z_0}\sin(2\beta z)\sin(2\omega t)$$

where we have used the trigonometric identities of $\cos(\zeta - \pi/2) = \sin\zeta$ and $2\sin\zeta\cos\zeta = \sin(2\zeta)$. Note that the instantaneous power carried by the standing wave oscillates in time at a rate twice that of the voltage and current and that its average over one period (i.e., $T = 2\pi/\omega$) is thus zero, as expected on the basis of the fact that the voltage and current are 90° out of phase.

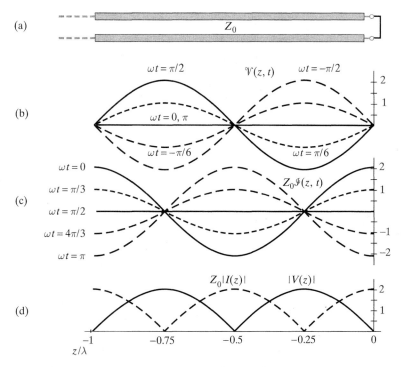

Figure 3.5   Voltage and current on a short-circuited line. (a) Schematic of a short-circuited line. (b) Instantaneous voltage $\mathscr{V}(z,t)$ versus z/λ at different times. (c) Corresponding instantaneous current times the characteristic impedance, namely $Z_0\mathscr{I}(z,t)$. (d) Magnitudes of the voltage and current phasors, showing $|V(z)|$ (solid line) and $Z_0|I(z)|$ (dashed line) as functions of z/λ. All of the plots shown are for $V^+ = 1$.

The line impedance seen looking toward the short circuit at any position z along the short-circuited line is

$$Z(z) = \frac{V(z)}{I(z)} = Z_0\frac{-2jV^+\sin(\beta z)}{2V^+\cos(\beta z)} = -jZ_0\tan(\beta z)$$

where $z < 0$. The input impedance of a short-circuited line segment of length l can then be found by evaluating the preceding equation at the source end, where $z = -l$:

$$\boxed{Z_{sc} = jZ_0\tan(\beta l) = jX_{sc}} \qquad (3.17)$$

We note from (3.17) that the input impedance of a short-circuited line of length l is purely reactive. As illustrated in Figure 3.6, the input impedance depends on line length l, or more generally, on the electrical length, which is defined as the ratio of the physical length of the line to the wavelength, that is, l/λ, where $\lambda = 2\pi/\beta$. The input impedance can be varied by varying the length or the frequency, or both, and can be capacitive (negative X_{sc}) or inductive (positive X_{sc}). It makes physical sense that the

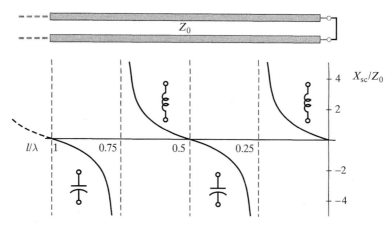

Figure 3.6 Input impedance of a short-circuited line. The normalized reactance X_{sc}/Z_0 of a short-circuited line segment of length l is shown as a function of electrical length l/λ.

input impedance is inductive for very short line lengths ($l < \lambda/4$); a shorted two-wire line of relatively short length resembles a small loop of wire. The fact that any reactive input impedance can be realized by simply varying the length of a short-circuited line (and the open-circuited line as will be discussed in the next subsection) is very useful in tuning- and impedance-matching applications at microwave frequencies, as discussed in Section 3.5.

Example 3.2: Inductance of a short television antenna lead-in wire. Consider a television antenna lead-in wire of length $l = 10$ cm having a characteristic impedance of $Z_0 = 300\Omega$, shorted at one end. Find the input impedance of this line if it is to be used at 300 MHz.

> **Solution:** We can first determine the electrical length of the line. We assume the conductors to be mostly surrounded by air (although they might in fact be held together by some plastic material) so that the phase velocity $v_p \simeq c$. Since the wavelength at 300 MHz is $\lambda = v_p/f \simeq 3 \times 10^8/(300 \times 10^6) = 1$ m, the electrical length is $(l/\lambda) = 0.1$. The input impedance of the shorted line can then be found directly from (3.17):

$$Z_{sc} = jZ_0 \tan(\beta l) = j(300)\tan\left(\frac{2\pi}{\lambda}l\right) \simeq j218\Omega$$

This is an inductive input impedance. At a frequency of 300 MHz, an inductive reactance of $X_{sc} = \omega L_{sc} = 218\Omega$ corresponds to a lumped inductor having an inductance of $L_{sc} = 218/(2\pi \times 300 \times 10^6) \simeq 0.116 \ \mu H$.

According to (3.17) and Figure 3.6, the input impedance of a short-circuited line of length $l = \lambda/4$ is infinite; that is, the line appears as if it is an open circuit. In practice, however, the input impedance of such a line is limited by its distributed conductance.

Note that considering the circuit model of the line (Figure 2.5) the per unit length shunt conductance G presents a resistance proportional to $(G)^{-1}$ across the input terminals of a line, even when the input impedance looking toward the load is infinite. Although G was assumed to be zero for our lossless analysis, it nevertheless is a nonzero value and thus limits the input impedance of the line to a finite value. Similarly, equation (3.17) and Figure 3.6 indicate that the input impedance of a short-circuited line of length $l = \lambda/2$ is zero, in other words, that the line appears as if it is a short circuit. In practice, however, the minimum value of input impedance of a short-circuited half-wavelength long line is determined by its per unit length series resistance R, which, although small, is nevertheless a nonzero value.

3.2.2 Open-Circuited Line

Figure 3.7 shows an open-circuited transmission line ($Z_L = \infty$). The analysis of open-circuited lossless transmission lines is very similar to that of short-circuited lines. Open-circuited termination forces the load current I_L to be zero, so that using (3.15) we have

$$I_L = [I(z)]_{z=0} = \left[\frac{V^+}{Z_0}e^{-j\beta z} - \frac{V^-}{Z_0}e^{j\beta z}\right]_{z=0} = \frac{V^+ - V^-}{Z_0} = 0$$

leading to

$$V^- = V^+ \quad \text{or} \quad \frac{V^-}{V^+} = +1$$

Note that the load voltage appearing across the open circuit can be found from (3.14) using $V^- = V^+$. We have

$$V_L = [V(z)]_{z=0} = [V^+ e^{-j\beta z} + V^- e^{j\beta z}]_{z=0} = (V^+ + V^-) = 2V^+$$

Anywhere else along the line we have

$$V(z) = V^+(e^{-j\beta z} + e^{j\beta z}) = 2V^+ \cos(\beta z)$$

$$I(z) = \frac{V^+}{Z_0}(e^{-j\beta z} - e^{j\beta z}) = -2\frac{V^+}{Z_0}j\sin(\beta z) = 2\frac{V^+}{Z_0}(e^{-j\pi/2})\sin(\beta z)$$

Figure 3.7 Open-circuited line. The input impedance Z_{oc} of an open-circuited line can be capacitive or inductive depending on the length of the line.

The instantaneous space–time expressions for the voltage and current[7] are

$$\mathcal{V}(z,t) = \mathcal{R}e\{V(z)e^{j\omega t}\} = 2|V^+|\cos(\beta z)\cos(\omega t)$$

$$\mathcal{I}(z,t) = \mathcal{R}e\{I(z)e^{j\omega t}\} = 2\frac{|V^+|}{Z_0}\sin(\beta z)\cos\left(\omega t - \frac{\pi}{2}\right)$$

where we have assumed $V^+ = |V^+|e^{j\phi^+}$, with $\phi^+ = 0$, without any loss of generality.

As for the short-circuited line, the current and voltage on an open-circuited line are in time quadrature (i.e., out of phase by 90°) so that the average power carried is again zero. Both $\mathcal{V}(z,t)$ and $\mathcal{I}(z,t)$ are purely standing waves. Their absolute amplitude patterns are shown in Figure 3.8b, in the same format as in Figure 3.5d.

The line impedance seen looking toward the open circuit at any position z along the line is

$$Z(z) = \frac{V(z)}{I(z)} = jZ_0\cot(\beta z)$$

where $z < 0$. The input impedance of an open-circuited line of length l (i.e, at $z = -l$) is then given by

$$\boxed{Z_{\text{oc}} = Z(z = -l) = -jZ_0\cot(\beta l) = jX_{\text{oc}}} \qquad (3.18)$$

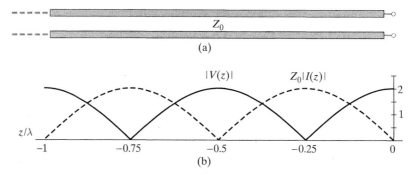

Figure 3.8 Voltage and current on an open-circuited line. (a) Schematic of an open-circuited line. (b) Magnitudes of the voltage and current phasors, showing $|V(z)|$ (solid line) and $Z_0|I(z)|$ (dashed line) as functions of z/λ, for $V^+ = 1$.

[7]Note for the current that

$$\mathcal{R}e\{I(z)e^{j\omega t}\} = \mathcal{R}e\left\{\frac{V^+}{Z_0}[e^{-j\beta z} - e^{j\beta z}]e^{j\omega t}\right\} = \mathcal{R}e\left\{2\frac{|V^+|}{Z_0}\sin(\beta z)e^{j\phi^+}e^{-j(\pi/2)}e^{j\omega t}\right\}$$

$$= \mathcal{R}e\left\{2\frac{|V^+|}{Z_0}\sin(\beta z)\left[\cos\left(\omega t - \frac{\pi}{2} + \phi^+\right) + j\sin\left(\omega t - \frac{\pi}{2} + \phi^+\right)\right]\right\}$$

$$= \frac{2|V^+|}{Z_0}\sin(\beta z)\cos\left(\omega t - \frac{\pi}{2} + \phi^+\right)$$

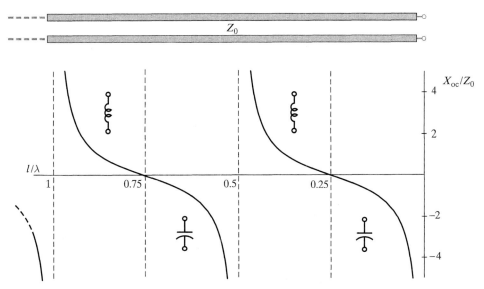

Figure 3.9   Input impedance of an open-circuited line. The normalized reactance X_{oc}/Z_0 of an open-circuited line segment of length l is shown as a function of electrical length l/λ.

As for a short-circuited line, the input impedance of an open-circuited line of length l is also purely reactive. The normalized reactance X_{oc}/Z_0 is plotted in Figure 3.9 as a function of electrical length l/λ. A capacitive or inductive reactance can be obtained simply by adjusting the line length l for a fixed wavelength λ (or frequency ω) or by adjusting the wavelength λ (or frequency ω) for fixed line length l. The fact that for $l < \lambda/4$ the impedance is capacitive makes physical sense, since a relatively short-length open-circuited line consists of two conductors with some separation between them, resembling an ordinary lumped capacitor.

Example 3.3: Capacitance of a short television antenna lead-in wire. Consider a television antenna lead-in wire of length $l = 20$ cm having a characteristic impedance of $Z_0 = 300\Omega$ and nothing connected at its end (i.e., open-circuited). Find the input impedance of this line if it is to be used at 300 MHz.

Solution: Once again we assume the phase velocity to be $v_p \simeq c$ and first determine the electrical length of the line. Since the wavelength at 300 MHz is $\lambda = v_p/f \simeq 3 \times 10^8/(300 \times 10^6) = 1$ m, the electrical length is $(l/\lambda) = 0.2$. The input impedance of the open-circuited line can then be found directly from (3.18):

$$Z_{oc} = -jZ_0 \cot(\beta l) = -j(300) \cot\left(\frac{2\pi}{\lambda}l\right) \simeq -j97.5\Omega$$

This is a capacitive input impedance. At a frequency of 300 MHz, a capacitive reactance of $X_{oc} = -(\omega C_{oc})^{-1} \simeq -97.5\Omega$ corresponds to a lumped capacitor of capacitance $C_{oc} = [(97.5)(2\pi \times 300 \times 10^6)]^{-1} \simeq 5.44$ pF.

According to (3.18) and Figure 3.9, the input impedance of an open-circuited line of length $l = \lambda/4$ is zero; that is, the line appears as if it is short-circuited. In practice, however, the minimum value of input impedance is determined by its per unit length series resistance R, which, although small, is nevertheless a nonzero value. Similarly, equation (3.18) and Figure 3.9 both indicate that the input impedance of an open-circuited line of length $l = \lambda/2$ is infinite, that is, that the line appears as if it is an open circuit. In practice, however, the input impedance of such a line is limited by its nonzero per unit length distributed conductance G.

3.2.3 Open- and Short-Circuited Lines as Reactive Circuit Elements

An important application of transmission lines involves their use as capacitive or inductive tuning elements in microwave circuits at frequencies between a few gigahertz to a few tens of gigahertz. In this frequency range, lumped inductors and capacitors become exceedingly small and difficult to fabricate. Furthermore, the wavelength is small enough that the physical sizes and separation distances of ordinary circuit components are no longer negligible. On the other hand, transmission line sections of appropriate sizes can be constructed with relative ease. For frequencies higher than $\sim$100 GHz, the physical size of transmission lines is too small, although novel transmission line implementations can operate[8] at frequencies as high as 500 GHz, corresponding to submillimeter wavelengths.

That transmission lines behave as reactive circuit elements is quite evident from Figures 3.6 and 3.9. Consider, for example, the input impedance as a function of frequency of a short-circuited line of length l such that $l = \lambda_0/4$. At a frequency of f_0, for which $\lambda = \lambda_0$, this line presents an infinite impedance (i.e., appears as an open circuit) at its input terminals. For frequencies slightly smaller than f_0, namely $f < f_0$ so that $\lambda > \lambda_0$, the length of the line is slightly shorter than $\lambda/4$, so it presents a very large inductive impedance (Figure 3.6). For frequencies slightly greater than f_0, the electrical length of the line is slightly larger than $\lambda/4$, so it has a very large capacitive impedance. Such behavior is similar to that of a lumped circuit consisting of a parallel combination of an inductor and a capacitor.

A similar analysis of a short-circuited line of length l such that $l = \lambda_0/2$ indicates that a half-wavelength line behaves as a lumped circuit consisting of a series combination of an inductor and a capacitor. As can be seen from Figure 3.6, the magnitude of the input impedance of a short-circuited line of length $l = \lambda_0/2$ is very small in the vicinity of its frequency f_0, and the input impedance is inductive for $f > f_0$ and capacitive for $f < f_0$.

Corresponding observations can also be made for open-circuited line segments, for which the input impedance is given as a function of electrical length in Figure 3.9. Lumped circuit counterparts of various transmission line segments are summarized in Figure 3.10.

---

[8]Linda P. B. Katehi, Novel transmission lines for the submillimeter-wave region, *Proc. IEEE,* 80(11), p. 1771, November 1992.

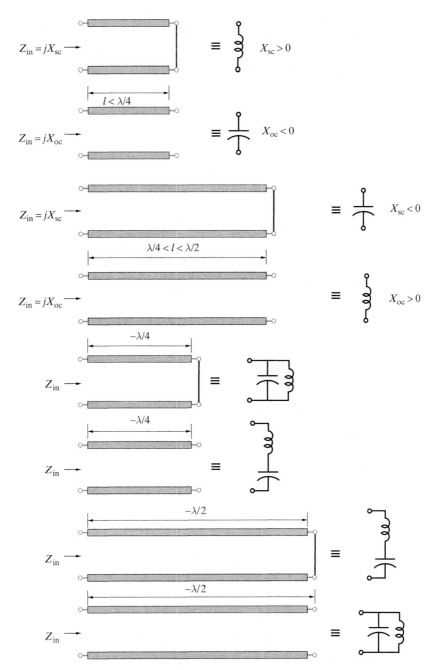

Figure 3.10 Lumped-circuit models of various open- and short-circuited line segments.

Example 3.4: Transmission line inductor. A short-circuited coaxial line with $v_p = 2.07 \times 10^8$ m-s^{-1} is to be designed to provide a 15 nH inductance for a microwave filter operating at 3 GHz. (a) Find the shortest possible length l if the characteristic impedance of the line is $Z_0 = 50\Omega$ and (b) find the lumped element value of the short-circuited line designed in part (a) at 4 GHz.

Solution:

(a) Equating the input impedance of a short-circuited line of length l to the impedance of a lumped inductor, we have

$$Z_{sc} = jZ_0 \tan\left(\frac{2\pi}{\lambda}l\right) = j\omega L_{sc}$$

where $\lambda = v_p/f = (2.07 \times 10^{10})/(3 \times 10^9) = 6.9$ cm. For $Z_0 = 50\Omega$, we can write

$$l = \frac{6.9}{2\pi} \tan^{-1}\left(\frac{2\pi \times 3 \times 10^9 \times 15 \times 10^{-9}}{50}\right) \simeq 1.53 \text{ cm}$$

(b) At 4 GHz, $\lambda = 2.07 \times 10^{10}/(4 \times 10^9) = 5.175$ cm. Thus, the input impedance of the short-circuited 50Ω coaxial line of length ~ 1.53 cm is

$$Z_{sc} \simeq j(50) \tan\left(\frac{2\pi \times 1.53}{5.175}\right) \simeq -j167.4\Omega$$

Therefore, at 4 GHz, the short-circuited 50Ω coaxial line designed in part (a) represents a lumped capacitor of element value given by

$$-\frac{j}{2\pi \times 4 \times 10^9 C} = -j167.4\Omega \quad \rightarrow \quad C_{sc} \simeq 0.238 \text{ pF}$$

The results are summarized in Figure 3.11.

Figure 3.11 A short-circuited 50Ω line as an inductor or capacitor. (a) Designed to provide a 15 nH inductance at 3 GHz. (b) Equivalent to a ~ 0.238 pF capacitance at 4 GHz.

3.3 LINES TERMINATED IN AN ARBITRARY IMPEDANCE

Most sinusoidal steady-state applications involve transmission lines terminated in arbitrary complex load impedances. The load to be driven may be an antenna, the feed-point impedance of which depends on the antenna characteristics and operating frequency in a complicated manner and is in general quite different from the characteristic impedance of the transmission line that connects it to a source. For efficient transmission of the energy from the source to the load, it is often necessary to match the load to the line, using various techniques to be discussed in Section 3.5. In this section, we consider the fundamental behavior of line voltage, current, and impedance for arbitrarily terminated transmission lines. Consider a transmission line of length l terminated in an arbitrary complex load impedance Z_L and excited by a sinusoidal voltage source, the phasor of which is represented by V_0, as shown in Figure 3.12. The line is uniform and lossless (i.e., $Z_0 = $ const., $R = 0$, and $G = 0$), so the voltage and current phasors at any position $z < 0$ along the line are in general given by equations (3.14) and (3.15), respectively. The boundary condition at the load end ($z = 0$) is simply

$$V_L = Z_L I_L \quad \rightarrow \quad V(z)|_{z=0} = Z_L I(z)|_{z=0} \quad \rightarrow \quad Z_L = Z_0 \frac{V^+ + V^-}{V^+ - V^-}$$

The ratio of the phasors of the reverse and forward waves at the load position ($z = 0$) is the *load voltage reflection coefficient,* defined as $\Gamma_L \equiv V^-/V^+$ such that

$$\boxed{Z_L = Z_0 \frac{1 + \Gamma_L}{1 - \Gamma_L} \quad \rightarrow \quad \Gamma_L = \rho e^{j\psi} = \frac{Z_L - Z_0}{Z_L + Z_0}} \tag{3.19}$$

where we explicitly recognize that Γ_L is in general a complex number with a magnitude ρ and phase angle ψ, where $0 \leq \rho \leq 1$. Note that, as shown before, $\Gamma_L = -1$ (i.e., $V^- = -V^+$) for a short-circuited line, $\Gamma_L = +1$ (i.e., $V^- = V^+$) for an open-circuited line, and $\Gamma_L = 0$ (i.e., $V^- = 0$) for a matched load (i.e., $Z_L = Z_0$).

The voltage and current phasors given by equations (3.14) and (3.15) can now be written in terms of Γ_L as

$$V(z) = V^+(e^{-j\beta z} + \Gamma_L e^{j\beta z}) = V^+ e^{-j\beta z}[1 + \Gamma(z)] \tag{3.20}$$

$$I(z) = \frac{V^+}{Z_0}(e^{-j\beta z} - \Gamma_L e^{j\beta z}) = \frac{V^+}{Z_0} e^{-j\beta z}[1 - \Gamma(z)] \tag{3.21}$$

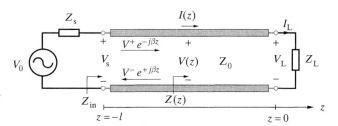

Figure 3.12   A terminated line. A lossless transmission line excited by a sinusoidal source terminated in a complex load impedance Z_L.

where

$$\Gamma(z) \equiv \frac{V^- e^{j\beta z}}{V^+ e^{-j\beta z}} = \Gamma_L e^{j2\beta z} = \rho e^{j(\psi + 2\beta z)}$$

is the *voltage reflection coefficient* at any position z along the line, defined as the ratio of the voltage phasors of the reverse and forward propagating waves at that position. The voltage reflection coefficient is a complex number with a constant magnitude ρ (equal to the magnitude of Γ_L) and a phase angle ψ varying with position z. We can view the quantity $\Gamma(z) \equiv \Gamma_L e^{j2\beta z}$ as a generalized reflection coefficient defined not only at the load but also at any point z along the line. Noting that the voltage along the line is given by

$$V(z) = \underbrace{V^+ e^{-j\beta z}}_{\text{forward wave}} + \underbrace{\Gamma_L V^+ e^{j\beta z}}_{\text{reflected wave}}$$

we can see that the quantity $\Gamma(z) = \Gamma_L e^{j2\beta z}$ is indeed the ratio of the reflected wave at point z to the forward wave at that same position.

Note that the line impedance seen looking toward the load at any position z along the line can be written in terms of $\Gamma(z)$ as

$$Z(z) = \frac{V(z)}{I(z)} = Z_0 \frac{1 + \Gamma(z)}{1 - \Gamma(z)} \qquad (3.22)$$

3.3.1 Voltage and Current Standing-Wave Patterns

To understand the nature of the voltage on the line, it is useful to examine the complete time function $\mathcal{V}(z, t)$. For this, we start our discussion with the case of a real (resistive) load impedance (i.e., $\psi = 0$ or π) and rewrite (3.20) as

$$V(z) = V^+ [e^{-j\beta z} \pm \rho e^{j\beta z}]$$
$$= V^+ [(1 \pm \rho) e^{-j\beta z} \pm \rho (-e^{-j\beta z} + e^{j\beta z})]$$

which, by using $\mathcal{V}(z, t) = \mathcal{R}e\{V(z) e^{j\omega t}\}$, gives

$$\mathcal{V}(z, t) = \underbrace{|V^+|(1 \pm \rho) \cos(\omega t - \beta z + \phi^+)}_{\text{propagating wave}} \pm \underbrace{|V^+|(2\rho) \sin(\beta z) \cos\left(\omega t + \phi^+ + \frac{\pi}{2}\right)}_{\text{standing wave}}$$

where we have taken $V^+ = |V^+| e^{j\phi^+}$. In other words, the voltage on the line consists of a standing wave plus a propagating wave.

According to (3.20), the magnitude of the voltage phasor (i.e., $|V(z)|$) alternates between the maximum and minimum values of $V_{\max}$ and $V_{\min}$ given by

$$V_{\max} = |V(z)|_{\max} = |V^+|(1 + |\Gamma_L|) = |V^+|(1 + \rho)$$
$$V_{\min} = |V(z)|_{\min} = |V^+|(1 - |\Gamma_L|) = |V^+|(1 - \rho)$$

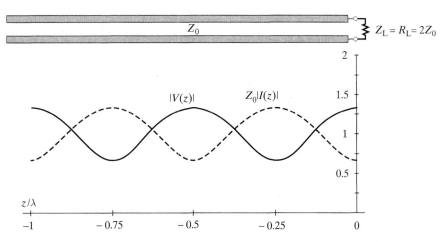

Figure 3.13   Standing-wave patterns for $Z_L = R_L = 2Z_0$. Magnitudes of the voltage and current phasors (i.e., $|V(z)|$ and $Z_0|I(z)|$) are shown as functions of electrical distance z/λ away from the load, for $V^+ = 1$.

Similarly, from equation (3.21), the magnitude of the current phasor (i.e., $|I(z)|$) alternates between the maximum and minimum values of

$$I_{max} = |I(z)|_{max} = \frac{|V^+|}{Z_0}(1 + \rho)$$

$$I_{min} = |I(z)|_{min} = \frac{|V^+|}{Z_0}(1 - \rho)$$

where I_{max} occurs at the same position as V_{min}, and I_{min} occurs at the same position as V_{max}. For example, Figure 3.13 shows the variations of both voltage and current magnitudes (represented by $|V(z)|$ and $Z_0|I(z)|$) as functions of position with respect to wavelength along the line, for the case of a purely resistive load, with $Z_L = R_L = 2Z_0$. As is apparent from Figure 3.13, the distance between successive voltage maxima (or minima) is $\lambda/2$. Note that for the case shown, with $Z_L = R_L > Z_0$, the reflection coefficient Γ_L is purely real with $\psi = 0$ and $0 \leq \rho \leq 1$.

Example 3.5: A Yagi antenna array driven by a coaxial line. To increase the geographic coverage area of a broadcast station, four Yagi antennas, each having a feed-point impedance of 50Ω, are stacked in parallel on a single antenna tower and connected to the transmitter by a 50Ω coaxial line, as shown in Figure 3.14a. (a) Calculate the load reflection coefficient Γ_L. (b) Calculate V_{max}, V_{min}, I_{max}, and I_{min} along the line, assuming $V^+ = 1$ V. (c) Sketch $|V(z)|$ and $|I(z)|$ as functions of z/λ, taking the position of the antenna array terminals to be at $z = 0$.

Solution:

(a) The total load impedance seen by the coaxial line is a parallel combination of the four 50Ω impedances, resulting in

$$Z_L = \frac{50}{4} = 12.5\Omega$$

The load reflection coefficient is then given by

$$\Gamma_L = \frac{Z_L - Z_0}{Z_L + Z_0} = \frac{12.5 - 50}{12.5 + 50} = -0.6 = 0.6e^{j180°}$$

so that $\rho = 0.6$ and $\psi = 180°$.

(b)

$$V_{max} = |V^+|(1 + \rho) = 1.6 \text{ V} \qquad\qquad V_{min} = |V^+|(1 - \rho) = 0.4 \text{ V}$$

$$I_{max} = \frac{|V^+|}{Z_0}(1 + \rho) = 32 \text{ mA} \qquad\qquad I_{min} = \frac{|V^+|}{Z_0}(1 - \rho) = 8 \text{ mA}$$

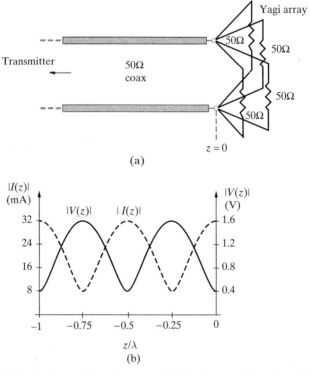

Figure 3.14 Yagi array driven by a coaxial line. (a) Array of a stack of four Yagi antennas fed by a coaxial line. (b) Voltage and current standing-wave patterns.

(c) The voltage reflection coefficient at any position z along the line is given by

$$\Gamma(z) = \rho e^{j(\psi + 2\beta z)} = 0.6 e^{j\pi(1 + 4z/\lambda)}$$

Using (3.20) and (3.21) with $V^+ = 1$ V, the magnitudes of the line voltage and current (i.e., $|V(z)|$ and $|I(z)|$) are plotted in Figure 3.14b as functions of electrical distance z/λ.

Standing-wave patterns such as those in Figure 3.13 are important in practice because, although the rapid temporal variations of the line voltages and currents are not easily accessible, the locations of the voltage minima and maxima and the ratio of the voltage maxima to minima are often readily measurable. A key parameter that is commonly used to characterize the termination of a transmission line is the standing-wave ratio (SWR), or S, defined as

$$S = \frac{V_{max}}{V_{min}} = \frac{I_{max}}{I_{min}} = \frac{1 + \rho}{1 - \rho} \quad \longrightarrow \quad \rho = \frac{S - 1}{S + 1} \qquad (3.23)$$

Note that S varies in the range $1 \le S \le \infty$.

Example 3.6: UHF blade antenna. A UHF blade antenna installed in the tail-cap of a small aircraft is used for communication over the frequency band 225–400 MHz. The following table provides the measured values of the feed-point impedance of the antenna at various frequencies:[9]

f (MHz)	$Z_L(\Omega)$
225	$22.5 - j51$
300	$35 - j16$
400	$45 - j2.5$

A 50Ω coaxial line is used to connect the communication unit to the antenna. Calculate the load reflection coefficient Γ_L and the standing-wave ratio S on the line at (a) 225 MHz, (b) 300 MHz, and (c) 400 MHz.

Solution:

(a) At 225 MHz, the load reflection coefficient is given by (equation (3.19))

$$\Gamma_L = \frac{Z_L - Z_0}{Z_L + Z_0} = \frac{22.5 - j51 - 50}{22.5 - j51 + 50} \simeq \frac{57.9 e^{-j118°}}{88.6 e^{-j35.1°}} \simeq 0.654 e^{-j83.2°}$$

The standing-wave ratio S at 225 MHz can then be obtained using (3.23) as follows:

$$S = \frac{1 + \rho}{1 - \rho} \simeq \frac{1 + 0.654}{1 - 0.654} \simeq 4.78$$

[9]R. L. Thomas, *A Practical Introduction to Impedance Matching*, Artech House, Inc., Dedham, Massachusetts, 1976.

(b) Using the same equations, at 300 MHz, we have $\Gamma_L =\simeq 0.254 e^{-j122°}$ and $S \simeq 1.68$.

(c) At 400 MHz, we have $\Gamma_L =\simeq 0.0588 e^{-j152°}$ and $S \simeq 1.13$.

We see that the reflections on the coaxial line are quite significant near 225 MHz but are much reduced near 400 MHz.

Another quantity that can sometimes be measured in experimental settings is z_{min}, or the distance from the load to the *first* minimum of the voltage standing-wave pattern.[10] From (3.20) we have

$$V(z) = V^+(e^{-j\beta z} + \Gamma_L e^{j\beta z}) = V^+ e^{-j\beta z}(1 + \rho e^{j\psi} e^{j2\beta z})$$

Since $|V(z)| = V_{min}$ when $e^{j(\psi+2\beta z)} = -1$, or

$$\psi + 2\beta z_{min} = -(2m+1)\pi \qquad m = 0, 1, 2, 3, \ldots$$

where $-\pi \le \psi < \pi$ and $z_{min} \le 0$. For any given frequency, measuring the wavelength (by measuring the distance between successive minima) provides a means to determine the phase velocity $v_p = f\lambda$.

At the location of the first minimum we have

$$\psi + 2\beta z_{min} = -\pi \quad \longrightarrow \quad \psi = -\pi - 2\beta z_{min} \tag{3.24}$$

We see that z_{min} is directly related to the phase ψ of the reflection coefficient Γ_L, whereas S determines its magnitude through (3.23). Once Γ_L is known, the load can be fully specified (assuming the characteristic impedance Z_0 is known), or Z_0 can be found (if Z_L is known). Thus, the two measurable quantities, S and z_{min}, completely characterize the transmission line terminated in an arbitrary load impedance.

It is often useful to rewrite (3.20) and (3.21) in terms of the load voltage V_L and load current I_L. Using the fact that $V_L = V(z)|_{z=0}$ and $I_L = I(z)|_{z=0}$, and after some manipulation, we have

$$V(z) = V_L \cos(\beta z) - jI_L Z_0 \sin(\beta z) \tag{3.25}$$

$$I(z) = I_L \cos(\beta z) - j\frac{V_L}{Z_0} \sin(\beta z) \tag{3.26}$$

The voltages and currents at the source end ($z = -l$) can be found from (3.25) and (3.26) by substituting $z = -l$. Note that for a general complex load impedance Z_L, the load voltage V_L and current I_L are in general complex, so that equations (3.25) and (3.26) do not necessarily constitute a decomposition of $V(z)$ and $I(z)$ into their real and imaginary parts.

---

[10]In practice, it may often be difficult to actually measure the first minimum; however, if the location of *any* of the minima can be measured, the location of the first minimum can be deduced by using the fact that successive minima are separated by $\lambda/2$.

In general, the voltage and current standing-wave patterns on a terminated line depend on the nature of the load. Typically what is plotted is $|V(z)|$, as was shown in Figure 3.13 for a purely resistive load $R_L = 2Z_0$. In the general case, with a complex load Z_L, the reflection coefficient Γ_L is complex, with $\psi \neq 0$. From (3.20) we have

$$V(z) = V^+[\cos(\beta z) - j\sin(\beta z) + \rho\cos(\psi + \beta z) + j\rho\sin(\psi + \beta z)]$$

and

$$\boxed{|V(z)| = |V^+|\sqrt{[\cos(\beta z) + \rho\cos(\psi + \beta z)]^2 + [-\sin(\beta z) + \rho\sin(\psi + \beta z)]^2}} \quad (3.27)$$

which is the quantity plotted in various figures as the voltage standing-wave pattern. For $\psi = 0$ or π (i.e., load is purely resistive) (3.27) reduces to

$$|V(z)| = |V^+|\sqrt{(1 \pm \rho)^2\cos^2(\beta z) + (-1 \pm \rho)^2\sin^2(\beta z)} \quad (3.28)$$

where the lower signs correspond to the case for $\psi = \pi$. Similar expressions can also be written for $|I(z)|$. Voltage and current standing wave patterns for different types of load impedances are shown in Figures 3.15 and 3.16. The interpretation of some of these patterns will become clearer after the discussion of line impedance in the following subsection.

In general, for purely resistive loads ($Z_L = R_L + j0$), the load position is a point of a voltage maximum or minimum, depending on whether $R_L > Z_0$ or $R_L < Z_0$, respectively. This behavior is apparent from Figure 3.15 and can also be seen by considering (3.20) and (3.21). For $R_L > Z_0$, $0 < \Gamma_L \leq 1$ and $|V(z = 0)| = V_{max} = |V^+|(1 + \rho)$, whereas for $R_L < Z_0$, $-1 \leq \Gamma_L < 0$ and $|V(z = 0)| = V_{min} = |V^+|(1 - \rho)$.

The standing-wave patterns in Figure 3.16 for $Z_L = Z_0 \pm jZ_0$ illustrate specific cases of the general behavior for loads with a reactive (capacitive or inductive) component. In general, the sign of the reactance (positive or negative) can be determined by inspection of the voltage standing-wave pattern. For $Z_L = R_L + jX_L$, X_L is negative (i.e., the load is capacitive) when the first minimum is at a distance smaller than one quarter of wavelength from the load (i.e., $-z_{min} < \lambda/4$) and X_L is positive (inductive) when the first minimum is at a distance greater than one quarter of a wavelength from the load (i.e., $\lambda/4 < -z_{min} < \lambda/2$), as illustrated in Figures 3.17a and b.

The behavior illustrated in Figures 3.17a and b can be understood upon careful examination of (3.19) and (3.20). For a general complex load impedance, (3.19) can be rewritten as

$$\Gamma_L = \rho e^{j\psi} = \frac{R_L + jX_L - Z_0}{R_L + jX_L + Z_0} = \frac{(R_L - Z_0) + jX_L}{(R_L + Z_0) + jX_L}$$

The phase angle ψ of the reflection coefficient is $0 < \psi < \pi$ if the load impedance is inductive ($X_L > 0$) and $-\pi < \psi < 0$ if the load impedance is capacitive ($X_L < 0$). The magnitude of the voltage along the line can be written from (3.20) as

$$|V(z)| = |V^+||1 + \Gamma(z)| = |V^+||1 + \rho e^{j(\psi + 2\beta z)}|$$

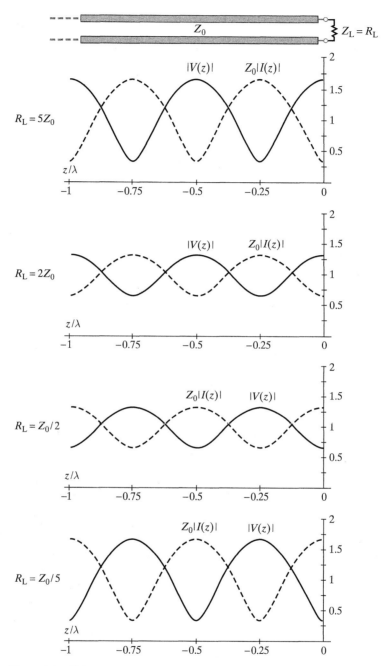

Figure 3.15 Voltage and current standing-wave patterns for different purely resistive loads. Magnitudes of the voltage and current phasors (i.e., $|V(z)|$ and $Z_0|I(z)|$) are shown as functions of electrical distance from the load position z/λ, for $V^+ = 1$ and for $R_L = 5Z_0, 2Z_0, Z_0/2$, and $Z_0/5$.

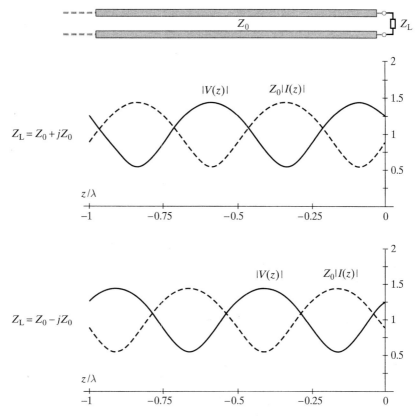

Figure 3.16   Voltage and current standing-wave patterns for complex loads. Magnitudes of the voltage and current phasors (i.e., $|V(z)|$ and $Z_0|I(z)|$) are shown as functions of electrical distance from the load z/λ for $V^+ = 1$ and for $Z_L = Z_0 + jZ_0$ (inductive load) and $Z_L = Z_0 - jZ_0$ (capacitive load).

where $z \leq 0$. Noting that $|V^+|$ is a constant, consider the second term and its variation with z (note that z decreases as one moves away from the load (at $z = 0$) along the transmission line). This term is the magnitude of the sum of two numbers, one being the real number 1 and the other being a complex number, $\Gamma(z) = \rho e^{j(\psi + 2\beta z)}$, which has a constant magnitude ρ ($0 \leq \rho \leq 1$, as determined by Z_L and Z_0) and a phase angle $\psi + 2\beta z$ that decreases with decreasing z (corresponding to clockwise rotation of this complex number on a circle with radius ρ centered at the origin on the complex plane). The two different cases of capacitive and inductive load are shown respectively in Figures 3.17c and d. For an inductive load, we see from Figure 3.17d that as we move away from the load (i.e., starting at $z = 0$ and rotating clockwise), $|V(z)|$ (which is proportional to $|1 + \Gamma(z)|$) first increases, reaches a maximum (at $\psi + 2\beta z = 0$), and then decreases, consistent with the variation of $|V(z)|$ for the inductive load as shown in Figure 3.17b. Similarly, for a capacitive load, we see from Figure 3.17c that as z decreases (starting with $z = 0$), $|V(z)|$ first decreases, reaches a minimum (at $\psi + 2\beta z = -\pi$), and then increases, consistent with Figure 3.17a.

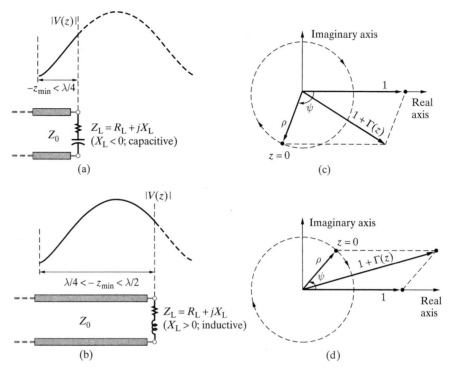

Figure 3.17 Variation of the voltage standing-wave pattern in the vicinity of the load for inductive or capacitive loads. In descriptive terms, starting from the load (i.e., $z = 0$), the standing-wave voltage at first increases (decreases) as one moves away from the load (i.e., clockwise in the diagrams shown) for an inductive (capacitive) load. (a) $|V(z)|$ for a capacitive load. (b) $|V(z)|$ for an inductive load. (c) $|V(z)| = |1 + \Gamma(z)|$ as the sum of two complex numbers, 1 and $\Gamma(z)$, for a capacitive load ($-\pi < \psi < 0$). (d) $|V(z)| = |1 + \Gamma(z)|$ as the sum of two complex numbers, 1 and $\Gamma(z)$, for an inductive load ($0 < \psi < \pi$). Note once again that $V^+ = 1$.

3.3.2 Transmission Line Impedance

An important property of a transmission line is its ability to transform impedances. In Section 3.2, we saw that the input impedance of a short- or open-circuited transmission line segment can be made equal to any arbitrary reactive impedance by simply adjusting its electrical length (l/λ). The input impedance of a transmission line terminated in an arbitrary load impedance Z_L is similarly dependent on the electrical length of the line or the distance from the load at which the impedance is measured. For the transmission line shown in Figure 3.12, the impedance seen looking toward the load Z_L at any position z along the line ($-l \le z \le 0$) given by (3.22) can be rewritten as

$$Z(z) = \frac{V(z)}{I(z)} = Z_0 \frac{e^{-j\beta z} + \Gamma_L e^{j\beta z}}{e^{-j\beta z} - \Gamma_L e^{j\beta z}} = Z_0 \frac{Z_L - jZ_0 \tan(\beta z)}{Z_0 - jZ_L \tan(\beta z)} \qquad (3.29)$$

using (3.19), (3.20), and (3.21). Expression (3.29) for the line impedance is often written as

$$Z(z) = Z_0 \frac{Z_L \cos(\beta z) - j Z_0 \sin(\beta z)}{Z_0 \cos(\beta z) - j Z_L \sin(\beta z)}$$

Note that since $\beta = 2\pi/\lambda$, the impedance varies periodically with electrical distance (z/λ) along the line, with the same impedance value attained at intervals in z of $\pm \lambda/2$. At the load, where $z = 0$, we have

$$Z(z = 0) = Z_0 \frac{1 + \Gamma_L}{1 - \Gamma_L} = Z_L$$

as expected. In particular, the input impedance seen by the source at the source end, where $z = -l$, is

$$\boxed{Z_{in} = [Z(z)]_{z=-l} = \frac{V_s}{I_s} = Z_0 \frac{Z_L + j Z_0 \tan(\beta l)}{Z_0 + j Z_L \tan(\beta l)}} \qquad (3.30)$$

For example, for a short-circuited line ($Z_L = 0$), the input impedance is

$$Z_{in} = j Z_0 \tan(\beta l)$$

and for an open-circuited line ($Z_L = \infty$), it is

$$Z_{in} = -j Z_0 \cot(\beta l)$$

as was shown in Sections 3.2.1 and 3.2.2.

Example 3.7: Input impedance of a line. Find the input impedance of a 75-cm long transmission line where $Z_0 = 70\Omega$, terminated with a $Z_L = 140\Omega$ load at 50, 100, 150, and 200 MHz. Assume the phase velocity v_p to be equal to the speed of light in free space.

Solution:

(a) At $f = 50$ MHz, we have $\lambda \simeq (3 \times 10^8)/(5 \times 10^7) = 6$ m, so the electrical length of the line is $l/\lambda \simeq 0.75/6 = 0.125$. Noting that $\beta l = 2\pi l/\lambda$, we then have, from (3.30),

$$Z_{in} \simeq 70 \frac{140 + j70 \tan(2\pi \times 0.125)}{70 + j140 \tan(2\pi \times 0.125)} = 56 - j42 \Omega$$

since $\tan(2\pi \times 0.125) = 1$. Note that the input impedance of the line at 50 MHz is capacitive.

(b) At $f = 100$ MHz, we have $\lambda \simeq 3$ m and $l/\lambda \simeq 0.75/3 = 0.25$. From (3.30) we have

$$Z_{in} \simeq 70 \frac{140 + j70 \tan(2\pi \times 0.25)}{70 + j140 \tan(2\pi \times 0.25)} = \frac{(70)^2}{140} = 35 \Omega$$

since $\tan(2\pi \times 0.25) = \infty$. Note that the input impedance of the line at 100 MHz is purely resistive.

(c) At $f = 150$ MHz, $\lambda \simeq 2$ m and $l/\lambda \simeq 0.75/2 = 0.375$. Again using (3.30), we have $Z_{in} \simeq 56 + j42\Omega$, which is an inductive impedance.

(d) At $f = 200$ MHz, we have $\lambda \simeq 1.5$ m and $l/\lambda \simeq 0.75/1.5 = 0.5$. Using (3.30), we have $Z_{in} \simeq 140\Omega$. Note that the input impedance of the line at 200 MHz is purely resistive and is exactly equal to the load impedance.

Normalized line impedance.   In transmission line analysis, it is often convenient and common practice to normalize all impedances to the characteristic impedance Z_0 of the transmission line. Denoting the normalized version of any impedance by using a bar at the top, we can rewrite (3.29) to express the normalized line impedance $\bar{Z}(z)$ in terms of the normalized load impedance $\bar{Z}_L$

$$\bar{Z}(z) = \frac{\bar{Z}_L - j \tan(\beta z)}{1 - j\bar{Z}_L \tan(\beta z)} = \frac{\bar{Z}_L \cos(\beta z) - j \sin(\beta z)}{\cos(\beta z) - j\bar{Z}_L \sin(\beta z)} \tag{3.31}$$

where

$$\bar{Z}_L = \frac{Z_L}{Z_0} = \frac{1 + \Gamma_L}{1 - \Gamma_L} = \frac{1 + \rho e^{j\psi}}{1 - \rho e^{j\psi}}$$

Using (3.23) and (3.24), we can further write

$$\bar{Z}_L = \frac{Z_L}{Z_0} = \frac{1 + jS \tan(\beta z_{min})}{S + j \tan(\beta z_{min})} = \frac{\cos(\beta z_{min}) + jS \sin(\beta z_{min})}{S \cos(\beta z_{min}) + j \sin(\beta z_{min})} \tag{3.32}$$

which expresses the normalized load impedance $\bar{Z}_L$ in terms of the measurable quantities S and z_{min}.

Sometimes, it is useful to express the real and imaginary parts of the load impedance $Z_L = R_L + jX_L$ explicitly in terms of z_{min} and S:

$$\bar{R}_L = \frac{R_L}{Z_0} = \frac{S}{S^2 \cos^2(\beta z_{min}) + \sin^2(\beta z_{min})}$$

$$\bar{X}_L = \frac{X_L}{Z_0} = \frac{(S^2 - 1) \cos(\beta z_{min}) \sin(\beta z_{min})}{S^2 \cos^2(\beta z_{min}) + \sin^2(\beta z_{min})}$$

The relationship between the polarity of X_L (i.e., inductive versus capacitive) and the distance to the first minimum, as depicted in Figure 3.17, can also be deduced by careful consideration of the preceding equation for $\bar{X}_L$.

Example 3.8: Unknown load. Determine an unknown load Z_L from S and z_{min} measurements. The following measurements are carried out on a 100Ω transmission line terminated with an unknown load Z_L, as shown in Figure 3.18. The voltage standing-wave ratio S is 5, the distance between successive voltage minima is 25 cm, and the distance from Z_L to the first voltage minimum is 8 cm. (a) Determine the load reflection coefficient Γ_L. (b) Determine the unknown load impedance Z_L. (c) Determine the location of the first voltage maximum with respect to the load.

Figure 3.18   **Transmission line terminated in an unknown impedance.**

Solution:

(a) Using (3.23) and (3.24), we have

$$\rho = \frac{S-1}{S+1} = \frac{5-1}{5+1} \simeq 0.667$$

$$\frac{\lambda}{2} = 25 \text{ cm} \quad \rightarrow \quad \lambda = 50 \text{ cm}$$

$$\psi = -\pi - 2\beta z_{\min} = -\pi + 2\left(\frac{2\pi}{50}\right)(8) = -0.36\pi \text{ rad} \quad \text{or} - 64.8°$$

$$\Gamma_{\mathrm{L}} = \rho e^{j\psi} \simeq 0.667 e^{-j64.8°}$$

(b) We have

$$(\beta z_{\min}) = \frac{-\pi + 0.36\pi}{2} = -0.32\pi \text{ rad} \quad \rightarrow \quad \tan(\beta z_{\min}) \simeq -1.58$$

and so using (3.32),

$$Z_{\mathrm{L}} = Z_0 \frac{1 + jS \tan(\beta z_{\min})}{S + j \tan(\beta z_{\min})} \simeq 100 \frac{1 - j5(1.58)}{(5 - j1.58)} \simeq 63.4 - j137.6\Omega$$

(c) The location of the first voltage maximum is $\lambda/4$ away from the location of the voltage minimum. Thus, we have

$$z_{\max} = z_{\min} - \lambda/4 = -8 - 12.5 = -20.5 \text{ cm}$$

Transmission line admittance.   In Sections 3.5 and 3.6, when we discuss impedance matching and the Smith chart, it will be useful at times to work with the line *admittance* rather than the impedance. From (3.29), we can find the expression for line admittance as

$$Y(z) = \frac{1}{Z(z)} = Y_0 \frac{1 - \Gamma_{\mathrm{L}} e^{+j2\beta z}}{1 + \Gamma_{\mathrm{L}} e^{+j2\beta z}} = Y_0 \frac{Y_{\mathrm{L}} - jY_0 \tan(\beta z)}{Y_0 - jY_{\mathrm{L}} \tan(\beta z)} \qquad (3.33)$$

where $Y_0 = (Z_0)^{-1}$. Similar to (3.31), the normalized line admittance $\bar{Y}(z) = Y(z)/Y_0$ can be written as

$$\bar{Y}(z) = \frac{\bar{Y}_{\mathrm{L}} - j \tan(\beta z)}{1 - j\bar{Y}_{\mathrm{L}} \tan(\beta z)} = \frac{\bar{Y}_{\mathrm{L}} \cos(\beta z) - j \sin(\beta z)}{\cos(\beta z) - j\bar{Y}_{\mathrm{L}} \sin(\beta z)} \qquad (3.34)$$

The load reflection coefficient Γ_L can also be written in terms of admittances as

$$\Gamma_L = \frac{Y_0 - Y_L}{Y_0 + Y_L}$$

Line impedance for resistive loads.   The variation with z of the real and imaginary parts of the normalized line impedance is illustrated in Figure 3.19a for the case of a resistive load with $Z_L = R_L = 2Z_0$. The voltage and current standing-wave patterns (from Figure 3.13) are also shown for reference in Figure 3.19b. Note from Figure 3.19a that, as viewed from different positions at a distance z from the load, the real part of the normalized line impedance varies between $\bar{R}(z) = 2$ and $\bar{R}(z) = 0.5$, with the distance between successive maxima being $\lambda/2$. The line impedance is purely real

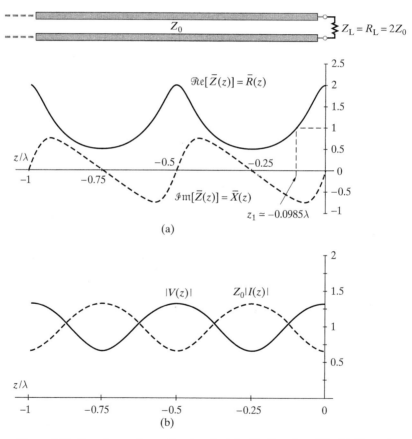

(a)

(b)

Figure 3.19   Impedance along a line terminated with $Z_L = R_L = 2Z_0$. (a) The real and imaginary parts of the normalized line impedance $\bar{Z}(z)$ are shown as functions of z/λ. (b) Magnitudes of the voltage and current phasors (i.e., $|V(z)|$ and $Z_0|I(z)|$) are shown as functions of electrical distance from the load z/λ for $V^+ = 1$.

at the load and at distances of integer multiples of $\lambda/4$ from the load. These positions also correspond to the positions of voltage maxima and minima along the line. For $-z < \lambda/4$, the line impedance $Z(z)$ is capacitive (i.e., its imaginary part is negative; $X(z) < 0$), reminiscent of the behavior of the open-circuited line[11] (see Figure 3.9). For $\lambda/4 < -z < \lambda/2$, the line impedance $Z(z)$ is inductive ($X(z) > 0$), and continues to alternate between capacitive and inductive impedance at intervals of $\lambda/4$.

An interesting aspect of the result in Figure 3.19a is the fact that $\Re e\{\bar{Z}(z)\} = 1$ at $z_1 \simeq -0.0985\lambda$.[12] If the imaginary part of the line impedance at that position could somehow be canceled (as we shall see in Section 3.5), the line would appear (from all positions at locations $z < -0.0985\lambda$) as if it were matched (i.e., terminated with an impedance Z_0). For example, such cancellation can in principle be achieved by introducing a purely reactive series impedance that is opposite in sign to the reactive part of $Z(z)$ at that position, as will be discussed in Section 3.5. The following example illustrates the determination of the point at which $\Re e\{\bar{Z}(z)\} = 1$ for a specific complex load impedance.

Example 3.9: An inverted-V antenna. A 50Ω coaxial line filled with teflon ($v_p \simeq 21$ cm-(ns)$^{-1}$) is connected to an inverted-V antenna represented by Z_L, as shown in Figure 3.20. At $f = 29.6$ MHz, the feed-point impedance of the antenna is approximately measured to be $Z_L \simeq 75 + j25\Omega$.[13] Find the two closest positions to the antenna along the line where the real part of the line impedance is equal to the characteristic impedance of the line (i.e., Z_0).

Solution: The line impedance at any position z is given by (3.29):

$$Z(z) = 50\frac{(75+j25) - j50\zeta}{50 - j(75+j25)\zeta} = 50\frac{3+j(1-2\zeta)}{(2+\zeta) - j3\zeta}$$

where $\zeta = \tan(\beta z)$, $\beta = 2\pi/\lambda$, and $\lambda = v_p/f \simeq (2.1 \times 10^8)/(29.6 \times 10^6) \simeq 7.09$ m. Multiplying both the numerator and the denominator with the complex conjugate of the denominator, we can extract the real part of $Z(z)$ as

$$\Re e\{Z(z)\} = \Re e\left\{50\frac{3+j(1-2\zeta)}{(2+\zeta) - j3\zeta} \cdot \frac{(2+\zeta) + j3\zeta}{(2+\zeta) + j3\zeta}\right\} = 50\frac{3(2+\zeta) - 3\zeta(1-2\zeta)}{(2+\zeta)^2 + (3\zeta)^2}$$

50Ω coax
$v_p \simeq 21$ cm/ns

Z_L

$z = 0$

$Z(z)$

Figure 3.20   A coaxial line connected to an antenna.

---

[11] Note that this makes sense because in Figure 3.19a the load resistance is larger than the characteristic impedance ($R_L > Z_0$), which is also the case for the open circuit.

[12] The value of z_1 can be read roughly from Figure 3.19a or accurately evaluated from (3.31) by letting $\bar{Z}(z) = 1 + j\bar{X}(z)$.

[13] R. Dean Straw (Ed.), *The ARRL Antenna Book,* 17th ed., American Radio Relay League, pp. 27–29, Newington, Connecticut, 1994–1996.

The value of z for which we have $\mathcal{R}e\{Z(z)\} = Z_0 = 50\Omega$ can then be found as

$$\mathcal{R}e\{Z(z)\} = Z_0 \quad \rightarrow \quad 50\frac{3(2+\zeta) - 3\zeta(1 - 2\zeta)}{(2+\zeta)^2 + (3\zeta)^2} = 50$$

$$\rightarrow \quad 2\zeta^2 + 2\zeta - 1 = 0 \quad \rightarrow \quad \zeta_1, \zeta_2 \simeq -1.37, 0.366$$

Using

$$\tan(\beta z) = \tan\left(\frac{2\pi z}{\lambda}\right) \simeq \tan\left[\frac{2\pi z}{7.09}\right] = \zeta$$

and noting that $z < 0$, we find

$$z_1 \simeq -0.149\lambda \simeq -1.06 \text{ m} \quad \text{and} \quad z_2 \simeq -0.444\lambda \simeq -3.15 \text{ m}$$

as the locations at which the real part of the line impedance is equal to the characteristic impedance of the line.

Some aspects of the behavior of the real and imaginary parts of the line impedance shown in Figure 3.19 for $\bar{Z}_L = 2$ can be generalized. For example, the line impedance seen at the positions of voltage maxima (minima) is always purely real and has maximum (minimum) magnitude. To see this, consider the voltage along the line at the position of a voltage maximum (i.e., $z = z_{max}$) given by

$$V(z = z_{max}) = V^+ e^{-j\beta z_{max}}[1 + \rho e^{j(\psi + 2\beta z_{max})}] = V^+ e^{-j\beta z_{max}}(1 + \rho)$$

with a maximum magnitude of

$$|V(z = z_{max})|_{max} = V_{max} = |V^+|(1 + \rho)$$

occurring at

$$\psi + 2\beta z_{max} = -m2\pi \qquad m = 0, 1, 2, 3, \dots$$

where $-\pi \leq \psi < \pi$, $z_{max} \leq 0$, and where $m = 0$ does not apply if $-\pi \leq \psi < 0$. At the same position, the current is equal to

$$I(z = z_{max}) = \frac{V^+}{Z_0} e^{-j\beta z_{max}}(1 - \rho)$$

with a minimum magnitude given by

$$|I(z = z_{max})|_{min} = I_{min} = \frac{|V^+|}{Z_0}(1 - \rho)$$

so that the line impedance $Z(z_{max}) = V(z_{max})/I(z_{max})$ is clearly purely resistive and has a maximum magnitude given by

$$\boxed{|Z(z_{max})|_{max} = R_{max} = \frac{V_{max}}{I_{min}} = Z_0\frac{1 + \rho}{1 - \rho} = SZ_0} \qquad (3.35)$$

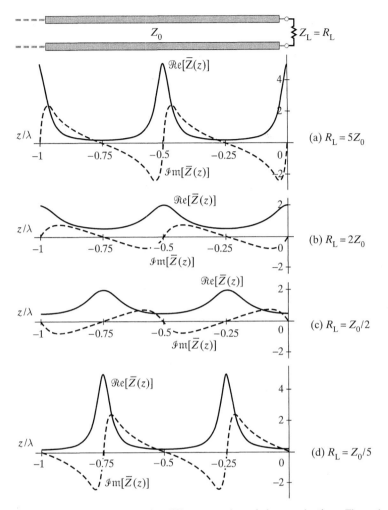

Figure 3.21 Line impedance for different purely resistive terminations. The real and imaginary parts of the normalized line impedance $\bar{Z}(z)$ are shown as functions of electrical distance z/λ along the line for Z_L equal to (a) $5Z_0$, (b) $2Z_0$, (c) $Z_0/2$, and (d) $Z_0/5$.

whereas at the voltage minima ($z_{min} = z_{max} - \lambda/4$), the line impedance $Z(z_{min})$ is purely resistive with a minimum magnitude given by

$$\boxed{|Z(z_{min})|_{min} = R_{min} = \frac{V_{min}}{I_{max}} = Z_0\frac{1-\rho}{1+\rho} = \frac{Z_0}{S}} \qquad (3.36)$$

Also, for purely resistive terminations ($Z_L = R_L$), the load is at a position of either minimum or maximum for the voltage and therefore, the load impedance R_L is either equal to the minimum ($R_L = Z_0/S$ when $R_L < Z_0$) or maximum ($R_L = SZ_0$ when $R_L > Z_0$)

magnitude for the line impedance. Note that we have

$$S = \frac{1+\rho}{1-\rho} = \begin{cases} \frac{R_L}{Z_0} & \text{for} \quad R_L > Z_0 \\ \frac{Z_0}{R_L} & \text{for} \quad R_L < Z_0 \end{cases}$$

Line impedance for complex load impedances. For general load impedances that are not purely resistive (i.e., $Z_L = R_L + jX_L$), the behavior of the line impedance $Z(z) = R(z) + jX(z)$ is similar to that for purely resistive loads, in that its real part $R(z)$ varies between a maximum value of SZ_0 and a minimum value of Z_0/S, and the imaginary part $X(z)$ alternates sign at intervals of $\lambda/4$. (SZ_0 occurs at the positions of the voltage maxima when the line impedance is purely resistive and therefore is also the maximum magnitude of the line impedance. Z_0/S occurs at the positions of the voltage minima when the line impedance is also purely resistive and therefore is also the minimum magnitude of the line impedance.) However, the maxima and minima of the magnitudes of either the voltage or the line impedance are not at

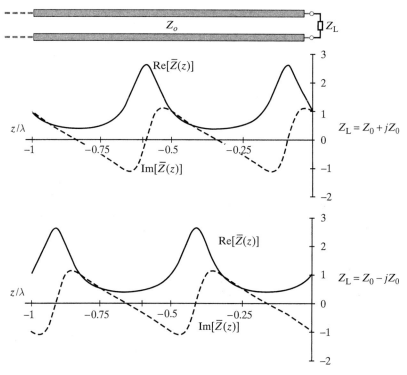

Figure 3.22 Line impedance for two different complex load impedances. The real and imaginary parts of the normalized line impedance $\bar{Z}(z)$ are shown as functions of electrical distance z/λ along the line for (a) $Z_L = Z_0 + jZ_0$ and (b) $Z_L = Z_0 - jZ_0$.

the load position. Figures 3.21 and 3.22 show plots of the real and imaginary parts of the normalized line impedance $\bar{Z}(z)$ as functions of z/λ for selected load impedances.

Example 3.10: Reflection coefficient, standing-wave ratio, and maximum and minimum resistances. A radio transmitter is connected to an antenna having a feed-point impedance of $Z_L = 70 + j100\Omega$ with a 50Ω coaxial line, as shown in Figure 3.23. Find (a) the load reflection coefficient, (b) the standing-wave ratio, and (c) the two positions closest along the line to the load where the line impedance is purely real, and their corresponding line impedance values.

Figure 3.23   Transmission line terminated in an antenna. In general, the feed-point impedance of an antenna is complex.

Solution:

(a) The load reflection coefficient is

$$\Gamma_L = \frac{Z_L - Z_0}{Z_L + Z_0} = \frac{70 + j100 - 50}{70 + j100 + 50} \simeq 0.653 e^{j38.9°}$$

(b) The standing-wave ratio is

$$S = \frac{1 + \rho}{1 - \rho} \simeq \frac{1 + 0.653}{1 - 0.653} \simeq 4.76$$

(c) The maximum voltage position is the position z_{max} at which the line impedance is purely real and the magnitude of the line impedance is a maximum; note that the position of the first voltage maximum is closer to the load than that of the first voltage minimum because the load impedance is inductive. To find the maximum voltage position, use $\psi + 2\beta z_{max} = 0 \rightarrow z_{max} \simeq -0.054\lambda$:

$$R_{max} = SZ_0 \simeq (4.76)(50) = 238\Omega$$

The minimum voltage position is the position z_{min} at which $|Z(z)|$ is minimum; note that this is the next closest position where the line impedance $Z(z)$ is real. To find the minimum voltage position, use

$$\psi + 2\beta z_{min} = -\pi \rightarrow z_{min} \simeq -0.304\lambda$$

Note that as expected, $z_{min} = z_{max} - \lambda/4$. We then have

$$R_{min} = Z_0/S \simeq 50/(4.76) \simeq 10.5\Omega$$

3.3.3 Calculation of V^+

Up to now, we have primarily focused on the line impedance and the variation of the voltage and current along the line without particular attention to the source end of the line. The source that excites the transmission line shown in Figure 3.12 is a voltage source with an open-circuit phasor voltage V_0 and a source impedance Z_s. Using equation (3.20), the voltage V_s at the source end of the line ($z = -l$) is

$$V_s = V(z = -l) = V^+ e^{j\beta l}(1 + \Gamma_L e^{-j2\beta l})$$

As seen from the source end, the transmission line can be represented by its input impedance, Z_{in}. We can thus also express the source-end voltage phasor V_s in terms of the source parameters V_0 and Z_s by noting the division of voltage between Z_{in} and Z_s, namely,

$$V_s = \frac{Z_{in}}{Z_{in} + Z_s} V_0$$

By equating the two preceding expressions for V_s, we can solve for the constant V^+:

$$V^+ = \frac{Z_{in} V_0}{(Z_{in} + Z_s)e^{j\beta l}(1 + \Gamma_L e^{-j2\beta l})}$$

Note that the knowledge of V^+ and the wavelength $\lambda = 2\pi/\beta$ completely specifies the transmission line voltage and current as given in (3.20) and (3.21), as for any given transmission line with characteristic impedance Z_0 and load Z_L (and hence Γ_L).

Example 3.11: Coaxial line feeding an antenna. A sinusoidal voltage source of $\mathcal{V}_0(t) = 10\cos(5\pi \times 10^7 t)$ V and $R_s = 20\Omega$ is connected to an antenna with feed-point impedance $Z_L = 100\Omega$ through a 3-m long, lossless coaxial transmission line filled with polyethylene ($v_p = 20$ cm-(ns)$^{-1}$) and with a characteristic impedance of $Z_0 = 50\Omega$, as shown in Figure 3.24a. Find (a) the voltage and current phasors, $V(z)$ and $I(z)$, at any location on the line and (b) the corresponding instantaneous expressions $\mathcal{V}(z,t)$ and $\mathcal{I}(z,t)$.

Solution:

(a) At $f = \omega/(2\pi) = 25$ MHz, the wavelength in a polyethylene-filled coaxial line is

$$\lambda = v_p/f = (20 \text{ cm-(ns)}^{-1})/(25 \text{ MHz}) = 8 \text{ m}$$

The electrical length of the 3-m line is then $l/\lambda = 3/8 = 0.375$, so we have $\beta l = 2\pi(0.375) = 3\pi/4$ and $\tan(\beta l) = -1$. The input impedance seen at the source end can be calculated using (3.30):

$$Z_{in} = 50\frac{100 + j50(-1)}{50 + j100(-1)} = 40 + j30\,\Omega$$

Equivalent Circuit
(b)

Figure 3.24   Coaxial line feeding an antenna. (a) Circuit configuration. (b) Thévenin equivalent circuit seen from the source end.

Using the equivalent circuit shown in Figure 3.24b, we have

$$V_s = \frac{Z_{in}}{R_s + Z_{in}} V_0 = \frac{40 + j30}{60 + j30}(10) \simeq \frac{5e^{j36.9°}}{3\sqrt{5}e^{j26.6°}}(10) \simeq 7.45e^{j10.3°} \text{ V}$$

We can also write an expression for V_s by evaluating $V(z)$ at $z = -3$ m as

$$V_s = V(z = -3 \text{ m}) = V^+ e^{j3\pi/4}(1 + \Gamma_L e^{-j3\pi/2})$$

where Γ_L is the load reflection coefficient given by (3.19):

$$\Gamma_L = \frac{Z_L - Z_0}{Z_L + Z_0} = \frac{100 - 50}{100 + 50} = \frac{1}{3}$$

Equating the two expressions for V_s, we can determine the complex constant V^+

$$V_s = V^+ e^{j3\pi/4}\left(1 + \frac{j}{3}\right) \simeq 7.45e^{j10.3°} \quad \rightarrow \quad V^+ \simeq 7.07e^{-j143°} \text{ V}$$

so that the voltage phasor at any position z from the load is given as

$$V(z) \simeq 7.07e^{-j143°}e^{-j\pi z/4}\left(1 + \tfrac{1}{3}e^{j\pi z/2}\right) \text{ V}$$

and the corresponding current phasor is

$$I(z) \simeq 0.141e^{-j143°}e^{-j\pi z/4}\left(1 - \tfrac{1}{3}e^{j\pi z/2}\right) \text{ A}$$

(b) Using $\mathcal{V}(z,t) = \mathcal{R}e\{V(z)e^{j\omega t}\}$ and $\mathcal{I}(z,t) = \mathcal{R}e\{I(z)e^{j\omega t}\}$, we find

$$\mathcal{V}(z,t) \simeq 7.07\cos\left(5\pi\,10^7 t - \frac{\pi z}{4} - 143°\right) + 2.36\cos\left(5\pi\,10^7 t + \frac{\pi z}{4} - 143°\right)\ \text{V}$$

and

$$\mathcal{I}(z,t) \simeq 0.141\cos\left(5\pi\,10^7 t - \frac{\pi z}{4} - 143°\right) - 0.0471\cos\left(5\pi\,10^7 t + \frac{\pi z}{4} - 143°\right)\ \text{A}$$

3.4 POWER FLOW ON A TRANSMISSION LINE

In practice, the primary purpose of most steady-state sinusoidal transmission line applications is to maximize the time-average power delivered to a load. The power and energy
flow on a transmission line can be determined from the line voltage and line current; the
product of the instantaneous current $\mathcal{I}(z,t)$ and instantaneous voltage $\mathcal{V}(z,t)$ at any point
z is by definition the power that flows into the line[14] at that point. In most applications,
the quantity of interest is not the rapidly varying instantaneous power but its average
over one sinusoidal period T_p, namely, the time-average power, which is given by

$$P_{\text{av}}(z) = \frac{1}{T_p}\int_0^{T_p}\mathcal{V}(z,t)\mathcal{I}(z,t)\,dt$$

where $T_p = 2\pi/\omega$. The time-average power can also be calculated directly from the
voltage and current phasors:

$$P_{\text{av}}(z) = \frac{1}{2}\mathcal{R}e\{V(z)[I(z)]^*\}$$

Consider the general expressions for the voltage and current phasors along a lossless
uniform transmission line:

$$V(z) = \quad V^+ e^{-j\beta z} + \Gamma_L V^+ e^{j\beta z}$$

$$I(z) = \quad \underbrace{\frac{V^+}{Z_0}e^{-j\beta z}}_{\text{forward wave}} - \underbrace{\Gamma_L\frac{V^+}{Z_0}e^{j\beta z}}_{\text{reverse wave}}$$

We denote the time-average power carried by the forward and backward traveling
waves as P^+ and P^-, respectively, and we evaluate them directly from the phasors. The
time-average power carried by the forward wave is

$$P^+ = \frac{1}{2}\mathcal{R}e\left\{\frac{(V^+ e^{-j\beta z})(V^+ e^{-j\beta z})^*}{Z_0}\right\} = \frac{V^+(V^+)^*}{2Z_0} = \frac{|V^+|^2}{2Z_0}$$

[14]Note that the product $\mathcal{V}(z,t)\mathcal{I}(z,t)$ represents power flow into the line, rather than out of the line, due
to the defined polarity of the current $I(z)$ and voltage $V(z)$ in Figure 3.12.

Note that although V^+ is in general complex, $V^+(V^+)^* = |V^+|^2$ is a real number.[15] The power carried by the reverse-propagating wave is

$$P^- = \frac{1}{2}\mathcal{R}e\left\{\frac{(\Gamma_L V^+ e^{j\beta z})(-\Gamma_L V^+ e^{j\beta z})^*}{Z_0}\right\} = -\Gamma_L\Gamma_L^* \frac{V^+(V^+)^*}{2Z_0} = -\rho^2\frac{|V^+|^2}{2Z_0}$$

The fact that P^- is negative simply indicates that the backward wave carries power in the opposite direction with respect to the defined polarity of $I(z)$ and $V(z)$ in Figure 3.12. The net total power in the forward direction is then given by

$$P_{av} = P^+ + P^- = \frac{|V^+|^2}{2Z_0} - \rho^2\frac{|V^+|^2}{2Z_0} = \frac{|V^+|^2}{2Z_0}(1 - \rho^2) \tag{3.37}$$

Thus, the net time-average power flow on a transmission line is maximized when the load reflection coefficient Γ_L is zero, which, according to (3.19), occurs when $Z_L = Z_0$. Note that when $\Gamma_L = 0$ and thus $Z_L = Z_0$, the standing-wave ratio $S = 1$, as is evident from (3.23). Since the transmission line is assumed to be lossless, all of the net power flowing in the $+z$ direction is eventually delivered to the load.

The same result can also be obtained by examining the power dissipation in the load, which is given by[16]

$$P_L = \frac{1}{2}\mathcal{R}e\{V_L I_L^*\} = \frac{1}{2}|I_L|^2\mathcal{R}e\{Z_L\} = \frac{1}{2}|I_L|^2 R_L$$

Noting that we have

$$V_L = V(z)|_{z=0} = V^+ + \Gamma_L V^+ = V^+(1 + \Gamma_L)$$

$$I_L = I(z)|_{z=0} = \frac{V^+}{Z_0} - \Gamma_L\frac{V^+}{Z_0} = \frac{V^+}{Z_0}(1 - \Gamma_L)$$

and substituting in the preceding expression for P_L, we find

$$P_L = \frac{1}{2}\mathcal{R}e\{V_L I_L^*\} = \frac{1}{2}\mathcal{R}e\left\{\frac{V^+(1 + \Gamma_L)[V^+(1 - \Gamma_L)]^*}{Z_0}\right\}$$

$$= \frac{1}{2}\mathcal{R}e\left\{\frac{V^+(V^+)^*[1 + (\Gamma_L - \Gamma_L^*) - \Gamma_L\Gamma_L^*]}{Z_0}\right\} = \frac{|V^+|^2}{2Z_0}(1 - \rho^2) \tag{3.38}$$

---

[15] If $V^+ = A + jB$, then $V^+(V^+)^* = (A + jB)(A - jB) = A^2 + B^2 = |A + jB|^2 = |V^+|^2$.

[16] Note that P_L can also be written in terms of the load voltage V_L and load admittance Y_L as

$$P_L = \frac{1}{2}|V_L|^2\mathcal{R}e\{Y_L\} = \frac{1}{2}|V_L|^2 G_L$$

where $Y_L = Z_L^{-1} = G_L + jB_L$.

which is identical to the total net forward power P_{av} as derived in (3.37). We see that $P_{av} = P_L$ as expected, since, for the case of a lossless line as assumed here, *all* of the net power traveling toward the load must be dissipated in the load.

The same result can further be obtained by evaluating $P(z)$ at any point along the line using the total voltage and current phasors (rather than separating them into forward and reverse traveling wave components). In other words,

$$P_{av}(z) = \frac{1}{2}\mathscr{R}e\{V(z)[I(z)]^*\}$$

where $V(z)$ and $I(z)$ are given by (3.20) and (3.21), respectively. (This derivation is left as an exercise for the reader.)

In summary, the total net power propagating in the $+z$ direction is

$$\boxed{P_{av}(z) = \frac{|V^+|^2}{2Z_0}(1 - \rho^2)} \tag{3.39}$$

and the following observations concerning power flow on a lossless transmission line can be made:

- For a given V^+, maximum power is delivered to the load when $Z_L = Z_0$, $\Gamma_L = 0$ (i.e., $\rho = 0$), and $S = 1$. Noting that Z_0 is a real number, this condition is realized when the load is purely resistive, that is, when $Z_L = R_L = Z_0$. When $R_L = Z_0$, the load is said to be *matched* to the line and all of the power P^+ is delivered to the load. Detailed discussion of impedance matching is given in Section 3.5.

- To deliver a given amount of power (say, P_L) when the line is not matched (i.e., $S > 1$) requires higher wave power in the incident wave with correspondingly higher voltages ($P^+ = |V^+|^2/(2Z_0)$). The higher voltages are undesirable as they may cause breakdown[17] of the insulation between the two conductors of the line.

- The power efficiency achieved by matching can be assessed by considering the ratio of the power P_L that is dissipated in a given load to the forward wave power P^+ that would be delivered to the load if the line were matched:

$$\frac{P_L}{P^+} = 1 - \rho^2 = \frac{4S}{(1+S)^2}$$

The variation of P_L/P^+ with S is plotted in Figure 3.25. We see that $P_L/P^+ = 1$ for $S = 1$ and monotonically decreases to zero as S gets larger. Note that for voltage standing-wave ratios $S < 1.5$, which are relatively easy to achieve in practice, more than 90 percent of the power in the forward wave is delivered to the load. In other words, it is not necessary to strive for S very near unity to attain maximum power transfer to

---

[17]Electrical breakdown of insulating materials will be discussed in Section 4.10.

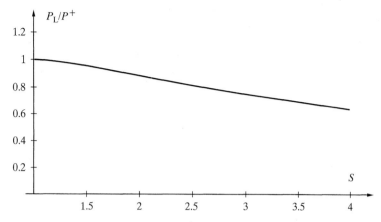

Figure 3.25   Power efficiency as a function of standing-wave ratio S.

the load. Usually the more important issues are ensuring that the value of S is not so large as to make the line performance highly sensitive to frequency (see Section 3.5), and that the design of the line can accommodate large reactive voltages and currents that accompany a large value of S.

The degree of mismatch between the load and the line is sometimes described in terms of *return loss,* which is defined as the decibel value of the ratio of the power carried by the reverse wave to the power carried by the forward wave, given as

$$\text{Return loss} = -20 \ \log_{10} \rho = 20 \ \log_{10} \frac{S+1}{S-1}$$

If the load is perfectly matched to the line ($\rho = 0$), the return loss is infinite, which simply indicates that there is no reverse wave. If the load is such that $\rho = 1$ (i.e., a short-circuited or open-circuited line, or a purely reactive load), then the return loss is 0 dB. In practice, a well-matched system has a return loss of 15 dB or more, corresponding to a standing-wave ratio of ~ 1.43 or less.

Example 3.12: A 125-MHz VHF transmitter-antenna system. A VHF transmitter operating at 125 MHz and developing $V_0 = 100e^{j0°}$ V with and source resistance of $R_s = 50\Omega$ feeds an antenna with a feed-point impedance[18] of $Z_L = 100 - j60$ through a 50Ω, polyethylene-filled coaxial line that is 17 m long. The setup is shown in Figure 3.26a. (a) Find the voltage $V(z)$ on the line. (b) Find the load voltage V_L. (c) Find the time-average power absorbed by the VHF antenna. (d) Find the power absorbed by the source impedance R_s.

---

[18]R. Dean Straw (Ed.), *The ARRL Antenna Compendium,* Vol. 4, p. 56, The American Radio Relay League, Newington, Connecticut, 1995–1996.

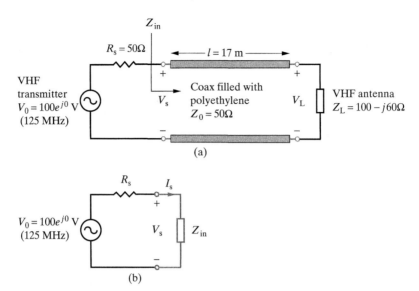

Figure 3.26 A 125-MHz VHF transmitter-antenna system. (a) Circuit diagram.
(b) Equivalent circuit seen by the source.

Solution:

(a) First we note that for a polyethylene-filled coaxial line, the wavelength at 125 MHz
is (using Table 3.1 and assuming v_p at 125 MHz is the same as that at 300 MHz)
$\lambda = v_p/f = (2 \times 10^8 \text{ m/s})/(125 \text{ MHz}) = 1.6$ m. The length of the line is then

$$l = 17 \text{ m} = 10.625\lambda = 10.5\lambda + 0.125\lambda$$

Noting that $\tan(\beta l) = \tan[(2\pi/\lambda)(0.125\lambda)] = \tan(\pi/4) = 1$, and the input impedance
of the line seen from the source end is then (equation (3.30)):

$$Z_{in} = 50\frac{(100 - j60) + j50\tan(\pi/4)}{50 + j(100 - j60)\tan(\pi/4)} \simeq 22.6 - j25.1\Omega$$

The equivalent circuit at the source end is as shown in Figure 3.26b. The source-end
voltage V_s is then

$$V_s = \frac{Z_{in}}{R_s + Z_{in}}V_0 \simeq \frac{22.6 - j25.1}{50 + 22.6 - j25.1}100e^{j0°} \simeq 44.0e^{-j28.9°} \text{ V}$$

But we can also evaluate V_s from the expression for the line voltage $V(z)$ as

$$V_s = V^+e^{-j\beta z}(1 + \Gamma_L e^{j2\beta z})|_{z=-17 \text{ m}} \simeq V^+e^{-j3\pi/4}(0.770 - j0.425)$$

where we have used the facts that $e^{j21.25\pi} = e^{j1.25\pi} = e^{-j0.75\pi}$, $e^{-j42.5\pi} = e^{-j0.5\pi}$, and that the load reflection coefficient Γ_L is

$$\Gamma_L = \frac{100 - j60 - 50}{100 - j60 + 50} = \frac{50 - j60}{150 - j60} \simeq 0.483 e^{-j28.4°}$$

Equating the two expressions for V_s, we can determine the unknown voltage V^+ as

$$V_s \simeq V^+ e^{-j3\pi/4}(0.770 - j0.425) \simeq 44.0 e^{-j28.9°} \quad \rightarrow \quad V^+ = 50 e^{j3\pi/4} \text{ V}$$

Thus the expression for the line voltage is

$$V(z) \simeq 50 e^{j3\pi/4} e^{-j5\pi z/4}(1 + 0.483 e^{-j(28.4° - 5\pi z/2)}) \text{ V}$$

(b) The voltage at the load end of the line is

$$V_L = V(z = 0) \simeq 50 e^{j3\pi/4}(1 + 0.483 e^{-j28.4°})$$

$$\simeq 50 e^{j3\pi/4}(1.43 - j0.230) \simeq 72.2 e^{-j126°} \text{ V}$$

(c) Using the value of V_L, the time-average power delivered to the VHF antenna can be calculated as

$$P_L = \frac{1}{2}|I_L|^2 R_L = \frac{1}{2}\left|\frac{V_L}{Z_L}\right|^2 R_L \simeq \frac{(72.2)^2(100)}{2(1.36 \times 10^4)} \simeq 19.2 \text{ W}$$

where $|Z_L|^2 = |100 - j60|^2 = (100)^2 + (60)^2 = 1.36 \times 10^4$. Note that P_L can also be found using the source-end equivalent circuit (Figure 3.26b). Since the line is lossless, all of the time-average power input to the line at the source end must be absorbed by the antenna. In other words,

$$P_L = \frac{1}{2}\left|\frac{V_s}{Z_{in}}\right|^2 R_{in} \simeq \frac{1}{2}\frac{(44.0)^2}{[(22.6)^2 + (25.1)^2]}(22.6) \simeq 19.2 \text{ W}$$

(d) Noting that $R_s = 50\,\Omega$ and $Z_{in} \simeq 22.6 - j25.1\,\Omega$, the current at the source end (again considering the lumped equivalent circuit shown in Figure 3.26b) is

$$I_s = \frac{V_0}{R_s + Z_{in}} \simeq \frac{100 e^{j0°}}{50 + 22.6 - j25.1} \simeq 1.30 e^{j19.1°} \text{ A}$$

Using the value of I_s, the time-average power dissipated in the source resistance R_s is then

$$P_{R_s} = \frac{1}{2}|I_s|^2 R_s \simeq \frac{1}{2}(1.30)^2(50) \simeq 42.3 \text{ W}$$

Therefore, the total power supplied by the VHF transmitter is $P_{total} = P_{R_s} + P_L \simeq 61.5$ W.

Example 3.13: Parallel transmission lines. Three lossless transmission lines are connected in parallel, as shown in Figure 3.27. Assuming sinusoidal steady-state excitation with a source to the left of the main line, find the reflection coefficient on the main line and the percentage of the total net forward power that is absorbed by the two loads Z_{L2} and Z_{L3} for the following cases: (a) $Z_{01} = Z_{02} = Z_{03} = Z_{L2} = Z_{L3} = 100\Omega$, (b) $Z_{01} = 50\Omega$, $Z_{02} = Z_{03} = Z_{L2} = Z_{L3} = 100\Omega$, and (c) $Z_{01} = Z_{02} = Z_{03} = 100\Omega$, $Z_{L2} = Z_{L3} = 50 + j50\Omega$.

Solution:

(a) $Z_{01} = Z_{02} = Z_{03} = Z_{L2} = Z_{L3} = 100\Omega$

Since the two parallel branches are both matched, the input impedance seen at the terminals of each branch is independent of its line length and is simply 100Ω. Thus, the line impedance Z_j seen from the main line is the parallel combination of two 100Ω impedances, or $Z_j = 50\Omega$. The reflection coefficient at the junction (as seen from the main line) is then

$$\Gamma_j = \frac{Z_j - Z_{01}}{Z_j + Z_{01}} = \frac{50 - 100}{50 + 100} = -\frac{1}{3}$$

In other words, the power efficiency, defined as the percentage of total power that is delivered to the loads versus that of the forward wave on the main line, is

$$\frac{P_L}{P^+} = (1 - |\Gamma_j|^2) \times 100 \simeq 88.9\%$$

Since each line presents the same impedance at the junction, each load absorbs half of the total power delivered, or approximately 44.4% of the total power of the incident wave.

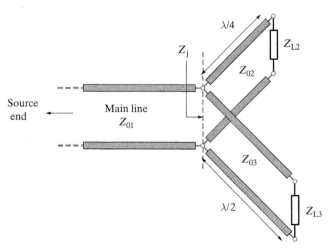

Figure 3.27   Parallel transmission lines. A main line with characteristic impedance Z_{01} drives two other lines of lengths $\lambda/4$ and $\lambda/2$, with characteristic impedances Z_{02} and Z_{03}, respectively.

(b) $Z_{01} = 50\Omega$, $Z_{02} = Z_{03} = Z_{L2} = Z_{L3} = 100\Omega$

The line impedance at the junction seen from the main line is once again $Z_j = 50\Omega$. However, the reflection coefficient is now zero since

$$\Gamma_j = \frac{Z_j - Z_{01}}{Z_j + Z_{01}} = \frac{50 - 50}{50 + 50} = 0$$

In other words, 100% of the power of the forward wave in this case is delivered, with each load absorbing 50%.

(c) $Z_{01} = Z_{02} = Z_{03} = 100\Omega$, $Z_{L2} = Z_{L3} = 50 + j50\Omega$

We now need to evaluate the input impedances of each of the two parallel sections at the junction as seen from the main line. Note that the length of line 2 is $\lambda/4$, so that using (3.30), its input impedance is

$$Z_{in_2} = Z_{02}\frac{Z_{L2} + jZ_{02}\tan(\frac{2\pi}{\lambda}\frac{\lambda}{4})}{Z_{02} + jZ_{L2}\tan(\frac{2\pi}{\lambda}\frac{\lambda}{4})} = \frac{Z_{02}^2}{Z_{L2}} = \frac{(100)^2}{50 + j50} = 100 - j100 \ \Omega$$

whereas line 3 has length $\lambda/2$ and thus presents Z_{L3} at the junction; in other words, $Z_{in_3} = Z_{L3} = 50 + j50\Omega$. The line impedance Z_j at the junction as seen from the main line is then the parallel combination of Z_{in_2} and Z_{in_3}, namely,

$$Z_j = \frac{(100 - j100)(50 + j50)}{100 - j100 + 50 + j50} = 60 + j20 \ \Omega$$

and the reflection coefficient at the end of the main line is

$$\Gamma_j = \frac{Z_j - Z_{01}}{Z_j + Z_{01}} = \frac{60 + j20 - 100}{60 + j20 + 100} \simeq 0.277e^{j\,146°}$$

The percentage of the incident power delivered to the two loads is thus

$$\frac{P_L}{P^+} = (1 - |\Gamma_j|^2) \times 100 \simeq 92.3\%$$

Since the two impedances Z_{in_2} and Z_{in_3} appear at the junction in parallel, they share the same voltage. In other words, we have

$$P_{L_2} = \frac{1}{2}\left|\frac{V_j}{Z_{in_2}}\right|^2 \mathcal{R}e\{Z_{in_2}\} \quad \text{and} \quad P_{L_3} = \frac{1}{2}\left|\frac{V_j}{Z_{in_3}}\right|^2 \mathcal{R}e\{Z_{in_3}\}$$

We can thus calculate the ratio of the powers delivered to the two loads as

$$\frac{P_{L_2}}{P_{L_3}} = \frac{|Z_{in_3}|^2 \mathcal{R}e\{Z_{in_2}\}}{|Z_{in_2}|^2 \mathcal{R}e\{Z_{in_3}\}} = \frac{[(50)^2 + (50)^2](100)}{[(100)^2 + (100)^2](50)} = 0.5$$

Therefore we have

$$P_{L_2} \simeq \frac{1}{3} \times 92.3\% \simeq 30.8\% \quad \text{and} \quad P_{L_3} \simeq \frac{2}{3} \times 92.3\% \simeq 61.5\%$$

Example 3.14: Cascaded transmission lines. An antenna of measured feed-point impedance of $72 + j36\Omega$ at 100 MHz is to be driven by a transmitter through two cascaded coaxial lines with the following characteristics:

$$Z_{01} = 120\Omega \qquad l_1 = 3.75 \text{ m} \quad \text{air-filled} \qquad\qquad v_{p1} = c \simeq 30 \text{ cm-(ns)}^{-1}$$

$$Z_{02} = 60\Omega \qquad l_2 = 1.75 \text{ m} \quad \text{polyethylene-filled} \qquad v_{p2} \simeq 20 \text{ cm-(ns)}^{-1}$$

where we have used Table 3.1 for the phase velocity v_{p2} for the polyethylene-filled coaxial line assuming it to be approximately the same at 100 MHz. (a) Assuming both lines to be lossless and assuming a source voltage of $V_0 = 100e^{j0}$ V and resistance of $R_s = 50\Omega$ for the transmitter, find the time-average power delivered to the load. (b) Repeat part (a) with $l_1 = 4.5$ m. The setup is shown in Figure 3.28a.

Solution:

(a) At 100 MHz, the wavelengths for the two coaxial lines are $\lambda_1 = v_{p1}/f \simeq 3 \times 10^8/10^8 = 3$ m and $\lambda_2 = v_{p2}/f = 2 \times 10^8/10^8 = 2$ m, respectively. The lengths of the lines are then

$$l_1 = 3.75 \text{ m} \simeq 1.25\lambda_1 = \lambda_1 + 0.25\lambda_1$$

$$l_2 = 1.75 \text{ m} = 0.875\lambda_2 = 0.5\lambda_2 + 0.375\lambda_2$$

Note that the corresponding phase constants are $\beta_1 = 2\pi/\lambda_1$ and $\beta_2 = 2\pi/\lambda_2$. The impedance Z_x seen looking toward the load at the interface between the two coaxial

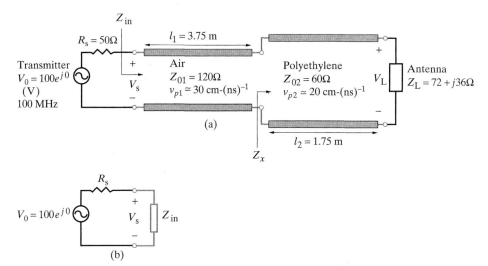

Figure 3.28 Cascaded transmission lines. (a) Circuit diagram showing an air-filled line of impedance 120Ω cascaded with a polyethylene-filled line of impedance 60Ω. (b) Equivalent circuit seen by the source.

lines is, using (3.30),

$$Z_x = Z_{02} \frac{Z_L + jZ_{02}\tan(\beta_2 l_2)}{Z_{02} + jZ_L\tan(\beta_2 l_2)} = 60\frac{(72 + j36) + j60(-1)}{60 + j(72 + j36)(-1)} = 36 + j12\,\Omega$$

The input impedance is then

$$Z_{\text{in}} = Z_{01}\frac{Z_x + jZ_{01}\tan(\beta_0 l_1)}{Z_{01} + jZ_x\tan(\beta_0 l_1)} = \frac{Z_{01}^2}{Z_x} \simeq \frac{(120)^2}{36 + j12} = 360 - j120\,\Omega$$

With reference to the equivalent circuit in Figure 3.28b, we have

$$V_s = \frac{Z_{\text{in}}}{R_s + Z_{\text{in}}}V_0 = \frac{360 - j120}{410 - j120}(100) \simeq 88.8e^{-j2.12°}\ \text{V}$$

Thus the power delivered to the antenna is

$$P_L = P_{\text{in}} = \frac{1}{2}\left|\frac{V_s}{Z_{\text{in}}}\right|^2 \mathcal{R}e\{Z_{\text{in}}\} = \frac{1}{2}\frac{(88.8)^2}{(1.44 \times 10^5)}(360) \simeq 9.86\ \text{W}$$

(b) With $l_1 = 4.5$ m $= 1.5\lambda_1$, we have

$$Z_{\text{in}} = 36 + j12 \quad \rightarrow \quad V_s = \frac{36 + j12}{86 + j12}(100) \simeq 43.7e^{j10.5°}\ \text{V}$$

so that

$$P_L \simeq \frac{1}{2}\frac{(43.7)^2}{(1440)}(36) \simeq 23.9\ \text{W}$$

which is a significant improvement in power delivered, achieved simply by making the first line segment longer. This result indicates that the amount of power delivered to a load sensitively depends on the electrical lengths of the transmission lines.

3.5 IMPEDANCE MATCHING

We have already encountered the concept of impedance matching in previous sections, in connection with standing waves on transmission lines. It was shown that if the characteristic impedance Z_0 of the line is equal to the load impedance Z_L, the reflection coefficient $\Gamma_L = 0$, and the standing-wave ratio is unity. When this situation exists, the characteristic impedance of the line and the load impedance are said to be *matched,* that is, they are equal. In most transmission line applications, it is desirable to match the load impedance to the characteristic impedance of the line in order to reduce reflections and standing waves that jeopardize the power-handling capabilities of the line and also distort the information transmitted. Impedance matching is also desirable in order to drive a given load most efficiently (i.e., to deliver maximum power to the load), although maximum efficiency also requires matching the generator to the line at the source end. In the presence of sensitive components (low-noise amplifiers, etc.), impedance matching improves the signal-to-noise ratio of the system and in other cases generally reduces amplitude and phase errors. In this section, we examine different methods of achieving impedance matching.

3.5.1 Matching Using Lumped Reactive Elements

The simplest way to match a given transmission line to a load is to connect a lumped reactive element in parallel (series) at the point along the line where the real part of the line admittance (impedance) is equal to the line characteristic admittance (impedance).[19] This method is useful only at relatively low frequencies for which lumped elements can be used. The method is depicted in Figure 3.29, which shows a shunt (parallel) lumped reactive element connected to the line at a distance l from the load.

Since the matching element is connected in parallel, it is more convenient to work with line admittance rather than line impedance. The normalized admittance $\bar{Y}(z)$ seen on the line looking toward the load from any position z is given by (3.33):

$$\bar{Y}(z) = \frac{Y(z)}{Y_0} = \frac{1 - \Gamma_L e^{j2\beta z}}{1 + \Gamma_L e^{j2\beta z}}$$

In Figure 3.29, matching requires that $\bar{Y}_2 = \bar{Y}_1 + \bar{Y}_s = 1$. Thus we first need to choose the position l along the line such that $\bar{Y}(z = -l) = \bar{Y}_1 = 1 - j\bar{B}$, that is, $\Re e\{\bar{Y}_1\} = 1$. Then we choose the lumped shunt element to be purely reactive with an appropriate value such that $\bar{Y}_s = j\bar{B}$, which results in $\bar{Y}_2 = \bar{Y}_1 + \bar{Y}_s = 1$. Substituting $z = -l$, we have

$$\bar{Y}_1 = \bar{Y}(z = -l) = \frac{1 - \Gamma_L e^{-j2\beta l}}{1 + \Gamma_L e^{-j2\beta l}} = 1 - j\bar{B}$$

from which we can solve for the position l of the lumped reactive element, its type (i.e., capacitor or inductor), and its normalized susceptance $-\bar{B}$. Since the load reflection

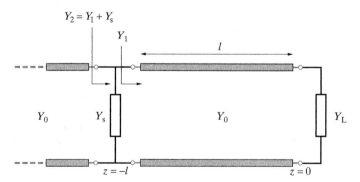

Figure 3.29 **Matching by a lumped shunt element.** The shunt element Y_s is connected at a distance l from the load such that the line admittance $Y_2(z = -l) = Y_1(z = -l) + Y_s = Y_0$.

---

[19]This possibility was noted earlier in Section 3.3.2 in connection with Figure 3.19, where $\Re e\{\bar{Z}(z_1)\} = 1$ at $z_1 \simeq -0.0985\lambda$. Note that, in general, the real part of the line admittance is equal to unity at some other point $z_2 \neq z_1$.

coefficient Γ_L is, in the general case, a complex number given by

$$\Gamma_L = \rho e^{j\psi} = \frac{Z_L - Z_0}{Z_L + Z_0}$$

we have

$$\bar{Y}_1 = \frac{1 - \rho e^{j\psi} e^{-j2\beta l}}{1 + \rho e^{j\psi} e^{-j2\beta l}} = \frac{1 - \rho e^{j\theta}}{1 + \rho e^{j\theta}} \tag{3.40}$$

where $\theta = \psi - 2\beta l$. By multiplying the numerator and the denominator by the complex conjugate of the denominator,[20] we obtain

$$\bar{Y}_1 = \underbrace{\frac{1 - \rho^2}{1 + 2\rho \cos \theta + \rho^2}}_{\mathcal{R}e\{\bar{Y}_1\}=1} - j \underbrace{\frac{2\rho \sin \theta}{1 + 2\rho \cos \theta + \rho^2}}_{-\mathcal{I}m\{\bar{Y}_1\}=\bar{B}} \tag{3.41}$$

Since $\mathcal{R}e\{\bar{Y}_1\} = 1$, we can write

$$\frac{1 - \rho^2}{1 + 2\rho \cos \theta + \rho^2} = 1$$

which yields

$$\theta = \psi - 2\beta l = \cos^{-1}(-\rho) \tag{3.42}$$

In other words, the distance l from the load at which $\bar{Y}(z = -l) = 1 - j\bar{B}$ is given by

$$\boxed{l = \frac{\psi - \theta}{2\beta} = \frac{\psi - \cos^{-1}(-\rho)}{2\beta} = \frac{\lambda}{4\pi}[\psi - \cos^{-1}(-\rho)]} \tag{3.43}$$

Note that, in general, $\theta = \cos^{-1}(-\rho)$ (with $\rho > 0$) has two solutions, one in the range $\pi/2 \leq \theta_1 \leq \pi$ and the other in the range $-\pi \leq \theta_2 \leq -\pi/2$. Also, if (3.43) results in negative values for l, then the corresponding physically meaningful solution can be found by simply adding $\lambda/2$.[21] To find $\bar{B}$, we substitute $\cos \theta = -\rho$ and $\sin \theta = \pm\sqrt{1 - \rho^2}$ (where the plus sign corresponds to $\pi/2 \leq \theta_1 \leq \pi$ and the minus sign corresponds to $-\pi \leq \theta_2 \leq -\pi/2$) in the imaginary part of $\bar{Y}_1$ given by (3.41), resulting in

$$\bar{B} = -\mathcal{I}m\{\bar{Y}_1\}|_{\cos \theta = -\rho} = \pm\frac{2\rho\sqrt{1 - \rho^2}}{1 - 2\rho^2 + \rho^2} = \pm\frac{2\rho}{\sqrt{1 - \rho^2}} \tag{3.44}$$

---

[20]$[1 + \rho e^{j\theta}]^* = 1^* + (\rho e^{j\theta})^* = 1 + \rho e^{-j\theta}$

[21]For negative l, we have $-2\pi < 2\beta l < 0$. By adding $\lambda/2$, we have $2\beta(l + \lambda/2) = 2\beta l + 2(2\pi/\lambda)(\lambda/2) = 2\beta l + 2\pi > 0$, which lies between 0 and 2π.

where the plus and minus signs correspond to a shunt capacitor ($\bar{B}_1 > 0$) and a shunt inductor ($\bar{B}_2 < 0$), respectively.[22] This susceptance also determines the value of the lumped reactive element $\bar{Y}_s$, which must be connected in parallel to the line in order to cancel out the reactive part of $\bar{Y}_1$. In particular, we should have $\bar{Y}_s = +j\bar{B}$ so that the total admittance $\bar{Y}_2$ seen from the left side of Y_s in Figure 3.29 is

$$\bar{Y}_2 = \bar{Y}_1 + \bar{Y}_s = 1 - j\bar{B} + j\bar{B} = 1$$

When matching with series lumped reactive elements, similar equations can be derived for the distance l away from the load at which $\bar{Z}_1 = \bar{Z}(z = -l) = 1 - j\bar{X}$,

$$\boxed{l = \frac{\psi - \cos^{-1}\rho}{2\beta} = \frac{\lambda}{4\pi}(\psi - \cos^{-1}\rho)} \tag{3.45}$$

and the normalized reactance of the series lumped element that would provide matching is given by

$$\bar{X} = \pm\frac{2\rho}{\sqrt{1 - \rho^2}}$$

Example 3.15: Matching with a single reactive element. An antenna having a feed-point impedance of 110Ω is to be matched to a 50Ω coaxial line with $v_p = 2 \times 10^8$ m-s^{-1} using a single shunt lumped reactive element, as shown in Figure 3.30. Find the position (nearest to the load) and the appropriate value of the reactive element for operation at 30 MHz using (a) a capacitor and (b) an inductor.

Solution: The load reflection coefficient is

$$\Gamma_L = \frac{Z_L - Z_0}{Z_L + Z_0} = \frac{110 - 50}{110 + 50} = 0.375$$

Using (3.44), the reactive admittance (or susceptance) at the position of the shunt element is

$$\bar{B} = \pm\frac{2 \times 0.375}{\sqrt{1 - (0.375)^2}} \simeq \pm 0.809$$

(a) For $\bar{B}_1 \simeq +0.809$, the shunt element must be a capacitor. The nearest position of the capacitor with respect to the load can be found as

$$l_1 = -\frac{\theta_1}{2\beta} \simeq -\frac{\lambda}{4\pi}\underbrace{\cos^{-1}(-0.375)}_{\pi/2 \leq \theta_1 \leq \pi} \simeq -\frac{\lambda}{4\pi}(1.955) \simeq -0.156\lambda$$

---

[22]Note that $\theta = \cos^{-1}(-\rho)$ is an angle that is either in the second quadrant, $\pi/2 \leq \theta_1 \leq \pi$ (when $\sin\theta_1 > 0$, requiring a capacitive element based on the polarity of the imaginary part of $\bar{Y}_1$ as given in (3.41)) or the third quadrant, $-\pi \leq \theta_2 \leq -\pi/2$ (when $\sin\theta_2 < 0$, requiring an inductive element).

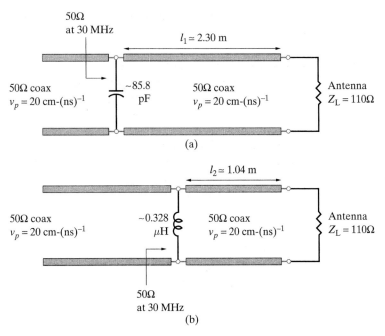

Figure 3.30   Matching with a single reactive element. The two solutions determined in Example 3.15. (a) Using a shunt capacitor. (b) Using a shunt inductor.

Since $l_1 < 0$, we add $\lambda/2$ to get $l_1 \simeq -0.156\lambda + 0.5\lambda = 0.344\lambda$. Using $\lambda = v_p/f = 2 \times 10^8/(30 \times 10^6) \simeq 6.67$ m, the actual position of the shunt capacitor is $l_1 \simeq 0.344 \times 6.67 \simeq 2.30$ m. To determine the capacitance C_s, we use

$$(j\omega C_s)(Z_0) = j\bar{B}_1 \quad \rightarrow \quad [j(2\pi \times 30 \times 10^6 C_s)(50)] \simeq +j0.809$$

$$\rightarrow \quad C_s \simeq 85.8 \text{ pF}$$

(b) For $\bar{B}_2 \simeq -0.809$, the shunt element must be an inductor. Similarly, the nearest position of the inductor is

$$l_2 = -\frac{\theta_2}{2\beta} \simeq -\frac{\lambda}{4\pi} \underbrace{\cos^{-1}(-0.375)}_{-\pi \leq \theta_2 \leq -\pi/2} \simeq -\frac{\lambda}{4\pi}(-1.955) \simeq 0.156\lambda$$

Using $\lambda \simeq 6.67$ m, the actual position of the shunt inductor is $l_2 \simeq 0.156 \times 6.67 \simeq 1.04$ m. To determine the inductance L_s, we use

$$[-j/(\omega L_s)](Z_0) = j\bar{B}_2 \quad \rightarrow \quad [-j/(2\pi \times 30 \times 10^6 L_s)](50) \simeq -j0.809$$

$$\rightarrow \quad L_s \simeq 0.328 \; \mu\text{H}$$

3.5.2 Matching Using Series or Shunt Stubs

In Section 3.2, we saw that short- or open-circuited transmission lines can be used as reactive circuit elements. At microwave frequencies, it is often impractical or inconvenient to use lumped elements for impedance matching. Instead, we use a common matching technique that uses single open- or short-circuited stubs (i.e., transmission line segments) connected either in series or in parallel, as illustrated in Figure 3.31. In practice, the short-circuited stub is more commonly used for coaxial and wave-guide applications because a short-circuited line is less sensitive to external influences (such as capacitive coupling and pick-up) and radiates less than an open-circuited line segment. However, for microstrips and striplines, open-circuited stubs are more common in practice because they are easier to fabricate. For similar practical reasons, the shunt (parallel) stub is more

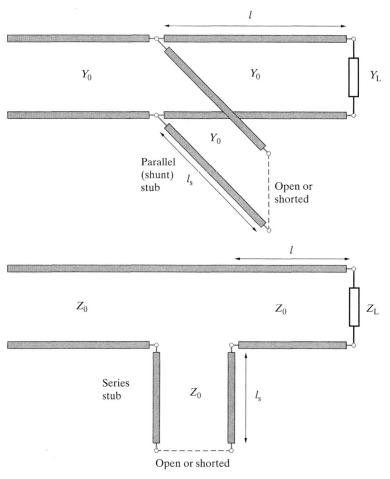

Figure 3.31   Matching by shunt or series open- or short-circuited stubs.

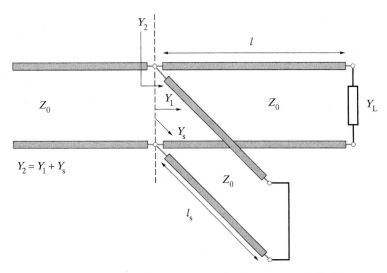

Figure 3.32 Matching with a single parallel (shunt) short-circuited stub.

convenient than the series stub; the discontinuity created by breaking the line may disturb the voltage and current in the case of the series stub.

The principle of matching with stubs is identical to that discussed in Section 3.5.1 for matching using shunt lumped reactive elements. The only difference here is that the matching admittance Y_s is introduced by using open- or short-circuited line segments (or stubs) of appropriate length l_s, as shown in Figure 3.32. In the following, we exclusively consider the case of matching with a short-circuited shunt stub, as illustrated in Figure 3.32. The corresponding analysis for open-circuited stubs is similar in all respects and is left as an exercise for the reader.

With the required location l and the normalized admittance $\bar{B}$ of the stub as determined from (3.43) and (3.44), we need only to find the length of the stub l_s necessary to present a normalized admittance of $\bar{Y}_s = +j\bar{B}$ at the junction. For this purpose, we can use expression (3.17) from Section 3.2 for the normalized input impedance of a short-circuited line of length l_s and set the corresponding normalized admittance equal to $+j\bar{B}$. Recalling that for a short-circuited line $\bar{Z}_{in} = j\tan(\beta l_s)$, we have

$$\bar{Y}_s = \frac{1}{j\tan(\beta l_s)} = +j\bar{B}$$

or

$$\tan(\beta l_s) = -\frac{1}{\bar{B}} \tag{3.46}$$

The value of $\bar{B}$ determined from (3.44) can be used in (3.46) to find the length l_s of the short-circuited stub. Note that in (3.46), we have assumed the characteristic impedance of the short-circuited stub to be equal to that of the main line.

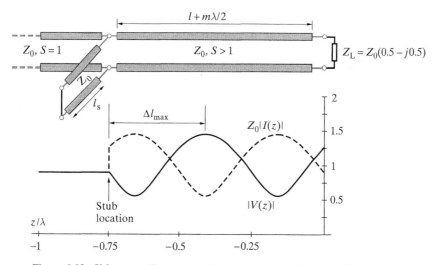

Figure 3.33 **Voltage standing-wave pattern on a transmission line with single-stub matching.** The standing-wave ratio S is unity to the left of the stub. The stub location is at a distance of $\Delta l_{\max}$ from the nearest voltage maximum toward the load end. The particular case shown is for $\bar{Z}_L = 0.5 - j0.5$. Note that, as usual, we have assumed $V^+ = 1$.

In practice, single-stub matching can be achieved even if the load impedance Z_L is not explicitly known, by relying on measurements of S to determine ρ and measurements of the location of the voltage minimum or maximum to determine ψ. To see this, consider that the stub location l can be measured relative to the position $z_{\max}$ of the nearest voltage maximum toward the load end so that $l - \Delta l_{\max} = |z_{\max}| < \lambda/2$, as shown in Figure 3.33. Using (3.42), we can then write

$$\theta = \psi - 2\beta l = \psi + 2\beta z_{\max} - 2\beta \Delta l_{\max}$$

$$= -m2\pi - 2\beta \Delta l_{\max} \qquad m = 0, 1, 2, \ldots$$

noting that $m = 0$ does not apply if $-\pi \le \psi < 0$. Using the preceding expression for θ, we have

$$\Delta l_{\max} = -\frac{1}{2\beta}[\theta + m2\pi] = -\frac{1}{2\beta}[\cos^{-1}(-\rho) + m2\pi], \quad m = 0, 1, 2, \ldots \qquad (3.47)$$

Thus, ρ can be directly determined from the measured standing-wave ratio, and (3.47) determines the stub location with respect to the measured location of the voltage maximum, as shown in Figure 3.33.

Figure 3.33 also illustrates the fact that, although the proper choice of the stub location l and its length l_s achieves matching so that the standing-wave ratio on the main line on the source side of the stub is unity (and so $V(z) = Z_0 I(z)$), a standing wave does exist on the segment of the main line between the stub and the load. Note also that some current is shunted to the matching stub, resulting in the discontinuity in the plot of $Z_0|I(z)|$ at the stub position along the main transmission line.

Example 3.16: Single-stub matching. Design a single-stub system to match a load consisting of a resistance $R_L = 200\Omega$ in parallel with an inductance $L_L = 200/\pi$ nH to a transmission line with characteristic impedance $Z_0 = 100\Omega$ and operating at 500 MHz. Connect the stub in parallel with the line.

Solution: At 500 MHz, the load admittance is given by

$$Y_L = \frac{1}{R_L} - j\frac{1}{\omega L_L} = \frac{1}{200} - j\frac{1}{[2\pi(500 \times 10^6)(200/\pi)(10^{-9})]} = 0.005 - j0.005 \text{ S}$$

The reflection coefficient at the load is

$$\Gamma_L = \frac{Y_0 - Y_L}{Y_0 + Y_L} = \frac{0.01 - (0.005 - j0.005)}{0.01 + (0.005 - j0.005)} = \frac{1+j}{3-j} \simeq 0.447 e^{j63.4°}$$

so that $\rho \simeq 0.447$ and $\psi \simeq 63.4°$. Using (3.43), we have

$$l \simeq \frac{1.11 - \cos^{-1}(-0.447)}{2(2\pi/\lambda)} \simeq \left(\frac{1.11 \mp 2.034}{4\pi}\right)\lambda \quad \rightarrow \quad \begin{array}{l} l_1 \simeq -0.073\lambda \\ l_2 = 0.25\lambda \end{array}$$

where the first solution with a negative value of l can be realized by simply adding 0.5λ so that the stub position is between the load and the source. Thus, the stub position for the first solution is $l_1 \simeq -0.073\lambda + 0.5\lambda = 0.426\lambda$. Note that we have used $\theta = \cos^{-1}(-0.447) \simeq \pm117° \simeq \pm2.034$ radians. Using (3.44), the normalized susceptance of the input admittance of the short-circuited shunt stub needed is

$$\bar{B} \simeq \pm\frac{2(0.447)}{\sqrt{1 - (0.447)^2}} \simeq \pm1$$

where the plus sign corresponds to the stub position l_1 and the minus sign corresponds to l_2. From (3.46), the length l_{s1} of the short-circuited stub at a distance l_1 from the load is

$$l_{s1} = \frac{\lambda}{2\pi} \tan^{-1}\left(-\frac{1}{\bar{B}}\right) \simeq \frac{\lambda}{2\pi} \tan^{-1}(-1) = -0.125\lambda \quad \rightarrow \quad 0.375\lambda$$

and similarly, the stub length l_{s2} needed at position l_2 is

$$l_{s2} = \frac{\lambda}{2\pi} \tan^{-1}(1) = 0.125\lambda$$

Both of the alternative solutions are shown in Figure 3.34. Since Solution 2 gives a stub position closer to the load and a shorter stub length, it would usually be preferred over Solution 1. In general, standing waves jeopardize power-handling capabilities of a line and also lead to signal distortion. Thus, it is desirable to minimize the lengths of line over which the standing-wave ratio is large. In the case shown in Figure 3.34, more of the line operates under matched ($S = 1$) conditions for Solution 2.

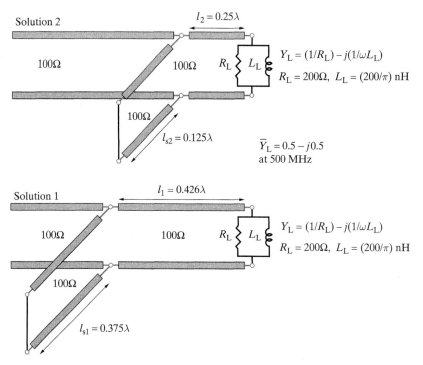

Figure 3.34   Two alternative single-stub matching solutions. Solution 2 would in general be preferred since it has shorter segments of line (and a shorter stub) over which the standing-wave ratio differs from unity.

The frequency dependence of the various designs is important in practice. A comparison of the frequency responses of the two alternative solutions for the previous example is given in Figure 3.35. Note that the load admittance as a function of frequency $f = \omega/(2\pi)$ is given by

$$Y_L(f) = \frac{1}{200} - j\frac{10^9}{400f}$$

Assuming that the phase velocity along the line is equal to the speed of light c, we have $\beta l = 2\pi f l/c$. The line admittance seen just to the right of the short-circuited stub is given as a function of frequency:

$$Y_1(f) = Y_0 \frac{Y_L + jY_0 \tan(2\pi f l/c)}{Y_0 + jY_L \tan(2\pi f l/c)}$$

If we also assume $\beta l_s = 2\pi f l_s/c$, the total admittance seen from the source side of the short-circuited stub (see Figure 3.32) is

$$Y_2(f) = Y_s + Y_1 = \frac{-jY_0}{\tan(2\pi f l_s/c)} + Y_0 \frac{Y_L + jY_0 \tan(2\pi f l/c)}{Y_0 + jY_L \tan(2\pi f l/c)}$$

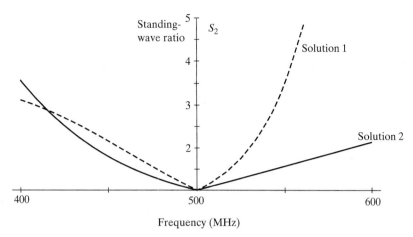

Figure 3.35   Frequency sensitivity of single-stub matching. The standing-wave ratio S_2 versus frequency for the two alternative solutions given in Figure 3.34.

The load reflection coefficient Γ_2 and the standing-wave ratio S_2 on the main line to the left of the short-circuited stub are then given as

$$\Gamma_2(f) = \frac{Y_0 - Y_2(f)}{Y_0 + Y_2(f)}; \quad S_2(f) = \frac{1 + |\Gamma_2(f)|}{1 - |\Gamma_2(f)|}$$

The quantity $S_2(f)$ is plotted in Figure 3.35 as a function of frequency between 400 and 600 MHz. Note that the bandwidths[23] of the two designs are dramatically different. Which solution to choose depends on the particular application in hand, although in most cases minimizing reflections (e.g., $S_2 < 2$) over a wider frequency range is desirable. Note that, for the case shown, Solution 2 provides matching over a substantially broader range of frequencies than does Solution 1.

3.5.3 Quarter-Wave Transformer Matching

A powerful method for matching a given load impedance to a transmission line that is used to drive it is the so-called quarter-wave transformer matching. This method takes advantage of the impedance inverting property of a transmission line of length $l = \lambda/4$, namely, the fact that the input impedance of a line of length $l = \lambda/4$ is given by

$$Z_{in}\big|_{l=\lambda/4} = Z_0 \left.\frac{Z_L + jZ_0 \tan(\beta l)}{Z_0 + jZ_L \tan(\beta l)}\right|_{l=\lambda/4} = \frac{Z_0^2}{Z_L}$$

---

[23]Defined as the frequency range over which the standing-wave ratio S_2 is lower than a given amount, for example, $S_2 < 2$. The maximum value of S_2 used depends on the application in hand.

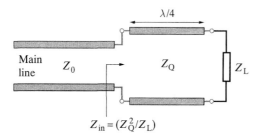

Figure 3.36 Quarter-wave transformer. A load Z_L to be driven by a transmission line of characteristic impedance Z_0 connected to the line via a quarter-wavelength-long line of characteristic impedance Z_Q.

or in terms of normalized impedances, we have

$$\bar{Z}_{in}\Big|_{l=\lambda/4} = \frac{Z_{in}}{Z_0}\Big|_{l=\lambda/4} = \frac{Z_0}{Z_L} = \frac{1}{\bar{Z}_L} \qquad (3.48)$$

hence the term "impedance inverter." Thus, a quarter-wave section transforms impedance in such a way that a kind of inverse of the terminating impedance appears at its input.

Consider a quarter-wavelength transmission line segment of characteristic impedance Z_Q, as shown in Figure 3.36. In the general case of a complex load impedance $Z_L = R_L + jX_L$, we have

$$Z_{in} = \frac{Z_Q^2}{R_L + jX_L} \qquad \longrightarrow \qquad Y_{in} = \frac{R_L}{Z_Q^2} + j\frac{X_L}{Z_Q^2}$$

so that a load consisting of a series resistance (R_L) and an inductive reactance ($X_L > 0$) appears at the input of the quarter-wave section as an admittance consisting of a conductance R_L/Z_Q^2 (or a resistance Z_Q^2/R_L) *in parallel* with a capacitive susceptance $X_L/Z_Q^2 > 0$. Similarly, if the load were capacitive ($X_L < 0$), it would appear as a conductance R_L/Z_Q^2 in parallel with an inductive susceptance $X_L/Z_Q^2 < 0$.

Purely resistive loads. The practical utility of the impedance-inverting property of a quarter-wavelength transmission line becomes apparent when we consider a purely resistive load. Any arbitrary purely resistive load impedance $Z_L = R_L$ is transformed into a purely resistive input impedance of Z_Q^2/R_L. Thus, by appropriately choosing the value of the characteristic impedance Z_Q of the quarter-wavelength line, its input impedance can be made equal to the characteristic impedance Z_0 (a real value for a lossless line) of the main line that is to be used to drive the load. This property of the quarter-wave line can be used to match two transmission lines of different characteristic impedances or to match a load impedance to the characteristic impedance of a transmission line. Note that the matching section must have a characteristic impedance of

$$Z_Q = \sqrt{R_1 R_2} \qquad (3.49)$$

where R_1 and R_2 are the two purely resistive impedances to be matched. Note that in the case shown in Figure 3.36, $R_1 = Z_0$ and $R_2 = R_L$. Alternatively, R_2 could be the characteristic impedance of another transmission line that may need to be matched to the main line (Z_0) using the quarter-wave section.

Example 3.17: Quarter-wave transformer for a monopole antenna. Design a quarter-wavelength section to match a thin monopole antenna of length 0.24λ[24] having a purely resistive feed-point impedance of $R_L = 30\Omega$ to a transmission line having a characteristic impedance of $Z_0 = 100\Omega$.

> **Solution:** With reference to Figure 3.37 and according to (3.49), the $\lambda/4$ section must match two impedances—$R_2 = R_L$ and $R_1 = Z_0 = 100\Omega$—and thus must have a characteristic impedance Z_Q of
>
> $$Z_Q = \sqrt{R_1 R_2} = \sqrt{Z_0 R_L} = \sqrt{(100)(30)} \simeq 54.8\Omega$$
>
> The standing-wave ratio is unity beyond the quarter-wave section. However, note that $S \simeq 1.83$ within the $\lambda/4$ section.

Figure 3.37   Quarter-wave transformer. The load is a monopole antenna of length 0.24λ, which has a purely resistive impedance of 30Ω.

Complex load impedances.

In using quarter-wave transformers to match a complex load impedance to a lossless transmission line (i.e., where Z_0 is real), it is necessary to insert the quarter-wave segment at the point along the line where the line impedance $\bar{Z}(z)$ is purely resistive. As discussed in previous sections, this point can be the position of either the voltage maximum or minimum. In most cases, it is desirable to choose the point closest to the load in order to minimize the length of the transmission line segment on which $S \neq 1$ because the presence of standing waves jeopardizes power-handling capabilities of the line, tends to reduce signal-to-noise ratio, and may lead to distortion of the signal transmitted.

---

[24]The reactive part of the impedance of such monopole antennas with length just shorter than a quarter-wavelength is nearly zero. Monopole or dipole antennas with purely resistive input impedances are referred to as resonant antennas and are used for many applications. (See Section 14.06 of E. Jordan and K. Balmain, *Electromagnetic Waves and Radiating Systems,* Prentice-Hall, New York, 1968.)

Example 3.18: Thin-wire half-wave dipole antenna. A thin-wire half-wave dipole antenna[25] has an input impedance of $Z_L = 73 + j42.50\Omega$. Design a quarter-wave transformer to match this antenna to a transmission line with characteristic impedance $Z_0 = 100\Omega$.

Solution: We start by evaluating the reflection coefficient at the load

$$\Gamma_L = \rho e^{j\psi} = \frac{Z_L - Z_0}{Z_L + Z_0} = \frac{73 + j42.5 - 100}{73 + j42.5 + 100} \simeq 0.283 e^{j\,109°}$$

so that $\psi \simeq 109° \simeq 1.896$ radians. Note that the standing-wave ratio is

$$S = \frac{1 + \rho}{1 - \rho} \simeq \frac{1 + 0.283}{1 - 0.283} \simeq 1.79$$

From previous sections, we know that, for an inductive load, the first voltage maximum is closer to the load than the first voltage minimum (see Figure 3.17). The first voltage maximum is at

$$\psi + 2\beta z_{max} = 0 \quad \longrightarrow \quad z_{max} = -\frac{\psi}{2\beta} \simeq -\frac{1.896\lambda}{4\pi} \simeq -0.151\lambda$$

Thus the quarter-wave section should be inserted at $z \simeq -0.151\lambda$, as shown in Figure 3.38.
Noting that the normalized load impedance is $\bar{Z}_L = 0.73 + j0.425$ and that we have $\tan[\beta(0.151\lambda)] \simeq 1.392$, the normalized line impedance $\bar{Z}(z)$ seen toward the load at $z \simeq -0.151\lambda$ is

$$\bar{Z}(z)|_{z\simeq-0.151\lambda} = \frac{\bar{Z}_L - j\tan(\beta z)}{1 - j\bar{Z}_L\tan(\beta z)}\bigg|_{z\simeq-0.151\lambda} \simeq \frac{0.73 + j0.425 + j1.392}{1 + j(0.73 + j0.425)(1.392)} \simeq 1.79$$

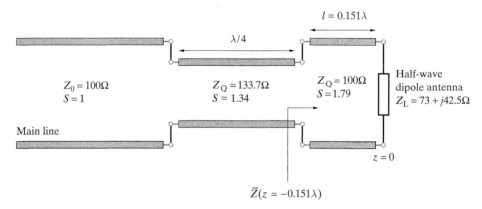

Figure 3.38   Quarter-wave matching of the half-wave dipole antenna in Example 3.18.

[25] See, for example, Section 14.06 of E. C. Jordan and K. Balmain, *Electromagnetic Waves and Radiating Systems,* Prentice-Hall, New York, 1968.

Note that $\bar{Z}(z \simeq -0.151\lambda) \simeq 1.79 = S$, as expected on the basis of the discussion in Section 3.3.2 (i.e., $\bar{R}_{\max} = S \simeq 1.79$). The characteristic impedance Z_Q of the quarter-wave section should thus be

$$Z_Q = \sqrt{Z(z \simeq -0.151\lambda)Z_0} \simeq \sqrt{(100)(1.79)(100)} \simeq 133.7\Omega$$

Note that in general, as in this specific example, we have $\bar{Z}(z = z_{\max}) = S$, and thus $Z_Q = Z_0\sqrt{S}$. Note also that the standing-wave ratio in the quarter-wave section is $S \simeq 1.34$, as can be calculated by using $Z(z \simeq -0.151\lambda)$ and Z_Q.

Frequency sensitivity of quarter-wave matching.

The frequency sensitivity of a quarter-wave transformer is a serious limitation since the design is perfect (i.e., provides $S = 1$) only at the frequency for which the length of the transformer segment is exactly $\lambda/4$. The bandwidth of the transformer can be assessed by plotting the standing-wave ratio S versus frequency, as was done in Section 3.5.2 for the single-stub tuning example and as is shown in the following example.

Example 3.19: Multiple-stage quarter-wave transformers. A resistive load of $R_L = 75\Omega$ is to be matched to a transmission line with characteristic impedance $Z_0 = 300\Omega$. The frequency of operation is $f_0 = 300$ MHz (which corresponds to $\lambda_0 = v_p/f_0 \simeq 3 \times 10^8/(300 \times 10^6) = 1$ m, assuming an air-filled coaxial line). Design multiple-cascaded quarter-wave matching transformers and at 300 MHz compare their frequency responses between 200 and 400 MHz.

Solution: Three different designs are shown in Figure 3.39. Note that the choice of the impedance Z_{Q1} for the single-transformer case is straightforward. For the double-transformer case, the condition for exact quarter-wave matching is $Z_{Q2} = Z_{Q1}\sqrt{Z_0/R_L}$, allowing for different choices of Z_{Q2} and Z_{Q1} as long as the condition is satisfied. The design shown in Figure 3.39b is one that provides a standing-wave ratio in the first and second quarter-wave sections of $S \simeq 1.40$ and $S \simeq 1.43$, respectively. For the triple transformer case the condition for exact quarter-wave matching can be shown to be $Z_{Q1}Z_{Q3} = Z_{Q2}\sqrt{Z_0R_L} = 150Z_{Q2}$. Once again, many different combinations of Z_{Q1}, Z_{Q2}, Z_{Q3} satisfy this condition, and other performance criteria (such as minimizing S in the quarter-wave sections) must be used to make particular design choices. One design approach[26] is to require the characteristic impedance of the second quarter-wave segment to be the geometric mean of the two impedances to be matched, namely $Z_{Q2} = \sqrt{(300)(75)} = 150\Omega$. The choices of Z_{Q1} and Z_{Q3} shown in Figure 3.39 is one that provides a relatively low value of $S <\sim 1.26$ in all the transmission line segments.

Figure 3.40 compares the three different designs in terms of their frequency response. As in Figure 3.35, we base this comparison on the behavior of the total standing-wave ratio $S(f)$ as a function of frequency. First we note that since the wavelength at 300 MHz is 1 m (assuming a transmission line with air as the material surrounding the conductors), all of the quarter-wave segments have physical lengths of 0.25 m each. To evaluate $S(f)$, we can start

---

[26]There are various established methods for the design of multisection quarter-wavelength transformers. The approach described here is an ad-hoc one.

Figure 3.39   Multiple-stage quarter-wave matching. Matching with triple, double, and single $\lambda/4$ sections are illustrated respectively in panels (a), (b), and (c).

at the load end and transform impedances as we move toward the source end in accordance with

$$Z_i(f) = Z_{0i} \frac{Z_{i-1} + jZ_{0i} \tan[2\pi f (0.25\lambda_0/v_p)]}{Z_{0i} + jZ_{i-1} \tan[2\pi f (0.25\lambda_0/v_p)]}$$

where $\lambda_0 = v_p/f_0 = 1$ m, $v_p = c$, and Z_i and Z_{0i} are, respectively, the input impedance and the characteristic impedance of the ith quarter-wave transformer over which the impedance is being transformed, and Z_{i-1} is the input impedance of the $(i-1)$th quarter-wave transformer seen looking toward the load.

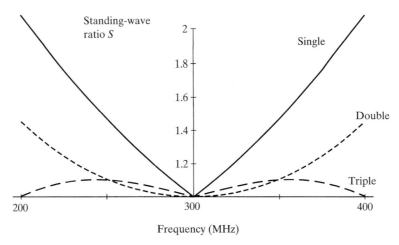

Figure 3.40   Quarter-wave transformer bandwidth. Standing-wave ratio S versus frequency for the three different quarter-wave transformers shown in Figure 3.39.

It is clear from Figure 3.40 that the bandwidth performance improves with the use of multiple segments. The improvement between a single and double quarter-wave section is very significant, with a tolerable standing-wave ratio of $S <\sim 1.2$ being achieved over a much larger range in a double quarter-wave section. The triple transformer provides for $S <\sim 1.1$ over the entire range of 200–400 MHz. In practice, little improvement is obtained by cascading more than four sections.[27]

In the limit of adding more and more sections, we would approach an infinitely long, smooth, gradually tapered transmission line with virtually no reflections. This is illustrated in Figure 3.41a. In practice, it is usually sufficient to make the tapered section of length $\sim\lambda$ or more.

A mechanical analogy. Impedance matching to achieve maximum energy transfer is an essential aspect of not just electrical but also other types of physical systems. One analogy is the transfer of energy in an elastic head-on collision between a mass M_1 moving with a speed v_M and a stationary mass M_2. In the absence of losses, and based on the conservation of momentum, and given the elastic nature of the collision, we know that if $M_2 = M_1$, then all the energy resident in M_1 is transferred to M_2 (i.e., M_1 stops and M_2 moves away at velocity v_M). If the $M_2 = M_1$ condition is not met, only a fraction of the total energy is transferred to M_2, and M_1 either reflects and moves in the reverse direction (if $M_1 < M_2$) or continues its motion at reduced speed (if $M_1 > M_2$).

---

[27]S. Guccione, Nomograms for speed design of $\lambda/4$ transformers, *Microwaves,* August 1975.

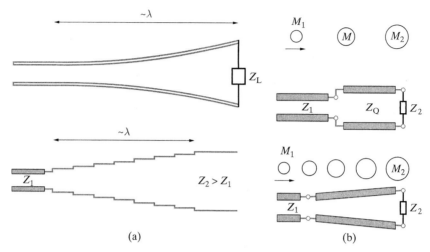

Figure 3.41   Tapered impedance transformer and a mechanical analogy. (a) A gradual taper provides wide bandwidth; many small reflections from a series of small incremental steps add with different phases to produce very small net total reflection. (b) Mechanical analog of impedance matching to provide better power absorption at a termination.

The transfer of energy between M_1 and M_2 can be improved by the insertion of a third mass between them, as shown in Figure 3.41b. The energy transfer is optimum if the third mass M is the geometric mean of the other two, that is, if $M = \sqrt{M_1 M_2}$. Further improvement in the energy transfer can be achieved by using several bodies with masses varying monotonically between M_1 and M_2.

3.6 THE SMITH CHART

Many transmission-line problems can be solved easily with graphical procedures, using the so-called *Smith chart*.[28] The Smith chart is also a useful tool for visualizing transmission-line matching and design problems. Many aspects of the voltage, current, and impedance patterns discussed in previous sections can also be interpreted and visualized by similar means using the Smith chart. One might think that graphical techniques are not as useful in this age of powerful computers and calculators, but it is interesting to note that some commonly used pieces of laboratory test equipment have displays that imitate the Smith chart, with the line impedance and standing-wave ratio results presented on such displays. In this section, we describe the Smith chart and provide examples of its use in understanding transmission-line problems.

---

[28]P. H. Smith, *Electronics,* January 1939. Also see J. E. Brittain, The Smith Chart, *IEEE Spectr.,* 29(8), p. 65, August 1992.

3.6.1 Mapping of Complex Impedance to Complex Γ

The Smith chart is essentially a conveniently parameterized plot of the normalized line impedance $\bar{Z}(z)$ of a transmission line and the generalized voltage reflection coefficient $\Gamma(z)$ as a function of distance from the load. To understand the utility of the Smith chart, we need to understand the relationship between $\bar{Z}(z)$ and $\Gamma(z)$.

From (3.22), we can write the normalized line impedance as

$$\bar{Z}(z) = \frac{1 + \Gamma(z)}{1 - \Gamma(z)} \tag{3.50}$$

where we note that $\Gamma(z) = \rho e^{j(\psi + 2\beta z)}$ is the voltage reflection coefficient at any position z along the line. Denoting $\Gamma(z)$ simply as Γ, while keeping in mind that it is a function of z, we can rewrite (3.50) as

$$\boxed{\bar{Z} = \frac{1 + \Gamma}{1 - \Gamma}} \tag{3.51}$$

where $\bar{Z} = \bar{R} + j\bar{X}$ and $\Gamma = u + jv$ are both complex numbers, so (3.51) represents a mapping between two complex numbers. Note that if the load is given, then we know Γ_L, and therefore Γ (and thus $\bar{Z} = \bar{R} + j\bar{X}$, from (3.51)), at any position at a distance z from the load. The Smith chart conveniently displays values of $\bar{Z}$ (or $\bar{R}, \bar{X}$) on the Γ (or u, v) plane for graphical calculation and visualization. From (3.51) we have

$$\bar{Z} = \bar{R} + j\bar{X} = \frac{1 + (u + jv)}{1 - (u + jv)} = \frac{[1 + (u + jv)][1 - (u - jv)]}{(1 - u)^2 + v^2} \tag{3.52}$$

Equating real parts in (3.52) and rearranging, we have

$$\boxed{\left(u - \frac{\bar{R}}{1 + \bar{R}}\right)^2 + v^2 = \left(\frac{1}{1 + \bar{R}}\right)^2} \tag{3.53}$$

which is the equation of a circle in the uv plane centered at $u = \bar{R}/(1 + \bar{R})$, $v = 0$, and having a radius of $1/(1 + \bar{R})$. Examples of such circles are shown in Figure 3.42a. Note that $\bar{R} = 1$ corresponds to a circle centered at $u = \frac{1}{2}$, $v = 0$, having a radius $\frac{1}{2}$, and passing through the origin in the uv plane.

Similarly, by equating the imaginary parts in (3.52) and rearranging, we find

$$\boxed{(u - 1)^2 + \left(v - \frac{1}{\bar{X}}\right)^2 = \frac{1}{\bar{X}}^2} \tag{3.54}$$

which is the equation for a circle in the uv plane centered at $u = 1$, $v = 1/\bar{X}$, and having a radius of $1/\bar{X}$. Examples of such circular segments are shown in Figure 3.42b. Note that $\bar{X} = \pm 1$ corresponds to a circle centered at $u = 1$, $v = \pm 1$, having a radius of 1, and tangent to the v axis at $v = \pm 1$.

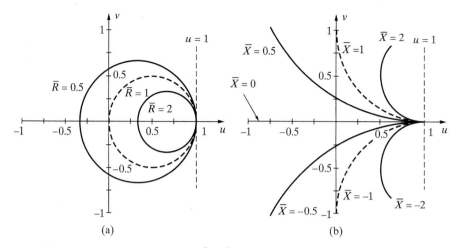

Figure 3.42   Contours of constant $\bar{R}$ or $\bar{X}$. (a) The circles in the uv plane are centered at $[\bar{R}(1+\bar{R})^{-1},0]$, with radius $(1+\bar{R})^{-1}$. Note the $\bar{R}=1$ circle (dashed lines) passes through the origin (i.e., $u,v=0$); this circle is centered at $(\frac{1}{2},0)$ with radius $\frac{1}{2}$. (b) The circles in the uv plane are centered at $(1,\bar{X}^{-1})$, with radius $\bar{X}^{-1}$. Note that for $\bar{X}=\pm1$ we have two circles (dashed lines) with unity radii and centered at $(1,\pm1)$.

The voltage reflection coefficient $\Gamma = u + jv$ is defined on the complex uv plane so that the locus of points of constant $|\Gamma| = |\Gamma_L| = \rho$ are circles centered at the origin. Once ρ is known (the value of which is set by Z_L and Z_0), motion along the line (i.e., variation of z) corresponds to motion around this Γ circle of fixed radius ρ. To see this, consider

$$\Gamma = \Gamma_L e^{j2\beta z} = \rho e^{j\psi} e^{j2\beta z} = \rho e^{j(\psi + 2\beta z)} \tag{3.55}$$

As illustrated in Figure 3.43, motion away from the load (i.e., decreasing z) corresponds to clockwise rotation of Γ around a circle in the uv plane. Since $\beta = 2\pi/\lambda$, a complete rotation of $2\beta z = -2\pi$ occurs when z decreases by $-\lambda/2$, which is why a complete cycle of line impedance (or admittance) is repeated every $\lambda/2$ length along the line. A circle

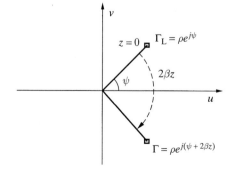

Figure 3.43   Complex reflection coefficient Γ. The complex number Γ is shown in the uv plane, together with its variation with z.

of constant radius ρ (corresponding to a given load) also corresponds to a fixed-voltage standing-wave ratio S, since $S = (1 + \rho)(1 - \rho)^{-1}$.

Although all Γ values along the line terminated with Z_L lie on the circle of radius ρ, each value of Γ corresponds (through (3.51)) to a different value of $\bar{Z} = \bar{R} + j\bar{X}$, which is the normalized line impedance seen at that position. On a typical Smith chart, shown in Figure 3.44, the contours of constant $\bar{R}$ or $\bar{X}$ are plotted and labeled on the uv plane so that the line impedance at any position along the constant ρ (or S) circle can be easily read from the chart. A summary of various Smith chart contours and key points is provided in Figure 3.45.

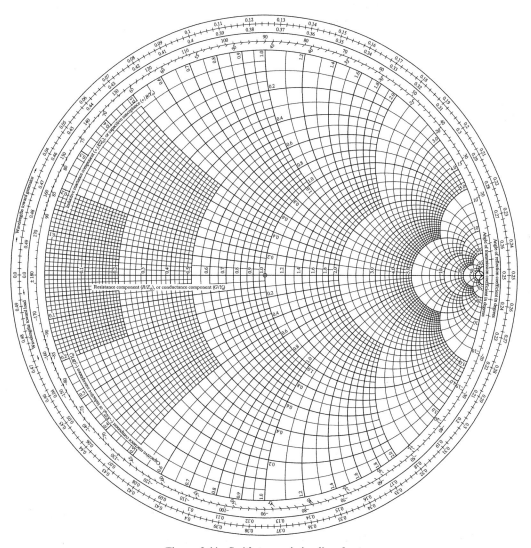

Figure 3.44   Smith transmission line chart.

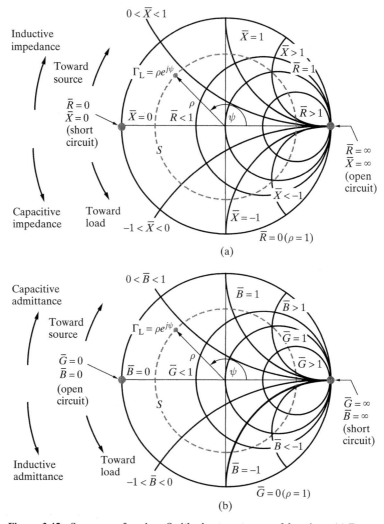

Figure 3.45 Summary of various Smith chart contours and locations. (a) For use as an impedance chart. (b) For use as an admittance chart.

As shown in Figures 3.45 and 3.46, the horizontal radius to the left of the chart center (i.e., the negative u axis) is the direction where $\Gamma = u + jv = \rho e^{(\psi + 2\beta z)} = \rho e^{-j\pi}$ or $\psi + 2\beta z = -(2m + 1)\pi$ where $m = 0, 1, 2, \ldots$; in this case, the magnitude of the line voltage is a minimum. Thus, every crossing of the negative u axis as one moves along the constant ρ circle corresponds to a minimum in the line voltage (and thus a maximum in the line current). The distance from the load to the first voltage minimum, namely z_{min}, can thus be found simply by equating $2\beta z$ to the negative of the angle from the load point (i.e., $\Gamma = \Gamma_L$) to this axis measured in the clockwise direction

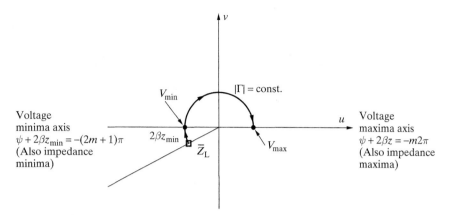

Figure 3.46   Location of voltage minima and maxima on the Smith chart. For the load shown in this figure, the first V_{max} is found using $m = 1$.

(i.e., $2\beta z_{min} = -(\pi + \psi)$, as was discussed in Section 3.3.1). Similarly, the crossings of the horizontal radius to the right (i.e., the positive u axis) represent voltage maxima. Note that when $\psi + 2\beta z = -m2\pi$ $(m = 0, 1, 2, \ldots)$ where $|V(z)| = V_{max}$, we have

$$\bar{Z} = \frac{1 + \rho}{1 - \rho} = \bar{R}_{max} = S$$

Hence if S is known (instead of ρ), the S circle (which is the same as the ρ circle) can be constructed with its center at the chart center and passing through the same point on the positive u axis as the $\bar{R} = S$ circle. This circle is then the locus of all impedances appearing at various positions along the transmission line, normalized to the characteristic impedance Z_0 of the line.

Once we realize that the upper (lower) half of the impedance Smith chart shown in Figure 3.45a corresponds to inductive (capacitive) reactances, that the negative u axis corresponds to a voltage minimum, and that moving away from the load corresponds to moving clockwise along a constant S (or constant ρ) circle around the chart, the interpretation of voltage standing-wave patterns for inductive versus capacitive loads (as depicted in Figure 3.17) becomes very clear. When we start anywhere in the upper half of the chart (i.e., inductive load) and move toward the source, we would encounter the voltage maxima (i.e., positive u axis) before the voltage minima so that the voltage magnitude would always first increase as we move away from an inductive load. The reverse would be true for a capacitive load. Many other aspects of the voltage, current, and impedance patterns discussed in previous sections can also be interpreted and visualized similarly using the Smith chart.

In cases for which it is more convenient to work with admittances than impedances, the Smith chart can be effectively used as an admittance chart. For this purpose, we note that since

$$\Gamma_L = \frac{Y_0 - Y_L}{Y_0 + Y_L} = -\frac{Y_L - Y_0}{Y_L + Y_0}$$

we have

$$\bar{Y}(z) = \bar{G} + j\bar{B} = \frac{1 - \Gamma(z)}{1 + \Gamma(z)} \tag{3.56}$$

instead of (3.51). In this case, the $\bar{R}$ and $\bar{X}$ circles can be treated as $\bar{G}$ and $\bar{B}$ circles, respectively. However, note that the upper (lower) half of the chart now corresponds to capacitive (inductive) susceptances, which are represented by positive (negative) values of $\bar{B}$. A summary of various Smith chart contours and key points for its use as an admittance chart is provided in Figure 3.45b.

3.6.2 Examples of the Use of the Smith Chart

We now consider some applications of the Smith chart. The examples selected illustrate the relatively easy determination of line impedance for given resistive, reactive, and complex loads; determination of unknown load impedance based on measurements of standing-wave ratio and location of voltage minimum; single-stub impedance matching; and quarter-wave transformer matching.

Example 3.20: Input impedance with pure resistive load. Find the input impedance of a lossless transmission line with the following parameters: $Z_0 = 100\,\Omega$, $Z_L = 50 + j0\,\Omega$, line length $l = 86.25$ cm, wavelength $\lambda = 1.5$ m.

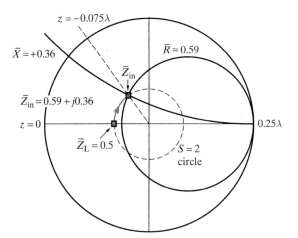

Figure 3.47   Graphical solution for Example 3.20.

Solution: We first note that the electrical length of the line is 0.575λ. Since impedance goes through a full cycle every 0.5λ, the input impedance of this line would be identical to one with length $0.075\lambda = (0.575 - 0.5)\lambda$. The normalized load impedance is $\bar{Z}_L = Z_L/Z_0 = 0.5 + j0$. We enter the Smith chart at the point where the $\bar{R} = 0.5$ circle crosses the horizontal axis (note that the imaginary part of the load impedance is zero). We draw a circle passing through this point and centered at the origin; this is the constant ρ circle. We move along this circle by 0.075λ (from 0 mark to -0.075λ mark) away from the load (i.e., clockwise) and read the impedance as $\bar{Z}_{in}(z = -0.075\lambda) \simeq 0.59 + j0.36$. Since $\bar{Z}$ is the normalized impedance, the actual line impedance is $Z_{in} \simeq 59 + j36\Omega$. The details of the graphical solution are shown in Figure 3.47.

Example 3.21: Input impedance with a pure reactive load. Find the input impedance of a lossless transmission line given the following parameters: $Z_0 = 50\Omega$, $Z_L = 0 - j75\Omega$, line length $l = 1.202\lambda$ (i.e., $\lambda + 0.202\lambda$).

Solution: The normalized load impedance is $\bar{Z}_L = -j1.5$. For a purely reactive load, $R_L = 0$, so that $\rho = 1$ and $S = \infty$. We enter the chart at the point on the outermost circle (which corresponds to $\bar{R} = 0$), which is intersected by the $\bar{X} = -1.5$ circle. The length scale at that point reads $\sim 0.344\lambda$. The angle of Γ_L, or ψ, may be read to be $\sim -67°$. We now move along the outer circle (which in this case is our constant ρ circle) a distance of 0.202λ to the point $\sim 0.046\lambda$. The impedance at that point is $\bar{Z}_{in} \simeq j0.3$, or $Z_{in} \simeq j15\Omega$. The details of the graphical solution are given in Figure 3.48.

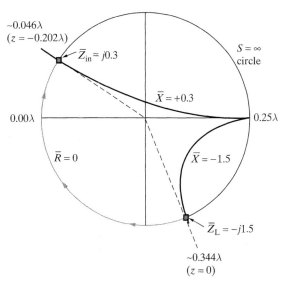

Figure 3.48   Graphical solution for Example 3.21.

Example 3.22: Input impedance with a complex load. Find the input impedance of a lossless transmission line given the following parameters: $Z_0 = 100\Omega$, $Z_L = 100 + j100\Omega$, line length $l = 0.676\lambda$ (i.e., $0.5\lambda + 0.176\lambda$).

Solution: The normalized load impedance is $\bar{Z}_L = 1.0 + j1.0$. We find the point on the chart corresponding to $\bar{R} = 1.0$ and $\bar{X} = 1.0$ (i.e., the intersection point of the $\bar{R} = 1.0$ and $\bar{X} = 1.0$ circles) and draw a circle passing through this point and centered at the origin. The intersection of this constant ρ circle with the right horizontal axis is at $\bar{R} \simeq 2.62$, which is also the value of S. To find the input impedance, we simply move along this circle (clockwise from the load position) a distance of 0.176λ and read $\bar{Z}_{in} = 1.0 - j1.0$. The input impedance of the line is then $Z_{in} = 100 - j100\Omega$. The details of the graphical solution are given in Figure 3.49.

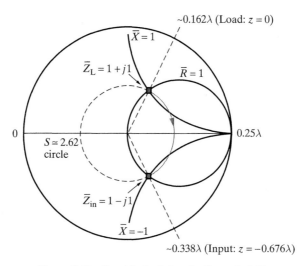

Figure 3.49   Graphical solution for Example 3.22.

Example 3.23: Unknown load impedance. Find the normalized load impedance on a transmission line with the following measured parameters: standing-wave ratio $S = 3.6$ and first voltage minimum $z_{min} = -0.166\lambda$.

Solution: We draw the constant ρ circle corresponding to $S = 3.6$ (i.e., intersecting the positive u axis at $\bar{R} = 3.6$). The point corresponding to z_{min} is that at which this circle crosses the negative u axis. We start at this point of intersection of the constant S circle with the left horizontal axis and move toward the load (i.e., counterclockwise) a distance of 0.166λ to find the normalized load impedance. This gives $\bar{Z}_L \simeq 0.89 - j1.3$. The details of the graphical solution are shown in Figure 3.50.

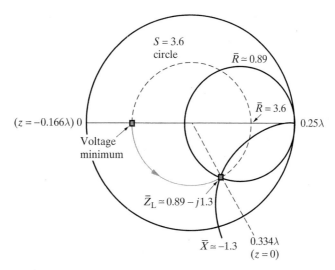

Figure 3.50   Graphical solution for Example 3.23.

Example 3.24: Single-stub impedance matching. Given a characteristic impedance $Z_0 = 80\Omega$ and a load impedance $Z_L = 160 - j80\Omega$, match the line to the given load by using a short-circuited shunt stub, as shown in Figure 3.32.

> **Solution:**   Refer to Figure 3.51 and to the discussion in Section 3.5 on impedance matching. Note that in view of the shunt connection of the stub, it is more convenient to deal with admittances. For this purpose, we use the Smith chart as an admittance chart. We require
>
> $$\bar{Y}_1 = 1 - j\bar{B} \quad \text{and} \quad \bar{Y}_s = +j\bar{B}$$
>
> where $\bar{Y}_1$ is the admittance seen looking toward the load at the position l where the stub is to be connected, and $\bar{Y}_s$ is the input admittance of the short-circuited stub of length l_s.
>   The normalized load impedance is $\bar{Z}_L = 2.0 - j1.0$. We enter the Smith chart at the point marked $\bar{Z}_L$ corresponding to the intersection of the resistance $\bar{R} = 2$ circle with the reactance $\bar{X} = -1.0$ circle, noting that negative reactances are in the lower half of the chart. The circle centered at the origin and passing through this point is our constant S (or constant ρ) circle along which the complex reflection coefficient Γ (or the line impedance) varies as we move away from the load. Noting that the Smith chart can be used equally for impedances and admittances, we choose to work with admittances in order to easily handle a parallel connected stub. The normalized load admittance can be found either directly (i.e., $\bar{Y}_L = (\bar{Z}_L)^{-1} = (2 - j)^{-1} = 0.4 + j0.2$) or by moving around the constant S circle by $180°$, as shown in Figure 3.51. The normalized load admittance is thus $\bar{Y}_L = 0.4 + j0.2$. Note that when we change an impedance to an admittance on the Smith chart and work from there, all of the $\bar{R}$ and $\bar{X}$ circles can now be used as $\bar{G}$ and $\bar{B}$ circles.
>   We now move along the constant S circle up to its point of intersection with the conductance $\bar{G} = 1$ circle (P_1). The amount that we need to move determines the stub position

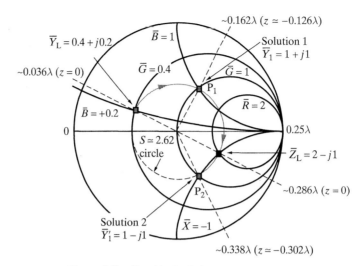

Figure 3.51   Graphical solution for Example 3.24.

at a distance l from the load. For this example, we find $l \simeq 0.126\lambda$. At the intersection point, the line impedance is $\bar{Y}_1 \simeq 1.0 + j1.0$, so that for matching we must have $\bar{Y}_s = -j1.0$.

To determine the length of a short-circuited stub that would present an admittance of $-j1.0$, we start from the point on the chart corresponding to a short circuit (i.e., $\bar{Y} = \infty$ on the right horizontal axis or the positive u axis). We move clockwise until we intersect the circle corresponding to a susceptance $\bar{B} = -1.0$. This determines the length of the stub to be $l_s = 0.125\lambda$.

Note that we might have taken the second intersection of the constant S circle with the $\bar{G} = 1$ circle (P_2), shown in Figure 3.51 as $\bar{Y}_1'$. This would have given $l \simeq 0.302\lambda$ and $\bar{Y}_1' \simeq 1.0 - j1.0$, requiring a stub impedance of $\bar{Y}_s = +j1.0$, which would be presented by a stub of length $l_s = 0.375\lambda$.

Example 3.25: Quarter-wave transformer matching. Given a transmission line with a characteristic impedance $Z_0 = 120\Omega$ and load impedance $Z_L = 72 + j96\Omega$, match the line to the given load using a quarter-wave transformer.

Solution: Refer to Figure 3.52, and the discussions in Section 3.5. We first move along the line a distance of l_1 such that the impedance Z_1 seen looking toward the line is purely resistive. Noting that the normalized load impedance is $\bar{Z}_L = (72 + j96)/120 = 0.6 + j0.8$, we enter the Smith chart at the point where the $\bar{R} = 0.6$ and $\bar{X} = 0.8$ circles intersect. We then move clockwise (away from load) along the constant S circle to its intersection with the horizontal axis, corresponding to the reactive part of the line impedance being zero. As shown in Figure 3.52, we need to move by 0.125λ, which means that the quarter-wave transformer can be placed at $l_1 = 0.125\lambda$. At that point, the line impedance normalized to a characteristic impedance of 120Ω is $\bar{Z}_1 = S_1 = 3.0$.

Noting that the quarter-wave transformation will occur on a line with characteristic impedance Z_Q, we now have to normalize the impedance to $Z_Q = \sqrt{Z_0 S_1 Z_0} \simeq 208\Omega$. The line impedance at $z = -0.125\lambda$, normalized to Z_Q, is then $\bar{Z}_1' = \bar{Z}_1(120/Z_Q) \simeq 1.73$.

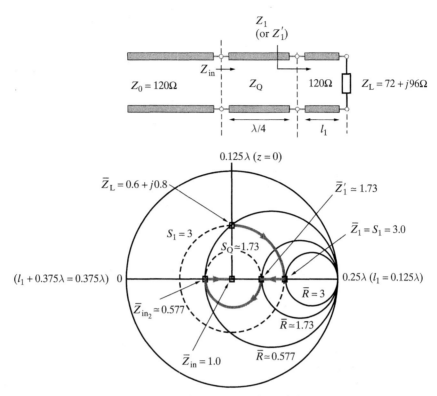

Figure 3.52   Quarter-wave transformer matching; graphical solution for Example 3.25. The path that we follow on the chart from the load impedance point to the origin (matching, i.e., $Z_{in} = Z_0$) is indicated by a thick dark line.

Following along the path on the Smith chart as shown in Figure 3.52, we thus move from $\bar{Z}_1$ to $\bar{Z}_1'$. The transformation along the quarter-wave segment is equivalent to a clockwise (away from load) rotation of $180°$, which brings us to $\bar{Z}_{in_2} = 1/S \simeq 0.577$. Note that this rotation is along the circle of $S_Q \simeq 1.73$, which is the standing-wave pattern within the transformer. We now note that $\bar{Z}_{in_2} \simeq 0.577$ is an impedance normalized to Z_Q, whereas the characteristic impedance of the line to be matched is 120Ω. Re-normalizing back to 120Ω, we find the input impedance of the line looking into the quarter-wave segment to be $\bar{Z}_{in} \simeq \bar{Z}_1'(208/120) \simeq 1.0$; this brings us to the center of the chart, which represents a matched line.

3.6.3 Voltage and Current Magnitudes from the Smith Chart

Note that for a lossless transmission line, we have

$$V(z) = V^+ e^{-j\beta z}(1 + \underbrace{\rho e^{j\psi} e^{j2\beta z}}_{\Gamma})$$

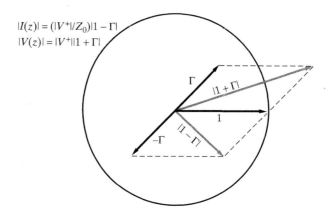

Figure 3.53 Line voltage and current from the Smith chart. We have assumed $V^+ = 1$.

so that the magnitude of the line voltage at any position z is given by

$$|V(z)| = |V^+||1 + \Gamma|$$

Since $|V^+|$ is just a scaling constant, the relative value of the line voltage can be obtained from the Smith chart by measuring the amplitude of the complex number $(1 + \Gamma)$. Note that at each position z along the line Γ is a new complex number as determined by a point on the uv plane (i.e., on the Smith chart). The length $|1 + \Gamma|$ can be determined graphically as shown in Figure 3.53. Note that as we move along the line away from the load, the generalized reflection coefficient vector Γ rotates clockwise, and the voltage vector $(1 + |\Gamma|)$ rotates clockwise like a crank. By inspection of Figure 3.53, we can see that the maximum and minimum values of the line voltage (i.e., maximum and minimum values of its magnitude) are $(1 + |\Gamma|)$ and $(1 - |\Gamma|)$, respectively, so that the standing-wave ratio is $S = (1 + |\Gamma|)/(1 - |\Gamma|) = (1 + \rho)/(1 - \rho)$, as was previously established in Section 3.3.1.

3.7 SINUSOIDAL STEADY-STATE BEHAVIOR OF LOSSY LINES

Our analyses so far have been based on the assumption that there is no power loss in the transmission line itself. A consequence of this assumption was the rather nonphysical result that the line current at a distance of a quarter wavelength from a short circuit is zero, and that the input impedance of a quarter-wavelength short-circuited line is infinite. In reality, every line consumes some power, partly because of the resistive losses (R) in the conductors and partly because of leakage losses (G) through the insulating medium surrounding the conductors. For lines with small losses, the effects of the losses on characteristic impedance, line voltage, line current, and input impedance are usually negligible, so that the lossless analysis is valid. However, in other cases, the losses and the resultant attenuation of signals cannot be ignored. The typical conditions under which losses cannot be neglected are (1) transmission of signals over long distances, (2) high-frequency applications since resistive losses increase with frequency, and (3) use of quarter-wavelength or half-wavelength long transmission line segments as circuit elements, when neglecting losses leads to nonphysical results such as zero current and/or infinite input impedance. In the third case, losses become the determining factor when the electrical quantities of interest tend toward zero or infinity. Thus, input impedance of a quarter-wavelength long open-circuited transmission line is in fact not zero but is a small nonzero value as determined by the losses. Similarly, the input impedance of a quarter-wavelength long short-circuited transmission line is not infinite, but a large finite value determined by the losses.

The sinusoidal steady-state behavior of lossy lines can be formulated in a manner quite similar to that of lossless lines. We can start with the most general form of the transmission line equations that were obtained in Section 2.2:

$$\frac{\partial \mathscr{V}(z,t)}{\partial z} = -\left(R + L\frac{\partial}{\partial t}\right)\mathscr{I}(z,t) \tag{3.57a}$$

$$\frac{\partial \mathscr{I}(z,t)}{\partial z} = -\left(G + C\frac{\partial}{\partial t}\right)\mathscr{V}(z,t) \tag{3.57b}$$

Under sinusoidal steady-state conditions, it is more convenient to work with the voltage and current equations written in terms of the phasor quantities $V(z)$ and $I(z)$, such that $\mathscr{V}(z,t) = \mathscr{R}e\{V(z)e^{j\omega t}\}$ and $\mathscr{I}(z,t) = \mathscr{R}e\{I(z)e^{j\omega t}\}$. We have

$$-\frac{dV(z)}{dz} = (R + j\omega L)I(z) \tag{3.58a}$$

$$-\frac{dI(z)}{dz} = (G + j\omega C)V(z) \tag{3.58b}$$

Taking the derivative of (3.58a) and substituting from (3.58b), we find

$$\frac{d^2V(z)}{dz^2} = (RG)V(z) + (LG + RC)(j\omega)V(z) + (j\omega)^2(LC)V(z)$$

$$= (R + j\omega L)(G + j\omega C)V(z)$$

$$\frac{d^2V(z)}{dz^2} - \gamma^2 V(z) = 0 \tag{3.59}$$

where

$$\gamma = \sqrt{(R + j\omega L)(G + j\omega C)} = \alpha + j\beta \tag{3.60}$$

is the propagation constant. Note that the propagation constant γ is in general a complex number, and its real and imaginary parts, α and β, are known as the *attenuation constant* and the *phase constant, respectively.*[29] For any given values of R, L, G, and C and the frequency f, the values of α and β can be directly calculated from (3.60).

Equation (3.59) is a second-order differential equation similar to the one we encountered for the lossless line case. Its general solution is

$$V(z) = V^+ e^{-\gamma z} + V^- e^{+\gamma z} \tag{3.61}$$

where V^+ and V^- are complex constants to be determined by the boundary conditions.

The current phasor $I(z)$ can be determined by simply substituting (3.61) into (3.58a). Thus, we have

$$(R + j\omega L)I(z) = -\frac{dV(z)}{dz}$$

$$I(z) = \frac{-1}{R + j\omega L} \frac{dV(z)}{dz} = +\frac{\gamma}{R + j\omega L}(V^+ e^{-\gamma z} - V^- e^{+\gamma z})$$

$$I(z) = \sqrt{\frac{G + j\omega C}{R + j\omega L}}(V^+ e^{-\gamma z} - V^- e^{+\gamma z}) \tag{3.62}$$

$$= \frac{1}{Z_0}(V^+ e^{-\gamma z} - V^- e^{+\gamma z})$$

where we have defined Z_0 as the *characteristic impedance,* namely,

$$Z_0 \equiv \sqrt{\frac{R + j\omega L}{G + j\omega C}} = |Z_0| e^{j\phi_z} \tag{3.63}$$

Compared with Z_0 for the lossless line case, we see that for the lossy case Z_0 is in general a complex number. Note once again that Z_0 depends on the physical line constants R, L, G, and C (which in turn depend on the physical makeup and dimensions of the line as well as the properties of the surrounding media) but now also on the frequency of operation $\omega = 2\pi f$. For future reference, the general solutions of the transmission line equations for the voltage and current phasors are

$$V(z) = V^+ e^{-\gamma z} + V^- e^{+\gamma z} \tag{3.64a}$$

$$I(z) = \frac{1}{Z_0}[V^+ e^{-\gamma z} - V^- e^{+\gamma z}] \tag{3.64b}$$

---

[29]Note that β for the lossy line case is not equal to that for the lossless case, which was defined earlier as $\beta = \omega\sqrt{LC}$. In the general lossy case, β is a function of R, L, G, and C and has a more complex dependence on the frequency f.

3.7.1 Infinitely Long or Matched Line

To understand the behavior of the time harmonic solutions for a lossy transmission line, we first consider the case of an infinitely long or a matched-terminated line. By analogy with the lossless case, we can see that the second terms in (3.64a) and (3.64b), those multiplied by the constant V^-, are zero in these cases, since no reflected wave exists. Accordingly, the voltage and current phasors are

$$V(z) = V^+ e^{-\gamma z}; \quad I(z) = \frac{1}{Z_0} V^+ e^{-\gamma z} \quad \rightarrow \quad \frac{V(z)}{I(z)} = Z_0$$

Note that everywhere on the line the ratio of the voltage to current phasors is the characteristic impedance Z_0, once again underscoring the physical meaning of the characteristic impedance.

It is instructive to write the voltage and current phasors explicitly in terms of the real and imaginary parts of γ. In other words, we have

$$V(z) = V^+ e^{-\alpha z} e^{-j\beta z} \tag{3.65a}$$

$$I(z) = \frac{V^+}{Z_0} e^{-\alpha z} e^{-j\beta z} \tag{3.65b}$$

Using (3.65), we can in turn obtain the space–time voltage and current functions as

$$\mathcal{V}(z,t) = \mathcal{R}e\{V^+ e^{-\alpha z} e^{-j\beta z}\} = V^+ e^{-\alpha z} \cos(\omega t - \beta z) \tag{3.66a}$$

$$\mathcal{I}(z,t) = \mathcal{R}e\left\{\frac{V^+}{Z_0} e^{-\alpha z} e^{-j\beta z}\right\} = \frac{V^+}{|Z_0|} e^{-\alpha z} \cos(\omega t - \beta z - \phi_z) \tag{3.66b}$$

where we have assumed V^+ to be real. The solutions for a lossy line are propagating waves with amplitudes exponentially decaying with increased distance. For physically realizable solutions, we must have $\alpha > 0$. Thus, in evaluating the propagation constant γ, the sign of the square root in (3.60) must be taken to be that which gives a value of $\alpha > 0$. To better visualize the behavior of the solutions, we show $\mathcal{V}(z,t)$ in Figure 3.54 as a function of position at fixed times and time at fixed positions. Also shown is the comparison between $\mathcal{V}(z,t)$ and $\mathcal{I}(z,t)$ as a function of space and time, clearly illustrating the phase difference ϕ_z between the two waveforms.

It can be shown from (3.65) that, for a matched or infinitely long line, the magnitude of the ratio of voltages or currents corresponding to two different positions separated by a length l is a constant. In other words, the magnitude of the ratio of the voltage at position z to the voltage at position $(z + l)$ is

$$\left|\frac{V(z)}{V(z+l)}\right| = \frac{e^{-\alpha z}}{e^{-\alpha(z+l)}} = e^{\alpha l} \tag{3.67}$$

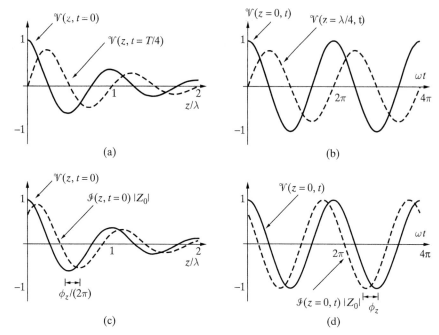

Figure 3.54 Voltage and current on a matched lossy line. (a) $\mathcal{V}(z,t) = V^+ e^{-\alpha z} \cos(\omega t - \beta z)$ versus z/λ for $t = 0$ and for $t = T_p/4$, where $\beta = 2\pi/\lambda$ and $T_p = 2\pi/\omega$. (b) $\mathcal{V}(z,t)$ versus ωt for $z = 0$ and $z = \lambda/4$. The attenuation constant was taken to be $\alpha = 1$ neper/λ. A comparison of voltage $\mathcal{V}(z,t)$ and current $\mathcal{I}(z,t)$ (c) at time $t = 0$ as a function of space and (d) at $z = 0$ as a function of time, for an assumed case where the R, L, G, and C values are such that $\phi_z = -\pi/4$. Note that we have assumed $V^+ = 1$.

Taking the natural logarithm of both sides, we have

$$\alpha l = \ln\left[\left|\frac{V(z)}{V(z+l)}\right|\right] \tag{3.68}$$

Note that αl is a dimensionless number since the units of α are in m^{-1} and l is in meters. However, to underscore the fact that αl expresses the attenuation on the line in terms of the natural (Naperian) logarithm of the magnitude of the ratio of voltages (or currents) at different positions, it is a common convention to express αl in units of *nepers* (np). Thus, in conventional usage, the unit of the attenuation constant α is nepers-m^{-1}.

In most engineering applications, a more commonly used unit for attenuation is the *decibel* (dB). The decibel is a unit derived from the *bel,* which in turn was named after Alexander Graham Bell and was used in early work on telephone systems. Specifically, the decibel is defined as

$$\text{Attenuation in decibels} \equiv 20 \log_{10} \left[\left| \frac{V(z)}{V(z+l)} \right| \right] \tag{3.69}$$

It is clear from (3.67) and (3.69) that a relation exists between attenuation expressed in decibels and that expressed in nepers. We have

$$\text{Attenuation in dB} = 20 \log_{10} e^{\alpha l} = (\alpha l) 20 \log_{10} e \simeq 8.686(\alpha l)$$

$$\simeq 8.686 \ (\text{attenuation in np}) \tag{3.70}$$

The advantage of a logarithmic unit such as the decibel or neper is that the total loss of several cascaded transmission lines (and other networks connected to them) can simply be found by adding the losses in the individual lines. As an example, if cascaded sections of a fiber-optic line have attenuations of 10 dB, 20 dB, and 5 dB, then the total attenuation of the signal in its passage through all three of the lines would be $10 + 20 + 5 = 35$ dB. Similarly, total gains of any number of cascaded amplifier stages in a system can also be easily calculated with the use of logarithmic units.

Example 3.26: Open-wire telephone line. An open-wire telephone line consists of two parallel lines made of copper with diameters $\sim$0.264 mm and spaced $\sim$20 cm apart on the crossarm of the wooden poles. Determine the propagation constant γ, its real and imaginary parts α and β, and the characteristic impedance Z_0. Assume it operates at 1.5 kHz.

Solution: Using the two-wire transmission line formulas given in Table 2.2, we find the transmission line parameters to be $R \simeq 24.4\Omega\text{-(km)}^{-1}$, $L \simeq 2.93$ mH-(km)$^{-1}$, and $C \simeq 3.80$ nF-(km)$^{-1}$. The value of G is assumed to be negligible. We have

$$\gamma = \sqrt{(R + j\omega L)(G + j\omega C)}$$

$$\simeq \sqrt{(24.4 + j2\pi \times 1.5 \times 10^3 \times 2.93 \times 10^{-3})(0 + j2\pi \times 10^3 \times 3.80 \times 10^{-9})}$$

$$\simeq \sqrt{(36.8e^{j48.6°})(3.58 \times 10^{-5}e^{j90°})} \simeq 0.0363e^{j69.3°} \ \text{(km)}^{-1}$$

$$\simeq 0.0128 + j0.0339 \ \text{(km)}^{-1}$$

$$\rightarrow \alpha \simeq 1.28 \times 10^{-2} \ \text{np-(km)}^{-1}; \quad \beta \simeq 3.39 \times 10^{-2} \ \text{rad-(km)}^{-1}$$

The phase velocity and wavelength are

$$v_p = \frac{\omega}{\beta} \simeq \frac{2\pi \times 1.5 \times 10^3}{3.39 \times 10^{-2}} \simeq 2.78 \times 10^5 \ \text{km-s}^{-1}; \qquad \lambda = \frac{2\pi}{\beta} \simeq 185 \ \text{km}$$

We find that the waves on an open-wire telephone cable propagate at a speed somewhat smaller than the speed of light in free space, namely, $c = 3 \times 10^8$ m-s^{-1}. The characteristic impedance is given by

$$Z_0 = \sqrt{\frac{R + j\omega L}{G + j\omega C}}$$

$$\simeq \sqrt{\frac{24.4 + j2\pi \times 1.5 \times 10^3 \times 2.93 \times 10^{-3}}{j2\pi \times 1.5 \times 10^3 \times 3.80 \times 10^{-9}}} \simeq 1014.5e^{-j20.7°} \simeq 949 - j359\,\Omega$$

Example 3.27: High-speed GaAs digital circuit coplanar strip interconnects. Transmission line properties of typical high-speed interconnects are experimentally investigated by fabricating and characterizing coplanar strip interconnects on semi-insulating GaAs substrates.[30] Measurements are carried out up to 18 GHz, from which the pertinent per-unit line parameters can be extracted. In one case, the values of the propagation constant γ and characteristic impedance Z_0 at 10 GHz are determined from the measurements to be $\gamma \simeq 1.2$ $(\text{np-(cm)}^{-1}) + j6$ (rad-(cm)^{-1}) and $Z_0 \simeq 105 - j25\,\Omega$, respectively. Using these values, calculate the per-unit length parameters $(R, L, G, \text{and } C)$ of the coplanar strip transmission line at 10 GHz.

Solution: The per-unit length parameters of the transmission line can readily be computed from γ and Z_0 using the relations

$$R + j\omega L = \gamma Z_0$$

$$G + j\omega C = \frac{\gamma}{Z_0}$$

Using the measured values of γ and Z_0 at 10 GHz, we have

$$R + j\omega L = (1.2 + j6)(105 - j25) \simeq 276 + j600\,\Omega$$

from which $R \simeq 276\,\Omega\text{-(cm)}^{-1}$ and $L \simeq 600/(2\pi \times 10^{10}) \simeq 9.55$ nH-(cm)$^{-1}$, respectively. Similarly, we have

$$G + j\omega C = \frac{1.2 + j6}{105 - j25} \simeq \frac{6.12e^{j78.9°}}{107.9e^{-j13.4°}} \simeq 0.0567e^{j92.1°} \simeq -0.0021 + j0.0567 \text{ S}$$

from which $G \simeq -0.0021$ S-(cm)$^{-1}$ and $C \simeq 0.0567/(2\pi \times 10^{10}) \simeq 0.902$ pF-(cm)$^{-1}$, respectively. The negative value of parameter G is nonphysical and is likely a result of measurement error.

---

[30]K. Kiziloglu, N. Dagli, G. L. Matthaei, and S. I. Long, Experimental analysis of transmission line parameters in high-speed GaAs digital circuit interconnects, *IEEE Trans. Microwave Theory Techn.*, 39(8), pp. 1361–1367, August 1991.

The average power delivered into the line at any given point z can be found using the phasor expressions (3.65) for voltage and current:

$$
\begin{aligned}
P_{av}(z) &= \frac{1}{2}\mathscr{R}e\{V(z)[I(z)]^*\} \\
&= \frac{1}{2}\mathscr{R}e\left\{V^+e^{-\alpha z}e^{-j\beta z}\frac{(V^+)^*}{Z_0^*}e^{-\alpha z}e^{+j\beta z}\right\} \\
&= \frac{|V^+|^2}{2|Z_0|}e^{-2\alpha z}\cos(\phi_z)
\end{aligned}
\tag{3.71}
$$

We find that the time-average power decreases with distance as $e^{-2\alpha z}$, with an effective attenuation constant that is twice that of the voltage and current. The difference in time-average powers evaluated at any two points z_1 and z_2 is the amount of power dissipated in the segment of the line between z_1 and z_2.

Low-loss lines.   An important practical case is that in which the losses along the line are small but not negligible. If the line is low loss, we can assume that $R \ll \omega L$ and $G \ll \omega C$, which means that the resistive losses and leakage losses in the surrounding medium are both small. In such cases, useful approximate expressions can be derived for the characteristic impedance Z_0 and the propagation constant γ.

We first consider Z_0:

$$
\begin{aligned}
Z_0 &= \sqrt{\frac{R+j\omega L}{G+j\omega C}} \\
&= \sqrt{\frac{L}{C}}\sqrt{\frac{1+R/(j\omega L)}{1+G/(j\omega C)}} \simeq \sqrt{\frac{L}{C}}\left[\frac{1+R/(j2\omega L)}{1+G/(j2\omega C)}\right]
\end{aligned}
$$

where we have used the fact that for $\zeta \ll 1$, $(1+\zeta)^{1/2} = 1+(\zeta/2)+\cdots \simeq 1+(\zeta/2)$. By neglecting the higher-order terms in the numerator and the denominator, and using $(1+\zeta)^{-1} \simeq 1-\zeta$ for $\zeta \ll 1$, we can write the characteristic impedance as

$$
\begin{aligned}
Z_0 &\simeq \sqrt{\frac{L}{C}}\left(1+\frac{R}{j2\omega L}\right)\left(1-\frac{G}{j2\omega C}\right) \\
&= \sqrt{\frac{L}{C}}\left[\left(1+\frac{RG}{4\omega^2 LC}\right)+j\frac{1}{2\omega}\left(\frac{G}{C}-\frac{R}{L}\right)\right]
\end{aligned}
\tag{3.72}
$$

In general, the second term in the real part of (3.72) is negligible since it involves the product of two small terms, namely, $R/(\omega L)$ and $G/(\omega C)$. Thus, the important effect of the losses on the transmission line is to introduce a small imaginary component to the characteristic impedance. In many cases, the imaginary part of Z_0 can be neglected so that the characteristic impedance is, to the first order, equal to that for the lossless line.

A similar simplification can also be obtained for the propagation constant γ, again using $(1 + \zeta)^{1/2} = 1 + (\zeta/2) + \cdots \simeq 1 + (\zeta/2)$ for $\zeta \ll 1$. We have

$$\gamma = [(R + j\omega L)(G + j\omega C)]^{1/2}$$

$$= \left[(j\omega L)(j\omega C) \left(1 + \frac{R}{j\omega L} \right) \left(1 + \frac{G}{j\omega C} \right) \right]^{1/2}$$

$$\simeq j\omega\sqrt{LC} \left[1 - j\frac{1}{2\omega} \left(\frac{R}{L} + \frac{G}{C} \right) \right]$$

The real and imaginary parts of γ for the low-loss line are thus

$$\alpha \simeq \frac{1}{2} \left[R\sqrt{\frac{C}{L}} + G\sqrt{\frac{L}{C}} \right] \tag{3.73a}$$

$$\beta \simeq \omega\sqrt{LC} \tag{3.73b}$$

As an example, coaxial lines used at high radio frequencies can be quite accurately represented by the above low-loss formulas of (3.73). Note that the phase constant β is the same as that in the lossless case, so the phase velocity $v_p = \omega/\beta = 1/\sqrt{LC}$, independent of frequency. The loss constant α also does not depend on frequency; it simply accounts for a decrease in the overall signal intensity as the wave propagates along the line. Thus the distortion of an information-carrying signal (consisting of a finite band of frequencies), due to the different speed and attenuation of its frequency components, is minimized for a low-loss line.

Example 3.28: Low-loss coaxial line. RG17A/U is a low-loss radio frequency coaxial line. The following data for the nominal parameters of this line are available: characteristic impedance $Z_0 = 50\Omega$, line capacitance $C \simeq 96.8$ pF-m^{-1}, and line attenuation ~ 3 dB-(100 m)$^{-1}$ at 100 MHz. Determine the inductance L and resistance R per unit length of this line, assuming that G is negligibly small. Determine the velocity of propagation.

Solution: Using (3.70), we can express the attenuation in np-m^{-1}. We have

$$3 \text{ dB-(100 m)}^{-1} = 0.03 \text{ dB-m}^{-1} \simeq \frac{0.03}{8.686} \simeq 3.45 \times 10^{-3} \text{ np-m}^{-1} = \alpha$$

Using the low-loss formulas, we have

$$\alpha \simeq \frac{1}{2} \left[\frac{R}{Z_0} + GZ_0 \right] = \frac{R}{2Z_0} \simeq 3.45 \times 10^{-3} \text{ np-m}^{-1}$$

which gives us $R \simeq 0.345\Omega\text{-m}^{-1}$ since $G \simeq 0$ and $Z_0 = 50\Omega$. The inductance can be determined from

$$Z_0 \simeq \sqrt{\frac{L}{C}} = 50\Omega \quad \rightarrow \quad L = Z_0^2 C \simeq (50)^2(96.8 \times 10^{-12}) = 0.242 \; \mu\text{H-m}^{-1}$$

We can check to see that the quantity $|R/(\omega L)| \simeq 2.27 \times 10^{-3}$, or is much smaller than 1, which is apparently why the characteristic impedance for this low-loss line is real. The phase velocity is given by

$$v_p = \frac{1}{\sqrt{LC}} = \frac{1}{\sqrt{Z_0^2 C^2}} = \frac{1}{Z_0 C} = \frac{1}{50 \times 96.8 \times 10^{-12}} \simeq 2.07 \times 10^8 \; \text{m-s}^{-1}$$

3.7.2 Terminated Lossy Lines

An important result of losses is that both the forward wave traveling toward the load and the reflected wave traveling away from the load are attenuated exponentially with distance. As an observer moves away from the load on a terminated lossless line, the standing-wave pattern remains the same. However, on a lossy line, the same observer finds that the attenuation of the reflected wave causes this wave to be less important as he or she moves farther from the load. In addition, since the magnitude of the forward wave becomes larger as the observer moves away from the load, the relative size of the reflected wave is doubly reduced in moving toward the source. The net result of this effect is that, regardless of its termination, the transmission line begins to appear more and more like an infinite (or matched) line when viewed farther and farther from the load.

We consider a terminated lossy transmission line as shown in Figure 3.55, with $z = 0$ taken to be the position of the load as in the case of lossless lines. In general, the expressions for voltage and current on a terminated lossy transmission line are

$$V(z) = V^+ e^{-\gamma z} + V^- e^{\gamma z} = V^+(e^{-\alpha z} e^{-j\beta z} + \Gamma_L e^{\alpha z} e^{j\beta z})$$

$$= V^+ e^{-\alpha z} e^{-j\beta z}[1 + \Gamma(z)] \qquad (3.74a)$$

$$I(z) = \frac{1}{Z_0} V^+ e^{-\alpha z} e^{-j\beta z}[1 - \Gamma(z)] \qquad (3.74b)$$

where $\Gamma(z)$ is the complex voltage reflection coefficient at any position z along the line defined as

$$\Gamma(z) \equiv \frac{V^- e^{\gamma z}}{V^+ e^{-\gamma z}} = \Gamma_L e^{2\alpha z} e^{j2\beta z}$$

and where Γ_L is the complex load reflection coefficient given as

$$\Gamma_L \equiv \frac{V^-}{V^+} = \rho e^{j\psi} = \frac{Z_L - Z_0}{Z_L + Z_0} \qquad (3.75)$$

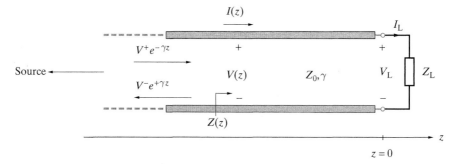

Figure 3.55   A terminated lossy transmission line.

The line impedance $Z(z)$ at any point z on the line is given by the ratio of the voltage and the current:

$$Z(z) = \frac{V(z)}{I(z)} = Z_0 \frac{e^{-\alpha z} e^{-j\beta z} + \Gamma_L e^{\alpha z} e^{j\beta z}}{e^{-\alpha z} e^{-j\beta z} - \Gamma_L e^{\alpha z} e^{j\beta z}} \qquad (3.76)$$

It is sometimes useful to rewrite the impedance as follows:

$$Z(z) = Z_0 \frac{1 + \Gamma_L e^{2\alpha z} e^{j2\beta z}}{1 - \Gamma_L e^{2\alpha z} e^{j2\beta z}} = Z_0 \frac{1 + \Gamma(z)}{1 - \Gamma(z)} \qquad (3.77)$$

We compare (3.77) for $Z(z)$ to equation (3.29) for the lossless line. We see that Γ_L in (3.29) is replaced with $\Gamma_L e^{2\alpha z}$, so that the magnitude of the reflection coefficient for the lossy case is effectively reduced exponentially between the observation point and the end of the line (i.e., the load position). As viewed from larger and larger distances from the load (i.e., as $z \to -\infty$), the effect of reflections becomes negligible, and the line impedance approaches Z_0, as if the line were an infinitely long or matched line. To understand this effect, consider the general voltage reflection coefficient at any point z, namely

$$\Gamma(z) = \Gamma_L e^{2\alpha z} e^{j2\beta z} = \rho e^{2\alpha z} e^{j(\psi + 2\beta z)} \qquad (3.78)$$

In Section 3.6, we noted that motion along the line away from the load corresponded to clockwise rotation of $\Gamma = u + jv$ in the uv plane (or on the Smith chart), while its magnitude $\Gamma = \rho$ remained constant. On lossy lines, (3.78) indicates that the same type of rotation occurs as determined by the $e^{j2\beta z}$ term, but that, in addition, the magnitude of $\Gamma(z)$, namely $|\Gamma| = \rho e^{2\alpha z}$, decreases as we move away from the load (i.e., as z decreases). Eventually, at some point, $|\Gamma(z)| \to 0$, and looking toward the load from the source side beyond this position, the line is indistinguishable from an infinitely long or matched line.

To examine the behavior of the line voltage, current, and impedance, we first consider a short-circuited line of length l, so that $Z_L = 0$ in Figure 3.55. In this case, the load reflection coefficient is $\Gamma_L = -1$, so the line voltage and current are

$$V(z) = V^+(e^{-\gamma z} - e^{\gamma z}) = -2V^+ \sinh(\gamma z) \tag{3.79a}$$

$$I(z) = \frac{V^+}{Z_0}(e^{-\gamma z} + e^{\gamma z}) = \frac{2V^+}{Z_0}\cosh(\gamma z) \tag{3.79b}$$

where we have used the defining expressions for the hyperbolic sine and cosine functions:

$$\sinh \zeta = \frac{e^\zeta - e^{-\zeta}}{2} \qquad \cosh \zeta = \frac{e^\zeta + e^{-\zeta}}{2}$$

Although the compact form of $V(z)$ in (3.79a) appears very similar to that for the lossless line (with sin replaced by sinh), the evaluation of $\sinh(\gamma z)$ is not trivial,[31] since γ is a complex quantity. Note that when $\alpha \to 0$, equations (3.79) for $V(z)$ and $I(z)$ reduce to their lossless equivalents, since $\sinh(\alpha z) \to 0$ and $\cosh(\alpha z) \to 1$.

Using (3.79a) and (3.79b), we can compactly write the input impedance of a short-circuited line of length l as

$$[Z_{\text{in}}]_{\text{sc}} = Z_0 \tanh(\gamma l)$$

Plots of magnitudes of the line voltage and current and the real and imaginary parts of the line impedance for a short-circuited line are provided in Figure 3.56, for two different values of the attenuation constant α, namely $\alpha = 0.5$ np/λ and 1.5 np/λ. For simplicity, we have assumed the phase of the characteristic impedance $\phi_z = 0$ in Figure 3.56. In general this phase angle is small, and leads to a phase difference between the voltage and current, as indicated in (3.66).

The resultant effects of the losses shown in Figure 3.56 become clear upon comparative examination of the lossless equivalents given in Figures 3.5 and 3.6. In the lossless case, the line voltage is zero at the load, and every half-wavelength thereafter, while the line current is a maximum at the same positions. The line impedance of the lossless line is zero at the load ($Z_L = 0$), inductive (i.e., $\mathcal{I}m\{Z(z)\} > 0$) in the range $-\lambda/4 < z < 0$, infinite (i.e., an open circuit) at $z = -\lambda/4$, capacitive in the range $-\lambda/2 < z < -\lambda/4$ back to zero at $z = -\lambda/2$, and repeating the same cycle thereafter.

For the lossy case, looking first at the relatively low-loss case of $\alpha = 0.5$ np/λ, we see that although the voltage and current exhibit generally similar cyclic behavior, the

---

[31]The hyperbolic sine of the complex number $\gamma = \alpha + j\beta$ can be expressed as

$$\sinh(\gamma z) = \sinh[(\alpha + j\beta)z]$$

$$= \sinh(\alpha z)\cosh(j\beta z) + \cosh(\alpha z)\sinh(j\beta z)$$

$$= \cos(\beta z)\sinh(\alpha z) + j\cosh(\alpha z)\sin(\beta z)$$

In practice, the evaluation of $\sinh[(\alpha + j\beta)z]$ would be straightforward using any reasonably sophisticated numerical evaluation tool (e.g., a software package or a scientific calculator); however, it is useful to note for insight the nature of the actual evaluation.

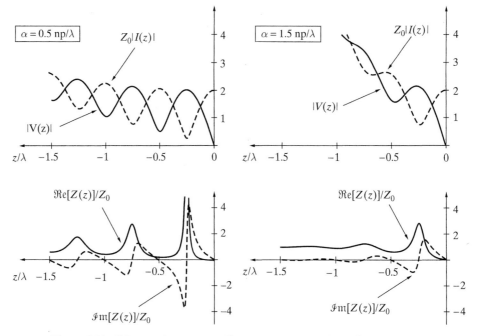

Figure 3.56 Voltage and current standing-wave patterns and impedance on a lossy short-circuited line. Results are shown for two different values of the attenuation constant, namely $\alpha = 0.5$ np/λ and 1.5 np/λ, where $\lambda = 2\pi/\beta$. For simplicity, we have assumed the characteristic impedance to be real, that is, $\phi_z = 0$, and $V^+ = 1$.

maximum and minimum values of both the line voltage and current increase with distance from the load. The line voltage is no longer zero at $z = -\lambda/2$. The differences between the values of the maxima and the minima also become smaller as z becomes increasingly negative, as is clearly evident from the relatively high-loss case of $\alpha = 1.5$ np/λ. At a sufficient distance away from the load, for example, for $z < -1.5\lambda$ in the case of $\alpha = 1.5$ np/λ, the magnitude of line voltage and current do not vary significantly over a distance of half-wavelength (i.e., the standing-wave ratio is unity), as if the line were infinitely long or matched.

The line impedance for the relatively low-loss ($\alpha = 0.5$ np/λ) case exhibits similar behavior to the lossless case. The impedance is inductive (i.e., $\mathcal{I}m\{Z(z)\} > 0$) in the approximate range $-\lambda/4 < z < 0$, attains a large real value (but not quite an open circuit) at $z = -\lambda/4$, is capacitive in the approximate range $-\lambda/2 < z < -\lambda/4$, but does not quite return to zero at $z = -\lambda/2$. The peak in the real part of the impedance at $z = -3\lambda/4$ is considerably smaller than that at $z = -\lambda/4$. In general, the maxima and minima of the imaginary part of $Z(z)$ both approach zero as z attains larger and larger negative values, while the maxima and minima of the real part of $Z(z)$ both approach Z_0. At sufficient distances from the load, for example, for $z < -1.5\lambda$ in the case of $\alpha = 1.5$ np/λ, the line impedance $Z(z) \simeq Z_0$, just as if the line were infinitely long or matched.

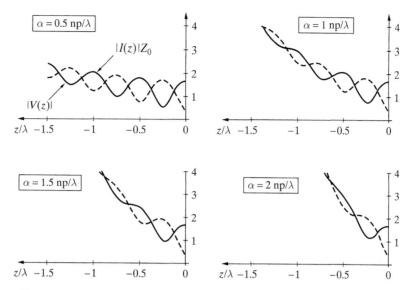

Figure 3.57   Voltage and current standing-wave patterns on a terminated lossy transmission line. The magnitudes of the voltage and current phasors (for current, the quantity plotted is $|I(z)|Z_0$) of a lossy line terminated in $Z_L = 5Z_0$ are shown for values of the attenuation constant $\alpha = 0.5, 1, 1.5,$ and 2 np/λ, where $\lambda = 2\pi/\beta$.

For a lossy line terminated in an open circuit ($Z_L = \infty$), expressions for $V(z)$, $I(z)$, and Z_{in} can be obtained in a manner analogous to the preceding discussion for a short-circuited line. This straightforward procedure is left as an exercise.

The general behavior of the line voltage, current, and impedance for other terminations is quite similar, as illustrated in Figures 3.57 and 3.58 for a resistive load impedance of $Z_L = 5Z_0$. Results are shown for four different values of the attenuation constant. For simplicity, we have once again assumed the phase of the characteristic impedance $\phi_z = 0$.

The time-average power at any point z along the line can be evaluated using the expressions (3.74) for $V(z)$ and $I(z)$. We have

$$
\begin{aligned}
P_{av}(z) &= \frac{1}{2}\mathcal{R}e\{V(z)[I(z)]^*\} \\
&= \frac{|V^+|^2 e^{-2\alpha z}}{2}\mathcal{R}e\left\{\frac{[1 + \Gamma_L e^{2\alpha z}e^{j2\beta z}][1 - \Gamma_L e^{2\alpha z}e^{j2\beta z}]^*}{Z_0^*}\right\} \\
&= \frac{|V^+|^2 e^{-2\alpha z}}{2}\mathcal{R}e\left\{\frac{1 - |\Gamma_L|^2 e^{4\alpha z} - \Gamma_L^* e^{2\alpha z}e^{-j2\beta z} + \Gamma_L e^{2\alpha z}e^{+j2\beta z}}{Z_0^*}\right\}
\end{aligned}
$$

Consider a terminated transmission line of length l. The time-average power at its input, namely at $z = -l$, is given by $P_{av}(z = -l)$, whereas that at the load is given by $P_{av}(z = 0)$. The difference between these quantities is the average power dissipated in

Figure 3.58   Line impedance standing on a terminated lossy transmission line. The real and imaginary parts of the line impedance (normalized to Z_0) of a lossy line terminated in $Z_L = 5Z_0$ are shown for values of the attenuation constant $\alpha = 0.5, 1, 1.5,$ and 2 np/λ, where $\lambda = 2\pi/\beta$.

the lossy transmission line. Thus, power lost in the line is given by

$$P_{lost} = P_{av}(z = -l) - P_{av}(z = 0)$$

Reflections in lossy lines can lead to substantially increased losses since each time a wave travels down the line it is further attenuated. If the load reflects part of the incident power, more power is dissipated in the lossy line than would have been dissipated if the line were matched (i.e., $\Gamma_L = 0$). If the power dissipated in a lossy line under matched conditions is P_{lost}^m, it can be shown[32] that the extra power dissipated as a result of reflections is

$$\frac{P_{lost} - P_{lost}^m}{P_{lost}^m} \simeq |\Gamma_L|^2(e^{2\alpha l} - 1)$$

The extra power dissipated due to mismatch can be substantial, especially when $|\Gamma_L| > 0.5$ and when the line is long.

Example 3.29: A high-speed microstrip interconnect. Consider a high-speed microstrip transmission line of length 20 cm used to connect a 1-V amplitude, 1-GHz, 50Ω sinusoidal voltage source to a digital logic gate having an input impedance of 1 kΩ, as shown in Figure 3.59. Based on measurements, the transmission line parameters of this interconnect at 1 GHz are

---

[32]See Section 6-3 of R. K. Moore, *Traveling Wave Engineering,* McGraw-Hill, New York, 1960.

(a)

(b)

Figure 3.59   A lossy high-speed microstrip interconnect. (a) The microstrip transmission line connected to a 1-kΩ load and driven by a 1-GHz source. (b) Thévenin equivalent circuit as seen by the source, where Z_{in} is the input impedance at the source end of the microstrip.

approximately given by $R = 5\,\Omega\text{-(cm)}^{-1}$, $L = 5$ nH-(cm)$^{-1}$, $C = 0.4$ pF-(cm)$^{-1}$, and $G = 0$, respectively. (a) Find the propagation constant γ and characteristic impedance Z_0 of the line. (b) Find the voltages at the source and the load ends of the line. (c) Find the time-average power delivered to the line by the source and the time-average power delivered to the load. What is the power dissipated along the line?

Solution:

(a) The propagation constant is given by

$$\gamma = \alpha + j\beta = \sqrt{(R + j\omega L)(G + j\omega C)}$$

$$= \sqrt{(500 + j2\pi \times 10^9 \times 500 \times 10^{-9})(j2\pi \times 10^9 \times 40 \times 10^{-12})}$$

$$\simeq 28.3e^{j85.5^\circ} \simeq 2.23 + j28.2$$

where $\alpha \simeq 2.23$ np-m^{-1} and $\beta \simeq 28.2$ rad-m^{-1}. The characteristic impedance is given by

$$Z_0 = \sqrt{\frac{R + j\omega L}{G + j\omega C}} \simeq \sqrt{\frac{500 + j3142}{j0.251}} \simeq \sqrt{\frac{3181e^{j80.96^\circ}}{0.251e^{j90^\circ}}}$$

$$\simeq 112.5e^{-j4.522^\circ}\ \Omega$$

(b) The reflection coefficient at the load end can be found as

$$\Gamma_L = \frac{Z_L - Z_0}{Z_L + Z_0} \simeq \frac{1000 - 112.5e^{-j4.522°}}{1000 + 112.5e^{-j4.522°}} \simeq 0.798e^{j1.03°}$$

The reflection coefficient at any position along the line is given by

$$\Gamma(z) = \Gamma_L e^{2\gamma z} \simeq 0.798e^{j1.029°} e^{4.458z} e^{j56.37z}$$

$$\simeq 0.798e^{4.458z} e^{j(56.37z+0.018)}$$

The input impedance of the line is given by

$$Z_{in} = Z(z)|_{z=-0.2 \text{ m}} = Z_0 \frac{1 + \Gamma(-0.2)}{1 - \Gamma(-0.2)}$$

We first find $\Gamma(-0.2)$ as

$$\Gamma(-0.2) \simeq 0.798e^{4.458(-0.2)} e^{j[56.37(-0.2)+0.018]} \simeq 0.327e^{j75.05°}$$

We now use $\Gamma(-0.2)$ to find Z_{in}. We have

$$Z_{in} \simeq 112.5e^{-j4.522°} \frac{1 + 0.327e^{j75.05°}}{1 - 0.327e^{j75.05°}}$$

$$\simeq (112.5e^{-j4.522°})(1.166e^{j35.31°}) \simeq 131.2e^{j30.79°}\Omega$$

Based on voltage division in the equivalent circuit of Figure 3.59b, the source-end voltage V_s is

$$V_s = \frac{Z_{in}}{Z_{in} + Z_s} V_0 \simeq \frac{131.2e^{j30.79°}}{112.7 + j67.16 + 50} \quad (1)$$

$$\simeq \frac{131.2e^{j30.79°}}{176e^{j22.43°}} \simeq 0.745e^{j8.361°} \text{ V}$$

Using (3.74a), V^+ can be written as

$$V^+ = \frac{V(z)}{e^{-\gamma z}[1 + \Gamma(z)]}$$

At $z = -0.2$ m, we have

$$V(z = -0.2) = V_s \simeq 0.745e^{j8.361°} \text{ V}$$

and

$$e^{-\gamma(-0.2)} \simeq e^{0.446} e^{-j37.01°} \quad \text{and} \quad \Gamma(-0.2) \simeq 0.327e^{j75.05°}$$

Using these values, the value of V^+ can be found as

$$V^+ \simeq \frac{0.745e^{j8.361°}}{e^{0.446} e^{-j37.01°}(1 + 0.327e^{j75.05°})} \simeq 0.423e^{j29.12°} \text{ V}$$

The voltage at the load end of the line is given as

$$V_L = V(z=0) = V^+[1+\Gamma_L] \simeq (0.423e^{j29.12°})(1+0.798e^{j1.029°}) \simeq 0.760e^{j29.57°} \text{ V}$$

(c) The time-average power delivered to the line is given by

$$P_s = \frac{1}{2}\left|\frac{V_s}{Z_{in}}\right|^2 R_{in} \simeq \frac{1}{2}\left|\frac{0.745}{131.2}\right|^2 (112.7) \text{ W} \simeq 1.82 \text{ mW}$$

Similarly, the time-average power delivered to the load can be found as

$$P_L = \frac{1}{2}\frac{|V_L|^2}{R_L} \simeq \frac{1}{2}\frac{(0.760)^2}{1000} \text{ W} \simeq 0.289 \text{ mW}$$

Thus, based on conservation of energy, the power dissipated in the lossy line is

$$P_{lost} = P_s - P_L \simeq 1.82 - 0.289 \simeq 1.53 \text{ mW}$$

3.8 SUMMARY

This chapter discussed the following topics:

- **Transmission line equations.** When a transmission line is excited by a sinusoidal source of angular frequency ω at steady state, the variations of the line voltage and current can be analyzed using the phasor form of the transmission line equations, which for a lossless line are

$$\frac{dV(z)}{dz} = -j\omega L I(z)$$

$$\frac{dI(z)}{dz} = -j\omega C V(z)$$

where L and C are the per-unit length distributed parameters of the line, and $V(z)$ and $I(z)$ are the voltage and current phasors, respectively, which are related to the actual space–time voltage and current expressions as follows:

$$\mathcal{V}(z,t) = \mathfrak{Re}\{V(z)e^{j\omega t}\}; \qquad \mathscr{I}(z,t) = \mathfrak{Re}\{I(z)e^{j\omega t}\}$$

- **Propagating-wave solutions, characteristic impedance, phase velocity, and wavelength.** The solutions of the lossless transmission line equations consist of a superposition of waves traveling in the $+z$ and $-z$ directions. The voltage and current phasors and the corresponding space–time functions have the form (assuming V^+ and V^- are both real)

$$V(z) = V^+ e^{-j\beta z} + V^- e^{+j\beta z}; \quad \mathcal{V}(z,t) = V^+ \cos(\omega t - \beta z) + V^- \cos(\omega t + \beta z)$$

$$I(z) = \frac{V^+}{Z_0}e^{-j\beta z} - \frac{V^-}{Z_0}e^{+j\beta z}; \quad \mathscr{I}(z,t) = \frac{V^+}{Z_0}\cos(\omega t - \beta z) - \frac{V^-}{Z_0}\cos(\omega t + \beta z)$$

The characteristic impedance Z_0 of the line is the ratio of the voltage to the current phasor of the wave propagating in the $+z$ direction (or the negative of the ratio of the voltage to the current phasor of the wave traveling in the $-z$ direction) and, for a lossless line, is given by $Z_0 = \sqrt{L/C}$. The phase velocity and the wavelength for a lossless line are given as

$$v_p = 1/\sqrt{LC}; \qquad \lambda = 2\pi/\beta = v_p/f$$

Note that the phase velocity of a lossless line is independent of frequency.

- **Input impedance of short- and open-circuited lines.** The line impedance of a transmission line seen looking toward the load at any position along the line is defined as

$$Z(z) = \frac{V(z)}{I(z)} = Z_0 \frac{V^+ e^{-j\beta z} + V^- e^{+j\beta z}}{V^+ e^{-j\beta z} - V^- e^{+j\beta z}}$$

The input impedances of short- or open-circuited transmission lines of length l are purely imaginary and are given by

$$Z_{\text{sc}} = jZ_0 \tan(\beta l) \qquad\qquad \text{short-circuited line}$$

$$Z_{\text{oc}} = -jZ_0 \cot(\beta l) \qquad\qquad \text{open-circuited line}$$

Since any arbitrary reactive impedance can be realized by simply adjusting the length l of open- or short-circuited stubs, these stubs are commonly used as reactive circuit elements for impedance matching and other applications.

- **Reflection coefficient.** It is common practice to treat steady-state transmission line problems by considering the wave traveling in the $+z$ direction (toward the load) as the incident wave and the wave traveling in the $-z$ direction (away from the load and toward the source) as the reflected wave. The ratio of the reflected to the incident voltage phasor at any position z along the line is defined as the reflection coefficient, represented by $\Gamma(z)$. The reflection coefficient at the load end of the line (where $z = 0$) is given by

$$\Gamma_{\text{L}} = \frac{V^-}{V^+} = (Z_{\text{L}} - Z_0)/(Z_{\text{L}} + Z_0) = \rho e^{j\psi}$$

The case of $Z_{\text{L}} = Z_0$ is referred to as a matched load, for which there is no reflected wave, since $\Gamma_{\text{L}} = 0$. The reflection coefficient $\Gamma(z)$ at any other location z (where $z < 0$) on a lossless line is given by

$$\Gamma(z) = \frac{V^- e^{j\beta z}}{V^+ e^{-j\beta z}} = \Gamma_{\text{L}} e^{j2\beta z} = \rho e^{j(\psi + 2\beta z)}$$

- **Standing-wave pattern.** The superposition of the incident and reflected waves constitutes a standing-wave pattern that repeats every $\lambda/2$ over the length of the line. The standing-wave ratio S is defined as the ratio of the maximum to minimum voltage (current) magnitude over the line and is given by

$$S = \frac{1 + \rho}{1 - \rho}$$

where $\rho = |\Gamma_L|$. The standing-wave ratio S has practical significance because it is easily measurable. The value of S varies in the range $1 \le S \le \infty$, where $S = 1$ corresponds to $\rho = 0$ (i.e., no reflection case) and $S = \infty$ corresponds to $\rho = 1$ (i.e., the load is either open or short circuit or purely reactive).

- **Transmission line as an impedance transformer.** The line impedance of a lossless transmission line terminated in an arbitrary load impedance, defined as the ratio of the total voltage to current phasor at position z, is in general complex and is a periodic function of z, with period of $\lambda/2$. The line impedance is purely real at locations along the line where the total voltage magnitude is a maximum or minimum.

- **Power flow.** The net time-average power propagating toward the load on a lossless transmission line is given by

$$P(z) = \frac{|V^+|^2}{2Z_0}(1 - \rho^2)$$

and is equal to the power P_L delivered to the load. For a given value of $|V^+|$, the power delivered to the load is maximized under matched conditions, or $\rho = 0$. The degree of mismatch between the load and the line can be described in terms of return loss, given as

$$\text{Return loss} = 20 \log_{10} \frac{S + 1}{S - 1}$$

- **Impedance matching.** In most applications it is desirable to match the load impedance to the line in order to reduce reflections and standing waves. In single-stub matching, a short- or open-circuited stub is placed in shunt or series at a location $z = -l$ along the line at which the normalized line admittance or the impedance is given as

$$\bar{Y}_1(z)|_{z=-l} = 1 - j\bar{B}; \qquad \bar{Z}_1(z)|_{z=-l} = 1 - j\bar{X}$$

The matching is then completed by choosing the length l_s of a short- or open-circuited stub so that it presents an admittance or impedance at $z = -l$ of $\bar{Y}_s = j\bar{B}$ or $\bar{Z}_s = j\bar{X}$. In quarter-wave matching, it is first necessary to determine the location l along the line at which the line impedance is purely real, that is, where

$$Z(z)|_{z=-l} = R + j0$$

Matching to a line of impedance Z_0 is then completed by using a quarter-wavelength long line of characteristic impedance $Z_Q = \sqrt{Z_0 R}$.

- **Smith chart.** The fact that the impedance $Z(z)$ and the reflection coefficient $\Gamma(z)$ on a lossless line are both periodic functions of position z along the line makes it possible to analyze and visualize the behavior of the line using a graphical display of $\Gamma(z)$, S, and $Z(z)$ known as the Smith chart. The Smith chart provides a convenient means of analyzing transmission line problems to determine values of impedance and reflection coefficient (or standing-wave ratio). The Smith chart is also a useful tool for matching network design.

- **Lossy transmission lines.** The solutions for voltage and current propagating in the z direction on a lossy transmission line have the form (assuming V^+ is real)

$$\mathscr{V}(z,t) = V^+ e^{-\alpha z} \cos(\omega t - \beta z)$$

$$\mathscr{I}(z,t) = \frac{V^+}{|Z_0|} e^{-\alpha z} \cos(\omega t - \beta z - \phi_z)$$

where α and β are the real and imaginary parts of the propagation constant $\gamma = \alpha + j\beta = \sqrt{(R+j\omega L)(G+j\omega C)}$, R, L, G, and C are the per-unit distributed parameters of the line, and ω is the angular frequency of the excitation. The characteristic impedance for a lossy line is in general complex and is given by

$$Z_0 = |Z_0| e^{j\phi_z} = \sqrt{\frac{R+j\omega L}{G+j\omega C}}$$

For terminated lines, the general expressions for line voltage and current are

$$V(z) = V^+ e^{-\alpha z} e^{-j\beta z} [1 + \Gamma(z)]$$

$$I(z) = \frac{1}{Z_0} V^+ e^{-\alpha z} e^{-j\beta z} [1 - \Gamma(z)]$$

where $\Gamma(z) = \Gamma_L e^{2\alpha z} e^{j2\beta z}$, with $\Gamma_L = (Z_L - Z_0)/(Z_L + Z_0)$ being the complex load voltage reflection coefficient. The line voltage and current exhibit a standing-wave pattern near the load, but the differences between the maxima and minima become smaller as distance from the load increases. At sufficient distances from the load, the magnitudes of the line voltage and current do not vary significantly with distance, as if the line were matched. The impedance of a lossy transmission line is given by

$$Z(z) = Z_0 \frac{1 + \Gamma(z)}{1 - \Gamma(z)} = Z_0 \frac{1 + \Gamma_L e^{2\alpha z} e^{j2\beta z}}{1 - \Gamma_L e^{2\alpha z} e^{j2\beta z}}$$

The real and imaginary parts of $Z(z)$ exhibit maxima and minima near the load, similar to that of a lossless line. However, at sufficient distances from the load, $Z(z)$ approaches Z_0, as if the line were matched.

PROBLEMS

3.1 **Transmission line capacitor.** (a) Design an open-circuited line with the shortest line that will provide the impedance of a 4 nF capacitor at 10 GHz. (b) Redesign the same capacitor using a short circuited 50Ω air transmission line. (c) Which design yield the shortest length and why?

3.2 **Transmission line inductor.** (a) Design an open-circuited 50Ω lossless microstrip transmission line with shortest electrical length that will provide the impedance of a 5 nH inductor at 5 GHz. (b) Repeat the same design using a short-circuited 50Ω microstrip line. (c) Which design resulted in a shortest length line and why?

3.3 **Input impedance of a transmission line.** Consider a short circuited 100Ω lossless transmission line as shown in Figure 3.60. Find the shortest electrical length of the line such that (a) $Z_{in} = j50\Omega$, (b) $Z_{in} = -j150\Omega$, (c) $Z_{in} = \infty$, and (d) $Z_{in} = 0$.

Figure 3.60   Short-circuited transmission line. Problem 3.3.

3.4 **Terminated transmission line.** A 50Ω lossless transmission line is terminated with a capacitive load having a load impedance of $Z_L = 100 - j100\Omega$, as shown in Figure 3.61. (a) Find the load reflection coefficient, Γ_L. (b) Find the standing wave ratio S on the line. (c) Find the input impedance of the line, Z_{in}, for four different line lengths: $l_1 = 0.125\lambda$, $l_2 = 0.25\lambda$, $l_3 = 0.375\lambda$, and $l_4 = 0.5\lambda$, respectively.

Figure 3.61   Terminated transmission line. Problem 3.4.

3.5 **Input impedance.** Find the input impedance Z_{in} of the two cascaded lossless transmission line system shown in Figure 3.62 for (a) $l_2 = 0.25\lambda$, (b) $l_2 = 0.5\lambda$, and (c) $l_2 = 0.125\lambda$.

3.6 **Source- and load-end voltages.** For the lossless transmission line circuit shown in Figure 3.63, calculate the phasor-form source-end and load-end voltages V_s and V_L.

3.7 **Capacitive termination** An air transmission line with $Z_0 = 50\Omega$ is terminated with a capacitive load impedance of $50 - j90\Omega$. If the line is operated at 100 MHz, calculate (a) the load reflection coefficient Γ_L, (b) the standing-wave ratio S on the line, (c) the distance from the load to the first voltage minimum, and (d) the distance from the load to the first voltage maximum.

Figure 3.62   Two cascaded transmission lines. Problem 3.5.

Figure 3.63   Source- and load-end voltages. Problem 3.6.

3.8 **A wireless communication antenna.** The following table provides the approximate values at various frequencies of the feed-point impedance of a circularly polarized patch antenna used in the wireless industry for making cellular phone calls in difficult environments, such as sport arenas and office buildings:

$f\,(\mathbf{MHz})$	$Z_L\,(\mathbf{\Omega})$
800	$21.5 - j\,19.4$
850	$38.5 + j\,2.24$
900	$43.8 + j\,9.74$
950	$55.2 - j\,10.2$
1000	$28.8 - j\,7.40$

If this antenna is directly fed by a $50\,\Omega$ transmission line, find and sketch the standing-wave ratio S as a function of frequency.

3.9 **Resistive line impedance.** A $100\,\Omega$ transmission line is terminated with a load impedance of $30 + j\,60\,\Omega$ at $z = 0$. Find the minimum electrical length l/λ of the line at which the line impedance (i.e., $Z(z = -l)$) is purely resistive. What is the value of the resistive line impedance?

3.10 **Real Z_{in} position.** Consider a $50\,\Omega$ transmission line terminated with a load impedance of $50 + j\,140\,\Omega$. (a) Find the load reflection coefficient Γ_L and the standing wave ratio S on the line. (b) Find the electrical position on the line nearest to the load where the line impedance is purely real. (c) Find the value of the line impedance at the position found in part (b).

3.11 **Cascaded transmission lines.** Two lossless transmission lines are cascaded as shown in Figure 3.64. (a) Find the standing-wave ratio S on each line. (b) Find the time-average power delivered to the load.

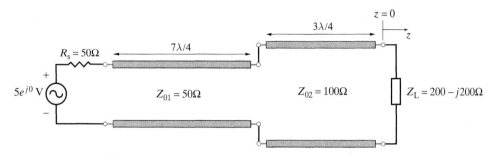

Figure 3.64   Inductive termination. Problem 3.11.

3.12 **Inductive termination.** A 75Ω lossless transmission line is terminated with an inductive load given by $45 + j60\Omega$, as shown in Figure 3.65. Calculate (a) the load reflection coefficient Γ_L, (b) the standing-wave ratio S on the line, (c) the percentage time-average incident power that is absorbed by the load, and (d) the V_{max} and V_{min} positions that are nearest to the load.

Figure 3.65   Inductive termination. Problem 3.12.

3.13 **Input impedance.** The input impedance of an 160Ω transmission line terminated with an unknown load Z_L and having an electrical length of $3/8$ at frequency f_1 is measured to be $64 - j48\Omega$. Find the new value of Z_{in} at frequency $f_2 = 2f_1$. Assume Z_L and Z_0 values to stay the same with frequency.

3.14 **Resistive load.** A lossless line is terminated with a resistive load of 120Ω. If the line presents an impedance of $48 + j36\Omega$ at a position $5\lambda/8$ away from the load, what is the characteristic impedance Z_0 of the line?

3.15 **Input impedance.** For the lossless transmission-line system shown in Figure 3.66, find Z_{in} for the following load impedances: (a) $Z_L = \infty$ (open circuit), (b) $Z_L = 0$ (short circuit), and (c) $Z_L = Z_0/2$.

Figure 3.66   Input impedance. Problem 3.15.

3.16 Input impedance. Find the input impedance Z_{in} of the two cascaded lossless transmission lines as shown in Figure 3.67, where $Z_P = j50\Omega$ is a lumped impedance connected at the junction.

Figure 3.67 Cascaded transmission line. Problem 3.16.

3.17 Input impedance. Repeat Problem 3.15 for the circuit shown in Figure 3.68.

Figure 3.68 Input impedance. Problem 3.17.

3.18 Unknown load. Consider a 50Ω transmission line terminated with an unknown load. If the standing-wave ratio on the line is measured to be 4.2 and the nearest voltage minimum point on the line with respect to the load position is located at 0.21λ, find the following: (a) The load impedance Z_L. (b) The nearest voltage maximum and the next voltage minimum positions with respect to the load. (c) The input impedance Z_{in} at each position found in part (b).

3.19 Input impedance. For the lossless transmission-line system shown in Figure 3.69, what is the ratio Z_{01}/Z_{02} if $Z_{in} = 450\Omega$?

Figure 3.69 Input impedance. Problem 3.19.

3.20 Unknown termination. Consider a transmission line with $Z_0 = 50\Omega$ terminated with an unknown load impedance Z_L. (a) Show that

$$Z_L = Z_0 \frac{1 - jS \tan(\beta l_{min})}{S - j \tan(\beta l_{min})}$$

where l_{min} is the length from the load to the first voltage minimum and S is the standing-wave ratio. (b) Measurements on a line with $Z_0 = 50\Omega$ having an unknown termination

Z_L show that $S = \sqrt{3}$, $l_{min} = 25$ mm, and that the distance between successive minima is 10 cm. Find the load reflection coefficient Γ_L and the unknown termination Z_L.

3.21 **Unknown load.** A 50Ω air transmission line with a standing wave ratio of 3 has its first and second voltage maximums nearest to the load located at 0.2 m and 0.4 m respectively. Calculate (a) the operating frequency f and (b) the unknown load impedance Z_L.

3.22 **Unkown load.** A 75Ω air transmission line with a standing wave ratio of 6 has a voltage maximum and an adjacent voltage minimum position located with respect to the load at 1.5 m and 3.5 m, respectively. Calculate (a) the operating frequency f and (b) the unknown load impedance Z_L.

3.23 **Distance to the first maximum.** Derive a formula similar to that in Problem 3.20 in terms of l_{max}, where l_{max} is the distance from the load to the first voltage maximum.

3.24 **Terminated transmission line.** A 50Ω air transmission line is terminated with an inductive load impedance given by $Z_L = 100 + j150\Omega$ and excited by a sinusoidal voltage source as shown in Figure 3.70. Calculate: (a) the load reflection coefficient Γ_L, (b) the standing wave ratio S, (c) the time-average power delivered to the load, and (d) the first two V_{max} and the first two V_{min} positions on the transmission line nearest to the load.

Figure 3.70   Terminated transmission line. Problem 3.24.

3.25 **Terminated transmission line.** A 50Ω, 10.5-m long air transmission line terminated with a load impedance of $Z_L = 70 + j10\Omega$ is excited by a sinusoidal voltage source, as shown in Figure 3.71. (a) Calculate the load reflection coefficient Γ_L and the standing wave ratio S on the line. (b) Find all the V_{max} and V_{min} positions (in actual lengths) on the line. (c) Find all the I_{max} and I_{min} positions on the line. (d) Find the input impedance Z_{in} seen at each V_{max} and V_{min} position. (e) Find the line impedance $Z(z)$ seen at the source end of the line and draw the equivalent lumped circuit with respect to the source end. (f) Find the phasor voltages V_s, V^+, V^-, and V_L. (g) Find the V_{max} and V_{min} values. (h) Find the I_{max} and I_{min} values. (i) Find the time-average powers P^+, P^-, P_{R_s}, P_L, and P_{source}. What percentage of the power carried by the incident wave reflects back towards the source? (j) Repeat parts (a) through (i) for a load impedance of $Z_L = 15 - j35\Omega$.

Figure 3.71   Terminated transmission line. Problem 3.25.

3.26 **Power dissipation.** For the lossless transmission line system shown in Figure 3.72, with $Z_0 = 100\Omega$, (a) calculate the time-average power dissipated in each load. (b) Switch the values of the load resistors (i.e., $R_{L1} = 200\Omega$, $R_{L2} = 50\Omega$), and repeat part (a).

Figure 3.72 **Power dissipation.** Problem 3.26.

3.27 **Power dissipation.** Consider the transmission line system shown in Figure 3.73. (a) Find the time-average power dissipated in the load R_L with the switch S open. (b) Repeat part (a) for the switch S closed. Assume steady state in each case.

Figure 3.73 **Power dissipation.** Problem 3.27.

3.28 **Power dissipation.** Repeat Problem 3.27 if the characteristic impedance of the transmission line on the source side is changed from 50Ω to $25\sqrt{2}\Omega$.

3.29 **Cascaded transmission lines.** Two cascaded lossless transmission lines are connected as shown in Figure 3.74. (a) Find the standing-wave ratio on each line. (b) Find the time-average powers delivered to impedances Z_1 and Z_2.

Figure 3.74 **Cascaded transmission lines.** Problem 3.29.

3.30 **Two antennas.** Two antennas having feed-point impedances of $Z_{L1} = 40 - j30\Omega$ and $Z_{L2} = 100 + j50\Omega$ are fed with a transmission line system, as shown in Figure 3.75. (a) Find S on the main line. (b) Find the time-average power supplied by the sinusoidal source. (c) Find the time-average power delivered to each antenna. Assume lossless lines.

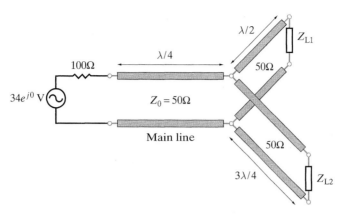

Figure 3.75   Two antennas. Problem 3.30.

3.31   Power dissipation. For the transmission line network shown in Figure 3.76, calculate the time-average power dissipated in the load resistor R_L.

Figure 3.76   Power dissipation. Problem 3.31.

3.32   Three identical antennas. Three identical antennas A1, A2, and A3 are fed by a transmission line system, as shown in Figure 3.77. If the feed point impedance of each antenna is $Z_L = 50 + j50\Omega$ find the average power delivered to each antenna.

Figure 3.77   Three identical antennas. Problem 3.32.

3.33   Power delivery. The transmission line system shown in Figure 3.78 is to be used at a frequency f such that $\omega = 2\pi f = 0.5 \times 10^9$ rad-s^{-1}. Determine the total time-average power supplied by the source and also the time-average power supplied to each of the three different 25Ω load resistances (connected to the ends of lines 2, 3, and 4).

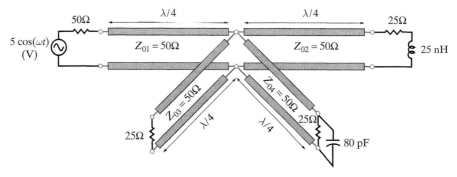

Figure 3.78 Power delivery. Problem 3.33.

3.34 Power delivery. Consider the transmission line system shown in Figure 3.79. (a) Calculate the time-average power dissipated in the load resistances $R_{L1}=50\Omega$ and $R_{L2}=50\Omega$ at $f=f_1$. (b) Calculate the time-average power dissipated in the load resistances $R_{L1}=50\Omega$ and $R_{L2}=50\Omega$ at $f=2f_1$. (c) Calculate the time-average power dissipated in the load resistances $R_{L1}=50\Omega$ and $R_{L2}=50\Omega$ at $f=1.5f_1$.

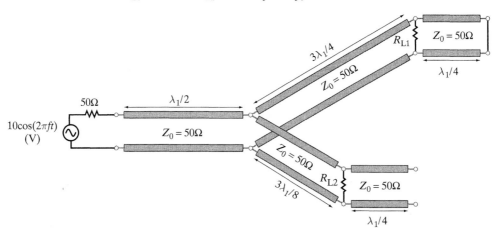

Figure 3.79 Power delivery. Problem 3.34.

3.35 Matching with a single lumped element. The transmission line matching networks shown in Figure 3.80 are designed to match a 100Ω load impedance to a 50Ω line. (a) For the network with a shunt element, find the minimum distance l from the load where the unknown shunt element is to be connected such that the input admittance seen at B–B' has a conductance part equal to 0.02 S. (b) Determine the unknown shunt element and its element value such that the input impedance seen at A–A' is matched to the line (i.e., $Z_{A-A'} = 50\Omega$) at 1 GHz. (c) For the matching network with a lumped series matching element, find the minimum distance l and the unknown element and its value such that a perfect match is achieved at 1 GHz. Assume $v_p = 30$ cm-(ns)$^{-1}$.

3.36 Matching with series stub. A load impedance of $135 - j90\Omega$ is to be matched to a 75Ω lossless transmission line system, as shown in Figure 3.81. If $\lambda = 20$ cm, what minimum length of transmission line l will yield a minimum length l_s for the series stub?

Figure 3.80   Matching with a single lumped element. Problem 3.35.

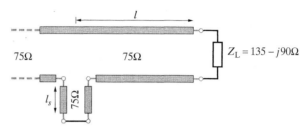

Figure 3.81   Matching with a series shorted stub. Problem 3.36.

3.37   Series stub matching. A series-shorted-stub matching network is designed to match a capacitive load of $R_L = 50\,\Omega$ and $C_L = 10/(3\pi)$ pF to a $100\,\Omega$ line at 3 GHz, as shown in Figure 3.82. (a) The stub is positioned at a distance of $\lambda/4$ away from the load. Verify the choice of this position and find the minimum electrical length of the stub to achieve a perfect match at the design frequency. (b) Calculate the standing-wave ratio S on the main line at 2 GHz. (c) Calculate S on the main line at 4 GHz.

Figure 3.82   Series stub matching. Problem 3.37.

3.38   Quarter-wave transformer. (a) Design a single-section quarter-wave matching transformer to match an $R_L = 20\,\Omega$ load to a line with $Z_0 = 80\,\Omega$ operating at 1.5 GHz. (b) Calculate the standing-wave ratio S of the designed circuit at 1.2 and 1.8 GHz.

3.39   Helical antenna. The feed-point impedance of an axial-mode helical antenna with a circumference C on the order of one wavelength is nearly purely resistive and is approximately given[33] by $R_L \simeq 140(C/\lambda)$, with the restriction that $0.8\lambda \le C \le 1.2\lambda$. Consider a helical antenna designed with a circumference of $C = \lambda_0$ for operation at a frequency f_0 and corresponding wavelength λ_0. The antenna must be matched for use with a 50Ω transmission line at f_0. (a) Design a single-stage quarter-wave transformer to realize the design objective. (b) Using the circuit designed in part (a), calculate the standing-wave ratio S on the 50Ω line at a frequency 15% above the design frequency. (c) Repeat part (b) at a frequency 15% below the design frequency.

3.40　**Quarter-wave matching.** Many microwave applications require very low values of S over a broad band of frequencies. The two circuits shown in Figure 3.83 are designed to match a load of $Z_L = R_L = 400\Omega$ to a line with $Z_0 = 50\Omega$, at 900 MHz. The first circuit is an air-filled coaxial quarter-wave transformer, and the second circuit consists of two air-filled coaxial quarter-wave transformers cascaded together. (a) Design both circuits. Assume $Z_{Q1}Z_{Q2} = Z_0Z_L$ for the second circuit. (b) Compare the bandwidth of the two circuits designed by calculating S on each line at frequencies 15% above and below the design frequency.

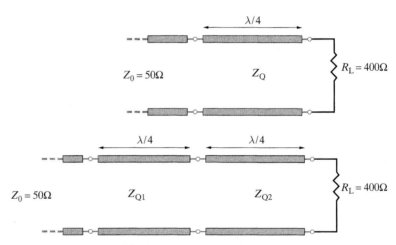

Figure 3.83　**Quarter-wave matching.** Problem 3.40.

3.41　**Quarter-wave matching.** A 75Ω coaxial line is connected directly to an antenna with a feed-point impedance of $Z_L = 156\Omega$. (a) Find the load-reflection coefficient and the standing-wave ratio. (b) An engineer is assigned the task of designing a matching network to match the feed-point impedance of the antenna (156Ω) to the 75Ω coaxial line. However, all he has available to use for this design is another coaxial line of characteristic impedance 52Ω. The engineer uses the quarter-wave matching technique to achieve the match. How?

3.42　**L-section matching networks.** A simple and practical matching technique is to use the lossless L-section matching network that consists of two reactive elements. (a) Two L-section matching networks marked A1 and A2, each consisting of a lumped inductor and a capacitor, as shown in Figure 3.84, are used to match a load impedance of $Z_L = 60 - j80\Omega$ to a 100Ω line. Determine the L section(s) that make(s) it possible to achieve the

---

[33] See Chapter 7 of J. D. Kraus, *Antennas*, 2nd. ed., McGraw-Hill, New York, 1988.

design goal, and calculate the appropriate values of the reactive elements at 800 MHz. (b) Repeat part (a) for the two L-section networks marked B1 and B2, consisting of two inductances and two capacitors, respectively.

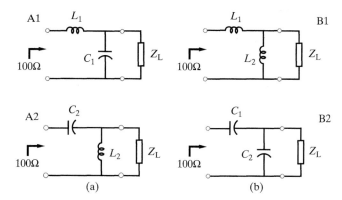

Figure 3.84   L-section matching networks. Problem 3.42.

3.43   Variable capacitor. A shunt stub filter consisting of an air-filled coaxial line terminated in a variable capacitor is designed to eliminate the FM radio frequencies (i.e., 88–108 MHz) on a transmission line with $Z_0 = 100\Omega$, as shown in Figure 3.85. If the stub length is chosen to be 20 cm, find the range of the variable capacitor needed to eliminate any frequency in the FM band. Assume the characteristic impedance of the stub to be also equal to 100Ω.

Figure 3.85   Variable capacitor.
Problem 3.43.

3.44   Impedance matching network design. Consider the transmission-line circuit as shown in Figure 3.86. As a design engineer, your task is to determine the electrical lengths of the two short-circuited stubs (each 50Ω) connected at the load position to match the load impedance $Z_L = 25 - j75\Omega$ to the line characteristic impedance $Z_0 = 50\Omega$ such that $l_1 + l_2$ is minimum.

3.45   Matching with lumped reactive elements. Two variable reactive elements are positioned on a transmission line to match an antenna having a feed-point impedance of $100 + j100\Omega$ to a $Z_0 = 100\Omega$ air-filled line at 5 GHz, as shown in Figure 3.87. (a) Determine the values of the two reactive elements to achieve matching. (b) If the reactive elements are to be replaced by shorted 50Ω air-filled stubs, determine the corresponding stub lengths.

3.46   Standing-wave ratio. For the transmission line shown in Figure 3.88, calculate S on the main line at (a) 800 MHz, (b) 880 MHz, and (c) 960 MHz.

3.47   Quarter-wave transformer design. Consider the transmission-line circuit, as shown in Figure 3.89. Design the quarter-wave matching network shown (find its electrical position l/λ from the load and its characteristic impedance Z_Q) to match the load impedance Z_L to Z_0 under the condition $Z_Q > Z_0$.

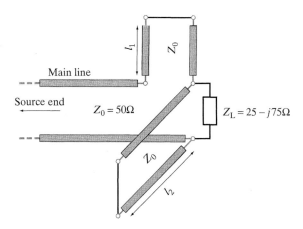

Figure 3.86   Impedance matching network design. Problem 3.44.

Figure 3.87   Matching with lumped reactive elements. Problem 3.45.

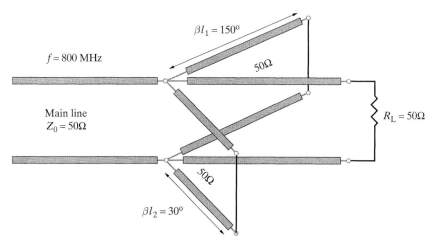

Figure 3.88   Standing-wave ratio. Problem 3.46.

Figure 3.89   Quarter-wave transformer design. Problem 3.47.

3.48 **Impedance matching.** A load of $Z_L = 36 + j40\Omega$ is to be matched to a main line of characteristic impedance $Z_0 = 100\Omega$ using a circuit as shown in Figure 3.90. (a) The open-circuited stub is positioned at a distance of $l_1 = \lambda/4$ away from the load. Determine the characteristic impedance Z_Q of this quarter-wave segment and the stub length l_s which would result in a match. (b) For the values found in part (a), calculate the standing wave ratio S on the quarter-wave line segment with the characteristic impedance Z_Q. Also find the magnitude and the phase of the V_1 at the open-circuited end of the stub.

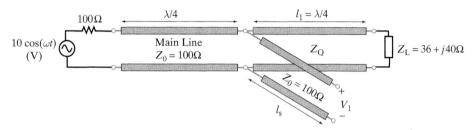

Figure 3.90   Impedance matching. Problem 3.48.

3.49 **Unknown feed-point impedance.** A 50Ω transmission line is terminated with an antenna that has an unknown feed-point impedance. An engineer runs tests on the line and measures the standing-wave ratio, wavelength, and a voltage minimum location away from the antenna feed point to be at 3.2 cm, 20 cm, and 74 cm, respectively. Use the Smith chart to find the feed-point impedance of the antenna.

3.50 **A lossy high-speed interconnect.** The per-unit line parameters of an IC interconnect at 5 GHz are extracted using a high-frequency measurement technique resulting in $R = 143.5\Omega\text{-(cm)}^{-1}$, $L = 10.1$ nH-(cm)$^{-1}$, $C = 1.1$ pF-(cm)$^{-1}$, and $G = 0.014$ S-(cm)$^{-1}$, respectively. [34] Find the propagation constant γ and the characteristic impedance Z_0 of the interconnect at 5 GHz.

3.51 **Characterization of a high-speed GaAs interconnect.** The propagation constant γ and the characteristic impedance Z_0 at 5 GHz of the GaAs coplanar strip interconnects considered in Example 3.27 are determined from the measurements to be $\gamma \simeq 1.1$ np-(cm)$^{-1} + j3$ rad-(cm)$^{-1}$ and $Z_0 \simeq 110 - j40\Omega$, respectively. Using these values, calculate the per-unit length parameters (R, L, G, and C) of the coplanar strip line at 5 GHz.

3.52 **A lossy high-speed interconnect.** Consider a high-speed microstrip transmission line of length 10 cm used to connect a 1-V amplitude, 2-GHz, 50Ω sinusoidal voltage source to an integrated circuit chip having an input impedance of 50Ω. The per-unit parameters of the microstrip line at 2 GHz are measured to be approximately given by $R = 7.5\Omega\text{-(cm)}^{-1}$, $L = 4.6$ nH-(cm)$^{-1}$, $C = 0.84$ pF-(cm)$^{-1}$, and $G = 0$, respectively. (a) Find the propagation constant γ and the characteristic impedance Z_0 of the line. (b) Find the voltages at the source and the load ends of the line. (c) Find the time-average power delivered to the line by the source and the time-average power delivered to the load. What is the power dissipated along the line?

[34]W. R. Eisenstadt and Y. Eo, *S* parameter-based IC interconnect transmission line characterization, *IEEE Trans. Comp. Hybrids Manufact. Technol.*, 15(4) pp. 483–489, August 1992.

4

The Static Electric Field

Following our discussion of transmission line behavior in Chapters 2 and 3, we begin in this chapter a more traditional study of electric and magnetic fields, following the historical development of ideas that led to the general laws of electromagnetics known as Maxwell's equations. The voltage and current waves on two-conductor transmission lines are but special cases of electromagnetic waves, which can propagate in empty space and in material media as well as being guided by a variety of conducting or insulating structures. Kirchhoff's voltage and current laws, which were used in Chapter 2 to derive the basic transmission line voltage and current equations, are special cases of Maxwell's equations, which in general are expressed in terms of electric and magnetic fields. This chapter is the first of four chapters in which we undertake the development of these fundamental laws of electromagnetics, culminating in the establishment of the full set of interconnected laws in Chapter 7. Our bases for the development of Maxwell's equations are experimentally established facts of nature, as verified by physical observations.

The interaction between electric and magnetic fields is at the root of electromagnetic wave propagation—the basis for all telecommunications, from the telegraph to today's satellite, wireless, and optical fiber networks. The interaction between electric and magnetic fields is also responsible for the behavior of electric circuit elements and networks as well as the workhorses of our industrial society: electrical motors and machinery. Our interest in electromagnetics stems from a need to understand the behavior of practical devices and systems, to describe such devices and systems mathematically, to predict their performance, and to design systems for particular applications. To achieve these goals requires an understanding of the physical bases of the fundamental laws of electromagnetics, which are developed in Chapters 4 through 7 and stated compactly as Maxwell's equations in Section 7.4. The branch of electromagnetics dealing with electric charges at rest, namely static electricity or *electrostatics,* involves the study of the first, and one of the most important, of these fundamental laws, known as Coulomb's law.[1] Coulomb's law is based on physical observation and is not logically or mathematically derivable from any other concept. The experimental basis of this law, the mathematical formulations that it leads to, and its broad range of applications and implications are the subjects of this chapter.

The study of electrostatics constitutes a relatively simple first step in our quest to understand the laws of electromagnetics. In electrostatics we deal only with electric charges that are at rest and that do not vary with time. Yet, mastering the behavior of static electric fields and the techniques for the solution of electrostatic problems is essential to the understanding of more complicated electromagnetic phenomena. Furthermore, many natural phenomena and the principles of some important industrial and technological applications are electrostatic in nature.

Lightning, corona discharge, and arcing are natural phenomena that involve very strong electrostatic fields that cause ionization in the surrounding medium. Lightning discharges involve currents of tens of kiloamperes that flow for a few hundred microseconds and represent release of energies of up to 10^{10} joules.[2] Applications of electrostatics in industry encompass diverse areas—cathode ray tubes (CRTs) and flat panel displays, which are widely used as display devices for computers and oscilloscopes; ink jet printers, which can produce good-quality printing at very fast speeds; and photocopy machines are all based on electrostatic fields. Electrostatic technologies are extensively used for sorting charged or polarized granular materials;[3] applications of this technology are reflected in hundreds of patents, extending from mineral beneficiation and seed conditioning to recycling of metals and plastics from industrial wastes. In Microelectromechanical Systems (MEMS), where small mechanical devices are fabricated together with integrated circuits, electrostatic forces are often utilized by the integrated digital electronics as a means to interact with the surrounding environment (see Section 4.14). Among other applications, electrostatic spraying and painting, electrostatic precipitators and filters, electrostatic transducers, and electrostatic recording are utilized in numerous industrial and household applications.

Our discussion of electrostatics in this chapter also serves to bring us closer to a full understanding of the underlying physical basis of the transmission line behavior discussed in Chapters 2 and 3. One of the important physical properties of a transmission line is its distributed capacitance, which comes about as a result of the separation of charge induced on the two conductors that constitute a transmission line. In Chapters 2 and 3, we took it for granted that a two-conductor system exhibits capacitance, and we presented formulas (in Table 2.2) for the distributed capacitances of a few common transmission line

---

[1]That electric charges attract or repel one another in a manner inversely proportional to the square of the distance between them. C. A. de Coulomb, Première mémoire sur l'électricité et magnétisme (First memoir on electricity and magnetism), *Histoire de l'Académie Royale des Sciences,* p. 569, 1785.

[2]See Sections 1.3 and 1.8 of M. A. Uman, *The Lightning Discharge,* Academic Press, San Diego, California, 1987.

[3]A. D. Moore (Ed.), *Electrostatics and Its Applications,* John Wiley & Sons, 1973; F. S. Knoll, J. E. Lawver, and J. B. Taylor, Electrostatic separation, in *Ullman's Encyclopedia of Industrial Chemistry,* 5th ed., VCH, Weinheim, New York, 1988, Vol. B2, pp. 20-1–20-11.

structures. In this chapter we define the concept of capacitance and discuss how the capacitance of different conductor configurations can be determined using Coulomb's law.

In addition to starting our discussion of the fundamental underpinnings of electromagnetic theory, we also present in this chapter the important concepts of vector algebra. Although some of the most basic aspects of vector algebra are provided in Appendix A, we introduce important concepts such as gradient and divergence along with the relevant physical laws. This approach ensures that the physical significance of the vector operations can be best understood in the electromagnetic context. Throughout this book, vector quantities are often written using boldface symbols (e.g., $\mathbf{G}$), although a bar above the symbol is also used (e.g., $\overline{\mathcal{G}}$).[4] In either case, the vector in question is understood in general to have three components, G_x, G_y, and G_z, and it is often written as $\mathbf{G} = \hat{\mathbf{x}}G_x + \hat{\mathbf{y}}G_y + \hat{\mathbf{z}}G_z$, where $\hat{\mathbf{x}}$, $\hat{\mathbf{y}}$, and $\hat{\mathbf{z}}$ are the unit vectors in the x, y, and z directions, respectively. The "hat" notation is always used to represent a unit vector, so we can also write the vector $\mathbf{G}$ as $\mathbf{G} = \hat{\mathbf{G}}G$, where $\hat{\mathbf{G}}$ is a unit vector in the direction of $\mathbf{G}$ and G is the magnitude of $\mathbf{G}$, $G = |\mathbf{G}|$. Points in three-dimensional space are identified by means of their position vectors, which point from the origin of the coordinate system to the point in question. For example, the position vector for point P located at (x_p, y_p, z_p) is $\mathbf{r}_p = \hat{\mathbf{x}}x_p + \hat{\mathbf{y}}y_p + \hat{\mathbf{z}}z_p$.

4.1 ELECTRIC CHARGE

Our experiences with electricity date back to ancient times[5] and have their roots in the observation that, for example, a piece of glass and a piece of resin (or rubber) attract one another if they are first rubbed together and then separated. Also, if a second piece of glass is rubbed with another piece of resin, the two pieces of glass or resin repel one another, while each glass piece attracts each piece of resin. Various manifestations of electrification by friction are encountered in our daily experiences.[6]

---

[4]This alternative notation is used starting in Chapter 7 in order to distinguish between real physical quantities and their corresponding complex *phasors,* necessary for sinusoidal (or time-harmonic) applications.

[5]For example, Thales of Miletus [640–540 B.C.] wrote that a piece of amber rubbed in silk attracts pieces of straw.

[6]In his text, *Electricity and Magnetism* (Cambridge Press, 1907), Sir James Jeans gives the following amusing account of Robert Symmer's [1759] observations: He was in the habit of wearing two pairs of stockings simultaneously: a worsted wool one for comfort and a silk pair for appearance. In pulling off his stockings, he noticed that they gave a crackling noise, and sometimes they even emitted sparks when taken off in the dark. On taking two stockings off together from the foot and then drawing the one from inside the other, he found that both became inflated to reproduce the shape of the foot and exhibited attractions and repulsions at a distance as much as a foot and a half. "When this experiment is performed with two black stockings in one hand, and two white in the other, it exhibits a very curious spectacle; the repulsion of those in the same colour, and attraction of those of different colours, throws them into an agitation that is not unentertaining, and makes them catch each at that of its opposite colour, and at a greater distance than one would expect. When allowed to come together they all unite in one mass. When separate, they resume their former appearance, and admit of the repetition of the experiment as often as you please, till their electricity, gradually wasting, stands in need of being recruited."

4.1.1 Electrification by Friction, Induction, and Conduction

These electrical phenomena of attraction and repulsion that come about due to *friction* and that are part of our daily experiences[7] are understood in terms of *electric charge*. Electric charge is said to be acquired by the material as a result of rubbing. In actual fact, when glass and resin, for example, are rubbed together, a small amount of charge is transferred from one to the other, causing each material to become non-neutral, that is, charged. The glass becomes positively charged, while the resin acquires negative charge. The different behavior of glass and resin indicates that there must be two different types of charge. Materials that behave upon electrification like glass are said to be *positively charged,* and those that behave like resin are said to be *negatively charged.* Other materials also acquire charge as a result of being rubbed, although this property may be less apparent for some than others.

Metallic materials, such as brass, do not retain electricity for a sufficient time (after they are rubbed to another object) for us to observe it. However, a brass rod with a glass handle becomes electrified to a marked degree on rubbing. Once it is electrified, such a brass rod with a glass handle loses all of its electricity if it comes in contact with water, flame, or the human body. Conversely, the same rod retains its power if it comes in contact with hard rubber, a piece of silk, or wood. We understand such behavior in terms of the classification of materials into *conductors* of electricity and *insulators*. Metals are excellent conductors of electricity. Solutions of salts, acids, and other electrolytes are also conductors. Examples of good insulators are oils, waxes, silk, glass, rubber, and plastics. Gases are ordinarily good insulators, but flames or other ionized gases are good conductors. Distilled water is almost a perfect insulator, but any other type of water contains impurities, which in general cause it to conduct reasonably well. Being made largely of water, the human body conducts reasonably well and is generally a bad insulator.

Although electrification by *friction* is part of our daily experiences, material bodies can also acquire electric charge through *induction*. If we suspend an uncharged metallic (e.g., brass) object by silk threads, as shown in Figure 4.1a, and bring near one end of it a positively (or negatively) charged rod, the nearby end of the suspended object becomes negatively (or positively) charged, while the other end becomes primarily positive (or negative). If the charged rod is removed, the suspended object loses its electrification. However, if the suspended object is cut in the middle before the inducing charge is removed, each of its parts remains electrified, one positive and the other negative, as shown in Figure 4.1b.

When a charged conductor is connected to another conductor (charged or not), for example by means of a metal wire, the total charge on both bodies is shared between the two. The second body is then said to have become electrified via *conduction* through the

---

[7]In our everyday lives, we experience frictional electrification every time we rub our feet across a wool carpet, pull clothes out of a dryer, or comb our hair with a plastic comb. On a dry summer day, it is not uncommon to receive a shock on touching a doorknob, hear an accompanying crackling noise, and sometimes see a spark. We see sparks in such cases because large amounts of charge create electric fields that cause local breakdown of air.

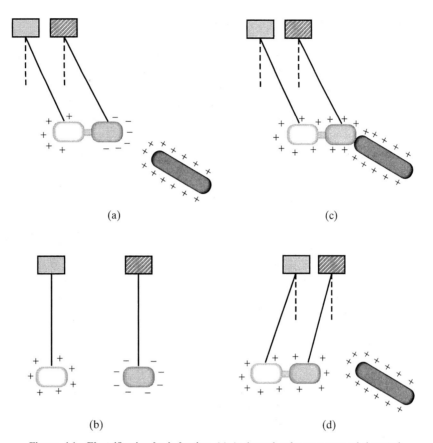

Figure 4.1 Electrification by induction. (a) A charged rod near a suspended neutral object induces opposite charge on the surface closer to it. Since the suspended object is initially neutral, an equal amount of positive charge remains on its far surface. Since the negative charge is closer to the charged rod, the attractive force is stronger than the repelling force, and the suspended object deflects toward the charged rod. (b) If we cut the suspended object in the middle while the charged rod is near it, each of its two parts remains electrified, one positive, one negative. This is an example of electrification by *induction*. (c) If the charged rod is brought in physical contact with the neutral body, the excess positive charge on the rod is shared between the two objects and the body is now positively charged. This is an example of electrification by *conduction*. (d) The two positively charged objects now repel one another.

metal wire. Such conduction would clearly not occur if the two bodies were connected with a silk thread. Once again we see a basis for classifying materials as conductors (e.g., metals) and insulators (e.g., silk). Electrification by conduction is essentially what occurs when we place the charged rod in contact with the neutral body as in Figure 4.1c; the charge on the rod is now shared between the two objects, which now repel one another, as shown in Figure 4.1d.

4.1.2 Faraday's Gold-Leaf Electroscope

Long series of experiments carried out by M. Faraday,[8] many years after the establishment of Coulomb's law, were instrumental in bringing about a physical understanding of electrical phenomena and demonstrated the principle of conservation of electric charge. In assessing the quantity of electrification or the quantity of charge associated with any body, it is useful[9] to think in terms of a gold-leaf electroscope, as shown in Figure 4.2.

Under normal conditions the gold leaves in Figure 4.2a hang flat side by side. When an electrified body (e.g., a charged rod) touches (conduction) or is brought near the brass rod of the gold-leaf electroscope (induction), the two gold leaves separate because of electrostatic repulsion (Figure 4.2b), so that the electroscope can be used to examine the degree to which a body is charged.

Now consider a metal vessel placed on top of the brass rod (Figure 4.2c), closed but having a lid attached to a silk thread so that it can be opened or closed without touching it. When a charged glass ball is inserted into the vessel and the lid is closed, opposite charges are induced on the inner surface of the vessel. Since the metal vessel and the gold leaves were originally neutral, unbalanced positive charge appears on the outside of the vessel, some of which is on the gold leaves, which repel one another and diverge. The separation of the gold leaves, which indicates the degree of electrification (i.e., amount of charge), remains exactly the same if we do the experiment by placing the electrified ball *at different positions* inside the vessel, as long as it does not come into contact with the vessel or other conductors. The separation of the gold leaves is also independent of any *changes in shape* of the glass piece (e.g., a thin rod versus a round ball or a straight rod versus a bent one) or any *changes in its state* (e.g., it might be heated[10] or cooled). It thus appears that the separation of the gold leaves is due only to a *quantity* of electricity, or *electric charge,* associated with the glass ball.

Now imagine two balls, one glass (A) and one rubber (B), electrified by rubbing against one another (Figure 4.2d). If we introduce A and B separately into the vessel, the gold leaves diverge by the same exact amount either way. If we introduce both A and B together into the vessel, we find that the leaves stay flat; that is, no electrification occurs outside the vessel. From this we conclude that the process of electrification by friction is *not* one that *creates* charge; rather, rubbing merely *transfers* charge[11] from one object to another, slightly disturbing the neutrality of each. If we insert two charged glass balls A and A′ into the vessel, the induced charge on the vessel would be the algebraic sum of what it would be when each ball was introduced separately. If, while A and A′ were inside the vessel, we connect them with a conducting wire, the induced charge on the vessel would not change, indicating once again that the *quantity of total charge remained constant.*

---

[8]M. Faraday, *Experimental Researches in Electricity,* Vol. 3, Art. 3249, Bernard Quatrich, London, 1855.

[9]We do not pretend here that Faraday's experiments can be carried out under modern conditions to verify electrical laws accurately or that a gold-leaf electroscope would be used today to measure charge. The discussion in this section should simply be viewed as a set of thought experiments that illustrate the properties and constitution of electricity remarkably well.

[10]Not with a flame, however; as was mentioned before, flames are good conductors because they consist of ionized matter.

[11]In actual fact, *electrons* are transferred, in this case from glass to rubber.

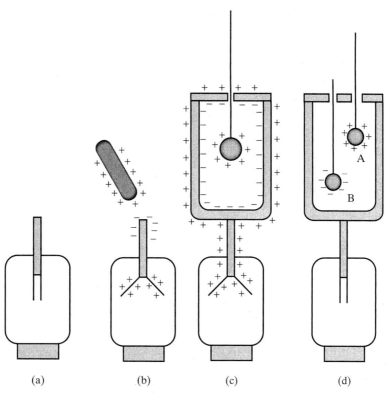

(a)   (b)   (c)   (d)

Figure 4.2   Faraday's gold-leaf electroscope. (a) An actual unit might consist of a glass vessel, through the top of which a metal (brass) rod is passed, to which are attached two gold leaves. (b) A charged object is brought near the brass rod; the gold leaves repel one another and diverge. (c) A closed metal vessel with a lid is placed on top of the brass rod so that a charged glass ball can be inserted within it. The gold leaves deflect by the same amount, regardless of the position of the charged ball within the vessel. (d) If two equally but oppositely charged bodies are inserted into the vessel, the gold leaves stay flat.

4.1.3 Electric Charge and the Atomic Structure

Our goal in this book is to develop and apply the fundamental laws of electrical and magnetic phenomena that are valid in the macroscopic, nonatomic realm. The physical experiments performed to verify Coulomb's law, and most applications of electrostatics, involve the use of ponderable macroscopic objects. The same is true for the other laws of electromagnetics that we shall study in later chapters. A brief review of modern notions concerning the atomic nature of matter is helpful in clarifying these macroscopic principles. Matter is composed of atoms with dimensions of order of one angstrom, or 10^{-8} cm, each containing a *nucleus* of dimensions $\sim 10^{-12}$ cm and electrons that move about the nucleus. It is the electrons' electrical influence on those of other atoms that effectively defines the atomic dimensions. Regardless of where they reside, electrons are all alike, each bearing a charge $q_e \simeq -1.602 \times 10^{-19}$ coulombs.[12] Most of the mass of an atom is contained in its nucleus,[13] which bears a positive charge of $N_a|q_e|$, where N_a is the atomic number[14] of the particular element. Thus, there essentially are two kinds of electrical charge in matter: positive and negative. Each atom or molecule in its normal state has as many electrons as it has units of positive charge in its nucleus or nuclei, and thus is electrically neutral.

Nearly all of the macroscopic effects of electricity arise from the fact that electrons may be separated from their atoms in some circumstances, leading to separation of positive and negative charges by distances appreciably greater than atomic dimensions. At present, the universe is believed to contain equal amounts of positive and negative electricity, with the result that an excess of one polarity of charge at one place implies the presence elsewhere of an equal but opposite charge. Furthermore, net amounts of electric charge can neither be created nor destroyed (the law of conservation of electric charge).

The atomic nature of matter is inherently quantized in that electric charges are made up of integral multiples of the electron charge q_e. However, any charged object of macroscopic size (i.e., much larger than atomic dimensions) typically holds so many electrons (or, if positively charged, would have a deficiency of so many electrons) that it is considered to have a "continuous" distribution of charge. Since we restrict our attention here to the macroscopic realm, all quantities that are dealt with, whether they be charges, fields, or potentials, should be understood to be macroscopic in nature. For example, the discussion of Coulomb's law in the next section describes the electrostatic force between two *point* charges residing at rest at two different locations in vacuum. In the context of our discussion, we implicitly understand that these point charges may actually occupy physical space of many atomic dimensions in extent, but that the size of these regions is nevertheless negligible on a macroscopic scale.

4.2 COULOMB'S LAW

Although the preceding discussion introduces us to the fundamentals of electrostatic principles, the law of action between electrified bodies needs to be specified for mathematical formulation of these concepts. This law of action, known as *Coulomb's law,* is the quantitative expression of the experimental observations discussed above. Coulomb's law states that the electric force between two point charges[15] Q_1 and Q_2 is proportional to the product of the two charges $Q_1 Q_2$ and inversely proportional to the square of the

---

[12]The most precise value of the charge of an electron is $q_e = -1.60217733(49) \times 10^{-19}$ C, as specified in *The CRC Handbook of Physics & Chemistry,* 76th ed., CRC Press, Boca Raton, Florida, 1995–1996.

[13]An electron is physically quite light, having a mass of $m_e \simeq 9.11 \times 10^{-31}$ kg, much smaller than the mass of a proton ($\sim 1.66 \times 10^{-27}$ kg) or a neutron, the two types of particles that together form the nucleus.

[14]The atomic number of an element is defined as the number of protons present in its nucleus.

[15]By "point charges" we mean charges that occupy a macroscopic region of space much smaller in extent than the distance between the charges.

distance between them. In vector form, the forces $\mathbf{F}_{12}$ and $\mathbf{F}_{21}$ felt by charges Q_2 and Q_1 are given as

$$\mathbf{F}_{12} = -\mathbf{F}_{21} = \hat{\mathbf{R}}\frac{kQ_1Q_2}{R^2} \qquad (4.1)$$

where $\mathbf{R} = \mathbf{r}_2 - \mathbf{r}_1$, with $\mathbf{r}_1$ and $\mathbf{r}_2$ being the position vectors[16] for points P_1 and P_2 (where charges Q_1 and Q_2 reside), respectively, shown in Figure 4.3. As for any other vector, we can write $\mathbf{R}$ as $\mathbf{R} = \hat{\mathbf{R}}R$, where $\hat{\mathbf{R}}$ is the unit vector in the direction of $\mathbf{R}$ and R is the magnitude of $\mathbf{R}$. Note that $\mathbf{R}$ is the vector pointing *from* point P_1 *to* P_2 and that $\hat{\mathbf{R}} = \mathbf{R}/R = (\mathbf{r}_2 - \mathbf{r}_1)/(|\mathbf{r}_2 - \mathbf{r}_1|)$ and $R = |\mathbf{r}_2 - \mathbf{r}_1|$. The force as given in (4.1) is repulsive (as shown in Figure 4.3) if the charges are alike in sign and is attractive if they are of opposite sign (i.e., $Q_1Q_2 < 0$, so that the directions of both $\mathbf{F}_{12}$ and $\mathbf{F}_{21}$ in Figure 4.3 are reversed). The proportionality constant k is equal to $(4\pi\epsilon_0)^{-1}$, with the 4π "rationalization" factor included so that a 4π factor does not appear in Maxwell's equations, which are more commonly used than (4.1).

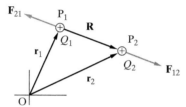

Figure 4.3 Coulomb's law. Forces on point charges Q_1 and Q_2 when the two charges are alike in sign.

Using the SI (*Le Système International d'Unités*, International System of Units) units of force, charge, and distance—respectively, newtons (N), coulombs (C), and meters (m)—the value of the quantity[17] ϵ_0, referred to as the permittivity of free space,[18] is $\epsilon_0 \simeq 8.854 \times 10^{-12}$ farads per meter (or F-m^{-1}). Since the proportionality constant k is needed in numerical evaluations, it is helpful to note that in SI units, $k = (4\pi\epsilon_0)^{-1} \simeq 9 \times 10^9$ m-F^{-1}. A coulomb can be defined as the quantity of charge that flows through a given cross section of a wire in 1 second when there is a steady current of 1 ampere (A) flowing in the wire.[19]

The following example introduces the application of Coulomb's law.

Example 4.1: Two point charges. Two point charges of $Q_1 = +37$ nC and $Q_2 = +70$ nC are located at points $(1, 3, 0)$ m and $(0, 0, 2)$ m, respectively, as shown in Figure 4.4. Find the force exerted on Q_2 by Q_1.

---

[16]As defined earlier, the position vector for any point P located in three-dimensional space is the vector pointing from the origin to the point P.

[17]It will later become evident that the precise value of this quantity is $\epsilon_0 = (4\pi \times 10^{-7}c^2)^{-1}$, where c is the speed of light in vacuum.

[18]The physical meaning of the dimensions of ϵ_0 will become clearer in later sections, when we discuss capacitance.

[19]The definition of an ampere will be discussed in Chapter 6 in connection with Ampère's law.

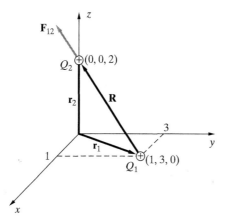

Figure 4.4 Two point charges. Configuration of two point charges for Example 4.1.

Solution: The repulsive force $\mathbf{F}_{12}$ on charge Q_2 exerted by Q_1 is

$$\mathbf{F}_{12} = \hat{\mathbf{R}}\frac{1}{4\pi\epsilon_0}\frac{Q_1 Q_2}{R^2}$$

where

$$R = |\mathbf{r}_2 - \mathbf{r}_1| = |2\hat{\mathbf{z}} - \hat{\mathbf{x}} - 3\hat{\mathbf{y}}| = [(-1)^2 + (-3)^2 + 2^2]^{1/2} = \sqrt{14} \text{ m}$$

since $\mathbf{r}_2 = 2\hat{\mathbf{z}}$ and $\mathbf{r}_1 = \hat{\mathbf{x}} + 3\hat{\mathbf{y}}$. The unit vector $\hat{\mathbf{R}}$ in direction of $\mathbf{F}_{12}$ is

$$\hat{\mathbf{R}} = \frac{\mathbf{R}}{R} = \frac{\mathbf{r}_2 - \mathbf{r}_1}{|\mathbf{r}_2 - \mathbf{r}_1|} = \frac{2\hat{\mathbf{z}} - \hat{\mathbf{x}} - 3\hat{\mathbf{y}}}{\sqrt{14}}$$

Therefore, we have

$$\mathbf{F}_{12} = \frac{9 \times 10^9 \text{ m-F}^{-1} \times 37 \times 10^{-9} \text{ C} \times 70 \times 10^{-9} \text{ C}}{14 \text{ m}^2}\underbrace{\left(\frac{-\hat{\mathbf{x}} - 3\hat{\mathbf{y}} + 2\hat{\mathbf{z}}}{\sqrt{14}}\right)}_{\hat{\mathbf{R}}}$$

$$\simeq 1.66 \times 10^{-6}\left(\frac{-\hat{\mathbf{x}} - 3\hat{\mathbf{y}} + 2\hat{\mathbf{z}}}{\sqrt{14}}\right) \simeq 445(-\hat{\mathbf{x}} - 3\hat{\mathbf{y}} + 2\hat{\mathbf{z}}) \text{ nN}$$

It is obvious that the repulsive force $\mathbf{F}_{21}$ on charge Q_1 exerted by Q_2 is $\mathbf{F}_{21} = -\mathbf{F}_{12}$.

To acquire a feel for the magnitude of the electric force as represented by (4.1), (e.g., in comparison with the gravitational force), consider the force between two protons in the nucleus of a helium atom, which are $\sim 10^{-15}$ m apart.[20] The gravitational force between the two protons is $\sim 1.84 \times 10^{-34}$ N, but the electrical force from (4.1) is ~ 230 N! The electrical force, in other words, is $\sim 10^{36}$ times larger. Were it not for the nuclear force that keeps the nucleus together, the initial acceleration that each of the protons would acquire due to the electrical force would be $\sim 10^{28}g$!

[20]Approximate radius of helium nuclei.

To appreciate the enormity of one coulomb of charge, note that two 1 C charges 1 m apart would exert an electrical force upon one another of $\sim 9 \times 10^9$ newtons! In normal air, a single 1 C charge would cause electrical breakdown[21] at distances of ~ 50 m.

If there are more than two charges present, (4.1) holds for every pair of charges, and the principle of superposition can be used to determine the net force on any one of the charges due to all others. For this purpose, the force due to each charge is determined as if it alone were present; the vector sum of these forces is then calculated to give the resultant force, as illustrated in Example 4.2.

Example 4.2: Three point charges. Consider two point charges of $Q_1 = +1$ μC and $Q_2 = +2$ μC located at $(1, 0)$ m and $(-1, 0)$ m, respectively, as shown in Figure 4.5. (a) What is the magnitude and direction of the electrical force felt by a third charge $Q_3 = +1$ nC when placed at $(0, 1)$ m? (b) At what point(s) must the third charge $Q_3 = +1$ nC be placed in order to experience no net force?

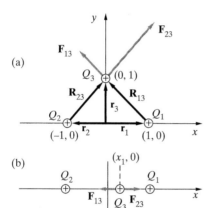

Figure 4.5   **Three point charges.** Configurations of the three point charges Q_1, Q_2, and Q_3 for Example 4.2.

Solution:

(a) Note from Figure 4.5a that $\mathbf{r}_3 = \hat{\mathbf{y}}$, $\mathbf{r}_1 = \hat{\mathbf{x}}$, and $\mathbf{r}_2 = -\hat{\mathbf{x}}$. Using Coulomb's law along with the superposition principle, the repulsive forces exerted on charge Q_3 by charges Q_1 and Q_2 are, respectively,

$$\mathbf{F}_{13} = \hat{\mathbf{R}}_{13} \frac{kQ_1Q_3}{R_{13}^2}; \qquad \mathbf{F}_{23} = \hat{\mathbf{R}}_{23} \frac{kQ_2Q_3}{R_{23}^2}$$

where

$$R_{13}^2 = |\mathbf{r}_3 - \mathbf{r}_1|^2 = |\hat{\mathbf{y}} - \hat{\mathbf{x}}|^2 = 2 \text{ m}^2; \qquad R_{23}^2 = |\mathbf{r}_3 - \mathbf{r}_2|^2 = |\hat{\mathbf{y}} + \hat{\mathbf{x}}|^2 = 2 \text{ m}^2$$

---

[21] Breakdown of air occurs when the electric field intensity in air exceeds 3×10^6 V-m^{-1}, at which time atmospheric electrons and ions are accelerated to high energies as a result of the Coulomb force. As these energized particles have inevitable collisions with neutral air molecules, many other electrons and ions are knocked out of the neutral air molecules, resulting in arcing and corona discharges.

and the unit vectors $\hat{\mathbf{R}}_{13}$ and $\hat{\mathbf{R}}_{23}$ are

$$\hat{\mathbf{R}}_{13} = \frac{\mathbf{r}_3 - \mathbf{r}_1}{|\mathbf{r}_3 - \mathbf{r}_1|} = \frac{\hat{\mathbf{y}} - \hat{\mathbf{x}}}{\sqrt{2}}; \qquad \hat{\mathbf{R}}_{23} = \frac{\mathbf{r}_3 - \mathbf{r}_2}{|\mathbf{r}_3 - \mathbf{r}_2|} = \frac{\hat{\mathbf{y}} + \hat{\mathbf{x}}}{\sqrt{2}}$$

Therefore we have

$$\mathbf{F}_{13} = \left(\frac{\hat{\mathbf{y}} - \hat{\mathbf{x}}}{\sqrt{2}}\right) \frac{9 \times 10^9 \text{ m-F}^{-1} \times 10^{-6} \text{ C} \times 10^{-9} \text{ C}}{2 \text{ m}^2} = \frac{4.5}{\sqrt{2}}(-\hat{\mathbf{x}} + \hat{\mathbf{y}}) \quad \mu\text{N}$$

and similarly

$$\mathbf{F}_{23} = \frac{9}{\sqrt{2}}(\hat{\mathbf{x}} + \hat{\mathbf{y}}) \quad \mu\text{N}$$

The total electrical force on charge Q_3 is thus

$$\mathbf{F}_3 = \mathbf{F}_{13} + \mathbf{F}_{23} = \frac{4.5}{\sqrt{2}}(\hat{\mathbf{x}} + 3\hat{\mathbf{y}}) \quad \mu\text{N}$$

(b) Since the forces exerted by the two charges on Q_3 must be equal and opposite in order to have no net force, Q_3 must be placed somewhere along the straight line between Q_1 and Q_2 (i.e., on the x axis), as shown in Figure 4.5b. Let us assume that Q_3 is located at $(x_1, 0)$ and find x_1. Referring to Figure 4.5b, we have

$$\mathbf{F}_3 = \mathbf{F}_{13} + \mathbf{F}_{23} = -\underbrace{\frac{9}{(1 - x_1)^2}}_{R_{13}^2}\hat{\mathbf{x}} + \underbrace{\frac{18}{(1 + x_1)^2}}_{R_{23}^2}\hat{\mathbf{x}} = 0 \quad \mu\text{N}$$

Solving for x_1, we find $x_1 \simeq 0.172$ m.

It is important to remember that Coulomb's law as stated above is defined for "point" charges. To see this, consider two spheres A and B at a distance R from one another; sphere A is charged by an amount $+Q_A$ while sphere B is uncharged, $Q_B = 0$. Direct application of Coulomb's law specifies the force between the two bodies to be

$$|\mathbf{F}| = \frac{1}{4\pi\epsilon_0} \frac{(+Q_A)(0)}{R^2} = 0$$

However, based on the discussion in connection with Figure 4.1, we know that the force is not zero, because sphere A will induce negative charge on the side of sphere B closer to it and will be attracted toward it. For this induction effect to be negligible, we must have $b \ll a$, where a is the radius of the charged sphere A and b is the radius of the smaller sphere B.

4.2.1 Coulomb and His Experiments

In 1785, Charles Augustin de Coulomb demonstrated the law of electric force by employing one of his inventions, called a torsion balance, to measure the repulsive force between two like charges. Using the apparatus shown in Figure 4.6, he was able to compensate

Figure 4.6   Coulomb's torsion balance apparatus. The glass cylinder ABCD, 12 inches in diameter by 12 inches high, was covered by a glass plate having two holes approximately one-twelfth of an inch in diameter, one at the center (at f) and the other at m. Cemented in the center was a vertical tube 24 inches in length, with a torsion micrometer at its top, consisting of a circumferential scale divided into degrees, a knob with a scale pointer, and a chuck or pincer to hold the torsion wire. The torsion wire was either silver, copper, or silk and attached at its lower end was a pincer, which held the torsion wire taut and linear and also provided support for a thin horizontal suspended straw or silk thread coated with sealing wax. At one end of the straw was a gilded elderwood pith-ball (one-sixth of inch in diameter), while at the other end was a small vertical paper disk, which served as a counterweight and to damp out oscillations. A second scale, ZOQ, marked in degrees, was attached outside the glass cylinder. Coulomb turned the micrometer at the top until the horizontal thread ag lined up with the zero marking on scale ZOQ. He then introduced (through the hole m) a thin insulated rod with a second identical pith-ball t, and made it touch the pith-ball a. Coulomb then charged an insulated pin and touched the pin to the two pith-balls, which, having acquired equal charge, separated by a certain distance, allowing Coulomb to measure the repulsive force between them. Figure and description adapted with permission from C. S. Gillmor, *Coulomb and the Evolution of Physics and Engineering in Eighteenth Century France,* Princeton University Press, Princeton, NJ, 1971. Copyright 1971 by Princeton University Press.

for the electric repulsion force by twisting the suspension head (thus exerting a force proportional to the angle of twist) to keep the two charged spheres apart at different distances and thus measure the force as a function of distance between them.[22],[23]

In the context of current scientific standards and practices, Coulomb's experimental data as presented in his Memoir would have left a lot to be desired as a demonstration of the inverse square law. Only three data points were presented to support Coulomb's hypothesis, and the values deviated from the inverse square law by as much as a few percent, probably due to the limited accuracy of the torsion balance measurements. However, scientists at the time expected an inverse square law for electrostatic forces (similar to Newton's gravitational law of attraction, then known for over 100 years), and Coulomb's results received wide acclaim, with none of his contemporaries expressing doubts. It is interesting to note that some recent attempts to reproduce Coulomb's experimental results have run into difficulties, raising questions as to whether Coulomb may have manipulated his data to obtain a result he intuitively expected.[24],[25]

It is also interesting that B. Franklin in 1755, J. Priestley in 1767, J. Robison in 1769, and H. Cavendish in 1773 conducted experiments of one kind or another that pointed to the inverse square law—in the case of Cavendish, with substantially better precision than Coulomb. However, these results were either not published or not well publicized and thus were unknown to the scientific community at the time of Coulomb's work.[26]

4.2.2 The Accuracy and Validity of Coulomb's Law

Since Coulomb's law is based solely upon physical observations (i.e., experimental fact), and in view of the fact that it provides one of the foundations of electromagnetics, it is important to assess the accuracy of the measurements by which it is established. It was mentioned above that Coulomb's measurements of the electric force were quite approximate, deviating by as much as a few percent from the r^{-2} law. Even with modern techniques, it is difficult to measure the *force* between two charged objects directly to a high degree of accuracy.[27] However, the accuracy of Coulomb's law can be assessed by verifying a fundamental consequence of the inverse square law, namely that the electric field inside a charged closed metallic surface (shell) is precisely zero. As we

---

[22]See Coulomb's Première mémoire sur l'électricité et magnétisme (First memoir on electricity and magnetism), *Histoire de l'Académie Royale des Sciences,* p. 569, 1785. For an English translation of excerpts, see W. F. Magie, *A Source Book in Physics,* McGraw-Hill Book Company, New York, 1935.

[23]For an excellent summary of Coulomb's experiments and related previous work of other researchers, see Chapter 3 of R. S. Elliott, *Electromagnetics,* IEEE Press, Piscataway, New Jersey, 1993.

[24]S. Dickman, Could Coulomb's experiment result in Coulomb's law? *Science,* p. 500, 22 October 1993.

[25]P. Heering, On Coulomb's inverse square law, *Am. J. Phys.* 60(11), p. 988, 1992.

[26]For an excellent account of these early works, see Chapter 3 of R. S. Elliott, *Electromagnetics,* IEEE Press, Piscataway, New Jersey, 1993.

[27]See A. N. Cleland and M. L. Roukes, A nanometre-scale mechanical electrometer, *Nature,* 392, p. 160, 1998.

shall see in Section 4.7, this result is based on Gauss's law (Section 4.5), which in turn is derived from Coulomb's law. Original experiments conducted by H. Cavendish in 1772, and later repeated[28] by J. C. Maxwell in 1879, indicated that the electrostatic force depends on $r^{-2+\zeta}$, with $|\zeta| < 10^{-5}$. This basic experiment, which involves the placement of an electrometer (an instrument that measures charge) inside a large sphere and the observation of whether any deflections occur when the sphere is charged to a high voltage, has been carried out carefully over the years. A null result is always obtained, which allows one to bound $|\zeta|$ using the known configuration and geometry of the apparatus and the sensitivity of the electrometer. Such an experiment essentially constitutes a *comparison* of the force law to the inverse square law and thus provides a high degree of precision. Plimpton and Lawton[29] found $|\zeta| < 2 \times 10^{-9}$. A more recent experiment by Williams, Faller, and Hill[30] places a limit of $|\zeta| < (2.7 \pm 3.1) \times 10^{-16}$.

In terms of the range of circumstances under which the inverse square law is valid, the various experiments described above show that Coulomb's law holds at distances of a few tens of centimeters. What about at smaller distances? E. Rutherford's analysis[31] of experiments involving the scattering of alpha particles[32] by atomic nuclei shows that the inverse square law holds at distances of order 10^{-11} cm. The accuracy of the inverse square law at such small distances is difficult to ascertain; however, measurements of Lamb and Retherford[33] of the relative positions of the energy levels of hydrogen indicate that $|\zeta| < 10^{-9}$ even at distances as small as 10^{-8} cm. At distances smaller than 10^{-11} cm, nuclear forces and strong interaction effects come into play and partially mask the effects of the inverse square law. Nevertheless, measurements in nuclear physics indicate that electrostatic forces still vary approximately as r^{-2} at distances of 10^{-13} cm. At even smaller distances, the question of the validity of Coulomb's law is still an open question.

A natural question that comes to mind here is the validity of Coulomb's law at large length scales. While electrostatic experiments at distance scales of more than few meters are difficult to carry out, magnetic field measurements can be used to assess the applicability of Coulomb's law at large distances.[34] Geophysical measurements of the earth's magnetic field thus provide a highly accurate assessment of the inverse square

[28] H. Cavendish, *Electrical Researches,* J. C Maxwell (Ed.), Cambridge University Press, Cambridge, UK, 1879, pp. 104–113.

[29] S. J. Plimpton and W. E. Lawton, A very accurate test of Coulomb's law of force between charges, *Phys. Rev.,* 50, p. 1066, 1936.

[30] E. R. Williams, J. E. Faller, and H. A. Hill, New experimental test of Coulomb's law: A laboratory upper limit on photon rest mass, *Phys. Rev. Lett.,* 26, p. 721, 1971.

[31] E. Rutherford, Scattering of α- and β-particles by matter, and the structure of the atom, *Phil. Mag.,* 21, p. 669, May 1911.

[32] α-particles are helium nuclei which move at speeds of $\sim 0.3c$ when they are emitted spontaneously by radioactive materials such as radium.

[33] W. E. Lamb, Jr. and R. C. Retherford, Fine structure of the hydrogen atom by a microwave method, *Phys. Rev.,* 72, p. 241, 1947; —, Fine structure of the hydrogen atom. Part IV, *Phys. Rev.,* 86, p. 1014, 1952, and references therein.

[34] In the context of special relativity theory, the fundamental laws of magnetostatics are related to Coulomb's law, so that a test of predictions of magnetostatics actually constitutes a test of the electrostatic inverse square law.

law. According to such measurements,[35] the inverse square law holds at distances of $\sim 10^9$ cm (10^4 km) to a degree of accuracy of $|\zeta| < 10^{-16}$, similar to that implied by the Williams, Faller, and Hill experiment.

In summary, Coulomb's law is known to be valid to a high degree of accuracy at distance scales ranging from 10^9 to 10^{-11} cm, or 20 orders of magnitude! In thinking about the inverse square law, it is helpful to realize the similarities between it and the law of gravitational attraction, which also varies[36] as r^{-2}, whose validity is established to a very high degree of accuracy because it explains the motions of heavenly bodies so well. Laplace was the first to show[37] that no function of distance except the inverse square law satisfies the condition that a uniform spherical mass shell exerts no gravitational force on a particle within it. Similarly, it can be shown on a purely mathematical basis[38] that no function of distance except the inverse square law is consistent with the fact that a spherical shell of charge exerts no electrostatic force on a charged particle within it.

In terms of the validity of aspects of Coulomb's law other than the r^{-2} dependence (for example, the assumptions that the electric force acts along the line connecting two point charges, that the force varies as the algebraic product of the two charges, and that the constant of proportionality is always $(4\pi\epsilon_0)^{-1}$), we rely on the accuracy of predictions based on these assumptions. Furthermore, since the charges used in various experiments were not really point charges but instead were distributed in space, the principle of superposition of forces was inherently assumed to hold. Once again, the validity of this particular assumption is verified only by virtue of the fact that predictions based on it are consistent with experiments.

4.3 THE ELECTRIC FIELD

Consider the region around a single point charge Q, as shown in Figure 4.7. Other charges (e.g., a test charge q) brought near this charge would experience a force whose magnitude depends on the distance between Q and q. The existence of this force can be described by saying that the charge Q produces an *electric field* in the region surrounding it. It is convenient to think of this region as being permeated by an electric field **E**, defined as the force per unit charge,

$$\boxed{\mathbf{E} = \lim_{q \to 0} \frac{\mathbf{F}}{q}} \qquad (4.2)$$

The SI unit of electric field is the newton per coulomb (N-C^{-1}). A more commonly used equivalent unit for the electric field is the volt per meter (V-m^{-1}). In order for (4.2)

---

[35]For an excellent review, see A. S. Goldbaher and M. M. Nieto, Terrestrial and extraterrestrial limits on the photon mass, *Rev. Mod. Phys.,* 43, p. 277, 1971.

[36]For an interesting discussion of inverse square law forces, see Chapter 2 of N. Feather, *Electricity and Matter,* Edinburgh University Press, Edinburgh, 1968.

[37]P. S. Laplace, *Mécanique céleste,* Vol. I., p. 163, Paris, A printing of Crapelet, 1799.

[38]For a discussion of Maxwell's original proof of this result, see Chapter 3 of R. S. Elliott, *Electromagnetics,* IEEE Press, Piscataway, New Jersey, 1993.

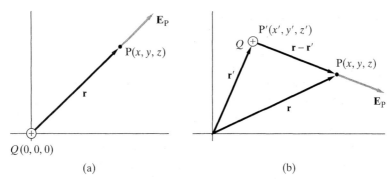

Figure 4.7 Electric field of single point charge. (a) Point charge at the origin. (b) Point charge at point P′(**r**′).

to be valid, the test charge q must be negligibly small (both in terms of its amount of charge and physical size) so that its force on Q does not cause Q to move and so that induction effects due to the presence of q are negligible.[39] In short, the test charge must both be small (in terms of quantity of charge) and must be a point charge[40] so that it does not alter the charge distribution from what it would be in the absence of the test charge.

The notion of a *field*, however abstract, is nevertheless the most useful way of thinking and working with electromagnetic problems and of describing the effects of charges on other charges at a distance.[41] We can simply proceed by defining a field to be the set of values assumed by a physical quantity at various points in a region of space at various instants of time. However, a few more words of discussion are in order in view of the abstract nature of this concept. The field concept is used in other areas of physics governed by "action-at-a-distance" laws similar to Coulomb's law. For example, Newtonian mechanics is based on the effects of gravitational "forces" on objects having mass, without reference to the specific nature of the sources of these forces. Similarly, it is convenient to think that the source charge produces "something" at surrounding points, and that this something then interacts with other charges brought to these points. Since we can use (4.2) to determine the magnitude of the electric field *whether or not* there is a charge there to be subject to a force, it is tempting to regard the field as an actual physical entity in its own right. Most of these ideas originated with M. Faraday, who believed that the presence of charges actually changed the physical properties of space and that the electric field was a manifestation of this altered state. For Faraday,

[39] If the electric field is set up by voltage differences between conductors, as it usually is, the presence of the test charge should not affect the distribution of charges on the conductors that set up the electric field.

[40] A useful analogy is in using a small float or fishing line to deduce the direction of current in a lake, and implicitly assuming that the float is infinitesimal in size. Clearly, we could not infer the direction of the current from the orientation of a float as large as an anchored ship since the presence of the ship would actually disturb the current flow.

[41] For an illuminating discussion, see Chapter 1 of Volume II of *The Feynman Lectures in Physics,* Addison Wesley, Reading, Massachusetts, 1964.

the electric field was a very real physical quantity.[42] If we view the electric field as a physical quantity, it becomes important that we are able to measure it. At first thought, it should be straightforward to simply bring a test charge q to point P and deduce the field from the electrical force that it experiences. However, the presence of this test charge q will in general subject the source charges to new forces and would thus change the field distribution that we set out to measure. It is in this context that the source charge q in (4.2) is required to be infinitesimally small.

For a positive charge Q located at the origin of our coordinate system and at rest in boundless free space as shown in Figure 4.7a, the electric field at any point P, identified by the position vector $\mathbf{r} = \hat{\mathbf{x}}x + \hat{\mathbf{y}}y + \hat{\mathbf{z}}z$, is

$$\boxed{\mathbf{E}_P = \hat{\mathbf{r}}\frac{Q}{4\pi\epsilon_0 r^2}} \tag{4.3}$$

where $\hat{\mathbf{r}} = \mathbf{r}/|\mathbf{r}|$ and $r = |\mathbf{r}| = [x^2 + y^2 + z^2]^{1/2}$. We note that the electric field at point P is a vector that points along the direction of the position vector $\mathbf{r}$ (i.e., outward from the positive charge Q), as shown in Figure 4.7a. Conversely, if the source charge were negative (i.e., $Q < 0$), the electric field would point in the $-\mathbf{r}$ direction (i.e., inward toward the negative charge).

In general, the source charge Q may not be located at the origin but at a point P′ described by the position vector $\mathbf{r}' = \hat{\mathbf{x}}x' + \hat{\mathbf{y}}y' + \hat{\mathbf{z}}z'$ marked in Figure 4.7b. The electric field at point P produced by this charge is then in the $\hat{\mathbf{R}}$ direction, where $\mathbf{R} = (\mathbf{r} - \mathbf{r}')$, and is given by

$$\boxed{\mathbf{E}_P = \hat{\mathbf{R}}\frac{Q}{4\pi\epsilon_0 R^2} = \hat{\mathbf{R}}\frac{Q}{4\pi\epsilon_0|\mathbf{r} - \mathbf{r}'|^2}} \tag{4.4}$$

Noting that $\hat{\mathbf{R}} \equiv \mathbf{R}/R = (\mathbf{r} - \mathbf{r}')/(|\mathbf{r} - \mathbf{r}'|)$, (4.4) can be written as

$$\mathbf{E}_P = \frac{Q}{4\pi\epsilon_0|\mathbf{r} - \mathbf{r}'|^2}\frac{\mathbf{r} - \mathbf{r}'}{|\mathbf{r} - \mathbf{r}'|} = \frac{Q}{4\pi\epsilon_0}\frac{\hat{\mathbf{x}}(x - x') + \hat{\mathbf{y}}(y - y') + \hat{\mathbf{z}}(z - z')}{[(x - x')^2 + (y - y')^2 + (z - z')^2]^{3/2}} \tag{4.5}$$

where the expanded form is written out in Cartesian coordinates. In (4.5), as well as throughout the rest of this book, we use primed (e.g., $\mathbf{r}'$) quantities to represent the position vector or coordinates of a source point, while unprimed (e.g., $\mathbf{r}$) quantities are used to represent the position vector or coordinates for a point at which the electric field (or another quantity) is determined.

Equation (4.4) implies that the electric field becomes arbitrarily large as the observation point P is brought closer to the point charge (i.e., as $\mathbf{r} \to \mathbf{r}'$ or $R \to 0$). The

---

[42]"Faraday, in his mind's eye, saw lines of force traversing all space where the mathematicians saw centres of force attracting at a distance: Faraday saw a medium where they saw nothing but distance: Faraday sought the seat of the phenomena in real actions going on in the medium, they (i.e., the mathematicians) were satisfied that they had found it (i.e., the seat of the phenomena) in a power of action at a distance impressed on electric fluids." From J. C. Maxwell, *A Treatise on Electricity and Magnetism,* Clarendon Press, Oxford, 1892.

resolution of this apparently unphysical result lies in our original premise (Section 4.1.3) of restricting our attention to effects on macroscopic scales, thus ensuring that the distance $R = |\mathbf{r} - \mathbf{r}'|$ between source and observation points is much larger than atomic dimensions.

Example 4.3: Electric field. Find the electric field at P(3, 1, 0) due to a point charge $Q = +80$ nC located at (2, 0, 2) m as shown in Figure 4.8. Assume all distances to be in meters.

Solution: Note from the geometry that $\mathbf{r} = 3\hat{\mathbf{x}} + \hat{\mathbf{y}}$ and $\mathbf{r}' = 2\hat{\mathbf{x}} + 2\hat{\mathbf{z}}$. Using (4.5) we have

$$\mathbf{E}_P = \frac{80 \text{ nC}}{(9 \times 10^9)^{-1}} \frac{\hat{\mathbf{x}} + \hat{\mathbf{y}} - 2\hat{\mathbf{z}}}{[(3-2)^2 + 1^2 + (-2)^2]^{3/2}} \simeq 49(\hat{\mathbf{x}} + \hat{\mathbf{y}} - 2\hat{\mathbf{z}}) \text{ V-m}^{-1}$$

Note from Figure 4.8 that $\mathbf{E}_P$ points predominantly downward, consistent with the fact that the magnitude of the z component is twice as large as that for the other two components.

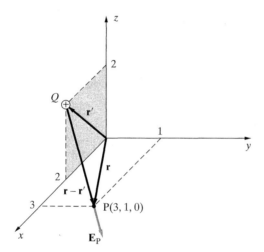

Figure 4.8   Electric field. Electric field $\mathbf{E}_P$ at point P(3, 1, 0) due to a positive point charge Q at (2, 0, 2) m.

4.3.1 Electric Field Resulting from Multiple Charges

The electric field concept is most useful when large numbers of charges are present, since each charge exerts a force on all the others. Assuming that for a particular test charge the forces resulting from the different charges can be linearly superposed,[43] we can evaluate

---

[43]We are able to assume linearity simply because, at the macroscopic level, its use produces accurate results in many different experiments and applications involving groups of charges and currents. We note here that some material media can behave in a highly nonlinear manner and that such behavior often leads to interesting applications or consequences, such as in magnetic materials, nonlinear optics, crystals, and electrostatic breakdown phenomena (e.g., lightning discharges). As far as fields in vacuum are concerned, linear superposition is remarkably valid even at the atomic level, where field strengths of order 10^{11} to 10^{17} V-m^{-1} exist at the orbits of electrons. For further discussion, see Section I.3 of J. D. Jackson, *Classical Electrodynamics,* Wiley, New York, 1975.

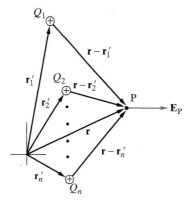

Figure 4.9 Electric field due to multiple charges. The electric field at the point P at position $\mathbf{r}$ is the sum of the electric fields due to the point charges $Q_1, Q_2, \ldots, Q_n$, located at source points $\mathbf{r}'_1, \mathbf{r}'_2, \ldots, \mathbf{r}'_n$, respectively.

the electric field at a given point simply as a vector sum of the field contributions resulting from all charges.

For a set of discrete point charges $Q_1, Q_2, \ldots, Q_n$ located at points with position vectors $\mathbf{r}'_1, \mathbf{r}'_2, \ldots, \mathbf{r}'_n$, as shown in Figure 4.9, the electric field at a point P with position vector $\mathbf{r}$ is given as

$$\mathbf{E}_P = \frac{1}{4\pi\epsilon_0} \sum_{k=1}^{n} \hat{\mathbf{R}}_k \frac{Q_k}{|\mathbf{r} - \mathbf{r}'_k|^2} \tag{4.6}$$

where $\hat{\mathbf{R}}_k$ is the unit vector pointing in the direction from the source point $P'_k(\mathbf{r}'_k)$ at which charge Q_k resides to the observation point $P(\mathbf{r})$. Noting that $\hat{\mathbf{R}}_k = (\mathbf{r} - \mathbf{r}'_k)/(|\mathbf{r} - \mathbf{r}'_k|)$, (4.6) can be written as

$$\mathbf{E}_P = \frac{1}{4\pi\epsilon_0} \sum_{k=1}^{n} \frac{Q_k(\mathbf{r} - \mathbf{r}'_k)}{|\mathbf{r} - \mathbf{r}'_k|^3} \tag{4.7}$$

Example 4.4: Two point charges. Consider two point charges $Q_1 = +5$ nC and $Q_2 = -5$ nC located at points $(1, 0, 0)$ m and $(-1, 0, 0)$ m as shown in Figure 4.10. (a) Calculate the electric field at points $P_1(0, 0, 0)$, $P_2(1, 1, 0)$, and $P_3(0, 0, 1)$, respectively. (b) Repeat for $Q_2 = +5$ nC. Assume all coordinates to be in meters.

Solution: Using (4.7), the electric field at point P with a position vector $\mathbf{r}$ resulting from point charges Q_1 and Q_2 located at positions $\mathbf{r}'_1 = \hat{\mathbf{x}}$ and $\mathbf{r}'_2 = -\hat{\mathbf{x}}$, respectively is given as

$$\mathbf{E}_P = \mathbf{E}_1 + \mathbf{E}_2 = \frac{Q_1(\mathbf{r} - \mathbf{r}'_1)}{4\pi\epsilon_0|\mathbf{r} - \mathbf{r}'_1|^3} + \frac{Q_2(\mathbf{r} - \mathbf{r}'_2)}{4\pi\epsilon_0|\mathbf{r} - \mathbf{r}'_2|^3}$$

(a) For point $P_1(0, 0, 0)$ we have $\mathbf{r} = 0$. Therefore,

$$\mathbf{E}_{P_1} = \frac{5 \times 10^{-9}}{(9 \times 10^9)^{-1}} \frac{(-\hat{\mathbf{x}})}{1^3} - \frac{5 \times 10^{-9}}{(9 \times 10^9)^{-1}} \frac{(+\hat{\mathbf{x}})}{1^3} = -90\hat{\mathbf{x}} \text{ V-m}^{-1}$$

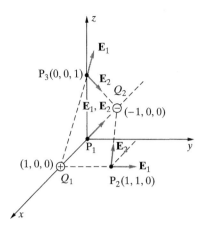

Figure 4.10 **Two point charges.** The electric field vectors $\mathbf{E}_1$ and $\mathbf{E}_2$, due to charges $Q_1 = +5$ nC and $Q_2 = -5$ nC, respectively, are shown at points P_1, P_2, and P_3.

For point $P_2(1, 1, 0)$, we have $\mathbf{r} = \hat{\mathbf{x}} + \hat{\mathbf{y}}$ so that

$$\mathbf{E}_{P_2} = \frac{5 \times 10^{-9}}{(9 \times 10^9)^{-1}} \frac{\hat{\mathbf{x}} + \hat{\mathbf{y}} - \hat{\mathbf{x}}}{1^3} - \frac{5 \times 10^{-9}}{(9 \times 10^9)^{-1}} \frac{\hat{\mathbf{x}} + \hat{\mathbf{y}} + \hat{\mathbf{x}}}{[2^2 + 1^2]^{3/2}}$$

$$= 45\hat{\mathbf{y}} - \frac{9}{\sqrt{5}}(2\hat{\mathbf{x}} + \hat{\mathbf{y}}) = -\frac{18}{\sqrt{5}}\hat{\mathbf{x}} + \frac{9(5\sqrt{5} - 1)}{\sqrt{5}}\hat{\mathbf{y}} \text{ V-m}^{-1}$$

For point $P_3(0, 0, 1)$ we have $\mathbf{r} = \hat{\mathbf{z}}$. Thus,

$$\mathbf{E}_{P_3} = \frac{5 \times 10^{-9}}{(9 \times 10^9)^{-1}} \frac{\hat{\mathbf{z}} - \hat{\mathbf{x}}}{[1^2 + 1^2]^{3/2}} - \frac{5 \times 10^{-9}}{(9 \times 10^9)^{-1}} \frac{\hat{\mathbf{z}} + \hat{\mathbf{x}}}{[1^2 + 1^2]^{3/2}} = -\frac{45}{\sqrt{2}}\hat{\mathbf{x}} \text{ V-m}^{-1}$$

(b) When Q_2 changes sign, all $\mathbf{E}_2$ vectors reverse direction. As a result, we have

$$\text{At } P_1 \qquad\qquad \mathbf{E}_{P_1} = 0$$

$$\text{At } P_2 \qquad\qquad \mathbf{E}_{P_2} = \frac{18}{\sqrt{5}}\hat{\mathbf{x}} + \frac{9(5\sqrt{5} + 1)}{\sqrt{5}}\hat{\mathbf{y}} \text{ V-m}^{-1}$$

$$\text{At } P_3 \qquad\qquad \mathbf{E}_{P_3} = \frac{45}{\sqrt{2}}\hat{\mathbf{z}} \text{ V-m}^{-1}$$

4.3.2 Electric Field Resulting from Continuous Charge Distributions

The atomic theory of matter has shown us that matter is quantized and that electric charges are made up of integral multiples of a certain minimum electric charge, which is known as the charge of an electron, namely $q_e \simeq -1.602 \times 10^{-19}$ C. Although we know that charge is so quantized, it is often useful to ignore that charges come in packages of electrons and protons and to instead think of continuous distributions of charge. Such a view is valid because the "graininess" of electrical charge is not significant over the macroscopic scale sizes that are considered here. We can consider charge as if it were "continuous" because we deal with very large numbers of charged particles that are extremely close (typically, confined to material bodies of macroscopic size). As an example, the current that flows

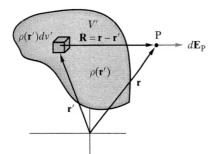

Figure 4.11 Electric field due to a continuous charge distribution. The field generated at point P (position vector $\mathbf{r}$), due to the continuous charge distribution $\rho(\mathbf{r}')$ in the source volume V', is found by integrating the contributions of infinitesimal volume elements such as that shown.

in a wire connected to an ordinary 110-volt, 100-watt light bulb is about 1 ampere (or coulomb per second), so that $\sim 6 \times 10^{18}$ elementary charges flow per second through the cross-sectional area of the wire. We can describe such "continuous" distributions of charge in terms of density of charge, much like representing matter in terms of its mass density rather than counting the number of molecules. The *volume charge density,* $\rho(x,y,z)$ can in general be a function of all three coordinates of space and has units of C-m^{-3}. In other cases, it might be more convenient to represent the charge distribution in terms of a *surface charge density* ρ_s or a *line charge density* ρ_l.

Consider electric charge to be distributed over a volume V' with a volume density $\rho(\mathbf{r}')$ as shown in Figure 4.11. Since a differential element of charge behaves like a point charge, the contribution of a charge element $dQ = \rho(\mathbf{r}')\,dv'$ located in a differential volume element[44] dv' to the electric field at point P in Figure 4.11 is

$$d\mathbf{E}_{\mathrm{P}} = \hat{\mathbf{R}}\frac{\rho(\mathbf{r}')\,dv'}{4\pi\epsilon_0|\mathbf{r}-\mathbf{r}'|^2} \tag{4.8}$$

The total electric field at point P due to all the charge in V' is then given as

$$\mathbf{E}_{\mathrm{P}} = \frac{1}{4\pi\epsilon_0}\int_{V'}\hat{\mathbf{R}}\frac{\rho(\mathbf{r}')\,dv'}{|\mathbf{r}-\mathbf{r}'|^2} \tag{4.9}$$

Note once again that (4.9) can alternatively be written as

$$\mathbf{E}_{\mathrm{P}} = \frac{1}{4\pi\epsilon_0}\int_{V'}\frac{(\mathbf{r}-\mathbf{r}')\rho(\mathbf{r}')\,dv'}{|\mathbf{r}-\mathbf{r}'|^3} \tag{4.10}$$

If the electric charge were instead distributed on a surface S' with a surface charge density ρ_s (having units of C-m^{-2}), then the integration would need to be carried out only over the surface. Thus,

$$\mathbf{E}_{\mathrm{P}} = \frac{1}{4\pi\epsilon_0}\int_{S'}\hat{\mathbf{R}}\frac{\rho_s(\mathbf{r}')\,ds'}{|\mathbf{r}-\mathbf{r}'|^2} \tag{4.11}$$

[44]Note that although dv' can be made very small compared with the macroscopic dimensions of the problem in hand, it is nevertheless taken to be large enough compared with atomic scales so that it still does contain many discrete charges.

or, alternatively,

$$\mathbf{E}_P = \frac{1}{4\pi\epsilon_0} \int_{S'} \frac{(\mathbf{r} - \mathbf{r}')\rho_s(\mathbf{r}')\,ds'}{|\mathbf{r} - \mathbf{r}'|^3} \tag{4.12}$$

Surface charge distributions arise commonly in electromagnetics, especially when metallic conductors are placed in electric fields. As we shall see later, freely mobile electrons in a metallic conductor redistribute quickly (within $\sim 10^{-19}$ s) under the influence of an applied electric field by establishing distributions of charge confined to infinitesimally (on macroscopic scales) thin regions at the conductor surfaces.

For a line charge distribution ρ_l (having units of C-m^{-1}) we have

$$\boxed{\mathbf{E}_P = \frac{1}{4\pi\epsilon_0} \int_{L'} \hat{\mathbf{R}}\,\frac{\rho_l(\mathbf{r}')\,dl'}{|\mathbf{r} - \mathbf{r}'|^2}} \tag{4.13}$$

or, alternatively,

$$\mathbf{E}_P = \frac{1}{4\pi\epsilon_0} \int_{L'} \frac{(\mathbf{r} - \mathbf{r}')\rho_l(\mathbf{r}')\,dl'}{|\mathbf{r} - \mathbf{r}'|^3} \tag{4.14}$$

where L' is the line along which the charge is distributed. Note that L' need not be a straight line, just as S' need not be a flat surface. The concept of a line charge distribution becomes convenient when the objects on which the charges reside have significant extent only in one dimension. A good example is a thin wire with negligible cross-sectional dimensions.

Example 4.5: Line charge of finite length. Consider a cylindrical rod of charge of length $2a$ centered at the origin and aligned with the z axis, as shown in Figure 4.12, where the diameter d of the rod is much smaller than any other dimension in the system, so we can represent[45] the charge distribution in terms of a line charge density ρ_l. Determine the electric field at a point $P(r, \phi, 0)$ equidistant from the end points of the line charge.

> **Solution:**   In view of the symmetry in the azimuthal (i.e., ϕ) direction, it is appropriate to use the cylindrical coordinate system, with r, ϕ, z. Using (4.13), the electric field at point P with $\mathbf{r} = \hat{\mathbf{r}}r$ due to a line element of charge $dQ = \rho_l\,dz'$ located at point $\mathbf{r}' = \hat{\mathbf{z}}z'$ and at a distance $|\mathbf{r} - \mathbf{r}'| = [r^2 + (z')^2]^{1/2}$ from point P is
>
> $$d\mathbf{E}_P = \frac{\rho_l\,dz'}{4\pi\epsilon_0}\frac{(\mathbf{r} - \mathbf{r}')}{[r^2 + (z')^2]^{3/2}} = \frac{\rho_l\,dz'}{4\pi\epsilon_0}\frac{1}{[r^2 + (z')^2]}\underbrace{\frac{\mathbf{r} - \mathbf{r}'}{|\mathbf{r} - \mathbf{r}'|}}_{\hat{\mathbf{R}}}$$
>
> This equation shows the field expressed in two slightly different ways; the expression on the right separates the magnitude of the vector from its direction (as expressed by the unit vector $\hat{\mathbf{R}}$, pointing from point $\mathbf{r}'$ to point $\mathbf{r}$). Note that the vector $d\mathbf{E}_P$ has components both in the radial (r) and z directions. However, in view of the symmetrical location of P, the contributions to the z component of charge elements located at positive z' locations cancel

---

[45]Note that the total charge on the cylindrical rod is given by $(2a)\rho_l$ or, in terms of volume charge density ρ, by $\rho(2a)\pi d^2/4$, so that $\rho_l = \rho\pi d^2/4$.

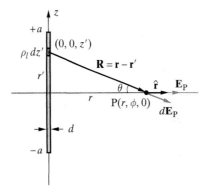

Figure 4.12 Line charge of finite length. A cylindrical rod of charge with linear charge density ρ_l C-m^{-1} produces the electric field $\mathbf{E}_P$ at point P equidistant from its endpoints.

out those of charge elements at negative z' locations. Thus, the net total electric field at point P is in the radial direction, and we need only to sum (integrate) the radial components of $d\mathbf{E}_P$. The magnitude of the radial component is

$$dE_r = |d\mathbf{E}_P|\cos\theta = \frac{\rho_l\, dz'}{4\pi\epsilon_0}\frac{1}{[r^2 + (z')^2]}\frac{r}{\sqrt{r^2 + (z')^2}}$$

Integrating[46] over z' we find

$$\mathbf{E}_P = \hat{\mathbf{r}}\frac{\rho_l r}{4\pi\epsilon_0}\int_{-a}^{+a}\frac{dz'}{[r^2 + (z')^2]^{3/2}} = \hat{\mathbf{r}}\frac{\rho_l a}{2\pi\epsilon_0 r\sqrt{r^2 + a^2}}$$

For a very long rod, such that $a \gg r$, we have

$$\mathbf{E}_P \simeq \hat{\mathbf{r}}\frac{\rho_l}{2\pi\epsilon_0 r}$$

which is a useful expression to remember. Note that the field of an infinitely long line charge falls off with distance as r^{-1}, although that for a point charge falls off with distance as r^{-2}.

Example 4.6: Ring of charge. Next we consider a ring of charge of radius a containing a total charge of amount Q, as shown in Figure 4.13. Assuming that the charge is uniformly distributed along the ring, and ignoring the thickness of the ring, the line charge density is $\rho_l = Q(2\pi a)^{-1}$. Evaluate the electric field at a point $P(0, \phi, z)$ along the axis of the ring at a distance z from its center.

Solution: Once again it is convenient to use cylindrical coordinates, and we take advantage of the symmetry of the problem to observe that the net total electric field at P is in the

---

[46]Using either integral tables or change of variables (e.g., $\zeta = r\cot\theta$) it can be shown that

$$\int\frac{d\zeta}{\sqrt{(\zeta^2 + b^2)^3}} = \frac{\zeta}{b^2\sqrt{\zeta^2 + b^2}} + \text{const.}$$

where b is a constant.

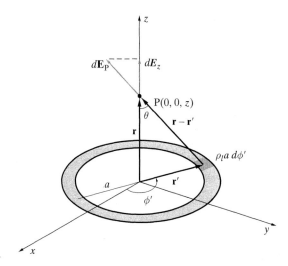

Figure 4.13 A ring of charge. Each elemental charge $\rho_l a\, d\phi'$ produces a differential electric field $d\mathbf{E}_P$ at point P.

z direction. Using (4.13), a given line element $dl' = a\, d\phi'$ along the ring, containing a total charge of $dQ = \rho_l(a\, d\phi') = Q(a\, d\phi')/(2\pi a)$, produces a differential field at P of

$$d\mathbf{E}_P = \frac{1}{4\pi\epsilon_0}\frac{Qa\, d\phi'}{2\pi a}\frac{(\mathbf{r}-\mathbf{r}')}{(a^2+z^2)^{3/2}} = \frac{1}{4\pi\epsilon_0}\frac{Qa\, d\phi'}{2\pi a}\frac{1}{(a^2+z^2)}\frac{\mathbf{r}-\mathbf{r}'}{|\mathbf{r}-\mathbf{r}'|}$$

where $\mathbf{r} - \mathbf{r}' = \hat{\mathbf{z}}z - \hat{\mathbf{r}}a$. Once again, we note that the electric field contributions of all elemental charges along the ring cancel out due to symmetry, except for the z components, so that the net total electric field at P is only in the z direction. The magnitude of the z component of the differential electric field is given by

$$dE_z = |d\mathbf{E}_P|\cos\theta = \frac{1}{4\pi\epsilon_0}\frac{Qa\, d\phi'}{2\pi a}\frac{1}{(a^2+z^2)}\frac{z}{\sqrt{a^2+z^2}}$$

so that the total electric field at point P is

$$\mathbf{E}_P = \hat{\mathbf{z}}\int_0^{2\pi}\frac{1}{4\pi\epsilon_0}\frac{Q\, d\phi'}{2\pi}\frac{1}{a^2+z^2}\frac{z}{\sqrt{a^2+z^2}} = \hat{\mathbf{z}}\frac{1}{4\pi\epsilon_0}\frac{Qz}{(2\pi)(a^2+z^2)^{3/2}}\int_0^{2\pi}d\phi'$$

$$= \hat{\mathbf{z}}\frac{1}{4\pi\epsilon_0}\frac{Qz}{(a^2+z^2)^{3/2}}$$

Note that for $z \gg a$, we have

$$\mathbf{E}_P \simeq \hat{\mathbf{z}}\frac{1}{4\pi\epsilon_0}\frac{Q}{z^2}$$

In other words, at large enough distances the ring produces the same electric field as a point charge located at the origin.

4.3.3 Is This All There Is to Electrostatics?

It is clear from the above that *if the positions of all the charges* (either discrete or continuous) *are specified,* finding the electric field simply involves carrying out the appropriate summations and integrals. In other words, all of electrostatics appears to boil down to Coulomb's law and integration operations. These integrals (equations (4.6), (4.9), (4.11), (4.13)) are often complicated and may involve all three dimensions; however, they are nevertheless straightforward operations that can be evaluated numerically, if not solved or approximated analytically. Although evaluating the integrals may be a straightforward numerical problem, *understanding* the electric field and its relationship to charge configurations is enhanced by introducing additional concepts, such as field lines, electric potential, and divergence, and by expressing Coulomb's law in other ways, such as the so-called Gauss's law. Familiarity with such concepts may help us find easier ways of evaluating these integrals or, if they were to be evaluated for us by a number-crunching machine, of interpreting the numerical results.

It should also be realized that in many electrostatic problems—in fact, in most of the ones that are of practical interest—the distribution of at least some of the charges *is not known.* We know only the general rules (i.e., boundary conditions) by which charges reorganize themselves under the influence of electric fields. This is especially the case when metallic conductors are present in the vicinity of charges: When a charged body is brought near a metallic conductor, the charge carriers within the conductor redistribute themselves (see Section 4.7) in a manner that depends on the electric field, which in turn depends on the eventual charge configuration. In other words, although we may know the locations of some of the charges in our system, we do not know the actual distribution of charges on the conductors and thus cannot evaluate the total electric field simply by direct integration.

4.3.4 Visualization of the Electric Field: Field Lines

Michael Faraday was probably the greatest scientist in history who was completely innocent of mathematics. He made up for this "deficiency" through his intuitive ability to pictorialize, an ability perhaps unequaled in scientific history. Through his pictorial and nonmathematical imagination, Faraday understood magnetic and electric fields in terms of lines of force. We owe our present concept of field lines, introduced and discussed below, largely to Faraday. Faraday's ideas and points of view had a significant influence on J. C. Maxwell, who couched Faraday's qualitative thinking in a mathematical framework. Since we separately discuss the related concept of *electric flux* in a later section, here we limit our attention simply to the graphical representation of the electric field distribution using field lines.

The field line concept is useful for visualization and graphical display of any vector quantity, which in general may vary as a function of space. The need for such a concept becomes apparent if we think of how to "plot" even the simplest electrostatic field: the field generated by a point charge at the origin. This simple field is in the spherically radial direction and varies with distance as r^{-2}. One way to "plot" the field so as to display both the direction and the magnitude information is as shown in Figure 4.14a, from which it is apparent that this method provides for rather coarse spatial resolution. Note that we can easily display the r^{-2} dependence of the magnitude of the electric field by means of a three-dimensional plot in Figure 4.14b; however, such a display does not

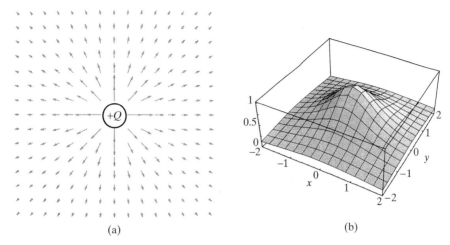

(a)   (b)

Figure 4.14   Electric field of a point charge. (a) A discretized vector plot of the electric field around a point charge. The sizes of the arrows at different positions represent the relative magnitude of the field at those positions. (b) The normalized magnitude $|\mathbf{E}(x, y, 1)|/|\mathbf{E}(0, 0, 1)|$ of the electric field due to a point charge $+Q$ at the origin is shown as a function of x, y at the $z = 1$ plane.

convey any information about the direction of the electric field. Graphical representation of the field becomes more important for more complicated field distributions.

Another way to plot the electric field distribution graphically is by drawing lines that are everywhere tangent to the electric field; such lines are called *field lines* or *lines of force*.[47] The magnitude information can be conveyed by the *density* of lines (i.e., number of lines per unit area through a surface perpendicular to the lines). Such displays are shown in Figures 4.15, 4.16, 4.17a, and 4.17b, respectively for a single point charge, two equal point charges of opposite polarity, two equal point charges of the same polarity, and two charges of different magnitudes and polarity. Note that the electric field lines emanate from positive charges and terminate at negative charges, some of which (e.g., in Figure 4.17b) may be located at infinity.

Note that in order for a field line plot to convey useful information, a sensible number of lines has to be utilized—enough to indicate any symmetry in the system and the *relative* magnitude of the field at different points. It would not be possible to display the magnitude of the field on any absolute scale (e.g., a given number of lines per unit charge), since the relative magnitude of the charge might lead to unreasonably sparse or too dense field line distributions.

With respect to Figure 4.15a, one may at first think that the total number of lines, for example, crossing a spherical surface of radius r, should decrease with distance as

---

[47]They are called lines of force because they literally represent the direction in which test charges initially placed at a point would move, or the direction of the *force* that the test charge would experience. In other words, a small charged test particle of negligible mass set free in the electric field would simply trace out a line of force, very much like a small float set free on a lake would trace out the line of water current.

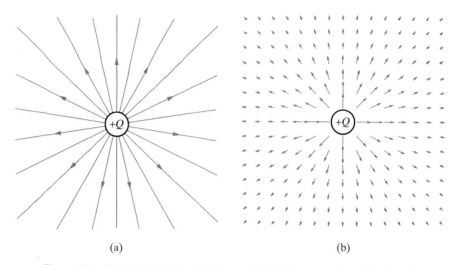

(a)   (b)

Figure 4.15   Electric field of a point charge. Graphical representation of the electric field of a point charge in terms of (a) field lines, (b) discrete vectors.

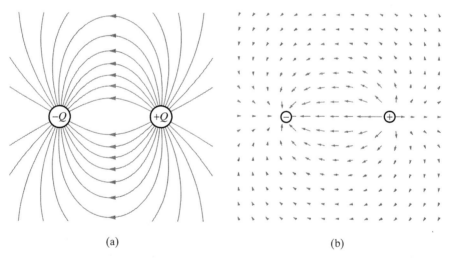

(a)   (b)

Figure 4.16   Electric field of two equal but opposite point charges. (a) Field line representation. (b) Representation using discrete vectors.

r^{-2} to represent relative magnitude of the field accurately. However, the area of such a spherical surface perpendicular to the lines increases as r^2, so the *density* of lines in Figure 4.15a is proportional to the magnitude of the field at any distance from the source charge, even though the total number of lines stays the same.

Since the electric field at any point has a definite direction, it follows that electric field lines cannot cross one another because that would imply two possible directions at the crossing point. Realizing this constraint also highlights an important deficiency of

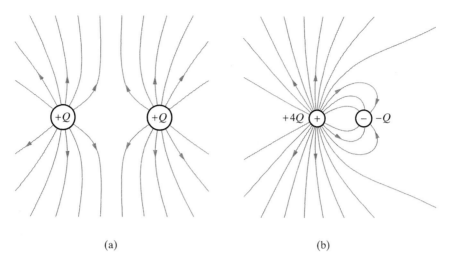

(a)   (b)

Figure 4.17   Field maps for two charges Q_1 and Q_2. (a) $Q_2 = Q_1 = Q$. (b) $Q_1 = -4Q_2 = 4Q$.

the field line concept. Although the electric field in the vicinity of two point charges is calculated (see Example 4.4) by linear superposition of the fields due to the individual charges, the field line plot of Figure 4.16 is not a simple superposition of two plots of the kind in Figure 4.15.[48] The principle of superposition, an important principle concerning electric fields, does not have an easy interpretation in the context of field lines. Although we can in principle visualize a vectorial superposition, we cannot easily deduce the field line distribution in Figure 4.16 from that in Figure 4.15.

4.4 THE ELECTRIC POTENTIAL

An electric field is essentially a field of force. If a body being acted upon by a force is moved from one point to another, work will be done on the body. In the absence of losses (or dissipation of energy), the energy (i.e., work) thus put in must be stored as either kinetic or potential energy. The electrostatic field is an example of such a *conservative* field, since if a charge is moved in a static field in the absence of friction, no energy is dissipated. In moving a small test charge about in a field, either we have to do work against electric forces or we find that these forces do work for us. If we have two charges of opposite sign, work must be done to separate them in opposition to the attractive electrostatic force between them. This energy is "stored" in the separated configuration of the two charges and can be recovered if the charges are allowed to return to their original relative positions. The stored energy is called *potential* (rather than kinetic) energy because it depends on the position (rather than motion) of the charges in a force field. The concept of *scalar electric potential* provides a quantitative

---

[48]Note that such a graphical superposition would create crossing field lines, which are not possible, as we just argued.

measure of the work or energy required to move charges from one point to another under the influence of an electrostatic field.

4.4.1 Work Required to Move a Test Charge: Electrostatic Potential

We assume that there is some distribution of charges that sustains an electric field. The work W that we need to do *against* the electrical force ($\mathbf{F}$) in carrying a test charge q along some path from point a to point b is

$$W_{a \to b} = -\int_a^b \mathbf{F} \cdot d\mathbf{l} \tag{4.15}$$

where $d\mathbf{l}$ is the differential vector displacement along the path, as shown in Figure 4.18a. Note that the units of W are joules (J). The minus sign in front of the integral is necessary because the work is done *against* the field. The dot product[49] accounts for the fact that it takes no work to move the test charge perpendicular to the field (no opposing force). Often we are more interested in work done (by an external force against the electrical force) in carrying a unit positive test charge (i.e., $q = 1$ C), in which case, using $\mathbf{F} = q\mathbf{E}$, we have

$$\boxed{\left[\frac{W}{q} \right]_{a \to b} = -\int_a^b \mathbf{E} \cdot d\mathbf{l}} \tag{4.16}$$

where the units of the quantity $[W/q]$ are joules per coulomb (J-C^{-1}) or volts (V), and the subscript $a \to b$ indicates that the work is done in moving from a to b. In evaluating line integrals such as (4.15), it is customary to take $d\mathbf{l}$ in the direction of the increasing coordinate value (e.g., the $+x$ direction if the integration is along the x axis) so that the manner in which the path of integration is traversed (i.e., $a \to b$ versus $b \to a$) is unambiguously determined by the limits of integration.

In general, a line integral of the type above depends on the path of integration. However, in a conservative field (i.e., one in which there is no mechanism for energy dissipation), the work done in moving from one point to another is independent of the path. If the integral depended on the path from a to b, we could extract energy out of the field by allowing the charge to move from a to b along the path for which $[W/q]$ is

---

[49]The dot product of two vectors $\mathbf{A}$ and $\mathbf{B}$ is a scalar denoted by $\mathbf{A} \cdot \mathbf{B}$ and is equal to the product of the magnitudes $|\mathbf{A}|$ and $|\mathbf{B}|$ of $\mathbf{A}$ and $\mathbf{B}$ and the cosine of the angle θ_{AB} between them. Namely,

$$\mathbf{A} \cdot \mathbf{B} \equiv |\mathbf{A}||\mathbf{B}| \cos\theta_{AB}$$

Noting that in rectangular coordinates we have $\mathbf{A} = \hat{\mathbf{x}}A_x + \hat{\mathbf{y}}A_y + \hat{\mathbf{z}}A_z$ and $\mathbf{B} = \hat{\mathbf{x}}B_x + \hat{\mathbf{y}}B_y + \hat{\mathbf{z}}B_z$, an alternative expression for the dot product is

$$\mathbf{A} \cdot \mathbf{B} = (\hat{\mathbf{x}}A_x + \hat{\mathbf{y}}A_y + \hat{\mathbf{z}}A_z) \cdot (\hat{\mathbf{x}}B_x + \hat{\mathbf{y}}B_y + \hat{\mathbf{z}}B_z) = A_x B_x + A_y B_y + A_z B_z$$

The dot product is sometimes referred to as the scalar product, since the result is a scalar quantity.

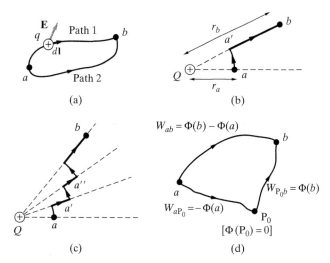

Figure 4.18   Work and voltage. The work done (by an external force against the electric field) in carrying a +1C charge from point a to b depends only on the end points, not on the path taken.

smaller and take it back from b to a along another. We could indefinitely extract work from the field[50] simply by repeating the process.

Now consider the field produced by a single point charge Q and two points a and b at distances of r_a and r_b, respectively, as shown in Figure 4.18b. Since the field is radial, it is clear that no work is done in moving a unit test charge along circular arc segments such as a–a' that are perpendicular to the field (i.e., $\mathbf{E} \cdot d\mathbf{l} = 0$ along the path). On the other hand, along path a'–b the field is in the direction of motion of the unit test charge, so $\mathbf{E} \cdot d\mathbf{l} = E \, dr$. Thus the work done by an external force against $\mathbf{E}$ in moving the unit test charge from point a to point b is

$$\left[\frac{W}{q} \right]_{a \to b} = \left[\frac{W}{q} \right]_{a \to a'} + \left[\frac{W}{q} \right]_{a' \to b} = -\int_a^{a'} \mathbf{E} \cdot d\mathbf{l} - \int_{a'}^b \mathbf{E} \cdot d\mathbf{l}$$

$$= 0 - \frac{Q}{4\pi\epsilon_0} \int_{r_{a'}}^{r_b} \frac{dr}{r^2} = -\frac{Q}{4\pi\epsilon_0} \left[-\frac{1}{r} \right]_{r_{a'}}^{r_b} = \frac{Q}{4\pi\epsilon_0} \left(\frac{1}{r_b} - \frac{1}{r_a} \right)$$

since $r_{a'} = r_a$. It is also clear that if we took some other path, one that has smaller steps in the radial direction interconnected with smaller circular arc segments (Figure 4.18c) for example, the result would be the same. Since any other path between a and b, however smooth it may be, can also be similarly divided into arc/radial segments, we conclude that the net work depends only on the end points of the path.

[50]There is nothing inherently wrong about extracting work from a field. Energy could be extracted from the field if the motion of the test charge produced forces that could influence the field, for example, by moving the charges that produce the field. However, in electrostatics, we assume that the charges producing the fields are fixed in their locations so that no work can be done on them.

Since $[W/q]_{a \to b}$ depends only on the end points, it can be represented as a differ-ence between two numbers. To see this, we can consider Figure 4.18d, which introduces a reference point P_0 and, for any point P, a function $\Phi(P)$ that is equal to the work done against the field in moving from P_0 to P. Since the work done in moving the test charge from P_0 to P depends only on the two end points, $\Phi(P)$ has only one value. Let us evaluate our line integral using a path that goes through P_0:

$$\left[\frac{W}{q}\right]_{a \to b} = -\int_a^b \mathbf{E} \cdot d\mathbf{l} = -\int_a^{P_0} \mathbf{E} \cdot d\mathbf{l} - \int_{P_0}^b \mathbf{E} \cdot d\mathbf{l} = \Phi(b) - \Phi(a)$$

since the work done in moving from a to P_0 is equal in magnitude but opposite in sign to that done in moving from P_0 to a.

The values of the function Φ constitute a relative *electrostatic potential* that can be assigned to every point in the field. This function Φ is a *scalar field* and is, in general, a function of x, y, and z. For convenience, the reference point P_0 is usually taken to be at infinity, so the *electrostatic potential* at any point P is

$$\boxed{\Phi(P) \equiv \left[\frac{W}{q}\right]_{\text{at P}} = -\int_\infty^P \mathbf{E} \cdot d\mathbf{l}} \tag{4.17}$$

Note that $\Phi(P)$ represents the amount of energy *potentially* available when a unit positive test charge is at point P; this energy can be extracted from the system by allowing the test charge to move to a distant point (i.e., to infinity). It is thus appropriate to view $\Phi(P)$ as *potential* energy per coulomb, with units of joules per coulomb (J-C^{-1}) or volts (V). Note that for a single point charge Q at the origin ($\mathbf{r}' = 0$) as the source of the electric field, the electrostatic potential (i.e., the work done to move a unit positive test charge from infinity to point P) is

$$\Phi(P) = \Phi(x, y, z) = -\int_\infty^r \left(\hat{\mathbf{r}}\frac{Q}{4\pi\epsilon_0|\mathbf{r}|^2}\right) \cdot (\hat{\mathbf{r}}\, dr) = \frac{Q}{4\pi\epsilon_0}\frac{1}{r} \tag{4.18}$$

where, for simplicity, the path of integration is chosen to be in the radial direction.

The work done in moving a unit test charge from point a to b is the *electrostatic potential difference* between the two points and is denoted by Φ_{ab}. Based on the above, this quantity is given by

$$\boxed{\left[\frac{W}{q}\right]_{a \to b} = \Phi_{ab} = -\int_a^b \mathbf{E} \cdot d\mathbf{l} = \Phi(b) - \Phi(a)} \tag{4.19}$$

where $\Phi(b)$ and $\Phi(a)$ are the electrostatic potentials respectively at points b and a.

One important corollary to the above discussion is that the line integral of the electrostatic field around any closed path is identically zero. This is readily seen from Figure 4.18d, if one imagines going around the closed path $a \to P_0 \to b \to a$. We thus have

$$\oint_{\text{any closed path}} \mathbf{E} \cdot d\mathbf{l} = 0 \tag{4.20}$$

This equation turns out to describe a general property of any *conservative* vector field. We will refer back to this property of the electrostatic field as we discuss the concept of *curl* of a vector field in Section 6.4.

An important point to note here is that we arrived at this property simply through symmetry arguments (e.g., we did use the fact that the force from a single charge was spherically symmetric and in the radial direction), and without using the fact that the electrostatic force depends on distance as r^{-2}. Thus, the above integral equation contains only part of the known laws of electrostatics. The additional property, known as Gauss's law, will be described in a later section. Together, these two integral properties represent electrostatics as contained in Coulomb's law.

Example 4.7: Work done in carrying charges. Determine the work done in carrying a $q = +2\ \mu$C point charge from point A$(0, -3, 0)$ to point B$(1, 0, 0)$ in a nonuniform electric field $\mathbf{E} = (3x + y)\hat{\mathbf{x}} + x\hat{\mathbf{y}} - 3\hat{\mathbf{z}}$ along (a) the shortest path (straight line), (b) an L-shaped path via the origin, and (c) along the parabola $y = 6x^2 - 3x - 3$, as shown in Figure 4.19.

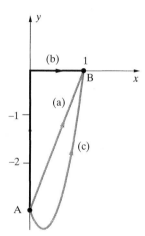

Figure 4.19   Example 4.7. The work done in moving a test charge from point A to point B is independent of the path.

Solution: In all three cases, we need to evaluate the line integral

$$W_{A \to B} = -q \int_A^B \mathbf{E} \cdot d\mathbf{l}$$

where

$$\mathbf{E} \cdot d\mathbf{l} = [(3x + y)\hat{\mathbf{x}} + x\hat{\mathbf{y}} - 3\hat{\mathbf{z}}] \cdot [\hat{\mathbf{x}}\, dx + \hat{\mathbf{y}}\, dy + \hat{\mathbf{z}}\, dz\,]$$

$$= (3x + y)dx + x\, dy - 3dz$$

following the usual convention of taking $d\mathbf{l}$ to be equal to $+\hat{\mathbf{r}}$ so that the way we integrate along the path is determined by the limits of integration. Therefore,

$$W_{A \to B} = (-2 \times 10^{-6}) \left[\int_0^1 (3x + y)dx + \int_{-3}^0 x\, dy - \int_0^0 3dz \right]$$

where the limits of each integral are provided with respect to its own variable of integration.

(a) Along the shortest path, we have $y = 3x - 3$. Using this equation, we can write

$$W_{A \to B} = (-2 \times 10^{-6}) \left[\int_0^1 (3x + 3x - 3)\, dx + \int_{-3}^0 \left[\left(\frac{y}{3} \right) + 1 \right] dy \right]$$

$$= (-2 \times 10^{-6}) \left\{ [3x^2 - 3x]_0^1 + \left[\frac{y^2}{6} + y \right]_{-3}^0 \right\} = -3 \ \mu J$$

(b) Along the L-shaped path, we have $x = 0$ from A to the origin and $y = 0$ from the origin to B. Thus, we can write

$$W_{A \to B} = (-2 \times 10^{-6}) \left[\int_0^1 3x\, dx \right] = (-2 \times 10^{-6}) \left[\frac{3x^2}{2} \right]_0^1 = -3 \ \mu J$$

(c) Along the parabolic path, we have $y = 6x^2 - 3x - 3$, so that

$$W_{A \to B} = (-2 \times 10^{-6}) \left[\int_0^1 (3x + 6x^2 - 3x - 3)\, dx + \int_{-3}^0 x\, dy \right]$$

Since along the parabola we have $dy = (12x - 3)\, dx$, and $x = 0$ when $y = -3$ and $x = 1$ when $y = 0$, we can rewrite the above as

$$W_{A \to B} = (-2 \times 10^{-6}) \left[\int_0^1 (6x^2 - 3)\, dx + \int_0^1 x(12x - 3)\, dx \right]$$

$$= (-2 \times 10^{-6}) \left[\int_0^1 (18x^2 - 3x - 3)\, dx \right]$$

$$= (-2 \times 10^{-6}) \left[6x^3 - \tfrac{3}{2}x^2 - 3x \right]_0^1 = -3 \ \mu J$$

As expected, the work done is the same regardless of the path we follow in moving the charge.

Example 4.8: Electrostatic potential in the vicinity of an infinitely long line charge. The electric field in the vicinity of an infinitely long line charge was derived in Example 4.5:

$$\mathbf{E} = \hat{\mathbf{r}} \frac{\rho_l}{2\pi \epsilon_0 r}$$

Determine the electrostatic potential Φ in its vicinity.

Solution: Since the field is in the r direction, for simplicity, we take a line integral along a radial path, with $d\mathbf{l} = \hat{\mathbf{r}}\, dr$, so we have

$$\Phi(r) = -\int_\infty^r \mathbf{E} \cdot d\mathbf{l} = -\frac{\rho_l}{2\pi \epsilon_0} \int_\infty^r \frac{\hat{\mathbf{r}}}{\zeta} \cdot (\hat{\mathbf{r}}\, d\zeta) = -\frac{\rho_l}{2\pi \epsilon_0} [\ln(\zeta)]_\infty^r = \infty \ !$$

We can understand this seemingly unphysical result by recalling the meaning of electrostatic potential at any point as the work that needs to be done to bring a unit positive test charge from infinity to that point. With an infinitely long line charge, the electric field apparently does not fall rapidly enough with distance to allow moving a test charge from ∞ to any point r without requiring infinite energy.

Note, however, that the difference in electrostatic potential between any two points $r = a$ and $r = b$, such that $0 < b < a$, is straightforward to evaluate; we find

$$\Phi_{a \to b} = \Phi(b) - \Phi(a) = -\frac{\rho_l}{2\pi\epsilon_0} \int_a^b \frac{dr}{r} = \frac{\rho_l}{2\pi\epsilon_0} \ln \frac{a}{b}$$

which is the amount of work that we need to do to move a unit positive test charge closer to the line charge from $r = a$ to $r = b$. Note that for $a < b$, we have $\ln(a/b) < 0$, so $\Phi_{a \to b} < 0$, indicating that the field does work (i.e., work is extracted from the field) when the unit positive test charge moves away from the positive line charge.

4.4.2 The Electrostatic Potential and the Electric Field

Since forces on charges are determined by the electric field **E**, the introduction of the additional concept of electrostatic potential may at first appear unnecessary. One advantage of the potential concept is that it is a scalar, which is relatively easier to evaluate in the vicinity of complicated charge distributions than the vector electric field, and from which the electric field can be readily calculated by simply taking derivatives, as described in the following paragraphs. Another advantage of the potential concept is that it allows us to link field concepts with the more familiar circuit concepts, which are couched in terms of potential (or voltage) differences between different points in a circuit and voltage drops across circuit elements.

We can now establish a differential relationship between the electrostatic potential Φ and the electric field **E**. Consider two points at x and $(x + \Delta x)$ with the same y and z coordinates. The differential amount of work done (dW) in moving a test charge q between these points is directly proportional to the potential difference between them. Hence,

$$(dW)_x = q\Phi(x + \Delta x, y, z) - q\Phi(x, y, z) = q\frac{\partial\Phi}{\partial x}\Delta x$$

If the two points also had slightly different y, z coordinates, we then have

$$dW = q\left(\frac{\partial\Phi}{\partial x}\Delta x + \frac{\partial\Phi}{\partial y}\Delta y + \frac{\partial\Phi}{\partial z}\Delta z\right)$$

But the work done by an external force against the electric field is also given by $dW = -q\mathbf{E} \cdot \Delta\mathbf{l}$, where $\Delta\mathbf{l}$ is the differential length element given by $\Delta\mathbf{l} = (\hat{\mathbf{x}}\Delta x + \hat{\mathbf{y}}\Delta y + \hat{\mathbf{z}}\Delta z)$. Thus we have

$$\frac{\partial\Phi}{\partial x}\Delta x + \frac{\partial\Phi}{\partial y}\Delta y + \frac{\partial\Phi}{\partial z}\Delta z = -\mathbf{E} \cdot \Delta\mathbf{l}$$

$$\left(\hat{\mathbf{x}}\frac{\partial\Phi}{\partial x} + \hat{\mathbf{y}}\frac{\partial\Phi}{\partial y} + \hat{\mathbf{z}}\frac{\partial\Phi}{\partial z}\right) \cdot (\hat{\mathbf{x}}\Delta x + \hat{\mathbf{y}}\Delta y + \hat{\mathbf{z}}\Delta z) = -\mathbf{E} \cdot \Delta\mathbf{l}$$

$$\left(\hat{\mathbf{x}}\frac{\partial\Phi}{\partial x} + \hat{\mathbf{y}}\frac{\partial\Phi}{\partial y} + \hat{\mathbf{z}}\frac{\partial\Phi}{\partial z}\right) \cdot \Delta\mathbf{l} = -\mathbf{E} \cdot \Delta\mathbf{l}$$

from which it follows that

$$\mathbf{E} = -\left(\hat{\mathbf{x}}\frac{\partial \Phi}{\partial x} + \hat{\mathbf{y}}\frac{\partial \Phi}{\partial y} + \hat{\mathbf{z}}\frac{\partial \Phi}{\partial z}\right) \qquad (4.21)$$

Equation (4.21) indicates that the electric field $\mathbf{E}$ at any given point will have its largest component in the direction opposite to that along which the spatial rate of change of electrostatic potential is the largest. In other words, the electric field at any given point is the negative *gradient*[51] of the electrostatic potential at that point.

It is often customary to represent the gradient of a scalar by using the *del* operator, sometimes also called the *grad* or *nabla* operator, represented by the symbol ∇. For this purpose, the del operator is defined in rectangular coordinates as

$$\nabla \equiv \hat{\mathbf{x}}\frac{\partial}{\partial x} + \hat{\mathbf{y}}\frac{\partial}{\partial y} + \hat{\mathbf{z}}\frac{\partial}{\partial z} \qquad (4.22)$$

so (4.21) can be compactly written as

$$\mathbf{E} = -\nabla \Phi \qquad (4.23)$$

As an example of gradients of potential functions that vary in two dimensions, consider the function shown in Figure 4.20. The function itself is illustrated using a three-dimensional surface plot, whereas the corresponding negative gradient (i.e., the electric field) is shown by means of a vector field plot. The plots illustrate that the electric field magnitudes (i.e., size of the arrows) are highest in regions of highest spatial slope of the potential functions and that the electric field has no components in directions along which there is no potential variation.

In problems for which a system of charges is specified and where it is necessary to determine the resultant electric field due to the charges, it is often simpler first to find the electrostatic potential $\Phi(x, y, z)$ and then determine $\mathbf{E}(x, y, z)$ by finding the negative gradient of the potential. This roundabout way of finding $\mathbf{E}$ is simpler because the electric field is a vector quantity, whereas Φ is a scalar and can be found as an algebraic (rather than vector) sum of the potentials due to each system of charge. In simple problems, there may be little advantage in using the potential method; however, in more complicated problems, the use of the scalar potential results in real simplification, as in the case of the electric dipole discussed in Section 4.4.3.

---

[51] The gradient of a scalar function at any given point is the maximum spatial rate of change (i.e., the steepest slope) of that function at that point. Gradient is thus naturally a vector quantity because the maximum rate of change with distance must occur in a given direction. The direction of the gradient vector is that in which the scalar function (e.g., temperature) changes most rapidly with distance. A good analogy here is the gravitational potential. A marble placed on the slope of a mountain rolls down (i.e., its velocity vector is oriented) in the direction opposite to the gradient of the gravitational potential, and its speed (i.e., the magnitude of its velocity) is determined by the spatial derivative in that direction.

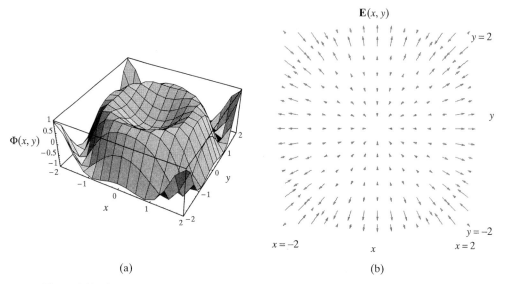

Figure 4.20   An example of a two-dimensional potential function and its gradient field. (a) The function shown is $\Phi(x,y) = \sin(x^2 + y^2)$. (b) The corresponding electric field is $\mathbf{E} = -\nabla\Phi = -\hat{\mathbf{x}}2x\cos(x^2 + y^2) - \hat{\mathbf{y}}2y\cos(x^2 + y^2)$. The units for the x and y axes are in (radians)$^{1/2}$ for both panels.

Gradient in other coordinate systems.   The preceding definition of the concept of the gradient of scalar potential in terms of the motion of a test charge in the presence of an electric field was presented in a rectangular coordinate system. We now consider gradient in other coordinate systems (see Appendix A).

If we consider the motion of the test charge in a cylindrical (r, ϕ, z) coordinate system, we realize that in moving by a differential amount $d\phi$ in the ϕ direction, we span a distance of $r\, d\phi$ (see Figure 4.21a), so the E_ϕ is $r^{-1}\,\partial\Phi/\partial\phi$, rather than just $\partial\Phi/\partial\phi$. Motion in the r and z directions spans differential distances of dr and dz, so that we have

$$\boxed{\mathbf{E} = -\nabla\Phi = -\left[\hat{\mathbf{r}}\frac{\partial\Phi}{\partial r} + \hat{\boldsymbol{\phi}}\frac{1}{r}\frac{\partial\Phi}{\partial\phi} + \hat{\mathbf{z}}\frac{\partial\Phi}{\partial z} \right]} \tag{4.24}$$

Similar considerations for a spherical coordinate system (r, θ, ϕ) indicate that in moving by an amount $d\theta$ in latitude we span a distance of $r\, d\theta$, whereas the distance we span in moving by an amount $d\phi$ in azimuth (longitude) depends on the latitude coordinate θ and is $r\sin\theta\, d\phi$, as shown in Figure 4.21b. Accordingly we have

$$\boxed{\mathbf{E} = -\nabla\Phi = -\left[\hat{\mathbf{r}}\frac{\partial\Phi}{\partial r} + \hat{\boldsymbol{\theta}}\frac{1}{r}\frac{\partial\Phi}{\partial\theta} + \hat{\boldsymbol{\phi}}\frac{1}{r\sin\theta}\frac{\partial\Phi}{\partial\phi} \right]} \tag{4.25}$$

Equations (4.24) and (4.25) also indicate that the definition of the del operator ∇ as given in (4.22) is valid only for a rectangular coordinate system.

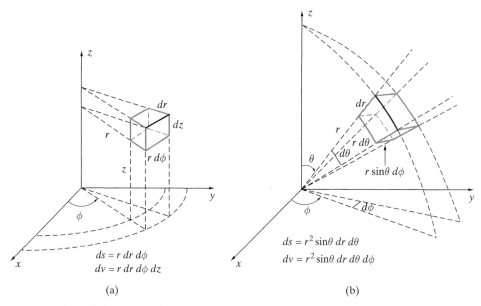

Figure 4.21 Gradient in other coordinate systems. (a) Cylindrical coordinate system. (b) Spherical coordinate system.

Example 4.9: Gradient in spherical coordinates. The potential distribution in the vicinity of a metal sphere placed in a uniform electric field is

$$\Phi(r,\theta) = E_0 \left[1 - \left(\frac{a}{r} \right)^3 \right] r \cos\theta$$

Find the corresponding electric field.

Solution: Using (4.25) we have

$$\mathbf{E} = -\nabla\Phi = -\frac{\partial\Phi}{\partial r}\hat{\mathbf{r}} - \frac{1}{r}\frac{\partial\Phi}{\partial\theta}\hat{\boldsymbol{\theta}} = -\hat{\mathbf{r}}E_0 \left[1 + 2\left(\frac{a}{r}\right)^3 \right]\cos\theta + \hat{\boldsymbol{\theta}}E_0 \left[1 - \left(\frac{a}{r}\right)^3 \right]\sin\theta$$

Note that the component of the electric field in the θ direction is generated as a result of the θ dependence of $\Phi(r,\theta)$, which has the simple $\cos\theta$ form, thus leading to a $\sin\theta$ dependence of the θ component of **E**.

Graphical representation of potential: Equipotential lines.   Just as it is convenient to graph and think of electric fields in terms of field lines, it is useful to visualize and graph scalar electric potential fields. The easiest way to represent potential is to draw surfaces on which Φ is a constant. Such surfaces are called *equipotentials*. Equipotentials for single and two point charges are shown in Figures 4.22 and 4.23,

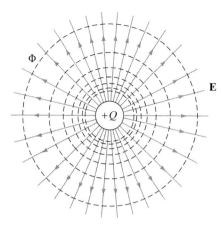

Figure 4.22 Equipotential lines. The dashed lines represent equipotential lines for a single positive point charge at the origin.

together with the electric field lines. We note from these figures that the equipotentials are at right angles to the field lines. This is a general property of equipotentials and field lines that follows from the fact that the electric field is the negative gradient of the potential. The gradient vector is in the direction of the most rapid change of potential and is therefore perpendicular to the equipotential surface. Another way of looking at it is to consider the fact that the potential is by definition constant over an equipotential surface, so any movement of charge over such a surface requires no work. Thus, equipotential surfaces must always be orthogonal to the electric field. Equipotential lines are analogous to contour elevation lines on maps, which connect the points of equal elevation.

Equations that describe the equipotential surfaces can be derived simply by setting the total electrostatic potential to a constant value. Using (4.18), for a single point charge Q located at the origin, the equation for the equipotential surfaces is

$$\frac{Q}{4\pi \epsilon_0 r} = \frac{Q}{4\pi \epsilon_0 \sqrt{x^2 + y^2 + z^2}} = C_1 \quad \rightarrow \quad x^2 + y^2 + z^2 = C_2$$

where C_1 and C_2 are constants. At the plane defined by $z = 0$, the equipotential surfaces are described by $x^2 + y^2 = C_2$ and are circles, as shown in Figure 4.22.

For the case of two point charges Q_1 and Q_2 located at points $(a, 0)$ and $(-a, 0)$, respectively, the superposition principle applies, so the equipotential surfaces are the set of curves corresponding to

$$\frac{Q_2}{r_2} + \frac{Q_1}{r_1} = \text{const.}$$

where r_1 and r_2 are, respectively, the distances to the point P located on the equipotential surface from the charges Q_1 and Q_2. The corresponding plots for $Q_2 = -Q_1 = Q$ and $Q_2 = -2Q_1 = 2Q$ are shown in Figure 4.23a. The orthogonality of the electric field lines and the equipotentials is clearly evident. For the case of charges of opposite sign (Figure 4.23a), note the clustering of the equipotentials at a point between the two charges where the polarity of the potential changes sign. The separating equipotential line corresponding to zero potential is indicated.

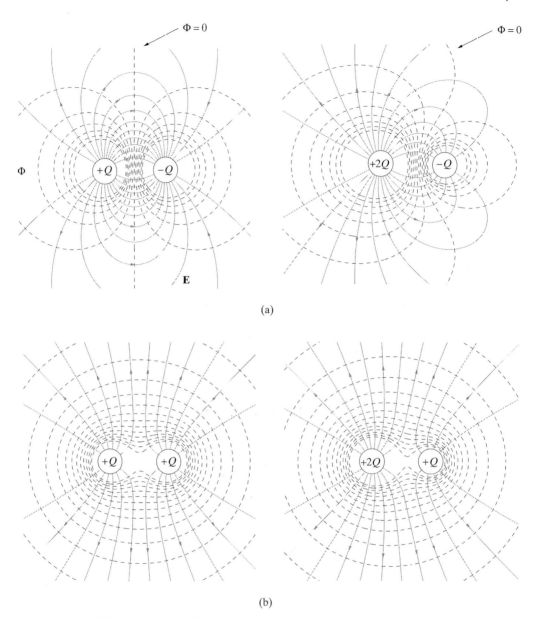

Figure 4.23 Equipotentials for two point charges. Equipotentials for two different pairs of charges are shown, together with the electric field lines for comparison. (a) Two charges of opposite polarity. (b) Two charges of the same polarity.

For the case of two equal positive charges of $+Q$ each, we note that the outermost equipotentials begin to look nearly circular. This is to be expected, since at far enough distances from the charges the field should be the same as that for a single point charge of $+2Q$ (left-hand panel of Figure 4.23b). The same is true for the $+2Q$ and $+Q$ pair shown in the right-hand panel of Figure 4.23b.

4.4.3 Electrostatic Potential Resulting from Multiple Point Charges

Using (4.18), the electric potential at point $\mathbf{r}$ in a system of n discrete point charges $Q_1, Q_2, \ldots, Q_n$ located at points $\mathbf{r}'_1, \mathbf{r}'_2, \ldots, \mathbf{r}'_n$ is, by superposition, the sum of potentials resulting from the individual charges:

$$\Phi = \frac{1}{4\pi\epsilon_0} \sum_{k=1}^{n} \frac{Q_k}{|\mathbf{r} - \mathbf{r}'_k|} \tag{4.26}$$

The most important example of a multiple charge distribution is the electrostatic dipole, considered next.

The electric dipole.   As an example, we consider the electrostatic dipole shown in Figure 4.24a, consisting of a pair of equal and opposite point charges $+Q$ and $-Q$ with a small separation d centered at the origin of the coordinate system. The distances from the charges to a point P are designated as r_+ and r_-, respectively. The potential at point P is obtained by superposing the potentials at P due to the individual charges, namely,

$$\Phi = \frac{Q}{4\pi\epsilon_0 r_+} - \frac{Q}{4\pi\epsilon_0 r_-} = \frac{Q}{4\pi\epsilon_0}\left(\frac{1}{r_+} - \frac{1}{r_-}\right)$$

where the distances r_+ and r_- follow from applying the law of cosines to the triangles AOP and BOP and are

$$r_\pm = \sqrt{r^2 + (d/2)^2 \mp rd\cos\theta} = r\sqrt{1 + (d/2r)^2 \mp (d/r)\cos\theta}$$

We are interested in the special case in which the charges are very close to one another so that $d \ll r_+, r_-$; in other words, the potential and electric field are observed at distances far away from the charges. Such closely spaced pairs of "dipole" charges are encountered often in physics and engineering. The fields produced by an electric dipole antenna can often be approximated as two charges separated by a small distance. More importantly, behavior of materials under the influence of electric fields can often be understood in terms of atomic dipoles established due to the relative displacement of electrons with respect to their nuclei, as discussed in Section 4.10.

The equation for the potential due to an electric dipole can be simplified further for the case $r \gg d$ (i.e., at a faraway point P). To obtain the simplified equation, we

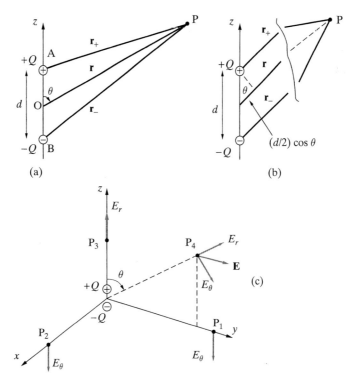

Figure 4.24 Electric dipole. The geometry of the problem of an electric dipole where
$r \gg d$. (a) The general case of a point P at a distance r. (b) Approximate geometry when
$r \gg d$. (c) The orientation of the electric field at different points P_i, all at large distances
(i.e., $r \gg d$).

expand r_+^{-1} and r_-^{-1} expressions in a Taylor series in terms of d/r and neglect the terms
of order $(d/r)^2$ and higher.[52] The Taylor series approximation is

$$\frac{1}{\sqrt{1+u}} = 1 - \frac{1}{2}u + \frac{3}{8}u^2 \cdots \simeq 1 - \frac{1}{2}u$$

as $u = \mp d/r \to 0$. Consequently, the r_+^{-1} and r_-^{-1} expressions are

$$\frac{1}{r_\pm} \simeq \frac{1}{r}\left(1 \pm \frac{d\cos\theta}{2r}\right).$$

---

[52]Note that the above result can be obtained in an even simpler manner by using the approximation
illustrated in Figure 4.24b. For $r \gg d$, we have from Figure 4.24b that $r_+ r_- \simeq r^2$ and $r_+ - r_- \simeq -d\cos\theta$,
so that

$$\Phi = \frac{Q|}{4\pi\epsilon_0 r_+} - \frac{Q}{4\pi\epsilon_0 r_-} = \frac{Q}{4\pi\epsilon_0}\frac{r_- - r_+}{r_+ r_-} \simeq \frac{Q}{4\pi\epsilon_0}\frac{d\cos\theta}{r^2}$$

However, by keeping higher-order terms in d/r, it is possible to determine the value of the potential nearer to
the dipole.

Thus, the potential is

$$\Phi(r,\theta) \simeq \frac{Q}{4\pi\epsilon_0 r}\left[\left(1 + \frac{d\cos\theta}{2r}\right) - \left(1 - \frac{d\cos\theta}{2r}\right)\right] = \frac{Qd\cos\theta}{4\pi\epsilon_0 r^2}$$

It is useful to write the potential in terms of an electric dipole moment $\mathbf{p} \equiv (Qd)\hat{\mathbf{z}}$, and noting that $\hat{\mathbf{z}} \cdot \hat{\mathbf{r}} = \cos\theta$,

$$\boxed{\Phi = \frac{1}{4\pi\epsilon_0}\frac{\mathbf{p}\cdot\hat{\mathbf{r}}}{r^2} = \frac{Qd\cos\theta}{4\pi\epsilon_0 r^2}} \qquad (4.27)$$

To find the electric field, we note that Φ does not have any ϕ dependence, so we have

$$\mathbf{E} = -\nabla\Phi = -\left[\hat{\mathbf{r}}\frac{\partial\Phi}{\partial r} + \hat{\boldsymbol{\theta}}\frac{1}{r}\frac{\partial\Phi}{\partial\theta}\right] = -\left[-\hat{\mathbf{r}}\frac{p\cos\theta}{2\pi\epsilon_0 r^3} - \hat{\boldsymbol{\theta}}\frac{p\sin\theta}{4\pi\epsilon_0 r^3}\right]$$

where $p = |p| = Qd$. This yields

$$\boxed{\mathbf{E} = \frac{p}{4\pi\epsilon_0 r^3}[\hat{\mathbf{r}}2\cos\theta + \hat{\boldsymbol{\theta}}\sin\theta]} \qquad (4.28)$$

as the electric field of an electric dipole at a distance r. We note that the above result is valid only at distances far from the dipole, since the approximation $r \gg d$ was used in deriving the potential. Far from the dipole the electric field varies as r^{-3}, in contrast with the r^{-2} dependence for the field of a point charge and the r^{-1} variation of the field of a line charge (Example 4.5).

A two-dimensional sketch of the electric field lines and equipotential lines for an electric dipole is shown in Figure 4.25. Note that close to the charges the dipole approximations (4.27) and (4.28) are not valid and the field line configuration is equal to that for two point charges, shown in Figure 4.23a.

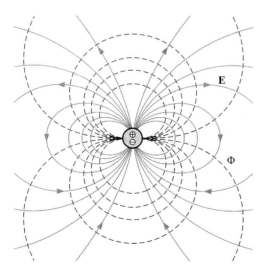

Figure 4.25 Electric field lines and equipotentials for a dipole. The dipole expressions (4.27) and (4.28) are not valid close to the charges, that is, in the shaded region at the center. In that region, the electric field configuration is as shown in Figure 4.23a.

4.4.4 Electrostatic Potential Resulting from Continuous Charge Distributions

Since potential is a scalar quantity, its calculation for a given charge configuration simply involves an algebraic sum of the contributions of *all* charges in the system, whether they are point, line, surface, or volume charges. In general, we have

$$\Phi = \frac{1}{4\pi\epsilon_0}\left[\sum_{k=1}^{n}\frac{Q_k}{|\mathbf{r}-\mathbf{r'}_k|} + \int_{L'}\frac{\rho_l(\mathbf{r'})\,dl'}{|\mathbf{r}-\mathbf{r'}|} + \int_{S'}\frac{\rho_s(\mathbf{r'})\,ds'}{|\mathbf{r}-\mathbf{r'}|} + \int_{V'}\frac{\rho(\mathbf{r'})\,dv'}{|\mathbf{r}-\mathbf{r'}|}\right] \qquad (4.29)$$

Example 4.10: Disk of charge. For a uniformly charged disk of radius a and surface charge density ρ_s, as illustrated in Figure 4.26, determine the potential Φ and the electric field $\mathbf{E}$ at a point P along its axis.

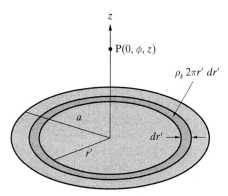

Figure 4.26 Disk of charge. Figure for Example 4.10.

Solution: We work in the cylindrical coordinate system (with the disk on the xy plane and centered at the origin). Consider an annular ring of radius r' and a differential width dr', as shown in Figure 4.26. The contribution of the surface charge residing on this ring to the electrostatic potential at point P along the z axis is

$$d\Phi_{\text{P}}(r=0) = \frac{\rho_s 2\pi r'\,dr'}{4\pi\epsilon_0\sqrt{z^2 + (r')^2}}$$

where the 2π term comes from an integration along the ring (i.e., ϕ varying from 0 to 2π). If we now integrate the foregoing expression over the surface of the entire disk, we find

$$\Phi(r=0) = \frac{\rho_s}{2\epsilon_0}\int_0^a \frac{r'\,dr'}{\sqrt{z^2 + (r')^2}} + C = \frac{\rho_s}{2\epsilon_0}[(a^2 + z^2)^{1/2} - |z|] + C$$

where C is a constant and we retain $|z| = \sqrt{z^2}$ in order to ensure that Φ decreases[53] as $z \to \pm\infty$ (or $|z| \to \infty$) and to ensure that $\Phi(z) = \Phi(-z)$, as is necessary in view of the

[53]At large distances from the disk the electrostatic potential should vary like that of a point charge.

symmetry in z. Since we view electrostatic potential at a point in reference to infinity, we need to have $C = 0$ so that $\Phi \to 0$ as $z \to \infty$. The electrostatic potential at point P is thus

$$\Phi(r = 0) = \frac{\rho_s}{2\epsilon_0}[\sqrt{a^2 + z^2} - |z|]$$

Note that for $z \gg a$ we can make the approximation $\sqrt{a^2 + z^2} \simeq z + a^2/(2z)$, so that we have

$$\Phi(r = 0) \simeq \frac{\rho_s}{2\epsilon_0}\left(z + \frac{a^2}{2z} - z\right) = \frac{(\rho_s \pi a^2)}{4\pi\epsilon_0 z} = \frac{1}{4\pi\epsilon_0}\frac{Q}{z} \qquad z > 0$$

where $Q = \rho_s \pi a^2$ is the total charge on the disk. For $z < 0$ we would have $\Phi \simeq -Q/(4\pi\epsilon_0 z)$ since for the positively charged disk (i.e., $\rho_s > 0$) the potential Φ must be positive. To see this, we need to remember that electrostatic potential at a given point is the energy required to bring a unit positive test charge from infinity to that point. We note that, as expected, at large distances the potential due to the disk charge Q behaves like that of a point charge.

To find the electric field, we must in general take the negative gradient of Φ, which means that we need to know the variation of Φ as a function of r, ϕ, z. However, because of the symmetry of the problem, the electric field at a point on the axis can only be in the z direction. Thus, the electric field at point P is related only to $\partial\Phi/\partial z$. In other words, even though Φ in the vicinity of point P may in general depend on r and ϕ, we are only concerned with its z dependence as determined above. Thus we have

$$[E_z]_\text{P} = -\frac{d\Phi}{dz} = \begin{cases} \dfrac{\rho_s}{2\epsilon_0}[1 - z(a^2 + z^2)^{-1/2}] & \text{for } z > 0 \\[2mm] \dfrac{\rho_s}{2\epsilon_0}[1 + z(a^2 + z^2)^{-1/2}] & \text{for } z < 0 \end{cases}$$

Note that $E_z(z) = -E_z(-z)$, as is expected on physical grounds.

Example 4.11: Finite-length line charge. In Example 4.5, we found the electric field at a point equidistant from the end points of a thin cylindrical rod charge, which had a component only in the r direction due to the symmetry involved. For points off the central axis, such as point $P(r, \phi, z)$ as shown in Figure 4.27, it is clear that the field would have both E_r and E_z components. This example illustrates the usefulness of the electrostatic potential concept by determining Φ at P via direct scalar integration and subsequently determining **E** from $\mathbf{E} = -\nabla\Phi$. Assume that $a \gg d$, so that the rod of charge can be represented by a line charge density ρ_l.

Solution: As shown in Figure 4.27, we consider an elemental charge $dQ = \rho_l\, dz'$ located at $(0, 0, z')$. The distance between this source element and point P is $|\mathbf{r} - \mathbf{r}'| = \sqrt{r^2 + (z - z')^2}$. The total potential Φ at P is then given by

$$\Phi = \int_{-a}^{+a} \frac{\rho_l\, dz'}{4\pi\epsilon_0[r^2 + (z - z')^2]^{1/2}} = -\frac{\rho_l}{4\pi\epsilon_0}\ln\left(\frac{z - a + [r^2 + (z - a)^2]^{1/2}}{z + a + [r^2 + (z + a)^2]^{1/2}}\right)$$

A three-dimensional plot of $\Phi(r, z)$ is shown in Figure 4.27. For the purpose of this display, the normalized value of the electrostatic potential has been limited to 3 units, since for $-a \leq z \leq a$, the potential increases without limit as $r \to 0$.

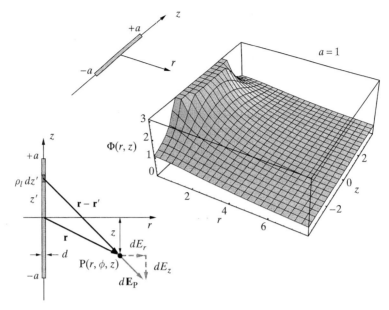

Figure 4.27 Electrostatic potential in the vicinity of a straight rod of charge. The amplitude range of the display is limited to 3 units since the potential becomes arbitrarily large as $r \to 0$.

The electric field components can now be found from $\mathbf{E} = -\nabla \Phi$; since Φ depends only on z and r, only E_r and E_z components exist and are given by

$$E_z = -\frac{\partial \Phi}{\partial z} = \frac{\rho_l}{4\pi \epsilon_0} \left(\frac{1}{[r^2 + (z-a)^2]^{1/2}} - \frac{1}{[r^2 + (z+a)^2]^{1/2}} \right)$$

$$E_r = -\frac{\partial \Phi}{\partial r} = \frac{\rho_l r}{4\pi \epsilon_0} \left(\frac{-1}{r^2 + (z-a)^2 + (z-a)[r^2 + (z-a)^2]^{1/2}} \right.$$
$$\left. + \frac{1}{r^2 + (z+a)^2 + (z+a)[r^2 + (z+a)^2]^{1/2}} \right)$$

It is useful to check the behavior of the field and potential in limiting cases. For $r \gg a$ or $z \gg a$, the potential approaches that of a point charge of $Q = \rho_l(2a)$, namely,

$$\lim_{r \to \infty} \Phi = \frac{\rho_l(2a)}{4\pi \epsilon_0 r} \qquad \text{and} \qquad \lim_{z \to \infty} \Phi = \frac{\rho_l(2a)}{4\pi \epsilon_0 z}$$

For $z = 0$, the solution boils down to that worked out in Example 4.5. For $r = 0$, and at points $|z| > a$, we have $E_r = 0$ and

$$E_z = \frac{\rho_l}{4\pi \epsilon_0} \left(\frac{1}{|z-a|} - \frac{1}{|z+a|} \right) = \frac{\pm \rho_l a}{2\pi \epsilon_0 (z^2 - a^2)}$$

where the "+" and "−" correspond to the cases of $z > a$ and $z < -a$, respectively.

The result in Figure 4.27 serves to give a better understanding of the relationship between the electric field and the electrostatic potential. Since $\mathbf{E} = -\nabla \Phi$, the direction of the electric field can be easily deduced from the three-dimensional surface plots of the potential. An analogy with gravitational potential is particularly useful here; if the quantity plotted, say, in Figure 4.27, were the gravitational potential, its gradient at any point is in the direction opposite to that in which a marble placed at that point rolls down (note that the top of the surface in Figure 4.27 looks flat only because of the limited amplitude range of the display). The electric field at any point specified by its r and z coordinates thus points in the direction of the most rapid change of slope of the potential distribution. The magnitude of the electric field is higher at points where the slope is higher—that is, closer to the disk charge.

4.5 ELECTRIC FLUX AND GAUSS'S LAW

The electric field line plots that we have seen suggest some sort of "flow," or *flux,* of electric energy emanating from positive charges and terminating at negative charges. If the field lines represented the velocity of fluid flow, we could indeed think of flux of fluid along the field lines. Although, in fact, electric field lines do not represent the flow of anything material, it is helpful to think of them as describing the flux of *something* that, like a fluid, is conserved and that emanates from the charges into the surrounding space.[54] In such a picture, each field line represents a unit of flux of that *something,* from each unit of positive charge to a unit of negative charge.

4.5.1 Electric Displacement and Flux Density D

Michael Faraday's experiments, carried out in 1837, well after Coulomb, and many years after his own work on induced electromotive force,[55] have established our present concepts of electric displacement and flux density. A schematic description of Faraday's experiments is shown in Figure 4.28.

Basically, Faraday used two concentric spheres. First, the inside sphere was charged by a known quantity of electricity (say, Q); then a larger outer sphere (uncharged) was placed around it[56] with a dielectric (insulating material) in between. The outer sphere was then grounded[57] momentarily; then the two spheres were separated and the charge remaining on the outer sphere was measured. Faraday found that this charge was equal in magnitude and opposite in sign to that on the inner sphere, *regardless of the size of the sphere* and *for all types of dielectric materials* filling the space between the spheres. He thus concluded that there was a sort of "displacement" from the charge on the inner sphere

---

[54]One possibility is to think of small "bullets" shot out of charges and to require that none of the bullets could disappear once they were produced; however, one needs to be very careful in carrying such a model any further than simply a thought exercise [R. P. Feynman, R. B. Leighton, and M. Sands, *The Feynman Lectures on Physics,* Addison Wesley, Reading, Massachusetts, 1964].

[55]See Chapter 7 for discussion of Faraday's law.

[56]Actually the outer sphere consisted of two hemispheres that could be firmly clamped together. Faraday prepared shells of dielectric material to occupy the entire volume between the spheres.

[57]In other words, connected to the ground, so that it can access as much charge as it required. In this context, "ground" is an uncharged object at zero potential that has essentially an infinite number of electrons (and positive nuclei) and that can accept or provide any amount of excess charge from or to any object connected to it.

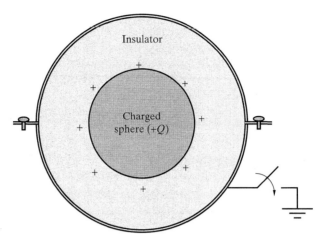

Figure 4.28   Faraday's experiments on electric displacement. The apparatus used in Faraday's experiments consisted of two concentric spheres separated by an insulating material.

through the insulator to the outer sphere, the amount of this displacement depending only on the magnitude of the charge Q. This displacement is the closest we can come to identifying the *something* that must flow along the field lines, and we adopt this notion as our definition of the *electric flux*. In SI units, the electric displacement (or electric flux) is equal in magnitude to the charge that produces it; namely, it has units of coulombs (C).

Consider now the case of an isolated point charge as shown in Figure 4.29a; all other bodies, such as the outer sphere of Figure 4.28, can be considered to be at infinity. The *density* of electric displacement, or *flux density* vector **D**, at any point on a spherical surface S of radius r centered at the isolated point charge Q is defined as

$$\mathbf{D} \equiv \hat{\mathbf{r}}\frac{Q}{4\pi r^2} \quad \text{C-m}^{-2} \qquad (4.30)$$

Note that the electric displacement per unit area, **D**, depends on the orientation of the area; hence, it is a vector quantity. In simple materials, we can write

$$\boxed{\mathbf{D} = \epsilon\mathbf{E}} \qquad (4.31)$$

with the dielectric constant ϵ being a simple constant[58] so that the vector **D** is in the same direction as the electric field **E**. Note, however, that **D** is in units of C-m^{-2}, while **E** is in V-m^{-1}. In view of the simple proportionality of **D** and **E**, our previous plots showing electric field lines also represent electric flux lines. In the following, we confine

---

[58]We discuss dielectric materials and their behavior under applied electric fields in Section 4.10. At this point, it suffices to note that, in general, the relationship between **D** and **E** can be quite complex. In addition to the possibility of the value of ϵ not being a constant and depending instead on the magnitude of the electric field ($|\mathbf{E}|$), x, y, z, and t, in so-called *anisotropic* materials ϵ might also depend on the *direction* of the applied electric field; for example, an electric field in the x direction (i.e, E_x) may produce an electric flux density in the y direction (i.e., D_y).

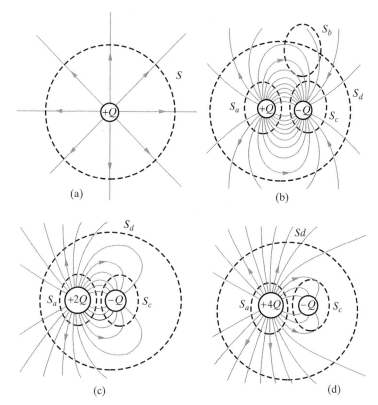

Figure 4.29   Electric field lines around point charges. (a) A positive point charge, (b) two opposite equal point charges, (c) two point charges $+2Q$ and $-Q$, and (d) two point charges $+4Q$ and $-Q$.

ourselves to the consideration of simple materials with constant ϵ unless otherwise stated. In fact, until dielectric materials are discussed in a later section, we exclusively consider free space, in which **D** and **E** differ through the constant ϵ_0, (i.e., $\mathbf{D} = \epsilon_0\mathbf{E}$).

Thinking of each unit of flux as a unit of charge enables us to have a deeper understanding of the distribution of electric field lines around charge configurations. Consider, for example, the flux lines around two opposite charges shown again in Figure 4.29b, and the closed surfaces S_a, S_b, S_c, and S_d. If a closed surface encloses a charge of $+Q$ (as is the case for S_a), all flux lines emanating from charge Q must go through it. Thus, by counting all the flux lines passing through this closed surface we should be able to know the amount of charges enclosed, independent of the properties of the material surrounding the charges. For this purpose, we would adopt the convention that lines outward (inward) represent positive (negative) enclosed charge. Since equal numbers of lines enter and leave from surface S_b, the charge enclosed by it is zero, whereas surface S_c encloses negative charge, equal in magnitude to that enclosed by surface S_a, since *all* lines of flux coming out of surface S_a eventually enter surface S_c. Similarly, the large surface S_d encloses both charges $+Q$ and $-Q$, so equal numbers of field lines enter and leave, and the net total flux through it is zero, just as the net amount of charge enclosed in it is zero.

Consider the other two cases shown in Figures 4.29c and d. In the case of Figure 4.29c, the positive charge is twice as large as the negative charge. Accordingly, twice as many (i.e., 24, versus 12 for the flux line density shown in Figure 4.29c) lines cross through surface S_a as do through surface S_c. Also, the large surface S_d now encloses a net charge of $2Q - Q = +Q$, so that it is crossed by the flux lines emanating from the $+2Q$ charge that do not connect back to the $-Q$ charge (i.e., 12 out of the total 24 lines[59] in Figure 4.29c), but instead terminate on charges located at infinity. We thus see that, indeed, the number of flux lines is an accurate measure of the amount of charge. Similar observations can be made in connection with Figure 4.29d, in which case the positive charge is four times as large as the negative charge; accordingly, four times as many flux lines (i.e., 24 in Figure 4.29d) cross through surface S_a as do through surface S_c (i.e., 6 in Figure 4.29d). The net number of outward flux lines (i.e., 18, Figure 4.29d) through S_d is determined by the net enclosed charge ($+4Q - Q = +3Q$).

4.5.2 Gauss's Law

Consider now the region of space around a single positive point charge and an arbitrary closed surface S as shown in Figure 4.30a. If the electric field ($\mathbf{E}$) is indeed like a flow, the net flow out of this truncated conical box should be zero. Let us find the surface integral of the component of the electric flux density vector $\mathbf{D}$ normal (perpendicular) to the surface. Using (4.30), we have $\mathbf{D} \cdot d\mathbf{s} = 0$ on the side faces, whereas on the radial faces we have $\mathbf{D} \cdot d\mathbf{s} = \pm|\mathbf{D}|ds$, plus on the outer surface S_b and minus on the inner surface S_a. Since the magnitude of the electric flux density $\mathbf{D}$ due to the positive point charge decreases as r^{-2}, and the surface areas of the radial faces increase as r^2, the total flow (or flux) inward through surface S_a completely cancels the flow outward though surface S_b. Since the total flow through the entire closed surface S is the integral of $\mathbf{D} \cdot d\mathbf{s}$ over S, we have $\oint_S \mathbf{D} \cdot d\mathbf{s} = 0$. Note that we would not have arrived at this conclusion if the electrostatic force law depended on distance in any other way than r^{-2}.

Considering Figure 4.30b, we conclude that the result we have just obtained is true even if the two end surfaces S_a and S_b were tilted with respect to the radial distance. Note that when the surface is tilted at an angle θ, the area is increased by a factor $(\cos \theta)^{-1}$, but the component of $\mathbf{D}$ normal to the surface is decreased by the factor $\cos \theta$, so the product $\mathbf{D} \cdot d\mathbf{s}$ is unchanged. Thus, the net total electric flux out of the closed surface is still zero.

It is thus clear that the net flux out of *any* closed surface (that does not enclose any charge) must be zero, because, as shown in Figure 4.30c, any such surface can be constructed out of infinitesimal truncated cones such as those in Figures 4.30a and b.

---

[59]We note that the particular number of lines (e.g., 12) is determined entirely by the resolution of the field line plot. In the case of Figure 4.29, we have chosen a particular density of field lines that conveys the essential aspects of the field distributions. Too high a density of lines (e.g., 100 lines emanating from $+Q$ instead of 12) would clutter the figure, while too few lines (e.g., 2 instead of 12) would not be sufficient to convey the underlying symmetries and to illustrate the field intensity quantitatively in the regions surrounding the charges.

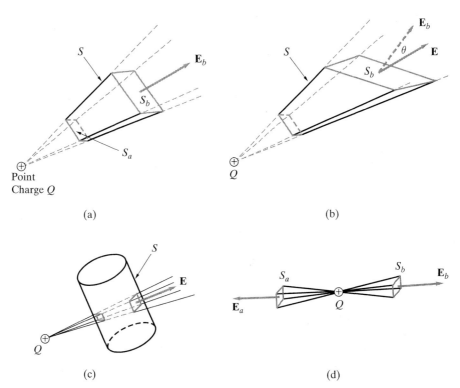

(a)   (b)

(c)   (d)

Figure 4.30   The flux of E out of a surface is zero. This figure is adapted with permission from R. P. Feynman, R. B. Leighton, and M. Sands, *The Feynman Lectures on Physics,* Addison-Wesley, 1964. Copyright 1964 by California Institute of Technology.

Now consider the surface shown in Figure 4.30d. We see that the magnitudes of the electric flux through the two surfaces S_a and S_b are once again equal (because the flux density decreases as r^{-2} but the area increases as r^2); however, they now have the same sign, so the net flux out of this surface is not zero.

To determine the net outward (or inward) flux from a surface enclosing a point charge, consider the general surface S shown in Figure 4.31, shown for simplicity as a two-dimensional cross-section. Emanating from the point charge Q are cones of flux, which might cross the surface one or more times as shown. Also shown is another small spherical surface S' around the point charge Q. Note that the volume enclosed between S and S' has no enclosed charge, so that the total flux emanating from this volume is zero, by the arguments given above.[60] Thus, the net total flux through the larger surface S must be identical to that through S'. In other words, the flux emanating from the end (near F) of the truncated cone between points E and F is precisely that which leaves the surface S' in the same solid angular range. Note that the above argument holds for a

[60]Basically, the net flux emanating from any of the truncated cones, such as those between points A and B, B and C, or C and D, is zero, since these surfaces are just like those in Figures 4.30a and b and do not enclose any charge.

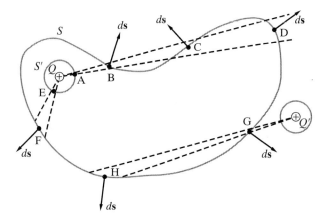

Figure 4.31   Gauss's law. The total flux emanating from the closed surface S' surrounding the charge Q is equal to that emanating from any other closed surface of arbitrary shape (i.e., S), as long as it encloses no other charges. This figure was inspired by Figures 8 and 9 of J. H. Jeans *The Mathematical Theory of Electricity and Magnetism*, 4th ed., Cambridge Press, 1920.

surface S' of any shape; we choose S' to be a sphere centered at the point charge Q for convenience so that we can easily calculate the flux through it.

Taking the radius of the sphere S' to be r, the electric flux density everywhere on its surface is given by (4.30). Since S' is a spherical surface with surface area $4\pi r^2$, we have

$$\oint_{S'} \mathbf{D} \cdot d\mathbf{s} = \left(\frac{Q}{4\pi r^2} \right) (4\pi r^2) = Q$$

As expected, the net flux out of the spherical surface S' is independent of the radius of the sphere. Since *all* of the flux out of S' must go out of the closed surface S, and since we imposed no restrictions whatsoever on the shape of S, we have arrived at a rather general result:

$$\int_{\text{any closed surface enclosing } Q} \mathbf{D} \cdot d\mathbf{s} = Q$$

To illustrate the point further, a second point charge Q' that lies outside the closed surface S is shown in Figure 4.31. It is clear from the foregoing discussion in connection with Figure 4.30 that any solid angular range (i.e., conical tube of flux) emanating from Q' cuts the closed surface S an even number of times (twice, four times, etc.), so the net flux from any of the pairs of truncated conical surfaces (such as G and H) is zero.

Although the result we have obtained is for a single point charge, it is obvious that it can be generalized to any free[61] charge distribution since any continuous or discrete

---

[61] In Faraday's experiments on electric displacement, he measured the total amount of mobile (or free) charges on the inner and outer conducting spheres. The flux density $\mathbf{D}$ is a measure of this free charge. We will see in Section 4.10 that in the presence of an electric field, the dielectric material that fills the space between the inner and outer spheres contains an induced bound polarization charge density.

distribution of charges can be subdivided into a smaller number of elements, each of which can each be treated as a point charge. Using the superposition principle, the total flux out of any closed surface S is the sum of that from each individual enclosed charge element. We are thus ready to write down the most general form of Gauss's law:

$$\oint_S \mathbf{D} \cdot d\mathbf{s} = \int_V \rho \, dv = Q_{enc} \tag{4.32}$$

where V is an arbitrary volume enclosed by the closed surface S, $\rho(x, y, z)$ is the volume density of free electric charge inside the volume V, and Q_{enc} is the total charge in volume V enclosed by the surface S.

This general result is the most convenient method at our disposal for expressing the electrostatic inverse square action law of Coulomb. *Gauss's law is a direct consequence of Coulomb's law:* that electrostatic force between point charges varies with distance as r^{-2}. It does not provide us with any additional information about the way static charges interact; it simply allows us to state Coulomb's law in a way that may be more useful in the solution of some electrostatic problems, especially those that exhibit some sort of symmetry. Gauss's contribution was actually not in originating the law but in providing the mathematical framework for this statement.

4.5.3 Applications of Gauss's Law

We now consider examples of the use of Gauss's law, as expressed in (4.32), to determine the electric flux density $\mathbf{D}$ (and hence $\mathbf{E} = \mathbf{D}/\epsilon$) in cases where the charge distribution is known. Easy solutions of this integral equation (where the unknown $\mathbf{D}$ is under the integral) are possible only if we can take advantage of the inherent symmetries of the problem to identify a so-called *Gaussian surface S* such that $\mathbf{D}$ is everywhere either tangential or normal to the closed surface, so that $\mathbf{D} \cdot d\mathbf{s}$ is either zero or simply $D_n \, ds$, with $D_n = $ const. on the surface. Gauss's law is thus useful only for symmetric charge distributions, which lead to symmetric distributions of $\mathbf{E}$ and $\mathbf{D}$. The following examples illustrate the use of Gauss's law for a few different charge distributions.

Example 4.12: Line charge. Consider an infinitely long line charge in free space with a line charge density ρ_l coulombs per meter, as shown in Figure 4.32. Find the electric field at a radius r from the line charge.

Solution: To find the field at a radius r, we can use Gauss's law (equation (4.32)) and integrate the flux density $\mathbf{D}$ over a coaxial cylindrical surface S of length l and radius r, as shown in Figure 4.32. Due to symmetry, the electric field has a component only in the r direction, so there is no contribution to the integral over the closed surface from the end surfaces of the cylinder ($E_z = 0$). Thus, we can write

$$\oint_S \mathbf{D} \cdot d\mathbf{s} = \oint \epsilon_0 \mathbf{E} \cdot d\mathbf{s} = \epsilon_0 \int_0^{2\pi} E_r l r \, d\phi = Q_{enc} = \rho_l l \quad \rightarrow \quad E_r = \frac{\rho_l}{2\pi \epsilon_0 r}$$

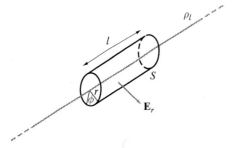

Figure 4.32 Line charge. A cylindrical Gaussian surface S is appropriate since by symmetry E_r is constant on the side surface of the cylinder.

When there is appropriate symmetry, the use of Gauss's law is clearly much simpler than direct use of Coulomb's law to find the electric field, as in Example 4.5.

Example 4.13: Spherical cloud of charge. Consider a spherical cloud of charge with radius a, as shown in Figure 4.33a. The electric charge is uniformly distributed over the spherical volume, with volume charge density ρ, so that the total charge in the cloud is $Q = \frac{4}{3}\pi a^3 \rho$. (a) Find the electric field $\mathbf{E}(r)$, for both $r < a$ and $r > a$. (b) Find the electric potential $\Phi(r)$.

Solution: (a) We first observe that, due to the spherical symmetry, the electric flux density $\mathbf{D}$ and hence $\mathbf{E}$ are both in the radial direction (i.e., $\mathbf{E} = \hat{\mathbf{r}}E_r$) and depend only on r. In

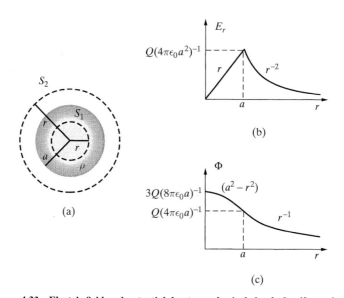

Figure 4.33 Electric field and potential due to a spherical cloud of uniform charge.
(a) The spherical cloud of charge and Gaussian surfaces S_1 and S_2. (b) Electric field.
(c) Potential.

other words, $\mathbf{D} = $ constant on any spherical surface that is concentric with the spherical charge cloud. We consider two spherical Gaussian surfaces S_1 and S_2, respectively, inside and outside the cloud of charge, and apply Gauss's law (equation (4.32)) to both surfaces:

$$\oint_{S_1 \text{ or } S_2} \epsilon_0 \mathbf{E} \cdot d\mathbf{s} = \epsilon_0 E_r 4\pi r^2 = \begin{cases} \rho \dfrac{4}{3}\pi r^3 = Q\left(\dfrac{r}{a}\right)^3 & r < a \quad \text{on } S_1 \\[2mm] \rho \dfrac{4}{3}\pi a^3 = Q & r \geq a \quad \text{on } S_2 \end{cases}$$

Thus, we have

$$\mathbf{E} = \begin{cases} \hat{\mathbf{r}} \dfrac{Q}{4\pi\epsilon_0 r^2} & r \geq a \\[3mm] \hat{\mathbf{r}} \dfrac{Qr}{4\pi\epsilon_0 a^3} & r < a \end{cases}$$

The variation of E_r is plotted as a function of r in Figure 4.33b.

(b) Using the definition of electric potential as given in (4.17), and taking the path of integration along the radial direction, we have

$$\Phi = -\int_\infty^r (\hat{\mathbf{r}} E_r) \cdot (\hat{\mathbf{r}}\, dr) = \begin{cases} -\displaystyle\int_\infty^r \dfrac{Q\, dr}{4\pi\epsilon_0 r^2} & r > a \\[4mm] -\displaystyle\int_\infty^a \dfrac{Q\, dr}{4\pi\epsilon_0 r^2} - \displaystyle\int_a^r \dfrac{Qr\, dr}{4\pi\epsilon_0 a^3} & r \leq a \end{cases}$$

Straightforward evaluation of these integrals yields

$$\Phi = \begin{cases} \dfrac{Q}{4\pi\epsilon_0 r} & r > a \\[4mm] \dfrac{Q}{4\pi\epsilon_0 a} + \dfrac{Q}{4\pi\epsilon_0 a^3}\dfrac{a^2 - r^2}{2} & r \leq a \end{cases}$$

The variation of the potential Φ is plotted in Figure 4.33c. The relationship between $\mathbf{E}$ and Φ, namely, $\mathbf{E} = -\nabla\Phi$ (in this case, $E_r = -d\Phi/dr$), is apparent.

Example 4.14: Infinite sheet of charge. Consider the planar sheet of charge of infinite extent[62] located in the xy plane (i.e., $z = 0$), as shown in Figure 4.34. The surface charge density is given as ρ_s. Determine the electric field $\mathbf{E}$ and the electrostatic potential Φ.

Solution: We observe from the symmetry of the problem that the electric field can only be in the z direction: $\mathbf{E} = \hat{\mathbf{z}}E_z$ for $z > 0$ and $\mathbf{E} = -\hat{\mathbf{z}}E_z$ for $z < 0$. We choose to work with cylindrical (r, ϕ, z) coordinates using a cylindrical pillbox-type Gaussian surface S of radius r, as shown in Figure 4.34, although rectangular (x, y, z) coordinates and a rectangular

---

[62]This appears as a perfect example of an academic problem of no practical interest; after all, an infinite sheet of charge certainly cannot exist. However, it turns out that several practical results (e.g., evaluating the capacitance of planar conductor configurations) can actually be based on the result derived here, as on other "symmetrical" charge configurations we study in this section.

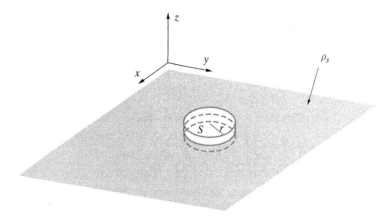

Figure 4.34 An infinite planar sheet of charge.

pillbox surface would also work just fine. Noting that the total charge enclosed by this surface is $Q = \rho_s \pi r^2$, we apply (4.32):

$$\underbrace{\int_{\text{top}} (\hat{\mathbf{z}} E_z) \cdot (\hat{\mathbf{z}}\, ds)}_{=E_z (\pi r^2)} + \underbrace{\int_{\text{bottom}} (-\hat{\mathbf{z}} E_z) \cdot (-\hat{\mathbf{z}}\, ds)}_{=E_z (\pi r^2)} + \underbrace{\int_{\text{side surfaces}} (\pm \hat{\mathbf{z}} E_z) \cdot (\hat{\mathbf{r}}\, ds)}_{=0 \text{ since } \hat{\mathbf{z}} \cdot \hat{\mathbf{r}} = 0} = \frac{\rho_s \pi r^2}{\epsilon_0}$$

hence

$$2E_z (\pi r^2) = \frac{\rho_s \pi r^2}{\epsilon_0} \quad \rightarrow \quad E_z = \frac{\rho_s}{2\epsilon_0} \quad \rightarrow \quad \mathbf{E} = \begin{cases} \hat{\mathbf{z}} \dfrac{\rho_s}{2\epsilon_0} & z > 0 \\[2mm] -\hat{\mathbf{z}} \dfrac{\rho_s}{2\epsilon_0} & z < 0 \end{cases}$$

This is a simple but an important result, which will be used in deriving the approximate expression for the capacitance of a parallel-plate capacitor. It is interesting to note that we could have arrived at this result by considering the expression for E_z above a disk of charge of radius a (Example 4.10) and simply taking the limit for $a \to \infty$. Note that the electric field does not decay as $z \to \infty$ because of the infinite extent of the sheet of charge. For a finite sheet of charge, the electric field decreases at very large distances; however, use of Gauss's law to calculate the fields for a finite sheet is not convenient because we cannot find a closed surface over which the electric field is constant.

If we were to attempt to find the electrostatic potential as we did in the previous example, we would run into a "problem," since, for any point above the sheet of charge ($z > 0$) and for a path of integration along the z axis, we have

$$\Phi(z) = -\int_{\infty}^{z} (\hat{\mathbf{z}} E_z) \cdot (\hat{\mathbf{z}}\, d\zeta) = -\int_{\infty}^{z} \frac{\rho_s}{2\epsilon_0} d\zeta = -\left[\frac{\rho_s}{2\epsilon_0} \zeta \right]_{\infty}^{z} = \infty \quad !$$

We can understand this seemingly unphysical result by recalling the meaning of electrostatic potential at any point as the work that needs to be done to bring a unit positive test charge from infinity to that point. With an infinite sheet of charge, the electric field is constant everywhere, so that one is faced with opposing a constant force over an infinite distance (from ∞ to z), requiring infinite energy. Note, however, that the difference in electrostatic potential between any two points $z = a$ and $z = b$, denoted as Φ_{ab}, is straightforward to evaluate; we find

$$\Phi_{ab} = \Phi(b) - \Phi(a) = -\int_a^b (\hat{\mathbf{z}} E_z) \cdot (\hat{\mathbf{z}}\, dz) = -\left[z\frac{\rho_s}{2\epsilon_0}\right]_a^b = \frac{\rho_s(a - b)}{2\epsilon_0}$$

which is the amount of work needed to move a unit positive test charge from $z = a$ to $z = b$. Note that for $a > b > 0$, Φ_{ab} is positive, meaning that work must be externally provided in bringing a positive test charge closer to the positively charged sheet.

Example 4.15: Two coaxial cylindrical shells of charge. Consider the charge configuration[63] shown in Figure 4.35, consisting of two coaxial, concentric, and infinitely long cylindrical shells of surface charges with radii a and b, such that $b > a$. The total amount of charge per unit length on each cylinder is equal in magnitude and opposite in sign, with the surface charge densities on the inner and outer shells being ρ_s and $-\rho_s(a/b)$, respectively. Find the electric field and the electrostatic potential everywhere.

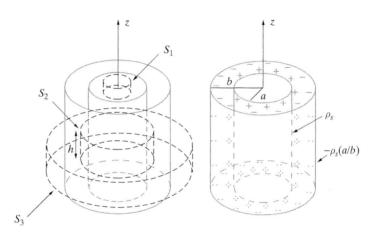

Figure 4.35   Two coaxial cylindrical sheets of charge.

Solution: Because of cylindrical symmetry, the electric field has only a radial component, $\mathbf{E} = \hat{\mathbf{r}} E_r$. It is most convenient to choose cylindrical Gaussian surfaces S_1, S_2, and S_3, respectively, at different radii, $r < a$, $a < r < b$, and $r > b$. It is clear that the net charge enclosed by the surfaces S_1 and S_3 is zero, and since the electric field must be constant over the side surfaces (due to symmetry), from Gauss's law we have $E_r = 0$. We now consider the surface S_2 and apply Gauss's law, noting that the surface elements on the top, bottom,

---

[63]This type of a charge distribution is in fact what occurs for a coaxial line, consisting of concentric coaxial conductors. The surface charge densities as assumed in this problem are established on the surfaces of the conducting cylinders when a potential difference is applied between them. The field and potential as determined in this example will be directly useful in the treatment of coaxial capacitances and coaxial transmission lines.

and side surfaces are $d\mathbf{s} = \hat{\mathbf{z}}r\,dr\,d\phi$, $d\mathbf{s} = -\hat{\mathbf{z}}r\,dr\,d\phi$, and $d\mathbf{s} = \hat{\mathbf{r}}r\,d\phi\,dz$, respectively, and that the charge enclosed by this surface is $Q_{enc} = \rho_s(2\pi ah)$. Using (4.32):

$$\underbrace{\int_{top}(\hat{\mathbf{r}}E_r)\cdot(\hat{\mathbf{z}}r\,dr\,d\phi)}_{=0} + \underbrace{\int_{bottom}(\hat{\mathbf{r}}E_r)\cdot(-\hat{\mathbf{z}}r\,dr\,d\phi)}_{=0} + \int_{side}(\hat{\mathbf{r}}E_r)\cdot(\hat{\mathbf{r}}r\,d\phi\,dz) = \frac{Q_{enc}}{\epsilon_0}$$

$$E_r\int_{side}r\,d\phi\,dz = \frac{Q_{enc}}{\epsilon_0} \quad\rightarrow\quad E_r(2\pi rh) = \frac{\rho_s 2\pi ah}{\epsilon_0} \quad\rightarrow\quad E_r = \frac{\rho_s a}{\epsilon_0 r}$$

where E_r was taken outside the integral since it is constant on the side surface of S_2. In other words, the electric field is

$$\mathbf{E} = \begin{cases} \hat{\mathbf{r}}\dfrac{\rho_s a}{\epsilon_0 r} & a < r < b \\[2mm] 0 & r < a,\, r > b \end{cases}$$

The electrostatic potential in the region between the cylinders (i.e., $a \le r \le b$) can be found as usual by integrating the electric field (equation (4.17)) as

$$\Phi(r) = -\int_\infty^r [\hat{\mathbf{r}}E_r(\zeta)]\cdot(\hat{\mathbf{r}}\,d\zeta) = -\int_b^r \frac{\rho_s a\,d\zeta}{\epsilon_0 \zeta} = \frac{\rho_s a}{\epsilon_0}\left[-\ln\zeta\right]_b^r = -\frac{\rho_s a}{\epsilon_0}\ln\frac{r}{b}$$

Often it is more convenient to write $\Phi(r)$ in terms of the potential difference Φ_{ba} between the conductors. In the notation used in Section 4.4, we have $\Phi_{ba} = \Phi(a) - \Phi(b)$ and

$$\Phi(r) = \Phi_{ba}\frac{\ln(r/b)}{\ln(a/b)} \qquad a \le r \le b$$

a result that will be quite useful in later sections. Note that the quantity Φ_{ba} is positive in this case since the inner shell (at $r = a$) is the one that has the positive charge. In other words, we would need to do work in order to move a positive test charge from $r = b$ to $r = a$.

4.6 DIVERGENCE: DIFFERENTIAL FORM OF GAUSS'S LAW

The concept of the divergence of a vector field is important for understanding the sources of fields. In the electrostatic context, divergence of the electric flux density $\mathbf{D}$ is directly related to the source charge density ρ. In this section, we discuss the definition and meaning of divergence and develop the relationship between $\mathbf{D}$ and ρ by applying Gauss's law to a very small (infinitesimal) volume.

If one envisions a closed surface S enclosing a source of any vector field $\mathbf{A}$, then the strength or magnitude of the source is given by the net outward flow or flux of $\mathbf{A}$ through the closed surface S, or $\oint_S \mathbf{A}\cdot d\mathbf{s}$. However, a more suitable measure of the *concentration* or *density* of the source is the outward flux per unit volume, or

$$\frac{\oint_S \mathbf{A}\cdot d\mathbf{s}}{V}$$

where V is the volume enclosed by the surface S. By reducing V to a differential volume, we obtain a point relation that gives the source density per unit volume at that point.

Since the source density in general varies from point to point, this measure of the source density is a scalar field and is called the *divergence* of the vector field **A**. The fundamental definition of divergence of a vector field **A** at any point is then

$$\text{div } \mathbf{A} \equiv \lim_{V \to 0} \frac{\oint_S \mathbf{A} \cdot d\mathbf{s}}{V} \qquad (4.33)$$

where S is the closed surface enclosing the volume V.

For the electrostatic field, we know from Gauss's law that the total outward flux through a surface S is equal to the total charge enclosed. For a differential volume element V, we can assume the volume charge density to be constant at all points within the infinitesimal volume, so the total free charge enclosed is ρV. Accordingly, we have

$$\text{div } \mathbf{D} \equiv \lim_{V \to 0} \frac{\oint_S \mathbf{D} \cdot d\mathbf{s}}{V} = \rho \qquad (4.34)$$

so that the divergence of the electric flux density at any point is equal to the volume charge density at that point.

We now derive the proper differential expression for div **D**. Although the divergence of a vector is clearly independent of any coordinate system, for simplicity we use a rectangular coordinate system and consider a small differential cubical volume element Δv placed in an electric flux density field **D** as shown in Figure 4.36. To determine the net outward flux from this volume, we can separately consider each pair of parallel faces: left-right, front-back, and bottom-top. We assume that the electric flux density **D** at the center of the cube is given by $\mathbf{D}(x_0, y_0, z_0) = \hat{\mathbf{x}}D_{x0} + \hat{\mathbf{y}}D_{y0} + \hat{\mathbf{z}}D_{z0}$. We first consider the net flux through the left and right faces due to the y component (i.e., D_y) of **D**. Note that on the left face we have $y = y_0 - \Delta y/2$, whereas on the right face $y = y_0 + \Delta y/2$.

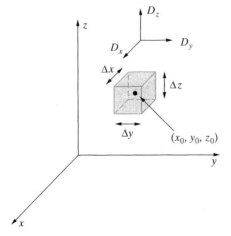

Figure 4.36   Divergence. A cubical volume element $\Delta v = \Delta x \, \Delta y \, \Delta z$ used to derive the differential expression for the divergence of **D**.

On the left face of the cubical volume element shown in Figure 4.36, $d\mathbf{s} = -\hat{\mathbf{y}}\,dx\,dz$, so that the flux leaving from the left face is, to the first order,[64]

$$\left[\hat{\mathbf{y}}D_y\left(x_0, y_0 - \frac{\Delta y}{2}, z_0\right)\right] \cdot (-\hat{\mathbf{y}}\Delta z\,\Delta x) = -(\Delta z\,\Delta x)D_y\left(x_0, y_0 - \frac{\Delta y}{2}, z_0\right)$$

$$\simeq -(\Delta z\,\Delta x)\left(D_{y0} - \frac{\Delta y}{2}\frac{\partial D_y}{\partial y}\right)$$

That from the right face[65] is similarly

$$+(\Delta z\,\Delta x)\left(D_{y0} + \frac{\Delta y}{2}\frac{\partial D_y}{\partial y}\right)$$

Thus, the net outward flux between these two faces is

$$\Delta z\,\Delta x\,\Delta y\frac{\partial D_y}{\partial y}$$

The net outward flux between the other two pairs of faces (front-back and bottom-top) can be similarly evaluated to find the net total outward flux from the cubical surface:

$$\underbrace{\Delta z\,\Delta x\,\Delta y}_{\Delta v}\left(\frac{\partial D_x}{\partial x} + \frac{\partial D_y}{\partial y} + \frac{\partial D_z}{\partial z}\right)$$

Since divergence was defined as the total outward flux per unit volume, we have

$$\boxed{\operatorname{div}\mathbf{D} = \frac{\partial D_x}{\partial x} + \frac{\partial D_y}{\partial y} + \frac{\partial D_z}{\partial z}} \qquad (4.35)$$

Using the definition of the del operator ∇ as a vector given in (4.22), the divergence of a vector $\mathbf{D}$ is typically written in a rectangular coordinate system as

$$\nabla \cdot \mathbf{D} = \left(\hat{\mathbf{x}}\frac{\partial}{\partial x} + \hat{\mathbf{y}}\frac{\partial}{\partial y} + \hat{\mathbf{z}}\frac{\partial}{\partial z}\right) \cdot (\hat{\mathbf{x}}D_x + \hat{\mathbf{y}}D_y + \hat{\mathbf{z}}D_z)$$

$$= \frac{\partial D_x}{\partial x} + \frac{\partial D_y}{\partial y} + \frac{\partial D_z}{\partial z}$$

An important result of our discussion of the concept of divergence of a vector field can now be stated. This result is the differential version of Gauss's law, namely

$$\boxed{\nabla \cdot \mathbf{D} = \rho} \qquad (4.36)$$

---

[64]At a more formal level, we can expand D_y in a Taylor's series around $y = y_0$ so that

$$D_y(x_0, y, z_0) = D_y(x_0, y_0, z_0) + (y - y_0)\frac{\partial D_y}{\partial y} + (y - y_0)^2\frac{\partial^2 D_y}{2!\partial y^2} + \text{higher-order terms}$$

Note that for a differential volume element, with $\Delta y = (y - y_0) \to 0$, the higher-order terms involving $(\Delta y)^k$ are negligible compared with the first-order term multiplied by Δy.

[65]Note that we have $d\mathbf{s} = +\hat{\mathbf{y}}\,dx\,dz$.

Equation (4.36) (or its integral form (4.32)) is one of the four fundamental equations of electromagnetics, the collection of which are referred to as Maxwell's equations. Note that this equation is simply a restatement of Gauss's law, which in turn is a convenient expression of the fundamental experimentally established inverse square law known as Coulomb's law.

Example 4.16: Divergence-free (zero-divergence) fields. Show that the following vector fields are divergence-free: (a) $\mathbf{A}_1 = -\hat{\mathbf{x}}x + \hat{\mathbf{y}}y$. (b) $\mathbf{A}_2 = \hat{\mathbf{x}}\frac{1}{2}x^2 - \hat{\mathbf{y}}xy$. (c) $\mathbf{A}_3 = \hat{\mathbf{x}}\sin y - \hat{\mathbf{y}}\cos x$. (d) $\mathbf{A}_4 = \hat{\mathbf{x}}\sin x \cos y - \hat{\mathbf{y}}\cos x \sin y$.

> **Solution:**   (a) $\nabla \cdot \mathbf{A}_1 = \partial A_x/\partial x + \partial A_y/\partial y = -1 + 1 = 0$.   (b) $\nabla \cdot \mathbf{A}_2 = x - x = 0$.
> (c) $\nabla \cdot \mathbf{A}_3 = 0 - 0 = 0$. (d) $\nabla \cdot \mathbf{A}_4 = \cos x \cos y - \cos x \cos y = 0$.
> These vector fields are shown in Figure 4.37. One common feature in the plots of all four vector fields is the fact that field vectors do not seem to converge toward or emerge from any "source" points.

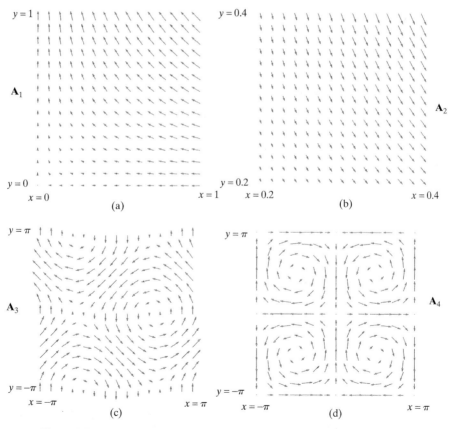

Figure 4.37 Examples of divergence-free fields. (a) $\mathbf{A}_1 = -\hat{\mathbf{x}}x + \hat{\mathbf{y}}y$. (b) $\mathbf{A}_2 = \hat{\mathbf{x}}\frac{1}{2}x^2 - \hat{\mathbf{y}}xy$. (c) $\mathbf{A}_3 = \hat{\mathbf{x}}\sin y - \hat{\mathbf{y}}\cos x$. (d) $\mathbf{A}_4 = \hat{\mathbf{x}}\sin x \cos y - \hat{\mathbf{y}}\cos x \sin y$.

Example 4.17: Vector fields with nonzero divergence. Find the divergence of the following vector fields: (a) $\mathbf{A}_5 = \hat{\mathbf{x}}(x^2/2) + \hat{\mathbf{y}}xy$. (b) $\mathbf{A}_6 = \hat{\mathbf{x}}\sin x - \hat{\mathbf{y}}\cos y$. (c) $\mathbf{A}_7 = \hat{\mathbf{x}}\sin x \cos y + \hat{\mathbf{y}}\cos x \sin y$. (d) $\mathbf{A}_8 = \hat{\mathbf{x}}\cos(xy) + \hat{\mathbf{y}}\sin(xy)$.

Solution: (a) $\nabla \cdot \mathbf{A}_5 = 2x$. (b) $\nabla \cdot \mathbf{A}_6 = \cos x + \sin y$. (c) $\nabla \cdot \mathbf{A}_7 = 2\cos x \cos y$. (d) $\nabla \cdot \mathbf{A}_8 = -y\sin(xy) + x\cos(xy)$.

Both the fields and their divergences are plotted in Figure 4.38. The emergence (or convergence) of the field vectors from source points is a common characteristic for $\mathbf{A}_5$, $\mathbf{A}_6$, $\mathbf{A}_7$, and $\mathbf{A}_8$. Note that the locations from which the field vectors appear to diverge are those at which the divergences are positive maxima, whereas those points at which the fields converge are negative minima. For $\mathbf{A}_5$, the divergence is steadily increasing with increasing x, representing a distributed source. In the case of $\mathbf{A}_8$, the arrows all have the same length since the magnitude of the vector is a constant: $|\mathbf{A}_8| = \sqrt{\cos^2(xy) + \sin^2(xy)} = 1$.

Divergence in other coordinate systems.

Although we use the notation $\nabla \cdot \mathbf{D}$ to indicate the divergence of a vector $\mathbf{D}$, it should be noted that a vector del operator as defined in (4.22) is useful only in a rectangular coordinate system. In other coordinate systems we still denote the divergence of $\mathbf{D}$ by $\nabla \cdot \mathbf{D}$, but note that the specific scalar derivative expressions to be used will need to be derived from the physical definition of divergence using a differential volume element appropriate for that particular coordinate system.

To derive the differential expression for divergence in the cylindrical coordinate system, we consider the cylindrical cuboid[66] volume element shown in Figure 4.39b. Note that the volume of this element is $\Delta v = r\,\Delta r\,\Delta\phi\,\Delta z$. The flux of the vector field $\mathbf{F}$ through the face marked as S_1 is

$$\int_{S_1} \mathbf{F} \cdot d\mathbf{s} = \int_{S_1} F_r\, ds \simeq F_r\left(r + \frac{\Delta r}{2}, \phi, z\right)\left(r + \frac{\Delta r}{2}\right)\Delta\phi\,\Delta z$$

while that through face 2 is

$$\int_{S_2} \mathbf{F} \cdot (-\hat{\mathbf{r}}\,ds) = -\int_{S_2} F_r\, ds \simeq -F_r\left(r - \frac{\Delta r}{2}, \phi, z\right)\left(r - \frac{\Delta r}{2}\right)\Delta\phi\,\Delta z$$

Adding these two and dividing by the volume $\Delta v = r\,\Delta r\,\Delta\phi\,\Delta z$ gives the net flux per unit volume out of the cube due to the r component of the vector field, namely

$$\frac{1}{\Delta v}\int_{S_1+S_2} \mathbf{F} \cdot d\mathbf{s} \simeq \frac{1}{r\,\Delta r}\left[\left(r + \frac{\Delta r}{2}\right)F_r\left(r + \frac{\Delta r}{2}, \phi, z\right)\right.$$
$$\left. - \left(r - \frac{\Delta r}{2}\right)F_r\left(r - \frac{\Delta r}{2}, \phi, z\right)\right]$$

---

[66] This terminology and discussion are adapted from H. M. Schey, *div grad curl and All That, an Informal Text on Vector Calculus*, W. W. Norton, New York, 1992.

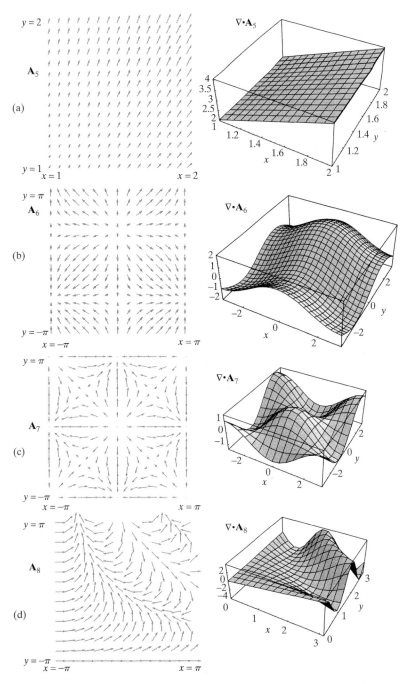

Figure 4.38   Examples of vector fields with nonzero divergence. (a) $\mathbf{A}_5 = \hat{\mathbf{x}}(x^2/2) + \hat{\mathbf{y}}xy$.   (b) $\mathbf{A}_6 = \hat{\mathbf{x}}\sin x - \hat{\mathbf{y}}\cos y$.   (c) $\mathbf{A}_7 = \hat{\mathbf{x}}\sin x\cos y + \hat{\mathbf{y}}\cos x\sin y$.   (d) $\mathbf{A}_8 = \hat{\mathbf{x}}\cos(xy) + \hat{\mathbf{y}}\sin(xy)$. The units for the x and y axes are in radians for all panels.

which in the limit as $\Delta r \to 0$ (and thus $\Delta v \to 0$) becomes

$$\frac{1}{r}\frac{\partial}{\partial r}(rF_r)$$

Evaluating the contributions from the other two pairs of faces in a similar fashion, we find the expression for divergence in cylindrical coordinates:

$$\text{div } \mathbf{F} = \nabla \cdot \mathbf{F} = \frac{1}{r}\frac{\partial}{\partial r}(rF_r) + \frac{1}{r}\frac{\partial F_\phi}{\partial \phi} + \frac{\partial F_z}{\partial z} \qquad (4.37)$$

Consideration of the flux in and out of the different opposing faces of the spherical cuboid shown in Figure 4.39d leads to the divergence expression in spherical coordinates:

$$\text{div } \mathbf{F} = \nabla \cdot \mathbf{F} = \frac{1}{r^2}\frac{\partial}{\partial r}(r^2 F_r) + \frac{1}{r \sin\theta}\frac{\partial}{\partial \theta}(\sin\theta F_\theta) + \frac{1}{r \sin\theta}\frac{\partial F_\phi}{\partial \phi} \qquad (4.38)$$

Example 4.18: Electrostatic field between two coaxial and concentric cylindrical shells of charge. In Example 4.15 we derived the electric flux density in the region between two concentric cylindrical shells of charge to be

$$\mathbf{D} = \epsilon_0 \mathbf{E} = \hat{\mathbf{r}}\frac{\rho_s a}{r} \qquad a < r < b$$

Find the divergence of the electric flux density.

> **Solution:**   Since there are no sources (charges) in the region between the shells ($a < r < b$), the divergence of $\mathbf{D}$ must be identically zero. This is expected on physical grounds; the flux density varies with radial distance as r^{-1}, but the volume of an element subtended by an angular range $\Delta\phi$, a height range Δz, and a radial range Δr is $r \, \Delta r \, \Delta\phi \, \Delta z$ and hence varies as r, exactly compensating the r^{-1} variation of $\mathbf{D}$. Analytically, we have
>
> $$\nabla \cdot \mathbf{D} = \frac{1}{r}\frac{\partial}{\partial r}\left(r\frac{\rho_s a}{r}\right) = 0$$

Example 4.19: Divergence of the dipole field. As an example of a field that has both r and θ components, consider the electric field of the electric dipole as derived in Section 4.4.3. For a dipole consisting of two opposite charges $\pm Q$ separated by a distance d, the electric flux density ($\mathbf{D} = \epsilon\mathbf{E}$) at faraway points ($r \gg d$) was shown to be

$$\mathbf{D} = \frac{Qd}{4\pi r^3}[\hat{\mathbf{r}}2\cos\theta + \hat{\boldsymbol{\theta}}\sin\theta]$$

Find the divergence of the electric flux density at faraway points.

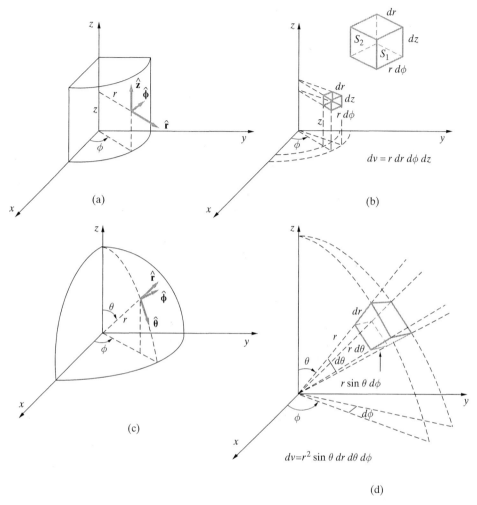

Figure 4.39 Other coordinate systems. (a, b) Unit vectors and volume element for cylindrical coordinates; (c, d) same for spherical coordinates.

Solution: The divergence of this flux density is

$$\text{div } \mathbf{D} = \frac{Qd}{4\pi} \left[\frac{1}{r^2} \frac{\partial}{\partial r} \left(r^2 \frac{2}{r^3} \cos \theta \right) + \frac{1}{r \sin \theta} \frac{\partial}{\partial \theta} \left(\sin \theta \frac{\sin \theta}{r^3} \right) \right]$$

$$= \frac{Qd}{4\pi} \left[\frac{2 \cos \theta}{r^2} \frac{\partial}{\partial r} \left(\frac{1}{r} \right) + \frac{1}{r^4 \sin \theta} \frac{\partial}{\partial \theta} (\sin^2 \theta) \right]$$

$$= \frac{Qd}{4\pi} \left[\frac{2 \cos \theta}{r^2} \left(-\frac{1}{r^2} \right) + \frac{1}{r^4 \sin \theta} 2 \sin \theta \cos \theta \right] = 0$$

Once again the divergence is zero as expected, since there are no charges at points far away from the source dipole.

4.6.1 The Divergence Theorem

For the electrostatic field, we can combine the integral and differential forms of Gauss's law to find

$$
\left.\begin{array}{c}
\displaystyle\oint_S \mathbf{D} \cdot d\mathbf{s} = \int_V \rho\, dv \\[2mm]
\nabla \cdot \mathbf{D} = \rho
\end{array}\right\} \quad \longrightarrow \quad \boxed{\oint_S \mathbf{D} \cdot d\mathbf{s} = \int_V \nabla \cdot \mathbf{D}\, dv} \qquad (4.39)
$$

This integral equation is entirely consistent with the interpretation of the divergence of a vector field as the outflow of flux per unit volume at a given point.

Consider a volume V, subdivided into smaller cells and surrounded by a surface S. The outward flux from any of the subcells is given by $(\nabla \cdot \mathbf{D})\Delta v$, where $(\nabla \cdot \mathbf{D})$ is the value of the divergence (i.e., source density) at that point. All of this outward flux enters the adjoining cells, unless the cell contains a portion of the outer surface. The divergence integrated over the volume V would give the total outward flux from that volume, which should thus be equal to the outward flux through the surface S enclosing the volume. Accordingly, we expect the above result to hold true for any arbitrary vector field $\mathbf{G}$, rather than just the electrostatic flux density $\mathbf{D}$. This general result, restated mathematically below for an arbitrary vector $\mathbf{G}$, is known as the *divergence theorem* or *Gauss's theorem* or *Green's theorem* and is fundamentally important and useful in many branches of engineering and physics.

$$
\boxed{\oint_S \mathbf{G} \cdot d\mathbf{s} = \int_V \nabla \cdot \mathbf{G}\, dv} \qquad (4.40)
$$

4.7 METALLIC CONDUCTORS

Until now, we have considered charges to be situated in vacuum, without any matter nearby. We have assumed the charges to be stationary and have found that the calculation of electric fields and potentials due to these charges in surrounding regions can basically be reduced to carrying out (albeit often complicated) integration operations. However, when material media are in the presence of electric fields, complications arise, because the way in which charges are distributed is not initially known and is determined by the properties of matter. The most important property of matter in this context is *conduction*, which is its ability to allow motion of charges over macroscopic distances. Materials greatly vary in this property, all the way from good conductors (metals) to very poor conductors (insulators, or dielectrics).

In this section, we briefly discuss metallic conductors. Most electrostatic problems consist of configurations of metallic bodies that are connected to primary sources (e.g., a battery) so that potential differences exist between conductors. To be able to determine the resulting charge distributions, we need to have a basic understanding of the electrical properties of metallic conductors.

Metallic materials are in general good conductors. A good conductor is a material within which charges can move freely. Since we are dealing with electrostatics, we assume *a priori* that the electric charges have reached[67] their equilibrium positions and are fixed in space. Under such static conditions, there must be zero electric field inside the metallic conductor, since otherwise charges would continue to flow, contrary to our original premise. The electrostatic potential Φ must be the same throughout the metallic conductor for the same reason. It thus follows that if such a conductor is charged, for example by connecting it to ground (see Figure 4.28) or by bringing it near another charged metallic conductor, the charges rearrange themselves such that the net electric field due to all the charges becomes zero inside the conductor. If a metallic conductor is placed in an external electric field $\mathbf{E}_0$, charge once again flows temporarily within it to produce a second field $\mathbf{E}_1$, which precisely cancels out the existing electric field inside the conductor, so that $\mathbf{E}_0 + \mathbf{E}_1 = 0$. Regardless of the shape of the conductor, the field everywhere within it becomes zero once the charges come to rest. In most good conductors (metals) the establishment of this equilibrium occurs in $\sim 10^{-19}$ s, as will be shown in Section 5.5.[68]

There are two important consequences of the fact that the electrostatic field inside a metallic conductor is identically zero. First, the entire space occupied by the conductor must be an equipotential volume. Secondly, all the charge, if any, on a conductor must reside entirely on its surface,[69] since if any charge did exist at any point within the body, then by Gauss's law a nonzero field would have to exist in the vicinity of a small Gaussian surface surrounding that point. At equilibrium, the surface charge is distributed in such a way that the total electric field inside the conductor and tangential to its surface is zero.[70] Note that the requirement for the tangential field to be zero also follows from the fact that the conductor is an equipotential, to which electric field lines must be orthogonal.

4.7.1 Macroscopic versus Microscopic Fields

The fundamental equations of electrostatics (Coulomb's law, Gauss's law), or for that matter the more general Maxwell's equations, which describe all electromagnetic

---

[67]We note that if a charge distribution is suddenly disturbed, it would take a finite time (see Section 5.5) for the charges to redistribute themselves.

[68]The value $\sim 10^{-19}$ s for the rearrangement time of excess charge is particular to copper. It is given by the ratio ϵ/σ for the material, where σ is the electrical conductivity (to be introduced in Chapter 5) and ϵ is the electrical permittivity (to be discussed in Section 4.10). For other metallic conductors, ϵ/σ is similarly very small (of order 10^{-18} to 10^{-19} s).

[69]We consider the surface as having strictly infinitesimal thickness. Actually, the excess or deficiency of electrons will move around in a layer of the order of atomic dimensions, or an angstrom (10^{-8} cm) in thickness, producing on the average a volume density of charge in a layer too thin to be of macroscopic importance, but nevertheless containing many electrons.

[70]The charged metallic conductor can be thought of as a body that is internally electrically neutral but has a charge distribution over its exterior surface. The interior of the conductor is equivalent to vacuum (see footnote 23 in Section 5.5) once equilibrium is established, since the net effect of all the microphysics of the material has been to distribute the surface charge properly. This distribution is such that if the conducting body were somehow removed, leaving the surface charge distribution intact in vacuum, the field $\mathbf{E}$ at all points previously occupied by the conductor would still remain zero.

phenomena, are written in terms of *macroscopic* quantities. For example, when we state that (4.36) is Gauss's law applied to a "point," we refer to a point that is very small compared with the physical dimensions of our system but very large compared with atomic dimensions, so that it actually contains many electrons, the volume charge density of which is ρ. A "point" in a macroscopic sense might have dimensions of 1 micron (or $1 \ \mu m = 10^{-6} \ m = 10^{-4} \ cm$), whereas a microscopic point would have atomic dimensions of the order of one angstrom or 10^{-8} cm. Thus, a macroscopic "point" in the form of a cube of size $1 \ \mu m \times 1 \ \mu m \times 1 \ \mu m$ may well contain 10^{12} atoms, assuming that each atom has a radius of about one angstrom.

Matter consists of primarily empty space sparsely filled with atoms and molecules. An atom is itself mostly empty space, with the nucleus and electrons occupying a tiny fraction of its volume. Quantitatively, and in purely classical (nonquantum) terms, we understand an atom as consisting of a heavy nucleus of approximate radius $\sim 10^{-13}$ cm. Negatively charged electrons, with physical dimensions also of order $\sim 10^{-13}$ cm, travel around the nucleus in orbits of approximate radii $\sim 10^{-8}$ cm. At room temperature and pressure, the spacing between the centers of atoms varies from $\sim 10^{-8}$ cm in solids to $\sim 10^{-6}$ cm in gases. To determine the range of dimensions over which a macroscopic description is appropriate, we can rely on the fact that reflection and refraction of visible light (wavelength $\sim 10^{-5}$ cm) are properly described by fundamental equations using macroscopic parameters, whereas diffraction of X rays (wavelength $\sim 10^{-8}$ cm) reveals the atomic nature of matter.[71] Based on this, it is possible to take a dimension of $10^{-6} \ cm = 10^{-2} \ \mu m = 100$ angstroms to be approximately the lower limit of the macroscopic domain.

Metals are conductors because electrons in them are free to move when electric fields of macroscopic extent are applied. In a classical picture, metallic structure can be thought of as an array of positive ions, which can vibrate about their lattice sites, and electrons as small ($< 10^{-12}$ cm) points or spheres that are free to wander throughout the body under the influence of thermal agitation. The vibrations of the ions and the movement of the electrons are both random thermal effects, and local electric fields exist at any atom site, varying with the motions of nearby ions and electrons. These microscopic fields can have very large intensities but vary extremely rapidly in both space and time. Instantaneous field strengths of order $\sim 10^{9}$ to $\sim 10^{15}$ V-cm^{-1} may exist at the orbits of electrons in atoms, and at the edge of a heavy nucleus the field intensities may be of order $\sim 10^{19}$ V-cm^{-1}. The spatial variations occur over distances of order $\sim 10^{-8}$ cm or less, and the temporal fluctuations occur with periods ranging from $\sim 10^{-13}$ to $\sim 10^{-17}$ seconds. Macroscopic measuring devices generally average over intervals in space and time that are much larger than the scales over which these microscopic

---

[71]For further discussion of macroscopic versus microscopic fields and charges, see the following references, listed in increasing level of complexity: J. D. Jackson, *Classical Electrodynamics,* Wiley, New York, 1975, Section 6.7; W. B. Cheston, *Elementary Theory of Electric and Magnetic Fields,* Wiley, New York, 1964, Section 2.7; H. A. Lorenz, *The Theory of Electrons and Its Applications to the Phenomena of Light and Radiant Heat,* Dover, New York, 1952, Chapter 1.

fluctuations occur. Consequently, these variations are *averaged* out,[72] resulting in smooth (in space) and slowly varying (in time) macroscopic quantities. In a conductor, the *average* value of the electric field is zero unless there is a drift of the electron cloud (current). No such macroscopic drifts can occur in electrostatic equilibrium in the absence of an externally applied electric field. The influence of a macroscopic applied electric field $\mathbf{E}$ is to produce a net electron drift velocity $\mathbf{v}_d$ proportional to $-\mathbf{E}$, which leads to conduction currents, to be studied in Chapter 5.

In terms of quantum theory, the electric field $\mathbf{E}$ can be thought to set up a continuous process of electrons jumping into empty conduction band levels,[73] while at the same time collisions with thermally agitated atomic nuclei, lattice boundaries, and imperfections produce backward jumps that counteract the effect of $\mathbf{E}$, so a steady state of electron flow is established, with $\mathbf{v}_d$ proportional[74] to $-\mathbf{E}$. We also note that charges that freely move in a conductor are strictly electrons; when a conductor surface is positively charged, the conductor has a net deficiency of electrons, versus an excess of electrons in the case of a negatively charged conductor. It is important to note, however, that this excess or deficiency of electrons is extremely small compared with the total number of electrons in the metallic material, as will be illustrated in Example 4.20.

4.7.2 Electric Field at the Surface of a Metallic Conductor

We can formally show that the electric field tangential to a conductor surface should be zero by using the conservative property of the electrostatic field. Figure 4.40a shows an interface between a conductor and free space, and a rectangular contour abcda having width Δw and height Δh, with sides bc and da parallel to the interface. Noting that the field inside the conductor is zero, letting $\Delta h \to 0$, and also noting that the line integral of the electrostatic field around any closed path is identically zero (4.20), we have

$$\oint_{abcda} \mathbf{E} \cdot d\mathbf{l} = \int_{da} E_t \, dl = E_t \, \Delta w = 0 \quad \to \quad \boxed{E_t = 0} \tag{4.41}$$

Thus, the tangential component of the electric field at the surface of a conductor is identically zero. Electric field lines terminate perpendicularly on conductor surfaces. In Chapter 7, when we examine Faraday's law, we shall see that, for time-varying fields, the line integral of the electric field around a closed loop is not zero (i.e., $\oint \mathbf{E} \cdot d\mathbf{l} \neq 0$)

---

[72]Only a spatial averaging of the fields is necessary, since in any macroscopically significant region there are so many nuclei and electrons that temporal fluctuations are completely washed out by averaging over space. To appreciate this, note that any macroscopic amount of ordinary matter (with a lowest one-dimensional extent of say $\sim 10^{-6}$ cm or volume of $\sim 10^{-18}$ cm^{-3}) has of order $\sim 10^6$ nuclei and electrons. For further discussion, see Section 6.7 of J. D. Jackson, *Classical Electrodynamics*, Wiley, New York, 1975.

[73]In metals, the lowest energy bands are only partially filled, with many vacant levels. For a brief elementary discussion, see Section 9.6 of R. L. Sproull, *Modern Physics, The Quantum Physics of Atoms, Solids, and Nuclei*, Wiley, New York, 1963. For more detail, see V. F. Weisskopf, *Am. J. Phys.*, 11:1943.

[74]This proportionality of $\mathbf{v}_d$ and $-\mathbf{E}$ is the basis for Ohm's law, which we will discuss in Chapter 5.

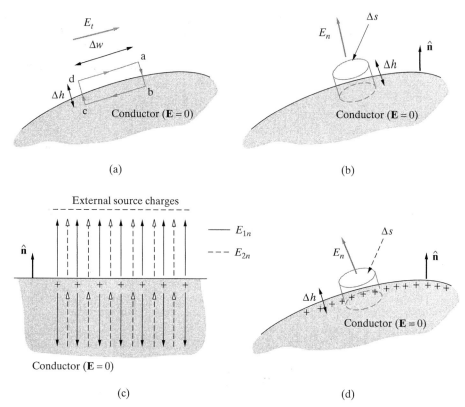

Figure 4.40   Field at the surface of a conductor. (a) Contour abcda used to show that $E_t = 0$. (b) Application of Gauss's law to relate normal electric field to surface charge. (c) Electric fields at the conductor surface; E_{1n} is the field due to the surface charge ρ_s, whereas E_{2n} is the field due to the external "source" charges of opposite polarity. (d) The fields at the surface of the conductor arise from two systems of charges.

but instead is given by the negative time rate of change of the magnetic flux passing through the surface enclosed by the closed loop. Nevertheless, the boundary condition on the conductor surface (i.e., the fact that $E_t = 0$) will be shown to be still valid in general.

Using Gauss's law it is possible to determine a relationship between the surface charge density on a conductor at equilibrium and the electrostatic field at the surface. Consider a very small portion of any charged conducting surface, as shown in Figure 4.40b. An infinitesimal pillbox-shaped Gaussian surface is visualized as shown, with half of it above and the rest below the surface. In applying Gauss's law, we note that no flux crosses the lower surface, since the field inside the conductor is zero. No flux leaves through the sides either, since this would require that **E** have a component tangential to the surface. Furthermore, we lose no generality by considering the limit $\Delta h \to 0$, so that the total surface area of the closed surface is primarily due to the top and bottom faces.

On the top surface, a normal component of $\mathbf{E}$ exists.[75] The net outward flux from this top surface is $\epsilon_0 \mathbf{E} \cdot \Delta \mathbf{s}$ and must equal the charge within it. Denoting the surface charge density on the conductor surface as ρ_s, and the electric field normal to the surface as E_n, we have

$$E_n \, \Delta s = \frac{\rho_s}{\epsilon_0} \Delta s \quad \longrightarrow \quad E_n = \hat{\mathbf{n}} \cdot \mathbf{E} = \frac{\rho_s}{\epsilon_0}$$

where $\hat{\mathbf{n}}$ is the unit vector outwardly perpendicular to the conductor surface, as shown. Note that since $\mathbf{D} = \epsilon_0 \mathbf{E}$, this condition can also be written as

$$\boxed{\hat{\mathbf{n}} \cdot \mathbf{D} = D_n = \epsilon_0 E_n = \rho_s} \tag{4.42}$$

This more general relationship is also valid if the conductor is surrounded by an insulating material other than vacuum, except that in (4.42), ϵ_0 would have to be replaced with a different constant ϵ to account for the electrical behavior of the particular material (see Section 4.10).

If we think of the conductor surface as an infinite plane surface with charge density ρ_s, the result just obtained (i.e., $E_n = \rho_s/\epsilon_0$) seems to contradict the electric field that we found for an infinite sheet of charge in Example 4.14, which was $E_n = E_z = \rho_s/(2\epsilon_0)$. The reason for this difference is that the field at the surface of a charged conductor actually arises from two systems of charges: the local surface charge and the "source" charges, which are remote from the conductor and that are physically required by the "charged" state of the conductor.[76] In the simplest geometry, we could think of these distant charges as lying uniformly on a distant parallel planar surface, as shown in Figure 4.40c. By examination of Figure 4.40c, we can see that the field solved in Example 4.14 represented only the partial field caused by the surface charge (i.e., E_{1n}) and that the total field can be found only if the external source charges of opposite polarity are included. In other words, the conducting surface can have net induced positive charge only if these negative external source charges are present, typically because a potential difference (e.g., with a battery) is applied between the conducting surface and some other conductor (on which the external negative charges reside). The field E_{2n} due to the external charges cancels out E_{1n} inside the conductor and doubles it outside the conductor, so the net field normal to the surface of the conductor is $E_n = E_{1n} + E_{2n} = \rho_s/\epsilon_0$.

We can also understand the difference between the field near a sheet of charge and that at the surface of the conductor by considering Gauss's law with a cylindrical pillbox surface, as shown in Figure 4.40d. Since the field inside the conductor is zero, and since we take $\Delta h \to 0$, the only nonzero contribution to the surface integral is from the top surface. Thus, we have $\epsilon_0 E_n \, \Delta s = \rho_s \, \Delta s$, which gives $E_n = \rho_s/\epsilon_0$. However, in the case of the sheet of charge (Example 4.14), the field is nonzero on both the top and

---

[75]Presumably this field is the reason the conductor is charged; that is, the electric field emanates from (or converges on) the positive (negative) surface charges and connects to (emanates from) other charges of negative (positive) sign, which constitute the source of the "charging" of the conductor.

[76]Considering the common situation in which the conductor is charged by applying a potential difference between it and another conductor, these distant charges are the charges of opposite sign induced on the other conductor.

bottom sides of a similar pillbox surface, so we have $2\epsilon_0 E_n \, \Delta s = \rho_s \, \Delta s$, which gives $E_n = \rho_s/(2\epsilon_0)$. Both of these points of view (i.e., as discussed in relation to Figures 4.40c and d) are equally valid.

Example 4.20: Surface charge on a conductor. It is interesting to consider how small an excess or deficiency of electrons can be expected to occur on the surface of a metallic conductor in practice. Normal air breakdown occurs (see Section 4.10.3) when the magnitude of the electric field in air reaches approximately $E_{max} = 3 \times 10^6$ V-m^{-1}. (a) Find the charge density ρ_s on the surface of a metallic conductor that would produce such a field. (b) How many excess electrons would such a surface density represent in a square pillbox volume of surface area 1 μm $\times$ 1 μm and thickness one angstrom? Note that a surface density at which the field immediately outside the conductor would be at the breakdown level for air represents essentially the maximum surface charge that can be placed on a conductor in air before one sees corona breakdown.[77]

Solution:

(a) The maximum surface charge density can be found as

$$\rho_s = \epsilon_0 E_{max} = 8.854 \times 10^{-12} \text{ F-m}^{-1} \times 3 \times 10^6 \text{ V-m}^{-1} \simeq 2.66 \times 10^{-5} \text{ C-m}^2$$

(b) Now consider the surface region of area $10^{-6} \times 10^{-6}$ m^2 and a thickness of 10^{-10} m. Such a region has a volume $10^{-6} \times 10^{-6} \times 10^{-10} = 10^{-22}$ m^3, with on average about $\sim 10^8$ free electrons and nuclei, assuming one free electron per atom and atomic dimensions of ~ 1 angstrom$^3 = 10^{-30}$ m^3. The number of excess electrons represented by the charge density ρ_s determined above, in the same volume just described, can be found as

$$\frac{\rho_s A}{|q_e|} \simeq \frac{2.66 \times 10^{-5} \text{ C-m}^{-2}}{1.6 \times 10^{-19} \text{ C-e}^{-2}} (10^{-6} \times 10^{-6} \text{ m}^2) \simeq 1.66 \times 10^2 \quad \ll 10^8 \text{ electrons!}$$

Thus, even the maximum excess charge that can be placed on a metallic conductor represents a truly minute change in the total number of free electrons available in the conductor.

4.7.3 Induced Charges on Conductors

The boundary conditions $E_t = 0$ and $D_n = \rho_s$ at the surface of a conductor imply that if a conductor is placed in an externally applied electric field, then (1) the field distribution

---

[77]The process of corona breakdown can be understood as follows: The intense electric field accelerates the electrons and ions in air to high velocities. These high-speed particles collide with neutral air molecules and knock electrons out of the molecules. As a result, vast numbers of additional ions and electrons are produced, leading to *avalanche breakdown*. The air in the vicinity of the conductor becomes much more conducting, with the result that the charged conductor quickly loses most of its charge. The air might even glow (exhibit corona effect) because of light emitted from the air molecules during these collisions. Further discussion of breakdown effects is given in Section 4.10.3.

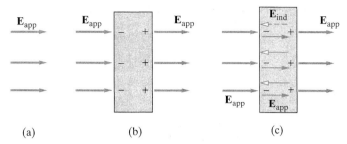

Figure 4.41   A conducting slab in an applied electric field.

will be distorted so that the electric field lines are normal to the conductor surface and (2) a surface charge will be induced on the conductor to support the electric field. We shall study examples of field distributions in the vicinity of conductors in Section 4.8. In this section, we consider some simple geometries to illustrate the behavior of the induced charge and its relationship to the electric field at a conceptual level.

Consider a horizontally oriented applied electric field E_{app}, into which we introduce a metallic slab, perpendicular to the electric field as shown in Figures 4.41a and b. Note that since E_t is already zero, the introduction of the conducting slab does not seem to distort the existing field configuration;[78] it appears as if the field enters the slab from the left and then emerges on the right side (Figure 4.41b). What actually happens is that the field pointing into the conductor on the left side induces negative charges in accordance with $D_n = \rho_s$. This negative surface charge consists of a very small fraction of the abundant supply of free electrons available in the metal. Since the metallic slab was initially neutral, movement of some negative charges toward the left side leaves an equal number of positive charges (a deficiency of electrons). During the charge rearrangement time (typically $\sim 10^{-19}$ sec), electrons move around in such a way as to make the electric field inside the conductor identically zero. As a result, the positive charges (that are, in effect, produced by the fact that some electrons move to the left side) remain on the right-hand surface of the conductor, where they are supported by the electric field that emerges from the slab in accordance with $D_n = \rho_s$.

Another way of thinking about the behavior of the conductor is illustrated in Figure 4.41c. The induced negative and positive surface charge layers establish an induced electric field E_{ind} within the conductor, which balances the applied electric field E_{app} such that the total electric field in the metal is zero. In other words, the electric field induced in the metal is $E_{ind} = -E_{app}$. When the conductor is placed in the electric field, the free charges quickly establish a charge distribution to set up this internal electric field in order to cancel the applied electric field.

[78] In contrast, if the conducting body were spherical, the field lines would have to bend in the vicinity of the sphere so as to be orthogonal on it. See Section 4.7 of D. K. Cheng, *Field and Wave Electromagnetics,* 2nd ed., Addison-Wesley, New York, 1989.

Example 4.21: Spherical metallic shell. Consider a positive point charge Q at the center of a spherical metallic shell[79] of an inner radius a and outer radius b as shown in Figure 4.42a. Find the (a) electric field and (b) electrostatic potential everywhere as a function of radial distance r.

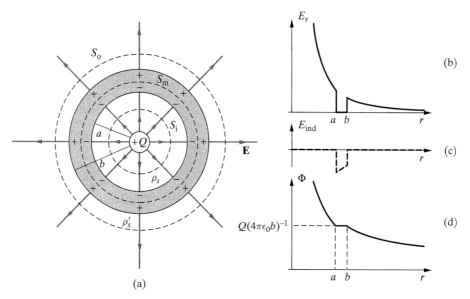

Figure 4.42 **Point charge at the center of a spherical conducting shell.** (a) The geometry. (b) The total electric field distribution. (c) The induced electric field distribution. (d) The electric potential distribution.

Solution:

(a) In view of the spherical symmetry, we apply Gauss's law to find **E** for three different spherical Gaussian surfaces, namely S_o with radius $r \geq b$, S_m with radius $a \leq r < b$, and S_i with radius $r < a$, as shown in Figure 4.42a with dashed lines. Note that due to spherical symmetry, in all three regions, we have $\mathbf{E} = \hat{r} E_r$. Applying Gauss's law ((4.32)) on the innermost surface S_i, we have

$$\oint_{S_i} \mathbf{E} \cdot d\mathbf{s} = E_r (4\pi r^2) = \frac{Q}{\epsilon_0} \quad \rightarrow \quad E_r = \frac{Q}{4\pi \epsilon_0 r^2} \qquad r < a$$

On the Gaussian surface S_m, which lies inside the metallic shell, the electrostatic field is identically zero as has been discussed. As a result of Gauss's law, the total charge enclosed inside S_m must be zero, so a negative surface charge of total amount $-Q$ (so that $Q_{enclosed} = Q - Q = 0$) is induced on the inner surface ($r = a$) of the spherical

[79]Note that the charge may have been placed through a small hole in the shell or the shell may have consisted of two separate hemispheres enclosed on the charge and secured together, like that in Faraday's experimental setup as shown in Figure 4.28.

conducting shell where the electric field lines emanating from the positive point charge located at the center can terminate. As a result of the spherical symmetry, the induced charge distribution is uniform, and the surface charge density can be found from

$$\rho_s (4\pi a^2) = -Q \quad \rightarrow \quad \rho_s = -\frac{Q}{4\pi a^2}$$

Note that since the conducting shell is initially neutral, a total amount of charge $+Q$ is also induced on the outer surface of the conductor, with a different surface charge density, given by $\rho'_s = Q/(4\pi b^2)$.

Application of Gauss's law to the outermost surface S_o yields

$$\epsilon_0 \oint_{S_o} \mathbf{E} \cdot d\mathbf{s} = E_r (4\pi r^2) = \underbrace{Q + (-Q) + Q}_{Q_{\text{enclosed in } S_o}} = Q \rightarrow E_r = \frac{Q}{4\pi \epsilon_0 r^2} \quad r \geq b$$

(b) The electrostatic potential Φ with respect to a point at infinity can be found by taking the line integral of $\mathbf{E}$ (equation (4.17)). For points outside the spherical shell, we have

$$\Phi = -\int_\infty^r E_r \, dr = -\int_\infty^r \frac{Q}{4\pi \epsilon_0 r^2} \, dr = \frac{Q}{4\pi \epsilon_0 r} \qquad r \geq b$$

Since the conductor is an equipotential, the electrostatic potential Φ remains constant in the region $a \leq r < b$, namely

$$\Phi = -\int_\infty^b E_r \, dr - \underbrace{\int_b^r E_r \, dr}_{=0 \text{ since } E_r = 0} = \frac{Q}{4\pi \epsilon_0 b} \qquad a \leq r < b$$

For points in the region enclosed by the spherical metallic shell (i.e., $r < a$), we have

$$\Phi = -\int_\infty^r E_r \, dr = -\int_\infty^b \frac{Q}{4\pi \epsilon_0 r^2} \, dr - \int_a^r \frac{Q}{4\pi \epsilon_0 r^2} \, dr$$

$$= \frac{Q}{4\pi \epsilon_0} \left[\frac{1}{b} + \frac{1}{r} - \frac{1}{a} \right] \qquad r < a$$

The variations of E_r and Φ with radial distance are shown in Figures 4.42b and d. Also shown, in Figure 4.42c, is the variation of the induced field E_{ind}, which is set up within the conducting shell and cancels out the external field produced by the point charge Q at the center.

Further insight into the behavior of induced charges and metallic conductors in an electric field can be obtained by considering a variation of the geometry of Example 4.21 as shown in Figure 4.43, where the point charge Q is moved to a point off from the center of the spherical conducting shell. Although an analytical solution of this problem is much too involved to be considered here, we can qualitatively describe the resulting configuration. The electric field distribution in the region inside the shell is now distorted to ensure the termination of electric field lines perpendicularly to the inner surface of the shell. The negative charge induced in the inner surface of the shell is now distributed

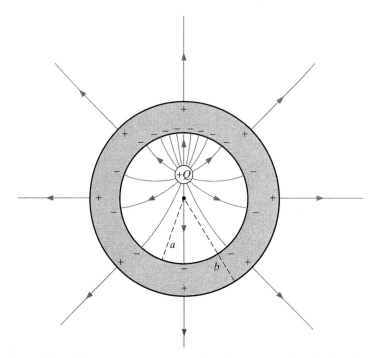

Figure 4.43   Off-centered point charge enclosed by a spherical conducting shell.

nonuniformly, in accordance with $D_n = \rho_s$, being denser at points of the surface closer to the charge Q, where the electric field magnitude is higher. The positive charge layer on the outer surface of the shell remains uniformly distributed, so the field at all points external to the shell remains the same, as if the exterior field originated from a point charge located at the center of the shell. In other words, the electric field produced by the charge induced in the inner surface of the conducting shell exactly cancels out the effect of the point charge for all radii larger than a.

We now can also better understand the charge displacement experiments carried out by Faraday, as depicted in Figure 4.28. The outer sphere in Faraday's experiments behaves exactly like the spherical shell considered here, with negative (positive) charge induced on its inner (outer) surface. When the outer sphere in Figure 4.28 is grounded, electrons move from the ground to the outer surface to neutralize it so that the outer sphere now has excess charge of amount exactly equal to $-Q$. With the ground connection open, and with the inner sphere removed, this $-Q$ charge stays on the conductor and is what Faraday measured.

4.7.4 Electrostatic Shielding and Tests of the Inverse Square Law

We now discuss an important consequence of Gauss's law: that the electric field inside an empty cavity completely enclosed by a conductor is identically zero, regardless of the

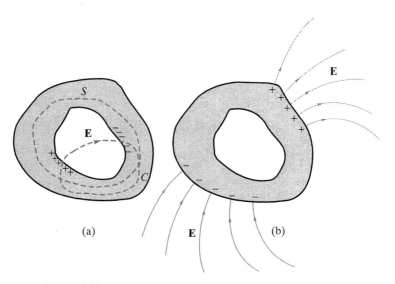

Figure 4.44   Electrostatic field inside an arbitrarily shaped cavity.

shape of the cavity and the electric field distribution outside the cavity shell. Consider an arbitrarily shaped cavity such as that shown in Figure 4.44. For any Gaussian surface S lying entirely in the conductor as shown, the field is zero at all points on it, so there is no net electric flux through it. In other words, the total charge inside S is zero. Unless the cavity is spherical, as in the case of Figure 4.42, the fact that the net total charge on the inner conductor surface is zero does not necessarily imply a uniform distribution of charge. Thus, we cannot rule out the possibility of an uneven distribution of charge (positive in one region and negative in another) on this surface.

In fact, there can be absolutely no charge on the inner surface, as can be seen by the following arguments. On a qualitative basis, we observe that if equal and opposite charges existed in different parts of the inner surface, electrons would quickly move along the surface to neutralize the positive charges. We can also use the conservative property of the electrostatic field (i.e., the fact that the line integral of **E** along any closed contour is identically zero) to show that there can be no charge separation along the inner conductor surface. Assume that there is a clustering of positive charges on some part of the surface. Based on the use of Gauss's law in connection with surface S in Figure 4.44, we must then have an equal number of negative charges somewhere else. Between the positive and negative charges, there must exist an electric field **E**, starting on the positive charges and ending on the negative charges (Figure 4.44a). Now consider a contour C that crosses the cavity along an electric field line and closes on itself via a path through the conductor. Since the line integral through the metal part is zero ($\mathbf{E} = 0$), we would then have

$$\oint_C \mathbf{E} \cdot d\mathbf{l} \neq 0$$

which contradicts the fact that the line integral of the electrostatic field along any closed contour is identically zero. Thus, we conclude that *there can be no fields inside the metallic cavity, nor any charges on its inner surface.* Note that our proof of this was not in any way dependent on any induced charges on the outer surface of the conductor. Thus, our conclusion is valid even in those cases when the cavity enclosure is charged or immersed in an external electric field, as shown in Figure 4.44b. In the latter case, the induced charges simply redistribute to support the external electric field, much like in the case described in connection with Figure 4.41.

The above result shows that if a cavity is completely enclosed by a metallic conductor, no static distribution of charges or fields outside can produce any fields inside. This result is the basis for the common practice of *shielding* electrical equipment by placing it in metal enclosures. Note that even extremely thin[80] conducting shells would give effective electrostatic shielding.[81] However, as we shall see in Chapter 8, this is not necessarily the case for time-varying fields. For such fields, the cavity is shielded if the thickness of the metallic enclosure is larger than the so-called "skin depth." Nevertheless, it should be noted that at, for example, 3 MHz, the skin depth for copper is $\sim$0.0382 mm, which means that, macroscopically speaking, the cavity shell can still be quite thin.

That the electrostatic field inside a metallic enclosure is zero is a direct consequence of Gauss's law, which in turn is based on Coulomb's inverse square law. Thus, this general conclusion was deduced from the single fact of the law of inverse square. This "prediction" of the inverse square law is valid for any shape of the conductor and allows for highly accurate verification of Coulomb's law. Even before Coulomb, the experiments carried out by H. Cavendish in 1772, and later repeated by J. C. Maxwell in 1879, indicated[82] that the instruments used would have shown charge on the inner sphere if the inverse square law were instead of the form $r^{-2+\varsigma}$. In this way, Maxwell was able to show that $|\varsigma| < 10^{-5}$. Later experimental verifications of Coulomb's law were also based on the same principle, with a more recent one by Williams, Faller, and Hill[83] establishing a bound of $|\varsigma| < (2.7 \pm 3.1) \times 10^{-16}$.

It might occur to the reader that the assumption that the electrostatic force law is of the form $r^{-2+\varsigma}$ might not be sufficiently general. What if the functional form of the force law as a function of distance were some other function $f(r)$? The experimental evidence that the force law depends on some power of inverse distance is the fact that conductors with similar geometric shapes have similar electrical properties when their

---

[80]Assuming macroscopic thickness of $> 10^{-6}$ m, or 1 micron.

[81]It is interesting to note here that shielding of static magnetic fields is quite a different problem from shielding of electric fields. Magnetic fields can effectively penetrate through conducting materials and thus cannot be confined to metallic cavities. Nevertheless, effective shielding of magnetic fields can be obtained by using high-permeability materials, which can confine and orient magnetic field lines away from selected regions. For a brief discussion at an appropriate level, see Section 10.2 of M. A. Plonus, *Applied Electromagnetics,* McGraw-Hill, 1978.

[82]H. Cavendish, *Electrical Researches,* J. C. Maxwell (Ed.), Cambridge University Press, Cambridge, UK, 1879, pp. 104–113. For an excellent description of these and other earlier experiments, see R. S. Elliott, *Electromagnetics.* IEEE Press, Piscataway, New Jersey, Chapter 3, 1993.

[83]E. R. Williams, J. E. Faller, and H. A. Hill, A new experimental test of Coulomb's law: A laboratory upper limit on photon rest mass, *Phys. Rev. Lett.,* 26, p. 721, 1971.

dimensions are scaled. For example, the electric field line configurations between two spheres remain similar when we scale all physical dimensions. Also, the inverse square law can be mathematically derived from the single fact that the electric field inside a spherical conducting shell is zero.[84]

4.7.5 Forces on Metallic Conductors

A differential element of charge $dQ = \rho_s \, ds$ on the surface of a metallic conductor experiences the electrostatic field of all the other charges in the system (i.e., external charges as in Figure 4.40c). Since the charge $dQ = \rho_s \, ds$ is bound to the conductor and is prevented from leaving it by atomic forces,[85] the force acting on $dQ = \rho_s \, ds$ is transmitted to the solid conductor itself. To determine the force on the conducting body, we need only to evaluate the forces on the surface charges. For this purpose, any small area on the surface of a conductor can be treated as approximately planar, and we can separately consider the electric field E_{2n} due to the external sources and E_{1n} due to the surface charges (see Figure 4.40c). Since the latter cannot exert a force on the charge producing itself, the force on the surface charge element $dQ = \rho_s \, ds$ is entirely due to the external field $E_{2n} = \rho_s/(2\epsilon_0)$. Since the electrostatic force is $F = qE$, the differential force on this charge element is

$$dF = E_{2n} \, dQ = \frac{\rho_s}{2\epsilon_0} \rho_s \, ds$$

so that the differential force per unit area is

$$\frac{dF}{ds} = \frac{\rho_s^2}{2\epsilon_0} = \frac{\epsilon_0 E_n^2}{2}$$

where $E_n = E_{1n} + E_{2n} = \rho_s/\epsilon_0$ is the total electric field normal to the surface of the conductor.

Note that the above result was derived by assuming the surface charge to reside in an infinitely thin plane sheet, with the external charges also planar, as implied in Figure 4.40c. In practice, the surface charge resides in a layer of finite thickness, and the surface may have roughness on a microscopic scale. Nevertheless, the force per unit area for conductor surfaces of any other type of geometrical shape and in cases where the surface charge resides in a region of finite thickness, is given by $\epsilon_0 E_n^2/2$.

The above result implies that any charged conductor surface is under the influence of a force that pulls it in the direction normal to the surface. Note that since the force is proportional to ρ_s^2, an induced surface charge of any polarity, or equivalently incident electric field on the surface of any direction, results in a pull on the surface. The magnitude of the pull on different parts of the surface varies in accordance with the surface charge

---

[84]See Chapter 3 of R. S. Elliott, *Electromagnetics,* IEEE Press, Piscataway, New Jersey, 1993.

[85]For a discussion of why electrons cannot easily escape from a metal surface, see Section 2.3 of J. D. Jackson, *Classical Electrodynamics,* Wiley, 1975, and Section 2.12 of M. A. Plonus, *Applied Electromagnetics,* McGraw-Hill, New York, 1978.

density as ρ_s^2 or the normal electric field as E_n^2. If the conductor surface is deformable, one can use the electrostatic force density together with the mechanical properties to determine the equilibrium shape that the conductor surface attains. For rigid conductor surfaces, the total force F_{total} on the conducting body can be found by integrating the force per unit area over the surface:

$$F_{total} = \int \frac{dF}{ds} \, ds = \int \frac{\epsilon_0 E_n^2}{2} \, ds$$

To appreciate the magnitude of the electrostatic force on a conductor surface, consider an aluminum surface region of size $10^{-2} \times 10^{-2}$ m^2 with a thickness of 10^{-3} m (1 mm). At an air breakdown electric field of 3×10^6 V-m^{-1}, the electrical force on this conductor per unit area would be

$$\frac{dF}{ds} = \frac{\epsilon_0 E_n^2}{2} = \frac{8.854 \times 10^{-12} \text{ F-m}^{-1} \times (3 \times 10^6 \text{ V-m}^{-1})^2}{2} \simeq 39.8 \text{ N-m}^{-2}$$

Thus, the total electrical force experienced by this thin 1 cm $\times$ 1 cm square shaped piece of aluminum would be $39.8 \times 10^{-2} \times 10^{-2} \simeq 4 \times 10^{-3}$ newtons. For comparison, we can consider the weight (or gravitational force) of such a piece of aluminum. Noting that the mass density of aluminum is 2.7×10^3 kg-m^{-3} and the gravitational constant is $g \simeq 9.8$ m-s^{-2}, the piece of aluminum with a volume of $10^{-2} \times 10^{-2} \times 10^{-3} = 10^{-7}$ m^3 weighs $(2.7 \times 10^3)(10^{-7})(9.8) \simeq 2.65 \times 10^{-3}$ newtons, which is almost equal to the electrostatic force. In other words, the electrostatic force is barely able to lift such a piece of aluminum. Note, however, that the electrostatic force does not depend on the thickness of the material and is the same for any 1 cm $\times$ 1 cm piece of material having a macroscopic thickness of larger than, say 10^{-6} m. Thus, the electrostatic force produced at the surface of aluminum by an electric field at the air breakdown level is larger than the gravitational force for conductor thicknesses of less than or equal to $\sim$1 mm.

Example 4.22: Can electrostatic forces break aluminum? The tensile strength for commercial hard-drawn aluminum is 2.9×10^8 N-m^{-2}. Find (a) the charge density ρ_s required to apply enough force to break aluminum, (b) the corresponding number of excess electrons in a surface region of area 1 μm $\times$ 1 μm and thickness one angstrom, (c) the corresponding amount of total surface charge for an aluminum sphere of radius 10 cm, and (d) the corresponding value of the electric field E_n just outside the surface.

Solution:

(a) Since the force per unit area is $\rho_s^2/(2\epsilon_0)$, the surface charge density corresponding to 2.9×10^8 N-m^{-2} is

$$\rho_s = \sqrt{2 \times 8.854 \times 10^{-12} \times 2.9 \times 10^8} \simeq 7.17 \times 10^{-2} \quad \text{C-m}^{-2}$$

(b) Now consider the surface region of area A $= 10^{-6} \times 10^{-6}$ m^2. The total number of excess electrons in such an area is

$$\frac{\rho_s A}{|q_e|} = \frac{(7.2 \times 10^{-2} \text{ C-m}^{-2})(10^{-6} \times 10^{-6} \text{ m}^2)}{1.6 \times 10^{-19} \text{ C}} \simeq 4.48 \times 10^5 \text{ electrons}$$

which is a small number compared with the total number of available free electrons in such a volume, on the order of $\sim 10^8$, as determined in Example 4.20. It thus appears that a relatively small number of excess electrons is required for the electrostatic force to break aluminum.

(c) On a sphere of radius $a = 10$ cm, the charge density of 7.2×10^{-2} C-m^{-2} corresponds to a total charge of $(7.2 \times 10^{-2}$ C-m$^{-2})(4\pi a^2) \simeq 9.0 \times 10^{-3}$ coulombs. While this amount of charge may not seem to be large, we shall see later[86] that placing such a charge on an aluminum sphere of 10 cm radius requires the application of a potential difference (between the sphere and the ground) of $\sim 8.1 \times 10^8$ volts, or ~ 800 megavolts!

(d) The electric field E_n just outside the surface corresponding to a $\rho_s \simeq 7.2 \times 10^{-2}$ C-m^{-2} is

$$E_n = \frac{\rho_s}{\epsilon_0} \simeq \frac{7.2 \times 10^{-2}}{8.854 \times 10^{-12}} \simeq 8.1 \times 10^9 \text{ V-m}^{-1}$$

which is well above the breakdown field for air of 3×10^6 V-m^{-1}. Thus, it is completely impractical to attain charge densities large enough to break metallic conductors such as aluminum electrostatically.[87]

4.8 POISSON'S AND LAPLACE'S EQUATIONS

We noted in Section 4.3 that *if the positions of all the charges* (discrete or continuous) *are known,* the solution of electrostatic problems boils down to carrying out appropriate integration operations. We also noted that, often the locations of the charges are *not known,* and that we know only the general rules (i.e., boundary conditions) by which charges organize themselves under the influence of fields. We can appreciate this point better now that we have seen (Example 4.20) that slight rearrangements of free charge in conducting bodies can create very large electric fields. In practical problems, often the constraints on the field distributions are not expressed in terms of the charge distributions but rather are given in terms of the electrostatic potentials at which the various conducting bodies are held. In such cases, it is convenient to determine the electric field distributions by solving the equations written directly in terms of the electrostatic potential Φ. These equations, easily derived from Gauss's law as will be shown, are Poisson's and Laplace's equations and are the subjects of this section.

In Section 4.6, we derived the differential equation (4.36), Gauss's law at a point, which implies in turn that

$$\boxed{\nabla \cdot \mathbf{D} = \rho \quad \rightarrow \quad \nabla \cdot \mathbf{E} = \frac{\rho}{\epsilon_0}} \tag{4.43}$$

which is valid for free space, for which ϵ_0 is a simple constant, so that it can be safely taken out of the spatial derivative. We also know from Section 4.4 that the electrostatic

---

[86] Any configuration of conductors has a finite *capacity* to hold charge per unit applied voltage, as we shall study in a later section on capacitance.

[87] However, with a similar result for magnetic fields, this type of consideration becomes extremely important in the design of high-field magnetic coils.

field is the negative gradient of the electrostatic potential Φ. Thus, we have

$$\nabla \cdot (-\nabla \Phi) = \frac{\rho}{\epsilon_0}$$

$$\nabla \cdot (\nabla \Phi) = -\frac{\rho}{\epsilon_0}$$

$$\left(\hat{x}\frac{\partial}{\partial x} + \hat{y}\frac{\partial}{\partial y} + \hat{z}\frac{\partial}{\partial z}\right) \cdot \left(\hat{x}\frac{\partial}{\partial x} + \hat{y}\frac{\partial}{\partial y} + \hat{z}\frac{\partial}{\partial z}\right)\Phi = -\frac{\rho}{\epsilon_0}$$

$$\left(\frac{\partial^2}{\partial x^2} + \frac{\partial^2}{\partial y^2} + \frac{\partial^2}{\partial z^2}\right)\Phi = -\frac{\rho}{\epsilon_0}$$

$$\nabla^2 \Phi = -\frac{\rho}{\epsilon_0} \qquad (4.44)$$

Equation (4.44) is known as *Poisson's equation.* For a charge-free region of space, where $\rho = 0$, it reduces to *Laplace's equation,* namely

$$\boxed{\nabla^2 \Phi = 0} \qquad (4.45)$$

The notation ∇^2 is referred to as the *Laplacian* and represents the operation

$$\nabla^2(\cdot) \equiv \nabla \cdot \nabla(\cdot) = \nabla \cdot [\nabla(\cdot)] \qquad (4.46)$$

If the electric field is known everywhere in a region where metallic conductors are present, the charges on the surfaces of the conductors can be determined using (4.42). Conversely, if the surface charge densities on the metallic conductors is given, the electric field $\mathbf{E}$ can be easily calculated by using (4.11) or (4.43). However, in typical electrostatic problems, neither $\rho_s(x, y, z)$ nor $\mathbf{E}(x, y, z)$ is likely to be known *a priori*. The much more common electrostatic problem involves a given geometrical configuration of charged conductors[88] in space otherwise free of charge. The objective is usually to solve for the electrostatic potential $\Phi(x, y, z)$ that satisfies Laplace's equation along with the boundary conditions on charged conductors. The boundary conditions are usually in the form of $\Phi_i(x, y, z) = \text{const.}$, on the different metallic conductor surfaces $S_i(x, y, z)$. In addition, the interior of the metallic conducting bodies must also be at the same potential as their surfaces. In other words, the problem specification consists of the geometry of the conductor configuration and the potentials of the different conductors, and the goal is to obtain the scalar potential $\Phi(x, y, z)$ by solving the partial differential equation (4.44). Problems of this type are appropriately called *boundary value problems.* An important property of such problems is expressed by the so-called uniqueness theorem[89]. According

---

[88]Conductors are usually charged by applying a potential difference between two or more conductors or between a conductor and ground.

[89]For a derivation, see Appendix B. Many other proofs, basically all along the same lines, can be found in other texts. For example, see Chapter 8 of R. F. Harrington, *Introduction to Electromagnetic Engineering*, McGraw-Hill, New York, 1958.

to this theorem, if the potential[90] is specified on all surfaces enclosing the region of interest (e.g., the boundaries may consist of conductors held at known potentials), if a solution $\Phi(x, y, z)$ that satisfies (4.44) can be found, then it is the only solution, regardless of the method by which it is found.

Relatively simple analytical solutions of (4.45) can be found only in a few cases involving symmetries of one form or another. A variety of powerful methods of mathematical physics have been used to find solutions for more complicated geometries in terms of infinite sums of cylindrical and spherical harmonics.[91] A host of powerful numerical methods has also been employed and specifically developed[92] for the solution of (4.45). As digital computer speed and memory capacity continue to increase, numerical solutions are becoming increasingly attractive, utilizing a variety of powerful techniques for the solution of partial differential equations subject to boundary conditions.[93] In this section we confine our attention to examples of cases that lend themselves to a direct analytical solution.

Example 4.23: Two parallel plates. Consider two parallel conducting planes of infinite extent in the x and y directions and separated by a distance d in the z direction, as shown in Figure 4.45. A potential difference is applied between the plates, such that the bottom plate, located in the xy plane, is at zero potential while the top plate, located at $z = d$, is at potential V_0 (i.e., $\Phi(x, y, 0) = 0$ and $\Phi(x, y, d) = V_0$). The region between the plates is free space and contains no free charges. (a) Find $\Phi(x, y, z)$ and $\mathbf{E}(x, y, z)$ in the region between the plates. (b) Find the surface charge densities induced on the two plates.

Solution:

(a) Laplace's equation (4.45) in rectangular coordinates is

$$\nabla^2 \Phi = \frac{\partial^2 \Phi}{\partial x^2} + \frac{\partial^2 \Phi}{\partial y^2} + \frac{\partial^2 \Phi}{\partial z^2} = 0$$

It is clear from the symmetry of the problem that the potential does not vary with x or y but instead depends only on z. Hence Laplace's equation reduces to

$$\nabla^2 \Phi = \frac{\partial^2 \Phi}{\partial z^2} = 0$$

[90] As shown in Appendix B, another consequence of the uniqueness theorem is that if the directional derivative normal to the bounding surface is specified, then the electrostatic field is uniquely determined. This condition is satisfied if the boundaries consist of conductors with known total charge.

[91] W. R. Smythe, *Static and Dynamic Electricity,* revised printing, Hemisphere Publishing Corp., Washington, DC, 1989; J. Van Bladel, *Electromagnetic Fields,* revised printing, Hemisphere Publishing Corp., Washington, DC, 1985; R. E. Collin, *Field Theory of Guided Waves,* revised printing, IEEE Press, Piscataway, New Jersey, 1991.

[92] R. Mittra, (Ed.) *Computer Techniques for Electromagnetics,* Pergamon, Oxford, p. 7, 1973; K. Umashankar and A. Taflove, *Computational Electromagnetics,* Artech House, Boston, Massachusetts, 1993.

[93] These techniques include the so-called finite-element methods, finite-difference methods, Fourier transformations, and the method of moments. For a description of these techniques and their applications see P. P. Silvester and R. L. Ferrari, *Finite Elements for Electrical Engineers,* Cambridge University Press, Cambridge, UK, 1983, and R. F. Harrington, *Field Computation by Moment Methods,* IEEE Press, Piscataway, New Jersey, 1993.

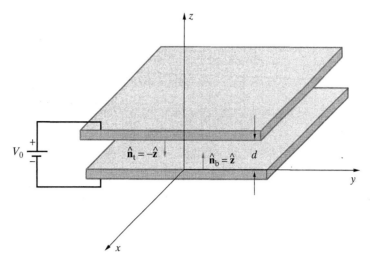

Figure 4.45   Two parallel plates. Two infinite-extent parallel plates of potential difference V_0 and separation distance d (Example 4.23).

which can be directly solved by double integration,

$$\Phi(z) = C_1 z + C_2$$

where C_1 and C_2 are the integration constants to be determined by the boundary conditions given in the problem specification. Namely,

$$\left.\begin{array}{r} \Phi(z=0) = 0 \\ \Phi(z=d) = V_0 \end{array}\right\} \quad \rightarrow \quad \begin{array}{l} C_2 = 0 \\ C_1 = \dfrac{V_0}{d} \end{array}$$

Therefore, the potential $\Phi(z)$ is

$$\Phi(z) = \frac{V_0}{d} z \tag{4.47}$$

which varies linearly from 0 to V_0 in the region between the plates. The corresponding electric field can be found by taking the negative gradient of $\Phi(z)$, namely,

$$\mathbf{E} = -\nabla\Phi(z) = -\hat{\mathbf{z}}\frac{V_0}{d} \tag{4.48}$$

which indicates that the electric field is constant in the region between the plates and is directed toward the plate having the lower potential.

(b) The surface charge densities induced on the bottom ($\rho_{s,b}$) and top ($\rho_{s,t}$) conducting plates can be found as

$$\rho_{s,b} = \epsilon_0(\hat{\mathbf{n}}_b \cdot \mathbf{E}) = \epsilon_0\left[\hat{\mathbf{z}} \cdot \left(-\hat{\mathbf{z}}\frac{V_0}{d}\right)\right] = -\epsilon_0\frac{V_0}{d}$$

$$\rho_{s,t} = \epsilon_0(\hat{\mathbf{n}}_t \cdot \mathbf{E}) = \epsilon_0\left[-\hat{\mathbf{z}} \cdot \left(-\hat{\mathbf{z}}\frac{V_0}{d}\right)\right] = +\epsilon_0\frac{V_0}{d}$$

Note that the unit vectors $\hat{n}_b$ and $\hat{n}_t$ are defined to be the *outward* normals of the two plates as shown in Figure 4.45.

Example 4.24: Coaxial cylinders. The geometry of this problem is very similar to that in Example 4.15, where the surface charge densities on each of two coaxial cylindrical surfaces were specified and the electric field was directly evaluated using Gauss's law. In this case, illustrated in Figure 4.46, the problem specification is that the surfaces are conducting cylinders with a potential difference applied between the two cylinders such that

$$\Phi(r = a) = V_0 \qquad \text{and} \qquad \Phi(r = b) = 0$$

Determine the potential $\Phi(r,\phi,z)$ by direct solution of (4.45) and subsequently the electric field from $\mathbf{E}(r,\phi,z) = -\nabla\Phi(r,\phi,z)$. We are interested only in the electric field and potential in the region between the cylinders, which is taken to be free space.

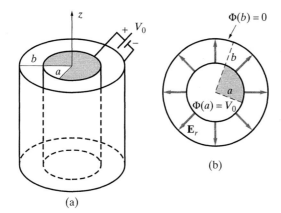

(a)

(b)

Figure 4.46 Two concentric cylinders. (a) Two concentric cylindrical conductors of radii a and b $(a < b)$. (b) Cross-sectional view with the electric field lines shown.

Solution: Note that cylindrical coordinates are most appropriate in view of the cylindrical conductor boundaries. Equation (4.45) written in cylindrical coordinates (as can be found using the definition of the nabla operator as given in (4.46)) is

$$\nabla^2\Phi = \frac{1}{r}\frac{\partial}{\partial r}\left(r\frac{\partial\Phi}{\partial r}\right) + \frac{1}{r^2}\frac{\partial^2\Phi}{\partial\phi^2} + \frac{\partial^2\Phi}{\partial z^2} = 0$$

Assuming that the conductors are sufficiently long, we do not expect any dependence of the potential on z, except near the ends, which we shall neglect for our purposes here. Also, there cannot be any dependence of Φ on ϕ, because of the cylindrical symmetry. Thus, we have

$$\frac{1}{r}\frac{\partial}{\partial r}\left(r\frac{\partial\Phi}{\partial r}\right) = 0$$

This equation can be solved by two successive direct integrations, yielding

$$\Phi(r) = C_1 \ln r + C_2$$

where C_1 and C_2 are the integration constants, which may be determined from the boundary conditions given in the problem specification. Namely, we have

$$\Phi(r = a) = V_0 = C_1 \ln a + C_2$$

$$\Phi(r = b) = 0 = C_1 \ln b + C_2$$

From these two equations, we can solve for C_1 and C_2 to find

$$C_1 = \frac{V_0}{\ln(a/b)} \quad \text{and} \quad C_2 = -\frac{V_0 \ln b}{\ln(a/b)}$$

which fully specifies the potential $\Phi(r)$ as

$$\Phi(r) = \frac{V_0}{\ln(a/b)}(\ln r - \ln b) = \frac{V_0}{\ln(a/b)} \ln \frac{r}{b}$$

which is identical to the electrostatic potential $\Phi(r)$ found in Example 4.15, when we note that $\Phi_{ba} = V_0$. Note that the two solutions are identical only because the cylindrical surface charge densities in the specification of Example 4.15 were appropriately chosen.

The electric field in the region between the conducting cylinders can be found by taking the negative gradient of the potential. Since $\Phi(r)$ is only a function of r, we have

$$\mathbf{E} = \hat{\mathbf{r}} E_r = -\hat{\mathbf{r}} \frac{\partial \Phi}{\partial r} = \hat{\mathbf{r}} \frac{V_0}{r \ln(b/a)}$$

The corresponding surface charge density ρ_{si} on the inner conductor is

$$\rho_{si} = \epsilon_0 \hat{\mathbf{n}} \cdot \mathbf{E}(r = a) = \epsilon_0 \hat{\mathbf{r}} \cdot \hat{\mathbf{r}} \frac{V_0}{a \ln(b/a)} = \frac{V_0 \epsilon_0}{a \ln(b/a)}$$

where $\hat{\mathbf{n}}$ is the outward normal. On the outer conductor we have

$$\rho_{so} = \epsilon_0 \hat{\mathbf{n}} \cdot \mathbf{E}(r = b) = \epsilon_0 (-\hat{\mathbf{r}}) \cdot \hat{\mathbf{r}} \frac{V_0}{b \ln(b/a)} = \frac{-V_0 \epsilon_0}{b \ln(b/a)} = -\rho_{si}\left(\frac{a}{b}\right)$$

just as was specified in the statement of Example 4.15.

Example 4.25: Potential around a conducting sphere. Consider a spherical conductor that is maintained at a potential of $\Phi = V_0$ by means of a battery connection as shown in Figure 4.47. Noting that spherical coordinates are appropriate, find $\Phi(r, \theta, \phi)$ and $\mathbf{E}(r, \theta, \phi)$. Assume that the region around the sphere is free space and that no other charges or conductors are nearby, so that we have $\Phi \to 0$ as $r \to \infty$.

Solution: Laplace's equation in spherical coordinates is

$$\nabla^2 \Phi = \frac{1}{r^2} \frac{\partial}{\partial r}\left(r^2 \frac{\partial \Phi}{\partial r}\right) + \frac{1}{r^2 \sin\theta} \frac{\partial}{\partial \theta}\left(\sin\theta \frac{\partial \Phi}{\partial \theta}\right) + \frac{1}{r^2 \sin^2\theta} \frac{\partial^2 \Phi}{\partial \phi^2} = 0$$

In view of the spherical symmetry, Φ cannot be a function of θ or ϕ, so we have

$$\frac{1}{r^2} \frac{\partial}{\partial r}\left(r^2 \frac{\partial \Phi}{\partial r}\right) = 0 \quad \to \quad \frac{\partial}{\partial r}\left(r^2 \frac{\partial \Phi}{\partial r}\right) = 0$$

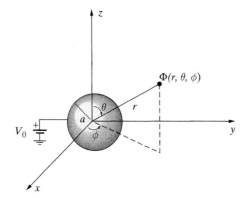

Figure 4.47   Potential around a conducting sphere. The sphere of radius a is held at a potential (with respect to ground) of V_0. Also shown are the spherical coordinates r, θ, ϕ.

Integrating twice with respect to r, we find

$$r^2 \frac{\partial \Phi}{\partial r} = C_1 \quad \rightarrow \quad \Phi(r) = -\frac{C_1}{r} + C_2$$

where C_1 and C_2 are integration constants to be determined by the boundary conditions. One boundary condition is that $\Phi(r = a) = V_0$, which gives

$$-\frac{C_1}{a} + C_2 = V_0 \quad \rightarrow \quad \Phi(r) = \frac{V_0 a}{r} + C_2 \left(1 - \frac{a}{r}\right)$$

To determine C_2, we need a second boundary condition. For a sphere of finite size, we would expect the potential $\Phi(r)$ to vanish as $r \to \infty$, which in turn implies that $C_2 = 0$. Our solution is then

$$\Phi(r) = \frac{V_0 a}{r} \quad \rightarrow \quad \mathbf{E} = -\hat{\mathbf{r}} \frac{\partial \Phi}{\partial r} = \hat{\mathbf{r}} \frac{V_0 a}{r^2}$$

This solution can be compared with what can be found using Gauss's law, if we were to assume that raising the sphere to a potential V_0 deposits a charge Q on it. In such a case, the electrostatic potential for $r > a$ would be

$$\Phi(r) = \frac{Q}{4\pi \epsilon_0 r}$$

For the two solutions to be identical, we must have

$$Q = V_0 (4\pi \epsilon_0 a)$$

We shall see in the next section that the ratio of Q/V_0 represents the capacity of the sphere to hold charge per unit applied voltage.

4.9 CAPACITANCE

The relation between induced charges on a set of conductors and the resulting potentials in their vicinity depends only on the geometric arrangement of the conductors. If we double the surface charge densities at every point on the conductors (presumably by doubling the applied potential difference between the conductors), the configuration (i.e., spatial variation) of the electric field remains unchanged, but the magnitude of the field is doubled, and hence the work required to take a unit positive test charge from one conductor to another (i.e., potential difference between them) is also doubled.

Consider a single isolated conductor of any shape. If we place a charge Q on it, the charge is distributed in some equilibrium pattern on the surface of the conductor. If we then increase the surface charge density ρ_s everywhere by the same factor, we still have an equilibrium arrangement, since the electrostatic potential is everywhere greater by the same factor. This constant charge-to-potential ratio of an isolated conductor is its *capacitance.* In other words, we define

$$C \equiv \frac{Q}{\Phi} \tag{4.49}$$

where Φ is the potential of the isolated conductor. The units of capacitance are coulombs per volt (C-V^{-1}), or farads (F). In particular, for a spherical conductor of radius a, far away from any other charges, conductors, or ground planes, we have $\Phi(r = a) = \Phi = Q/(4\pi\epsilon_0 a)$ (see Example 4.25), so that we have

$$C = 4\pi\epsilon_0 a \tag{4.50}$$

For example, a sphere of radius $a = 10$ cm would have a capacitance of

$$C = 4\pi \underbrace{(8.854 \times 10^{-12})}_{\text{farads-m}^{-1}} \underbrace{(10 \times 10^{-2})}_{\text{m}} \simeq 11.1 \text{ pF}$$

In Example 4.20 it was found that a surface charge density of $\rho_s \simeq 2.66 \times 10^{-5}$ C-m^{-2} on a metallic conductor would cause electrical breakdown of the surrounding air. Such a charge density uniformly distributed over the surface of a sphere of radius $a = 10$ cm corresponds to a total charge of $Q = \rho_s 4\pi a^2 \simeq 3.34 \times 10^{-6}$ C. Using the capacitance value of $C \simeq 11.1$ pF, we find that placement of that much charge on a 10-cm radius sphere requires $\Phi = Q/C \simeq (3.34 \times 10^{-6})/(11.1 \times 10^{-12}) \simeq 3 \times 10^5$ V, or $\sim$300 kilovolts!

In the case of an isolated conductor, the electric flux that leaves the conductor can be thought to terminate at infinity. If the amount of charge on the conductor is increased, the flux pattern remains the same, but the flux density **D** proportionally increases. If we have a pair of conductors and if these conductors are given equal and opposite charges (by applying a potential difference $\Phi_{12} = \Phi_2 - \Phi_1$ between them[94]), then all of the flux

---

[94]When a battery is connected between the two conductors, each conductor becomes an extended battery terminal. Within $\sim 10^{-19}$ s following the connection, electrons flow from the conductor connected to the positive terminal through the battery and connecting wires and to the conductor connected to the negative terminal, and equilibrium is established. As mentioned before, the time involved here ($\sim 10^{-19}$ s) is the time in which charge is rearranged in a material, which is in the range of 10^{-18} to 10^{-19} s for most metallic conductors. The subject of redistribution of charge is discussed in Section 5.5.

emanating from the positively charged conductor terminates on the negatively charged one. If the potential difference is increased, the magnitude of charge on both conductors, as well as the flux density **D**, also increases, without a change in the distribution of flux. Once again, we see that the amount of charge and the potential difference are proportional to one another, and we define their ratio to be the *capacitance* of the two-conductor configuration:

$$C \equiv \frac{Q}{\Phi_2 - \Phi_1} = \frac{Q}{\Phi_{12}} \qquad (4.51)$$

In (4.51), Q is the amount of charge on one of the conductors, usually taken to be the positive one. Note that the case of the single isolated conductor can be considered as a special case with the second conductor being at "infinity," so that $\Phi_2 = 0$. In summary, capacitance is the measure of the ability of a conductor configuration to hold charge per unit applied voltage between the conductors (i.e., their capacity to store charge).

In the following examples, we evaluate the capacitance of some simple two-conductor configurations, including the parallel-plate and coaxial transmission line configurations that were considered in Section 2.7, the capacitance expressions for which were given in Table 2.2. Although the capacitance expressions for these configurations are derived on the basis of electrostatic formulations, they are nevertheless quite valid up to optical frequencies.[95]

Example 4.26: Parallel-plate capacitor. Consider two parallel conducting plates separated by free space as shown in Figure 4.48. Determine the capacitance C.

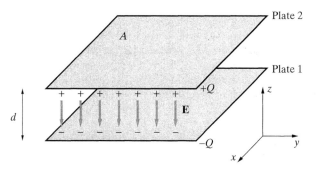

Figure 4.48   Parallel-plate capacitance.

Solution: We start by assuming that the plates are charged so that the upper plate has a total charge of $+Q$ while the lower plate has $-Q$. The charge on both plates is uniformly distributed on the inner surfaces, supporting an electric field pattern as shown. Note that

---

[95]The basic reason for this is the extremely small ($\sim 10^{-19}$ s) time within which charges can redistribute in metallic conductors, as mentioned in Section 4.7 and the previous footnote and further discussed in Section 5.5.

when the smallest linear dimension of the plates is sufficiently larger than the plate separation d, we can neglect "fringing" fields on the sides and assume the electric field between the plates to be identical to that in Example 4.23. If the area of the plates is A, the surface charge density is $\rho_s = Q/A$, positive on the upper plate and negative on the lower one.

Based on the results of Section 4.7.2 and Example 4.23, the electric field between the plates is

$$\mathbf{E} = -\hat{\mathbf{z}}\frac{\rho_s}{\epsilon_0}$$

The potential difference between the plates can be found from

$$\Phi_{12} = -\int_0^d \left(-\hat{\mathbf{z}}\frac{\rho_s}{\epsilon_0}\right) \cdot (\hat{\mathbf{z}}\, dz) = -\left[-\frac{\rho_s}{\epsilon_0}z\right]_0^d = \frac{\rho_s d}{\epsilon_0} = \frac{Qd}{\epsilon_0 A}$$

The capacitance is then

$$C = \frac{Q}{\Phi_{12}} = \frac{Q}{Qd/(\epsilon_0 A)} = \frac{\epsilon_0 A}{d}$$

We note that the capacitance is a function only of the geometry of the configuration (i.e., A and d). If the region between the conductors is an insulating material other than air, then the constant ϵ_0 is replaced by another constant characteristic to the material (Section 4.10). Thus, capacitance is also a function of the material medium around the conductors.

Example 4.27: Coaxial capacitor. Consider the coaxial capacitor, consisting of two concentric cylinders separated by free space, as shown in Figure 4.35 in connection with Example 4.15. Find the capacitance per unit length.

Solution: We can find the capacitance of this configuration by the usual method: (1) Assume a uniform surface charge density ρ_s on the inner conductor (i.e., a total charge of $Q = \rho_s(2\pi a h)$, where h is the height of the cylinder); (2) find $\mathbf{E}$ in terms of ρ_s; (3) integrate $\mathbf{E}$ to find the potential difference between the inner conductor ($r = a$) and the outer conductor ($r = b$), denoted Φ_{ba}, in terms of ρ_s; and (4) take the ratio of Q and Φ_{ba}. However, in this case we already have the solution of the problem worked out in Example 4.15, where two different expressions were found for $\Phi(r)$:

$$\Phi(r) = -\frac{\rho_s a}{\epsilon_0}\ln\frac{r}{b} \quad \text{and} \quad \Phi(r) = \Phi_{ba}\frac{\ln(r/b)}{\ln(a/b)}$$

Equating the two expressions for $\Phi(r)$ we find

$$-\frac{\rho_s a}{\epsilon_0}\ln\frac{r}{b} = \Phi_{ba}\frac{\ln(r/b)}{\ln(a/b)} \quad \rightarrow \quad -\frac{Q}{2\pi\epsilon_0 h} = -\frac{\Phi_{ba}}{\ln(b/a)}$$

Hence we have

$$C = \frac{Q}{\Phi_{12}} = \frac{2\pi\epsilon_0 h}{\ln(b/a)} \quad \rightarrow \quad C_u = \frac{C}{h} = \frac{2\pi\epsilon_0}{\ln(b/a)} \simeq \frac{55.6}{\ln(b/a)} \text{ pF-m}^{-1}$$

which agrees with the expression listed in Table 2.2 as the per-unit-length capacitance of a coaxial line.

As a numerical example, consider a practical coaxial line such as RG-11, which has an inner-conductor diameter of ~ 1.21 mm and the inner diameter of its outer conductor of ~ 7.24 mm. Assuming the conductors to be separated by free space, the capacitance per meter of this line is $C_u \simeq 55.6/[\ln(7.24/1.21)] \simeq 31$ pF-m^{-1}.

Example 4.28: Capacitance of concentric spheres. Consider a spherical capacitor consisting[96] of two concentric conducting spheres as shown in Figure 4.49a. Find the capacitance.

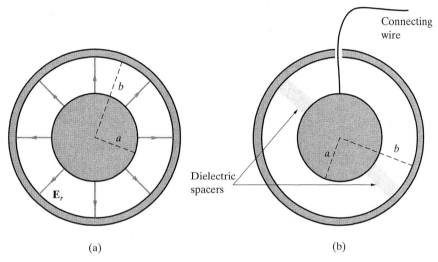

Figure 4.49   A spherical capacitor. (a) The geometry of a spherical capacitor consisting of two concentric conductor spheres of radii a and b. (b) In practice, the inner conductor would need to be supported by insulating spacers and can be accessed by means of a wire through a small hole as shown. The surface charges on the inner and outer spheres are not shown to avoid cluttering the figure.

Solution: If we assume a charge $+Q$ on the inner sphere and $-Q$ on the outer one, the electric field between the two spheres is

$$\mathbf{E} = \hat{\mathbf{r}}E_r = \hat{\mathbf{r}}\frac{Q}{4\pi\epsilon_0 r^2} \qquad a \leq r \leq b$$

consistent with Gauss's law. The potential difference between the spheres, denoted Φ_{ba} (note that Φ_{ba} is positive, since the inner sphere is positively charged) is given by

$$\Phi_{ba} = -\int_b^a \frac{Q}{4\pi\epsilon_0 r^2}\, dr = -\frac{Q}{4\pi\epsilon_0}\int_b^a \frac{dr}{r^2} = \frac{Q}{4\pi\epsilon_0}\left(\frac{1}{a} - \frac{1}{b}\right)$$

[96]Note that, in practice, the inner sphere would need to be held in place by some insulating spacers and also a small opening through the outer sphere is needed so that a wire can be connected to the inner sphere, by means of which a potential difference can be applied between the two spheres, as shown in Figure 4.49b.

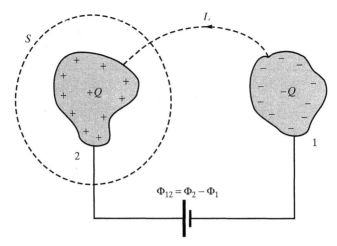

Figure 4.50   **Two arbitrary charged conductors forming a capacitor.**

so that the capacitance is

$$C = \frac{Q}{\Phi_{ba}} = 4\pi\epsilon_0 \frac{ab}{b-a}$$

Note that if we let $b \to \infty$, we get the capacitance of an isolated sphere discussed earlier. In such a case, the flux lines originating on the positively charged inner conductor terminate at "infinity."

In general, regardless of the conductor configuration, and with reference to Figure 4.50, we can define capacitance as

$$C \equiv \frac{Q}{\Phi_{12}} = \frac{\oint_S \mathbf{D} \cdot d\mathbf{s}}{-\int_L \mathbf{E} \cdot d\mathbf{l}} \tag{4.52}$$

where S is any surface enclosing the positively charged conductor and L is any path from the negative (lower potential) to the positive (higher potential) conductor.

The general procedure for determining the capacitance of any two-conductor configuration can be summarized as follows: (1) Choose a coordinate system appropriate to the geometrical layout and shapes of the two conductors. (2) Assume charges $+Q$ and $-Q$ on the conductors. (3) Find $\mathbf{E}$ from Gauss's law, by direct integration or by other methods; the result will be proportional to Q. (4) Integrate $\mathbf{E}$ along any path between the two conductors to determine Φ_{12}; result will be proportional to the assumed charge Q. (5) Find C by taking the ratio Q/Φ_{12}; note that Q will cancel out and the result will depend only on the geometry of the conductors.

Example 4.29: Capacitance of the two-wire line. A rather practical transmission line configuration used in telephony, radio engineering, and electric power distribution is the two-wire line, consisting of two parallel cylindrical conductors, each of radius a, separated by a distance d,

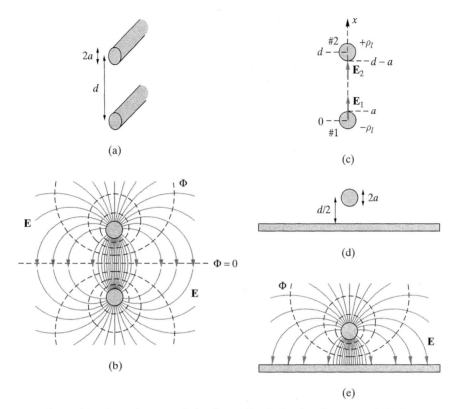

Figure 4.51   Two-wire transmission line and a single wire above ground. (a) The two-wire line. (b) Electric field configuration. (c) Representation in terms of two line charges. (d) Single wire above ground. (e) Field configuration for single wire above ground.

as shown in Figure 4.51a. General solution of the capacitance for this configuration is more involved than that for the coaxial cable, since, in general, the charge distribution on the conductor surfaces is not uniform, because of the Coulomb attraction between the induced surface charges on the two conductors. This so-called "proximity effect" causes the charge density to be larger (smaller) on the sides of each conductor facing toward (away from) the other. The resultant distribution of the electric field lines and equipotentials is shown in Figure 4.51b. Closed-form analytical expressions for the electric field and potential can be found using various methods. Find a simplified expression of the capacitance of the two-wire line, valid for cases where $d \gg a$. Note that this approximation is indeed quite valid in most cases, including the well-known examples of the power transmission lines (for which a is a few cm, while d is many meters) and the TV antenna wire ($a < 1$ mm and $d \sim 1$ cm).

Solution: For $d \gg a$, the nonuniformity of the charge distribution on the conductor surfaces can be neglected and thus the electric field outside the conductors is the same as what would be produced by two line charges of opposite polarity (ρ_l and $-\rho_l$). From Example 4.11 we know that the electric field at a distance r from a line charge of linear charge density ρ_l

is given as $E_r = \rho_l/(2\pi\epsilon_0 r)$. Taking the $+\rho_l$ and $-\rho_l$ line charges to be located at $r = d$ and $r = 0$, respectively, the electric field as a function of distance x along the x axis (i.e., $y = 0$ and $z = 0$) is given as

$$E_x(x,0,0) = \frac{-\rho_l}{2\pi\epsilon_0 x} + \frac{\rho_l}{2\pi\epsilon_0(x-d)}$$

To find the capacitance using its general definition (4.52), we need to integrate the electric field *along any path* from one conductor to the other. It is obvious from Figure 4.51c that one convenient integration path[97] is along the x axis, from $x = a$ to $x = d - a$. We find

$$\Phi_{12} = -\frac{1}{2\pi\epsilon_0}\int_a^{d-a}\left[\frac{-\rho_l}{x} + \frac{+\rho_l}{x-d}\right]dx = \frac{\rho_l}{2\pi\epsilon_0}\int_a^{d-a}\left[\frac{1}{x} + \frac{1}{d-x}\right]dx$$

$$= +\frac{\rho_l}{2\pi\epsilon_0}[\ln x - \ln(d-x)]_a^{d-a} = \frac{\rho_l}{2\pi\epsilon_0}\left[\ln\left(\frac{x}{d-x}\right)\right]_a^{d-a}$$

$$= \frac{\rho_l}{2\pi\epsilon_0}\left[\ln\left(\frac{d-a}{d-(d-a)}\right) - \ln\left(\frac{a}{d-a}\right)\right] = \frac{\rho_l}{2\pi\epsilon_0}2\ln\left(\frac{d-a}{a}\right)$$

$$\simeq \frac{\rho_l}{\pi\epsilon_0}\ln\frac{d}{a}$$

since $d \gg a$. Noting that the total charge on one of the conductors of unit length is $Q = \rho_l$, the *per unit length* capacitance C_u of the two-wire line is

$$C_u = \frac{\rho_l}{|\Phi_{12}|} \simeq \frac{\pi\epsilon_0}{\ln(d/a)} \simeq \frac{27.8}{\ln(d/a)} \quad \text{pF-m}^{-1}$$

As an example, consider the capacitance per kilometer between two Drake-type steel-reinforced aluminum conductors (ACSR) of a 115-kV power transmission line, separated by $d = 3$ m. For Drake ACSR, we have $a \simeq 1.407$ cm, so that the approximation $d \gg a$ is valid. Using the above, we find $C_u \simeq 5.19$ nF-(km)$^{-1}$.

The above expression is valid for $d \gg a$. A more accurate analysis for the general case yields

$$C_u = \frac{\pi\epsilon_0}{\ln[d/2a + \sqrt{(d/2a)^2 - 1}]} \simeq \frac{27.8}{\ln[d/2a + \sqrt{(d/2a)^2 - 1}]} \quad \text{pF-m}^{-1}$$

which is the expression listed in Table 2.2 for the per-unit-length capacitance of a two-wire line.

It is clear from (4.52) that the capacitance of a two-conductor configuration is primarily dependent on the electric field distribution. This realization allows us to simply infer the capacitance of some conductor configurations from others for which we have derived expressions. A good example is the case of a single cylindrical conductor above a ground plane, as shown in Figure 4.51d. We can simply obtain this configuration by placing an infinitely large conducting sheet (i.e., ground) coincident with the flat equipotential

---

[97]Note that the electric field at points other than along the x axis in general has components in both the x and y directions, as can be determined by a vector superposition of $\mathbf{E}_1$ and $\mathbf{E}_2$. However, since Φ_{12} can be found by integrating the total electric field along *any* path between the two conductors, it is convenient to choose this path along the x axis, where the electric field has only one nonzero component (i.e., E_x).

plane of the two-wire line configuration of Figure 4.51b. Such a conducting plane does not influence the existing field distribution, since the electric field lines are already perpendicular to it at all points. For a given value of ρ_l, the electric field everywhere is the same for the single-wire above-ground case as that for the two-wire line. However, if the charge density of ρ_l corresponded to a potential difference of V_0 between the two wires, the potential difference between a single conductor and the ground plane (for the same ρ_l) is $V_0/2$. In other words, the capacitance per unit length of the configuration shown in Figure 4.51d is

$$C_{u,\text{single wire above ground}} = 2C_{u,\text{two-wire line}} \simeq \frac{2\pi\epsilon_0}{\ln(d/a)} \tag{4.53}$$

Another way to analyze the problem of the single wire above a ground plane is the so-called method of images, which involves the replacement of boundary surfaces by appropriately placed "image" charges instead of a formal solution of Laplace's equation. In the case in hand, the conducting plane can be replaced by an image line charge of line charge density $-\rho_l$ at a distance $d/2$ below the plane so that $\Phi = 0$ on the ground plane. The electric field at any given point is then simply a superposition of the field due to the original line charge (single wire) and the image line charge, which of course is entirely equivalent to the case of the two-wire line. The method of images is a powerful method for finding the fields produced by charges in the presence of dielectric and conducting boundaries with certain symmetries.[98]

4.10 DIELECTRIC MATERIALS

An insulator is a substance whose electrons and ions cannot move over macroscopic distances under the influence of an applied field. Although no material is a perfect insulator, in this section we deal with materials within which the motion of charges over macroscopic distances is negligible. Insulators are said to be *polarized* when the presence of an applied field displaces the electrons within a molecule away from their average positions. When we consider the polarizability of insulators, we refer to them as *dielectrics*.[99] Although all substances are polarizable, the effects of polarizability can be readily observed only when the material does not conduct electricity—that is, when it is an insulator.

---

[98] See, for example, Section 4.4 of D. K. Cheng, *Field and Wave Electromagnetics,* 2nd ed., Addison-Wesley, Reading, Massachusetts, 1989.

[99] The word *dielectric* was coined by Faraday, who extensively studied the behavior of different insulators placed between electrodes and found that the capacitances of spherical capacitors filled with dielectrics were higher. His results and insights were documented in *Experimental Researches in Electricity,* Vol. 1, Section 1252–1270, B. Quatrich, London, 1839. For a historical overview and for an excellent and thorough discussion of dielectric materials see R. S. Elliott, *Electromagnetics,* IEEE Press, Piscataway, New Jersey, 1993. Another excellent reference on the topic of dielectrics is A. von Hippel (Ed.), *Dielectric Materials and Applications,* Artech House, Boston, 1995.

4.10.1 Polarizability

In Section 4.7 we saw (e.g., Example 4.20) that redistribution of charges in a metallic conductor involves the transfer of a very small percentage of the amount of free charge in a conductor over macroscopic distances. Polarization of dielectrics, on the other hand, involves the displacement of one or more electrons per atom over subatomic or microscopic distances. The atoms of dielectric materials have their outermost electron shells almost completely filled. As a consequence, the electrons are tightly bound, and only a negligible number of electrons are available for conduction of electric current. A measure of the conducting ability of a material is the charge rearrangement time, which as we shall see in Chapter 5 is $\sim 10^{-19}$ s for metallic conductors. For a good dielectric, such as fused quartz, the charge rearrangement time is ~ 50 days. Thus, free charge deposited inside quartz stays in place for all practical purposes and can be considered as bound charge. The difference in the electrical behavior of dielectrics versus conductors is essentially due to the different numbers of free versus bound charges, with semiconductors exhibiting intermediate behavior.

When a dielectric material is placed in an electric field, the electrons respond by shifting with respect to the nuclei in the direction opposite to the applied field, essentially establishing many small electric dipoles, as shown in Figure 4.52. Note that electric dipoles aligned with the field are produced in both polar and nonpolar materials. *Nonpolar* materials consist of molecules that do not possess a permanent dipole moment; the external field both induces the dipoles and orients them as shown in Figure 4.52a.

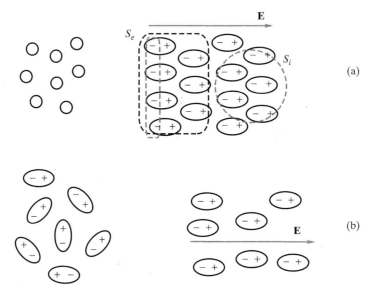

Figure 4.52 Dipoles oriented in an applied field. (a) Nonpolar molecules with no permanent dipole moment; the external field induces the dipoles and orients them along the field. (b) Polar molecules have permanent dipole moments randomized by thermal agitation; the applied field aligns the dipoles.

The molecules of *polar* materials (e.g., NaCl) have a permanent dipole moment, but the individual molecular dipoles are usually randomly oriented due to thermal agitation, as shown in Figure 4.52b; when an applied field is present, the individual dipoles tend to align in the direction of the field.

It is important to note that polarization does not produce net charge inside the dielectric. Any interior volume of macroscopic dimensions[100] (e.g., that enclosed by the surface S_i in Figure 4.52a) contains equal amounts of positive and negative charge. However, a net amount of surface polarization charge does appear on the surface of the polarized dielectric, as shown in Figure 4.52a for the volume enclosed by the surface S_e. We see that a layer of charge adjacent to the boundary remains unneutralized and appears as polarization charge with bound surface charge density ρ_{sp}. We shall see that the amount of surface charge is a direct indication of the degree of polarization of a material and that ρ_{sp} is proportional to the magnitude of the electric field.

Electronic polarizability.   Figure 4.53a shows a very simplified model of the atom, consisting of a positively charged nucleus surrounded by a negatively charged spherical electron cloud.[101] With an applied steady field the nucleus and the electron cloud are displaced (as illustrated in Figure 4.53b) until their mutual attractive force is just equal to the force due to the applied field. We can thus calculate the amount of displacement r by equating the two forces.

We can treat the electron cloud with radius a as a spherical cloud of charge much like that in Example 4.13, with uniform volume charge density $\rho = 3q_e/(4\pi a^3)$, as shown in Figure 4.53a. Note that a is the effective atomic radius. The electric field due to such a spherical cloud of total charge Q at a distance r from its center was derived in Example 4.13 to be $E_r = Qr/(4\pi\epsilon_0 a^3)$ for $r \leq a$, which for the case in hand (replacing Q by $|q_e|$) gives

$$E_r = \frac{|q_e|r}{4\pi\epsilon_0 a^3} \qquad r \leq a$$

When the atom is placed in an external electric field, the electron cloud is displaced with respect to the nucleus as shown in Figure 4.53b. Although the shape of the electron cloud may also be elongated as shown in Figure 4.53c, we assume that this effect is quite small and consider the displaced electron cloud to be spherical as shown in Figure 4.53b. Since the charge on the nucleus is $+|q_e|$, the Coulomb restoring force F_r of the spherical

---

[100]It is important to understand that S_i has *macroscopic* dimensions; otherwise, Figure 4.52 may give the impression that there is net surface charge on the outer edges of S_i just as in the case of S_e. However, since S_i has macroscopic dimensions, it contains a very large number of individual molecules and is thus neutral. Since the precise location of the boundary of S_i is defined only on a macroscopic scale, it does not make sense to think of it as cutting through a molecule of microscopic dimensions.

[101]Note that the nucleus has a diameter of order $\sim 10^{-15}$ m, while that of the electron cloud is $\sim 10^{-10}$ m, so that the nucleus can be considered a point source. Nevertheless, it should be realized that the atomic model shown in Figure 4.53 is highly simplified and corresponds to the early classical model of the atom. In the Bohr model of the atom, the electron orbits at a fixed discrete radius, whereas the modern quantum mechanical model specifies the location of the electron on the basis of a wave function that describes the probability of the electron being located at different radii.

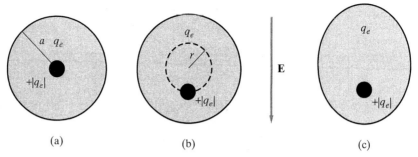

Figure 4.53 Simple model of an atom. (a) The simplified model of an atom with no applied field. (b) Displacement of the nucleus by an applied electric field. (c) The spherical electron cloud may in fact be somewhat elongated as a result of the applied field.

charge cloud acting on the nucleus when it is displaced by an amount r (where $r \le a$) is[102]

$$F_r = E_r(+|q_e|) = \frac{q_e^2 r}{4\pi\epsilon_0 a^3}$$

Equating this force on the nucleus to that due to the external electric field E (which is equal in magnitude but opposite to the direction of E_r), we find the electric dipole moment per atom as

$$p \equiv |q_e| r = 4\pi\epsilon_0 a^3 E$$

or, in vector form,

$$\mathbf{p} = 4\pi\epsilon_0 a^3 \mathbf{E} \quad \text{C-m-atom}^{-1} \tag{4.54}$$

The proportionality between the dipole moment per atom $\mathbf{p}$ and the electric field $\mathbf{E}$ holds only for small applied fields, and thus for small displacements. If the material has N atoms per unit volume, then the dipole moment per unit volume is $\mathbf{P} = N\mathbf{p}$, referred to as the *polarization vector*. We often write $\mathbf{p} = \alpha_e \mathbf{E}$, where $\alpha_e = 4\pi\epsilon_0 a^3$ is called the *electronic polarizability* of the material. Although the atomic model shown in Figure 4.53 is extremely primitive, the result (α_e) is accurate to within a factor of four or so for many simple atoms. With atomic radii a for typical materials being of order 1 angstrom, the numerical value of α_e is of order $\sim 10^{-40}$ F-m^2.

Ionic polarizability.   In some materials, two different atoms may join together as a molecule by forming a chemical bond. We can think of such molecules as consisting of positively and negatively charged ions, with the Coulomb forces between them serving as the binding force. Examples of such materials are hydrochloric acid (HCl), carbon dioxide (CO_2), and water (H_2O). Depending on whether or not the electrons are transferred or shared, the bond can be *ionic* or *covalent*. In both cases, the material may possess a

---

[102]We note that this is a highly simplified argument. Not only is the classical model of the atom not precisely valid, but the displaced electron cloud is in general not circular in shape (Figure 4.53c), and the charge density within the sphere is not uniform.

permanent dipole moment, and the application of an electric field to any such molecule also displaces the positive ions with respect to the negative ones and thereby induces a dipole moment. A similar analysis as given above[103] shows that this process leads to the creation of an average dipole moment per molecule given by

$$\mathbf{p} = \alpha_i \mathbf{E} \quad \text{where} \quad \alpha_i = \frac{64\epsilon_0 d_a}{7(n+1) - 16}$$

where d_a is the average interatomic spacing between the centers of the two ions constituting the molecule; n is a number between 5 and 12, depending on the material[104]; and α_i is referred to as the *ionic polarizability*. For typical values of d_a, the ionic polarizability has values of $\sim 10^{-40}$ F-m^2, comparable to α_e.

Orientational polarizability.   Some polyatomic molecules, such as H_2O, may be at least partially ionic and may consist of *polar* molecules, which carry a permanent dipole moment. With no electric field, the individual dipole moments point in random directions, so the net dipole moment is zero, as shown in the left-hand panel of Figure 4.52b. When an electric field is applied, such materials exhibit the electronic and ionic polarization effects just discussed, but in addition, the electric field tends to line up the individual dipoles to produce an additional net dipole moment per unit volume. If all the dipoles in a material were to line up, the polarization would be very large; however, at ordinary temperatures and relatively small electric field levels, the collisions of the molecules in their thermal motion allow only a small fraction to line up with the field. The resulting effective polarization can be shown[105] to be

$$\alpha_o = \frac{p^2}{3k_\mathrm{B}T}$$

where T is the absolute temperature in kelvins (K), $k_\mathrm{B} \simeq 1.38 \times 10^{-23}$ J-K^{-1} is the Boltzmann's constant, and $p = |q_e|d$ is the molecular dipole moment. For $q_e \simeq -1.602 \times 10^{-19}$ C and $d = 1$ angstrom (10^{-10} m), we have $p = |q_e|d \simeq 1.6 \times 10^{-29}$ C-m, so that at room temperature $\alpha_o \simeq 2 \times 10^{-38}$ m^3, comparable with α_e and α_i. The quantity α_o is referred to as the *orientational polarizability* of a material.

Polarization per unit volume.   The total polarization of a material may arise as a result of electronic, ionic, and orientational polarizability, leading in general to a dipole moment per unit volume of

$$\mathbf{P} = N \underbrace{\left(\alpha_e + \alpha_i + \frac{p^2}{3k_\mathrm{B}T} \right)}_{\alpha_\mathrm{T}} \mathbf{E} \qquad \text{C-m}^{-2} \tag{4.55}$$

---

[103]For a discussion at an appropriate level, see R. S. Elliott, *Electromagnetics,* IEEE Press, Piscataway, New Jersey, 1993, Section 6.7.

[104]L. Pauling, *The Nature of the Chemical Bond,* Cornell University Press, Ithaca, New York, 1948, p. 339.

[105]R. S. Elliott, *Electromagnetics,* IEEE Press, Piscataway, New Jersey, 1993, Section 6.7.

where α_o is explicitly written to emphasize its temperature dependence. Note that while α_e and α_i depend only on atomic and molecular configurations, α_o is inversely proportional to temperature, as well as dependent on the atomic properties.

In general, the electric field $\mathbf{E}$ in (4.55) should not be taken as the external field that exists in the absence of the dielectric, because the presence of a large number density N of neighboring dipoles also contributes to the polarization of a given atom or molecule. The electric field $\mathbf{E}$ in (4.55) should thus be taken to be the total local electric field $\mathbf{E}_{\mathrm{loc}}$, consisting of the external electric field $\mathbf{E}$ plus the molecular field $\mathbf{E}_{\mathrm{mol}}$. The local field $\mathbf{E}_{\mathrm{loc}}$ is the field that actually exists at the molecule. For gases, the separation of the molecules is sufficiently large that $\mathbf{E}_{\mathrm{loc}}$ is closely approximated by the applied field $\mathbf{E}$. For solid dielectrics, the effect of the fields of adjacent molecules cannot be neglected. The molecular field $\mathbf{E}_{\mathrm{mol}}$ and thus the local field $\mathbf{E}_{\mathrm{loc}}$ can be calculated by removing the molecule in question, maintaining all the other molecules in their time-average polarized positions, and calculating the space-averaged electrostatic field in the cavity previously occupied by the molecule. Such an analysis[106] shows that the effective local field acting to polarize the molecule is

$$\mathbf{E}_{\mathrm{loc}} = \mathbf{E} + \frac{\mathbf{P}}{3\epsilon_0}$$

so the total polarization per unit volume $\mathbf{P}$ from (4.55) is given as

$$\mathbf{P} = N\alpha_{\mathrm{T}}\mathbf{E}_{\mathrm{loc}} = N\alpha_{\mathrm{T}}\left(\mathbf{E} + \frac{\mathbf{P}}{3\epsilon_0}\right)$$

where α_{T} is the total polarizability. The above equation can be solved for $\mathbf{P}$ as

$$\mathbf{P} = \frac{N\alpha_{\mathrm{T}}}{1 - N\alpha_{\mathrm{T}}/(3\epsilon_0)}\mathbf{E} \tag{4.56}$$

The concepts of polarizability and the local electric field were introduced only so that we can relate the macroscopic behavior of the fields in a dielectric to the underlying microscopic causes. In practice, it is useful to lump together the various factors contributing to the polarization in the following expression:

$$\boxed{\mathbf{P} = \epsilon_0\chi_e\mathbf{E}} \tag{4.57}$$

where

$$\chi_e = \frac{N\alpha_{\mathrm{T}}/\epsilon_0}{1 - N\alpha_{\mathrm{T}}/(3\epsilon_0)}$$

is a dimensionless quantity called the *electric susceptibility* and $\mathbf{E}$ is the macroscopic field present everywhere in the dielectric.

---

[106]This calculation was first carried out in 1906 by H. A. Lorentz, *Theory of Electrons,* Dover, New York, 1952. For reference, see Section 6.5 and Appendix K of R. S. Elliott, *Electromagnetics,* IEEE Press, Piscataway, New Jersey, 1993; Section 3.9 of R. Plonsey and R. E. Collin, *Principles and Applications of Electromagnetic Fields,* McGraw-Hill, New York, 1961; Chapter 13 of C. Kittel, *Introduction to Solid State Physics,* Wiley, New York, 1971.

A brief discussion of macroscopic versus microscopic fields, along the lines of Section 4.7.1, is in order here. The quantity $\mathbf{P}$, polarization per unit volume, was defined to be $N\mathbf{p}$, where $\mathbf{p}$ is the atomic or molecular dipole moment. In this sense, $\mathbf{P}$ may appear to be a microscopic quantity. If that were the case, its use in (4.57) in connection with the macroscopic electric field $\mathbf{E}$ may be inappropriate. However, we should remember here that matter consists primarily of empty space interspersed with atoms and molecules, as discussed in Section 4.7.1. We note once again that the radii of nuclei are of order $\sim 10^{-15}$ m, whereas the spacing between adjacent atoms is $\sim 10^{-10}$ m. If we looked at the variation of $\mathbf{P}$ on a microscopic spatial scale, we would see it change drastically in magnitude and direction from one atomic nucleus to another, with zero values in the intervening empty space. Fortunately, a description of $\mathbf{P}$ or the electric field $\mathbf{E}$ on such a scale is not of interest to us. When we consider macroscopic dimensions of, say, 1 micron (10^{-6} m), the net polarization $\mathbf{P}$ and the external electric field $\mathbf{E}$ represent the spatial averages over many ($\sim 10^{12}$ atoms for a cube of size 1 μm) atoms, so these quantities vary smoothly over macroscopic distances and can properly be used as macroscopic quantities.

4.10.2 The Permittivity Concept

We have seen in the previous section that the effect of an applied electric field on the atoms or molecules of a dielectric material is to create a dipole moment distribution $\mathbf{P}$. This distribution sets up induced secondary fields, so the net field in the presence of the dielectric is modified from its free-space value. We now proceed to the consideration of fields generated by the induced dipole moments in a dielectric material. For this purpose, we use the methods studied earlier in Section 4.4 to find electrostatic potentials and fields from given distributions of charge. The pertinent expression for the potential is

$$\Phi(\mathbf{r}) = \frac{1}{4\pi\epsilon_0} \int_{V'} \frac{\rho(\mathbf{r}')\,dv'}{|\mathbf{r} - \mathbf{r}'|} \tag{4.58}$$

and the electric field can be found from the potential using $\mathbf{E} = -\nabla\Phi$.

When a dielectric is introduced in an external electric field set up by an arbitrary charge distribution,[107] a dipole moment distribution $\mathbf{P}$ is induced in the volume occupied by the dielectric. This distribution constitutes a secondary electric field source and must be included to determine the complete electric field in the presence of the dielectric.

The electrostatic potential at a faraway point of a z-directed dipole with dipole moment $\mathbf{p} = p\hat{\mathbf{z}}$ located at the origin was found in Section 4.4.3 to be[108]

$$\Phi(\mathbf{r}) = \frac{p\cos\theta}{4\pi\epsilon_0 r^2} = \frac{\mathbf{p}\cdot\hat{\mathbf{r}}}{4\pi\epsilon_0 r^2} = \frac{1}{4\pi\epsilon_0}\left[-\mathbf{p}\cdot\nabla\left(\frac{1}{r}\right)\right]$$

---

[107] Typically the external field is set up by applying a potential difference between two conductors, such as two parallel plates, and we are considering the introduction of the dielectric material between the plates.

[108] Note that in spherical coordinates we have

$$\nabla\left(\frac{1}{r}\right) = \nabla r^{-1} = \hat{\mathbf{r}}\frac{\partial r^{-1}}{\partial r} + \hat{\boldsymbol{\theta}}\frac{1}{r}\frac{\partial r^{-1}}{\partial\theta} + \hat{\boldsymbol{\phi}}\frac{1}{r\sin\theta}\frac{\partial r^{-1}}{\partial\phi} = -\frac{1}{r^2}\hat{\mathbf{r}} + 0 + 0 = -\frac{1}{r^2}\hat{\mathbf{r}}$$

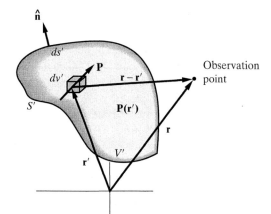

Figure 4.54 A volume distribution of polarization. Note that $\mathbf{P}(\mathbf{r}')$ is the distribution of volume polarization, and $\mathbf{P}$ is the polarization vector.

The above is obviously valid for any orientation of the dipole moment and can be adapted to an arbitrary position $\mathbf{r}'$ of the dipole. Thus, if $|\mathbf{r} - \mathbf{r}'|$ is the distance from the source point (location of the dipole) $\mathbf{r}'$ to the observation point $\mathbf{r}$, the electrostatic potential at a faraway point is

$$\Phi(\mathbf{r}) = \frac{1}{4\pi\epsilon_0}\left[-\mathbf{p}\cdot\nabla\left(\frac{1}{|\mathbf{r}-\mathbf{r}'|}\right)\right] = \frac{1}{4\pi\epsilon_0}\left[\mathbf{p}\cdot\nabla'\left(\frac{1}{|\mathbf{r}-\mathbf{r}'|}\right)\right]$$

where ∇' represents the del operator applied to the source coordinates $(\mathbf{r}')$. Note that since $R = |\mathbf{r} - \mathbf{r}'|$, we have $\nabla R^{-1} = -\nabla' R^{-1}$.

Now consider a large number of infinitesimal dipoles distributed throughout a given volume V' as shown in Figure 4.54. Since $\mathbf{P}$ is the dipole moment per unit volume, the dipole moment in a volume element dv' is $\mathbf{P}\,dv'$ and the potential at the observation point is given by

$$\Phi(\mathbf{r}) = \frac{1}{4\pi\epsilon_0}\left[\int_{V'}\mathbf{P}\cdot\nabla'\left(\frac{1}{|\mathbf{r}-\mathbf{r}'|}\right)dv'\right]$$

We can now use a vector identity,

$$\nabla\cdot(a\mathbf{A}) = a\nabla\cdot\mathbf{A} + \mathbf{A}\cdot\nabla a$$

where $\mathbf{A}$ is any vector and a is any scalar quantity, to rewrite the potential as

$$\Phi(\mathbf{r}) = \frac{1}{4\pi\epsilon_0}\left[\int_{V'}\nabla'\cdot\left(\frac{\mathbf{P}}{|\mathbf{r}-\mathbf{r}'|}\right)dv' - \int_{V'}\frac{\nabla'\cdot\mathbf{P}}{|\mathbf{r}-\mathbf{r}'|}dv'\right]$$

We can further use the divergence theorem (see Section 4.6) to express $\Phi(\mathbf{r})$ as

$$\Phi(\mathbf{r}) = \frac{1}{4\pi\epsilon_0}\left[\oint_{S'}\frac{\mathbf{P}\cdot\hat{\mathbf{n}}}{|\mathbf{r}-\mathbf{r}'|}ds' - \int_{V'}\frac{\nabla'\cdot\mathbf{P}}{|\mathbf{r}-\mathbf{r}'|}dv'\right] \tag{4.59}$$

where $\hat{\mathbf{n}}$ is the outward normal to the surface S' that encloses the volume V', as shown in Figure 4.54. Equation (4.59) allows for a simple and physically based interpretation of electrostatic potential due to a volume distribution of electric dipoles. By comparison with (4.58), we see that a volume distribution of dipoles may be represented as an equivalent volume (ρ_p) and surface (ρ_{sp}) distribution of charges given by

$$\rho_p = -\nabla \cdot \mathbf{P} \qquad (4.60)$$

and

$$\rho_{sp} = \mathbf{P} \cdot \hat{\mathbf{n}} \qquad (4.61)$$

These equivalent volume and surface charge densities can be understood on physical grounds. At points in space where $\nabla \cdot \mathbf{P} \neq 0$, we have creation of a net dipole moment per unit volume, which means that charge density from adjacent dipoles does not completely cancel. The surface charge density ρ_{sp} occurs because the dipoles with one end on the surface cannot be neutralized since they do not have an adjacent dipole layer on that end (see Figure 4.52a).

If in addition there are also free charges[109] within the dielectric medium as represented by volume charge density ρ, the total electrostatic potential is due to the superposition of the effects of bound polarization charges ρ_p and ρ_{sp} and the free charge ρ. In other words, we have

$$\Phi(\mathbf{r}) = \frac{1}{4\pi\epsilon_0}\left[\oint_{S'} \frac{\rho_{sp}}{|\mathbf{r}-\mathbf{r}'|}\,ds' + \int_{V'} \frac{\rho+\rho_p}{|\mathbf{r}-\mathbf{r}'|}\,dv'\right] \qquad (4.62)$$

In general, the polarization charges may react back on the free charges and modify the distribution ρ. In many cases, however, such effects are small.

The introduction of Gauss's law in Section 4.5 was framed in the context of Faraday's experiments with free charges. If we instead keep a separate accounting of the free and bound charges, Gauss's law becomes

$$\epsilon_0 \nabla \cdot \mathbf{E} = \rho + \rho_p \qquad (4.63)$$

Noting that $\rho_p = -\nabla \cdot \mathbf{P}$ we have

$$\nabla \cdot (\epsilon_0 \mathbf{E}) + \nabla \cdot \mathbf{P} = \rho \qquad \rightarrow \qquad \nabla \cdot (\epsilon_0 \mathbf{E} + \mathbf{P}) = \rho$$

This suggests that we can use a displacement density or flux density vector defined as

$$\mathbf{D} = \epsilon_0 \mathbf{E} + \mathbf{P} \qquad (4.64)$$

which permits us to write the familiar form of the differential form of Gauss's law (equation (4.36)) as

$$\nabla \cdot \mathbf{D} = \rho \qquad (4.65)$$

Note that in materials for which (4.57) holds we have

$$\mathbf{D} = \epsilon_0 \mathbf{E} + \epsilon_0 \chi_e \mathbf{E} = (1 + \chi_e)\epsilon_0 \mathbf{E} = \epsilon \mathbf{E} \qquad (4.66)$$

---

[109]Free charge includes all charge that is not the result of polarization, such as free-flowing electrons in a conductor.

where the parameter $\epsilon = (1 + \chi_e)\epsilon_0$ is called the *electrical permittivity* or the *dielectric constant* of the material. Typically we write

$$\epsilon = \epsilon_r \epsilon_0 \tag{4.67}$$

where ϵ_r is a dimensionless quantity called the *relative permittivity* or the *relative dielectric constant.*

Note that (4.66) holds only for a class of dielectrics under certain conditions. Since (4.66) originates in (4.57), it depends primarily on the *linear* relationship between the polarization **P** and the electric field **E**. An important example of nonlinear behavior occurs when the applied electric field is intense enough to pull electrons completely out of their bound locations, causing dielectric breakdown, discussed in Section 4.10. Equation (4.66) also requires that the material be *isotropic;* that is, χ_e must be independent of the direction of **E**. In *anisotropic* materials (e.g., crystals), electric field in one direction can produce polarization (and thus **D**) in another direction. Accordingly, the relation between **D** and **E** must be expressed as a matrix, more commonly called a tensor.[110] Equation (4.66) is valid for both *homogeneous* (where ϵ is the same everywhere in the material) and for *inhomogeneous* materials as long as we allow for ϵ to vary throughout the material as a function of position, expressed as $\epsilon(x, y, z)$.

Some authors refer to the charge density in (4.65) as the "free" charge density and use the symbol ρ_f to distinguish this quantity from the bound polarization charge density ρ_p. However, the distinction between bound charge distributions and all other charge density terms is usually clear from context alone. Since the definition of **D** includes the polarization vector **P**, the physical effects of bound charges, as determined through $-\nabla \cdot \mathbf{P}$, are already included in the specification of **D** and ϵ.

Comparing (4.65) with (4.63), we see the simplification introduced by the notion of electric flux density **D**. From Coulomb's law and the definition of the electric field, **E** is determined by *all* charges, including both free charges and bound charges, the latter resulting from a polarization density **P**. Hence, we see that the charge densities (volume, surface, or line) in equations (4.9), (4.11), (4.13), and thus (4.29) refer to the total charge densities, which include bound charges. By extension, the general form of Gauss's law (equation (4.63)) and Poisson's equation (equation (4.44)) also depend on both the bound and free charges. With the introduction of **D** and the concept of electrical permittivity, we can simply express Gauss's law in terms of free charges only, represented by ρ (4.65). Similarly, for linear media (where $\mathbf{D} = \epsilon\mathbf{E}$), by rewriting (4.43) as $\nabla \cdot \mathbf{E} = \rho/\epsilon$, Poisson's equation can be rewritten as

$$\nabla^2 \Phi = -\frac{\rho}{\epsilon} \tag{4.68}$$

where, again, ρ now excludes any bound polarization charges.

---

[110]For a brief discussion, see Section 3.3 of D. H. Staelin, A. W. Morgenthaler, and J. A. Kong, *Electromagnetic Waves,* Prentice-Hall, New Jersey, 1994. For further discussion of electromagnetics of anisotropic media, see Section 13.8 of S. Ramo, J. R. Whinnery, and T. Van Duzer, *Fields and Waves in Communication Electronics,* 3rd ed., Wiley, New York, 1994.

While there are versions of Gauss's law and Poisson's equation that involve $\mathbf{D}$ and the free charge distribution only, explicit equations for $\mathbf{E}$ and Φ in terms of the total charge distribution, such as equations (4.9), (4.11), (4.13), and (4.29), that depend on the fact that $\nabla \times \mathbf{E} = 0$ for static electric fields, do not have equivalents for $\mathbf{D}$. In general, even for static fields, we may have $\nabla \times \mathbf{D} \neq 0$, since $\nabla \times \mathbf{P} \neq 0$ may be true.

The introduction of the permittivity concept greatly simplifies our analysis and understanding of electrostatic phenomena in the presence of dielectric materials. The polarizability and local molecular electric field concepts were introduced in Section 4.10.1 so that we can relate microscopic physical behavior to macroscopic fields. For most future problems we shall describe materials by means of their permittivity ϵ, since this constant is easily measured. If ϵ is known, we can find the polarization from the relation

$$\boxed{\mathbf{P} = \epsilon_0 \chi_e \mathbf{E} = \mathbf{D} - \epsilon_0 \mathbf{E} = (\epsilon - \epsilon_0)\mathbf{E}} \tag{4.69}$$

The total polarizability α_T is thus

$$\epsilon - \epsilon_0 = \epsilon_0 \chi_e = \frac{N \alpha_T}{1 - N\alpha_T/(3\epsilon_0)} \quad \rightarrow \quad \alpha_T = \frac{3\epsilon_0(\epsilon - \epsilon_0)}{N(\epsilon + 2\epsilon_0)}$$

Permittivities of selected materials. The relative permittivities of some selected materials are listed in Table 4.1. The values given are at room temperature and for static or low-frequency applied electric fields; the electrical permittivity of most materials depends on frequency and can in general be quite different from the low-frequency values, especially in the vicinity of resonances (see Section 11.2.1). The value of ϵ_r for most materials listed lies in the range 1–25, with the exception of distilled water, titanium dioxide, and barium titanate. The high permittivity of distilled water arises from the partial orientation of permanent dipole moments of its polar molecules. However, other polar materials such as mica and quartz do not have unusually large values of ϵ_r, so the substantially higher permittivity of distilled water is largely a result of the specific microscopic molecular structure of this ubiquitous material. Titanium dioxide is also a polar material but is also an anisotropic crystalline substance whose permittivity depends on the direction of the applied electric field. The value of ϵ_r for TiO_2 is 89 when the applied field is in the direction of one of its crystal axes and 173 when the applied field is perpendicular to this axis. Other materials such as quartz also exhibit anisotropic behavior but with a narrower range of permittivities (e.g., 4.7–5.1 for quartz).

Titanates (combinations of TiO_2 with other oxides) also belong to a class of materials known as *ferroelectric* materials, which exhibit spontaneous polarization just as ferromagnetic materials (e.g., iron) exhibit permanent magnetization (see Chapter 6). Some of these materials have relative permittivities at room temperature of 500–6000; a good example is barium titanate (BaO·TiO_2 or $BaTiO_3$),[111] with $\epsilon_r \simeq 1200$.

---

[111] $BaTiO_3$ is also anisotropic, with $\epsilon_r \simeq 160$ along its principal ferroelectric axis and $\epsilon_r \simeq 5000$ along the perpendicular axis at room temperature. For an aggregation of randomly oriented $BaTiO_3$ crystals, the permittivity can be taken to be $\epsilon_r \simeq 1200$.

TABLE 4.1 RELATIVE PERMITTIVITY AND DIELECTRIC STRENGTH OF SELECTED MATERIALS

Material	Relative Permittivity (ϵ_r) (at Room Temperature)	Dielectric Strength (MV-m^{-1}) (at Room Temp. and 1 atm)
Air	1	~3
Alumina (Al_2O_3)	~8.8	
Amber	2.7	
Bakelite	~4.8	25
Barium titanate ($BaTiO_3$)	1200	7.5
Freon	1	~8
Fused quartz (SiO_2)	3.9	~1000
Gallium arsenide (GaAs)	13.1	~40
Germanium (Ge)	16	~10
Glass	~4–9	~30
Glycerin	50	
Ice	3.2	
Mica (ruby)	5.4	200
Nylon	~3.6–4.5	
Oil	2.3	15
Paper	1.5–4	15
Paraffin wax	2.1	30
Plexiglass	3.4	
Polyethylene	2.26	
Polystyrene	2.56	20
Porcelain	~5–9	11
Rubber	~2.4–3.0	25
Rutile (TiO_2)	100	
Silicon (Si)	11.9	~30
Silicon nitride (Si_3N_4)	7.2	~1000
Sodium chloride (NaCl)	5.9	
Styrofoam	1.03	
Sulphur	4	
Tantalum pentoxide (Ta_2O_5)	~25	
Teflon (PTFE)	2.1	
Vaseline	2.16	
Water (distilled)	81	
Wood (balsa)	1.4	

Such materials can be used to construct compact general-purpose capacitors with very small dimensions but relatively high capacitances, in the range of 5 pF–0.1 μF. The permittivity of the titanates depends sensitively on temperature and also on the applied electric field strength. The latter property can be very useful in nonlinear circuit applications. Other crystals that have similar structure to $BaTiO_3$ and show similar behavior include $SrTiO_3$, $NaTaO_3$, and $LaFeO_3$. Other examples of ferroelectric materials include lithium selenite $LiH_3(SeO_3)_2$ and related salts, ammonium cadmium sulfate $(NH_4)_2Cd_2(SO_4)_3$, and lead metaniobate $Pb(NbO_3)_2$.

Further discussion of electrical properties of different dielectric materials can be found elsewhere.[112]

4.10.3 Dielectric Breakdown

The linear relationship (4.66) between **E** and **D** holds only for relatively small values of the applied electric field **E**. If the applied field is sufficiently high, a dielectric material is suddenly transformed from a good insulator into an extremely good conductor, causing substantial current to flow. *Dielectric breakdown* is the term used for this sudden loss of the insulating property of a dielectric under the influence of an electric field. The electric field strength at which dielectric breakdown occurs is referred to as the *dielectric strength* and is denoted by E_{BR}. The values of E_{BR} for selected materials are given in Table 4.1.

The dielectric strength of a given material may vary by several orders of magnitude depending on the conditions under which it is used. For gases, E_{BR} is proportional to pressure. For solids, microstructural defects, impurities, the shape of the dielectric, the manner in which it was prepared, and its environment are all factors that may strongly affect E_{BR}. Under normal conditions, dielectric breakdown does not permanently affect gaseous or liquid dielectrics. In solids, on the other hand, the breakdown leads to the formation of highly conductive channels, leaving behind a characteristic damage to the texture of the material, typically in the form of a channel of molten material, jagged holes, or a tree-like decomposition pattern of carbonized matter.

We now comment briefly on the nature of dielectric breakdown in gaseous and solid dielectrics.

Breakdown in gases.   Consider the parallel-plate capacitor of Figure 4.48 to be filled with a gas and connected to a constant voltage applied across its plates, and arrangements made so that the current flowing through the dielectric region between the plates can be measured. As the applied voltage is slowly increased, the current initially rises to a small value of a small fraction of a microampere and levels off to a saturation level. As the voltage is increased further, the current remains nearly constant until a certain critical voltage, $V_{BR} = E_{BR}d$, where V_{BR} is called the breakdown voltage and d is the separation of the plates, is reached. When a voltage greater than V_{BR} is applied, the current through the dielectric quickly rises (typically in a time interval of the order of $\sim 10^{-8}$ s) to the maximum value allowed by the power supply.

This behavior was first analyzed by Townsend,[113] who introduced the concept of *electron avalanche breakdown.* Avalanche breakdown occurs when the applied field is large enough to accelerate a free electron (a few of which are always present in any gas due to cosmic ray ionization) to sufficiently large energies so that it can liberate a new electron by impact ionization of a neutral gas molecule, leaving behind a positively charged gas ion. Both the initial electron and the newly produced electron can then be subsequently accelerated by the field again and produce more electrons by impact ionization.

---

[112]See A. von Hippel (Ed.), *Dielectric Materials and Applications,* Artech House, 1995.

[113]J. S. Townsend, *Electricity in Gases,* Oxford University Press, Oxford, UK, 1914.

This process thus develops into an avalanche of impact ionization, rapidly creating many free electrons and positive ions in the gas, and accounts for the highly conducting behavior of the gas. The acceleration of the initial electron to energies at which it can impact-ionize the neutrals must occur between two successive collisions of the electron with gas molecules. Since the density of gas molecules (i.e., the average distance between them) is larger at higher pressures, the dielectric strength for gases is proportional to pressure;[114] in other words, pressurized gases can, in general, withstand stronger electric fields. This property of gases, and the fact that a gas recovers its insulating properties if the applied field is removed after breakdown, accounts for the frequent use of pressurized gases as dielectrics, especially in high-voltage applications.

In most cases, the electric field distribution in a gas is not uniform, as was implicitly assumed to be the case for the gas-filled parallel-plate capacitor just described. For example, in a single-wire-above-ground configuration as shown in Figure 4.51e, the electric field is strongest on the surface of the conductor facing the ground. Similarly, the electric field in a coaxial line is strongest on the inner conductor, decreasing with distance as r^{-1}. In general, on any charged conductor the electric field is higher in the vicinity of sharp points or conductor surfaces with the lowest radius of curvature. These high fields can sometimes locally exceed the dielectric strength of air and produce a breakdown in the vicinity of the high-field region although, further away from the conductor, the field is not sufficiently strong to sustain the breakdown. This type of a local discharge is referred to as a *corona discharge*. Consider, for example, a single-wire power transmission line above ground as shown in Figure 4.51e, with a typical height above ground of tens of meters. As the air breaks down in the vicinity of the conductor, it is ionized and behaves like a conductor. At night, the ionized air is apparent as a reddish (or sometimes blue-yellow) glow that crowns the conductor, hence the term "corona." To maximize power transmission, power lines typically operate very close to their highest voltage ratings. On days of bad weather, increased humidity and other effects may lower the dielectric strength of air, leading to corona discharges. Corona also comes about because of accumulation of rust and dirt on the lines, which create local inhomogeneities, resulting in large local electric fields in their vicinity. Corona discharge on power transmission lines must be avoided because it substantially increases losses and emits electromagnetic waves that can interfere with certain types of nearby communication systems. The reader may have experienced the effects of corona discharge while driving under a transmission line with the radio on.

It was mentioned above that corona discharge occurs when the field is above the breakdown level in the vicinity of a conductor but not away from it. If the applied voltage is increased even further, a continuously ionized path in the form of a bright luminous arc is formed, extending to the nearest point of opposite polarity. An intense current (typically hundreds to thousands of amperes) flows through the gas, and an *arc*

---

[114]This important result was first demonstrated by F. Paschen in 1889 and is valid at moderate pressures. At higher pressure levels, the breakdown voltage rises with increased pressure since too many collisions take place and an excessive amount of energy is wasted in various excitation processes. For a discussion and references, see A. von Hippel (Ed.), *Dielectric Materials and Applications,* Artech House, 1995; A. H. Beck, *Handbook of Vacuum Physics,* Macmillan, Pergamon Press, New York, 1965.

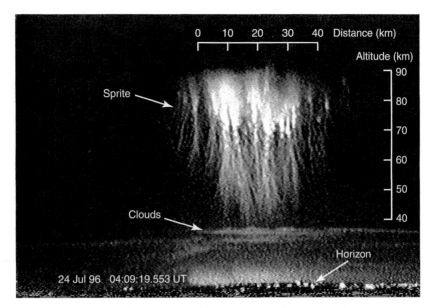

Figure 4.55   Sprites. Spectacular luminous glows occurring in clear air at 50–90-km altitudes above large active thunderstorms are called sprites. This image was taken from Fort Collins, Colorado, using a low-light-level television camera pointed toward the east. The event occurred above a thunderstorm in northern Kansas.

discharge is said to occur. The giant arcs of lightning are the best known examples of an arc discharge.[115]

Rather spectacular examples of dielectric breakdown phenomena in air are the recently discovered luminous glows that occur at high altitudes above thunderstorms, referred to as *sprites*.[116] An example of sprites observed in the midwestern United States is shown in Figure 4.55. Sprites are believed to occur as a result of dielectric breakdown of air under the influence of quasi-static electric fields that appear at high altitudes following intense cloud-to-ground lightning discharges.

Breakdown in solids.   The processes by which electrical breakdown occurs in liquids and solids are quite complex and depend on the particular periodic lattice structure of the material, the number of electrons in its conduction band, as well as imperfections due to the presence of foreign atoms. The externally measured voltage–current characteristics are quite similar to those observed for gases. When a pure, homogeneous solid dielectric is placed between suitable electrodes across which an increasing

[115]M. A. Uman, *The Lightning Discharge,* Academic Press, Orlando, 1987.

[116]D. D. Sentman, E. M. Wescott, D. L. Osborne, D. L. Hampton, and M. J. Heavner, Preliminary results from the Sprites94 aircraft campaign, 1: Red sprites, *Geophysical Research Letters,* 22(10), pp. 1205–1208, 1995. For two descriptive articles, see Heaven's new fires, *Discover,* pp. 100–107, July 1997, and Lightning between earth and space, *Scientific American,* pp. 56–59, August 1997.

voltage is applied, a small (microampere) current flows and levels off to a saturation value until the applied voltage exceeds a certain critical voltage. At this point, the current through the material suddenly (within $\sim 10^{-8}$ s) increases to the maximum value allowed by the supply. The dielectric strength of solids may vary by several orders of magnitude depending on the purity of the material, the shape of the material, the manner in which it was manufactured, the ambient temperature, and the duration of the applied field. In general, it is not possible to determine E_{BR} for solids on purely theoretical grounds; practical experimental tests under precisely defined conditions are generally required to specify the dielectric strength of the material reliably in the context of a particular application.

Three basic mechanisms of breakdown can be identified. So-called *intrinsic breakdown* is electronic in nature and depends on the presence of conduction-band electrons capable of migration through the lattice. The underlying process is an avalanche process similar to what occurs in gases; however, the microphysics can be understood only on the basis of the band theory of electronic structure and by including electron–lattice interactions—topics well beyond the scope of this book. The electric field levels at which intrinsic breakdown occurs typically represent the theoretical upper limits for E_{BR} for a solid dielectric, reached only by extremely pure and homogeneous materials. More typically, impurities and structural defects cause the material to break down at lower E_{BR} values. The second type of breakdown is called *thermal breakdown* and arises from the fact that small leakage currents flowing through a dielectric lead to resistive losses, which are dissipated as heat in the material. If the local heat is generated faster than it can be dissipated, the temperature of the material rises and breakdown occurs, typically by melting or decomposition. The dielectric strength E_{BR} due to this breakdown mechanism depends on the duration of the application of the field as well as the ambient temperature. The third type of breakdown mechanism is often referred to as *discharge breakdown* and depends on the presence of voids or cracks within the material. The presence of air in such cracks or voids produces a reduction in the dielectric strength. Partial discharges in the vicinity of such imperfections can in time lead to arcing, localized melting, and chemical transformations, creating conducting channels. Even in relatively pure and homogeneous materials, microscopic cracks can initially occur because of temperature or mechanical stresses, oxidation, or other aging (progressive degradation) effects.

Example 4.30: Parallel-plate capacitor with sandwiched dielectrics. A parallel-plate capacitor is constructed using two separate dielectric materials, stacked between two conductor plates, as shown in Figure 4.56. (a) Find the capacitance of this configuration. (b) If glass ($\epsilon_{r1} = 6$ and $(E_{BR})_1 = 30$ MV-m^{-1}) and mica ($\epsilon_{r2} = 5$ and $(E_{BR})_2 = 200$ MV-m^{-1}) are chosen to be used with 2 cm^2 cross-sectional areas and equal thicknesses of 5 mm each (i.e., $d_1 = d_2 = 5$ mm), find the capacitance and voltage rating of the capacitor, using a safety factor of 10.

   Solution:

   (a) As usual in capacitance problems, we start by assuming that a potential difference is applied between the plates and that charges $+Q$ and $-Q$ are induced on the lower and upper plates, respectively. The surface charge densities on the plates are $\rho_{s1} = Q/A$ and

Figure 4.56   Dielectric strength. Parallel-plate capacitor with two sandwiched dielectric layers.

$\rho_{s2} = -Q/A$, respectively, where A is the area of the plates. Taking a Gaussian surface S_1 as shown, we have

$$[\epsilon_1 E_1]A = \rho_{s1}A \quad \rightarrow \quad E_1 = \frac{\rho_{s1}}{\epsilon_1}$$

where we arbitrarily assume the positive direction for the electric field to be upward. Similarly, using the Gaussian surface S_2, we can show that $E_2 = -\rho_{s2}/\epsilon_2 = \epsilon_1 E_1/\epsilon_2$. The magnitude of the total potential drop between the plates is

$$|\Phi| = E_1 d_1 + E_2 d_2 = \frac{Q}{A}\left(\frac{d_1}{\epsilon_1} + \frac{d_2}{\epsilon_2}\right)$$

and the capacitance is then

$$C = \frac{Q}{|\Phi|} = \frac{A\epsilon_0}{(d_1/\epsilon_{r1}) + (d_2/\epsilon_{r2})}$$

Using the permittivity values given with $d_1 = d_2 = 5$ mm, we find $C \simeq 0.966$ pF.

(b) The voltage rating of the capacitor is determined by the dielectric layer that breaks down first. We have $(E_{BR})_1 = 30$ MV-m^{-1} and $(E_{BR})_2 = 200$ MV-m^{-1}. For a safety factor of 10, we must have

$$(E_1)_{max} = \frac{(E_{BR})_1}{10} = 3 \text{ MV-m}^{-1} \quad \text{and} \quad (E_2)_{max} = \frac{(E_{BR})_2}{10} = 20 \text{ MV-m}^{-1}$$

But since $E_{2y} = \epsilon_1 E_{1y}/\epsilon_2 = 6E_{1y}/5$, the voltage rating of this capacitor is determined by the glass layer since the glass layer breaks down first. Thus, the voltage rating is

$$|\Phi|_{max} = (E_1)_{max}d_1 + \epsilon_{r1}(E_1)_{max}\frac{d_2}{\epsilon_{r2}} = (E_1)_{max}\left[d_1 + \frac{\epsilon_{r1}d_2}{\epsilon_{r2}}\right]$$

$$= (3 \text{ MV-m}^{-1})(5 \times 10^{-3} \text{ m})\left[1 + \frac{6}{5}\right] = 33 \text{ kV}$$

Note that if the dielectric were filled uniformly with mica, the breakdown voltage with a safety factor of 10 would be $|\Phi|_{max} = (E_2)_{max}(d_1 + d_2) = (20 \text{ MV-m}^{-1})(0.01 \text{ m}) = 0.2$ MV. Thus, the presence of the glass layer lowers the voltage rating of the capacitor.

4.11 ELECTROSTATIC BOUNDARY CONDITIONS

In dealing with electrostatic problems, it is often necessary to relate the polarization charges induced in dielectric media to electric fields produced by external charges (e.g., on conducting bodies). This is facilitated by relations such as (4.66), which are called *constitutive relations*. In most electrostatic applications, we also deal with interfaces between two or more different dielectric media. The manner in which **D** and **E** behave across such interfaces is described by the boundary conditions. The boundary conditions are derived from fundamental laws of electrostatics, which we reiterate below.

Starting with the experimental facts as expressed in Coulomb's law, namely that the electric force between charges is a function of the distance between them[117] and that its magnitude is proportional to the inverse square of the distance, we have derived the fundamental laws of electrostatics[118] as

$$\nabla \times \mathbf{E} = 0 \qquad \oint_C \mathbf{E} \cdot d\mathbf{l} = 0 \qquad \text{Conservative field}$$

$$\nabla \cdot \mathbf{D} = \rho \qquad \oint_S \mathbf{D} \cdot d\mathbf{s} = \int_V \rho \, dv \qquad \text{Coulomb's law}$$

The dielectric properties of different physical media are accounted for by the *permittivity* ϵ of the material, and so, as stated earlier, the charge density ρ excludes any bound polarization charge density. For linear (polarization **P** is linearly proportional to **E**) and isotropic (polarization **P** does not depend on the *direction* of **E**) media, **D** and **E** are related via (4.66), namely $\mathbf{D} = \epsilon \mathbf{E}$, which is valid even for inhomogeneous materials where ϵ is a function of position expressed as $\epsilon(x, y, z)$.

A very common type of inhomogeneity occurs in practice at the interface between two electrically different (i.e., characterized by different permittivity values) materials. To establish a basis for solving such problems, we now study the behavior of **D** and **E** at the boundary between two different materials and derive the conditions that these vectors must satisfy at such interfaces. These conditions are referred to as the boundary conditions. The boundary conditions must be derived using the integral forms of the fundamental electrostatic laws, because the differential forms apply only at a point.

For the normal component of the electric flux density, we consider a pillbox-shaped closed differential surface as shown in Figure 4.57a. Since the contributions from the sides can be made infinitesimally small by taking $\Delta h \to 0$, we have

$$\oint_S \mathbf{D} \cdot d\mathbf{s} = (\hat{\mathbf{n}} \cdot \mathbf{D}_1)\Delta s - (\hat{\mathbf{n}} \cdot \mathbf{D}_2)\Delta s = \rho_s \Delta s \quad \to \quad \boxed{D_{1n} - D_{2n} = \rho_s} \qquad (4.70)$$

where the minus sign in front of the second term on the left-hand side is due to the unit vector normal to the surface of the pillbox being $-\hat{\mathbf{n}}$, and ρ_s is the free surface charge density at the interface. Hence, the normal component of the electric flux density **D** is discontinuous through a surface that has free surface-charge density present; however, at

---

[117]It is important to remember that our arguments concerning the conservative nature of the electrostatic force—that $\oint_C \mathbf{E} \cdot d\mathbf{l} = 0$, or that $\int_C \mathbf{E} \cdot d\mathbf{l}$ is path-independent (Section 4.4)—were based only on the fact that the force from a single charge was radial and spherically symmetric. The fact that the force was proportional to r^{-2} was not important in this regard; any other r dependence would give the same result. Gauss's law, on the other hand, is based entirely on the fact that the electric force is proportional to r^{-2}.

[118]We do not discuss the definition and physical meaning of the notation $\nabla \times \mathbf{E}$ until Chapter 6, but it follows from $\oint_C \mathbf{E} \cdot d\mathbf{l}$ by Stokes's theorem, also to be introduced and discussed in Chapter 6.

interfaces between two dielectrics, the free surface charge density ρ_s is usually zero unless a surface charge density is externally placed at the interface. For $\rho_s = 0$, (4.70) becomes

$$\boxed{\hat{\mathbf{n}} \cdot \mathbf{D}_1 = \hat{\mathbf{n}} \cdot \mathbf{D}_2 \quad \text{or} \quad D_{1n} = D_{2n}} \tag{4.71}$$

In other words, the normal component of $\mathbf{D}$ is continuous across a charge-free dielectric boundary. The normal component of $\mathbf{E}$ is then not continuous across the boundary, since we have

$$\epsilon_1(\hat{\mathbf{n}} \cdot \mathbf{E}_1) = \epsilon_2(\hat{\mathbf{n}} \cdot \mathbf{E}_2) \quad \rightarrow \quad \epsilon_1 E_{1n} = \epsilon_2 E_{2n} \tag{4.72}$$

The discontinuity in $\hat{\mathbf{n}} \cdot \mathbf{E}$ or E_n can be understood physically as follows. The change in the normal component of the polarization vector $\mathbf{P}$ at the boundary can be related to the surface polarization charge by integrating (4.60) over the differential pillbox shown in Figure 4.57a as $\Delta h \rightarrow 0$. Noting that $\mathbf{P} \cdot d\mathbf{s} = \mathbf{P} \cdot \hat{\mathbf{n}} ds = P_n ds$, we have

$$\int_V -\nabla \cdot \mathbf{P} dv = -\oint_S \mathbf{P} \cdot d\mathbf{s} = -(P_{1n} - P_{2n})\Delta s = \int_V \rho_p dv$$

Since $\Delta h \rightarrow 0$, the volume integral of ρ_p selects only the surface polarization charge density at the boundary:

$$\int_V \rho_p dv = \rho_p \Delta h \Delta s = \rho_{sp} \Delta s$$

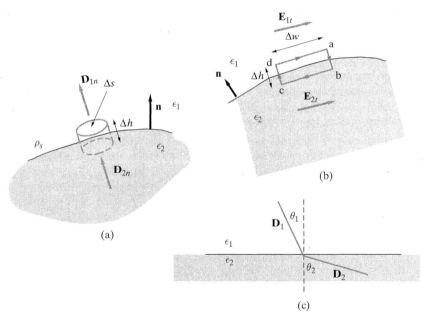

(a)

(b)

(c)

Figure 4.57   The boundary between two electrically different media. (a) A differential pillbox-shaped closed surface. (b) A differential closed contour. (c) $\mathbf{D}$ vectors at the boundary for the case $\epsilon_2 > \epsilon_1$.

Simplifying, we can write

$$-(P_{1n} - P_{2n}) = \rho_{sp} \qquad (4.73)$$

But, using (4.64), we have

$$P_{1n} = D_{1n} - \epsilon_0 E_{1n}$$
$$P_{2n} = D_{2n} - \epsilon_0 E_{2n}$$

so that

$$-(P_{1n} - P_{2n}) = -\underbrace{(D_{1n} - D_{2n})}_{=0 \text{ from } (4.71)} + \epsilon_0(E_{1n} - E_{2n})$$

Using (4.73), we find the discontinuity in the normal component of the electric field at the interface in terms of the bound polarization surface charge as

$$E_{1n} - E_{2n} = \frac{\rho_{sp}}{\epsilon_0}$$

Thus, at the surface of the dielectric the normal component of the electric field **E** is in general discontinuous at the boundary by an amount ρ_{sp}/ϵ_0, which can be interpreted as if ρ_{sp} is a surface layer of free charge at that location. The surface polarization charge represents the amount of charge on the ends of dipoles in medium 2 that are not canceled by the charge of opposite polarity on the ends of the dipoles in medium 1 (see Figure 4.52a).

In practice, surface polarization charge does not need to be taken into account explicitly, since the usual methodology involves finding suitable solutions for **E** and **D** in the two dielectric regions and then adjusting the magnitudes of the solutions to satisfy the boundary conditions. Using the different permittivities in the two regions fully accounts for all the requirements placed on the fields due to the different microphysical electrical behavior of the materials.

For the tangential components of the electric field, we consider the line integral of the electric field **E** around a closed differential contour such as that shown in Figure 4.57b. Since the contributions from sides ab and cd can be made infinitesimally small by taking $\Delta h \to 0$, we have

$$\int_d^a \mathbf{E}_1 \cdot d\mathbf{l} + \int_b^c \mathbf{E}_2 \cdot d\mathbf{l} = (E_{1t} - E_{2t})\Delta w = 0 \quad \to \quad \boxed{E_{1t} = E_{2t}} \qquad (4.74)$$

Thus the tangential component of the electric field is continuous across the interface between the two materials. Since $\mathbf{D} = \epsilon \mathbf{E}$, the tangential component of the electric flux density **D** is not continuous across a boundary. Instead we have

$$\frac{D_{1t}}{\epsilon_1} = \frac{D_{2t}}{\epsilon_2}$$

A consequence of (4.70) and (4.74) is the change in orientation of electric flux lines across material interfaces, as illustrated in Figure 4.57c. Consider an electric field

E_1 that is oriented at an angle θ_1 from the normal in medium 1. From (4.70) and (4.74) we have

$$E_1 \sin \theta_1 = E_2 \sin \theta_2$$

$$\epsilon_1 E_1 \cos \theta_1 = \epsilon_2 E_2 \cos \theta_2$$

where E_1 and E_2 are the magnitudes of $\mathbf{E}_1$ and $\mathbf{E}_2$ and where we have assumed $\rho_S = 0$. Therefore, we can find

$$\tan \theta_2 = \frac{\epsilon_2}{\epsilon_1} \tan \theta_1$$

and

$$E_2 = E_1 \sqrt{\left(\frac{\epsilon_1}{\epsilon_2} \cos \theta_1\right)^2 + \sin^2 \theta_1}$$

The above relationships indicate that electric field lines are *bent further away from the normal* to the boundary in the medium with the *higher* permittivity.

Example 4.31: Spherical dielectric shell. A positive point charge Q is at the center of a spherical dielectric shell of an inner radius a and an outer radius b. The relative dielectric constant of the shell is ϵ_r. Determine $\mathbf{E}, \Phi, \mathbf{D}, \mathbf{P}$, and the polarization charge densities ρ_p and ρ_{sp}.

Solution: This example is very similar to Example 4.21 except that the conducting shell has now been replaced by a dielectric shell (see Figure 4.58). In view of the spherical symmetry, we apply Gauss's law to find $\mathbf{E}$ and $\mathbf{D}$ in three regions: (a) $r \geq b$, (b) $a \leq r < b$, and (c) $r < a$. The electric potential Φ is then determined from the negative line integral of $\mathbf{E}$, and polarization $\mathbf{P}$ follows from the relation

$$\mathbf{P} = \mathbf{D} - \epsilon_0 \mathbf{E} = \epsilon_0(\epsilon_r - 1)\mathbf{E}$$

Note that the $\mathbf{E}$, $\mathbf{D}$, and $\mathbf{P}$ vectors each have only radial components.

(a) For the region $r \geq b$, $\mathbf{E}$ and Φ are exactly the same as in Example 4.21. We have

$$E_r = \frac{Q}{4\pi \epsilon_0 r^2} \qquad r \geq b \qquad \Phi = \frac{Q}{4\pi \epsilon_0 r} \qquad r \geq b$$

From equations (4.64) and (4.66) we obtain

$$P_r = 0 \qquad\qquad\qquad\qquad\qquad r \geq b$$

$$D_r = \epsilon_0 E_r = \frac{Q}{4\pi r^2} \qquad\qquad\qquad r \geq b$$

(b) For the region $a \leq r < b$ (i.e., inside the dielectric shell), $\mathbf{E}$ can be found using Gauss's law (equation (4.32)) using the spherical Gaussian surface S_d as

$$E_r = \frac{Q}{4\pi \epsilon_0 \epsilon_r r^2} = \frac{Q}{4\pi \epsilon r^2} \qquad a \leq r < b$$

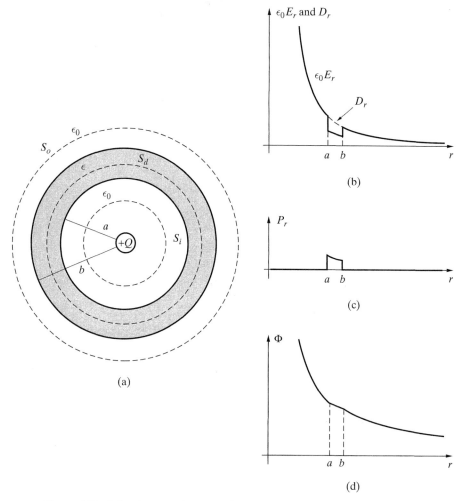

Figure 4.58 Point charge at the center of a spherical dielectric shell. (a) The geometry. (b) D_r and $\epsilon_0 E_r$ versus r. (c) P_r versus r. (d) Φ versus r. Note that $D_r = P_r + \epsilon_0 E_r$, as can be seen from the plots (b) and (c).

and **D** follows as (equation (4.66))

$$D_r = \epsilon E_r = \frac{Q}{4\pi r^2} \qquad a \leq r < b$$

The polarization vector **P** is (equation (4.69))

$$P_r = D_r - \epsilon_0 E_r = \frac{Q}{4\pi r^2}\left(1 - \frac{1}{\epsilon_r}\right) \qquad a \leq r < b$$

The electrostatic potential is (equation (4.17))

$$\Phi = -\int_{\infty}^{b} \frac{Q}{4\pi\epsilon_0 r^2} dr - \int_{b}^{r} \frac{Q}{4\pi\epsilon r^2} dr = \frac{Q}{4\pi\epsilon_0 b} + \left[\frac{Q}{4\pi\epsilon r}\right]_{b}^{r}$$

$$= \frac{Q}{4\pi\epsilon_0 r} \left[\frac{r}{b}\left(1 - \frac{1}{\epsilon_r}\right) + \frac{1}{\epsilon_r}\right] \qquad a \le r < b$$

(c) For the region $r < a$, the application of Gauss's law on surface S_i yields the same expressions for E_r, D_r, and P_r as in region $r \ge b$:

$$E_r = \frac{Q}{4\pi\epsilon_0 r^2} \qquad\qquad r < a$$

$$D_r = \frac{Q}{4\pi r^2} \qquad\qquad r < a$$

$$P_r = 0 \qquad\qquad r < a$$

To find Φ, we must add to Φ at $r = a$ the negative line integral of E_r in the region $r < a$:

$$\Phi = \Phi(a) - \int_{a}^{r} \frac{Q}{4\pi\epsilon_0 r^2} dr = \frac{Q}{4\pi\epsilon_0 r}\left[1 + \left(\frac{r}{b} - \frac{r}{a}\right)\left(1 - \frac{1}{\epsilon_r}\right)\right] \qquad r < a$$

The variations of $\epsilon_0 E_r$ and D_r versus r are plotted in Figure 4.58b. The difference $(D_r - \epsilon_0 E_r)$ is P_r and is also shown in Figure 4.58c. We note that D_r is a continuous curve exhibiting no sudden changes in going from one medium to another and that P_r exists only in the dielectric region. The plot for Φ is also shown in Figure 4.58d. The polarization charge densities associated with the dielectric shell are

$$\rho_p = -\nabla \cdot \mathbf{P} = -\frac{1}{r^2}\frac{\partial}{\partial r}\left[r^2 \frac{Q}{4\pi r^2}\left(1 - \frac{1}{\epsilon_r}\right)\right] = 0$$

inside the dielectric shell,

$$\rho_{sp}|_{r=a} = \mathbf{P} \cdot (-\hat{\mathbf{r}})|_{r=a} = -\frac{Q}{4\pi a^2}\left(1 - \frac{1}{\epsilon_r}\right)$$

on the inner surface of the shell, and

$$\rho_{sp}|_{r=b} = \mathbf{P} \cdot \hat{\mathbf{r}}|_{r=b} = \frac{Q}{4\pi b^2}\left(1 - \frac{1}{\epsilon_r}\right)$$

on the outer surface of the shell, respectively.

These results indicate that there is no net polarization volume charge inside the dielectric shell. However, polarization surface charges of negative and positive polarity exist on the inner and the outer surfaces, respectively. These surface charges produce an electric field inside the dielectric shell that is directed radially inward, thus reducing the **E** field inside the dielectric shell produced by the point charge $+Q$ located at its center.

4.12 ELECTROSTATIC ENERGY

When we lift a flowerpot and place it on a windowsill, the work we do (i.e., the energy we expend) against gravity is stored in the form of potential energy. Any object with nonzero mass acquires gravitational potential energy when raised through a certain height. When we compress a spring, the work we expend to do so is stored in the form of elastic potential energy. The gravitational potential energy can be recovered by lowering the raised object, and the elastic potential energy can be recovered by releasing the spring. Bringing two like-charges together from infinite separation against their electrostatic repulsion also requires work; *electrostatic energy* is thus stored in such a configuration of charges, and this energy may be recovered by allowing the charges to recede to an infinite separation. In Section 4.4, we defined the concept of electrostatic potential of a given point as the work that needs to be done to move a unit positive test charge from infinity to that point. In this section we further consider the relationship between electrostatic energy, electric potential, charges, and electric fields.

The absolute potential due to a single charge Q in a simple dielectric medium is given as (equation (4.18))

$$\Phi = \frac{Q}{4\pi\epsilon R}$$

where R is the distance between the point at which we measure the potential and the position of Q. In the presence of this potential distribution (or the corresponding electric field), the energy required to bring another charge q from infinity to a distance R from Q is

$$W_e = q\Phi = \frac{Qq}{4\pi\epsilon R} \tag{4.75}$$

where $W_e = q\Phi$ follows from the definition of potential as the work per unit charge. This energy is what is required to hold these two charges at a distance R from one another, and is referred to as the *potential energy* of the charge configuration. This electric energy associated with an assembly of charges is only a function of the final position of the charges and can be determined by calculating the work required to gather the charges together. That any assemblage of point or continuous charge distributions "stores" energy can be seen by considering the motion of the charges if they are to be released from their fixed (i.e., externally imposed) points. The large Coulomb forces that exist between the charges cause the charge assembly to literally fly apart.[119] The net kinetic energy gained by the charges after they have been separated equals the initial potential energy stored in the assemblage.

4.12.1 Electrostatic Energy in Terms of Charges and Potential

We begin by considering a series of separated point charges at infinity and then calculate the energy required to bring these charges, one by one, from infinity to form the

---

[119]Assuming that no other forces act on the particles—that is what we mean by having them "released" from their fixed locations.

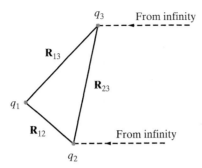

Figure 4.59   An assemblage of charges. Point charges q_i gathered together by bringing them one by one from infinity.

assemblage of charges as illustrated in Figure 4.59. The energy required to bring the first charge q_1 from infinity to the region of interest is $W_1 = 0$, since no other charges are present to exert any force on q_1. The energy required to bring the second charge q_2 to a distance R_{12} from q_1 is

$$W_2 = q_2\Phi_{12}$$

where Φ_{12} is defined as the electric potential at the location of q_2 due to the presence of charge q_1; that is,

$$\Phi_{12} = \frac{q_1}{4\pi\epsilon R_{12}}$$

In Φ_{12}, the second subscript denotes the location where the potential is measured, whereas the first subscript denotes the source charge q_1 creating the potential at the position of q_2. If a third charge q_3 is brought from infinity to a distance R_{13} from q_1 and R_{23} from q_2, the energy required is

$$W_3 = q_3(\Phi_{13} + \Phi_{23})$$

The energy required to bring a fourth charge is

$$W_4 = q_4(\Phi_{14} + \Phi_{24} + \Phi_{34})$$

and thus the total energy required to gather together a series of charges is

$$W_e = \underbrace{q_2\Phi_{12}}_{W_2} + \underbrace{q_3(\Phi_{13} + \Phi_{23})}_{W_3} + \underbrace{q_4(\Phi_{14} + \Phi_{24} + \Phi_{34})}_{W_4} + \cdots \tag{4.76}$$

where the subscript "e" indicates that W_e is electrostatic energy, to be distinguished from magnetostatic energy W_m, discussed in Chapter 7. We can develop a compact series expression for W_e by recognizing that W_2 may be written alternatively as

$$W_2 = q_2\Phi_{12} = \frac{q_2 q_1}{4\pi\epsilon R_{12}} = q_1\frac{q_2}{4\pi\epsilon R_{12}} = q_1\Phi_{21}$$

which indicates that we have

$$q_i\Phi_{ki} = q_k\Phi_{ik}$$

The above is true simply because no matter what order the charges are assembled in, the partial energy associated with any two charges, say, the ith and the kth, equals $q_i q_k/(4\pi\epsilon R_{ki})$, and this term occurs only once during the assembly process.

It is then possible to rewrite (4.76) in its alternative form as

$$W_e = q_1\Phi_{21} + q_1\Phi_{31} + q_2\Phi_{32} + q_1\Phi_{41} + q_2\Phi_{42} + q_3\Phi_{43} + \cdots \qquad (4.77)$$

corresponding to assembling the charges in reverse order. We can now add (4.76) and (4.77) to find

$$2W_e = q_1(\Phi_{21} + \Phi_{31} + \Phi_{41} + \cdots) + q_2(\Phi_{12} + \Phi_{32} + \Phi_{42} + \cdots)$$
$$+ q_3(\Phi_{13} + \Phi_{23} + \Phi_{43} + \cdots) + \cdots$$

which may be conveniently written as

$$W_e = \tfrac{1}{2}\sum_{i=1}^{N} q_i\Phi_i \qquad (4.78)$$

where

$$\Phi_i = \Phi_{1i} + \Phi_{2i} + \Phi_{3i} + \cdots + \Phi_{(i-1)i} + \Phi_{(i+1)i} + \cdots + \Phi_{Ni} \qquad (4.79)$$

is the total potential at the ith location due to all the other $(N-1)$ charges present, except for the ith charge. In other words, Φ_i includes the contributions to the total potential at the ith point of all charges other than the one that is at that point. Equation (4.78) represents the total energy of an assemblage of N charges in terms of a simple summation over the charges.

We can write (4.78) alternatively as

$$W_e = \tfrac{1}{2}\sum_{i=1}^{N}\sum_{j=1}^{N} q_i\Phi_{ji} \qquad (i\neq j) \qquad (4.80)$$

where the quantity inside the double summation is the electrostatic potential at the location of the ith charge produced by the jth charge multiplied by the value of the ith charge.

It might have been tempting to write (4.80) by inspection, without carrying out the indicated steps as was done above. However, any such intuitive argument that leads from the definition of the electrostatic potential to the electrostatic potential energy is more than likely to ignore the $\tfrac{1}{2}$ factor that stands outside the summation in (4.80). This factor is a consequence of the fact that the potential energy associated with any two charges is a property of both charges but is not a property that can be assigned to each charge separately.

Example 4.32: Energy required to assemble three point charges. Determine the energy required to bring three charges from infinity to three points on the x axis: $(-1,0,0)$, $(0,0,0)$, and $(+1,0,0)$. Assume free space.

Solution: The total energy of the configuration can be calculated using (4.76). Assume that charge q_1 is brought to $(0,0,0)$, charge q_2 to $(-1,0,0)$, and q_3 to $(+1,0,0)$. We have

$$W_e = q_2\Phi_{12} + q_3\Phi_{13} + q_3\Phi_{23} = q_2\frac{q_1}{4\pi\epsilon_0 R_{12}} + q_3\frac{q_1}{4\pi\epsilon_0 R_{13}} + q_3\frac{q_2}{4\pi\epsilon_0 R_{23}}$$

where we have $R_{12} = 1$, $R_{13} = 1$ and $R_{23} = 2$. If $q_1 = q_2 = q_3 = q$, we then have

$$W_e = \frac{5q^2}{8\pi\epsilon_0}$$

Note that the same result could have been obtained using (4.78), which is simply another way of writing (4.76):

$$W_e = \frac{1}{2}q_1\left(\frac{q_2}{4\pi\epsilon_0 R_{12}} + \frac{q_3}{4\pi\epsilon_0 R_{13}}\right) + \frac{1}{2}q_2\left(\frac{q_1}{4\pi\epsilon_0 R_{12}} + \frac{q_3}{4\pi\epsilon_0 R_{23}}\right)$$
$$+ \frac{1}{2}q_3\left(\frac{q_1}{4\pi\epsilon_0 R_{13}} + \frac{q_2}{4\pi\epsilon_0 R_{23}}\right)$$

which upon adding terms with values substituted also gives $W_e = 5q^2/(8\pi\epsilon_0)$.

4.12.2 Electrostatic Energy in Terms of Fields

It is interesting to consider *where* the energy associated with the charge distribution is stored. This question is analogous to one we might ask in mechanics when we have two masses attached to the ends of a compressed spring. The stored energy might be considered to reside in the masses or in the stressed state of the spring. The first viewpoint is the same as that expressed by (4.78), where the stored energy is linked to the charges and the electric potentials at their positions. This view ascribes physical reality only to the charges and their spatial distributions. On the other hand, if the field concept is to be complete, it should be capable of expressing the stored energy without recourse to a description of the charges producing it. The energy should be describable in terms of the "elastic" quality of the electric field, just as the spring stores the energy in the preceding mechanical example.

Another mechanical example is the flowerpot on the windowsill, which possesses potential energy because if it is pushed over the edge, it gains kinetic energy. Is the potential energy stored in the windowsill, in the flowerpot, or in the gravitational field?

It turns out that the energy associated with a charge distribution can indeed be expressed solely in terms of the fields **E** and **D**. We can generalize (4.78) to the case of a continuous distribution of charges in a volume V by making the following transformations and taking the limit as $N \to \infty$,

$$q_i \to \rho(r)dv, \qquad \Phi_i \to \Phi(r) \quad \text{and} \quad \sum_{i=1}^{N} \to \int_V$$

which gives

$$\boxed{W_e = \frac{1}{2}\int_V \rho\Phi\, dv} \tag{4.81}$$

where Φ is now a continuous function of position and is the potential distribution due to all charges. It is no longer necessary to exclude the contribution due to $\rho\,\delta v$ located at the point where Φ is evaluated because in the limit $\delta v \to 0$, this contribution is zero anyway. Equation (4.81) can be interpreted as the volume integral over V of an electric energy density $w_e = \frac{1}{2}\rho\Phi$, which exists in regions where $\rho \neq 0$.

From Section 4.10, we know that a charge distribution can in general include both free charges and bound polarization charges. Let us first consider the case where the 'ρ' in (4.81) represents only the free charge density (this scenario is certainly the more practical consideration since free charges are the ones that we can directly control). In this case, as the free charge distribution is assembled, the resulting electric field induces polarization in the dielectric, represented by $\mathbf{P}$. The induced polarization weakens the electric field in the dielectric due to the shielding effect[120] of the effective bound polarization charge densities. Thus, considering only the free and induced polarization charge distributions, it appears that the emergence of the polarization charge reduces the energy required to assemble the free charge distribution. However, while bringing the free charges into their final configurations, no net work is done on the bound charges, since they are bound to their final positions. By calculating (4.81) for free charges only, we implicitly include an extra energy term: the energy necessary to polarize the dielectric material itself.[121] Writing out all the energy terms explicitly,

$$W_e = W_f + W_b + W_p \tag{4.82}$$

where W_f is the energy associated with the free charges, W_b is the energy associated with the bound charges, and W_p is the energy required to polarize the dielectric material. Since $W_b = -W_p$, by calculating W_f, we actually account for all energy terms, including the energy required to polarize the dielectric material. Thus, if we use the free charge density as the 'ρ' term in (4.81), we calculate both W_f and W_e. Hence ρ should indeed be solely the free charge density. If we instead include both the bound and free charge densities in (4.81), we calculate only $W_f + W_b$ and leave out the final term W_p in the equation for the total stored energy of the charge system.

Since the charge distribution in question is the free charge represented by ρ distributed in volume V, to convert (4.81) into an expression involving $\mathbf{D}$ and $\mathbf{E}$, we can substitute $\nabla \cdot \mathbf{D}$ for ρ:

$$W_e = \frac{1}{2}\int_V (\nabla \cdot \mathbf{D})\Phi\,dv$$

By using the vector identity

$$\nabla \cdot (\Phi\mathbf{D}) \equiv (\nabla \cdot \mathbf{D})\Phi + \mathbf{D} \cdot \nabla\Phi$$

we find

$$W_e = \frac{1}{2}\int_V (\nabla \cdot \Phi\mathbf{D} - \mathbf{D} \cdot \nabla\Phi)\,dv$$

---

[120]As illustrated in Figure 4.52, since the induced polarization is oriented in the same direction as $\mathbf{E}$ for isotropic materials, the electric field from the induced effective bound charge densities has a destructive effect on the electric field generated by externally applied charges inside the dielectric (see also Example 4.31).

[121]This energy may be thought of as a potential energy equivalent to stretching or compressing a mechanical spring. It is the energy required to "stretch" the bound charges from their equilibrium states.

Since for electrostatic fields $\mathbf{E} = -\nabla\Phi$, we have

$$W_e = \tfrac{1}{2} \oint_S \Phi\mathbf{D} \cdot \hat{\mathbf{n}}\, ds + \tfrac{1}{2} \int_V \mathbf{D} \cdot \mathbf{E}\, dv$$

where S is the surface that encloses volume V, and $\hat{\mathbf{n}}$ is the outward unit normal vector of the differential surface element ds. The surface S, and thus the volume V, must be large enough to enclose all the charges in the region of interest. Since there are no other restrictions on the choice of the surface S, we can extend it to infinity. By doing so, the surface integral term in W_e reduces to zero, since the dependencies of the various quantities at large distances from the charges (at infinity the entire charge distribution looks just like a point charge) are

$$\Phi \sim \frac{1}{r}, \qquad D \sim \frac{1}{r^2} \qquad \text{and} \qquad ds \sim r^2$$

We thus have

$$\lim_{r\to\infty} \Phi\mathbf{D} \cdot \hat{\mathbf{n}}\, ds \quad \to \quad \lim_{r\to\infty} \frac{1}{r}\frac{1}{r^2}r^2 \to 0$$

so the electrostatic energy W_e reduces to simply the volume integral term,

$$\boxed{W_e = \tfrac{1}{2} \int_V \mathbf{D} \cdot \mathbf{E}\, dv} \tag{4.83}$$

Since we let S extend to infinity, we must have $V \to \infty$, meaning that the integral in (4.83) has to be carried out over all space in which the electric field is nonzero. Based on the above, we can define the *volume energy density* for the electrostatic field as

$$\boxed{w_e = \tfrac{1}{2}\mathbf{D} \cdot \mathbf{E} = \tfrac{1}{2}\epsilon E^2} \tag{4.84}$$

We have thus found a way of expressing the stored energy of a charge distribution solely in terms of the electric field.

Note that the two alternative expressions we have found for electrostatic energy density, namely, $\tfrac{1}{2}\rho\Phi$ from (4.81) and $\tfrac{1}{2}\epsilon E^2$ from (4.84) are quite different. The former implies that energy exists where free charges exist and is zero where $\rho = 0$. However, the latter indicates that electrostatic energy exists wherever the fields exist. Both points of view have merit, and it is neither necessary[122] nor possible to determine which one is more "correct." The dilemma here is the same as that in the mechanical examples discussed earlier: The two masses at the ends of a compressed spring and the flowerpot

---

[122]For an interesting discussion, see Vol. II, Section 8 of *The Feynman Lectures on Physics,* Addison-Wesley, Reading, Massachusetts, 1964. A more advanced discussion is given in Chapter 4 of W. B. Cheston, *Elementary Theory of Electric and Magnetic Fields,* Wiley, 1964. It is possible to formulate the subject of electrostatics such that the electrostatic energy and the fields produced by a system of charges are fundamental quantities and the force between charged particles is a derived concept—an approach more fruitful for treatment of quantum phenomena. In the context of the experimentally based approach adopted here, the electric force, as defined in Coulomb's law, is taken to be the fundamental quantity.

on the windowsill. It is not possible to "localize" energy or to decide whether it is associated with the charge or the field. Thus, the quantities $\frac{1}{2}\rho\Phi$ and $\frac{1}{2}\epsilon E^2$ represent energy density only to the extent that their volume integral over space yields the total potential energy.

We can also write W_e in terms of the polarization $\mathbf{P} = \mathbf{D} - \epsilon_0\mathbf{E}$ as

$$W_e = \tfrac{1}{2}\int_V \epsilon_0 E^2 dv + \tfrac{1}{2}\int_V \mathbf{P}\cdot\mathbf{E}\, dv \tag{4.85}$$

In other words, to establish a given electric field $\mathbf{E}$ in a dielectric medium rather than in free space, additional energy needs to be supplied (by the source that sustains the field). This additional energy is given by the second integral in the last equation above and is the energy required to polarize the dielectric. We can also interpret (4.85) in the context of (4.82). The first term, $\frac{1}{2}\int_V \epsilon_0 E^2 dv$, calculates the energy required to assemble all of the charges (free and polarization charge densities) as if they were all free charges assembled in free space. Using the notation in (4.82), we have $W_f + W_b = \frac{1}{2}\int_V \epsilon_0 E^2 dv$. Hence the final term in (4.85) is the energy required to polarize the dielectric material: $W_p = \int_V \mathbf{P}\cdot\mathbf{E}\, dv$.

Energy stored in a capacitor.   Equation (4.81) or (4.84) can be used to derive the expression for energy stored in a capacitor, which we know from circuit theory to be $W_e = \frac{1}{2}CV^2$. Consider any configuration of two conductors constituting a general capacitor structure as shown in Figure 4.50. For simplicity, assume conductor 1 to be at potential $-V_0/2$ and conductor 2 to be at potential $V_0/2$, with reference to zero potential at infinity. Since the charge density is zero everywhere inside the conductors and is nonzero only on their surfaces, (4.81) reduces to

$$W_e = \tfrac{1}{2}\left[\int_{S_{c1}} \rho_{s1}\Phi_1\, ds + \int_{S_{c2}} \rho_{s2}\Phi_2\, ds\right]$$

where S_{c1} and S_{c2} are the surfaces of the two conductors. Using $\Phi_1 = -V_0/2$ and $\Phi_2 = V_0/2$, we have

$$W_e = \tfrac{1}{2}\left[\tfrac{-V_0}{2}\int_{S_{c1}} \rho_{s1}\, ds + \tfrac{V_0}{2}\int_{S_{c2}} \rho_{s2}\, ds\right]$$
$$= \tfrac{1}{2}\left[\tfrac{-V_0}{2}(-Q) + \tfrac{V_0}{2}(Q)\right] = \tfrac{1}{2}V_0 Q = \tfrac{1}{2}CV_0^2 \tag{4.86}$$

where Q and $-Q$ are the total surface charges residing on conductors 2 and 1, respectively, and where we have used $C = Q/V_0$.

Example 4.33: Energy of a parallel-plate capacitor. Consider the dielectric filled parallel-plate capacitor shown in Figure 4.60. The parallel plates, separated by a distance d and each with area A, are held at a potential difference V_0. Calculate the total energy stored in the capacitor.

Solution: Neglecting fringing effects and using the coordinate system shown in Figure 4.60, the electric field between the capacitor plates is given by (see Example 4.23)

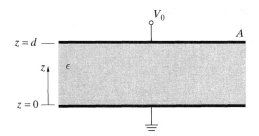

Figure 4.60 Energy in a capacitor. Parallel-plate capacitor filled with a dielectric.

$\mathbf{E} = -\hat{\mathbf{z}}V_0/d$. Using (4.83), the total energy stored, which equals the energy required to assemble the free charges on the conducting plates (W_f), is

$$W_e = \frac{1}{2}\int_V \epsilon \mathbf{E}\cdot\mathbf{E}dv = \frac{1}{2}A\int_0^d \epsilon\frac{V_0^2}{d^2}dz = \left(\frac{1}{2}A\epsilon_0\frac{V_0^2}{d}\right)\epsilon_r$$

where $\epsilon = \epsilon_r\epsilon_0$. Compared with the same capacitor filled with free space, using (4.69), the power supply responsible for maintaining the constant voltage V_0 supplies an additional energy in the amount (see (4.85) and subsequent discussion)

$$W_p = \frac{1}{2}\int_V \mathbf{P}\cdot\mathbf{E}dv = \frac{1}{2}A\int_0^d \epsilon_0(\epsilon_r-1)\frac{V_0^2}{d^2}dz = \left(\frac{1}{2}A\epsilon_0\frac{V_0^2}{d}\right)(\epsilon_r-1)$$

We can also directly calculate the total energy W_b required to assemble the final distribution of bound charges as if they were free charges assembled in a vacuum. From the induced polarization $\mathbf{P}$, there are two layers of bound surface charge: a negative layer at the top of the dielectric (at $z = d$) and a positive layer at the bottom of the dielectric (at $z = 0$). Using (4.61), these surface charge densities are

$$\rho_{sp}|_{z=d} = \mathbf{P}\cdot(\hat{\mathbf{z}})|_{z=d} = -\frac{V_0}{d}\epsilon_0(\epsilon_r-1)$$

$$\rho_{sp}|_{z=0} = \mathbf{P}\cdot(-\hat{\mathbf{z}})|_{z=0} = \frac{V_0}{d}\epsilon_0(\epsilon_r-1)$$

Using (4.19), the electric potential $\Phi(z)$ with respect to the lower plate is given by

$$\Phi(z) = -\int_{z'=0}^z -\hat{\mathbf{z}}\frac{V_0}{d}\cdot\hat{\mathbf{z}}dz' = \frac{V_0}{d}z$$

Hence, substituting ρ_p for ρ in (4.81) and converting the integration into a surface integral, we have

$$W_b = \frac{1}{2}\int_S \rho_p\Phi ds = \frac{1}{2}A\rho_{sp}(z=d)\Phi(z=d) = -\left(\frac{1}{2}A\epsilon_0\frac{V_0^2}{d}\right)(\epsilon_r-1)$$

where we used $\Phi(z=0) = 0$. As expected, the energy W_b associated with the bound charges is equal and opposite to the extra energy W_p required to polarize the dielectric. We note that the energy required to assemble the free and bound charges, excluding the energy required to polarize the dielectric, is

$$W_f + W_b = \left(\frac{1}{2} A \epsilon_0 \frac{V_0^2}{d} \right)$$

We could have arrived at this result more directly by simply evaluating (again with reference to (4.85) and subsequent discussion)

$$W_f + W_b = \frac{1}{2} \int_V \epsilon_0 E^2 dv$$

Example 4.34: Energy of a sphere of charge. Consider a sphere of radius b having uniform charge of volume density ρ in free space. Determine the electrostatic energy of this charge configuration using (a) (4.81) and (b) (4.84).

Solution: The potential $\Phi(r)$ and the electric field $E_r(r)$ were determined in Example 4.13:

$$E_r = \frac{1}{3} \frac{r\rho}{\epsilon_0} \qquad r < b \qquad \Phi(r) = -\frac{1}{6} \frac{r^2 \rho}{\epsilon_0} + \frac{1}{2} \frac{\rho b^2}{\epsilon_0}$$

$$E_r = \frac{1}{3} \frac{\rho b^3}{\epsilon_0 r^2} \qquad r \geq b \qquad \Phi(r) = \frac{1}{3} \frac{\rho b^3}{\epsilon_0 r}$$

where we have made the appropriate substitutions of $a \to b$ and $\rho \to Q/[(4\pi/3)a^3]$.

(a) Using (4.81), the stored energy is

$$W_e = \frac{1}{2} \int_V \rho \Phi(r)\, dv = \frac{1}{2} 4\pi \int_0^b \rho \Phi(r) r^2\, dr = \frac{2\pi \rho^2 \left[\dfrac{b^5}{6} - \dfrac{b^5}{30} \right]}{\epsilon_0} = \frac{4\pi}{15} \frac{\rho^2 b^5}{\epsilon_0}$$

Note that ρ is finite only where $r < b$, so that we only need to evaluate Φ for $r < b$.

(b) Since $D_r = \epsilon_0 E_r$, the stored energy from (4.84) is

$$W_e = \frac{1}{2} \int_0^\infty \mathbf{D} \cdot \mathbf{E}\, dv = \frac{1}{2} 4\pi \int_0^\infty D_r E_r r^2\, dr = \frac{2\pi \rho^2}{\epsilon_0} \frac{1}{9} \left[\int_0^b r^2 r^2 dr + \int_b^\infty \frac{b^6}{r^4} r^2 dr \right]$$

$$= \frac{2\pi \rho^2}{\epsilon_0} \frac{1}{9} \left[\frac{b^5}{5} + b^5 \right] = \frac{4\pi \rho^2 b^5}{15 \epsilon_0}$$

As expected, this result is identical to the stored energy as calculated from (4.81).

Example 4.35: Electrostatic energy around a charged conducting sphere. Consider a conducting sphere of radius a carrying a total charge of Q in free space. Determine the electrostatic energy stored around this sphere using (a) (4.86) and (b) (4.84).

Solution: The electrostatic potential for such a conducting sphere was determined in Example 4.25. The variations with r of the electric field $\mathbf{E}(r)$ and the potential $\Phi(r)$ are

$$\Phi(r) = \begin{cases} \dfrac{Q}{4\pi\epsilon_0 r} & r \geq a \\[2mm] \dfrac{Q}{4\pi\epsilon_0 a} & r < a \end{cases} \qquad \mathbf{E}(r) = \begin{cases} \hat{\mathbf{r}}\,\dfrac{Q}{4\pi\epsilon_0 r^2} & r \geq a \\[2mm] 0 & r < a \end{cases}$$

The capacitance of a charged conducting sphere is given by (4.50) as $C = 4\pi\epsilon_0 a$.

(a) We first calculate the stored energy by considering the sphere as a capacitor. Noting that the sphere is at a potential $V_0 = \Phi(r = a)$, the energy stored is given by

$$W_e = \frac{1}{2}QV_0 = \frac{1}{2}Q\left(\frac{Q}{4\pi\epsilon_0 a}\right) = \frac{Q^2}{8\pi\epsilon_0 a}$$

(b) We can also determine the stored energy by integrating the electric field over the entire space within which it exists (i.e., it is nonzero), by using (4.84). We have

$$W_e = \frac{1}{2}\int_V \epsilon_0 E^2 \, dv = \frac{1}{2}\int_0^{2\pi}\int_0^{\pi}\int_a^{\infty} \epsilon_0 \left(\frac{Q}{4\pi\epsilon_0 r^2}\right)^2 r^2 \sin\theta \, dr \, d\theta \, d\phi$$

$$= \frac{1}{2}(4\pi)\int_a^{\infty} \epsilon_0\left(\frac{Q}{4\pi\epsilon_0}\right)^2 \frac{1}{r^2} \, dr = \frac{Q^2}{8\pi\epsilon_0}\left[-\frac{1}{r}\right]_a^{\infty} = \frac{Q^2}{8\pi\epsilon_0 a}$$

which is the same as that found in (a).

4.13 ELECTROSTATIC FORCES

The force experienced by a conducting body placed in an electric field was discussed in Section 4.7.5. The magnitude of this force per unit area at a point on the conductor on which an electric field E_n is incident is given by $\epsilon_0 E_n^2/2 = \rho_s^2/(2\epsilon_0)$, where ρ_s is the surface charge density at the same point. The fact that there is a force of attraction between the plates of a parallel-plate capacitor is thus to be expected, because of the presence on the plates of charges of opposite sign. In calculating the electrostatic force on a charged conducting body, it is in principle possible to simply integrate the force per unit area [i.e., $\rho_s^2/(2\epsilon_0)$] over its entire surface. In fact, electrostatic forces on and between charged conductors are special cases of the subject of electrostatic force on matter in general, which can be formulated entirely in terms of electric fields that produce stress in the media which they permeate.[123]

---

[123]Faraday was the first to speak of lines of force as elastic bands that transmit tension and compression. Maxwell also spent considerable time on this concept and placed Faraday's notions into clear mathematical focus. For brief discussions of the concept of electromagnetic stress see C. C. Johnson, *Field and Wave Electrodynamics,* Section 1.19, McGraw-Hill, New York, 1965; W. B. Cheston, *Elementary Theory of Electric and Magnetic Fields,* Sections 10.7–10.8, Wiley, New York, 1964. For more complete coverage, see L. M. Landau and E. M. Lifshitz, *Electrodynamics of Continuous Media,* Addison-Wesley, Reading, Massachusetts, 1960; J. A. Stratton, *Electromagnetic Theory,* McGraw-Hill, New York, 1941.

In this book we confine our attention to selected special cases involving simple field geometries. To calculate the electrostatic forces, we consider how the energy of the system varies due to a small virtual change in geometry. This method, referred to as the *principle of virtual work,* provides an entirely satisfactory means of determining the electrostatic forces for the cases considered. Any component of a force on an object at rest has a definite meaning only when we have defined the way in which the object might move in the direction indicated, at least infinitesimally—that is, if we can consider the force in question to perform at least an infinitesimal amount of work. The object may of course be in equilibrium under a balance of electrical and mechanical forces; by *motion* we mean motion against the mechanical restraints. If, for example, the possible motion is in the x direction, the work done by an external agent F_{ext} for a displacement dx would be $F_{ext}\, dx$. The calculation of this *virtual* work thus provides a means to calculate the opposing internal electrical force $F_x = -F_{ext}$.

In the next section we illustrate the application of the principle of virtual work to determine the electrostatic force on a charged capacitor plate. We then demonstrate how the principle of virtual work can be used to determine the force on a dielectric slab.

4.13.1 Electrostatic Forces on Charged Conductors

When calculating electrostatic forces using the principle of virtual displacement, the conductors can be kept isolated (i.e., constant charge) or be maintained at a constant potential (i.e., by externally connected batteries) during their virtual displacement. The electrostatic force should not depend on this choice, since it could in principle be calculated from $\rho_s^2/(2\epsilon_0)$ as stated above. However, the change in the energy of the system due to a displacement dx is different based on whether or not energy is available from the batteries. It is thus necessary to consider both possibilities separately. In this section we illustrate the application of the principle of virtual work to determine the electrostatic force on a capacitor plate using first, a constant-charge constraint, and then next, a constant-potential constraint.

Constant-Charge. Consider a parallel-plate capacitor in free space with a total charge of $+Q$ on one plate and $-Q$ on the other, as shown in Figure 4.61a. We can

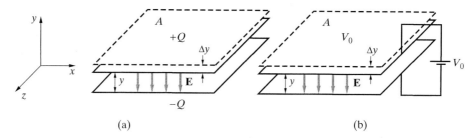

(a)   (b)

Figure 4.61 Virtual displacement of one plate in a parallel-plate capacitor.
(a) Charge kept constant. (b) Potential difference between plates kept constant.

visualize the upper plate as being displaced upward by an amount Δy as a consequence of applying an external force F_{ext}. From conservation of energy, this mechanical work must reappear as energy elsewhere, and in this case the stored energy of the electric field is the only other energy term that could possibly be involved. The energy stored in the electric field must therefore have increased, and the electrostatic force F_y that must be overcome by the external force to move the plates further away by Δy is related to the increase in the stored energy of the system as

$$-F_y \, \Delta y = F_{\text{ext}} \, \Delta y = \Delta W_e|_Q \tag{4.87}$$

where $\Delta W_e|_Q$ denotes the change in stored energy with the charge held constant. The increase in stored energy can be determined from

$$\Delta W_e = W_e|_{\text{after}} - W_e|_{\text{before}} = \frac{1}{2} \frac{Q^2}{C_{\text{after}}} - \frac{1}{2} \frac{Q^2}{C_{\text{before}}}$$

$$= \frac{1}{2} \frac{Q^2(y + \Delta y)}{\epsilon_0 A} - \frac{1}{2} \frac{Q^2 y}{\epsilon_0 A} = \frac{Q^2 \, \Delta y}{2\epsilon_0 A} = \frac{\epsilon_0}{2} E_y^2 A \, \Delta y$$

where we use the fact that $\epsilon_0 E A = Q$, as required by Gauss's law, and $C = \epsilon_0 A/y$, for the capacitance of the parallel plates separated by an amount y and of area A (Example 4.26), and also neglected fringing fields near the edges of the plates. Consequently, we have

$$-F_y \, \Delta y = \frac{\epsilon_0}{2} E^2 A \, \Delta y \quad \rightarrow \quad \boxed{F_y = -\frac{\epsilon_0 E^2 A}{2}} \tag{4.88}$$

Note that the magnitude of the force per unit area, as calculated using the principle of virtual work, is the same as the magnitude that was deduced in Section 4.7.5 from field considerations, namely $\epsilon_0 E_n^2/2 = \rho_s^2/(2\epsilon_0)$. For the case of this simple geometry, we could have calculated the total force F_y on the capacitor plate by simply integrating $\epsilon_0 E_n^2/2$ over the area of the plates.

In the parallel-plate capacitor example considered above, the virtual displacement increased the separation distance between the capacitor plates by Δy. Since the capacitance varies inversely with the separation distance, the change in stored energy as a result of this virtual displacement is linearly proportional to Δy. As a result, the common Δy terms cancel in (4.88). In the general case, the change in energy may not be a linear function of the virtual displacement term, and so we need to instead consider a differential change dy in separation distance:

$$dW_e|_Q = -F_y \, dy \quad \rightarrow \quad F_y = -\frac{dW_e|_Q}{dy} \tag{4.89}$$

We note that in the constant-charge configuration, where there are no external sources of energy, the internal electrostatic force acts in a direction to minimize the internally stored electrostatic energy.[124] We can simplify (4.89) by direct substitution of $W_e|_Q = Q^2/(2C)$ to obtain

$$F_y = -\frac{dW_e|_Q}{dC}\frac{dC}{dy} = \frac{Q^2}{2C^2}\frac{dC}{dy} \tag{4.90}$$

We return now to the parallel-plate capacitor example depicted in Figure 4.61a. Substituting $C = \epsilon_0 A/y$ into (4.90) and again noting that $\epsilon_0 EA = Q$, we have

$$F_y = \frac{Q^2}{2C^2}\left(\frac{-\epsilon_0 A}{y^2}\right) = -\frac{\epsilon_0 E^2 A}{2}$$

which matches our result above.

Constant voltage.   Figure 4.61b depicts a parallel-plate capacitor held at a constant potential V_0. In this configuration, when the plates are separated by an increment Δy due to an external force F_{ext}, the energy associated with the removal of charge by the battery from the plates to maintain the same potential must also be considered. The total energy increase in the electric field of the capacitor is equal to the mechanical work supplied by the external force moving the plates against the internal electrical force, plus the energy supplied by the battery to deposit an additional ΔQ charge against the voltage V_0:

$$\Delta W_e|_V = F_{\text{ext}}\Delta y + V_0\Delta Q \tag{4.91}$$

where $W_e|_V$ represents the energy stored in a capacitor at a fixed potential. Thus, in a constant-voltage configuration, the internal electrostatic force F_y is related to the change in stored energy as

$$F_y\Delta y = -\Delta W_e|_V + V_0\Delta Q \tag{4.92}$$

In terms of differentials,

$$F_y dy = -dW_e|_V + V_0 dQ \quad \rightarrow \quad F_y = -\frac{dW|_V}{dy} + V_0\frac{dQ}{dy} \tag{4.93}$$

Using (4.86) and $Q = CV_0$, (4.93) becomes

$$F_y = -\frac{d}{dy}\left(\frac{1}{2}CV_0^2\right) + V_0^2\frac{dC}{dy} = \frac{V_0^2}{2}\frac{dC}{dy} \tag{4.94}$$

Substituting $V_0 = Q/C$ into (4.94) yields (4.90), which was derived using the constant-charge configuration.

---

[124]This result may be generalized to a force acting in three dimensions by considering a virtual displacement in the x, y, and z directions:

$$\mathbf{F} = -\frac{\partial W_e|_Q}{\partial x}\hat{\mathbf{x}} - \frac{\partial W_e|_Q}{\partial y}\hat{\mathbf{y}} - \frac{\partial W_e|_Q}{\partial z}\hat{\mathbf{z}} = -\nabla W_e|_Q$$

4.13.2 Electrostatic Forces in the Presence of Dielectrics

If metallic conductors are in the presence of a dielectric of finite extent (i.e., instead of being immersed in a dielectric permeating all space), then any virtual displacement of the dielectric is also expected to change the stored energy, depending on the geometries involved. Since change in stored energy under virtual displacement indicates the presence of mechanical force, it follows that dielectric bodies in electrostatic fields will experience a net force. The magnitude of this force can be calculated using the principle of virtual work, in a manner similar to the case of conducting bodies. We illustrate this technique in the following example.

Example 4.36: Force on a dielectric slab. Consider a uniform dielectric slab of permittivity ϵ partially inserted between the plates of a parallel-plate capacitor of width w, and separation d, as shown in Figure 4.62 and having depth L (not shown). The capacitor plates are held at a constant potential difference V_0. Determine the electrostatic force acting on the dielectric slab.

Figure 4.62   Force on a dielectric slab between capacitor plates.

Solution: Suppose that the portion of the dielectric between the plates is of length x. In this case, the deflection coordinate is in the x direction. Assuming a constant-voltage configuration as shown in Figure 4.62, equation (4.94) becomes

$$F_x = \frac{V_0^2}{2}\frac{dC}{dx} = \frac{E_y^2 d^2}{2}\frac{dC}{dx} \tag{4.95}$$

where we used $V_0 = E_y d$. To evaluate the total capacitance C, we can take the usual approach of evaluating the total charge Q on the capacitor plates under the assumed potential difference V_0. Using Gauss's law (equation (4.32)), the total charge residing on the capacitor plate adjacent to the dielectric slab is $\epsilon E_y L x$. Similarly, the total charge on the plate adjacent to free space is $\epsilon_0 E_y L(w - x)$. Hence the total capacitance is

$$C = \frac{Q}{V_0} = \frac{\epsilon E_y L x}{E_y d} + \frac{\epsilon_0 E_y L(w - x)}{E_y d} + C_{\text{fringing}} = \frac{\epsilon L x}{d} + \frac{\epsilon_0 L(w - x)}{d} + C_{\text{fringing}}$$

where we have noted the effect of the fringing field on the capacitance explicitly. As long as the ends of the slab are not too close to the plate edges, the fringing term does not change much with the virtual displacement, so we can simplify dC/dx to

$$\frac{dC}{dx} = \frac{L\epsilon}{d} - \frac{L\epsilon_0}{d} + \frac{d(C_{\text{fringing}})}{dx} \simeq \frac{L}{d}(\epsilon - \epsilon_0) \tag{4.96}$$

Substituting (4.96) into (4.95) gives

$$F_x = \frac{E_y^2 d^2}{2} \frac{L(\epsilon - \epsilon_0)}{d} = \frac{1}{2} E_y^2 (\epsilon - \epsilon_0) L d$$

Hence the direction of the force is such as to draw the dielectric slab further into the region between the capacitor plates.

It is interesting to calculate the work W required (i.e., the energy that needs to be supplied by the batteries) to place the dielectric slab completely inside the capacitor. The total work to move the slab from $x = 0$ to $x = w$ can be rewritten as

$$W = \int_0^w F_x \, dx = \frac{1}{2} E_y^2 (\epsilon - \epsilon_0) L w d = \frac{1}{2} \int_V \epsilon \mathbf{E} \cdot \mathbf{E} \, dv - \frac{1}{2} \int_V \epsilon_0 \mathbf{E} \cdot \mathbf{E} \, dv$$

$$= \frac{1}{2} \int_V (\epsilon_0 \mathbf{E} \cdot \mathbf{E} + \mathbf{P} \cdot \mathbf{E}) \, dv - \frac{1}{2} \int_V \epsilon_0 \mathbf{E} \cdot \mathbf{E} \, dv = \frac{1}{2} \int_V \mathbf{P} \cdot \mathbf{E} \, dv$$

which is the additional energy needed to polarize the inserted dielectric, as was discussed in Section 4.12.

The physical origin of electrostatic forces on dielectrics.

Although we were able to calculate the force on the dielectric slab using the principle of virtual work, the physical origin of the force on a dielectric is interesting in its own right. This question relates to our most basic everyday experiences with electrostatics, which were discussed at the start of this chapter. Why does a charged (e.g., by rubbing) object pick up small dielectric objects? Why did Thales of Miletus observe that pieces of amber rubbed in silk attracted small pieces of straw? At first thought, one might conclude that the amber had one kind of charge and that the straws had the opposite charge. But the straws were not rubbed onto anything and thus were electrically neutral. Although they do not have any net charge, they were nevertheless attracted to the charged amber. What was the physical reason for this attraction?

The answer lies in the fact that the electric field produced by the charged amber (or any charged object of finite size) is nonuniform. As shown in Figure 4.63, a dielectric placed in an electric field is polarized. There are polarization charges of both signs, which experience attraction and repulsion forces due to the electric field. However, there is net attraction, because the electric field nearer to the source (at the top in Figure 4.63) is stronger than that farther away. If the electric field were uniform (constant intensity everywhere), there would be no net attraction. Indeed, if pieces of straw are placed between the plates of a large parallel-plate capacitor, they are not attracted to either of the plates, regardless of the intensity of the electric field. The spatial variation (or the nonuniformity) of the electric field is the fundamental reason why dielectrics experience a force in the presence of electric fields. A dielectric always tends to move towards regions of stronger electric fields, as illustrated in Figure 4.63.

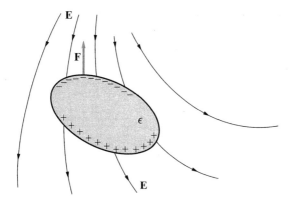

Figure 4.63 A dielectric object in an electric field. The force experienced by the object is proportional to the gradient of the square of the magnitude of the electric field. The dielectric tends to move toward regions of higher electric field.

As we have seen in Example 4.36, the force experienced by the dielectric depends on the square of the field. In fact, it can be shown that for small objects the force on a dielectric is proportional to the gradient of the square of the electric field, $-\nabla E^2$. This is because the induced polarization charges are proportional to the electric fields, and for given distributions of charges the electrostatic force is proportional to the electric field. As a result, the object experiences a net force only if the square of the field varies in space. The constant of proportionality that determines the magnitude of the force depends on the dielectric constant of the object as well as its size and shape. The calculation of the electrostatic force on a dielectric object can in general be a quite difficult problem to solve.

The case of the dielectric slab between parallel plates considered in Example 4.36 is thus a very special case, where the calculation of the force is facilitated by the principle of virtual work. In actual fact, the force on the dielectric slab is produced by the nonuniform fringing fields (see Figure 4.62), whose intensity decreases with distance away from the plates. The direct calculation of the force would require the accurate determination of these fringing fields, evaluation of the gradient of the square of magnitude of the electric field, and its integration over the entire body of the dielectric. It is clear that the principle of virtual work comes in rather handy in this case.

4.14 MICROELECTROMECHANICAL SYSTEMS (MEMS)

Microelectromechanical systems (MEMS[125]) are miniature devices that integrate electronic circuits with mechanical machines that interact with the surrounding physical environment. The field of MEMS has its roots in the integrated circuit (IC) industry. One of the primary advantages of MEMS technology is the ability to integrate mechanical systems with electronic circuits—a process called monolithic integration. This ability

---

[125]One of the most widely referenced early papers on MEMS was published by K.E. Petersen: "Silicon as a Mechanical Material," *Proceedings of the IEEE*, 1982, 70(5), p. 420–457. For a modern reference on many of the broad applications covered by MEMS technology at an appropriate level, see C. Liu *Foundations of MEMS*, 2nd ed., Pearson, 2012.

reduces the overall manufacturing costs and allows for greater uniformity among the constituent electrical and mechanical components.

Since the transistor was invented in 1947 at the AT&T Bell Laboratories, the blistering pace of innovation and advancements in integrated circuits has steadily pushed forward the capabilities of microfabrication and miniaturization. Against enormous technical challenges, Moore's Law—an observation by Gordon Moore, one of the co-founders of Intel, that the density of integrated transistors doubles every 12–18 months—remains valid even today. This continuously advancing fabrication technology forms the backdrop that enabled the functional integration of mechanical devices with analog and digital electronics. Today, three-dimensional mechanical structures can be machined out of bulk or thin film silicon, often on the same silicon die.

A large portion of commercial applications of MEMS technology is centered around sensors and actuators. In general, sensors provide a way to monitor some physical environmental variable, such as pressure, temperature, or incident radiation. Actuators, on the other hand, take an input control signal, such as a voltage or current and produce a force or torque in order to generate mechanical movement. Sensors and actuators are both examples of transducers, which transform power from one energy domain (such as mechanical or thermal) to another (such as electrical).

Our study of forces resulting from charge separation on a capacitor provides the theoretical basis for a common class of MEMS transducers: electrostatic sensors and actuators. Electrostatic transducers convert power between electrical and mechanical domains. As a basic example, consider again the parallel-plate capacitor. If the distance between the two plates changes as a result of an external force, the overall capacitance also changes. This principle forms the basis for electrostatic sensing of position. Conversely, if more or less charge were deposited on the capacitor, perhaps by varying the voltage or by using a current source, the attractive force on the plates would change. If there was some mechanical means of movement, the surface would also change position, thus forming a simple actuator. The length scales of MEMS devices generally range from 1 μm to 1 cm. Due to their small size, the gravitational force is usually negligible compared with the electrostatic force, and can be ignored.

Electrostatic sensors and actuators are found in a wide array of modern commercial products. We review two such products, one involving electrostatic sensors and the other electrostatic actuators, in order to illustrate some of the possibilities enabled by MEMS technology: automobile air-bags and the Digital Light Processing (DLP) chip developed by Texas Instruments for video projection systems.

Air-bag systems in automobiles are designed to detect rapid changes in acceleration (such as the rapid deceleration during a collision) in order to deploy the bag. This industry was one of the first to use accelerometers designed using MEMS technology. In the typical configuration, two electrodes consisting of interdigitated fingers fabricated on a silicon substrate are used (see Figure 4.69 in Example 4.38). This pair of electrodes has an associated capacitance, determined by the size of the fingers and the distance between them. One electrode is held fixed, while the other is suspended in equilibrium and is connected to a mechanical spring with spring constant k. If the mass of the suspended electrode is m and the device is accelerating at a rate a, there is a force $F = ma$ on the suspended electrode. This force pushes the movable electrode toward the fixed electrode, changing the distance between them, thereby changing the overall capacitance. This small change in capacitance is then detected by electronics that are fabricated on the same piece of silicon, which then relays a signal to deploy the air bag. Integrating the mechanical structure (the suspended electrode) with the sensing electronics helps to reduce the cost of this important safety component by eliminating the need for manual integration of the electrical and mechanical components.

The DLP projection system,[126] which was invented by Dr. Larry Hornbeck of Texas Instruments in 1987, uses an array of torsion hinge-mounted microscopic mirrors to reflect a light source onto a projection screen. Each mirror, which measures approximately $16 \times 16 \ \mu m^2$, is responsible for one pixel of the projected image. The monolithic integration of the mirrors and logic circuitry enables individual control over a large, dense array: up to 2 million mirrors can be micromachined on a single semiconductor chip. The mirror itself is mounted on a plate, called a yoke, that is free to rotate around a torsion hinge by a deflection angle of $\pm 10°$. The deflection angle is kept precise by using mechanical stops. Electrodes are placed on either side of the torsion hinge. When the yoke and the electrodes are uncharged, the plate is stable at zero deflection, held in place by the mechanical restoring force of the torsion hinge. When the mirror and yoke are charged up to a bias potential, then applying a voltage differential across the electrodes causes an electrostatic attraction toward the electrode with a larger net opposite charge. To change the orientation of the mirror, the mirror bias voltage is removed, allowing the restoring force of the torsion rod to bring the mirror flat. The bias voltage is then applied again, causing the mirror to rotate to the new state of the electrodes. While each mirror has only two states, "on" (reflecting light toward the screen) or "off" (reflecting light away from the screen), fast switching of the state allows shades of grey by controlling the percent of time the mirror is in the "on" state. Each mirror can oscillate in the kHz range to create an illusion of a grey scale.

In the following sections, we revisit the attractive force between the two plates in a capacitor and show how this force may be used to design an electrostatic actuator. We consider both a charge-controlled and a voltage-controlled actuator.

4.14.1 Parallel-Plate Electrostatic Actuators

Consider the model of a parallel-plate electrostatic actuator shown in Figure 4.64. The capacitor plates each have an area A and are separated by a gap distance d. The bottom plate is anchored, and the top plate moves with a mechanical restoring force dictated by the spring constant k. The origin is chosen at the position of the top plate when there is no charge on the capacitor and no force from the spring, with y increasing as the top plate moves toward the bottom plate. We define d_0 to be the gap distance with $y = 0$, so $d(y) = d_0 - y$. The capacitance is

$$C = \frac{A\epsilon}{d} = \frac{A\epsilon}{d_0 - y} \tag{4.97}$$

---

[126]For a modern overview reference, see J. Hornbeck, Combining digital optical MEMS, CMOS and algorithms for unique display solutions. *Electron Devices Meeting, 2007. IEDM 2007. IEEE International.*, 2007.

Figure 4.64 Model of a parallel-plate electrostatic actuator. The bottom plate of a parallel-plate actuator is held fixed, while the top plate is suspended by a spring with spring constant k and is free to move in the vertical direction. The two plates have area A and are separated by distance d. Two control mechanisms are shown: (a) Charge control, where the current source controls the charge Q on each plate and (b) Voltage control, where the voltage source controls the potential difference Φ across the capacitor plates.

In this coordinate system, $\mathbf{F}_{\text{spring}} = -ky\hat{\mathbf{y}}$. We assume that the restoring force of the spring remains linear throughout the range of motion.

As before, we assume an ideal parallel plate capacitor, ignoring fringing fields at the edge of the capacitor plates and assuming the electric field is constant between the plates. This assumption introduces little error when the linear dimensions of the plates are much larger than the gap length d. We also consider the general case where the permittivity of the material between the two plates is uniform and equal to ϵ. From (4.88), and noting the coordinate system in Figure 4.64, we can write the electrostatic force pulling the two plates together as

$$\mathbf{F} = \frac{\epsilon E^2 A}{2}\hat{\mathbf{y}} \tag{4.98a}$$

$$= \frac{\epsilon \Phi^2 A}{2d^2}\hat{\mathbf{y}} \tag{4.98b}$$

$$= \frac{Q^2}{2A\epsilon}\hat{\mathbf{y}} \tag{4.98c}$$

For (4.98b) we used the fact that[127] $\Phi = Ed$, as shown in Example 4.23. Equation (4.98c) can be derived from (4.98b) using the capacitance of the parallel plate capacitor, $C = Q/\Phi = \epsilon A/d$.

There are two ways to control the electrostatic force between the two plates: by controlling the charge Q, through a current source, or by controlling the potential difference Φ, through a voltage source. These two configurations are shown in Figure 4.64a and b. When the force is expressed in terms of the charge Q, it is independent of the

---

[127]In this section, we use Φ instead of V_0 for the potential difference between the capacitor electrodes since this value is not fixed in a transducer. The time dependence of $\Phi(t)$ is implicit: it changes either in response to an external agent changing the capacitance (sensor) or due to an internal agent controlling the charge or voltage (actuator).

distance d between the two plates (equation (4.98c)), whereas the force depends on the gap distance when expressed in terms of the potential difference Φ (equation (4.98b)). We consider these two configurations separately.

Charge control. We first consider the case of moving the actuator by controlling the amount of charge on the capacitor plates. Since the electrostatic force is independent of the gap distance d when expressed in terms of the charge Q, the equations dictating the gap distance as a function of the charge are easy to derive. This independence of gap distance leads to some desirable properties in terms of the mechanical considerations of the design.

The total force acting on the upper plate is the vector sum of the electrostatic and spring forces: $\mathbf{F} + \mathbf{F}_{\text{spring}}$. At equilibrium, the total force must be zero. Therefore, at equilibrium we have

$$\frac{Q^2}{2A\epsilon} = ky \quad \rightarrow \quad y_{\text{eq}} = \frac{Q^2}{2kA\epsilon} \tag{4.99}$$

The gap between the two plates is thus

$$d = d_0 - y_{\text{eq}} = d_0 - \frac{Q^2}{2kA\epsilon} \tag{4.100}$$

The gap of the charge-controlled actuator monotonically decreases as the charge increases and is therefore stable over the range of valid movement. Eventually, assuming the full range of motion is possible, the gap becomes zero when enough charge is placed on the capacitor. The total amount of charge Q_0 needed to close the gap entirely is

$$d = 0 \quad \rightarrow \quad Q_0 = \sqrt{d_0 2k\epsilon A} \tag{4.101}$$

The stable deflection as a function of the deposited charge is a valuable attribute of the charge-controlled actuator, since it enables a full range of motion as a function of the charge Q. However, the above formulation ignored the parallel parasitic capacitance C_p between the mobile electrode and ground. Since capacitances involved in such applications are typically small, on the order of femto-Farads, parasitic capacitances, which alter the amount of charge that must be deposited by the current source, provide a practical limitation to a charge-control design approach. A parasitic capacitance that is at least half as large as the actuator capacitance at equilibrium can cause a so-called charge pull-in effect, where a positive feedback loop occurs after a critical amount of charge is deposited on the electrodes and the top plate snaps down.[128]

Example 4.37: Potential difference across a charge-controlled actuator. Find the maximum potential difference across a charge-controlled actuator. What is the corresponding gap distance?

---

[128]For a detailed discussion of this effect, see J. Seeger and B. Boser. Charge control of parallel-plate, electrostatic actuators and the tip-in instability, *Journal of Microelectromechanical Systems* 12(5), pp. 656–671, 2003.

Solution: Since the electric field is constant, the potential difference across the plates is given by (using (4.100))

$$\Phi = Ed = \frac{Q}{\epsilon A} d = \frac{Q}{\epsilon A}\left(d_0 - \frac{Q^2}{2kA\epsilon}\right) \tag{4.102}$$

To find the maximum potential difference, we differentiate (4.102) with respect to Q, set the result to 0, and first solve for the maximum charge:

$$\frac{d\Phi}{dQ} = \frac{d_0}{\epsilon A} - \frac{3Q^2}{2kA^2\epsilon^2} = 0 \quad \rightarrow \quad Q_{max} = \sqrt{\frac{d_0 2kA\epsilon}{3}} = \frac{Q_0}{\sqrt{3}} \tag{4.103}$$

where Q_0 is given by (4.101). Substituting Q_{max} into (4.102), we have

$$\Phi_{max} = \frac{Q_{max}}{\epsilon A}\left(d_0 - \frac{Q_{max}^2}{2kA\epsilon}\right) = \frac{\sqrt{d_0 2kA\epsilon/3}}{\epsilon A}\left(d_0 - \frac{d_0 2kA\epsilon/3}{2kA\epsilon}\right)$$

$$= \sqrt{\frac{8}{27}\frac{d_0^3 k}{A\epsilon}} \tag{4.104}$$

The deflection distance at this charge level is (using (4.99))

$$y_{max} = \frac{Q_{max}^2}{2kA\epsilon} = \frac{d_0 2kA\epsilon/3}{2k\epsilon A} = \frac{d_0}{3} \tag{4.105}$$

and so the gap width at the maximum voltage point is $d_{max} = d_0 - d_0/3 = 2d_0/3$.

The normalized voltage and deflection distance are plotted as a function of the deposited charge in Figure 4.65.

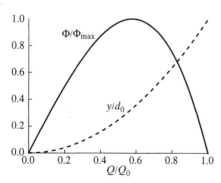

Figure 4.65 Plot of normalized deflection and potential difference. The potential difference across the charge-controlled capacitor and the deflection distance y are plotted against the normalized charge Q/Q_0. The maximum potential difference occurs at $Q/Q_0 = 1/\sqrt{3} \simeq 0.577$. At this point, the separation between the upper and lower plate is $d = 2d_0/3$.

Voltage control. Unlike the charge control configuration, the voltage control actuator design is not susceptible to parasitic capacitances (since a parallel capacitance C_p does not alter the potential difference across the actuator electrodes) and is therefore a popular choice in practice. However, this ease of implementation comes at the cost of a reduced range of motion.

For a voltage-controlled actuator, we begin with (4.98b), which expresses the electrostatic force on the plates in terms of the potential difference. As before, we seek to find the equilibrium position of the upper plate by setting the total force to zero. This equilibrium is achieved when

$$\frac{\epsilon \Phi^2 A}{2d^2} = ky \quad \rightarrow \quad y_{eq} = \frac{\Phi^2 A \epsilon}{2kd^2} = \frac{\Phi^2 A \epsilon}{2k(d_0 - y_{eq})^2} \tag{4.106}$$

Because the electrostatic force is inversely proportional to the square of the gap separation d and therefore the deflection distance y, the solution for the equilibrium position cannot be derived as easily as with the charge-controlled configuration. The resulting equation for y_{eq} is a cubic function, and so there may be more than one solution in the allowable range of motion.

The characteristics of the solutions can be more readily seen by solving for the potential difference in (4.106) as a function of displacement:

$$\Phi = \sqrt{\frac{y_{eq} 2k}{A\epsilon}}(d_0 - y_{eq}) \tag{4.107}$$

The potential difference, normalized by the maximum value (4.104) found in the charge-controlled example, is plotted in Figure 4.66. As seen, for each potential difference, there are two equilibrium values y that balance out the electrostatic and spring forces.

While the forces cancel at each solution satisfying (4.106), the solution also needs to be stable. Let F_{tot} be the net force in the $\hat{y}$ direction. At a solution point that satisfies (4.106), $F_{tot} = 0$. At a stable solution, the incremental net force ΔF resulting from a small downward deflection Δy would bring the top plate back up to the equilibrium position. For small deflections,

$$\Delta F = \frac{\partial F_{tot}}{\partial y} \Delta y \tag{4.108}$$

Thus, the derivative of the total force at a stable equilibrium point must satisfy

$$\left. \frac{\partial F_{tot}}{\partial y} \right|_{y=y_{eq}} < 0 \tag{4.109}$$

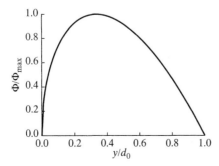

Figure 4.66 Potential difference versus displacement of a voltage-controlled parallel plate actuator. For each potential difference, there are two solutions for the deflection y. Φ_{max} is given by (4.104).

For the voltage-controlled parallel-plate actuator using (4.98b), the net force in the $\hat{y}$ direction is

$$F_{\text{tot}} = \frac{\Phi^2 A\epsilon}{2(d_0 - y)^2} - ky \tag{4.110}$$

and so for stability, by applying (4.109), we need

$$\frac{\Phi^2 A\epsilon}{(d_0 - y)^3} - k < 0 \tag{4.111}$$

Solving this inequality for y and using (4.106), the stable deflection range is

$$y < \frac{d_0}{3}. \tag{4.112}$$

Referring to Figure 4.66 and based on (4.105), we see that the voltage indeed reaches a peak at $y/d_0 = 1/3$. So, based on (4.112), the solutions to the left of this point, where $y < d_0/3$, are stable; those to the right are unstable. This graph also indicates that the maximum potential difference is the same as that found in the charge-controlled actuator, which can also be verified by substituting $y = d_0/3$ into (4.107). If the potential difference is increased further, the actuator enters the unstable region. The instability is due to a destabilizing positive feedback from the voltage source. When the actuator deflects further (d decreases), the capacitance, given by $C = \epsilon A/d$, increases. As a result, the voltage source adds more charge to maintain the same voltage, causing the actuator to deflect further. This instability can also be explained by noting that, when expressed in terms of potential difference, the electrostatic force is inversely proportional to the gap distance squared (equation (4.98b)). In the unstable region, as the plates move even closer together, the electrostatic force increases at a faster rate than the restoring spring force. As a result, the upper plate snaps in until it hits the other plate or until some other mechanical failure occurs. The maximum potential difference given by 4.104 is therefore also called the "Pull-in" voltage:

$$V_{\text{PI}} \equiv \sqrt{\frac{8}{27} \frac{d_0^3 k}{A\epsilon}} \tag{4.113}$$

We now proceed with a full graphical solution to (4.106). The family of solutions may be plotted quite generally by normalizing the electrostatic and spring forces by $d_0 k$. The normalized spring force is $F_{\text{spring}}/(d_0 k) = -y/d_0$, and the normalized electrostatic force using (4.98b) and (4.113) is

$$\frac{F_y}{d_0 k} = \frac{\Phi^2 A\epsilon}{2 d_0 k (d_0 - y)^2} = \frac{\Phi^2 A\epsilon}{2 d_0^3 k \left(1 - \dfrac{y}{d_0}\right)^2} = \Phi^2 \frac{4}{27} \left(\frac{27 A\epsilon}{8 d_0^3 k}\right) \frac{1}{\left(1 - \dfrac{y}{d_0}\right)^2}$$

$$= \frac{4}{27} \frac{\Phi^2}{V_{\text{PI}}^2} \frac{1}{\left(1 - \dfrac{y}{d_0}\right)^2}$$

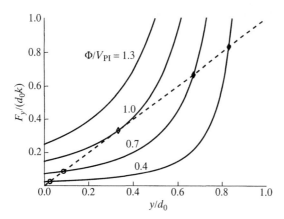

Figure 4.67 Normalized electrostatic and spring forces of a voltage-controlled parallel plate actuator. The family of solid lines trace out the normalized electrostatic force $F_y/(d_0 k)$ as a function of normalized displacement y/d_0, parameterized by the ratio of the applied potential difference to the pull-in voltage. The dashed line traces out the negative of the normalized spring force, $-F_{spring}/(d_0 k) = y/d_0$. Solutions occur at the intersection points. Stable solutions, where $y < d_0/3$, are labeled with a circle; unstable solutions are labeled with a diamond.

Figure 4.67 shows a graphical solution to $F_{tot} = 0$ for several Φ/V_{PI} ratios. When $\Phi/V_{PI} < 1$, there are two solutions: a stable solution, where $y < d_0/3$, and an unstable solution with $y > d_0/3$. At the stable point, the restoring spring force grows faster than the increased electrostatic force when y is increased, causing the plate to return to equilibrium if deflected by an incremental amount Δy. Beyond $y = d_0/3$, though, the opposite is true, leading to the pull-in effect.

In practical MEMS devices, the true pull-in voltage and maximum deflection distance will deviate from the values calculated above. The above formulation neglected the contribution of the fringe field to the overall capacitance. Inclusion of this fringe capacitance would alter the expression for the electrostatic force and therefore the maximum pull-in voltage. Also, we have assumed a linear restoring force for the mechanical spring. In practice, the restoring force deviates from this idealized model, particularly if the displacement is large.

Another practical limit on the maximum applied potential difference is dielectric breakdown. In Section 4.10.3, we noted that in gaseous dielectrics at moderate pressures, the breakdown voltage increases with increasing pressure. This relationship is a special case of a more general result that relates the gas pressure, gap distance, and breakdown voltages, known as Paschen's law:

$$V_{BR} = \frac{Apd}{\ln(pd) + B} \tag{4.114}$$

where V_{BR} is the breakdown voltage, d is the distance between the electrodes, and p is the pressure of the gas. The constants A and B are set for a given gas composition. For air at atmospheric pressure, $A = 4.36 \times 10^7$ V·atm^{-1}·m^{-1} and $B = 12.8$. Differentiating with respect to pd and setting the result to 0, the minimum breakdown voltage occurs at

$$pd = e^{1-B} \tag{4.115}$$

As we discussed in Section 4.10.3, dielectric breakdown occurs when there is sufficiently strong electric fields to dislodge an electron and sustain an avalanche process that eventually forms a conductive path between the two electrodes. At higher pd values, a dislodged electron makes many more collisions with gas molecules. Since each collision randomizes the electron's direction, there is a reduced chance of the electron sustaining acceleration by the electric field and gaining enough energy to cause further ionizing collisions needed to initiate the avalanche breakdown process. This is the regime we discussed in Section 4.10.3, and it describes the breakdown voltage for moderately sized capacitors. At lower pd values, when $pd < e^{1-B}$, there are fewer impacts with gas molecules between the electrode and anode, and so a higher voltage is required to initiate an avalanche process. Hence, at very small distances, the breakdown voltage may be increased by reducing the capacitor spacing d even at a standard atmospheric pressure.

4.14.2 Constant-Gap Electrostatic Microactuators

There are several approaches to mitigate the limited range of a voltage-controlled actuator. The gap can be made larger, so that the 1/3 range before the pull-in threshold is sufficient for the given application. However, the maximum gap is typically determined by the fabrication technology used, and so this option may not be available. Another approach involves using feedback circuitry to stabilize the actuator in the unstable region, or adding a mechanical stop before the deflection reaches the pull-in point.

These undesirable properties all stem from the fact that the electrostatic force is a function of displacement. A common configuration that removes this dependency is to move the top electrode in the transverse direction. In this case, the gap d between the plates is held constant, and the deflection coordinate x determines the overlap area between the two electrodes, as shown in Figure 4.68.

With this geometrical configuration, the capacitance increases approximately linearly with the overlap length $l - x$. We can write the total capacitance as the sum of two terms:

$$C_{\text{tot}} = C_0 + C(x) \tag{4.116}$$

C_0 is the capacitance due to fringing fields, and $C(x)$ is the capacitance due to the fields between the overlapping region of the plates. We assume that the fringing fields do not

Figure 4.68   Constant-gap voltage-controlled actuator. The separation between the capacitor plates is kept at a constant d, but the overlap distance $l - x$ varies. The spring is at equilibrium when $x = x_0$.

change appreciably with a change in x, and so C_0 is independent of the overlap area. The capacitance due to the overlap area between the electrodes is

$$C(x) = \frac{A_{\text{overlap}}}{\epsilon}d = \frac{(l - x)t\epsilon}{d} \tag{4.117}$$

Using (4.95), the electrostatic force can be derived by taking the derivative with respect to the deflection coordinate x:

$$F_x = -F_{\text{ext}} = \frac{\Phi^2}{2}\frac{dC}{dx} = -\frac{\Phi^2}{2}\frac{t\epsilon}{d} \tag{4.118}$$

As with the charge-controlled parallel plate actuator, the force is independent of the displacement. Thus, there is no snap-down effect in the x direction. However, there is still an electrostatic force acting in the direction of the gap length d. In practice, the structure should be made as stiff as possible in this vertical direction so d does not change.

If the spring is at equilibrium when $x = x_0$, the spring force is $\mathbf{F}_{\text{spring}} = k(x_0 - x)\hat{\mathbf{x}}$. As before, the equilibrium point is found by setting the electrostatic force equal to the negative of the spring force:

$$-F_{\text{spring}} = -k(x_0 - x) = -\frac{\Phi^2}{2}\frac{t\epsilon}{d} \quad \rightarrow \quad x_{\text{eq}} = x_0 - \frac{\Phi^2}{2}\frac{t\epsilon}{kd} \tag{4.119}$$

Example 4.38: Electrostatic comb drive. A common geometrical configuration for a constant-gap transducer is to use a sequence of interdigitated fingers, as illustrated in Figure 4.69. Because of the resemblance of the capacitor plates to the teeth of a comb, such devices are referred to as comb drives. Calculate the equilibrium position x_{eq} of the comb drive depicted in Figure 4.69, assuming the spring force is zero when $x = 0$. Assume the capacitance C_0 due to fringe fields does not depend on x and that the capacitor is filled with free space.

    Solution: The capacitance due to the overlap of the interleaved plates can be derived by viewing this configuration as N parallel plate capacitors stacked in a sequence, each with

Figure 4.69 Model of an electrostatic comb drive. Each plate is kept at a constant distance d from its neighbors. The overlap distance is x and the width of each plate is t. In this voltage-controlled configuration, a voltage source is used to maintain a potential difference Φ between the movable and the stationary plates.

an area xt and a separation d. Since each plate faces two plates (one above and the other below) from the opposing electrode, the total capacitance is

$$C(x) = \frac{2tx\epsilon_0 N}{d} \qquad (4.120)$$

Using (4.95), the electrostatic force can be derived by taking the derivative with respect to the deflection coordinate x:

$$F_x = \frac{\Phi^2}{2} \frac{dC}{dx} = \frac{\Phi^2 N \epsilon_0 t}{d} \qquad (4.121)$$

The electrostatic force acts in the $+\hat{\mathbf{x}}$ direction, pulling the two sets of fingers together. The equilibrium position is found by equating the spring force $F_{\text{spring}} = -kx$ with the negative of the electrostatic force:

$$kx_{\text{eq}} = \frac{\Phi^2 N \epsilon_0 t}{d} \quad \rightarrow \quad x_{\text{eq}} = \frac{\Phi^2 N \epsilon_0 t}{kd} \qquad (4.122)$$

4.15 SUMMARY

This chapter discussed the following topics:

* **Coulomb's law.** The electric force between two point charges Q_1 and Q_2 is given by

$$\mathbf{F} = \hat{\mathbf{R}} \frac{Q_1 Q_2}{4\pi\epsilon_0 R^2}$$

where $\hat{\mathbf{R}}$ is the unit vector pointing from one charge to the other and R is the distance between them. The force is repulsive between like charges and attractive between charges of opposite polarity.

* **Electric field.** Any physical region of space in the vicinity of point charges Q_k, a line charge distribution $\rho_l(\mathbf{r})$, a surface charge distribution $\rho_s(\mathbf{r})$, and/or a volume charge distribution $\rho(\mathbf{r})$, is said to be permeated by an electric field. The electric field is defined as the force per unit positive charge and is given by

$$\mathbf{E}(\mathbf{r}) = \frac{1}{4\pi\epsilon_0} \left[\sum_{k=1}^{n} \hat{\mathbf{R}}_k \frac{Q_k}{|\mathbf{r} - \mathbf{r}_k'|^2} + \int_{L'} \hat{\mathbf{R}} \frac{\rho_l(\mathbf{r}')dl'}{|\mathbf{r} - \mathbf{r}'|^2} + \int_{S'} \hat{\mathbf{R}} \frac{\rho_s(\mathbf{r}')ds'}{|\mathbf{r} - \mathbf{r}'|^2} + \int_{V'} \hat{\mathbf{R}} \frac{\rho(\mathbf{r}')dv'}{|\mathbf{r} - \mathbf{r}'|^2} \right]$$

where $\hat{\mathbf{R}}$ is the unit vector pointing from the charge density source position $\mathbf{r}'$ to the observation point P at $\mathbf{r}$.

* **Electrostatic potential.** Electrostatic potential at any point P is defined as the work required to move a unit positive test charge from infinity to the point P in the presence of an electrostatic field. The electrostatic potential at any point $P(\mathbf{r})$ is related to the electric field as

$$\Phi(\mathbf{r}) = -\int_{\infty}^{r} \mathbf{E} \cdot d\mathbf{l}$$

so that the electric field is the negative gradient of potential:

$$\mathbf{E} = -\nabla \Phi$$

Electrostatic potential can also be directly evaluated from a known distribution of point charges, a line charge distribution, a surface charge distribution, and/or a volume charge distribution:

$$\Phi(\mathbf{r}) = \frac{1}{4\pi\epsilon_0} \left[\sum_{k=1}^{n} \frac{Q_k}{|\mathbf{r} - \mathbf{r}'_k|} + \int_{L'} \frac{\rho_l(\mathbf{r}')dl'}{|\mathbf{r} - \mathbf{r}'|} + \int_{S'} \frac{\rho_s(\mathbf{r}')ds'}{|\mathbf{r} - \mathbf{r}'|} + \int_{V'} \frac{\rho(\mathbf{r}')dv'}{|\mathbf{r} - \mathbf{r}'|} \right]$$

- **Electric flux and Gauss's law.** The electric flux density vector $\mathbf{D}$ is a measure of electric displacement and is a function only of the electric charge producing it, independent of the type of medium surrounding the charge. In free space, $\mathbf{D} = \epsilon_0 \mathbf{E}$, so for a point charge located at the origin we have

$$\mathbf{D} = \hat{\mathbf{r}} \frac{Q}{4\pi r^2}$$

Gauss's law states that the total electric flux out of any closed surface S is a constant and is equal to the total charge in the volume V enclosed by S:

$$\oint_S \mathbf{D} \cdot d\mathbf{s} = \int_V \rho \, dv = Q_{\text{enc}}$$

where ρ is the free volume charge density (which excludes any bound polarization charge density) in volume V, and Q_{enc} is the total charge in volume V. Gauss's law is a direct consequence of Coulomb's law, namely that the electrostatic force varies with distance as r^{-2}. It is particularly useful for the analysis of symmetric charge distributions.

- **Divergence.** Divergence is a measure of the source density per unit volume of a vector field. Qualitatively, divergence of a vector field is a scalar quantity and is nonzero only at those points where new flux lines emerge or terminate. For the electrostatic field, we have

$$\nabla \cdot \mathbf{D} = \rho$$

which is the differential version of Gauss's law. Thus, the electrostatic field has nonzero divergence only at points where charges exist. The divergence theorem is valid in general for any vector field $\mathbf{G}$:

$$\oint_S \mathbf{G} \cdot d\mathbf{s} = \int_V \nabla \cdot \mathbf{G} \, dv$$

where V is the volume enclosed by the surface S.

- **Metallic conductors.** Metallic conductors have ample free electrons, which can move rapidly over macroscopic distances. When a conductor is charged or placed in an external electric field, electrons rapidly rearrange so that the net electric field inside the conductor is identically zero. Thus, the entire body of a charged metallic conductor is an equipotential volume to which electric field lines are always orthogonal. The tangential electric field at the surface of a metallic conductor is identically zero, while the normal component of the electric flux density is equal to the surface charge density, namely,

$$\hat{\mathbf{n}} \cdot \mathbf{D} = \epsilon_0 \hat{\mathbf{n}} \cdot \mathbf{E} = \rho_s$$

In general, slight rearrangement of free charge in metallic conductors can create very large electric fields.

- **Laplace's equation.** When electrostatic fields are set up by applying potential differences between metallic conductors, the electric potential and field distributions at points free of charges can be found uniquely by solving Laplace's equation, namely,

$$\nabla^2 \Phi = 0$$

to find Φ, and $\mathbf{E} = -\nabla\Phi$ to find $\mathbf{E}$. Laplace's equation can be analytically solved only for simple geometries; however, a host of numerical methods are available for its solution in more general cases. The more general form of Laplace's equation, which applies in the presence of charges, is Poisson's equation, namely $\nabla^2 \Phi = -\rho/\epsilon_0$.

- **Capacitance.** The capacitance of a configuration of two conductors is a measure of its ability to hold charge per unit applied voltage between them. The capacitance is defined as

$$C \equiv \frac{Q}{\Phi_{12}} = \frac{\oint_S \mathbf{D} \cdot d\mathbf{s}}{-\int_L \mathbf{E} \cdot d\mathbf{l}}$$

where S is any surface enclosing the positively charged conductor, L is any path from the negative to the positive conductor, and Φ_{12} is the potential difference between them. To find the capacitance of a configuration, we typically assume charges $+Q$ and $-Q$ on the two conductors, find the electric field $\mathbf{E}$ from Gauss's law or Coulomb's law and integrate it along a path L to determine Φ_{12}.

- **Dielectric materials.** When a dielectric material is placed in an external electric field $\mathbf{E}$, a dipole moment distribution $\mathbf{P}$ is created, which is called polarization and is given by $\mathbf{P} = \epsilon_0 \chi_e \mathbf{E}$, where χ_e is the electric susceptibility, dependent on the particular microscopic atomic, molecular, and orientational properties of the material. The net effect of this induced dipole moment distribution within the dielectric is that the total electric flux density is now given by $\mathbf{D} = \epsilon_0 \mathbf{E} + \epsilon_0 \chi_e \mathbf{E}$, a fact that is typically accounted for by assigning to each material an electric permittivity $\epsilon = \epsilon_0(1 + \chi_e)$ such that $\mathbf{D} = \epsilon \mathbf{E}$.

- **Electrostatic boundary conditions.** Experimentally established laws of electrostatics dictate that the normal component of electric flux density is continuous across the interface between two materials, except when there is free surface charge present, which typically occurs at the surface of metallic conductors. The tangential component of the electric field is always continuous across any interface. In summary, we have

$$D_{1n} - D_{2n} = \rho_s \quad \text{and} \quad E_{1t} = E_{2t}$$

- **Electrostatic energy.** Any configuration of charges stores electrostatic energy of an amount equal to the work that was required to bring the charges together. The total energy stored in a distribution of free charge is given by

$$W_e = \frac{1}{2} \int_V \rho \Phi \, dv$$

where V is the volume over which the charge is distributed. Alternatively, the electrostatic energy can be viewed as residing in the fields and can be found from

$$W_e = \frac{1}{2} \int_V \epsilon E^2 \, dv$$

where V is the entire volume over which the electric field is nonzero. The volume energy density of the electrostatic field is $w_e = \frac{1}{2}\epsilon E^2$.

- **Electrostatic forces.** Both conducting and dielectric materials experience forces when in the presence of electric fields. A relatively easy means of calculating the electrostatic force on such objects is to use the principle of virtual work. The physical origin of electrostatic forces on metallic conductors is the attraction or repulsion between induced charges, whereas dielectrics experience a force only when placed in a nonuniform field.

- **MEMS.** Microelectromechanical Systems integrate mechanical devices with electronic circuits using a unified fabrication process. A common class of MEMS devices, which are found in a broad array of commercial applications, are electrostatic transducers. These devices electronically control or monitor the charge or voltage on the electrodes of a capacitor and use the resulting electrostatic force between these electrodes to interact with the environment. As an example we considered parallel-plate actuators under either charge or voltage control. Charge-controlled parallel-plate actuators are stable throughout the entire range of motion, but parasitic capacitances limit their practical use. Voltage-controlled parallel-plate actuators are stable as long as the deflection is less than one-third of the neutral gap width.

The relationships among the total volume charge density (which in general may contain both a free charge density and a bound charge density ρ_p due to a polarization vector **P**), electric field **E**, and the electrostatic potential Φ described above may be summarized in Table 4.2. Each row lists equations for the respective quantity of interest. The equations are organized according to the independent quantity listed on the top of each column.

TABLE 4.2 SUMMARY OF ELECTROSTATIC EQUATIONS

	$\rho^\dagger$	$\mathbf{E}$	Φ
$\rho^\dagger =$		$\nabla \cdot \epsilon_0 \mathbf{E}$	$-\epsilon_0 \nabla^2 \Phi$
$\mathbf{E} =$	$\dfrac{1}{4\pi\epsilon_0} \displaystyle\int_{V'} \hat{\mathbf{R}} \dfrac{\rho(\mathbf{r}')dv'}{|\mathbf{r}-\mathbf{r}'|^2}$		$-\nabla\Phi$
$\Phi =$	$\dfrac{1}{4\pi\epsilon_0} \displaystyle\int_{V'} \dfrac{\rho(\mathbf{r}')dv'}{|\mathbf{r}-\mathbf{r}'|}$	$-\displaystyle\int_\infty^r \mathbf{E} \cdot d\mathbf{l}$	

$^\dagger\rho$ represents the total volume charge density, which in general may include the volume bound polarization charge density $-\nabla \cdot \mathbf{P}$.

In dielectric materials where we can write $\mathbf{D} = \epsilon\mathbf{E}$, we can express the free charge density directly in terms of the **E**-field and electrostatic potential, as summarized in Table 4.3. The free charge density ρ excludes the bound polarization charge density $-\nabla \cdot \mathbf{P}$, so that $\nabla \cdot \mathbf{D} = \rho$.

TABLE 4.3 FREE CHARGE DENSITY

	$\rho^\dagger$	$\mathbf{E}$	Φ
$\rho^\dagger =$		$\nabla \cdot \epsilon\mathbf{E}$	$-\epsilon\nabla^2\Phi$

$^\dagger\rho$ represents the free volume charge density, which excludes the volume bound polarization charge density $-\nabla \cdot \mathbf{P}$.

PROBLEMS

4.1 **Two point charges.** Two identical point charges 1 m apart from each other in free space are experiencing a repulsion force of 1 N each. What is the magnitude of each charge?

4.2 **Two point charges.** Two small identical spheres have charges of $+20$ nC and -25 nC, respectively. (a) What is the force between them if they are apart by 10 cm? (b) The two spheres are brought into contact and then separated again by 10 cm. What is the force between them now?

4.3 **Two suspended charges.** Two small, identical, electrically charged conducting spheres of mass 2.5 g and charge $+150$ nC each are suspended by weightless strings of length 12 cm each, as shown in Figure 4.70. Calculate the deflection angle θ.

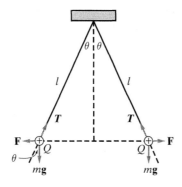

Figure 4.70 Two suspended charges.
Problem 4.3.

4.4 **Zero force.** (a) Three point charges of $Q_1 = +40$ nC, $Q_2 = -20$ nC, and $Q_3 = +10$ nC are all situated on the x axis in such a way that the net force on charge Q_2 due to Q_1 and Q_3 is equal to zero. If Q_1 and Q_2 are located at points $(-2,0,0)$ and $(0,0,0)$, respectively, what is the location of charge Q_3? (b) Q_3 is moved to a different position on the x axis such that the force on itself due to Q_1 and Q_2 is equal to zero. What is the new position of Q_3?

4.5 **Three charges.** Three identical charges of charge Q are located at the vertices of an equilateral triangle of side length a. Determine the force on one of the charges due to the other two.

4.6 **Three point charges.** Three point charges of values $+15$ nC, $+10$ nC, and $+20$ nC are located at points $(0,0,0)$, $(1,0,0)$, and $(0,1,0)$, respectively. (a) Find the force on each charge due to the other charges. Which charge experiences the largest force? (b) Repeat part (a) for -100 nC as the second charge.

4.7 **Two point charges.** Two point charges of $+Q_1$ and $+Q_1$ are located at $(0,0,0)$ and at $(0,a,0)$ respectively. Find the electric field at $P_1(0,0,3a)$.

4.8 **Zero field from three charges.** Two point charges of $+Q$ each are located at $(10,0)$ cm and $(-10,0)$ cm, while a third one of charge $-2Q$ is at $(0,-10)$ cm. Find the coordinates of the point where the electric field is zero.

4.9 **Three point charges.** Two point charges of $+10$ nC each are located at points $(0.46,0,0)$ and $(-0.46,0,0)$, respectively. Where should a third point charge of $+15$ nC be placed such that $\mathbf{E} = 0$ at point $(0,1,0)$?

4.10 **Four point charges on a square.** Four point charges of $+20$ nC each are located at the corners of a square of side length 6 cm located on the xy plane and centered at the origin. Find and sketch the electric potential on the z axis.

4.11 **Six point charges on a hexagon.** Six hexagon identical point charges of $+20$ nC each are situated in space at the corners of a regular hexagon whose sides of length 6 cm. (a) Find potential at the center of the hexagon.

4.12 **Two straight-line charges.** Consider two uniformly charged wires, each of length 1 m and a total charge $+100$ nC, with their ends separated by 1 m, as shown in Figure 4.71. (a) Find the electric potential Φ at the point P, midway between the two wires. (b) Find the electric field $\mathbf{E}$ at point P.

Figure 4.71   Two straight-line charges. Problem 4.12.

4.13   Seven point charges on a cube. Seven identical point charges of +20 nC each occupy seven of the eight corners of a cube 3 cm on each side. Find the electric potential at the unoccupied corner.

4.14   Two line charges. A uniform line charge of $\rho_{l1} = -8\pi \times 8.85$ pC-m^{-1} is located between the points $(-5, 0)$ and $(-2, 0)$ m and another such line of positive charge (i.e., $\rho_{l2} = 8\pi \times 8.85$ pC-m^{-1}) between $(5, 0)$ and $(2, 0)$ m, as shown in Figure 4.72. (a) Find the electric potential Φ at $(1, 0)$ m.

Figure 4.72   Two line charges. Problem 4.14.

4.15   Circular ring of charge. A total charge of Q_1 is distributed uniformly along a half-circular ring as shown in Figure 4.73. Two point charges, each of magnitude Q_2, are situated as shown. The surrounding medium is free space. (a) Find Q_2 in terms of Q_1 so that the potential Φ at the center of the ring is zero. (b) Find Q_2 in terms of Q_1 so that the electric field $\mathbf{E}$ at the center of the ring is zero.

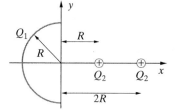

Figure 4.73   Circular ring of charge. Problem 4.15.

4.16   Semicircular line charge. A thin line charge of density ρ_l is in the form of a semicircle of radius a lying on the xy plane with its center located at the origin, as shown in Figure 4.74. Find the electric field at the origin for the cases in which (a) the line charge density $\rho_l = \rho_0$ is a constant and (b) the line charge density varies along the semicircular ring as $\rho_l = \rho_0 \sin \phi$.

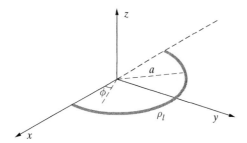

Figure 4.74 Semicircular line charge.
Problem 4.16.

4.17 Charge on a hemisphere. The curved surface of a hemisphere of radius a centered at the origin carries a total charge of Q uniformly distributed over its curved surface, as shown in Figure 4.75. (a) Find the electric potential on the z axis. (b) Find the electric field on the z axis. (c) Repeat parts (a) and (b) if the charge Q is uniformly distributed throughout the volume of the hemisphere.

Figure 4.75 Charge on a hemisphere.
Problem 4.17.

4.18 Sheet of charge with hole. An infinite sheet of uniform charge density ρ_s is situated coincident with the xy plane at $z = 0$. The sheet has a hole of radius a centered at the origin. Find (a) the electric potential Φ and (b) the electric field $\mathbf{E}$ at points along the z axis.

4.19 Spherical charge distribution. A charge density of

$$\rho(r) = Ke^{-br}$$

where K and b are constants, exists in a spherical region of space defined by $0 < r < a$. (a) Find the total charge in the spherical region. (b) Find the electric field at all points in space. (c) Find the electric potential at all points in space. (d) Show that the potential found in part (c) satisfies the equation $\nabla^2 \Phi = -\rho(r)/\epsilon_0$ for both $r < a$ and $r > a$.

4.20 The electron charge density in a hydrogen atom. According to quantum mechanics, the electron charge of a hydrogen atom in its ground state is distributed like a cloud surrounding its nucleus, extending in all directions with steadily decreasing density such that the total charge in this cloud is equal to q_e (i.e., the charge of an electron). This electron charge distribution is given by

$$\rho(r) = \frac{q_e}{\pi a^3} e^{-2r/a}$$

where a is the *Bohr radius,* $a \simeq 0.529 \times 10^{-10}$ m. (a) Find the electric potential and the electric field due to the electron cloud only. (b) Find the total electric potential and the electric field in the atom, assuming that the nucleus (proton) is localized at the origin.

4.21 **Spherical shell of charge.** The space between two concentric spheres of radii a and b ($a < b$) in free space is charged to a volume charge density given by

$$\rho(r) = \frac{K}{r^2}; \quad a \leq r \leq b$$

where K is a constant. (a) Find the total charge in the shell. (b) Find the electric field at all points in space. (c) Find the electric potential at all points in space. (d) What happens if $b \to a$?

4.22 **Space charge between parallel plates.** The space between two perfectly conducting parallel plates is filled with space charge (i.e., free charge) of density given as:

$$\rho = \rho_0 \sin\left(\frac{\pi z}{d}\right)$$

where d is the separation distance of the plates. Other than the space charge, the medium between the plates is air, with permittivity ϵ_0. The upper and lower plates, located at $z = 0$ and $z = d$ are kept at potentials of $\Phi = 0$ and $\Phi = V_0$, respectively. Determine expressions for the potential $\Phi(z)$ and the electric field $\mathbf{E}(z)$ between the plates.

4.23 **Spherical charge distribution.** A spherical charge distribution exists in free space in the region $0 < r < a$ given by

$$\rho(r) = \rho_0\left(1 - \frac{r^2}{a^2}\right)$$

(a) Find the total charge. (b) Determine $\mathbf{E}$ everywhere. (c) Determine Φ everywhere. (d) Sketch both $|\mathbf{E}|$ and Φ as a function of r.

4.24 **Spherical charge with a cavity.** A spherical region of radius b in free space is uniformly charged with a charge density of $\rho = K$, where K is a constant. The sphere contains an uncharged spherical cavity of radius a. The centers of the two spheres are separated by a distance d such that $d + a < b$. Find the electric field inside the cavity.

4.25 **Charge on a hollow metal sphere.** A hollow metal sphere of 10 cm diameter is given a total charge of 1 μC. Find the electric field and the electric potential at the center of the sphere.

4.26 **Electric field.** An electric field in empty space is given as:

$$\mathbf{E}(x, y, t) = \hat{\mathbf{x}} \sin x \, \cos y \, \sin \omega t + \hat{\mathbf{y}} \cos x \, \sin y \, \sin \omega t \qquad \text{V-m}^{-1}$$

Determine the energy required to move a positive unit test charge in the presence of this electric field from the point $(0, 0)$ to $(1, 1)$ along the shortest straight line in the x-y plane. Determine the energy required to move the same test charge from point $(1, 1)$ back to $(0, 0)$ but via a different path, by first going fro $(1, 1)$ to $(1, 0)$ and then from $(1, 0)$ to $(0, 0)$.

4.27 **A 1-farad capacitor.** To get an idea about the physical size of a 1-F capacitor, consider a parallel-plate capacitor with the two metal plates separated by 1 mm thickness of air. Calculate the area of the metal plates needed so that the capacitance is 1 F.

4.28 **Gate oxide capacitance of a MOS transistor.** A basic MOS transistor consists of a gate conductor and a semiconductor (which is the other conductor), separated by a gate dielectric. Consider a MOS transistor using silicon dioxide (SiO_2) ($\epsilon_r = 3.9$) as the gate oxide. The gate oxide capacitance can be approximated as a parallel-plate capacitor. The gate oxide capacitance per unit area is given by

$$C_{ox} = \frac{\epsilon_{ox}}{t_{ox}}$$

where ϵ_{ox} and t_{ox} are the permittivity and the thickness of the gate dielectric. (a) If the thickness of the SiO_2 layer is 2×10^{-6} cm, find the gate oxide capacitance per unit area. (b) If the length and the width of the gate region are $L = 5 \times 10^{-4}$ cm and $W = 2 \times 10^{-3}$ cm, respectively, find the total gate capacitance.

4.29 **RG 6 coaxial cable.** A coaxial cable (RG 6) designed for interior use, such as connecting a TV set to a VCR, has a per-unit-length capacitance listed as 17.5 pF/ft. If the relative dielectric constant of the insulator material in the cable is $\epsilon_r \simeq 1.64$, find the ratio of the inner and outer radii of the insulator.

4.30 **Radius of a high-voltage conductor sphere.** Consider an isolated charged metallic conductor sphere in a dielectric medium at an electric potential of 500 kV. Calculate the minimum radius of the sphere such that dielectric breakdown will not occur if the surrounding dielectric is (a) air ($E_{\text{BR}} = 3$ MV-m^{-1}), (b) a gaseous dielectric such as sulfur hexafluoride (SF_6) ($\epsilon_r \simeq 1$ and $E_{\text{BR}} = 7.5$ MV-m^{-1}), (c) a liquid dielectric such as oil ($\epsilon_r = 2.3$ and $E_{\text{BR}} = 15$ MV-m^{-1}), and (d) a solid dielectric such as mica ($\epsilon_r = 5.4$ and $E_{\text{BR}} = 200$ MV-m^{-1}).

4.31 **Parallel-plate capacitor.** A parallel-plate capacitor is constructed from two aluminum foils of 1 cm^2 area each placed on both sides of rubber ($\epsilon_r = 2.5$ and $E_{\text{BR}} = 25$ MV-m^{-1}) of thickness 2.5 mm. Find the voltage rating of the capacitor using a safety factor of 10.

4.32 **Energy in a capacitor.** A 9-V battery is connected across a parallel-plate air-filled capacitor. The battery is subsequently removed, and a block of solid dielectric ($\epsilon = 2\epsilon_0$) is inserted between the plates. (a) What is the voltage across the capacitor after the introduction of the dielectric? (b) Compare the total electrostatic energy stored in this capacitor before and after the introduction of the dielectric. Comment and explain any differences. Neglect all fringing effects.

4.33 **Coaxial capacitor.** Consider a coaxial capacitor as shown in Figure 4.76. Given $a = 5$ mm, $l = 3$ cm, and the voltage rating of the capacitor to be 2 kV with a safety factor of 10, what is the maximum capacitance that can be designed using (a) oil ($\epsilon_r = 2.3$ and $E_{\text{BR}} = 15$ MV-m^{-1}) and (b) mica ($\epsilon_r = 5.4$ and $E_{\text{BR}} = 200$ MV-m^{-1}).

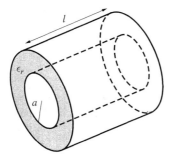

Figure 4.76   Coaxial capacitor. Problem 4.33.

4.34 **Coaxial capacitor with two dielectrics.** A coaxial capacitor consists of two conducting coaxial surfaces of radii a and b ($a < b$). The space between is filled with two different dielectric materials with relative dielectric constants ϵ_{1r} and ϵ_{2r}. (a) Find the capacitance of this configuration.

Figure 4.77 Coaxial capacitor with two dielectrics. Problem 4.34.

4.35 **Capacitor with spacers.** The cross-sectional view of an air-filled coaxial capacitor with spacers made out of material with permittivity ϵ is shown in Figure 4.78. (a) Find the capacitance of this coaxial line in terms of ϵ, a, b, and ϕ.

Figure 4.78 Coaxial capacitor with spacers. Problem 4.35.

4.36 **Earth capacitor.** Consider the earth to be a large conducting sphere. (a) Find its capacitance (the earth's radius is $\sim 6.371 \times 10^6$ m). (b) Find the total charge and energy stored on the earth (take the electric field on the surface of the earth to be 100 V-m^{-1}). (c) Find the maximum charge and energy that can be stored on the earth.

4.37 **Coaxial capacitor with variable ϵ.** A coaxial capacitor of inner radius a and outer radius b is filled with a dielectric material whose relative permittivity varies as $\epsilon_r = 10\,a/r$ over the region from $r = a$ to $r = b$. Find the capacitance per unit length and compare with the capacitance of the same coaxial cable when filled with air.

4.38 **Planar charge.** A surface charge distribution $\rho_s(x, z)$ exists on the x-z plane, with no charge anywhere else (i.e., $\rho = 0$ for $|y| > 0$). Which of the following potential functions are valid solutions for the electrostatic potential in the half-space $y > 0$, and what is the corresponding charge distribution $\rho_s(x, z)$ on the x-z plane?

$$\Phi_1 = e^{-y}\cosh x \qquad \Phi_2 = e^{-y}\cos x \qquad \Phi_3 = e^{-\sqrt{2}y}\cos x \sin x$$

$$\Phi_4 = \sin x \sin y \sin z$$

4.39 **Parallel power lines.** An isolated pair of parallel power lines a distance of d_1 apart have a potential difference of V_{AB} and are located a distance h above a pair of telephone wires, as shown in Figure 4.79. The parameter values are $d_1 = 2$ m, $a = 3$ cm, $V_{AB} = 400$ V, $h = 50$ cm, and $d_2 = 20$ cm. (a) Find the direction and magnitude of the electric field at points 1 and 2. Take the midpoint between the power lines as the origin of your coordinate system. (b) Determine the potential difference Φ_{12} between points 1 and 2.

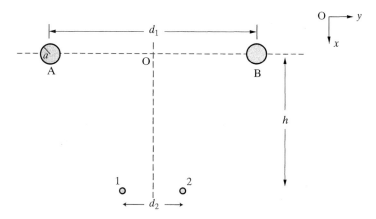

Figure 4.79 **Parallel power lines above telephone lines.** Problem 4.39.

4.40 **Field under high-voltage line.** Many 60 Hz high-voltage transmission lines operate at an rms alternating voltage of 885 kV. (a) What is the peak electric field at ground level under such a line if the wire is 10 m above the ground? (b) What is the peak potential difference between the head and feet of a 6-ft tall person? (c) Is the field sufficient to ignite a standard (110 V) fluorescent lamp of 2 ft length?

4.41 **Capacitance of the two-wire line with dielectric sleeve.** A two-wire line consists of two metallic conductors of radius a enclosed by dielectric (permittivity ϵ) sleeves of radii b separated by a distance d as shown in Figure 4.80, with $d \gg a$. The space surrounding the dielectric sleeves is air, with permittivity ϵ_0. (a) Determine the capacitance per unit length of this two-wire line. (b) The values of wire radius and separation distance are given to be $a = 1$ cm and $d = 20$ cm, respectively, while the radius of the dielectric sleeve is $b = 5$ cm. If the dielectric sleeve is made of mica $\left(\epsilon = 5.4\epsilon_0, E_{BR}^{mica} = 200 \text{ MV-m}^{-1}\right)$, determine the maximum operating voltage of the two-wire line capacitor. Note that the breakdown field for air is $E_{BR}^{air} = 3 \text{ MV-m}^{-1}$.

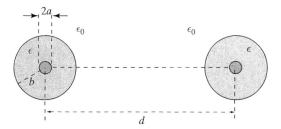

Figure 4.80 **Two-wire transmission line.** Problem 4.41.

4.42 **Thundercloud fields.** A typical thundercloud can be modeled as a capacitor with horizontal plates with 10 km^2 area separated by a vertical distance of 5 km. Just before a large lightning discharge, the upper plate may have a total positive charge of up to 300 C, with the lower plate having an equal amount of negative charge. (a) Find the electrostatic energy stored in the cloud just before a discharge. (b) What is the potential difference between the top and bottom plates? (c) What is the average electric field within the cloud? How does this value compare to the dielectric breakdown field of dry air (3 MV-m^{-1})?

4.43 **Two conducting spheres.** Consider a pair of small conducting spheres with radii a, b, small compared with the separation distance d between their centers (i.e., $a, b \ll d$). (a) Determine the electrostatic energy stored by this configuration, assuming that the spheres with radii a and b carry charges of Q and $-Q$, respectively. Your answer should depend on d. State all assumptions. (b) Repeat part (a) assuming that the spheres with radii a and b carry charges of $+Q$ and $+2Q$, respectively.

4.44 **Voltage-controlled actuator.** Consider a voltage-controlled parallel-plate actuator, as shown in Figure 4.64b, with plate area $A = 250 \times 250$ μm^2, equilibrium separation distance $d_0 = 2$ μm, and effective spring constant $k = 0.01$ N-m^{-1}. The capacitor is filled with free space. (a) What is the usable voltage range of this actuator? (b) What is the spacing d between the top and bottom capacitor plates when a voltage of 0.05 V is applied?

4.45 **Air-bag accelerometer.** The comb-drive accelerometer depicted in Figure 4.81 is to be used in an integrated circuit that is responsible for deploying an air-bag in the event of an automobile crash. Assuming a sufficient overlap between the plates such that the effects of fringing fields remain unchanged and effective spring constant $k = 0.01$ N-m^{-1}, determine the change in capacitance during a deceleration of $40g$, where $g \approx 9.8$ m-s^{-2}. Assume the capacitor is filled with free space, there are $N = 200$ capacitor plate pairs, and the following dimensional parameters: $L = 100$ μm, $w = 2$ μm, $t = 5$ μm, and $d = 2$ μm. Assume all of the mass in the movable half of the comb drive is in the teeth (the capacitor plates that comprise the movable electrode), and that the teeth are made of polycrystalline silicon, which has a mass density $\rho = 2.33$ g-cm^{-3}.

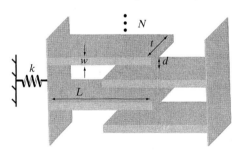

Figure 4.81   Comb drive accelerometer.
Problem 4.45.

5

Steady Electric Currents

The special properties of metallic conductors were noted in Chapter 4, namely that such materials hold a large number of free electronic charges that are readily able to move over macroscopic distances under the influence of applied electric fields. Although metals are particularly good conductors, many other materials also contain free charges that can move over macroscopic distances and are considered to be conductors. In this chapter, we introduce a measure of the ability of materials to conduct electricity, namely *conductivity,* and study the conditions that lead to steady flow of electric charge in conducting materials. Note that the steady flow of electric charge we discuss in this chapter is quite different from the transient redistribution of charge that occurs when a conductor is brought into a region of electric field or when charge is placed on a conductor. For steady currents to exist, we need not only a steady electric field to impart a velocity to the free charges, but also a mechanism that prevents the charge from piling up, which would tend to reduce the field. The subject of steady electric currents links field theory to several important circuit concepts, namely Ohm's law, Kirchhoff's voltage and current laws, and Joule's law. In addition, using the concept of steady current flow and the principle of conservation of charge, we can derive the fundamental equation of continuity, which relates current and charge density.

In this chapter, we define the concept of resistance and discuss how the resistance of different configurations can be evaluated using fundamental laws of electrostatics. The performance of transmission lines with losses (Section 3.7) are dependent on the line parameters R and G, representing resistive losses in the conductors and leakage losses, respectively, in the dielectric media between them. Our study of steady electric currents enables us to calculate the transmission line parameter G used in Chapters 2 and 3. The resistive losses in the conductors, however, are determined by both the bulk conductivity and the frequency of operation, which we discuss in Chapters 8 and 10.

5.1 CURRENT DENSITY AND THE MICROSCOPIC VIEW OF CONDUCTION

We describe the flow of electric charge by introducing the concepts of current and current density. *Current* is a flow of charge, measured by the rate at which charge passes through any specified surface area, for example, the cross section of a wire. Current is a scalar quantity, and its unit, coulombs per second, is called the *ampere*. A current I is said to flow when the charge that passes through the surface area in time Δt is given by $\Delta Q = I \Delta t$. Since different amounts of charge may flow through different parts of a given cross-sectional area, it is often more appropriate to describe current in terms of *current density*, denoted as $\mathbf{J}$ and having units of amperes per square meter, or coulombs-s^{-1}-m^{-2}. The current density $\mathbf{J}(x, y, z)$ at any given point (x, y, z) in a conductor is a vector quantity having the direction of the flow of positive charge (which is opposite to the direction of electron flow) and a magnitude equal to the current per unit area that is normal to the direction of flow. The current through any small area Δs represented by a vector $\Delta \mathbf{s} = \hat{\mathbf{n}} \Delta s$ can be calculated as $\Delta I = \mathbf{J} \cdot \Delta \mathbf{s}$. For example, the current through the area element Δs in Figure 5.1 is the same as that through the area $\Delta s \cos \theta$ perpendicular to $\mathbf{J}$, or $J \Delta s \cos \theta$. To determine the current through a larger area S, we need to integrate $\mathbf{J}$ over S, namely

$$I = \int_S \mathbf{J} \cdot d\mathbf{s} \tag{5.1}$$

Although electric current is a macroscopic quantity, it results from the motion of many microscopic charges. At a simple level entirely sufficient for our purposes, we can think of the conducting material as composed of a lattice of fixed positive ions and an electron gas that is free to move about. The free electrons in a metallic conductor are in fact the conduction-band electrons, which are very loosely bound to their atoms and are essentially free to wander through the lattice. For example, atoms of copper have

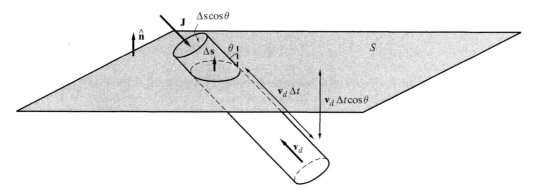

Figure 5.1 Current density J. Since current is defined as the rate of charge flow through a specified surface Δs, it is given by the projection of the selected surface perpendicular to the direction of the flow of conduction electrons.

29 electrons, 28 of which are in tightly bound shells. The remaining outermost electron is essentially free to move about the crystalline structure of copper. Ordinarily, these free electrons (each only loosely bound to an atom) are in a state of random motion because of their thermal energy. Without any external electric field applied, they move about randomly colliding with the atoms of the lattice and with one another, as illustrated in Figure 5.2a. Between collisions, however, the electrons in a metal at room temperature move at rather high thermal speeds of v_{th} of $\sim 10^5$ to $\sim 10^6$ m-s^{-1}. This rapid random motion is the source of the fluctuating currents known as *Johnson noise*,[1] which can be measured across any resistor. The mean free time t_c, defined as the average time between collisions, is typically of order $\sim 10^{-14}$ s, at room temperature. This value corresponds to a *mean free path* $\bar{l}$, defined as the average distance traveled between collisions, of $\sim 10^{-8}$ m. After each collision, the electrons acquire a new velocity nearly independent of their previous velocity. Because of the entirely random nature of this motion in the absence of any applied electric field, there is no net motion in any given direction, and the net charge density averaged over macroscopic spatial dimensions (i.e., $\gg 10^{-8}$ m) and time scales (i.e., $\gg 10^{-14}$ s) remains approximately zero.

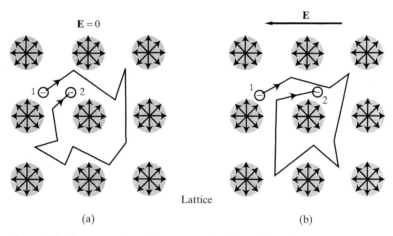

Figure 5.2 Representation of electron motion through the vibrating lattice structure of a solid. The shaded areas represent lattice sites, at which individual atoms or molecules are located; the atoms vibrate randomly in all directions as indicated by arrows. (a) Without an externally applied electric field and (b) with an externally applied electric field. The field produces a small *drift* component toward the right (electrons move opposite to the field). The drift shown here is greatly exaggerated. In fact, for typical conduction current levels of a few amperes or less, the drift velocity in metals is $<10^{-4}$ m-s^{-1}, whereas the electron velocity between collisions due to thermal motion is $\sim 10^5$ m-s^{-1}.

[1] The average size of random fluctuations in the voltage across a resistor depends on the resistor and the temperature. For a 400-kΩ resistor at room temperature, the amplitude of fluctuations with frequency < 20 kHz is ~ 10 μV. For more information, see A. L. King, *Thermophysics,* Freeman, San Francisco, 1962, pp. 212–214.

Conduction, or net flow of free charge, occurs due to the relatively slow drift of electrons (compared with the rapid thermal motions of individual electrons) caused by an applied electric field. When an electric field E is established within a conductor (typically by applying a potential difference between the two ends of the conductor), the free electrons are accelerated opposite to the direction of the field. At first, one might think that the electrons would experience a steady acceleration[2] given by $(|q_e|E/m_e)$; however, before they can acquire any appreciable speed, the electrons collide with the lattice and acquire a new random velocity. Since the applied electric field essentially has to start accelerating all over again every $\sim 10^{-14}$ s, it can alter the random thermal velocities of the electrons only slightly, but in a systematic manner. This relatively slight systematic drift (shown on a highly exaggerated scale in Figure 5.2b) of the free electrons is the basis of conduction. Following a brief initial transient, the electrons acquire a steady-state average *drift velocity* $\mathbf{v}_d$, determined by the balance between the accelerating force of the applied field and the scattering effect of the collisions with the lattice. These collisions are also the mechanism by which some of the energy of the electrons, and thus of the electric field, is dissipated as heat.

If there are n_c conduction-band electrons[3] per cubic meter traveling with an average drift velocity $\mathbf{v}_d$, each travels a distance $v_d \Delta t$ in a time Δt, and all those in a cylinder of slant height $v_d \Delta t$ and base Δs (see Figure 5.1) pass through Δs in a time Δt. The volume of this cylinder is $v_d \Delta t \Delta s \cos\theta$, so that $n_c v_d \Delta t \Delta s \cos\theta$ electrons cross Δs, involving a total charge of $|q_e| n_c v_d \Delta t \Delta s \cos\theta$. Thus, the magnitude of the current density $\mathbf{J}$, equal to the total charge per unit time per unit area, is given as

$$|\mathbf{J}| = \frac{|q_e| n_c v_d \Delta t \Delta s \cos\theta}{\Delta t \Delta s \cos\theta} = |q_e| n_c v_d \quad \rightarrow \quad \mathbf{J} = n_c q_e \mathbf{v_d}$$

Note that since q_e is negative, $\mathbf{J}$ is in the direction opposite to the electron drift velocity $\mathbf{v}_d$.

We can crudely estimate the average drift velocity by assuming that each free electron has the same thermal speed v_{th}, the same mean free time t_c between collisions, and thus the same mean free path between collisions of $\bar{l} = v_{th} t_c$. The electric field gives each electron an acceleration of $q_e E / m_e$, so that the change in the velocity of a free electron before the next collision is $\Delta v_{th} \simeq q_e \mathbf{E} t_c / m_e$. The drift velocity averaged over the entire population of electrons should be of the same order as Δv_{th}, although the actual value will depend on the lattice and band structure and other material properties, such as the energy distribution of free electrons. For metals, the average drift velocity is

---

[2]A steady acceleration would mean that the current in the conductor would increase indefinitely with time, in contradiction with Ohm's law, which predicts a constant current for a constant applied potential (i.e., a constant electric field).

[3]For copper, the density of conduction electrons is $n_c = 8.45 \times 10^{28}$ el-m^{-3}, which is also the number of copper atoms per unit volume, since copper has one conduction-band electron per atom. Note that the number of atoms per unit volume is approximately the same for all solids.

given as[4]

$$\mathbf{v}_d \simeq \frac{q_e \mathbf{E} t_c}{m_e} \quad \text{or} \quad \mathbf{v}_d \simeq \frac{q_e \mathbf{E} \bar{l}}{m_e v_{th}}$$

The proportionality constant between $\mathbf{v}_d$ and $\mathbf{E}$, namely $\mu_e = (|q_e| t_c / m_e)$, is referred to as the *mobility* of the conduction electrons in the particular material. In general, the drift velocity v_d is very small compared with v_{th}, since the electric field makes only a slight change in the velocity distribution that existed before the field was applied. Therefore, the average values of t_c and $\bar{l}$, which depend on the thermal velocity of the electrons, do not change appreciably due to the presence of the applied field. This means that the asymptotic value of the drift velocity is constant for a given applied electric field, and thus the current is constant. In most cases, the magnitude of drift velocity is quite small—less than $\sim 10^{-4}$ m-s^{-1} in metals (see Example 5.1)—for reasonable conduction currents.

The slowly[5] drifting electrons constitute a current of density

$$\mathbf{J} = n_c q_e \mathbf{v}_d = \frac{n_c q_e^2 \bar{l}}{m_e v_{th}} \mathbf{E} \quad \rightarrow \quad \boxed{\mathbf{J} = \sigma \mathbf{E}} \tag{5.2}$$

where $\sigma = n_c q_e^2 \bar{l} / (m_e v_{th})$ is identified as the *conductivity* of the material. The conductivity is the macroscopic quantity that is hereafter used to account for the microscopic behavior of conductors. The unit of conductivity is Siemens-m^{-1} or S-m^{-1}, sometimes also referred to as mhos-m^{-1}. Values of σ for common materials vary over ~ 24 orders of magnitude, ranging from excellent conductors (e.g., 6.17×10^7 S-m^{-1} for silver) to excellent insulators (e.g., $\sim 10^{-17}$ S-m^{-1} for fused quartz). Equation (5.2) holds in cases

---

[4]The *drift velocity* is the solution of an equation of motion with a damping term, namely

$$m_e \frac{d\mathbf{v}_d}{dt} + m_e \frac{\mathbf{v}_d}{t_c} = q_e \mathbf{E}$$

where the term t_c behaves essentially as a coefficient of friction. Note that the solution of this equation is

$$\mathbf{v}_d = \frac{q_e t_c}{m_e} \mathbf{E}[1 - e^{-(t/t_c)}]$$

so that for a time period t longer than t_c, the drift velocity asymptotically approaches $\mathbf{v}_d = (q_e t_c / m_e)\mathbf{E}$ as t increases (e.g., this asymptotic value is practically accurate when $t \geq 5t_c$). For further discussion, see R. L. Sproull, *Modern Physics: The Quantum Physics of Atoms, Solids, and Nuclei*, 3rd ed., John Wiley, New York, 1980, Sec. 8.5. For an excellent elementary review of conduction in metals, see V. F. Weisskopf, On the theory of electric resistance of metals, *Am. J. Phys.*, 11(1), pp. 1–12, 1943.

[5]Note that the drift velocity is indeed quite small ($\sim 10^{-5}$ to $\sim 10^{-4}$ m-s^{-1}) considering that an applied signal on a transmission line propagates at the speed of light. It is clear that the propagation of a voltage/current disturbance along a wire does not depend on how much time it would take for a particular tagged electron to travel the distance through the wire. In fact, such an applied signal propagates as an electromagnetic wave outside the wire.

when the electric field is constant or variable in time, as long as the field remains approximately constant for times much longer than the mean free time between collisions (i.e., $t_c \simeq 10^{-14}$ s). In some contexts, it might be more useful to work with *resistivity*, which is simply the inverse of conductivity (i.e., σ^{-1}) and has units of Ω-m.

The manner in which the conductivity depends on temperature differs drastically between metals and semiconductors and dielectrics. In the case of metals, *all* of the conduction-band electrons are already free to move at room temperature, and the bound electrons are not released from their shells even for very large temperatures or electric fields. The dominant effect of increased temperature for metals is the fact that the lattice atoms undergo larger vibrations (larger excursions around their central locations), thus increasing the probability of collisions and impeding the motion of the carriers. As a result, the conductivity σ of metallic conductors decreases with increasing temperature. Semiconductors and dielectrics, on the other hand, have relatively fewer free electrons available for conduction at room temperature.[6] As the temperature increases, the lattice vibrations become stronger, and, depending on the type of material, some bound electrons are knocked loose and become available for conduction. Although the increased lattice vibrations still tend to impede the flow of current carriers, the net dominant effect in semiconductors and dielectrics is increasing conductivity with increasing temperature, due to the larger number of conduction electrons.[7]

For pure metallic elements, the conductivity depends rather strongly on temperature, decreasing by $\sim$0.1–0.5% per degree Celsius, depending on the type of material. Specifically, if $\sigma_{20°}$ is the conductivity of a metal at 20°C, its conductivity at a Celsius temperature T (in °C) is given by[8]

$$\frac{1}{\sigma_T} \simeq \frac{1}{\sigma_{20°}}[1 + \alpha_\sigma(T - 20)]$$

where the coefficient α_σ is called the temperature coefficient and has the value 0.001–0.005 for most metals. Table 5.1 provides values of conductivity σ and temperature

---

[6]For semiconductors, the conductivity is limited more by the small number of current carriers than by the size of t_c. By properly accounting for the energy distribution of conduction electrons, it can be shown that a more accurate expression for the conductivity of a semiconductor is

$$\sigma = \frac{4 n_c q_e^2 \bar{l}}{3\sqrt{2\pi m_e k_B T}}$$

where k_B is Boltzmann's constant ($k_B \simeq 1.38 \times 10^{-23}$ J-K^{-1}) and T is the absolute temperature in °K. A more common expression for the conductivity of a semiconductor is

$$\sigma = |q_e|(\mu_e N_e + \mu_p N_p)$$

where N_e and N_p are the densities of electrons and holes, respectively, and μ_e and μ_p are the mobilities of the electrons and holes, respectively; see Problem 5.9.

[7]For further discussion, see Chapter 8 of R. L. Sproull, *Modern Physics: The Quantum Physics of Atoms, Solids, and Nuclei*, 3rd ed., John Wiley, New York, 1980.

[8]See Chapter 5 of W. T. Scott, *The Physics of Electricity and Magnetism*, 2nd ed., John Wiley, New York, 1966, and Section 8.5 of R. L. Sproull, ibid.

TABLE 5.1   CONDUCTIVITIES AND TEMPERATURE COEFFICIENTS OF SELECTED MATERIALS

Material	Conductivity σ (S-m^{-1}) (at 20°C)	Temperature Coefficient α_σ [(°C)$^{-1}$]
Aluminum	3.82×10^7	0.0039
Bismuth	8.70×10^5	0.004
Brass (66 Cu, 34 Zn)	2.56×10^7	0.002
Carbon (graphite)	7.14×10^4	−0.0005
Constantan (55 Cu, 45 Ni)	2.26×10^6	0.0002
Copper (annealed)	5.80×10^7	0.0039
Dry, sandy soil	$\sim 10^{-3}$	
Distilled water	$\sim 10^{-4}$	
Fresh water	$\sim 10^{-2}$	
Germanium (intrinsic)	~ 2.13	−0.048
Glass	$\sim 10^{-12}$	−0.07
Gold	4.10×10^7	0.0034
Iron	1.03×10^7	0.0052–0.0062
Lead	4.57×10^6	0.004
Marshy soil	$\sim 10^{-2}$	
Mercury (liquid)	1.04×10^6	0.00089
Mica	$\sim 10^{-15}$	−0.07
Nichrome (65 Ni, 12 Cr, 23 Fe)	1.00×10^6	0.00017
Nickel	1.45×10^7	0.0047
Niobium	8.06×10^6	
Platinum	9.52×10^6	0.003
Polystyrene	$\sim 10^{-16}$	
Porcelain	$\sim 10^{-14}$	
Quartz (fused)	$\sim 10^{-17}$	
Rubber (hard)	$\sim 10^{-15}$	
Seawater	~ 4	
Silicon (intrinsic)	$\sim 4.35 \times 10^{-4}$	
Silver	6.17×10^7	0.0038
Sodium	2.17×10^7	
Stainless steel	1.11×10^6	
Sulfur	$\sim 10^{-15}$	
Tin	8.77×10^6	0.0042
Titanium	2.09×10^6	
Tungsten	1.82×10^7	0.0045
Y Ba$_2$Cu$_3$O$_7$ (at < 80K)	$\sim 10^{20}$	
Wood	10^{-11}–10^{-8}	
Zinc	1.67×10^7	0.0037

coefficient α_σ for selected materials. Note that the temperature coefficients for nonmetals are typically negative, representing increasing conductivity with increasing temperature. The preceding relation assumes that $(\sigma_T)^{-1}$ varies linearly with T; in fact, quadratic terms can be used for greater accuracy.

Superconductivity. A dramatic deviation from the nearly linear dependence of resistance on temperature occurs in some materials that exhibit a phenomenon called *superconductivity,* discovered[9] by Kamerlingh Onnes in 1911. Below a critical temperature T_c, the conductivity of these materials becomes nearly infinite ($>10^{20}$ S-m^{-1} have been measured), although at slightly higher temperatures, the normal temperature-dependent resistance is observed. Most superconductors are metallic elements, compounds, or alloys and exhibit transitions into superconducting states at critical temperatures approaching absolute zero (0 K or $-273°$C). Some examples of metallic elements that exhibit superconductivity are aluminum ($T_c = 1.2$ K), lead ($T_c = 7.2$ K), and niobium ($T_c = 9.2$ K). Interestingly, some metals do not become superconducting; for example, copper is not superconducting even at 0.05 K.

Until recently, the Nb$_3$Al-Nb$_3$Ge ($T_c = 21$ K) alloy was believed to have the highest critical temperature, and alloys and compounds of niobium (e.g., niobium with a tin coating) are widely used as superconducting wire and tape in magnets operating at liquid-helium temperatures. A practical use for superconductivity has been in generating intense magnetic fields for research projects.[10] Niobium-tin compounds can support current densities of over 10^9 A-m^{-2}. In 1986, it was discovered that some oxides become superconducting at temperatures above the boiling temperature of liquid nitrogen (77 K). For example, yttrium-barium-copper oxide, or YBa$_2$Cu$_3$O$_7$, has $T_c = 80$ K, so its superconductivity can be utilized by cooling with liquid nitrogen. Since then, materials with T_c values as high as 125 K have been discovered.[11]

5.2 CURRENT FLOW, OHM'S LAW, AND RESISTANCE

Consider a piece of current carrying material such as the conductor shown in Figure 5.3. If at any time current is directed toward a boundary between the conductor and the

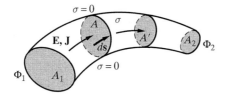

Figure 5.3 An arbitrary-shaped conductor. The surfaces A_1 and A_2 are maintained at potentials Φ_1 and Φ_2, respectively and are assumed to be equipotential surfaces (i.e., they are coated by material with conductivity much larger than that of the conductor).

[9]See R. de Bruyn Outober, Heike Kamerlingh Onnes's discovery of superconductivity, *Scientific American,* pp. 98–103, March 1997.

[10]For more information on the practical uses of superconductivity, see the August 1989 issue of the *Proceedings of the IEEE.*

[11]V. Z. Kresin and S. A. Wolf, *Fundamentals of Superconductivity,* Plenum, New York, 1990.

surrounding dielectric, charge accumulates at the boundary and produces a steadily increasing electric field. To maintain a constant electric field and a steady current flow, we must thus have $\mathbf{J}$ and $\mathbf{E}$ parallel to the conductor boundaries. Consider any two surfaces A and A', both of which are perpendicular to $\mathbf{J}$, with their circumferences defined by the conductor boundaries. The currents passing through these surfaces, namely $I = \int_A \mathbf{J} \cdot d\mathbf{s}$ and $I' = \int_{A'} \mathbf{J} \cdot d\mathbf{s}$, respectively, must be equal, since otherwise charge accumulates indefinitely in the region between the two surfaces. Because the two surfaces were chosen arbitrarily, the current must be the same through any other surface that cuts across the wire.

For homogeneous media (i.e., where σ is not a function of position), the distribution of current flow lines is the same as that of the electric field, and the charge density ρ inside the conductor is zero[12] so that the current lines do not begin or end anywhere in the conductor. The electric field within the conductor is presumed to be set up and maintained by means of a potential difference applied between the cross-sectional surfaces at the ends of the conductor, namely A_1 and A_2 in Figure 5.3. For simplicity, we take these end surfaces to be equipotential surfaces.[13] Note that no current can flow outside the conductor ($\sigma = 0$) and that charge cannot indefinitely accumulate at the boundary between the conductor and the surrounding dielectric. Thus, the electric field lines, and thus the lines of current flow, *are always parallel to the side edges of the wire,* regardless of how curved or twisted the conductor may be. Since the conservative property of the electric field requires (see Section 4.11) that the tangential component of $\mathbf{E}$ is continuous across any interface, the electric field just outside the wire must have the same tangential component as that just inside. In general, the electric field outside the wire also has a component perpendicular to the side surface of the wires, which is supported by surface charge induced on the surface of the wire.[14] Further discussion of the structure of the electric field at the boundary between a conductor and an insulator is found in Section 5.6.

The flow of current (i.e., $\mathbf{J}$) at each point in the material is in the direction of the electric field $\mathbf{E}$. The potential difference Φ_{12} between the two equipotential surfaces A_1 and A_2 at potentials Φ_1 and Φ_2, respectively, is

$$\Phi_{12} = \Phi_2 - \Phi_1 = -\int_L \mathbf{E} \cdot d\mathbf{l} = -\int_L \frac{1}{\sigma} \mathbf{J} \cdot d\mathbf{l}$$

---

[12]It will be shown later in Section 5.4 that, based on the principle of conservation of charge, the divergence of steady current must be zero, so that current lines cannot originate or terminate anywhere in the conductor. For homogeneous media, $\nabla \cdot \mathbf{J} = 0$ implies that $\nabla \cdot \mathbf{E} = 0$, which in turn means that volume charge density is zero. However, if σ varies with position, we may then have volume charge density $\rho \neq 0$, with the electric field terminating on volume charges (see footnote 22 in Section 5.4).

[13]This could be achieved by coating the end surfaces with a material whose conductivity is much greater than that of the conductor. For example, if the conductor itself is made of graphite, the end surfaces may be copper- or silver-coated.

[14]This perpendicular field typically exists because of an applied potential difference between the conductor and other conductors in a circuit. For example, if the conductor in question is the inner conductor of a coaxial line, the perpendicular field is radially outward between the inner and outer conductors such that the negative line integral of this field from the inner conductor (of radius a) to the outer one (of radius b) is equal to the voltage difference Φ_{ab}. Also see Section 5.6 and Figure 5.12.

where the path of integration L is any path starting at any point on A_1 and ending at any point on A_2. Note that in order for Φ_{12} to be positive, A_1 must be at a lower potential than A_2 (i.e., $\Phi_2 > \Phi_1$). Note also that, as argued in the previous paragraph, the same total current must pass through all of the surfaces A_1, A_2, A, and A' in Figure 5.3. This current is given by

$$I = \int_A \mathbf{J} \cdot d\mathbf{s} = \int_A \sigma \mathbf{E} \cdot d\mathbf{s}$$

If the potential difference Φ_{12} between A_1 and A_2 is increased, the electric flux lines do not change shape, but $\mathbf{E}$ proportionally increases everywhere within the conductor. Since the current is linearly related to the electric field, assuming that the increased field and the resultant increased current are not sufficient to lead to a substantial temperature increase that might modify the conductivity σ, $\mathbf{J}$ increases everywhere in the same proportion as $\mathbf{E}$, as does the total current I. The ratio of the potential difference to the total current is thus a constant, defined as the *resistance,* and is given by

$$\boxed{R \equiv \frac{\Phi_{12}}{I} = \frac{-\int_L \mathbf{E} \cdot d\mathbf{l}}{\int_A \sigma \mathbf{E} \cdot d\mathbf{s}}} \tag{5.3}$$

where L is any path from the end surface A_1 to A_2 (assuming $\Phi_2 > \Phi_1$), and where R remains constant for different Φ_{12} and I, but depends on the flow pattern between the end surfaces A_1 and A_2, and on the value of the conductivity σ. The relation between Φ_{12}, I, and R is the well known[15] *Ohm's law:*

$$\boxed{\Phi_{12} = IR} \tag{5.4}$$

Equation (5.3) constitutes a general definition for resistance akin to that for capacitance given in (4.52). The duality between $\mathbf{J} = \sigma \mathbf{E}$ and $\mathbf{D} = \epsilon \mathbf{E}$ is further discussed in Section 5.7.

As a simple example of the application of (5.3), we consider a rectangular conductor of conductivity σ, cross-sectional area A, and length l, as shown in Figure 5.4. Assuming the two end surfaces (A_1 and A_2) to be coated with highly conducting material and maintained at potentials of Φ_1 and Φ_2, respectively, the electric field within the material is given[16] by $\mathbf{E} = -\hat{\mathbf{z}}E_0$, where E_0 can be found as

$$\Phi_{12} = \Phi_2 - \Phi_1 = -\int_{z_1}^{z_2} \mathbf{E} \cdot d\mathbf{z} = +\int_{z_1}^{z_2} E_0 \hat{\mathbf{z}} \cdot \hat{\mathbf{z}} dz \quad \rightarrow \quad E_0 = \frac{\Phi_2 - \Phi_1}{z_2 - z_1} = \frac{\Phi_{12}}{l}$$

---

[15]For a discussion of Ohm's law and contributions of G. S. Ohm, see M. S. Gupta, Georg Simon Ohm and Ohm's Law, *IEEE Transactions on Education*, 23(3), pp. 156–162, August 1980.

[16]Note that we have assumed (on the basis of symmetry) the electric field to be constant everywhere inside the rectangular conductor.

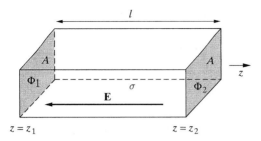

Figure 5.4   A rectangular conductor of uniform cross section. The end surfaces are coated with highly conducting material and are thus equipotential surfaces.

so that the current is given by

$$I = \int_A \sigma \mathbf{E} \cdot d\mathbf{s} = \frac{\sigma \Phi_{12} A}{l} \qquad \rightarrow \qquad R = \frac{\Phi_{12}}{I} = \frac{\Phi_2 - \Phi_1}{I} = \frac{l}{\sigma A} \qquad (5.5)$$

The preceding expression is not specific to a rectangular block and is in fact valid for the resistance of a conductor of length l having a uniform cross-sectional area A of any shape.

Example 5.1: Copper wire. Consider a copper wire 1 km long and 1 mm in radius. (a) Find the resistance R. (b) If the wire carries a current of 1 A, determine the duration of time in which the electrons drift across the length of the wire.

Solution:

(a) For copper, we have $\sigma = 5.8 \times 10^7$ S-m^{-1}. For $l = 10^3$ m and a 1-mm radius, which gives $A = \pi \times 10^{-6}$ m^2, using (5.5) we have

$$R = \frac{10^3 \text{ m}}{5.8 \times 10^7 \text{ S-m}^{-1} \times \pi \times 10^{-6} \text{ m}^2} \simeq 5.49\Omega$$

(b) A current of 1 A and a cross-sectional area $A = \pi \times 10^{-6}$ m^2 implies a current density of $\mathbf{J} \simeq 3.18 \times 10^5$ A-m^{-2}. Since $\mathbf{J} = n_c q_e \mathbf{v}_d$ and $n_c = 8.45 \times 10^{28}$ el-m^{-3}, we have $|\mathbf{v}_d| \simeq 2.35 \times 10^{-5}$ m/s. In other words, the electrons can drift across a distance of 1 m in about $\sim$12 hours. A particular tagged electron can traverse the entire 1-km length of this copper wire in approximately $\sim$492 days! On the other hand, we know from transmission-line analysis that an applied signal (e.g., a change in current through or voltage across a line) travels to the other end of the line at nearly the speed of light in air (i.e., approximately $\sim$3.33 μs for a 1-km wire). The transmission-line signal propagates outside the conductors (or wires) as an electromagnetic wave, not via the drifting motion of electrons within the conductors that constitute the line.

Example 5.2: A microstrip trace. Consider a printed circuit board microstrip trace, as shown in Figure 5.5. If the metal trace is made of copper with thickness $t = 34.3$ μm, find the trace resistance per centimeter for a trace width w of (a) 0.25 mm and (b) 0.5 mm.

Copper trace w

Insulator

Figure 5.5   A microstrip trace.
The cross-sectional view of a microstrip trace.

Solution: Using (5.5) where $l = 1$ cm $= 0.01$ m, $\sigma = 5.8 \times 10^7$ S-m^{-1}, and $A = wt = 3.43 \times 10^{-5}w$ m^2:

(a) For $w_1 = 2.5 \times 10^{-4}$ m, we have $R_1 \simeq 20.1$ mΩ-cm^{-1}, and

(b) for $w_2 = 2w_1 = 5 \times 10^{-4}$ m, we have $R_2 = R_1/2 \simeq 10.05$ mΩ-cm^{-1}.

Example 5.3: Curved bar. The resistance of a rectangular block of metal was discussed in this section and found to be given by $R = l/(\sigma A)$, where σ is the conductivity, l is the length, and A is the cross-sectional area. As a more complicated geometry, consider the resistor in the form of a curved bar, which is made of the rectangular block bent to form the arc of a circle, with the two end surfaces remaining flat, as shown in Figure 5.6a. The edges are coated with perfectly conducting material so that they constitute electrodes at uniform potential. Find the resistance R between the electrodes.

Solution: This problem has a cylindrical symmetry with respect to the O-O′ axis, as shown. The equipotential surfaces are the intersections of the curved rectangular bar and the planes

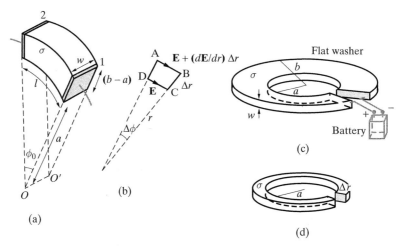

Figure 5.6   A curved-bar resistor. (a) A rectangular bar bent into a circular arc of rectangular cross section. (b) A differential angular element along the arc. (c) A flat washer is equivalent to a curved bar of arc length $\phi_0 \simeq 2\pi$. (d) The flat washer can be viewed as a parallel combination of elemental resistances of radial width Δr, as shown.

passing through the O-O' axis. The current flow lines $\mathbf{J}$ are circular arcs centered on the O-O' axis (i.e., $\mathbf{J}$ is in the $\hat{\phi}$ direction). We may suppose from symmetry that the $\mathbf{E}$ vector is the same at all points inside the bar that are equidistant from the O-O' axis. However, the electric field must vary with radial distance from the axis (r) across the curved bar, since the inward portions of the end faces are closer to each other than the outward portions are (i.e., $\mathbf{E}$ must be a function of r, or $\mathbf{E} = \mathbf{J}/\sigma = \hat{\phi}E_\phi(r)$). To determine how the electric field varies with radial distance, we can use the fact that $\oint_C \mathbf{E} \cdot d\mathbf{l} = 0$ (equation (4.20)). Consider a closed contour ABCD, as shown in Figure 5.6b. We have

$$\oint_C \mathbf{E} \cdot d\mathbf{l} = \left(E_\phi + \frac{dE_\phi}{dr}\Delta r\right)(r + \Delta r)\Delta\phi - E_\phi r \Delta\phi = 0$$

where the first term is the contribution from path segment AB and the second term is that from path segment CD. We thus have

$$E_\phi \Delta r + r\frac{dE_\phi}{dr}\Delta r + \frac{dE_\phi}{dr}(\Delta r)^2 = 0 \qquad \rightarrow \qquad \frac{dE_\phi}{E_\phi} = -\frac{dr}{r}$$

where the last term drops out as $\Delta r \rightarrow 0$. Therefore

$$\ln E_\phi = -\ln r + K_1 \qquad \rightarrow \qquad E_\phi = \frac{K}{r}$$

where K_1 and K are constants. The functional form of E_ϕ is thus determined. Note that we could alternatively have found $\Phi(r)$ using Laplace's equation, namely $\nabla^2\Phi = 0$, and subsequently determined $\mathbf{E}$ from $\mathbf{E} = -\nabla\Phi$. The voltage between the end faces is thus

$$\Phi_{12} = -\int_1^2 \mathbf{E} \cdot d\mathbf{l} = -\int_{\phi_0}^0 \frac{K}{r}\hat{\phi} \cdot \hat{\phi} r \, d\phi = \int_0^{\phi_0} \frac{K}{r} r \, d\phi = K\phi_0 = \frac{Kl}{a}$$

where we have implicitly assumed the end face at the lower potential to be at $\phi = \phi_0$, so that $\mathbf{E} = \hat{\phi}(K/r)$. The total current between the end faces is given by

$$I = \int_S \mathbf{J} \cdot d\mathbf{s} = \int_S \sigma\mathbf{E} \cdot d\mathbf{s} = \sigma w K \ln\frac{b}{a}$$

Hence the resistance is given by

$$R = \frac{\Phi_{12}}{I} = \frac{l}{\sigma w a \ln(b/a)}$$

Note that the preceding expression for R applies for any arc length l of the curved bar. For example, consider the flat washer shown in Figure 5.6c, with resistance between the edges of a saw-cut through the radius of the washer (i.e., the shaded surface and the one facing it). In this case, and assuming the arc length of the cut region is very small ($\ll 2\pi a$), we simply replace the length l in the previous expression with $l \simeq 2\pi a$. Note from Figure 5.6d that the resistance of the flat washer between the indicated faces can also be thought of as being the parallel combination of elemental flat washer resistances, each of radial thickness Δr.

Example 5.4: Buried electrode. A ground connection is made by burying halfway a perfectly conducting spherical electrode of radius a in the earth, as shown in Figure 5.7. Assuming the earth's conductivity to be σ, find the resistance of the conductor to distant points in the ground (i.e., between the electrode and a concentric, perfectly conducting hemisphere of infinite radius).

> **Solution:** We can set up a spherical coordinate system as shown in Figure 5.7, where the azimuth angle ϕ is not shown. Since the electrode is a perfect conductor, the potential will be constant at every point within and on the surface of this electrode. We can arbitrarily take the reference point for potential to be at infinity. Due to symmetry, all field quantities (and thus the current) are independent of the azimuth angle ϕ. The electric field is also independent of θ, although current flows only in the lower hemisphere. We then have $\Phi(r) = K/r$, where K is a constant, as an admissible general solution of Laplace's equation for this configuration. Assuming that the spherical electrode is held at a certain potential V_0, we must have $\Phi(r) = V_0$ at $r = a$, so that $K = aV_0$. Thus, $\Phi(r) = aV_0/r$.
>
> The corresponding electric field can be found from $\mathbf{E} = -\nabla\Phi$, which gives $\mathbf{E}(r) = \hat{\mathbf{r}}aV_0/r^2$. Thus, the current density $\mathbf{J}$ is also in the radial direction and is given by

$$\mathbf{J} = \sigma\mathbf{E} = \hat{\mathbf{r}}\frac{\sigma aV_0}{r^2}$$

Note that this current satisfies all the boundary conditions since it is entirely tangential to the earth-free space interface at $\theta = \pi/2$. The total current I that crosses any concentric hemisphere of radius r is

$$I = \int_S \mathbf{J}\cdot d\mathbf{s} = \frac{\sigma a\Phi_0}{r^2}\frac{4\pi r^2}{2} = 2\pi\sigma aV_0$$

Therefore the total resistance between the electrode and distant points is

$$R = \frac{V_0}{I} = \frac{V_0}{2\pi\sigma V_0 a} = \frac{1}{2\pi\sigma a}$$

Note that because we took the reference point for potential to be at infinity, the voltage drop V_0 is that between the electrode and the hemisphere at infinity.

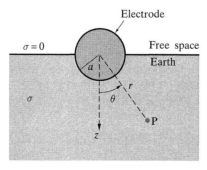

Figure 5.7 Buried spherical electrode. The electrode is made of perfectly conducting material and is thus an equipotential surface.

Figure 5.8   Thin-film contact resistance. Two concentric circular electrodes on a thin film of conducting material ($\sigma \neq 0$) lying on an insulator.

Example 5.5: A thin semiconductor film contact. Consider a thin semiconductor film of thickness h with concentric perfectly conducting electrodes (or contacts) of radii a and b, respectively ($h \leq a, h \leq b$). (See Figure 5.8.) The thin film of semiconductor material is deposited on an insulator, and the conductivity of this material is σ S-m^{-1}. Find the resistance between the electrodes.

Solution: We consider a cylindrical coordinate system with the z axis pointing up and passing through the center of the inner electrode. Due to symmetry, the electric field in the region between the electrodes is independent of ϕ and only has a component in the r direction. Considering circular electrodes of radii a and b, where $a \geq h, b \geq h$, and assuming the electrodes to be perfect conductors, the total radial current I flowing between the electrodes is given by

$$I = \int_S \sigma \mathbf{E} \cdot d\mathbf{s} = \int_0^h \int_0^{2\pi} \sigma E_r \hat{\mathbf{r}} \cdot \hat{\mathbf{r}} r \, d\phi \, dz = \sigma E_r (2\pi r h) \quad \rightarrow \quad E_r = \frac{I}{2\pi \sigma h r}$$

Taking the electrode at $r = b$ to be at zero potential and the electrode at $r = a$ to be at a potential V_0, we find that the potential at a radius r (where $a \leq r \leq b$) is

$$\Phi(r) = -\int_b^r \mathbf{E} \cdot d\mathbf{l} = -\int_b^r \frac{I}{2\pi \sigma h \zeta} d\zeta = \frac{I}{2\pi \sigma h} \ln \frac{b}{r}$$

Substituting $r = a$, we find the potential difference and the total resistance between the electrodes:

$$\Phi_{ba} = V_0 = \frac{I}{2\pi \sigma h} \ln \frac{b}{a} \quad \rightarrow \quad R = \frac{\Phi_{ba}}{I} = \frac{1}{2\pi \sigma h} \ln \frac{b}{a}$$

5.3 ELECTROMOTIVE FORCE AND KIRCHHOFF'S VOLTAGE LAW

Until now, we have not discussed the requirements for the maintenance of steady currents, although it is clear that we must start with an electric field (i.e., $\mathbf{J} = \sigma \mathbf{E}$). It turns out that electrostatic fields produced by stationary charges cannot by themselves maintain a steady current, since they are conservative in nature (i.e., $\oint_C \mathbf{E} \cdot d\mathbf{l} = 0$). Thus electromotive force produced by another energy source, such as a chemical battery, is necessary to sustain steady currents. In this section, we discuss the maintenance of steady currents and electromotive force.

Suppose we wish to establish current in a rectangular conducting bar such as that shown in Figure 5.4. Since current needs to be driven by an electric field, we might think of inserting the conductor in an electrostatic field, such as that between the plates of a parallel-plate capacitor (see Figure 5.9a). At the instant the bar is inserted, the conductor is in a uniform electric field, and thus a current $I = \sigma E A$ flows. But this current cannot continue since it is essentially a redistribution of charge inside the conductor. Within a very short time scale (e.g., $\sim 10^{-19}$ s for copper) negative charge accumulates at the top of the bar, leaving behind positive charge at the bottom, and these separated charges set up a field within the conductor that completely cancels the external field and reduces the total field within the conductor to zero,[17] so that the current simply stops. It is thus apparent that an electrostatic field cannot sustain steady current flow.

We observed in connection with Figure 5.3 that steady currents cannot originate or terminate within the conductor, since this would imply nonstop removal or accumulation of charge. Accordingly, steady current flow requires a closed circuit. There may be branches in the circuit where three or more wires are connected, but any given flow line must close on itself without branching. Let us now consider one such closed circuit, as shown in Figure 5.9b, and the motion of the current carriers. For the sake of this argument, we neglect the wire resistances and assume that current can flow through the wires without any dissipation of energy. For steady-state current to exist, the current carriers (e.g., electrons) must make a complete circuit (closed loop), starting from point 1 and ending at point 1, repeating their motion ad infinitum. However, since current through the resistor (rectangular bar) produces energy loss in the form of heat (due to collisions), the electrons lose energy as they travel through this part of the circuit; they cannot gain this energy back from the conservative electrostatic field because the round-trip motion leads to zero net energy (i.e., $\oint_C \mathbf{E} \cdot d\mathbf{l} = 0$). Thus, another source of energy is necessary to maintain steady electric current flow through the circuit by continuously supplying the energy dissipated in the resistor as heat. In other words, as we go around the circuit starting at a given point, we observe a continual decrease in electrostatic potential. Thus, there *must* be some portion of the circuit that does not behave like a

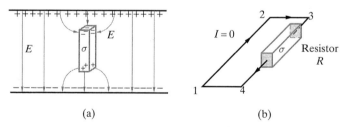

(a)   (b)

Figure 5.9 Steady currents cannot be maintained without an external energy source. (a) The placement of a rectangular conducting bar in a uniform field leads to rapid ($\sim 10^{-19}$ s) redistribution of the charge in the bar, but the field does not support steady current flow in the bar. (b) Current in a closed circuit cannot be maintained since the current carriers lose their energy in traveling through the resistor, and they cannot gain this energy back from the conservative electrostatic field.

[17]This internal field is the induced field E_{ind} that was discussed in Section 4.7.3 in connection with Figure 4.41.

conductor (i.e., does not obey Ohm's law), but where potential *rises* along the direction of the current.

The sources of the external energy required to maintain steady currents and to create rises in potential can be chemical batteries, electric generators, thermocouples, photovoltaic cells, or other devices. These electrical energy sources, when connected in a circuit, provide the driving force for current carriers. This force manifests itself in the form of an external electric field, which we designate as $\mathbf{E}_{\text{emf}}$. (The subscript "emf" will be defined shortly.) Consider the simple electric circuit shown in Figure 5.10a. The circuit consists of a rectangular conducting bar that acts as a resistor R and a battery. A steady current I is assumed to be present at all points in the circuit. We interpret the internal action of the battery from the standpoint of producing a nonconservative electric field $\mathbf{E}_{\text{emf}}$. In general, an electrostatic field (designated by $\mathbf{E}$) is also created by the battery as a result of accumulation, at the battery terminals or elsewhere in the circuit, of stationary charge (e.g., such as that on a capacitor). The total electric field is then

$$\mathbf{E}_{\text{total}} = \mathbf{E}_{\text{emf}} + \mathbf{E}$$

Note also that although $\mathbf{E}$ is conservative and can be derived from the gradient of a scalar potential due to stationary charges (i.e., $\mathbf{E} = -\nabla\Phi$), $\mathbf{E}_{\text{emf}}$ is not related to any stationary charges. At all points in the circuit we have

$$\frac{\mathbf{J}}{\sigma} = \mathbf{E}_{\text{emf}} + \mathbf{E}$$

(where $|\mathbf{J}| = I/A$), although $\mathbf{E}_{\text{emf}}$ is actually zero at all points outside the battery. If we now integrate around a complete circuit (closed loop), we find

$$\oint_C \mathbf{E} \cdot d\mathbf{l} + \oint_C \mathbf{E}_{\text{emf}} \cdot d\mathbf{l} = \oint_C \frac{1}{\sigma}\mathbf{J} \cdot d\mathbf{l} = \int_a^b \frac{I}{\sigma A}\,dl$$

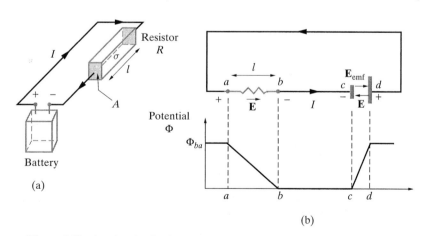

Figure 5.10   **A series circuit with a battery and resistor.** (a) Battery driving a circuit with a resistor. (b) Variation of potential around the circuit.

The first term is zero, since the line integral around a closed loop of an electrostatic field produced by stationary charges is always zero. However, the second term, involving the line integral of $\mathbf{E}_{emf}$, is not zero but is instead equal to a voltage called the *electromotive force,* or *emf,* of the circuit, typically denoted by $\mathscr{V}_{emf}$. The right-hand side of the preceding equation equals the total voltage drop around the circuit, which is simply equal to IR, assuming that the internal resistance of the battery is zero[18] and that the wire resistances are also zero (i.e., no voltage drops occur across the wires). Thus we have

$$\mathscr{V}_{emf} = I \int_a^b \frac{dl}{\sigma A} = IR \qquad \text{or} \qquad R = \int_a^b \frac{dl}{\sigma A}$$

If current were not allowed to flow (e.g., by disconnecting part of the circuit so that it is no longer a closed circuit), or if we consider a battery under open circuit conditions, an electrostatic field exists everywhere because of the accumulation of charge at the battery terminals. Within the battery, the electrostatic field is neutralized by the nonconservative E_{emf} field. From the point of view of a positive test charge, it is possible to acquire energy in moving from the positive to the negative terminal (i.e., in the direction of decreasing potential) external to the battery, but by completing the circuit through the battery, where there is zero field, the acquired energy is not returned to the field. The test charge can thus make a complete circuit with a net accumulation of energy.[19] In an actual circuit such as that in Figure 5.10a, the acquired energy is dissipated in the resistor as heat. Nevertheless, the test particle is capable of making repeated round trips, hence constituting a steady electric current.

In general, for a closed circuit containing many resistors and emf sources, we have

$$\boxed{\sum \mathscr{V}_{emf} = I \sum R} \tag{5.6}$$

This relation is known as *Kirchhoff's voltage law.* It states that the algebraic sum of the emfs around a closed circuit equals the algebraic sum of the voltage drops (i.e., IR) over the resistances around the circuit. Kirchhoff's voltage law applies not only for an isolated closed network but also for any single mesh (closed path) of a larger network. Although the preceding discussion relates to stationary electric fields and currents, Kirchhoff's voltage law also applies to time-varying situations, as long as the dimensions of the circuit are much smaller than a wavelength. In such cases, the algebraic sum of the *instantaneous* emfs equals the algebraic sum of the *instantaneous IR* drops around the circuit.

Further insight into the nature of the electromotive force and Kirchhoff's voltage law can be gained by examining the variation of potential around the circuit, as shown in Figure 5.10b. Taking the points b, c to be at zero potential, a positively charged current carrier starting at point a and traveling in the direction of the current undergoes a decrease

---

[18]If the battery has an internal resistance R_s, the total voltage drop around the circuit is simply $IR + IR_s$.

[19]Accumulation of kinetic energy in many such round trips through a nonconservative field in the absence of resistive losses is the basis for particle accelerators such as the cyclotron.

in potential as it travels through the resistor and then an equivalent increase as it travels through the battery. Note that the potential difference Φ_{ab} is related to the electrostatic field **E**, so that

$$\Phi_{ab} = \Phi_b - \Phi_a = -\int_a^b \mathbf{E} \cdot d\mathbf{l}$$

whereas the potential difference Φ_{cd} is simply equal to the electromotive force $\mathscr{V}_{\text{emf}}$ of the battery.

5.4 THE CONTINUITY EQUATION AND KIRCHHOFF'S CURRENT LAW

We now consider one of the fundamental principles of electromagnetics, namely the continuity equation. This important equation is a mathematical statement of the conservation of charge and is also the basis of Kirchhoff's current law. Since current consists of the flow of charge and since charge is conserved,[20] we must have a buildup of charge within a region if there is net current flow into it. If we consider an arbitrary volume V bounded by a surface S, as shown in Figure 5.11a, we have

$$-\oint_S \mathbf{J} \cdot d\mathbf{s} = \frac{\partial}{\partial t} \int_V \rho \, dv$$

where the left-hand term gives the net inflow of current (hence the negative sign, since $d\mathbf{s}$ is by convention defined as being outward) and the right-hand side represents the net rate of increase of total free charge enclosed by the surface S. Note from Figure 5.11a that **J** is defined as outward from the volume.

Using the divergence theorem (equation (4.40)), we have $\oint_S \mathbf{J} \cdot d\mathbf{s} = \int_V \nabla \cdot \mathbf{J} \, dv$, and thus the preceding equation can be rewritten as

$$\int_V \left(\nabla \cdot \mathbf{J} + \frac{\partial \rho}{\partial t} \right) dv = 0$$

---

[20]Conservation of charge is one of the basic laws of physics. Charge is indestructible; it cannot be lost or created. Electric charge can move from place to place but can never appear from nowhere. The principle of conservation of charge is as important to physics as the constancy of the speed of light. The validity of this principle is verified by every experiment to date and is a direct consequence of Maxwell's equations. Like the speed of light, charge is the same for every frame of reference. Although other quantities such as mass, energy, and electric and magnetic fields all change with frames of reference, charge does not (see Section 11-8 of A. M. Portis, *Electromagnetic Fields,* John Wiley, 1978). The conservation of electric charge, based on experiments with electrical bodies and the transfer of electrification, was put forth in 1746 by William Watson and in 1747 by Benjamin Franklin. The first satisfactory experimental proof was provided by M. Faraday in *Phil. Mag.*, Vol. xxii, 1843, p. 200. For an excellent discussion of the history, see Chapters II and VI of E. Whittaker, *A History of the Theories of Aether and Electricity,* Thomas Nelson and Sons Ltd., London, 1951.

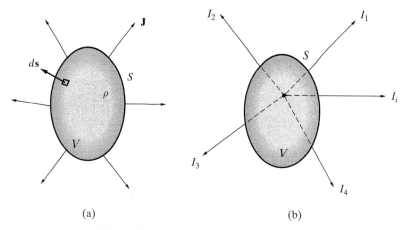

Figure 5.11   **Current flow out of a volume** *V*.

Since this relation must be true regardless of the choice of the volume *V*, the integrand itself must be zero, or

$$\boxed{\nabla \cdot \mathbf{J} + \frac{\partial \rho}{\partial t} = 0}$$

(5.7)

Equation (5.7) is commonly referred to as the *continuity equation* and is essentially a differential form of the law of conservation of charge. For steady currents, we must have $\partial \rho / \partial t = 0$ so that

$$\nabla \cdot \mathbf{J} = 0$$

In other words, for stationary currents, the current density $\mathbf{J}$ is *solenoidal*.[21] We noted earlier that steady currents must flow in closed loops in order to avoid continuous accumulation of charge.

Since Ohm's law must hold in a conducting medium, we have $\mathbf{J} = \sigma \mathbf{E}$. If σ does not vary with position (i.e., if the medium is homogeneous), we must have $\nabla \cdot \mathbf{J} = \sigma \nabla \cdot \mathbf{E}$ for both steady and time-varying cases. In general, the continuity equation can then be written in terms of the electric field

$$\nabla \cdot \mathbf{E} + \frac{1}{\sigma}\frac{\partial \rho}{\partial t} = 0$$

For the static case ($\partial \rho / \partial t = 0$), the electric field in a homogeneous conducting medium must have zero divergence; that is, $\nabla \cdot \mathbf{E} = 0$. Since from Gauss's law we have

---

[21] A vector field is said to be a solenoidal field if its divergence is zero everywhere. The zero divergence indicates that there are no sources or sinks in the field for the lines of flux to originate from or terminate on. Accordingly, the flux lines of a solenoidal vector field always close on themselves.

$\nabla \cdot \mathbf{E} = \rho/\epsilon$, we conclude that the volume density of free charge ρ must be zero in a homogeneous conducting medium.[22]

Using the divergence theorem, the solenoidal nature of steady currents can be expressed in integral form as

$$\nabla \cdot \mathbf{J} = 0 \quad \rightarrow \quad \oint_S \mathbf{J} \cdot d\mathbf{s} = 0$$

The preceding relation is for steady currents and applies to any closed surface. The volume enclosed by this surface may be entirely inside a conducting medium, or it may only be partially filled with conductors. The conductors may form a network inside the volume, or they may all meet at a point. If the steady current is carried into the volume by different wires meeting at a node, as shown in Figure 5.11b, then the preceding relation implies that the *algebraic sum of all the currents at the junction is zero,* since a junction of wires can be neither a sink nor a source for charges. This condition is *Kirchhoff's current law,* which can be expressed mathematically as

$$\boxed{\sum I = 0} \tag{5.8}$$

Like Kirchhoff's voltage law, Kirchhoff's current law, as just derived for stationary fields and currents, also applies to time-varying situations, as long as the dimensions of the circuit (i.e., the length of the current-carrying wires) are much smaller than a wavelength (see Chapter 1).

5.5 REDISTRIBUTION OF FREE CHARGE

We have noted on several occasions that charge placed at a point inside a conducting body moves to the surface and redistributes itself in such a way that zero field exists within and tangent to the conductor surface. With the continuity equation now in hand, we can quantitatively evaluate the length of time required for this process. We shall show that there can be no permanent distribution of free charge within a homogeneous region

---

[22] In an inhomogeneous medium, with σ and ϵ both functions of position, we have

$$\nabla \cdot \mathbf{D} = \epsilon \nabla \cdot \mathbf{E} + \mathbf{E} \cdot \nabla \epsilon = \rho \quad \rightarrow \quad \nabla \cdot \mathbf{E} = \frac{\rho - \mathbf{E} \cdot \nabla \epsilon}{\epsilon}$$

and

$$\nabla \cdot \mathbf{J} = \sigma \nabla \cdot \mathbf{E} + \mathbf{E} \cdot \nabla \sigma = \frac{\sigma \rho}{\epsilon} - \frac{\sigma}{\epsilon} \mathbf{E} \cdot \nabla \epsilon + \mathbf{E} \cdot \nabla \sigma$$

In other words, $\nabla \cdot \mathbf{J} = 0$ does not imply that $\rho = 0$. Volume charge density can exist even under conditions of steady current flow (i.e., where $\nabla \cdot \mathbf{J} = 0$) in regions of variable conductivity, and its magnitude is proportional to the gradient of the conductivity, or $\rho = -(\epsilon/\sigma)\mathbf{E} \cdot \nabla \sigma + \mathbf{E} \cdot \nabla \epsilon$. Such regions of static free charge may exist in thin layers at the boundary between two materials of different conductivity. A good example is a parallel-plate capacitor with two different lossy materials sandwiched between its plates (see Problem 5.28).

of nonzero conductivity, and that the rearrangement or relaxation time is determined by the permittivity-to-conductivity ratio of the material (i.e., ϵ/σ).

Consider a homogeneous, isotropic, time-invariant, and linear conducting region where conductivity σ and permittivity ϵ are simple constants and are not functions of position, direction, time, or the applied electric field. Using $\nabla \cdot \mathbf{D} = \rho$ (Gauss's law) and the continuity equation, and assuming that the relations between $\mathbf{D}$ and $\mathbf{E}$, and between $\mathbf{J}$ and $\mathbf{E}$ are linear, we have

$$\nabla \cdot \mathbf{J} + \frac{\partial \rho}{\partial t} = \nabla \cdot (\sigma \mathbf{E}) + \frac{\partial \rho}{\partial t} = \sigma \nabla \cdot \mathbf{E} + \frac{\partial \rho}{\partial t} = 0$$

On the other hand, we have

$$\nabla \cdot \mathbf{D} = \nabla \cdot (\epsilon \mathbf{E}) = \epsilon \nabla \cdot \mathbf{E} = \rho \quad \rightarrow \quad \nabla \cdot \mathbf{E} = \frac{\rho}{\epsilon}$$

Combining the two preceding equations we find

$$\frac{\partial \rho}{\partial t} + \frac{\sigma}{\epsilon} \rho = 0$$

Rewriting and integrating, we have

$$\frac{\sigma}{\epsilon} \partial t = -\frac{\partial \rho}{\rho} \quad \rightarrow \quad \frac{\sigma}{\epsilon} \int_0^t \partial t = -\int_{\rho_0}^{\rho} \frac{\partial \zeta}{\zeta} \quad \rightarrow \quad \frac{\sigma}{\epsilon} t = -\ln\left(\frac{\rho}{\rho_0}\right)$$

$$\rightarrow \quad \rho(x,y,z,t) = \rho_0(x,y,z) e^{-(\sigma/\epsilon)t}$$

where $\rho_0(x,y,z)$ is the initial value of the free charge density at $t = 0$. The initial charge distribution throughout the conductor decays exponentially with time at every point, completely independent of any applied electric fields. If the charge density is initially zero, it remains zero at all times thereafter.

The time $\tau_r = \epsilon/\sigma$ is is referred to as the *relaxation time* and is the time required for the charge at any point to decay to $1/e$ of its original value. The relaxation time is extremely short for good conductors and relatively large for insulators or dielectrics. In fact, whether a material is considered a conductor or an insulator is decided on the basis of the relaxation time. When τ_r is extremely short compared with measurable times or times of interest in a given application, the material is considered to be a conductor; whereas when τ_r is very long, the material behaves like an insulator. For most metals, τ_r is indeed far too short to measure or observe; for example, for copper we have $\sigma = 5.8 \times 10^7$ S-m^{-1} and $\epsilon = \epsilon_0$,[23] so that $\tau_r^{\text{copper}} \simeq 10^{-19}$ s. In fact, τ_r is very small for all but the poorest of conductors. Even for distilled water, $\tau_r^{\text{H}_2\text{O}} \simeq 10^{-5}$ s. On the

---

[23] The dielectric constant ϵ for a metallic conductor is not easily measurable, since any polarization effect is completely overshadowed by conduction. Nevertheless, based on measurements of reflectivity of metals and the fact that atomic resonances for metals lie in the ultraviolet and x-ray ranges, metallic conductors can be treated as if their dielectric constant is ϵ_0 at frequencies up to and including the visible range (i.e., $\sim 10^{15}$ Hz).

other hand, for a good dielectric τ_r is very large; for example, $\tau_r^{amber} \simeq 4 \times 10^3$ s, $\tau_r^{mica} \simeq$ 10–20 hours, and $\tau_r^{quartz} \simeq 50$ days.

Consider a thought experiment in which we suppose that at $t = 0$, charge is concentrated within a small spherical region located near the center of a very large conducting sphere. In every other region of the conductor, the charge density is initially zero. Starting at $t = 0$, the charge within the small spherical region begins to fade away exponentially, but since charge anywhere in the conductor can only *decrease* in time (because of the expression $\rho = \rho_0 e^{-(\sigma/\epsilon)t}$ derived previously), no charge can appear anywhere *within* the conductor. Where then does the charge in the small spherical region go? Because charge is conserved, the exponentially vanishing charge near the center must begin to appear at the surface of the conducting sphere, no matter how great the radius of the conducting sphere may be. However, the surface charge must make its appearance at the exact instant that the interior charge begins to decay, because the total charge in the system is constant.[24]

Although we are concerned in this chapter with electrostatic fields, the concept of relaxation time is also used to determine the electrical nature (i.e., conductor versus insulator) of materials for time-varying fields. At any given frequency of operation f, a material is considered a good conductor if τ_r is much shorter than the period $T = 1/f$, that is, if $\tau_r \ll T$. Conversely, the material is considered an insulator if $\tau_r \gg T$. We can now see that some materials that are considered to be good conductors at certain frequencies tend to become insulators at sufficiently higher frequencies. For example, seawater ($\sigma = 4$ S-m^{-1}) is considered to be a good conductor at frequencies up to $\sim$100 MHz but is an insulator for frequencies above $\sim$10 GHz. Further discussion of conducting or dielectric properties of materials at different frequencies is provided in Sections 8.3 and 11.2.1.

5.6 BOUNDARY CONDITIONS FOR STEADY CURRENT FLOW

Most applications of steady current flow involve considerations of the interfaces between current-carrying conductors and dielectrics (or insulators) or between two conducting materials of different conductivity. The manner in which the current density **J** and the electric field **E** behave across such interfaces is governed by the *boundary conditions,* which are formulated and discussed in this section. We separately consider conductor–dielectric and conductor–conductor interfaces.

---

[24]This interesting observation at first appears to indicate that this phenomenon may be used to transmit signals with infinite velocity. However, in order to convey such a signal at a given moment ($t = 0$), we must, until this moment, prevent charge on the small sphere from being dispersed. This can be done by means of an insulating envelope (e.g., a thin membrane) that might be withdrawn at the given moment to initiate the signal. Before this could happen, however, an induced charge equal and opposite to the charge on the small sphere would appear on the external surface of the envelope, and at the same time an induced charge equal to the original charge in the small sphere would be produced on the surface of the large sphere. The withdrawal of the insulating envelope would merely cause the charges on its two sides to unite (M. Abraham and R. Becker, *Electricity and Magnetism,* Blackie & Son Limited, Glasgow, pp. 260, 272, 1944).

5.6.1 Current and Electric Field at Conductor – Dielectric Interfaces

In Section 4.7, we showed that the surfaces of metallic conductors were equipotentials and that, therefore, the tangential component of electric field on a metallic surface must necessarily be zero (i.e., $E_t = 0$). On the other hand, when a conductor with finite conductivity carries a current, an electric field given by $\mathbf{E} = \mathbf{J}/\sigma$ exists within the medium. At the boundary of any conductor–dielectric interface, the current must flow tangentially to the boundary surface, as shown in Figure 5.12a, since current cannot flow across the boundary into the dielectric region with zero conductivity. Thus, on the conductor side of the interface, we have $E_t = J_t/\sigma$. By the continuity of the tangential electric field at a boundary, the tangential field on the dielectric side must also be E_t.

 If a potential difference is applied between the two ends of a good (but not perfect) conductor,[25] current flows, and the body of the conductor is no longer an equipotential.

Figure 5.12 Conductor–dielectric interface. (a) Current must be tangential to the surface since it cannot flow into the dielectric region. (b) Battery connected to the two ends of a conducting bar. (c) Two metal strips separated by air or another dielectric are maintained at a potential difference of V_0. (d) The connection of a load across the metal strips leads to current flow, which in turn means that the electric fields on the surfaces of the metal strips are not entirely vertical.

[25]In practice this can be achieved by connecting the wire across a battery, which is rather unwise in most cases since the resistance of a piece of conducting wire is typically extremely small (much smaller than the internal resistance of the battery), and the battery is shorted.

Assuming that the two ends of the conducting bar are coated with perfectly conducting material so that they are equipotentials, the potential varies uniformly along such a current-carrying wire, as shown in Figure 5.12b. Assuming that the electric field E is uniform, the potential difference $d\Phi$ over any differential length dl of the wire is $E\,dl$. The integral $\int E\,dl$ carried out over the full length l of the wire is equal to the applied potential difference V_0 and is also given by $V_0 = IR$, where R is the resistance of the wire and I is the total current. The field is the same just inside and just outside of the wire and is entirely tangential (i.e., parallel to the axis of the wire).

Note, however, that in practice a conducting wire is very rarely used in the configuration implied in Figure 5.12b, since for typical applied potentials (e.g., $V_0 = 1$ V from a battery) the current is rather large, and the wire shorts out the battery. As an example, consider a silver wire of radius 1 mm and length 10 cm. The resistance of such a wire is approximately $\sim 5 \times 10^{-4}\Omega$, which means that a current of $I \simeq 2000$ amperes flows when this single wire is connected across a 1-V battery! In reality, however, any battery has an internal resistance much larger than the resistance of the wire, so that most of the electromotive force of the battery would appear across its internal resistance, resulting in all of the battery power being dissipated internally. This would lead to rapid generation of internal heat, which could cause the battery housing to melt down or burst.

In practice, a common use of conducting materials involves two or more conducting objects that are separated by dielectrics (or free space) and that are at different potentials.[26] In such cases, a surface charge distribution ρ_s exists on the conductor surfaces, and a component of electric field normal to the conductor–dielectric boundary exists, given by $E_n = \rho_s/\epsilon$. The amount of charge on any small portion of the surface of a conductor is equal to the average potential of that portion multiplied by its capacitance to ground or to the appropriate nearby conductor. An example of such a situation is shown in Figure 5.12c, where two metal strips are maintained at potential difference of V_0 by a battery connection. Just inside the conductors, we must have $E_n = 0$ since the the current density **J** (and hence **E**) must be tangential to the boundary, as argued in Section 5.2. If no path is available between the metal strips for current flow (e.g., Figure 5.12c), then the current $\mathbf{J} = 0$, and hence we also have $E_t = 0$, so that the total field inside both conductors is zero.

If a load is now connected between the two conducting strips so that current can flow, the electric field lines no longer terminate on the conductors at right angles but are tilted, as shown in Figure 5.12d. Typically, the tangential electric field E_t inside the conductors is much smaller than the normal electric field E_n due to the external circuit connections. This circumstance is implied by the size of the arrows representing E_t and E_n in Figure 5.12d; however, the size of E_t and thus the slant of the electric field lines are nevertheless greatly exaggerated. In reality, the tilt angle θ of the field lines is much smaller than a degree, since $E_t \ll E_n$. Note that the small value of E_t is consistent with the fact that the strips are relatively good metallic conductors, in which

---

[26]The difference in potentials is due either to intentional application of a potential difference, such as in the case of the outer and inner conductors of a coaxial line, or to the unavoidable proximity of other conductors at different potentials.

very small fields can establish sizable currents. As an example, consider the strips to be copper ($\sigma = 5.8 \times 10^7$ S-m^{-1}) wires of 2-mm diameter and 1 cm length, the separation of the conductors to be 1 cm, $V_0 = 1$ V, and the load to be an $R_L = 1\Omega$ resistor. Since the wire resistance is entirely negligible compared with the 1Ω load, the current is determined by R_L alone and is $I = 1$ A. For a cylindrical wire of 1 mm radius, the current density is then $J = I/(\pi 10^{-6}) \simeq 3.2 \times 10^5$ A-m^{-2}, implying a tangential electric field of $E_t = J/\sigma \simeq 5.5 \times 10^{-3}$ V-m^{-1}. To determine E_n, we can use the expression derived in Section 4.9 for the capacitance per unit length of the two-wire line, namely, $C_u = (\pi \epsilon_0)/\ln(d/a)$, where in our case $d = 1$ cm and $a = 1$ mm. Substituting values, we find $C_u \simeq 12$ pF-m^{-1}. Noting that capacitance per unit length is $C_u = \rho_l/\Phi_{12}$ and with $\Phi_{12} = V_0 = 1$ V, in this case the equivalent line charge density on the conductors is $\rho_l \simeq 1.2 \times 10^{-11}$ C-m^{-1}. Again from Chapter 4, we know that the electric field at a distance r from a line charge density of ρ_l is $E_r = \rho_l/(2\pi \epsilon_0 r)$. Using the value of ρ_l just found, we find that the electric field midway between the wires (i.e., at $r = 0.5$ cm) due to one of the conductors is of order $E_n \simeq 43$ V-m^{-1}. It is thus obvious that $E_n \gg E_t$. In this case, we have $\theta = \tan^{-1}(E_t/E_n) \simeq 0.007°$,[27] so that the tilt of the electric field lines due to the current flow is indeed extremely small.

The charges that reside on the surfaces of wires carrying steady currents, which are induced via the potential differences between the wires and other conductors, thus play a fundamental role in the production of steady currents. The distributed capacitances (also called stray capacitances) by which these charges are induced are the sources of electric fields much larger than the field that supports the electron drift that constitutes the current. The surface charges must of course be in motion as are those inside the wire; however, the amount of charge (i.e., charge density ρ) at any place must remain constant, being maintained by steady currents.

5.6.2 Bending of Current Flow at Interfaces between Two Conductors

The boundary conditions for current flow across an interface between two conducting materials can be derived in a manner entirely analogous to the derivation of the electro-static boundary conditions that was undertaken in Section 4.11. Figure 5.13 shows the interface between two media of conductivities σ_1 and σ_2. To determine the conditions for the component of the steady currents normal to the interface, we consider a cylindrical Gaussian surface, as shown in Figure 5.13a, and use the fact that

$$\int_V (\nabla \cdot \mathbf{J})\, dv = \oint_S \mathbf{J} \cdot d\mathbf{s} = 0$$

[27]Note that the electric field due to the other conductor is in the same direction and essentially doubles the electric field. Also note that for the case of the two-wire configuration, the electric field varies with distance between the conductors, being much larger near the wires. Thus, the tilt of the lines is in fact substantially smaller than the $\sim 0.007°$ calculated here.

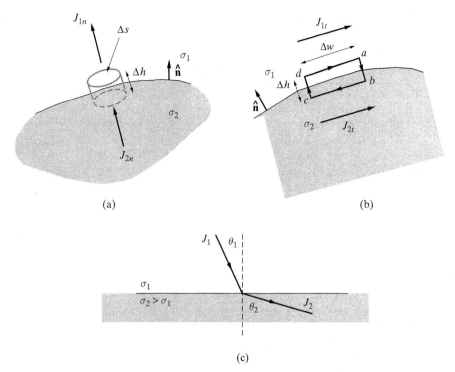

Figure 5.13   The boundary between two different conductors. (a) A differential pillbox-shaped closed surface. (b) A differential closed contour. (c) The bending of current flow at the boundary for $\sigma_2 > \sigma_1$.

Assuming the height Δh of the cylinder to approach zero ($\Delta h \to 0$), the only nonzero contribution to the surface integral comes from the top and bottom surfaces. In other words,

$$(J_{1n} - J_{2n}) \, \Delta s = 0 \qquad \to \qquad \boxed{J_{2n} = J_{1n}} \tag{5.9}$$

which implies that

$$\sigma_1 E_{1n} = \sigma_2 E_{2n}$$

If medium 2 is a perfect dielectric ($\sigma_2 = 0$), we must have $J_{2n} = 0$, meaning that no current can flow perpendicular to the surface; that is,

$$J_{1n} = 0 \qquad \to \qquad E_{1n} = 0$$

Thus, the current flow at the surface of a conductor with a perfect insulator must be parallel to the surface, as mentioned earlier.

For the tangential components of the current, we consider the tangential component of the electric field around a closed differential contour, as shown in Figure 5.13b. The fundamental equation we use here is the same as that used in Section 4.11, namely, that the electrostatic field is conservative, or in other words, $\oint_C \mathbf{E} \cdot d\mathbf{l} = 0$. Naturally, the condition we find when we let $\Delta h \to 0$ is then also the same as in Section 4.11, namely, that $E_{1t} = E_{2t}$. In terms of the tangential components of the current, we then have

$$\boxed{\frac{J_{1t}}{\sigma_1} = \frac{J_{2t}}{\sigma_2}} \tag{5.10}$$

Together, (5.9) and (5.10) indicate that upon crossing a boundary, the current line bends by an amount proportional to the ratio of the conductivities. Noting the definition of θ_1 and θ_2 from Figure 5.13c, we can write

$$\tan \theta_1 = \frac{J_{1t}}{J_{1n}} \qquad \tan \theta_2 = \frac{J_{2t}}{J_{2n}} = \frac{\sigma_2 J_{1t}}{\sigma_1 J_{1n}}$$

or

$$\tan \theta_2 = \frac{\sigma_2}{\sigma_1} \tan \theta_1$$

Note that if medium 1 is a good conductor and medium 2 is a low-loss dielectric (i.e., $\sigma_1 \gg \sigma_2$), the current enters medium 2 at a right angle to the boundary for practically all angles of incidence from medium 1 (i.e., $J_2 \simeq J_{2n}$). This corresponds to the requirement (noted in Section 4.7) that the electric field is normally incident to the surface of a good conductor.

In general, the condition (5.9) requires that the normal component of the electric field be discontinuous across the interface. Using the boundary condition (4.70), we also have

$$\epsilon_1 E_{1n} - \epsilon_2 E_{2n} = \rho_s$$

indicating that the discontinuity of the normal component of electric flux density necessitates a surface charge layer ρ_s at the boundary. Using $\sigma_1 E_{1n} = \sigma_2 E_{2n}$, we can write

$$\rho_s = \left(\epsilon_1 \frac{\sigma_2}{\sigma_1} - \epsilon_2 \right) E_{2n} = \left(\epsilon_1 - \epsilon_2 \frac{\sigma_1}{\sigma_2} \right) E_{1n}$$

Note that the surface charge ρ_s will vanish only for the special case of $\epsilon_2/\epsilon_1 = \sigma_2/\sigma_1$. Alternatively, we can write ρ_s as

$$\rho_s = J_n \left(\frac{\epsilon_1}{\sigma_1} - \frac{\epsilon_2}{\sigma_2} \right) = J_n (\tau_{r1} - \tau_{r2})$$

where τ_{r1} and τ_{r2} are the charge rearrangement (or relaxation) times in media 1 and 2, respectively. If both media are metallic conductors, we have $\epsilon_1 \simeq \epsilon_2 \simeq \epsilon_0$, so that we have

$$\rho_s = \epsilon_0 J_n \left(\frac{1}{\sigma_1} - \frac{1}{\sigma_2} \right)$$

5.7 DUALITY OF J AND D: THE RESISTANCE–CAPACITANCE ANALOGY

In most materials, both the current density **J** and the electric flux density **D** are linearly proportional to the electric field. As a result, there exists a dual relationship between **J** and **D** in regions where nonconservative fields are not present, that is, in regions outside the interior of the batteries and in the absence of time-varying fields. In this section, we analyze some of these dual relationships.

If current enters and leaves a conducting medium via two "perfect" conductors (also referred to as electrodes), the equivalent resistance is given in terms of the potential difference $\Phi_{12} = \Phi_2 - \Phi_1$ between the perfect conductors as $R = \Phi_{12}/I$, where I is the total current leaving the positively charged electrode. Taking an arbitrary surface completely enclosing the positive electrode, this current is given by

$$I = \oint_S \mathbf{J} \cdot d\mathbf{s} = \sigma \oint_S \mathbf{E} \cdot d\mathbf{s} = \frac{\sigma}{\epsilon} \oint_S \mathbf{D} \cdot d\mathbf{s} = \frac{\sigma Q}{\epsilon}$$

where Q is the total charge induced on the positive electrode due to the capacitance between it and the other (negative) electrode. Noting that this capacitance is given by $C = Q/\Phi_{12}$, we thus have

$$R = \frac{\Phi_{12}}{I} = \frac{\epsilon \Phi_{12}}{\sigma Q} = \frac{\epsilon}{\sigma C}$$

Note that the definition of C depends on the existence of static charge on the electrodes. This charge is proportional to Φ_{12} and is independent of whether a current also exists. Note, however, that *all* of the electric flux, and thus current, that leaves one electrode enters the other.

Consider now a pair of electrodes that can be placed either in a dielectric (or free space) or in a conducting medium in which there are no boundaries that would disturb the field and the current patterns. The electric field pattern must be the same in both cases because charge is distributed in the same way on the conductors in each case, since $\nabla \cdot \mathbf{E} = 0$ holds in the intervening space. Thus, if we can find the capacitance for the case when the electrodes are immersed in a dielectric, we can then determine the resistance when the conducting medium is present simply by using the formula $R = \epsilon/(\sigma C)$. Note that since C is proportional to the permittivity ϵ (see, e.g., the general definition of capacitance given by (4.52)), it is evident that the resistance R does not depend on the value of ϵ. Similarly, since R is inversely proportional to σ (see (5.3)), the capacitance C does not depend on σ.

This duality relationship can be easily verified by considering the case of a slab of low-conductivity material of thickness d sandwiched between and in contact with parallel planar electrodes of area A, as shown in Figure 5.14b. The resistance of this arrangement is $R = d/(\sigma A)$, and the capacitance of the same configuration when the conducting slab is replaced by a dielectric (Figure 5.14a) is $C = \epsilon A/d$, consistent with $RC = \epsilon/\sigma$.

The duality relationships between linear homogeneous dielectric and conducting media are summarized in Table 5.2 and in Figure 5.14.

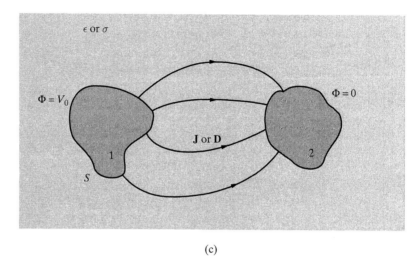

(c)

Figure 5.14   Duality between J and D. The fact that the configuration of the electric field lines is determined only by $\oint_C \mathbf{E} \cdot d\mathbf{l} = 0$ brings about the duality between resistance and capacitance between two arbitrary, perfectly conducting bodies (electrodes). (a) A dielectric medium sandwiched between two electrodes. (b) A conducting medium between two electrodes. (c) Two electrodes of arbitrary shape surrounded by a homogeneous medium, either a dielectric or a conductor.

In homogeneous, linear, isotropic, and time-invariant materials, an important consequence of the duality is

$$RC = \frac{\epsilon}{\sigma} = \tau_{\mathrm{r}} \qquad (5.11)$$

which is consistent with our previous discussions of τ_{r} as the relaxation time constant of a material.[28] The relation (5.11) can be very useful in deriving expressions for resistance of electrode configurations for which we already have the capacitance or vice versa.

---

[28]Note that $\tau_{\mathrm{r}} = \epsilon/\sigma$ is akin to the RC time constant of a first-order capacitive circuit, indicating that the behavior of a lossy dielectric medium under different applied voltage/current conditions can be modeled by such an RC equivalent circuit.

TABLE 5.2 DUALITY BETWEEN DIELECTRIC
AND CONDUCTING MEDIA

Conducting Media	Dielectric Media
$\oint_C \mathbf{E} \cdot d\mathbf{l} = 0$	$\oint_C \mathbf{E} \cdot d\mathbf{l} = 0$
$\mathbf{J} = \sigma \mathbf{E}$	$\mathbf{D} = \epsilon \mathbf{E}$
$\nabla \cdot \mathbf{J} = 0$	$\nabla \cdot \mathbf{D} = 0$
$\mathbf{J} = \sigma \mathbf{E} = -\sigma \nabla \Phi$	$\mathbf{D} = \epsilon \mathbf{E} = -\epsilon \nabla \Phi$
$J_{1n} = J_{2n}$	$D_{1n} = D_{2n}$
$\sigma_1^{-1} J_{1t} = \sigma_2^{-1} J_{2t}$	$\epsilon_1^{-1} D_{1t} = \epsilon_2^{-1} D_{2t}$
$R = \dfrac{-\int_L \mathbf{E} \cdot d\mathbf{l}}{\oint_S \sigma \mathbf{E} \cdot d\mathbf{s}}$	$C = \dfrac{\oint_S \epsilon \mathbf{E} \cdot d\mathbf{s}}{-\int_L \mathbf{E} \cdot d\mathbf{l}}$

For example, we had derived in Example 4.28 that the capacitance of two concentric spheres is

$$C = \frac{4\pi \epsilon ab}{b - a}$$

Using (5.11), we can immediately determine that, if the region between two concentric spheres is filled with a homogeneous conducting material, the resistance between the two spheres is

$$R = \frac{\epsilon}{\sigma C} = \frac{b - a}{4\pi \sigma ab}$$

Note that for $b \to \infty$, this resistance becomes one half of the resistance of the buried hemispherical electrode found in Example 5.4. This result is to be expected, since the geometry of the electric field (and thus that of the current) is exactly the same. But a full spherical electrode at the same potential V_0 has twice as much current flow. In other words, since half of the space in Example 5.4 is free space, half as much current flows for the same potential difference V_0, thus resulting in twice as much resistance.

Example 5.6: Resistance of a coaxial shell. A cross section of a coaxial line consists of an inner conductor of radius a and an outer shell of radius b separated with a dielectric with ϵ, as shown in Figure 5.15. When the coaxial line is used as a transmission line, the current flows along the inner conductor and returns in the outer shell (or vice versa). The capacitance (per unit length) of this configuration was found in Example 4.27 to be $C_u = (2\pi \epsilon)/\ln(b/a)$. When the space between radii a and b is filled with imperfect dielectric with conductivity σ, a radial leakage current tends to flow between the inner conductor and the outer shell. Find the resistance that determines this leakage current.

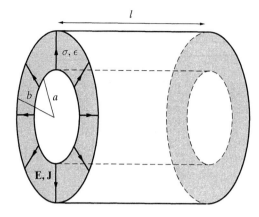

Figure 5.15 Resistance of a coaxial shell.
When the space between the inner and outer conductors of a coaxial line is filled with lossy material, leakage current **J** flows between the two conductors, as the arrows show.

Solution: Since the electric field line configuration is radial in both cases, we can use the duality of **D** and **J** to find the resistance R over a length l. Using (5.11),

$$R = \frac{1}{C}\frac{\epsilon}{\sigma} = \frac{1}{C_u l}\frac{\epsilon}{\sigma} = \frac{1}{2\pi\sigma l}\ln\frac{b}{a}$$

Note that the leakage resistance of the coaxial shell is identical to that of the thin circular semiconductor contact discussed in Example 5.5 (with $h = l$). This result is to be expected, since the contact in Example 5.5 is simply a coaxial shell that has very small height $h \ll a, b$, instead of having a length l comparable to or larger than the inner and outer radii a and b.

We note that the resistance is inversely proportional to the length l of the coaxial shell. This is an expected result, since at a given voltage there is more leakage current for a longer shell length (the leakage resistances of two 1-m long segments of coaxial line connected together are connected in parallel, since they share the same voltage). We can define the resistance per unit length $R_u \equiv Rl$, which in this case has units Ω-m (rather than Ω-m^{-1}) since the total leakage resistance is inversely proportional to the length. The per-unit-length leakage conductance G provided in Table 2.2 for a coaxial line structure is the inverse of R_u, where a conductivity of $\sigma = 1.17 \times 10^{-4}$ S-m^{-1} was assumed for polyethylene at 3 GHz.

Example 5.7: Measuring soil conductivity. Conductivity of soil is an important parameter in a number of applications such as the design of AM broadcast systems, grounding electrical equipment, and locating buried objects in the ground. A simple experiment is proposed[29] to determine the conductivity of soil by using two parallel cylindrical metal electrodes, as shown in Figure 5.16a, that are buried in the soil. The soil and the electrodes are placed in a nonconducting container with diameter several times greater than the spacing between the two electrodes, as shown in Figure 5.16b. The electrodes are connected to an ohmmeter, which is used to measure the ohmic resistance R between the ends of the electrodes. The corresponding conductivity of the soil can then be calculated from the value of R using the geometric dimensions of the electrodes. A student

[29] A. D. Wunsch, A backyard experiment for electromagnetics I, *IEEE Trans. Educ.*, 34(1), pp. 142–144, February 1991.

Figure 5.16   Soil conductivity measurement. (a) Two cylindrical electrodes. (b) Soil sample and electrodes in a plastic bucket several times larger than the separation between the electrodes.

is given a pair of copper rods, each of which is 5 mm in diameter and 15 cm in length, and an ordinary ohmmeter to conduct this experiment as an assignment in her backyard at home. The student conducts three sets of measurements at three different electrode spacings, with results summarized as follows:

d (cm)	R (kΩ)
2	1.47
4	2.10
8	3.04

Find the average conductivity of the soil.

Solution: In Chapter 4, Example 4.29, the per-unit-length capacitance of a two-wire line with diameter $2a$ and separation d was shown to be approximately given by

$$C_{\mathrm{u}} = \frac{\pi \epsilon}{\ln(d/a)}$$

for the case when $d \gg a$. Using the duality relationship $RC = \epsilon/\sigma$ and noting that the total capacitance for a two-wire line of length l is $C_{\mathrm{u}}l$, we can write the resistance of the two electrodes embedded in the soil using the duality relationship (5.11) as

$$R = \frac{\epsilon}{C\sigma} = \frac{\ln(d/a)}{\pi \sigma l}$$

From this, we can write the soil conductivity in terms of the measured resistance R between the two electrodes, each of length l, as

$$\sigma = \frac{\ln(d/a)}{\pi R l}$$

where, in our case, $a = 0.25$ cm and $l = 0.15$ m. Thus, for the first measurement, with $d_1 = 2$ cm, we have

$$\sigma_1 = \frac{\ln(2 \text{ cm}/0.25 \text{ cm})}{\pi(1.47 \times 10^3 \ \Omega)(0.15 \text{ m})} \simeq 3.00 \times 10^{-3} \text{ S-m}^{-1}$$

Similarly, for $d_2 = 4$ cm, we have $\sigma_2 \simeq 2.80 \times 10^{-3}$ S-m^{-1}, and for $d_3 = 8$ cm, we have $\sigma_3 \simeq 2.42 \times 10^{-3}$ S-m^{-1}. Therefore, the average conductivity based on the three sets of measurements is

$$\sigma = \frac{\sigma_1 + \sigma_2 + \sigma_3}{3} \simeq 2.74 \times 10^{-3} \text{ S-m}^{-1}$$

5.8 JOULE'S LAW

Based on our understanding of the microscopic picture of conduction discussed in Section 5.1, steady current flow in a material must necessarily lead to dissipation of power, as individual current carriers (electrons) are accelerated by the applied field and transfer their energy to the material medium via collisions. In this section, we discuss the power dissipation resulting from steady current flow and express it in terms of such macroscopic parameters as the electric field $\mathbf{E}$ and current density $\mathbf{J}$.

The energy required to maintain steady current flow can be determined by considering the definition of electrostatic potential. As current flows along a piece of wire of any shape, positive charge moves to lower values of potential, so that work is continually being done by the electric field. For each coulomb of charge that moves through one volt drop in potential, one *joule* of work is performed. The work done to move a charge Q through a potential difference $\Delta\Phi$ is given by $W = Q\Delta\Phi$. The rate at which the work is done is the *power P* expended, which is given by

$$\frac{dW}{dt} = P = \Delta\Phi \frac{dQ}{dt} = \Delta\Phi I$$

or, since $\Delta\Phi = IR$ by Ohm's law, we have

$$\boxed{P = I^2 R} \tag{5.12}$$

in joules/second, or *watts*. Relation (5.12) is the well-known *Joule's law*.[30] Since in steady flow no kinetic energy is gained (i.e., the electron drift velocity $\mathbf{v}_d$ is constant in

---

[30]In 1841, J. P. Joule conducted a series of experiments by coiling wires of different lengths, cross sections, and composition into thin glass tubes, and then immersing the assemblies in separate containers filled with measured quantities of water. When the same intensity of steady current was passed through the different coils, the water was found to heat up to an equilibrium temperature that differed among the several containers, but in such a way that the change in temperature was proportional to the resistances of the coils. This led Joule to conclude that "...when a given quantity of voltaic electricity is passed through a metallic conductor for a given length of time, the quantity of heat evolved by it is always proportional to the resistance which it presents, whatever may be the length, thickness, shape, or kind of that metallic conductor" (J. P. Joule, On the heat evolved by metallic conductors of electricity, *Phil. Mag.,* 19, pp. 260–265, August 1841.) For further discussion, see Chapter 8 of R. S. Elliott, *Electromagnetics,* IEEE Press, Piscataway, New Jersey, 1993.

time) and no charges are redistributed, this power must all become heat; such dissipation of electrical energy is referred to as *Joule heating*. Note that even though we might reverse the polarity of voltage or the direction of current, P remains positive; thus, Joule heating is irreversible.

The expression (5.12) cannot be applied at a particular point. However, power dissipated per unit volume can be defined *at* a point by formulating a differential version of Joule's law. Consider a differential volume element, as shown in Figure 5.17a. The potential difference between the two ends of such an element is given by

$$d\Phi = \mathbf{E} \cdot d\mathbf{l} = \frac{\mathbf{E} \cdot \mathbf{J}}{J} \, dl$$

where $\mathbf{E}$ is the electric field within the element and $d\mathbf{l} = (\mathbf{J}/J) \, dl$. The total current through the differential resistor element $dI = J \, ds$. Thus, according to Joule's law, the power dissipated in this differential element is

$$dP = dI \, d\Phi = \underbrace{\mathbf{E} \cdot \mathbf{J}}_{(\text{V-m}^{-1})(\text{A-m}^{-2})} \overbrace{dl \, ds}^{\text{m}^3} \quad \rightarrow \quad (\text{volts})(\text{amperes}) = \text{watts}$$

Thus, the differential form of Joule's law is

$$\frac{dP}{dv} = \mathbf{E} \cdot \mathbf{J}$$

so that $\mathbf{E} \cdot \mathbf{J}$ represents the power dissipated per unit volume. In other words, the electric field gives up $(\mathbf{E} \cdot \mathbf{J})$ watts-m^{-3} to the steady current flow of $\mathbf{J}$. This energy is converted into heat in the conducting material.

By considering an arbitrarily shaped resistor, as shown in Figure 5.17b, we can now derive a general expression for the power dissipated. With surfaces A_1, A_2 kept, at potentials Φ_1, Φ_2, respectively, (presumably because they are coated with material of

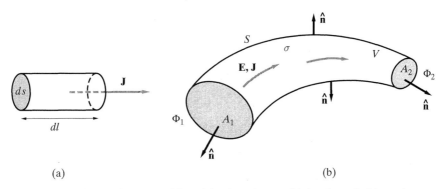

(a)   (b)

Figure 5.17   Joule's law. (a) A differential resistor element. (b) A resistor of arbitrary shape.

conductivity much greater than σ), and noting that the current $\mathbf{J}$ flows normal to these end surfaces and parallel to the sides of the conductor, we have

$$P = I^2 R = \int_V \mathbf{E} \cdot \mathbf{J}\, dv = \int_V \mathbf{E} \cdot \sigma \mathbf{E}\, dv \quad \rightarrow \quad R = \frac{P}{I^2} = \frac{\int_V \mathbf{E} \cdot \mathbf{J}\, dv}{I^2}$$

where R is the total resistance between the terminal faces of the resistor in terms of the total power P dissipated. This expression for resistance is to be compared with equation (5.3), which expresses R in terms of only the electric field.

5.9 SURFACE AND LINE CURRENTS

Thus far we have discussed the volume current density $\mathbf{J}$, which specifies the current per unit area perpendicular to the direction of current flow (A-m^{-2}). Physically, all conduction current is caused by the drift of charge carriers flowing in a three-dimensional conducting region, sustained by an electric field in the material. However, in several practical cases, we can approximate the current (the physical transport of charges) to be confined to either two or one dimensions.

5.9.1 Surface Currents

Surface currents are analogous to surface charge; they are currents flowing over a two-dimensional surface. Consider the band of current flowing in a tube of differential width dl and height Δw as depicted in Figure 5.18a. If the volume current density is confined entirely to the thickness Δw, and if Δw is much smaller than other relevant feature sizes, we may as a matter of convenience assume that the current flows along the surface only. In such cases, we define a surface current $\mathbf{J}_s$ as

$$\mathbf{J}_s = \frac{dI}{dl}\hat{\mathbf{u}} \tag{5.13}$$

Figure 5.18   Surface and Line Currents. (a) Surface current density. (b) Line current.

where $\hat{\mathbf{u}}$ is a unit vector pointing in the current flow direction, and dI is the total current in the band of differential width dl. That is, $\mathbf{J}_s$ is the current per unit width that is perpendicular to the current flow direction, in units A-m^{-1}. By definition, $\mathbf{J}_s$ is nonzero only on the surface of the material.

Surface currents are a convenient construct when one dimension of a conductor is much smaller than all other feature sizes of a problem. For example, a thin sheet of a conducting material with a relatively small thickness effectively confines the charge carriers to flow in only two dimensions. In such a case, it is convenient to express the movement of charge carriers in terms of a surface current density.

Surface currents also arise in good conductors. In Chapter 4 we saw that surface charge resides on the surface of good conductors. However, we also observed that the surface charge exists in a macroscopically infinitesimal (but nevertheless a few atomic dimensions thick, so that we can still think in terms of "density" of charge) surface layer. In practice, the depth to which a surface current on a good conductor is confined depends on the conductivity σ and the frequency of operation f. Quantitatively, the thickness of the surface current layer is given by the *skin depth* δ, which, as we discuss in Chapter 8, is inversely proportional to the square root of both frequency and conductivity. For copper, for example, we have $\delta \simeq 0.07f^{-1/2}$ (in meters), so that at 1 MHz, $\delta \simeq$ 0.07 mm, which is small but not infinitesimal on a microscopic scale. Nevertheless, in most applications, a current density $\mathbf{J}$ (A-m^{-2}) confined to a region of thickness $\delta \simeq 0.07$ mm can be approximated as a surface current of $\mathbf{J}_s = \mathbf{J}\delta$ (A-m^{-1}), for all practical purposes. The difficulty arises when one considers extremely low frequencies or purely static (dc) fields. In such cases, the skin depth $\delta \to \infty$, and the current density $\mathbf{J}$ exists throughout the body of the conductor. The only exception would be the case of perfect conductors, or superconductors, for which $\sigma = \infty$ and the skin depth $\delta \to 0$ even for static fields.

5.9.2 Line Currents

Consider a conducting filament with cross section A, as shown in Figure 5.18b. If the cross section A is much smaller than all other relevant dimensions in the problem, then it is convenient to assume all the current to flow in a single dimension along the central axis of the filament. If the total current passing through an arbitrary cross section is I, then we may ascribe a *line current* vector $\mathbf{I} = I\hat{\mathbf{u}}$ (in units of Amperes) to each point along the line, where $\hat{\mathbf{u}}$ is the current-direction unit vector parallel to the central axis of the conductor.

A typical example is a current-carrying thin conducting wire. If the cross-sectional dimensions of the wire are much smaller than all other relevant physical dimensions involved in the problem, then it is reasonable to make the simplifying assumption that all of the current is confined to a single dimension. In this case, the current can be approximated as a total current I flowing through the conductor, and the unit vector $\hat{\mathbf{u}}$ that describes the current-flow direction of the physical wire at each point along the wire.

In the limit of $\sigma \to \infty$, we refer to the current-carrying filament as a *source current*. In this limit, the electric field inside the filament is 0, and the electric potential is the same at each point along the wire. In reality, all materials have a finite σ, but the notion of a source current is a useful construct in many applications. For example, we may have a copper wire connected to a current source, in which case we may assume that the current density in the wire, denoted as $\mathbf{J}_{\text{source}}$, is known. We can also use $\mathbf{J}_{\text{source}}$ as a tool for solving electromagnetic problems by assuming this current density exists and is maintained by some unspecified external agency.

5.10 SUMMARY

This chapter discussed the following topics:

- **Current density and conduction.** A constant electric field $\mathbf{E}$ in a conducting medium leads to steady current flow, with the current density $\mathbf{J}$ given by

$$\mathbf{J} = n_c q_e \mathbf{v}_d = \frac{n_c q_e^2 \bar{l}}{m_e v_{\text{th}}} \mathbf{E} \quad \rightarrow \quad \mathbf{J} = \sigma \mathbf{E}$$

  where parameters such as the number of conduction band electrons n_c, the mean free path $\bar{l}$, and thermal velocity v_{th} depend on the material properties; q_e and m_e are the charge and mass of the electron; and σ is the conductivity of the material. Microscopically, conduction occurs as the electrons, which are in a state of random agitation, slowly drift in the direction of the applied electric field with velocity $\mathbf{v}_d$, where $|\mathbf{v}_d| \ll v_{\text{th}}$.

- **Resistance and Ohm's law.** When the space between two perfectly conducting electrodes is filled with a conducting material, the resistance between the electrodes is given as

$$R = \frac{\Phi_{12}}{I} = \frac{-\int_L \mathbf{E} \cdot d\mathbf{l}}{\int_A \sigma \mathbf{E} \cdot d\mathbf{s}}$$

  where Φ_{12} is the potential difference between the two electrodes, I is the total current through them, A is the cross-sectional area of any of the electrodes, and L is any path from the electrode 1 at a lower potential to electrode 2 at a higher potential. This relation between Φ_{12} and I is also known as Ohm's law.

- **Kirchhoff's voltage law.** Steady current flow can be maintained only by an external energy source (e.g., a battery) supplying electromotive force. In general, for any closed circuit containing many resistors and emf sources, we have

$$\sum \mathcal{V}_{\text{emf}} = I \sum R$$

- **Continuity equation and Kirchhoff's current law.** The fundamental principle of conservation of electric charge indicates that the net inflow of current through a closed surface must be equal to the net rate of charge increase in the enclosed volume. The differential form of this fundamental principle is known as the continuity equation:

$$\nabla \cdot \mathbf{J} + \frac{\partial \rho}{\partial t} = 0$$

For steady currents, we must thus have

$$\nabla \cdot \mathbf{J} = 0 \quad \text{or} \quad \oint_S \mathbf{J} \cdot d\mathbf{s} = 0$$

indicating that steady current must flow in closed loops. A more commonly known version of this result is Kirchhoff's current law, which states that the algebraic sum of all currents at a junction is zero.

- **Redistribution of free charge.** The ratio of permittivity of a material to its conductivity is known as the relaxation time $\tau_r = \epsilon/\sigma$, which defines the time period over which free charge is redistributed in the material. For metals, this time is extremely small, being of the order $\sim 10^{-19}$ s; for good dielectrics τ_r can be on the order of hours or days.

- **Boundary conditions for steady currents.** The component of current density normal to the boundary between two different conducting materials is continuous across the interface. The continuity of the tangential component of the electric field across any such interface indicates that the tangential component of current density is not continuous. In summary, we have

$$J_{1n} = J_{2n} \quad \text{and} \quad \frac{J_{1t}}{\sigma_1} = \frac{J_{2t}}{\sigma_2}$$

- **The resistance–capacitance analogy.** The resistance R and capacitance C between two perfectly conducting electrodes in a linear, homogeneous, and isotropic medium is related by

$$RC = \frac{\epsilon}{\sigma} = \tau_r$$

Many of the fundamental relationships for current density $\mathbf{J}$ are dual to those for the electric flux density $\mathbf{D}$, as summarized in Table 5.2.

- **Power dissipation and Joule's law.** The flow of a steady current in a material leads to dissipation of power in the form of heat. The rate at which work is done and power expended is $P = I^2 R$, where I is the total current flowing through a resistance R. At a differential level, the volume density of power dissipation, in units of watts-m^{-3}, is represented by $\mathbf{E} \cdot \mathbf{J}$.

- **Surface and line currents.** If the flow of charge carriers in a conductor is confined to an infinitesimal thickness, we may describe the current density in terms of a surface current density $\mathbf{J}_s$, defined as the current dI per differential width dl perpendicular to the flow direction $\hat{\mathbf{u}}$:

$$\mathbf{J}_s = \frac{dI}{dl}\hat{\mathbf{u}} \tag{5.14}$$

If the current is confined to a small cross section, such as in a conducting wire, then it is convenient to define a line current $\mathbf{I} = I\,\hat{\mathbf{u}}$, defined by the total current I and the axial direction of the current represented by the unit vector $\hat{\mathbf{u}}$.

PROBLEMS

5.1 **Current in a wire.** If a current of 1 A is flowing in a conductor wire, find the number of electrons that pass through its cross section each second.

5.2 **Current in a copper wire.** If a copper wire of cross-sectional area 1 mm^2 is carrying a current of 1 A, find the average drift velocity of the conduction electrons in the wire. The density of conduction electrons in copper is $n_c = 8.45 \times 10^{28}$ el-m^{-3}.

5.3 **Electric field in an aluminum wire.** An aluminum wire of conductivity $\sigma = 3 \times 10^7$ S-m^{-1} is carrying a uniform current of density 100 A-cm^{-2}. (a) Find the electric field in the wire. (b) Find the electric potential difference per meter of wire.

5.4 **Silver versus porcelain.** The conductivity of silver is $\sigma = 6.17 \times 10^7$ S-m^{-1} and that of porcelain is about $\sigma \simeq 10^{-14}$ S-m^{-1}. (a) If a voltage of 10 V is applied across a silver plate of 1 cm thickness, find the current density through it. Assume a uniform electric field. (b) Repeat part (a) for a porcelain plate of same thickness and compare your results.

5.5 **Electron mobility in copper.** The average drift velocity of free electrons in a metal is proportional to the applied electric field approximately given by

$$v_d \simeq \frac{q_e t_c}{m_e} E = \mu_e E$$

where μ_e is a proportionality factor called the mobility of the electron. Since $v_d = \mu_e E$, the mobility describes how strongly the motion of an electron is influenced by an applied electric field. The conductivity of copper is $\sigma = 5.8 \times 10^7$ S-m^{-1} at room temperature and is due to the mobility of electrons that are free (one per atom) to move under the influence of an electric field. (a) Find the electron mobility in copper at room temperature and compare it with the mobility values of pure silicon and germanium (see Problem 5.9). (b) Find the average drift velocity of the electrons in the direction of the current flow in a copper wire of 1 mm diameter carrying a current of 1 A.

5.6 **Resistance of a short, small-diameter conductor.** A technique has been developed to measure the dc resistance at room temperature of conductors that are short in length and small in diameter.[31] To demonstrate this system, the resistance of commercial bare copper wire ($\sigma = 5.85 \times 10^7$ S-m^{-1} at 20°C) was measured in four different diameters. The manufacturer's reported diameters for these wires were 25.4 μm, 20.3 μm, 12.7 μm, and 7.6 μm.

---

[31]C. A. Thompson, Apparatus for resistance measurement of short, small-diameter conductors, *IEEE Trans. Instrumentation and Measurement,* 43(4), pp. 675–677, August 1994.

The resistances per unit length of these wires were measured to be 0.340, 0.527, 1.22, and 2.56, all in Ω-cm^{-1}. (a) Calculate the actual diameter of each wire using the measured resistance values. (b) Find the percentage of error in the reported diameter values by comparing them to the calculated values. (Note that the uniformity and the surface condition of each wire were examined with a scanning electron microscope. All of the wires had uniform diameters, and only the 7.6-μm diameter wire showed significant surface cracking and roughness. So a larger error for the 7.6-μm diameter wire is most likely due to its degraded surface condition.)

5.7 **Resistance of a copper wire.** Consider a copper wire of 1 mm diameter. Find its resistance per unit length if the wire temperature is (a) $-20°$C, (b) $20°$C, and (c) $60°$C. For copper, the conductivity is $\sigma = 5.8 \times 10^7$ S-m^{-1} at $20°$C.

5.8 **Resistance of a copper wire.** Consider a copper wire at $20°$C. To what temperature must it be heated in order to double its resistance?

5.9 **Intrinsic semiconductor.** A semiconductor is said to be intrinsic (pure) when it is free of any dopant impurity atoms. The only mobile carriers are those caused by thermal excitation, a process that creates an equal number of holes and electrons. The intrinsic charge carrier concentration N_i is a strong function of temperature. Under most conditions, it is given by

$$N_i^2 = N_c N_v e^{-E_g/k_B T}$$

where N_c and N_v are related to the density of allowed states near the edges of the conduction and valence bands, E_g is the energy gap between the conduction and the valence bands, k_B is Boltzmann's constant $k_B \simeq 1.38 \times 10^{-23}$ J-K^{-1}, and T is the temperature in K. The conductivity of a semiconductor sample is given by

$$\sigma = |q_e|(\mu_e N_e + \mu_p N_p)$$

where $|q_e| \simeq 1.6 \times 10^{-19}$ C; μ_e and μ_p are the electron and the hole mobilities, which are both functions of temperature; and N_e and N_p are the electron and hole concentrations, which, in the case of a pure semiconductor, are both equal to the intrinsic charge carrier concentration; that is, $N_e = N_p = N_i$. (a) For a bulk sample of pure silicon (Si) at 300K, $N_c = 2.8 \times 10^{19}$ cm^{-3}, $N_v = 1.04 \times 10^{19}$ cm^{-3}, $E_g = 1.124$ eV (where 1 eV $\simeq$ 1.602×10^{-19} J), $\mu_e = 1450$ cm^2-(V-s)$^{-1}$, and $\mu_p = 450$ cm^2-(V-s)$^{-1}$. Calculate the intrinsic carrier concentration (in cm^{-3}) and the conductivity (in S-m^{-1}) of the Si sample at 300K. (b) Repeat part (a) for a bulk sample of pure germanium (Ge). For Ge at 300K, $N_c = 1.04 \times 10^{19}$ cm^{-3}, $N_v = 6.0 \times 10^{18}$ cm^{-3}, $E_g = 0.66$ eV, $\mu_e = 3900$ cm^2-(V-s)$^{-1}$, and $\mu_p = 1900$ cm^2-(V-s)$^{-1}$.

5.10 **Extrinsic semiconductor.** Semiconductors in which conduction results primarily from carriers contributed by impurity atoms are said to be extrinsic (impure). The impurity atoms, which are intentionally introduced to change the charge carrier concentration, are called dopant atoms. In doped semiconductors, only one of the components of the conductivity expression in Problem 5.9 is generally significant because of the very large ratio between the two carrier densities N_e and N_p, the product of which is always equal to the square of the intrinsic charge carrier concentration, that is, $N_e N_p = N_i^2$. In addition, the mobility values of the charge carriers vary with different doping levels. (a) Phosphorus donor atoms with a concentration of $N_d = 10^{16}$ cm^{-3} are added uniformly to a pure sample of Si (which is called n-type Si since the electrons of the phosphorus atoms are the majority carriers, that is,

$N_e \simeq N_d \gg N_p$). Find the conductivity of the n-type Si sample at 300K. For the mobility of majority charge carriers, use $\mu_e = 1194$ cm^2-(V-s)$^{-1}$. (b) Repeat part (a) if boron acceptor atoms with a concentration of $N_a = 10^{16}$ cm^{-3} are added to a pure sample of Si (which is called p-type Si since the holes are now the majority carriers, that is, $N_p \simeq N_a \gg N_e$). For the mobility of majority charge carriers, use $\mu_p = 444$ cm^2-(V-s)$^{-1}$. (c) Compare your results in parts (a) and (b) with the conductivity of the intrinsic Si and comment on the differences.

5.11 **A silicon resistor.** A silicon bar 1 mm long and 0.01 mm^2 in cross-sectional area is doped with $N_d = 10^{17}$ cm^{-3} arsenic (As) atoms. Find the resistance of the bar and compare with the resistance of the same bar made of pure silicon (see Problem 5.9). For the mobility of the free arsenic electrons, use $\mu_e = 731$ cm^2-(V-s)$^{-1}$.

5.12 **A silicon resistor.** A sample of p-type silicon is 8 μm long and has a cross-sectional area of 2.5 μm $\times$ 2.5 μm. If the measurements show that the average hole concentration and the resistance of the sample are $N_d = 10^{16}$ cm^{-3} and $R = 18$ kΩ, respectively, find the hole mobility μ_p.

5.13 **Sheet resistance R_{sq}.** Consider the resistance of a uniformly doped n-type Si layer of length l, width w, and thickness t, as shown in Figure 5.19. This resistor can be divided into square sheets of dimension w on each side, as shown. The resistance of any one of these square sheets is called the sheet resistance, denoted by the symbol R_{sq}, in units of Ω-(sq)$^{-1}$. (a) Show that the sheet resistance is given by

$$R_{sq} = \frac{1}{|q_e|\mu_e N_e t}$$

(b) Show that the total resistance of the Si layer is given by

$$R = \frac{l}{w} R_{sq}$$

(c) Calculate the total resistance of a Si layer that has a length of 50 μm, width of 5 μm and a sheet resistance of 150Ω-(sq)$^{-1}$.

Figure 5.19   Sheet resistance. Problem 5.13.

5.14 **Integrated-circuit (IC) resistor.** Consider an integrated-circuit (IC) resistor of length l, width w, and thickness t that is made of Si doped with 10^{16} cm^{-3} phosphorus atoms (i.e., $N_e \simeq 2.5 \times 10^{16}$ cm^{-3}). Given $t = 2$ μm, find the aspect ratio, w/l, such that the resistance of the IC resistor is 20 kΩ at 300K. Take $\mu_e = 1000$ cm^2-(V-s)$^{-1}$.

5.15 An ion-implanted IC resistor. An ion-implanted n-type resistor layer with an average doping concentration $N_d = 4 \times 10^{17}$ cm^{-3} is designed with 9 μm thickness and 4 μm width to provide a resistance of 3 kΩ for an IC chip. Find the sheet resistance and the required length of the resistor. Assume the electron mobility as $\mu_e = 450$ cm^2-(V-s)$^{-1}$.

5.16 Diffused IC resistor. An IC resistor is frequently fabricated by diffusing a thin layer of p-type impurity into an n-type isolation island, as shown in Figure 5.20. If contacts are made near the two ends of the p-type region and a voltage is applied, a current will flow parallel to the surface in this region. It is not possible to use $R = l/(\sigma A)$, however, to calculate the resistance of this region because the impurity concentration in it is not uniform. The impurity concentration resulting from the diffusion process is maximum near the surface ($x = 0$) and decreases as one moves in the x direction. (a) Show that the sheet resistance of the p-type layer is given by

$$R_{\mathrm{sq}} = \left[\int_0^t \sigma(x)\, dx \right]^{-1}$$

(b) Assuming the conductivity of the p-type layer decreases linearly from σ_0 at the surface ($x = 0$) to $\sigma_1 \ll \sigma_0$ at the interface ($x = t$) with the n-type wafer, find the sheet resistance, R_{sq}. (c) Repeat part (b) if the conductivity decreases exponentially from σ_0 at $x = 0$ to $\sigma_1 \ll \sigma_0$ at $x = t$.

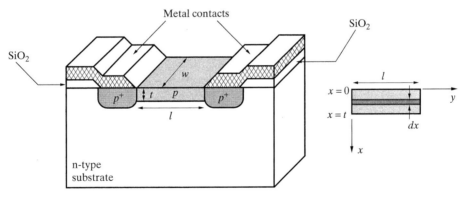

Figure 5.20 Diffused IC resistor. Problem 5.16.

5.17 Resistance of a copper-coated steel wire. A steel wire ($\sigma = 1.11 \times 10^6$ S-m^{-1}) 6 mm in diameter is to be coated with cu ($\sigma = 5.8 \times 10^7$ S-m^{-1}) in order to reduce its resistance per unit length by 50%. Find the thickness of cu coating needed.

5.18 Resistance of an aluminum conductor, steel-reinforced (ACSR) wire. A power utility company uses a steel-reinforced aluminum conductor (ACSR) wire of 3 cm diameter as an extra-high-voltage (EHV) transmission line. It is made of aluminum ($\sigma = 3.82 \times 10^7$ S-m^{-1}) with an inner core of stainless steel ($\sigma = 1.11 \times 10^6$ S-m^{-1}) along the center axis such that the steel content of the ACSR wire is only 10%. (a) Find the resistance per kilometer of the wire at 20°C. (b) If a current of 1000 A flows in the wire, find the current that flows in each metal. (Neglect the changes in the conductivity values due to changes in temperature because of the current flow.) (c) If the ACSR wire between two EHV towers is 300 m long, find the voltage drop between the two towers. (d) Repeat parts (a), (b), and (c) if the steel content of the ACSR wire is increased to 25%.

5.19 **Conductivity of lunar soil.** A simulated version of the lunar soil obtained from the Apollo 11 site on the moon is investigated as a possible material to be used for electrical insulation in high-voltage power systems in space.[32] A sample of this soil is placed as an insulator between the electrodes of a parallel-plate capacitor with 1 mm separation and 10 cm^2 area, and dc voltages are applied across the electrodes. Leakage currents of 10 nA, 20 nA, 30 nA, 40 nA, and 50 nA are recorded for electric field strengths of 2.5 kV-(mm)$^{-1}$, 3 kV-(mm)$^{-1}$, 3.6 kV-(mm)$^{-1}$, 4 kV-(mm)$^{-1}$, and 4.2 kV-(mm)$^{-1}$, respectively. (a) Using Ohm's law, find the conductivity of the lunar soil in each case. Note that the conductivity of this soil varies with the applied field. (b) Using a curve-fitting technique, the conductivity of the soil is approximated as

$$\sigma \simeq \frac{4^{(2E-5)/3}}{10^{14}E}$$

where σ is in S-(mm)$^{-1}$ and E is in kV-(mm)$^{-1}$. Using this expression, find the value of σ and its percentage difference from its corresponding value found in part (a) for each case.

5.20 **Resistance of a semicircular ring.** The ends of a semicircular conductor ring of rectangular cross section are connected to a dc battery, as shown in Figure 5.21. (a) Write the total resistance R of the conductor using the result of Example 5.3. (b) Find R by assuming the conductor to be a straight conductor with a length $l = \pi(a+b)/2$. (c) Using $\sigma = 7.4 \times 10^4$ S-m^{-1} (graphite) and $b = 1.5a = 10t = 3$ cm, find R using both expressions obtained in parts (a) and (b) and compare the results. (d) Show that for $a \gg (b - a)$, the expression of part (a) reduces to the expression of part (b).

Battery

Figure 5.21   Semicircular ring.
Problem 5.20.

5.21 **Resistance of a hemispherical conductor.** A perfectly conducting hemispherical conductor is buried in the earth to achieve a good ground connection (see Figure 5.7). If the diameter of the conductor is 125 cm and the conductivity of the ground is 10^{-4} S-m^{-1}, find the resistance of the conductor to distant points in the ground.

[32]H. Kirkici, M. F. Rose, and T. Chaloupka, Experimental study on simulated lunar soil: high voltage breakdown and electrical insulation characteristics, *IEEE Trans. Dielect. and Elect. Insul.*, 3(1), pp. 119–125, February 1996.

5.22 **Resistance of a toroidal conductor.** Consider a segment of a toroidal (doughnut-shaped) resistor with a horizontal cross section, as shown in Figure 5.22. Show that the resistance between the flat ends having a circular cross section is given by

$$R = \frac{\phi_0}{\sigma \pi (\sqrt{b} - \sqrt{a})^2}$$

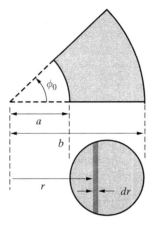

Figure 5.22   Toroidal conductor. Problem 5.22.

5.23 **Leakage resistance.** A 2-mm-diameter copper wire is enclosed in an insulating sheath of 4-mm outside diameter. If the wire is buried in a highly conducting ground, what is the leakage resistance per kilometer of the sheath to the ground? The sheath can be assumed to have a conductivity of 10^{-8} S-m^{-1}.

5.24 **Pipeline resistance.** Two parallel steel pipelines have centers spaced 10 m apart. The pipes are half buried in the ground, as shown in Figure 5.23. The diameter of the pipes is 1 m. The ground in which the pipes lie is marshy soil ($\sigma \simeq 10^{-2}$ S-m^{-1}). Find the resistance per kilometer between the two pipes.

Figure 5.23   Pipeline resistance. Problem 5.24.

5.25 **Energy dissipation in lossy coaxial cable.** Consider the coaxial cable shown in Figure 5.24, with inner conductor radius of a, outer conductor radius of b, and filled with two different conductive materials, having the same thickness, $w_1 = w_2 = (b - a)/2$. Available in hand are two different types of lossy material, with conductivities of σ and 3σ. (a) Assuming a constant voltage V_{ab} to be applied between the inner and outer conductors, which of the two lossy materials should be used as the inner one in order to minimize losses (per unit length)? In other words, is it better to have $\sigma_1 = \sigma$, $\sigma_2 = 3\sigma$ or vice versa? By what factor does the loss per unit length increase if the two materials are flipped? (b) Determine the values that w_1 and w_2 must have in order for the loss per unit length to be the same regardless of the position of the materials. In other words, find values of w_1 and w_2 (in terms of a and b) such that the loss per unit length is the same for (i) $\sigma_1 = \sigma$ and $\sigma_2 = 3\sigma$, or (ii) $\sigma_1 = 3\sigma$ and $\sigma_2 = \sigma$.

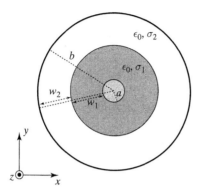

Figure 5.24 Lossy coaxial cable.
Problem 5.25.

5.26 Ground current. A vertical lightning rod discharges 10^5 amperes into the ground. Find the voltage produced by the discharge between two points 1 m apart on a radial line if the point nearest the rod is at a distance of (a) 3 m, and (b) 6 m from the rod. Assume the ground to consist of a 20-cm layer of conducting soil with $\sigma \simeq 10^{-2}$ S-m^{-1}.

5.27 Inhomogeneous medium. Consider an inhomogeneous medium in which both ϵ and σ are functions of position. Show that a steady current $\mathbf{J}$ flowing though such a medium would necessarily establish a charge distribution given by $\rho = -[\nabla\epsilon - (\epsilon/\sigma)\nabla\sigma] \cdot \nabla\Phi$. Note that the field arising from the flow of current can still be derived from a scalar potential, but the potential no longer satisfies Laplace's equation, since $\rho \neq 0$.

5.28 Leaky capacitor. Consider the parallel-plate capacitor with spacing $2d$ and plate area A, as shown in Figure 5.25. The region between the plates is filled with two lossy dielectric slabs, each of thickness d and with parameters ϵ_1, σ_1, and ϵ_2, σ_2, respectively. A potential V_0 is applied across the plates. (a) Find the steady-state value of the electric field in each of the two materials. (b) Find an expression for the surface charge density at the interface between the two materials.

Figure 5.25 Leaky capacitor.
Problem 5.28.

5.29 **Lossy dielectric between spherical electrodes.** Consider a spherical capacitor with the region between the inner and outer electrodes filled with lossy material with permittivity ϵ_0 and conductivity

$$\sigma_r = \sigma_0 \, \frac{e^{-r/b}}{r^3}$$

where b is the radius of the outer electrode and σ_0 is a constant. The inner electrode (with radius a) is held at a constant potential of V_0 while the outer electrode (with radius $b > a$) is at zero potential. (a) Determine the charge Q on the inner and outer electrodes, and the charge distribution $\rho(r)$ in the region between the electrodes. (b) Determine the total space-distributed charge in the region between the electrodes.

5.30 **Inhomogeneous medium.** A spherical electrode of radius a is surrounded by a concentric spherical shell of radius b. The space between the two conductors is filled with material whose conductivity varies linearly with distance from the common center of the spheres (i.e., $\sigma = Kr$, where K is a proportionality constant). If a potential difference of V volts is maintained between the spheres, with the inner electrode grounded (i.e., at zero potential), what is the electric potential at a distance r ($a < r < b$) from the center?

5.31 **Leakage resistance.** A very long wire of radius a is suspended horizontally near the bottom of a deep lake. Assume the lake to have a plane bottom that is a very good conductor. The wire is parallel to the bottom and is at a height $h \gg a$ above it. If the conductivity of the water is σ, find the resistance (per unit length) between the wire and the planar bottom of the lake.

5.32 **Two conducting spheres.** Two metallic conducting spheres of radii a_1 and a_2 are buried deep in poorly conducting ground of conductivity σ and permittivity ϵ. The distance b between the spheres is much larger than both a_1 and a_2. Determine the resistance between the two spheres.

6

The Static Magnetic Field

In the preceding chapters, we discussed various aspects of electrostatics or the configurations of static electric fields and their relationship to stationary charges and conducting or dielectric boundaries. Our discussions in Chapter 4 involved charges that were assumed to be at rest, and the various effects studied were consequences of Coulomb's law—the only experimental fact that we needed to introduce. In Chapter 5 we considered motion of charges under the influence of the Coulomb force resulting from an applied static electric field, constituting steady electric current. We now focus our attention on a series of new phenomena caused by the steady motion of charges. These phenomena constitute the subject of *magnetostatics* and can be understood as the consequences of a new experimental fact: namely, that current-carrying wires exert forces on one another.

In a certain sense, *magnetostatics* is an unfortunate name for the class of phenomena discussed in this chapter. First, the phenomena are caused by charges in motion, making the "static" part of the name a misnomer. Secondly, the "magnet" part of the name immediately suggests phenomena involving actual magnets. In reality, the physical phenomena involving natural magnets are highly complex, and it is very difficult, if not impossible, to construct a theory of magnetostatics based on experiments with magnets. It is much easier to formulate the theory of magnetostatics based on experiments involving moving charges or electric currents, a task which we undertake in this chapter.

As we noted in Chapter 4, our earliest experiences with electricity date back to ancient times and involve the attraction of objects by other objects. Interestingly enough, the subject that we now call *magnetism* also began with the observation that certain natural minerals, readily found near the ancient city of Magnesia in Anatolia, which is now part of Turkey, could attract other materials.[1] For centuries, magnetism was thought

---

[1] That lodestone attracts iron was first noted in print by Roman poet and philosopher Lucretius [99?–55? B.C.] in his philosophical and scientific poem titled *De rerum natura (On the Nature of Things)*. Lodestone is a form of iron oxide called magnetite, found in the shape of elongated fragments. The name *lodestone* comes from Middle English *lode,* "course," because a compass needle made of this material can be used for navigation. In the same manner, the lodestar is the pole star, which marks north in the sky. Also see G. L. Verschuur, *Hidden Attraction, The Mystery and History of Magnetism,* Oxford University Press, New York, 1993.

to be independent of electricity. Phenomena involving magnets were studied extensively because lodestone was readily available, and performing experiments with magnets was relatively simple.[2] However, construction of a theory of magnetostatics based upon results of such experiments requires that we introduce highly artificial concepts and fictions such as a "magnetic pole." After many years of experimentation, and the accidental discovery in 1820 (see the next paragraph) of the experimental fact that an electric current can exert a force on a compass needle, it was realized that the simple understanding of magnetostatic phenomena lay in the relatively difficult experiments involving moving charges rather than in the relatively simple experiments involving magnets. We adopt this modern viewpoint of magnetostatics, founded on Ampère's experimentally based law of force between current-carrying wires.

That wires carrying electric currents produce magnetic fields was first discovered[3] apparently by accident by H. C. Oersted in 1820. Within a few weeks after hearing of Oersted's findings, André-Marie Ampère announced[4] that current-carrying wires exert forces on one another, and J.-B. Biot and F. Savart repeated Oersted's experiments and put forth[5] a compact law of static magnetic fields generated by current elements in a circuit, which is known as the Biot–Savart law. The most thorough treatment of the subject was undertaken by Ampère in the following three years, culminating in an extensive memoir[6] in 1825. This work established Ampère's law, which describes the law of force between two current elements and is analogous to Coulomb's force law in electrostatics. Ampère's studies also led him to postulate that magnetism itself was due to circulating currents on an atomic scale, thus closing the gap between the magnetic fields produced by currents and those produced by natural magnets.

This exciting period of development of the experimentally based underpinnings of magnetostatics was one of the most interesting in the history of science and was not without controversy.[7] After the development of the special theory of relativity, it was

[2]An example is the work of Pierre de Maricourt, a native of Picardy [his *Epistola Petri Peregrini de Maricourt de magnete* was written in 1629], who experimented with lodestone and needles and introduced the concept of magnetic poles. Also notable was the work of William Gilbert [Gulielmi Gilberti, *de Magnete, Magneticisque corporibus, et de magno magnete tellure*, London, 1600], who was the personal physician of Queen Elizabeth. Gilbert studied the properties of magnets and realized that magnets set themselves in definite orientations with respect to the earth because the earth is itself a giant spherical magnet. He was thus the first to discover the earth's magnetic field. Gilbert was rather carried away with this new discovery, however, and conjectured that magnetic forces also accounted for the earth's gravity and the motions of the planets.

[3]H. C. Oersted, *Experiments on the Effect of Current of Electricity on the Magnetic Needle*, privately distributed pamphlet dated July 21, 1820. English translation in *Ann. Philos.*, 16, p. 273, 1820.

[4]A.-M. Ampère, Memoir on the mutual action of two electric currents, *Ann. Chimie et Phys.*, 15, p. 59, 1820.

[5]J.-B. Biot and F. Savart, *Ann. Chimie et Phys.*, 15, p. 222, 1820.

[6]A.-M. Ampère, On the mathematical theory of electrodynamic phenomena uniquely deduced from experiment, *Mem. Acad.*, pp. 175–388, 1825. About 50 years later, J. C. Maxwell described Ampère's work as being "one of the most brilliant achievements in science." Also see L. P. Williams, André-Marie Ampère, *Scientific American*, pp. 90–97, January 1989.

[7]For an excellent account of the history and a thorough treatment of underlying fundamentals see R. S. Elliott, *Electromagnetics*, IEEE Press, Piscataway, New Jersey, 1993.

realized that magnetostatic theory may be derived from Coulomb's law.[8] Nevertheless, in this text we adopt an experimentally based formulation of Maxwell's equations, so Ampère's law of magnetic force and the Biot–Savart law of magnetic induction are considered quantitative statements of experimental facts.

The study of magnetostatics constitutes the second major step in our quest to understand the foundations of the laws of electromagnetics. In magnetostatics we deal with magnetic fields produced by steady currents, which are themselves constant in time and therefore do not allow inductive coupling between circuits or the coupling between electric and magnetic fields. Yet, mastering the behavior of static magnetic fields and the techniques of solution of magnetostatic problems is essential to the understanding of the more complicated electromagnetic phenomena. Furthermore, many natural phenomena and the principles of some important industrial and technological applications are based in magnetostatics. In this connection, it suffices to say that magnetic recording was a more than \$20 billion industry[9] in 1984. Today, annual sales of just hard disk drives, which record data by magnetizing a thin film of ferromagnetic material (see Section 6.8.3) on a disk, are still over \$30 billion worldwide.

Our coverage of magnetostatics in this chapter also brings us one step closer to a full understanding of the underlying physical basis of the transmission line behavior discussed in Chapters 2 and 3. An important physical property of a transmission line is its distributed inductance, which comes about because of the magnetic fields generated by the transmission line currents. In Chapters 2 and 3, we took it for granted that any two-conductor system has some inductance and we relied on formulas (in Table 2.2) to determine the distributed inductances of a few common transmission line structures. In this chapter, we define the physical basis of the concept of inductance and discuss how the inductance of different types of current-carrying conductors can be determined using fundamental laws of magnetostatics.

6.1 AMPÈRE'S LAW OF FORCE

Before we formally write down the magnetic force law, it is useful to review the new experimental facts in simple terms. Figure 6.1 illustrates the direction of the force between current-carrying wires as first experienced by Oersted, its dependence on the orientation of the wires, and the direction of the current flowing in them. The experimental facts indicate that two parallel wires carrying like (i.e., in the same direction) current attract

---

[8]L. Page, A derivation of the fundamental relations of electrodynamics from those of electrostatics, *Am. J. Sci.,* 34, p. 57, 1912; see Section 4.2 of R. S. Elliott, *Electromagnetics,* IEEE Press, Piscataway, New Jersey, 1993, for a clear treatment. It should be realized, however, that such derivations implicitly make other assumptions in addition to using special relativity; see Section 12.2 of J. D. Jackson, *Classical Electrodynamics,* 2nd ed., Wiley, 1975.

[9]See, for example, R. M. White (Ed.), *Introduction to Magnetic Recording,* IEEE Press, Piscataway, New Jersey, 1984.

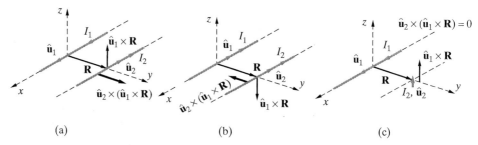

Figure 6.1 Magnetic force between straight current-carrying wires. (a) Two infinitely long straight parallel wires carrying current in opposite directions repel one another. (b) Two straight parallel wires carrying current in the same direction attract. (c) When a small wire element carrying a current I_2 is oriented perpendicular to the wire carrying current I_1, it feels no magnetic force, regardless of the direction of its current.

one another while those carrying opposite current repel, and that when a small[10] current-carrying wire element is oriented perpendicular to another current-carrying wire (see Figure 6.1c) it feels no magnetic force. This set of experimental facts is represented by expressing the force between two current-carrying wires as a double cross product,[11] namely

$$\mathbf{F}_{12} = kI_2\hat{\mathbf{u}}_2 \times (I_1\hat{\mathbf{u}}_1 \times \mathbf{R})$$

where $\mathbf{F}_{12}$ is the force exerted on wire 2 by wire 1, $\mathbf{R}$ is the vector from wire 1 to wire 2, $\hat{\mathbf{u}}_1$ and $\hat{\mathbf{u}}_2$ are unit vectors along wires 1 and 2 in the direction of currents I_1 and I_2, respectively, and k is a proportionality constant. Note from Figure 6.1a that the vector $(I_1\hat{\mathbf{u}}_1 \times \mathbf{R})$ is in the z direction, but the cross product of $\hat{\mathbf{u}}_2 \times (\hat{\mathbf{u}}_1 \times \mathbf{R})$ is in the y direction, indicating repulsion of the wires carrying oppositely directed current. Note that we could have just as well written an expression for the force $\mathbf{F}_{21}$ due to wire 2 at the location of wire 1, which points in the $-y$ direction. The magnitude of the magnetic force is experimentally determined to be inversely proportional to the square of the distance between wires.

[10]The reason we consider a small element in Figure 6.1c is that a long wire oriented in the z direction and carrying current I_2 would in fact experience a torque (see Section 6.10 on magnetic forces and torques). More precisely, the portion of the long wire above the x-y plane would feel a force in the $+x$ direction, while that below the x-y plane would experience a force in the $-x$ direction.

[11]The cross product of two vectors $\mathbf{A}$ and $\mathbf{B}$ is a vector, denoted by $(\mathbf{A} \times \mathbf{B})$, with its magnitude equal to the product of the magnitudes of the two vectors times the sine of the angle $\psi_{\mathbf{AB}}$ between them and its direction following that of the thumb of the right hand, when the fingers rotate from $\mathbf{A}$ to $\mathbf{B}$ through the angle $\psi_{\mathbf{AB}}$. Namely,

$$\mathbf{A} \times \mathbf{B} \equiv \hat{\mathbf{n}}|\mathbf{A}||\mathbf{B}|\sin\psi_{\mathbf{AB}}$$

where $\hat{\mathbf{n}}$ is normal to both $\mathbf{A}$ and $\mathbf{B}$ and its direction is in the direction of advance of a right-handed screw as $\mathbf{A}$ is turned toward $\mathbf{B}$. In rectangular coordinates, and noting that $\hat{\mathbf{x}} \times \hat{\mathbf{y}} = \hat{\mathbf{z}}$, $\hat{\mathbf{y}} \times \hat{\mathbf{z}} = \hat{\mathbf{x}}$, and $\hat{\mathbf{z}} \times \hat{\mathbf{x}} = \hat{\mathbf{y}}$, we can use the distributive property of the cross product to write

$$\mathbf{A} \times \mathbf{B} = (\hat{\mathbf{x}}A_x + \hat{\mathbf{y}}A_y + \hat{\mathbf{z}}A_z) \times (\hat{\mathbf{x}}B_x + \hat{\mathbf{y}}B_y + \hat{\mathbf{z}}B_z)$$
$$\mathbf{A} \times \mathbf{B} = \hat{\mathbf{x}}(A_yB_z - A_zB_y) + \hat{\mathbf{y}}(A_zB_x - A_xB_z) + \hat{\mathbf{z}}(A_xB_y - A_yB_x)$$

The cross product is sometimes referred to as the vector product, since the result is a vector quantity.

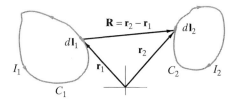

Figure 6.2 Ampère's law of force. Two separate circuits C_1 and C_2, carrying currents of I_1 and I_2, respectively, exert a force on one another.

The simplest form of Coulomb's law (Section 4.2) dealt with forces between individual point charges. By analogy, we might expect that we should similarly consider forces between elemental-length current-carrying wires. In practice, however, steady currents must necessarily flow in complete circuits, so our fundamental experimental law must describe the total force between two complete circuits. Consider two idealized[12] complete circuits C_1 and C_2, consisting of two very thin conducting loops (wires) carrying filamentary currents of I_1 and I_2, respectively. With respect to an arbitrary origin, as shown in Figure 6.2, the position vectors describing points on the two loops are taken to be $\mathbf{r}_1$ and $\mathbf{r}_2$ as indicated. We examine the force exerted on the circuit C_2 by the circuit C_1, neglecting for the purposes of this discussion the forces between current elements within the same loop. The vector distance from an elemental length $d\mathbf{l}_1$ along C_1 to another $d\mathbf{l}_2$ on C_2 is thus $(\mathbf{r}_2 - \mathbf{r}_1) = \mathbf{R} = R\hat{\mathbf{R}}$, $\hat{\mathbf{R}}$ is the unit vector directed *from*[13] $d\mathbf{l}_1$ *to* $d\mathbf{l}_2$, and $R = |\mathbf{R}| = |\mathbf{r}_2 - \mathbf{r}_1|$ is the distance between the two current elements. With $d\mathbf{l}_1$ at $\mathbf{r}_1 = \hat{\mathbf{x}}x_1 + \hat{\mathbf{y}}y_1 + \hat{\mathbf{z}}z_1$ and $d\mathbf{l}_2$ at $\mathbf{r}_2 = \hat{\mathbf{x}}x_2 + \hat{\mathbf{y}}y_2 + \hat{\mathbf{z}}z_2$, we have $R = [(x_2 - x_1)^2 + (y_2 - y_1)^2 + (z_2 - z_1)^2]^{1/2}$, and $\hat{\mathbf{R}} = (\mathbf{r}_2 - \mathbf{r}_1)/|\mathbf{r}_2 - \mathbf{r}_1| = \mathbf{R}/R$.

In his extensive experiments, Ampère found that the total vector force $\mathbf{F}_{12}$ exerted on C_2 by C_1 (both of which are located in free space) due to the mutual interaction of the currents I_1 and I_2 can be expressed[14] as

$$\mathbf{F}_{12} = \frac{\mu_0}{4\pi} \oint_{C_2} \oint_{C_1} \frac{I_2 d\mathbf{l}_2 \times (I_1 d\mathbf{l}_1 \times \hat{\mathbf{R}})}{R^2} \tag{6.1}$$

Equation (6.1) is referred to as *Ampère's law of force* and constitutes the foundation of magnetostatics. In MKS units, $\mathbf{F}_{12}$ is measured in newtons, the currents I_1 and I_2 in amperes, and the lengths $d\mathbf{l}_1$, $d\mathbf{l}_2$, and R in meters. The proportionality constant is $\mu_0/(4\pi)$ because of our use of MKS units and includes the 4π "rationalization" factor so

[12]The circuits are idealized in the sense that the batteries that would have to be sustaining the currents are not shown. It is assumed that such batteries are some distance away and that the conducting leads from them are twisted closely together; one of Ampère's first experiments showed that two oppositely directed currents close together produced no effect on another current.

[13]That is, from the "source" point to the "field" point or "observation" point, since we are aiming to write an expression for the force experienced by the circuit C_2 due to the presence of C_1. A comparison between Figure 6.2 and Figure 4.3 illustrates the similarity between the fundamental force laws of magnetostatics and electrostatics.

[14]It may appear incredible that a formula with the generality implied in (6.1) could have been established from a few experiments on circuits of special and simple shapes as was done by Ampère. Indeed, (6.1) represents a generalization from results found in special arrangements of current loops; however, it should be noted that (6.1) continues to be valid (for steady currents), providing consistent results for every experiment involving a certain arrangement of loops that has been carried out since the time of Ampère.

that a 4π factor does not appear in Maxwell's equations. In the MKS system of units, μ_0 is *defined* to have precisely the value $\mu_0 = 4\pi \times 10^{-7}$ henrys per meter $\simeq 1.26\ \mu\text{H-m}^{-1}$. This constant is called the *permeability of free space;* for practical purposes μ_0 is also the permeability of air.[15] From (6.1), we see that the dimensions of μ_0 are force-(current)$^{-2}$, or N-A^{-2}. Thus, we have 1 H $= 1$ N-m-A$^{-2} = 1$ J-A^{-2}. Since the newton and the meter are determined independently, the above choice of the value of μ_0 constitutes a definition of the unit of electric current or ampere (and hence also the unit electric charge or coulomb).

Example 6.1: An extra-high-voltage dc transmission line. An extra-high-voltage direct-current (dc) overhead transmission line consists of two very long parallel wires $a = 10$ m apart and each located $h = 35$ m above ground as shown in Figure 6.3. If the wires carry an equal current of $I = 3000$ A each flowing in opposite directions, find the repulsion force per unit length on each line. Neglect the effects of the ground plane.

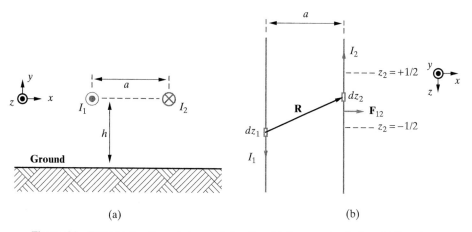

Figure 6.3   Extra-high-voltage dc transmission line. (a) Cross-sectional view. (b) Top view.

Solution: The wires are assumed to be in the z direction, with $I_1\, d\mathbf{l}_1 = \hat{\mathbf{z}}I\, dz_1$ and $I_2\, d\mathbf{l}_2 = -\hat{\mathbf{z}}I\, dz_2$ as shown. Using Ampère's law of force we can calculate the net force exerted by one of the wires on a unit length section of the other wire as

$$\mathbf{F}_{12} = \frac{\mu_0 I^2}{4\pi} \int_{z_2=-1/2}^{1/2} \int_{z_1=-\infty}^{\infty} \frac{(-dz_2)\hat{\mathbf{z}} \times (dz_1\hat{\mathbf{z}} \times \hat{\mathbf{R}})}{R^2}$$

where $R = \sqrt{(z_2 - z_1)^2 + a^2}$ is the magnitude of $\mathbf{R}$ shown in Figure 6.3b and $\hat{\mathbf{R}} = [\hat{\mathbf{z}}(z_2 - z_1) + \hat{\mathbf{x}}a]/R$ and z_1 and z_2 are the integration variables over wires 1 and 2, respectively. Note that $\mathbf{F}_{12}$ is the force exerted on a 1-m long segment of wire 2 (thus the limits of the z_2 integral are from $-\frac{1}{2}$ to $+\frac{1}{2}$) by the entire wire 1 (thus the limits of $\pm\infty$ for the z_1 integral).

---

[15]The physical meaning of permeability and its dimensions will become clearer in Section 6.8, when we discuss magnetic materials and inductance. At this point we may note that permeability has the same significance for magnetostatics as permittivity has for electrostatics.

Substituting for R and $\hat{\mathbf{R}}$, we have

$$\mathbf{F}_{12} = \frac{\mu_0 I^2}{4\pi} \int_{z_2=-1/2}^{1/2} (-dz_2)\hat{\mathbf{z}} \times \hat{\mathbf{y}} \int_{z_1=-\infty}^{\infty} \frac{a\, dz_1}{[(z_2 - z_1)^2 + a^2]^{3/2}}$$

where we have used $\hat{\mathbf{z}} \times \hat{\mathbf{R}} = \hat{\mathbf{y}} a[(z_2 - z_1)^2 + a^2]^{-1/2}$. Performing the integration[16] over z_1 we find

$$\mathbf{F}_{12} = \frac{\mu_0 I^2 a}{4\pi} \int_{z_2=-1/2}^{1/2} (-dz_2)\hat{\mathbf{z}} \times \hat{\mathbf{y}} \left[\frac{-(z_2 - z_1)}{a^2\sqrt{a^2 + (z_2 - z_1)^2}} \right]_{z_1=-\infty}^{\infty}$$

$$= \hat{\mathbf{x}} \frac{\mu_0 I^2 a}{2\pi a^2} \int_{z_2=-1/2}^{1/2} dz_2 = \hat{\mathbf{x}} \frac{\mu_0 I^2}{2\pi a}$$

Using the numerical values given as $I = 3000$ A and $a = 10$ m, we find the repulsion force to be

$$\mathbf{F}_{12} = \hat{\mathbf{x}} \frac{(4\pi \times 10^{-7} \text{ H-m}^{-1})(3000 \text{ A})^2}{2\pi(10 \text{ m})} = 0.18 \text{ N-m}^{-1}$$

This lateral repulsion force can be compared with the weight of a line segment of 1 m length. Assuming the wires to have outside diameters of a few centimeters, their mass would be a few kg, and their weight a few tens of newtons. Thus, the per-unit-length magnetostatic repulsion force is quite small compared with the per-unit-length weight of the wires. However, it should be noted that the force between the two wires is proportional to the square of the current, so it can reach substantial values when much larger currents (tens to hundreds of kA) may flow due to accidental shorts (see Problem 6.3).

Ampère's force law is another example of a force law describing "action at a distance," analogous in this sense to Coulomb's law for electrostatics. Just as it was useful to divide Coulomb's law into two parts by using the concept of an electric field as an intermediary to describe the interaction between charges, we can use Ampère's force law to define an appropriate field that may be regarded as the means by which currents exert forces on one another. The so-called[17] *magnetostatic induction field* or *magnetic flux density* **B** can be defined by rewriting (6.1) as follows:

$$\mathbf{F}_{12} = \oint_{C_2} I_2\, d\mathbf{l}_2 \times \underbrace{\frac{\mu_0}{4\pi} \oint_{C_1} \frac{I_1\, d\mathbf{l}_1 \times \hat{\mathbf{R}}}{R^2}}_{\mathbf{B}_{12}} \qquad (6.2)$$

---

[16] This is a rather common integral already encountered in Example 4.5. Using integral tables, or simply a change of variables, it can be shown that

$$\int \frac{d\zeta}{[\zeta^2 + b^2]^{3/2}} = \frac{\zeta}{b^2\sqrt{\zeta^2 + b^2}} + \text{const.}$$

[17] For historical reasons, the term "magnetic field" is generally used for a different vector, which will be defined and discussed in Section 6.8. In retrospect, it would have been more proper to refer to **B** as the magnetic field. The terms "magnetostatic induction field" or "magnetic flux density" are in fact quite inappropriate descriptions of the nature of the **B** field (also see Section 6.2.2). For this reason, and whenever possible, we shall try to refer to **B** simply as the "**B** field."

where

$$\mathbf{B}_{12} = \frac{\mu_0}{4\pi} \oint_{C_1} \frac{I_1 \, d\mathbf{l}_1 \times \hat{\mathbf{R}}}{R^2} \tag{6.3}$$

where $\mathbf{B}_{12}$ is the $\mathbf{B}$ field at point $\mathbf{r}_2$ (i.e., at the location of $d\mathbf{l}_2$ at a distance R from $d\mathbf{l}_1$) due to the current I_1 in circuit C_1. Equation (6.2) evaluates the force on circuit C_2 in terms of the interaction of its current I_2 with the field $\mathbf{B}_{12}$, which is set up by the current I_1 in circuit C_1. The current–field interaction takes place over the entire circuit C_2, while the field $\mathbf{B}_{12}$ depends only on the current and configuration of the circuit C_1, which sets up the field. In the MKS system of units, the $\mathbf{B}$ field has units of[18] tesla (T) or weber per meter^2. Since one tesla is relatively large as a practical value, $\mathbf{B}$ is often also given in the CGS units of gauss (G), where $1 \, T = 10^4$ G.

For reference purposes, the earth's magnetic field on the surface is ~ 0.5 G, the $\mathbf{B}$ fields of permanent magnets range from a few to thousands of gauss, while the $\mathbf{B}$ field at the surface of neutron stars is believed to be 10^{12} Gauss. The most powerful permanent magnets,[19] such as samarium-cobalt or neodymium-iron-boron magnets, have fields of 3000–4000 gauss, and several of them could easily lift an entire refrigerator. World-record magnetic fields of nearly a million gauss have been achieved with pulsed electromagnets of novel designs.[20] Fields of such intensity represent a burst of energy comparable to an exploding stick of dynamite; the resultant forces imposed on the current-carrying wire surpass the tensile strength of copper and the resistive losses generate enough heat to melt the copper wire. Table 6.1 provides additional examples of $\mathbf{B}$ field values in different applications, ranging from high-technology applications such as superconducting quantum interference devices[21] to relatively low-technology (but no less important) applications such as cow magnets.[22]

---

[18] Nikola Tesla (1856–1943) was a brilliant electrical engineer who invented the induction motor and many other useful electromagnetic devices and made alternating current practical. See M. Cheney, *Tesla: Man Out of Time,* Prentice-Hall, Englewood Cliffs, New Jersey, 1981. Wilhelm E. Weber was a professor of physics at Göttingen and a close colleague of J. K. F. Gauss. For most of his professional life, Weber worked in collaboration with Gauss on the study of magnetic phenomena.

[19] P. Campbell, *Permanent Magnet Materials and Their Design,* Cambridge University Press, 1994.

[20] G. Boebinger, A. Passner, and J. Bevk, Building world-record magnets, *Scientific American,* pp. 58–66, June 1995.

[21] Superconducting quantum interference devices (SQUIDs) are the basis for scanning SQUID microscopes, which can image magnetic fields at the surface of samples under study with an unprecedented sensitivity. SQUID imaging of fields as low as 10^{-11} gauss in a 1 cm^2 area was reported by J. Clarke, SQUIDs, *Scientific American,* pp. 46–53, August 1994. At the level of sensitivity of these devices, nearly everything is magnetic. Applications of SQUIDs range from diagnosis of brain tumors to tests of relativity. For further information, see J. Kirtley, Imaging magnetic fields, *IEEE Spectr.,* pp. 40–48, December 1996.

[22] The cow magnet is an alnico (Al-Ni-Co-Fe) cylinder three inches long and a half-inch in diameter; it has rounded ends and is usually coated with plastic for protection against corrosion and breakage. Grazing cows are encouraged to swallow the cow magnet as a protection against the so-called "hardware disease," which occurs when a grazing cow indiscriminately swallows sharp steel objects, such as bits of wire used to bale hay, which can cause damage to the walls of her intestines. The cow magnet remains in the cow's stomach, attracting to it any steel objects she later swallows, and can be retrieved and reused when the cow is slaughtered. For more on cow magnets and other interesting applications, see J. D. Livingston, *Driving Force: The Natural Magic of Magnets,* Harvard University Press, Cambridge, Massachusetts, 1996.

TABLE 6.1   TYPICAL **B** FIELD VALUES IN SELECTED APPLICATIONS

Application	B Field (Gauss)
Sensitivity of a scanning SQUID microscope*	10^{-11}
Human brain	10^{-8}
Intergalactic and interstellar magnetic fields	10^{-6}
Human heart	10^{-4}–10^{-3}
Earth's magnetic field	0.5
Refrigerator memo magnets	10–800
Electron beam of CRT (computer or TV)	50–100
Magnetic read switch	100–200
1-horsepower electric motor	1000–2000
Cow magnets*	2000
Powerful permanent magnets*	3000–4000
Magnetic resonance imaging (MRI)	10^3–10^5
High-energy particle accelerators	10^5
Pulsed electromagnets*	10^5–10^6
White dwarf stars	10^8
Neutron stars	10^{12}

*See references in text.

The vector field **B** can be calculated at any field point **r** using (6.3) even if there is no current element there to experience a magnetostatic force. When a current-carrying wire is placed in a region permeated by a **B** field, it experiences a magnetostatic force given by $\mathbf{F} = \int Id\mathbf{l} \times \mathbf{B}$, as is evident from (6.2), which can be calculated as illustrated in Example 6.2. As noted in Section 4.3, the notion of a "field," however abstract, not only is useful and convenient in thinking and working with action-at-a-distance phenomena but may also be regarded as an actual physical entity in its own right.[23] For our purposes in the rest of this chapter, it is more useful[24] to continue our discussions of the implications of (6.1) using the so-called *Biot–Savart law,* which quantifies the **B** field produced at a point by currents in its vicinity.

Example 6.2: A semicircular loop in a B field. A semicircular loop of wire of 1 m diameter carrying a current of $I = 10$ A lies in a uniform **B** field of magnitude $B_0 = 1.5$ T, as shown in Figure 6.4. Determine the total magnetostatic force experienced by the loop.

Solution: Using (6.2), the total magnetostatic force on the straight portion of the current loop is

$$\mathbf{F}_{\text{str}} = \int_{-a}^{a} (-I\,dx)\hat{\mathbf{x}} \times \hat{\mathbf{z}}B_0 = \hat{\mathbf{y}}IB_0 \int_{-a}^{a} dx$$

$$= \hat{\mathbf{y}}2IB_0a = \hat{\mathbf{y}}2(10 \text{ A})(1.5 \text{ T})(0.5 \text{ m}) = \hat{\mathbf{y}}15 \text{ N}$$

[23]Which it certainly was for M. Faraday; see Footnote 42 in Section 4.3.

[24]Direct use of (6.1) to determine the force on a circuit is feasible only in relatively simple geometries. It is usually easier to determine the force on a particular circuit (C_2) through the concept of the **B** field produced by the current of another circuit (C_1).

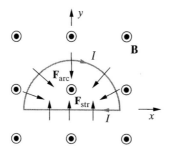

Figure 6.4 **A semicircular current loop in a B field.** The **B** field is directed out of the page, as represented with a circle having a central dot.

Similarly, the total force experienced by the semicircular arc portion of the current loop is

$$\mathbf{F}_{arc} = \int_{\phi=0}^{\pi} (-Ia \, d\phi)\hat{\boldsymbol{\phi}} \times \hat{\mathbf{z}} B_0 = -IB_0 a \int_0^{\pi} \hat{\mathbf{r}} \, d\phi = -IB_0 a \int_0^{\pi} (\hat{\mathbf{x}} \cos\phi + \hat{\mathbf{y}} \sin\phi) \, d\phi$$

$$= -IB_0 a [\hat{\mathbf{x}} \sin\phi - \hat{\mathbf{y}} \cos\phi]_0^{\pi} = -\hat{\mathbf{y}} 2IB_0 a = -\hat{\mathbf{y}} 15 \text{ N}$$

The fact that the force is in the y direction makes sense, since the net effect of the x components of the force on the semicircular arc portion cancel out due to symmetry, as is evident in Figure 6.4. Thus, the net total magnetostatic force exerted on the semicircular loop by the **B** field is

$$\mathbf{F} = \mathbf{F}_{str} + \mathbf{F}_{arc} = \hat{\mathbf{y}} 2IB_0 a - \hat{\mathbf{y}} 2IB_0 a = 0$$

6.2 THE BIOT–SAVART LAW AND ITS APPLICATIONS

The Biot–Savart law is the most basic law of magnetostatics; it describes how the **B** field at a given point is produced by the moving charges (i.e., currents) in the neighborhood of that point. This basic law has been experimentally verified by innumerable measurements of the fields produced by many types of current distributions. A highly accurate assessment of the inverse square dependence is provided by geophysical measurements of the earth's magnetic field, which indicate[25] that not only is the $r^{-2+\zeta}$ dependence valid at distances of $\sim 10^4$ km but that $|\zeta| < 10^{-16}$.

We consider electric currents that are either steady or slowly varying.[26] Using superposition,[27] we can consider any current to consist of infinitesimal current elements. More complex circuits and current distributions can always be expressed as a sum of such elemental currents.

---

[25] For an excellent review, see A. S. Goldbaher and M. M. Nieto, Terrestrial and extraterrestrial limits on the photon mass, *Rev. Mod. Phys.*, 43, pp. 277–295, 1971.

[26] The currents may vary with time if the time interval over which appreciable variation occurs is much longer compared with r/c, where r is the maximum distance from any of the elementary source currents to the observation point where the magnetic field is to be computed, and c is the velocity of light in free space. Any given circuit that the current flows through undoubtedly has its internal time constant (due, e.g., to its distributed capacitance and/or inductance); the time intervals over which the currents vary appreciably must also be much longer than this time constant.

[27] As in the case of electric fields due to charges, we are able to use superposition only because it is verified by experiment.

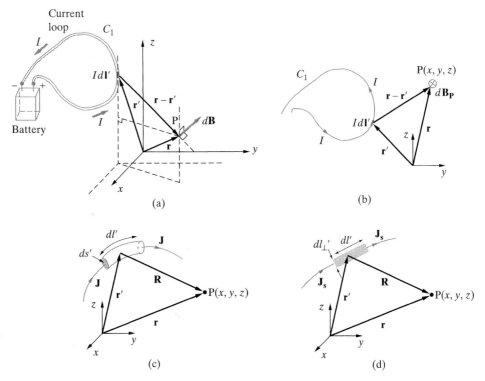

Figure 6.5   Magnetic field of a current element. (a) Note that the differential current element is necessarily part of a circuit or loop of current, driven by a battery or other source as shown. (b) A two-dimensional view with $d\mathbf{l}'$ and $\mathbf{R} = (\mathbf{r} - \mathbf{r}')$ on the plane of the paper, in which case $d\mathbf{B}_\mathrm{P}$ is into or out of the paper. (c) An arbitrary volume distribution of current can be analyzed in terms of differential tubular current density elements. (d) An arbitrary surface distribution of current can be analyzed in terms of differential surface current density elements.

The **B** field vector at any point P identified by the position vector $\mathbf{r}$, as shown in Figure 6.5a, due to a differential line current element $I\,d\mathbf{l}'$ located at position $\mathbf{r}'$, follows from (6.3) as

$$d\mathbf{B}_\mathrm{p} = \frac{\mu_0 I\,d\mathbf{l}' \times \hat{\mathbf{R}}}{4\pi R^2} \tag{6.4}$$

where $\hat{\mathbf{R}}$ is the unit vector pointing *from* the location of the current element *to* the field point P, and $R = |\mathbf{r} - \mathbf{r}'|$ is the distance between them. Note that we resort to our practice of using primed quantities to represent the position vector or coordinates of source points, while unprimed quantities are used to represent the position vector or coordinates of points at which the **B** field is evaluated. The cross product in (6.4) indicates that $d\mathbf{B}$ is perpendicular to both $d\mathbf{l}'$ and $\hat{\mathbf{R}}$; the orientation of $d\mathbf{B}$ is determined by the right-hand rule. As indicated in Figure 6.6b, the *right-hand rule* states that if the fingers close up as

Figure 6.6 The right-hand rule. (a) Determination of the direction of the **B** field from the direction of current. (b) Determination of the direction of the cross product of two vectors, in this case $I\, d\mathbf{l}' \times \hat{\mathbf{R}}$.

they move from $I\, d\mathbf{l}'$ to $\hat{\mathbf{R}}$, then the thumb points in the direction of their cross product, namely in the direction of $d\mathbf{B}$. As indicated in Figure 6.6a, the right-hand rule also states that if the thumb of the right hand is pointed in the direction of the current, the curled right-hand fingers encircling the current element point in the direction of the **B** field. The latter form of the right-hand rule is more useful in interpreting the direction of the **B** field near current-carrying conductors. In relation to Figure 6.5b, where the current loop C_1 is assumed to be entirely confined to the y-z plane, note that we have adopted the customary convention of representing the direction of vectors that are perpendicular to the page (in this case the **B** field) by a small circle with a cross at the center where the vector points into the page. Vectors pointing out of the page are represented with circles having a central dot, as was done in Figure 6.4.

Note that, in general, the current element $Id\mathbf{l}'$ is part of a thin filamentary[28] closed current loop C_1, with more complex circuits and current distributions represented as a superposition of many such closed filamentary loops of current. The total magnetic flux density at point P due to the filamentary current loop C_1 is given by the sum (superposition) of the contributions from its individual differential current elements $I\, d\mathbf{l}'$, namely

$$\boxed{\mathbf{B} = \frac{\mu_0}{4\pi} \oint_{C_1} \frac{I\, d\mathbf{l}' \times \hat{\mathbf{R}}}{R^2}} \tag{6.5}$$

Equations (6.4) and (6.5) are statements of the *Biot–Savart law.*

In practice, we do not always have currents flowing in thin conductors, and it is thus necessary to generalize the definition of **B** so that it applies for any arbitrary volume distribution of current. Since a steady electric current is divergence-free ($\nabla \cdot \mathbf{J} = 0$, see

---

[28]The diameter of the conductor is much smaller than any other dimensions in the system, so we can assume that all the current flows along the axis of the conductor (see Section 5.9.2).

Section 5.4), all current flow lines form closed loops. Consider a differential length dl' of a single current-flow tube of differential cross-sectional area ds' as shown in Figure 6.5c, carrying a current with a current density $\mathbf{J}(\mathbf{r}')$. Since we can associate the direction with the current density vector $\mathbf{J}(\mathbf{r}')$ (rather than with an arc length dl'), a given current-flow tube element of length dl' and area ds' located at $\mathbf{r}'$ produces a field $d\mathbf{B}$ at point P given by

$$d\mathbf{B}_\mathrm{P} = \frac{\mu_0}{4\pi} \frac{\mathbf{J}(\mathbf{r}')\, ds'\, dl' \times \hat{\mathbf{R}}}{R^2}$$

since the total current flowing in this differential tube is $\mathbf{J}(\mathbf{r}')ds'$. The total current contained in a volume V' thus produces a $\mathbf{B}$ field given by

$$\mathbf{B}(\mathbf{r}) = \frac{\mu_0}{4\pi} \int_{V'} \frac{\mathbf{J}(\mathbf{r}') \times \hat{\mathbf{R}}}{R^2}\, dv' \tag{6.6}$$

where the integration is to be carried out over the source coordinates x', y', and z', and dv' is an element of volume V' given by $dv' = ds'\, dl'$.

Similarly, if we have a surface current density $\mathbf{J}_s$, the current flow patch over a *surface area* of length dl' and width $dl_\perp'$ (perpendicular to $\mathbf{J}_s$) located at $\mathbf{r}'$ produces a differential field $d\mathbf{B}$ at point P at $\mathbf{r}$ given by

$$d\mathbf{B}_\mathrm{P} = \frac{\mu_0}{4\pi} \frac{\mathbf{J}_s(\mathbf{r}')dl_\perp'dl' \times \hat{\mathbf{R}}}{R^2}$$

since the current flowing in a strip of width $dl_\perp'$ is $\mathbf{J}_s(\mathbf{r}')dl_\perp'$. [Note again that $R = |\mathbf{r} - \mathbf{r}'|$ and $\hat{\mathbf{R}} = (\mathbf{r} - \mathbf{r}')/R$.] Integrating the differential field over the entire surface, and using the differential surface element $ds' = dl_\perp'dl'$, the total current flowing over the surface S' produces a $\mathbf{B}$ field given by

$$\mathbf{B}(\mathbf{r}) = \frac{\mu_0}{4\pi} \int_{S'} \frac{\mathbf{J}_s(\mathbf{r}') \times \hat{\mathbf{R}}}{R^2}\, ds' \tag{6.7}$$

Example 6.3: Finite-length straight wire. Consider a straight, current-carrying filamentary conductor of length $2a$, as shown in Figure 6.7a. Find the $\mathbf{B}$ field at a point $P(r, \phi, 0)$ equidistant from the end points of the conductor.

Solution: Since we have azimuthal symmetry (i.e., no dependence on ϕ), a cylindrical coordinate system with the wire oriented along the z-axis and centered at the origin is the most appropriate. Using (6.4), the differential magnetic field $d\mathbf{B}_\mathrm{P}$ at point P (note that $\mathbf{r} = \hat{\mathbf{r}}r$) due to current element $I\, d\mathbf{l}'$, located at $\mathbf{r}' = \hat{\mathbf{z}}z'$ at a distance $R = |\mathbf{r} - \mathbf{r}'| = \sqrt{r^2 + (z')^2}$ from point P is

$$d\mathbf{B}_\mathrm{P} = \frac{\mu_0}{4\pi} \frac{I\, dz'\hat{\mathbf{z}} \times \hat{\mathbf{R}}}{R^2} = \hat{\boldsymbol\phi}\, \frac{\mu_0 I}{4\pi} \frac{\sin\alpha\, dz'}{[r^2 + (z')^2]} = \hat{\boldsymbol\phi}\, \frac{\mu_0 I}{4\pi} \frac{r\, dz'}{[r^2 + (z')^2]^{3/2}}$$

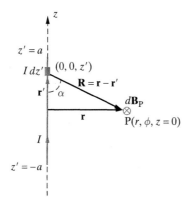

Figure 6.7 Finite-length current-carry-ing wire. Coordinate system and the source and observation points.

where we have used the fact that the magnitude of the cross product of two vectors is equal to the product of the magnitudes of the two vectors times the sine of the angle between them; that is, $\hat{\mathbf{z}} \times \hat{\mathbf{R}} = \hat{\boldsymbol{\phi}}|\hat{\mathbf{z}}||\hat{\mathbf{R}}| \sin\alpha = \hat{\boldsymbol{\phi}} \sin\alpha$. Note that

$$\sin\alpha = r[r^2 + (z')^2]^{-1/2}$$

with α as shown in Figure 6.7. We can find the total $\mathbf{B}_P$ field at point P by integrating[29] $d\mathbf{B}_P$ over the length of the wire,

$$\mathbf{B}_P = \hat{\boldsymbol{\phi}} \frac{\mu_0 I r}{4\pi} \int_{z'=-a}^{+a} \frac{dz'}{[r^2 + (z')^2]^{3/2}} = \hat{\boldsymbol{\phi}} \frac{\mu_0 I r}{4\pi} \left[\frac{z'}{r^2 \sqrt{r^2 + (z')^2}} \right]_{z'=-a}^{+a}$$

$$= \hat{\boldsymbol{\phi}} \frac{\mu_0 I a}{2\pi r \sqrt{r^2 + a^2}}$$

For an infinitely long conductor, or at small distances from a finite-length conductor (i.e., $r \ll a$), we have

$$\mathbf{B}_P \simeq \hat{\boldsymbol{\phi}} \frac{\mu_0 I}{2\pi r}$$

analogous to $\mathbf{E} \simeq \hat{\mathbf{r}} \rho_l/(2\pi\epsilon_0 r)$ for an infinite line charge (see Example 4.5).

As a numerical example, the $\mathbf{B}$ field at a distance of 1 cm from a long straight wire carrying a current of $I = 5$ A is $B \simeq \mu_0 I/(2\pi r) = (4\pi \times 10^{-7})(5)/[2\pi(0.01)] = 10^{-4}$ T or 1 gauss.

Example 6.4: Square loop of current. Use the result derived in Example 6.3 to find the $\mathbf{B}$ field at the center of a square loop of current. Consider the square loop of side length $2a$ carrying a steady current I as shown in Figure 6.8.

Solution: First consider one side of the loop, say DE. Using the result of Example 6.3, the $\mathbf{B}$ field due to the current flowing along side DE is given by

$$\mathbf{B}_1 = -\hat{\mathbf{z}} \frac{\mu_0 I}{2\sqrt{2}\pi a}$$

---

[29]Either by substitution of variables or by looking up integral tables, just as in the case of Example 4.5.

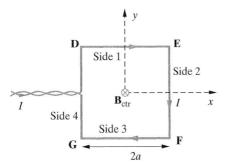

Figure 6.8   Square loop of current. A current I flows in a square wire loop.

From symmetry, the contributions by the other sides of the square loop are the same, since by the right-hand rule they all produce a magnetic field in the same direction. Thus, the total **B** field at the center is

$$\mathbf{B}_{\text{ctr}} = 4\mathbf{B}_1 = -\hat{\mathbf{z}}\frac{\sqrt{2}\mu_0 I}{\pi a}$$

Example 6.5: Two infinitely long parallel wires. Consider two infinitely long parallel wires, each carrying a current I in the z direction, one passing through the point $(x = 0, y = -a)$ and the other through $(x = 0, y = a)$, as shown in Figure 6.9. Find **B** at $\mathrm{P}_1(x = 0, y = 0)$, $\mathrm{P}_2(x = b, y = 0)$, and $\mathrm{P}_3(x = 0, y = b)$.

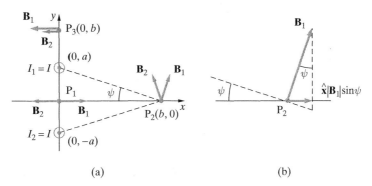

(a)   (b)

Figure 6.9   Two parallel wires. (a) Each wire is infinitely long carrying a current I in the z direction. (b) Expanded geometry of the vicinity of point P_2.

Solution: Using the result of Example 6.3, namely that the **B** field at a distance r from an infinitely long straight wire carrying a current I in the z direction is $\mathbf{B} = \hat{\boldsymbol{\phi}}\mu_0 I/(2\pi r)$, we have, using superposition,

$$\mathbf{B}_{\mathrm{P}_1} = \mathbf{B}_1 + \mathbf{B}_2 = \hat{\mathbf{x}}\frac{\mu_0 I}{2\pi a} - \hat{\mathbf{x}}\frac{\mu_0 I}{2\pi a} = 0$$

Similarly, at P$_2$, we have

$$\mathbf{B}_{P_2} = \frac{\mu_0 I}{2\pi\sqrt{a^2 + b^2}}(\hat{\mathbf{x}}\sin\psi + \hat{\mathbf{y}}\cos\psi) + \frac{\mu_0 I}{2\pi\sqrt{a^2 + b^2}}(-\hat{\mathbf{x}}\sin\psi + \hat{\mathbf{y}}\cos\psi)$$

$$= \hat{\mathbf{y}}\frac{\mu_0 I b}{\pi(a^2 + b^2)}$$

where we have used $\cos\psi = b(a^2 + b^2)^{-1/2}$. At P$_3$, we have

$$\mathbf{B}_{P_3} = (-\hat{\mathbf{x}})\frac{\mu_0 I}{2\pi(b - a)} + (-\hat{\mathbf{x}})\frac{\mu_0 I}{2\pi(b + a)} = (-\hat{\mathbf{x}})\frac{\mu_0 I b}{\pi(b^2 - a^2)}$$

Note that this result is valid regardless of whether $b > a$ or $a > b$.

Example 6.6: Circular loop of current. Consider a circular loop of radius a, carrying a current I, and situated as shown in Figure 6.10. Find **B** at a point P on the axis of the loop.

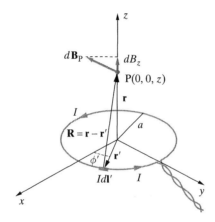

Figure 6.10 Circular loop of current. A circular loop of wire of radius a carrying a current I producing a **B** field along its axis with only a z component.

Solution: Using the cylindrical coordinate system and (6.4), the differential field at point P due to a current element $I\,d\mathbf{l}' = Ia\,d\phi'\hat{\boldsymbol{\phi}}$ located at $\mathbf{r}'$ is given by

$$d\mathbf{B}_P = \frac{\mu_0 I\,d\mathbf{l}' \times \hat{\mathbf{R}}}{4\pi R^2} = \frac{\mu_0 Ia\,d\phi'\hat{\boldsymbol{\phi}} \times (\hat{\mathbf{z}}z - \hat{\mathbf{r}}a)}{4\pi(a^2 + z^2)^{3/2}}$$

since in cylindrical coordinates we have $\mathbf{r} = \hat{\mathbf{z}}z$, $\mathbf{r}' = \hat{\mathbf{r}}a$, $R = \sqrt{a^2 + z^2}$, and $\hat{\mathbf{R}} = (\mathbf{r} - \mathbf{r}')/R$. (Note that in rectangular coordinates, the vectors $\mathbf{r}$ and $\mathbf{r}'$ are given as $\mathbf{r} = z\hat{\mathbf{z}}$ and $\mathbf{r}' = \hat{\mathbf{x}}a\cos\phi' + \hat{\mathbf{y}}a\sin\phi'$). The cross product $\hat{\boldsymbol{\phi}} \times \hat{\mathbf{z}} = \hat{\mathbf{r}}$ gives the component of $d\mathbf{B}_P$ in the r direction, which cancels out upon integration over the whole loop, due to symmetry. The magnitude of the z component (dB_z), which results from the cross product $\hat{\boldsymbol{\phi}} \times (-\hat{\mathbf{r}}) = \hat{\mathbf{r}} \times \hat{\boldsymbol{\phi}} = \hat{\mathbf{z}}$, is

$$dB_z = \frac{\mu_0 Ia^2\,d\phi'}{4\pi(a^2 + z^2)^{3/2}}$$

so that the total **B** at point P due to the whole loop is

$$\mathbf{B_P} = \hat{\mathbf{z}} \int_{\phi'=0}^{2\pi} \frac{\mu_0 I a^2}{4\pi (a^2 + z^2)^{3/2}} \, d\phi' = \hat{\mathbf{z}} \frac{\mu_0 I a^2}{4\pi (a^2 + z^2)^{3/2}} \int_{\phi'=0}^{2\pi} d\phi' = \hat{\mathbf{z}} \frac{\mu_0 I a^2}{2(a^2 + z^2)^{3/2}}$$

Furthermore, at the center of the loop ($z = 0$), the **B** field simplifies to

$$\mathbf{B}_{\text{ctr}} = \hat{\mathbf{z}} \frac{\mu_0 I}{2a}$$

Example 6.7: Finite-length wire B at an off-axis point. The problem arrangement is identical to that for Example 6.3, except that we now evaluate the **B** field at an arbitrary point $P(r, \phi, z)$, as shown in Figure 6.11.

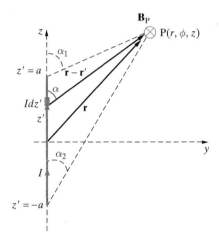

Figure 6.11 Magnetic field at an arbitrary point near a finite-length current-carrying wire.

Solution: From the Biot–Savart law (equation (6.4)), we have

$$d\mathbf{B_P} = \frac{\mu_0}{4\pi} \frac{I \, dz' \, \hat{\mathbf{z}} \times \hat{\mathbf{R}}}{R^2} = \hat{\boldsymbol{\phi}} \frac{\mu_0}{4\pi} \frac{I \sin\alpha \, dz'}{[r^2 + (z - z')^2]}$$

where we note that $\mathbf{r} = \hat{\mathbf{r}} r + \hat{\mathbf{z}} z$ and $\mathbf{r}' = \hat{\mathbf{z}} z'$ so that $R = \sqrt{r^2 + (z - z')^2}$, and $\hat{\mathbf{R}} = (\mathbf{r} - \mathbf{r}')/R = [\hat{\mathbf{r}} r + \hat{\mathbf{z}}(z - z')]/R$. From Figure 6.11, we have

$$\sin\alpha = \frac{r}{\sqrt{r^2 + (z - z')^2}}$$

Substituting into the expression for $d\mathbf{B_P}$, we find

$$d\mathbf{B_P} = \hat{\boldsymbol{\phi}} \frac{\mu_0}{4\pi} \frac{I r \, dz'}{[r^2 + (z - z')^2]^{3/2}}$$

Integrating over the full length of the wire, we find the total **B** field at point P as

$$\mathbf{B}_P = \hat{\boldsymbol{\phi}}\frac{\mu_0 I r}{4\pi}\int_{z'=-a}^{+a}\frac{dz'}{[r^2+(z-z')^2]^{3/2}} = \hat{\boldsymbol{\phi}}\frac{\mu_0 I r}{4\pi}\left[\frac{z-z'}{r^2\sqrt{r^2+(z-z')^2}}\right]_{z'=+a}^{-a}$$

$$= \hat{\boldsymbol{\phi}}\frac{\mu_0 I}{4\pi r}\left[\frac{z+a}{\sqrt{r^2+(z+a)^2}} - \frac{z-a}{\sqrt{r^2+(z-a)^2}}\right]$$

where the integration can be carried out using either integral tables or substitution of variables, just as in the case of Example 4.5. We now note that

$$\frac{z+a}{\sqrt{r^2+(z+a)^2}} = \cos\alpha_2 \quad \text{and} \quad \frac{z-a}{\sqrt{r^2+(z-a)^2}} = \cos\alpha_1$$

so that we have

$$\mathbf{B}_P = \hat{\boldsymbol{\phi}}\frac{\mu_0 I}{4\pi r}(\cos\alpha_2 - \cos\alpha_1)$$

Note that for a point on the bisecting axis of the wire (i.e., at $z=0$), we have

$$\cos\alpha_2 = -\cos\alpha_1 = \frac{a}{\sqrt{r^2+a^2}}$$

and the result in Example 6.3 is obtained, namely

$$\mathbf{B}_P = \hat{\boldsymbol{\phi}}\frac{\mu_0 I a}{2\pi r\sqrt{r^2+a^2}}$$

Example 6.8: B field of a strip of sheet current. Consider an infinitely long perfectly conducting sheet of width d with a surface current density of $\mathbf{J}_s = \hat{\mathbf{z}}J_{s0}$ amperes per meter, as shown in Figure 6.12. Determine the **B** field produced by this strip at an arbitrary point P(x, y, z).

Solution: First, we can decompose the sheet into incremental filamentary currents, each located at $y=0$, $x=x'$, carrying a current of $J_{s0}\,dx'$, and producing a field $d\mathbf{B}$ as found in Example 6.3. This field would have components dB_x and dB_y given by

$$dB_x = \frac{-\mu_0 J_{s0}}{2\pi}\frac{y\,dx'}{y^2+(x-x')^2} \quad \text{and} \quad dB_y = \frac{\mu_0 J_{s0}}{2\pi}\frac{(x-x')\,dx'}{y^2+(x-x')^2}$$

Note that we used the basic result from Example 6.3 that for a strip of width dx' along the z axis, carrying a total current of $J_{s0}dx'$, we have

$$d\mathbf{B} = \hat{\boldsymbol{\phi}}\frac{\mu_0 J_{s0}dx'}{2\pi\sqrt{x^2+y^2}} = [-\hat{\mathbf{x}}\sin\phi + \hat{\mathbf{y}}\cos\phi]\frac{\mu_0 J_{s0}dx'}{2\pi\sqrt{x^2+y^2}}$$

$$= -\hat{\mathbf{x}}\frac{y\mu_0 J_{s0}dx'}{2\pi(x^2+y^2)} + \hat{\mathbf{y}}\frac{x\mu_0 J_{s0}dx'}{2\pi(x^2+y^2)} = \hat{\mathbf{x}}dB_x + \hat{\mathbf{y}}dB_y$$

since $\cos\phi = x(x^2+y^2)^{-1/2}$ and $\sin\phi = y(x^2+y^2)^{-1/2}$ as shown in Appendix A. Replacing x with $(x-x')$ gives the dB_x and dB_y components due to a strip located at $x=x'$

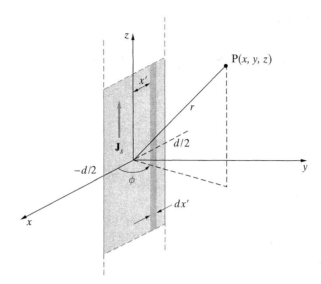

Figure 6.12   Surface current J_{s0} on an infinitely long sheet of width d. Note that the ribbonlike sheet lies in the x-z (i.e., $y = 0$) plane.

instead of along the z axis. The total **B** field is found by integrating over x' from $-d/2$ to $d/2$. After manipulation of integrals[30] the components of the field are found to be

$$B_x(x,y,z) = \frac{-\mu_0 J_{s0}}{2\pi}\left[\tan^{-1}\left(\frac{x+d/2}{y}\right) - \tan^{-1}\left(\frac{x-d/2}{y}\right)\right]$$

$$B_y(x,y,z) = \frac{\mu_0 J_{s0}}{4\pi}\ln\left[\frac{y^2+(x+d/2)^2}{y^2+(x-d/2)^2}\right]$$

Note that if the sheet current is infinitely wide (i.e., $d \to \infty$) in addition to being infinitely long (i.e., surface current flowing on the whole $y = 0$ plane), then the y components of the **B** field would cancel because of symmetry. Thus, the total **B** field in the x direction is given as

$$B_x = \begin{cases} -\dfrac{\mu_0 J_{s0}}{2} & y > 0 \\[3mm] \dfrac{\mu_0 J_{s0}}{2} & y < 0 \end{cases}$$

6.2.1 Magnetic Field Inside a Solenoid

The result found in Example 6.6 can be used directly to evaluate the **B** field for a commonly used practical circuit element, namely a solenoid. A solenoid is a coil consisting of many turns of an insulated wire wound on a cylindrical support with a circular

---

[30]The essential integrals are

$$\int \frac{dx'}{y^2+(x-x')^2} = -\frac{1}{y}\tan^{-1}\left(\frac{x-x'}{y}\right) \quad \text{and} \quad \int \frac{(x-x')\,dx'}{y^2+(x-x')^2} = -\frac{1}{2}\ln[y^2+(x-x')^2]$$

cross section, as shown in Figure 6.13a. The cross section of the solenoid is shown in Figure 6.13b. Fairly strong fields can be produced by winding wires in the form of a solenoid so that the **B** fields due to the successive turns are additive.

Because of the coil being very tightly wound in a spiral, and since the wire diameter is typically much smaller than the length l of the solenoid, it is possible to treat the problem in terms of an equivalent surface current of density $J_s = (NI)/l$ A-m^{-1}, where N is the total number of turns and I is the current flowing in the wire, as shown in Figure 6.13c. A three-dimensional view of the equivalent sheet current $\mathbf{J}_s$ is also shown in Figure 6.13d. The current in an elemental length dz' of the solenoid is then given by

$$J_s \, dz' = \frac{NI}{l} \, dz'$$

We can view each dz' length of the solenoid as a thin circular current loop as in Example 6.6 and use the result derived there to find the **B** field at any point z along the axis of the solenoid. Using the result of Example 6.6, the differential field $d\mathbf{B}$ at any point z along the axis of the solenoid due to such a loop of differential thickness dz' located at a distance z' from the center $(z = 0)$ of the solenoid is given by

$$d\mathbf{B} = \hat{\mathbf{z}} \frac{\mu_0 \left(\dfrac{NI}{l} \, dz' \right) a^2}{2[a^2 + (z - z')^2]^{3/2}}$$

To find the total field **B** of the solenoid at any point z along its axis, we integrate $d\mathbf{B}$ to sum the contributions of all such loops. We find

$$\mathbf{B}(z) = \hat{\mathbf{z}} \frac{\mu_0 N I a^2}{2l} \int_{-l/2}^{+l/2} \frac{dz'}{[a^2 + (z - z')^2]^{3/2}}$$

$$= \hat{\mathbf{z}} \frac{\mu_0 N I a^2}{2l} \int_{z+l/2}^{z-l/2} \frac{-du}{(a^2 + u^2)^{3/2}} = \hat{\mathbf{z}} \frac{\mu_0 N I a^2}{2la^2} \left[\frac{u}{\sqrt{a^2 + u^2}} \right]_{z-l/2}^{z+l/2}$$

$$= \hat{\mathbf{z}} \frac{\mu_0 N I}{2l} \left[\frac{(z + l/2)}{\sqrt{a^2 + (z + l/2)^2}} - \frac{(z - l/2)}{\sqrt{a^2 + (z - l/2)^2}} \right]$$

where we have made the substitution $u = z - z'$, so that $du = -dz'$. To find the total field **B** at the center of the solenoid (the point marked "ctr" in Figure 6.13c), we substitute $z = 0$, yielding

$$\mathbf{B}_{\text{ctr}} = \hat{\mathbf{z}} \frac{\mu_0 N I}{\sqrt{(2a)^2 + l^2}} = \hat{\mathbf{z}} \frac{\mu_0 N I}{l \sqrt{1 + (2a/l)^2}}$$

Similarly, the total field **B** at the ends of its axis (i.e., $z = \pm l/z$) is given by

$$\mathbf{B}_{\text{end}} = \hat{\mathbf{z}} \frac{\mu_0 N I}{2l \sqrt{1 + (a/l)^2}}$$

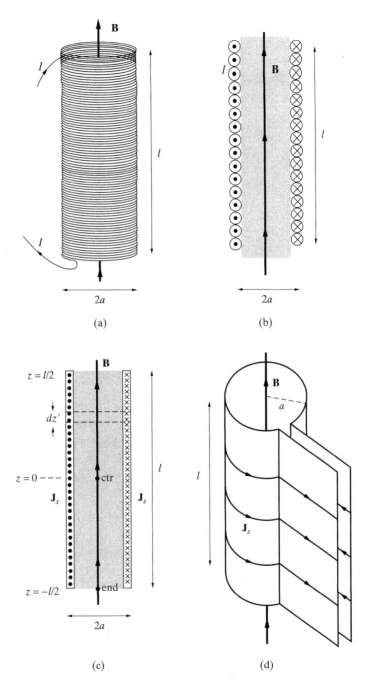

Figure 6.13   Solenoid. (a) Closely wound solenoid coil. (b) Cross section of the solenoid showing many fewer than the actual number of turns. (c) Equivalent surface current sheet of density $\mathbf{J}_s$ in A-m^{-1} representing the current flowing through the coil of the solenoid. (d) Three-dimensional view of equivalent surface current sheet of the solenoid.

For a very long solenoid or one with very small radius (i.e., $l \gg a$), the **B** field at the center and endpoints of the solenoid are approximately

$$\mathbf{B}_{\text{ctr}} \simeq \hat{\mathbf{z}}\frac{\mu_0 NI}{l}$$

$$\mathbf{B}_{\text{end}} \simeq \hat{\mathbf{z}}\frac{\mu_0 NI}{2l}$$

Example 6.9: A solenoid-type magnetic filter. A magnetic filter involves a solenoid-type magnet used to separate very fine metallic particles from food products such as flours and corn starch. The solenoid has a 50 cm diameter and 1 m length. If the solenoid has 1000 turns of uniformly wound wire with a current of 100 A flowing through it, calculate the magnetic field (a) at its center and (b) at the end of its axis. Assume an air-core solenoid.

Solution: Using $a = 25$ cm and $l = 1$ m, we have

(a)
$$\mathbf{B}_{\text{ctr}} = \hat{\mathbf{z}}\frac{(4\pi \times 10^{-7} \text{ H-m}^{-1})(1000 \text{ turns})(100 \text{ A})}{(1 \text{ m})\sqrt{1 + (0.5)^2}} \simeq 0.112 \text{ T} = 1120 \text{ gauss}$$

and

(b)
$$\mathbf{B}_{\text{end}} = \hat{\mathbf{z}}\frac{(4\pi \times 10^{-7} \text{ H-m}^{-1})(1000 \text{ turns})(100 \text{ A})}{(2 \text{ m})\sqrt{1 + (0.25)^2}} \simeq 0.061 \text{ T} = 610 \text{ gauss}$$

That $\mathbf{B}_{\text{end}}$ is approximately one-half of $\mathbf{B}_{\text{ctr}}$ can be seen by considering a very long solenoid of length $2l$ and with $2N$ turns. The field at the center of this solenoid is $\mu_0(2N)I/(2l)$, half of which is equal to $\mathbf{B}_{\text{end}}$ (by superposition, we expect the field to be halved when the contributions of the elemental loops from one half of the solenoid are removed). Thus, for a long solenoid, the **B** field at the ends of its axis is one-half that of the field at its center. The variation of the **B** field along the axis of the solenoid between its two ends is shown in Figure 6.14, for two different values of the ratio of the solenoid radius to its length. We see that for a long solenoid ($a/l = 0.1$), the magnetic field is effectively confined to the region inside the solenoid and is nearly uniform except close to the edges. However, for a shorter solenoid ($a/l = 0.5$), there is more leakage of the **B** field outside the solenoid, and also the field is not as uniform within the solenoid. It is interesting to note, however, that the **B** field at the end (i.e., at $z = l/2$) of the solenoid is approximately equal for both values of a/l.

In the foregoing discussion, we confined ourselves to the evaluation of the **B** field at points along the axis of the solenoid. For a long and slim solenoid (i.e., $l \gg a$), the field is nearly uniform within the solenoid even at points off its axis, as we shall show in the next section using Ampère's law.

6.2.2 The Duality of the B and E Fields

The **B** field is in many ways analogous to the electric field **E** that we worked with in Chapter 4, rather than to the electric flux density **D**. In this section we comment on

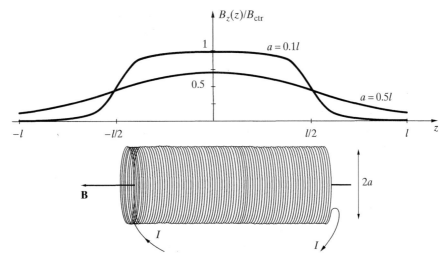

Figure 6.14   The variation of the B field along the axis of a solenoid. Normalized magnitude of B_z (i.e., $B(z)/B_{ctr}^\infty$ where $B_{ctr}^\infty = \mu_0 NI/l$) is plotted versus z for two different ratios of radius versus length, namely $a/l = 0.1$ and 0.5.

this duality. The basic experimental fact of magnetostatics, as elucidated[31] by Ampère, is that current-carrying wires exert forces on one another. The Biot–Savart law is one statement of this experimental fact, as discussed at the beginning of Section 6.2. In terms of Ampère's law of force, the magnetic force $\mathbf{F}_m$ felt by a current element $I d\mathbf{l}$ in the presence of a magnetic field $\mathbf{B}$ is

$$\mathbf{F}_m = I \, d\mathbf{l} \times \mathbf{B}$$

Noting that current is charge in motion, a charge q moving at a velocity $\mathbf{v}$ is equivalent to an element of current $I \, d\mathbf{l} = q\mathbf{v}$, and hence, in the presence of a magnetic field $\mathbf{B}$, would experience a force $\mathbf{F}_m$ given by

$$\boxed{\mathbf{F}_m = q\mathbf{v} \times \mathbf{B}} \qquad (6.8)$$

This force is called the *Lorentz magnetic force,* and (6.8) is sometimes[32] taken as the defining equation for $\mathbf{B}$.

   In many ways, the quantity $\mathbf{B}$ is analogous to the electric field $\mathbf{E}$, as can be seen by comparing the electric and magnetic force expressions given below:

$$\mathbf{F}_e = q\mathbf{E} = q \int_{V'} \frac{\rho(\mathbf{r}')\hat{\mathbf{R}} \, dv'}{4\pi\epsilon_0 R^2}$$

---

[31] A.-M. Ampère, On the mathematical theory of electrodynamic phenomena uniquely deduced from experiment, *Mem. Acad.,* pp. 175–388, 1825.

[32] The Lorentz force acting on a particle with charge q is the vector sum of the magnetic and electric Lorentz forces; that is, $\mathbf{F} = q\mathbf{E} + q\mathbf{v} \times \mathbf{B}$. Also see Section 6.10.1.

$$\mathbf{F}_m = q\mathbf{v} \times \mathbf{B} = q\mathbf{v} \times \int_{V'} \frac{\mathbf{J}(\mathbf{r}') \times \hat{\mathbf{R}}\, dv'}{4\pi \mu_0^{-1} R^2}$$

We can see that both $\mathbf{E}$ and $\mathbf{B}$ act on a charge q to produce force, both are related to their respective sources (ρ for $\mathbf{E}$ and $\mathbf{J}$ for $\mathbf{B}$) in a similar manner, and both are medium-dependent (through ϵ_0^{-1} and μ_0) quantities. In view of this similarity, it is unfortunate that $\mathbf{B}$ is usually referred to as the *magnetic flux density,* which conveys the incorrect impression[33] of a similarity between $\mathbf{B}$ and the electric flux density $\mathbf{D}$. In this text, we may succumb to this convention and on occasion refer to $\mathbf{B}$ as the *magnetic flux density,* although in most cases we simply refer to it as the "$\mathbf{B}$ field."

6.3 AMPÈRE'S CIRCUITAL LAW

In electrostatics, problems involving symmetries of one form or another can often be solved more easily using Gauss's law than by direct application of Coulomb's law. For static magnetic fields, the Biot–Savart law and Ampère's force law are the point relationships that are analogous to Coulomb's law, whereas *Ampère's circuital law* serves the same purpose as Gauss's law. Ampère's circuital law, hereafter referred to simply as *Ampère's law,* is a mathematical consequence[34] of the Biot–Savart law, much as Gauss's law was that of Coulomb's law. Ampère's law can simply be stated as

$$\boxed{\oint_C \mathbf{B} \cdot d\mathbf{l} = \int_S \mu_0 \mathbf{J} \cdot d\mathbf{s} = \mu_0 I} \tag{6.9}$$

where C is the contour that encloses the surface S, as shown in Figure 6.15. In words, (6.9) states that the line integral of $\mathbf{B}$ around any closed contour is equal to μ_0 times the total net current I passing through the surface S enclosed by the contour C. This law is

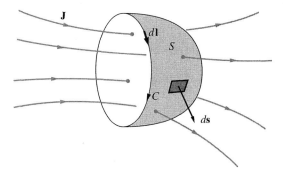

Figure 6.15 Illustration of Ampère's Law. The direction of the line integral around the contour C is related to the direction of the surface element $d\mathbf{s}$ used for the surface integral by right-hand rule. With the fingers of the right hand pointed in the direction of the line element $d\mathbf{l}$, the thumb points in the direction of the surface element vector $d\mathbf{s}$.

[33]The magnetostatic quantity that is the correct analogue of the electric flux density vector $\mathbf{D}$ is in fact a different vector $\mathbf{H}$, which we introduce later in Section 6.8. Here, it suffices to mention that $\mathbf{H}$ is a medium-independent quantity defined by the relation $\mathbf{H} \equiv \mu^{-1}\mathbf{B}$, where μ is the magnetic permeability of the medium also defined and introduced in Section 6.8. Also see Footnote 71.

[34]A formal derivation of Ampère's law from the Biot–Savart law is provided in Appendix C.

particularly useful in solving magnetostatic problems having some degree of symmetry (usually cylindrical symmetry), when a contour C over which the $\mathbf{B}$ field is constant can be identified, in which case the integration on the left-hand side of (6.9) can be readily carried out.

Example 6.10: Infinitely long cylindrical conductor. Consider an infinitely long, straight cylindrical conductor of radius a carrying a steady current I, as shown in Figure 6.16a. Assume[35] the current density to be uniform over the cross section of the conductor. Find the magnetic field both inside and outside the conductor.

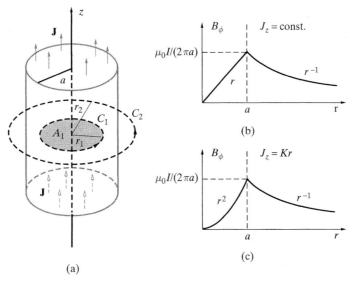

Figure 6.16   Infinitely long cylindrical conductor. (a) The cylindrical conductor carrying a current with current density $\mathbf{J}$. (b) Magnetic flux density B_ϕ versus distance r for the case of uniform current distribution. (c) B_ϕ versus r when $J_z = Kr$.

Solution: We choose to use the cylindrical coordinate system with the current I flowing in the z direction. In view of the cylindrical symmetry, it is convenient to use Ampère's law to find $\mathbf{B}$. According to the Biot–Savart law and the right-hand rule, $\mathbf{B}$ is in the $\hat{\boldsymbol{\phi}}$ direction, so that we have $\mathbf{B}(r) = B(r)\hat{\boldsymbol{\phi}}$. We take two circular contours, C_1 and C_2, respectively, with radii $r_1 < a$ and $r_2 > a$, as shown in Figure 6.16a. Applying Ampère's law along path C_1 lying inside the conductor, we have

$$\oint_{C_1} \mathbf{B} \cdot d\mathbf{l} = \mu_0 \int_{A_1} \mathbf{J} \cdot d\mathbf{s}$$

---

[35]We shall see in Chapter 8 that the current can be uniformly distributed across the cross section only for steady currents. In most applications involving time-varying fields, the current is confined mostly to a narrow region at the outer edge of the conductor.

$$\int_{\phi=0}^{2\pi} (B_\phi \hat{\phi}) \cdot (\hat{\phi} r_1 \, d\phi) = \mu_0 \int_0^{2\pi} \int_0^{r_1} \frac{I}{\pi a^2} \hat{z} \cdot \hat{z} r \, dr \, d\phi = \mu_0 I_1$$

where I_1 is the portion of the total current I through the area $A_1 = \pi r_1^2$ enclosed by contour C_1. From the above we have

$$B_\phi r_1 \int_0^{2\pi} d\phi = \mu_0 \frac{I}{\pi a^2} A_1 = \mu_0 \frac{I}{\pi a^2} \pi r_1^2 \quad \rightarrow \quad B_\phi = \frac{\mu_0 I r_1}{2\pi a^2}, \qquad r_1 \le a$$

Similarly, by applying Ampère's law along contour C_2 lying outside the conductor (i.e., $r_2 > a$), we find

$$B_\phi (2\pi r_2) = \mu_0 I \quad \rightarrow \quad B_\phi = \frac{\mu_0 I}{2\pi r_2}, \qquad r_2 > a$$

In summary, the **B** field at any position r is

$$\mathbf{B} = \begin{cases} \hat{\phi} \dfrac{\mu_0 I r}{2\pi a^2} & r \le a \\[3mm] \hat{\phi} \dfrac{\mu_0 I}{2\pi r} & r > a \end{cases}$$

which is plotted in Figure 6.16b.

Example 6.11: Nonuniform distribution of current. Consider an infinitely long, straight cylindrical conductor of radius a carrying a steady current I, as in the previous example, except that the current is distributed in such a way that the current density J is proportional to r, the distance from the axis (assume $J_z = Kr$, where K is a constant). Find the **B** field both inside and outside the conductor.

Solution: Using (5.1), the total current I flowing through the conductor can be found as

$$I = \int_0^{2\pi} \int_0^a Kr\hat{z} \cdot \hat{z} r \, dr \, d\phi \quad \rightarrow \quad K = \frac{3I}{2\pi a^3}$$

So, the current density is given by $\mathbf{J} = \hat{z} 3Ir / (2\pi a^3)$. Applying Ampère's law along path C_1 inside the conductor, we can write

$$\oint_{C_1} \mathbf{B} \cdot d\mathbf{l} = \mu_0 \int_{A_1} \mathbf{J} \cdot d\mathbf{s} \quad \rightarrow \quad \int_0^{2\pi} B_\phi \hat{\phi} \cdot (\hat{\phi} r_1 \, d\phi) = \left(\frac{3\mu_0 I}{2\pi a^3} \right) \left(\int_0^{r_1} 2\pi r^2 \, dr \right)$$

$$B_\phi (2\pi r_1) = \left(\frac{3\mu_0 I}{2\pi a^3} \right) \left(\frac{2\pi r_1^3}{3} \right) \quad \rightarrow \quad B_\phi = \frac{\mu_0 I r_1^2}{2\pi a^3}, \qquad r_1 \le a$$

Similarly, by applying Ampère's law along contour C_2 lying outside the conductor, we find

$$B_\phi (2\pi r_2) = \mu_0 I \quad \rightarrow \quad B_\phi = \frac{\mu_0 I}{2\pi r_2}, \qquad r_2 > a$$

In summary, the **B** field at any position r is

$$\mathbf{B} = \begin{cases} \hat{\phi}\dfrac{\mu_0 I r^2}{2\pi a^3} & r \leq a \\[3mm] \hat{\phi}\dfrac{\mu_0 I}{2\pi r} & r > a \end{cases}$$

Figure 6.16c shows the variation of B_ϕ as a function of r.

Example 6.12: Coaxial line. Consider an infinitely long coaxial transmission line carrying a uniformly distributed current I in the inner conductor of radius a and $-I$ in the outer conductor of inner and outer radii b and c such that $c > b > a$, whose cross-sectional view is shown in Figure 6.17a. Find the **B** field everywhere.

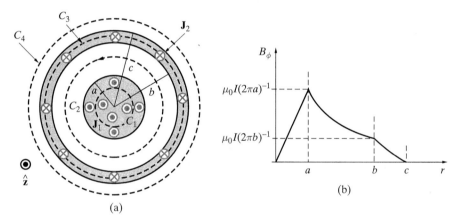

Figure 6.17   Coaxial line. (a) Cross-sectional view. (b) Variation of the **B** field as a function of r.

Solution: Following a similar approach as in Example 6.10, we can apply Ampère's law for $r_1 \leq a$ to write

$$\oint_{C_1} \mathbf{B} \cdot d\mathbf{l} = \oint_{C_1} (B_\phi \hat{\phi}) \cdot (\hat{\phi} r_1 \, d\phi) = \mu_0 \int_{A_1} \mathbf{J}_1 \cdot d\mathbf{s} = \mu_0 I_1$$

where I_1 is the portion of the current I passing through the area A_1. Noting that $\mathbf{J}_1 = \hat{z} I / (\pi a^2)$, we have

$$B_\phi(2\pi r_1) = \mu_0 \frac{I}{\pi a^2} \pi r_1^2 \quad \rightarrow \quad B_\phi = \frac{\mu_0 I r_1}{2\pi a^2} \qquad r_1 \leq a$$

which is identical to that found in Example 6.10 for $r_1 < a$. For $a < r_2 < b$, we have

$$B_\phi = \frac{\mu_0 I}{2\pi r_2} \qquad a < r_2 \leq b$$

also identical to that in Example 6.10. For $b < r_3 \leq c$, using $\mathbf{J}_2 = -\hat{\mathbf{z}} I / [\pi (c^2 - b^2)]$ we have

$$B_\phi (2\pi r_3) = \mu_0 \left[\int_0^{2\pi} \int_0^a \frac{I}{\pi a^2} r \, dr \, d\phi + \int_0^{2\pi} \int_b^{r_3} \frac{-I}{\pi (c^2 - b^2)} r \, dr \, d\phi \right]$$

$$= \mu_0 \left[I - \frac{r_3^2 - b^2}{c^2 - b^2} I \right] = \mu_0 I \frac{c^2 - r_3^2}{c^2 - b^2}$$

$$\rightarrow \quad B_\phi = \frac{\mu_0 I}{2\pi r_3} \frac{c^2 - r_3^2}{c^2 - b^2} \qquad b < r_3 \leq c$$

For $r_4 > c$, we have

$$B_\phi = \frac{\mu_0}{2\pi r_4} \left[\underbrace{\int_0^{2\pi} \int_0^a \frac{I}{\pi a^2} r \, dr \, d\phi}_{I} + \underbrace{\int_0^{2\pi} \int_b^c \frac{-I}{\pi (c^2 - b^2)} r \, dr \, d\phi}_{-I} \right] = 0$$

since the net current linking the area enclosed by contour C_4 is zero. In summary, the **B** field at any position r is given by

$$\mathbf{B} = \begin{cases} \hat{\boldsymbol{\phi}} \dfrac{\mu_0 I r}{2\pi a^2} & r \leq a \\[2ex] \hat{\boldsymbol{\phi}} \dfrac{\mu_0 I}{2\pi r} & a < r \leq b \\[2ex] \hat{\boldsymbol{\phi}} \dfrac{\mu_0 I}{2\pi r} \dfrac{c^2 - r^2}{c^2 - b^2} & b < r \leq c \\[2ex] 0 & r > c \end{cases}$$

Figure 6.17b shows the variation of B_ϕ as a function of r.

Example 6.13: Conductor with hole. A long cylindrical conductor of radius b and carrying a uniform current I in the z direction has a cylindrical hole of radius a along its entire length, as shown in Figure 6.18. The center of the hole is offset from the center of the conductor by a distance d. Find **B** inside the hole.

Solution: We can consider the region of the hole, within which the current density is zero, to result from two equal currents flowing in opposite directions. Accordingly, we can represent the problem in hand as a superposition of a cylindrical conductor of radius b carrying a current with density **J** without a hole and another of radius a with a current density of the same magnitude but flowing in the opposite direction (i.e., $-\mathbf{J}$). This representation is indicated in Figure 6.18.

Note that the current density **J** is

$$\mathbf{J} = \hat{\mathbf{z}} \frac{I}{\pi (b^2 - a^2)}$$

Applying Ampère's law, the **B** field at point $P_1 (r_1, \phi_1, z_1 = 0)$ in the hole region due to the current in the larger conductor (radius b) with no hole is

$$\mathbf{B}_1 = \hat{\boldsymbol{\phi}} \frac{\mu_0 I r_1}{2\pi (b^2 - a^2)} = \frac{\mu_0 I}{2\pi (b^2 - a^2)} (\hat{\mathbf{z}} \times \mathbf{r}_1) = \frac{\mu_0}{2} (\mathbf{J} \times \mathbf{r}_1)$$

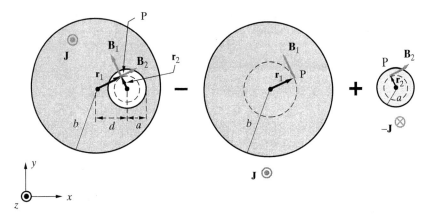

Figure 6.18 Cylindrical conductor with a hole. The conductor with a hole is viewed as a superposition of two solid cylindrical conductors of radii a and b, carrying equal but opposite current density **J**.

where we used the result derived in Example 6.10 and noted $\hat{\mathbf{z}} \times \mathbf{r}_1 = \hat{\boldsymbol{\phi}} r_1$, where $\mathbf{r}_1 = \hat{\mathbf{r}} r_1$. Similarly, the **B** field at P_1 due to the current in the smaller conductor (radius a) is

$$\mathbf{B}_2 = \frac{\mu_0 I}{2\pi(b^2 - a^2)}[(-\hat{\mathbf{z}}) \times \mathbf{r}_2] = \frac{\mu_0}{2}[(-\mathbf{J}) \times \mathbf{r}_2]$$

where $\mathbf{r}_2 = \hat{\mathbf{r}} r_2$. Therefore, the total **B** field at P_1 is

$$\mathbf{B} = \mathbf{B}_1 + \mathbf{B}_2 = \frac{\mu_0}{2}[\mathbf{J} \times (\mathbf{r}_1 - \mathbf{r}_2)] = \frac{\mu_0}{2}[\mathbf{J} \times (\hat{\mathbf{x}} d)]$$

where we have noted from Figure 6.18 that $(\mathbf{r}_1 - \mathbf{r}_2) = \hat{\mathbf{x}} d$. This result shows that the magnetic field anywhere inside the hole is constant, and its direction is at right angles to the line joining the centers of the two circles. In the case shown, with the center of the hole on the x axis, the **B** field everywhere in the hole is constant and is in the y direction. Rewriting in terms of I, we then have

$$\mathbf{B} = \hat{\mathbf{y}} \frac{\mu_0 I d}{2\pi(b^2 - a^2)}$$

We can check the above result in two limiting cases. First, if the hole is directly at the center of the conductor (i.e., $d = 0$), we have $\mathbf{B} = 0$, as expected based on Ampère's law (i.e., no current enclosed). Secondly, if the radius a of the hole approaches zero ($a \to 0$), then **B** reduces to the form obtained in Example 6.10.

Example 6.14: The infinitely long solenoid. In Section 6.2.1 we considered the magnetic field inside a tightly wound solenoid of finite length and derived an expression for the **B** field along its axis using the Biot–Savart law. In this example, we consider a very long solenoid and find the **B** field inside it using Ampère's law.

Solution: Consider the cross section of a closely wound solenoid as shown in Figure 6.19. For an infinitely long solenoid, the **B** field everywhere inside must be parallel to the axis

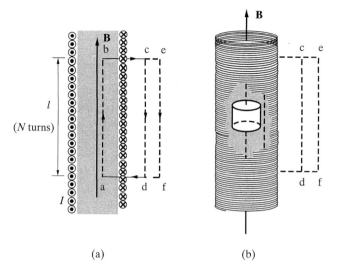

Figure 6.19   Cross section of a solenoid.

(i.e., $\mathbf{B} = \hat{\mathbf{z}}B_z$, with $B_r = 0$ and $B_\phi = 0$) of the solenoid,[36] and the field outside the solenoid must be zero.[37] Applying Ampère's law to a closed contour indicated in Figure 6.19 as abcd, selected to be of length l and containing N turns of the solenoid coil, gives

$$\oint_{abcd} \mathbf{B} \cdot d\mathbf{l} = \int_{ab} + \underbrace{\int_{bc} + \int_{cd} + \int_{da}}_{=0}$$

Note that the integral along cd is zero because $\mathbf{B} = 0$ outside the solenoid, and the integrations along bc and da are zero since $\mathbf{B} = \hat{\mathbf{z}}B_z$ is perpendicular to $d\mathbf{l}$. Thus, we have

$$\int_{ab} \mathbf{B} \cdot d\mathbf{l} = \mu_0 NI \quad \rightarrow \quad Bl = \mu_0 NI \quad \rightarrow \quad B = \frac{\mu_0 NI}{l}$$

which is identical to the result found in Section 6.2.1 for $\mathbf{B}_{ctr}$ for the case of a very long solenoid. This result indicates also that the $\mathbf{B}$ field is constant across the cross section of the solenoid; note that we did not in any way restrict the location of segment ab in Figure 6.19 to be on the axis of the solenoid.

In the context of the above formulation, it is easy to show that the z component of the $\mathbf{B}$ field at either end of a long solenoid would be half of that at its center. To see this,

---

[36]The reason for this is the fact that the integral of the $\mathbf{B}$ field over any closed surface is identically zero, as will be shown later in Section 6.7. Considering an axial cylinder of unit length inside the solenoid as shown in Figure 6.19b, the integral $\oint \mathbf{B} \cdot d\mathbf{s}$ over its surface is simply $2\pi r B_r$, since the integrals over the two end faces cancel. Thus, B_r must be zero in order for $\oint \mathbf{B} \cdot d\mathbf{s}$ to be zero. Note that $B_\phi = 0$ because all of the current flow is in the $\hat{\boldsymbol{\phi}}$ direction and we know from the Biot–Savart law that $\mathbf{B}$ must be perpendicular to the direction of current flow.

[37]If the field outside was not zero, then the line integral of $\mathbf{B}$ around the contour cdfe would not be zero (as it should be, since it does not enclose any current) unless the field were the same along cd and ef (no contribution along the ce and df paths since $B_r = 0$), which would in turn imply that a constant field exists everywhere outside the solenoid, contrary to experimental observation.

we can cut a very long solenoid into two. If the same current I is maintained in the two parts, the magnetic flux density in the newly created ends must drop to half of its original values; otherwise the field would not combine to its original value when the two ends were reconnected. Note, however, that the radial component B_r is not zero at the ends of a finite length solenoid.

Example 6.15: The toroid. If the long solenoid of Example 6.14 is bent into a circle and closed on itself, we obtain a toroidal (doughnut-shaped) coil. A toroidal coil with a rectangular cross section is shown in Figure 6.20a. Find the **B** field inside the toroid.

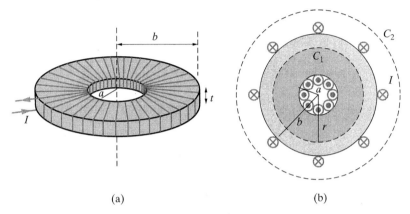

(a)   (b)

Figure 6.20   A toroid. (a) Three-dimensional view showing the windings. (b) Cross-sectional view showing two Amperian contours C_1 and C_2.

Solution: Let the toroid consist of N turns of wire carrying a current I, uniformly and tightly distributed around its circumference as shown in Figure 6.20a. Because of cylindrical symmetry, the magnetic field **B** has only a ϕ component. We apply Ampère's law to any circular contour C_1 of radius r such that $a < r < b$ lying inside the toroid, as shown. By symmetry the field **B** must be the same everywhere along such a contour. Noting that the total current crossing through the area enclosed by the contour C_1 is NI, we have

$$\oint_{C_1} \mathbf{B} \cdot d\mathbf{l} = \oint_{C_1} (B_\phi \hat{\boldsymbol{\phi}}) \cdot (\hat{\boldsymbol{\phi}} r \, d\phi) = B_\phi (2\pi r) = \mu_0 NI \quad \rightarrow \quad B_\phi = \frac{\mu_0 NI}{2\pi r}$$

For any circular path C_2 that lies outside the toroid ($r > b$), the total net current crossing through the enclosed area of C_2 is $N(I - I) = 0$, so that $\oint_{C_2} \mathbf{B} \cdot d\mathbf{l} = 0$, yielding $\mathbf{B} = 0$. Note that if we consider a contour with radius $r < a$, we can conclude by the same arguments that the magnetic field is also zero in the hole of the toroid. Thus, the magnetic field of a closely wound toroid is confined to the interior of its windings.

For a numerical example, consider the air-core toroid reactor vessel of the Tokamak Fusion Test Reactor[38] used for high-power plasma fusion experiments at Princeton University. This toroid has a mean radius of 2.2 m and consists of 20 coils, each with 44 turns per

---

[38]For a description of fusion reactors and the Tokamak facility referred to here, see Chapter 9 of F. F. Chen, *Introduction to Plasma Physics*, Plenum Press, 1974.

coil, uniformly wound around it. With each coil carrying a current of 73.3 kA, the **B** field at its mean radius can be found using the expression derived above. Namely,

$$B_\phi = \frac{(4\pi \times 10^{-7}\ \text{H-m}^{-1})(20 \times 44\ \text{turns})(73.3 \times 10^3\ \text{A})}{2\pi(2.2\ \text{m})} \simeq 5.86\ \text{T}$$

6.4 CURL OF THE MAGNETIC FIELD: DIFFERENTIAL FORM OF AMPÈRE'S LAW

In electrostatics, we discussed the concept of the divergence of a vector field, and used it to express the differential form of Gauss's law as $\nabla \cdot \mathbf{D} = \rho$. A second very important property of a vector field is its circulation. We can observe circulation as we watch water drain out of a bathtub or sink. For a perfectly symmetrical circulating fluid with angular velocity ω, a measure of the circulation at any radius r may be defined as the product of angular velocity and circumference [i.e., $\omega(2\pi r)$]. We can extend this concept of circulation to any vector field **G** by defining total or net circulation about any arbitrary closed path C as

$$\text{circulation} \equiv \oint_C \mathbf{G} \cdot d\mathbf{l}$$

which is called the *circulation* of **G** around C. Since any closed loop C encloses a surface S, the net circulation per unit surface area is

$$\frac{\oint_C \mathbf{G} \cdot d\mathbf{l}}{S}$$

Since a given vector field **G** may have different amounts of circulation at different points, we now consider a differential surface element $\Delta\mathbf{s} = \hat{\mathbf{s}}\Delta s$ located at some point in space, and surrounded by a contour C. We expect the circulation per unit area of the vector field about C to depend on the orientation $\hat{\mathbf{s}}$ of the surface enclosed by C. The *curl* of a vector field **G** at the location of this surface element is defined as an axial vector (i.e., perpendicular to Δs) whose magnitude is the maximum (as the direction of the surface element $\Delta\mathbf{s} = \hat{\mathbf{s}}\Delta s$ is varied) net circulation of vector **G** per unit area as the area of the surface element approaches zero. In other words,

$$\text{curl } \mathbf{G} \equiv \left[\lim_{\Delta s \to 0} \frac{\oint_C \mathbf{G} \cdot d\mathbf{l}}{\Delta s} \right]_{\text{max}} \hat{\mathbf{s}} \qquad (6.10)$$

Note that curl is clearly a directional quantity; by convention, the direction of curl is chosen to be the direction of the surface element $\Delta\mathbf{s} = \hat{\mathbf{s}}\Delta s$ that gives the maximum value for the magnitude of the net circulation per unit area in the defining expression (6.10). Curl **G** is thus a vector; for example, the x component of the vector (curl **G**) represents the line integral of **G** per unit area along an infinitesimally small (i.e., $\Delta s \to 0$) closed path lying in the y-z plane.

For the **B** field, we know from Ampère's law that the circulation of **B** along any closed contour C (i.e., line integral of **B** along C) is equal to the total current passing through the surface area S enclosed by C times μ_0. For a differential surface element, we can assume that the current density is constant at all points over the surface, so the total current is $|\mathbf{J}|\Delta s$, if the surface is chosen to be orthogonal to the direction of the current flow so that the net circulation is maximized. Accordingly, we have

$$\text{curl } \mathbf{B} \equiv \left[\lim_{\Delta s \to 0} \frac{\oint_C \mathbf{B} \cdot d\mathbf{l}}{\Delta s} \right]_{\max} \hat{\mathbf{s}} = \mu_0 \mathbf{J}$$

6.4.1 Curl in the Rectangular Coordinate System

We now utilize the definition of curl to obtain a convenient expression in terms of partial derivatives. For this, we simply need to evaluate curl $\mathbf{B}$ at a general point in a given coordinate system. With reference to Figure 6.21, we can evaluate the component of curl $\mathbf{B}$ in the x direction by conducting a line integral along the path abcda, which, as shown, lies entirely in the y-z plane. Note that this is an infinitesimally small path with dimensions Δy and Δz, so the magnitude of $\mathbf{B}$ at different sides can be simply related to the value of the field at the center point x_0, y_0, z_0. From Ampère's law, we have

$$\oint_{\text{abcda}} \mathbf{B} \cdot d\mathbf{l} = B_{1y}\Delta y + B_{2z}\Delta z - B_{3y}\Delta y - B_{4z}\Delta z = \mu_0 J_x \Delta y \Delta z$$

$$\underbrace{(B_{1y} - B_{3y})}_{-(\partial B_y/\partial z)\Delta z}\Delta y + \underbrace{(B_{2z} - B_{4z})}_{(\partial B_z/\partial y)\Delta y}\Delta z = \mu_0 J_x \Delta y \Delta z$$

$$\left[\frac{\partial B_z}{\partial y} - \frac{\partial B_y}{\partial z}\right]\Delta y \Delta z = \mu_0 J_x \Delta y \Delta z$$

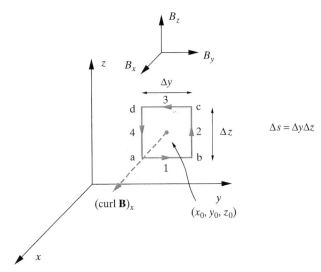

Figure 6.21   Evaluation of the x component of curl **B**.

from which

$$(\text{curl }\mathbf{B})_x = \lim_{\Delta s \to 0} \frac{\oint_{abcda} \mathbf{B} \cdot d\mathbf{l}}{\Delta s} = \frac{\partial B_z}{\partial y} - \frac{\partial B_y}{\partial z} = \mu_0 J_x$$

where the partial derivatives are to be evaluated at the center point (x_0, y_0, z_0). By taking line integrals along other contours lying in the x-z and x-y planes, we can show that

$$\text{curl }\mathbf{B} = \hat{\mathbf{x}}\left(\frac{\partial B_z}{\partial y} - \frac{\partial B_y}{\partial z}\right) + \hat{\mathbf{y}}\left(\frac{\partial B_x}{\partial z} - \frac{\partial B_z}{\partial x}\right) + \hat{\mathbf{z}}\left(\frac{\partial B_y}{\partial x} - \frac{\partial B_x}{\partial y}\right)$$

$$= \mu_0(\hat{\mathbf{x}}J_x + \hat{\mathbf{y}}J_y + \hat{\mathbf{z}}J_z) = \mu_0\mathbf{J}$$

in rectangular coordinates. Using the *del* operator we can express the curl of $\mathbf{B}$ as

$$\boxed{\text{curl }\mathbf{B} = \nabla \times \mathbf{B} = \mu_0\mathbf{J}} \tag{6.11}$$

which is the differential form of Ampère's law. A convenient method for remembering the expression for $\nabla \times \mathbf{B}$ in a rectangular coordinate system is to use the determinant form

$$\nabla \times \mathbf{B} = \begin{vmatrix} \hat{\mathbf{x}} & \hat{\mathbf{y}} & \hat{\mathbf{z}} \\ \dfrac{\partial}{\partial x} & \dfrac{\partial}{\partial y} & \dfrac{\partial}{\partial z} \\ B_x & B_y & B_z \end{vmatrix}$$

Example 6.16: Circulation in a fluid. Further physical insight into the nature of the curl operation can be gained by considering a simple example. Consider the vector field representing the velocity on a water surface. Suppose this velocity is given by

$$\mathbf{v} = \hat{\mathbf{y}}(v_0 x)$$

as shown in Figure 6.22. Find the curl of this vector field.

Solution: Consider the closed contour marked abcda in Figure 6.22. We can easily see that there is a net circulation $\oint_{abcda} \mathbf{v} \cdot d\mathbf{l}$ around this path, since the velocity field is stronger on side ab than it is on side cd, and there is no contribution to the line integral from the sides bc or ad. The line integral of the velocity vector around the contour abcda is thus given by

$$\oint_{abcda} \mathbf{v} \cdot d\mathbf{l} = v_0\left(x + \frac{\Delta x}{2}\right)\Delta y - v_0\left(x - \frac{\Delta x}{2}\right)\Delta y = v_0\Delta x \Delta y$$

since the velocity $\mathbf{v}$ does not vary with y. Since $\Delta s = \Delta x \Delta y$, and $\hat{\mathbf{s}} = \hat{\mathbf{z}}$, we have

$$\text{curl }\mathbf{v} = \left[\lim_{\Delta s \to 0} \frac{\oint_C \mathbf{v} \cdot d\mathbf{l}}{\Delta s}\right]\hat{\mathbf{s}} = \hat{\mathbf{z}}v_0 \qquad \text{or} \qquad \nabla \times \mathbf{v} = \hat{\mathbf{z}}v_0$$

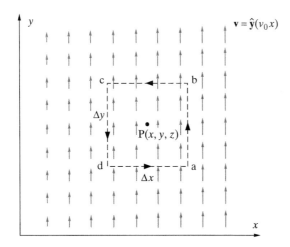

Figure 6.22   Velocity field on a water surface.

which is the same result that one would obtain by using the differential expression, namely,

$$\nabla \times \mathbf{v} = \hat{\mathbf{z}}\left(\frac{\partial v_y}{\partial x} - \frac{\partial v_x}{\partial y}\right) = \hat{\mathbf{z}}(v_0 + 0) = \hat{\mathbf{z}}v_0$$

We note that the velocity flow itself is entirely in the y direction and does not bend or curve around in any way. However, if we placed an infinitesimal wooden chip with its length in the x direction at the point on the surface where we have evaluated the curl, the chip would tend to rotate, since the speed of the flow is greater on side ab than on side cd. As viewed from above, the wooden chip rotates in the counterclockwise direction, which, according to our definitions is perpendicular to the plane of circulation and is in the direction defined by the fingers of the right hand when the thumb points in the direction of the curl vector (i.e., the right-hand rule).

6.4.2 Curl in Other Coordinate Systems

Although we use the notation $\nabla \times \mathbf{A}$ to indicate the curl of a vector $\mathbf{A}$, expression of curl as a cross product with a vector *del* operator (having the form $\nabla = \hat{\mathbf{x}}(\partial/\partial x) + \hat{\mathbf{y}}(\partial/\partial y) + \hat{\mathbf{z}}(\partial/\partial z)$ for rectangular coordinates) is useful only in the rectangular coordinate system. In other coordinate systems, we still denote the curl of $\mathbf{A}$ by $\nabla \times \mathbf{A}$, but note that the specific derivative expressions for each component of curl will need to be derived from the physical definition of curl as given in (6.10), using a differential surface element appropriate for that particular coordinate system.

To derive the differential expression for curl in the spherical coordinate system, we consider the cuboid volume element shown in Figure 6.23b. To determine the r component of the curl, we consider the curvilinear contour ABCDA. Note that this contour is perpendicular to $\hat{\mathbf{r}}$, and the sense of rotation is related to $\hat{\mathbf{r}}$ by the right-hand rule if we go around the contour in the A $\rightarrow$ B $\rightarrow$ C $\rightarrow$ D $\rightarrow$ A direction. The lengths

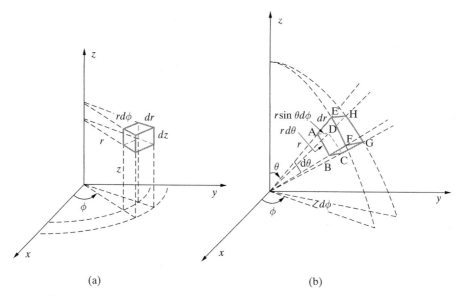

Figure 6.23 Derivation of curl in other coordinate systems. (a) Cylindrical coordinates. (b) Spherical coordinates.

of the sides are $AB = r \, d\theta$, $CD = r \, d\theta$, $BC = r \sin(\theta + d\theta) \, d\phi$, and $DA = r \sin \theta \, d\phi$. The line integral is given by

$$\oint_{ABCDA} \mathbf{A} \cdot d\mathbf{l} = A_\theta r \, d\theta + \left(A_\phi + \frac{\partial A_\phi}{\partial \theta} \, d\theta \right) r \sin(\theta + d\theta) \, d\phi$$

$$- \left(A_\theta + \frac{\partial A_\theta}{\partial \phi} \, d\phi \right) r \, d\theta - A_\phi r \sin \theta \, d\phi$$

$$= -\frac{\partial A_\theta}{\partial \phi} r \, d\theta \, d\phi + \frac{\partial}{\partial \theta} (A_\phi \sin \theta) r \, d\theta \, d\phi$$

where we have retained small quantities up to second order. The area of the contour ABCDA is $\Delta s = r^2 \sin \theta \, d\theta \, d\phi$, so

$$[\nabla \times \mathbf{A}]_r = \frac{\oint_{ABCDA} \mathbf{A} \cdot d\mathbf{l}}{\Delta s} = \frac{1}{r \sin \theta} \left[\frac{\partial}{\partial \theta} (A_\theta \sin \theta) - \frac{\partial A_\phi}{\partial \phi} \right]$$

The other two components can be obtained by considering the other sides of the cuboid in Figure 6.23b, namely, the contour ABFEA for the ϕ component and the contour BCGFB for the θ component. The complete expression for the curl in spherical coordinates is

$$\text{curl } \mathbf{A} = \nabla \times \mathbf{A} = \hat{\mathbf{r}} \frac{1}{r \sin \theta} \left[\frac{\partial}{\partial \theta} (A_\phi \sin \theta) - \frac{\partial A_\theta}{\partial \phi} \right]$$

$$+ \hat{\boldsymbol{\theta}} \frac{1}{r} \left[\frac{1}{\sin \theta} \frac{\partial A_r}{\partial \phi} - \frac{\partial}{\partial r} (r A_\phi) \right] + \hat{\boldsymbol{\phi}} \frac{1}{r} \left[\frac{\partial}{\partial r} (r A_\theta) - \frac{\partial A_r}{\partial \theta} \right]$$

Similar reasoning in a cylindrical coordinate system using a volume element such as that shown in Figure 6.23a leads to the curl expression in cylindrical coordinates:

$$\text{curl } \mathbf{A} = \nabla \times \mathbf{A} = \hat{\mathbf{r}} \left[\frac{1}{r} \frac{\partial A_z}{\partial \phi} - \frac{\partial A_\phi}{\partial z} \right] + \hat{\boldsymbol{\phi}} \left[\frac{\partial A_r}{\partial z} - \frac{\partial A_z}{\partial r} \right] + \hat{\mathbf{z}} \frac{1}{r} \left[\frac{\partial}{\partial r} (rA_\phi) - \frac{\partial A_r}{\partial \phi} \right]$$

Example 6.17: Curl of the B field due to an infinitely long cylindrical conductor. In Example 6.10, the **B** field of an infinitely long straight cylindrical current-carrying conductor of radius a was found to be

$$\mathbf{B} = \begin{cases} \hat{\boldsymbol{\phi}} \dfrac{\mu_0 I r}{2\pi a^2} & r \le a \\[2ex] \hat{\boldsymbol{\phi}} \dfrac{\mu_0 I}{2\pi r} & r > a \end{cases}$$

Find $\nabla \times \mathbf{B}$.

Solution: Since **B** is expressed in cylindrical coordinates, we use the cylindrical coordinate expression for curl. Note that the only nonzero component of **B** is B_ϕ, which varies only with r. Thus, the curl of **B** is

$$\nabla \times \mathbf{B} = \hat{\mathbf{z}} \frac{1}{r} \frac{\partial}{\partial r} (rB_\phi) = \hat{\mathbf{z}} \frac{1}{r} \left[B_\phi + r \frac{\partial B_\phi}{\partial r} \right]$$

so that for $r \le a$, we have

$$\nabla \times \mathbf{B} = \hat{\mathbf{z}} \frac{1}{r} \left[\frac{\mu_0 I r}{2\pi a^2} + r \frac{\mu_0 I}{2\pi a^2} \right] = \hat{\mathbf{z}} \frac{\mu_0 I}{\pi a^2} = \mu_0 \mathbf{J}$$

where **J** is the current density vector in the conductor given by $\mathbf{J} = \hat{\mathbf{z}} I / (\pi a^2)$. Similarly, for $r > a$, we have

$$\nabla \times \mathbf{B} = \hat{\mathbf{z}} \frac{1}{r} \left[\frac{\mu_0 I}{2\pi r} - r \frac{\mu_0 I}{2\pi r^2} \right] = 0$$

as expected, since $\mathbf{J} = 0$ for $r > a$.

Example 6.18: Curl of the electrostatic dipole field. As an example of a field that has both r and θ components, consider the electric field of the electric dipole as derived in Section 4.4.3. For a dipole consisting of two opposite charges $\pm Q$ separated by a distance d, the electric field at faraway points ($r \gg d$) was shown to be (equation (4.28))

$$\mathbf{E} = \frac{Qd}{4\pi \epsilon_0 r^3} [\hat{\mathbf{r}} 2 \cos \theta + \hat{\boldsymbol{\theta}} \sin \theta]$$

Find the curl of the electric field at faraway points.

Solution: The curl of **E** in spherical coordinates is given by

$$\nabla \times \mathbf{E} = \hat{\boldsymbol{\phi}} \frac{1}{r} \left[\frac{\partial}{\partial r} (rE_\theta) - \frac{\partial E_r}{\partial \theta} \right] = \hat{\boldsymbol{\phi}} \frac{1}{r} \left[-\frac{2 \sin \theta}{r^3} + \frac{2 \sin \theta}{r^3} \right] = 0$$

Example 6.19: Curl-free fields. Show that the following vector fields, shown in Figure 6.24, are curl-free: (a) $\mathbf{A}_1 = -\hat{\mathbf{x}}x + \hat{\mathbf{y}}y$. (b) $\mathbf{A}_2 = \hat{\mathbf{x}}\sin x - \hat{\mathbf{y}}\cos y$. (c) $\mathbf{A}_3 = \hat{\mathbf{x}}\sin x \cos y + \hat{\mathbf{y}}\cos x \sin y$. (d) $\mathbf{A}_4 = 1\hat{\mathbf{r}} = \hat{\mathbf{x}}(x/\sqrt{x^2+y^2}) + \hat{\mathbf{y}}(y/\sqrt{x^2+y^2})$. (e) $\mathbf{A}_5 = \hat{\mathbf{r}}2\phi r + \hat{\boldsymbol{\phi}}r$. (f) $\mathbf{A}_6 = \hat{\boldsymbol{\phi}}r^{-1}$.

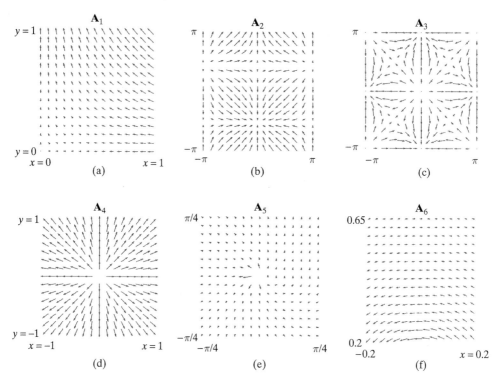

Figure 6.24   Examples of curl-free vector fields. For simplicity of presentation, two-dimensional vector fields are considered, with $A_z = 0$ and with no variation of any component with z. (a) $\mathbf{A}_1 = -\hat{\mathbf{x}}x + \hat{\mathbf{y}}y$. (b) $\mathbf{A}_2 = \hat{\mathbf{x}}\sin x - \hat{\mathbf{y}}\cos y$. (c) $\mathbf{A}_3 = \hat{\mathbf{x}}\sin x \cos y + \hat{\mathbf{y}}\cos x \sin y$. (d) $\mathbf{A}_4 = 1\hat{\mathbf{r}} = \hat{\mathbf{x}}(x/\sqrt{x^2+y^2}) + \hat{\mathbf{y}}(y/\sqrt{x^2+y^2})$. (e) $\mathbf{A}_5 = \hat{\mathbf{r}}2\phi r + \hat{\boldsymbol{\phi}}r$. (f) $\mathbf{A}_6 = \hat{\boldsymbol{\phi}}r^{-1}$.

Solution: Since these vector fields are all two dimensional, they can have curl only in the z direction, given by

$$[\nabla \times \mathbf{A}]_z = \left(\frac{\partial A_y}{\partial x} - \frac{\partial A_x}{\partial y} \right)$$

in rectangular coordinates, and

$$[\nabla \times \mathbf{A}]_z = \frac{1}{r} \left[\frac{\partial (rA_\phi)}{\partial r} - \frac{\partial A_r}{\partial \phi} \right]$$

in cylindrical coordinates. We thus have

$$\text{(a)}\quad [\nabla \times \mathbf{A}_1]_z = \frac{\partial (y)}{\partial x} - \frac{\partial (-x)}{\partial y} = 0 + 0 = 0$$

(b) $[\nabla \times \mathbf{A}_2]_z = \dfrac{\partial(-\cos y)}{\partial x} - \dfrac{\partial(\sin x)}{\partial y} = 0 + 0 = 0$

(c) $[\nabla \times \mathbf{A}_3]_z = -\sin x \sin y - (-\sin x \sin y) = 0$

(d) $[\nabla \times \mathbf{A}_4]_z = -\dfrac{xy}{(x^2 + y^2)^{3/2}} - \left(-\dfrac{xy}{(x^2 + y^2)^{3/2}}\right) = 0$

(e) $[\nabla \times \mathbf{A}_5]_z = \dfrac{1}{r}\left[\dfrac{\partial(r^2)}{\partial r} - \dfrac{\partial(2\phi r)}{\partial \phi}\right] = \dfrac{1}{r}[2r - 2r] = 0$

(f) $[\nabla \times \mathbf{A}_6]_z = \dfrac{1}{r}\left[\dfrac{\partial(rr^{-1})}{\partial r} - 0\right] = 0$

6.4.3 Skilling's Paddle Wheel

The term "curl" suggests an association with motion in curved lines. That such motion is not necessary in order to have nonzero curl was illustrated well in Example 6.16, where we found that straight-line motion may also have nonzero curl. If the velocity field considered there represents the water flow in a canal, every particle of water may indeed move in a straight line, but nevertheless there is curl. One of the best ways of thinking about the curl of a vector field is to visualize (see Figure 6.25) a small paddle wheel[39] to be dipped into the flow field. The paddle wheel rotates only if the curl of the vector field is nonzero; the speed with which it would rotate is a measure of the magnitude of curl, while the direction it rotates as viewed from above (clockwise or counterclockwise) is an indication of the direction (down or up, respectively) of the curl vector. Another way to relate the direction of rotation to the curl vector is to note that if the paddle wheel turned a right-hand screw, it would drive the screw in the direction of the curl vector, as shown in Figure 6.25a.

Applying the paddle wheel concept to a case of straight-line flow as illustrated in Figure 6.25b (similar to that in Example 6.16), we clearly see that the flow indeed has nonzero curl. The paddle wheel turns in the clockwise direction, since the stream is more rapid on its upper blades than on its lower ones. Figure 6.25c shows water flow in another canal, in which the flow bends around a corner but is nevertheless curl-free. It is possible for water flow to turn around the corner with zero curl, provided that it flows faster along the inner margin of the channel by just the right amount. An enlarged view of the paddle wheel at the corner is shown in Figure 6.25c, with arrows indicating the force of water on each of its blades. Because of the curvature of the flow lines, more than half of its blades are driven clockwise. However, the velocity of water is greatest on the inner side, and, although fewer blades are driven counterclockwise, they are acted upon more forcefully. It is thus conceivable that the curvature and the variation of velocity be so related that the wheel does not rotate. Curved motion with zero curl is in fact characteristic of a truly

[39]This extremely useful concept is now widely used in many textbooks but was first introduced by H. H. Skilling, *Fundamentals of Electric Waves,* John Wiley & Sons, Inc., New York, 1942. Skilling's paddle wheel is mounted on frictionless bearings so that it can freely turn when dipped into the "flow"; it is also "small" so that it does not interfere with the flow.

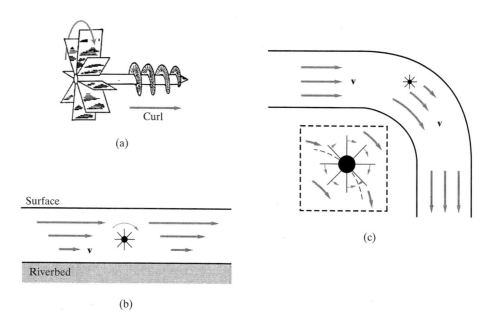

Figure 6.25 Skilling's paddle wheel. (a) If the paddle wheel turned a screw, its rotation would drive it in the direction of the curl. (b) Straight-line flow is not necessarily curl-free. (c) Flow fields that bend and go around corners can nevertheless be curl-free. (Figure adapted from H. H. Skilling, *Fundamentals of Electric Waves,* 2nd ed., John Wiley & Sons, Inc., New York, 1948.) H. H. Skilling was a pioneering electrical engineering educator and a prolific writer of many outstanding textbooks on electrical circuits and networks, electromechanics, transmission lines, and electromagnetics. He was the Chairman of the Electrical Engineering Department at Stanford University during 1941–1964.

frictionless fluid. In practice, it is often desirable to have a surface so that air or water can flow with minimum curl, since motion with nonzero curl develops eddies that waste energy.

Figure 6.24 shows some actual vector fields that exhibit flow lines that bend but are nevertheless curl-free. Two clear examples are $\mathbf{A}_1$ and $\mathbf{A}_6$. Since all of the vector fields shown in Figure 6.24 are curl-free, the reader is encouraged to visualize Skilling's paddle wheel placed in these fields and to consider whether it rotates.

Example 6.20: Vector fields with nonzero curl. Find the curl of the following vector fields: (a) $\mathbf{A}_7 = \hat{\mathbf{x}}\frac{x^2}{2} - \hat{\mathbf{y}}xy$. (b) $\mathbf{A}_8 = -\hat{\mathbf{x}}\sin y + \hat{\mathbf{y}}\cos x$. (c) $\mathbf{A}_9 = \hat{\mathbf{x}}\sin x \cos y - \hat{\mathbf{y}}\cos x \sin y$. (d) $\mathbf{A}_{10} = \hat{\mathbf{x}}\cos(xy) + \hat{\mathbf{y}}\sin(xy)$.

Solution:

$$\text{(a)} \quad [\nabla \times \mathbf{A}_7]_z = \hat{\mathbf{z}}\left[\frac{\partial(-xy)}{\partial x} - \frac{\partial(x^2/2)}{\partial y}\right] = -\hat{\mathbf{z}}y$$

(b) $[\nabla \times \mathbf{A}_8]_z = \hat{\mathbf{z}}\left[\dfrac{\partial(\cos x)}{\partial x} - \dfrac{\partial(-\sin y)}{\partial y}\right] = \hat{\mathbf{z}}(-\sin x + \cos y)$

(c) $[\nabla \times \mathbf{A}_9]_z = \hat{\mathbf{z}}\left[\dfrac{\partial(-\cos x \sin y)}{\partial x} - \dfrac{\partial(\sin x \cos y)}{\partial y}\right] = \hat{\mathbf{z}}(2\sin x \sin y)$

(d) $[\nabla \times \mathbf{A}_{10}]_z = \hat{\mathbf{z}}\left[\dfrac{\partial[\sin(xy)]}{\partial x} - \dfrac{\partial[\cos(xy)]}{\partial y}\right] = \hat{\mathbf{z}}[y\cos(xy) + x\sin(xy)]$

Both the vector fields and their curls are plotted in Figures 6.26 and 6.27. Note that the curl of all of the vectors is exclusively in the z direction. For the curl of these vectors to have an x or y component, either their z component must vary with y or x, respectively, or the y or x components must vary with z. Since $A_z = 0$ for all vectors, and since A_x and A_y are independent of z, their curl does not have x or y components. The reader is encouraged to visualize Skilling's paddle wheel placed in these fields. The places where the magnitude of the curl peaks are where the paddle wheel would rotate the fastest. Note that $\mathbf{A}_7$ is similar in appearance to the curl-free $\mathbf{A}_1$ and represents a flow that turns around a corner much like that shown in Figure 6.25c. However, the water velocity (i.e., $\mathbf{A}_7$) is apparently not larger on the inner side of the curve by the right amount, leading to nonzero curl.

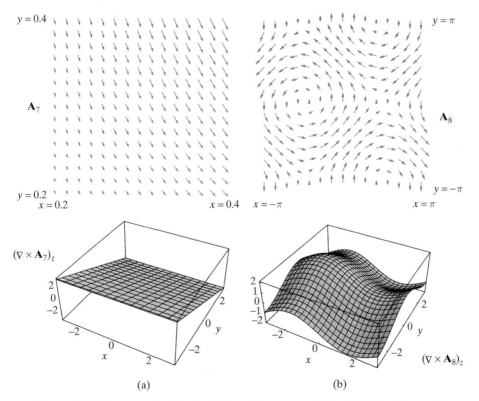

Figure 6.26 Two examples of vector fields with nonzero curl. For simplicity of presentation, two-dimensional vector fields are considered, with $A_z = 0$ and with no variation of any component with z. (a) $\mathbf{A}_7 = \hat{\mathbf{x}}(x^2/2) - \hat{\mathbf{y}}xy$. (b) $\mathbf{A}_8 = -\hat{\mathbf{x}}\sin y + \hat{\mathbf{y}}\cos x$.

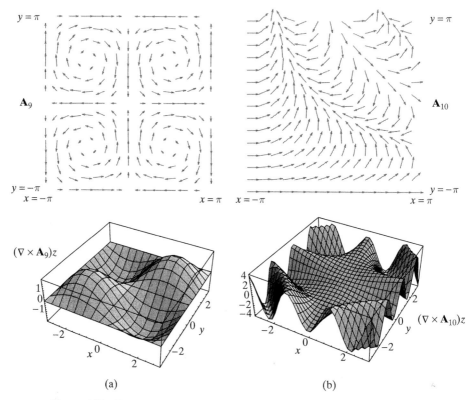

Figure 6.27 Two examples of vector fields with nonzero curl. For simplicity of presentation, two-dimensional vector fields are considered, with $A_z = 0$ and with no variation of any component with z. (a) $\mathbf{A}_9 = \hat{\mathbf{x}} \sin x \cos y - \hat{\mathbf{y}} \cos x \sin y$. (b) $\mathbf{A}_{10} = \hat{\mathbf{x}} \cos(xy) + \hat{\mathbf{y}} \sin(xy)$.

6.4.4 Stokes's Theorem

Now that we have discussed circulation and curl of a vector field, we are ready to cite an important theorem widely used in electromagnetics and other fields. This theorem, known as Stokes's theorem, relates the line integral of a vector around a closed contour to the integral of its curl over the surface enclosed by the contour. For the magnetostatic field, we can combine the integral and differential forms of Ampère's law to find

$$\left. \begin{array}{c} \oint_C \mathbf{B} \cdot d\mathbf{l} = \int_S \mu_0 \mathbf{J} \cdot d\mathbf{s} \\ \nabla \times \mathbf{B} = \mu_0 \mathbf{J} \end{array} \right\} \quad \longrightarrow \quad \boxed{\oint_C \mathbf{B} \cdot d\mathbf{l} = \int_S \nabla \times \mathbf{B} \cdot d\mathbf{s}}$$

where C is the contour that encloses the surface S. This integral equation is entirely consistent with the interpretation of the curl of a vector field as the circulation per unit surface area at a given point. Since the curl is the net circulation per unit surface area, it

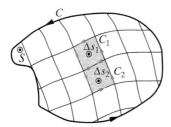

Figure 6.28 Stokes's theorem. The sum of the closed line integrals (i.e., circulations) about the perimeter of every small area element is the same as the closed line integral along the perimeter of S enclosed by C due to the cancellations at the edges of the interior paths.

is to be expected that the total circulation around the boundary of the surface (i.e., along the contour C) can be obtained by integrating the curl over the enclosed surface. This relationship between the curl of a vector field and its line integral is true in general for *any* vector field. In other words, for any vector field **G** we have

$$\oint_C \mathbf{G} \cdot d\mathbf{l} = \int_S (\nabla \times \mathbf{G}) \cdot d\mathbf{s} \qquad (6.12)$$

It is quite easy to see that (6.12) is true by considering a general surface as shown in Figure 6.28. Taking the surface to be planar for simplicity of discussion, we can envision it to be subdivided into differential surface elements Δs_i bounded by infinitesimal contours C_i as shown. We have

$$\oint_C \mathbf{G} \cdot d\mathbf{l} = \sum_i \int_{C_i} \mathbf{G} \cdot d\mathbf{l}$$

since the terms arising from the common boundaries between any two elements cancel out under the summation. The component of curl **G** normal to one of the surface elements is (curl **G**) $\cdot \hat{\mathbf{s}}_i$, where $\hat{\mathbf{s}}_i$ is the unit vector in the direction of the surface element Δs_i. From the definition of curl as given in (6.10) we have

$$(\nabla \times \mathbf{G}) \cdot \hat{\mathbf{s}}_i = \left[\lim_{\Delta s_i \to 0} \frac{\oint_{C_i} \mathbf{G} \cdot d\mathbf{l}}{\Delta s_i} \right]_{\max}$$

where $\hat{\mathbf{s}}_i$ is the unit vector in the direction of $\Delta \mathbf{s}_i$. Thus we may write

$$\oint_C \mathbf{G} \cdot d\mathbf{l} = \lim_{\Delta s_i \to 0} \sum_i [(\nabla \times \mathbf{G}) \cdot \hat{\mathbf{s}}_i] \Delta s_i$$

In the limit, the summation becomes an integral, $\Delta \mathbf{s}_i$ becomes $d\mathbf{s}$, and we have

$$\oint_C \mathbf{G} \cdot d\mathbf{l} = \int_S (\nabla \times \mathbf{G}) \cdot d\mathbf{s}$$

thus proving the theorem.

One of the important consequences of Stokes's theorem is that the curl of a conservative vector field is zero. As was discussed in Chapter 4, a vector field is said to be *conservative* when $\oint_C \mathbf{G} \cdot d\mathbf{l} = 0$, a terminology that stems from an analogy with force fields. Let **G** be the force exerted on a particle at a given point. If the vector field does not include any friction or dissipation, then no net energy is required to move the particle around a closed path, so the energy is conserved. When $\oint_C \mathbf{G} \cdot d\mathbf{l} = 0$ for any arbitrary contour C, Stokes's theorem requires that $\nabla \times \mathbf{G} = 0$. For the electrostatic field, we then have

$$\oint_C \mathbf{E} \cdot d\mathbf{l} = 0 \quad \text{is equivalent to} \quad \nabla \times \mathbf{E} = 0$$

In other words, the electrostatic field is curl-free, as was indicated in passing in Section 4.11.

Example 6.21: Water drain in a sink. Consider water flow going down the drain in a sink, represented by a simplified vector field given by $\mathbf{v} = \hat{\boldsymbol{\phi}}\omega_0 r$, where ω_0 is a constant and r is the distance from the z axis (the axis of rotation) as shown in Figure 6.29. Verify Stokes's theorem using a flat circular (radius a) surface, shown as shaded in Figure 6.29, bounded by the contour marked C.

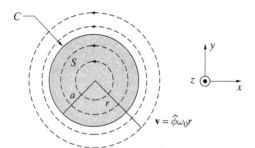

Figure 6.29   Velocity field of water drain.

Solution: First, we take the line integral of the velocity vector **v** around the $r = a$ contour as

$$\oint_C \mathbf{v} \cdot d\mathbf{l} = \int_0^{2\pi} \omega_0 a \hat{\boldsymbol{\phi}} \cdot \hat{\boldsymbol{\phi}} a \, d\phi = \omega_0 a^2 \int_0^{2\pi} d\phi = 2\omega_0 a^2 \pi$$

Second, we take the curl of the velocity field as

$$\nabla \times \mathbf{v} = \hat{\mathbf{z}} \frac{1}{r} \frac{\partial}{\partial r}(rv_\phi) = \hat{\mathbf{z}} \frac{1}{r} \frac{\partial}{\partial r}(\omega_0 r^2) = \hat{\mathbf{z}} 2\omega_0$$

which we use to take the surface integral on the circular surface enclosed by $r = a$ as

$$\int_S (\nabla \times \mathbf{v}) \cdot d\mathbf{s} = \int_0^a \int_0^{2\pi} (2\omega_0 \hat{\mathbf{z}}) \cdot (\hat{\mathbf{z}} r \, dr \, d\phi) = 4\omega_0 \pi \int_0^a r \, dr = 2\omega_0 a^2 \pi$$

which is the same result obtained from the line integral, as expected on the basis of Stokes's theorem.

6.5 VECTOR MAGNETIC POTENTIAL

We have seen the general form of the Biot–Savart law to be

$$\mathbf{B}(\mathbf{r}) = \frac{\mu_0}{4\pi} \int_{V'} \frac{\mathbf{J}(\mathbf{r}') \times \hat{\mathbf{R}}}{R^2} \, dv' \tag{6.6}$$

where $\mathbf{R} = (\mathbf{r} - \mathbf{r}') = \hat{\mathbf{R}} R$ and $R = |\mathbf{r} - \mathbf{r}'|$. Note that $\hat{\mathbf{R}}$ is the unit vector directed *from* the source point *to* the observation point. We can manipulate (6.6) by noting[40] that $\hat{\mathbf{R}}/R^2 = -\nabla(1/R)$, in which case the integrand becomes $-\mathbf{J} \times \nabla(1/R)$. We then note that the *del* operator ∇ operates on the variables x, y, z and thus does not affect quantities such as $\mathbf{J}(\mathbf{r}')$ that are functions only of the source coordinates x', y', z'. Thus, we can take $\mathbf{J}(\mathbf{r}')$ inside the ∇ operator without affecting the integral as

$$\mathbf{B}(\mathbf{r}) = \frac{\mu_0}{4\pi} \int_{V'} \nabla \times \frac{\mathbf{J}}{R} \, dv' = \nabla \times \frac{\mu_0}{4\pi} \int_{V'} \frac{\mathbf{J}(\mathbf{r}')}{R} \, dv' \tag{6.13}$$

where we have interchanged the $\nabla \times$ and the integration operations, since the former operates on the coordinates x, y, z, whereas the integration is carried out over the source coordinates x', y', z'.

It appears from (6.13) that we can express the magnetic field $\mathbf{B}(\mathbf{r})$ as the curl of another vector $\mathbf{A}$:

$$\mathbf{B} = \nabla \times \mathbf{A} \tag{6.14}$$

where the vector $\mathbf{A}$ is given by

$$\mathbf{A}(\mathbf{r}) = \frac{\mu_0}{4\pi} \int_{V'} \frac{\mathbf{J}(\mathbf{r}')}{R} \, dv' \tag{6.15}$$

The vector $\mathbf{A}$ is referred to as the *vector magnetic potential* and is in some ways analogous to the scalar electric potential Φ. In some cases, it is easier to evaluate $\mathbf{A}$ from (6.15) and then find $\mathbf{B}$ from $\mathbf{B} = \nabla \times \mathbf{A}$, rather than to evaluate $\mathbf{B}$ directly from (6.6). The analogy between the scalar potential Φ and $\mathbf{A}$ can be seen by noting from (4.29) that Φ is related to the source charge distribution ρ in a similar manner. In other words, we have

$$\Phi(\mathbf{r}) = \frac{1}{4\pi \epsilon_o} \int_{V'} \frac{\rho(\mathbf{r}')}{R} \, dv' \tag{6.16}$$

---

[40]To see this we can expand $(1/R)$ into its components and manipulate as follows:

$$\frac{1}{R} = \frac{1}{[(x - x')^2 + (y - y')^2 + (z - z')^2]^{1/2}}$$

so that

$$\nabla \left(\frac{1}{R} \right) = \hat{\mathbf{x}} \frac{\partial}{\partial x} \left(\frac{1}{R} \right) + \hat{\mathbf{y}} \frac{\partial}{\partial y} \left(\frac{1}{R} \right) + \hat{\mathbf{z}} \frac{\partial}{\partial z} \left(\frac{1}{R} \right)$$

$$= -\frac{\hat{\mathbf{x}}(x - x') + \hat{\mathbf{y}}(y - y') + \hat{\mathbf{z}}(z - z')}{[(x - x')^2 + (y - y')^2 + (z - z')^2]^{3/2}} = -\frac{\mathbf{R}}{R^3} = -\hat{\mathbf{R}} \frac{1}{R^2}$$

The same result was obtained using spherical coordinates in Footnote 108 in Chapter 4.

which is a scalar integral similar to that for each component of (6.15). In other words,

$$A_{x,y \text{ or } z}(\mathbf{r}) = \frac{\mu_0}{4\pi} \int_{V'} \frac{J_{x,y \text{ or } z}(\mathbf{r}')}{R} \, dv' \tag{6.17}$$

indicating that, for example, A_x is related to its source J_x in the same manner as Φ is related to ρ.

Recall from Chapter 4 that (6.16) is the solution to Poisson's equation $\nabla^2 \Phi = -\rho/\epsilon_0$, assuming a reference of $\Phi = 0$ at infinity. By observing the similarity between (6.16) and (6.17), one can immediately establish an analogous vector differential equation expressing $\mathbf{A}$ in terms of $\mathbf{J}$:

$$\nabla^2 \mathbf{A} = -\mu_0 \mathbf{J} \tag{6.18}$$

We can derive (6.18) more directly by substituting $\nabla \times \mathbf{A}$ into the differential form of Ampère's law:

$$\nabla \times \mathbf{B} = \nabla \times (\nabla \times \mathbf{A}) = \nabla(\nabla \cdot \mathbf{A}) - \nabla^2 \mathbf{A} = \mu_0 \mathbf{J} \tag{6.19}$$

Note that (6.19) yields (6.18) if we have $\nabla \cdot \mathbf{A} = 0$. Since $\mathbf{B}$ is only specified in terms of the curl of $\mathbf{A}$, we are free in the definition of $\mathbf{A}$ to add any function with zero curl, that is, the gradient of any scalar function.[41] Introducing $\mathbf{A} = \mathbf{A}' + \nabla\phi$, where ϕ is a scalar function, the value of $\mathbf{B}$ is the same if either vector $\mathbf{A}$ or $\mathbf{A}'$ is taken as the vector potential, since $\nabla \times \nabla\phi \equiv 0$. However, since

$$\nabla \cdot \mathbf{A} = \nabla \cdot \mathbf{A}' + \nabla^2 \phi,$$

for any $\mathbf{A}'$, we can always find a function ϕ such that[42] $\nabla \cdot \mathbf{A}' = -\nabla^2 \phi$. Thus, for simplicity, we choose $\nabla \cdot \mathbf{A} = 0$.[43] With this choice of $\nabla \cdot \mathbf{A}$, (6.19) reduces to (6.18), and we can thus express $\mathbf{A}$ directly in terms of $\mathbf{J}$ as given in (6.15).

---

[41] The curl of the gradient of a scalar function ϕ is always zero: $\nabla \times \nabla\phi \equiv 0$.

[42] To see why, note that $\nabla^2 \phi = -\nabla \cdot \mathbf{A}'$ is mathematically equivalent to Poisson's equation if we replace $\nabla \cdot \mathbf{A}'$ with ρ/ϵ_0. Thus we may simply set

$$\phi = \frac{1}{4\pi} \int \frac{\nabla \cdot \mathbf{A}'}{R} \, dv'$$

to ensure $\nabla \cdot \mathbf{A}' = -\nabla^2 \phi$ and thus $\nabla \cdot \mathbf{A} = 0$.

[43] This particular choice for the divergence of $\mathbf{A}$ is known as the *Coulomb gauge*. In the Coulomb gauge, the divergence and curl of $\mathbf{A}$ and $\mathbf{B}$ have a close resemblance:

$$\nabla \cdot \mathbf{B} = 0 \qquad \nabla \times \mathbf{B} = \mu_0 \mathbf{J}$$
$$\nabla \cdot \mathbf{A} = 0 \qquad \nabla \times \mathbf{A} = \mathbf{B}$$

We already have a solution to $\mathbf{B}$ in terms of $\mathbf{J}$: the Biot–Savart law, given by (6.6). Hence we can write down the equation for $\mathbf{A}$ in terms of $\mathbf{B}$ with the substitution $\mathbf{B} \rightarrow \mathbf{A}$ and $\mu_0 \mathbf{J} \rightarrow \mathbf{B}$:

$$\mathbf{A} = \frac{1}{4\pi} \int_{V'} \frac{\mathbf{B}(\mathbf{r}') \times \hat{\mathbf{R}}}{|\mathbf{r} - \mathbf{r}'|^2} \, dv'$$

where $\mathbf{r}$ is the observation point of the field $\mathbf{A}$, $\mathbf{r}'$ is the coordinate of the $\mathbf{B}$-field inside the integral, and $\hat{\mathbf{R}}$ is the unit vector pointing from $\mathbf{r}'$ to $\mathbf{r}$. Note however that this explicit expression for $\mathbf{A}$ in terms of $\mathbf{B}$ is not very useful in practice.

Note that for currents flowing in thin filamentary wires we have $\mathbf{J}\,dv' = JS'\,dl' = I\,d\mathbf{l}'$ (where $d\mathbf{l}'$ and S' are the differential length and the cross-sectional area of the filament), so (6.15) becomes

$$\mathbf{A} = \frac{\mu_0 I}{4\pi} \oint_{C'} \frac{d\mathbf{l}'}{R} \tag{6.20}$$

where C' is the closed contour over which the filamentary current I flows. When currents are restricted to surface currents, such as in a perfect conductor, we have $\mathbf{J}\,dv' = \mathbf{J}_s\,ds'$, so (6.15) becomes

$$\mathbf{A(r)} = \frac{\mu_0}{4\pi} \int_{S'} \frac{\mathbf{J}_s(\mathbf{r}')}{R}\,ds' \tag{6.21}$$

where S' is the surface over which the surface current $\mathbf{J}_s$ flows.

Examples 6.22 and 6.23, and the discussion of the magnetic dipole in the next section, demonstrate the use of the vector magnetic potential $\mathbf{A}$ to determine $\mathbf{B}$ indirectly. However, for simple cases, such as for a finite-length straight wire, using $\mathbf{A}$ to find the $\mathbf{B}$ field might actually lead to a more complicated solution. On the other hand, in more complicated cases of practical significance, such as the magnetic dipole, the use of the vector magnetic potential results in significant simplification. The concept of vector magnetic potential is commonly used in most antenna problems to relate source current distributions (on antenna wires) to radiated fields.

An interesting difference between the electric potential Φ and the vector magnetic potential $\mathbf{A}$ should be noted here. Electrostatic problems more often involve the determination of electric fields from specified potentials rather than from charges. Thus, solving for the potential Φ (usually by solving Poisson's equation) is actually more practical than finding the electric fields directly from charge distributions. In magnetostatics, on the other hand, we usually have a known distribution of currents from which the $\mathbf{B}$ field can be directly determined. The vector potential $\mathbf{A}$ is thus an auxiliary quantity that is used primarily for simplicity. Although $\mathbf{A}$ plays a similar role in magnetostatics as Φ does in electrostatics, it has not been given any simple physical meaning. In classical terms, it is possible to view $\mathbf{A}$ as an entity that is not a "real" field,[44] since in any region where $\mathbf{B}$ is zero, a moving charged particle experiences no magnetic force (since $\mathbf{F}_m = q\mathbf{v} \times \mathbf{B} = 0$) even if $\mathbf{A} \neq 0$. However, there are phenomena in quantum mechanics that show that $\mathbf{A}$ is in fact a "real" field.[45] In terms of examples studied in this chapter, it can be shown[46] that although the $\mathbf{B}$ fields outside an infinitely long solenoid and an

---

[44]R. P. Feynman defines a "real" field as a mathematical function that is introduced to avoid the idea of action at a distance. In other words, a "real" field is a set of numbers we specify in such a way that what happens *at a point* depends only on the number *at that point*.

[45]For an interesting discussion, see Chapter II–15 of R. P. Feynman, R. B. Leighton, and M. Sands, *The Feynman Lectures in Physics*, Addison Wesley, 1964. Also see M. D. Semon and J. R. Taylor, Thoughts on the magnetic vector potential, *Amer. J. Phys.*, 64(11), pp. 1361–9, November 1996, and references therein.

[46]See, for example, Section 7.11 of W. T. Scott, *The Physics of Electricity and Magnetism*, 2nd ed., John Wiley & Sons, Inc., New York, 1966.

ideal toroid are identically zero, the **A** fields are not. For an infinitely long solenoid of
radius a, the vector potentials inside and outside the solenoid are given by

$$A_\phi = \begin{cases} \dfrac{\mu_0 N I a^2}{2r} & r > a \\[2ex] \frac{1}{2}\mu_0 N I r & r \leq a \end{cases}$$

where r is the radial distance from the solenoid axis. Note that inside the solenoid ($r \leq a$),
$\nabla \times \mathbf{A}$ indeed yields the **B** field for an infinitely long solenoid, as found in Example 6.14.
For $r > a$, with **A** having only a $\hat{\boldsymbol{\phi}}$ component proportional to r^{-1}, the curl of **A** is
zero, since in cylindrical coordinates we have $\nabla \times \mathbf{A} = \hat{\mathbf{z}}(1/r)[\partial(rA_\phi)/\partial r] = 0$. This
nonvanishing feature of **A** in regions where $\mathbf{B} = 0$ has important physical significance.[47]

Example 6.22: Finite-length straight wire. Calculate the **B** field on the bisecting plane (i.e., the
$z = 0$ plane) due to a current I flowing in a straight wire of length $2a$ oriented along the z axis and
centered at $z = 0$ as shown in Figure 6.30. This problem constitutes a repeat of Example 6.3, except
that we shall first find the vector magnetic potential **A** and then determine **B** using $\mathbf{B} = \nabla \times \mathbf{A}$.

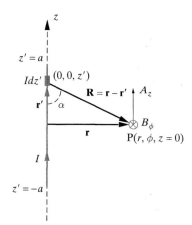

Figure 6.30 Vector potential **A** around a
wire of length $2a$.

Solution: Taking the current I to flow in the z direction and noting that $\int_{S'} J\, ds' = I$,
where S' is the cross-sectional area of the wire, the vector potential **A** on the $z = 0$ plane
is given by (6.15):

$$\mathbf{A} = \frac{\mu_0}{4\pi} \int_{V'} \frac{\mathbf{J}}{R}\, dv' = \frac{\mu_0}{4\pi} \int_{-a}^{a} \frac{\hat{\mathbf{z}} I\, dz'}{R}$$

---

[47]For example, if a wire is looped around an infinitely long solenoid, Faraday's law (to be discussed in
Chapter 7) requires that an electromotive force be induced on the loop if the current in the solenoid varies
with time. We might then ask: How can a **B** field that never touches a wire have an effect on the charges in it?
The apparent conceptual difficulty is resolved when we realize that the induced emf can be directly determined
using **A**, which is nonzero.

From the geometry of the problem (Figure 6.30), and since $\mathbf{r}' = \hat{\mathbf{z}}z'$, $\mathbf{r} = \hat{\mathbf{r}}r$, we have $R = \sqrt{r^2 + (z')^2}$, so that the $\mathbf{A}$ field at a distance r from the wire on the $z = 0$ plane is given as

$$\mathbf{A} = \hat{\mathbf{z}}A_z = \frac{\mu_0 I}{4\pi} \int_{-a}^{a} \frac{dz'}{\sqrt{r^2 + (z')^2}} = \hat{\mathbf{z}}\frac{\mu_0 I}{2\pi} \left[\ln\left(\frac{z'}{r} + \sqrt{\left(\frac{z'}{r}\right)^2 + 1}\right) \right]_0^a$$

$$= \hat{\mathbf{z}}\frac{\mu_0 I}{2\pi} \ln\left(\frac{a}{r} + \sqrt{\frac{a^2}{r^2} + 1}\right)$$

where the integration is carried out either by change of variables or by using integral tables. By using $\mathbf{B} = \nabla \times \mathbf{A}$ in cylindrical coordinates, and noting that there is no dependence on ϕ, we have

$$\mathbf{B} = \nabla \times \mathbf{A} = \hat{\boldsymbol{\phi}} \left(-\frac{\partial A_z}{\partial r} \right)$$

so that on the bisecting plane, where $z = 0$, we have

$$B_\phi = -\frac{\mu_0 I}{2\pi} \frac{\partial}{\partial r} \ln\left(\frac{a}{r} + \sqrt{\frac{a^2}{r^2} + 1}\right) = \frac{\mu_0 I}{2\pi r} \frac{a/r}{\sqrt{1 + (a/r)^2}} = \frac{\mu_0 I}{2\pi r} \frac{a}{\sqrt{r^2 + a^2}}$$

Note that this result is identical to that obtained in Example 6.3. We can observe that for $r \ll a$, B_ϕ varies as $1/r$, consistent with what we obtained for an infinitely long wire. For $r \gg a$, on the other hand, we have B_ϕ varying as $1/r^2$, consistent with the field of a very short current element, as is expected from the Biot–Savart law.

Example 6.23: Field and potential at the center of a spherical shell. Consider a spherical shell of inner and outer radius a and b, respectively, and conducting a uniform current density $\mathbf{J} = \hat{\boldsymbol{\phi}}J_0$ (where J_0 is a constant), which is always tangent to circles of constant latitude, as shown in Figure 6.31. Find the vector potential $\mathbf{A}$ and the $\mathbf{B}$ field at the center of the sphere (i.e., at $z = 0$).

Solution: From (6.15), we have

$$\mathbf{A}(\mathbf{r} = 0) = \frac{\mu_0}{4\pi} \int_{V'} \frac{\mathbf{J}(\mathbf{r}')}{R} \, dv' = \frac{\mu_0}{4\pi} \int_{V'} \frac{\mathbf{J}(\mathbf{r}')}{|\mathbf{r}'|} \, dv'$$

where V' is the volume of the shell, and the origin of the coordinate system is located at the center of the sphere—that is, at the field point (in other words, $\mathbf{r} = 0$). Note that the magnitude of $\mathbf{J}$ is uniform throughout the volume V' of the spherical shell. Each current density element $\mathbf{J}\,dv'$ at a given point on the shell produces a differential vector magnetic potential $d\mathbf{A}$ at the center of the sphere, which is in the same direction as the current element. However, in view of the spherical symmetry, for each such element there exists another one

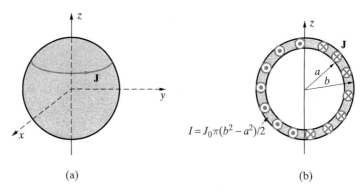

Figure 6.31   A spherical current-conducting shell. (a) The density **J** is everywhere tangent to circles of constant latitude. The polar axis of the sphere is along the z axis. (b) Vertical cross-sectional view along the polar axis.

located diametrically opposite and having oppositely directed current flow. Thus, the vector potential at the center of the sphere is $\mathbf{A}(\mathbf{r} = 0) = 0$. On the other hand, $\mathbf{B}(\mathbf{r} = 0)$ does not vanish. From (6.6), we have

$$\mathbf{B}(\mathbf{r} = 0) = \frac{\mu_0}{4\pi} \int_{V'} \frac{\mathbf{J}(\mathbf{r}') \times \hat{\mathbf{R}}}{R^2} \, dv'$$

where $\mathbf{J}(\mathbf{r}') = \hat{\boldsymbol{\phi}} J_0$ is the current flowing between $r = a$ and $r = b$, and $\hat{\mathbf{R}}$ is the unit vector pointing from a source point $\mathbf{r}'$ in the spherical shell to the origin (i.e., the observation point in this case) and is the inward normal to the spherical surface. From symmetry considerations, only the component of $\mathbf{J} \times \hat{\mathbf{R}}$ that is parallel to the polar axis of the sphere contributes to the **B** field at $\mathbf{r} = 0$. Noting that this component is proportional to $\sin \theta'$, where θ' is the spherical coordinate of the source point, we have

$$\mathbf{B}(\mathbf{r} = 0) = \frac{\mu_0}{4\pi} J_0 \int_{V'} \hat{\mathbf{z}} \frac{\sin \theta'}{R^2} \, dv'$$

where $R^2 = (r')^2$. Note also that because the spherical coordinate volume element is given as $dv' = (r')^2 \sin \theta' \, d\theta' \, d\phi' \, dr'$, the volume integral can be easily carried out as

$$B_z(\mathbf{r} = 0) = \frac{\mu_0 J_0}{4\pi} \int_a^b dr' \int_0^\pi \sin^2(\theta') \, d\theta' \int_0^{2\pi} d\phi' = \frac{\mu_0 J_0 \pi^2}{4\pi}(b - a)$$

If the shell carries a total current I given by $I = (\pi/2) J_0 (b^2 - a^2)$, we can write

$$B_z(\mathbf{r} = 0) = \frac{\mu_0 I}{2(b + a)}$$

In the limit $b \to a$, with I fixed, we approach an infinitely thin shell conducting a finite total current I, and the **B** field at the center is

$$\mathbf{B}(\mathbf{r} = 0) = \hat{\mathbf{z}} \frac{\mu_0 I}{4a} \, .$$

The spherical shell carrying a uniform current provides an interesting example of a configuration for which $\mathbf{A} = 0$ but $\mathbf{B} \neq 0$. This circumstance is not at all surprising when we realize that curl is essentially a vector differentiation; thus, the case in hand is analogous to a one-dimensional function that crosses zero with a nonzero slope.

Example 6.24: Saddle-shaped gradient coils. Magnetic resonance imaging (MRI) scanners use gradient coils to produce transverse gradients (nonuniformities) in the main field of the scanner in order for imaging to be possible. The magnetic fields of the gradient coils strengthen the main field (typically a constant field of a required strength and uniformity produced by superconducting or other types of coils or permanent magnets) in one region while weakening it in another, resulting in a total field whose strength varies in a continuous fashion in a desired direction. Two types[48] of gradient head coils are shown in Figures 6.32a and b. Typically, the gradient fields are much weaker than the main field, so much less powerful electromagnets can be used. Saddle-shaped coils are used in MRI scanners to produce transverse gradients. The optimum geometry[49] of saddle coils consists of two semicircular loops of wire connected by straight wires. The semicircular portions are active in creating the gradient; the straight portions serve only to connect the loops and do not contribute. A pair of identical saddle-shaped coils wound on the surface of a cylinder with a circular cross section oriented in the z direction is shown in Figure 6.32c. Find the $\mathbf{B}$ field at the center of the imaging system (i.e., at the origin of the coordinate system in Figure 6.32c).

Solution: First we consider the $\mathbf{B}$ field due to the arcs of the two coils. The vector magnetic potential at a point along the axis due to an element of current on the circular arc is given by the integrand of (6.20):

$$d\mathbf{A} = \frac{\mu_0 I}{4\pi} \frac{d\mathbf{l}'}{R}$$

where $d\mathbf{l}' = \hat{\boldsymbol{\phi}} a\, d\phi' = (-a \sin\phi')\, d\phi'\hat{\mathbf{x}} + (a\cos\phi')\, d\phi'\hat{\mathbf{y}}$ and $R = |\mathbf{r} - \mathbf{r}'| = |\hat{\mathbf{z}}(z - z') - \hat{\mathbf{r}}a| = \sqrt{a^2 + (z - z')^2}$. Since the contributions from all four circular arcs add together, the total $\mathbf{B}$ field at the center (i.e., $|z - z'| = l/2$) is

$$\mathbf{B}_{\text{arcs}} = \nabla \times \mathbf{A} = \left\{ \nabla \times \left[\frac{\mu_0 I}{\pi} \int_{-\phi_0/2}^{\phi_0/2} \frac{(-a\sin\phi')\hat{\mathbf{x}} + (a\cos\phi')\hat{\mathbf{y}}}{\sqrt{a^2 + (z-z')^2}}\, d\phi' \right] \right\}_{z-z'=l/2}$$

$$= \left\{ \nabla \times \left[\hat{\mathbf{y}} \frac{\mu_0 I}{\pi} \frac{2a\sin(\phi_0/2)}{\sqrt{a^2 + (z-z')^2}} \right] \right\}_{z-z'=l/2}$$

$$= \hat{\mathbf{x}} \frac{\mu_0 I}{\pi} \frac{al\sin(\phi_0/2)}{[a^2 + (l/2)^2]^{3/2}}$$

[48] For more information, see W. H. Oldendorf, M.D. and W. Oldendorf, Jr., *Basics of Magnetic Resonance Imaging*, Martinus Nijhoff Publishing, 1988: also see W. H. Oldendorf, *MRI Primer*, Raven Press, New York, 1991.

[49] D. M. Ginsberg and M. J. Melchner, Optimum geometry of saddle shaped coils for generating a uniform magnetic field, *Rev. Sci. Instrum.*, 41(1), pp. 122–123, January 1970.

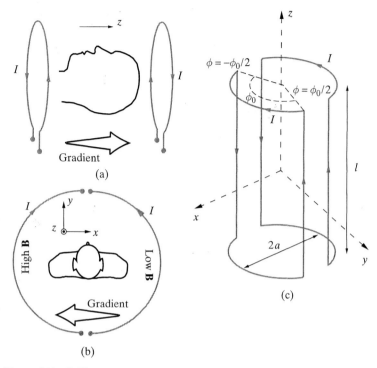

Figure 6.32   Saddle-shaped gradient coils. (a) A longitudinal z axis gradient is created by two circular coils. Current is passed in opposite directions through the two coils. The main **B** field (in the $+z$ direction) is reinforced near the coil on the right and is diminished near the coil on the left, resulting in a magnetic gradient between the two coils. (b) A transverse gradient is created using opposed semicircular loops in which current flows in opposite directions. The field from one semicircular loop adds to the main field (in the $+z$ direction), while the field from the other subtracts from it, forming a gradient. (c) The geometry of the saddle coils and the coordinate system.

We next consider the **B** field due to the vertical portions of the coils. Using the results of Example 6.3, and realizing that the x components of the vertical wires add together while the y components cancel, the **B** field due to four vertical parts of the coils is

$$\mathbf{B}_{\text{vert}} = \hat{\mathbf{x}} 4 \frac{\mu_0 I\,(l/2)\sin(\phi_0/2)}{2\pi a \sqrt{a^2 + (l/2)^2}} = \hat{\mathbf{x}} \frac{\mu_0 Il}{\pi a} \frac{\sin(\phi_0/2)}{\sqrt{a^2 + (l/2)^2}}$$

which is directed in the same direction as $\mathbf{B}_{\text{arcs}}$. Thus, the total **B** field at the center of the pair of saddle coils shown in Figure 6.32c is

$$\mathbf{B} = \mathbf{B}_{\text{arcs}} + \mathbf{B}_{\text{vert}} = \hat{\mathbf{x}} \frac{\mu_0 Il \sin(\phi_0/2)}{\pi \sqrt{a^2 + (l/2)^2}} \left[\frac{a}{a^2 + (l/2)^2} + \frac{1}{a} \right]$$

Determination of the **B** field at points off the center point is also straightforward but requires numerical evaluation of the resulting integrals. By examining the variation of the **B** field with position in the vicinity of the center point, it can be shown[50] that optimum homogeneity is obtained for $l = 2a$ and $\phi_0 = 120°$.

6.6 THE MAGNETIC DIPOLE

A small current-carrying loop constitutes the magnetic equivalent of the electric dipole. In this section we derive an expression for **B** at a large distance from such a small loop, which is referred to as the *magnetic dipole*. The concept of a magnetic dipole is useful in understanding the behavior of magnetic materials (Section 6.8), being analogous to the electric dipole, which was used in Section 4.10 to determine the response to an external electric field of dielectric materials.

We consider a small circular loop of radius a carrying a steady current I with its axis coincident with the z axis and its center located at the origin, as shown in Figure 6.33a. The radius a of the loop is assumed to be much smaller than r (i.e., $a \ll r$), where r is the distance from the origin of the point P at which we wish to determine the **B** field. Under these conditions, the loop is considered to be a magnetic dipole. We would like to find the magnetic vector potential **A** at point P and use it to obtain the **B** field.

The exact expression for the vector **A** at point P (r, θ, ϕ) is given by

$$\mathbf{A} = \frac{\mu_0 I}{4\pi} \oint_C \frac{d\mathbf{l}'}{R}$$

where $d\mathbf{l}' = \hat{\boldsymbol{\phi}} a\, d\phi'$; noting that $\mathbf{r}' = \hat{\mathbf{x}} a \cos \phi' + \hat{\mathbf{y}} a \sin \phi'$, we have

$$R = |\mathbf{R}| = |\mathbf{r} - \mathbf{r}'| = \sqrt{(x - a \cos \phi')^2 + (y - a \sin \phi')^2 + z^2}$$

Substituting $x = r \sin \theta \cos \phi, y = r \sin \theta \sin \phi, z = r \cos \theta$, we can write

$$
\begin{aligned}
R = |\mathbf{R}| &= \sqrt{(r \sin \theta \cos \phi - a \cos \phi')^2 + (r \sin \theta \sin \phi - a \sin \phi')^2 + r^2 \cos^2 \theta} \\
&= \sqrt{a^2 + r^2 - 2ar \sin \theta (\cos \phi \cos \phi' + \sin \phi \sin \phi')} \\
&= \sqrt{a^2 + r^2 - 2ar \sin \theta \cos(\phi - \phi')} \\
&= r\sqrt{1 + \frac{a^2}{r^2} - \frac{2a}{r} \sin \theta \cos(\phi - \phi')}
\end{aligned}
$$

---

[50]D. I. Hoult and R. E. Richards, The signal-to-noise ratio of the nuclear magnetic resonance experiment, *J. Magn. Reson.,* 24, pp. 71–85, 1976.

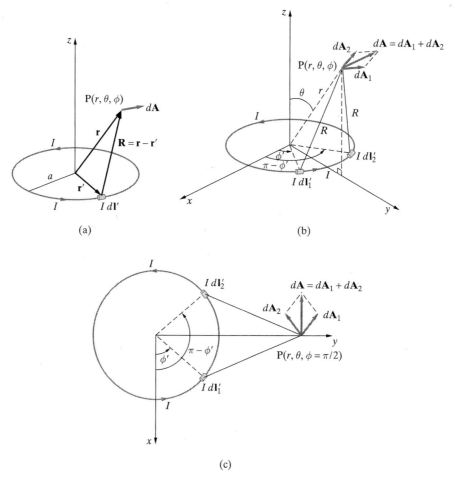

Figure 6.33 A circular current loop in free space. (a) A circular loop carrying a steady current I. (b) We lose no generality by taking the observation point to be in the y-z plane ($\phi' = \pi/2$). (c) Top view, showing the vector potentials produced by two current elements symmetrical with respect to the y axis.

It is clear from symmetry that the potential is independent of the azimuth angle ϕ. We can consider straightforward evaluation of $\mathbf{A}(r,\theta)$ by substituting the expression for $\mathbf{R}$ and $d\mathbf{l}$ in the integral expression for $\mathbf{A}$. Thus, we have

$$\mathbf{A}(r,\theta) = \frac{\mu_0 I}{4\pi} \int_0^{2\pi} \frac{\hat{\boldsymbol{\phi}} a \, d\phi'}{\sqrt{a^2 + r^2 - 2ar \sin\theta \cos(\phi - \phi')}}$$

This integral does not have a closed-form solution and is not particularly revealing; however, direct numerical evaluation of $\mathbf{A}$ using the integral expression is straightforward, since the integrand is a well-behaved function.

In our case, we are interested in the fields at points far away from the loop, for which case useful approximate expressions can be derived for **A**. Since $r \gg a$, we can use the Taylor series expansion[51] of R^{-1} in r to write

$$\frac{1}{R} \simeq \frac{1}{r}\left[1 + \frac{a}{r}\sin\theta\cos(\phi - \phi')\right]$$

Hence, it follows that

$$\mathbf{A} \simeq \frac{\mu_0 I a}{4\pi}\int_0^{2\pi}\hat{\boldsymbol{\phi}}\frac{1}{r}\left[1 + \frac{a}{r}\sin\theta\cos(\phi - \phi')\right]d\phi'$$

Without any loss of generality, the point P can be located directly above the y axis as shown in Figure 6.33b (i.e., $\phi = \pi/2$). The variable direction of the unit vector $\hat{\boldsymbol{\phi}}$ in the integrand is handled by considering two current elements, $I\,d\mathbf{l}_1$ and $I\,d\mathbf{l}_2$, which are symmetrically located with respect to the y axis. From the geometry, we have

$$I\,d\mathbf{l}_1 = (-\hat{\mathbf{x}}\sin\phi' + \hat{\mathbf{y}}\cos\phi')Ia\,d\phi'$$

and

$$I\,d\mathbf{l}_2 = (-\hat{\mathbf{x}}\sin\phi' - \hat{\mathbf{y}}\cos\phi')Ia\,d\phi'$$

the sum of which results in a net differential vector magnetic potential $d\mathbf{A}$ at point P that is $-x$-directed (or ϕ-directed; note that at point P with $\phi = \pi/2$ we have $-\hat{\mathbf{x}} = \hat{\boldsymbol{\phi}}$). This is more clearly seen from the top view shown in Figure 6.33c. Thus, the total vector **A** for $r \gg a$ becomes

$$\mathbf{A} = \hat{\boldsymbol{\phi}}\frac{\mu_0 I a}{4\pi}2\int_{-\pi/2}^{\pi/2}\sin\phi'\frac{1}{r}\left[1 + \frac{a}{r}\sin\theta\sin\phi'\right]d\phi'$$

$$= \hat{\boldsymbol{\phi}}\frac{\mu_0 I a}{2\pi}\left[\frac{1}{r}\int_{-\pi/2}^{\pi/2}\sin\phi'\,d\phi' + \frac{a}{r^2}\sin\theta\int_{-\pi/2}^{\pi/2}\sin^2\phi'\,d\phi'\right]$$

where we have also noted that with $\phi = \pi/2$, $\cos(\phi - \phi') = \sin\phi'$. Note that the first integral is zero, so evaluating the second integral yields

$$\mathbf{A} = \hat{\boldsymbol{\phi}}\frac{\mu_0 I a^2 \sin\theta}{4r^2}$$

We can now find **B** from (6.14). Using spherical coordinates, since $A_r = A_\theta = 0$ and $\partial A_\phi/\partial\phi = 0$, it follows that

$$\mathbf{B} = \hat{\mathbf{r}}B_r + \hat{\boldsymbol{\theta}}B_\theta$$

where

$$B_r = \frac{1}{r\sin\theta}\frac{\partial}{\partial\theta}(A_\phi\sin\theta) = \frac{\mu_0 I a^2 \cos\theta}{2r^3}$$

[51] See Section 4.4.3 for the electric dipole, where a similar expansion was shown in detail.

and

$$B_\theta = -\frac{1}{r}\frac{\partial}{\partial r}(rA_\phi) = \frac{\mu_0 I a^2 \sin\theta}{4r^3}$$

It is interesting to compare these expressions with those obtained for the electric field due to an electric dipole. For the electric dipole (from Section 4.4.3), we have

$$\mathbf{E} = \hat{\mathbf{r}}\frac{p\cos\theta}{2\pi\epsilon_0 r^3} + \hat{\boldsymbol{\theta}}\frac{p\sin\theta}{4\pi\epsilon_0 r^3}$$

where $p = Qd$ is the electric dipole moment. For the magnetic dipole we have

$$\mathbf{B} = \hat{\mathbf{r}}\frac{\mu_0 I a^2 \cos\theta}{2r^3} + \hat{\boldsymbol{\theta}}\frac{\mu_0 I a^2 \sin\theta}{4r^3}$$

As these expressions are identical in form, we can associate the magnetic dipole with a dipole moment $\mathbf{m}$ such that $\mathbf{m} = \hat{\mathbf{z}}|\mathbf{m}| = \hat{\mathbf{z}}I(\pi a^2)$, where πa^2 is the total area of the circular loop. For a magnetic dipole with N turns, the magnetic dipole moment is $\mathbf{m} = \hat{\mathbf{z}}NI\pi a^2$. The $\mathbf{B}$ field for the magnetic dipole can be rewritten as

$$\mathbf{B} = \hat{\mathbf{r}}\frac{\mu_0|\mathbf{m}|\cos\theta}{2\pi r^3} + \hat{\boldsymbol{\theta}}\frac{\mu_0|\mathbf{m}|\sin\theta}{4\pi r^3} \tag{6.22}$$

Although (6.22) was derived for a circular loop, it is also valid for small, current-carrying loops of other symmetric shapes (e.g., a square loop), with $\mathbf{m} = \hat{\mathbf{z}}NIA$, where A is the area of the loop.

The similarity of the fields for the magnetic and electric dipoles is illustrated in Figure 6.34. The top panels show the normalized electric and magnetic dipole field lines as given, by (4.28) and (6.22), respectively, which of course have identical shapes but are valid only at large distances from the dipoles. The bottom panels show the close-up views, illustrating that the electric field lines terminate on the two charges, whereas the magnetic field lines close on themselves.

Note that magnetic dipole moment $\mathbf{m}$ is a vector, which has a magnitude $|\mathbf{m}|$ and a direction along the axis of the magnetic dipole loop determined by the direction of the current flow using the right-hand rule. The vector $\mathbf{A}$ due to a magnetic dipole can be written in terms of its dipole moment $\mathbf{m}$ as

$$\mathbf{A} = \frac{\mu_0\mathbf{m}\times\hat{\mathbf{r}}}{4\pi r^2} = \hat{\boldsymbol{\phi}}\frac{\mu_0|\mathbf{m}|\sin\theta}{4\pi r^2}$$

Noting that $\nabla r^{-1} = -\hat{\mathbf{r}}/r^2$ (see Footnote 40 in Section 6.5), this result can also be written as

$$\mathbf{A} = \frac{\mu_0}{4\pi}\left(\mathbf{m}\times\frac{\hat{\mathbf{r}}}{r^2}\right) = \frac{\mu_0}{4\pi}\left(\nabla\times\frac{\mathbf{m}}{r}\right)$$

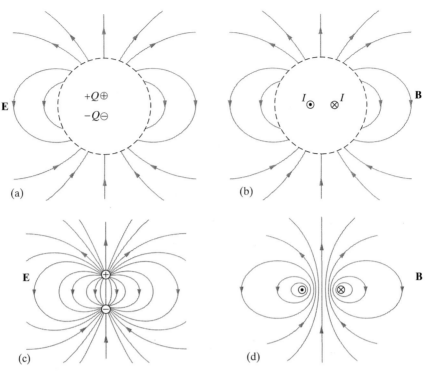

Figure 6.34   The field lines of electric and magnetic dipoles. (a) Field lines as defined by (4.28) for the **E** field at large distances from an electric dipole. (b) **B** field lines at large distances as given by (6.22) for a magnetic dipole. (c) Close-up view showing the termination of the **E** field on the charges. (d) Close-up view showing the self-closure of the **B** field lines.

Using the above, we can also write an expression directly relating **B** and **m**. Using (6.14), we have

$$\mathbf{B} = \nabla \times \mathbf{A} = \frac{\mu_0}{4\pi} \nabla \times \left(\nabla \times \frac{\mathbf{m}}{r} \right)$$

which can be simplified[52] to the form

$$\mathbf{B} = \frac{\mu_0}{4\pi} \nabla \left[\mathbf{m} \cdot \nabla \left(\frac{1}{r} \right) \right] \tag{6.23}$$

[52]Using the fact that $\nabla^2 (1/r) = 0$ for $r \neq 0$ and the identity

$$\nabla \times \nabla \times \mathbf{G} \equiv \nabla (\nabla \cdot \mathbf{G}) - \nabla^2 \mathbf{G}$$

Example 6.25: Field of a long uniform solenoid. As an application of the magnetic dipole results derived above, consider the **B** field of a long uniform solenoid as shown in Figure 6.35, having N turns per meter and carrying a current I. The length of the solenoid is l, and its radius is a. Determine the **B** field at point $P(x, y, z)$ such that $r \gg a$, $R_1 \gg a$, and $R_2 \gg a$.

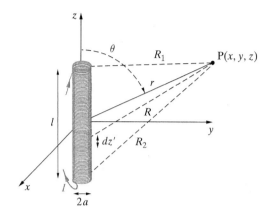

Figure 6.35   B field of a solenoid.

Solution: We can consider the solenoid as consisting of many magnetic dipoles stacked on top of one another, and since $r \gg a$ and $R_1, R_2 \gg a$, we can use the magnetic dipole expressions derived above for distant observation points. For a segment of length dz', we have Ndz' turns, each with current I, so that the magnetic dipole moment of that differential segment is $\mathbf{m} = \hat{\mathbf{z}} NIdz'\pi a^2$. Although we can first find the vector potential **A** and then use $\mathbf{B} = \nabla \times \mathbf{A}$ to find the **B** field, we can also use the direct relationship between **B** and **m** as given above. Using the expression for the **B** field of a magnetic dipole shown in (6.23), the differential contribution to the field at P due to the differential segment of the solenoid is

$$d\mathbf{B} = \frac{\mu_0}{4\pi} \nabla \left[\mathbf{m} \cdot \nabla \left(\frac{1}{R} \right) \right] = \frac{\mu_0}{4\pi} \nabla \left[NIdz'\pi a^2 \hat{\mathbf{z}} \cdot \nabla \left(\frac{1}{R} \right) \right]$$

where R is the distance between the elemental magnetic dipole and point P as shown in Figure 6.35. Note that $\nabla(R)^{-1} = -\hat{\mathbf{R}}/R^2 = -\mathbf{R}/R^3$, as derived in Footnote 40 in Section 6.5. To find the total **B** field at P we simply need to integrate over the length of the solenoid. In other words,

$$\mathbf{B} = \frac{\mu_0}{4\pi} \nabla \int_{-l/2}^{l/2} \nabla \left(\frac{1}{R} \right) \cdot \hat{\mathbf{z}} NI \pi a^2 \, dz'$$

Noting that $R = |\mathbf{r} - \mathbf{r}'| = \hat{\mathbf{x}}x + \hat{\mathbf{y}}y + \hat{\mathbf{z}}(z - z')$, we have

$$\mathbf{B} = \frac{-NIa^2\mu_0}{4} \nabla \int_{-l/2}^{l/2} -\frac{\hat{\mathbf{R}}}{R^2} \cdot \hat{\mathbf{z}} \, dz'$$

$$= \frac{-NIa^2\mu_0}{4} \nabla \int_{-l/2}^{l/2} -\frac{(z - z') \, dz'}{[x^2 + y^2 + (z - z')^2]^{3/2}}$$

$$= \frac{-NIa^2\mu_0}{4} \nabla [x^2 + y^2 + (z - z')^2]^{-1/2} \Big|_{-l/2}^{l/2}$$

$$= \frac{-NIa^2\mu_0}{4} \left[\nabla \left(\frac{1}{R} \right) \right]_{-l/2}^{l/2} = \frac{-NIa^2\mu_0}{4} \nabla \left[\frac{1}{R_1} - \frac{1}{R_2} \right]$$

where R_1 and R_2 are, respectively, the distances to the observation point from the top and bottom of the solenoid.

The evaluation of the gradient in the above expression for **B** is difficult in the general case, because of the complicated dependence of R_1 and R_2 on the spatial coordinates. However, when the point P is very far away from the solenoid (i.e., $r \gg l$), it can be shown that $R_1^{-1} \simeq r^{-1}[1 - (l/2r)\cos\theta]$ and $R_2^{-1} \simeq r^{-1}[1 + (l/2r)\cos\theta]$, so we have

$$\mathbf{B} = \frac{-NIa^2\mu_0}{4} \nabla \left(\frac{l\cos\theta}{r^2} \right) = \frac{\mu_0}{4\pi} \underbrace{(NI\pi a^2 l)}_{|\mathbf{m}_{\text{slnd}}|} \left(\hat{\mathbf{r}} \frac{2\cos\theta}{r^3} + \hat{\boldsymbol{\theta}} \frac{\sin\theta}{r^3} \right) \tag{6.24}$$

We note that the above expression for **B** is identical to (6.22), if we replace $|\mathbf{m}|$ with the dipole moment of the solenoid, namely $|\mathbf{m}_{\text{slnd}}| = NI\pi a^2 l$. Thus, as far as the field at large distances is concerned, the solenoid produces the same type of field structure as the elementary magnetic dipole, as if it had a total dipole magnetic moment of $NI\pi a^2 l$.

6.7 DIVERGENCE OF B, MAGNETIC FLUX, AND INDUCTANCE

As we discussed in Section 4.5, it is often advantageous to think of a vector field as representing the flow of something. Faraday's experiments on electric displacement were the basis for the concepts of electric flux and electric flux density **D**, which was shown (Section 4.6) to originate from or terminate at sources (or electric charges), as described by the relation $\nabla \cdot \mathbf{D} = \rho$. In this section, we first discuss the concept of magnetic flux and the divergence of the **B** field. In fact, in engineering practice we more often think in terms of magnetic flux than electric flux. Much of our present view of magnetic flux has its origins in Faraday's visualizations of the magnetic force as stretching out in all directions from the electric current that produces it, filling all space as a magnetic field. Through this magnetic flux, which permeates all space, seemingly isolated circuits are inductively coupled; hence, we next consider the important concept of inductance.

6.7.1 Divergence of B and Magnetic Flux

We saw in Section 6.5 that the magnetic field **B** can be derived from the curl of an auxiliary vector potential function **A**. This result leads at once to an important physical property for the **B** field. The divergence of the curl of any vector is identically equal to zero, and hence $\nabla \cdot \nabla \times \mathbf{A} = 0$, from which it follows that the divergence of **B** is identically zero; that is,

$$\boxed{\nabla \cdot \mathbf{B} = 0} \tag{6.25}$$

This property of $\mathbf{B}$ is a mathematical consequence of the formulation of Ampère's experimentally based force law or its alternative statement in the Biot–Savart law. In Section 6.8, we show that the effects of material bodies on the distribution of the magnetic field can be accounted for by their equivalent volume and surface magnetization currents, which also produce magnetic fields, as given by (6.6). Thus, even in the presence of material media it is possible to derive $\mathbf{B}$ from the curl of a vector potential $\mathbf{A}$, and $\nabla \cdot \mathbf{B} = 0$ is valid in general.

As was discussed in Chapter 5 in connection with steady currents for which $\nabla \cdot \mathbf{J} = 0$, the fact that $\nabla \cdot \mathbf{B} = 0$ requires that the flux lines of $\mathbf{B}$ are always continuous and form closed loops. Thus, (6.25) implies that there are no sources of magnetic fields. (In this connection, "source" is used in a mathematical sense; literally, electric current is the source of magnetic fields.) In a mathematical sense, (6.25) means that there exist no *magnetic charges* (or free magnetic poles), corresponding to electric charges, from which $\mathbf{B}$ field lines can emerge or onto which they can terminate. If the divergence of $\mathbf{B}$ were not zero, magnetic field lines could originate at sources (at which $\nabla \cdot \mathbf{B} \neq 0$), just as electric field lines originate at charges ($\nabla \cdot \mathbf{D} = \rho$). However, at the present time, there is no experimental evidence whatsoever for the existence of free magnetic poles[53] or "magnetic charge." On the contrary, all available experimental data can be interpreted on the basis of $\nabla \cdot \mathbf{B} = 0$. All magnets have both a North and a South pole, and the field $\mathbf{B}$ is continuous through the magnet. For this reason the magnetostatic field $\mathbf{B}$ is fundamentally different from the electrostatic field $\mathbf{E}$.

Our discussion of the concepts of flux, flux tubes, and flux lines, given in Section 4.5 for electric fields, also applies to the $\mathbf{B}$ field. In thinking about the $\mathbf{B}$ field, it is often useful to consider it as representing the flow of something. It is in this context that the $\mathbf{B}$ field is often thought of as representing a magnetic flux density, although the implied analogy between $\mathbf{B}$ and the electric flux density $\mathbf{D}$ is somewhat misleading, as discussed in Section 6.2.2. The magnetic flux through a differential area ds is given by the dot product of $\mathbf{B}$ with $d\mathbf{s}$, namely $\mathbf{B} \cdot d\mathbf{s}$. The dot product selects the normal component of $\mathbf{B}$ through the surface ds, as shown in Figure 6.36a, so the total magnetic flux Ψ passing through the surface S is given by

$$\Psi = \int_S \mathbf{B} \cdot d\mathbf{s} \tag{6.26}$$

---

[53]The lack of magnetic poles or charges is the one outstanding asymmetry between electrostatics and magnetostatics. If magnetic poles did exist, say with a volume density ρ_m, the divergence of the $\mathbf{B}$ field would then be $\nabla \cdot \mathbf{B} = \mu_0 \rho_m$. If the magnetic charges moved with a velocity $\mathbf{v}$, the quantity $\rho_m \mathbf{v}$ would constitute a magnetic conduction current density $\mathbf{J}_m$, which in turn would produce an electric field, just as electric conduction current density $\mathbf{J}$ produces a $\mathbf{B}$ field. P. A. M. Dirac [1902–1984] carried out an extensive theoretical investigation of magnetic poles and found no fundamental reason for isolated magnetic charges (*magnetic monopoles*) not to exist. However, he found that a quantum of magnetic charge would be so great that if pairs of opposite sign combined together, it would take highly energetic cosmic rays to separate them. See P. A. M. Dirac, The theory of magnetic poles, *Phys. Rev.*, 74, p. 817, 1948. For a brief discussion of experimental attempts to detect magnetic monopoles, see pp. 136–137 of M. Schwartz, *Principles of Electrodynamics*, Dover, 1972. Also see Sections 6.12 and 6.13 of J. D. Jackson, *Classical Electrodynamics*, Wiley, 2nd ed., New York, 1975.

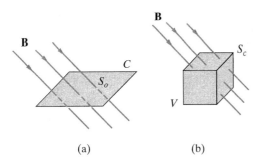

Figure 6.36 Magnetic flux through a surface S. (a) An open surface S_o enclosed by a contour C and (b) a closed surface S_c enclosing a volume V.

in units of webers (Wb). The magnetic flux passing through the surface S bounded by the contour C is said to *link* the contour C and is commonly referred to as the flux linkage.

For a closed surface S, however, as in Figure 6.36b, just as much flux leaves the surface as enters because of the continuous nature of the flux lines. Hence the integral of $\mathbf{B} \cdot d\mathbf{s}$ over a closed surface is always equal to zero. Mathematically, this result follows from (6.25) using the divergence theorem. Since $\nabla \cdot \mathbf{B} = 0$, we have

$$\int_V \nabla \cdot \mathbf{B} \, dv = \oint_S \mathbf{B} \cdot d\mathbf{s} = 0 \qquad (6.27)$$

where the volume V is enclosed by the surface S. We have thus established the fact that the integral of the $\mathbf{B}$ field on any closed surface is zero, a fact that was utilized in Example 6.14 to deduce that the $\mathbf{B}$ field everywhere inside an infinitely long solenoid is directed along the axis of the solenoid.

The magnetic flux that links a contour C may also be expressed in terms of the vector potential $\mathbf{A}$. Since $\mathbf{B} = \nabla \times \mathbf{A}$, we have

$$\Psi = \int_S \mathbf{B} \cdot d\mathbf{s} = \int_S \nabla \times \mathbf{A} \cdot d\mathbf{s}$$

The latter integral may be transformed to a contour integral by using Stokes's theorem:

$$\Psi = \int_S \nabla \times \mathbf{A} \cdot d\mathbf{s} = \oint_C \mathbf{A} \cdot d\mathbf{l}$$

This latter integral is sometimes more convenient to evaluate than $\int_S \mathbf{B} \cdot d\mathbf{s}$ when determining the magnetic flux linked by a contour C.

6.7.2 Inductance

In Chapter 4 we pointed out that electrostatic problems are often difficult because the distribution of the electric charges is not known and is determined by the configuration of metallic conductors, thus requiring the solution of Poisson's equation. The concept of capacitance was introduced as a measure of the distribution of the electric field in the vicinity of conductor configurations. In practice, calculation of magnetostatic fields is relatively simpler than that of electrostatic fields, because the magnetostatic problem usually involves a known distribution of currents, from which the **B** field can be found using the Biot–Savart law. However, practical problems are nevertheless complicated by the magnetic properties of the surrounding medium. Fortunately, a detailed description of the **B** field is rarely desirable; more often than not, some overall measure of the field is sufficient. *Inductance* is this single measure of the distribution of the magnetic field near a current-carrying conductor.

Capacitance was introduced as a measure of the ability of a conductor configuration to hold charge per unit applied voltage, or store electrical energy, and was shown to be a property of the physical arrangement of the conductors. In a similar vein, inductance is another property of the physical layout of conductors and is a measure of the ability of a conductor configuration to *link magnetic flux,* or store magnetic energy. For our purposes, we define *flux linkage* as the integral of the magnetic field over the area enclosed by a closed circuit.

Before defining inductance, we must introduce the concept of flux linkage. Consider two neighboring closed loops C_1 and C_2 as shown in Figure 6.37. If a current I_1 flows around the closed loop C_1, a magnetic field $\mathbf{B}_1$ is produced, and some of this magnetic field links C_2 (i.e., will pass through the area S_2 enclosed by C_2).[54] This magnetic flux produced by the current I_1 flowing around C_1 but linked by the area S_2 enclosed by C_2 can be designated as

$$\Psi_{12} = \int_{S_2} \mathbf{B}_1 \cdot d\mathbf{s}_2$$

using the same subscript indexing as in previous cases, where the first subscript indicates the source of the quantity while the second indicates the location at which it is observed. In practice, inductances typically consist of coils with multiple turns. If C_1 consists of

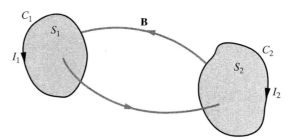

Figure 6.37   Two magnetically coupled circuits.

[54]We choose to not give a precise definition of the term "links," because intuition gives a better feel for this particular concept than any formal definition. Linking of one circuit by another can be thought of as the magnetic flux produced by one circuit "passing through" or connecting with another circuit.

multiple turns N_1, then the total flux produced is N_1 times larger (see Section 6.2.1); namely, $\Lambda_{12} = N_1 \Psi_{12}$.

The *mutual inductance* L_{12} between the two coils is thus defined as

$$L_{12} \equiv \frac{N_2 \Lambda_{12}}{I_1} = \frac{N_2 N_1 \Psi_{12}}{I_1}$$

Noting that in general C_2 has N_2 turns, the magnetic flux produced by C_1 is thus linked N_2 times by C_2.

Some of the magnetic flux produced by I_1 links the area S_1 enclosed by the closed contour C_1, so that we can also define the *self-inductance* of loop C_1 as

$$L_{11} \equiv \frac{N_1 \Lambda_{11}}{I_1} = \frac{N_1^2 \Psi_{11}}{I_1}$$

where $\Lambda_{11} = N_1 \Psi_{11}$ is the the total flux linked by a single turn of C_1, and Ψ_{11} is the magnetic flux produced by a single turn of C_1 and linked by a single turn of C_1.

Note that if a current I_2 flows around the closed loop C_2, it generates a magnetic flux that links C_1, so that the mutual inductance between the two loops is given by

$$L_{21} = \frac{N_1 N_2 \Psi_{21}}{I_2}$$

where Ψ_{21} is the flux generated by the current I_2 flowing around a single turn of C_2 and linking the area enclosed by a single turn of C_1.

To show that $L_{12} = L_{21}$, we can rely on the representation of the **B** field by a vector potential. Consider the expression for the mutual inductance L_{12}:

$$L_{12} = \frac{N_2}{I_1} \int_{S_2} \mathbf{B}_1 \cdot d\mathbf{s}_2$$

where $\mathbf{B}_1$ is the total field produced by C_1 (with N_1 turns) at the surface S_2 enclosed by C_2 (with N_2 turns). Noting that $\mathbf{B}_1 = \nabla \times \mathbf{A}_1$ we can rewrite L_{12} as

$$L_{12} = \frac{N_2}{I_1} \int_{S_2} \mathbf{B}_1 \cdot d\mathbf{s}_2 = \frac{N_2}{I_1} \int_{S_2} (\nabla \times \mathbf{A}_1) \cdot d\mathbf{s}_2 = \frac{N_2}{I_1} \oint_{C_2} \mathbf{A}_1 \cdot d\mathbf{l}_2$$

where we used Stokes's theorem. The vector potential $\mathbf{A}_1$ is related to its source current I_1 through (6.15), so we have

$$\mathbf{A}_1 = \frac{\mu_0 N_1 I_1}{4\pi} \oint_{C_1} \frac{d\mathbf{l}_1}{R}$$

where $R = |\mathbf{r} - \mathbf{r}'|$, with $\mathbf{r}$ and $\mathbf{r}'$ being the positions of the observation and source points, respectively. We can thus write

$$L_{12} = \frac{\mu_0 N_1 N_2}{4\pi} \oint_{C_1} \oint_{C_2} \frac{d\mathbf{l}_1 \cdot d\mathbf{l}_2}{R} \qquad (6.28)$$

which indicates that $L_{12} = L_{21}$, since the dot product is commutative and the line integrals can be interchanged.

Equation (6.28) is known as the *Neumann formula* for mutual inductance. It underscores the fact that the mutual inductance is only a function of the geometrical arrangement of the conductors. We shall see in Section 6.8 that if the medium the circuits are located in is a magnetic material, the constant μ_0 needs to be replaced by another constant to be defined as magnetic permeability μ. Thus, inductance also depends on the magnetic properties of the medium that the circuits are located in, as represented by its permeability μ. In general, carrying out the double line integral is quite involved. In most cases, we take advantage of the symmetries inherent in the problem to determine the flux linkage or stored magnetic energy so that we can find the mutual inductance without resorting to (6.28). In some cases, however, it actually is easier to use (6.28); we shall use this formula in Example 6.31 to determine the self-inductance of a current-carrying loop.

For any given conductor configuration, the self-inductance of the closed loop C can be evaluated in a manner very similar to the procedure used to find the capacitance in electrostatics. We can first assume a current I to flow in the closed loop, from which we can determine **B** using Ampère's law or the Biot–Savart law or the vector potential **A**. This magnetic field is proportional to the current I. We can then find the flux linked by the circuit by conducting an integral, namely

$$\Psi = \int_S \mathbf{B} \cdot d\mathbf{s}$$

where S is the area enclosed by the closed loop C. Given that the number of turns in the loop is N, and that all the turns produce and link the same flux, the self-inductance is then given by $L = (N^2\Psi)/I$. Note that I will cancel out so that L depends only on the geometrical arrangement (i.e., shape and dimensions) of the circuit C and the magnetic properties of the medium as mentioned in the preceding paragraph.

Example 6.26: Self-inductance of a long solenoid. Find the self inductance of a solenoidal coil of length l, radius a, and total number of turns equal to N.

> **Solution:** In Section 6.2.1, it was shown that for a long solenoid the intensity of the **B** field at the end points of the axis is half that at the center because of the flux leakage near the ends. However, this leakage is mainly confined to the ends of the solenoid and can be neglected. So, a good approximation is to assume that the **B** field is constant over the entire interior of the solenoid, being equal to its value at the center given by the result derived in Example 6.14, namely
>
> $$\mathbf{B} \simeq \mathbf{B}_{\text{ctr}} \simeq \hat{\mathbf{z}}\frac{\mu_0 NI}{l}$$
>
> As a result, the total flux linkage of every individual turn of the solenoidal coil is
>
> $$\Psi = B_z A = \frac{\mu_0 NI}{l}A$$

where A is the cross-sectional area of the solenoid (equal to $A = \pi a^2$ for a solenoid with a circular cross section of radius a). Since there are N turns, the total flux linkage of all N turns is

$$\Lambda = N\Psi = \frac{\mu_0 N^2 IA}{l}$$

Thus, the self-inductance of a long solenoid is

$$L = \frac{\Lambda}{I} = \frac{\mu_0 N^2 A}{l}$$

To get a feel about the orders of magnitude involved, let us calculate the self-inductance of a solenoid of 100 turns wound uniformly over a cylindrical wooden core ($\mu = \mu_0$) of length 10 cm and diameter 1 cm. Substituting these values yields

$$L = \frac{(4\pi \times 10^{-7} \text{ H-m}^{-1}) \times (100)^2 \times \pi (0.5 \times 10^{-2})^2 \text{ m}^2}{10 \times 10^{-2} \text{ m}} \simeq 9.87 \ \mu\text{H}$$

Using a linear ferromagnetic core material (Section 6.8) with $\mu = 1000\mu_0$ instead of the wooden core increases the self-inductance of the 100-turn solenoid by a factor of 1000, yielding $L \simeq 9.87$ mH.

Example 6.27: Inductance of a toroid. Find the inductance of a toroidal coil having N turns, similar to that shown in Figure 6.20 but having a circular cross section.

Solution: For a toroidal coil, we assume[55] that the mean radius of the toroid is much greater than the diameter of the coil; that is, $r_m = (a + b)/2 \gg b - a$. With this assumption, the magnetic field **B** is approximately uniform throughout the inner part of the toroid (also referred to as the core) and is given by

$$\mathbf{B} \simeq \hat{\phi} \frac{\mu_0 NI}{2\pi r_m} \qquad a \le r \le b$$

Since the same flux links all the turns, the total flux linkage is

$$\Lambda = N \int_S \mathbf{B} \cdot d\mathbf{s} = NB_\phi \int_S ds = NB_\phi A = \frac{\mu_0 N^2 IA}{2\pi r_m} \quad \rightarrow \quad L = \frac{\Lambda}{I} = \frac{\mu_0 N^2 A}{2\pi r_m}$$

where we noted that $A = \pi[(b - a)/2]^2$ is the cross-sectional area of the toroid.

For example, the inductance of a 1000-turn air-core toroid having a mean radius of $r_m = 5$ cm and a core diameter of $(b - a) = 1$ cm is approximately given by

$$L = \frac{(4\pi \times 10^{-7}) \times (1000)^2 \times \pi (0.5 \times 10^{-2})^2}{2\pi (5 \times 10^{-2})} \simeq 0.314 \text{ mH}$$

Again, the inductance of the toroid could be increased significantly by using a ferromagnetic type of core material.

---

[55]This assumption is by no means necessary in order to allow a tractable solution. See Problems 6.40 and 6.41.

Example 6.28: Coaxial inductor. In Example 4.27, we found the capacitance per unit length of an infinitely long coaxial line. Now we find the inductance per unit length of a coaxial line as shown in Figure 6.38. For simplicity, assume that the current flows only in thin layers[56] at $r = a$ and $r = b$, respectively, where $r = b$ is the inner radius of the outer conductor.

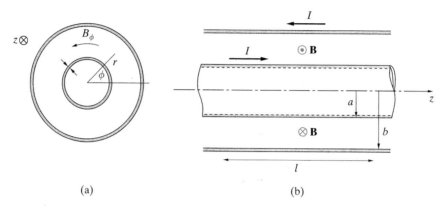

(a)   (b)

Figure 6.38   Coaxial transmission line with a thin-walled inner conductor.
(a) Cross-sectional view. (b) Side view.

Solution: The current in the inner conductor is I and the outer conductor is of the same magnitude with opposite direction. Using the results found in Example 6.12, the **B** field is nonzero only between the two conductors, given by

$$B_\phi = \frac{\mu_0 I}{2\pi r} \qquad a \leq r \leq b$$

The total flux linkage of the coaxial line of length l can be found as

$$\Psi = \int_S \mathbf{B} \cdot d\mathbf{s} = \int_0^l \int_a^b B_\phi \hat{\boldsymbol{\phi}} \cdot (\hat{\boldsymbol{\phi}} \, dr \, dz) = \frac{\mu_0 I l}{2\pi} \int_a^b \frac{dr}{r} = \frac{\mu_0 I l}{2\pi} \ln\left(\frac{b}{a}\right)$$

Hence, the inductance of the coaxial line of length l is

$$L = \frac{\Psi}{I} = \frac{\mu_0 l}{2\pi} \ln\left(\frac{b}{a}\right)$$

or, the inductance *per unit length* of the coaxial line is

$$L_u = \frac{L}{l} = \frac{\mu_0}{2\pi} \ln\left(\frac{b}{a}\right) = 0.2 \ln\left(\frac{b}{a}\right) \quad \mu\text{H-m}^{-1}$$

which is identical to the expression given in Table 2.2 for the inductance per unit length of a coaxial line.

---

[56]This is effectively the case when the walls of the conductors are thin. In addition, at high frequencies the current is effectively confined to a thin layer (skin depth) at $r = a$ for the inner conductor and at $r = b$ for the outer, even if the conductors are not thin (the inner conductor is typically not hollow).

Consider a practical coaxial line such as RG-214 (typically used for high-frequency (MHz range), high-power transmission) having an inner conductor diameter of $\sim$2.256 mm and the inner diameter of its outer conductor of $\sim$7.24 mm, filled with solid polyethylene (which has the same magnetic properties as air). Assuming the currents to be flowing in thin layers (excellent assumption at MHz frequencies, as we shall see in Chapters 8 and 10), the inductance per meter of this line is

$$L_u \simeq 0.2 \ln \frac{7.24}{2.256} \simeq 0.233 \ \mu\text{H-m}^{-1}$$

Example 6.29: Inductance of the two-wire line. In Example 4.29, the capacitance per unit length of a two-wire line was calculated. Find the inductance per unit length of the two-wire line, as shown in Figure 6.39.

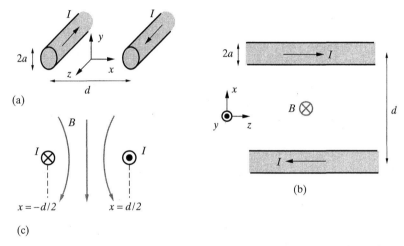

Figure 6.39   Two-wire transmission line. (a) Two parallel conductors, each carrying a current I, in opposite senses with radii a and separation distance d. (b) Horizontal cross-sectional view. (c) Vertical cross-sectional view.

Solution: For $d \gg a$, the steady currents in each conductor are assumed to be confined along their axes. Hence, the **B** field in the region of the $y = 0$ plane that extends from $x = -d/2 + a$ to $x = d/2 - a$ is approximately given by the sum of the fields of two infinitely long parallel wires as

$$B_y \simeq -\frac{\mu_0 I}{2\pi}\left(\frac{1}{x + d/2} - \frac{1}{x - d/2}\right)$$

The total flux linkage through the area on the $y = 0$ plane between the conductors each of length l can be found from

$$\Psi = \int_S \mathbf{B} \cdot d\mathbf{s} = \frac{\mu_0 I l}{2\pi} \int_{-d/2+a}^{d/2-a} \left[\frac{1}{x + d/2} - \frac{1}{x - d/2}\right] dx$$

$$= \frac{\mu_0 I l}{2\pi}\left[\ln(x + d/2) - \ln(x - d/2)\right]_{-d/2+a}^{d/2-a} = \frac{\mu_0 I l}{\pi} \ln\left(\frac{d - a}{a}\right)$$

Since $d \gg a$, we can write

$$\Psi \simeq \frac{\mu_0 I l}{\pi} \ln \frac{d}{a}$$

Therefore, the inductance per unit length of two-wire line is

$$L_u = \frac{L}{l} = \frac{\Psi}{Il} = \frac{\mu_0}{\pi} \ln \frac{d}{a} = 0.4 \ln \frac{d}{a} \quad \mu\text{H-m}^{-1}$$

The above result is for the case with $d \gg a$. A more accurate analysis for the general case shows that

$$L_u = \frac{\mu_0}{\pi} \ln \left[\frac{d}{2a} + \sqrt{\left(\frac{d}{2a}\right)^2 - 1} \right] = 0.4 \ln \left[\frac{d}{2a} + \sqrt{\left(\frac{d}{2a}\right)^2 - 1} \right] \quad \mu\text{H-m}^{-1}$$

which is the same as given in Table 2.2.

As a numerical example, consider a $Z_0 = 300\Omega$ air-spaced two-wire transmission line, which is commonly used in connecting a TV to an antenna. For this line, the ratio $d/(2a)$ is about 6, and the corresponding line inductance per unit length is approximately given as

$$L_u = \frac{4\pi \times 10^{-7}}{\pi} \ln[6 + \sqrt{6^2 - 1}] \simeq 0.991 \ \mu\text{H-m}^{-1}$$

Example 6.30: Two circular coils. Two circular coils with centers on a common axis as shown in Figure 6.40 have N_1 and N_2 turns, each of which is closely wound, and radii a and b, respectively. The coils are separated by a distance d, which is assumed to be much larger than both radii (i.e., $d \gg a, b$). Find the mutual inductance between the coils.

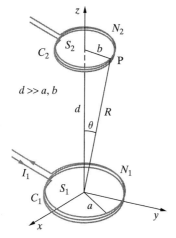

Figure 6.40   **Two circular coils of turns N_1 and N_2.**

Solution: Assume a current I_1 flowing in coil 1. Using the results obtained for a magnetic dipole, the magnetic vector potential $\mathbf{A}_1$ due to coil 1 at an arbitrary point P on coil 2 can be written as

$$\mathbf{A}_1 = \hat{\boldsymbol{\phi}} \frac{\mu_0 N_1 I_1 (\pi a^2)}{4\pi R^2} \sin\theta = \hat{\boldsymbol{\phi}} \frac{\mu_0 N_1 I_1 a^2}{4R^2} \frac{b}{R}$$

$$= \hat{\boldsymbol{\phi}} \frac{\mu_0 N_1 I_1 a^2 b}{4(d^2+b^2)^{3/2}} \simeq \hat{\boldsymbol{\phi}} \frac{\mu_0 N_1 I_1 a^2 b}{4d^3}$$

since $d \gg b$. The total flux linkage through the area S_2 of coil 2 due to the field $\mathbf{B}_1$ produced by coil 1 can be written as

$$\Lambda_{12} = N_2 \int_{S_2} \mathbf{B}_1 \cdot d\mathbf{s}_2 = N_2 \int_{S_2} (\nabla \times \mathbf{A}_1) \cdot d\mathbf{s}$$

Using Stokes's theorem, we can rewrite Λ_{12} as

$$\Lambda_{12} = N_2 \oint_{C_2} \mathbf{A}_1 \cdot d\mathbf{l} = N_2 \oint_{C_2} \frac{\mu_0 N_1 I_1 a^2 b}{4d^3} \hat{\boldsymbol{\phi}} \cdot (\hat{\boldsymbol{\phi}} b\, d\phi) = \frac{\mu_0 N_1 N_2 I_1 a^2 b}{4d^3} (2\pi b)$$

Therefore, the mutual inductance L_{12} is

$$L_{12} = \frac{\Lambda_{12}}{I_1} = \frac{\mu_0 \pi N_1 N_2 a^2 b^2}{2d^3}$$

6.7.3 Inductances of Some Practical Coils

The determination of the inductance of coils involving geometries more complicated than those considered in Examples 6.26 through 6.30 is a straightforward matter in principle and requires no new concepts but often involves difficult mathematical analyses. Much early work on electromagnetics involved the design of coils with optimum geometries for various applications.[57] In this section, we briefly discuss two types of practical coils: the finite-length solenoid and a single loop of wire.

---

[57] J. C. Maxwell himself spent considerable effort on the choice of the geometrical arrangement that would produce the maximum inductance for a fixed length of wire. See Article 706 in Vol. 2 of J. C. Maxwell, *A Treatise on Electricity and Magnetism*, Dover, 1954 (reprinted from 1873 original). The answer, by the way, turns out to be the Brooks inductor; see P. N. Murgatroyd, The Brooks inductor: A study of optimal solenoid cross-sections, *IEEE Proc.*, B 133, pp. 309–314, 1986; P. N. Murgatroyd and A. D. Hinley, The well tempered coil winder, *IEEE Trans. Educ.*, 37(4), pp. 329–331, 1994. An excellent summary of inductance formulas complete enough to satisfy the ordinary needs of engineers and physicists and accurate to better than 0.5% is given in Section 2 of F. E. Terman, *Radio Engineers' Handbook*, McGraw-Hill, New York, 1943. More extensive collections of formulas are provided in E. B. Rosa and F. W. Grover, Formulas and tables for the calculation of mutual and self inductance, *Bur. Standards Bull.*, 8, pp. 1–237, January 1, 1912, and in F. W. Grover, *Inductance Calculations, Working Formulas and Tables*, Dover, 1946.

Inductance of solenoids of finite length. The expression for the self-inductance of a solenoid as derived in Example 6.26 is valid only for an infinitely long solenoid and is a good approximation for solenoids with large length-to-radius ratio (i.e., $l \gg a$). For solenoids of finite length, it is clear that the inductance should be different, since the **B** field is obviously not constant throughout the solenoid as was assumed in Example 6.26. In fact, based on the discussion in Section 6.2.1, the axial **B** field at the ends of the solenoid is equal to one-half of that at the center, because of the leakage of the **B** field outside the solenoid. As a result of leakage, the **B** field also varies over the cross section of the solenoid, whereas in the infinitely long solenoid the **B** field is constant over the cross section as was shown in Example 6.14. The exact distribution of the **B** field for a finite-length solenoid is difficult to evaluate, because of the nature of the integrals involved; however, the shape of the **B** field lines is expected to be as shown in Figure 6.41a.

For solenoids of intermediate length, empirical or semiempirical formulas for inductance are available. The most famous of these is the Nagaoka formula,[58] which applies a correction factor to the formula for an infinitely long solenoid. According to this,

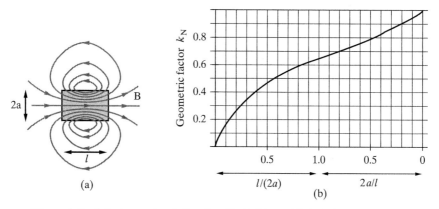

Figure 6.41 Inductance of a finite solenoid. (a) Lines of **B** field for a solenoid of finite length. Adapted from Figure 565 in E. Durand, *Électrostatique et Magnétostatique,* Masson et Cie, Paris, 1953. (b) The Nagaoka correction factor for the inductance of a solenoid of intermediate length.

[58]Prof. H. Nagaoka, Professor of Physics at the Imperial University, Tokyo, wrote a series of papers on self and mutual inductance of concentric coaxial solenoids and other coils between 1903 and 1912, starting with H. Nagaoka, On the potential and lines of force of a circular current, *Phil. Mag.,* 6, p. 19, 1903. Other papers included: *J. Coll. Sci., Tokyo,* 27, art. 6, 1909, and *Tokyo Math. Phys. Soc.,* 6, p. 10, 1911. Also see E. B. Rosa and F. W. Grover, Formulas and tables for the calculation of mutual and self inductance, *Bur. Standards Bull.,* 8, pp. 1–237, January 1, 1912. Nagaoka was also interested in atomic structure and proposed in 1904 that electrons encircled positively charged atoms much like the planets encircling the sun, rather than the atom being a sphere of positively charged matter with electrons placed on its surface, as was suggested by J. J. Thomson, who discovered the electron. Within two years, E. Rutherford showed that there was indeed a central positively charged nucleus.

the self-inductance of a short solenoid is less than that of a long one by a factor of k_N,

$$L = k_N \frac{\mu_0 N^2 A}{l}$$

where k_N is a dimensionless number less than unity. Numerical values of k_N are plotted in Figure 6.41 for values of $l/(2a)$ from zero to unity and for values of $2a/l$ from unity to zero (where a is the radius of the cross section of the solenoid, so $A = \pi a^2$), thus covering the complete range from no solenoid to an infinitely long solenoid. A simple approximate formula valid to within less than a few percent for solenoidal coils with $l > 0.8a$ is[59]

$$L \simeq \frac{10\mu_0 \pi a^2 N^2}{9a + 10l}$$

Inductance of a single loop of wire. The extreme case of a short solenoid is a single-conductor loop of arbitrary shape. In general, in evaluating the self-inductance of such a loop, we have to account for the internal and external inductances of the wire separately. For most circuits the total magnetic flux generated by a current can be partitioned into the portion lying outside the conductor plus the flux that lies wholly inside the conductor. The storage of magnetic energy and flux linkages associated with the internal flux lead to an *internal inductance*, while those associated with the fluxes outside the conductor are represented by the *external inductance*. Up to now, we have implicitly assumed circuits consisting of filamentary currents, thus neglecting internal inductance. For most high-frequency applications, internal inductance is negligible, since the magnetic fields (and thus magnetic fluxes) are confined to a very thin region on the surface of the current-carrying conductors. Determination of internal inductance using the linking of magnetic flux requires the introduction of the concept of partial flux linkages and unnecessarily complicated analysis.[60] It is generally much easier to evaluate the internal inductance by considering the magnetic energy stored inside the conductor, as will be done in Example 6.32, after we briefly introduce the concept of magnetic energy. We show in Example 6.32 that the dc internal inductance per unit length of an infinitely long wire (or the inner conductor of a coaxial line) is $\mu/(8\pi)$, where μ is the magnetic permeability (defined in Section 6.8) of the material the wire is made of. It is shown in Section 6.8 that $\mu \simeq \mu_0$ for most materials, including highly conducting metals such as copper and aluminum, and that μ is substantially different from μ_0 only for magnetic materials such as iron. Thus, the internal self-inductance per unit length of a wire is in most cases simply equal to $\mu_0/(8\pi)$.

For a thin wire of total length l bent into an arbitrary loop as shown in Figure 6.42, the magnetic field near the surface is very nearly the same as that for an infinitely long

---

[59]H. A. Wheeler, Simple inductance formulas for radio coils, *Proc. I. R. E.,* 16, pp. 1398–1400, 1928.

[60]For examples of partial flux linkage analyses, see Section 5-11 of C. T. A. Johnk, *Engineering Electromagnetic Fields and Waves*, John Wiley & Sons, New York, 1988, and Section 8.5 of R. Plonsey and R. E. Collin, *Principles and Applications of Electromagnetic Fields*, McGraw-Hill, New York, 1961.

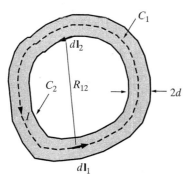

Figure 6.42   A conductor of finite radius d bent into a closed loop.

wire, provided the radius of curvature of the wire is much greater than the radius of the wire at all points. In other words, we may treat the wire locally as though it were part of an infinitely long wire. Thus, the internal self-inductance of any loop of mean length l is[61] $\mu_0 l/(8\pi)$. Thus, the internal self-inductance of a thin wire ($d \ll a$) circular loop (Figure 6.43) of radius a is $\mu_0(2\pi a)/(8\pi) = \mu_0 a/4$.

To evaluate the external self-inductance of an arbitrarily shaped loop, we can apply the Neumann formula (6.28) for the mutual inductance between two loops. With respect to Figure 6.42, consider the contour C_2 to coincide with the interior edge of the conducting loop. The external self-inductance is due to the flux produced by the current in the conductor that links the contour C_2. To evaluate the magnetic flux linking the interior contour C_2, we may assume with negligible error that the total current I flowing in the conductor is concentrated in an infinitely thin filamentary contour C_1 along the center of

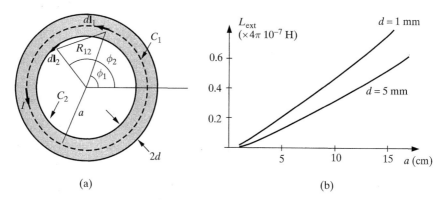

(a)   (b)

Figure 6.43   A circular conducting loop. (a) The circular loop showing various quantities referred to in the text. (b) Plots of the external self-inductance L_{ext} versus loop radius a for two different values of d, namely $d = 1$ mm and 5 mm.

---

[61]Note, however, that at frequencies in the radio range or higher the internal self-inductance is much smaller than the external one, since the magnetic flux is confined only to a small region on the surface, resulting in much smaller magnetic energy storage within the conductor.

the conductor. The external self-inductance can now be evaluated by finding the mutual inductance between contours C_1 and C_2. We have

$$L_{\text{ext}} = L_{12} = \frac{\mu_0}{4\pi} \oint_{C_1} \oint_{C_2} \frac{d\mathbf{l}_1 \cdot d\mathbf{l}_2}{R_{12}} \qquad (6.29)$$

where R_{12} is the distance between the differential elements $d\mathbf{l}_1$ and $d\mathbf{l}_2$.

Example 6.31: Self-inductance of a circular loop. Consider a conductor of radius d bent into a circular shape of mean radius a, as shown in Figure 6.43a. Find the external self-inductance of this circular loop.

Solution: To determine the external inductance, we use the expression (6.29) for L_{ext}. From Figure 6.43a, the magnitudes of $d\mathbf{l}_1$ and $d\mathbf{l}_2$ are given by

$$|d\mathbf{l}_1| = a\,d\phi_1 \qquad |d\mathbf{l}_2| = (a - d)\,d\phi_2 \simeq a\,d\phi_2$$

where we have assumed $a \gg d$ (i.e., a thin wire loop). The angle between $d\mathbf{l}_1$ and $d\mathbf{l}_2$ is $(\phi_2 - \phi_1)$, and we have

$$d\mathbf{l}_1 \cdot d\mathbf{l}_2 = a^2 \cos(\phi_2 - \phi_1)\,d\phi_1\,d\phi_2$$

Using the law of cosines, the distance R_{12} between the two differential elements is given by

$$R_{12}^2 = a^2 + (a - d)^2 - 2a(a - d)\cos(\phi_2 - \phi_1)$$

Thus, the external self-inductance is given by

$$L_{\text{ext}} = \frac{\mu_0 a^2}{4\pi} \int_0^{2\pi} \int_0^{2\pi} \frac{\cos(\phi_2 - \phi_1)\,d\phi_1\,d\phi_2}{\sqrt{a^2 + (a - d)^2 - 2a(a - d)\cos(\phi_2 - \phi_1)}}$$

We can now integrate over ϕ_2 first, after changing variables to replace $(\phi_2 - \phi_1)$ by ϕ and $d\phi_2$ by $d\phi$; thus,

$$L_{\text{ext}} = \frac{\mu_0 a^2}{4\pi} \int_0^{2\pi} \int_0^{2\pi} \frac{\cos\phi\,d\phi\,d\phi_1}{\sqrt{d^2 + 2a(a - d)(1 - \cos\phi)}}$$

where we have also rewritten the denominator in a more compact form. Note that the limits of integration for ϕ are kept as 0–2π, since the origin for ϕ_1 is arbitrary because of the circular symmetry. The integration over ϕ_1 simply brings in a multiplication with 2π, so we have

$$L_{\text{ext}} = \frac{\mu_0 a^2}{2} \int_0^{2\pi} \frac{\cos\phi\,d\phi}{\sqrt{d^2 + 2a(a - d)(1 - \cos\phi)}}$$

The above integral does not have a compact analytical form, but it can be expressed in terms of elliptic integrals. However, it is even simpler to evaluate L_{ext} numerically for any set of parameters, since the integrand is a well-behaved function. Plots of L_{ext} for selected parameters are shown in Figure 6.43b. For d in the range of a few mm, and radius a being a few to tens of cm, the external self-inductance is of order μH or less. For the same range of parameters, the internal self-inductance is typically much smaller.

For a thin wire circular loop ($d \ll a$), the result of the above integral reduces to

$$L_{\text{ext}} \simeq \mu_0 a \left(\ln \frac{8a}{d} - 2 \right)$$

On the scale of Figure 6.43b, the values of L_{ext} given by this approximate expression are indistinguishable from those evaluated directly from the integral.

As a matter of practical interest, it is useful to consider the inductance of single loops of shapes other than a circle as in Example 6.31. Fundamentally, we expect the inductance to depend more on the area of the loop than on the particular shape; however, for a given length of conductor, the circular shape represents the best utilization for the purpose of producing and linking maximum magnetic flux. A simple formula[62] for a single-conductor thin-wire loop of fixed length l is

$$L \simeq \frac{\mu_0 l}{2\pi} \left(2.303 \log_{10} \frac{4l}{d} - \theta_s \right)$$

where θ_s is a shape factor. The values of θ_s for different shapes are 2.451, 2.636, 2.853, and 3.197, respectively, for a circle, a regular hexagon, a square, and an equilateral triangle.

6.7.4 Energy in a Magnetic Field

Just as configurations of charges store electrostatic energy, configurations of stationary currents store magnetic energy. In electrostatics we were able to find expressions for energy stored in electric fields by considering the work necessary to bring charges from infinity to their locations. Such an approach is not appropriate in magnetostatics, since *magnetic forces do no work*. Consider a charge q propagating at a velocity $\mathbf{v}$ (a charge q moving with velocity $\mathbf{v}$ constitutes an element of current $I d\mathbf{l}$) under the influence of a magnetic field $\mathbf{B}$. The work done on the charge by the magnetic field is given by

$$dW_{\text{m}} = \mathbf{F}_{\text{m}} \cdot d\mathbf{l} = (q\mathbf{v} \times \mathbf{B}) \cdot (\mathbf{v} dt) = 0$$

Since the magnetic force is always perpendicular to the motion, the work done by this force must be zero.

To calculate the energy in the magnetic fields, we must instead consider the *creation* of steady-state current configurations, since the associated magnetic fields

[62]See Section 2 of F. E. Terman, *Radio Engineers' Handbook*, McGraw-Hill, Inc., New York, 1943.

involve an initial transient period during which currents are increased from zero to their final values. According to Faraday's law (Chapter 7), such time-varying fields generate electromotive forces against which work must be done. The work to establish the current distribution is thus done during this initial transient period, so we must consider time-varying fields in order to evaluate the total energy stored in a configuration of stationary currents.

As we are not yet equipped with Faraday's law and the concept of induced electric fields, we delay a complete consideration of energy in magnetic fields to Chapter 7. However, having just studied the concept of inductance, we provide below a preliminary discussion of stored magnetic energy, with expressions drawn from Chapter 7.

Analogous to the energy density $w_e = \frac{1}{2}\epsilon_0 E^2$ (J-m^{-3}) stored in electric fields, the magnetic energy density stored in a magnetic field configuration in free space is given by

$$w_m = \frac{1}{2\mu_0}B^2$$

in units of J-m^{-3}. The total magnetic energy W_m stored in a given steady-current configuration can be found by integrating the associated **B** field *over the entire volume V of its existence*, namely

$$W_m = \frac{1}{2}\int_V \frac{B^2}{\mu_0}\, dv' \tag{6.30}$$

Given any configuration of current-carrying conductors, we know from circuit theory that we can alternatively express the stored magnetic energy using the inductance[63] L of the configuration. Namely,

$$W_m = \tfrac{1}{2}LI^2 \tag{6.31}$$

where I is the current in the circuit. This relation between W_m and inductance is formally derived in Section 7.3. We have thus arrived at an alternative method of calculating the inductance of a circuit; namely, we can determine W_m from (6.30) and then find L from

$$\boxed{L = \frac{2W_m}{I^2}} \tag{6.32}$$

Example 6.32: Inductance of a coaxial line. Consider a coaxial line with a solid inner conductor of radius a and an outer conductor of thickness $(c - b)$ as was shown in Figure 6.17, and discussed in Example 6.12. Find the inductance per unit length of this coaxial line using (6.32).

---

[63]Note that the inductance we refer to in this context is the self-inductance L_{11} of a circuit C_1, in which a current I_1 is assumed to flow. In this context, we drop the subscript 1 and simply use L and I.

Solution: The magnetic field distribution for this case was found in Example 6.12 to be

$$B_\phi = \begin{cases} \dfrac{\mu_0 I r}{2\pi a^2} & r \leq a \\[2mm] \dfrac{\mu_0 I}{2\pi r} & a < r \leq b \\[2mm] \dfrac{\mu_0 I}{2\pi r}\dfrac{c^2 - r^2}{c^2 - b^2} & b < r \leq c \\[2mm] 0 & r > c \end{cases}$$

The total stored magnetic energy can be found by separately evaluating the energy stored in all three regions in which the field is nonzero. Considering a unit-length coaxial line, for $r \leq a$ we have

$$W_{m1} = \frac{1}{2\mu_0}\int_0^a B_{1\phi}^2 2\pi r\, dr = \frac{\mu_0 I^2}{4\pi a^4}\int_0^a r^3\, dr = \frac{\mu_0 I^2}{16\pi}$$

in units of J-m^{-1}. Similarly, for $a < r \leq b$ we have

$$W_{m2} = \frac{1}{2\mu_0}\int_a^b B_{2\phi}^2 2\pi r\, dr = \frac{\mu_0 I^2}{4\pi}\int_a^b \frac{1}{r}\, dr = \frac{\mu_0 I^2}{4\pi}\ln\frac{b}{a}$$

and for $b < r \leq c$

$$W_{m3} = \frac{1}{2\mu_0}\int_b^c B_{3\phi}^2 2\pi r\, dr = \frac{\mu_0 I^2}{4\pi(c^2 - b^2)^2}\int_b^c \frac{(c^2 - r^2)^2}{r}\, dr$$

$$= \frac{\mu_0 I^2}{4\pi(c^2 - b^2)^2}\int_b^c \left(\frac{c^4}{r} - 2c^2 r + r^3\right)dr = \frac{\mu_0 I^2}{4\pi(c^2 - b^2)^2}\left[c^4\ln r - c^2 r^2 + \frac{r^4}{4}\right]_b^c$$

$$= \frac{\mu_0 I^2}{4\pi}\left[\left(\frac{c^2}{c^2 - b^2}\right)^2 \ln\frac{c}{b} - \frac{c^2}{c^2 - b^2} + \frac{1}{4}\left(\frac{c^2 + b^2}{c^2 - b^2}\right)\right]$$

The self-inductance of the coaxial line per unit length is then found to be

$$L_u = \frac{2(W_{m1} + W_{m2} + W_{m3})}{I^2}$$

$$= \underbrace{\frac{\mu_0}{2\pi}\ln\frac{b}{a}}_{\text{External inductance}} + \underbrace{\frac{\mu_0}{8\pi} + \frac{\mu_0}{2\pi}\left[\left(\frac{c^2}{c^2 - b^2}\right)^2 \ln\frac{c}{b} - \frac{c^2}{c^2 - b^2} + \frac{1}{4}\left(\frac{c^2 + b^2}{c^2 - b^2}\right)\right]}_{\text{Internal inductance}}$$

The first term (due to W_{m2}) is identical to the self-inductance of the thin-walled coaxial line calculated in Example 6.28. The last two terms (due to W_{m1} and W_{m3}), represent the inductance due to the magnetic energy stored *inside* the inner and outer conductors, respectively. These inductances can be important only at near-dc frequencies, and if the wires are not completely nonmagnetic[64] (i.e., $\mu > \mu_0$). At any appreciable frequency, it will

---

[64]For a quantitative discussion of cases in which internal inductance is important, see Chapter 7 of H. H. Skilling, *Electric Transmission Lines*, McGraw-Hill, 1951.

be shown in Chapter 8 that the current is effectively confined to a thin layer at $r = a$ on the inner and at $r = b$ on the outer conductor, resulting in negligible internal stored magnetic energy and, thus, negligible internal self-inductance. Therefore, the per unit length self-inductance of the coaxial line is for all practical purposes given by

$$L_{\mathrm{u}} \simeq L_{\mathrm{u,ext}} = \frac{\mu_0}{2\pi} \ln\left(\frac{b}{a}\right)$$

as was found using considerations of flux linkage in Example 6.28.

6.8 MAGNETIC FIELDS IN MATERIAL MEDIA

Our discussion of magnetostatics has so far been concerned with the physical effects produced by steady-state currents in vacuum or free space. In this section, we introduce and discuss the additional effects associated with the presence of material media. Our goals in this respect, as well as the level of discussion provided, are very similar to those in Section 4.10 on dielectric materials. We must necessarily start with a description of the material behavior at a microscopic level but then perform suitable space and time averages to represent the microscopic effects in terms of macroscopic field quantities.

Materials are typically classified according to their most important characteristics. Glass is a *dielectric* because its dielectric constant is appreciably different from that of free space (i.e., $\epsilon \simeq 4$–9). The conductivity of glass is extremely small ($\sim 10^{-12}$ S-m^{-1} at 20°C), and its behavior in the presence of magnetic fields is not measurably different than that of free space. Copper, on the other hand, is a *conductor* because its conductivity is very large (5.7×10^8 S-m^{-1}), while its electric permittivity and magnetic properties are essentially the same as those of free space. The class of materials known as *magnetic* are those that exhibit magnetic properties. The most common example of such materials is iron, which is also a conductor of electricity ($\sigma_{\mathrm{iron}} \simeq 10^7$ S-m^{-1}) but is better known for its striking magnetic behavior. Some other materials, such as nickel, cobalt, and a few special alloys, also exhibit similarly strong and clearly evident magnetic properties. Materials with such striking magnetic behavior are known as *ferromagnetic* materials; they exhibit permanent magnetic dipole moments and retain their magnetization after the external field is removed or when no macroscopic current flows in them or in their vicinity. The physics of ferromagnetic materials is inherently complex and is beyond the scope of this book. However, all substances exhibit some magnetic effects, although at levels substantially lower (typically 10^3–10^6 times less) than in ferromagnetic materials. In the following we confine our attention to a description of ordinary magnetism, which provides a basis for our formulation of magnetostatics in the presence of ordinary (i.e., nonferromagnetic) material media.

In Section 4.10 we stated that in isotropic materials electric polarization **P** was in the same direction as the electric field **E**, and in Section 4.13 we showed that dielectrics are always attracted toward regions of higher electric field. Unlike this electrical effect in matter, some materials are attracted toward regions of higher magnetic fields, whereas others are repelled from such regions. Although the magnetic effect in ordinary materials is quite small, these two different tendencies of some materials (i.e., attraction or repulsion) can be easily demonstrated[65] using a strong electromagnet with a sharply pointed pole to create a gradient in the magnetic field. Substances such as bismuth are repelled away from the high-field region; they are said to acquire magnetization in a direction *opposite* (or antiparallel) to the external field and are called *diamagnetic* materials. Other materials, such as aluminum, are attracted toward the high-field region; they are said to be magnetized in the same direction (i.e., parallel) as the external field and are thus referred to as *paramagnetic* materials.

The magnetic behavior of materials on a macroscopic scale is inherently due to the fact that tiny currents exist in all materials on an atomic (microscopic) scale. Every medium is composed of atoms, and these atoms may be considered as consisting of electrons moving in orbits about fixed nuclei. Both these orbital motions and the inherent spins of the electrons about their axes constitute microscopic electric currents. On a macroscopic scale, these current loops are infinitesimal in size, so we can treat them as small magnetic dipoles. In ordinary materials the microscopic magnetic dipole moments are randomly oriented and cancel each other out when averaged over a finite volume. However, an external magnetic field leads to a net alignment of these magnetic dipoles, causing the medium to be magnetically polarized.

6.8.1 Microscopic Basis of Diamagnetism and Paramagnetism

Both diamagnetic and paramagnetic behavior of materials are inherently quantum mechanical phenomena that cannot be properly described in classical terms.[66] However, especially for diamagnetic materials, we can provide some semiquantitative classical arguments to illustrate the basic processes that lead to the formation of induced magnetic dipole moments in a direction opposite to the applied field. Paramagnetism is due almost entirely to the spin magnetic dipole of the electrons, so it is more difficult to discuss its bases in quantitative classical terms.

Diamagnetism can be qualitatively understood by considering the effects of an external magnetic field on electrons orbiting about their nuclei. For simplicity, we consider the case for which the applied magnetic field is aligned with the axis of revolution of an electron as shown in Figure 6.44. In the classical model, the electron orbits around the positively charged nuclei in such a way that the outward centrifugal force is balanced by the Coulomb attraction force. If the radius of the orbit is a and the angular velocity is ω_0 in the absence of the external field, we must have

$$m_e \omega_0^2 a = \frac{q_e^2}{4\pi\epsilon_0 a^2} \quad \rightarrow \quad \omega_0^2 = \frac{q_e^2}{4\pi\epsilon_0 m_e a^3}$$

The magnetic dipole moment of this unperturbed orbit is the current times the area. The current is the charge per unit time that passes any point on the orbit, or simply the charge

---

[65] For further discussion, see R. P. Feynman, R. B. Leighton, and M. Sands, *The Feynman Lectures on Physics,* Addison-Wesley, Reading, Massachusetts, 1964.

[66] See Section 34-35 of *The Feynman Lectures on Physics,* Addison-Wesley, Reading, Massachusetts, 1964.

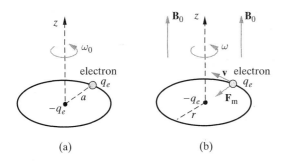

Figure 6.44 Perturbation of electron orbit by applied B field. The nuclei are positively charged by an amount $-q_e \simeq 1.6 \times 10^{-19}$ C. (a) No external **B** field. (b) The presence of the external **B** field produces a magnetic force $\mathbf{F}_{\mathrm{m}}$.

q_e divided by the time taken for one revolution, or $2\pi/\omega_0$. The unperturbed magnetic dipole moment of the electron orbit is thus

$$\mathbf{m}_{\mathrm{u}} = \hat{\mathbf{z}} I \left(\pi a^2\right) = \hat{\mathbf{z}} q_e \frac{\omega_0}{2\pi} \pi a^2 = \hat{\mathbf{z}} \frac{q_e \omega_0 a^2}{2}$$

Since the charge of the electron $q_e = -1.6 \times 10^{19}$ C, the current flow is in the direction opposite to that of the electron, so the dipole magnetic moment is oriented in the $-\hat{\mathbf{z}}$ direction.

In the presence of the external field $\mathbf{B} = \hat{\mathbf{z}} B_0$, the electron motion is influenced by an additional force given by $\mathbf{F}_{\mathrm{m}} = q_e \mathbf{v} \times \mathbf{B}$, where $\mathbf{v}$ is the linear velocity of the electron, whose magnitude is given by $|\mathbf{v}| = \omega r$, where ω is the new angular velocity and r is the new orbital radius, both of which may be different from their values in the absence of the external field. The force balance equation that is now required for the electron to orbit is

$$m_e \omega^2 r = \frac{q_e^2}{4\pi \epsilon_0 r^2} - q_e r \omega B_0$$

Assuming[67] that the orbital radius remains constant, so that we have $r \simeq a$, we have

$$\omega^2 = \omega_0^2 - \frac{q_e B_0}{m_e} \omega$$

Since we expect the applied field $\mathbf{B} = \hat{\mathbf{z}} B_0$ to produce only a small perturbation (i.e., $(\omega - \omega_0) \ll \omega_0$), we can write

$$\omega^2 - \omega_0^2 = (\omega - \omega_0)(\omega + \omega_0) \simeq 2\omega_0(\omega - \omega_0)$$

[67]It is difficult to justify this assumption in the context of the classical model discussed here. However, an alternative way to study the effect of the external magnetic field on the electron orbit is to assume the field **B** to be slowly turned on. The associated time rate of change of the magnetic field would then induce an electric field via Faraday's law (see Chapter 7), which would accelerate the electron. It can be shown that the associated increase in the kinetic energy of the electron is exactly equal to that required to sustain circular motion at the same radius. For further discussion, see Chapter 9 of W. B. Cheston, *Elementary Theory of Electric and Magnetic Fields,* John Wiley & Sons, New York, 1964.

so that the net effect of the applied field is to accelerate the electron to a new angular speed given by

$$\omega \simeq \omega_0 - \frac{q_e B_0}{2m_e}$$

thus increasing the magnetic dipole moment of the electron orbit by an amount

$$\Delta \mathbf{m} = -\hat{\mathbf{z}} \frac{q_e^2 a^2 B_0}{4m_e}$$

We note that this induced differential magnetic moment $\Delta \mathbf{m}$ is in the direction opposite, or *antiparallel,* to $\mathbf{B}$, hence the term *dia*magnetism. Note that if the applied field were in the $-\hat{\mathbf{z}}$ direction, the orbiting electron would then be slowed down by the $q_e \mathbf{v} \times \mathbf{B}$ force, so that the induced magnetic dipole moment would now be in the $+\hat{\mathbf{z}}$ direction—still opposite to the applied field.

In the absence of an applied field, the electron orbits are randomly oriented with respect to one another, and the orbital dipole moments cancel out when averaged over any finite volume. With the applied field, however, each atom acquires[68] a small additional dipole moment, and all these incremental moments are antiparallel (i.e., in the opposite direction) to the applied field. If N is the effective number of electron orbits per unit volume, the induced magnetic dipole moment per unit volume, also called *magnetization* and denoted by $\mathbf{M}$, is given by

$$\mathbf{M} = -\frac{N q_e^2 a^2 \mathbf{B}}{4m_e} \tag{6.33}$$

where $\mathbf{M}$ is expressed in A-m^{-1}. This quantity is entirely analogous to the polarization vector $\mathbf{P}$ for dielectric materials, as defined in (4.69). In order to keep analogous notation with (4.57), we might be tempted to write

$$\mathbf{M} = \frac{\chi_m \mathbf{B}}{\mu_0} \tag{6.34}$$

where χ_m is the magnetic susceptibility. However, the magnetic susceptibility as traditionally defined relates the magnetization $\mathbf{M}$ to the magnetic field intensity $\mathbf{H}$, which we introduce in the next section. With this definition, the relation between the magnetic susceptibility and the $\mathbf{B}$ field is

$$\mathbf{M} = \left(\frac{\chi_m}{1 + \chi_m}\right) \frac{\mathbf{B}}{\mu_0} \tag{6.35}$$

---

[68] We should note that the geometry considered in Figure 6.44 is rather idealized, with $\mathbf{B}$ being along the axis of revolution. The interaction with the applied field of other electron orbits is more complicated. Nevertheless, the net effect is that while the electrons in an atom move nearly in the same orbits in the presence of an applied field as they do in zero field, superimposed upon their rapid orbital motion of the electron is a slow uniform precessional motion of the entire atom about the field direction.

Substitution in (6.33) of typical values for N of 10^{28} to 10^{29} atoms-m^{-3} in a solid and a of order one angstrom shows that $\chi_m/(1 + \chi_m)$ is of order $\sim -10^{-6}$ to -10^{-5}, indicating that the diamagnetic effect is very small and negligible in most practical situations (i.e., $|\chi_m/(1 + \chi_m)| \ll 1$). Also, when $|\chi_m/(1 + \chi_m)| \ll 1$, (6.34) is an excellent approximation, since $\chi_m/(1 + \chi_m) \approx \chi_m$.

Based on the foregoing discussion, all materials should exhibit diamagnetic behavior. However, many materials exist where the magnetization **M** is parallel to the applied **B**-field, a good example being aluminum. Such materials are called *paramagnetic* materials, and are characterized by a positive χ_m. Paramagnetism is due almost entirely to the spin magnetic dipole moment of the electron. It is not really possible to understand paramagnetism in classical terms; however, we can note that the primary cause of the effect is that the atoms of some materials have a permanent magnetic dipole moment associated with them. In the absence of an external field, the atomic and molecular dipoles are randomly oriented, and the net resultant magnetic dipole moment, averaged over a finite volume, vanishes. However, in the presence of an applied magnetic field the individual electron spin dipoles are aligned with the field. The macroscopic moment per unit volume of magnetization **M** produced by this orientation is parallel to **B** for linear materials, thus corresponding to positive susceptibility χ_m. Typical values of χ_m for paramagnetic materials is of order 10^{-5} to 10^{-4} at room temperature. For example, oxygen gas is paramagnetic, with $\chi_m = 2.1 \times 10^{-6}$ at atmospheric pressure and room temperature. Many metals are also paramagnetic, such as aluminum, with $\chi_m = 2.1 \times 10^{-5}$ at room temperature. The magnetic susceptibility χ_m is typically inversely proportional to absolute temperature.[69] In materials that exhibit paramagnetic behavior, diamagnetic effects are also in general present. However, the diamagnetic reaction is completely masked by the paramagnetic effect, for reasons that cannot be described without constructing a microscopic theory of the material—a task well beyond the scope of this book.

6.8.2 Magnetic Field Intensity and the Permeability Concept

We saw in the previous section that the net effect of an applied magnetic field on a magnetic material is to create a macroscopic magnetization field **M**. We now proceed to consider the magnetic fields generated by this induced macroscopic magnetic dipole moment. In other words, we now set out to replace the material medium possessing an **M**-field in its interior by an equivalent distribution of currents residing in vacuum. The procedure we follow for this purpose is quite analogous to that used in Section 4.10.2 for the consideration of fields generated by induced electric dipole moments in dielectric materials.

We start by recalling a result derived in Section 6.6, namely that the vector potential **A** at a point **r** due to an infinitesimal magnetic dipole **m** located at point **r**$'$ is given by

$$\mathbf{A}(\mathbf{r}) = \frac{\mu_0}{4\pi} \frac{\mathbf{m} \times \hat{\mathbf{R}}}{R^2}$$

[69]See Chapter 15 of C. Kittel, *Introduction to Solid State Physics,* Wiley, New York, 1971.

where $\hat{\mathbf{R}}$ is the unit vector pointing from the source point $\mathbf{r}'$ to the observation point $\mathbf{r}$ and is given by $\hat{\mathbf{R}} = \mathbf{R}/R$ where $\mathbf{R} = (\mathbf{r} - \mathbf{r}')$ and $R = |\mathbf{r} - \mathbf{r}'|$. If the material medium under consideration is subdivided into differential volumes dv', each possessing a dipole moment $\mathbf{M}(\mathbf{r}')dv'$, the vector potential due to all elemental magnetic dipoles in a given volume V' of the material can be written as

$$\mathbf{A}(\mathbf{r}) = \frac{\mu_0}{4\pi} \int_{V'} \frac{\mathbf{M}(\mathbf{r}') \times \hat{\mathbf{R}}}{R^2} \, dv' = \frac{\mu_0}{4\pi} \int_{V'} \left(\mathbf{M}(\mathbf{r}') \times \nabla' \frac{1}{R} \right) dv'$$

where we have used the identity $\nabla' R^{-1} = -\nabla R^{-1} = (\hat{\mathbf{R}}/R^2)$, as established in Footnote 40 in Section 6.5, and where ∇' represents the del operator applied only to the source coordinates $(\mathbf{r}')$. We now use a vector identity, namely,

$$\nabla' \times \left[\frac{\mathbf{M}}{R} \right] \equiv \frac{1}{R} \nabla' \times \mathbf{M} - \mathbf{M} \times \nabla' \frac{1}{R}$$

Substituting this identity in the expression for $\mathbf{A}(\mathbf{r})$, we find

$$\mathbf{A}(\mathbf{r}) = \frac{\mu_0}{4\pi} \int_{V'} \frac{\nabla' \times \mathbf{M}(\mathbf{r}')}{R} \, dv' - \frac{\mu_0}{4\pi} \int_{V'} \nabla' \times \left[\frac{\mathbf{M}(\mathbf{r}')}{R} \right] dv'$$

The second integral in this expression can be transformed into an integral over a closed surface using an identity reminiscent of Stokes's theorem:

$$\int_V \nabla \times \mathbf{G} \, dv \equiv - \oint_S \mathbf{G} \times \hat{\mathbf{n}} \, ds$$

where $\hat{\mathbf{n}}$ is the unit outward normal to the surface S enclosing the volume V. We thus have

$$\mathbf{A}(\mathbf{r}) = \frac{\mu_0}{4\pi} \int_{V'} \frac{\nabla' \times \mathbf{M}(\mathbf{r}')}{R} \, dv' + \frac{\mu_0}{4\pi} \oint_{S'} \frac{\mathbf{M}(\mathbf{r}') \times \hat{\mathbf{n}}}{R} \, ds' \qquad (6.36)$$

We can now compare (6.36) with expression (6.15) and (6.21), which give the vector potential from a distribution of true volume and surface current densities, respectively. This comparison suggests that the term $\nabla' \times \mathbf{M}$ in (6.36) can be interpreted as an equivalent *volume magnetization current* density $\mathbf{J}_{\mathrm{m}}$ (in A-m^{-2}) and that, similarly, the term $\mathbf{M} \times \hat{\mathbf{n}}$ can be interpreted as an equivalent *surface magnetization current* density $\mathbf{J}_{sm}$ (in A-m^{-1}). In other words, if we define

$$\boxed{\mathbf{J}_{\mathrm{m}} = \nabla \times \mathbf{M}} \qquad (6.37)$$

and

$$\boxed{\mathbf{J}_{sm} = \mathbf{M} \times \hat{\mathbf{n}}} \qquad (6.38)$$

Figure 6.45   Equivalent currents in a magnetized material. Even if the magnetization was uniform throughout the material so that the volume density of the polarization current ($\mathbf{J}_m = \nabla \times \mathbf{M}$) is zero, elementary current loops result in a surface polarization current $\mathbf{J}_{sm} = \mathbf{M} \times \hat{\mathbf{n}}$, which exists at the surface of the material.

we can rewrite (6.36) as

$$\mathbf{A} = \frac{\mu_0}{4\pi} \left(\int_{V'} \frac{\mathbf{J}_m(\mathbf{r}')}{R} \, dv' + \oint_{S'} \frac{\mathbf{J}_{sm}(\mathbf{r}')}{R} \, ds' \right)$$

The quantities $\mathbf{J}_m$ and $\mathbf{J}_{sm}$ are more than mathematical artifacts; they represent net effective currents, albeit inaccessible. A bound volume current $\mathbf{J}_m$ arises when adjacent molecular current loops do not completely cancel one another because the magnetization is inhomogeneous (i.e., $\mathbf{M}$ is not constant throughout the material). That a bound surface current $\mathbf{J}_{sm}$ exists at the outer boundary of a magnetic material can be seen by thinking of a uniformly magnetized medium in terms of tiny adjacent molecular currents, as shown in Figure 6.45. Although all internal currents are canceled by a contiguous current in the opposite direction, a net current is left uncanceled at the exterior edge of the material. This surface magnetization current vanishes only when the magnetization $\mathbf{M}$ is perpendicular to the outer surface such that $\mathbf{M} \times \hat{\mathbf{n}}$ is zero. The quantities $\mathbf{J}_m$ and $\mathbf{J}_{sm}$ are analogous to the equivalent bound volume (ρ_p) and surface (ρ_{sp}) polarization charge densities for a dielectric, as discussed in Section 4.10.

Example 6.33: Field of a cylindrical bar magnet. Determine the **B** field produced by a cylindrical bar magnet at a point $P(x, y, z)$ as shown in Figure 6.46. The length of the magnetic material is l and its radius is a. The magnet is assumed to be permanently magnetized with a uniform magnetization per unit volume of $\mathbf{M} = \hat{\mathbf{z}} M_0$.

Solution: Since the magnetization is in the z direction, the magnetization currents must flow in the $\hat{\boldsymbol{\phi}}$ direction. Since **M** is uniform, the volume density of the magnetization current $\mathbf{J}_m$ is zero. However, an uncanceled surface magnetization current flows circumferentially along the outer boundary of the cylindrical bar as shown. This current is given by

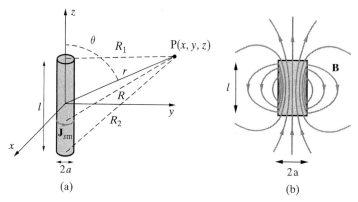

Figure 6.46 B field of a cylindrical bar magnet. (a) Coordinate system. Note the similarity with the solenoid geometry shown in Figure 6.35. (b) Approximate sketch of magnetic field lines in the y-z plane. Note the similarity with Figure 6.41a.

$\mathbf{J}_{sm} = \mathbf{M} \times \hat{\mathbf{n}} = M_0 \hat{\mathbf{z}} \times \hat{\mathbf{r}} = M_0 \hat{\boldsymbol{\phi}}$. By comparing the geometry of Figure 6.46 to that of Figure 6.35 in Section 6.6, it is clear that the field produced at point P by the bar magnet will be the same as that from an equivalent solenoid having an effective surface current of M_0 amperes per meter. In other words, a closely wound solenoid of N turns per meter carrying a current of I is equivalent to a bar magnet of magnetization M_0 if $NI = M_0$. The total magnetic dipole moment of the bar magnet can be found by integrating its magnetization density $\mathbf{M}$ over its volume; we find $|\mathbf{m}_{\text{bar}}| = \pi a^2 l M_0$ to be compared with the total moment of the solenoid of $|\mathbf{m}_{\text{slnd}}| = NI \pi a^2 l$. Substitution of $\mathbf{m}_{\text{bar}}$ instead of $\mathbf{m}_{\text{slnd}}$ in (6.24) gives the expression for the magnetic field of the bar magnet at distant ($r \gg l$) points.

While (6.24) is based on the dipole approximation (i.e., a distant observation point, or $r \gg a$ and $r \gg l$), it is possible to use the magnetization current sheet model of the permanent magnet to evaluate the $\mathbf{B}$ field everywhere, both inside and outside the magnet. Methods for such evaluations are described elsewhere.[70] However, an approximate sketch of the magnetic field lines around a cylindrical bar magnet is shown in Figure 6.46b. Note the similarity between the $\mathbf{B}$ field distribution for a bar magnet and that for a finite length solenoid, shown in Figure 6.41a.

Just as we differentiated between the free charge density and the bound polarization charge density ρ_p in Chapter 4, we need to separately account for a free current density (due to the net transport of charge carriers across a given cross-sectional area) and the bound magnetization current density given by (6.37). In Chapter 5, we considered one type of free current density: the conduction current density, which is sustained by an electric field in a conducting material through $\mathbf{J} = \sigma \mathbf{E}$. However, conduction current is

---

[70]For a comparative discussion of the $\mathbf{B}$ fields for a cylindrical rod magnet and a finite-length solenoid, see Section 9.4 of M. A. Plonus, *Applied Electromagnetics,* McGraw-Hill, 1978. For a discussion of the formal procedure used to determine the fields around a permanent magnet, see Section 6.6 of J. Van Bladel, *Electromagnetic Fields,* Revised Publishing, Hemisphere Publishing Corp., New York, 1985.

not the only mechanism through which we can have free current. For example, one could imagine a cloud of suspended electrons blowing in the wind. This current may not be driven in accordance with Ohm's law, but it nevertheless represents a net movement of charges, and so can be defined a free current density.

We now consider the $\mathbf{B}$ field inside (i.e., $\mathbf{J}_{sm} = 0$) a magnetic material body in which a free current distribution $\mathbf{J}$ exists. This current produces a macroscopic $\mathbf{B}$ field, which magnetizes the material, resulting in the magnetization current density $\mathbf{J}_m$ given by (6.37). Since the total effective current density is then the vector sum of $\mathbf{J}$ and $\mathbf{J}_m$, we can write the differential form of Ampère's law as

$$\nabla \times \mathbf{B} = \mu_0(\mathbf{J} + \mathbf{J}_m)$$

Noting that we have $\mathbf{J}_m = \nabla \times \mathbf{M}$, we can rewrite this as

$$\nabla \times \left(\frac{\mathbf{B}}{\mu_0} - \mathbf{M} \right) = \mathbf{J}$$

Therefore, the vector $[(\mathbf{B}/\mu_0) - \mathbf{M}]$ has as its source only the free current $\mathbf{J}$. Thus, to eliminate the necessity of dealing with the magnetization vector $\mathbf{M}$, we introduce a new vector $\mathbf{H}$, defined as

$$\boxed{\mathbf{H} \equiv \frac{\mathbf{B}}{\mu_0} - \mathbf{M}} \tag{6.39}$$

The vector $\mathbf{H}$ is called the *magnetic field intensity* vector, and it is directly analogous to the electric flux density vector $\mathbf{D}$ in electrostatics in that both $\mathbf{D}$ and $\mathbf{H}$ are medium-independent and are directly related to their sources. The dimensions of $\mathbf{H}$ are amperes per meter. We can also express Ampère's law in terms of $\mathbf{H}$ as follows:

$$\oint_C \mathbf{H} \cdot d\mathbf{l} = \int_S \mathbf{J} \cdot d\mathbf{s} = I \quad \text{or} \quad \nabla \times \mathbf{H} = \mathbf{J} \tag{6.40}$$

where I is the total net free current passing through the surface S enclosed by C. Note that the medium-independent nature of $\mathbf{H}$ is once again apparent. Taking the divergence of (6.39), we obtain

$$\nabla \cdot \mathbf{H} = \frac{1}{\mu_0} \nabla \cdot \mathbf{B} - \nabla \cdot \mathbf{M} = -\nabla \cdot \mathbf{M}$$

since $\nabla \cdot \mathbf{B} = 0$. Note that $\nabla \cdot \mathbf{H} = -\nabla \cdot \mathbf{M}$ is in general nonzero and can be thought to correspond mathematically to an equivalent magnetic charge.

For most materials other than those which are ferromagnetic, the magnetization $\mathbf{M}$ is directly proportional to the applied external field $\mathbf{B}$, so that, based on (6.39), $\mathbf{M}$ is also directly proportional to $\mathbf{H}$. It is thus customary to write

$$\mathbf{M} = \chi_m \mathbf{H} \tag{6.41}$$

where χ_m is the dimensionless proportionality constant called the *magnetic susceptibility* of the material that we introduced in (6.35). The value of χ_m provides a measure of how susceptible a material is to becoming magnetized by an applied magnetic field. χ_m is negative for diamagnetic materials and positive for paramagnetic materials. For ferromagnetic materials, χ_m in general depends on the applied magnetic field and the past history of magnetization of the material. Substituting (6.41) into (6.39), we find

$$\boxed{\mathbf{B} = \mu_0(\mathbf{H} + \mathbf{M}) = \mu_0(1 + \chi_m)\mathbf{H} = \mu_0\mu_r\mathbf{H} = \mu\mathbf{H}} \qquad (6.42)$$

where the quantity $\mu = \mu_0(1 + \chi_m) = \mu_0\mu_r$ is called the *magnetic permeability* and is entirely analogous to the electric permittivity ϵ in electrostatics. Note that μ_r is the relative permeability, analogous to ϵ_r in electrostatics. Just like electric permittivity ϵ, the permeability μ is a macroscopic parameter[71] that is used to account for the microscopic effects of material bodies in the presence of external magnetic fields. In practice, the permeability μ for a given material can be determined experimentally. When its value is known, μ can be used in (6.42) to relate $\mathbf{B}$ and $\mathbf{H}$ directly, thus eliminating the necessity of taking the magnetization $\mathbf{M}$ into account explicitly.

It is important to note that (6.42) only holds for a class of materials and under certain conditions. In particular, since (6.42) originates in (6.41), it depends primarily on the linear relationship between $\mathbf{M}$ and $\mathbf{H}$. Magnetic materials, especially ferromagnetic materials discussed in the next section, often exhibit permanent magnetization and highly nonlinear behavior, in which case (6.41) (and thus (6.42)) is not valid. Equation (6.42) also requires that the material be isotropic—that is, that χ_m be independent of the direction of $\mathbf{B}$. In anisotropic materials, such as ferrites (see Section 6.8.3), a $\mathbf{B}$ field in one direction can produce magnetization (and thus $\mathbf{M}$) in another direction. Accordingly, the relation between $\mathbf{B}$ and $\mathbf{H}$ must be expressed[72] as a type of matrix, more commonly called a *tensor*.

We recall from Chapter 4 that the introduction of the electric flux density vector $\mathbf{D}$ enabled us to simplify Gauss's law by focusing on the free charge density. Similarly, introduction of the quantity $\mathbf{H}$, together with the permeability μ, allows us to simplify Ampère's law by using only the free (e.g., conduction) current density. The Biot–Savart law (6.6) and Ampère's law as stated in (6.9) both depend on the *total* current density, which in general consists of the free current density $\mathbf{J}$ and the magnetization current density $\mathbf{J}_m$. However, in calculating the $\mathbf{H}$ field through (6.40), we only need to account for the free current, which in practice is usually externally driven (such as a current flowing through a wire).

While the magnetic field intensity vector $\mathbf{H}$ is analogous to electric flux density vector $\mathbf{D}$, in practice $\mathbf{H}$ is used much more often then $\mathbf{D}$. In a laboratory, we typically

---

[71]We should note here that in the interior of any material the values of $\mathbf{M}$ and $\mathbf{B}$ may vary rapidly with both time and space. However, in any practical problem that involves the measurement of fields, the field is sampled over regions of space much larger than atomic dimensions. Thus, the quantities $\mathbf{B}$, $\mathbf{M}$, and $\mathbf{H}$ in the above expressions are all macroscopic quantities.

[72]See Section 13.12 of S. Ramo, J. R. Whinnery, and T. Van Duzer, *Fields and Waves in Communication Electronics,* 3rd ed., John Wiley & Sons, Inc., New York, 1994.

control conduction currents through wires and voltages on metal conductors. Thus, by reading our equipment dials we can directly calculate the line integrals of the electric field **E** using (4.19) and the **H** field using (6.40). If instead of controlling voltages, there existed an easier means to place a known amount of charge on each conductor, then **D** instead of **E** would be the more widely used quantity in electrostatics.

In retrospect, it is quite unfortunate that the duality of **E** and **B** (see Section 6.2.2) and that of **D** and **H** were not appreciated during the early development of the theory of electromagnetics, since the use of the reciprocal of μ_0 in relating **H** to **B** would have been more appropriate. Also, the common references to **H** as the "magnetic field strength" and to **B** as the "magnetic flux density" are rather misleading.[73]

For paramagnetic and diamagnetic materials the value of μ differs only insignificantly from μ_0. For one of the strongest diamagnetic materials, bismuth, we have $\mu = 0.999833\mu_0$, whereas for the paramagnetic tungsten $\mu = 1.000078\mu_0$. The permeability of air is $\mu = 1.00000037\mu_0$. For copper and water, we have $\mu = 0.999991\mu_0$. Thus, it is very common to assume the relative permeability $\mu_r = 1$ for diamagnetic and paramagnetic materials. However, for ferromagnetic materials, discussed in the next section, μ_r is in general much larger than unity and in some special cases can be as large as 10^6! The relative permeability (μ_r) values for selected materials are given in Table 6.2.

6.8.3 Ferromagnetic Materials

The value of μ is generally much larger than μ_0 for *ferromagnetic* materials, which exhibit large magnetizations even in the presence of very weak magnetic fields. Only the three elements iron, nickel, and cobalt are ferromagnetic at room temperature and above. Almost all ferromagnetic alloys and compounds contain one or more of these three elements or manganese, which belongs to the same group of transition elements in the periodic table. The ease of magnetization of these materials results from quantum mechanical effects that tend to align neighboring atomic spins parallel to each other even in the absence of an applied magnetic field. In ferromagnetic materials the diamagnetic and paramagnetic effects also occur, but the contributions of these effects to the total magnetization is negligible even in relatively large fields. The permanent magnetization of cobalt,

[73]The inappropriateness of the term "magnetic field strength" for **H** is discussed in Section 2 of A. Sommerfeld, *Electrodynamics*, Academic Press, 1952, reproduced here with permission of Academic Press, Inc. In Sommerfeld's words (p. 11): "We may indicate finally a subdivision of physical entities into entities of intensity and entities of quantity. **E** and **B** belong to the first class, **D** and **H**, to the second. The entities of the first class are answers to the question 'how strong,' those of the second class, to the question 'how much.' In the theory of elasticity, for example, the stress is an entity of intensity, the corresponding strain, one of quantity; in the theory of gases, pressure and volume form a corresponding pair of entities. In **D** the quantity character is clearly evident as the quantity of electricity (i.e., flux) that has passed through; in **H** the situation is slightly obscured by the fact there are no isolated magnetic poles. We are in general inclined to regard the entities of intensity as cause, the corresponding entities of quantity as their effect." On p. 45 of his text, Sommerfeld further states, "The unhappy term 'magnetic field' for **H** should be avoided as far as possible. It seems to us that this term has led into error none less than Maxwell himself." For an interesting discussion of the question of the relative physical merits of **B** versus **H** in connection with permanent magnets, see pp. 136–139 of M. Abraham and R. Becker, *Electricity and Magnetism*, Blackie & Son Limited, Glasgow, 1944.

TABLE 6.2 RELATIVE PERMEABILITY OF SELECTED MATERIALS

Material	Relative Permeability (μ_r)
Air	1.00000037
Aluminum	1.000021
Bismuth	0.999833
Cobalt	250.
Copper	0.9999906
Iron (Purified: 99.96% Fe)	280,000.
Iron (Motor grade: 99.6%)	5,000.
Lead	0.9999831
Manganese	1.001
Manganese-zinc ferrite	1,200.
Mercury	0.999968
Nickel	600.
Nickel-zinc ferrite	650.
Oxygen	1.000002
Palladium	1.0008
Permalloy: 78.5% Ni, 21.5% Fe	70,000.
Platinum	1.0003
Silver	0.9999736
Supermalloy: 79% Ni, 15% Fe, 5% Mo, 0.5% Mn	1,000,000.
Tungsten	1.00008
Water	0.9999912

nickel, and iron leads to the magnetic permeabilities of $\mu_{Co} = 250\mu_0$, $\mu_{Ni} = 600\mu_0$, and $\mu_{Fe} = 6000\mu_0$. For purified iron, μ ranges from $10{,}000\mu_0$ to $200{,}000\mu_0$; for supermalloy (79% Ni, 15% Fe, 5% Mo, 0.5% Mn) μ can be up to $\sim10^6\mu_0$. Typically, the value of μ for a ferromagnetic material is not unique, because of strong nonlinearities. In practice, the relation between **B** and **H** for ferromagnetic materials is represented graphically in the form of a curve known as the hysteresis curve. The magnetic properties of ferromagnetic materials are strongly temperature-dependent. In the absence of an applied field, they exhibit spontaneous magnetization below a temperature T_C, known as the Curie temperature, and are strongly paramagnetic above that temperature, susceptibilities (χ_m) decreasing with increasing temperature. The Curie temperatures for iron, nickel, and cobalt are 1043 K, 631 K, and 1393 K, respectively.

Some elements that are neighbors of the ferromagnetic elements on the periodic table, such as chromium and manganese, also have strong quantum mechanical coupling forces between the atomic dipole moments. However, for these elements the coupling leads to antiparallel alignment of electron spins between adjacent atoms, so the net magnetic moment is zero. Elements with this property are called *antiferromagnetic*. Antiferromagnetism is also strongly temperature-dependent. At temperatures above the Curie temperature, the spin directions suddenly become random, and the material exhibits paramagnetic properties.

A number of oxides containing iron, nickel, or cobalt exhibit a magnetic behavior between ferromagnetism and antiferromagnetism and are called *ferrimagnetic* materials. The magnetic moments of these molecules alternate from atom to atom in an unequal manner, resulting in a net magnetic moment, but one that is much smaller than those of ferromagnetic materials. The most common ferrimagnetic materials are Fe_3O_4 and the family of ferrites described by the chemical formula $KO \cdot Fe_2O_3$, where K is any divalent metal such as Fe, Co, Ni, Mn, Mg, Cu, Zn, Cd, or a mixture of these. These are ceramic-like compounds with very low conductivities (for ferrites $\sigma = 10^{-4}$ to 1 S-m^{-1} as opposed to 10^7 S-m^{-1} for ferromagnetic Fe) and thus exhibit very low eddy current losses at high frequencies. Thus, ferrites are commonly used in high-frequency and microwave applications as cores for FM antennas, transformers, and phase shifters. An important property of ferrites is that they are *anisotropic* in the presence of magnetic fields, meaning that the **H** and **B** vectors are in different directions. Microwave devices that utilize the anisotropy of ferrites include isolators, gyrators, and some directional couplers.

Further discussion of ferromagnetic materials is beyond the scope of this book.[74]

6.8.4 The Meissner Effect in Superconductors

It was mentioned in Section 5.1 that some materials become superconducting below a critical temperature, with their conductivity becoming nearly infinite ($> 10^{20}$). This phenomenon is accompanied by a type of *repulsion* or *exclusion* of the **B** field from the interior of the material. In effect, the material becomes perfectly diamagnetic, with $\mathbf{B} = 0$ within the metal independent of the existence and nature of current carriers in the neighborhood of the metal, as if its permeability $\mu = 0$. This phenomenon was discovered in 1933 by W. Meissner and R. Ochsenfeld and is known as the *Meissner effect*. The description of the Meissner effect as a complete repulsion of the **B** field is an oversimplification. In fact, for a metal below the critical temperature the **B** field penetrates into the surfaces to optical distances, 50–200 nm. This result is true for dc fields as well as time-varying fields and may be understood by modeling the superconductor as a dense collisionless plasma.[75] The Meissner effect completes our understanding of a "perfect" conductor as one in which no static or time-varying electric or magnetic fields exist. A perfect conductor is usually understood to be one in which no electric field (at least none that varies more slowly than the relaxation time τ_r, which is typically of order $\sim 10^{-19}$ s for metallic conductors) can exist, since otherwise the abundantly available free electrons would be in perpetual motion. Maxwell's equations imply that there must then be no time-varying magnetic fields inside a perfect conductor either. However, a

---

[74]For an excellent and relatively concise discussion at an appropriate level, see Chapters 9 and 10 of M. A. Plonus, *Applied Electromagnetics*, McGraw-Hill, New York, 1978.

[75]T. Van Duzer and C. W. Turner, *Principles of Superconductive Devices and Circuits*, Elsevier, New York, 1981. Plasma is the fourth state of matter, consisting of an ionized collection of positive and negatively charged particles. Plasmas are inherently diamagnetic, since motions of both positively charged and negatively charged particles tend to reduce the magnetic field, as illustrated for a single electron in Figure 6.44.

truly static magnetic field should in principle be completely uncoupled from the electric field and should not be affected by the conductivity of the material, even if $\sigma = \infty$. To the degree that superconductors are the physical materials that can be viewed most nearly as "perfect" conductors, the Meissner effect ensures that the interiors of perfect conductors are completely free of electric or magnetic fields.[76]

6.9 BOUNDARY CONDITIONS FOR MAGNETOSTATIC FIELDS

In some magnetostatic problems we deal with interfaces between two or more different types of materials. The manner in which **B** or **H** behaves across such interfaces is described by the boundary conditions. The boundary conditions are derived from fundamental laws of magnetostatics, which we reiterate below.

Starting with the experimental fact that electric currents create magnetic fields, as expressed by the Biot–Savart law, we have derived the following fundamental differential and integral laws of magnetostatics:

$$\nabla \times \mathbf{H} = \mathbf{J} \qquad \oint_C \mathbf{H} \cdot d\mathbf{l} = \int_S \mathbf{J} \cdot d\mathbf{s} \qquad \text{Ampère's law}$$

$$\nabla \cdot \mathbf{B} = 0 \qquad \oint_S \mathbf{B} \cdot d\mathbf{s} = 0 \qquad \text{No magnetic charges}$$

Note that Ampère's law was first derived in terms of **B** and used as such during most of this chapter. However, on the basis of our discussions in Section 6.8, we now understand that the more general version of Ampère's law relates the **H** field directly to the free current sources (i.e., **J**), independent of the material media. The magnetic properties of different physical media are accounted for via the *permeability* μ of the material. For linear (magnetization **M** proportional to **B**) and isotropic (magnetization **M** in the same direction as **B**) media, we have

$$\mathbf{H} = \mu^{-1}\mathbf{B} \qquad \text{or} \qquad \mathbf{B} = \mu\mathbf{H} \tag{6.43}$$

Note that (6.43) is valid even for inhomogeneous materials as long as we allow μ to be a function of position, that is, $\mu(x, y, z)$.

A very common type of inhomogeneity encountered in practice occurs at the interface between two magnetically different materials. To establish a basis for solving such problems, we now study the behavior of **H** and **B** in crossing the boundary between two different materials and derive the conditions that **B** and **H** must satisfy at such interfaces. These conditions are referred to as the *boundary conditions;* they are similar in nature to the boundary conditions for electrostatic fields studied in Section 4.11.

---

[76]For further reading on this interesting subject, see Section 13.4 of S. Ramo, J. R. Whinnery, and T. Van Duzer, *Fields and Waves in Communication Electronics,* 3rd ed., John Wiley & Sons, Inc., New York, 1994; also see Section 9.10 of W. B. Cheston, *Elementary Theory of Electric and Magnetic Fields,* John Wiley & Sons, Inc., New York, 1964.

The boundary conditions must necessarily be derived using the integral forms of the fundamental magnetostatic laws, since the differential forms apply only at a point.

The methodology we use for derivation is identical to that used in Section 4.11 for electrostatics. For the normal component of the magnetic field, we consider the surface of a differential pillbox, as shown in Figure 6.47a. Since the contributions from the side surfaces can be made infinitesimally small by taking $\Delta h \to 0$, using (6.27) we have

$$\oint_S \mathbf{B} \cdot d\mathbf{s} = \mathbf{B}_1 \cdot \hat{\mathbf{n}} - \mathbf{B}_2 \cdot \hat{\mathbf{n}} = 0 \quad \to \quad \boxed{B_{1n} = B_{2n}} \tag{6.44}$$

Hence the normal components of the magnetic field $\mathbf{B}$ are always continuous across an interface between two different materials. Since $\mathbf{B} = \mu\mathbf{H}$, the magnetic field intensity $\mathbf{H}$ is not continuous across a boundary:

$$\mu_1 H_{1n} = \mu_2 H_{2n}$$

For the parallel components of the magnetic field, we consider the line integral of $\mathbf{H}$ around a closed contour abcda such as that shown in Figure 6.47b. In general there may be a free current in a layer of vanishing thickness flowing along the boundary between two materials, as occurs with time-varying fields and when one of the materials

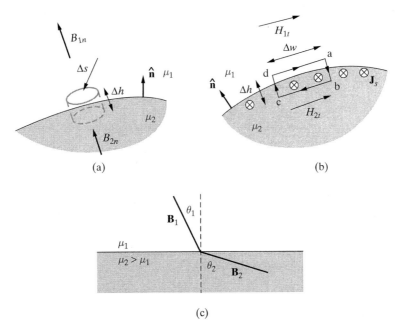

(a)   (b)

(c)

Figure 6.47 The boundary between two magnetic media. (a) A differential pillbox-shaped closed surface. (b) A differential closed contour. (c) $\mathbf{B}$ vectors at the boundary for the case $\mu_2 > \mu_1$.

is an excellent conductor. From Figure 6.47b, since the contributions from the sides can be made infinitesimally small by taking $\Delta h \to 0$, applying (6.40), we have

$$\int_d^a \mathbf{H}_1 \cdot d\mathbf{l} - \int_b^c \mathbf{H}_2 \cdot d\mathbf{l} = \int_0^{\Delta w} \int_{-\Delta h/2}^{\Delta h/2} J \, dh \, dw$$

$$H_{1t} \Delta w - H_{2t} \Delta w = J \Delta w \Delta h$$

$$H_{1t} - H_{2t} = J \Delta h = J_s$$

Since the relationship between the surface current density $\mathbf{J}_s$ and $\mathbf{H}$ must satisfy the right-hand rule, this boundary condition can be compactly expressed as

$$\boxed{\hat{\mathbf{n}} \times (\mathbf{H}_1 - \mathbf{H}_2) = \mathbf{J}_s} \tag{6.45}$$

where $\hat{\mathbf{n}}$ is the *outward unit normal pointing from medium 2* at the interface and $\mathbf{J}_s$ is normal to the page pointing inward.

In the absence of any free surface currents at the interface (i.e., $\mathbf{J}_s = 0$), the tangential component of the $\mathbf{H}$ field is continuous across the interface between two materials:

$$\boxed{H_{1t} = H_{2t}} \quad (\mathbf{J}_s = 0) \tag{6.46}$$

The tangential component of $\mathbf{H}$ is also continuous across the interface when the conductivities of both media are finite, since in such cases currents are defined by volume current densities instead of surface currents, so $J \Delta h \to 0$ as $\Delta h \to 0$ and so $J_s = 0$.

When one of the media (say, medium 2) is a perfect conductor, the magnetic field in its interior is zero (see the discussion in Section 6.8.4), so we have $\mathbf{H}_2 = 0$ and thus from (6.45),

$$\boxed{\hat{\mathbf{n}} \times \mathbf{H}_1 = \mathbf{J}_s} \quad \text{(Perfect conductor)} \tag{6.47}$$

as the basic boundary condition on the surface of a perfect conductor. Note that another consequence of the fact that $\mathbf{B} = 0$ inside a perfect conductor is that $\mathbf{B}_{2n} = 0$, which implies (based on (6.44)) that $\mathbf{B}_{1n} = 0$. In other words, magnetic fields cannot terminate normally on a perfect conductor. The magnetic field immediately outside a perfect conductor must always be tangential to the surface.

Even in the absence of free surface currents, since $\mathbf{B} = \mu \mathbf{H}$, the tangential component of $\mathbf{B}$ is not continuous across a boundary between two magnetically different materials. Instead, from (6.46) we have

$$\frac{B_{1t}}{\mu_1} = \frac{B_{2t}}{\mu_2} \tag{6.48}$$

The discontinuity of the tangential $\mathbf{B}$ field can be understood in terms of the bound magnetization current. Integrating (6.37) over the differential surface element bounded by the closed contour abcda shown in Figure 6.47b, and noting that $\Delta h \to 0$ we have

$$\int_S \nabla \times \mathbf{M} \cdot d\mathbf{s} = \oint_C \mathbf{M} \cdot d\mathbf{l} = (M_{1t} - M_{2t})\Delta w = \int_S \mathbf{J}_m \cdot d\mathbf{s}$$

Since $\Delta h \to 0$, the surface integral of $\mathbf{J}_m$ selects only the surface magnetization current density at the boundary:

$$\int_S \mathbf{J}_m \cdot d\mathbf{s} = J_m \Delta h \Delta w = J_{sm}\Delta w$$

Combining the two previous expressions gives

$$M_{1t} - M_{2t} = J_{sm} \tag{6.49}$$

To write (6.49) in terms of the **B** field, we first note from (6.39) that

$$M_{1t} = \frac{B_{1t}}{\mu_0} - H_{1t}$$

$$M_{2t} = \frac{B_{2t}}{\mu_0} - H_{2t}$$

so that

$$M_{1t} - M_{2t} = \frac{B_{1t}}{\mu_0} - \frac{B_{2t}}{\mu_0} - \underbrace{(H_{1t} - H_{2t})}_{=0 \text{ from } (6.46)}$$

Substituting this result into (6.49) gives

$$B_{1t} - B_{2t} = \mu_0 J_{sm}$$

Noting that the magnetic field and surface current density vectors are related via the right-hand rule, we can write this result more compactly as

$$\hat{\mathbf{n}} \times (\mathbf{B}_1 - \mathbf{B}_2) = \mu_0 \mathbf{J}_{sm}$$

where $\hat{\mathbf{n}}$ is the outward unit normal pointing from medium 2 at the interface. At the surface between two magnetically different media, the tangential component of the **B** field is discontinuous by the amount $\mu_0 J_{sm}$, as if there exists a surface current density at the boundary. This surface current arises due to unequal magnetization surface current densities at the interface between the two different media (see Figure 6.45). This result is analogous to the discontinuity of the normal electric field due to effective surface charge density at the interface between two dielectrics, as discussed in Section 4.11.

A consequence of (6.44) and (6.48) is the bending of magnetic field lines across material interfaces, as illustrated in Figure 6.47c. Consider a magnetic field **B** that is oriented at an angle θ_1 from the normal in medium 1. From (6.44) and (6.46) we have

$$B_1 \cos \theta_1 = B_2 \cos \theta_2$$

$$\mu_2 B_1 \sin \theta_1 = \mu_1 B_2 \sin \theta_2$$

where $B_{1,2}$ are the magnitudes of $\mathbf{B}_{1,2}$. From these results we can find

$$\tan \theta_2 = \frac{\mu_2}{\mu_1} \tan \theta_1$$

and

$$B_2 = B_1 \sqrt{\left(\frac{\mu_2}{\mu_1} \sin \theta_1 \right)^2 + \cos^2 \theta_1}$$

The above relationships indicate that magnetic flux lines are *bent further away from the normal* in the medium with the *higher* magnetic permeability. Most materials have magnetic permeabilities of $\mu \simeq 1$, except for ferromagnetic materials, for which μ can be very large. If medium 2 is a ferromagnetic material and medium 1 is vacuum or air (or any other nonmagnetic material), $\mu_2 \gg \mu_1$, so $\tan \theta_1 \ll \tan \theta_2$, and the magnetic flux lines are for all practical purposes normal to the surface on the air side, provided θ_2 is not near 90°. This sharp bending of magnetic field lines is evident in Figure 6.46b, noting that the bar magnet must be a high-permeability material.

6.10 MAGNETIC FORCES AND TORQUES

We started this chapter with Ampère's force law, describing the manner in which current-carrying wires exert forces on one another, as the experimental basis of magnetostatics. The discovery that current-carrying wires experience physical forces in the presence of magnetic fields (created by other current-carrying wires or permanent magnets) immediately raised the possibility of using these forces to do work and indeed led to significant enhancements in industrial technology, including the development of direct- (i.e., steady-) current electric motors and sensitive instruments for electrical measurements (e.g., voltmeters and ammeters). Much of the machinery of our present-day industrial and technological environment is actually based on alternating currents and fields, which operate on the basis of Faraday's law of electromagnetic induction, covered in Chapter 7. However, forces and torques produced by static magnetic fields and direct-current motors and generators also account for a range of applications involving critical speed or position control requirements, including automobile power windows, food blenders, hand power tools, fans, electric cars, elevators, and hoists. In this section we briefly discuss magnetostatic forces and torques, which form the basic principles of direct-current rotating machines.

6.10.1 Magnetic Force on Moving Charges and the Hall Effect

Magnetic fields exert forces on matter via their influence on charged particles. In any region of space where a magnetic field $\mathbf{B}$ is present, the magnetic force $\mathbf{F}_m$ on a charge q moving with a velocity $\mathbf{v}$ is given by

$$\mathbf{F}_m = q\mathbf{v} \times \mathbf{B} \tag{6.50}$$

which is equivalent to $\mathbf{F}_m = I \, d\mathbf{l} \times \mathbf{B}$ when we note that a charge q moving with velocity $\mathbf{v}$ constitutes an element of current $I d\mathbf{l}$, as previously noted in Section 6.2.2.

Equation (6.50) describes the Lorentz magnetic force, which is the magnetic component of the total Lorentz force acting on a charged particle in the presence of electric and magnetic fields, namely,

$$\boxed{\mathbf{F} = q[\mathbf{E} + \mathbf{v} \times \mathbf{B}]} \tag{6.51}$$

Equation (6.51) is the well-known Lorentz[77] force equation. The solution of the Lorentz force equation is required when determining the motion of electrons in electrostatic and magnetic deflection systems (such as in cathode-ray tubes, in microwave devices such as the klystron and magnetron, and in particle accelerators such as the cyclotron) and in other applications involving ionized gases or plasmas.

While the Lorentz force equation (6.51) is commonly used to study the motion of freely mobile electrons and positively charged particles such as those in an ionized gas, it is equally applicable to the dynamics of current carriers in solids, such as electrons and holes in semiconductors.[78] An interesting and important manifestation of the motion of electrons and holes under the influence of the Lorentz force is the Hall effect, which is described briefly below.

The Hall effect.   In 1879, Edwin H. Hall, at that time conducting an experiment as a student in Baltimore, discovered[79] a new effect of a magnetic field on electric currents. Hall placed a strip of gold leaf mounted on glass, forming part of an electric circuit through which a current passed, between the poles of an electromagnet, the plane

---

[77]H. A. Lorentz was the first to study the motion of electrons in the presence of electric and magnetic fields. In particular, he applied equation (6.51) to put forth an explanation for the Zeeman effect, involving the splitting of spectral lines in the presence of a magnetic field [P. Zeeman, Doublets and triplets in the spectrum produced by external magnetic forces, *Phil. Mag.* 43 (5), p. 226, 1897]. Zeeman was a student of Lorentz, and the two later shared the Nobel Prize in physics in 1902. Lorentz's famous theory of electrons is described in the extensive compilation of lectures that he gave at Columbia University in the Spring of 1906: H. A. Lorentz, *The Theory of Electrons and Its Applications to the Phenomena of Light and Radiant Heat*, Dover Publications, Inc., 1952.

[78]A. S. Grove, *Physics and Technology of Semiconductor Devices*, John Wiley & Sons, Inc., New York, 1967.

[79]E. H. Hall, *Phil. Mag.* ix, p. 225, and x, p. 301, 1880.

of the strip being perpendicular to the magnetic flux lines. The two ends of a sensitive voltmeter were then placed in contact with different parts of the strip, until two points at the same potential were found, so that the voltmeter reading was zero. When the magnetic field (i.e., the current of the electromagnet) was turned on or off, a deflection of the galvanometer needle was observed, indicating a potential difference between the voltmeter leads. In this way, Hall showed that the magnetic field produces a new electromotive force in the strip of gold, at right angles to the primary electromotive force and to the magnetic force, and proportional to the product of these two forces.[80] Physically, we can regard this effect as an additional electromotive force generated by the action of a magnetic field on the current. This phenomenon, called the Hall effect, is extremely useful in practice in the determination of charge densities in materials, especially in semiconductors.

Consider a rectangular semiconductor bar as shown in Figure 6.48. Assume that this is a p-type semiconductor, so that the majority charge carriers are holes, each with charge $|q_e|$ where $q_e \simeq -1.6 \times 10^{-19}$ C is the charge of an electron. A potential difference of V_0 applied between the two side faces sets up an applied electric field E_0 in the x direction, which causes a current $I_0 = J_x A = w \, d\sigma E_0$ to flow in the x direction, where σ is the conductivity of the semiconductor. In general, the conductivity for a semiconductor is given by

$$\sigma = |q_e|(\mu_e N_e + \mu_p N_p)$$

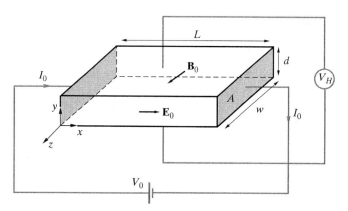

Figure 6.48   Illustration of the Hall effect.

[80]In his excellent historical account, *A History of the Theories of Aether and Electricity,* Thomas Nelson and Sons, Ltd., 1951, E. T. Whittaker notes that in the early 1870s Oliver Lodge had done experiments on the flow of electricity in a metallic sheet and had come very close to discovering the Hall effect. However, he was deterred from making the crucial test upon reading a passage in Article 501 in Vol. ii of Maxwell's *Electricity and Magnetism:* "It must be carefully remembered, that the mechanical force which urges a conductor carrying a current across the lines of magnetic force acts, not on the electric current, but on the conductor which carries it."

where N_e and N_p are, the densities of electrons and holes, respectively, and μ_e and μ_p are the mobilities of the electrons and holes, respectively. As was noted in Section 5.1 in connection with (5.2), mobility is the ratio between the drift velocity of the charge carriers and the applied electric field. For a p-type semiconductor, we have $N_p \gg N_e$, so that $\sigma \simeq |q_e|\mu_p N_p$, and the drift velocity is $v_d = \mu_p E_0$, where $\mu_p \simeq \sigma/(|q_e|N_p)$. If a magnetic field B_0 is applied in the z direction, the holes in the p-type semiconductor deflect in the $-y$ direction. Using vector notation, the total force on a single hole due to the electric and magnetic fields is given by (6.51), namely,

$$\mathbf{F} = |q_e|(\mathbf{E} + \mathbf{v}_d \times \mathbf{B}) = |q_e|(\hat{\mathbf{x}}E_0 + \hat{\mathbf{x}}v_d \times \hat{\mathbf{z}}B_0)$$

$$= |q_e|(\hat{\mathbf{x}}E_0 - \hat{\mathbf{y}}v_d B_0)$$

According to this result, the holes in the semiconductor experience a magnetic force of $-\hat{\mathbf{y}}(|q_e|v_d B_0)$ and thus an acceleration in the $-y$ direction. However, the displacement of the holes in the $-y$ direction leads to the creation of an electric field E_y between the positively charged region into which the holes move and the negatively charged atoms that are left behind. Thus, this restoring electric field points in the y direction and at steady state must have a magnitude $v_d B_0$ to balance the magnetic force $|q_e|v_d B_0$. The formation of this electric field E_y sets up a potential difference V_H between the top and bottom ends of the bar, given by

$$V_H = E_y d = v_d B_0 d$$

and is called the Hall voltage.

Substituting for $v_d = \mu_p E_0$ and $\mu_p = \sigma/(|q_e|N_p)$ and noting that $\mathbf{J} = \sigma\mathbf{E}$, the electric field E_y can be written in terms of the current density as

$$E_y = v_d B_0 \simeq \frac{\sigma E_0}{|q_e|N_p} B_0 = R_H J_x B_0$$

where $R_H = (|q_e|N_p)^{-1}$ is called the Hall coefficient and N_p is the hole concentration in the p-type semiconductor. The hole concentration N_p is given as

$$N_p = \frac{1}{|q_e|R_H} = \frac{J_x B_0}{|q_e|E_y} = \frac{[I_o/(wd)]B_0}{|q_e|(V_H/d)} = \frac{I_o B_0}{|q_e|w V_H}$$

Since I_0, B_0, w, $|q_e|$, and V_H are either known or can be measured, the Hall effect can be used to determine (quite accurately) the hole concentration in the p-type semiconductor. Note that the same is true for an n-type semiconductor, with the only difference being that the charge carriers are electrons.

The Hall effect can also be used to determine the conductivity of the semiconductor material. If the resistance of the semiconductor bar is measured ($R = V_0/I_0$), the conductivity is given by

$$\sigma = \frac{L}{Rwd} = \frac{LI_0}{wdV_0}$$

However, since the conductivity of the p-type bar is $\sigma \simeq |q_e|\mu_p N_p$, we can write

$$\mu_p \simeq \frac{\sigma}{|q_e|N_p}$$

so that by measuring the hole concentration N_p and the conductivity of the bar (inferred from the resistance of the bar), the mobility of the holes in the p-type material can be determined. Such measurements are very important in the analysis of semiconductor materials.

6.10.2 Magnetic Force on Current-Carrying Conductors

It is evident from (6.2) and (6.3), and the discussion in Section 6.2.2, that when a filamentary conductor carrying current I is placed in a region permeated by a magnetic field, it experiences a total net magnetic force given by

$$\mathbf{F}_\mathrm{m} = \int_C d\mathbf{F}_\mathrm{m} = \int_C I\,d\mathbf{l} \times \mathbf{B} \tag{6.52}$$

where C is the contour defined by the filamentary conductor, which in many engineering applications is the length of a wire. Note that $d\mathbf{F}_\mathrm{m} = I\,d\mathbf{l} \times \mathbf{B}$ is the incremental magnetic force felt by an infinitesimal wire element $I\,d\mathbf{l}$.

In Example 6.1, we used Ampère's force law (6.1) to determine the repulsive magnetic force between two infinitely long wires carrying oppositely directed current. In the following Example 6.34, we use (6.52) to determine the magnetic attraction force between two infinitely long wires carrying equally directed current. Example 6.35 considers the magnetic force exerted on a rigid square loop by an infinitely long current-carrying wire.

Example 6.34: Magnetic force between two long wires. Two long wires in free space at a distance a from one another carry currents I_1 and I_2, both flowing in the z direction, as shown in Figure 6.49. Find an expression for the magnetic force per unit length between the wires in terms of I_1, I_2, and a.

Figure 6.49   Force between two long wires.

Solution: Based on the result of Example 6.3, the magnetic field $\mathbf{B}_1$ due to the current in wire 1 at any point P at a distance r outside the wire is $\mathbf{B}_1 = \hat{\boldsymbol{\phi}}\mu_0 I_1/(2\pi r)$, where r and

ϕ are the cylindrical coordinate quantities, with wire 1 assumed to be located at the origin and oriented in the z direction. The **B** field at the location of wire 2 is then simply

$$[\mathbf{B}_1]_{\text{at wire } 2} = \hat{\mathbf{y}}\frac{\mu_0 I_1}{2\pi a}$$

since $r = a$ and $\hat{\boldsymbol{\phi}} = \hat{\mathbf{y}}$ for $\phi = 0$. The differential force on an elemental length of wire 2 is then given by

$$d\mathbf{F}_{12} = \hat{\mathbf{z}}(I_2\,dl_2) \times \hat{\mathbf{y}}\left(\frac{\mu_0 I_1}{2\pi a}\right)$$

which, when integrated over unit length, gives the magnetic force per unit length as

$$\mathbf{F}_{12} = -\hat{\mathbf{x}}\frac{\mu_0 I_1 I_2}{2\pi a}$$

The force is attractive as expected. The reader is encouraged to evaluate the magnetic force per unit length $\mathbf{F}_{21}$ exerted on wire 1 by wire 2 and show that it is given by $\mathbf{F}_{21} = +\hat{\mathbf{x}}\mu_0 I_1 I_2/(2\pi a)$.

Example 6.35: Magnetic force on a current-carrying loop. A rigid rectangular loop carrying current I_2 is located near an infinitely long wire carrying current I_1 as shown in Figure 6.50. Find the magnetic force exerted by the long wire on the loop.

Figure 6.50   Force on a rectangular loop near a long wire. Coordinate system, dimensions, and loop orientation. The directions of the magnetic forces on each side of the loop are indicated.

Solution: Note that the lead wires of the loop are twisted as in Example 6.4, so that the currents on the two wires (and thus the magnetic forces that they experience) effectively cancel. Thus, we need to consider only the force on the square loop. Once again using the result of Example 6.3, the **B** field produced by the long wire at any point in the plane of the loop and at a distance r from the long wire is

$$\mathbf{B}_1 = -\hat{\mathbf{x}}\frac{\mu_0 I_1}{2\pi r}$$

Assuming that the loop is rigid (i.e., it maintains its shape even under the influence of the magnetic force; note that this can be achieved by simply winding the loop out of strong enough wire or just winding it on a nonmagnetic frame such as a wooden frame), the total magnetic force exerted on it is given by

$$\mathbf{F}_{\text{loop}} = I_2 \oint_C d\mathbf{l}_2 \times \mathbf{B}_1$$

$$= I_2 \left[\int_{AB} d\mathbf{l}_2 \times \mathbf{B}_1 + \int_{BC} d\mathbf{l}_2 \times \mathbf{B}_1 + \int_{CD} d\mathbf{l}_2 \times \mathbf{B}_1 + \int_{DA} d\mathbf{l}_2 \times \mathbf{B}_1 \right]$$

$$= \frac{\mu_o I_1 I_2}{2\pi} \left\{ \int_A^B \left[\hat{\mathbf{z}} \, dz \times \frac{-\hat{\mathbf{x}}}{a} \right] + \int_B^C \left[\hat{\mathbf{y}} \, dy \times \frac{-\hat{\mathbf{x}}}{y} \right] \right.$$

$$\left. + \int_C^D \left[\hat{\mathbf{z}} \, dz \times \frac{-\hat{\mathbf{x}}}{a+b} \right] + \int_D^A \left[\hat{\mathbf{y}} \, dy \times \frac{-\hat{\mathbf{x}}}{y} \right] \right\}$$

The directions of the forces on each side of the loop are shown in Figure 6.50. Note that the forces on the BC and DA sides are equal and opposite and thus cancel out. The forces on the AB and CD sides are also in opposing directions but their magnitudes are different, in view of their different distances from the long wire. The net total force on the loop is thus given by

$$\mathbf{F}_{\text{loop}} = -\hat{\mathbf{y}} \frac{\mu_0 I_1 I_2 d}{2\pi} \left(\frac{1}{a} - \frac{1}{a+b} \right)$$

Thus the loop is pulled toward the long wire. Note that if the polarity of either (but not both) of the two currents I_1 or I_2 were reversed, the loop would be pushed away from the long wire.

6.10.3 Torque on a Current-Carrying Loop

The orientation of the loop in Example 6.35 is such that the plane of the loop and thus all of its wires are entirely perpendicular to the **B** field generated by the long wire. If the loop is instead immersed in a region with a **B** field in the y direction, as shown in Figure 6.51a, its two sides AB and CD experience magnetic forces, respectively, into and out of the page, as shown, while the forces on the BC and DA sides are zero, since $d\mathbf{l}$ is parallel to **B**. Assuming again that the loop is rigid, it thus experiences a tendency to twist or turn around the z–z' axis shown in dashed lines. For given magnitudes of the forces $\mathbf{F}_{AB}$ and $\mathbf{F}_{CD}$, the loop has a greater tendency to rotate if the AB and CD sides are farther away from the z–z' axis. The quantitative measure of the tendency of a force to cause or change rotational motion is torque. Mathematically, when a force **F** acts on a body at a distance **r** from a reference axis, the torque **T** acting on a body with respect to that reference axis is defined as

$$\mathbf{T} = \mathbf{r} \times \mathbf{F}$$

The SI units for torque are newton-meters (N-m). The direction of torque is perpendicular to both **r** and **F** and defines the direction of rotation in the right-hand-rule sense. In other

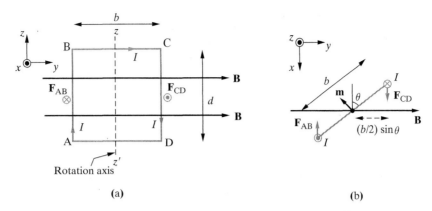

Figure 6.51   Torque on a loop in a B field. (a) Loop immersed in a uniform magnetic field **B**. The rotation axis is z–z'. (b) View from the top.

words, with the thumb of the right hand pointed in the direction of torque, rotation is in the direction of the fingers.

Let us return to the loop in Figure 6.51a and assume that it is mounted such that it is free to rotate about the z–z' axis and that the **B** field is uniform and is given by $\mathbf{B} = \hat{\mathbf{y}}B_0$. The total magnetic force on the loop can be calculated in the same manner as in Example 6.35. Noting once again that the forces on the BC and DA sides are zero since $d\mathbf{l}$ is parallel to **B**, we have

$$\mathbf{F}_{\text{loop}} = \mathbf{F}_{\text{AB}} + \mathbf{F}_{\text{CD}}$$

$$= I\left[\int_A^B d\mathbf{l} \times \mathbf{B} + \int_C^D d\mathbf{l} \times \mathbf{B}\right] = I\left[\int_A^B \hat{\mathbf{z}}\, dz \times \hat{\mathbf{y}}B_0 + \int_C^D \hat{\mathbf{z}}\, dz \times \hat{\mathbf{y}}B_0\right] = 0$$

Note that the net total force acting on the loop is zero. However, since $\mathbf{F}_{\text{AB}}$ and $\mathbf{F}_{\text{CD}}$ act at different points, they produce a torque, as illustrated in Figure 6.51b. The magnitudes of $\mathbf{F}_{\text{AB}}$ and $\mathbf{F}_{\text{CD}}$ are equal, namely, $|\mathbf{F}_{\text{AB}}| = |\mathbf{F}_{\text{CD}}| = B_0 I d$, and each of these two forces produces torque in the same direction, at distances of $r = (b/2)\sin\theta$ with respect to the rotation axis. The resultant torque is thus given as

$$\mathbf{T} = \mathbf{r}_{\text{AB}} \times \mathbf{F}_{\text{AB}} + \mathbf{r}_{\text{CD}} \times \mathbf{F}_{\text{CD}}$$

$$= \left(-\hat{\mathbf{y}}\frac{b}{2}\sin\theta\right) \times (-\hat{\mathbf{x}}B_0 I d) + \left(\hat{\mathbf{y}}\frac{b}{2}\sin\theta\right) \times (\hat{\mathbf{x}}B_0 I d) = -\hat{\mathbf{z}}B_0 I b d \sin\theta$$

The sense of rotation of the coil as shown in Figure 6.51b is clockwise, consistent with the torque being in the $-z$ direction (i.e., if the right-hand thumb is pointed in the $-z$ direction, the fingers rotate clockwise).

The total torque acting on the loop is thus found to be proportional to the product (bd), which is the area A of the loop. Thus we can rewrite **T** as

$$\mathbf{T} = -\hat{\mathbf{z}} B_0 I A \sin\theta$$

where A is the area of the loop. As long as the loop is symmetrically positioned with respect to the z–z' axis of rotation and has the same area A, the torque acting on it can be shown to be independent of its shape; that is, a circular, hexagonal, or rectangular loop experiences the same torque as a square loop.

Using the definition of the magnetic dipole from Section 6.6, we can write $|\mathbf{m}| = IA$ and note that the direction of the dipole moment **m** is related to the current flow via the right-hand rule and is as shown in Figure 6.51b. Considering the relationship between **T**, **m**, and **B**, we can see that the torque can be compactly expressed as

$$\boxed{\mathbf{T} = \mathbf{m} \times \mathbf{B}} \tag{6.53}$$

Equation (6.53) is quite general and describes the torque acting on a loop when it is in a region permeated by a uniform magnetic field **B**. The torque described by (6.53) is what aligns microscopic magnetic dipoles in paramagnetic materials and causes them to be magnetized as discussed in Section 6.8.1.

Direct-current machines.   Equation (6.53) indicates that rotational motion can be brought about from an arrangement of magnetostatic fields and steady electric current-carrying wires. Electrical machines that operate on this principle are called *direct-current* (or dc) *motors* and *generators*. Noting that the subject of electrical machinery and its applications is wide-ranging and well beyond the scope of this book, we briefly comment here on the basic principles of direct-current machines.

Careful examination of Figure 6.51 indicates that the torque acting on the loop tends to move it toward an equilibrium position at which the torque is zero. It is easy to see from Figure 6.51 that the torque is zero when **m** is parallel to **B**—that is, when the loop is perpendicular to the magnetic field. At this position there is no net force or torque acting on the loop. It turns out, however, that because of the finite mass of any coil, it approaches the equilibrium position with inertia and rotates through or overshoots it. As soon as the coil moves past the equilibrium position, the direction of torque is reversed (since $\sin\theta$ in Figure 6.51b is now negative) and it tends to rotate back toward the equilibrium position. Because of the damping effect of the surrounding air and mechanical friction, the coil eventually stops at the equilibrium position.

Even without continuous rotational motion, the fact that the torque produced is proportional to the current passing through the loop facilitates sensitive electrical measurements. The operation of voltmeters and ammeters is based on an arrangement such as that shown in Figure 6.52a, although an electromagnet may more often be used instead of a permanent magnet.

In order to obtain continuous rotation of the loop, it is necessary to prevent the reversal of the direction of torque as the loop moves past the equilibrium position. One possible way of achieving this is sketched in Figure 6.52b. Noting that the direction

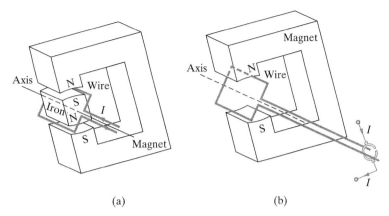

(a)   (b)

Figure 6.52   Simple dc motors. (a) Using a permanent magnet and soft iron material. In the case shown, there is no continuous motion, since the torque changes direction as the loop goes past the equilibrium position. (b) Using a permanent magnet and a commutator arrangement. By reversing the current every half cycle of rotation, continuous motion can be achieved.

of the torque is dependent on the direction of current flowing in the loop, the direction of current is changed as soon as the loop reaches its equilibrium position, using a so-called *commutator* arrangement. The commutator in a dc machine changes the direction of current once per rotation so that the torque always remains in the same direction.

The simplistic dc motor depicted in Figure 6.52 is not useful, since the magnitude of the torque varies periodically with time, although its sense remains constant. Smoother operation can be achieved by designing the magnetic field structure to provide a nearly constant **B** field at all points during one rotational cycle.

6.11 SUMMARY

This chapter discussed the following topics:

- **Ampère's force law.** The magnetostatic force F_{12} exerted by a circuit C_1 carrying a current I_1 on another circuit C_2 carrying current I_2 is given by

$$\mathbf{F}_{12} = \frac{\mu_0}{4\pi} \oint_{C_2} \oint_{C_1} \frac{I_2 \, d\mathbf{l}_2 \times (I_1 \, d\mathbf{l}_1 \times \hat{\mathbf{R}})}{R^2}$$

  where $\hat{\mathbf{R}}$ is the unit vector pointing from current element $I_1 \, d\mathbf{l}_1$ to $I_2 \, d\mathbf{l}_2$ and R is the distance between them. Ampère's force law is experimentally based and constitutes the foundation of magnetostatics.

- **The magnetic field.** Any physical region of space in the vicinity of steady electric currents is said to be permeated by a magnetic field. The relationship between the **B** field and its source current is provided by the Biot–Savart law, which is given by

$$\mathbf{B} = \frac{\mu_0}{4\pi} \left[\oint_{C'} \frac{I \, d\mathbf{l}' \times \hat{\mathbf{R}}}{R^2} + \int_{S'} \frac{\mathbf{J}_s(\mathbf{r}') \times \hat{\mathbf{R}}}{R^2} \, ds' + \int_{V'} \frac{\mathbf{J}(\mathbf{r}') \times \hat{\mathbf{R}}}{R^2} \, dv' \right]$$

where I is the current flowing in a filamentary contour C, $\mathbf{J}_s$ is a surface current density, and $\mathbf{J}$ is a volume current density. The unit vector $\hat{\mathbf{R}}$ points *from* the location of the current element *to* the observation point P and $R = |\mathbf{r} - \mathbf{r}'|$ is the distance between the current element and P. The Biot–Savart law is a direct consequence of Ampère's force law and is thus experimentally based.

- **Ampère's circuital law.** Ampère's circuital law states that the line integral of **B** over any closed contour C is a constant and is equal to μ_0 times the total net current I passing through the surface S enclosed by the contour. In other words,

$$\oint_C \mathbf{B} \cdot d\mathbf{l} = \int_S \mu_0 \mathbf{J} \cdot d\mathbf{s} = \mu_0 I$$

where C is the contour that encloses the surface S and I is the total current passing through the surface S. Ampère's circuital law is a direct consequence of the Biot–Savart law and is particularly useful for the analysis of problems involving a high degree of symmetry, such that contours over which the **B** field is constant can be identified.

- **Curl.** Curl is a measure of the circulation per unit area of a vector field. It is defined as an axial vector whose magnitude is the maximum net circulation of the vector field and whose direction is normal to the surface element that results in maximum circulation. For the magnetostatic field we have

$$\nabla \times \mathbf{B} = \mu_0 \mathbf{J}$$

which is the differential version of Ampère's law. Thus, the magnetostatic field has nonzero curl only at points where currents exist. Stokes's theorem is valid in general for any vector field,

$$\oint_C \mathbf{G} \cdot d\mathbf{l} = \int_S (\nabla \times \mathbf{G}) \cdot d\mathbf{s}$$

where S is the surface enclosed by the contour C.

- **Vector magnetic potential.** A direct consequence of the experimentally based Biot–Savart law is that the **B** field can be written as the curl of another vector **A** related to the source current **J** as

$$\mathbf{A}(\mathbf{r}) = \frac{\mu_0}{4\pi} \int_{V'} \frac{\mathbf{J}(\mathbf{r}')}{R} \, dv'$$

In some cases it is easier to evaluate **A** directly from the source currents using the above expression and subsequently find the **B** field using $\mathbf{B} = \nabla \times \mathbf{A}$.

- **The magnetic dipole.** A small current-carrying loop of radius a constitutes a magnetic dipole. The **B** field at large distances $(r \gg a)$ produced by a small loop of radius a lying in the x-y plane and carrying current I is given by

$$\mathbf{B} = \hat{\mathbf{r}} \frac{\mu_0 |\mathbf{m}| \cos\theta}{2\pi r^3} + \hat{\boldsymbol{\theta}} \frac{\mu_0 |\mathbf{m}| \sin\theta}{4\pi r^3}$$

  where $\mathbf{m} = \hat{\mathbf{z}} I (\pi a^2)$ is the magnetic dipole moment.

- **Divergence of B and magnetic flux.** A direct consequence of the fact that we can write $\mathbf{B} = \nabla \times \mathbf{A}$ is that the divergence of the **B** field is identically zero. In other words,

$$\nabla \cdot \mathbf{B} = 0 \qquad \rightarrow \qquad \oint_S \mathbf{B} \cdot d\mathbf{s} = 0$$

  which in turn means that the flux lines of the **B** field are always continuous and form closed loops.

- **Inductance.** The inductance of a circuit is a measure of its ability to link magnetic flux per unit current. The mutual inductance L_{12} between two coils C_1 and C_2 having, N_1 and N_2, respectively, turns is given by

$$L_{12} = \frac{N_1 N_2 \int_{S_2} \mathbf{B}_1 \cdot d\mathbf{s_2}}{I_1}$$

  where S_2 is the surface enclosed by circuit C_2, I_1 is the current in loop C_1, and $\mathbf{B}_1$ is the field produced by a single turn of coil 1. In general, $L_{12} = L_{21}$. The self-inductance of a circuit C having N turns is given by

$$L = \frac{N^2 \int_S \mathbf{B} \cdot d\mathbf{s}}{I}$$

  where S is the surface enclosed by the circuit C and **B** is the field produced by a single turn of C. To find the inductance of a circuit, we typically assume current I to flow and find the **B** field using the Biot–Savart law or Ampère's law and integrate it over the surface S to determine the flux linked by the circuit. A more general formula for the mutual inductance between two circuits C_1 and C_2 is given by the Neumann formula, namely,

$$L_{12} = \frac{\mu_0 N_1 N_2}{4\pi} \oint_{C_1} \oint_{C_2} \frac{d\mathbf{l}_1 \cdot d\mathbf{l}_2}{R}$$

  where N_1 and N_2 are the number of turns in circuits C_1 and C_2, respectively.

- **Magnetic materials.** When a material is placed in an external **H** field, a magnetization distribution **M** is created, which is given by $\mathbf{M} = \chi_m \mathbf{H}$, where χ_m is the magnetic susceptibility, dependent on the particular microscopic properties of the material. The net effect of this magnetization within the material is that the total magnetic field is now given by $\mathbf{B} = \mu_0(1 + \chi_m)\mathbf{H}$, a fact that is typically accounted for by assigning to each material a magnetic permeability $\mu = \mu_0(1 + \chi_m)$ such that $\mathbf{B} = \mu\mathbf{H}$. For most materials, the value of χ_m is quite small, so that $\mu \simeq \mu_0$. For ferromagnetic materials, however, the values of μ can be very large, ranging from a few hundred up to a million.

- **Magnetostatic boundary conditions.** Experimentally established laws of magnetostatics dictate that the normal component of the $\mathbf{B}$ field is continuous across the interface between two materials with different permeability. The tangential component of the magnetic field $\mathbf{H}$ is always continuous across any interface except when there is surface current density $\mathbf{J}_s$ present, which typically occurs at the surfaces of metallic conductors. In summary, we have

$$B_{1n} = B_{2n} \quad \text{and} \quad H_{1t} = H_{2t}$$

  except at a perfect conductor surface, when we have

$$\hat{\mathbf{n}} \times \mathbf{H}_1 = \mathbf{J}_s$$

  where $\hat{\mathbf{n}}$ is the outward normal unit vector from the perfect conductor surface.

- **Magnetic forces and torques.** Our starting point for magnetostatics was the experimental fact that current-carrying wires exert forces on one another. An alternative way of expressing this result is to state that a circuit C carrying current I placed in a $\mathbf{B}$ field experiences a magnetic force given by

$$\mathbf{F}_{\mathrm{m}} = \int_C I \, d\mathbf{l} \times \mathbf{B}$$

  When a loop of wire of area A, carrying current I and having a magnetic moment $\mathbf{m}$ such that $|\mathbf{m}| = IA$ is placed in a uniform magnetic field $\mathbf{B}$, it experiences a torque given by

$$\mathbf{T} = \mathbf{m} \times \mathbf{B}$$

The relationships among the total volume current density (which in general may contain both a free current density and a bound magnetization current density $\mathbf{J}_{\mathrm{m}}$ due to a magnetization vector $\mathbf{M}$), the $\mathbf{B}$ field, and the vector magnetic potential $\mathbf{A}$ described earlier are summarized in Table 6.3. Each row lists equations for the respective quantity of interest. The equations are organized according to the independent quantity listed on the top of each column.

In magnetic materials with $\mathbf{H} = \mu^{-1}\mathbf{B}$, we can write the free current density directly in terms of the $\mathbf{B}$-field and magnetic vector potential, as summarized in Table 6.4. The free current density $\mathbf{J}$ includes all sources of current excluding the bound magnetization current density, so that $\nabla \times \mathbf{H} = \mathbf{J}$.

TABLE 6.3 SUMMARY OF MAGNETOSTATIC EQUATIONS

	$\mathbf{J}^{\dagger}$	$\mathbf{B}$	$\mathbf{A}$
$\mathbf{J}^{\dagger} =$		$\dfrac{1}{\mu_0} \nabla \times \mathbf{B}$	$-\dfrac{1}{\mu_0} \nabla^2 \mathbf{A}$
$\mathbf{B} =$	$\dfrac{\mu_0}{4\pi} \displaystyle\int_{V'} \dfrac{\mathbf{J}(\mathbf{r}') \times \hat{\mathbf{R}}}{|\mathbf{r} - \mathbf{r}'|^2}\, dv'$		$\nabla \times \mathbf{A}$
$\mathbf{A}^{\ddagger} =$	$\dfrac{\mu_0}{4\pi} \displaystyle\int_{V'} \dfrac{\mathbf{J}(\mathbf{r}')}{|\mathbf{r} - \mathbf{r}'|}\, dv'$	$\dfrac{1}{4\pi} \displaystyle\int_{V'} \dfrac{\mathbf{B}(\mathbf{r}') \times \hat{\mathbf{R}}}{|\mathbf{r} - \mathbf{r}'|^2}\, dv'$	

$^{\dagger}\mathbf{J}$ represents the total volume current density, which in general may include the volume bound magnetization current density $\nabla \times \mathbf{M}$.

‡These formulas assume the so-called Coulomb gauge, that is, that we choose $\mathbf{A}$ such that $\nabla \cdot \mathbf{A} = 0$ (see Section 6.5 and Footnote 43).

TABLE 6.4 FREE CURRENT DENSITY

	$\mathbf{J}^{\dagger}$	$\mathbf{B}$	$\mathbf{A}$
$\mathbf{J}^{\dagger} =$		$\dfrac{1}{\mu} \nabla \times \mathbf{B}$	$-\dfrac{1}{\mu} \nabla^2 \mathbf{A}$

$^{\dagger}\mathbf{J}$ represents the free volume current density, which excludes the volume bound magnetization current density $\nabla \times \mathbf{M}$.

PROBLEMS

6.1 **Force between two infinitely long wires.** A dc transmission line consists of two infinitely long, parallel wires separated by a distance of 50 cm. The two wires carry the same current I in opposite directions. Find I if the force per unit length experienced by each wire is 0.5 N-m^{-1}.

6.2 **Forces between two wires.** Consider a two-wire transmission line consisting of two parallel conductors of 1 mm diameter separated by 1 cm. If a potential difference of 10 V is applied between the conductors, equal charges of opposite sign would be induced on the two conductors, resulting in an electrostatic force of attraction between them. Is there a value of line current I for which this electric attraction force might be balanced by the magnetic force acting on the wires?

6.3 **Bundle clash in a transmission line.** Each phase of a three-phase alternating current transmission line consists of a bundle of two parallel wires each 4 cm in diameter and 50 cm apart, carrying current in the same direction. Under normal operation, the currents flowing in the wires are of the order of hundreds of amperes, and the magnetic force of attraction between the wires in the bundle is relatively small. An out-of-control airplane accidentally strikes the power transmission line, causing short-circuit loading, as a result of which the peak current in each wire in the bundle reaches a value of 100 kA. Calculate the peak magnetic force of attraction per unit length on each wire and comment on the possibility of a bundle clash. Note that similar short-circuit loading of power lines can also occur due, for example, to a tree falling on the line or a snake climbing up a power-line pole.

6.4 **Magnetic force between current elements.** Two elements of current are oriented in air in the same plane and positioned with respect to each other as illustrated in Figure 6.53a. (a) Find the magnitude and direction of the force exerted upon each element by the other. (b) Repeat (a) for the configuration of the current elements shown in Figure 6.53b.

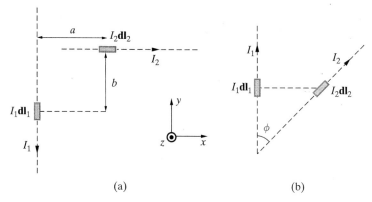

(a)   (b)

Figure 6.53   **Force between current elements.** Problem 6.4.

6.5 **Square loop.** Does a square loop carrying current have a tendency to expand or contract?

6.6 **Force on a current-carrying wire.** A single conductor of a transmission line extends in the east-west direction and carries a current of 1 kA. The earth's **B** field is directed essentially straight north at the point in question and has a value of $\sim$0.5 T. What is the magnitude of the force per meter on the current-carrying conductor?

6.7 **Rigid rectangular loop.** A rigid rectangular loop carrying a current of 5 A is located in the x-y plane with its four corners at $(0,0)$, $(1,0)$, $(1,2)$, and $(0,2)$. Determine the magnetic force exerted on each side of the loop if the region in which the loop is located is permeated with a **B** field given by (a) $\mathbf{B} = \hat{\mathbf{x}}3$ T and (b) $\mathbf{B} = \hat{\mathbf{z}}3$ T.

6.8 **Two parallel wires.** Consider two infinitely long parallel wires, each carrying a steady current of I in the z direction, one passing through the point $(4,0,0)$ and the other through $(0,4,0)$, as shown in Figure 6.54. (a) Find **B** at the origin. (b) Find **B** at point $(1,1,0)$. (c) Find **B** at $(2,2,0)$. (d) Repeat parts (a), (b), and (c) if the direction of the current in the wire on the x axis is switched.

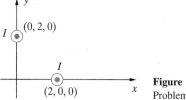

Figure 6.54   **Two parallel wires.** Problem 6.8.

6.9   **An infinitely long L-shaped wire.** Consider a single wire extending from infinity to the origin along the y axis and back to infinity along the x axis and carrying a current I. Find **B** at the following points: (a) $(-2a,0,0)$, (b) $(0,+2a,0)$, and (c) $(0,0,a/2)$.

6.10   **B field in cylindrical conductor.** Consider a cylindrical conductor of radius a, having a **B** field given by

$$\mathbf{B}(r) = \hat{\boldsymbol{\phi}} \, B_0 \left[\frac{r}{2a} - \frac{1}{3}\left(\frac{r}{a}\right)^2 \right] \qquad r \le a$$

(a) Determine the current density $\mathbf{J}(r)$ and total current I through the conductor. (b) What is the **B** field for the region $r > a$?

6.11   **Irregular loop.** Find **B** at point P due to the wire carrying a current I in free space, as shown in Figure 6.55.

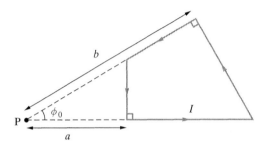

Figure 6.55   Irregular loop.
Problem 6.11.

6.12   **An N-sided regular loop.** A regular polygon-shaped loop, with N sides, carries a current I. (a) Show that the magnetic field at the center of the loop has a magnitude given by $B = [\mu_0 NI/(\pi d)]\tan(\pi/N)$, where d is the diameter of the circle passing through the corners of the polygon. (b) Also show that as $N \to \infty$, B approaches that found in Example 6.6.

6.13   **Square loop of current.** A current I flows in a square loop of side length a as shown in Figure 6.56. Find the **B** field at the point $P(a/4, a/4)$.

Figure 6.56   Square loop of current.
Problem 6.13.

6.14   **A wire with two circular arcs.** Consider a loop of wire consisting of two circular and two straight segments carrying a current I, as shown in Figure 6.57. Find **B** at the center P of the circular arcs.

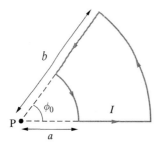

Figure 6.57   Wire with two circular arcs.
Problem 6.14.

6.15   Wire with four 90° bends. An infinitely long wire carrying current $I = 1$ A has four sharp 90° bends 1 m apart as shown in Figure 6.58. Find the numerical value (in Wb-m^{-2}) and direction (i.e., the vector expression) of the **B** field at point $P(1, 0, 0)$.

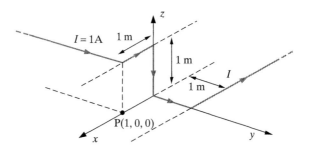

Figure 6.58   Wire with four 90° bends. Problem 6.15.

6.16   Infinitely long copper cylindrical wire carrying uniform current. An infinitely long solid copper cylindrical wire 2 cm in diameter carries a total current of 1000 A, distributed uniformly throughout its cross section. Find **B** at points inside and outside the wire.

6.17   Helmholtz coils. Two thin circular coaxial coils each of radius a, having N turns, carrying current I, and separated by a distance d, as shown in Figure 6.59, are referred to as *Helmholtz coils* for the case when $d = a$. This setup is well known for producing an approximately uniform magnetic field in the vicinity of its center of symmetry. (a) Find **B** on the axis of symmetry—the z axis—of the Helmholtz coils. (b) Show that $dB_z/dz = 0$ at the point P midway between the two coils. (c) Show that both $d^2B_z/dz^2 = 0$ and $d^3B_z/dz^3 = 0$ at the midway when $d = a$. (Note that d is called the Helmholtz spacing, which corresponds to

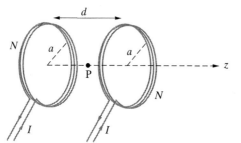

Figure 6.59   Helmholtz coils.
Problem 6.17.

the coil separation for which the second derivative of B_z vanishes at the center.) (d) Show that B_z at the midpoint P between the Helmholtz coils is

$$B_z \simeq 0.8992 \times 10^{-6} \frac{NI}{a} \quad \text{T}$$

(e) Find B_z at the center of each loop and compare it with the value at the midpoint between the coils.

6.18 Helmholtz coils. Design a pair of Helmholtz coils, separated by 0.5 m, to produce a magnetic field of 150 μT midway between the two coils. Take $I = 2$ A.

6.19 Helmholtz coils. Consider the Helmholtz coils discussed in Problem 6.17. Calculate and sketch B_z along the axis of symmetry between the two coils if one of the coils is connected backward (i.e., its current is switched).

6.20 Square Helmholtz coils. Consider a pair of square coils, similar to Helmholtz coils, each of same side a, having N turns, and carrying current I, separated by a distance d. Show that the Helmholtz spacing d (i.e., the coil separation for which $d^2 B_z / dz^2 = 0$ at the center) is equal[81] to $d \simeq 0.5445a$. (Note that the Helmholtz spacing for circular coils is $d = a$, where a is the radius of each coil.) To simplify your solution, take each side to be of unity length and use a simple iterative procedure.

6.21 A circular and a square coil. Consider a pair of coils, similar to Helmholtz coils, separated by a distance d, where one of the coils is circular in shape with radius a and the other square with side $2a$, each having N turns and carrying current I, and $d = a$. For $a = 25$ cm, $I = 1$ A, and $N = 20$, calculate and plot the magnitude of the **B** field along the axis of symmetry between the two coils.

6.22 Magnetic field of a solenoid. For an air-core solenoid of $N = 400$ turns, length $l = 50$ cm, radius $a = 2$ cm, and having a current of 1 A, find the **B** field (a) at the center and (b) at the ends of the solenoid.

6.23 Magnetic field of a surface current distribution. A circular disk of radius a centered at the origin with its axis along the z axis carries a surface current flowing in a circular direction around its axis given by

$$\mathbf{J}_s = \hat{\phi} K r \quad \text{A-m}^{-2}$$

where K is a constant. Find **B** at a point P on the z axis.

6.24 B inside the solenoid. An air-core solenoid of 800 turns, 40 cm length, and 5 cm^2 cross-sectional area is carrying a current of 1.5 A. (a) Find **B** along the axis of the solenoid. (b) Find $\mathbf{B}_{\text{ctr}}$ at the center of the solenoid. (c) Sketch $|\mathbf{B}|$ along the axis of the solenoid.

6.25 B inside the solenoid. A long solenoid consists of a tightly wound coil around a magnetic core ($\mu_r = 300$) carrying a current of 10 A. (a) If the **B** field magnitude at the center of the core is $B_{\text{ctr}} \simeq 1.5$ T, find the number of turns of wire wound around the core per centimeter. (b) Repeat part (a) if this was an air-core solenoid.

6.26 Split washer. The thin flat washer shown in Figure 6.60 has an inner radius of $a = 10$ mm and an outer radius of $b = 50$ mm. A uniform voltage V_0 is applied between the edges of the radial slot, resulting in an angular current of $I = 120$ A. Find the **B** field at the center of

[81] A. H. Firester, Designs of square Helmholtz coil systems, *Rev. Sci. Instr.*, 37, pp. 1264–1265, 1966.

the washer. Note that the current density is a function of the radius from the center. Neglect the width of the slot. Note that the application of the voltage between the edges is as shown in Figure 5.6c.

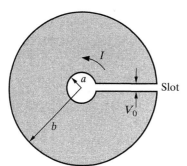

Figure 6.60   Split washer. Problem 6.26.

6.27   Long wire encircled by small loop. A long straight wire oriented along the z axis carries a current of 1 A. A loop of 5 cm radius carrying a current of 100 A is located in the x-y (i.e., $z = 0$) plane, with its center at the origin, concentric with the wire, so that the axis of the loop coincides with the z axis. Find the **B** field at a point on the x-y plane at a distance of 1 m from the origin.

6.28   Flux through a rectangular loop. A long, straight wire carrying a current I and a rectangular loop of wire are separated by a distance a as shown in Figure 6.61. (a) If the sides of the rectangular loop parallel and perpendicular to the straight wire are a and $2a$, respectively, find the magnetic flux Ψ that links the rectangular loop due to the straight current-carrying wire. (b) The rectangular loop is rotated by 90° around its perpendicular symmetry axis. Find the percentage change in Ψ linking the loop.

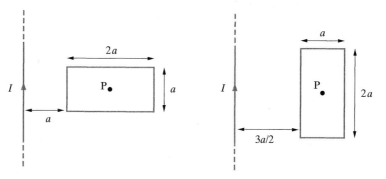

Figure 6.61   Flux through rectangular loop. Problem 6.28.

6.29   Flux through a triangular loop. A long, straight wire carrying a current I and a triangular loop of wire are as shown in Figure 6.62. (a) Find the magnetic flux Ψ that links the triangular loop in terms of a, b, and ϕ_0. (b) Find Ψ if $I = 120$ A, $b = 4a = 40$ cm, and $\phi_0 = 45°$, respectively.

Figure 6.62 Flux through a triangular loop. Problem 6.29.

6.30 **A toroidal coil around a long, straight wire.** A long, straight wire carrying a current of 100 A coincides with the principal axis of symmetry of a 150-turn rectangular toroid of inner and outer radii $a = 6$ cm and $b = 9$ cm, thickness $t = 3$ cm, core material with $\mu_r = 250$, respectively. No current flows in the toroid. Find the total magnetic flux Ψ linking the toroid due to the current in the long, straight wire.

6.31 **Vector potential.** Determine an expression for the magnetic vector potential $\mathbf{A}$ inside $(r < a)$ and outside $(r \geq a)$ an infinite cylindrical conductor such as that in Example 6.10 (Figure 6.16), oriented in the z direction, having radius a and carrying a current I uniformly distributed over its cross-sectional area [i.e., $\mathbf{J} = \hat{\mathbf{z}} I /(\pi a^2)$]. Note that the vector potential $\mathbf{A}$ cannot be uniquely determined, so you are asked to simply find one of the possible expressions for $\mathbf{A}$ in both regions. [Hint: You might want consider using the $\mathbf{B}$ field expressions given in Example 6.10.]

6.32 **Infinitely long wire with a cylindrical hole.** Consider an infinitely long cylindrical conductor wire of radius b, the cross section of which is as shown in Figure 6.63. The wire contains an infinitely long cylindrical hole of radius a parallel to the axis of the conductor. The axes of the two cylinders are apart by a distance d such that $d + a < b$. If the wire carries a total current I, (a) find the vector magnetic potential $\mathbf{A}$ and (b) use $\mathbf{A}$ to find the $\mathbf{B}$ field in the hole. Compare your result with that found in Example 6.13.

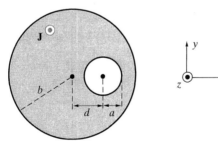

Figure 6.63 Wire with hole. Problem 6.32.

6.33 **Finite-length straight wire: A at an off-axis point.** Consider the straight, current-carrying filamentary conductor of length $2a$, as shown in Figure 6.11. (a) Find the magnetic vector potential $\mathbf{A}$ at an arbitrary point $P(r, \phi, z)$. (b) Find the $\mathbf{B}$ field using $\mathbf{B} = \nabla \times \mathbf{A}$ to verify the result of Example 6.7.

6.34 **Two infinitely long wires.** Consider two infinitely long parallel wires oriented in the z direction each carrying a current I in opposite directions, respectively. Show that the magnetic vector potential $\mathbf{A}$ is give by $\mathbf{A} = \hat{\mathbf{z}}[\mu_0 I/(2\pi)]\ln(b/a)$, where a and b are the distances from the observation point to the wires.

6.35 **Wire with oval-shaped hole.** Consider a wire with the cross-sectional shape shown in Figure 6.64, having an oval-shaped axial hole. The wire segments on both sides of the hole carry uniform current densities $\mathbf{J}_1$ and $\mathbf{J}_2$ of equal magnitude J_0 and opposite sign. Find the magnetic field $\mathbf{B}$ at points P_1 and P_2 in the hole.

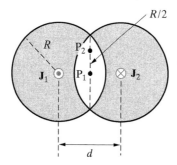

Figure 6.64 Wire with oval-shaped hole.
Problem 6.35.

6.36 **Inductance of a short solenoid.** Find inductance of a short air-core solenoid of 400 turns, 10 cm length & 5.5 cm diameter. Use the Nagaoka formula.

6.37 **Inductance of a finite-length solenoid.** The secondary coil of a high-voltage electric generator is an air-core solenoid designed with the following parameters: $l = 1.8$ m, $a = 18$ cm, and $N = 800$ turns. Calculate the self-inductance of the coil by treating it (a) as an infinitely long solenoid, and (b) as a finite-length solenoid. Compare your results.

6.38 **Inductance of a rectangular toroid.** Consider the toroidal coil of rectangular cross section shown in Figure 6.20. (a) Show that the inductance of this coil on an air core is in general given by

$$L = 2 \times 10^{-9} t N^2 \ln\left(\frac{b}{a}\right) \quad \text{H}$$

where N is the number of turns and t is the vertical thickness of the core in cm. (Do not assume $r_m \gg b - a$.) (b) If the dimensions of this toroid are $a = 1.2$ cm, $b = 2$ cm, and $t = 1.5$ cm, $N = 1000$, find the inductance using the expression in part (a). (c) Find the inductance using the approximate expression derived in Example 6.27, assuming $r_m \gg b - a$ and compare it with the result of part (b).

6.39 **Inductance of a rectangular toroid.** Consider an air-core rectangular toroid as shown in Figure 6.20 with the values of its dimensions given by $a = 3.8$ mm, $b = 6.4$ mm, and $t = 4.8$ mm, respectively. (a) Find the total number of turns to be wound on this core such that its total inductance is around 1 mH. (b) Repeat part (a) for a powdered nickel-iron core (assume $\mu_r = 200$) having the same geometric dimensions.

6.40 **Inductance of a rectangular toroid.** A 50-mH toroid inductor is to be designed using a molypermalloy powder core with $\mu_r = 125$, $a = 7.37$ mm, $b = 13.5$ mm, and $t = 11$ mm. Find the approximate number of turns N required.

6.41 **Inductance of a circular toroid.** Consider a toroid of circular cross section of radius a and mean radius r_m as shown in Figure 6.65. Show that the inductance of this coil is given by

$$L = \mu_r \mu_0 N^2 [r_m - (r_m^2 - a^2)^{1/2}]$$

where μ_r is the relative permeability of the core material.

6.42 **Inductance of a two-wire line.** Determine the inductance per unit length of a two-wire transmission line in air as shown in Figure 6.39, designed for an amateur radio transmitter, with conductor radius $a = 1$ mm and spacing $d = 6$ cm. (b) Repeat part (a) if the conductor spacing is doubled (i.e., $d = 12$ cm).

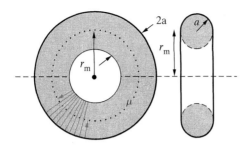

Figure 6.65   Inductance of a circular toroid. Problem 6.41.

6.43   Inductance of a thin circular loop of wire. (a) Find the inductance of a single circular loop of wire with $d = 4$ mm and $a = 10$ cm. (b) Repeat part (a) with $d = 0.4$ mm and $a = 10$ cm. Refer to Figure 6.43.

6.44   Mutual inductance between two solenoidal coils. An air-core solenoid 25 cm long and 2.5 cm diameter is wound of 1000 turns of closely spaced, insulated wire. A separate coil of 100 turns, 3 cm long, with about the same diameter is located at the center of the longer coil (where **B** is approximately constant). Find the mutual inductance between the two coils.

6.45   Mutual inductance between a wire and a circular loop. Find the mutual inductance between an infinitely long, straight wire and a circular wire loop, as shown in Figure 6.66.

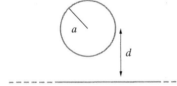

Figure 6.66   Wire and circular loop. Problem 6.45.

6.46   Long wire and loop. An infinitely long wire carrying current I passes just under a circular loop of radius a, also carrying the same current I as shown in Figure 6.67. (a) Find the

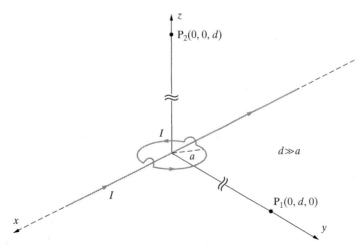

Figure 6.67   Long wire and loop. Problem 6.46.

magnitude and direction of the **B** field at point P_1. Note that $d \gg a$. (b) Repeat (a) for point P_2. (c) What is the mutual inductance between the wire and the loop?

6.47 **Inductance and energy of a solenoid.** An air-core solenoid 25 cm long and 1 cm diameter is wound of 800 turns of closely spaced, insulated wire. (a) Find the inductance of the solenoid. (b) Find the energy stored in the solenoid if it is carrying a current of 15 A.

6.48 **Two square coils.** Two square coils each of side length a are both located on the x-y plane such that the distance between their centers is d. (a) Find a simple expression for the mutual inductance between the two loops for the case $d \gg a$. (b) Find the mutual inductance between the two coils for the case $d = 4a$.

6.49 **Fixed-length piece of wire.** You have a fixed length of copper wire of 1 mm diameter and 10 cm length. You can bend this wire in any shape or form to obtain as large an external self-inductance as you can. (a) What is the value of the maximum inductance, and what arrangement would work best? You cannot use any magnetic materials. (b) If you had the tools necessary to melt the copper wire and form it into a longer wire of diameter 0.1 mm instead, how would your answer to (a) change? Also compare and comment on the resistance of the two different wires.

6.50 **Magnetic energy.** A conductor consists of a cylinder with radius b with a hole of radius a ($a < b$) drilled coaxially through its axis. The current density is uniform and corresponds to a total current of I. Find the magnetic energy stored inside per unit length of the wire.

6.51 **Explosion in a power transformer.** A current transformer used for 500 kV transmission lines has a single primary coil which is connected to the high voltage line via two parallel wires that are 20 cm apart and carrying the same current in opposite directions. A fault occurs on the line and causes the current of each wire of the transformer to reach a peak value of 100 kA which results in an explosion in the oil/paper insulation structure of the wires. Calculate the per-unit-force on each wire and explain what happened.

6.52 **Force between a long wire and a loop.** A long wire extending along the z axis is situated near a rigid loop as shown in Figure 6.68. Find an expression for the force and torque (about the origin) experienced by the loop.

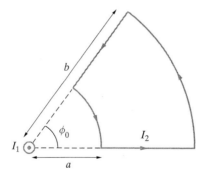

Figure 6.68 Force between a long wire and a loop. Problem 6.52.

7

Time-Varying Fields and Maxwell's Equations

Oersted's accidental discovery in 1820 of the close connection between electricity and magnetism led to almost immediate advances in technology. The fact that wires carrying current experience forces and torques when in the presence of a magnetic field was put to good use in direct-current electric motors (Section 6.10) and in the design of galvanometers, sensitive instruments for measuring electrical current. These instruments were crucial in facilitating scientific experimentation. However, a much broader range of industrial and technological applications of electricity and magnetism was yet to come, with Faraday's discovery in 1831 of electromagnetic induction—the fact that magnetic fields that change with time produce electric fields. This experimentally based fact constitutes the last of our three experimental pillars of electromagnetics[1] and is the subject of this chapter.

Electrostatics (Chapter 4) deals with the effects of stationary charges, the spatial distributions of which are determined by the presence of conducting bodies and applied potentials. The governing laws of electrostatics, derived from Coulomb's force law, can be summarized as follows:

$$\nabla \times \mathbf{E} = 0 \qquad \oint_C \mathbf{E} \cdot d\mathbf{l} = 0$$

$$\nabla \cdot \mathbf{D} = \rho \qquad \oint_S \mathbf{D} \cdot d\mathbf{s} = \int_V \rho \, dv$$

where ρ is the free volume charge density, and the fields $\mathbf{E}$ and $\mathbf{D}$ are related via $\mathbf{D} = \epsilon \mathbf{E}$, with ϵ being a scalar constant for linear, isotropic, homogeneous, and time-invariant media.

---

[1]The first pillar of electromagnetics is the fact that electric charges attract or repel one another in a manner inversely proportional to the distance between them (Coulomb's law). The second pillar is the fact that current-carrying wires create magnetic fields and exert forces on one another (Ampère's law of force or alternatively the Biot–Savart law).

Magnetostatics (Chapter 6) deals with the physical effects produced by charges in motion (i.e., steady currents). The governing laws of magnetostatics, derived from Ampère's force law (expressed also in the Biot–Savart Law), can be summarized as follows:

$$\nabla \times \mathbf{H} = \mathbf{J} \qquad\qquad \oint_C \mathbf{H} \cdot d\mathbf{l} = \int_S \mathbf{J} \cdot d\mathbf{s}$$

$$\nabla \cdot \mathbf{B} = 0 \qquad\qquad \oint_S \mathbf{B} \cdot d\mathbf{s} = 0$$

where $\mathbf{J}$ is the free volume current density, and the fields $\mathbf{B}$ and $\mathbf{H}$ are related via $\mathbf{H} = \mu^{-1}\mathbf{B}$ where μ is a constant scalar for linear, isotropic, homogeneous, and time-invariant media.

Up to now, we have exclusively considered static cases and paid little attention to electric and magnetic effects produced by charges, currents, or circuits whose positions or intensities vary with time.[2] Under the static assumption, the electric and magnetic fields are unrelated, and the field vectors $\mathbf{E}$, $\mathbf{D}$ and $\mathbf{B}$, $\mathbf{H}$ form independent pairs.[3] However, following Oersted's findings that electric currents produce magnetic fields, M. Faraday and many others started thinking about the possibility that magnetic fields may in turn produce electric fields, and that the two sets of preceding equations may be coupled.

After many years of various trials,[4] Faraday carried out the classic experiment[5] of induction on August 29, 1831, and showed that $\mathbf{E}$ and $\mathbf{B}$ were indeed related, but only for the case in which the fields change in time, that is, under nonstatic conditions. At about the same time, J. Henry of Albany Academy in New York studied similar effects and arrived at similar conclusions. However, Faraday's results were published earlier and had the greatest impact in the scientific world.[6] As a result, the law of electromagnetic induction is commonly referred to as Faraday's law.

Faraday's law of electromagnetic induction, namely that magnetic fields that change with time induce[7] electromotive force, is the third and final experimental

---

[2]Occasionally, it was necessary to apply Gauss's law to time-dependent phenomena with little or no justification, for example, in the derivation of the continuity equation (Section 5.5).

[3]In a conducting medium ($\sigma \neq 0$), a static electric field causes a steady current to flow ($\mathbf{J} = \sigma\mathbf{E}$), which in turn gives rise to a static magnetic field. However, at steady state, the electric field can be completely determined from the static charges or potential distributions. The magnetic field is simply a consequence of the current and does not affect the electric field.

[4]For a brief historical account, see R. Elliott, *Electromagnetics,* IEEE Press, New Jersey, 1993. Another excellent review of the history and development of electromagnetics, with Chapter VI devoted to Faraday, is given by E. Whittaker, *History of the Theories of the Aether and Electricity,* T. Nelson and Sons, Ltd., London, 1951. It is interesting, for example, that a number of scientists, including Ampère, only narrowly missed discovering the law of induction.

[5]M. Faraday, *Experimental Researches in Electricity,* Taylor, London, Vol. I., 1839, pp. 1–109.

[6]Henry's lack of promptness in announcing his results, combined with the fact that the New World was remote from European centers, caused his results to be viewed as merely confirmation of Faraday's findings.

[7]The word "induce" is used quite literally in this context; to induce means to cause the formation of. The word "induction," on the other hand, has several meanings in nontechnical parlance, including the initiation of a person into military service. In our context, the word *induction* refers to the process by which the electromotive force is brought into existence by the time-varying magnetic field.

fact[8] that forms the basis for Maxwell's equations and is arguably the most important of our three experimental pillars of electromagnetics. Without it, electromagnetic waves would not propagate through free space, and wireless communication would not be possible. Faraday's discoveries also form the basic underpinnings of modern electrical technology; without electromagnetic induction, the workhorses of our industrial world—electric motors and generators—would not be possible.

Our coverage of Faraday's law in this chapter completes our understanding of the underlying physical basis of the transmission line behavior discussed in Chapters 2 and 3. The fact that temporal variation of a voltage applied between the two conductors of a transmission line induces electromotive force, which in turn affects the current flow, is the very basis of the distributed inductance of the line. Completing our formulation of Maxwell's equations later in the chapter also allows us to exhibit the similarity between the fundamental equations that relate electric and magnetic fields for electromagnetic waves and the transmission line equations that relate line voltages and currents. Natural solutions of both the electric/magnetic field equations and the voltage/current equations are in the form of waves that propagate with a finite speed, leading to the distributed circuit effects studied in Chapters 2 and 3, and the propagation, reflection, and guiding of electromagnetic waves covered in Chapters 8–10.

As we discuss time-varying magnetic fields and Faraday's law in the first three sections of this chapter, we are concerned with variations that are slow enough so that radiation effects are negligible. This is the so-called *quasi-static approximation,* which implies that the system of conductors carrying currents has dimensions much smaller than a wavelength. A wide range of problems in laboratory physics and engineering lie within the domain of the quasi-static approximation. The criteria for the validity of the quasi-static approximation are the same as those discussed in Chapter 1 in connection with determining the applicability of the lumped analysis and the necessity of a distributed circuit treatment. Basically, the quasi-static approximation restricts us to considering circuits of such sizes and rates of change of current that the electromagnetic disturbance propagates over much of the useful parts of the circuit before the current has changed significantly. Under such conditions, we implicitly assume that at any given instant, the magnetic fields everywhere in the circuit are strictly proportional to the current at that instant. In general, the quasi-static approximation amounts to calculating all fields as if

---

[8]We note once again that the complete set of Maxwell's equations can be derived via transformation of Coulomb's law, using special theory of relativity. This was first shown in 1912 [L. Page, A derivation of the fundamental relations of electrodynamics from those of electrostatics, *Am. J. Sci.,* 34, pp. 57–68, 1912]; a clear treatment is given in Chapter 5 of R. S. Elliott, *Electromagnetics,* IEEE Press, New Jersey, 1993. However, such derivations must necessarily make other implicit assumptions; see Section 12.2 of J. D. Jackson, *Classical Electrodynamics,* Wiley, New York, 2nd ed., 1975. Even before the introduction of the special theory of relativity, and about ten years after Faraday's discovery, H. von Helmholtz and Lord Kelvin recognized and suggested [Helmholtz, *Über die Erhaltung der Kraft,* 1847; Kelvin, *Trans. Brit. Assoc.,* 1848 and *Phil. Mag.,* December 1851] that it is possible to deduce the existence of induced currents from energy considerations. For a brief discussion, see Article 481 of G. H. Livens, *The Theory of Electricity,* Cambridge University Press, Cambridge, UK, 1918. Nevertheless, and in view of the profound importance of the law of electromagnetic induction in forming the basis of Maxwell's equations, in this book we consider Faraday's law to be based on experimental fact.

they were stationary, without having to account for the travel time of fields from one point of the circuit to another. Formally, the quasi-static approximation is valid so far as $\partial \rho / \partial t$ is negligible compared with $\nabla \cdot \mathbf{J}$. When free-charge density varies rapidly enough that $\partial \rho / \partial t$ is comparable to $\nabla \cdot \mathbf{J}$, time variations of electric fields produce magnetic fields, leading to propagation of electromagnetic waves, as is discussed in Section 7.4.

7.1 FARADAY'S LAW

Michael Faraday was completely innocent of all mathematics, but at the same time he was one of the most imaginative and visual scientific thinkers of all time. Since Oersted demonstrated that an electric current could deflect a compass needle (i.e., generate a magnetic field), Faraday was convinced that a magnetic field or a magnet should be capable of producing a current. In 1831, Faraday set up an apparatus[9] consisting of an iron ring in the shape of a toroid on which were wound two coils of wire as sketched in Figure 7.1. The primary coil was connected through a switch to a voltaic cell; the ends of the secondary coil were joined with a wire that ran above a compass. Any current induced in the secondary coil would thus deflect the compass needle to the left or right of its normal position, depending on its flow direction. Upon closing the switch, current flowed in the primary coil, producing a magnetic field in the iron ring that passed through the secondary coil. Faraday observed a momentary deflection of the compass needle detecting a brief surge of current induced in the secondary coil. However, the compass needle quickly settled back to zero, indicating that the induced current existed only during the initial transient. In other words, no current flowed in the secondary coil once the current in the primary coil became steady and the field in the toroid reached

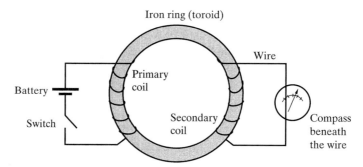

Figure 7.1   Faraday's iron-ring experiment. A sketch showing the elements of the apparatus used by Faraday that led to the discovery of the law of electromagnetic induction (also called Faraday's law). The iron ring provides a path for the magnetic flux coupling the two circuits.

[9]See H. Kondo, Michael Faraday, *Scientific American,* 10, pp. 91–98, 1953 and Chapter 6 of G. L. Verschuur, *Hidden Attraction, The Mystery and History of Magnetism,* Oxford University Press, New York, 1993.

its final value. When the switch was opened, terminating the current in the primary coil and thus eliminating the magnetic field in the toroid, another momentary deflection was observed, opposite in polarity with respect to the one observed when the switch was closed. Based on these extraordinary observations, Faraday deduced that a magnetic field could be generated by a steady current, but that a current could be induced only by a *changing* magnetic field.[10] This meant that the mathematical description of this phenomenon, which came to be known as electromagnetic induction, or simply induction, had to depend on the time variation of the magnetic flux.

Later, Faraday investigated the possibility of generating a steady current in the secondary coil and realized that a *continuously moving* magnetic field would be necessary. For this purpose he made a copper disk rotate with its edge between the poles of a magnet and found that current flowed from the center toward the edge of the disk (or vice versa). This was the world's first dynamo or electric generator. Faraday also used the motion of a magnet in and out of a helical coil to create a current, as sketched in Figure 7.2. He also determined that a more efficient way to produce currents was to move coils of wire in a magnetic field, the principle on which electric generators are built.

These experiments enabled Faraday to formulate his famous law of electromagnetic induction, or *Faraday's law,* which states that when the magnetic flux enclosed by a loop

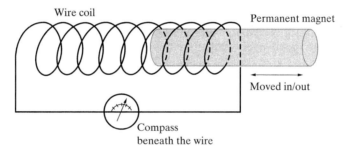

Figure 7.2   A magnet moving in a coil. A sketch of one of Faraday's experiments used to induce a current in a helical coil of wire by moving a magnet into and out of the coil—the principle of the electric generator. The induced current is detected by the deflection of the compass needle located under the wire combining the ends of the coil.

[10]The transitory nature of the electromotive force produced, due to its dependence on the *change* in magnetic flux, contributed to the elusive nature of the effect. An outstanding example of this was the experiments of Jean Daniel Colladon in Geneva in 1825. Colladon made a large helix of insulated copper wire 8–10 cm long and 4–5 cm in diameter, and he connected the ends to a sensitive galvanometer. He then brought a permanent magnet up to one end of the helix, expecting to produce a permanent current flow to be registered by the galvanometer. However, being the careful scientist he was, he wanted to place the galvanometer far from the magnet so that it would not be affected by the magnetic field of the magnet itself. His solution was to connect the galvanometer to the helix with 50 m of wire and to place it under a glass jar in another room well away from the magnet. There is little doubt that Colladon's galvanometer registered the temporary current caused by his movement of the magnet; however, by the time he walked to the other room, the galvanometer needle had returned to its zero reading and Colladon missed his chance of discovering electromagnetic induction. (From W. A. Atherton, The history of electromagnetic induction, *Am. J. Phys.,* 48(9), pp. 781–782, September 1980.)

of wire changes with time, a current is produced in the loop, indicating that a voltage, or an *electromotive force* (emf), is induced. The variation of the magnetic flux can result from a time-varying magnetic field linking a stationary and fixed loop of wire or from a loop of wire moving in a static magnetic field or from both, that is, a loop of wire moving in a time-varying magnetic field.

The mathematical form of Faraday's law is

$$\mathcal{V}_{ind} = -\frac{d\Psi}{dt}$$ (7.1)

where $\mathcal{V}_{ind}$ is the induced voltage across the terminals of the loop C, and Ψ is the total magnetic flux linking[11] the closed loop C, as shown in Figure 7.3. Faraday's law states that the induced voltage $\mathcal{V}_{ind}$ around a closed loop C is equal to the negative of the time rate of change of the magnetic flux linking C. Note that Ψ is the total magnetic flux linking the contour C, given by (6.26):

$$\Psi = \int_S \mathbf{B} \cdot d\mathbf{s}$$

where S is the area of the surface enclosed by contour C, as shown in Figure 7.3. Having an induced voltage $\mathcal{V}_{ind}$ appear across the terminals of the loop means that current will flow if a resistance is connected across the terminals. In this sense, this induced voltage is similar to the electromotive force as defined and discussed in Chapter 5. The electromotive force is the agent that "pushes" the electrons that constitute the current. Faraday thus discovered that changing magnetic fields produce an electromotive force or emf, which acts just like the emf generated in other voltage sources, such as a chemical battery, a piezoelectric crystal (emf produced in response to mechanical pressure), or a

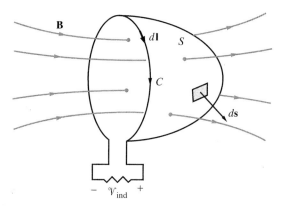

Figure 7.3   Illustration of Faraday's law. Voltage $\mathcal{V}_{ind}$ is induced between the terminals of loop C due to a time-varying magnetic field **B** that links the area S enclosed by C, where the direction of the differential length dl on the contour and the direction of the differential area ds on the surface S are related by the right-hand rule.

[11]With loops of N turns, where each turn links the same amount of flux, we can also write $\Lambda = N\Psi$, where Λ is the total flux linkage of N turns and Ψ is the flux linked by each turn. We then have $\mathcal{V}_{ind} = -N \, d\Psi/dt = -d\Lambda/dt$.

thermocouple (emf resulting from a temperature gradient). The term "electromotive force" is somewhat misleading because an emf is actually not a force but rather a line integral of a force per unit charge (i.e., an electric field). More precisely, *emf is defined as the tangential force per unit charge along the wire integrated over its length, around the complete circuit.*

Having a nonzero emf across the terminals of the loop in turn means that the line integral of the electric field around the loop is not zero, since the force that moves the current carriers is ultimately due to an electric field. Note that this electric field is certainly not an electrostatic field, because otherwise its integral around the closed circuit must be zero. Nor can the source of the force be a magnetostatic field, since the stationary charges cannot experience magnetic forces. The field that provides the force to drive the current is an entirely new kind of electric field produced by the new physical effect embodied in Faraday's law.[12] In words, Faraday's law states that *a changing magnetic field induces an electric field.* The induced voltage $\mathcal{V}_{ind}$ can thus be expressed as

$$\mathcal{V}_{ind} = \oint_C \mathbf{E} \cdot d\mathbf{l} \tag{7.2}$$

where **E** is the *induced* electric field. Combining (7.1) and (7.2), the mathematical statement of Faraday's law is

$$\boxed{\oint_C \mathbf{E} \cdot d\mathbf{l} = -\frac{d}{dt} \int_S \mathbf{B} \cdot d\mathbf{s}} \tag{7.3}$$

where S is the surface enclosed by the contour C. According to (7.3), an electric field induced by a changing magnetic field exists in space regardless of whether there are conducting wires present. If conducting wires are present, an *induced* current flows through these wires. Faraday's law is the basic principle on which electric generators operate; mechanical energy is supplied to change the magnetic flux Ψ that links the coil C (e.g., by rotating coil C and thus the surface S) and thus to produce induced voltage across its terminals through (7.3). Faraday himself invented the first direct-current generator in 1831 and the prototype of the modern electric generator in 1851.

The minus sign in (7.3) indicates that the induced emf is in a direction that opposes the change in the magnetic flux that caused the emf in the first place. In other words, the induced voltage leads to current flow[13] in a direction that produces an opposing magnetic

---

[12]This new kind of physical force that drives electric current is also called an electric field to recognize the fact that it moves charges in the same manner as an electrostatic field. Keep in mind, however, that this new field is quite different from an electrostatic field, whereas the line integral of an electrostatic field around a closed path is identically zero, that of an induced field is obviously not zero.

[13]In typical circuits, which have nonzero resistance, the induced currents may be tiny fractions of the original currents that produce the magnetic fluxes linking the circuit. In other words, the magnetic fields produced by the induced currents are usually negligible, so that although the circuit opposes the changing flux by establishing a current flow, the resulting currents typically fall far short of fully compensating for the flux change.

flux Ψ. This statement of experimental fact[14] is known as *Lenz's law*. In terms of the correct polarities of the two integrals in (7.3), it is important to note that the orientation of the two integrals cannot be independently chosen. In other words, if $d\mathbf{l}$ is chosen such that one finds the first integral by going counterclockwise around the contour C, then the direction of $d\mathbf{s}$ in the second integral must be outward from the page assuming that the area S enclosed by loop C lies in the same plane as the loop. The right-hand rule applies here; with the fingers of the right hand pointing in the direction of $d\mathbf{l}$ along C, the thumb points in the direction of $d\mathbf{s}$.

The relationships between the polarity of the induced voltage $\mathcal{V}_{\text{ind}}$, the relative orientations of the induced electric field $\mathbf{E}$, and the other quantities (e.g., $d\mathbf{l}$) in (7.3) are depicted in Figure 7.4. We take the contour C to be enclosing the surface S, as shown in Figure 7.4a, and we take the orientation of the surface element $d\mathbf{s}$ to be in the z direction (out of the page), in which case $d\mathbf{l}$ must be counterclockwise, as shown in Figure 7.4a. For the purposes of this discussion, we take the magnetic field $\mathbf{B}$ to be pointing out of the page, uniformly present everywhere, and steadily increasing in time (i.e., $\partial B/\partial t > 0$).

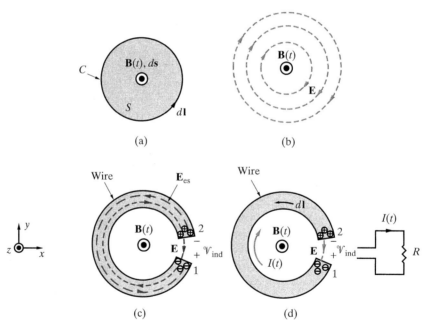

Figure 7.4 **The polarity of the induced voltage and the electric field.** The case illustrated is for $\partial B/\partial t > 0$.

[14]We note here that although Lenz's law is taken to be an experimental fact, it is in fact a consequence of the conservation of energy. In other words, if the polarity of the induced current were opposite to that dictated in Lenz's law, magnetic energy would not be conserved, as is discussed later in Section 7.3. We expect this result on a purely heuristic basis; if the polarity of the induced currents were such as to enhance the effect that produced the currents, the magnetic flux would increase without limit.

The orientation of the induced electric field **E** is shown in Figure 7.4b. Note that this field is induced regardless of whether conducting wires are present. Since $\partial B/\partial t > 0$ and **B** and $d\mathbf{s}$ are both out of the page, the right-hand side of (7.3) is negative. The fact that $d\mathbf{s}$ is out of the page requires that $d\mathbf{l}$ must be in the counterclockwise direction, so the induced electric field **E** must encircle the **B** field in the clockwise direction in order for the left-hand side of (7.3) to also be negative.

Now consider what happens when a conducting wire loop is present, with a small gap across which the induced voltage appears as shown in Figure 7.4c. Note that the thickness of the wire is exaggerated for convenience. At the instant this wire loop is introduced into the system, the induced electric field causes the free electrons inside the conductor to move to one end and leave the other end positively charged, as shown in Figure 7.4c. Assuming that the wire is a good conductor such as copper, this rearrangement of charge within the wire occurs in $\sim 10^{-19}$ s and produces an electrostatic field $\mathbf{E}_{es}$ due to the separation of charge, which cancels the induced field, so that the net electric field inside the conductor is zero. The potential difference between the ends 2 and 1 is

$$\Phi_{12} = \Phi_2 - \Phi_1 = -\int_1^2 \mathbf{E} \cdot d\mathbf{l}$$

This relation is consistent with the definition of electric potential in Section 4.4, where Φ_{12} is the work done in moving a unit positive test charge from point 1 to point 2 (e.g., see (4.19)) where $[W/q]_{a \to b} = \Phi_{ab} = \Phi(b) - \Phi(a) = -\int_a^b \mathbf{E} \cdot d\mathbf{l}$. Note that if we integrate the total electric field around the contour C in the counterclockwise direction, the only contribution to the integral comes from across the gap (since the total electric field is zero within the wire), where $\mathbf{E} \cdot d\mathbf{l}$ is negative, which in turn results in a positive value of Φ_{12}. If a load resistance R is connected across the gap (Figure 7.4d), current flows in the clockwise direction, tending to produce magnetic flux opposing the increase in **B**; this is consistent with the orientation of the induced electric field, which drives the current carriers. If the loop did not have a gap, then current would still flow in the clockwise direction, in accordance with the wire resistance and self-inductance of the loop. In effect, the external resistor R can then be thought of as the internal resistance of the wire.

Turning now to the choice of the polarity of $\mathcal{V}_{ind}$, we note from (7.3) and the definition of $\mathcal{V}_{ind} = \oint \mathbf{E} \cdot d\mathbf{l}$ that for $\partial B/\partial t > 0$, and for $d\mathbf{s}$ chosen to be in the same direction as **B** as is the case in Figure 7.4, $\mathcal{V}_{ind}$ as determined by (7.3) is negative. Thus, the polarity of $\mathcal{V}_{ind}$ in Figure 7.4 must be chosen to be positive at end 1 and negative at end 2, so that we have $\mathcal{V}_{ind} = -\Phi_{12}$, that is, a negative value, consistent with the result obtained from the equation (7.3). More generally, in applying (7.3) to any loop with a gap, the polarity of $\mathcal{V}_{ind}$ must be defined to be positive on the terminal at which the $d\mathbf{l}$ element, the polarity of which is determined by the right hand rule once $d\mathbf{s}$ is chosen, points outward from that terminal.

For further elaboration of Lenz's law, consider the circuit shown in Figure 7.5a, which consists of a magnetic core material[15] having two separately wound coils of wire

---

[15]The magnetic material enhances the mutual inductance, and thus the inductive coupling, between the two coils.

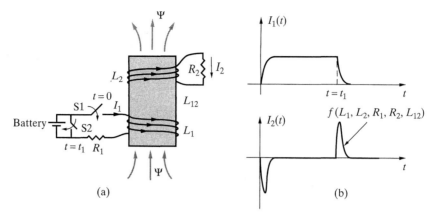

Figure 7.5 Illustration of Lenz's law. (a) A solenoidal core with two separately wound coils of wire. A battery is connected to or disconnected from the lower coil through switches S1 and S2. (b) The variation of the currents I_1 and I_2 as a function of time for the case when the switch S1 is closed at $t = 0$ and S2 is closed at $t = t_1$.

(similar to Faraday's experiment shown in Figure 7.1, except the circuit in Figure 7.5 is a solenoid instead of a toroid). The lower coil is connected to a battery through the switch S1. When the switch S1 is closed (say at time $t = 0$), a current I_1 starts to flow in the lower coil. As a result of the increase in the magnetic flux linking the upper coil, a current I_2 is induced in the direction opposite to I_1, opposing the change in the magnetic flux linking the upper coil. Note that, in general, the current I_1 does not begin to flow immediately after switch S1 is closed but is instead governed by a time constant determined by the inductance and resistance of the lower coil, as well as on the mutual inductance between the two coupled coils. Similarly, the establishment of the current I_2 is governed by a time constant dependent on R_2, L_2, as well as on the mutual inductance between the coils. If the wire were a perfect conductor, the induced current I_2 would flow in the upper coil as long as current I_1 flows in the lower coil. However, in a practical coil, the resistive losses of the wire cause the current I_2 to decay exponentially with a time constant determined by the self-inductance and resistance (L_1, R_1, L_2, and R_2) of each coil, as well as the mutual inductance L_{12} between the two coupled coils. Thus, at steady state, the current I_2 becomes zero and a constant magnetic flux passes through the upper coil due to the current I_1 flowing in the lower coil. Later, when the switch S2 is closed (say at time $t = t_1$), the upper coil tries to keep the flux constant by inducing a current I_2 in the upper coil that is in the same direction as the current I_1 in the lower coil in order to maintain constant flux, until I_1 gradually decreases to zero. Both I_1 and the induced current I_2 die off with time (in a manner defined by the solution of the differential equation describing the coupled circuit consisting of L_1, L_2, R_1, R_2, and L_{12}, that is, not necessarily a simple exponential decay) due to the resistive losses in their respective loops. Figure 7.5b shows the variation of the two currents I_1 and I_2 as a function of time as just discussed.

Example 7.1: Emf induced in a coil. Consider a single-turn coil of wire of radius a, as shown in Figure 7.6a. The region the coil lies in is permeated by a magnetic field $\mathbf{B}(t) = \hat{\mathbf{z}}B_z(t) = \hat{\mathbf{z}}B_0\sin(\omega t)$. Find the induced voltage between the two open end terminals of the coil.

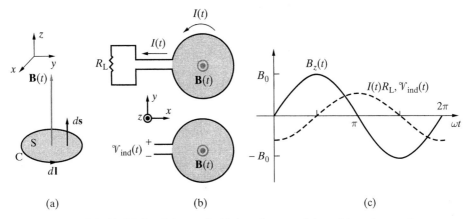

(a)   (b)   (c)

Figure 7.6   Emf induced in a coil. (a) A stationary coil located in a time-varying magnetic field. (b) The best way to think about the direction of the induced current and the polarity of the induced voltage $\mathcal{V}_{ind}$ is to view the coil as a voltage source. (c) The variation of the magnetic field B_z, the induced emf $\mathcal{V}_{ind} = IR_L$ as a function of time. The direction of induced current flow is determined by Lenz's law.

Solution: Using (7.1), the induced voltage is given by

$$\mathcal{V}_{ind}(t) = -\frac{d}{dt}\int_S \mathbf{B}\cdot d\mathbf{s} = -\frac{d}{dt}\int_S [\hat{\mathbf{z}}B_0\sin(\omega t)]\cdot(\hat{\mathbf{z}}\,ds)$$

$$= -\frac{d}{dt}[B_0\sin(\omega t)](\pi a^2) = -\omega\pi a^2 B_0\cos(\omega t)$$

To appreciate the polarity of $\mathcal{V}_{ind}$, it is best to plot the variation of $\mathcal{V}_{ind}(t)$ over one cycle together with $B_z(t)$. In fact, it is more illuminating to think in terms of the current $I(t)$ that would flow in the coil if an external load were connected to the loop, as shown in Figure 7.6b.

  The plots of $B_z(t)$, $I(t)$, and $\mathcal{V}_{ind}(t)$ in Figure 7.6b clearly illustrate Lenz's law in effect; when the magnetic field $B_z(t)$ exhibits its largest rate of increase (at $t = 0$), the current $I(t)$ is maximally negative (i.e., flows clockwise so that its associated magnetic field is in the $-z$ direction and *opposes* the increase in $B_z(t)$). Negative (i.e., clockwise) current flows throughout the interval $0 < \omega t < \pi/2$, but with the magnitude of the current decreasing as the rate of change of $B_z(t)$ decreases, with $I(t)$ eventually equal to zero at $\omega t = \pi/2$ when $B_z(t)$ has reached its peak value B_0, so that $\partial B_z/\partial t = 0$. Similar manifestations of Lenz's law can be observed during the rest of the cycle. Note that, in this context, the coil is best viewed as a voltage source (or battery) for a load R_L connected to it.

  If the coil consists of N turns, the surface S is enclosed N times, where each turn induces a voltage. These voltages all add in series, so that the total voltage induced across the terminals of the coil is N times greater and is given by

$$\mathcal{V}_{\text{ind}}(t) = -N\omega\pi a^2 B_0 \cos(\omega t)$$

The induced voltage is thus proportional to the rate of change of the **B** field (ω), the number of turns (N), and the intensity of the field (B_0) linking each turn with the area of the coil.

Example 7.2: Pair of lines and a rectangular loop. Consider a rectangular loop in the vicinity of a pair of infinitely long wires, as shown in Figure 7.7. The two wires carry a current I flowing in opposite directions, the magnitude of which increases at a rate $dI/dt > 0$. Find the electromotive force induced in the loop (i.e., the $\mathcal{V}_{\text{ind}}$ that would be measured across the small opening if the loop were broken) for the two different placements of the rectangular loop, as shown in Figure 7.7.

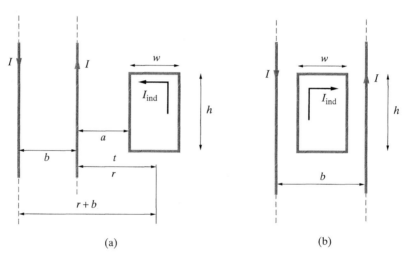

(a)   (b)

Figure 7.7   Pair of parallel wires and a loop. (a) The loop outside the region between the wires. (b) The loop in between the two wires equidistant from each wire.

Solution: We know from Example 6.3 that a current I in an infinitely long straight wire produces a field $\mathbf{B} = \hat{\boldsymbol{\phi}} B_\phi$ at a distance r of

$$B_\phi = \frac{\mu_0 I}{2\pi r}$$

(a) Since the **B** field due to both infinitely long straight wires is given by the same expression, the total flux linking the loop can be written as

$$\Psi = \frac{\mu_0 I}{2\pi} \left(\int_a^{a+w} \frac{h\,dr}{r} - \int_{b+a}^{b+a+w} \frac{h\,dr}{r} \right) = \frac{\mu_0 h I}{2\pi} \ln\left[\frac{(b+a)(a+w)}{a(b+a+w)} \right]$$

The induced emf in the rectangular loop is then given by

$$\mathcal{V}_{ind} = -\frac{\mu_0 h}{2\pi} \ln\left[\frac{(b+a)(a+w)}{a(b+a+w)}\right]\frac{dI}{dt} = \frac{\mu_0 h}{2\pi} \ln\left[\frac{a(b+a+w)}{(b+a)(a+w)}\right]\frac{dI}{dt}$$

Note that the polarity of the emf is such that the induced current I_{ind} in the loop flows in the *counterclockwise* direction when dI/dt is positive.

(b) When the loop is between the wires (Figure 7.7b), placed equidistant from both straight wires, the magnetic fluxes linking the loop due to the current flowing in each wire are equal in magnitude and in the same direction. Thus, using (6.26), the total flux linking the rectangular loop is

$$\Psi = 2\frac{\mu_0 I h}{2\pi} \int_{(b-w)/2}^{(b+w)/2} \frac{dr}{r} = \frac{\mu_0 I h}{\pi} \ln\left[\frac{b+w}{b-w}\right]$$

The induced emf is

$$\mathcal{V}_{ind} = -\frac{d\Psi}{dt} = -\frac{\mu_0 h}{\pi} \ln\left[\frac{b+w}{b-w}\right]\frac{dI}{dt}$$

Note that this time, the polarity of the induced voltage is such that the induced current I_{ind} in the loop flows in the *clockwise* direction when dI/dt is positive.

Example 7.3: Jumping ring. A conducting ring of radius a is placed on top of a solenoidal coil wound around an iron core, as shown in Figure 7.8. Describe what happens to the ring when the switch is closed.

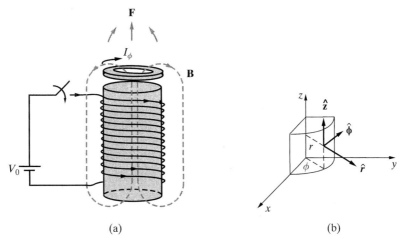

Figure 7.8 Jumping ring. (a) A conducting ring placed on top of a solenoidal coil. (b) A cylindrical coordinate system.

Solution:

Before the switch is closed, the conducting ring has zero flux linkage. When the current is first turned on in the solenoid by closing the switch, the flux produced by the solenoid links the ring in the upward direction, inducing a current in the ring opposite in direction to that in the coil (i.e., Lenz's law). Since currents in opposite directions repel, the repulsion force lifts the conducting ring off the top of the solenoid. Using (6.52), this force is given by

$$\mathbf{F} = \int_C I \; d\mathbf{l} \times \mathbf{B} = \int_0^{2\pi} -I_\phi(\hat{\boldsymbol{\phi}} a \; d\phi) \times (\hat{\mathbf{r}} B_r + \hat{\mathbf{z}} B_z)$$

$$= -\hat{\boldsymbol{\phi}} 2\pi a I_\phi \times \hat{\mathbf{r}} B_r = 2\pi a I_\phi B_r \hat{\mathbf{z}}$$

where I_ϕ is the magnitude of the induced current I (note that, by Lenz's law, this current must be in the direction $-\hat{\boldsymbol{\phi}}$, as shown), a is the radius of the ring, and B_r and B_z are the radial and vertical components of the **B** field. Note that the radial component of the force due to the B_z component cancels out because of the circular symmetry, so the resultant force is in the z direction. Note also that the actual dynamics of the motion of the ring are quite complicated indeed. The magnitude of the induced current I_ϕ depends on the mutual inductance between the solenoid and the ring; once the ring lifts off and begins to move away, the magnitude of the **B** field (and thus the magnetic force) experienced by the ring changes; the time rate of change of the magnetic flux (and hence the duration of the induced current I) depends on the inductance of the solenoidal coil, its internal resistance, and so on.

Example 7.4: Two concentric coils. Consider two circular coils of radii $a = 50$ cm and $b = 5$ cm, number of turns $N_1 = 25$ and $N_2 = 100$, and both centered at the origin and located on the same plane, as shown in Figure 7.9. (a) Find the mutual inductance of the two coils. (b) If the larger coil is carrying an alternating current of $I(t) = 10\sin(377t)$ A, find the induced emf across the terminals of the smaller coil.

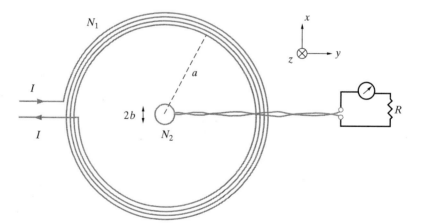

Figure 7.9   Two concentric coils. The circuit is arranged such that emf is to be measured across the terminals of the smaller coil.

Solution:

(a) In Example 6.6, we found out that the **B** field produced at the center of a circular current-carrying wire is given by

$$\mathbf{B}_{\text{ctr}} = \hat{\mathbf{z}}\frac{\mu_0 I}{2a}$$

The **B** field at the center of a circular wire of N_1 turns is

$$\mathbf{B}_{\text{ctr}} = \hat{\mathbf{z}}\frac{\mu_0 N_1 I}{2a}$$

Since the radius of the smaller loop $b \ll a$, we can assume the **B** field to be constant over its flat surface. The total magnetic flux linking the small loop due to the current flowing in the large loop is thus $\Lambda_{12} = N_2 B_{\text{ctr}}\pi b^2$. The mutual inductance is then

$$L_{12} = \frac{\Lambda_{12}}{I} = \frac{N_2 N_1 \pi b^2 \mu_0}{2a}$$

(b) Substituting the values given, we find

$$\mathbf{B}_{\text{ctr}} = \hat{\mathbf{z}}\frac{(4\pi \times 10^{-7}\ \text{H-m}^{-1})(25)[10\sin(377t)\ \text{A}]}{2 \times (0.5\ \text{m})} \simeq \hat{\mathbf{z}}\, 3.14 \times 10^{-4}\sin(377t)\ \text{T}$$

The total flux linked by a single turn of the small loop can be found from (6.26) as

$$\Psi_{12} = \int_S \mathbf{B}_{\text{ctr}} \cdot d\mathbf{s} \simeq |\mathbf{B}_{\text{ctr}}| \int_S ds = \frac{\mu_0 N_1 I}{2a}(\pi b^2)$$

$$= [\pi \times 10^{-4}\sin(377t)\ \text{T}][\pi \times (5 \times 10^{-2})^2]$$

$$\simeq 2.47 \times 10^{-6}\sin(377t)\ \text{Wb}$$

Therefore, the induced emf across the terminals of the smaller loop of $N_2 = 100$ turns is found as

$$\mathcal{V}_{\text{ind}}(t) = -N_2\frac{d\Psi_{12}}{dt}$$

$$\simeq -100 \times 2.47 \times 10^{-6} \times 377\cos(377t) \simeq -93\cos(377t)\ \text{mV}$$

Note that we are not concerned with the polarity of $\mathcal{V}_{\text{ind}}$ since the terminals of the small loop are not identified. However, Lenz's law tells us that when the current $I(t)$ increases during part of its alternating cycle, the induced current in the small loop must flow in the counterclockwise direction in order to oppose the increase of the **B** field, which is in the z direction.

7.2 INDUCTION DUE TO MOTION

When conductors move in the presence of magnetic fields, an induced voltage is produced, in addition to the induced voltage due to the variation of magnetic fields with time. Induction due to motion is the basis for most of the engines of our industrial society, including the giant generators of electricity in hydroelectric dams, large motors used in steel mills, and tiny motors in dentist's drills or children's toys.[16] In general, when a loop travels through space, its motion through a magnetic field may alter the amount of flux through the loop even though the magnetic field may be constant at all points in space. If, in addition to motion, the magnetic field itself varies with time, then electromotive force is induced due to both motion and the time variation of the magnetic field.

7.2.1 Motion in a Constant Magnetic Field

The magnitude of the voltage induced due to motion of a conductor in a constant magnetic field can be found from the Lorentz force, which was defined in Chapter 6 with (6.8). The Lorentz force due to a magnetic field is the force $\mathbf{F}_m$ that a $\mathbf{B}$ field exerts on a charged particle q moving with a velocity $\mathbf{v}$. In other words,

$$\mathbf{F}_m = q\mathbf{v} \times \mathbf{B}$$

Note that this force acts in a direction perpendicular to both $\mathbf{v}$ and $\mathbf{B}$, and is in fact just another form of the Ampère's force law in expression (6.2), or $\mathbf{F}_m = Id\mathbf{l} \times \mathbf{B}$, since $q\mathbf{v}$ can be viewed as a current element (see Section 6.2.2).

The fact that the charges in the wire experience a force of $q\mathbf{v} \times \mathbf{B}$ indicates that an electromotive force (emf) is induced across the ends of the wire. To see this, note that in Section 7.1, we defined emf as the tangential force per unit charge integrated over its length. Thus, in the case shown in Figure 7.10, the induced emf is given by the integral of $\mathbf{F}_m$ over the length of the bar. If a resistance were connected across the ends of the moving wire, current flows, with the current carriers "pushed" by the magnetic force,

---

[16]For an excellent qualitative and enlightening discussion of motors and generators and their applications, see Chapter 16 of R. P. Feynman, R. B. Leighton, and M. Sands, *The Feynman Lectures on Physics, Volume II,* Addison-Wesley, 1964. As reproduced here with permission, Feynman discusses the engineering miracle of hot lights (electricity) produced from cold water (a hydroelectric dam) hundreds of miles away, all done with specially arranged pieces of copper and iron: "Hundreds of little wheels, turning in response to the turning of the big wheel at Boulder Dam. Stop the big wheel, and all the little wheels stop; the lights go out. They are really connected. Yet there is more. The same phenomena that take the tremendous power of the river and spread it through the countryside, until a few drops of the river are running the dentist's drill, come again into the building of extremely fine instruments... for the detection of incredibly small amounts of current... for the transmission of voices, music, and pictures... for computers... for automatic machines of fantastic precision."

Figure 7.10 A conducting bar moving in a static magnetic field. The charges in the conducting bar redistribute themselves in such a way that an induced emf is generated across the bar.

namely $(q\mathbf{v} \times \mathbf{B})$. Note that we do not need to invoke any *induced* electric field, since, due to the motion of the bar, the current carriers can be driven by a magnetic force.[17] The electromotive force between points 1 and 2 (Figure 7.10) is thus given by

$$\mathcal{V}_{\text{ind}} = \int_2^1 (\mathbf{v} \times \mathbf{B}) \cdot d\mathbf{l} \qquad (7.4)$$

since tangential force per unit charge in the wire is $(\mathbf{v} \times \mathbf{B})$.

This expression is the mathematical formulation of Faraday's observation that moving conductors in the vicinity of magnets (or moving magnets in the vicinity of conductors) induce electromotive force. The induced voltage as given by (7.4) is sometimes referred to as the *motional emf*.

Example 7.5: Moving metal bar in uniform static magnetic field. Consider the arrangement shown in Figure 7.11, in which a metal bar can slide over conducting rails. The magnetic field is static (i.e., constant with respect to time) and uniform (i.e., the same everywhere) and is out of the page as shown and given by $\mathbf{B} = \hat{\mathbf{z}}B_0$. The bar and the rails constitute a conducting wire loop, the area of which expands as the bar moves to the right with a constant velocity $\mathbf{v} = \hat{\mathbf{x}}v_0$. Find the total emf induced in the loop.

---

[17]An electrostatic field is in fact set up within the conductor as a result of separation of charge. When a conductor moves with velocity $\mathbf{v}$ in a static magnetic field $\mathbf{B}$, as shown in Figure 7.10, the $q\mathbf{v} \times \mathbf{B}$ force causes the electrons inside the conductor to flow to one end of the conductor and leave the other end positively charged. This brief current flows until equilibrium is reached (very quickly, over a time of $\sim 10^{-19}$ s for most metals), when the coulomb force of attraction $\mathbf{F}_{es}$ due to charge separation cancels the force due to the magnetic field. In other words, we have $\mathbf{F} = \mathbf{F}_{es} + \mathbf{F}_m = q\mathbf{E}_{es} + q\mathbf{v} \times \mathbf{B} = 0$, or $\mathbf{E}_{es} = -(\mathbf{v} \times \mathbf{B})$, where $\mathbf{E}_{es}$ is the electrostatic field within the conductor. Note that the induced electric field $\mathbf{E}$, which gives $\mathcal{V}_{\text{ind}}$ when integrated along any external path between the two ends of the bar, is related to $(\mathbf{v} \times \mathbf{B})$ via (7.4), or

$$\mathcal{V}_{\text{ind}} = \int_2^1 \mathbf{E} \cdot d\mathbf{l} = \int_2^1 (\mathbf{v} \times \mathbf{B}) \cdot d\mathbf{l}$$

so that $\mathbf{E} = \mathbf{v} \times \mathbf{B}$, or $\mathbf{E} = -\mathbf{E}_{es}$. Note that unlike $\mathbf{E}_{es}$, the induced field $\mathbf{E}$ is nonconservative.

To an observer moving with the bar, the charges in the bar appears to be stationary, so the fact that they experience a force (equal to $q\mathbf{v} \times \mathbf{B}$) can be interpreted as being due to an electric field given by $(\mathbf{v} \times \mathbf{B})$, as long as $|v| \ll c$. The electrostatic field set up by the displaced charges may be observed in both a stationary frame of reference and a moving frame attached to the conductor.

Figure 7.11   A moving metal bar on stationary conducting rails. The magnetic field **B** is constant, uniform, and points out of the page.

Solution: The emf induced between terminals A and B can be calculated using Faraday's law. Since the magnetic field does not vary with time, it is appropriate to use (7.4) to find $\mathcal{V}_{\text{ind}}$. In applying (7.4), we need to integrate around the entire loop, part of which is the moving bar. In other words, we have

$$\mathcal{V}_{\text{ind}} = \int_C (\mathbf{v} \times \mathbf{B}) \cdot d\mathbf{l}$$

where C is the contour BCDA. However, since all other parts of the loop are stationary ($\mathbf{v} = 0$), the only contribution to the right-hand side is from the moving bar (i.e., the CD segment of the loop). Thus, we have

$$\mathcal{V}_{\text{ind}} = \int_{\text{CD}} (\mathbf{v} \times \mathbf{B}) \cdot d\mathbf{l} = \int_{\text{CD}} (\hat{\mathbf{x}} v_0 \times \hat{\mathbf{z}} B_0) \cdot (\hat{\mathbf{y}} \, dy) = -v_0 B_0 l$$

Note that the polarity of the induced emf is such that terminal B is positive with respect to terminal A. In other words, if a load resistance is connected between terminals A and B, a current of $I = v_0 B_0 l / R_{\text{L}}$ flows in the *clockwise* direction through the loop to oppose the increasing magnetic flux linking the loop (i.e., Lenz's law). The electrical power dissipated in the load resistor R_{L} is

$$P_{\text{elec}} = I^2 R_{\text{L}} = \frac{(v_0 B_0 l)^2}{R_{\text{L}}}$$

Based on energy conservation, this electrical power must originate from the mechanical energy used to move the metal bar at a velocity **v**. Energy is essentially transferred from the mechanical source to the load resistor. To quantitatively evaluate the energy transfer, we can consider the magnetic force experienced by the moving bar. In the presence of a magnetic field, a current-carrying wire element experiences a magnetic force given by

$$d\mathbf{F}_{\text{m}} = I \, d\mathbf{l} \times \mathbf{B}$$

so that the total magnetic force experienced by the metal bar is

$$\mathbf{F}_{\text{m}} = I \int_{\text{C}}^{\text{D}} -\hat{\mathbf{y}} \, dy \times \hat{\mathbf{z}} B_0 = -\hat{\mathbf{x}} I B_0 l$$

where the induced current in the loop flows in the clockwise direction, that is, $I \, d\mathbf{l} = -I \hat{\mathbf{y}} \, dy$. The magnetic force experienced by the bar is thus in the $-x$ direction, *opposing* the motion.

In order to sustain the motion of the bar at a constant velocity $\mathbf{v} = \hat{\mathbf{x}}v_0$, the mechanical force must be exactly equal and opposite to this magnetic force. The mechanical power is given by

$$P_{\text{mech}} = -\mathbf{F}_{\text{m}} \cdot \mathbf{v} = Iv_0B_0l = \left(\frac{v_0B_0l}{R_{\text{L}}}\right)v_0B_0l = \frac{(v_0B_0l)^2}{R_{\text{L}}}$$

since $\mathbf{F}_{\text{m}}$ and $\mathbf{v}$ are parallel. Thus, we see that, as expected, $P_{\text{mech}} = P_{\text{elec}}$. The system shown in Figure 7.11 is thus a simple example of an electric generator, where the applied mechanical energy is converted into electrical energy.

In this example, the Lorenz magnetic force is responsible for generating the current. When an external mechanical force is applied to the bar, the resulting motion in the x direction generates a force on the positive charge carriers in the $-y$ direction through (6.50). An interesting question to pose here is, what does this external force pull against? When the bar acquires a velocity v, the net motion of the positive charge carriers is no longer strictly in the $-y$ direction but also has a component in the x direction. Since the magnetic Lorenz force is always perpendicular to the velocity, the net force on the positive charge carriers acquire a component in the $-x$ direction as the bar begins to move. The external mechanical force must overcome this horizontal component of the $q\mathbf{v} \times \mathbf{B}$ force to maintain a constant velocity. Thus, while the magnetic field does no work, it redirects the horizontal force generated by the external agent to a vertical force that is responsible for moving the charge carriers around the circuit.

7.2.2 Moving Conductor in a Time-Varying Magnetic Field

We have seen that an electromotive force is induced through (7.3), when the magnetic field linking a stationary loop varies with time, or through (7.4), when a conductor moves through a static magnetic field. The general case involves motion of a loop C in addition to a time variation of the magnetic field. In such a case, the induced voltage is given by[18]

$$\boxed{\mathcal{V}_{\text{ind}} = -\int_S \frac{\partial \mathbf{B}}{\partial t} \cdot d\mathbf{s} + \oint_C (\mathbf{v} \times \mathbf{B}) \cdot d\mathbf{l}} \tag{7.5}$$

where the contour C encloses the surface S. Care must be taken so that $d\mathbf{l}$ and $d\mathbf{s}$ are related via the right-hand rule, as discussed in connection with (7.3). Note that the velocity of the different parts of the loop need not be the same, since C may be changing shape as well as undergoing translation (linear motion) and/or rotation.

The motional emf term (i.e., the second term) of (7.5) can in many cases be thought of in terms of Faraday's original changing-flux concept, as expressed in (7.1). To see this, consider Figure 7.12. In a time interval dt, an element $d\mathbf{l}$ of C sweeps out an area $d\mathbf{s} = (\mathbf{v} \times d\mathbf{l})\,dt$. The change in magnetic flux $d\Psi_1$ caused by the motion of the element $d\mathbf{l}$ is equal to the integral of $\mathbf{B}$ through the swept-out area, or $d\Psi_1 = \mathbf{B} \cdot d\mathbf{s} = \mathbf{B} \cdot (\mathbf{v} \times d\mathbf{l})dt$.

---

[18]If the contour C enclosing the surface S is stationary, then the time derivative outside the integral in (7.3) can be absorbed into the integrand as a partial time derivative of $\mathbf{B}$; see Section 7.2.3.

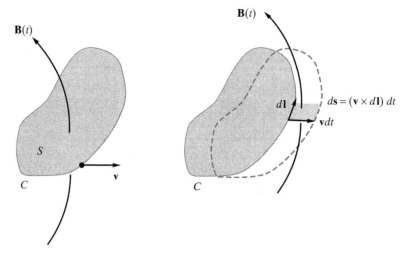

Figure 7.12   A conducting loop C moving in a time-varying magnetic field.

The total change in magnetic flux due to the displacement of the entire contour C is then given as $d\Psi = \oint_C d\Psi_1$ or

$$d\Psi = \oint_C \mathbf{B} \cdot (\mathbf{v} \times d\mathbf{l})\, dt \quad \rightarrow \quad -\frac{d\Psi}{dt} = -\oint_C \mathbf{B} \cdot (\mathbf{v} \times d\mathbf{l}) = \oint_C (\mathbf{v} \times \mathbf{B}) \cdot d\mathbf{l}$$

where $\oint_C (\mathbf{v} \times \mathbf{B}) \cdot d\mathbf{l}$ is the motional emf part of (7.5).

Based on the preceding expression, it appears that the most general form of Faraday's law,[19] including the cases involving moving circuits, can be expressed as $\mathcal{V}_{\text{ind}} = -d\Psi/dt$, or

$$\boxed{\mathcal{V}_{\text{ind}} = \oint_C \mathbf{E} \cdot d\mathbf{l} = -\frac{d}{dt} \int_S \mathbf{B} \cdot d\mathbf{s}} \tag{7.6}$$

where $\mathbf{E}$ is the induced electric field.

Although (7.6) applies in general, there are situations in which a closed contour cannot be identified.[20] An example in which it is not clear how the flux linkage can

---

[19]By comparing equation (7.5) with (7.6), we see that the general form of Faraday's law actually consists of two distinct mechanisms for generating an emf. Based on the first term in (7.5), a time-changing magnetic field generates an emf. This is a fundamental fact based on experimental observations and cannot be derived using any previous result from Chapters 4 to 6. The second term in (7.5) is a restatement of Ampere's law; the emf is magnetic in nature. The fact that the general form of Faraday's law, as stated in (7.1), (7.3), and (7.6), simultaneously describes these two distinct phenomena is quite remarkable. Furthermore, there is a deeper reason for such a convergence: interestingly enough, this curious coincidence is what drove Einstein to develop the special theory of relativity.

[20]For examples and a discussion, see Chapter II-17 of R. P. Feynman, R. B. Leighton, and M. Sands, *The Feynman Lectures on Physics,* Definitive Edition, Addison-Wesley, San Francisco, 2006.

be evaluated is the single moving conductor of Figure 7.10. In such cases, the correct
physics is always revealed by (7.5).

**Example 7.6: Elementary alternating current generator: A rotating conductor loop in a static
magnetic field.** This example illustrates the basic operating principle of an elementary alternating
current generator, which typically involves an armature coil mechanically rotated in a steady
magnetic field at a constant rate. Consider the single-turn rectangular coil shown in Figure 7.13,
rotating with an angular velocity ω_0 about its axis (z axis) in the presence of a static uniform
magnetic field $\mathbf{B} = \hat{\mathbf{x}}B_0$. Determine the voltage induced between terminals 1 and 2.

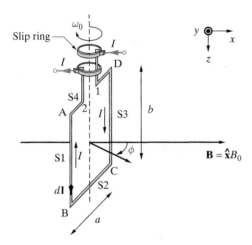

Figure 7.13 Elementary current generator. A rectangular coil is rotated
at a constant angular velocity in a uniform
magnetic field. The slip rings maintain electrical contact with the terminals of the coil
as it rotates.

Solution: Since $\partial\mathbf{B}/\partial t = 0$, the induced emf is due solely to the motion of the conductor
(i.e., motional emf), and we can use either (7.4) or (7.6) to determine $\mathcal{V}_{\text{ind}}$. We first select
the polarity of $\mathcal{V}_{\text{ind}}$ to be such that the terminal marked 2 is positive. In other words,
$\mathcal{V}_{\text{ind}} \equiv \Phi_{12} = \Phi_2 - \Phi_1$.

We proceed by using the generalized Faraday's law, namely (7.6). At any instant of
time t, the magnetic flux Ψ through the coil is

$$\Psi = \int_S \mathbf{B} \cdot d\mathbf{s} = \int_S B_0\hat{\mathbf{x}} \cdot \hat{\mathbf{r}} \, ds = B_0 A \cos\phi = ab B_0 \cos(\omega_0 t)$$

where $A = ab$ is the area of the loop. The induced voltage $\mathcal{V}_{\text{ind}}$ is then

$$\mathcal{V}_{\text{ind}} = -\frac{d\Psi}{dt} = \omega_0 ab B_0 \sin(\omega_0 t)$$

The same result can also be obtained by using (7.4). Note that the velocity of an
electron along the vertical sides (S1 and S3) of the coil is $\mathbf{v} = \hat{\boldsymbol{\phi}}\omega_0(a/2)$ and the velocity $\mathbf{v}$
and the field $\mathbf{B} = \hat{\mathbf{x}}B_0$ are at an angle $\phi = \omega_0 t$, so that we have $|\mathbf{v} \times \mathbf{B}| = |v|B_0 \sin(\omega_0 t) =$
$(a\omega_0/2)B_0 \sin(\omega_0 t)$. Note that the directions of $\mathbf{v} \times \mathbf{B}$ are opposite to each other on sides
AB and CD. The sides BC and DA do not contribute to the line integral since $\mathbf{v} \times \mathbf{B}$ and

$d\mathbf{l}$ are orthogonal to each other, and thus $(\mathbf{v} \times \mathbf{B}) \cdot d\mathbf{l} = 0$. Assuming that the loop wire is perfectly conducting, the only voltage drop is across terminals 1 and 2. Thus, we have

$$\mathcal{V}_{ind} = \Phi_{12} = \int_{2}^{1} \mathbf{E} \cdot d\mathbf{l} = \int_{2}^{1} (\mathbf{v} \times \mathbf{B}) \cdot d\mathbf{l}$$

$$= \int_{A}^{B} \hat{\mathbf{z}} \frac{a\omega_0}{2} [B_0 \sin(\omega_0 t)] \cdot (\hat{\mathbf{z}} \, dz) + \int_{C}^{D} -\hat{\mathbf{z}} \frac{a\omega_0}{2} [B_0 \sin(\omega_0 t)] \cdot (\hat{\mathbf{z}} \, dz)$$

$$= b \frac{a\omega_0}{2} B_0 \sin(\omega_0 t) + b \frac{a\omega_0}{2} B_0 \sin(\omega_0 t) = ba\omega_0 B_0 \sin(\omega_0 t)$$

which is the same as the previous result obtained from (7.6).

In general, depending on the impedance of the external load connected across the generator terminals 1 and 2, the current I can be out of phase with respect to $\mathcal{V}_{ind}$. Assuming the current that flows is $I(t) = I_0 \cos(\omega_0 t + \theta)$, the instantaneous electric power delivered to the external load by the generator is

$$P_{elec} = \mathcal{V}_{ind} I(t) = [\omega_0 ab B_0 \sin(\omega_0 t)][I_0 \cos(\omega_0 t + \theta)]$$

The source of this electrical power is the mechanical power that maintains the rotation of the coil at ω_0. To determine the mechanical power needed, we note that as it rotates in a constant magnetic field, the coil experiences a torque given by (6.53), namely $\mathbf{T} = \mathbf{m} \times \mathbf{B}$, where $\mathbf{m}$ is the magnetic dipole moment such that $\mathbf{m} = \hat{\mathbf{r}} ab I(t) = \hat{\mathbf{r}} ab I_0 \cos(\omega_0 t + \theta)$. Thus, the torque acting on the coil is

$$\mathbf{T} = [\hat{\mathbf{r}} ab I_0 \cos(\omega_0 t + \theta)] \times \hat{\mathbf{x}} B_0 = -\hat{\mathbf{z}} [ab I_0 \cos(\omega_0 t + \theta)](B_0) \sin \phi$$

$$= -\hat{\mathbf{z}} [ab I_0 \cos(\omega_0 t + \theta)](B_0) \sin(\omega_0 t)$$

Note that the sign of the torque is such that it opposes the rotation of the coil. In order to maintain the rotation of the coil at a fixed rate ω_0, the externally applied mechanical force must overcome this electromagnetic opposition. Thus, noting that mechanical power in rotational motion is the product of torque and angular velocity, we have

$$P_{mech} = \omega_0 T = \omega_0 ab B_0 I_0 \sin(\omega_0 t) \cos(\omega_0 t + \theta)$$

Note that $P_{elec} = P_{mech}$, as was the case for the translational motion electrical generator of Example 7.5.

Example 7.7: Expanding circular loop. A circular conducting rubber loop expands at a constant radial velocity $\mathbf{v} = \hat{\mathbf{r}} v_0$ in a region (see Figure 7.14) permeated by a uniform magnetic field that is perpendicular to the loop and varies in time as $\mathbf{B} = \hat{\mathbf{z}} B_0 t$. The radius of the loop at any time t is $r_t = v_0 t$. Neglect the thickness of the conducting rubber wire. Find the induced emf $\mathcal{V}_{ind}$. How does the result change if the magnetic field is oriented in the $-z$ direction (i.e., $\mathbf{B} = -\hat{\mathbf{z}} B_0 t$)?

Solution: We can solve the problem using (7.6) or (7.5). We first use equation (7.6). Since the magnetic field is in the $+z$ direction, it is convenient to choose $d\mathbf{s}$ also in the $+z$ direction so that the integrand in (7.6) is positive. This in turn means that the contour integral in (7.6) should be taken in the counterclockwise direction, since the contour C and the surface S

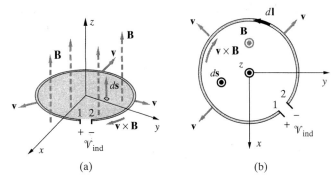

Figure 7.14 An expanding circular loop. (a) Coordinate system. (b) Top view with the magnetic field pointing out of the page ($+z$ direction).

in (7.6) are related via the right-hand rule. Therefore, the polarity of $\mathcal{V}_{ind}$ is defined to be positive at terminal 1, at which $d\mathbf{l}$ is outward (see Figure 7.4 and accompanying discussion).

Noting that we then have $d\mathbf{s} = r\,dr\,d\phi\,\hat{\mathbf{z}}$ and $\mathbf{B} = \hat{\mathbf{z}}B_0t$, from (7.6) we have

$$\mathcal{V}_{ind} = -\frac{d}{dt}\int_{S(t)} \mathbf{B}\cdot d\mathbf{s} = -\frac{d}{dt}\int_0^{r_t=v_0t}\int_0^{2\pi} B_0tr\,dr\,d\phi$$

$$= -\frac{d}{dt}\left[2\pi B_0t\left(\frac{1}{2}r^2\right)\right]_0^{v_0t} = -\frac{d}{dt}[B_0\pi v_0^2 t^3]$$

$$= -3B_0\pi v_0^2 t^2 = -3B_0\pi\left(\frac{r_t^2}{t^2}\right)t^2 = -3B_0\pi r_t^2$$

We now use (7.5) to determine $\mathcal{V}_{ind}$. We integrate around the loop in the counterclockwise direction, with $d\mathbf{l}$ in (7.5) chosen by convention to be in the $+\hat{\boldsymbol{\phi}}$ direction. We have

$$\mathbf{v}\times\mathbf{B} = v_0\hat{\mathbf{r}}\times\hat{\mathbf{z}}B_0t = -B_0v_0t\hat{\boldsymbol{\phi}} = -B_0r_t\hat{\boldsymbol{\phi}} \qquad \text{and} \qquad d\mathbf{l} = r_t\,d\phi\hat{\boldsymbol{\phi}}$$

so that

$$\oint_{C(t)} (\mathbf{v}\times\mathbf{B})\cdot d\mathbf{l} = \int_0^{2\pi} -B_0r_t^2\,d\phi = -2\pi B_0r_t^2$$

and

$$-\int_{S(t)}\frac{\partial\mathbf{B}}{\partial t}\cdot d\mathbf{s} = -\int_{S(t)}\frac{\partial}{\partial t}[B_0t\hat{\mathbf{z}}]\cdot\hat{\mathbf{z}}r\,dr\,d\phi = -\int_0^{v_0t}\int_0^{2\pi} B_0r\,d\phi\,dr$$

$$= \left[-2\pi B_0\left(\frac{1}{2}r^2\right)\right]_0^{v_0t} = -\pi B_0v_0^2 t^2 = -\pi B_0r_t^2$$

or

$$\mathcal{V}_{ind} = -\pi B_0r_t^2 - 2\pi B_0r_t^2 = -3\pi B_0r_t^2$$

which is the same as the result we found using (7.6).

If the magnetic field points in the $-z$ direction and we maintain the same orientation for $d\mathbf{l}$ (and thus $d\mathbf{s}$), we have $d\mathbf{s}$ and $\mathbf{B}$ in the opposite directions, so that the contribution of the first term in (7.5) is the negative of that found in the preceding expression, or $+\pi B_0r_t^2$.

However, since $\mathbf{B} = -\hat{\mathbf{z}}B_0 t$ and $\mathbf{v} = \hat{\mathbf{r}}v_0$, we have $(\mathbf{v} \times \mathbf{B}) = +B_0 v_0 t \hat{\boldsymbol{\phi}}$, so that $(\mathbf{v} \times \mathbf{B})$ and $d\mathbf{l}$ are now also in opposite directions. Thus, the contribution from the second term in (7.5) also changes sign and is now $+2\pi B_0 r_t^2$. Thus, the induced emf is

$$\mathcal{V}_{\text{ind}} = \left(+\pi B_0 r_t^2\right) + \left(+2\pi B_0 r_t^2\right) = 3\pi B_0 r_t^2$$

As expected, the magnitude of the induced emf is not affected by the direction of the magnetic field. However, the polarity of $\mathcal{V}_{\text{ind}}$ is of course reversed.

7.2.3 Differential Form of Faraday's Law

The most general form of Faraday's law, namely (7.6), can be expressed in differential form by using Stokes's theorem. We have

$$\oint_C \mathbf{E} \cdot d\mathbf{l} = \int_S (\nabla \times \mathbf{E}) \cdot d\mathbf{s} = -\frac{d}{dt} \int_S \mathbf{B} \cdot d\mathbf{s}$$

which immediately yields a differential equation involving $\mathbf{E}$ and $\mathbf{B}$ *if* it is legitimate to carry out the following operation:

$$\frac{d}{dt} \int_S \mathbf{B} \cdot d\mathbf{s} = \int_S \frac{\partial \mathbf{B}}{\partial t} \cdot d\mathbf{s}$$

The interchange of integration over the variables in S and the differentiation with respect to time is legitimate if C is an arbitrary contour fixed in space, that is, if contour C has a fixed geometric shape and is stationary in space. With this interpretation of C, the integrands in the preceding equation can be equated since S is arbitrary. Thus, the differential form of Faraday's law in a stationary frame of reference[21] is

$$\boxed{\nabla \times \mathbf{E} = -\frac{\partial \mathbf{B}}{\partial t}} \tag{7.7}$$

equation (7.7) is one of the fundamental laws of electromagnetics, namely Maxwell's equations. The fields $\mathbf{E}$ and $\mathbf{B}$ in (7.7) must be measured in the *same* coordinate system since the contour C must be fixed when converting from the integral form of Faraday's law to the differential form. In other words, the spatial derivatives of $\mathbf{E}$ are related to the time derivative of $\mathbf{B}$ at a point according to (7.7) only if measurements of these derivatives are made by the same observer.

---

[21] It can be shown that Faraday's law for a moving medium is

$$\nabla \times \mathbf{E}' = -\frac{d\mathbf{B}}{dt} = -\left[\frac{\partial \mathbf{B}}{\partial t} + (\mathbf{v} \cdot \nabla)\mathbf{B}\right]$$

where $\mathbf{E}'$ is measured in the moving frame of reference and the velocity $\mathbf{v}$ is assumed to be small relative to the velocity of light in free space. See Appendix A for an expanded form of $(\mathbf{v} \cdot \nabla)\mathbf{B}$. However, a stationary observer sees an electric field $\mathbf{E} = \mathbf{E}' - \mathbf{v} \times \mathbf{B}$ for which we have $\nabla \times \mathbf{E} = -\partial \mathbf{B}/\partial t$. For further discussion, see Chapter 9 of W. K. H. Panofsky and M. Phillips, *Classical Electricity and Magnetism*, 2nd. ed., Addison-Wesley, Reading, Massachusetts, 1962.

7.2.4 Faraday's Law in Terms of the Vector Potential A

Faraday's law can be alternately stated in terms of the magnetic vector potential **A**. Since
$\mathbf{B} = \nabla \times \mathbf{A}$, we have from (7.6)

$$\mathcal{V}_{\text{ind}} = \frac{d\Psi}{dt}$$

$$\oint_C \mathbf{E} \cdot d\mathbf{l} = -\frac{d}{dt} \int_S (\nabla \times \mathbf{A}) \cdot d\mathbf{s}$$

$$\oint_C \mathbf{E} \cdot d\mathbf{l} = -\frac{d}{dt} \oint_C \mathbf{A} \cdot d\mathbf{l} \tag{7.8}$$

where the last step follows from Stokes's theorem. Equation (7.8) is generally valid, even
if **A** varies with time while the contour C is in motion. When contour C and surface S
are stationary, we can rewrite (7.8) as

$$\oint_C \mathbf{E} \cdot d\mathbf{l} = -\oint_C \frac{\partial \mathbf{A}}{\partial t} \cdot d\mathbf{l} \tag{7.9}$$

Since (7.9) must be valid for *any* path C, we must have

$$\mathbf{E} = -\frac{\partial \mathbf{A}}{\partial t}$$

at each point, as long as **E** represents the induced electric field. It is important here
again to distinguish between induced electric field and the electric field associated with
stationary electric charges. For electrostatic fields $\mathbf{E}_{\text{es}}$ associated with electric charges, we
have $\oint \mathbf{E}_{\text{es}} \cdot d\mathbf{l} = 0$; for such fields, we can define an electrostatic potential Φ such that
$\mathbf{E}_{\text{es}} = -\nabla\Phi$. Induced electric fields, on the other hand, are generated precisely because
$\oint \mathbf{E} \cdot d\mathbf{l} \neq 0$. If stationary electrical charges are present in addition to electromagnetic
induction, their effects can be described by a scalar potential Φ, producing an additional
electric field $\mathbf{E}_{\text{es}} = -\nabla\Phi$. Since $\mathbf{E}_{\text{es}}$ is conservative and its curl is thus zero, the total
electric field $\mathbf{E}_{\text{t}} = \mathbf{E}_{\text{es}} + \mathbf{E}$ still obeys (7.7). In other words, we have

$$\nabla \times \mathbf{E}_{\text{t}} = -\frac{\partial \mathbf{B}}{\partial t}$$

$$\nabla \times \mathbf{E}_{\text{t}} = -\frac{\partial}{\partial t}(\nabla \times \mathbf{A})$$

$$\nabla \times \left(\mathbf{E}_{\text{t}} + \frac{\partial \mathbf{A}}{\partial t} \right) = 0$$

Since any vector with zero curl can be expressed as the gradient of a scalar (see
Appendix A), we can write

$$\mathbf{E}_{\text{t}} + \frac{\partial \mathbf{A}}{\partial t} = -\nabla\Phi \qquad \rightarrow \qquad \boxed{\mathbf{E}_{\text{t}} = -\nabla\Phi - \frac{\partial \mathbf{A}}{\partial t}} \tag{7.10}$$

showing that the total electric field $\mathbf{E}_t$ is produced either by electromagnetic induction $(-\partial \mathbf{A}/\partial t)$ or stationary electric charges $(-\nabla \Phi)$ or both. Note that for the static case, with $\partial \mathbf{A}/\partial t = 0$, (7.10) reduces to $\mathbf{E}_t = -\nabla \Phi$, so that $\mathbf{E}_t$ can be determined from Φ alone, and $\mathbf{B}$ can be determined from $\mathbf{A}$ alone. For time-varying fields, the total electric field $\mathbf{E}_t$ depends on both Φ and $\mathbf{A}$.

7.3 ENERGY IN A MAGNETIC FIELD

Since current-carrying loops exert forces and torques on each other, a finite amount of work must be expended in establishing or modifying a given configuration of current loops. It is thus to be expected that a system of conductors carrying steady-state currents stores magnetic energy (W_m) in a manner analogous to a configuration of electric charges storing electrostatic energy (W_e). In Chapter 6, we deferred the discussion of methods of evaluation of W_m until after we were equipped with Faraday's law, because the establishment of the current configurations necessarily involves a transient period during which currents and associated fields are increased from zero to their final values. We now consider the magnetic energy of current-carrying loops and express the stored energy solely in terms of the fields produced by the currents. Once this is done, we can ascribe energy to the magnetic field in the same manner that energy was associated with the electrostatic field in Section 4.12.

In electrostatics, when we considered the establishment of charge configurations, the calculation of the work expended depended on the existence of two well-defined physical quantities, namely the electric charge q and the electrostatic potential Φ. In magnetostatics, magnetic charges do not exist, and further, it is not possible, in general, to define a magnetostatic potential that is a single-valued function.[22] Thus, the development of an expression for W_m must differ markedly from that used to calculate W_e.

We first consider placing all current-carrying loops at infinite distances apart from each other so that they do not initially exert forces or torques on one another. Then, one by one, the loops can be brought to their final positions and orientations while we calculate the work expended. Although such a procedure appears straightforward, we need to make two observations: (1) each current-carrying loop or circuit contains finite energy; that is, the loop would change its shape if it were not for the mechanical stresses in the wires and so on that hold it together (this is the *self-energy* of a current-carrying loop), and (2) a certain amount of work is necessary to keep the currents flowing in the loops as they are moved into their positions.[23] Alternatively, we can think of current-carrying

---

[22]Although it is true that we can always find a vector $\mathbf{A}$ such that $\mathbf{B} = \nabla \times \mathbf{A}$, the vector $\mathbf{A}$ is not uniquely specified until we also define its divergence $\nabla \cdot \mathbf{A}$. As we discussed in Section 6.5, for simplicity, we chose $\nabla \cdot \mathbf{A} = 0$, which is called the Coulomb gauge. In electromagnetic wave problems, however, it is customary to use the Lorentz gauge, or Lorentz condition, where the divergence of $\mathbf{A}$ is taken to be $\nabla \cdot \mathbf{A} = -(\mu\epsilon)\,\partial \Phi/\partial t$. This particular choice uncouples the equations for the magnetic vector potential $\mathbf{A}$ and the scalar electric potential Φ, making the formulation and solution of antenna radiation problems relatively easier.

[23]The work referred to here includes only the *reversible* work that must be expended because of the induced electromotive force (via Faraday's law); it does not include the Joule heating (i.e., ohmic losses or I^2R losses). We are also not concerned here with the mechanical work required to move the coils physically.

loops as being initially at their fixed positions and orientations but carrying zero current. As we establish current in the loops one by one, we have to exert electrical energy to oppose the induced electromotive forces. Consider for a moment that we have only a single circuit with a constant current I flowing in it. Faraday's law states that an electromotive force $\mathcal{V}_{ind}$ is induced around it if the magnetic flux Ψ through the circuit changes (e.g., if we bring the loop closer to other current-carrying loops). In order to keep the current I constant, the sources of the current must do work. The amount of work[24] that must be done in a time interval Δt is given by $\Delta W = -I\mathcal{V}_{ind}\Delta t = I\Delta\Psi$, where we have used $\mathcal{V}_{ind} = -\Delta\Psi/\Delta t$ (Faraday's law). The sources can supply this work by providing an applied voltage of $-\mathcal{V}_{ind}$ for the time interval Δt, while the current I flows.

We now proceed to use Faraday's law to determine the self-energy of a single current loop, the magnetic energy of a system of current-carrying loops, and, finally, the magnetic energy in terms of the **B** and **H** fields.

7.3.1 Self-Energy of a Current-Carrying Loop

We consider the self-energy of a single filamentary loop carrying a current I. To determine the self-energy of such a loop, consider the magnetic flux Ψ threading through a loop carrying a current I. Using (6.14), the magnetic flux Ψ can be written in terms of the **B** field or the vector potential **A** as

$$\Psi = \int_S \mathbf{B} \cdot d\mathbf{s} = \int_S (\nabla \times \mathbf{A}) \cdot d\mathbf{s} = \oint_C \mathbf{A} \cdot d\mathbf{l}$$

where S is the area enclosed by the loop defined by the contour C, and where we have invoked Stokes's theorem. From Section 6.5, we know that the vector potential is related to the current I as (see (6.20))

$$\mathbf{A}(\mathbf{r}) = \frac{\mu_0 I}{4\pi} \oint_C' \frac{d\mathbf{l}'}{R}$$

where $R = |\mathbf{r} - \mathbf{r}'|$. Thus, we have

$$\Psi = \oint_C \mathbf{A} \cdot d\mathbf{l} = I \left[\frac{\mu_0}{4\pi} \oint_C \oint_C' \frac{d\mathbf{l} \cdot d\mathbf{l}'}{R} \right] = IL$$

In view of the definition of inductance as magnetic flux linked per unit current, the preceding term in square brackets is the *self-inductance L* of the loop. Note that this quantity depends solely on the geometry of the loop and is thus a constant for a loop with fixed geometry. Note also that the term in square brackets is a special case of the Neumann formula for inductance that was derived in Section 6.7 for the mutual inductance of two circuits.

---

[24]The time rate of change in energy of a particle with velocity **v** acted on by a force **F** is $\mathbf{F} \cdot \mathbf{v}$. When changing flux induces an additional electric field **E**, the energy of each conduction electron of charge q_e and drift velocity $\mathbf{v}_d$ increases by $|q_e \mathbf{v}_d \cdot \mathbf{E}|$ per unit time. The power required to sustain the current (i.e., $-I\mathcal{V}_{ind}$) corresponds to a summation over all of the electrons in the circuit.

The self-inductance of a circuit acts as a sort of electric inertia. Since $\Psi = IL$ and L is a constant determined by geometry, whenever the current in the circuit varies, there is a corresponding variation in magnetic flux, which by Faraday's law gives rise to an electromotive force tending to prevent the variation. The self-energy of a loop carrying current can thus be determined by considering the fact that the establishment of a current I requires energy in order to oppose this induced electromotive force. If the current in the loop is increased by an amount ΔI, the change in the magnetic flux threading through the surface S is $\Delta\Psi = L\,\Delta I$. The rate at which magnetic energy is added to the circuit when the current is increased by ΔI is

$$\frac{dW_m}{dt} = -I\mathcal{V}_{ind} = I\frac{\Delta\Psi}{\Delta t} = IL\frac{dI}{dt}$$

when we consider a differential change in current such that $\Delta I \to 0$. Consequently, the self-energy W_{ms} of the loop, defined as the total energy added to the loop to increase its current from zero to a final value of I, is

$$W_{ms} = \int_0^I \left(\frac{dW_{ms}}{dt}\right) dt = L\int_0^I I'\,dI' = \frac{1}{2}LI^2$$

where I' is simply used to denote the dummy variable of integration. We have thus arrived at the general result that we know from circuit theory, namely that the magnetic energy stored in an inductance L carrying current I is given by $W_m = \frac{1}{2}LI^2$.

7.3.2 Magnetic Energy of a System of Current-Carrying Loops

Consider two conducting loops C_1 and C_2, as shown in Figure 7.15, in which we would like to establish currents I_1 and I_2, respectively. Work must be done to increase the currents from zero to their final values. This work will result in energy being stored in the magnetic field surrounding the conductors. To evaluate this energy, we can start by assuming that both of the loops initially have zero current. We can then increase the current in C_1 from zero to I_1, while keeping the current in C_2 zero. This action will have two consequences: (1) an emf will be induced in C_1 due to the self-inductance of C_1 and (2) an emf will be induced in C_2 due to the mutual inductance between C_1 and C_2. The work that must be supplied to overcome the first emf is the self-energy of C_1. An additional amount of energy must be supplied to C_2 in order to maintain zero current in it while the current in C_1 is being changed. Once the current I_1 is established in C_1, the

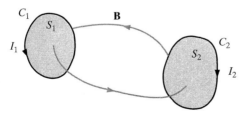

Figure 7.15 Two magnetically coupled current-carrying loops.

next step is to increase the current in C_2 from zero to I_2. Once again, this action will have two consequences: (1) an emf will be induced in C_2 due to the self-inductance of C_2, requiring that energy be supplied that will eventually be stored as the self-energy of C_2 and (2) an emf will be induced in C_2 due to the mutual inductance between C_2 and C_1, so that additional energy must be supplied in order to maintain the current in C_1 as I_1.

We now analyze the preceding sequence of events more quantitatively. Let us first maintain the current i_2 in C_2 at zero while we increase the current i_1 in C_1 from zero to I_1. When we change the current in C_1 by an amount di_1 in a time interval dt, the magnetic field $\mathbf{B}_1$ due to the current in C_1 changes at a rate of $d\mathbf{B}_1/dt$ and will thus produce, based on (7.1), an induced voltage $\mathcal{V}_{\text{ind}_1} = -d\Psi_{11}/dt$ in C_1 and an induced voltage $\mathcal{V}_{\text{ind}_2} = -d\Psi_{12}/dt$ in C_2. Thus, in order to increase the current in C_1 by an amount di_1 in a time interval dt, we must apply a voltage $-\mathcal{V}_{\text{ind}_1}$ in C_1. At the same time, we must apply a voltage $-\mathcal{V}_{\text{ind}_2}$ in C_2 in order to maintain[25] its current at zero. The work done in a time interval dt by an applied voltage of $-\mathcal{V}_{\text{ind}_1}$ is

$$dW_1 = -\mathcal{V}_{\text{ind}_1} i_1 \, dt = i_1 \, d\Psi_{11} = L_{11} i_1 \, di_1$$

where we have used the definition of the self-inductance of loop C_1 as $L_{11} i_1 = \Psi_{11}$ and noted that $L_{11} \, di_1 = d\Psi_{11}$ because L_{11} is constant. The voltage $-\mathcal{V}_{\text{ind}_2}$ applied in C_2 does no work because its current i_2 is kept at zero. The total work done in increasing i_1 from zero to I_1 is thus

$$W_1 = \int_0^{I_1} L_{11} i_1 \, di_1 = \tfrac{1}{2} L_{11} I_1^2$$

which is the self-energy of the single loop C_1. Since the current i_2 in C_2 is kept at zero, the physical presence of the loop C_2 has no effect on loop C_1.

The next step is to keep $i_1 = I_1$ and increase i_2 by an amount di_2 in a time interval dt. This results in the induced voltages

$$\mathcal{V}_{\text{ind}_2} = -\frac{d\Psi_{22}}{dt} = -L_{22}\frac{di_2}{dt} \quad \text{in } C_2$$

$$\mathcal{V}_{\text{ind}_1} = -\frac{d\Psi_{21}}{dt} = -L_{21}\frac{di_2}{dt} \quad \text{in } C_1$$

To maintain i_1 at a constant value I_1, we must apply a voltage $-\mathcal{V}_{\text{ind}_1}$, which in time dt does work of an amount

$$dW_{21} = -\mathcal{V}_{\text{ind}_1} I_1 \, dt = I_1 L_{21} \, di_2$$

Note that dW_{21} can be negative (work taken out of the system), depending on whether $\mathcal{V}_{\text{ind}_1}$ tends to increase or decrease i_1. Similarly, the work done by the voltage $-\mathcal{V}_{\text{ind}_2}$ that must be applied to change i_2 by an amount di_2 is

$$dW_2 = -\mathcal{V}_{\text{ind}_2} i_2 \, dt = L_{22} i_2 \, di_2$$

---

[25]Note that having a voltage of $\mathcal{V}_{\text{ind}_2} = -d\Psi_{12}/dt = -L_{12} \, di_2/dt$ induced in circuit C_2 means that the time derivative of i_2 is nonzero and that i_2 will thus change from its value of zero, unless we keep the net total voltage in circuit C_2 zero by applying $-\mathcal{V}_{\text{ind}_2}$.

The total work done in raising i_2 from zero to I_2 while keeping i_1 at I_1 is then

$$W_{21} + W_{22} = I_1 L_{21} \int_0^{I_2} di_2 + L_{22} \int_0^{I_2} i_2 \, di_2 = I_1 I_2 L_{21} + \tfrac{1}{2} I_2^2 L_{22}$$

The total work done on the system to establish the currents I_1 and I_2 in loops C_1 and C_2, respectively, is the sum of W_1 and $(W_{21} + W_{22})$, which represents the energy W_m stored in the magnetic field surrounding the two loops. This energy is thus given by

$$W_m = W_1 + W_{21} + W_{22} = \tfrac{1}{2} L_{11} I_1^2 + L_{21} I_1 I_2 + \tfrac{1}{2} L_{22} I_2^2 = \tfrac{1}{2} \sum_{i=1}^{2} \sum_{j=1}^{2} L_{ij} I_i I_j$$

The final result written in terms of a double summation can easily be generalized to a system of N loops by simply extending the summations for both i and j up to N. This equation for W_m provides an interpretation of the coefficients of inductance L_{ij} as the coefficients in the quadratic expression for the energy stored in a magnetic field. Note that L_{ij} $(i \neq j)$ may be either positive or negative, depending on the direction in which the mutual magnetic flux links the respective circuits. The sign of L_{ij} depends on the choices of positive directions for currents I_i and I_j. If the directions of I_i and I_j are such that they produce magnetic flux in the same direction, then L_{ij} is positive, since the stored magnetic energy is increased. If the magnetic fluxes produced by I_i and I_j tend to cancel one another, L_{ij} is negative.

Example 7.8: Magnetic energy stored by parallel coaxial coils. Consider the parallel coaxial coils considered in Example 6.30 and shown in Figure 6.40. If $a = 2$ cm, $b = 1$ cm, $d = 10$ cm, and $N_1 = N_2 = 10$, calculate the total energy stored in this system if currents of $I_1 = I_2 = 1$ mA flow in both coils. Assume that the thickness of the wire that the coils are made of is 0.4 mm.

> **Solution:** The self-inductance of a single circular loop of wire was determined in Example 6.31. For wire thickness $2t$ much less than the loop radius a, the self-inductance is given by
>
> $$L \simeq \mu_0 a \left(\ln \frac{8a}{t} - 2 \right)$$
>
> The self-inductance of a loop with N turns is simply N^2 times larger, assuming that each turn produces and links the same magnetic flux (such is the case if the coils are wound tightly enough and if the number of turns is not large, as shown in Figure 6.40). Using this formula, the self-inductances of coils 1 and 2, having radii of $a = 2$ cm and $b = 1$ cm, respectively, are

$$L_{11} \simeq N^2 \mu_0 a \left(\ln \frac{8a}{t} - 2 \right)$$

$$= 10^2 (4\pi \times 10^{-7})(2 \times 10^{-2} \text{ m}) \left[\ln \left(\frac{8 \times 2 \times 10^{-2} \text{ m}}{0.2 \times 10^{-3} \text{ m}} \right) - 2 \right] \simeq 12 \ \mu\text{H}$$

and $L_{22} \simeq 5 \ \mu\text{H}$.

The mutual inductance between two parallel coaxial coils as shown in Figure 6.40 was evaluated in Example 6.30 to be

$$L_{12} \simeq \frac{\mu_0 \pi N_1 N_2 a^2 b^2}{2d^3}$$

for the case when $d \gg a, b$, which is the case in hand with $d = 0.1$ m. Using this expression for L_{12}, we find

$$L_{12} \simeq \frac{(4\pi \times 10^{-7})\pi (10)(10)(0.02)^2 (0.01)^2}{2(0.1)^3} \simeq 0.008 \ \mu\text{H}$$

Thus, we see that the mutual inductance between the two coils is much less than their respective self-inductances. The main reason for this is that the coils are relatively far apart, with $d = 10$ cm. Note that L_{12} is inversely proportional to d^3, so that it is larger by approximately a factor of 1000 if the coils are 1 cm (instead of 10 cm) apart. However, care should be exercised here since the expression for L_{12} derived in Example 6.30 is only valid for the case $d \gg a, b$.

The sign of the mutual inductance L_{12} depends on the definition of the polarity of the currents I_1 and I_2 in the two circuits. If these are defined such that positive values of I_1 and I_2 produce magnetic flux in the same direction, then L_{12} is positive. The total magnetic energy stored in the system of two parallel coaxial coils is thus

$$W_{\text{m}} = \tfrac{1}{2}L_{11}I_1^2 \pm L_{12}I_1 I_2 + \tfrac{1}{2}L_{22}I_2^2$$
$$\simeq \tfrac{1}{2}(12 \times 10^{-6} \text{ H})(10^{-3} \text{ A})^2 \pm (8 \times 10^{-9} \text{ H})(10^{-3} \text{ A})(10^{-3} \text{ A})$$
$$+ \frac{1}{2}(5 \times 10^{-6} \text{ H})(10^{-3} \text{ A})^2$$
$$\simeq (8.5 \pm 0.008) \quad \text{pJ}$$

Note that the mutual coupling, numerically small in this case, can either slightly decrease or slightly increase the stored energy, depending on the relative directions of the two currents I_1 and I_2.

Example 7.9: A solenoid with a secondary winding. Find the total magnetic energy stored by the solenoid shown in Figure 7.16, which has a secondary winding wound concentrically over the main winding. Neglecting leakage flux from the ends, determine the total magnetic energy stored in this system if currents of $I_1 = I_2 = 1$ mA are flowing through the two coils and other parameters are $N_1 = 1000$, $N_2 = 100$, $l_1 = 10$ cm, $l_2 = 5$ cm, and $a = 1$ cm. Assume a nonmagnetic core ($\mu = \mu_0$).

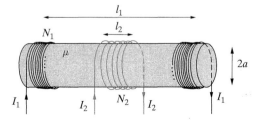

Figure 7.16   A solenoid with a secondary winding. The primary winding (N_1 turns) in fact covers the full length (l_1) of the core, but the middle portion is not shown to avoid clutter in the diagram.

Solution: From Example 6.26, we know that the self-inductance of a solenoid of length l and having N turns with a cross-sectional area of A is $L = \mu_0 N^2 A/l$. Thus, we have

$$L_{11} = \frac{\mu_0 N_1^2 \pi a^2}{l_1} = \frac{(4\pi \times 10^{-7})(1000)^2[\pi(0.01)^2]}{0.10} \simeq 3.95 \text{ mH}$$

and similarly, we find $L_{22} \simeq 0.079$ mH. The mutual inductance between the two coils is given by $L_{12} = N_2 \Lambda_{12}/I_1$. Once again invoking the result of Example 6.26, the flux produced by the primary coil is

$$\Lambda_{12} = \frac{\mu_0 N_1 I_1 \pi a^2}{l_1}$$

Since this flux is linked N_2 times by the secondary coil, we have

$$L_{12} = \frac{N_2 \Lambda_{12}}{I_1} = \frac{\mu_0 N_1 N_2 \pi a^2}{l_1}$$

Upon substitution of the parameter values, this yields $L_{12} \simeq 0.395$ mH. Thus, we see that the mutual inductance between the two coils is larger than the self-inductance of the secondary coil.

The total magnetic energy stored in this system is given by

$$W_m = \tfrac{1}{2} L_{11} I_1^2 \pm L_{12} I_1 I_2 + \tfrac{1}{2} L_{22} I_2^2$$

$$\simeq \tfrac{1}{2}[(3.95 \times 10^{-3})(10^{-3})^2 + (0.079 \times 10^{-3})(10^{-3})^2] \pm (0.395 \times 10^{-3})(10^{-3})(10^{-3})$$

$$\simeq (2.01 \pm 0.395) \quad \text{nJ}$$

If the currents I_1 and I_2 are positive with directions as shown in Figure 7.16, they reinforce one another so that L_{12} is positive and $W_m \simeq 2.41$ nJ. Otherwise, the magnetic flux produced by I_1 is reduced as a result of that produced by I_2, so that the total energy stored is $W_m \simeq 1.62$ nJ.

7.3.3 Magnetic Energy in Terms of the Fields

As we did for electrostatics in Section 4.12, it is possible to express the stored magnetic energy of a configuration of current-carrying loops solely in terms of the **B** and **H** fields. To this end, we start with the general expression derived in Section 7.3.2 for the energy of N current-carrying loops, namely

$$W_m = \tfrac{1}{2} \sum_{i=1}^{N} \sum_{j=1}^{N} L_{ij} I_i I_j$$

From Section 6.7, we recall the Neumann formula for the mutual inductance between two circuits C_i and C_j as

$$L_{ij} = \frac{\mu_0}{4\pi} \oint_{C_i} \oint_{C_j} \frac{d\mathbf{l}_i \cdot d\mathbf{l}_j}{R_{ij}}$$

where R_{ij} is the distance between the differential lengths $d\mathbf{l}_i$ and $d\mathbf{l}_j$. Using this, we have

$$
W_{\mathrm{m}} = \frac{1}{2} \sum_{i=1}^{N} I_i \left[\sum_{j=1}^{N} I_j L_{ij} \right] = \frac{1}{2} \sum_{i=1}^{N} I_i \sum_{j=1}^{N} I_j \left[\frac{\mu_0}{4\pi} \oint_{C_i} \oint_{C_j} \frac{d\mathbf{l}_i \cdot d\mathbf{l}_j}{R_{ij}} \right]
$$

$$
= \frac{1}{2} \sum_{i=1}^{N} I_i \oint_{C_i} \left[\frac{\mu_0}{4\pi} \sum_{j=1}^{N} I_j \oint_{C_j} \frac{d\mathbf{l}_j}{R_{ij}} \right] \cdot d\mathbf{l}_i
$$

Recalling from (6.20) in Section 6.5 the definition of the vector potential $\mathbf{A}$, we recognize that the expression in the square brackets is the vector potential $\mathbf{A}$, at a point on the loop identified by subscript i, due to (or produced by) currents flowing in all the loops. Therefore, we can write

$$
W_{\mathrm{m}} = \frac{1}{2} \sum_{i=1}^{N} \oint_{C_i} I_i d\mathbf{l}_i \cdot \mathbf{A}
$$

In terms of the more general free volume current density $\mathbf{J}$, and noting that $\oint_C I \, d\mathbf{l}$ is equivalent to $\int_V \mathbf{J} \, dv$, the preceding expression can be further generalized as

$$
\boxed{W_{\mathrm{m}} = \frac{1}{2} \int_V \mathbf{J} \cdot \mathbf{A} \, dv} \tag{7.11}
$$

where the volume V is the entire region of space within which $\mathbf{J} \cdot \mathbf{A} \neq 0$.

   Equation (7.11) is analogous to (4.81) associated with electrostatics. This expression identifies the quantity $\frac{1}{2} \mathbf{J} \cdot \mathbf{A}$ as the volume density of magnetostatic energy. Using the relationships between $\mathbf{A}$, $\mathbf{B}$ and $\mathbf{J}$, $\mathbf{H}$, we can write W_{m} entirely in terms of the $\mathbf{B}$ or $\mathbf{H}$ fields. To do this, we proceed in a manner identical to that used in electrostatics. If we assume $\mathbf{J}$ in (7.11) to be the free volume current density (thus excluding the bound magnetization current $\nabla \times \mathbf{M}$), we can substitute $\mathbf{J} = \nabla \times \mathbf{H}$ (Ampère's law). Using the vector identity (see Appendix A)

$$
\nabla \cdot (\mathbf{H} \times \mathbf{A}) \equiv (\nabla \times \mathbf{H}) \cdot \mathbf{A} - (\nabla \times \mathbf{A}) \cdot \mathbf{H}
$$

equation (7.11) can be rewritten as

$$
W_{\mathrm{m}} = \frac{1}{2} \int_V \mathbf{J} \cdot \mathbf{A} \, dv = \frac{1}{2} \int_V (\nabla \times \mathbf{H}) \cdot \mathbf{A} \, dv
$$

$$
= \frac{1}{2} \int_V (\nabla \times \mathbf{A}) \cdot \mathbf{H} \, dv + \frac{1}{2} \int_V \nabla \cdot (\mathbf{H} \times \mathbf{A}) \, dv
$$

$$
= \frac{1}{2} \int_V \mathbf{B} \cdot \mathbf{H} \, dv + \frac{1}{2} \oint_S (\mathbf{H} \times \mathbf{A}) \cdot d\mathbf{s}
$$

where S is the surface that encloses the volume V and where we have used Stokes's theorem to rewrite the second term as a surface integral. Note that the surface S and the volume V must be large enough to enclose all currents in the region of interest. Since there are no other restrictions on the choice of the surface S, we can extend it to infinity. In this case, the surface integral reduces to zero because dependencies of the various quantities at large distances from the currents (i.e., at points from which the entire distribution of currents looks just like an elemental current) are

$$|\mathbf{A}| \sim \frac{1}{r}, \quad |\mathbf{H}| \sim \frac{1}{r^2}, \quad \text{and} \quad |d\mathbf{s}| \sim r^2$$

so that

$$\lim_{r \to \infty} (\mathbf{H} \times \mathbf{A}) \cdot d\mathbf{s} \quad \to \lim_{r \to \infty} \frac{1}{r} \frac{1}{r^2} r^2 \to 0$$

In other words, the magnetostatic energy W_{m} reduces to simply the volume integral term, or

$$\boxed{W_{\mathrm{m}} = \tfrac{1}{2} \int_V \mathbf{H} \cdot \mathbf{B} \, dv} \tag{7.12}$$

Note that since we let S extend to infinity, we must have $V \to \infty$, meaning that the integral in (7.12) has to be carried out over *all* space in which the magnetic field is nonzero. Based on (7.12), we can define the *volume energy density* for the magnetostatic field as

$$\boxed{w_{\mathrm{m}} = \tfrac{1}{2} \mathbf{H} \cdot \mathbf{B}} \tag{7.13}$$

For linear[26] and isotropic materials, where the permeability μ is a simple constant, we have $\mathbf{H} \cdot \mathbf{B} = \mu H^2$, so that the magnetic energy density is $w_{\mathrm{m}} = \tfrac{1}{2} \mu H^2$ in units of J-m^{-3}. Alternatively, magnetic energy density can be expressed in terms of the $\mathbf{B}$ field as $w_{\mathrm{m}} = \tfrac{1}{2}(B^2/\mu)$. The relationship between magnetic energy density and the magnetic field $\mathbf{H}$ is entirely analogous to the corresponding expression for the electrostatic field, namely $w_{\mathrm{e}} = \tfrac{1}{2} \epsilon E^2$, found in Section 4.12.

As in the electrostatic case, we note that the two alternative expressions for magnetic energy density, namely $\tfrac{1}{2} \mathbf{J} \cdot \mathbf{A}$ and $\tfrac{1}{2} \mathbf{H} \cdot \mathbf{B}$, are quite different. The former implies that magnetic energy exists at places where currents exist and is zero where $\mathbf{J} = 0$. However, the latter indicates that magnetic energy exists wherever the fields exist. As before, both points of view have merit, and it is neither necessary nor possible to determine which one is more "correct." It is not possible to localize energy or to decide whether it is associated with the current or the field. Thus, the quantities $\tfrac{1}{2} \mathbf{J} \cdot \mathbf{A}$ and $\tfrac{1}{2} \mathbf{H} \cdot \mathbf{B}$ represent magnetic energy density only to the extent that their volume integral over space is the total stored magnetic energy.

---

[26] Even equation (7.12) is not the most general form of magnetic energy applicable for nonlinear materials. The work done by external sources in establishing a magnetic field of $\mathbf{B}$ in material media, which by definition is the magnetic energy stored in that field, is given by

$$W_{\mathrm{m}} = \int_V \left[\int_0^{\mathbf{B}} \mathbf{H} \cdot d\mathbf{B} \right] dv$$

where the integral is to be carried out over all space. For further discussion, see Section 2.15 of J. A. Stratton, *Electromagnetic Theory,* McGraw-Hill, New York, 1941. On the other hand, equations (7.12) and (7.13) apply to both linear and anisotropic materials, in which the permeability is a tensor, such as in the case of ferrites. See Section 5 of A. Sommerfeld, *Electrodynamics,* Academic Press, New York, 1952.

Example 7.10: Magnetic energy near long, straight, current-carrying wires. Compare the magnetic energy stored per unit length by two separate infinitely long, straight wires of different radii a and b such that $b > a$; each wire has a circular cross section and carries a uniformly distributed current I.

Solution: The **B** field at any position r in the vicinity of a wire carrying uniformly distributed current and having a circular cross section of radius a was determined in Example 6.10 to be

$$\mathbf{B} = \begin{cases} \hat{\boldsymbol{\phi}} \dfrac{\mu_0 I r}{2\pi a^2} & r \le a \\[3mm] \hat{\boldsymbol{\phi}} \dfrac{\mu_0 I}{2\pi r} & r > a \end{cases}$$

To determine the magnetic energy stored per unit length, we can integrate $\frac{1}{2}(B_\phi^2/\mu_0)$ over a cylindrical volume of 1-m axial length. In other words,

$$W_{\mathrm{m}} = \frac{1}{2\mu_0} \int_V B_\phi^2 \, dv = \frac{1}{2\mu_0} \int_0^\infty B_\phi^2 2\pi r \, dr$$

$$= \frac{1}{2\mu_0} \frac{\mu_0^2 I^2}{(2\pi a^2)^2} \int_0^a r^2 2\pi r \, dr + \frac{1}{2\mu_0} \frac{\mu_0^2 I^2}{(2\pi)^2} \int_a^\infty \frac{1}{r^2} 2\pi r \, dr$$

$$= \frac{\mu_0 I^2}{4\pi a^4} \left[\frac{r^4}{4} \right]_{r=0}^{r=a} + \frac{\mu_0 I^2}{4\pi} [\ln r]_{r=a}^{r=\infty} = \frac{\mu_0 I^2}{16\pi} + \infty$$

which indicates that the energy residing in the magnetic field surrounding a 1-m length of such a wire is infinite. This result is physically not surprising, since the very premise of sustaining a current I in an infinitely long wire is nonphysical. In reality, any such current carried in one direction by a long straight wire has to return via another wire or set of wires. This return current would generate **B** fields in the opposite direction, thus reducing the net total **B** field so that the stored magnetic energy is not divergent.

An interesting aspect of the preceding expression for W_{m} is the fact that the first term, which represents the magnetic energy stored within the wire in the region $r \le a$, is independent of the radius a of the wire. This internal magnetic energy given by $(W_{\mathrm{m}})_{\mathrm{int}} = \mu_0 I^2/(16\pi)$ is the quantity represented by the internal inductance of a wire discussed in Section 6.7. In other words, using (6.32) we have

$$L_{\mathrm{int}} = \frac{2(W_{\mathrm{m}})_{\mathrm{int}}}{I^2} = \frac{\mu_0}{8\pi}$$

Although the total magnetic energy stored per unit length for a straight wire is divergent, the difference between the energy stored by two different wires of radii a and b is finite. Since the internal stored energy is the same for both radii, the difference between the total magnetic energies stored per unit length for the two different wires is given by

$$W_{\mathrm{m}}^a - W_{\mathrm{m}}^b = \frac{1}{2\mu_0} \frac{\mu_0^2 I^2}{(2\pi)^2} \int_a^b \frac{1}{r^2} 2\pi r \, dr = \frac{\mu_0 I^2}{4\pi} \ln \frac{b}{a}$$

7.3.4 Magnetic Forces and Torques in Terms of Stored Magnetic Energy

In Section 4.13, we saw that electrostatic forces can be expressed in terms of spatial derivatives of stored electrostatic energy. The underlying principle used was called the principle of virtual displacement, which involves consideration of how the energy of a system changes for a small virtual change in geometry. In an analogous manner, magnetic forces and torques can also be derived from stored magnetic energy using the principle of virtual displacement. In this manner, one can usually calculate the lifting force of magnets.[27] In this section, we briefly discuss this topic with analogy to Section 4.13.

In Section 4.13, we noted that the relationship between electrostatic force and stored energy depended on the premise under which we allowed the virtual displacement to occur. Specifically, it was possible to consider virtual displacement under the constraint of constant charge or constant voltage. For magnetostatics, we can consider a system of current-carrying loops with constant currents or constant magnetic flux linkages. In the following, we consider only the case of constant current.

Consider a solenoid with a movable, permeable core, where $\mu > \mu_0$, as shown in Figure 7.17. A constant-current generator is attached to the coil, and the core of the solenoid is allowed to move a distance Δx, as shown. Assume that the movement of the core by an amount Δx changes the magnetic flux linked by the coil from Ψ_1 to Ψ_2. Note that the magnetic energy stored in this system is given by $W_m = \frac{1}{2}LI^2$, or alternatively $W_m = \frac{1}{2}I\Psi$, since $L = \Psi/I$. Thus, the change in flux corresponds to a change in stored magnetic energy of

$$\Delta W_m = \tfrac{1}{2}I\Psi_2 - \tfrac{1}{2}I\Psi_1 = \tfrac{1}{2}I\,\Delta\Psi$$

This additional energy must be provided by the generator in order to maintain the current I constant. The generator can do this by canceling out the induced emf by supplying a voltage of $-\mathcal{V}_{ind}$, in which case the amount of work done by the generator is

$$\Delta W_g = I(-\mathcal{V}_{ind})\,\Delta t = I\frac{\Delta\Psi}{\Delta t}\,\Delta t = I\,\Delta\Psi = 2\,\Delta W_m$$

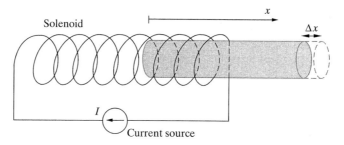

Figure 7.17   Solenoid with a movable core.

[27]For a discussion at an appropriate level, see Chapter 10 of M. A. Plonus, *Applied Electromagnetics,* McGraw-Hill, New York, 1978.

where Δt is the time duration over which the virtual displacement is assumed to occur. In order for energy to be conserved, we must have

$$F_x \,\Delta x + \Delta W_m = \Delta W_g \quad \rightarrow \quad F_x \,\Delta x + \Delta W_m = 2\,\Delta W_m \quad \rightarrow \quad F_x = \frac{\Delta W_m}{\Delta x}$$

where $F_x \,\Delta x$ is the virtual work done to move the core a distance Δx. In other words, F_x is the magnetic force exerted on the core by the coil, and F_x must be overcome in order to move the core. We have thus arrived at a result analogous to that found for electrostatic forces in Section 4.13.

For the case shown in Figure 7.17, under the constant-current constraint, the **H** field everywhere inside the coil remains constant. With the magnetic energy density being $\frac{1}{2}\mu H^2$, and for $\mu > \mu_0$ (i.e., paramagnetic or ferromagnetic material), the total stored magnetic energy decreases when the core moves to the right, so that less of it is within the coil. Thus, $F_x = dW_m/dx$ is negative, and the core experiences a force that tends to pull it within the coil, in a manner analogous to the case of the dielectric slab between parallel plates analyzed in Example 4.36. Note that if the material were diamagnetic, that is, if $\mu < \mu_0$, then the direction of the force would be reversed, tending to push the core out of the coil. In practice, however, μ is significantly different from μ_0 only for ferromagnetic materials, so that magnetic forces are significant only when the core is made out of such materials; in this case, the magnetic force tends to pull the core in, under constant-current conditions.

More generally, it can be shown that, under the constraint of constant currents, the magnetic force acting on a circuit is given by

$$[\mathbf{F}]_{I=\text{const.}} = \nabla W_m$$

If the circuit is constrained to rotate about the z axis, then the z component of the torque acting on it is given by

$$[T_z]_{I=\text{const.}} = \frac{\partial W_m}{\partial \phi}$$

Magnetic forces in terms of mutual inductance.   The principle of virtual displacement can also be used to determine magnetic forces and torques between current-carrying circuits. Consider, for example, two parallel coaxial loops carrying currents I_1 and I_2 and with the z axis passing through their centers, as shown in Figure 6.40. Based on what we found in Section 7.3.2, and also discussed in Example 7.8, the magnetic energy stored in this system is given by

$$W_m = \tfrac{1}{2}L_{11}I_1^2 \pm L_{12}I_1I_2 + \tfrac{1}{2}L_{22}I_2^2$$

If the currents I_1 and I_2 are maintained (i.e., kept constant) in both circuits (presumably by driving them with current sources), then the mutual inductance between the two coils is the only quantity that varies as a result of any relative motion of the two coils. Thus, an infinitesimal displacement by an amount Δz of either circuit with respect to the other results in a change of energy given by

$$\Delta W_{\mathrm{m}} = I_1 I_2 \frac{\Delta L_{12}}{\Delta z} \, \Delta z = I_1 I_2 \, \Delta L_{12}$$

Note that ΔW_{m} is a positive quantity when I_1 and I_2 have the same sign and when ΔL_{12} is positive. Thus, the force between the two circuits, given by the gradient of W_{m}, tends to pull the circuits in the direction in which L_{12} increases most rapidly. In other words, the two circuits attract each other whenever their currents are in the same direction, since their mutual inductance obviously increases when they move closer. When I_1 and I_2 are in opposite directions, ΔW_{m} is negative and the circuits repel one another.

In general, it can be shown that under constant-current conditions, the magnetic force between two circuits is proportional to the gradient of mutual inductance or

$$\mathbf{F} = I_1 I_2 \, \nabla L_{12}$$

and if the circuit is constrained to rotate about the z axis, then the z component of the torque acting on it is given by

$$T_z = I_1 I_2 \, \frac{\partial L_{12}}{\partial \phi}$$

7.4 DISPLACEMENT CURRENT AND MAXWELL'S EQUATIONS

In a series of brilliant papers between 1856 and 1865, culminating in his classic paper,[28] James Clerk Maxwell formulated the complete classical theory of electromagnetics.[29] He provided a mathematical framework for Faraday's results, clearly elucidated the different behavior of conductors and insulators under the influence of fields, imagined and introduced the concept of displacement current, and inferred the electromagnetic nature of light. His theoretical framework predicted the existence of electromagnetic waves in the absence of any experimental evidence. His hypotheses were to be confirmed 23 years later (in 1887) in the experiments of Heinrich Hertz.[30]

The crucial step in Maxwell's development of the theory of electromagnetics was his introduction of the notion of a displacement current. In the following section, we discuss this concept.

7.4.1 Displacement Current and the General Form of Ampère's Law

Before Maxwell developed his theory of electromagnetics, Ampère's circuital law was known as the relationship between a static magnetic field $\mathbf{H}$ and the conduction-current

---

[28]For an excellent account with passages quoted from Maxwell's papers, see Chapter 5 of R. S. Elliott, *Electromagnetics,* IEEE Press, New Jersey, 1993.

[29]J. C. Maxwell, A dynamical theory of the electromagnetic field, *Phil. Trans. Royal Soc.* (London), 155, p. 450, 1865.

[30]H. Hertz, On the finite velocity of propagation of electromagnetic actions, *Sitzb. d. Berl. Akad. d. Wiss.,* February 2, 1888; for a collected English translation of this and other papers by H. Hertz, see H. Hertz, *Electric Waves,* MacMillan, London, 1893.

density[31] $\mathbf{J}_c$. From Chapter 6, the differential form of this relationship is

$$\nabla \times \mathbf{H} = \mathbf{J}_c$$

Another fundamental relation concerning conduction-current density was derived in Chapter 5 on the basis of conservation of charge. This relation, known as the continuity equation and given in (5.7), is repeated here:

$$\nabla \cdot \mathbf{J}_c + \frac{\partial \rho}{\partial t} = 0 \tag{7.14}$$

In magnetostatics, we simply have $\nabla \cdot \mathbf{J}_c = 0$, since $\partial \rho / \partial t = 0$ by definition. We can then relate Ampère's law to the continuity equation by taking the divergence of the former,

$$\nabla \cdot (\nabla \times \mathbf{H}) = \nabla \cdot \mathbf{J}_c \quad \rightarrow \quad 0 = \nabla \cdot \mathbf{J}_c$$

where we have observed (see Appendix A) that the divergence of the curl of any vector is identically zero. We see that Ampère's law is indeed consistent with the principle of conservation of charge under static conditions. However, in the more general case when the field and source quantities are allowed to vary with time, Ampère's law, in its form used in Chapter 6, is clearly inconsistent with the fundamental principle of conservation of charge.

Faced with this inconsistency, Maxwell developed the time-varying form of Ampère's circuital law by postulating an additional current term $\mathbf{J}_d$ such that

$$\nabla \times \mathbf{H} = \mathbf{J}_c + \frac{\partial \mathbf{D}}{\partial t} = \mathbf{J}_c + \mathbf{J}_d \tag{7.15}$$

where $\mathbf{J}_c$ is the free conduction current density and

$$\boxed{\mathbf{J}_d = \frac{\partial \mathbf{D}}{\partial t}} \tag{7.16}$$

is the *displacement-current density*. Note that the divergence of (7.15) gives

$$\nabla \cdot (\nabla \times \mathbf{H}) = 0 = \nabla \cdot \mathbf{J}_c + \nabla \cdot \frac{\partial \mathbf{D}}{\partial t} = \nabla \cdot \mathbf{J}_c + \frac{\partial}{\partial t} (\nabla \cdot \mathbf{D})$$

Substituting for $\nabla \cdot \mathbf{D} = \rho$ from Gauss' law, we have

$$0 = \nabla \cdot \mathbf{J}_c + \frac{\partial \rho}{\partial t}$$

Thus, we see that (7.15) is consistent with the continuity equation (7.14) and hence with the principle of conservation of charge.

---

[31]The current density term in $\nabla \times \mathbf{H} = \mathbf{J}$ includes all forms of charge transport except for the bound magnetization current density $\nabla \times \mathbf{M}$. In this section, we use the label $\mathbf{J}_c$ to distinguish this free current density (which in practice often is conduction current) from the displacement current term $\mathbf{J}_d$.

Equation (7.15) states that the net circulation (or curl) of **H** is the sum of the conduction-current density $\mathbf{J_c}$ plus the displacement-current density $\mathbf{J_d} = \partial \mathbf{D}/\partial t$. Note that this new term has the dimensions of a current density and indicates that $\nabla \times \mathbf{H}$ may be generated (or sustained) at a point in space (even in the absence of $\mathbf{J_c}$) when a time-varying electric field is present. This relationship between **D** and **H** is analogous to Faraday's law (7.7), which states that $\nabla \times \mathbf{E}$ may be generated at a point in space when a time-varying magnetic field is present. The combined action of the $\nabla \times \mathbf{H}$ and $\nabla \times \mathbf{E}$ equations is precisely what leads to the propagation of electromagnetic waves, as discussed in Section 7.4.2 in connection with Figure 7.20. Without the displacement current, electromagnetic waves do not exist. It is not surprising that this term was not discovered earlier (e.g., in Ampère's experiments); in current-carrying wires, and especially at low frequencies, the magnitude of $\mathbf{J_d}$ is much smaller than that of $\mathbf{J_c}$ (see the discussion that follows). On the other hand, for electromagnetic waves propagating in free space, where there cannot be any conduction current, the displacement current density $\mathbf{J_d} = \partial \mathbf{D}/\partial t$ is the only current term.

The displacement current density $\mathbf{J_d}$ consists of two terms corresponding to the time derivative of both the electric field and polarization vectors of **D** in (4.64):

$$\mathbf{J_d} = \frac{\partial \mathbf{D}}{\partial t} = \epsilon_0 \frac{\partial \mathbf{E}}{\partial t} + \frac{\partial \mathbf{P}}{\partial t} \tag{7.17}$$

The first term states that a time-varying electric field generates a magnetic field. It is not surprising that the time rate of change of electric field **E** behaves like a current. In any given configuration of conductors, any increase (decrease) in **E** implies a buildup (decrease) of charge, which requires current flow. Consider a parallel-plate capacitor having plates separated by free space, as shown in Figure 7.18a. A current I_c charges up the left-hand plate to a surface charge density ρ_s, which means that current moves

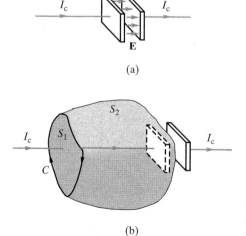

(a)

(b)

Figure 7.18 Displacement current concept. (a) Flow of current I_c charges up the left-hand plate, thus increasing the electric field. (b) Two different surfaces S_1 and S_2, enclosed by the same contour C. Note that S_2 is shaped like a sack, with C being the circumference of its opening.

away from the right-hand plate to create an opposing charge density $-\rho_s$. The resulting electric field between the capacitor plates is $E = \rho_s/\epsilon_0$. If the current flow is continuous, ρ_s (and thus E) increases with time. Thus, current flow into the capacitor is associated with a time rate of change of $\mathbf{E}$ (i.e., $\partial \mathbf{E}/\partial t$) between the capacitor plates. In effect, the displacement current, given by $\epsilon_0 \partial \mathbf{E}/\partial t$, "takes over" the conduction current in the free-space region between the capacitor plates.

The physical nature of the second term, $\partial \mathbf{P}/\partial t$ is quite different. This term is called the *polarization current density*:

$$\mathbf{J}_\mathrm{p} = \frac{\partial \mathbf{P}}{\partial t} \tag{7.18}$$

As $\mathbf{P}$ changes with time, the electrons in a dielectric are displaced with respect to their parent nuclei and move to and fro in accordance with $\partial \mathbf{E}/\partial t$. Although the charge never leaves the parent molecule, the back-and-forth motion constitutes a true alternating current. Indeed, we can check if this current satisfies the continuity equation (5.7) for bound charges:

$$\nabla \cdot \mathbf{J}_\mathrm{p} = \nabla \cdot \frac{\partial \mathbf{P}}{\partial t} = \frac{\partial}{\partial t}\left(\nabla \cdot \mathbf{P}\right) = -\frac{\partial \rho_\mathrm{p}}{\partial t} \tag{7.19}$$

where ρ_p is the bound polarization volume charge density. Hence in time-varying fields, the polarization current accounts for the creation of the bound charge in regions where $\nabla \cdot \mathbf{P} \neq 0$.

To further illustrate the need for a displacement current, consider an arbitrary contour C_1 encircling both the wire and two different surfaces S_1 and S_2, as shown in Figure 7.18b. Considering surface S_1, we have from Ampère's law

$$\oint_{C_1} \mathbf{H} \cdot d\mathbf{l} = \int_{S_1} \mathbf{J}_\mathrm{c} \cdot d\mathbf{s} = I_\mathrm{c}$$

where I_c is the conduction current in the wire, which passes through S_1. When we consider surface S_2, however, we have a problem. Without the displacement current term, we have no conduction current through the surface S_2 so that we would have to conclude $\oint_{C_1} \mathbf{H} \cdot d\mathbf{l} = 0$, which contradicts what we found for surface S_1 enclosed by the same contour C_1! This difficulty is resolved by including the displacement-current term. Let A be the area of the parallel-plate capacitor and a be the separation between its plates. The capacitance is then given by $C = \epsilon A/a$. When a displacement current I_d flows through the capacitor, the potential Φ across its plates is determined by the well-known voltage–current relationship for a capacitor, namely

$$I_\mathrm{d} = C\,\frac{d\Phi}{dt}$$

assuming that the material between the capacitor plates is a perfect dielectric with $\sigma = 0$ (i.e., $I_\mathrm{c} = 0$). But the potential difference across the plates is $\Phi = Ea$, where E is the electric field between the plates. Recalling from Example 4.26 that $C = \epsilon A/a$, we have

$$I_\mathrm{d} = Ca\,\frac{dE}{dt} = \frac{Ca}{\epsilon A}\frac{d(\epsilon A E)}{dt} = \frac{d(\epsilon E A)}{dt} = \frac{d}{dt}\left(\int_{S_2} \mathbf{D} \cdot d\mathbf{s}\right) = \int_{S_2} \mathbf{J}_\mathrm{d} \cdot d\mathbf{s}$$

Note that if we neglect any fringing effects so that the electric field between the capacitor plates is uniform over the area A and zero elsewhere, the integral of $\mathbf{D} \cdot d\mathbf{s}$ over any arbitrary surface is simply equal to $\epsilon E A$. Therefore, we see that the choice of either surface S_1 or S_2 gives the same result for $\oint_{C_1} \mathbf{H} \cdot d\mathbf{l}$, illustrating that $I_c = I_d$ and that consistent results are obtained by including the displacement-current term in Ampère's law.

While we can construct a strong plausibility case for its existence, the real physical bases of the displacement current are the three experimental pillars of electromagnetics, namely, Coulomb's law (equation (4.1)), the Biot–Savart law (equation (6.5)), and Faraday's law (equation (7.1)). After all, Maxwell introduced the $\partial \mathbf{D}/\partial t$ term in order to reconcile these physical experimental facts with the principle of conservation of charge (i.e., the continuity equation (5.7)). The conclusive experimental proof of the presence of displacement current was provided by Hertz's experiments in 1887, when he demonstrated that electromagnetic waves (whose existence, as we discuss in the next section, depend on the displacement current) can indeed travel through free space.

Example 7.11: A parallel-plate capacitor. Consider a parallel-plate capacitor consisting of two metal plates of 50-cm^2 area each, separated by a porcelain layer of thickness $a = 1$ cm (for porcelain, $\epsilon_r = 5.5$ and $\sigma = 10^{-14}$ S-m^{-1}). If a voltage $\Phi(t) = 110\sqrt{2}\cos(120\pi t)$ V is applied across the capacitor plates, find (a) J_c, (b) J_d, and (c) the total current I through the capacitor.

Solution: Using (4.48), the electric field in the porcelain layer is

$$E(t) = \frac{\Phi(t)}{a} = \frac{110\sqrt{2}\cos(120\pi t) \text{ V}}{0.01 \text{ m}} = (1.1\sqrt{2}) \times 10^4 \cos(120\pi t) \text{ V-m}^{-1}$$

The conduction (J_c) and the displacement-current (J_d) densities are then

(a)
$$J_c = \sigma E(t) = (10^{-14} \text{ S-m}^{-1})[(1.1\sqrt{2}) \times 10^4 \cos(120\pi t) \text{ V-m}^{-1}]$$
$$= 1.1\sqrt{2} \times 10^{-10} \cos(120\pi t) \text{ A-m}^{-2}$$

(b)
$$J_d = \frac{dD}{dt} = \epsilon \frac{dE(t)}{dt} = \epsilon_r \epsilon_0 \frac{dE(t)}{dt}$$
$$= (5.5 \times 8.85 \times 10^{-12} \text{ F-m}^{-1})[-1.1\sqrt{2} \times 10^4 \times 120\pi \sin(120\pi t) \text{ V-(ms)}^{-1}]$$
$$\simeq -(2.86 \times 10^{-4}) \sin(120\pi t) \text{ A-m}^{-2}$$

Note that the conduction current is in phase with the electric field, whereas the displacement current is 90° out of phase. Also note that the amplitude of the conduction current is $\sim 5 \times 10^5$ times smaller than the displacement current, as expected since porcelain is an excellent insulator.

(c) The total current I can be found by integrating the total current density over the cross-sectional area of the capacitor plates as

$$I = \int_A (\mathbf{J}_c + \mathbf{J}_d) \cdot d\mathbf{s} = (J_c + J_d)A$$

since $\mathbf{J}_c$ and $\mathbf{J}_d$ are both approximately uniform and perpendicular to the plates throughout the dielectric region. Note also that $|\mathbf{J}_c|_{max} \ll |\mathbf{J}_d|_{max}$, so that for all practical purposes we can write

$$I \simeq J_d A \simeq [-2.86 \times 10^{-4} \sin(120\pi t) \text{ A-m}^{-2}](50 \times 10^{-4} \text{ m}^2)$$

$$\simeq -1.43 \times 10^{-6} \sin(120\pi t) \text{ A} = -1.43 \sin(120\pi t) \quad \mu\text{A}$$

Note that we could have easily obtained this result by using the familiar current–voltage relationship of an ideal capacitor, that is, $I(t) = C\, d\Phi(t)/dt$, where $C = \epsilon A/a \simeq 24.3$ pF.

It is instructive to estimate the ratio of conduction to displacement-current density in a metallic conductor. For an alternating electric field $E_0 \cos(\omega t)$ within a metallic conductor of conductivity σ, the conduction-current density is $J_c = \sigma E_0 \cos(\omega t)$, whereas the displacement-current density is $J_d = \partial D/\partial t = -\omega\epsilon E_0 \sin(\omega t)$, where ϵ is the dielectric constant of the material. Thus, we have

$$\frac{|J_c|_{max}}{|J_d|_{max}} = \frac{\sigma}{\omega\epsilon} \tag{7.20}$$

The dielectric constant ϵ for a metallic conductor is not easily measurable, since any polarization effect is completely overshadowed by conduction. Nevertheless, based on measurements of reflectivity of metals and the fact that atomic resonances for metals lie in the ultraviolet and x-ray ranges, metallic conductors can be treated as if their dielectric constant is ϵ_0 at frequencies up to and including the visible range (i.e., $\sim 10^{15}$ Hz). Taking $\sigma \simeq 10^7$ S-m^{-1} for a metallic conductor, we then have $|J_c|_{max}/|J_d|_{max} \simeq 10^{17}/f$, where $f = \omega/(2\pi)$ is the frequency of operation. Thus, the displacement current within a metallic conductor is completely negligible compared with the conduction current at frequencies below the optical range of up to $\sim 10^{15}$ Hz. Note that this calculation concerns the displacement current inside a metallic conductor. For propagation of electromagnetic waves in free space, however, where $\sigma = 0$, the only current term that exists is the displacement current.

Example 7.12: Conduction to displacement current ratio for humid soil. Consider a certain type of humid soil with the following properties: $\sigma \simeq 10^{-2}$ S-m^{-1}, $\epsilon_r = 30$, and $\mu_r = 1$. Find the ratio of the amplitudes of the conduction and displacement currents at 1 kHz, 1 MHz, and 1 GHz.

Solution: Using $|J_c|_{max}/|J_d|_{max} = \sigma/(\omega\epsilon)$, we have

$$\frac{\sigma}{\omega\epsilon} = \frac{1}{2\pi f} \frac{10^{-2} \text{ S-m}^{-1}}{30 \times 8.85 \times 10^{-12} \text{ F-m}^{-1})} \simeq \frac{6 \times 10^6}{f}$$

Thus, we have $\sigma/(\omega\epsilon) \simeq 6000$, 6, and 0.006, respectively, at 1 kHz, 1MHz, and 1 GHz.

7.4.2 Maxwell's Equations

We are now ready to examine the complete set of four Maxwell's equations:

$$\oint_C \overline{\mathscr{E}} \cdot d\mathbf{l} = -\int_{S_C} \frac{\partial \overline{\mathscr{B}}}{\partial t} \cdot d\mathbf{s} \qquad \nabla \times \overline{\mathscr{E}} = -\frac{\partial \overline{\mathscr{B}}}{\partial t} \qquad (7.21a)$$

$$\oint_{S_V} \overline{\mathscr{D}} \cdot d\mathbf{s} = \int_V \tilde{\rho} \, dv \qquad \nabla \cdot \overline{\mathscr{D}} = \tilde{\rho} \qquad (7.21b)$$

$$\oint_C \overline{\mathscr{H}} \cdot d\mathbf{l} = \int_{S_C} \overline{\mathscr{J}} \cdot d\mathbf{s} + \int_{S_C} \frac{\partial \overline{\mathscr{D}}}{\partial t} \cdot d\mathbf{s} \qquad \nabla \times \overline{\mathscr{H}} = \overline{\mathscr{J}} + \frac{\partial \overline{\mathscr{D}}}{\partial t} \qquad (7.21c)$$

$$\oint_{S_V} \overline{\mathscr{B}} \cdot d\mathbf{s} = 0 \qquad \nabla \cdot \overline{\mathscr{B}} = 0 \qquad (7.21d)$$

where C is a closed contour that encloses the surface S_C, and S_V is a closed surface that encloses the volume V, as depicted in Figure 7.19. At this point, we introduce the notation $\overline{\mathscr{E}}$ (rather than $\mathbf{E}$) for the various vector quantities in order to distinguish[32] real field quantities as functions of time and space $\overline{\mathscr{E}}(x, y, z, t)$ from complex phasors representing time-harmonic quantities $\mathbf{E}(x, y, z)$ (see Section 7.4.4). Similarly, the free charge density $\tilde{\rho}(x, y, z, t)$ is distinguished from its phasor $\rho(x, y, z)$. Equations are written for a coordinate system fixed in space, which is why the integral equations have $\partial/\partial t$ instead of d/dt. The charge density $\tilde{\rho}$ includes all free charges; the bound polarization charge density $\tilde{\rho}_{\mathrm{p}}$ is accounted for by our use of the macroscopic parameter ϵ. The current density $\overline{\mathscr{J}}$ in (7.21c) can be any free current, such as a source current (see Section 5.9.2) or a conduction current that flows due to the finite conductivity of the medium (i.e., $\overline{\mathscr{J}} = \sigma \overline{\mathscr{E}}$); the magnetization current $\overline{\mathscr{J}}_{\mathrm{m}}$ is accounted for by our use of the macroscopic parameter μ. The complete specification of the field quantities must also involve the

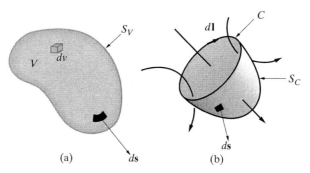

(a) $d\mathbf{s}$ (b)

Figure 7.19   Contour, surface, and volume. (a) A closed surface S_V enclosing a volume V. (b) A closed contour C enclosing a surface S_C.

---

[32]Note that until now $\mathbf{E}$ was used to represent the real electric field vector as a function of time and space, that is, $\mathbf{E}(x, y, z, t)$. From here on, however, $\mathbf{E}(x, y, z)$ is a phasor, and is related to the real field quantity $\overline{\mathscr{E}}$ as $\overline{\mathscr{E}}(x, y, z, t) = \mathfrak{Re}\{\mathbf{E}(x, y, z)e^{j\omega t}\}$, as discussed in Section 7.4.4.

constitutive relations, which in simple media (i.e., linear, isotropic, homogeneous media) are $\overline{\mathcal{H}} = \mu^{-1}\overline{\mathcal{B}}$ and $\overline{\mathcal{D}} = \epsilon\overline{\mathcal{E}}$.

It is possible to eliminate the vectors $\overline{\mathcal{D}}$ and $\overline{\mathcal{H}}$ from Maxwell's equations by substituting for them as follows:

$$\overline{\mathcal{D}} = \epsilon_0\overline{\mathcal{E}} + \overline{\mathcal{P}} \quad \text{and} \quad \overline{\mathcal{H}} = \frac{\overline{\mathcal{B}}}{\mu_0} - \overline{\mathcal{M}}$$

where $\overline{\mathcal{P}}$ is the polarization vector in a dielectric in units of coulombs-m^{-2}, and $\overline{\mathcal{M}}$ is the magnetization vector in a magnetic medium in units of ampere-m^{-1}. These two quantities account for the presence of matter at the points considered. Maxwell's equations then take the following form:

$$\nabla \times \overline{\mathcal{E}} = -\frac{\partial \overline{\mathcal{B}}}{\partial t}$$

$$\nabla \cdot \overline{\mathcal{E}} = \frac{1}{\epsilon_0}(\tilde{\rho} - \nabla \cdot \overline{\mathcal{P}})$$

$$\nabla \times \overline{\mathcal{B}} = \epsilon_0\mu_0\frac{\partial \overline{\mathcal{E}}}{\partial t} + \mu_0\left(\overline{\mathcal{J}} + \frac{\partial \overline{\mathcal{P}}}{\partial t} + \nabla \times \overline{\mathcal{M}}\right)$$

$$\nabla \cdot \overline{\mathcal{B}} = 0$$

The preceding equations are completely general but are expressed in a way that stresses the contributions of the medium. Note that the presence of matter has the effect of adding the bound volume charge density $-\nabla \cdot \overline{\mathcal{P}}$ (Section 4.10), the polarization current density $\partial\overline{\mathcal{P}}/\partial t$ (equation (7.18)), and the equivalent volume magnetization current density $\nabla \times \overline{\mathcal{M}}$ (Section 6.8).

Careful examination of (7.21) provides a qualitative understanding of the plausibility of electromagnetic wave propagation in space. Consider, for example, (7.21a) and (7.21c) in a nonconducting medium (i.e., $\sigma = 0$) and in the absence of sources (i.e., $\overline{\mathcal{J}}, \tilde{\rho} = 0$). According to (7.21a), any magnetic field $\overline{\mathcal{B}}$ that varies with time generates an electric field along a contour C surrounding it, as shown in Figure 7.20. On the other hand, according to (7.21c), this electric field $\overline{\mathcal{E}}$, which would typically vary in time because $\overline{\mathcal{B}}$ was taken to be varying with time, in turn generates a magnetic field along a contour C surrounding itself. This process indefinitely continues, as shown in

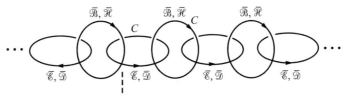

Figure 7.20 Plausibility of electromagnetic wave propagation as dictated by equations (7.21a) and (7.21c). Starting with a time-varying magnetic field at the location of the dashed line, electric and magnetic fields are successively generated at surrounding regions.

Figure 7.20. It thus appears that if we start with a magnetic field at one point in space and vary it with time, Maxwell's equations dictate that magnetic fields and electric fields are created at surrounding points—that is, that the disturbance initiated by the changing magnetic field propagates away from its point of origin, indicated by the dashed line in Figure 7.20. Note that although Figure 7.20 shows propagation toward the left and right, the same sequence of events takes place in all directions.

7.4.3 Transverse Electromagnetic (TEM) Waves

Maxwell's equations are listed in Section 7.4.2 in their most general forms, when the various field quantities may be full vectors with all three components finite (i.e., $\mathcal{E}_x$, $\mathcal{E}_y$, and $\mathcal{E}_z$ all nonzero) and may vary with all three spatial coordinates as well as with time (i.e., $\overline{\mathcal{E}} = \overline{\mathcal{E}}(x, y, z, t)$). However, in a variety of useful practical cases, Maxwell's equations reduce to simpler forms. Consider, for example, an electric field that has only one nonzero component (say, $\mathcal{E}_x$) and that varies only with one spatial coordinate (say, z). In other words, $\overline{\mathcal{E}} = \hat{\mathbf{x}}\mathcal{E}_x(z, t)$. In this case, the integral form of (7.21a) reduces to

$$\oint_C \hat{\mathbf{x}}\mathcal{E}_x \cdot d\mathbf{l} = -\int_{S_C} \frac{\partial \overline{\mathcal{B}}}{\partial t} \cdot d\mathbf{s} \tag{7.22}$$

Consider, for example, the contour C to be the square-shaped contour abcda in the x-z plane that encloses the surface S_1, as shown in Figure 7.21. Noting that the side lengths are $\Delta x = \Delta z$, we evaluate the left-hand side of (7.22) as

$$\int_{\text{abcda}} \hat{\mathbf{x}}\mathcal{E}_x \cdot d\mathbf{l} = \mathcal{E}_x(z, t) \, \Delta x + 0 - \mathcal{E}_x(z + \Delta z, t) \, \Delta x + 0 = -\frac{\partial \mathcal{E}_x}{\partial z} \Delta z \, \Delta x$$

assuming Δz and Δx are infinitesimally small. Since $\overline{\mathcal{B}}$ can be taken to be constant over such a differential surface element, the right-hand side of (7.22) reduces to

$$-\frac{\partial}{\partial t} \int_{S_1} \overline{\mathcal{B}} \cdot d\mathbf{s} = -\frac{\partial}{\partial t} \int_{S_1} \overline{\mathcal{B}} \cdot (-\hat{\mathbf{y}} \, ds) = \frac{\partial \mathcal{B}_y}{\partial t} \Delta x \, \Delta z$$

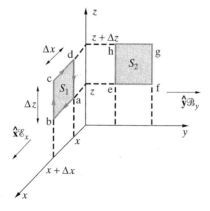

Figure 7.21 Transverse electromagnetic wave fields. Contours abcda and efghe used in the discussion of the one-dimensional version of Maxwell's equations for $\overline{\mathcal{E}} = \hat{\mathbf{x}}\mathcal{E}_x(z, t)$.

Note that only the $\mathcal{B}_y$ component contributes to the integral since the surface S_1 enclosed by contour abcda lies in the x-z plane. Also note that the surface element $d\mathbf{s}$ points in the $-y$ direction according to the right-hand rule; with the fingers of the right hand describing the sense of line integration around contour C, the thumb points in the direction of the differential surface element. Equating both sides once again, we have

$$-\frac{\partial \mathcal{E}_x}{\partial z} \Delta z\ \Delta x = \frac{\partial \mathcal{B}_y}{\partial t} \Delta x\ \Delta z \quad \rightarrow \quad \frac{\partial \mathcal{E}_x}{\partial z} = -\frac{\partial \mathcal{B}_y}{\partial t}$$

which is actually the reduced form of the differential form of equation (7.21a). Note that our choice of an electric field of the form $\overline{\mathcal{E}} = \hat{\mathbf{x}}\mathcal{E}_x(z,t)$ requires that the magnetic field $\overline{\mathcal{B}}$ have only a nonzero y component, so that $\overline{\mathcal{B}} = \hat{\mathbf{y}}\mathcal{B}_y(z,t)$.

   Assuming that the physical regions in consideration are far from the sources, so that $\overline{\mathcal{J}}, \tilde{\rho} = 0$, and assuming free space conditions so that $\overline{\mathcal{B}} = \mu_0 \overline{\mathcal{H}} = \hat{\mathbf{y}}\mu_0 \mathcal{H}_y(z,t)$ and $\overline{\mathcal{D}} = \epsilon_0 \overline{\mathcal{E}} = \hat{\mathbf{x}}\epsilon_0 \mathcal{E}_x(z,t)$, we now consider, in a similar fashion, the integral form of equation (7.21c) for the contour efghe in the y-z plane. We derive the corresponding differential form as

$$-\frac{\partial \mathcal{H}_y}{\partial z} = \epsilon_0 \frac{\partial \mathcal{E}_x}{\partial t}$$

Electromagnetic waves that have single, orthogonal vector electric and magnetic field components (e.g., $\mathcal{E}_x$ and $\mathcal{H}_y$), both varying with a single coordinate of space (e.g., z), are known as *uniform plane waves* or *transverse electromagnetic (TEM) waves*. These types of waves will be considered explicitly and extensively in Chapters 8 and 10. At this point, we note that the voltage and current waves on lossless transmission lines as discussed in Chapters 2 and 3 were also TEM waves. To see this, consider the similarity between the fundamental transmission line equations and the preceding derived expressions for $\mathcal{E}_x$ and $\mathcal{H}_y$:

$$\frac{\partial \mathcal{E}_x}{\partial z} = -\mu_0 \frac{\partial \mathcal{H}_y}{\partial t} \qquad \longleftrightarrow \qquad \frac{\partial \mathcal{V}}{\partial z} = -L \frac{\partial \mathcal{I}}{\partial t}$$

$$\frac{\partial \mathcal{H}_y}{\partial z} = -\epsilon_0 \frac{\partial \mathcal{E}_x}{\partial t} \qquad \longleftrightarrow \qquad \frac{\partial \mathcal{I}}{\partial z} = -C \frac{\partial \mathcal{V}}{\partial t}$$

   This correspondence between transmission line equations and the equations relating the electric and magnetic fields of transverse electromagnetic waves is more than a simple analogy. As we shall see in Chapter 10, the transmission line voltage and current waves that we worked with in Chapters 2 and 3 are indeed transverse electromagnetic waves and can also be analyzed solely in terms of electric and magnetic fields. Their treatment in terms of distributed circuits concepts and in terms of voltage and current waves is in fact a simplification of their more general behavior, which is best expressed in terms of field quantities. By the same token, the fact that the equations relating $\mathcal{E}_x$ to $\mathcal{H}_y$ and $\mathcal{V}$ to $\mathcal{I}$ are virtually identical indicates that we can expect transverse electromagnetic waves to exhibit the full range of behavior of transmission lines in terms of both transient and steady-state waves as analyzed, respectively, in Chapters 2 and 3. The association between transmission line behavior and transverse electromagnetic wave behavior will be further discussed in Chapter 10.

7.4.4 Time-Harmonic Maxwell's Equations

Numerous practical applications (e.g., broadcast radio and TV, radar, optical, and microwave applications) involve transmitting sources that operate in such a narrow band of frequencies that the behavior of all the field components is very similar to that of the central single-frequency sinusoid (i.e., the carrier). Most generators also produce sinusoidal voltages and currents, and hence electric and magnetic fields that vary sinusoidally with time. In many applications, the transients involved at the time the signal is switched on (or off) are not of concern, so the steady-state sinusoidal approximation is most suitable. For example, for an AM broadcast station operating at a carrier frequency of 1 MHz, any turn-on transients would last only a few μs and are of little consequence to the practical application. For all practical purposes, the signal propagating from the transmitting antenna to the receivers can be treated as a sinusoid, with its amplitude modulated within a narrow bandwidth (e.g., ± 5 kHz) around the carrier frequency. Since the characteristics of the propagation medium do not vary significantly over this bandwidth, we can describe the propagation behavior of the AM broadcast signal by studying a single sinusoidal carrier at a frequency of 1 MHz.

The time-harmonic (sinusoidal steady-state) forms of Maxwell's equations[33] are listed here with their more general versions:

$$\nabla \times \overline{\mathscr{E}} = -\frac{\partial \overline{\mathscr{B}}}{\partial t} \qquad\qquad \nabla \times \mathbf{E} = -j\omega\mathbf{B} \qquad (7.23a)$$

$$\nabla \cdot \overline{\mathscr{D}} = \tilde{\rho} \qquad\qquad \nabla \cdot \mathbf{D} = \rho \qquad (7.23b)$$

$$\nabla \times \overline{\mathscr{H}} = \overline{\mathscr{J}} + \frac{\partial \overline{\mathscr{D}}}{\partial t} \qquad\qquad \nabla \times \mathbf{H} = \mathbf{J} + j\omega\mathbf{D} \qquad (7.23c)$$

$$\nabla \cdot \overline{\mathscr{B}} = 0 \qquad\qquad \nabla \cdot \mathbf{B} = 0 \qquad (7.23d)$$

Note that in (7.23a)–(7.23d), $\overline{\mathscr{E}}$, $\overline{\mathscr{D}}$, $\overline{\mathscr{H}}$, and $\overline{\mathscr{B}}$ are real (measurable) quantities that can vary with time, whereas the vectors $\mathbf{E}$, $\mathbf{D}$, $\mathbf{H}$, and $\mathbf{B}$ are complex phasors that do not vary with time. In general, we can obtain the former from the latter by multiplying by $e^{j\omega t}$ and taking the real part. For example,

$$\overline{\mathscr{E}}(x,y,z,t) = \mathscr{R}e\{\mathbf{E}(x,y,z)e^{j\omega t}\}$$

Note that the same is true for all of the quantities. For example,

$$\tilde{\rho}(x,y,z,t) = \mathscr{R}e\{\rho(x,y,z)e^{j\omega t}\}$$

---

[33]The actual derivation of the time-harmonic form, for example, for (7.21a) is as follows:

$$\nabla \times \overline{\mathscr{E}} = -\frac{\partial \overline{\mathscr{B}}}{\partial t} \quad\rightarrow\quad \nabla \times [\underbrace{\mathscr{R}e\{\mathbf{E}(x,y,z)e^{j\omega t}\}}_{\overline{\mathscr{E}}}] = -\frac{\partial}{\partial t}[\underbrace{\mathscr{R}e\{\mathbf{B}(x,y,z)e^{j\omega t}\}}_{\overline{\mathscr{B}}}]$$

$$\rightarrow\quad \mathscr{R}e\{e^{j\omega t}\nabla \times \mathbf{E}\} = \mathscr{R}e\{-j\omega e^{j\omega t}\mathbf{B}\} \quad\rightarrow\quad \nabla \times \mathbf{E} = -j\omega\mathbf{B}$$

Example 7.13: Aircraft VHF communication signal. The electric field component of an electromagnetic wave in air used by an aircraft to communicate with the air traffic control tower can be represented by

$$\overline{\mathscr{E}}(z,t) = \hat{\mathbf{y}}0.02\cos(7.5 \times 10^8 t - \beta z) \quad \text{V-m}^{-1}$$

Find the corresponding wave magnetic field $\overline{\mathscr{H}}(z,t)$ and the constant β.

Solution: In view of the sinusoidal nature of the electric field signal, it is appropriate to work with phasors. The phasor form of the electric field is

$$\mathbf{E}(z) = \hat{\mathbf{y}}E_y(z) = \hat{\mathbf{y}}0.02e^{-j\beta z} \quad \text{V-m}^{-1}$$

Using (7.23a), we can write

$$\mathbf{H}(z) = -\frac{1}{j\omega\mu_0}\nabla \times \mathbf{E}(z) = \hat{\mathbf{x}}\frac{1}{j\omega\mu_0}\frac{\partial E_y}{\partial z}$$

$$= -\hat{\mathbf{x}}\frac{\beta}{\omega\mu_0}E_y = -\hat{\mathbf{x}}\frac{0.02\beta}{\omega\mu_0}e^{-j\beta z}$$

where $\omega = 7.5 \times 10^8$ rad-s^{-1} and $\mu_0 = 4\pi \times 10^{-7}$ H-m^{-1}. Substituting the expression for $\mathbf{H}(z)$ into (7.23c), we find

$$\mathbf{E}(z) = \frac{1}{j\omega\epsilon_0}\nabla \times \mathbf{H}(z) = \hat{\mathbf{y}}\frac{1}{j\omega\epsilon_0}\frac{\partial H_x}{\partial z} = \hat{\mathbf{y}}\frac{0.02\beta^2}{\omega^2\mu_0\epsilon_0}e^{-j\beta z}$$

where $\epsilon_0 = 8.85 \times 10^{-12}$ F-m^{-1}. But this expression for $\mathbf{E}(z)$ must be the same as the electric field phasor expression we started with. Thus, we must have

$$\frac{0.02\beta^2}{\omega^2\mu_0\epsilon_0} = 0.02 \quad \rightarrow \quad \beta = \omega\sqrt{\mu_0\epsilon_0} = \frac{7.5 \times 10^8 \text{ rad-s}^{-1}}{3 \times 10^8 \text{ m-s}^{-1}} = 2.5 \text{ rad-m}^{-1}$$

where we have used the fact that $(\mu_0\epsilon_0)^{-1/2} = c$, the speed of light in free space.
     The corresponding magnetic field $\mathbf{H}$ is then

$$\mathbf{H}(z) = -\hat{\mathbf{x}}\frac{(0.02)(2.5)}{(7.5 \times 10^8)(4\pi \times 10^{-7})}e^{-j2.5z} \simeq -\hat{\mathbf{x}}53.1 \times e^{-j2.5z} \quad \mu\text{A-m}^{-1}$$

and the instantaneous magnetic field $\overline{\mathscr{H}}(z,t)$ is

$$\overline{\mathscr{H}}(z,t) = \mathscr{R}e\{\mathbf{H}(z)e^{j\omega t}\} = \mathscr{R}e\{-\hat{\mathbf{x}}53.1e^{-j2.5z}e^{j\omega t}\}$$

$$= -\hat{\mathbf{x}}53.1\cos(7.5 \times 10^8 t - 2.5z) \quad \mu\text{A-m}^{-1}$$

7.5 REVIEW OF MAXWELL'S EQUATIONS

A brief review of Maxwell's equations[34] and their underlying foundations is now in order, since their development was spread over Chapters 4 through 7. We have established these

[34]J. C. Maxwell, *A Treatise in Electricity and Magnetism*, Clarendon Press, Oxford, 1892, Vol. 2, pp. 247–262.

four fundamental equations of electromagnetics on the basis of three separate experimentally established facts, namely, Coulomb's law,[35] Ampère's law[36] (or the Biot–Savart law), Faraday's law,[37] and the principle of conservation of electric charge. The validity of Maxwell's equations is based on their consistency with all of our experimental knowledge to date concerning electromagnetic phenomena. The physical meaning of the equations is better perceived in the context of their integral forms, which are listed below together with their differential counterparts:

(1) Faraday's law is based on the experimental fact that time-changing magnetic flux induces electromotive force:

$$\oint_C \overline{\mathscr{E}} \cdot d\mathbf{l} = -\int_{S_C} \frac{\partial \overline{\mathscr{B}}}{\partial t} \cdot d\mathbf{s} \qquad \nabla \times \overline{\mathscr{E}} = -\frac{\partial \overline{\mathscr{B}}}{\partial t} \qquad (7.21a)$$

where the contour C is that which encloses the surface S_C, as shown in Figure 7.19b and where the sense of the line integration over the contour C (i.e., $d\mathbf{l}$) must be consistent with the direction of the surface vector $d\mathbf{s}$ in accordance with the right-hand rule.

(2) Gauss's law is a mathematical expression of the experimental fact that electric charges attract or repel one another with a force inversely proportional to the square of the distance between them (i.e., Coulomb's law):

$$\oint_{S_V} \overline{\mathscr{D}} \cdot d\mathbf{s} = \int_V \tilde{\rho} \, dv \qquad \nabla \cdot \overline{\mathscr{D}} = \tilde{\rho} \qquad (7.21b)$$

where the surface S_V encloses the volume V, as shown in Figure 7.19a. The free volume charge density is represented with $\tilde{\rho}$ to distinguish it from the phasor form ρ used in the time-harmonic form of Maxwell's equations.

(3) Maxwell's third equation is a generalization of Ampère's law, which states that the line integral of the magnetic field over any closed contour must equal the total current enclosed by that contour:

$$\oint_C \overline{\mathscr{H}} \cdot d\mathbf{l} = \int_{S_C} \overline{\mathscr{J}} \cdot d\mathbf{s} + \int_{S_C} \frac{\partial \overline{\mathscr{D}}}{\partial t} \cdot d\mathbf{s} \qquad \nabla \times \overline{\mathscr{H}} = \overline{\mathscr{J}} + \frac{\partial \overline{\mathscr{D}}}{\partial t} \qquad (7.21c)$$

where the contour C is that which encloses the surface S_C, as shown in Figure 7.19b. Maxwell's third equation expresses the fact that time-varying electric fields produce magnetic fields. The first term of this equation (also referred

[35]Electric charges attract or repel one another in a manner inversely proportional to the square of the distance between them; C. A. de Coulomb, Première mémoire sur l'électricité et magnétisme (First memoir on electricity and magnetism), *Histoire de l'Académie Royale des Sciences*, 1785, p. 569.

[36]Current-carrying wires create magnetic fields and exert forces on one another, with the intensity of the magnetic field (and thus the force) depending on the inverse square of the distance; A.-M. Ampère, *Recueil d'observations électrodynamiques*, Crochard, Paris, 1820–1833.

[37]Magnetic fields that change with time induce electromotive force; M. Faraday, *Experimental Researches in Electricity*, Taylor and Francis, London, Vol. I, 1839, pp. 1–109.

to as the conduction-current term) is Ampère's law, which is a mathematical statement of the experimental findings of Oersted, whereas the second term, known as the displacement-current term, was introduced theoretically by Maxwell in 1862 and verified experimentally many years later in Hertz's experiments.[38]

(4) Maxwell's fourth equation is based on the fact that there are no magnetic charges (i.e., magnetic monopoles) and that, therefore, magnetic field lines always close on themselves:

$$\oint_{S_V} \overline{\mathscr{B}} \cdot d\mathbf{s} = 0 \qquad \nabla \cdot \overline{\mathscr{B}} = 0 \qquad (7.21\mathrm{d})$$

where the surface S_V encloses the volume V, as shown in Figure 7.19a. As we have seen in Section 6.7, this equation can actually be derived from the Biot–Savart law, so it is not completely independent.[39]

The two constitutive relations $\overline{\mathscr{D}} = \epsilon \overline{\mathscr{E}}$ and $\overline{\mathscr{B}} = \mu \overline{\mathscr{H}}$ (more properly expressed as $\overline{\mathscr{H}} = \mu^{-1} \overline{\mathscr{B}}$, because in the magnetostatic case $\overline{\mathscr{H}}$ is the medium-independent quantity) govern the manner by which the electric and magnetic fields, $\overline{\mathscr{E}}$ and $\overline{\mathscr{B}}$, are related to the medium-independent quantities, $\overline{\mathscr{D}}$ and $\overline{\mathscr{H}}$, in material media, where, in general, $\epsilon \neq \epsilon_0$ and $\mu \neq \mu_0$. In general, the free current density $\overline{\mathscr{J}}$ consists of conduction currents ($\overline{\mathscr{J}}_c = \sigma \overline{\mathscr{E}}$), source currents ($\overline{\mathscr{J}}_{\text{source}}$), or any other source of charge transfer excluding the polarization and the bound magnetization current densities. The free charge density $\tilde{\rho}$ represents all charge density excluding the bound polarization charge density. The free current density $\overline{\mathscr{J}}$ and the free volume charge density $\tilde{\rho}$ represent the sources from which magnetic and electric fields originate, respectively.

Note that ϵ, μ, and σ are macroscopic parameters that describe the relationships among macroscopic field quantities, but they are based on the microscopic behavior of the atoms and molecules in response to the fields. These parameters are simple constants only for *simple* material media, which are linear, homogeneous, time-invariant, and isotropic. Otherwise, for complex material media that are nonlinear, inhomogeneous, time-variant, or anisotropic, ϵ, μ, and σ may depend on the magnitudes of $\overline{\mathscr{E}}$ and $\overline{\mathscr{B}}$

---

[38] H. Hertz, On the finite velocity of propagation of electromagnetic actions, *Sitzb. d. Berl. Akad. d. Wiss.*, Feb 2, 1888; for a collected English translation of this and other papers by H. Hertz, see H. Hertz, *Electric Waves,* MacMillan, London, 1893.

[39] It is also interesting that (7.21d) can be derived from (7.21a) by taking the divergence of the latter and using the vector identity of $\nabla \cdot (\nabla \times \mathbf{G}) \equiv 0$, which is true for any vector $\mathbf{G}$. We find

$$\nabla \cdot (\nabla \times \overline{\mathscr{E}}) = -\nabla \cdot \left(\frac{\partial \overline{\mathscr{B}}}{\partial t} \right) \quad \rightarrow \quad 0 = -\frac{\partial (\nabla \cdot \overline{\mathscr{B}})}{\partial t} \quad \rightarrow \quad \text{const.} = \nabla \cdot \overline{\mathscr{B}}$$

The constant can then be shown to be zero by the following argument. If we suppose that the $\overline{\mathscr{B}}$ field was produced a finite time ago, that is, it has not always existed, then, if we go back far enough in time, we have $\overline{\mathscr{B}} = 0$ and therefore $\nabla \cdot \overline{\mathscr{B}} = 0$. Hence, it would appear that

$$\nabla \cdot \overline{\mathscr{B}} = 0 \quad \text{and} \quad \int_S \overline{\mathscr{B}} \cdot d\mathbf{s} = 0$$

(nonlinear), on spatial coordinates (x, y, z) (inhomogeneous), on time (time-variant), or on the orientations of $\overline{\mathcal{E}}$ and $\overline{\mathcal{H}}$ (anisotropic). For anisotropic media, ϵ, μ, or σ is generally expressed as a matrix (called a *tensor*) whose entries relate each component (e.g., the x, y, or z component) of $\overline{\mathcal{E}}$ (or $\overline{\mathcal{H}}$) to the other three components (e.g., the x, y, and z components) of $\overline{\mathcal{D}}$ or $\overline{\mathcal{J}}$ (or $\overline{\mathcal{B}}$). In ferromagnetic materials, the magnetic field $\overline{\mathcal{B}}$ is determined by the past history of the field $\overline{\mathcal{H}}$ rather than by its instantaneous value; such substances are said to exhibit *hysteresis*. Some hysteresis effects can also be seen in certain dielectric materials.

The continuity equation, which expresses the principle of conservation of charge in differential form, is contained in Maxwell's equations and in fact can be readily derived[40] by taking the divergence of (7.21c) and using (7.21b). For the sake of completeness, we give the integral and differential forms of the continuity equation:

$$-\oint_{S_V} \overline{\mathcal{J}} \cdot d\mathbf{s} = \frac{\partial}{\partial t} \int_V \tilde{\rho} \, dv \qquad \nabla \cdot \overline{\mathcal{J}} = -\frac{\partial \tilde{\rho}}{\partial t} \tag{7.24}$$

where the surface S_V encloses the volume V, as shown in Figure 7.19a.

Note that for all of the equations (7.21a) through (7.21d) and equation (7.24), the differential forms can be derived from the integral forms (or vice versa) by using either Stokes's or the divergence theorem, both of which are valid for any arbitrary vector field $\overline{\mathcal{G}}$. These theorems are

$$\oint_C \overline{\mathcal{G}} \cdot d\mathbf{l} = \int_{S_C} (\nabla \times \overline{\mathcal{G}}) \cdot d\mathbf{s} \qquad \text{(Stokes's theorem)} \tag{7.25}$$

where the contour C encloses the surface S_C, and

$$\oint_{S_V} \overline{\mathcal{G}} \cdot d\mathbf{s} = \int_V (\nabla \cdot \overline{\mathcal{G}}) \, dv \qquad \text{(divergence theorem)} \tag{7.26}$$

where the surface S_V encloses volume V.

Electromagnetic boundary conditions. The integral forms of equations (7.21) can be used to derive the relationships between electric- and magnetic-field components on both sides of interfaces between two different materials (i.e., different μ, ϵ, or σ).

The electromagnetic boundary conditions can be summarized as follows:

(1) It follows from applying (7.21a) to a contour C, as shown in Figure 7.22b, that the tangential component of the electric field $\overline{\mathcal{E}}$ is continuous across any interface:

$$\hat{\mathbf{n}} \times [\overline{\overline{\mathcal{E}}}_1 - \overline{\overline{\mathcal{E}}}_2] = 0 \qquad \rightarrow \qquad \mathcal{E}_{1t} = \mathcal{E}_{2t} \tag{7.27}$$

where $\hat{\mathbf{n}}$ is the unit vector perpendicular to the interface and outward from medium 2 (shown as $\hat{\mathbf{n}}$ in Figure 7.22b).

---

[40]That the continuity equation can be derived from equations (7.21b) and (7.21c) indicates that Maxwell's equations (7.21b) and (7.21c) are not entirely independent, if we accept conservation of electric charge as a fact.

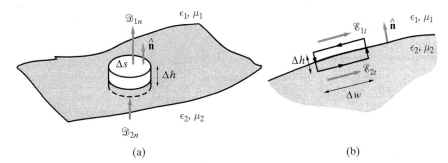

Figure 7.22   Interfaces between two different materials. The boundary conditions for the electromagnetic fields are derived by applying the surface integrals to the cylindrical surface as shown in (a) and the line integrals to the rectangular contour in (b).

(2) It can be shown by applying (7.21c) to the same contour C in Figure 7.22b that the tangential component of the magnetic field $\overline{\mathcal{H}}$ is continuous across any interface:

$$\hat{\mathbf{n}} \times [\overline{\mathcal{H}}_1 - \overline{\mathcal{H}}_2] = 0 \qquad \rightarrow \qquad \mathcal{H}_{1t} = \mathcal{H}_{2t} \tag{7.28}$$

except where surface currents ($\mathbf{J}_s$) may exist, such as at the surface of a perfect conductor (i.e., $\sigma = \infty$):

$$\hat{\mathbf{n}} \times \overline{\mathcal{H}}_1 = \overline{\mathcal{J}}_s$$

Noting that the field $\overline{\mathcal{H}}_2$ inside the perfect conductor is zero, as discussed in Section 6.8.4.

(3) It can be shown by applying (7.21b) to the surface of the cylinder shown in Figure 7.22a that the normal component of electric flux density $\overline{\mathcal{D}}$ is continuous across interfaces, except where surface charge ($\tilde{\rho}_s$) may exist, such as at the surface of a metallic conductor or at the interface between two lossy dielectrics ($\sigma \neq 0$):

$$\hat{\mathbf{n}} \cdot (\overline{\mathcal{D}}_1 - \overline{\mathcal{D}}_2) = \tilde{\rho}_s \qquad \rightarrow \qquad \mathcal{D}_{1n} - \mathcal{D}_{2n} = \tilde{\rho}_s \tag{7.29}$$

(4) A consequence of applying (7.21d) to the surface of the cylinder in Figure 7.22a is that the normal component of the magnetic field $\overline{\mathcal{B}}$ is continuous across interfaces:

$$\hat{\mathbf{n}} \cdot [\overline{\mathcal{B}}_1 - \overline{\mathcal{B}}_2] = 0 \qquad \rightarrow \qquad \mathcal{B}_{1n} = \mathcal{B}_{2n} \tag{7.30}$$

(5) It follows from applying (7.24) to the surface of the cylinder in Figure 7.22a that, at the interface between two lossy media (i.e., $\sigma_1 \neq 0$, $\sigma_2 \neq 0$), the normal component of the current density $\overline{\mathcal{J}}$ is continuous, except where time-varying surface charge may exist, such as at the surface of a perfect conductor or at the interface between lossy dielectrics ($\epsilon_1 \neq \epsilon_2$ and $\sigma_1 \neq \sigma_2$):

$$\hat{\mathbf{n}} \cdot (\overline{\mathcal{J}}_1 - \overline{\mathcal{J}}_2) = -\frac{\partial \tilde{\rho}_s}{\partial t} \qquad \rightarrow \qquad \mathcal{J}_{1n} - \mathcal{J}_{2n} = -\frac{\partial \tilde{\rho}_s}{\partial t} \tag{7.31}$$

Note that (7.31) is not completely independent of (7.27) through (7.30), since (7.24) is contained in (7.21a) through (7.21d), as mentioned previously. For the stationary case ($\partial/\partial t = 0$), (7.31) implies $\mathcal{J}_{1n} = \mathcal{J}_{2n}$, or $\sigma_1 \mathcal{E}_{1n} = \sigma_2 \mathcal{E}_{2n}$, which means that at the interface between lossy dielectrics (i.e., $\epsilon_1 \neq \epsilon_2$ and $\sigma_1 \neq \sigma_2$) there must in general be finite surface charge (i.e., $\tilde{\rho}_s \neq 0$) because otherwise (7.29) demands that $\epsilon_1 \mathcal{E}_{1n} = \epsilon_2 \mathcal{E}_{2n}$.

7.6 SUMMARY

This chapter discussed the following topics:

- **Faraday's law.** When the magnetic flux enclosed by a loop of wire changes with time, an electromotive force is induced, which is given by

$$\mathcal{V}_{\text{ind}} = \oint_C \mathbf{E} \cdot d\mathbf{l} = -\frac{d}{dt} \int_S \mathbf{B} \cdot d\mathbf{s}$$

  where S is the surface enclosed by the contour C and where the orientations of the two integrals are related by the right-hand rule. The polarity of the induced emf, determined by Lenz's law, opposes the change in magnetic flux that causes it. Faraday's law states that an electric field is induced by a time-varying magnetic field in space, regardless of whether wires are present. The direction of the induced electric field is also determined by Lenz's law.

- **Induction due to motion.** When conductors move in the presence of magnetic fields, an induced voltage is produced even if the magnetic fields do not vary in time. The general expression for the induced electromotive force is given by

$$\mathcal{V}_{\text{ind}} = -\int_S \frac{\partial \mathbf{B}}{\partial t} \cdot d\mathbf{s} + \oint_C (\mathbf{v} \times \mathbf{B}) \cdot d\mathbf{l}$$

  where the contour C encloses the surface S and where $d\mathbf{l}$ and $d\mathbf{s}$ are related via the right-hand rule. These two components represent the two physical mechanisms for generating an emf represented by the general form of Faraday's law.

- **Differential form of Faraday's law.** The differential form of Faraday's law in a stationary frame of reference is

$$\nabla \times \mathbf{E} = -\frac{\partial \mathbf{B}}{\partial t}$$

  which describes the fact that any time-varying magnetic field in space is encircled by an electric field $\mathbf{E}$.

- **Energy in a magnetic field.** The self-energy of a current-carrying loop is given by

$$W_{\text{ms}} = \tfrac{1}{2} \int_V \mathbf{J} \cdot \mathbf{A} \, dv$$

  where $\mathbf{A}$ is the magnetic vector potential and $\mathbf{J}$ is the free volume current density. This expression reduces to the familiar expression for the energy stored in circuit with inductance L, namely, $(\tfrac{1}{2})LI^2$.

The magnetic energy stored by two current-carrying loops is given by

$$W_{\mathrm{m}} = \tfrac{1}{2}L_{11}I_1^2 \pm L_{12}I_1I_2 + \tfrac{1}{2}L_{22}I_2^2$$

where L_{12} is the mutual inductance between the two circuits.

The magnetostatic energy stored in a distribution of magnetic fields is given by

$$W_{\mathrm{m}} = \tfrac{1}{2}\int_V \mathbf{H} \cdot \mathbf{B}\, dv$$

where V is the entire volume in which the magnetic field is nonzero. The volume energy density of the magnetostatic field for a linear and isotropic medium is $w_{\mathrm{m}} = \tfrac{1}{2}\mu H^2$.

- **Displacement current.** Maxwell developed the time-varying form of Ampère's law by postulating the displacement-current term such that

$$\nabla \times \mathbf{H} = \mathbf{J} + \frac{\partial \mathbf{D}}{\partial t}$$

The displacement current term makes Ampère's law consistent with the law of conservation of charge and indicates that time-varying electric fields produce magnetic fields. Together with Faraday's law, the $\partial \mathbf{D}/\partial t$ term makes it possible for electromagnetic waves to propagate through free space, a prediction confirmed in 1887 by Hertz's experiments.

- **Maxwell's equations.** The general form of Maxwell's equations are:

$$\nabla \times \overline{\mathscr{E}} = -\frac{\partial \overline{\mathscr{B}}}{\partial t}$$

$$\nabla \cdot \overline{\mathscr{E}} = \frac{1}{\epsilon_0}\left(\tilde{\rho} - \nabla \cdot \overline{\mathscr{P}}\right)$$

$$\nabla \times \overline{\mathscr{B}} = \mu_0\left(\overline{\mathscr{J}} + \nabla \times \overline{\mathscr{M}} + \frac{\partial \overline{\mathscr{P}}}{\partial t} + \epsilon_0\frac{\partial \overline{\mathscr{E}}}{\partial t}\right)$$

$$\nabla \cdot \overline{\mathscr{B}} = 0$$

The total volume charge density includes the free ($\tilde{\rho}$) and the bound polarization ($-\nabla \cdot \overline{\mathscr{P}}$) charge densities. The total current density includes the free volume current density $\overline{\mathscr{J}}$ (e.g., conduction and/or source currents) and the bound magnetization current density ($\nabla \times \overline{\mathscr{M}}$); the displacement current density $\overline{\mathscr{J}}_{\mathrm{d}}$ includes the polarization current density ($\partial \overline{\mathscr{P}}/\partial t$) and $\epsilon_0 \partial \overline{\mathscr{E}}/\partial t$.

It is possible to eliminate the polarization vector $\overline{\mathscr{P}}$ and magnetization vector $\overline{\mathscr{M}}$ from Maxwell's equations by introducing the electric displacement $\overline{\mathscr{D}}$ and $\overline{\mathscr{H}}$-field terms

$$\overline{\mathscr{D}} = \epsilon_0\overline{\mathscr{E}} + \overline{\mathscr{P}} \quad \text{and} \quad \overline{\mathscr{H}} = \frac{\overline{\mathscr{B}}}{\mu_0} - \overline{\mathscr{M}}$$

With these definitions, we may write Maxwell's equations in terms of the free volume charge density $\tilde{\rho}$ and free volume current density $\overline{\tilde{\mathcal{J}}}$ as

$$\nabla \times \overline{\mathcal{E}} = -\frac{\partial \overline{\mathcal{B}}}{\partial t}$$

$$\nabla \cdot \overline{\mathcal{D}} = \tilde{\rho}$$

$$\nabla \times \overline{\mathcal{H}} = \overline{\tilde{\mathcal{J}}} + \frac{\partial \overline{\mathcal{D}}}{\partial t}$$

$$\nabla \cdot \overline{\mathcal{B}} = 0$$

In addition, a complete specification of the field quantities involves the constitutive relations, which in simple media (i.e., linear, isotropic, homogeneous media) are $\overline{\mathcal{H}} = \mu^{-1}\overline{\mathcal{B}}$ and $\overline{\mathcal{D}} = \epsilon\overline{\mathcal{E}}$.

PROBLEMS

7.1 **Stationary rectangular loop.** Consider a fixed single-turn rectangular loop of area A with its plane perpendicular to a uniform magnetic field. Find the voltage induced across the terminals of the loop if the magnetic flux density is given by (a) $B(t) = 2B_0 t e^{-\alpha t}$ and (b) $B(t) = B_0 e^{-2\alpha t} \sin(\omega t)$.

7.2 **Stationary circular loop.** Consider a 20-turn circular loop of wire of 15 cm diameter with its plane perpendicular to a uniform magnetic field, as shown in Figure 7.6. If the magnetic flux density B is given by

$$B = 10\cos(120\pi t) \quad \text{G}$$

find the rms value of the induced current through the 10Ω resistor at the following time instants: (a) $t = 0$, (b) $t = 10$ ms, (c) $t = 100$ ms, and (d) $t = 1$ s.

7.3 **Two circular coils.** Two circular coils of 5 cm radius each have the same axis of symmetry and are 1 m apart from each other. Coil 1 has 20 turns and coil 2 has 150 turns. (a) If an alternating current of 100 A amplitude and 1 kHz frequency is passed through coil 1, find the amplitude of the induced voltage across the open terminals of coil 2. (b) Repeat part (a) for a frequency of 10 kHz.

7.4 **Two concentric coils.** Consider the two concentric coils shown in Figure 7.9. The magnetic flux density produced at the center of the two coils due to the current I flowing in the larger coil is given by

$$\mathbf{B} = \hat{z}2.5 \times 10^{-5}I \text{ T}$$

(a) If the larger coil has 30 turns, find its radius. (b) If the smaller coil has 75 turns and its radius is 1 cm, find the total flux linking the smaller coil. (c) If the current I in the larger coil is given by $I(t) = 10\cos(120\pi t)$ A, find the induced current through the resistor R assuming $R = 10\Omega$.

7.5 **Triangular loop and long wire.** An equilateral triangular loop is situated near a long current-carrying wire, as shown in Figure 7.23. The wire is part of a power line carrying

60-Hz sinusoidal current. An ac ammeter inserted in the loop reads a current of amplitude 1 mA. Assume the total resistive impedance of the loop to be 0.01Ω. Find the amplitude of the current I in the long wire.

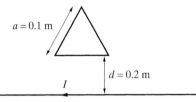

$a = 0.1$ m

$d = 0.2$ m

I

Figure 7.23   Triangular loop and long wire. Problem 7.5.

7.6   **Faraday's law.** Consider a circular loop (radius $a = 5$ cm) of wire lying in the x-y plane with its center at the origin, in the presence of a z-directed magnetic field

$$\mathbf{B}(r, t) = \hat{\mathbf{z}} B_0 \left[1 - (10 \text{ m}^{-1})r\right] \cos(2\pi f\, t)$$

where r is the polar coordinate and $B_0 = 10$ mT (milliwebers/m^2). The loop wire is made up of copper ($\sigma = 5.8 \times 10^7$ S/m) and has a cross-sectional area of 2 mm^2. It is known that the wire would melt if the total current I flowing through it exceeds 25 A ($I > 25$ A). Stating all assumptions, calculate the maximum frequency f for which this wire can be used in the presence of this magnetic field.

7.7   **Toroidal coil around a long, straight wire.** A long, straight wire carrying an alternating current of $I(t) = 100 \cos(377t)$ A coincides with the principal axis of symmetry of a 200-turn coil wrapped uniformly around a rectangular, toroidal-shaped iron core of inner and outer radii $a = 6$ cm and $b = 8$ cm, thickness $t = 3$ cm, and relative permeability $\mu_r = 1000$, as shown in Figure 7.24. Calculate the induced voltage $\mathcal{V}_{\text{ind}}$ between the terminals of the coil. (This type of coil, called a *current transformer*, is used to measure the current in a conductor wire that passes through it.)

200-turn coil

I Iron core

$+$

$\mathcal{V}_{\text{ind}}$ a t

$-$ b

$\mu_r = 1000$

I Wire

Figure 7.24   Toroidal coil around long wire. Problem 7.7.

7.8   **Current transformer.** A current transformer is used to measure the current in a high-voltage transmission line, as shown in Figure 7.25. The circular toroidal core has a mean diameter of 6 cm, circular cross section of 1 cm diameter, and relative permeability of $\mu_r = 200$. The winding consists of $N = 300$ turns. If the 60-Hz current of amplitude 1000 A flows through the high-voltage line, find the rms value of the open circuit voltage induced across the terminals of the toroid.

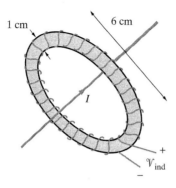

Figure 7.25 A current transformer. Problem 7.8.

7.9 **Sliding bar in a constant magnetic field.** A metal bar able to move on fixed rails is initially at rest on two stationary conducting rails that are $l = 50$ cm apart and connected to each other via a $V_0 = 10$ V voltage source in series with a $R = 10\Omega$ resistor, as shown in Figure 7.26. A constant magnetic field of $\mathbf{B}_0 = \hat{z}1.5$ T is turned on at $t = 0$. Assume the rails and the metal bar to be perfectly conducting and neglect the self-inductance of the loop formed by the rails and the bar. The initial resting position of the bar is at $y_0 = 75$ cm. (a) With no friction and assuming that the mass of the bar is 0.5 kg, calculate the magnitudes and directions of its initial acceleration and its final velocity. Explain your reasoning. (b) Describe what happens if the voltage source is turned off (i.e., $V_0 = 0$) at $t = t_0$, when the bar has reached a velocity v_0. (c) What happens if the voltage source remains at V_0 but the magnetic field is turned off (i.e., $\mathbf{B} = 0$) at $t = t_0$?

Figure 7.26 Sliding bar in a constant magnetic field. Problem 7.9.

7.10 **Oscillating bar in a time-varying magnetic field.** A metal wire is in contact with two long, parallel, stationary conducting rails connected with a resistance R at one end, as shown in Figure 7.27. The wire oscillates symmetrically around $x = x_0$ with a velocity $\mathbf{v}(t) = \hat{x}v_0 \cos(\omega_0 t)$ in the presence of a time-varying magnetic field that is perpendicular to the plane of the rails. The magnetic field is given by $\mathbf{B}(t) = \hat{z}B_0 \sin(\omega_0 t)$. Find the induced current through the resistance R.

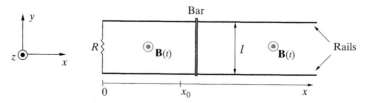

Figure 7.27 Oscillating bar in a time-varying magnetic field. Problem 7.10.

7.11 **Moving square loop.** A uniform magnetic field of $B_0 = 1$ T is confined to a square-shaped area 10 cm to a side as shown in Figure 7.28, with zero magnetic field everywhere else. A square loop of side length 5 cm moves through the loop with a velocity of $v = 10$ cm-s^{-1}. Find the electromotive force induced in the square loop and plot its value as a function of distance x, for $0 \leq x \leq 15$ cm. Note that the plane of the loop is orthogonal to the magnetic field.

Figure 7.28 Moving square loop. Problem 7.11.

7.12 **Rectangular loop and long wire.** A rectangular loop of wire lies in the plane of a long, straight wire carrying a steady current I, as shown in Figure 7.29. If the loop is moving radially away from the long wire with a velocity $\mathbf{v} = \hat{\mathbf{r}} v_0$ without changing its shape, find the voltage induced across the open terminals of the loop with its polarity indicated.

Figure 7.29 Rectangular loop and long wire. Problem 7.12.

7.13 **Rotating wire in a constant magnetic field.** A semicircular-shaped wire of radius a is rotated in a constant magnetic field B_0 at a constant angular frequency ω, as shown in Figure 7.30. If the wire forms a closed loop with a resistance R, find the induced current in R. Neglect the resistance of the wires.

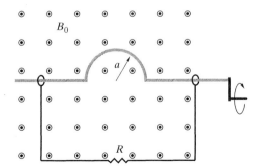

Figure 7.30 Rotating wire in constant magnetic field. Problem 7.13.

7.14 **Rotating rectangular loop in a constant magnetic field.** A single turn rectangular coil of dimensions 5 cm × 10 cm is rotated at 1800 rpm in a uniform magnetic field of 50 mT normal to the axis of rotation, as shown in Figure 7.13. Find the induced voltage across the open terminals of the loop.

7.15 **Rotating rectangular loop in a nonuniform magnetic field.** Consider the rectangular coil of 10 turns with dimensions 20 cm × 10 cm, rotating at an angular velocity of 2000 rpm about its axis, as shown in Figure 7.31, in the presence of a magnetic field. If the **B** field vector is in the cylindrical radial direction, which only varies with the azimuthal angle ϕ as $\mathbf{B} = \hat{\mathbf{r}} 3 \cos \phi$ mT, find the voltage $\mathcal{V}_{ab}$ induced across the open terminals of the coil.

Figure 7.31   Rotating rectangular loop.
Problem 7.15

7.16 **Extracting power from a power line.** Consider the possibility of putting your electromagnetics knowledge to practical use. You live out in the country and suddenly realize that a large power line passes by your farm. The power-line easement extends to 20 m on either side, so that the border of your property is at a ground distance of 20 m from the wire. You have at your disposal 200 m of number 6 gauge (4.1-mm diameter) copper wire, which can be deployed in the form of a 1-turn rectangular loop of side lengths a and b, as shown in Figure 7.32. (a) Assuming the power line to carry a sinusoidal (60-Hz) current of $I = 4000$ A, find the maximum amount of power (in watts) that you can extract from the power line. State all assumptions. (b) Could you extract more power by using your 200-m wire in the form of a multiturn loop? Note that this is a hypothetical problem, which, like many engineering problems, may have legal ramifications. The power company may claim that it owns the spillover energy; on the other hand, you can claim that the spillover energy is an unauthorized intrusion into your private property and that the least you can do is to take advantage of it.

Figure 7.32   Extracting power from a power line. Problem 7.16.

7.17 **Induction.** Two infinitely long wires carrying currents I_1 and I_2 cross (without electrical contact) at the origin, as shown in Figure 7.33. A small rectangular loop is placed next to the wires. (a) If $I_1 = \cos(\omega t)$ and $I_2 = \sin(\omega t)$, determine the polarity and magnitude of the induced voltage $\mathcal{V}_{ind}(t)$. Sketch $\mathcal{V}_{ind}(t)$ together with $I_1(t)$ and $I_2(t)$. (b) If I_1 and I_2 are both constant so that $I_1 = I_2 = I$, and if we move the loop away from the wires at a constant velocity $\mathbf{v}$, which direction should it be moved in order to produce the largest $|\mathcal{V}_{ind}|$? Find this value of $|\mathcal{V}_{ind}|$.

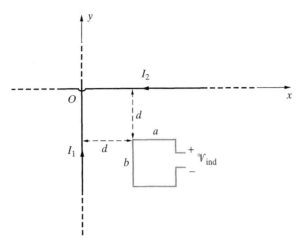

Figure 7.33 Induction. Problem 7.17.

7.18 **Induction.** Consider the square loop located between two infinitely long parallel wires carrying currents I_1 and I_2, where the sides of the loop are equidistant from each wire, as shown in Figure 7.34. (a) Starting at $t = 0$, the current I_1 decreases exponentially over time according to $I_1 = I_0 e^{-t}$, while the current I_2 stays constant at $I_2 = I_0$. Find an expression for the induced voltage $\mathcal{V}_{\text{ind}}(t)$ and sketch it as a function of time, paying particular attention to its polarity. (b) Repeat part (a) for $I_1 = I_0 \sin(\omega t)$. Your sketch of $\mathcal{V}_{\text{ind}}(t)$ should cover the range $0 < t < 2\pi/\omega$.

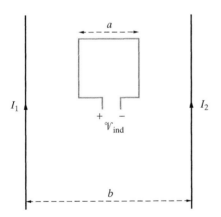

Figure 7.34 Induction. Problem 7.18.

7.19 **Wave propagation.** An electromagnetic pulse of length l traveling in the $+x$ direction with a constant velocity v_p has an electric field in the z direction, which is shown at $t = 0$ as a function of x. (See Figure 7.35.) An observer capable of measuring $\mathcal{E}_z$ is located at position $x = d$ (where $d > l$). (a) Sketch $\mathcal{E}_z$ as a function of x at $t = (d - l)/v_p$, $(d - l/2)/v_p$, $(d + l/2)/v_p$, and $(d + 3l)/v_p$. (b) Find $\mathcal{E}_z$ measured at $t = (d - l/3)/v_p$, $(d - 2l/3)/v_p$, and $(d + l)/v_p$.

Figure 7.35 Wave propagation. Problem 7.19.

7.20 Wave propagation. An observer located at $y = y_1$ measures the magnetic field of an electromagnetic pulse propagating with a constant velocity v_p in the $+y$ direction as

$$\overline{\mathscr{H}}(y_1, t) = \begin{cases} 0 & t < 0 \\ \hat{\mathbf{z}} H_0 t / t_1 & 0 < t < t_1 \\ \hat{\mathbf{z}} H_0 & t_1 < t < 3t_1 \\ 0 & t > 3t_1 \end{cases}$$

Sketch $\mathscr{H}_z$ versus y at the following times: (a) $t = 2t_1$ and (b) $t = 4t_1$. Indicate the position of the observer in each sketch.

7.21 Interference of two waves. Two transverse electromagnetic pulses A and B propagate in the $+z$ and $-z$ directions, respectively, each with a velocity v_p. At $t = 0$, the fronts of the two pulses are at a distance d from one another. The electric field waveform for pulse A is known, whereas the temporal shape of pulse B is unknown. If an observer located midway between the two waves measures a total electric field as shown in Figure 7.36, sketch the electric field waveform of pulse B. Assume that the electric fields of both pulses are in the x direction.

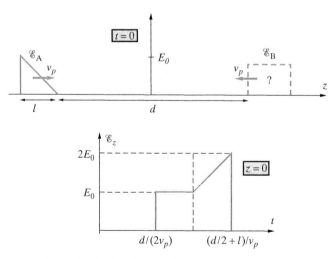

Figure 7.36 Interference of two waves. Problem 7.21.

7.22 **Wave propagation.** The electric field of an electromagnetic wave propagating in air has the shape of a Gaussian pulse, given by

$$\overline{\mathscr{E}}(z,t) = \hat{\mathbf{y}}50e^{-\pi(z-v_p t)^2}$$

in V-m^{-1}, where $v_p = 3 \times 10^8$ m-s^{-1}. An observer located at $z = 1$ km has a receiver that is capable of measuring a minimum electric field of 0.1 μV-m^{-1}. (a) How long does the observer continue to detect a signal? (b) At what time does the field measured by the observer reach a maximum? What is the maximum $\overline{\mathscr{E}}$ measured? (c) Sketch $\mathscr{E}_y(z,t)$ as a function of z at $t = 5$ μs with the position of the observer indicated. Does the observer detect a measurable signal at this time?

7.23 **Phasors.** Write the following real-time expressions in phasor form: (a) $\overline{\mathscr{E}}(z,t) = \hat{\mathbf{y}}\cos(\omega t - z)$ V-m^{-1}. (b) $\overline{\mathscr{H}}(x,t) = 0.1[\hat{\mathbf{y}}\cos(\omega t - 0.3x) + \hat{\mathbf{z}}0.5\sin(\omega t + 0.3x)]$ mA-m^{-1}. (c) $\overline{\mathscr{B}}(y,z,t) = \hat{\mathbf{x}}40\sin(3 \times 10^8 t + 0.8y - 0.6z + \pi/4)$ μT. (d) $\overline{\mathscr{E}}(x,y,t) = \hat{\mathbf{z}}E_0\sin(ax)\cos(\omega t + by)$.

7.24 **Phasors.** Express the following phasors as real-time quantities: (a) $\mathbf{E}(y) = \hat{\mathbf{z}}5e^{-j40\pi y}$ V-m^{-1}. (b) $\mathbf{B}(z) = \hat{\mathbf{x}}0.1e^{-j2\pi z} - \hat{\mathbf{y}}0.3je^{-j2\pi z}$ μT. (c) $\mathbf{E}(x) = \hat{\mathbf{z}}0.1(e^{-j18x} - 0.5e^{j18x})$ V-m^{-1}. (d) $\mathbf{H}(x,z) = \hat{\mathbf{y}}e^{-j48\pi x}e^{j64\pi z}$ mA-m^{-1}. (e) $\mathbf{J}(y) = \hat{\mathbf{x}}40e^{-0.1y(1+j)+j\pi/3}$ μA-m^{-2}.

7.25 **Displacement current in a capacitor.** Consider a parallel-plate capacitor with metal plates of 2-cm^2 area, each separated by mica ($\epsilon_r = 6$) 1 mm thick. If an alternating voltage of $V(t) = 15\cos(2\pi ft)$ V is applied across the capacitor plates, find the displacement current $\mathbf{J}_d$ through the capacitor at (a) $f = 10$ kHz and (b) $f = 1$ MHz.

7.26 **Propagation through lake water.** The electric field of an electromagnetic wave propagating in a deep lake (for lake water, assume $\sigma = 4 \times 10^{-3}$ S-m^{-1}, $\epsilon_r = 81$, and $\mu_r = 1$) transmitted by an electromagnetic probe submerged in the lake is approximately given by

$$\overline{\mathscr{E}} \simeq \hat{\mathbf{x}}10e^{-0.08z}\cos(2.7 \times 10^6 \pi t - 0.27z) \text{ mV-m}^{-1}$$

Find the peak values of the vectors of conduction-current density $\overline{\mathscr{J}}_c$ and displacement-current density $\overline{\mathscr{J}}_d$ at (a) $z = 0$, (b) $z = 10$ m, and (c) $z = 100$ m.

7.27 **Sea water.** For sea water, $\sigma = 4$ S-m^{-1}, $\epsilon_r = 81$, and $\mu_r = 1$. (a) Find the ratio of the magnitudes of the conduction-current density and the displacement-current density at 20 kHz, 2 MHz, 200 MHz, and 20 GHz. (b) Find the frequency at which the ratio is 4.

7.28 **Dry soil.** For dry soil, assume $\sigma \simeq 10^{-4}$ S-m^{-1}, $\epsilon_r \simeq 3$, and $\mu_r = 1$. Find the frequency at which the ratio of the magnitudes of the conduction-current and displacement-current densities is 2.

7.29 **Maxwell's equations.** Consider an electromagnetic wave propagating in a source-free non-conducting medium represented by $\overline{\mathscr{E}}$ and having components given by

$$\mathscr{E}_x = p_1(z - v_p t) + p_2(z + v_p t) \qquad \mathscr{E}_y = \mathscr{E}_z = 0$$

where p_1 and p_2 are any two arbitrary functions and $v_p = 1/\sqrt{(\mu\epsilon)}$. Show that $\overline{\mathscr{E}}$ satisfies all of Maxwell's equations, and find the components of the corresponding $\overline{\mathscr{H}}$.

7.30 **Maxwell's equations.** The magnetic field in a source free (no charges or currents), lossless (i.e., $\sigma = 0$), and simple non-magnetic dielectric medium (i.e., permittivity ϵ and permeability μ_0) is given as follows:

$$\overline{\mathscr{B}}(x,y,t) = \hat{\mathbf{x}}\sin x \cos y \cos 100t - \hat{\mathbf{y}}\cos x \sin y \cos 100t \qquad \text{Wb-m}^{-2}$$

Determine all components of the electric field $\overline{\mathscr{E}}(x, y, t)$. Do the magnetic and electric fields constitute a propagating electromagnetic wave?

7.31 AM radio waves. The electric-field and magnetic-field components of an AM radio signal propagating in air are given by

$$\overline{\mathscr{E}} = \hat{x} E_0 \cos(7.5 \times 10^6 t - \beta z)$$

$$\overline{\mathscr{H}} = \hat{y} \frac{E_0}{\eta} \cos(7.5 \times 10^6 t - \beta z)$$

Find the values of β and η such that these expressions satisfy all of Maxwell's equations.

7.32 Maxwell's equations. The magnetic field phasor of an electromagnetic wave in air is given by

$$\mathbf{H}(y) = \hat{z} 1.83 \times 10^{-4} e^{-j4y} \text{ A-m}^{-1}$$

(a) Find the angular frequency ω of the wave such that $\mathbf{H}$ satisfies all of Maxwell's equations. (b) Find the corresponding time-harmonic electric field $\mathbf{E}$. (c) Find the electric flux density $\mathbf{D}$ and the displacement-current density $\mathbf{J}_d$.

7.33 Superposition of two waves. The sum of the electric fields of two time-harmonic (sinusoidal) electromagnetic waves propagating in opposite directions in air is given as

$$\overline{\mathscr{E}}(z, t) = \hat{x} 100 \sin(\beta z) \sin(21 \times 10^9 \pi t) \text{ mV-m}^{-1}$$

(a) Find the constant β. (b) Find the corresponding $\overline{\mathscr{H}}$.

7.34 Electromagnetic wave in free space. An electromagnetic wave propagating in free space has an electric field given by

$$\overline{\mathscr{E}}(x, z, t) = \hat{y} 5 \cos(1.8 \times 10^9 \pi t - ax - 2.5az) \text{ V-m}^{-1}$$

where a is a constant. Find the value of a and the corresponding expression for the magnetic field $\overline{\mathscr{H}}$.

7.35 Coaxial lines. The electric field of a transverse electromagnetic wave guided within a lossless coaxial transmission line along the z axis is expressed in cylindrical coordinates as

$$\mathbf{E}(r, z) = \hat{r} \frac{E_0}{r} e^{-j\beta z} \qquad a \leq r \leq b$$

where $\beta = \omega\sqrt{\mu\epsilon}$; μ and ϵ are the permeability and permittivity of the dielectric material, respectively, and a and b are the inner and the outer radii of the coaxial line. (a) Find the corresponding $\mathbf{H}$. (b) Write the time-domain expressions for $\overline{\mathscr{E}}$ and $\overline{\mathscr{H}}$. (c) Sketch both $\mathscr{E}_r$ and $\mathscr{H}_\phi$ as functions of r over the range $a \leq r \leq b$ at position $z = 0$ and time $t = 0$. (d) Sketch both $\mathscr{E}_r$ and $\mathscr{H}_\phi$ as functions of z at position $r = a$ and time $t = 0$.

8

Waves in an Unbounded Medium

Having established the physical basis of the complete set of Maxwell's equations, we now proceed to discuss what is arguably their most important consequence: electromagnetic waves. In this chapter, we consider the propagation in an *unbounded, simple,* and *source-free* medium of a special type of electromagnetic waves known as *uniform plane waves.* Uniform plane waves are waves in which the amplitude and phase of the electric field (and the magnetic field) at any instant of time are constants over infinite planes orthogonal to the direction of propagation. The characteristics of uniform plane waves are particularly simple, so their study constitutes an excellent starting point in understanding more complicated electromagnetic waves. Furthermore, many practically important electromagnetic waves can be approximated as uniform plane waves, so our study of such waves is also of significant practical importance. Uniform plane waves are often also referred to as transverse electromagnetic (TEM) waves, since both the electric and magnetic fields of the wave are transverse to the propagation direction.

Many of the electromagnetic wave concepts that we study in this chapter directly correspond to voltage and current wave concepts that we studied in Chapters 2 and 3. The basic wave equations and their general solutions are identical, as are the concepts of propagation constant, wavelength, phase velocity, and attenuation constant. We take note of this correspondence at various points during the chapter. The similarity of voltage and current waves and uniform plane waves is, of course, not just a coincidence. The uniform two-conductor transmission lines considered in Chapters 2 and 3 are merely special cases of more general guiding structures that can efficiently transmit electromagnetic energy from one point to another, a correspondence that we discuss further in Chapter 10. Thus, the voltage and current waves of Chapters 2 and 3 are in fact uniform plane electromagnetic waves and can be studied entirely in terms of electric and magnetic fields, rather than voltages and currents. The relatively simple physical configuration of two-conductor transmission lines provided us the ability to analyze the propagation of electromagnetic waves on them by means of voltages and currents, which is why we were able to study transmission lines before introducing Maxwell's equations.

Our coverage in this chapter of uniform plane waves in an unbounded, simple and source-free medium starts a sequence of four chapters in each of which we consider the propagation of electromagnetic waves under different circumstances. In this chapter, we restrict our attention to waves for which all three of these conditions hold true. In Chapter 9, we remove the first condition and consider how waves reflect and refract in the presence of planar boundaries. In Chapter 10, we consider the case of guided wave propagation in regions bounded by conducting or dielectric structures. In Chapter 11, we return to an unbounded medium, but now allow it not to be simple, but rather have material properties that depend on the frequency of the electromagnetic wave.

8.1 PLANE WAVES IN A SIMPLE, SOURCE-FREE, AND LOSSLESS MEDIUM

We mentioned in Chapter 7 that two of Maxwell's equations, namely the two curl equations (7.21a) and (7.21c), which represent the facts that changing magnetic fields produce electric fields and that changing electric fields produce magnetic fields, necessarily lead to propagation of electromagnetic waves. In this section, we study the characteristics of such electromagnetic waves, as they propagate in unbounded, simple, source-free, and lossless media. Our starting point consists of Maxwell's equations, repeated below for convenience:

$$\nabla \times \overline{\mathscr{E}} = -\frac{\partial \overline{\mathscr{B}}}{\partial t} \tag{8.1a}$$

$$\nabla \cdot \overline{\mathscr{D}} = \tilde{\rho} \tag{8.1b}$$

$$\nabla \times \overline{\mathscr{H}} = \overline{\mathscr{J}} + \frac{\partial \overline{\mathscr{D}}}{\partial t} \tag{8.1c}$$

$$\nabla \cdot \overline{\mathscr{B}} = 0 \tag{8.1d}$$

The $\overline{\mathscr{J}}$ term in (8.1c) can in general be nonzero because of the presence of source currents $\overline{\mathscr{J}}_{\text{source}}$ (e.g., wires carrying current) and/or because of conduction current $\overline{\mathscr{J}}_c = \sigma\overline{\mathscr{E}}$, which flows in media with nonzero conductivity ($\sigma \neq 0$). The latter leads to loss of electromagnetic power, with volume density of power dissipation represented by $\overline{\mathscr{E}} \cdot \overline{\mathscr{J}}$. In this section, we consider electromagnetic wave propagation in source-free (i.e., $\overline{\mathscr{J}}_{\text{source}} = 0, \tilde{\rho} = 0$), simple,[1] and lossless (i.e., $\sigma = 0$ and thus $\overline{\mathscr{J}}_c = 0$) media, so that $\overline{\mathscr{J}} = \overline{\mathscr{J}}_{\text{source}} + \overline{\mathscr{J}}_c = 0$. Our goal is to describe the properties of different types of electromagnetic waves that can exist (i.e., can satisfy Maxwell's equations) in regions without any source currents and charges, regardless of where and when those fields may have originated—undoubtedly at faraway sources $\overline{\mathscr{J}}_{\text{source}}$ and/or $\tilde{\rho}$.

---

[1]In this context, a "simple" medium is a material medium that is linear, time-invariant, isotropic, and homogeneous, so that ϵ, μ, and σ are simple constants.

In general, the two coupled equations (8.1a) and (8.1c) can be combined to obtain equations in terms of only $\overline{\mathscr{E}}$ (or $\overline{\mathscr{H}}$). Taking the curl of (8.1a) and using (8.1c) gives

$$\nabla \times \nabla \times \overline{\mathscr{E}} = -\mu \frac{\partial}{\partial t} \nabla \times \overline{\mathscr{H}}$$

$$= -\mu \frac{\partial^2 \overline{\mathscr{D}}}{\partial t^2}$$

where we have taken into account the fact that $\overline{\mathscr{J}} = 0$ and assumed that μ is a constant (i.e., the medium is magnetically linear, isotropic, homogeneous, and time-invariant) and that $\overline{\mathscr{H}}$ is a continuous function of time and space, so that spatial and temporal derivatives can be interchanged.

We now use the vector identity

$$\nabla \times \nabla \times \overline{\mathscr{E}} \equiv \nabla(\nabla \cdot \overline{\mathscr{E}}) - \nabla^2 \overline{\mathscr{E}}$$

and the fact that under source-free conditions ($\tilde{\rho} = 0$) we have (equation (8.1b))

$$\nabla \cdot \overline{\mathscr{D}} = 0 \quad \rightarrow \quad \nabla \cdot (\epsilon \overline{\mathscr{E}}) = 0 \quad \rightarrow \quad \nabla \cdot \overline{\mathscr{E}} = 0$$

assuming that ϵ is constant (i.e., that the medium is electrically linear, isotropic, homogeneous, and time-invariant) to obtain

$$\boxed{\nabla^2 \overline{\mathscr{E}} - \mu\epsilon \frac{\partial^2 \overline{\mathscr{E}}}{\partial t^2} = 0} \tag{8.2}$$

We can obtain a similar equation for the magnetic field $\overline{\mathscr{H}}$ by taking the curl of (8.1c), using (8.1a) and making similar assumptions (i.e., a simple medium, being linear, isotropic, homogeneous, and time-invariant). We find

$$\boxed{\nabla^2 \overline{\mathscr{H}} - \mu\epsilon \frac{\partial^2 \overline{\mathscr{H}}}{\partial t^2} = 0} \tag{8.3}$$

Equations of the type (8.2) and (8.3) are encountered in many branches of science and engineering and have natural solutions in the form of propagating waves; they are thus referred to as *wave equations*. The solutions of these equations describe the characteristics of the electromagnetic waves as dictated by Maxwell's equations and the properties (μ, ϵ) of simple lossless material media in regions without any source currents and charges. It is important to note that the equations (8.2) and (8.3) are not independent (since they were both obtained from (8.1a) and (8.1c)), and that either (8.2) or (8.3) (together with (8.1a) or (8.1c)) can be used to solve for both $\overline{\mathscr{E}}$ and $\overline{\mathscr{H}}$. We follow the usual convention of solving for $\overline{\mathscr{E}}$ from the electric field equation (8.2) and then determining $\overline{\mathscr{H}}$ from (8.1a).

We note that in the general case when $\overline{\mathcal{E}}$ has three nonzero components ($\mathcal{E}_x$, $\mathcal{E}_y$, and $\mathcal{E}_z$), which may vary with the three Cartesian spatial coordinates (x, y, and z), (8.2) is actually a set of three scalar second-order partial differential equations:

$$\left[\frac{\partial^2}{\partial x^2} + \frac{\partial^2}{\partial y^2} + \frac{\partial^2}{\partial z^2}\right]\mathcal{E}_x - \mu\epsilon\frac{\partial^2\mathcal{E}_x}{\partial t^2} = 0 \tag{8.2a}$$

$$\left[\frac{\partial^2}{\partial x^2} + \frac{\partial^2}{\partial y^2} + \frac{\partial^2}{\partial z^2}\right]\mathcal{E}_y - \mu\epsilon\frac{\partial^2\mathcal{E}_y}{\partial t^2} = 0 \tag{8.2b}$$

$$\left[\frac{\partial^2}{\partial x^2} + \frac{\partial^2}{\partial y^2} + \frac{\partial^2}{\partial z^2}\right]\mathcal{E}_z - \mu\epsilon\frac{\partial^2\mathcal{E}_z}{\partial t^2} = 0 \tag{8.2c}$$

We do not need to consider the most general case in order to study the characteristics of propagating electromagnetic waves contained in the solution of (8.2). For simplicity, we limit ourselves here to the special case in which the electric field $\overline{\mathcal{E}}$ is independent of two dimensions (say x and y). Equation (8.2) then becomes

$$\frac{\partial^2\overline{\mathcal{E}}}{\partial z^2} - \mu\epsilon\frac{\partial^2\overline{\mathcal{E}}}{\partial t^2} = 0 \tag{8.4}$$

which is equivalent to three scalar equations, one for each component of $\overline{\mathcal{E}}$. With no loss of generality, we further restrict our attention to one of the three components, say $\mathcal{E}_x$, the equation for which is then

$$\boxed{\frac{\partial^2\mathcal{E}_x}{\partial z^2} - \mu\epsilon\frac{\partial^2\mathcal{E}_x}{\partial t^2} = 0} \tag{8.5}$$

Second-order partial differential equations of the type (8.5) commonly occur in many branches of engineering and science. For example, replacing $\mathcal{E}_x(z,t)$ with $\mathcal{V}(z,t)$ results in the wave equation that describes the voltage wave $\mathcal{V}(z,t)$ on a lossless transmission line (see equation (2.5)), whereas replacing $\mathcal{E}_x(z,t)$ with $u(z,t)$ gives the wave equation describing the variation of velocity $u(z,t)$ for acoustic waves in a fluid.

Following a similar approach to the solution of equation (2.7) provided by (2.8) as discussed in Chapter 2, the general solution of (8.5) is of the form

$$\mathcal{E}_x(z,t) = p_1(z - v_p t) + p_2(z + v_p t) \tag{8.6}$$

where $v_p = 1/\sqrt{\mu\epsilon}$ and p_1 and p_2 are arbitrary functions representing the shape (e.g., square pulse, sinusoid with a Gaussian envelope, exponentially decaying pulse) of the electric field excited by a remote source. Examples of functions[2] of $(z - v_p t)$ include $Ae^{-(z-v_p t)^2}$, $A\sqrt{z - v_p t}$, and $Ae^{-(z-v_p t)}\cos(z - v_p t)$. Note also that the functions p_1 and

[2]An important function of $(z - v_p t)$ that is often encountered and that we shall introduce in the next section and study in the rest of this book is the sinusoidal traveling-wave function, $A\cos[\omega(t - z/v_p)]$. Depending on the location of the observation point $z > 0$ along the z axis, this function replicates the sinusoidal variation $A\cos(\omega t)$ occurring at $z = 0$, except delayed by (z/v_p) seconds at the point z. Thus, (z/v_p) represents a time shift, or delay, which is a characteristic of the class of wave functions of the variable $(z - v_p t)$.

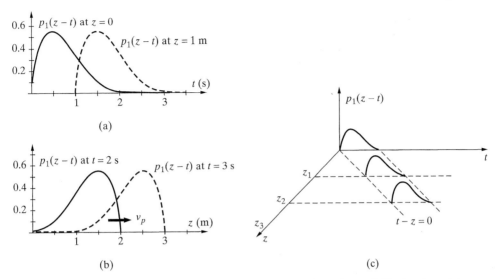

Figure 8.1   Variation in space and time of an arbitrary function $p_1(z - v_p t)$. The function $p_1(\zeta)$ shown above is $p_1(\zeta) = \zeta^{1/2}(\sin\zeta/\zeta)^6 u(\zeta)$, where $u(\zeta)$ is the unit step function (i.e., $u(\zeta) = 0$ for $\zeta < 0$ and $u(\zeta) = 1$ for $\zeta > 0$). For the purpose of the plots on the left-hand side, the speed of propagation is taken to be $v_p = 1$ m-s^{-1}. (a) $p_1(z - t)$ as a function of t at two positions $z = 0$ and $z = 1$ m. (b) $p_1(z - t)$ as a function of z at two time instants, showing how the pulse travels in the z directions as time progresses. (c) $p_1(z - t)$ as a function of z and t.

p_2 are not necessarily the same. The fact that (8.6) is a solution of (8.5) can be seen by simple substitution.[3]

That the solutions $p_1(z - v_p t)$ and $p_2(z + v_p t)$ represent waves propagating, respectively, in the $+z$ and $-z$ directions can be seen by observing their variation with z at different times t_i, as shown for one type of $p_1(z - v_p t)$ function in Figure 8.1. In this context, a wave is to be understood as some disturbance (e.g., an electric field variation) that occurs at one place at a given time and at other places at other times, with time delays proportional to the spatial separations from the first location. Since the wave travels with

---

[3]Consider

$$\mathscr{E}_x(z, t) = f(z - v_p t) = f(\zeta)$$

To see that this function $f(\cdot)$ of the variable $\zeta = (z - v_p t)$ is a solution of (8.5), we can express the time and space derivatives of $\mathscr{E}_x(z, t)$ in terms of the derivatives of $f(\zeta)$ with respect to ζ:

$$\frac{\partial \mathscr{E}_x}{\partial t} = \frac{\partial f}{\partial \zeta}\frac{\partial \zeta}{\partial t} = -v_p\frac{\partial f}{\partial \zeta} \quad \text{and} \quad \frac{\partial^2 \mathscr{E}_x}{\partial t^2} = -v_p\frac{\partial^2 f}{\partial \zeta^2}\frac{\partial \zeta}{\partial t} = v_p^2\frac{\partial^2 f}{\partial \zeta^2}$$

since $\partial\zeta/\partial t = -v_p$. Similarly, noting that $\partial\zeta/\partial z = 1$, we have

$$\frac{\partial \mathscr{E}_x}{\partial z} = \frac{\partial f}{\partial \zeta} \quad \text{and} \quad \frac{\partial^2 \mathscr{E}_x}{\partial z^2} = \frac{\partial^2 f}{\partial \zeta^2}$$

Substituting in (8.5) we find that the wave equation is indeed satisfied by any function $f(\cdot)$ of the variable $\zeta = (z - v_p t)$, where $v_p = 1/\sqrt{\mu\epsilon}$.

a velocity v_p, a time period of (z/v_p) elapses as the wave propagates from $z = 0$ to the position z. Thus, an observer at point $z > 0$ sees events that actually occurred (at $z = 0$) at an earlier time. For example, light waves from a supernova explosion may arrive at earth millions of years after their source has been extinguished. A wave is not necessarily a repetitive or oscillatory disturbance in time; a single pulse moving in space also constitutes a wave, such as a voltage pulse when a transmission line is briefly connected to a battery or a tsunami is generated in an undersea earthquake.

Note that at any time, say t_1, the function $p_1(z - v_p t_1)$ is simply a function of z, because $v_p t_1$ is a constant. At a later time t_2, the function is $p_1(z - v_p t_2)$, which has exactly the same type of dependence on z, but displaced to the right by an amount $v_p(t_2 - t_1)$. In other words, the disturbance (in this case the electric field) represented by $p_1(z - v_p t)$ has traveled in the positive z direction with a velocity v_p. Note that in free space, $v_p = 1/\sqrt{\mu_0 \epsilon_0} \simeq 3 \times 10^8$ m-s^{-1}, or the speed of light, a fact that led Maxwell to suggest that light is a form of electromagnetic radiation. The space-time dependence of a wave pulse such as $p_1(z - v_p t)$ is illustrated in Figure 8.1.

Note that the second term in (8.6), $p_2(z + v_p t)$, represents a wave traveling in the negative z direction. The general solution of (8.5) is thus a superposition of two waves: a forward wave traveling in the $+z$ direction (away from the source, if the source is assumed to be located far away in the $-z$ direction) and a reverse wave traveling in the $-z$ direction (back toward the source). The reverse wave term is nonzero only if there are discontinuities (surfaces with different ϵ and/or μ) that "reflect" some of the forward-traveling wave energy back to the source. The reflection and refraction of electromagnetic waves at planar interfaces are studied in Chapter 9.

Electromagnetic waves for which the field components have functional dependencies as given by (8.6) belong to a class of waves known as *plane waves*. This term originates from the fact that the surfaces over which the argument of the function is a constant (e.g., $z - v_p t = $ constant) are planes.[4] Waves for which the field amplitudes do not vary with position over the planes of constant phase are known as *uniform plane waves*. Generalized expressions for uniform plane waves are given in Section 8.6. Note that although we studied the solution of (8.4) only for the x component of $\overline{\mathscr{E}}$, identical behavior would be expected for the y component $\mathscr{E}_y$, since the differential equation governing it is the y component of (8.4), which is identical in form to (8.5). In general, both components may exist, in which case the total field is simply a linear superposition of the two separate solutions of the corresponding differential equations. Although it would seem that the $\mathscr{E}_z$ would also behave in a similar manner, a uniform plane wave propagating in the z direction in a simple medium cannot have a z component, as can be seen by examining equation (8.1b). We have

$$\nabla \cdot \overline{\mathscr{D}} = 0 \longrightarrow \nabla \cdot \overline{\mathscr{E}} = 0 \longrightarrow \frac{\partial \mathscr{E}_x}{\partial x} + \frac{\partial \mathscr{E}_y}{\partial y} + \frac{\partial \mathscr{E}_z}{\partial z} = 0$$

---

[4]Such surfaces are known as *surfaces of constant phase, phase surfaces*, or *phase fronts*. Although the concept of phase is more commonly associated with sinusoidal steady-state or time-harmonic waves, in which case the function $p_1(z - v_p t)$ has the form $\cos[\omega(t - z/v_p)]$, it is nevertheless equally valid for nonsinusoidal functions. Time-harmonic waves are discussed in Section 8.2.

The first two terms ($\partial \mathcal{E}_x / \partial x$ and $\partial \mathcal{E}_y / \partial y$) are zero (since the field components do not vary with x or y), which requires that $\partial \mathcal{E}_z / \partial z = 0$, which in turn means that $\mathcal{E}_z$ cannot vary with z. Substituting $\partial \mathcal{E}_z / \partial z = 0$ into the z component of (8.4), we find $\partial^2 \mathcal{E}_z / \partial t^2 = 0$, which means that $\mathcal{E}_z$ can at most be a linearly increasing function of time. Since a quantity that is constant in space and linearly increasing in time cannot contribute to wave motion, a uniform plane wave propagating in the z direction has no z component. Since $\nabla \cdot \mathcal{H} = 0$ (from (8.1d)), we can use the same derivation to show $\mathcal{H}_z = 0$ in a uniform plane wave propagating in the z direction.

8.1.1 Relation between $\overline{\mathcal{E}}$ and $\overline{\mathcal{H}}$

Although we have just discussed the solution of the wave equation for the electric field, we could have just as easily solved (8.3) for the magnetic field $\overline{\mathcal{H}}$ by making the same assumption (i.e., uniform plane wave) and restricting our attention to a single component of the magnetic field. However, once we have described a solution for the electric field of a uniform plane wave (say the component $\mathcal{E}_x(z)$), we are no longer free to choose arbitrarily one of the two components of its magnetic field (i.e., $\mathcal{H}_x$ or $\mathcal{H}_y$) to be nonzero. Because the wave equation (8.4) for the electric field was obtained using the entire set of Maxwell's equations, the magnetic field is already determined once we specify the electric field. The proper means for finding the magnetic field of a uniform plane wave is thus to derive it from the electric field using Maxwell's equations.

Let us again assume that we have a uniform plane wave having an electric field with only an x component propagating in a simple medium in the $+z$ direction:

$$\mathcal{E}_x(z, t) = p_1(z - v_p t)$$

Since $\mathcal{E}_y = 0, \mathcal{E}_z = 0, \mathcal{H}_z = 0$, $\mathcal{E}_x$ varies only with z and t, and any components of the magnetic field $\overline{\mathcal{H}}$ also vary only with z and t, we have from (8.1a)

$$\nabla \times \overline{\mathcal{E}} = -\frac{\partial \overline{\mathcal{B}}}{\partial t} \quad \rightarrow \quad \hat{\mathbf{y}} \frac{\partial \mathcal{E}_x}{\partial z} = -\mu \hat{\mathbf{x}} \frac{\partial \mathcal{H}_x}{\partial t} - \mu \hat{\mathbf{y}} \frac{\partial \mathcal{H}_y}{\partial t}$$

whereas from (8.1c) we have

$$\nabla \times \overline{\mathcal{H}} = \frac{\partial \overline{\mathcal{D}}}{\partial t} \quad \rightarrow \quad -\hat{\mathbf{x}} \frac{\partial \mathcal{H}_y}{\partial z} + \hat{\mathbf{y}} \frac{\partial \mathcal{H}_x}{\partial z} = \epsilon \hat{\mathbf{x}} \frac{\partial \mathcal{E}_x}{\partial t}$$

Equating the different components of the above two equations, we find

$$\frac{\partial \mathcal{E}_x}{\partial z} = -\mu \frac{\partial \mathcal{H}_y}{\partial t}; \qquad -\frac{\partial \mathcal{H}_y}{\partial z} = \epsilon \frac{\partial \mathcal{E}_x}{\partial t}$$

It thus appears that the magnetic field of a uniform plane wave with an electric field of $\overline{\mathcal{E}} = \hat{\mathbf{x}} p_1(z - v_p t)$ has only a y component, which is related to $\mathcal{E}_x$ by the above equations. For any given function $p_1(z - v_p t)$, $\mathcal{H}_y$ can be found using direct differentiation and integration, as we show in the next section for sinusoidal uniform plane waves, for which case $p_1(z - v_p t) = A \cos(z - v_p t)$. By substituting $\mathcal{E}_x = p_1(z - v_p t)$

into $\partial \mathscr{E}_x / \partial z = -\mu \partial \mathscr{H}_y / dt$, taking the derivative with respect to z, and then integrating with respect to time, we get

$$\mathscr{H}_y = \left(\sqrt{\frac{\epsilon}{\mu}} \right) p_1(z - v_p t) = \frac{1}{\eta} p_1(z - v_p t)$$

where $\eta \equiv \sqrt{\mu/\epsilon}$ is a quantity that has units of impedance (ohms) and is defined as the *intrinsic impedance* of the medium. Thus, for uniform plane waves in a simple lossless medium, the ratio of the electric and magnetic fields is η and is determined only by the material properties of the medium (i.e., μ, ϵ). The intrinsic impedance of free space[5] is $\eta = \sqrt{\mu_0/\epsilon_0} \simeq 120\pi \simeq 377\Omega$.

An important characteristic of uniform plane waves is that their electric and magnetic fields are perpendicular to one another. In simple and lossless media, the variation of $\overline{\mathscr{E}}$ and $\overline{\mathscr{H}}$ in space and time are identical [i.e., they are both proportional to $p_1(z - v_p t)$]. In other words, $\overline{\mathscr{E}}$ and $\overline{\mathscr{H}}$ propagate in unison along z, reaching their maxima and minima at the same points in space and at the same times. The orientation of $\overline{\mathscr{E}}$ and $\overline{\mathscr{H}}$ is such that the vector $\overline{\mathscr{E}} \times \overline{\mathscr{H}}$ is in the $+z$ direction, which is the direction of propagation of the wave. The orthogonality of the electric and magnetic fields and the propagation of a disturbance are illustrated in Figure 8.2.

Note that since the choice of the coordinate system cannot affect the physical relationship between $\overline{\mathscr{E}}$ and $\overline{\mathscr{H}}$, if we start with $\overline{\mathscr{E}}$ having only a y component (i.e., $\overline{\mathscr{E}} = \hat{\mathbf{y}} \mathscr{E}_y$), $\overline{\mathscr{H}}$ then comes out to be $\overline{\mathscr{H}} = -\hat{\mathbf{x}} \mathscr{H}_x$. The relationship between $\overline{\mathscr{E}} = \hat{\mathbf{y}} \mathscr{E}_y$ and $\overline{\mathscr{H}} = -\hat{\mathbf{x}} \mathscr{H}_x$ for a uniform plane wave propagating in the $+z$ direction is depicted in Figure 8.3.

The concept of uniform plane waves makes the solution of the wave equation (8.2) tractable, as we have seen. In a strict sense, uniform plane waves can be excited (created) only by sources infinite in extent (e.g., uniformly distributed over the entire xy plane to produce a uniform plane wave propagating in the $+z$ direction). However, uniform plane waves are in fact often excellent approximations in practice, especially when we observe (or receive) electromagnetic waves at large distances (compared with wavelength) from their sources (e.g., in broadcast radio and television applications). Electromagnetic fields emanating from a point source spread spherically, and since a very small portion of the surface of a large sphere is approximately planar, they can be considered as plane waves at large distances from their sources. Uniform plane waves are also important because the electromagnetic fields of an arbitrary wave (i.e., nonuniform and/or nonplanar wave) can be expressed[6] as a superposition of component plane waves, much like a Fourier decomposition. Further discussion of uniform plane waves and nonuniform or nonplanar waves are provided in Section 8.7. We shall also see examples of nonuniform waves in Chapters 9 and 10 when we study reflection, refraction, and guiding of electromagnetic waves.

---

[5]Using $\mu_0 = 4\pi \times 10^{-7}$ H-m^{-1} and $\epsilon_0 = 1/(\mu_0 c^2) \simeq 10^{-9}/(36\pi)$ F-m^{-1}.

[6]P. C. Clemmow, *The Plane Wave Spectrum Representation of Electromagnetic Fields,* Pergamon Press, Oxford, 1966.

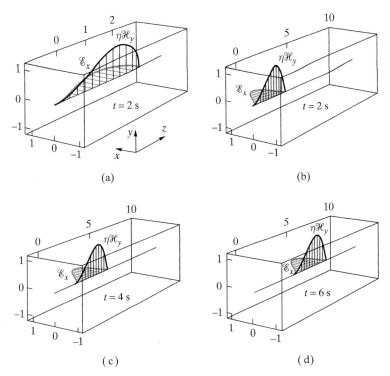

Figure 8.2   The propagation of a uniform plane electromagnetic disturbance in the z direction. The electric and magnetic fields are orthogonal at all times. For the purpose of this figure, the speed of propagation is taken to be $v_p = 1$ m-s^{-1}. (a) $\mathscr{E}_x(z,t)$ and $\eta\mathscr{H}_y(z,t)$ as a function of z at $t = 2$ s. (b) Same as (a) but shown on a compressed distance scale. (c) $\mathscr{E}_x(z,t)$ and $\eta\mathscr{H}_y(z,t)$ as a function of z at $t = 4$ s. (d) $\mathscr{E}_x(z,t)$ and $\eta\mathscr{H}_y(z,t)$ as a function of z at $t = 6$ s. The propagation of the pulse in the $+z$ direction is clearly evident. Note that the pulse shapes for the electric and magnetic fields are identical.

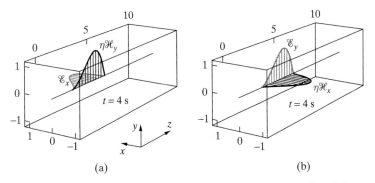

Figure 8.3   Electric and magnetic fields of a uniform plane wave. The relationship between $\overline{\mathscr{E}}$ and $\overline{\mathscr{H}}$ for a uniform plane wave is independent of the choice of a particular coordinate system. (a) $\mathscr{E}_x(z,t)$ and $\eta\mathscr{H}_y(z,t)$. (b) $\mathscr{E}_y(z,t)$ and $\eta\mathscr{H}_x(z,t)$.

8.2 TIME-HARMONIC UNIFORM PLANE WAVES IN A LOSSLESS MEDIUM

We now study uniform plane waves for which the temporal behavior is harmonic, or sinusoidal. It was pointed out in Section 7.4.4 that a large number of practical applications (e.g., broadcast radio and TV, radar, optical and microwave applications) involve sources (transmitters) that operate in such a narrow band of frequencies that the behavior of all of the field components is very similar to that of the central single-frequency sinusoid (i.e., the carrier).

The AM radio broadcast signal was mentioned in Section 7.4.4 as an example of a signal with a narrow bandwidth centered around a carrier frequency. Other examples include FM radio broadcasts, which utilize carrier frequencies of 88–108 MHz with a bandwidth of ~200 kHz, or UHF television broadcasts, where the signal bandwidth is ~6 MHz for carrier frequencies in the range 470–890 MHz. For all practical purposes, the propagation of the signals from the transmitters to receivers can be described by studying single sinusoids at the carrier frequency, since the characteristics of the propagation medium do not vary significantly over the signal bandwidth. In other words, while at the transmitter the actual signal is constituted by a superposition of Fourier components at different frequencies within the signal bandwidth, each of the frequency components propagates to the receiver in a manner identical to the propagation of the carrier.

In other applications, such as in the case of guided propagation of electromagnetic waves, the propagation characteristics may vary significantly over the bandwidth of the signal (especially when terminated transmission lines or waveguides are concerned). However, sinusoidal steady-state (or time-harmonic) analysis is still useful, because any arbitrary waveform can be represented as a linear superposition of its Fourier components whose behavior is well represented by the time-harmonic analysis.

To study the propagation of time-harmonic electromagnetic waves, we use the time-harmonic form of Maxwell's equations as given in (7.23). Note that these are written in terms of the vector phasor quantities represented with the boldface symbols ($\mathbf{E}$, $\mathbf{H}$, etc.,) which are related to the real space-time fields in the usual manner previously described in Section 7.4.4; for example,

$$\overline{\mathscr{E}}(x, y, z, t) = \mathscr{R}e \left\{ \mathbf{E} e^{j\omega t} \right\}$$

For lossless ($\sigma = 0$), source-free ($\mathbf{J}_{\text{source}} = 0, \rho = 0$), linear [i.e., $\epsilon, \mu \neq f(\mathbf{E}, \mathbf{H})$], homogeneous [i.e., $\epsilon, \mu \neq f(x, y, z)$], isotropic [i.e., $\epsilon, \mu \neq f(\text{direction})$], and time-invariant [i.e., $\epsilon, \mu \neq f(t)$] medium, we can derive the wave equation for the electric field by first taking the curl of (7.23a):

$$\nabla \times \nabla \times \mathbf{E} = -j\omega\mu(\nabla \times \mathbf{H})$$

Note that μ must be independent of direction and spatial coordinates in order to write $\nabla \times \mathbf{B} = \mu \nabla \times \mathbf{H}$. We now use a vector identity to replace the left-hand side of the preceding equation:

$$\nabla \times \nabla \times \mathbf{E} \equiv \nabla(\nabla \cdot \mathbf{E}) - \nabla^2 \mathbf{E}$$

Since ϵ is independent of direction and spatial coordinates, we have from (7.23b)

$$\nabla \cdot \mathbf{D} = 0 \longrightarrow \nabla \cdot (\epsilon \mathbf{E}) = \epsilon \nabla \cdot \mathbf{E} = 0 \longrightarrow \nabla \cdot \mathbf{E} = 0$$

Using (7.23c) further reduces the preceding equation to

$$-\nabla^2 \mathbf{E} = -j\omega\mu(j\omega\epsilon\mathbf{E})$$

or

$$\boxed{\nabla^2 \mathbf{E} - (j\beta)^2 \mathbf{E} = 0} \tag{8.7}$$

where

$$\boxed{\beta = \omega\sqrt{\mu\epsilon}} \tag{8.8}$$

is the *phase constant*, also often called the *wave number* or *propagation constant*, in units of radians-m^{-1}. Note that the propagation constant $\beta = \omega\sqrt{\mu\epsilon}$ for uniform plane waves in free space is in fact identical to the phase constant for voltage and current waves on uniform lossless transmission lines, namely $\beta = \omega\sqrt{LC}$, where L and C are, respectively, the per-unit-length inductance and capacitance of the line.[7] Equation (8.7) is known as the *vector wave equation* or *Helmholtz equation* and is actually a set of three scalar equations:

$$\left[\frac{\partial^2}{\partial x^2} + \frac{\partial^2}{\partial y^2} + \frac{\partial^2}{\partial z^2} \right] E_x - (j\beta)^2 E_x = 0 \tag{8.9a}$$

$$\left[\frac{\partial^2}{\partial x^2} + \frac{\partial^2}{\partial y^2} + \frac{\partial^2}{\partial z^2} \right] E_y - (j\beta)^2 E_y = 0 \tag{8.9b}$$

$$\left[\frac{\partial^2}{\partial x^2} + \frac{\partial^2}{\partial y^2} + \frac{\partial^2}{\partial z^2} \right] E_z - (j\beta)^2 E_z = 0 \tag{8.9c}$$

We now limit our attention to uniform plane waves by considering solutions of (8.7) for which $\mathbf{E}$ has only an x component and is only a function of z. In other words, we let

$$\mathbf{E}(x, y, z) = \hat{\mathbf{x}} E_x(z)$$

The wave equation (8.7) then reduces to

$$\boxed{\frac{d^2 E_x}{dz^2} - (j\beta)^2 E_x = 0} \tag{8.10}$$

which is a second-order ordinary differential equation encountered in sinusoidal steady-state applications in many branches of physics and engineering. For example, replacing E_x with V gives the wave equation for the voltage phasor V on a lossless transmission line.

---

[7]The reader is encouraged to use the various capacitance and inductance expressions tabulated in Table 2.2 and derived in Chapters 4 and 6 for lossless transmission lines to show that $LC = \mu\epsilon$.

The general solution of (8.10) is

$$E_x(z) = \underbrace{C_1 e^{-j\beta z}}_{E_x^+(z)} + \underbrace{C_2 e^{+j\beta z}}_{E_x^-(z)} \tag{8.11}$$

where C_1 and C_2 are constants to be determined by boundary conditions. Note that the quantity $E_x(z)$ is a complex phasor. The real (or instantaneous) electric field $\overline{\mathscr{E}}(z,t)$ can be found from $E_x(z)$ in (8.11) as

$$\mathscr{E}_x(z,t) = \mathfrak{Re}\{(C_1 e^{-j\beta z} + C_2 e^{+j\beta z})e^{j\omega t}\} \tag{8.12}$$

or

$$\mathscr{E}_x(z,t) = \underbrace{C_1 \cos(\omega t - \beta z)}_{\mathscr{E}_x^+(z,t)} + \underbrace{C_2 \cos(\omega t + \beta z)}_{\mathscr{E}_x^-(z,t)} \tag{8.13}$$

where C_1 and C_2 are assumed to be real constants. The first term $\mathscr{E}_x^+(z,t)$ represents a wave traveling in the $+z$ direction, as can be seen from the successive snapshots in time plotted as a function of z in Figure 8.4a. The z axis is normalized to wavelength, which is defined as $\lambda = 2\pi/\beta$. Note that the electric field at any given point in space varies sinusoidally in time, as illustrated in Figure 8.4b (where the t axis is normalized by period $T_p = 2\pi/\omega$).

By the same token, the second term $\mathscr{E}_x^-(z,t)$ in (8.13) represents a wave traveling in the $-z$ direction. Both terms represent waves traveling at a speed given by the phase velocity $v_p = \omega/\beta$, which is the velocity of travel of any point on the wave identified by a given phase (i.e., argument) of the sinusoid, as defined in Chapter 3. In other words, if we were to observe a fixed phase point on the wave, we have

$$\omega t - \beta z = \text{const.} \quad \longrightarrow \quad \frac{dz}{dt} = v_p = \frac{\omega}{\beta}$$

Note that $v_p = \omega/(\omega\sqrt{\mu\epsilon}) = 1/\sqrt{\mu\epsilon}$, which for free space ($\mu = \mu_0$, $\epsilon = \epsilon_0$) is the speed of light in free space, or $v_p = c$. Note also that substituting $\beta = 2\pi/\lambda$ in $v_p = \omega/\beta$ gives the familiar expression $v_p = f\lambda$, where $f = \omega/(2\pi)$. It is interesting to note that, for a simple lossless medium, the phase velocity is a function only of the medium parameters and is independent of frequency. This is an important property of uniform plane waves in a simple lossless medium, which does not hold true for waves propagating in a lossy medium or those that propagate in guiding structures, as we shall see later. However, it should be noted that ϵ and μ are in general functions of frequency even in a lossless medium and can be considered to be constant only over limited frequency ranges (see Section 11.2.1).

Although we have discussed the solution of the wave equation for the electric field, we could have just as easily derived a wave equation identical to (8.7) for the magnetic field **H**, and we could have solved it by making the same assumption (i.e., uniform plane wave) and restricting our attention to a single component of the magnetic field. However, once we have described a solution for the electric field of a uniform plane wave (say the component E_x), we are no longer free to choose one of the two components of its magnetic field (i.e., H_x or H_y) arbitrarily to be nonzero, because the wave equation for the electric field (i.e., (8.7)) was obtained using the entire set of equations (7.23) and

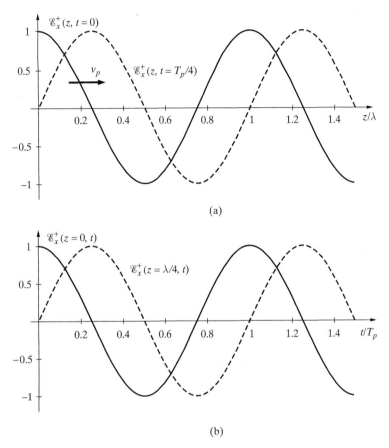

Figure 8.4 Wave behavior in space and time. (a) $\mathcal{E}_x^+(z,t) = \cos[(2\pi/T_p)t - (2\pi/\lambda)z]$ versus z/λ and for $t = 0$ and $t = T_p/4$. (b) $\mathcal{E}_x^+(z,t)$ versus t/T_p for $z = 0$ and $z = \lambda/4$.

the associated magnetic field is already determined once we specify its electric field. The proper means for finding the magnetic field of a uniform plane wave is thus to derive it from the electric field using (7.23a). We have

$$\nabla \times \mathbf{E} = \hat{\mathbf{y}}\frac{dE_x(z)}{dz} = -j\omega\mu[\hat{\mathbf{x}}H_x + \hat{\mathbf{y}}H_y + \hat{\mathbf{z}}H_z]$$

It thus appears that the corresponding **H** has only a y component, which for $E_x(z)$ as given in (8.11) can be found as

$$H_y(z) = \frac{1}{-j\omega\mu}\frac{d}{dz}(C_1 e^{-j\beta z} + C_2 e^{j\beta z})$$

$$= \frac{\beta}{\omega\mu}(C_1 e^{-j\beta z} - C_2 e^{j\beta z}) = \frac{1}{\eta}(C_1 e^{-j\beta z} - C_2 e^{j\beta z})$$

where we have used $\eta = \sqrt{\mu/\epsilon} = \omega\mu/\beta$. Note that the intrinsic impedance η is a real number for a simple lossless medium, but, as we will see in Section 8.3, it is a complex number for a lossy medium.

The instantaneous magnetic field $\overline{\mathcal{H}}(x,y,z,t)$ can be obtained from the phasor $\mathbf{H}(x,y,z)$ in the same way as in (8.12). Considering only the first term in (8.13) and the associated magnetic field, the electric and magnetic fields of a time-harmonic uniform plane wave propagating in the $+z$ direction are

$$\boxed{\begin{aligned} \mathcal{E}_x(z,t) &= C_1 \cos(\omega t - \beta z) \\ \mathcal{H}_y(z,t) &= \frac{1}{\eta} C_1 \cos(\omega t - \beta z) \end{aligned}} \qquad (8.14)$$

We find, as before (Section 8.1), that the electric and magnetic fields of a uniform plane wave are perpendicular to one another. The $\overline{\mathcal{E}}$ and $\overline{\mathcal{H}}$ fields are also both perpendicular to the direction of propagation, which is why uniform plane waves are also often referred to as *transverse electromagnetic,* or *TEM,* waves. In a simple lossless medium, the variation of $\overline{\mathcal{E}}$ and $\overline{\mathcal{H}}$ in space and time are identical (i.e., they are in phase); in other words, the $\overline{\mathcal{E}}$ and $\overline{\mathcal{H}}$ propagate in unison along z, reaching their maxima and minima at the same points in space and at the same times. The orientation of $\overline{\mathcal{E}}$ and $\overline{\mathcal{H}}$ is such that the vector $\overline{\mathcal{E}} \times \overline{\mathcal{H}}$ is in the $+z$ direction, which is the direction of propagation for the wave. Note that since the choice of the coordinate system cannot change the physical picture, if we start with $\overline{\mathcal{E}}$ having only a y component (i.e., $\overline{\mathcal{E}} = \hat{\mathbf{y}}\mathcal{E}_y$), then $\overline{\mathcal{H}}$ comes out to be $\overline{\mathcal{H}} = -\hat{\mathbf{x}}\mathcal{H}_x$, that is, $\overline{\mathcal{E}} \times \overline{\mathcal{H}}$ would still be in the direction of propagation (i.e., the z direction). The relationship between $\overline{\mathcal{E}}$ and $\overline{\mathcal{H}}$ for a time-harmonic uniform plane wave is depicted in Figure 8.5. As time progresses, the electric and magnetic fields propagate in the $+z$ direction, staying in phase at all points and at all times.

Equations (8.14) and Figure 8.5 completely describe the properties of uniform plane waves in a simple lossless unbounded medium. As long as μ and ϵ are simple constants, an electromagnetic wave in free space differs from that in the material medium (with $\epsilon \neq \epsilon_0$ and/or $\mu \neq \mu_0$) primarily in terms of its wavelength, given by $\lambda = (f\sqrt{\mu\epsilon})^{-1}$. We shall see in the next section that the characteristics of uniform plane waves in a lossy medium are substantially more influenced by the material properties.

Example 8.1: AM broadcast signal. The instantaneous expression for the electric field component of an AM broadcast signal propagating in air is given by

$$\overline{\mathcal{E}}(x,t) = \hat{\mathbf{z}}10 \, \cos(1.5\pi \times 10^6 t + \beta x) \quad \text{V-m}^{-1}$$

(a) Determine the direction of propagation and frequency f. (b) Determine the phase constant β and the wavelength λ. (c) Write the instantaneous expression for the corresponding magnetic field $\overline{\mathcal{H}}(x,t)$.

Solution:

(a) The wave propagates in the $-x$ direction. From $\omega = 2\pi f = 1.5\pi \times 10^6$ rad-s^{-1}, $f = 750$ kHz.

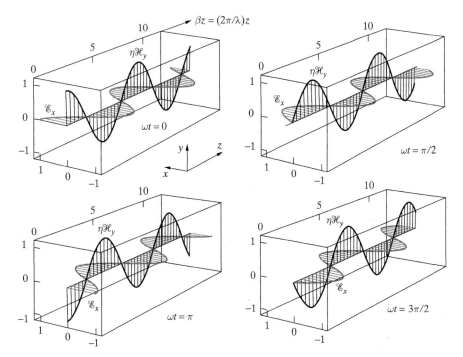

Figure 8.5 Electric and magnetic fields of a uniform plane wave in a lossless medium. Snapshots of $\mathcal{E}_x(z,t)$ and $\eta\mathcal{H}_y(z,t)$ for a sinusoidal uniform plane wave shown as a function of βz at $\omega t = 0$, $\pi/2$, π, and $3\pi/2$. The plots shown are for $C_1 = 1$.

(b) Using (8.8), the phase constant is given by

$$\beta = \omega\sqrt{\mu_0\epsilon_0} \simeq 1.5\pi \times 10^6/(3 \times 10^8) = 0.005\pi \ \text{rad-m}^{-1}$$

The wavelength in air is equal to that in free space, or

$$\lambda = 2\pi/\beta \simeq 2\pi/0.005\pi = 400 \ \text{m}$$

(c) The phasor electric field is given by

$$\mathbf{E}(x) = \hat{\mathbf{z}}10e^{+j\beta x} = \hat{\mathbf{z}}10e^{j0.005\pi x} \quad \text{V-m}^{-1}$$

To find the corresponding **H**, we use (7.23a) to find

$$-j\omega\mu_0\mathbf{H} = \nabla \times \mathbf{E} \quad \rightarrow \quad H_y(x) = -\frac{1}{j\omega\mu_0}\left[-\frac{\partial E_z(x)}{\partial x}\right] = \frac{1}{\eta}E_z(x)$$

$$\mathbf{H}(x) = \hat{\mathbf{y}}\frac{1}{\eta}E_z(x) \simeq \hat{\mathbf{y}}\frac{10 \ \text{V-m}^{-1}}{377\Omega}e^{j0.005\pi x}$$

Therefore,

$$\overline{\mathcal{H}}(x,t) \simeq \hat{\mathbf{y}}26.5 \times 10^{-3}\cos(1.5\pi \times 10^6 t + 0.005\pi x) \quad \text{A-m}^{-1}$$

Example 8.2: FM broadcast signal. An FM radio broadcast signal traveling in the y direction in air has a magnetic field given by the phasor

$$\mathbf{H}(y) = 2 \times 10^{-3} e^{-j0.68\pi y} (-\hat{\mathbf{x}} + \hat{\mathbf{z}}j) \quad \text{A-m}^{-1}$$

(a) Determine the frequency (in MHz) and wavelength (in m). (b) Find the corresponding $\mathbf{E}(y)$. (c) Write the instantaneous expression for $\overline{\mathscr{E}}(y,t)$ and $\overline{\mathscr{H}}(y,t)$.

Solution:

(a) Using (8.8), we have

$$\beta = \omega\sqrt{\mu_0\epsilon_0} \simeq 0.68\pi \quad \text{rad-m}^{-1}$$

from which we find

$$f = \frac{\omega}{2\pi} \simeq \frac{0.68\pi \text{ rad-m}^{-1} \times 3 \times 10^8 \text{ m-s}^{-1}}{2\pi} = 102 \text{ MHz}$$

and

$$\lambda = \frac{2\pi}{\beta} \simeq \frac{2\pi}{0.68\pi} \simeq 2.94 \text{ m}$$

(b) Using (7.23c)

$$\nabla \times \mathbf{H} = \hat{\mathbf{x}}\frac{\partial H_z}{\partial y} - \hat{\mathbf{z}}\frac{\partial H_x}{\partial y} = j\omega\epsilon_0\mathbf{E}$$

and performing the partial differentiation yields

$$\mathbf{E}(y) \simeq 0.754 e^{-j0.68\pi y}(-\hat{\mathbf{x}}j - \hat{\mathbf{z}}) \quad \text{V-m}^{-1}$$

(c) The instantaneous expressions for $\overline{\mathscr{E}}(y,t)$ and $\overline{\mathscr{H}}(y,t)$ are given by

$$\overline{\mathscr{E}}(y,t) = -\hat{\mathbf{x}}0.754\cos(2.04\pi \times 10^8 t - 0.68\pi y + \pi/2)$$
$$-\hat{\mathbf{z}}0.754\cos(2.04\pi \times 10^8 t - 0.68\pi y) \quad \text{V-m}^{-1}$$
$$\overline{\mathscr{H}}(y,t) = -\hat{\mathbf{x}}2 \times 10^{-3}\cos(2.04\pi \times 10^8 t - 0.68\pi y)$$
$$+\hat{\mathbf{z}}2 \times 10^{-3}\cos(2.04\pi \times 10^8 t - 0.68\pi y + \pi/2) \quad \text{A-m}^{-1}$$

Example 8.3: Uniform plane wave. Consider a uniform plane wave traveling in the z direction in a simple lossless nonmagnetic medium (i.e., $\mu = \mu_0$) with a y-directed electric field of maximum amplitude of 60 V-m^{-1}. If the wavelength is 20 cm and the velocity of propagation is 10^8 m-s^{-1}, (a) determine the frequency of the wave and the relative permittivity of the medium and (b) write complete time-domain expressions for both the electric and magnetic field components of the wave.

Solution:

(a) We know that $f = v_p/\lambda = 10^8/0.2 = 500$ MHz. We also know that $v_p = 1/\sqrt{\mu\epsilon} = 1/\sqrt{\mu_0\epsilon_r\epsilon_0} \simeq 3 \times 10^8/\sqrt{\epsilon_r} = 10^8$, from which we find the relative permittivity to be $\epsilon_r = 9$.

(b) The phase constant is $\beta = 2\pi/\lambda = 2\pi/0.2 = 10\pi$ rad-m^{-1}, and the intrinsic impedance is $\eta = \sqrt{\mu_0/\epsilon} = \sqrt{\mu_0/(\epsilon_r \epsilon_0)} \simeq 120\pi/3 \simeq 126\Omega$. Therefore, the instantaneous electric and magnetic fields are given by

$$\overline{\mathscr{E}}(z,t) = \hat{\mathbf{y}}60\cos(10^9\pi t - 10\pi z) \quad \text{V-m}^{-1} \text{ and } \overline{\mathscr{H}}(z,t) \simeq -\hat{\mathbf{x}}0.477\cos(10^9\pi t - 10\pi z) \quad \text{A-m}^{-1}$$

Example 8.4: UHF cellular phone signal. The magnetic field component of a UHF electromagnetic signal transmitted by a cellular phone base station is given by

$$\mathbf{H}(y) = \hat{\mathbf{x}}50e^{-j(17.3y - \pi/3)} \quad \mu\text{A-m}^{-1}$$

where the coordinate system is defined such that the z axis is in the vertical direction above a horizontal ground (xy plane or the $z = 0$ plane). (a) Determine the frequency f and the wavelength λ. (b) Write the corresponding expression for the electric field, $\mathbf{E}(y)$. (c) An observer located at $y = 0$ is using a vertical electric dipole antenna to measure the electric field as a function of time. Assuming that the observer can measure the vertical component of the electric field without any loss, what is the electric field at the time instants corresponding to $\omega t_1 = 0$, $\omega t_2 = \pi/2$, $\omega t_3 = \pi$, $\omega t_4 = 3\pi/2$, $\omega t_5 = 2\pi$ radians?

Solution:

(a) From $\beta = \omega/v_p = 2\pi f/c = 17.3$ rad-m^{-1}, we find $f \simeq 826$ MHz. The corresponding wavelength is $\lambda = 2\pi/\beta \simeq 36.3$ cm.

(b) From $\nabla \times \mathbf{H} = j\omega\epsilon\mathbf{E}$, we can solve for $\mathbf{E}$ as

$$\mathbf{E}(y) = \hat{\mathbf{z}}\eta H_x(y) \simeq \hat{\mathbf{z}}377 \times 50 \times 10^{-6}e^{-j(17.3y - \pi/3)} = \hat{\mathbf{z}}18.85e^{-j(17.3y - \pi/3)} \quad \text{mV-m}^{-1}$$

(c) The instantaneous electric field is given by

$$\overline{\mathscr{E}}(y,t) = \mathscr{R}e\{\mathbf{E}(y)e^{j\omega t}\} \simeq \hat{\mathbf{z}}18.85\cos(5.19 \times 10^9 t - 17.3y + \pi/3) \quad \text{mV-m}^{-1}$$

Substituting $y = 0$ and $t_1 = 0$ yields $\overline{\mathscr{E}}(0, t_1) = \hat{\mathbf{z}}9.425$ mV-m^{-1}. Similarly, $\overline{\mathscr{E}}(0, t_2) \simeq -\hat{\mathbf{z}}16.3$ mV-m^{-1}, $\overline{\mathscr{E}}(0, t_3) = -\hat{\mathbf{z}}9.425$ mV-m^{-1}, $\overline{\mathscr{E}}(0, t_4) \simeq \hat{\mathbf{z}}16.3$ mV-m^{-1}, and $\overline{\mathscr{E}}(0, t_5) = \hat{\mathbf{z}}9.425$ mV-m^{-1}.

8.2.1 The Electromagnetic Spectrum

The properties of uniform plane waves, as defined by the electric and magnetic field expressions given in (8.14), are identical over all of the electromagnetic spectrum that has been investigated experimentally, frequencies ranging from millihertz to 10^{24} Hz. Regardless of their frequency, all uniform plane electromagnetic waves propagate in unbounded free space with the same velocity, namely $v_p = c$, but with different wavelengths, as determined by $\lambda = c/f$. The speed of propagation of electromagnetic waves is also independent of frequency in a simple lossless material medium. (However, the material properties themselves are often functions of frequency, so the assumption of a simple lossless material medium does not hold true over the entire electromagnetic spectrum for any material medium.) Table 8.1 lists the various designated frequency and

TABLE 8.1 THE ELECTROMAGNETIC SPECTRUM AND RELATED APPLICATIONS

Frequency	Designation	Selected Applications	Wavelength (in Free Space)
$> 10^{22}$ Hz	Cosmic rays	Astrophysics	
10^{18}–10^{22} Hz	γ-rays	Cancer therapy, astrophysics	
10^{16}–10^{21} Hz	X-rays	Medical diagnosis	
10^{15}–10^{18} Hz	Ultraviolet	Sterilization	0.3–300 nm
3.95×10^{14}–7.7×10^{14} Hz	Visible light	Vision, astronomy, optical communications	390–760 nm
		Violet	390–455
		Blue	455–492
		Green	492–577
		Yellow	577–600
		Orange	600–625
		Red	625–760
30–300 THz	Near infrared	Heating, night vision, optical communications	1–10 μm
3–30 THz	Medium to far infrared	Scanning/identification of objects/materials	10–100 μm
0.3–3 THz	Millimeter	Astronomy	0.1–1 mm
30–300 GHz	EHF	Radar, remote sensing	0.1–1 cm
80–100		W-band	
60–80		V-band	
40–60		U-band	
27–40		K_a-band	
3–30 GHz	SHF	Radar, satellite comm.	1–10 cm
18–27		K-band	
12–18		K_u-band	
8–12		X-band	
4–8		C-band	
0.3–3 GHz	UHF	Radar, TV, GPS, cellular phone	10–100 cm
2–4		S-band	
2.45		Microwave ovens	
1–2		L-band, GPS system	
470–890 MHz		TV Channels 14–83	
30–300 MHz	VHF	TV, FM, police,	1–10 m
174–216		TV Channels 7–13	
88–108		FM radio	
76–88		TV Channels 5–6	
54–72		TV Channels 2–4	
3–30 MHz	HF	Short-wave, citizens' band	10–100 m
0.3–3 MHz	MF	AM broadcasting	0.1–1 km
30–300 kHz	LF	Navigation, radio beacons	1–10 km
3–30 kHz	VLF	Navigation, positioning, naval communications	10–100 km
0.3–3 kHz	ULF	Telephone, audio	0.1–1 Mm
30–300 Hz	SLF	Power transmission, submarine communications	1–10 Mm
3–30 Hz	ELF	Detection of buried metals	10–100 Mm
< 3 Hz		Geophysical prospecting	> 100 Mm

wavelength ranges of the electromagnetic spectrum and selected applications for each range. Maxwell's equations and the results derived from them thus encompass a truly amazing range of physical phenomena and applications that affect nearly every aspect of human life and our physical environment. In this section, we briefly comment on a few of the many applications listed in Table 8.1.

At frequencies in the ultraviolet range and higher, physicists are more accustomed to thinking in terms of the associated energy level of the photon (a quantum of radiation), which is given by hf, where $h \simeq 6.63 \times 10^{-34}$ J-s is *Planck's constant*. Cosmic rays, consisting typically of photons at energies 10 MeV or greater, are constantly present in our universe; they ionize the highest reaches of Earth's atmosphere and help maintain the ionosphere at night, in the absence of solar radiation. Short-duration bursts of γ-rays, which bathe our solar system (and our galaxy) about three times a day, are believed to be produced in the most powerful explosions in the universe, releasing (within a few seconds or minutes) energies of 10^{51} ergs—more energy than our sun will produce in its entire ten billion years of existence.[8] Brief flashes of γ-rays have been observed to be originating from Earth,[9] and represent the highest natural energy emission within Earth's atmosphere. These bursts of γ-rays are generated from electrons accelerated by strong electric fields associated with thunderstorms. The exact generation mechanism is still an active area of research; recent theoretical work has shown that sufficient electron acceleration for γ-ray production may result from strong electric fields within thunderclouds,[10] from intense electromagnetic impulses generated by individual lightning flashes,[11] or from strong electric fields above a thundercloud after a large lightning discharge to ground.[12]

It is also interesting to note that only a very narrow portion of the electromagnetic spectrum is perceptible to human vision, namely the visible range. We can also "feel" infrared as heat, and our bodies can be damaged by excessive amounts of microwave radiation, X-rays, and γ-rays. Applications of X-rays, ultraviolet, visible, and infrared light are far too numerous to be commented on here and include vision, lasers, optical fiber communications, and astronomy.

The part of the electromagnetic spectrum between the low-frequency end of infrared and the high-frequency end of the millimeter range is called terahertz radiation. The frequency in this region is measured in terahertz (THz), or 10^{12} Hz, corresponding to wavelengths in the submillimeter range. This frequency band is used by astronomers to study the chemical composition of celestial bodies. In addition, artificial generation of terahertz waves is an active area of research. Typically, optical lasers are modified to generate

[8]G. J. Fishman and D. H. Hartmann, Gamma-ray bursts, *Scientific American,* pp. 46–51, July 1997.

[9]G. J. Fishman, P. N. Bhat, R. Mallozzi, J. M. Horack, T. Koshut, C. Kouveliotou, G. N. Pendleton, C. A. Meegan, R. B. Wilson, W. S. Paciesas, S. J. Goodman, and H. J. Christian, Discovery of intense gamma-ray flashes of atmospheric origin, *Science,* 264, pp. 1313–1316, May 1994.

[10]J. R. Dwyer and D. M. Smith, A comparison between Monte Carlo simulations of runaway breakdown and terrestrial gamma-ray flash observations *Geophys. Res. Lett.,* 32(22), 2005.

[11]U. S. Inan and N. G. Lehtinen, Production of terrestrial gamma-ray flashes by an electromagnetic pulse from a lightning return stroke, *Geophys. Res. Lett.* 32(19), 2005.

[12]N. G. Lehtinen, T. F. Bell, and U. S. Inan, Monte Carlo simulation of runaway MeV electron breakdown with application to red sprites and terrestrial gamma ray flashes, *J. Geophys. Res. Space Phys.,* 104(A11), 1999.

radiation at the lower submillimeter frequency range, such as by mixing two signals or using a pulsed laser to dislodge electrons from a semiconductor that are then accelerated by an applied field.[13] Technological advancements in THz radiation sources have enabled a growing number of practical applications that leverage this frequency band. Unlike visible or infrared light, some frequencies of terahertz radiation can penetrate low-moisture biological tissue, clothing, and some polymers, enabling novel imaging technologies for a wide variety of medical, security, and quality control applications. The submillimeter band is also a candidate for use in high data rate wireless transmission applications.

Each decade of the electromagnetic spectrum below the millimeter range of frequencies is divided[14] into designated ranges, with the acronyms indicated in Table 8.1: extremely high frequency (EHF), super high frequency (SHF), ultra high frequency (UHF), very high frequency (VHF), high frequency (HF), medium frequency (MF), low frequency (LF), very low frequency (VLF), ultra low frequency (ULF), super low frequency (SLF), extremely low frequency (ELF).

The microwave band of frequencies is vaguely defined as the range from 300 MHz up to 1 THz, including the millimeter range. It is extensively utilized for radar, remote sensing, and a host of other applications too numerous to cite here.[15] In radar work, the microwave band is further subdivided into bands with alphabetical designations, which are listed in Table 8.1.

The VLF range of frequencies is generally used for global navigation and naval communications, using transmitters that utilize huge radiating structures.[16] Although VLF transmissions are used for global communications with surface ships and submarines near the water surface, even lower frequencies are required for communication with deeply submerged submarines. The Sanguine system operated by the U.S. Navy utilizes two sets of large (22.5 km length) orthogonal horizontal antennas, one located in Wisconsin and the other in Michigan at a distance of 240 km, and operates[17] at 72 to 80 Hz.

The lowest frequencies of the experimentally investigated electromagnetic spectrum are commonly used to observe so-called micropulsations, which are electromagnetic waves at frequencies 0.001–10 Hz generated as a result of large-scale currents flowing in the earth's auroral regions and by the interaction between the earth's magnetic field and the energetic particles that stream out of the sun in the form of the solar wind.[18]

---

[13]For a summary of some modern terahertz generation technologies, see Nagatsuma, T. Generating Millimeter and Terahertz Wave, *IEEE Microwave Magazine*, 2009.

[14]There is a certain arbitrariness in the designations of these frequency ranges. In geophysics and solar terrestrial physics, the designation ELF is used for the range 3 Hz–3 kHz, while ULF is used to describe frequencies typically below 3 Hz.

[15]See J. Thuery, *Microwaves: Industrial, Scientific and Medical Applications,* Artech House, Boston-Norwood, Massachusetts, 1992. Also see O. P. Gandhi, editor, *Biological Effects and Medical Applications of Electromagnetic Energy,* Prentice-Hall, 1990.

[16]See A. D. Watt, *VLF Radio Engineering,* Pergamon Press, New York, 1967; J. C. Kim and E. I. Muehldorf, *Naval Shipboard Communications Systems,* Prentice-Hall, Englewood Cliffs, New Jersey, 1995.

[17]C. H. Richard, *Sub vs. Sub,* Orion Books, New York, 1988; M. F. Genge and R. D. Carlson, Project ELF Electromagnetic Compatibility Assurance Program, *IEEE J. Oceanic Eng.,* 9(3), pp. 143–153, July 1984.

[18]J. A. Jacobs, *Geomagnetic Micropulsations,* Springer-Verlag, New York, Heidelberg, and Berlin, 1970; J. K. Hargreaves, *The Solar-Terrestrial Environment,* Cambridge University Press, 1992.

These natural signals are also often used for geophysical prospecting and magnetotelluric studies.[19] Another potentially very important practical use of this band may yet emerge, based on experimental evidence of detectable electromagnetic precursors produced many hours prior to major earthquakes.[20]

8.3 PLANE WAVES IN LOSSY MEDIA

Many of the more interesting electromagnetic applications involve the interactions between electric and magnetic fields and matter. The manner in which waves interact with matter is introduced in Chapter 11. For now, we represent the microscopic interactions of electromagnetic waves with matter in terms of the macroscopic parameters ϵ, μ, and σ. In general, most dielectric media exhibit small but nonzero conductivity or complex permittivity and can absorb electromagnetic energy, resulting in the attenuation of an electromagnetic wave as it propagates through the medium. The performance of most practical transmission lines, and of other devices that convey electromagnetic energy from one point to another, is limited by small losses in conductors or dielectrics. The inherently lossy nature of some media (e.g., seawater, animal tissue) determines the range of important applications (e.g., submarine communications and medical diagnostic implants). All media exhibit losses in some frequency ranges; for example, although air is largely transparent (lossless) over the radio and microwave ranges, it is a highly lossy medium at optical frequencies. The fact that upper atmospheric ozone is lossy at ultraviolet frequencies (i.e., it absorbs ultraviolet light) protects life on Earth from this harmful radiation.

When a material exhibits a nonzero conductivity σ, the electric field of a propagating wave causes a conduction current of $\mathbf{J}_c = \sigma\mathbf{E}$ to flow. This current, which is in phase with the wave electric field, leads to dissipation of some of the wave energy as heat within the material, with the power dissipated per unit volume being given by $\mathbf{E} \cdot \mathbf{J}$, as discussed in Section 5.8. This dissipation requires that the waves electric and magnetic fields attenuate with distance as they propagate in the lossy material, much like the attenuation of voltage and current waves propagating on a lossy transmission line as discussed in Section 3.7. To determine the characteristics of uniform plane waves in lossy media, we start with Maxwell's equations and follow a procedure quite similar to that used in the previous section.

In a conducting (or lossy) medium the $\mathbf{J}$ term in (7.23c) is nonzero even in the absence of external sources, since a conduction current of $\mathbf{J}_c = \sigma\mathbf{E}$ flows in response to the wave's electric field. We then have

$$\nabla \times \mathbf{H} = \sigma\mathbf{E} + j\omega\epsilon\mathbf{E} \tag{8.15}$$

[19]M. N. Nabighian, (Ed.), *Electromagnetic Methods in Applied Geophysics, Vol. 2: Application,* Parts A and B, Society of Exploration Geophysics, 1991.

[20]A. C. Fraser-Smith, A. Bernardi, P. R. McGill, M. E. Ladd, R. A. Helliwell, and O. G. Villard, Jr., Low-frequency magnetic field measurements near the epicenter of the M_s 7.1 Loma Prieta earthquake, *Geophys. Res. Lett.,* 17, pp. 1465–1468, 1990; also see B. Holmes, Radio hum may herald quakes, *New Scientist,* p. 15, 23/30 December 1995.

and, taking the curl of (7.23a), we have

$$\nabla \times \nabla \times \mathbf{E} \equiv \nabla(\nabla \cdot \mathbf{E}) - \nabla^2 \mathbf{E} = -j\omega \nabla \times \mathbf{B}$$

Using the fact that $\nabla \cdot \mathbf{E} = 0$ for a simple, source-free medium, and substituting from (7.23c), we find

$$\nabla^2 \mathbf{E} - j\omega\mu(\sigma + j\omega\epsilon)\mathbf{E} = 0$$

or

$$\boxed{\nabla^2 \mathbf{E} - \gamma^2 \mathbf{E} = 0} \tag{8.16}$$

where $\gamma = \sqrt{j\omega\mu(\sigma + j\omega\epsilon)}$ is known as the propagation constant (or wave number) and is in general a complex quantity, which can be expressed in terms of its real and imaginary parts as $\gamma = \alpha + j\beta$. Note that the complex propagation constant γ for uniform plane waves in an unbounded lossy medium is analogous to the complex propagation constant γ given in (3.60) and derived in Section 3.7 for propagation of voltage and current waves on a lossy transmission line.

Realizing that (8.16) represents three scalar equations similar to (8.7), we limit our attention once again to uniform plane waves propagating in the z direction (i.e., all field quantities varying only in z), with the electric field having only an x component, for which (8.16) becomes

$$\boxed{\frac{d^2 E_x(z)}{dz^2} - \gamma^2 E_x(z) = 0} \tag{8.17}$$

which, when we replace $E_x(z)$ with $V(z)$, is identical to equation (3.59) in Section 3.7 that describes the variation of the voltage phasor $V(z)$ on a lossy transmission line. The general solution of (8.17) is

$$E_x(z) = C_1 e^{-\gamma z} + C_2 e^{+\gamma z} = \underbrace{C_1 e^{-\alpha z} e^{-j\beta z}}_{E_x^+(z)} + \underbrace{C_2 e^{+\alpha z} e^{+j\beta z}}_{E_x^-(z)} \tag{8.18}$$

For $\alpha, \beta > 0$, the two terms $E_x^+(z)$ and $E_x^-(z)$ represent waves propagating in the $+z$ and $-z$ directions, respectively. Note that the constants C_1 and C_2 are in general complex numbers.

The instantaneous electric field can be found from (8.18) in the same way[21] as described in (8.12). We find

$$\mathcal{E}_x(z, t) = \underbrace{C_1 e^{-\alpha z} \cos(\omega t - \beta z)}_{\mathcal{E}_x^+(z,t)} + \underbrace{C_2 e^{+\alpha z} \cos(\omega t + \beta z)}_{\mathcal{E}_x^-(z,t)} \tag{8.19}$$

assuming C_1 and C_2 to be real. The nature of the waves described by the two terms of (8.19) is shown in Figure 8.6. We see, for example, that the wave propagating in the $+z$

---

[21] $\mathfrak{Re}\{C_1 e^{-\alpha z} e^{-j\beta z} e^{j\omega t}\} = C_1 e^{-\alpha z} \mathfrak{Re}\{e^{j(\omega t - \beta z)}\} = C_1 e^{-\alpha z} \cos(\omega t - \beta z)$, where C_1 is assumed to be a real constant.

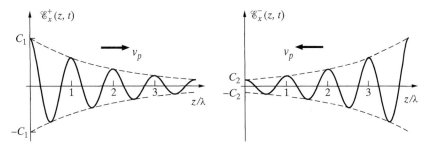

Figure 8.6 Snapshots of waves in a lossy medium. The two terms of equation (8.19) plotted as a function of z/λ at $\omega t = 0$.

direction has a decreasing amplitude with increasing distance at a fixed instant of time. In other words, the wave is attenuated as it propagates in the medium. The rate of this attenuation is given by the attenuation constant α, in units of np-m^{-1} (see Section 3.7.1 for a discussion of the unit nepers per meter [np-m^{-1}]). The quantity β (in rad-m^{-1}) determines the phase velocity and wavelength of the wave and is referred to as the phase constant. We shall see below that the wavelength of a uniform plane wave in lossy materials can be substantially different from that in free space. We note once again that all aspects of the solution given in (8.19) are entirely analogous to that given in (3.66) for voltage and current waves on a lossy transmission line. In this connection, the similarity between Figure 8.6 and Figure 3.54a should also be noted.

Note that the complex propagation constant γ is given by

$$\gamma = \alpha + j\beta = \sqrt{j\omega\mu(\sigma + j\omega\epsilon)} = j\omega\sqrt{\mu\epsilon}\sqrt{1 - j\frac{\sigma}{\omega\epsilon}} \qquad (8.20)$$

Explicit expressions for α and β can be found[22] by squaring and equating the real and imaginary parts. We find

$$\alpha = \omega\sqrt{\frac{\mu\epsilon}{2}}\left[\sqrt{1 + \left(\frac{\sigma}{\omega\epsilon}\right)^2} - 1\right]^{1/2} \qquad (8.21)$$

[22] We have

$$\gamma^2 = (\alpha + j\beta)^2 = j\omega\mu(\sigma + j\omega\epsilon) \quad \rightarrow \quad \underbrace{\alpha^2 - \beta^2 = -\omega^2\mu\epsilon}_{\text{Real part}} \quad \text{and} \quad \underbrace{2\alpha\beta = \omega\mu\sigma}_{\text{Imaginary part}}$$

Squaring and adding the two equations and taking the square root, we have

$$(\alpha^2 - \beta^2)^2 + (2\alpha\beta)^2 = \omega^4\mu^2\epsilon^2 + (\omega\mu\sigma)^2 \rightarrow \alpha^4 + 2\alpha^2\beta^2 + \beta^4 = \omega^4\mu^2\epsilon^2 + \omega^2\mu^2\sigma^2$$

$$\rightarrow (\alpha^2 + \beta^2)^2 = \omega^4\mu^2\epsilon^2\left[1 + \left(\frac{\sigma}{\omega\epsilon}\right)^2\right] \rightarrow \alpha^2 + \beta^2 = \omega^2\mu\epsilon\sqrt{1 + \left(\frac{\sigma}{\omega\epsilon}\right)^2}$$

Equation (8.21) follows by adding the expressions for $(\alpha^2 + \beta^2)$ and $(\alpha^2 - \beta^2)$.

and

$$\beta = \omega \sqrt{\frac{\mu\epsilon}{2}} \left[\sqrt{1 + \left(\frac{\sigma}{\omega\epsilon}\right)^2} + 1 \right]^{1/2} \tag{8.22}$$

Consider now only the wave propagating in the $+z$ direction, the first term of (8.18):

$$E_x^+(z) = C_1 e^{-\alpha z} e^{-j\beta z} \tag{8.23}$$

The wave magnetic field that accompanies this electric field can be found by substituting (8.23) into (7.23a), as we did for the lossless case, or by making an analogy with the lossless-media solution obtained in the previous section. Following the latter approach, we can rewrite (8.15) as

$$\nabla \times \mathbf{H} = \sigma\mathbf{E} + j\omega\epsilon\mathbf{E} = j\omega\epsilon_{\text{eff}}\mathbf{E}$$

where

$$\epsilon_{\text{eff}} \equiv \epsilon - j\sigma/\omega \tag{8.24}$$

It thus appears that the solutions we obtained in the previous section for a lossless medium can be used as long as we make the substitution $\epsilon \to \epsilon_{\text{eff}}$. On this basis, the propagation constant may be written in a more compact form as

$$\gamma = j\omega\sqrt{\mu\epsilon_{\text{eff}}} \tag{8.25}$$

The intrinsic impedance of a conducting medium, which is also a complex quantity, can be found as

$$\eta_c = |\eta_c| e^{j\phi_\eta} = \sqrt{\frac{\mu}{\epsilon_{\text{eff}}}} = \sqrt{\frac{\mu}{\epsilon - j\dfrac{\sigma}{\omega}}} = \frac{\sqrt{\dfrac{\mu}{\epsilon}}}{\left[1 + \left(\dfrac{\sigma}{\omega\epsilon}\right)^2\right]^{1/4}} e^{j(1/2)\tan^{-1}[\sigma/(\omega\epsilon)]} \tag{8.26}$$

and using (8.23) the associated magnetic field $\mathbf{H}$ is

$$\mathbf{H} = \hat{\mathbf{y}} H_y^+(z) = \hat{\mathbf{y}} \frac{1}{\eta_c} C_1 e^{-\alpha z} e^{-j\beta z}$$

The instantaneous magnetic field $\overline{\mathcal{H}}(z, t)$ can be found as

$$\mathcal{H}_y^+(z, t) = \mathcal{R}e\{H_y^+(z) e^{j\omega t}\} = \mathcal{R}e \left\{ \frac{1}{|\eta_c| e^{j\phi_\eta}} C_1 e^{-\alpha z} e^{-j\beta z} e^{j\omega t} \right\}$$

$$= \frac{1}{|\eta_c|} C_1 e^{-\alpha z} \mathcal{R}e\{e^{j(\omega t - \beta z)} e^{-j\phi_\eta}\} \tag{8.27}$$

$$\mathcal{H}_y^+(z, t) = \frac{C_1}{|\eta_c|} e^{-\alpha z} \cos(\omega t - \beta z - \phi_\eta)$$

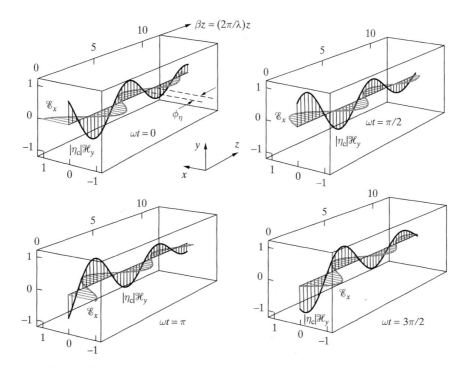

Figure 8.7   Time snapshots of the electric and magnetic fields in a conducting medium. The curves shown are for $C_1 = 1$, $\phi_\eta = 45°$, and $\alpha = 0.1\beta$, where $\beta = 2\pi/\lambda$.

Thus we see that the electric and magnetic fields of a uniform plane wave in a conducting medium do not reach their maxima at the same time (at a fixed point) or at the same point (at a fixed time); in other words, $\overline{\mathcal{E}}(z,t)$ and $\overline{\mathcal{H}}(z,t)$ are *not in phase*. The magnetic field lags the electric field by a phase difference equal to the phase of the complex intrinsic impedance, ϕ_η. This relationship is depicted in Figure 8.7.

We can see from (8.21), (8.22), and (8.26) that the rate of attenuation α, the phase constant β, and the intrinsic impedance η_c depend sensitively on $\sigma/(\omega\epsilon)$. For $\sigma \ll (\omega\epsilon)$, the attenuation rate is small, and the propagation constant and intrinsic impedance are only slightly different from that for a lossless medium; namely, $\beta \simeq \omega\sqrt{\mu\epsilon}$ and $\eta_c \simeq \sqrt{\mu/\epsilon}$. For $\sigma \gg (\omega\epsilon)$, the attenuation rate α is large, causing the uniform plane wave to decay rapidly with distance, and the intrinsic impedance is very small, approaching zero as $\sigma \to \infty$, as it does for a perfect conductor. A perfect conductor, in other words, is a medium with zero intrinsic impedance,[23] so the presence of an electric field of nonzero

---

[23] A perfect conductor is thus the ultimate lossy material. Although this may at first appear counterintuitive, the behavior of a perfect conductor as a medium with zero impedance is analogous to a short circuit. The voltage drop across a short circuit is always zero, while the current through it is determined by the external circuits it is connected to. Similarly, we shall see in Chapter 9 that when a uniform plane wave is incident on a perfect conductor, the wave is perfectly reflected, since no electromagnetic fields can exist inside the conductor, while the current that flows on the surface of the perfect conductor is determined by the magnetic field intensity of the incident uniform plane wave.

magnitude $C_1 \neq 0$ requires a magnetic field of an infinite magnitude. Thus, no time-harmonic electric or magnetic fields can exist in a perfect conductor.[24]

The quantity

$$\tan \delta_c \equiv \frac{\sigma}{\omega \epsilon} \tag{8.28}$$

is called the *loss tangent* of the medium and is a measure of the degree to which the medium conducts. The complex propagation constant can be expressed in terms of the loss tangent as

$$\gamma = j \omega \sqrt{\mu \epsilon} \sqrt{1 - j \tan \delta_c} \tag{8.29}$$

A medium is considered to be a good conductor if $\tan \delta_c \gg 1$; most metals are good conductors at frequencies of up to 100 GHz or so. A medium is considered to be a poor conductor, or a good insulator, if $\tan \delta_c \ll 1$.

Example 8.5: Microwave exposure of muscle tissue. Find the complex propagation constant γ and the intrinsic impedance η_c of a microwave signal in muscle tissue[25] at 915 MHz ($\sigma = 1.6$ S-m^{-1}, $\epsilon_r = 51$, $\mu_r = 1$).

Solution: The complex propagation constant is given by

$$\gamma = \alpha + j\beta$$

where α and β are given by (8.21) and (8.22). Using (8.28), the loss tangent of the muscle tissue at 915 MHz is

$$\tan \delta_c = \frac{\sigma}{\omega \epsilon} \simeq \frac{1.6 \text{ S-m}^{-1}}{2\pi \times 915 \times 10^6 \text{ rad-s}^{-1} \times 51 \times 8.85 \times 10^{-12} \text{ F-m}^{-1}} \simeq 0.617$$

So α and β are given by

$$\begin{Bmatrix} \alpha \\ \beta \end{Bmatrix} \simeq (2\pi \times 915 \times 10^6 \text{ rad-s}^{-1}) \frac{\sqrt{51}}{\sqrt{2} \times 3 \times 10^8 \text{ m-s}^{-1}} \left[\sqrt{1 + (0.617)^2} \mp 1 \right]^{1/2}$$

$$\simeq \begin{Bmatrix} 40.7 \text{ np-m}^{-1} \\ 142.7 \text{ rad-m}^{-1} \end{Bmatrix}$$

So the complex propagation constant is given by

$$\gamma \simeq 40.7 + j\,142.7$$

The complex intrinsic impedance can be found using (8.26) as

$$\eta_c \simeq \frac{377/\sqrt{51}}{[1 + (0.617)^2]^{1/4}} \exp \left[j \frac{1}{2} \tan^{-1}(0.617) \right] \simeq 48.7 e^{j\,15.8°} \ \Omega$$

---

[24]Note that static electric fields inside metallic conductors must also be zero and that static magnetic fields cannot exist in a perfect conductor because of the Meissner effect. See Sections 4.7 and 6.8.4.

[25]C. C. Johnson and A. W. Guy, Nonionizing electromagnetic wave effects in biological materials and systems, *Proc. IEEE*, 60(6), pp. 692–718, June 1972.

At this point it is important to remember that the material properties themselves (i.e., σ and ϵ) may well be functions of frequency (as discussed further in Sections 8.3.1 and 11.2). For typical good conductors, both σ and ϵ are nearly independent of frequency, at frequencies below the optical range, but for lossy dielectrics the material constants σ and ϵ tend to be functions of frequency. For some materials, the loss tangent $\tan \delta_c = \sigma/(\omega\epsilon)$ tends to vary less over the frequency range of interest. Properties of dielectrics are usually given in terms of ϵ and $\tan \delta_c$. The ratio $\sigma/(\omega\epsilon)$ is plotted as a function of frequency in Figure 8.8 for some common materials, assuming the parameters ϵ and σ to be constants. Table 8.2 gives the values of σ and ϵ used in Figure 8.8. The values above the microwave region (>10 GHz) are not likely to be accurate since σ and ϵ may vary significantly with frequency.

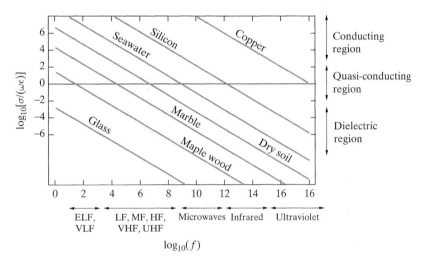

Figure 8.8   Loss tangent versus frequency. The loss tangent $\tan \delta_c = \sigma/(\omega\epsilon)$ for selected materials plotted as a function of frequency, assuming that the material constants σ and ϵ are as given in Table 8.2 and that they do not vary with frequency.

TABLE 8.2   RELATIVE PERMITTIVITY AND CONDUCTIVITY OF SELECTED MATERIALS

Medium	Relative Permittivity ϵ_r (Dimensionless)	Conductivity σ, (S-m^{-1})
Copper	1	5.8×10^7
Seawater	81	4
Doped silicon	12	10^3
Marble	8	10^{-5}
Maple wood	2.1	3.3×10^{-9}
Dry soil	3.4	10^{-4} to 10^{-2}
Fresh water	81	$\sim 10^{-2}$
Mica	6	10^{-15}
Flint glass	10	10^{-12}

The most common cases of practical interest involving the propagation of electromagnetic waves in lossy media concern lossy dielectrics, which exhibit complex permittivity, and good conductors. We consider these two cases separately in the next two sections.

8.3.1 Uniform Plane Wave Propagation in a Lossy Dielectrics

In most dielectrics that are good insulators, the direct conduction current (which is due to finite σ) is usually negligible. However, at high frequencies an alternating current that is in phase with the applied field is present, because the rapidly varying applied electric field alternately polarizes the bound electrons, thus doing work against molecular forces. As a result, materials that are good insulators at low frequencies can consume considerable energy when they are subjected to high-frequency fields. The heat generated as a result of such radio-frequency heating is used in molding plastics, in microwave cooking, and in microwave drying of a wide variety of materials.[26]

The microphysical basis of such effects are different for solids, liquids, and gases and are too complex to be summarized here.[27] When an external time-varying field is applied to a material, bound charges are displaced, giving rise to volume polarization density $\mathbf{P}$. At sinusoidal steady state, the polarization $\mathbf{P}$ varies at the same frequency as the applied field $\mathbf{E}$. At low frequencies, $\mathbf{P}$ is also in phase with $\mathbf{E}$, both quantities reaching their maxima and minima at the same points in the radio-frequency cycle. As the frequency is increased, however, the charged particles resist acceleration by the changing field, not just because of their mass but also because of the elastic and frictional forces that keep them attached to their molecules. This "inertia" tends to prevent the polarization $\mathbf{P}$ from keeping in phase with the applied field. The work that must be done against the frictional damping forces causes the applied field to lose power, and this power is deposited in the medium as heat. This condition of out-of-phase polarization that occurs at higher frequencies can be characterized by a complex and frequency-dependent electric susceptibility $\chi_e(\omega)$ and hence a complex permittivity ϵ_c (see Section 11.2). By convention, ϵ_c can be expressed in terms of its real and imaginary components as

$$\epsilon_c = \epsilon' - j\epsilon''$$

We can analyze the resulting effects by substituting the preceding equation into (7.23c):

$$\nabla \times \mathbf{H} = \sigma\mathbf{E} + j\omega\epsilon_c\mathbf{E} = (\sigma + \omega\epsilon'')\mathbf{E} + j\omega\epsilon'\mathbf{E} \qquad (8.30)$$

Comparing (8.30) to (8.15), it thus appears that the imaginary part of ϵ_c (namely, ϵ'') leads to an extra volume current density term that is in phase with the electric field, as

---

[26] J. Thuery, *Microwaves: Industrial, Scientific and Medical Applications,* Artech House, Boston-Norwood, Massachusetts, 1992.

[27] For a discussion at an appropriate level, see C. Kittel, *Introduction to Solid State Physics,* 5th ed., Wiley, New York, 1976.

if the material has an effective conductivity

$$\sigma_{\text{eff}} = \sigma + \omega\epsilon'' \tag{8.31}$$

The propagation constant γ can be reevaluated by assuming a permittivity, permeability, and conductivity for the medium given, respectively, by ϵ', μ, and σ_{eff} and using (8.20) with σ replaced with σ_{eff}. Alternatively, we can substitute σ with σ_{eff} in (8.24) to lump all lossy terms into an effective complex permittivity given by

$$\epsilon_{\text{eff}} = \epsilon' - j\frac{\sigma_{\text{eff}}}{\omega} = \epsilon' - j\left(\epsilon'' + \frac{\sigma}{\omega}\right) \tag{8.32}$$

Substituting (8.32) into equations (8.25) and (8.26) (where σ is again replaced with σ_{eff}) recovers the relevant expressions for the complex propagation constant and the intrinsic impedance, respectively.

At low frequencies, $\omega\epsilon''$ is small, because ω is small and ϵ'' is itself small, so the dielectric losses are largely negligible. However, at high frequencies $\omega\epsilon''$ increases and produces the same macroscopic effect as if the dielectric had an extra effective conductivity term $\omega\epsilon''$. When a steady current $\mathbf{J}_c = \sigma\mathbf{E}$ flows in a conducting material in response to a constant applied field $\mathbf{E}$, the electrical power per unit volume dissipated in the material is given by $\mathbf{E} \cdot \mathbf{J}_c = \sigma|\mathbf{E}|^2$ in units of W-m^{-3}. Similarly, when a dielectric is excited at frequencies high enough for $\omega\epsilon''$ to be appreciable, an alternating current density of $\omega\epsilon''\mathbf{E}$ flows (in addition to the displacement current density given by $j\omega\epsilon'\mathbf{E}$) in response to an applied alternating field $\mathbf{E}$, leading to an instantaneous power dissipation of $[(\omega\epsilon''\mathbf{E}) \cdot \mathbf{E}] = \omega\epsilon''|\mathbf{E}|^2$ in units of W-m^{-3}. When the electric field is time-harmonic, that is, $\mathscr{E}(z,t) = |\mathbf{E}|\cos(\omega t \pm \beta z)$, the time-average power dissipated in the material per unit volume at any arbitrary point (e.g., $z = 0$) is given by

$$P_{\text{av}} = T_p^{-1}\int_0^{T_p}\omega\epsilon''|\mathscr{E}(0,t)|^2\,dt = T_p^{-1}\int_0^{T_p}\omega\epsilon''|\mathbf{E}|^2\cos^2(\omega t)\,dt = \tfrac{1}{2}\omega\epsilon''|\mathbf{E}|^2$$

This dissipated power density is the basis for microwave heating of dielectric materials. Since the values of ϵ'' are often determined by measurement, it is not necessary to distinguish between losses due to nonzero conductivity σ and the dielectric losses discussed here. In such cases, both the frictional damping and any other ohmic losses (due to nonzero conductivity σ) are included in ϵ'', and hence (8.31) simply becomes $\sigma_{\text{eff}} = \omega\epsilon''$.

In general, both ϵ' and ϵ'' can depend on frequency in complicated ways, exhibiting several resonances over wide frequency ranges. The typical behavior in the vicinity of resonances is enhanced losses (i.e., large values of ϵ'') and a reduction of ϵ' to a new level.[28] Different materials exhibit a large variety of resonances, which originate in the basic energy-level structure of the atoms and molecules in these materials. For example, for water at room temperature, the resonance frequency is $\sim$24 GHz, whereas

---

[28]The frequency response of a typical dielectric is shown in Figure 11.4 of Section 11.2, with contributions from the different types of polarizations (orientational, ionic, and electronic) identified in terms of the frequency ranges in which they are significant.

for ice at $-20°C$ the relaxation frequency is ~ 1 kHz. Table 8.3 provides values of real and imaginary parts of the relative permittivity for selected materials.[29] Since these properties (especially ϵ'') are strongly dependent on frequency, and also to some degree on temperature, the frequency and temperature to which the given values correspond are specified. In some cases, the moisture content of the material is also noted, since the permittivity values also depend on this parameter.

Substituting (8.31) into (8.28), the loss tangent $\tan \delta_c$ becomes

$$\tan \delta_c = \frac{\sigma_{\text{eff}}}{\omega \epsilon'} = \frac{\sigma}{\omega \epsilon'} + \frac{\epsilon''}{\epsilon'} \tag{8.33}$$

As discussed earlier, in practice, it is generally not necessary to distinguish between losses due to σ or $\omega \epsilon''$, since ϵ' and σ_{eff} (or $\tan \delta_c$) are often determined by measurement. In such cases, we assume $\sigma = 0$ and (8.33) becomes $\tan \delta_c = \epsilon''/\epsilon'$.

The attenuation in a lossy dielectric is frequently expressed as the attenuation distance or penetration depth d over which the field intensity decreases by a factor of e^{-1}. Using (8.21) we can write

$$d \equiv \frac{1}{\alpha} = \frac{\sqrt{2}}{\omega \sqrt{\mu \epsilon'} [\sqrt{1 + \tan^2 \delta_c} - 1]^{1/2}} = \frac{c\sqrt{2}}{\omega \sqrt{\mu_r \epsilon_r'} [\sqrt{1 + \tan^2 \delta_c} - 1]^{1/2}} \tag{8.34}$$

where c is the speed of light in free space.

Example 8.6: Complex permittivity of distilled water. Consider distilled water at 25 GHz ($\epsilon_{cr} \simeq 34 - j9.01$). Calculate the attenuation constant α, propagation constant β, penetration depth d, and wavelength λ.

Solution: The loss tangent of distilled water at 25 GHz is $\tan \delta_c = \epsilon''/\epsilon' = 0.265$. From (8.21), the attenuation constant is then

$$\alpha \simeq \frac{(2\pi \times 25 \times 10^9 \text{ rad-m}^{-1})\sqrt{34}}{\sqrt{2} \times (3 \times 10^8 \text{ m-s}^{-1})}[\sqrt{1 + (0.265)^2} - 1]^{1/2} \simeq 401 \text{ np-m}^{-1}$$

---

[29]The values in Table 8.3 were taken from the following sources: *Reference Data for Radio Engineers,* 8th ed., Sams Prentice Hall Computer Publishing, Carmel, Indiana, 1993; J. Thuery, *Microwaves: Industrial, Scientific and Medical Applications,* Artech House, Inc., Boston-Norwood, Massachusetts, 1992; D. I. C. Wang and A. Goldblith, Dielectric properties of food, *Tech. Rep. TR-76-27-FEL,* MIT, Dept. of Nutrition and Food Science, Cambridge, Massachusetts, November 1976; S. O. Nelson, Comments on "Permittivity measurements of granular food products suitable for computer simulations of microwave cooking processes," *IEEE Trans. Instrum. Meas.,* 40(6), pp. 1048–1049, December 1991; A. D. Green, Measurement of the dielectric properties of cheddar cheese, *J. Microwave Power Electromagn. Energy,* 32(1), pp. 16–27, 1997; S. Puranik, A. Kumb-harkhane, and S. Mehrotra, Dielectric properties of honey-water mixtures between 10 MHz to 10 GHz using time domain technique, *J. Microwave Power Electromagn. Energy,* 26(4), pp. 196–201,1991; H. C. Rhim and O. Buyukozturk, Electromagnetic properties of concrete at microwave frequency range, *ACI Mater. J.,* 95(3), pp. 262–271, May–June 1998; J. M. Osepchuk, Sources and basic characteristics of microwave/RF radiation, *Bull. N. Y. Acad. Med.,* 55(11), pp. 976–998, December 1979; S. O. Nelson, Microwave dielectric properties of fresh onions, *Trans. Amer. Soc. Agr. Eng.,* 35(3), pp. 963–966, May–June 1992; R. F. Shiffmann, Understanding microwave reactions and interactions, *Food Product Design,* pp. 72–88, April 1993.

TABLE 8.3 DIELECTRIC PROPERTIES OF SELECTED MATERIALS

Material	f (GHz)	ϵ_r'	ϵ_r''	T(°C)
Aluminum oxide (AL_2O_3)	3.0	8.79	8.79×10^{-3}	25
Barium titanate ($BaTiO_3$)	3.0	600	180	26
Bread	2.45	4.6	1.20	
Bread dough	2.45	22.0	9.00	
Butter (salted)	2.45	4.6	0.60	20
Cheddar cheese	2.45	16.0	8.7	20
Concrete (dry)	2.45	4.5	0.05	25
Concrete (wet)	2.45	14.5	1.73	25
Corn (8% moisture)	2.45	2.2	0.2	24
Corn oil	2.45	2.5	0.14	25
Distilled water	2.45	78	12.5	20
Dry sandy soil	3.0	2.55	1.58×10^{-2}	25
Egg white	3.0	35.0	17.5	25
Frozen beef	2.45	4.4	0.528	−20
Honey (100% pure)	2.45	10.0	3.9	25
Ice (pure distilled)	3.0	3.2	2.88×10^{-3}	−12
Milk	3.0	51.0	30.1	20
Most plastics	2.45	2 to 4.5	0.002 to 0.09	20
Papers	2.45	2 to 3	0.1 to 0.3	20
Potato (78.9% moisture)	3.0	81.0	30.8	25
Polyethylene	3.0	2.26	7.01×10^{-4}	25
Polystyrene	3.0	2.55	8.42×10^{-4}	25
Polytetrafluoroethylene (Teflon)	3.0	2.1	3.15×10^{-4}	22
Raw beef	2.45	52.4	17.3	25
Snow (fresh fallen)	3.0	1.20	3.48×10^{-4}	−20
Snow (hard packed)	3.0	1.50	1.35×10^{-3}	−6
Some glasses (Pyrex)	2.45	~4.0	0.004 to 0.02	20
Smoked bacon	3.0	2.50	0.125	25
Soybean oil	3.0	2.51	0.151	25
Steak	3.0	4.0	12.0	25
White onion (78.7% moisture)	2.45	53.8	13.5	22
White rice (16% moisture)	2.45	3.8	0.8	24
Wood	2.45	1.2 to 5	0.01 to 0.5	25

In a similar manner, the phase constant β can be found from (8.22) to be $\beta \simeq 3079$ rad-m^{-1}. The penetration depth is $d = \alpha^{-1} \simeq 2.49$ mm, and the wavelength is $\lambda = 2\pi/\beta \simeq 2.04$ mm.

In other words, the depth of penetration into distilled water is of the order of one wavelength. Although the condition $\tan \delta_c \ll 1$ is not satisfied for this case, the value of β is nevertheless within less than one percent of that for a lossless dielectric with $\epsilon_r = 34$.

Example 8.7: Complex permittivity of bread dough. The dielectric properties of commercially available bread dough are investigated in order to develop more efficient microwave systems and sensors for household and industrial baking.[30] The material properties depend not only on the microwave frequencies used but also on the time of the baking process. Table 8.4 provides the dielectric constants of the bread baked for 10, 20 (not done yet), and 30 (baking complete) minutes, respectively, at two microwave frequencies, 600 MHz and 2.4 GHz. Calculate the depth of penetration for each case and compare the results.

TABLE 8.4 RELATIVE PERMITTIVITY $\epsilon_{cr} = \epsilon'_r - j\epsilon''_r$ OF BREAD DOUGH VERSUS FREQUENCY AND BAKING TIME

	10 Min	20 Min	30 Min
600 MHz	$23.1 - j11.85$	$16.78 - j6.66$	$8.64 - j2.51$
2.4 GHz	$12.17 - j4.54$	$9.53 - j3.12$	$4.54 - j1.22$

Solution: At $f = 600$ MHz, after baking for 10 minutes, the loss tangent of bread dough is

$$\tan \delta_c = \frac{11.85}{23.1} \simeq 0.513$$

from which the depth of penetration can be calculated as

$$d \simeq \frac{3 \times 10^8 \text{ m-s}^{-1}}{2\pi \times 6 \times 10^8 \text{ rad-s}^{-1}} \left[\frac{2}{23.1(\sqrt{1 + (0.513)^2} - 1)} \right]^{1/2} \simeq 6.65 \text{ cm}$$

Similarly, we can calculate the other values. The results are tabulated in Table 8.5.

TABLE 8.5 LOSS TANGENT $\tan \delta_c$ AND DEPTH OF PENETRATION d VERSUS FREQUENCY AND BAKING TIME

	10 Min		20 Min		30 Min	
	$\tan \delta_c$	d	$\tan \delta_c$	d	$\tan \delta_c$	d
600 MHz	0.513	6.65 cm	0.397	9.97 cm	0.291	18.8 cm
2.4 GHz	0.373	3.11 cm	0.327	3.99 cm	0.269	7.01 cm

As we see from Table 8.5, the loss tangent decreases with both increasing baking time and increasing frequency. The depth of penetration increases with baking time but decreases with frequency.

Low-loss dielectrics. A low-loss dielectric (or a good dielectric) is one for which $\tan \delta_c \ll 1$. In many applications, dielectric materials that are used are very nearly perfect but nevertheless do cause some nonzero amount of loss. A good example is the dielectric fillings of a coaxial line; the materials used for this purpose are typically nearly

---

[30]J. Zuercher, L. Hoppie, R. Lade, S. Srinivasan, and D. Misra, Measurement of the complex permittivity of bread dough by an open-ended coaxial line method at ultra high frequencies, *J. Microwave Power Electromagn. Energy,* 25(3), pp. 161–167, 1990.

perfect insulators, and an approximate evaluation of the small amount of losses is entirely adequate for most applications.

Therefore, for low-loss dielectrics, we do not need to use the exact expressions for α, β, and η_c given, respectively, by (8.21), (8.22), and (8.26). When $\tan \delta_c \ll 1$, equations (8.21) and (8.22) for α and β can be simplified by applying the binomial approximation[31] to (8.20)

$$\gamma = \alpha + j\beta = j\omega\sqrt{\mu\epsilon'}\left(1 - j\frac{\sigma_{\text{eff}}}{\omega\epsilon'}\right)^{1/2} \simeq j\omega\sqrt{\mu\epsilon'}\left[1 - j\frac{\sigma_{\text{eff}}}{2\omega\epsilon'} + \frac{1}{8}\left(\frac{\sigma_{\text{eff}}}{\omega\epsilon'}\right)^2\right]$$

where we replaced σ with σ_{eff}. If we assume $\sigma_{\text{eff}} = \omega\epsilon''$, we can then write

$$\boxed{\alpha \simeq \frac{\sigma_{\text{eff}}}{2}\sqrt{\frac{\mu}{\epsilon'}} = \frac{\omega\epsilon''}{2}\sqrt{\frac{\mu}{\epsilon'}}} \tag{8.35}$$

as the attenuation constant and

$$\boxed{\beta \simeq \omega\sqrt{\mu\epsilon'}\left[1 + \frac{1}{8}\left(\frac{\sigma_{\text{eff}}}{\omega\epsilon'}\right)^2\right] = \omega\sqrt{\mu\epsilon'}\left[1 + \frac{1}{8}\left(\frac{\epsilon''}{\epsilon'}\right)^2\right]} \tag{8.36}$$

as the phase constant. Note that since $\tan \delta_c \ll 1$, the phase constant β (and therefore the wavelength $\lambda = 2\pi/\beta$) in a low-loss dielectric are very nearly equal to those in a lossless medium. The attenuation constant imposed on the wave by the medium is usually very small. For example, for dry earth (take $\sigma = 10^{-5}$ S-m^{-1} and $\epsilon = 3\epsilon_0$) at 10 MHz, $\tan \delta_c \simeq 0.006$ and $\alpha \simeq 1.09 \times 10^{-3}$ np-m^{-1}, so the signal is attenuated down to $1/e$ of its value in $\sim$920 m. The wavelength $(2\pi/\beta)$, on the other hand, is $\sim$17.3 m, so the depth of penetration for a low-loss dielectric is typically many wavelengths. Such a picture is illustrated in Figure 8.9.

Note that a simplified expression for the phase velocity $(v_p = \omega/\beta)$ can be obtained by using (8.22) and the binomial expansion for $(1 + x)^k$ and noting that the exponent k is negative:

$$v_p \simeq \frac{1}{\sqrt{\mu\epsilon'}}\left[1 - \frac{1}{8}\left(\frac{\sigma_{\text{eff}}}{\omega\epsilon'}\right)^2\right] = \frac{1}{\sqrt{\mu\epsilon'}}\left[1 - \frac{1}{8}\left(\frac{\epsilon''}{\epsilon'}\right)^2\right]$$

Since $1/\sqrt{\mu\epsilon'}$ is the phase velocity in the lossless medium, we see that the small loss introduced by the low-loss dielectric also slightly reduces the velocity of the wave.

An approximate expression for the intrinsic impedance of a good dielectric can also be obtained, again using the binomial expansion. We find:

$$\eta_c = \sqrt{\frac{\mu}{\epsilon_{\text{eff}}}} = \sqrt{\frac{\mu}{\epsilon'}}\left(1 - j\frac{\sigma_{\text{eff}}}{\omega\epsilon'}\right)^{-1/2} \simeq \sqrt{\frac{\mu}{\epsilon'}}\left(1 + j\frac{\sigma_{\text{eff}}}{2\omega\epsilon'}\right) = \sqrt{\frac{\mu}{\epsilon'}}\left(1 + j\frac{\epsilon''}{2\epsilon'}\right)$$

[31]$(1 \pm x)^k = 1 \pm kx + \dfrac{k(k-1)}{2}x^2 \pm \cdots$

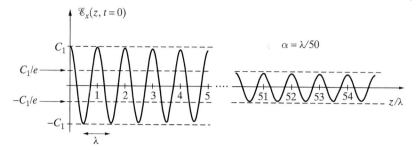

Figure 8.9 Low-loss dielectric. The electric field component of a uniform plane wave propagating in a low-loss (i.e., good) dielectric, in which the depth of penetration is typically much larger than a wavelength.

We see that the small amount of loss also introduces a small reactive (in this case inductive) component to the intrinsic impedance of the medium. This will in turn result in a slight phase shift between the electric and magnetic fields, since ϕ_η in (8.26) is nonzero. We note, however, that the phase shift is usually quite small; for dry earth, for example, at 10 MHz, $\phi_\eta \simeq 0.172°$.

Example 8.8: Uniform plane wave in gallium arsenide. A uniform plane wave of frequency 10 GHz propagates in a sufficiently large sample of gallium arsenide (GaAs, $\epsilon_r \simeq 12.9$, $\mu_r = 1$, $\tan \delta_c \simeq 5 \times 10^{-4}$), which is a commonly used substrate material for high-speed solid-state circuits. Find (a) the attenuation constant α in np-m^{-1}, (b) phase velocity v_p in m-s^{-1}, and (c) intrinsic impedance η_c in Ω.

Solution: Since $\tan \delta_c = 5 \times 10^{-4} \ll 1$, we can use the approximate expressions for a good dielectric.

(a) We have

$$\alpha \simeq \frac{\sigma}{2}\sqrt{\frac{\mu}{\epsilon}} = \frac{\omega\epsilon \tan \delta_c}{2}\sqrt{\frac{\mu}{\epsilon}}$$

$$= \frac{2\pi \times 10^{10} \times 5 \times 10^{-4}}{2}\sqrt{\mu\epsilon}$$

$$= \frac{2\pi \times 10^{10} \times 5 \times 10^{-4}}{2}\sqrt{\mu_r\epsilon_r}\sqrt{\mu_0\epsilon_0}$$

$$= \frac{2\pi \times 10^{10} \times 5 \times 10^{-4}}{2 \times 3 \times 10^8}\sqrt{12.9} \simeq 0.188 \text{ np-m}^{-1}$$

(b) Since phase velocity $v_p = \omega/\beta$ where $\beta \simeq \omega\sqrt{\mu\epsilon}$, we have $v_p \simeq 1/\sqrt{\mu\epsilon} \simeq 3 \times 10^8/\sqrt{12.9} \simeq 8.35 \times 10^7$ m-s^{-1}. Note that the phase velocity is ~3.59 times slower than that in air.

(c) The intrinsic impedance $\eta_c \simeq \sqrt{\mu/\epsilon} \simeq 377/\sqrt{12.9} \simeq 105\Omega$. Note that the intrinsic impedance is ~3.59 times smaller than that in air.

Example 8.9: Microwave heating of milk. The dielectric properties of milk with 7.3% moisture content at 20°C and 3 GHz are $\epsilon_r' = 51$ and $\tan \delta_c = 0.59$. Calculate (a) ϵ_r'' and (b) the average dissipated power per unit volume if the peak electric field inside the dielectric is 30 kV-m^{-1}.

Solution: (a) From $\tan \delta_c = 0.59 = \epsilon_r''/\epsilon_r' = \epsilon_r''/51$, we find $\epsilon_r'' \simeq 30.1$. (b) The average power dissipated per unit volume is

$$\tfrac{1}{2}\omega\epsilon'' E_{\text{peak}}^2 \simeq \tfrac{1}{2}(2\pi \times 3 \times 10^9 \text{ rad-s}^{-1})(30.1 \times 8.85 \times 10^{-12} \text{ F-m}^{-1})(30 \text{ kV-m}^{-1})^2$$

$$\simeq 2.26 \times 10^9 \text{ W-m}^{-3} = 2.26 \text{ W-mm}^{-3}$$

Complex permeability: Magnetic relaxation. Although our discussion above dealt exclusively with lossy dielectrics, an analogous effect also occurs in magnetic materials, although it is of less importance for electromagnetic wave applications. If a magnetic field is suddenly applied to a paramagnetic[32] material, the magnetization exhibits some inertia and does not immediately reach its static value but instead approaches it gradually. Similar inertial delay also occurs when the field is turned off. The inertia exhibited is attributable to the energy exchanges between the spinning electrons and the lattice vibrations as well as energy exchanges between neighboring spins.[33] In analogy with dielectric relaxation, such effects can be described by introducing a complex permeability μ such that

$$\mu_c = \mu' - j\mu'' \tag{8.37}$$

To represent the fact that this effect would lead to power dissipation in the medium, consider the time rate of change of magnetic energy density, or

$$\frac{\partial w_{\text{m}}}{\partial t} = \frac{\partial}{\partial t}\left(\frac{1}{2}\overline{\mathcal{H}} \cdot \overline{\mathcal{B}}\right) \qquad \rightarrow \qquad \tfrac{1}{2}j\omega(\mathbf{H} \cdot \mathbf{B})$$

With $\mathbf{B} = (\mu' - j\mu'')\mathbf{H}$, we have

$$\tfrac{1}{2}j\omega(\mathbf{H} \cdot \mathbf{B}) = \tfrac{1}{2}(j\omega\mu')H^2 + \tfrac{1}{2}\omega\mu''H^2$$

where $H = |\mathbf{H}|$. For a time-harmonic magnetic field, the real part of this quantity represents the time-average power dissipated in a medium. We note that the first term is purely imaginary; that is, this term simply represents the volume density of energy stored in the magnetic field per unit time. The second term, on the other hand, is purely real and represents the time-average power density of the magnetization loss.

---

[32]Paramagnetic materials are those for which the permeability μ is slightly above μ_0. See Chapter 9 of M. A. Plonus, *Applied Electromagnetism,* McGraw-Hill, New York, 1978 and Chapter 14 of P. S. Neelakanta, *Handbook of Electromagnetic Materials,* CRC Press, Boca Raton, 1995.

[33]For further details, see Section 7.16 of R. S. Elliott, *Electromagnetics,* IEEE Press, 1993, and Chapter 16 of C. Kittel, *Introduction to Solid State Physics,* 5th ed., Wiley, New York, 1976.

Substituting (8.37) and (8.32) into (8.25), the complex propagation constant is

$$\gamma = j\omega\sqrt{(\mu' - j\mu'')\epsilon_{\text{eff}}}$$

$$= j\omega\sqrt{(\mu' - j\mu'')\left(\epsilon' - j\epsilon'' - j\frac{\sigma}{\omega}\right)}$$

$$= \sqrt{j\omega\mu'\left[j\omega\epsilon' + (\sigma + \omega\epsilon'')\left(1 + j\frac{\mu''}{\mu'}\right) + \frac{\omega\epsilon'\mu''}{\mu'}\right]}$$

Comparing the last line with (8.20), we see that the effective conductivity is now given by

$$\sigma_{\text{eff}} = (\sigma + \omega\epsilon'')\left(1 + j\frac{\mu''}{\mu'}\right) + \frac{\omega\epsilon'\mu''}{\mu'} \tag{8.38}$$

Hence when $\mu'' \ll \mu'$, the additional losses due to magnetization damping can be approximately represented by an additional effective conductivity $\epsilon'\omega\mu''/\mu'$.

In principle, diamagnetic materials also exhibit time-varying relaxation behavior. Although the resultant effects are so small that one might think such phenomena are of little practical interest, resonant absorption in diamagnetic liquids is the cause of the phenomenon of nuclear magnetic resonance,[34] which in turn is the basis for magnetic resonance imaging (MRI) technology.[35]

8.3.2 Uniform Plane Wave Propagation in a Good Conductor

Another special case of wave propagation in lossy media that is of practical importance involves propagation in nearly perfect conductors. Many transmission lines and guiding systems are constructed using metallic conductors, such as copper, aluminum, or silver. In most cases these conductors are nearly perfect, so the losses are relatively small. Nevertheless, these small losses often determine the range of applicability of the systems utilized and need to be quantitatively determined. Fortunately, approximate determinations of the losses are perfectly adequate in most cases, using simplified formulas derived on the basis that the materials are good conductors; that is, $\sigma \gg \omega\epsilon$.

For $\tan\delta_c \gg 1$, the propagation constant γ in (8.20) can be simplified as follows:

$$\gamma = j\omega\sqrt{\mu\epsilon}\sqrt{1 - j\frac{\sigma}{\omega\epsilon}} \simeq j\omega\sqrt{\mu\epsilon}\sqrt{-j\frac{\sigma}{\omega\epsilon}} = \sqrt{j\omega\mu\sigma} = \sqrt{\omega\mu\sigma}\,e^{j45°}$$

---

[34]Resonant absorption occurs because of the highly frequency-dependent nature of the permeability μ, analogous to that shown in Figure 11.4 for ϵ_c. For further discussion and relevant references, see Chapter 16 of C. Kittel, *Introduction to Solid State Physics*, 5th ed., Wiley, New York, 1976.

[35]See M. A. Brown and R. C. Semelka, *MRI Basic Principles and Applications*, Wiley-Liss, New York, 1995.

where we have used the fact that $\sqrt{j} = e^{j\pi/4}$. The real and imaginary parts (i.e., the attenuation and phase constants) of γ for a good conductor are thus equal:

$$\alpha = \beta \simeq \sqrt{\frac{\omega\mu\sigma}{2}} \qquad (8.39)$$

The phase velocity and wavelength can be obtained from β, as $v_p = \sqrt{2\omega/(\mu\sigma)}$ and $\lambda = 2\sqrt{\pi/(f\mu\sigma)}$. Since both α and β are proportional to $\sigma^{1/2}$ and σ is large, it appears that uniform plane waves not only are attenuated heavily but also undergo a significant phase shift per unit length as they propagate in a good conductor. The phase velocity of the wave and the wavelength are both proportional to $\sigma^{-1/2}$ and are therefore significantly smaller than the corresponding values in free space.

For example, for copper ($\sigma = 5.8 \times 10^7$ S-m^{-1}) at 300 MHz, we have $v_p \simeq$ 7192 m-s^{-1}, and $\lambda \simeq 0.024$ mm, which are much smaller than the free-space values of $c \simeq 3 \times 10^8$ m-s^{-1} and $\lambda \simeq 1$ m at 300 MHz. At 60 Hz, the values for copper are even more dramatic, being $v_p \simeq 3.22$ m-s^{-1} and $\lambda \simeq 53.6$ mm, compared with the free-space wavelength of $\sim$5000 km. As an example of a nonmetallic conductor, for seawater ($\epsilon = 81\epsilon_0$, $\sigma = 4$ S-m^{-1}) at 10 kHz we have $v_p \simeq 1.58 \times 10^5$ m-s^{-1} and $\lambda \simeq 15.8$ m, compared with a free-space wavelength of 30 km. It is interesting to note here that in the context of undersea (submarine) communications, a half-wavelength antenna (i.e., a reasonably efficient radiator) operating in seawater at 10 kHz has to be only $\sim$7.91 m long, whereas a half-wavelength antenna for a ground-based very low frequency (VLF) transmitter radiating at 10 kHz in air would have to be $\sim$15 km long. In practice, VLF communication and navigation transmitters use electrically short (a few hundred meters high) vertical monopole antennas above ground planes and/or in top-loaded fashion.[36]

Using (8.26) we can express the intrinsic impedance for $\sigma \gg \omega\epsilon$ as

$$\eta_c = \sqrt{\frac{\mu}{\epsilon}} \frac{1}{\left(1 - j\dfrac{\sigma}{\omega\epsilon}\right)^{1/2}} \simeq \sqrt{\frac{j\omega\mu}{\sigma}} = \sqrt{\frac{\mu\omega}{\sigma}} e^{j45°} \qquad (8.40)$$

The magnitude of η_c for good conductors is typically quite small (proportional to $\sigma^{-1/2}$), and the medium presents an inductive impedance (i.e., $\mathscr{H}$ lags $\mathscr{E}$). The phase of the intrinsic impedance (ϕ_η) is very nearly 45° for all good conductors. For copper at 300 MHz, the magnitude of the intrinsic impedance is $\sim$6.39 $\times 10^{-3}\Omega$, compared with 377Ω in free space. Thus, for a given uniform plane-wave electric field intensity, the corresponding wave magnetic field in copper is $\sim$6 $\times 10^4$ times as large as that in free space. For seawater at 10 kHz, $|\eta_c| \simeq 0.14\Omega$, which is a factor of $\sim$2700 times smaller than the free-space value of 377Ω.

A useful parameter used in assessing the degree to which a good conductor is lossy and the degree to which electromagnetic waves can penetrate into a good conductor is

---

[36]A. D. Watt, *VLF Radio Engineering*, Pergamon Press, New York, 1967.

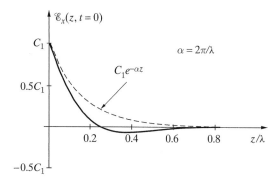

Figure 8.10 Good conductor. The electric field component of a uniform plane wave in a good conductor.

the so-called *skin depth* δ, defined as the depth in which the wave is attenuated to $1/e$ (or $\sim36.8\%$) of its original intensity (i.e., its value immediately below the surface if a wave is incident on the material from above). The skin depth[37] is given by

$$\delta \equiv \frac{1}{\alpha} \simeq \sqrt{\frac{2}{\omega\mu\sigma}} = \frac{1}{\sqrt{\pi f \mu \sigma}} \qquad (8.41)$$

For example, for copper at 300 MHz, $\delta \simeq 0.00382$ mm $= 3.82$ μm, which is a very small distance both in absolute terms and also as compared with the wavelength ($\lambda \simeq 0.024$ mm $= 24$ μm, see above). The nature of the wave propagation in a good conductor, as shown in Figure 8.10, is quite different from the case of that in a good dielectric shown in Figure 8.9. In a good conductor, since $\lambda = 2\pi\delta$, the wave attenuates rapidly with distance and reaches negligible amplitudes after traveling only a small fraction of a wavelength. Note that the skin depth in good conductors such as copper is very small even at audio or low radio frequencies (tens to hundreds of kHz). For example, at 30 kHz, we have $\delta \simeq 0.382$ mm, which means that any time-varying fields and current densities exist only in a thin layer on the surface. In typical transmission lines (e.g., a coaxial line), electromagnetic power propagates in the dielectric region between the conductors, typically at the velocity of light in free space (or at the phase velocity appropriate for the dielectric medium). The electromagnetic wave is guided by the conductor structure and penetrates into the conductor only to the degree that the conductor is imperfect. For example, in the case of a coaxial line, the propagation characteristics of the wave are not different in a solid inner metallic conductor versus one that consists of a thin cylindrical shell of metallic conductor that is hollow or filled with any other material (conductor or insulator).

Example 8.10: Comparison of skin depths. Calculate the skin depth in each of the following nonmagnetic ($\mu = \mu_0$) media at 10 kHz: seawater ($\epsilon_r = 81$, $\sigma = 4$ S-m^{-1}), wet earth ($\epsilon_r = 10$, $\sigma = 10^{-2}$ S-m^{-1}), and dry earth ($\epsilon_r = 3$, $\sigma = 10^{-4}$ S-m^{-1}).

---

[37]Note that the definition of the skin depth δ as being equal to $1/\alpha$ is identical to that of the penetration depth d in (8.34). However, the term "skin depth" is exclusively used for good conductors and emphasizes that the penetration is confined to a thin region on the skin of the conductor.

Solution: Using (8.28), the loss tangent for seawater can be found as

$$\tan \delta_c \simeq \frac{4 \text{ S-m}^{-1}}{(2\pi \times 10^4 \text{ rad-s}^{-1})(81 \times 8.85 \times 10^{-12} \text{ F-m}^{-1})} \simeq 8.88 \times 10^4$$

Similarly, for wet earth we find $\tan \delta_c \simeq 1.8 \times 10^3$, and for dry earth $\tan \delta_c \simeq 59.9$. Thus, at 10 kHz, all three media can be considered to be good conductors, and we can use the skin depth formula given by (8.41). For seawater at 10 kHz we then have

$$\delta \simeq \frac{1}{\sqrt{\pi(10^4 \text{ s}^{-1})(4\pi \times 10^{-7} \text{ H-m}^{-1})(4 \text{ S-m}^{-1})}} \simeq 2.52 \text{ m}$$

Similarly, the skin depths for wet and dry earth can be found to be, respectively, ~50.3 m and ~503 m. The relatively large skin depths facilitate the use of VLF frequencies (3–30 kHz) for various applications such as undersea communication, search for buried objects, and geophysical prospecting. Note that the skin depths are even larger for the ELF, SLF, and ULF ranges of frequencies (see Table 8.1), in the range 3 Hz–3 kHz. These relatively large penetration distances facilitate the use of these frequencies for geophysical prospecting.

Example 8.11: Propagation of radio waves through lake water versus seawater. Consider the propagation of a 10 MHz radio wave in lake water (assume $\sigma = 4 \times 10^{-3}$ S-m^{-1} and $\epsilon_r = 81$) versus seawater ($\sigma = 4$ S-m^{-1} and $\epsilon_r = 81$). Calculate the attenuation constant and the penetration depth in each medium and comment on the difference. Note that both media are nonmagnetic.

Solution: For lake water, using (8.28), the loss tangent at 10 MHz is given by

$$\tan \delta_c = \frac{\sigma}{\omega \epsilon_r \epsilon_0} \simeq \frac{4 \times 10^{-3} \text{ S-m}^{-1}}{2\pi \times 10^7 \text{ rad-s}^{-1} \times 81 \times 8.85 \times 10^{-12} \text{ F-m}^{-1}} \simeq 8.88 \times 10^{-2} \ll 1$$

Therefore, at 10 MHz, lake water behaves like a low-loss dielectric. Using (8.35), the penetration depth for lake water is

$$d = \frac{1}{\alpha} \simeq \frac{2}{\sigma}\sqrt{\frac{\epsilon_r \epsilon_0}{\mu_0}} \simeq \left[\frac{2}{4 \times 10^{-3} \text{ S-m}^{-1}}\right]\left[\frac{\sqrt{81}}{377\Omega}\right] \simeq 11.9 \text{ m} \simeq 12 \text{ m}$$

For seawater, the loss tangent at 10 MHz is given by

$$\tan \delta_c \simeq \frac{4 \text{ S-m}^{-1}}{2\pi \times 10^7 \text{ rad-s}^{-1} \times 81 \times 8.85 \times 10^{-12} \text{ F-m}^{-1}} \simeq 88.8 \gg 1$$

Therefore, at 10 MHz, seawater behaves like a good conductor. The skin depth for seawater is given by (8.41):

$$\delta \simeq \frac{1}{\sqrt{\pi \times 10^7 \text{ rad-s}^{-1} \times 4\pi \times 10^{-7} \text{ H-m}^{-1} \times 4 \text{ S-m}^{-1}}} = \frac{1}{4\pi} \text{ m} \simeq 8 \text{ cm}$$

The results clearly indicate that communication through seawater at 10 MHz is not feasible, since the skin depth is only 8 cm. However, 10 MHz electromagnetic waves do penetrate lake water, making communication possible at relatively shallow depths of tens of meters.

Example 8.12: VLF waves in the ocean. The electric field component of a uniform plane VLF electromagnetic field propagating vertically down in the z direction in the ocean ($\sigma = 4$ S-m^{-1}, $\epsilon_r = 81$, $\mu_r = 1$) is approximately given by

$$\overline{\mathscr{E}}(z,t) = \hat{x}E_0 e^{-\alpha z}\cos(6\pi \times 10^3 t - \beta z)$$

where E_0 is the electric field amplitude at $z = 0^+$ immediately below the air–ocean interface, at $z = 0$. Note that the frequency of the wave is 3 kHz. (a) Find the attenuation constant α and phase constant β. (b) Find the wavelength λ, phase velocity v_p, skin depth δ, and intrinsic impedance η_c, and compare them to their values in air. (c) Write the instantaneous expression for the corresponding magnetic field, $\overline{\mathscr{H}}(z,t)$. (d) A submarine located at a depth of 100 m has a receiver antenna capable of measuring electric fields with amplitudes of 1 μV-m^{-1} or greater. What is the minimum required electric field amplitude immediately below the ocean surface (i.e., E_0) in order to communicate with the submarine? What is the corresponding value for the amplitude of the magnetic field?

Solution: At $\omega = 6\pi \times 10^3$ rad-s^{-1}, using (8.28) the loss tangent of the ocean is given by

$$\tan \delta_c = \frac{\sigma}{\omega \epsilon_r \epsilon_0} \simeq \frac{4}{6\pi \times 10^3 \text{ rad-s}^{-1} \times 81 \times (8.85 \times 10^{-12} \text{ F-m}^{-1})} \simeq 2.96 \times 10^5 \gg 1$$

Therefore, the ocean acts as a good conductor at $f = 3$ kHz.

(a) Using the approximate expressions given by (8.39), we have

$$\alpha = \beta \simeq \sqrt{\frac{\omega \mu \sigma}{2}} = \sqrt{\frac{(6\pi \times 10^3 \text{ rad-s}^{-1})(4\pi \times 10^{-7} \text{ H-m}^{-1})(4 \text{ S-m}^{-1})}{2}}$$

$$\simeq 0.218 \text{ np-m}^{-1} \text{ or rad-m}^{-1}$$

(b) The wavelength in the ocean is

$$\lambda = \frac{2\pi}{\beta} \simeq \frac{2\pi}{0.218} \simeq 28.9 \text{ m}$$

Since in air $\lambda \simeq 100$ km at 3 kHz, the wavelength in the ocean is approximately 3464 times smaller than that in air. The phase velocity is

$$v_p = f\lambda \simeq 3 \times 10^3 \times 28.9 \simeq 8.66 \times 10^4 \text{ m-s}^{-1}$$

Compared with $c \simeq 3 \times 10^8$ m-s^{-1} in air, the phase velocity in the ocean at 3 kHz is approximately $\sim$3464 times smaller. The skin depth and the intrinsic impedance in the ocean are given by (8.41) and (8.40), respectively:

$$\delta = 1/\alpha \simeq 4.59 \text{ m}$$

$$\eta_c = \sqrt{\frac{(4\pi \times 10^{-7} \text{ H-m}^{-1})(6\pi \times 10^3 \text{ rad-s}^{-1})}{4 \text{ S-m}^{-1}}} e^{j45°} \simeq 7.70 \times 10^{-2} e^{j45°} \, \Omega$$

Compared with $\eta \simeq 377\Omega$ in air, the magnitude of η_c in the ocean is approximately 4900 times smaller. In addition, η_c in the ocean is a complex number with a phase angle of $\sim$45°.

(c) Using (8.27), the corresponding magnetic field is given by

$$\overline{\mathcal{H}}(z,t) = \hat{\mathbf{y}}\frac{E_0}{|\eta_c|}e^{-\alpha z}\cos(6\pi \times 10^3 t - \beta z - \phi_\eta)$$

$$\simeq 13E_0 e^{-0.218z}\cos(6\pi \times 10^3 t - 0.218z - \pi/4)\text{A-m}^{-1}$$

(d) At $z_1 = 100$ m, the amplitude of the electric field is given by

$$E_0 e^{-\alpha z_1} \simeq E_0 e^{-(0.218)(100)} \geq E_{min} = 1 \ \mu\text{V-m}^{-1}$$

Therefore, the minimum value for E_0 to establish communication with the submarine located at a depth of 100 m is

$$E_0 \simeq 2.83 \ \text{kV-m}^{-1}$$

The corresponding value for H_0 is

$$H_0 = \frac{E_0}{|\eta_c|} \simeq 13E_0 \simeq 36.8 \ \text{kA-m}^{-1}$$

8.4 ELECTROMAGNETIC ENERGY FLOW AND THE POYNTING VECTOR

Electromagnetic waves carry power through space, transferring energy and momentum from one set of charges and currents (i.e., the sources that generated them) to another (those at the receiving points). Our goal in this section is to derive (directly from Maxwell's equations) a simple relation between the rate of this energy transfer and the electric and magnetic fields $\overline{\mathcal{E}}$ and $\overline{\mathcal{H}}$. We can in fact attribute definite amounts of energy and momentum to each elementary volume of space occupied by the electromagnetic fields. The energy and momentum exchange between waves and particles is described by the Lorentz force equation (6.51). This equation represents the forces exerted on a charged particle (having a charge q and velocity $\tilde{\mathbf{v}}$) by electric and magnetic fields: $\overline{\mathcal{F}}_{elec} = q\overline{\mathcal{E}}$ and $\overline{\mathcal{F}}_{mag} = q\tilde{\mathbf{v}} \times \overline{\mathcal{B}}$, where $\overline{\mathcal{B}} = \mu_0\overline{\mathcal{H}}$. When fields exert forces on the charges, the charges respond to the force (i.e., they move) and gain energy, which must be at the expense of the energy lost by the fields. It is thus clear that electromagnetic fields store energy, which is available to charged particles when the fields exert forces (do work) on them. Since energy is transmitted to the particles at a specific rate, we can think of the fields as transmitting power.

8.4.1 Flow of Electromagnetic Energy and Poynting's Theorem

The densities of energy stored in static electric and magnetic fields in linear and isotropic media are,[38] respectively, $\frac{1}{2}\epsilon E^2$ and $\frac{1}{2}\mu H^2$. When these fields vary with time, the associated stored energy densities can also be assumed to vary with time. If we consider a given volume of space, electromagnetic energy can be transported into or out of it by electromagnetic waves, depending on whether the source of the waves is inside or

---

[38]These stored electrostatic and magnetostatic energies represent the work that needs to be done to establish the fields, or the charge and distributions that maintain them.

outside the volume. In addition, electromagnetic energy can be stored in the volume in the form of electric and magnetic fields, and electromagnetic power can be dissipated in it in the form of Joule heating (i.e., as represented by $\overline{\mathscr{E}} \cdot \overline{\mathscr{J}}$). Poynting's theorem concerns the balance between power flow into or out of a given volume and the rate of change of stored energy and power dissipation within it. This theorem provides the physical framework by which we can express the flow of electromagnetic power in terms of the fields $\overline{\mathscr{H}}$ and $\overline{\mathscr{E}}$.

We now proceed to find an expression for electromagnetic power in terms of the field quantities $\overline{\mathscr{E}}$ and $\overline{\mathscr{H}}$. Since we expect the power flow to be related to the volume density of power dissipation represented by $\overline{\mathscr{E}} \cdot \overline{\mathscr{J}}$, we can start with (8.1c):

$$\overline{\mathscr{J}} = \nabla \times \overline{\mathscr{H}} - \frac{\partial \overline{\mathscr{D}}}{\partial t}$$

and take the dot product of both sides with $\overline{\mathscr{E}}$ to obtain $\overline{\mathscr{E}} \cdot \overline{\mathscr{J}}$ in units of W-m^{-3}:

$$\overline{\mathscr{E}} \cdot \overline{\mathscr{J}} = \overline{\mathscr{E}} \cdot (\nabla \times \overline{\mathscr{H}}) - \overline{\mathscr{E}} \cdot \frac{\partial \overline{\mathscr{D}}}{\partial t}$$

and use the vector identity

$$\nabla \cdot (\overline{\mathscr{E}} \times \overline{\mathscr{H}}) \equiv \overline{\mathscr{H}} \cdot (\nabla \times \overline{\mathscr{E}}) - \overline{\mathscr{E}} \cdot (\nabla \times \overline{\mathscr{H}})$$

to find

$$\overline{\mathscr{E}} \cdot \overline{\mathscr{J}} = \overline{\mathscr{H}} \cdot (\nabla \times \overline{\mathscr{E}}) - \nabla \cdot (\overline{\mathscr{E}} \times \overline{\mathscr{H}}) - \overline{\mathscr{E}} \cdot \frac{\partial \overline{\mathscr{D}}}{\partial t} \qquad (8.42)$$

We now use (8.1a),

$$\nabla \times \overline{\mathscr{E}} = -\frac{\partial \overline{\mathscr{B}}}{\partial t}$$

and substitute into (8.42) to obtain

$$\overline{\mathscr{E}} \cdot \overline{\mathscr{J}} = -\overline{\mathscr{H}} \cdot \frac{\partial \overline{\mathscr{B}}}{\partial t} - \overline{\mathscr{E}} \cdot \frac{\partial \overline{\mathscr{D}}}{\partial t} - \nabla \cdot (\overline{\mathscr{E}} \times \overline{\mathscr{H}}) \qquad (8.43)$$

We now consider the two terms with the time derivatives. For a simple medium[39] (i.e., ϵ and μ simple constants) we can write

$$\overline{\mathscr{H}} \cdot \frac{\partial \overline{\mathscr{B}}}{\partial t} = \overline{\mathscr{H}} \cdot \frac{\partial (\mu \overline{\mathscr{H}})}{\partial t} = \frac{1}{2} \frac{\partial (\mu \overline{\mathscr{H}} \cdot \overline{\mathscr{H}})}{\partial t} = \frac{\partial}{\partial t} \left(\frac{1}{2} \mu |\overline{\mathscr{H}}|^2 \right)$$

[39]The more general form of Poynting's theorem, without these assumptions about the medium, is

$$\int_V \overline{\mathscr{E}} \cdot \overline{\mathscr{J}} \, dv = -\int_V \left(\overline{\mathscr{H}} \cdot \frac{\partial \overline{\mathscr{B}}}{\partial t} + \overline{\mathscr{E}} \cdot \frac{\partial \overline{\mathscr{D}}}{\partial t} \right) dv - \oint_S (\overline{\mathscr{E}} \times \overline{\mathscr{H}}) \cdot d\mathbf{s}$$

which is valid even for nonlinear media, as long as there are no hysteresis effects. For ferromagnetic materials, for which the relation between $\overline{\mathscr{B}}$ and $\overline{\mathscr{H}}$ is often multivalued due to hysteresis, an additional amount of energy is deposited within the material.

and

$$\overline{\mathscr{E}} \cdot \frac{\partial \overline{\mathscr{D}}}{\partial t} = \overline{\mathscr{E}} \cdot \frac{\partial(\epsilon \overline{\mathscr{E}})}{\partial t} = \frac{1}{2}\frac{\partial(\epsilon \overline{\mathscr{E}} \cdot \overline{\mathscr{E}})}{\partial t} = \frac{\partial}{\partial t}\left(\frac{1}{2}\epsilon|\overline{\mathscr{E}}|^2\right)$$

respectively. With these substitutions, (8.43) can be rewritten as

$$\overline{\mathscr{E}} \cdot \overline{\mathscr{J}} = -\frac{\partial}{\partial t}\left(\frac{1}{2}\mu|\overline{\mathscr{H}}|^2\right) - \frac{\partial}{\partial t}\left(\frac{1}{2}\epsilon|\overline{\mathscr{E}}|^2\right) - \nabla \cdot (\overline{\mathscr{E}} \times \overline{\mathscr{H}}) \qquad (8.44)$$

Integrating (8.44) over an arbitrary volume V, we have

$$\int_V \overline{\mathscr{E}} \cdot \overline{\mathscr{J}}\, dv = -\frac{\partial}{\partial t}\int_V \left(\frac{1}{2}\mu|\overline{\mathscr{H}}|^2 + \frac{1}{2}\epsilon|\overline{\mathscr{E}}|^2\right) dv - \int_V \nabla \cdot (\overline{\mathscr{E}} \times \overline{\mathscr{H}})\, dv \qquad (8.45)$$

Using the divergence theorem on the last term in (8.45) we find

$$\boxed{\int_V \overline{\mathscr{E}} \cdot \overline{\mathscr{J}}\, dv = -\frac{\partial}{\partial t}\int_V \left(\frac{1}{2}\mu|\overline{\mathscr{H}}|^2 + \frac{1}{2}\epsilon|\overline{\mathscr{E}}|^2\right) dv - \oint_S (\overline{\mathscr{E}} \times \overline{\mathscr{H}}) \cdot d\mathbf{s}} \qquad (8.46)$$

where the surface S encloses the volume V.

We can now interpret the various terms in (8.46) physically. The left-hand term is the generalization of Joule's law and represents the instantaneous power dissipated in the volume V. If $\overline{\mathscr{E}}$ is the electric field that produces $\overline{\mathscr{J}}$ in a lossy medium, this term represents the ohmic (I^2R) power loss in the medium. Note that in a simple isotropic medium (i.e., σ a simple constant), $\overline{\mathscr{E}}$ and $\overline{\mathscr{J}}$ are in the same direction. However, in general this is not true. For example, in Earth's ionosphere (which is an anisotropic medium because of the presence of Earth's magnetic field), an applied electric field in one direction can cause current flow not only in that direction but also in other directions. Even in such cases, however, $\overline{\mathscr{E}} \cdot \overline{\mathscr{J}}$ still represents the power dissipated per unit volume, although $\overline{\mathscr{J}}$ and $\overline{\mathscr{E}}$ are not parallel. Alternatively, there could be an energy source within the volume V, such as an antenna carrying current, in which case $\overline{\mathscr{E}} \cdot \overline{\mathscr{J}}$ is negative and represents power flow out of that region.

The first term on the right-hand side represents the rate at which the electromagnetic energy stored in volume V decreases (negative sign), with the terms $\frac{1}{2}\mu|\overline{\mathscr{H}}|^2$ and $\frac{1}{2}\epsilon|\overline{\mathscr{E}}|^2$ representing, respectively, the magnetic and electric energy densities. Note that, strictly speaking, the quantities $W_e = \frac{1}{2}\epsilon|\overline{\mathscr{E}}|^2$ and $W_m = \frac{1}{2}\mu|\overline{\mathscr{H}}|^2$ are known to represent electric and magnetic energy densities for static fields. However, it is generally assumed that these quantities also represent stored energy densities in the case of time-varying fields.[40]

[40]Such an assumption is entirely reasonable, since the energy density is defined at a given point. From another point of view, we can consider Poynting's theorem or (8.46) to be the definition of energy density for time-varying fields. The correct amount of total electromagnetic energy is always obtained by assigning an amount $\frac{1}{2}(\overline{\mathscr{B}} \cdot \overline{\mathscr{H}} + \overline{\mathscr{D}} \cdot \overline{\mathscr{E}}) = \frac{1}{2}(\epsilon|\overline{\mathscr{E}}|^2 + \mu|\overline{\mathscr{H}}|^2)$ to each unit of volume. Other expressions for static energy densities, such as $\frac{1}{2}\rho\Phi$ for electrostatic fields, where Φ is the electrostatic potential, are not applicable for time-varying fields. See Chapter 27 of Volume II of R. P. Feynman, R. B. Leighton, and M. Sands, *The Feynman Lectures on Physics,* the Definitive Edition Addison-Wesley, San Francisco, 2006, and Section 2.19 of J. A. Stratton, *Electromagnetic Theory,* McGraw-Hill, New York, 1941.

From conservation of energy, the last term in (8.46) must represent the flow of energy inward or outward through the surface S enclosing the volume V. Thus, the vector $\overline{\mathscr{S}} = \overline{\mathscr{E}} \times \overline{\mathscr{H}}$, which has dimensions of watts per square meter (W-m^{-2}), appears to be a measure of the rate of energy flow per unit area at any point on the surface S. The direction of power flow is perpendicular to both $\overline{\mathscr{E}}$ and $\overline{\mathscr{H}}$. In other words, the power density in an electromagnetic wave is given by

$$\boxed{\overline{\mathscr{S}} = \overline{\mathscr{E}} \times \overline{\mathscr{H}}}$$

Equation (8.46) is known as *Poynting's theorem,* and the vector $\overline{\mathscr{S}}$ is known as the *Poynting vector.* In words, we have, "Electromagnetic power flow into a closed surface at any instant equals the sum of the time rates of increase of the stored electric and magnetic energies plus the ohmic power dissipated (or electric power generated, if the surface encloses a source) within the enclosed volume."

Note that the interpretation of $\overline{\mathscr{S}}$ as the local power flux vector does not follow from (8.46) with mathematical rigor. In principle, any vector $\overline{\mathscr{Y}}$ for which the integral over the closed surface S is zero (as is true for any vector that is the curl of another vector)[41] can be added to $\overline{\mathscr{E}} \times \overline{\mathscr{H}}$ without changing the result. Similar concerns may be raised about the interpretation of the terms $\frac{1}{2}\epsilon|\overline{\mathscr{E}}|^2$ and $\frac{1}{2}\mu|\overline{\mathscr{H}}|^2$ as energy densities. In general, Poynting's theorem makes physical sense in terms of its integral form, enabling us to describe the net flow of electromagnetic power through a closed surface. However, we run into difficulty when we try to describe where the energy resides.[42] This problem is reminiscent of the potential energy of a raised object, which is given by the weight times the height above the ground, but for which we also find it difficult to determine *where* the energy actually resides. The fact that we cannot pinpoint the "location" of electromagnetic energy is reminiscent of the fact that it is neither necessary nor possible to "determine" the location of electric and magnetic energies stored in static fields (see Sections 4.12 and 7.3).

The work done per unit time per unit volume by the fields (i.e., $\overline{\mathscr{J}} \cdot \overline{\mathscr{E}}$) is a conversion of electromagnetic energy into mechanical energy or heat. Since matter ultimately consists of charged particles (electrons and nuclei), we can think of this rate of conversion as a rate of increase of energy of the charged particles per unit volume. In this sense, Poynting's theorem for the microscopic fields can be interpreted as a statement of conservation of energy of the combined system of particles and fields.

Example 8.13: Wire carrying direct current. Consider a long cylindrical conductor of conductivity σ and radius a, carrying a direct current I as shown in Figure 8.11. Find the power dissipated in a portion of the wire of length l, using (a) the left-hand side of (8.46) and (b) the right-hand side of (8.46). Note that the cross-sectional area of the wire is $A = \pi a^2$.

---

[41]Let $\overline{\mathscr{Y}} = \nabla \times \overline{\mathscr{X}}$. Then $\oint_S \overline{\mathscr{Y}} \cdot d\mathbf{s} = \int_V \nabla \cdot (\nabla \times \overline{\mathscr{X}})\, dv \equiv 0$, since $\nabla \cdot \nabla \times \overline{\mathscr{X}} \equiv 0$ for any vector $\overline{\mathscr{X}}$.

[42]For a qualitative discussion see Section II-27-4 of R. P. Feynman, R. B. Leighton, and M. Sands, *The Feynman Lectures on Physics,* Addison-Wesley, 1964.

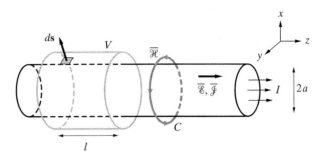

Figure 8.11   Long straight wire. The power dissipated in a cylindrical conductor carrying a direct current I is given by $I^2 R$.

Solution:

(a) The current density and electric field within the conductor are respectively $\overline{\mathscr{J}} = \hat{\mathbf{z}}(I/A)$ and $\overline{\mathscr{E}} = \hat{\mathbf{z}} I/(\sigma A)$. We can evaluate the left-hand side of (8.46) for a cylindrical volume V of arbitrary radius greater than a as shown in Figure 8.11. Noting that the current density is zero outside the wire, the limits of integration for r need extend only up to $r = a$, and we have

$$\int_V \overline{\mathscr{E}} \cdot \overline{\mathscr{J}} \, dv = \int_0^l \int_0^{2\pi} \int_0^a \left(\frac{I}{\sigma A} \hat{\mathbf{z}} \right) \cdot \left(\hat{\mathbf{z}} \frac{I}{A} \right) r \, dr \, d\phi \, dz$$

$$= \frac{I^2}{\sigma A^2} (\pi a^2 l) = \frac{I^2 l}{\sigma A} = I^2 R$$

where $R = l/(\sigma A)$ is the resistance of the cylindrical conductor. Thus, the left-hand side of (8.46) represents the total ohmic losses in the volume V.

(b) Since the current I and thus the associated electric and magnetic fields are static ($\partial/\partial t = 0$), the rate of increase of the electric and magnetic energies within the conductor is zero. Thus, the right-hand side of (8.46) reduces to the surface integral of the $\overline{\mathscr{E}} \times \overline{\mathscr{H}}$ term. Using Ampère's law with a contour of radius $r \geq a$, the magnetic field surrounding the conductor is

$$\overline{\mathscr{H}} = \hat{\boldsymbol{\phi}} \frac{I}{2\pi r}$$

At the surface of the conductor ($r = a$), we thus have $\overline{\mathscr{H}} = \hat{\boldsymbol{\phi}} I/(2\pi a)$, so that the Poynting vector is

$$\overline{\mathscr{S}} = \overline{\mathscr{E}} \times \overline{\mathscr{H}} = \frac{I}{\sigma A} \hat{\mathbf{z}} \times \hat{\boldsymbol{\phi}} \frac{I}{2\pi a} = -\hat{\mathbf{r}} \frac{I^2}{2\pi a \sigma A}$$

which is directed radially inwards toward the surface of the wire. The right-hand side of (8.46) is simply the negative of the integral of $\overline{\mathscr{S}}$ over the closed surface of the cylindrical wire (note that there is no contribution from the ends of the cylinder, since $\overline{\mathscr{S}}$ is parallel to those surfaces):

$$-\oint_S \overline{\mathscr{S}} \cdot d\mathbf{s} = -\int_0^l \int_0^{2\pi} \frac{I^2}{2\pi a \sigma A} (-\hat{\mathbf{r}}) \cdot \hat{\mathbf{r}} a \, d\phi \, dz = \frac{I^2 l}{\sigma A} = I^2 R$$

which is equal to the left-hand side of (8.46) evaluated in part (a), thus verifying the Poynting theorem. Note here that if the wire were a perfect conductor ($\sigma = \infty$), then the electric field inside the wire (and thus just outside the wire) must be zero. In such a case, $\overline{\mathscr{E}} \times \overline{\mathscr{H}} = 0$ and there is no power flow into the wire, consistent with the fact that there is no power dissipation in the wire.

Example 8.14: Power flow in a coaxial line. Consider a coaxial line delivering power to a resistor as shown in Figure 8.12. Assume the wires to be perfect conductors, so that there is no power dissipation in the wires, and the electric field inside them is zero. The configurations of electric and magnetic field lines in the coaxial line are as shown, the electric field being due to the applied potential difference between the inner and outer conductors. Find the power delivered to the resistor R.

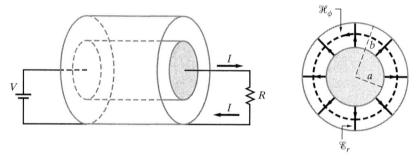

Figure 8.12 **Power flow in a coaxial line.** Coaxial wire delivering power to a resistor R.

Solution: From Ampère's law we have

$$\overline{\mathscr{H}} = \hat{\phi} \frac{I}{2\pi r} \qquad a \le r \le b$$

To find the electric field, we recall that the capacitance (per unit length) of a coaxial line is $C_u = (2\pi\epsilon)/\ln(b/a)$, so the applied voltage V induces a charge per unit length of $Q = C_u V$, from which we can find the $\overline{\mathscr{E}}$ field using Gauss's law (i.e., the integral form of equation (7.21b)) as

$$\overline{\mathscr{E}} = \hat{\mathbf{r}} \frac{Q}{2\pi\epsilon r} = \hat{\mathbf{r}} \frac{V}{r \ln(b/a)}$$

Therefore, the power delivered through the cross-sectional area of the coaxial line can be found by integrating the Poynting vector $\overline{\mathscr{S}} = \overline{\mathscr{E}} \times \overline{\mathscr{H}}$ over the area,

$$\oint_S \overline{\mathscr{E}} \times \overline{\mathscr{H}} \cdot d\mathbf{s} = \int_a^b \int_0^{2\pi} \frac{V}{r \ln(b/a)} \left(\frac{I}{2\pi r}\right) \hat{\mathbf{z}} \cdot (\hat{\mathbf{z}} r \, dr \, d\phi)$$

$$= \int_a^b \frac{V}{r \ln(b/a)} \left(\frac{I}{2\pi r}\right) 2\pi r \, dr = \frac{VI}{\ln(b/a)} \int_a^b \frac{dr}{r} = VI$$

as expected on a circuit theory basis. We note that the electromagnetic power is carried entirely outside the conductors. Even if the inner and outer conductors of the coaxial line were imperfect conductors, this simply leads to a component of the Poynting vector radially inward, as we saw in the previous example. No part of the axially directed Poynting flux (i.e., the power flux flowing along the coaxial line in the z direction) is carried within the conductors.

Example 8.15: Energy flow in a capacitor. Figure 8.13 shows a parallel-plate capacitor of capacitance $C = \epsilon A / d$, being charged by a current I flowing in the connecting wires. When it is charged, an electric field $\mathscr{E}_z$ exists (taking the z axis to be along the current-carrying wire) between the capacitor plates, with $\frac{1}{2}\epsilon\mathscr{E}_z^2$ representing the stored energy density. Investigate the manner in which the energy enters into the region between the capacitor plates.

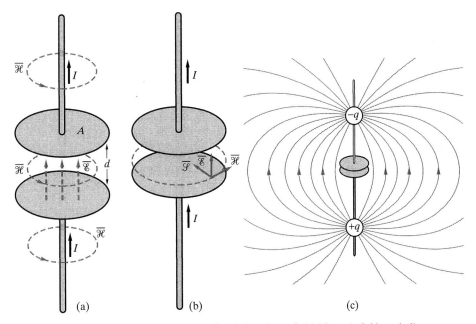

Figure 8.13   A parallel-plate capacitor being charged. (a) Magnetic fields encircling the conduction (outside the region between plates) or displacement (inside the region between the plates) currents. (b) Power evidently enters the storage region (i.e., the region between plates) through the sides via $\overline{\mathscr{E}} \times \overline{\mathscr{H}}$. (c) The intensification of the electric field lines near the capacitor as two point charges come closer to the plates.

Solution: Neglecting fringing fields, the electric field between the capacitor plates is uniform, so the total electrical energy stored between the plates is

$$W_{\mathrm{e}} = \int_V \tfrac{1}{2}\epsilon\mathscr{E}_z^2 \, dv = (\tfrac{1}{2}\epsilon\mathscr{E}_z^2)Ad$$

The time rate of change of this energy is given by

$$\frac{dW_{\mathrm{e}}}{dt} = Ad\frac{d}{dt}\left(\frac{1}{2}\epsilon\mathscr{E}_z^2\right) = \epsilon Ad\mathscr{E}_z\frac{d\mathscr{E}_z}{dt}$$

which means that there must be a flow of energy into the volume between the plates from somewhere. Can it be arriving through the wire? Apparently not, according to Poynting's theorem, since we note that $\overline{\mathscr{E}}$ is perpendicular to the plates, and thus $\overline{\mathscr{E}} \times \overline{\mathscr{H}}$ must necessarily be parallel to the plates.

A magnetic field $\overline{\mathscr{H}}$ encircles the z axis, supported by the conduction current along the wires and the displacement current in the region between the plates, as shown in Figure 8.13a. In the region between the plates, where $\overline{\mathscr{J}} = 0$, the magnetic field at any radial distance r is given by (7.21c) as

$$\oint_C \overline{\mathscr{H}} \cdot d\mathbf{l} = \int_S \frac{\partial(\epsilon\overline{\mathscr{E}})}{\partial t} \cdot d\mathbf{s} \quad \rightarrow \quad \mathscr{H}_\phi(2\pi r) = \epsilon(\pi r^2)\frac{\partial\mathscr{E}_z}{\partial t} \quad \rightarrow \quad \mathscr{H}_\phi = \frac{\epsilon r}{2}\frac{\partial\mathscr{E}_z}{\partial t}$$

Thus, Poynting's vector $\overline{\mathscr{P}} = \overline{\mathscr{E}} \times \overline{\mathscr{H}} = \mathscr{E}_z\hat{\mathbf{z}} \times \hat{\boldsymbol{\phi}}\mathscr{H}_\phi = \mathscr{E}_z\mathscr{H}_\phi(-\hat{\mathbf{r}})$ has only an r component, pointed inward, as shown in Figure 8.13b. The energy stored in the capacitor is apparently being supplied through the sides of the cylindrical region between the plates. The total power flow into the region can be found by integrating $\overline{\mathscr{P}}$ over the cylindrical side surface of the capacitor at $r = a$:

$$\oint_S \overline{\mathscr{P}} \cdot d\mathbf{s} = \int_0^{2\pi}\int_{z=0}^d \mathscr{E}_z\left(\frac{\epsilon a}{2}\frac{\partial\mathscr{E}_z}{\partial t}\right) a\, dz\, d\phi = \pi a^2 d\frac{\partial}{\partial t}\left(\frac{1}{2}\epsilon\mathscr{E}_z^2\right)$$

which is identical to the dW_e/dt expression we found above, where we note that $A = \pi a^2$. Thus, we see that the power supplied into the region from the sides is precisely equal to the time rate of change of stored energy.

This result is quite contrary to our intuitive expectations of the power being supplied to the capacitor via the wires. One way to think about this result[43] is illustrated in Figure 8.13c. Consider two opposite charges above and below the capacitor plates, approaching the capacitor plates via the wire. When the charges are far away from the plates, the electric field between the plates is weak but spread out in the region surrounding the capacitor. As the charges approach the plates, the electric field between the capacitor plates becomes stronger (i.e., the field energy moves toward the capacitor) and in the limit is eventually completely confined to the region between the plates. Thus, as shown, the electromagnetic energy flows into the region between the capacitor plates through the openings on the sides.

8.4.2 Electromagnetic Power Carried by a Uniform Plane Wave in a Lossless Medium

Consider the time-harmonic uniform plane wave propagating in a lossless medium that was studied in Section 8.2:

$$\mathscr{E}_x = E_0\cos(\omega t - \beta z)$$

$$\mathscr{H}_y = \frac{1}{\eta}E_0\cos(\omega t - \beta z)$$

---

[43]Taken from Section II-27-5 of R. P. Feynman, R. B. Leighton, and M. Sands, *The Feynman Lectures on Physics,* Addison-Wesley, 1964.

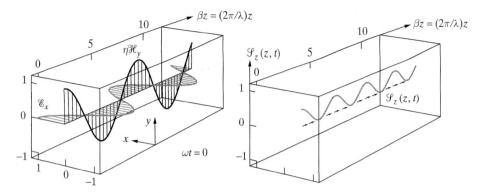

Figure 8.14 The Poynting vector of a uniform plane wave. (a) The wave electric and magnetic fields at $\omega t = 0$. (b) The Poynting vector at $\omega t = 0$.

where $\beta = \omega\sqrt{\mu\epsilon}$ and $\eta = \sqrt{\mu/\epsilon}$. The electric and magnetic fields of the wave are as shown in Figure 8.14a. The Poynting vector for this wave is given by

$$\overline{\mathscr{P}} = \overline{\mathscr{E}} \times \overline{\mathscr{H}} = \hat{\mathbf{z}}E_0 \left(\frac{E_0}{\eta}\right)\cos^2(\omega t - \beta z) = \hat{\mathbf{z}}\frac{E_0^2}{2\eta}\{1 + \cos[2(\omega t - \beta z)]\}$$

We see from Figure 8.14b that $\overline{\mathscr{P}}$ everywhere is directed in the $+z$ direction. The power carried by the wave has a constant component and a component varying at a frequency twice that of the fields, exhibiting two positive peaks per wavelength of the wave, as shown in Figure 8.14b. In most cases, the constant component, called the time-average Poynting vector $\mathbf{S}_{av}$, is the quantity of interest. The quantity $\mathbf{S}_{av}$ is defined as the average over one period ($T_p = 2\pi/\omega$) of the wave of $\overline{\mathscr{P}}$. For a time-harmonic uniform plane wave in a lossless medium, we have

$$\mathbf{S}_{av} = \frac{1}{T_p}\int_0^{T_p} \overline{\mathscr{P}}(z,t)\,dt$$

$$= \frac{1}{T_p}\int_0^{T_p} \hat{\mathbf{z}}\frac{E_0^2}{2\eta}\{1 + \cos[2(\omega t - \beta z)]\}\,dt \quad \rightarrow \quad \boxed{\mathbf{S}_{av} = \hat{\mathbf{z}}\frac{E_0^2}{2\eta}} \qquad (8.47)$$

According to Poynting's theorem as expressed in (8.46), and since the medium is assumed to be lossless and source-free, so $\overline{\mathscr{J}} = 0$, the integral of the Poynting vector $\overline{\mathscr{P}}$ over any closed surface should be balanced by the time rate of change of the sum of stored electric and magnetic energies in the volume enclosed by that surface. The electric and magnetic energy densities for this uniform plane wave are

$$W_e = \tfrac{1}{2}\epsilon|\mathscr{E}_x|^2 = \tfrac{1}{2}\epsilon E_0^2\cos^2(\omega t - \beta z) \qquad (8.48a)$$

$$W_m = \frac{1}{2}\mu|\mathscr{H}_y|^2 = \frac{1}{2}\frac{\mu}{(\sqrt{\mu/\epsilon})^2}E_0^2\cos^2(\omega t - \beta z) = \frac{1}{2}\epsilon E_0^2\cos^2(\omega t - \beta z) \qquad (8.48b)$$

We thus note that $W_e = W_m$, indicating that the instantaneous values of the electric and magnetic stored energy densities are equal at all points in space and at all times. The reader is encouraged to choose a suitable closed surface and show that the volume integral of the partial derivative with respect to time of $W_m + W_e$ is indeed equal to the surface integral of $\mathcal{S}$. Alternatively, one can consider the differential form of Poynting's theorem, as given in (8.44), which for $\overline{\mathscr{E}} \cdot \overline{\mathcal{S}} = 0$ reduces to

$$\frac{\partial}{\partial t}\left(\frac{1}{2}\mu|\overline{\mathcal{H}}|^2\right) + \frac{\partial}{\partial t}\left(\frac{1}{2}\epsilon|\overline{\mathscr{E}}|^2\right) = -\nabla \cdot (\overline{\mathscr{E}} \times \overline{\mathcal{H}})$$

and show its validity by simply substituting (8.48a) and (8.48b) and carrying out the differentiation with respect to time. The manipulations are left as an exercise for the reader.

The instantaneous variations of the stored energies are often not of practical interest. Instead, one is often interested in the time averages of these quantities over one period ($T_p = 2\pi/\omega$). Denoting these time averages as $\overline{W}_e$ and $\overline{W}_m$, we have

$$\overline{W}_e = \overline{W}_m = \frac{1}{T_p}\int_0^{T_p}\frac{1}{2}\epsilon E_0^2 \cos^2(\omega t - \beta z)\, dt$$

$$= \frac{1}{2}\epsilon E_0^2 \frac{1}{T_p}\int_0^{T_p}\left\{\frac{1}{2} + \frac{1}{2}\cos[2(\omega t - \beta z)]\right\} dt = \frac{1}{4}\epsilon E_0^2 \qquad (8.49)$$

We undertake a more general discussion of instantaneous and time-average electromagnetic power in the next section, applicable to uniform plane waves in lossy media.

Example 8.16: Solar radiation at Earth. If the time-average power density arriving at the surface of the earth due to solar radiation is about 1400 W-m^{-2} (an average over a wide band of frequencies), find the peak electric and magnetic field values on the earth's surface due to solar radiation. The radii of Earth and the sun are $\sim$6.37 Mm and $\sim$700 Mm, respectively, and the radius of Earth's orbit around the sun is $\sim$1.5 × 10^8 km.

Solution: Using (8.47) we can write

$$|\mathbf{S}_{av}| = \frac{1}{2}\frac{E_0^2}{\eta} = 1400 \text{ W-m}^{-2}$$

where $\eta \simeq 377\Omega$. Solving for E_0 we find $E_0 \simeq \sqrt{2 \times 377 \times 1400} \simeq 1027$ V-m^{-1}. The corresponding magnetic field H_0 is $H_0 = E_0/\eta \simeq 1027/377 \simeq 2.73$ A-m^{-1}.

Example 8.17: A spaceship in lunar orbit. A spaceship in lunar orbit (the Earth–Moon distance is $\sim$380 Mm) transmits plane waves with an antenna operating at 1 GHz and radiating a total power of 1 MW isotropically (i.e., equally in all directions). Find (a) the time-average power density, (b) the peak electric field value on Earth's surface, and (c) the time it takes for these waves to travel from the spaceship to Earth.

Solution:

(a) The time-average power density on Earth's surface a distance $R \simeq 380$ Mm away from the spaceship radiating a total power of $P_{tot} = 1$ MW isotropically is

$$|\mathbf{S}_{av}| = \frac{P_{tot}}{4\pi R^2} = \frac{10^6 \text{ W}}{4\pi (380 \times 10^6)^2 \text{ m}^2} \simeq 5.51 \times 10^{-13} \text{ W-m}^{-2}$$

(b) The corresponding peak electric field value E_0 can be found from (8.47):

$$|\mathbf{S}_{av}| = \frac{1}{2}\frac{E_0^2}{\eta} \quad \rightarrow \quad E_0 \simeq \sqrt{2 \times 377 \times 5.51 \times 10^{-13}} \simeq 2.04 \times 10^{-5} \text{ V-m}^{-1}$$

(c) The time it takes for the waves to travel from the spaceship to the earth is given by

$$t = \frac{R}{c} \simeq \frac{380 \times 10^6 \text{ m}}{3 \times 10^8 \text{ m-s}^{-1}} \simeq 1.27 \text{ s}$$

Example 8.18: FM broadcasting. A transmitter antenna for an FM radio broadcasting station operates around 92.3 MHz (KGON station in Portland, Oregon) with an average radiated power of 100 kW, as limited by Federal Communications Commission (FCC) regulations to minimize interference problems with other broadcast stations outside its coverage area. For simplicity, assume isotropic radiation, that is, radiation equally distributed in all directions, such as from a point source. (a) For a person standing 50 m away from the antenna, calculate the time-average power density of the radiated wave and compare it with the IEEE safety limit[44] for an uncontrolled environment at that frequency, namely 2 W-m^{-2} or 0.2 mW-(cm)$^{-2}$. (b) Find the minimum distance such that the time-average power density is equal to the IEEE safety standard value at that frequency. (c) If the antenna is mounted on a tower that is 200 m above the ground, repeat part (a) for a person standing 50 m away from the foot of the tower.

Solution: Assume the antenna to be a point source and neglect all effects from boundaries.

(a) The wavelength at 92.3 MHz is

$$\lambda \simeq \frac{3 \times 10^8}{92.3 \times 10^6} \cong 3.25 \text{ m}$$

Since 50 m $\gg$ 3.25 m, we can assume far field, as if the wave were a uniform plane wave. The total power of $P_{tot} = 100$ kW is radiated equally in all directions, so if we take a sphere of radius 50 m centered at the antenna, the time-average power density on the surface of the sphere is given by

$$|\mathbf{S}_{av}| = \frac{P_{tot}}{4\pi R^2} = \frac{100 \times 10^3}{4\pi (50)^2} \simeq 3.18 \text{ W-m}^{-2} = 0.318 \text{ mW-(cm)}^{-2}$$

Since 0.318 mW-(cm)$^{-2}$ > 0.2 mW-(cm)$^{-2}$, it is not safe to stand at a distance of 50 m from this FM broadcast antenna.

---

[44]IEEE c95.1-2005, *IEEE Standard for Safety Levels with Respect to Human Exposure to Radio Frequency Electromagnetic Fields,* 3 kHz to 300 GHz. For a related discussion, see M. Fischetti, The cellular phone scare, *IEEE Spectr.,* pp. 43–47, June 1993.

(b) From

$$\frac{P_{\text{tot}}}{4\pi R_{\text{min}}^2} = \frac{100 \times 10^3}{4\pi R_{\text{min}}^2} = 2 \text{ W-m}^{-2}$$

we find $R_{\text{min}} \simeq 63.1$ m.

(c) Considering a 200 m tower height and a person standing on the ground at a distance 50 m from the base of the tower, the distance between the antenna and the person is $\sqrt{(200)^2 + (50)^2}$ m, so that we have

$$|\mathbf{S}_{\text{av}}| = \frac{100 \times 10^3}{4\pi[(200)^2 + (50)^2]} \simeq 0.187 \text{ W-m}^{-2} < 2 \text{ W-m}^{-2}$$

Therefore, the radiation level on the ground is well below the IEEE safety standard.

Example 8.19: VHF/UHF broadcast radiation. A survey[45] conducted in the United States indicates that $\sim$50% of the population is exposed to average power densities of approximately 0.005 μW-(cm)$^{-2}$ due to VHF and UHF broadcast radiation. Find the corresponding amplitudes of the electric and magnetic fields.

Solution: By direct application of (8.47) we have

$$|\mathbf{S}_{\text{av}}| = \frac{1}{2}\frac{E_0^2}{\eta} = 0.005 \ \mu\text{W-(cm)}^{-2}$$

so

$$E_0 \simeq \sqrt{2 \times 377 \times 5 \times 10^{-9}/10^{-4}} \simeq 194 \text{ mV-m}^{-1}$$

$$H_0 = \frac{E_0}{\eta} \simeq \frac{194 \text{ mV-m}^{-1}}{377\Omega} \simeq 515 \ \mu\text{A-m}^{-1}$$

where we have noted that $\eta \simeq 377\Omega$ in air.

8.4.3 Instantaneous and Time-Average Power and the Complex Poynting Theorem

In most applications we are interested in the Poynting vector averaged over time rather than in its instantaneous value. Consider the expressions for the instantaneous electric and magnetic fields of a time-harmonic uniform plane wave in the general case of a lossy medium. From (8.19) and (8.27) we have

$$\overline{\mathscr{E}}(z,t) = \hat{\mathbf{x}}C_1 e^{-\alpha z} \cos(\omega t - \beta z)$$

and

$$\overline{\mathscr{H}}(z,t) = \hat{\mathbf{y}}\frac{C_1}{|\eta_c|}e^{-\alpha z} \cos(\omega t - \beta z - \phi_\eta)$$

---

[45]R. A. Tell and E. D. Mantiply, Population exposure to VHF and UHF broadcast radiation in the United States, *Proc. IEEE,* 68(1), pp. 6–12, January 1980.

Noting that $\hat{\mathbf{x}} \times \hat{\mathbf{y}} = \hat{\mathbf{z}}$, the instantaneous Poynting vector given by the cross product $\overline{\mathscr{E}} \times \overline{\mathscr{H}}$ is in the $+z$ direction:

$$\overline{\mathscr{S}}(z,t) = \overline{\mathscr{E}}(z,t) \times \overline{\mathscr{H}}(z,t) = \hat{\mathbf{z}} \frac{C_1^2}{|\eta_c|} e^{-2\alpha z} \cos(\omega t - \beta z) \cos(\omega t - \beta z - \phi_\eta) \qquad (8.50)$$

As expected, the electromagnetic power carried by a uniform plane wave propagating in the $+z$ direction flows in the $+z$ direction. We can simplify (8.50) to

$$\overline{\mathscr{S}}(z,t) = \overline{\mathscr{E}}(z,t) \times \overline{\mathscr{H}}(z,t) = \hat{\mathbf{z}} \frac{C_1^2}{2|\eta_c|} e^{-2\alpha z} [\cos(\phi_\eta) + \cos(2\omega t - 2\beta z - \phi_\eta)] \qquad (8.51)$$

We see that the instantaneous Poynting vector has a component that does not change with time and a component that oscillates at twice the rate at which electric and magnetic fields change in time. In most applications the time-average value of the power transmitted by a wave is more significant than the fluctuating component. The relationship between the electric and magnetic fields and the Poynting vector $\overline{\mathscr{S}}(z,t)$ is illustrated in Figure 8.15. Note that the instantaneous Poynting vector itself can be negative in certain regions; however, the time-average value is always positive for a uniform plane wave and represents real power flow in the $+z$ direction.

The time-average value can be obtained from (8.51) by integrating over one period T_p of the sinusoidal variation. We find

$$\mathbf{S}_{av}(z) = \frac{1}{T_p} \int_0^{T_p} \overline{\mathscr{S}}(z,t)\ dt = \hat{\mathbf{z}} \frac{C_1^2}{2|\eta_c|} e^{-2\alpha z} \cos(\phi_\eta) \qquad (8.52)$$

As an example, consider a uniform plane wave at 300 MHz in copper, for which $\eta_c \simeq 6.39 \times 10^{-3} e^{j45°}\ \Omega$. The time-average electromagnetic power carried by a wave with

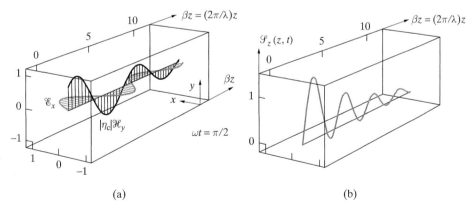

Figure 8.15 The Poynting vector and fields for a uniform plane wave in a lossy medium. (a) The wave electric and magnetic fields at $\omega t = \pi/2$ and (b) the Poynting vector $\overline{\mathscr{S}}(z, \omega t = \pi/2)$.

1 V-m^{-1} field amplitude at $z = 0$ (i.e., $C_1 = 1$ V-m^{-1}) is ~55.3 W-m^{-2}, compared with ~0.0013 W-m^{-2} in free space for the same C_1.

Finding the time-average Poynting flux $\mathbf{S}_{av}$ by an integration over one period as in (8.52) is in general not necessary for sinusoidal signals, since $\mathbf{S}_{av}$ can be directly obtained from the phasors $\mathbf{E}$ and $\mathbf{H}$, as shown below.

The Poynting theorem is expressed mathematically in (8.46) in terms of the instantaneous field quantities $\overline{\mathscr{E}}$ and $\overline{\mathscr{H}}$. The complex version of the theorem in terms of the phasors $\mathbf{E}$ and $\mathbf{H}$ cannot be obtained from (8.46) by substituting $\partial/\partial t \to j\omega$, since the terms in (8.46) involve vector products of field quantities. Thus, the complex Poynting theorem has to be derived from the phasor form of Maxwell's equations (7.23a) and (7.23c):

$$\nabla \times \mathbf{E} = -j\omega\mathbf{B} \qquad \nabla \times \mathbf{H} = \mathbf{J} + j\omega\mathbf{D}$$

We start from the complex form of the vector identity

$$\nabla \cdot (\mathbf{E} \times \mathbf{H}^*) = \mathbf{H}^* \cdot (\nabla \times \mathbf{E}) - \mathbf{E} \cdot (\nabla \times \mathbf{H}^*)$$

Substituting the two curl equations into this identity, we have

$$\nabla \cdot (\mathbf{E} \times \mathbf{H}^*) = \mathbf{H}^* \cdot (-j\omega\mathbf{B}) - \mathbf{E} \cdot (\mathbf{J}^* - j\omega\mathbf{D}^*) \tag{8.53}$$

Integrating (8.53) over a volume V and using the divergence theorem,

$$\int_V \nabla \cdot (\mathbf{E} \times \mathbf{H}^*)\, dv = \oint_S (\mathbf{E} \times \mathbf{H}^*) \cdot d\mathbf{s} = \int_V -[\mathbf{E} \cdot \mathbf{J}^* + j\omega(\mathbf{H}^* \cdot \mathbf{B} - \mathbf{E} \cdot \mathbf{D}^*)]\, dv \tag{8.54}$$

If the medium is isotropic and if all the losses occur through conduction currents $\mathbf{J} = \sigma\mathbf{E}$, equation (8.54) becomes

$$\oint_S (\mathbf{E} \times \mathbf{H}^*) \cdot d\mathbf{s} = -\int_V \sigma\mathbf{E} \cdot \mathbf{E}^*\, dv - j\omega \int_V (\mu\mathbf{H} \cdot \mathbf{H}^* - \epsilon\mathbf{E} \cdot \mathbf{E}^*)\, dv \tag{8.55}$$

In the case of lossy dielectrics, for which losses occur through effective conduction currents $\overline{\mathbf{J}} = \sigma_{\text{eff}}\overline{\mathbf{E}} = \omega\epsilon''\overline{\mathbf{E}}$, equation (8.54) is valid when we replace σ with $\sigma_{\text{eff}} = \omega\epsilon''$ and ϵ with ϵ'. Recognizing from (8.49) that $\overline{W}_e = \frac{1}{4}\epsilon\mathbf{E} \cdot \mathbf{E}^*$ and $\overline{W}_m = \frac{1}{4}\mu\mathbf{H} \cdot \mathbf{H}^*$ are the time averages of the energy densities $W_e = \frac{1}{2}\epsilon\overline{\mathscr{E}} \cdot \overline{\mathscr{E}}$ and $W_m = \frac{1}{2}\mu\overline{\mathscr{H}} \cdot \overline{\mathscr{H}}$, respectively, and noting that the volume density of time-average power dissipation in conduction currents is $P_c = \frac{1}{2}\sigma\mathbf{E} \cdot \mathbf{E}^*$, we can equate the real and imaginary parts of (8.55) as

$$\mathscr{R}e\left\{\oint_S (\mathbf{E} \times \mathbf{H}^*) \cdot d\mathbf{s}\right\} = -2\int_V P_c\, dv \tag{8.56}$$

$$\mathscr{I}m\left\{\oint_S (\mathbf{E} \times \mathbf{H}^*) \cdot d\mathbf{s}\right\} = -4\omega\int_V (\overline{W}_m - \overline{W}_e)\, dv \tag{8.57}$$

The instantaneous Poynting vector $\overline{\mathscr{P}}(z,t)$ can be written in terms of the phasors $\mathbf{E}(z)$ and $\mathbf{H}(z)$, as follows:

$$\overline{\mathscr{P}}(z,t) = \mathscr{R}e\{\mathbf{E}(z)e^{j\omega t}\} \times \mathscr{R}e\{\mathbf{H}(z)e^{j\omega t}\} = \tfrac{1}{2}\mathscr{R}e\{\mathbf{E}(z) \times \mathbf{H}^*(z) + \mathbf{E}(z) \times \mathbf{H}(z)e^{j2\omega t}\}$$

where we have used the fact that in general, for two different complex vectors $\mathbf{G}$ and $\mathbf{F}$, and noting that $(\mathbf{G} \times \mathbf{F}^*)^* = \mathbf{G}^* \times \mathbf{F}$, we have

$$\begin{aligned}
\mathscr{R}e\{\mathbf{G}\} \times \mathscr{R}e\{\mathbf{F}\} &= \tfrac{1}{2}(\mathbf{G} + \mathbf{G}^*) \times \tfrac{1}{2}(\mathbf{F} + \mathbf{F}^*) \\
&= \tfrac{1}{4}[(\mathbf{G} \times \mathbf{F}^* + \mathbf{G}^* \times \mathbf{F}) + (\mathbf{G} \times \mathbf{F} + \mathbf{G}^* \times \mathbf{F}^*)] \\
&= \tfrac{1}{2}\mathscr{R}e\{\mathbf{G} \times \mathbf{F}^* + \mathbf{G} \times \mathbf{F}\}
\end{aligned}$$

since in general we have $\mathscr{R}e\{\mathbf{G}\} = \tfrac{1}{2}(\mathbf{G} + \mathbf{G}^*)$ for any complex vector $\mathbf{G}$. The time-average power density, which was denoted $\mathbf{S}_{\text{av}}$ in (8.52), can then be obtained by integrating $\overline{\mathscr{P}}(z,t)$ over one period $T_p = 2\pi/\omega$. Since the time-average of the $\mathbf{E}(z) \times \mathbf{H}(z)e^{j2\omega t}$ term vanishes, we find that the time-average Poynting vector is given by

$$\boxed{\mathbf{S}_{\text{av}} = \tfrac{1}{2}\mathscr{R}e\{\mathbf{E} \times \mathbf{H}^*\}} \qquad (8.58)$$

This result allows us to evaluate $\mathbf{S}_{\text{av}}$ conveniently from the phasor field quantities. Note that

$$\mathbf{S} = \mathbf{E} \times \mathbf{H}^* \qquad (8.59)$$

is referred to as the *complex Poynting vector*. The imaginary part of the integral of $\mathbf{S}$ as given in (8.57) is equal to the difference between the average energies stored in the magnetic and the electric fields and represents reactive power flowing back and forth to supply the instantaneous changes in the net stored energy in the volume.

Example 8.20: VLF waves in the ocean. Consider VLF wave propagation in the ocean as discussed in Example 8.12. Find the time-average Poynting flux at the sea surface ($z = 0$) and at the depth of the submarine ($z = 100$ m).

> **Solution:** The phasor expressions for the minimum electric and magnetic fields to communicate with the submarine at 100 m depth are given by
>
> $$\mathbf{E}(z) \simeq \hat{\mathbf{x}}2.83e^{-0.218z}e^{-j0.218z} \text{ kV-m}^{-1}$$
>
> $$\mathbf{H}(z) \simeq \hat{\mathbf{y}}36.8e^{-0.218z}e^{-j0.218z}e^{-j\pi/4} \text{ kA-m}^{-1}$$
>
> Using (8.58), the time-average Poynting vector is given by
>
> $$\mathbf{S}_{\text{av}}(z) \simeq \hat{\mathbf{z}}\tfrac{1}{2}(2.83)(36.8)e^{-2(0.218z)}\cos(\pi/4) \simeq \hat{\mathbf{z}}36.8e^{-0.435z} \text{ MW-m}^{-2}$$
>
> At the sea surface (i.e., $z = 0^+$), $\mathbf{S}_{\text{av}} \simeq \hat{\mathbf{z}}36.8$ MW-m^{-2}. At the location of the submarine (i.e., $z = 100$ m), $\mathbf{S}_{\text{av}} \simeq \hat{\mathbf{z}}4.59 \times 10^{-12}$ W-m^{-2}. The power density required at the sea surface is extremely high, indicating that it is not feasible to use 3 kHz signals to communicate with a submarine at 100 m depth.

8.4.4 Radiation Pressure and Electromagnetic Momentum

In addition to carrying energy as discussed above, an electromagnetic wave can also exert pressure (i.e., force per unit area) or transport linear momentum in the direction of propagation. In other words, pressure is exerted on an object by shining an electromagnetic wave (e.g., a beam of light) on it.[46] Clearly, such forces must ordinarily be quite small,[47] since we do not notice them in our ordinary environment. However, ultrahigh-intensity lasers[48] can nowadays have power densities of $\sim 10^{24}$ W-m^{-2}, corresponding to gigantic light pressures of up to tens of 10^9 bars (1 bar is defined as 10^5 N-m^{-2}; ~ 1.013 bar is equal to 1 atmospheric pressure (atm), that is, the air pressure at sea level). A simple rotating device that uses radiation pressure of sunlight as its power source is *Crookes's radiometer*,[49] consisting of four vanes delicately mounted on a vertical axis. One side of the vanes is silvered, and the other is blackened, to reflect and absorb light, respectively, thus acquiring relative angular momentum, leading to rotational motion (also see Section 9.1.1). In astrophysics, consideration of radiation pressure is essential in understanding physical processes in stellar interiors and the mechanisms governing supernovae, nebulae, and black holes.[50]

The first suggestion that light could exert pressure dates back to Johannes Kepler, who first proposed in 1619 that it was the radiation pressure of sunlight that "blew" back a comet's tail so that it points away from the sun. Maxwell's new theories of electromagnetic waves and the wave nature of light allowed him in 1873 to predict formally the magnitude of momentum carried by electromagnetic waves and the radiation pressure exerted by them. Maxwell calculated the radiation pressure to be ordinarily quite small, but he ventured to say, "Such rays falling on a thin metallic disk, delicately suspended in vacuum, might perhaps produce an observable effect." The first experimental observation of radiation pressure was to be achieved about 30 years after Maxwell's

---

[46]G. E. Henry, Radiation pressure, *Scientific American,* pp. 99–108, June 1957; A. Ashkin, The pressure of laser light, *Scientific American,* pp. 63–71, February 1972.

[47]In 1905, J. H. Poynting, in his presidential address to the British Physical Society, said, "A very short experience in attempting to measure these light forces is sufficient to make one realize their extreme minuteness, which appears to put them beyond consideration in terrestrial affairs" [J. H. Poynting, Radiation pressure, *The London, Edinburgh, and Dublin Philosophical Magazine and Journal of Science,* 9(52), pp. 393–406, April 1905]. This view of electromagnetic light forces as being of no practical importance held true until the advent of the laser in 1960.

[48]G. A. Mourou, C. P. J. Barty, and M. D. Perry, Ultrahigh-intensity lasers: Physics of the extreme on a table top, *Physics Today,* pp. 22–28, January 1998.

[49]W. Crookes, *Phil. Trans.,* p. 501, 1873. Crookes apparently believed in 1873 that he had found the true radiation pressure with his newly invented radiometer and cautiously suggested that his experiments might have some bearing on the prevailing theory of the nature of light. He discovered certain gas pressures for which the combined gas and radiation forces neutralized, but since he did not discriminate between forces due to radiation and gas forces, his results were apparently capricious and his reasoning was somewhat confused. For further discussion, see E. F. Nichols and G. F. Hull, The pressure due to radiation, *Phys. Rev.,* 17(1), pp. 26–50, July 1903.

[50]See p. 252 of D. Goldsmith, *The Evolving Universe,* Benjamin Cummings, 1985, and Section 16.6 of A. J. McMahon, *Astrophysics and Space Science,* Prentice-Hall, Englewood Cliffs, New Jersey, 1965.

predictions, by measuring with careful instrumentation the force exerted on an object by a light beam.[51]

The radiation pressure is by definition the force per unit area exerted by a wave upon a material on which it is incident. If we assume that the material completely absorbs the electromagnetic energy of a wave traveling at the speed of light c and carrying a time-average Poynting vector $\mathbf{S}_{av}$ normally incident on it for a time duration of Δt, the momentum $\Delta \overline{\mathscr{G}}$ delivered to an elemental surface area ΔA of the object is

$$\Delta \overline{\mathscr{G}} = \frac{\mathbf{S}_{av} \Delta A \Delta t}{c} \qquad \text{kg-m-s}^{-1} \quad \text{(or N-s)}$$

From Newton's second law, and assuming that the time-average power flux is steady during the time interval Δt, the time-average force exerted on the elemental surface ΔA of the object is given by

$$\mathbf{F}_{av} = \frac{\mathbf{P}_{av}}{\Delta t} = \frac{\mathbf{S}_{av} \Delta A}{c} \quad \text{N}$$

so that the time-average force per unit area, or time-average pressure, is given by

$$\boxed{\mathbf{P}_{av} = \frac{\mathbf{S}_{av}}{c} \qquad \text{N-m}^{-2}} \tag{8.60}$$

The quantity $\mathbf{P}_{av}$ is essentially a surface force density or stress. Considerations of this type constitute the subject of electromagnetic stress or the so-called *Maxwell stress tensor,* a topic beyond the scope of this book.[52]

The pressure of electromagnetic radiation encountered in our daily environment is extremely small, as mentioned above. For example, the power density of all electromagnetic radiation from the sun at the Earth's surface is $|\mathbf{S}_{av}| \simeq 1400$ W-m^{-2}, so that the pressure is only $(|\mathbf{S}_{av}|/c) \simeq 4.67 \times 10^{-6}$ N-m^{-2}, exceedingly small indeed. The radiation pressure varies as the inverse square of the distance from the source, as does the Poynting vector $\mathbf{S}_{av}$. Thus, we can expect the radiation pressure to be much higher at

---

[51]The first such measurement was realized independently in both the United States and Russia [E. F. Nichols and G. F. Hull, *Phys. Rev.* 13, p. 307, October 1901; P. Lebedev, *Ann. Physik,* 6, p. 433, October 1901; E. F. Nichols and G. F. Hull, The pressure due to radiation, *Phys. Rev.* 17(1), p. 26–50, July 1903]. One of the experiments, as described in E. F. Nichols and G. F. Hull, Pressure due to radiation, *Amer. Acad. Proc.,* 38(20), pp. 559–599, April 1903, used a torsion balance in a carefully undertaken experiment to measure a pressure of $\sim 7 \times 10^{-6}$ N-m^{-2}, or a force on the ~ 1 cm^2 area of their silvered glass vanes of $\sim 7 \times 10^{-10}$ N. The Poynting flux in the light beam they used was $|\mathbf{S}_{av}| \simeq 0.1$ W-(cm)$^{-2}$. In comparison, typical laboratory laser beams in use today may have a total power of ~ 5 mW and a width of ~ 0.5 mm (or $|\mathbf{S}_{av}| \simeq 2$ W-(cm)$^{-2}$).

[52]Faraday was the first to speak of lines of force as elastic bands that transmit tension and compression. Maxwell also spent considerable time on this concept and placed Faraday's notions into clear mathematical focus. For brief discussions of the concept of electromagnetic stress see C. C. Johnson, *Field and Wave Electrodynamics,* Section 1.19, McGraw-Hill, New York, 1965; W. B. Cheston, *Elementary Theory of Electric and Magnetic Fields,* Section 10.7–10.8, Wiley, New York, 1964. For more complete coverage, see L. M. Landau and E. M. Lifshitz, *Electrodynamics of Continuous Media,* Addison-Wesley, Reading, Massachusetts, 1960; J. A. Stratton, *Electromagnetic Theory,* McGraw-Hill, New York, 1941.

the surface of the sun. Using a mean Earth–sun distance of $\sim 1.5 \times 10^{11}$ m, we find the radiation pressure on the surface of the sun to be only ~ 0.21 N-m^{-2}, equivalent to a weight (on Earth) of $\sim 2.2 \times 10^{-3}$ gm-(cm)$^{-2}$. Radiation pressure is unimportant even in the interior of the sun, but it is important in the dynamics of brighter stars.

Much higher electromagnetic power densities and thus radiation pressures can be attained in waveguides, where power densities of up to 10^9 W-m^{-2} are possible. With propagation at velocities very close to c, 10^9 W-m^{-2} corresponds to a radiation pressure of ~ 3.33 N-m^{-2}, equivalent to a weight of 0.034 gm-(cm)$^{-2}$. For a powerful laser, which may utilize narrow (~ 0.1 mm radius) pulsed beams of a thousand gigawatts (i.e., $|S_{av}| = 10^{12}/[\pi(0.1 \times 10^{-3})^2] \simeq 3.2 \times 10^{19}$ W-m^{-2}), the force exerted on an absorbing material across the size of the beam would be ~ 3300 newtons, or as much as the weight of a ~ 340 kg object.

The radiation pressure of a wave acts in the direction of wave propagation, and the force of the wave is transferred to an object upon which the wave is incident. If the incident wave energy is completely absorbed by the object (no reflection), a force of $S_{av}\Delta A/c$ acts on it. If the incident energy is entirely reflected, as it would be if the object was a perfect conductor, the force acting on the body is twice[53] that for absorption, namely $2S_{av}\Delta A/c$. We shall revisit this subject when we study reflection of plane waves from conducting surfaces in Chapter 9.

The question of the momentum of an electromagnetic wave can be more formally studied by using the Lorentz force equation. The force on an object is the rate at which momentum is transferred to that object. The force per unit volume on the charges and currents in a given region is the rate of transfer per unit volume. Using a development very similar to that used in the derivation of Poynting's theorem (see above), it can be shown[54] that the total rate of transfer is equal to the rate of decrease of momentum residing in the field. One finds that each unit of volume in the electromagnetic field appears to carry a momentum of

$$\boxed{\bar{g} = \overline{\mathfrak{D}} \times \overline{\mathfrak{B}} = \epsilon\mu(\overline{\mathscr{E}} \times \overline{\mathscr{H}})} \qquad (8.61)$$

which is dimensionally a *momentum per unit volume*. The total momentum of the electromagnetic field contained in a volume V is

$$\overline{\mathscr{G}} = \int_V \bar{g}\, dv \qquad \text{N-s}$$

An interesting implication of the fact that electromagnetic fields carry momentum is that Newton's third law and the principle of conservation of momentum are strictly

---

[53] Note that this is similar to the fact that twice as much momentum is delivered to an object when a perfectly elastic ball is bounced from it as when it is struck with a perfectly inelastic ball. We can also think of this circumstance as the conductor first absorbing the energy and then recoiling as it re-emits it.

[54] See W. T. Scott, *The Physics of Electricity and Magnetism*, Section 10.4, Wiley, New York, 1966. The procedure basically involves starting with the Lorentz force equation describing the electromagnetic force $\overline{\mathfrak{F}}$ on a material volume, namely $\overline{\mathfrak{F}} = \bar{\rho}\overline{\mathscr{E}} + \overline{\mathscr{J}} \times \overline{\mathfrak{B}}$, and eliminating $\bar{\rho}$ and $\overline{\mathscr{J}}$ using Maxwell's equations.

valid only when the momentum of an electromagnetic field is taken into account along with that of the matter with which it interacts.[55] A related issue is the fact that the magnetostatic force law (i.e., $q\bar{v} \times \overline{\mathcal{B}}$) is not by itself consistent with Newton's third law, in that if one considers two moving charges $q_1\bar{v}_1$ and $q_2\bar{v}_2$, the vector sum of the force exerted by one on the other and vice versa is not zero[56] (i.e., $\mathbf{F}_{12} + \mathbf{F}_{21} \neq 0$).

The fact that an electromagnetic wave carries momentum of amount $\mathbf{S}_{av}/c$ is also consistent with atomic physics. According to quantum theory, electromagnetic energy is transported in units of hf, where h is Planck's constant ($h \simeq 6.63 \times 10^{-34}$ joule-s) and f is the frequency. Thus, each photon of energy must carry a momentum of $hf/c = h/\lambda$, as has been verified in experiments, such as the Compton effect.[57]

The above discussion concerns *linear* momentum. Electromagnetic waves can also possess *angular* momentum.[58] The angular momentum about a given origin carried by an electromagnetic radiation is $\overline{\mathcal{L}} = \mathbf{r} \times \overline{\mathcal{G}}$, where $\mathbf{r}$ is the vector from the origin to the point at which the momentum is measured. When a beam of circularly polarized (see Section 8.5) light with Poynting flux $\mathbf{S}_{av}$ is incident on a completely absorbing object of area ΔA for a time Δt, the angular momentum transferred to the object is given by $\overline{\mathcal{L}} = (\mathbf{S}_{av}\Delta A\Delta t)/(2\pi f)$, where f is the frequency of light. The first experimental verification of the angular momentum of electromagnetic waves was the fact that a doubly refracting slab that produces circularly polarized light experiences a reaction torque.[59]

8.5 POLARIZATION OF ELECTROMAGNETIC WAVES

In previous sections we noted that the electric field vector of a uniform plane wave lies in the plane perpendicular to the direction of propagation, but we did not concern ourselves with the particular direction in which the electric field oscillates. In a practical communication environment, the electric field vector of a monochromatic (i.e., single-frequency) uniform plane wave propagating in the z direction can have both an x and a y component, which can oscillate independently, depending on the manner in which the wave was generated at its source. The resultant effect produced by two independent oscillations at right angles to one another can result in a variety of different behavior of the total electric field vector. The orientation and the behavior of the total electric field vector is determined by the *polarization* of the wave. Polarized electromagnetic waves

[55]See Section 8.5 of J. A. Stratton, *Electromagnetic Theory,* McGraw-Hill, New York, 1941.

[56]For further discussion, see Sections 6.1, 6.2, and 10.8 of W. B. Cheston, *Elementary Theory of Electric and Magnetic Fields,* Wiley, New York, 1962.

[57]A. H. Compton, Quantum Theory of the Scattering of X-Ray by Light Elements, *Phys. Rev.,* 21, pp. 483–502, 1923. A. H. Compton earned a Nobel Prize for this work in 1927. For a historical account, see A. H. Compton, The Scattering of X-Rays as Particles, *Am. J. Phys.,* pp. 817–820, 1961.

[58]A. M. Portis, *Electromagnetic Fields: Sources and Media,* Section 11–18, Wiley, 1978; R. P. Feynman, R. B. Leighton, and M. Sands, *The Feynman Lectures in Physics,* Section II-27-6, Definitive Edition, Addison-Wesley, San Francisco, 2006.

[59]R. A. Beth, Mechanical Detection and Measurement of the Angular Momentum of Light, *Phys. Rev.,* 50, p. 115–125, 1936.

and light are commonly encountered in our everyday environment and are also put to very good use in practice, as discussed in Section 8.5.5.

The polarization of a uniform plane wave refers to the behavior of its electric field vector as a function of time at a fixed point in space. More specifically, polarization describes the manner in which the shape and orientation of the locus of the tip of the electric field vector varies with time at a fixed plane in space. Note that polarization refers to the wave electric field by convention; the wave magnetic field can be obtained from the wave electric field and should thus behave in a corresponding manner.[60] For uniform plane waves, the wave electric field is confined to the plane perpendicular to the direction of propagation, so that the polarization of the wave necessarily describes the behavior of the electric field vector in this plane. For nonuniform plane waves, of the type we shall encounter in Chapters 9 and 10, the wave electric field vector may well have components in the direction of propagation. In such cases, polarization of the wave can be separately described in different planes, including that which is perpendicular to the direction of propagation, as well those which are parallel to it.

8.5.1 Linear Polarization

A uniform plane wave propagating in the z direction is said to be *linearly* polarized (LP) when it has only one component ($\mathscr{E}_x$ or $\mathscr{E}_y$) or when its two transverse components ($\mathscr{E}_x$ and $\mathscr{E}_y$) are *in phase*. In other words, we have

$$\boxed{\begin{aligned}\mathscr{E}_x(z,t) &= C_{1x}\cos(\omega t - \beta z + \zeta) \\ \mathscr{E}_y(z,t) &= C_{1y}\cos(\omega t - \beta z + \zeta)\end{aligned}} \quad \text{Linear polarization} \quad (8.62)$$

where C_{1x} and C_{1y} are constants, either of which may be zero, or in general both can be nonzero. When viewed at a fixed location (i.e., fixed z) in the plane perpendicular to

---

[60]The choice of the wave electric field vector (rather than the wave magnetic field vector) as the one to which we refer in delineating the orientation of the wave with respect to the plane of reference is somewhat arbitrary. After all, the electric field vector of a propagating wave cannot exist without the magnetic field vector and vice versa. However, when we pay attention to the physical effect of these vectors, we might think that $\overline{\mathscr{E}}$ is more qualified to represent the electromagnetic field than $\overline{\mathscr{H}}$ or $\overline{\mathscr{B}}$. To see this, consider the effect of the electromagnetic field on a small particle of charge q moving with a velocity $\tilde{\mathbf{v}}$. The total force experienced by the particle is

$$\overline{\mathscr{F}} = q(\overline{\mathscr{E}} + \tilde{\mathbf{v}} \times \overline{\mathscr{B}})$$

which is the Lorentz force. For a uniform plane wave propagating in the z direction, say with only $\mathscr{E}_x$ and $\mathscr{B}_y$ components, the magnitudes of the fields are related via Faraday's Law, or $\beta\mathscr{E}_x = \omega\mathscr{B}_y \longrightarrow \mathscr{B}_y = \sqrt{\mu\epsilon}\,\mathscr{E}_x = \mathscr{E}_x/v_p$, so that we have $|\overline{\mathscr{B}}| = |\overline{\mathscr{E}}|/v_p$. Thus, the Lorentz force acting on the particle is

$$\overline{\mathscr{F}} = q\left[\overline{\mathscr{E}} + \left(\frac{|\overline{\mathscr{E}}|}{v_p}\right)\frac{\tilde{\mathbf{v}} \times \overline{\mathscr{B}}}{|\overline{\mathscr{B}}|}\right]$$

The electric field vector can thus act on the particle even when the particle is at rest, whereas the magnetic force is comparatively smaller in magnitude since typically $|\tilde{\mathbf{v}}| \ll v_p$.

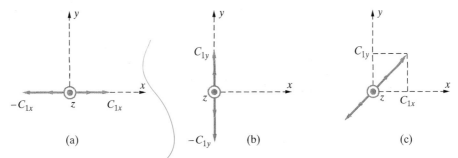

Figure 8.16   The electric field vector of a linearly polarized wave. (a) $C_{1x} \neq 0$ and $C_{1y} = 0$. (b) $C_{1x} = 0$ and $C_{1y} \neq 0$. (c) $C_{1x} \neq 0$ and $C_{1y} \neq 0$.

the propagation direction (i.e., the x-y plane), the tip of the total electric field vector of a linearly polarized wave vibrates in time along a straight line. The orientation of the straight line to which the electric field vector is confined is shown in Figure 8.16 for different combinations of values of C_{1x} and C_{1y}.

8.5.2 Circular Polarization

A uniform plane wave is *circularly* polarized when its two transverse components are out of phase by 90° (or $\pi/2$ radians) but have equal amplitudes. In other words:

$$\boxed{\begin{aligned} \mathscr{E}_x(z,t) &= C_1 \cos(\omega t - \beta z) \\ \mathscr{E}_y(z,t) &= C_1 \cos\left(\omega t - \beta z \pm \frac{\pi}{2}\right) \end{aligned}} \qquad \text{Circular polarization} \qquad (8.63)$$

As shown in Figure 8.17, the tip of the total electric field vector as observed at a fixed point in the x-y plane (chosen to be at the origin in this case) moves along a fixed circle as time progresses. For the case shown in Figure 8.17a, with the motion of the tip of the total electric field vector being counterclockwise when viewed looking toward the $-z$ direction, the wave is said to be *right-hand* circularly polarized (RHCP). An easy way to remember this convention is to use your right hand with the thumb pointing in the direction of propagation (in this case the $+z$ direction); if the electric field moves in the direction of your other fingers, then the wave is polarized in the right-handed sense.[61] If the total electric field vector moves in the opposite sense, the wave is said

---

[61]The right-hand rule described here is the IEEE convention. Amazingly enough, there is considerable disagreement on how to define the sense of polarization of a circularly polarized wave. Most physicists, as well as scientists and engineers specializing in optics, prefer to have the thumb point to *where the wave is coming from,* exactly opposite of the IEEE convention. A further source of confusion is the preference of physicists to use $e^{-j\omega t}$ instead of $e^{j\omega t}$, which of course reverses the sense of rotation. In view of these ambiguities, it is wise in any given case to examine the actual sense of rotation of the total instantaneous electric field $\overline{\mathscr{E}}$ carefully by examining the field vector orientation at a fixed point in space at two specific times (separated by less than $T_p/2$, where $T_p = 2\pi/\omega$), such as $\omega t = 0$ and $\omega t = \pi/2$, and compare with Figure 8.17.

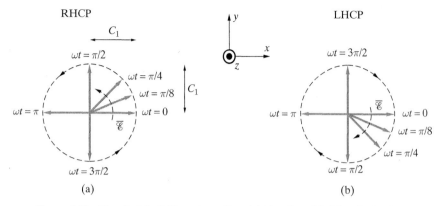

Figure 8.17   The electric field vectors of a right-hand and left-hand circularly polarized waves. (a) RHCP. (b) LHCP. In both cases, the positions of the total electric field vector at the origin are shown at different times as indicated. Note that the wave is assumed to be propagating in the $+z$ direction.

to be *left-hand* circularly polarized (LHCP). Note that for a left-hand polarized wave, $\mathscr{E}_y(z,t)$ would lead $\mathscr{E}_x(z,t)$ by 90° (i.e., $+\pi/2$ instead of $-\pi/2$ radians), as shown in Figure 8.17b.

That the locus of the total electric field vector traces a circle as time progresses can be seen by examining $\mathscr{E}_x(z,t)$ and $\mathscr{E}_y(z,t)$ given by (8.63) at any fixed observation point, say at $z = 0$, as in Figure 8.17. We then have

$$\mathscr{E}_x = C_1 \cos(\omega t) \qquad \mathscr{E}_y = C_1 \cos\left(\omega t - \frac{\pi}{2}\right) = C_1 \sin(\omega t)$$

for the RHCP wave. Thus, the magnitude of the total electric field vector can be written as

$$|\overline{\mathscr{E}}|^2 = \mathscr{E}_x^2 + \mathscr{E}_y^2 = C_1^2$$

which is the equation for a circle.

As observed at a fixed point, the magnitude of the total electric field vector $|\overline{\mathscr{E}}|$ remains constant in time but changes its direction. At a fixed instant of time, the orientation of the total electric field vector at different points in space can be found by vector addition of the $\mathscr{E}_x$ and $\mathscr{E}_y$ components, as illustrated in Figure 8.18. For a RHCP wave, the locus of the tip of the total electric field describes a helix in space, which advances in the propagation direction (i.e., the $+z$ direction) as time progresses in the same way as a right-handed screw.

Note that we can alternately express the electric field components of the circularly polarized wave in phasor form. We have

RHCP: $\quad E_x(z) = C_1 e^{-j\beta z} \qquad E_y(z) = C_1 e^{-j\beta z} e^{-j\frac{\pi}{2}} = -jC_1 e^{-j\beta z}$ (8.64a)

LHCP: $\quad E_x(z) = C_1 e^{-j\beta z} \qquad E_y(z) = C_1 e^{-j\beta z} e^{j\frac{\pi}{2}} = jC_1 e^{-j\beta z}$ (8.64b)

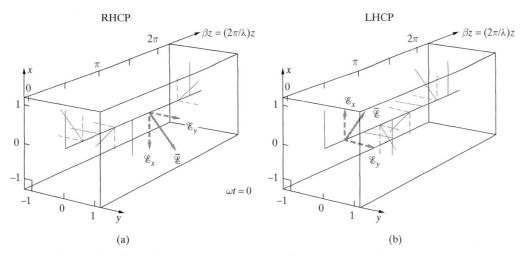

Figure 8.18 Circularly polarized wave in space. The total electric field of the wave at any given point in space and at any time is given by the vector addition of the $\mathscr{E}_x$ and $\mathscr{E}_y$ components. The direction of propagation is the $+z$ direction. (a) Right-hand circularly polarized (RHCP) wave. (b) Left-hand circularly polarized (LHCP) wave.

Polarization of the wave magnetic field. As noted before, our discussion of wave polarization concerns only the behavior of the total wave electric field vector, essentially as a matter of convention. However, each of the total wave electric field components must necessarily be accompanied by an associated magnetic field component, and the $\overline{\mathscr{H}}$-components exhibit time variations (as observed at a fixed point in space) consistent with the $\overline{\mathscr{E}}$-components. For a RHCP wave propagating in the $+z$ direction we have

$$\overline{\mathscr{E}}(z,t) = \hat{\mathbf{x}}\mathscr{E}_x + \hat{\mathbf{y}}\mathscr{E}_y = C_1\left[\hat{\mathbf{x}}\cos(\omega t - \beta z) + \hat{\mathbf{y}}\cos\left(\omega t - \beta z - \frac{\pi}{2}\right)\right] \qquad (8.65\text{a})$$

$$\overline{\mathscr{H}}(z,t) = \hat{\mathbf{x}}\mathscr{H}_x + \hat{\mathbf{y}}\mathscr{H}_y = \frac{C_1}{\eta}\left[-\hat{\mathbf{x}}\cos\left(\omega t - \beta z - \frac{\pi}{2}\right) + \hat{\mathbf{y}}\cos(\omega t - \beta z)\right] \qquad (8.65\text{b})$$

Note that $\mathscr{E}_x$ is associated with $\mathscr{H}_y$ and $\mathscr{E}_y$ with $\mathscr{H}_x$. If we were to observe the tip of the total wave magnetic field vector as a function of time, looking towards the propagation direction, we would see it rotate clockwise (i.e., RHCP) together with the total electric field vector, always remaining perpendicular to it and oriented such that $\overline{\mathscr{E}} \times \overline{\mathscr{H}}$ (i.e., the direction of electromagnetic power flow) is always in the $+z$ direction.

Example 8.21: Geopositioning satellite. A low earth orbit (LEO) geopositioning satellite orbiting at an altitude of 1000 km transmits a total power of $P_{\text{tot}} = 40$ kW isotropically at a downlink frequency of about 137.5 MHz, as shown in Figure 8.19a. The electric field of the received wave on the earth at a point immediately below the satellite is given by

$$\mathbf{E}(z) = C_1(-j\hat{\mathbf{x}} + \hat{\mathbf{y}})e^{-j\beta z}$$

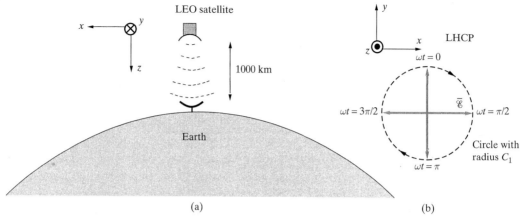

Figure 8.19   A satellite at low earth orbit and the polarization of the received signal. The direction of propagation is the $+z$ direction.

(a) Find the polarization of this wave. (b) Find the values of β and C_1. (c) Find the corresponding magnetic field, $\mathbf{H}(z)$.

Solution:

(a) The instantaneous expression for the total electric field is given by

$$\overline{\mathcal{E}}(z,t) = \mathcal{R}e\{\mathbf{E}(z)e^{j\omega t}\}$$
$$= \mathcal{R}e\{\hat{\mathbf{x}}C_1 e^{-j\pi/2}e^{-j\beta z}e^{j\omega t} + \hat{\mathbf{y}}C_1 e^{-j\beta z}e^{j\omega t}\}$$
$$= \hat{\mathbf{x}}C_1 \cos(\omega t - \beta z - \pi/2) + \hat{\mathbf{y}}C_1 \cos(\omega t - \beta z)$$
$$= \hat{\mathbf{x}}C_1 \sin(\omega t - \beta z) + \hat{\mathbf{y}}C_1 \cos(\omega t - \beta z)$$

We see that the two components of the wave electric field are equal in magnitude but 90° out of phase, meaning that this wave is circularly polarized. Further, if we sketch the total electric field vector as a function of time at $z = 0$, as seen in Figure 8.19b, we find that the wave is left-hand circularly polarized (i.e., LHCP).

(b) $\beta = 2\pi f/c \simeq 2\pi(137.5 \times 10^6)/(3 \times 10^8) \simeq 2.88$ rad-m^{-1}. Neglecting all losses and effects from boundaries, we have

$$|\mathbf{S}_{\text{av}}| = \frac{P_{\text{tot}}}{4\pi R^2} = \left(\frac{1}{2}\frac{C_1^2}{\eta}\right)$$

so

$$C_1 \simeq \sqrt{\frac{2 \times 377 \times 40 \times 10^3}{4\pi(10^6)^2}} \simeq 1.55 \times 10^{-3} \text{ V-m}^{-1} = 1.55 \text{ mV-m}^{-1}$$

(c) The phasor expression for the corresponding wave magnetic field can written as

$$\mathbf{H}(z) = \frac{C_1}{\eta}(-\hat{\mathbf{x}} - j\hat{\mathbf{y}})e^{-j\beta z} \simeq 4.11(-\hat{\mathbf{x}} - j\hat{\mathbf{y}})e^{-j2.88z} \quad \mu\text{A-m}^{-1}$$

8.5.3 Elliptical Polarization

The general case of elliptical polarization occurs when $\mathscr{E}_x$ and $\mathscr{E}_y$ have different magnitudes or if the phase difference between them is different from 90°. The case of 90° phase difference with different amplitudes is shown in Figure 8.20, for which we have the two transverse wave electric field components expressed in phasor form as

$$E_x(z) = C_{1x}e^{-j\beta z} \tag{8.66a}$$

$$E_y(z) = C_{1y}e^{-j\beta z}e^{j\zeta} \tag{8.66b}$$

where $\zeta = \pm\pi/2$, with the $+$ and $-$ representing respectively left- and right-hand polarization senses.

Even if the magnitudes of the two field components were equal, the wave is elliptically polarized if $0 < \zeta < 90°$. In the general case we can write the two transverse components of the wave electric field in time domain as

$$\boxed{\begin{aligned} \mathscr{E}_x(z,t) &= C_{1x}\cos(\omega t - \beta z) \\ \mathscr{E}_y(z,t) &= C_{1y}\cos(\omega t - \beta z + \zeta) \end{aligned}} \quad \text{Elliptical polarization} \tag{8.67}$$

The corresponding shape of the polarization ellipses for different values of ζ are given in Figure 8.21. The top panels represent left-hand elliptical polarization (LHEP), whereas the bottom panels show right-hand elliptical polarization (RHEP).

For the case when $C_{1x} = C_{1y}$, the full range of possible polarizations can be realized by simply varying the phase angle ζ. When the phase difference ζ varies from $\zeta = 0$ to $\zeta = \pi$, the polarization of a wave with $|\mathscr{E}_x| = |\mathscr{E}_y|$ varies from linear ($\zeta = 0$) tilted

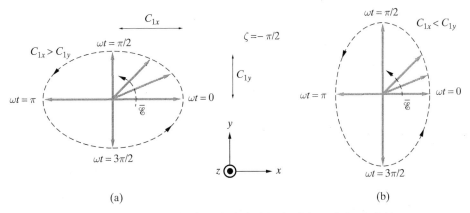

(a)   (b)

Figure 8.20 Elliptical polarization. The loci of the tip of the total electric field vector shown at the origin for an elliptically polarized wave propagating in the $+z$ direction represented by (8.67) with $\zeta = -\pi/2$. (a) The major axis of the ellipse is along the x axis when $C_{1x} > C_{1y}$. (b) The major axis of the ellipse is along the y axis when $C_{1x} < C_{1y}$. As in the case of circular polarization, it can be shown using the right-hand rule that this wave is right-hand elliptically polarized (RHEP).

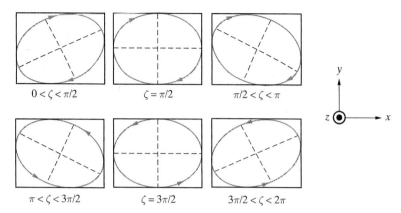

Figure 8.21 Elliptical polarization ellipses. The loci of the tip of the total electric field vector for an elliptically polarized wave as given by (8.67) for different ranges of values of ζ. It is assumed that $C_{1x} > C_{1y}$. Top panels are LHEP, whereas the bottom panels are RHEP waves. The straight-line loci for the cases of $\zeta = 0$ and π, corresponding to linear polarization, are not shown.

at $45°$ with respect to the x-axis to elliptical $(0 < \zeta < \pi/2)$ to circular $(\zeta = \pi/2)$ to elliptical $(\pi/2 < \zeta < \pi)$ and back to linear tilted at $-45°$ $(\zeta = \pi)$.

We can show that the tip of the total electric field vector does indeed trace an ellipse by eliminating $\omega t - \beta z$ between the two equations of (8.67) and rewriting them in the form[62]

$$\left(\frac{\mathscr{E}_x}{C_{1x}}\right)^2 + \left(\frac{\mathscr{E}_y}{C_{1y}}\right)^2 - 2\frac{\mathscr{E}_x}{C_{1x}}\frac{\mathscr{E}_y}{C_{1y}}\cos\zeta = \sin^2\zeta \tag{8.68}$$

This equation represents a tilted polarization ellipse, as shown in Figure 8.22.

The major and minor axes of this ellipse do not coincide with the x and y axes of the rectangular coordinate system; rather, they coincide with the x' and y' axes obtained by rotating the x and y axes by a tilt angle ψ, as shown in Figure 8.22, given by

$$\psi = \tfrac{1}{2}\tan^{-1}[\tan(2\alpha)\cos\zeta] \tag{8.69}$$

where

$$\tan\alpha = C_{1y}/C_{1x} \tag{8.70}$$

The principal semiaxes $A_{x'}$ and $A_{y'}$ (in other words, the half-lengths of the principal axes) can be written in terms of C_{1x}, C_{1y}, ζ, and ψ as

$$A_{x'} = \left(C_{1x}^2\cos^2\psi + C_{1y}^2\sin^2\psi + C_{1x}C_{1y}\sin(2\psi)\cos\zeta\right)^{1/2}$$

$$A_{y'} = \left(C_{1x}^2\sin^2\psi + C_{1y}^2\cos^2\psi - C_{1x}C_{1y}\sin(2\psi)\cos\zeta\right)^{1/2}$$

---

[62]To eliminate $\omega t - \beta z$, we can expand $\cos(\omega t - \beta z + \zeta)$, substitute $\cos(\omega t - \beta z) = \mathscr{E}_x(z,t)/C_{1x}$, and use $\sin(\cdot) = \sqrt{1 - \cos^2(\cdot)}$. See *Principles of Optics*, M. Born and E. Wolf, 5th ed., Pergamon Press, Oxford, 1975, pp. 24–30.

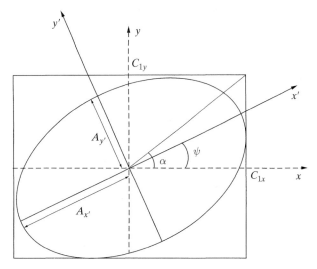

Figure 8.22   The tilted polarization ellipse.

From power considerations, $A_{x'}$, $A_{y'}$, C_{1x}, and C_{1y} are related by

$$A_{x'}^2 + A_{y'}^2 = C_{1x}^2 + C_{1y}^2$$

One can rewrite the equation of the polarization ellipse given by (8.68) with respect to the rotated axes x' and y' as

$$\left(\frac{\mathscr{E}_{x'}}{A_{x'}}\right)^2 + \left(\frac{\mathscr{E}_{y'}}{A_{y'}}\right)^2 = 1$$

where $\mathscr{E}_{x'}$ and $\mathscr{E}_{y'}$ are the components of the wave electric field in the rotated coordinate system.

Also of interest is the so-called axial ratio, AR, which is defined as the ratio of the major to the minor axes. If $A_{x'} > A_{y'}$, then $AR = A_{x'}/A_{y'}$, whereas if $A_{x'} < A_{y'}$, then $AR = (A_{x'}/A_{y'})^{-1}$, the value of which satisfies $1 \leq AR \leq \infty$.

Example 8.22: Elliptical polarization. An electromagnetic wave travels in the x direction with its magnetic field given by

$$\overline{\mathscr{H}}(x,t) = \hat{\mathbf{y}}12\cos(\omega t - \beta x) + \hat{\mathbf{z}}8\cos(\omega t - \beta x + 70°)$$

in mA-m^{-1}. Find the tilt angle, axial ratio, and rotation sense of the polarization ellipse.

Solution: From geometry, we have

$$\tan\alpha = \frac{C_{1y}}{C_{1z}} = \frac{8}{12} = \frac{2}{3}$$

from which

$$\tan(2\alpha) = \frac{2\tan\alpha}{1 - \tan^2\alpha} = 2.4$$

So, the tilt angle measured from the z axis is given by

$$\psi = \tfrac{1}{2}\tan^{-1}[\tan(2\alpha)\cos\zeta] = \tfrac{1}{2}\tan^{-1}[(2.4)\cos 70°] \simeq 19.7°$$

The principal semiaxes $A_{x'}$ and $A_{y'}$ are found as

$$A_{y'} \simeq \left(8^2\cos^2(19.7°) + 12^2\sin^2(19.7°) + 8 \times 12\sin(2 \times 19.7°)\cos(70°)\right)^{1/2} \simeq 12.5$$

$$A_{y'} \simeq \left(8^2\sin^2(19.7°) + 12^2\cos^2(19.7°) - 8 \times 12\sin(2 \times 19.7°)\cos(70°)\right)^{1/2} \simeq 7.23$$

So the axial ratio is

$$\text{AR} \simeq \frac{12.5}{7.23} \simeq 1.73$$

Figure 8.23 shows the polarization ellipse and the sense of rotation of the magnetic field vector as a function of time at the origin. Since the magnetic field vector rotates around the ellipse in the clockwise direction as shown, this is a LHEP wave.

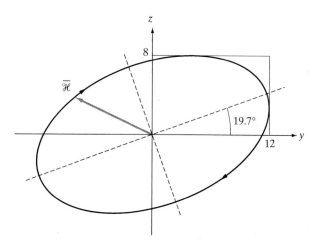

Figure 8.23   **Polarization ellipse of Example 8.22.**

Sometimes it is helpful to express an elliptically polarized wave as a sum of two oppositely rotating circularly polarized waves, as shown in Figure 8.24. To see this decomposition analytically, note that the total electric field phasor for a left-hand circularly polarized wave can be written as

$$\mathbf{E}_1 = \hat{\mathbf{x}}C_1 e^{-j\beta z} + \hat{\mathbf{y}}jC_1 e^{-j\beta z}$$

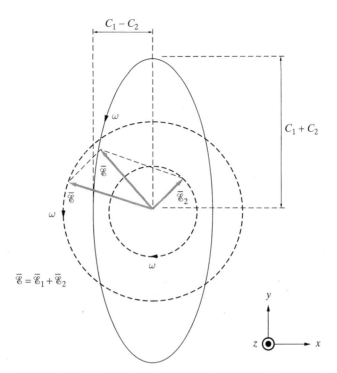

Figure 8.24   The decomposition of an elliptically polarized wave into two circularly polarized waves. Any elliptically polarized wave ($\overline{\mathscr{E}}$) can be decomposed into two counter-rotating circularly polarized components $\overline{\mathscr{E}}_1$ and $\overline{\mathscr{E}}_2$. In the limiting case when $\overline{\mathscr{E}}$ is linearly polarized (i.e., $\zeta = 0$ in (8.67)), $|\overline{\mathscr{E}}_1| = |\overline{\mathscr{E}}_2|$. Otherwise, the sense of rotation of $\overline{\mathscr{E}}$ is in the direction of the larger of $\overline{\mathscr{E}}_1$ and $\overline{\mathscr{E}}_2$.

whereas that for a RHCP wave is

$$\mathbf{E}_2 = \hat{\mathbf{x}}C_2 e^{-j\beta z} - \hat{\mathbf{y}}jC_2 e^{-j\beta z}$$

the sum of the electric fields for these two waves is then

$$\mathbf{E} = \hat{\mathbf{x}}(C_1 + C_2)e^{-j\beta z} + \hat{\mathbf{y}}j(C_1 - C_2)e^{-j\beta z}$$

which is an elliptically polarized wave (since $\mathscr{E}_x$ and $\mathscr{E}_y$ have different magnitudes and are 90° out of phase). Note that for elliptically polarized waves in which the major and minor axes are not aligned with the x (or y) and y (or x) axes, a similar type of decomposition can be used after rotating the coordinate system to align the polarization axes with the x (or y) and y (or x) axes.

Example 8.23: Two linearly polarized waves. Two linearly polarized waves propagating in the same direction at the same frequency are given by

$$E_1(y) = \hat{x} C_1 e^{-j\beta y}$$

$$E_2(y) = \hat{z} C_2 e^{-j\beta y} e^{j\theta}$$

where C_1, C_2, and θ are constants and $\beta = \omega\sqrt{\mu\epsilon}$. Find the polarization of the sum of these waves for the following cases: (a) $\theta = 0°$, (b) $\theta = \pi/2$, (c) $\theta = \pi/2$ and $C_1 = C_2$, and (d) $\theta = \pi$.

Solution: The real-time expression for the sum of these two waves can be found as

$$\overline{\mathscr{E}}(y,t) = \mathscr{R}e\{[E_1(y) + E_2(y)]e^{j\omega t}\}$$

$$= \mathscr{R}e[\hat{x} C_1 e^{-j\beta y} e^{j\omega t} + \hat{z} C_2 e^{-j\beta y} e^{j\theta} e^{j\omega t}]$$

$$= \hat{x} C_1 \cos(\omega t - \beta y) + \hat{z} C_2 \cos(\omega t - \beta y + \theta)$$

(a) When $\theta = 0°$, the two components of the total electric field are in phase, therefore, resulting in a linearly polarized (LP) wave oscillating in the direction as shown in Figure 8.25a.

(b) When $\theta = \pi/2$, the two components are 90° out of phase. Since the amplitudes are in general different, the total wave is elliptically polarized. Figure 8.25b shows the variation of the total electric field at $y = 0$ as a function of time (for $C_2 > C_1$) from which we can conclude that the wave is right-hand elliptically polarized (RHEP).

(c) When $\theta = \pi/2$ and $C_1 = C_2$, the wave is circularly polarized (i.e., RHCP), the time variation of which is shown at $y = 0$ in Figure 8.25c.

(d) When $\theta = \pi$, the two components are 180° out of phase. Rewriting the total electric field yields

$$\overline{\mathscr{E}}(y,t) = \hat{x} C_1 \cos(\omega t - \beta y) + \hat{z} C_2 \cos(\omega t - \beta y + \pi)$$

$$= \hat{x} C_1 \cos(\omega t - \beta y) - \hat{z} C_2 \cos(\omega t - \beta y)$$

At any time instant t, the direction of the total electric field is given by the unit vector as

$$\hat{u}_\epsilon = \frac{\overline{\mathscr{E}}(y,t)}{|\overline{\mathscr{E}}(y,t)|}$$

$$= \hat{x}\frac{C_1}{\sqrt{C_1^2 + C_2^2}} - \hat{z}\frac{C_2}{\sqrt{C_1^2 + C_2^2}}$$

which does not change with time. Therefore, the wave is linearly polarized, as shown in Figure 8.25d.

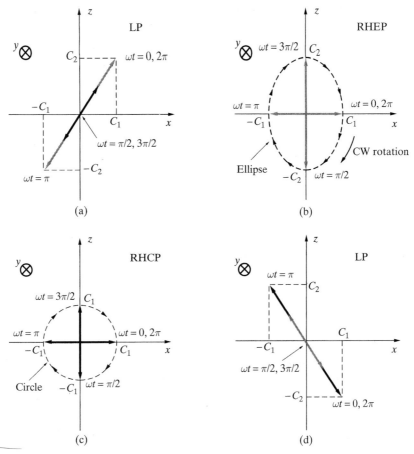

Figure 8.25 Two linearly polarized waves. The variation of the field vector at $y = 0$ is shown as a function of time at five different time instants, namely $\omega t = 0$, $\pi/2$, π, $3\pi/2$ and 2π, for the cases when (a) $\theta = 0°$, (b) $\theta = \pi/2$, (c) $\theta = \pi/2$ and $C_1 = C_2$, and (d) $\theta = \pi$. The direction of propagation is the y direction. Also note that when $C_1 \neq C_2$, it is assumed that $C_1 < C_2$.

8.5.4 Poynting Vector for Elliptically Polarized Waves

Consider the electric field phasor for an elliptically polarized wave given by

$$\mathbf{E}(z) = \mathbf{E}_1(z) + \mathbf{E}_2(z) = \hat{\mathbf{x}} C_{1x} e^{-j\beta z} + \hat{\mathbf{y}} C_{1y} e^{-j(\beta z - \zeta)}$$

with its accompanying magnetic field, which follows as

$$\mathbf{H}(z) = \mathbf{H}_1(z) + \mathbf{H}_2(z) = -\hat{\mathbf{x}} \frac{C_{1y}}{\eta} e^{-j(\beta z - \zeta)} + \hat{\mathbf{y}} \frac{C_{1x}}{\eta} e^{-j\beta z}$$

where $\mathbf{E}_1 = \hat{\mathbf{x}}(\cdots)$ is associated with $\mathbf{H}_1 = \hat{\mathbf{y}}(\cdots)$, whereas $\mathbf{E}_2 = \hat{\mathbf{y}}(\cdots)$ is associated with $\mathbf{H}_2 = -\hat{\mathbf{x}}(\cdots)$. The time-average Poynting vector (8.58) for this wave can be shown to be equal to

$$\mathbf{S}_{av} = \hat{\mathbf{z}}\frac{1}{2\eta}\left\{C_{1x}^2 + C_{1y}^2\right\} \tag{8.71}$$

It should be noted here that $\mathbf{S}_{av}$ is independent of ζ. For a lossy medium the time-average Poynting vector of an elliptically polarized wave can be shown to be

$$\mathbf{S}_{av} = \hat{\mathbf{z}}\frac{1}{2|\eta_c|}\left[C_{1x}^2 + C_{1y}^2\right]\cos\phi_\eta e^{-2\alpha z} \tag{8.72}$$

where $\alpha = \mathcal{R}e\{\gamma\}$.

Example 8.24: Elliptical polarization. Calculate the total time-average power carried by the electromagnetic wave given in Example 8.22.

    Solution: The total time-average power per unit area is equal to the time-average Poynting vector for this wave given by

$$\mathbf{S}_{av} = \frac{1}{2}\eta[C_{1y}^2 + C_{1z}^2] \simeq \frac{1}{2}(377)[8^2 + (12)^2] \simeq 39.2 \ \text{mW-m}^{-2}$$

8.5.5 Polarization in Practice

Polarization of electromagnetic waves and thus light can be observed in our everyday environment and is also put to very good use in practice. Direct sunlight is virtually unpolarized, but the light in the rainbow exhibits distinct polarization, as does the blue sky light itself.[63] Polarization studies were crucial to the early investigations of the nature of light[64] and to the determination that X rays were electromagnetic in nature.[65] Much useful information about the structure of atoms and nuclei is gained from polarization studies of the emitted electromagnetic radiation. Polarized light also has many practical applications in industry and in engineering science. In the following, we describe some of the more common engineering applications at radio frequencies.

---

    [63]The color of the sky is blue because of the scattering of sunlight from tiny molecules of air, a process which leads to polarized light. Sometimes, small clouds, hardly visible in air, can be seen more clearly in a reflection in water (e.g., a lake), because the light from the clouds is not polarized and is thus more efficiently reflected by the water than the blue sky light, which is polarized. Reflection and refraction of electromagnetic waves will be studied in Chapter 9. For more on the color of the sky and many other interesting phenomena, see M. Minnaert, *The Nature of Light & Colour in the Open Air*, Dover, 1954.

    [64]Polarization was discovered by E. L. Malus [*Nouveau Bulletin des Sciences, par la Soc. Philomatique*, p. 266, 1809]; also see C. Huygens, *Traité de la Lumière*, 1690.

    [65]C. G. Barkla, Polarization in Secondary Röntgen Radiation, *Roy. Soc. Proc. Ser. A*, 77, pp. 247–255, 1906.

Linearly polarized waves are commonly utilized in radio and TV broadcast applications. AM broadcast stations operate at relatively lower frequencies and utilize large antenna towers and generate *vertically* polarized signals, with the wave electric field perpendicular to the ground. Although the primary polarization of the AM signal is vertical, there is typically also a component of the electric field in the direction of propagation, due to ground-losses. For maximum reception of an AM radio signal, the receiver antenna should be parallel to the electric field, or perpendicular to the ground (i.e., vertical). Most AM radio receivers use ferrite-rod antennas instead of wire antennas to detect the wave magnetic field, which is horizontal. For maximum reception, the ferrite-rod antenna should be parallel to the magnetic field. In North America, TV broadcast signals are typically horizontally polarized, which is why typical rooftop antennas are horizontal. Most FM radio broadcast stations in the United States utilize circularly polarized waves, so that the orientation of an FM receiving antenna is not critical as long as the antenna lies in a plane orthogonal to the direction from which the signal arrives.

Polarization of an electromagnetic signal is initially determined by the source antenna that launches the signal.[66] However, as a signal propagates through some medium, its polarization may change. A good example of this occurs in propagation of waves through Earth's ionosphere, when a rotation of the plane of polarization occurs, known as Faraday rotation. Circularly polarized waves are utilized in some communication and radar systems, primarily to relax the requirement for the receiving antenna to be carefully aligned with the wave electric vector. Many international communication satellites also utilize circularly polarized signals. In radar applications, circular polarization is used because such a wave reflects from a metal target with the opposite sense (i.e., RHCP becomes LHCP), providing a method to distinguish metal targets (i.e., aircraft) from clouds and clutter.

Use of polarized transmissions also provides a means of multiple simultaneous usage of the same frequency band. Most North American communication satellites use linear polarization, alternating between horizontal and vertical polarization between adjacent satellites.[67]

8.6 ARBITRARILY DIRECTED UNIFORM PLANE WAVES

In previous sections we used expressions for uniform plane wave electric and magnetic fields for which the direction of propagation was chosen to coincide with one of the coordinate axes, namely, the z axis. This choice obviously results in no loss of generality in our considerations of waves in an unbounded medium in this chapter. In Chapter 9, however, we consider electromagnetic wave propagation in the presence of planar boundaries, resulting in their reflection and refraction. When uniform plane waves

---

[66] A very interesting antenna that can selectively generate right-hand or left-hand circularly polarized signals is the *helical* antenna, discovered by J. D. Kraus. See J. D. Kraus, The Helical Antenna, *Proc. IRE*, 37(3), pp. 263–272, 1949. For detailed discussion of the helical antenna and other antennas see J. D. Kraus, *Antennas*, McGraw-Hill, New York, 1988.

[67] W. Sinnema, *Electronic Transmission Technology*, Prentice-Hall, 1988.

are normally incident on an infinite planar boundary, the direction of propagation of the wave can be chosen to be along one of the principal axes without any loss of generality. However, the general case of reflection and refraction of waves at infinite planar boundaries involves waves that are incident on the boundaries at an arbitrary angle, in which case we shall need to work with uniform plane waves propagating in arbitrary directions. In this section, we derive general expressions for such waves and introduce the notation to be used. We also briefly discuss nonuniform plane waves, because such waves are encountered in later chapters.

8.6.1 Uniform Plane Waves in an Arbitrary Direction

Before we proceed with deriving some general relationships concerning uniform plane waves, it is useful to note that the physical nature of the waves we consider here is still the same as those studied in Sections 8.1 through 8.5. In other words, the waves considered here are uniform plane waves in which the field vectors are confined to an infinite plane (which is the surface of constant phase) and exhibit spatial variations only in the direction perpendicular to that plane.

Consider the vector wave equation (8.7) for time-harmonic fields in a lossless medium, repeated here for convenience:

$$\nabla^2 \mathbf{E} - (j\beta)^2 \mathbf{E} = 0 \tag{8.73}$$

where $\beta = \omega\sqrt{\mu\epsilon}$. We know that one solution of (8.73) is

$$\mathbf{E}(z) = \mathbf{E}_0 e^{-j\beta z} = [\hat{\mathbf{x}}E_{0x} + \hat{\mathbf{y}}E_{0y}]e^{-j\beta z}$$

A more general solution representing a wave propagating in an arbitrary direction $\mathbf{r} = \hat{\mathbf{x}}x + \hat{\mathbf{y}}y + \hat{\mathbf{z}}z$ is

$$\mathbf{E}(x,y,z) = \mathbf{E}_0 e^{-j\beta_x x - j\beta_y y - j\beta_z z} = [\hat{\mathbf{x}}E_{0x} + \hat{\mathbf{y}}E_{0y} + \hat{\mathbf{z}}E_{0z}]e^{-j\beta_x x - j\beta_y y - j\beta_z z} \tag{8.74}$$

It can be shown by direct substitution in (8.73) that (8.74) satisfies the wave equation as long as we have

$$\beta^2 = \beta_x^2 + \beta_y^2 + \beta_z^2 = \omega^2\mu\epsilon$$

We can rewrite (8.74) in a more compact form by using vector notation for the exponent:

$$\boxed{\mathbf{E}(\mathbf{r}) = \mathbf{E}_0 e^{-j\beta\hat{\mathbf{k}}\cdot\mathbf{r}}} \tag{8.75}$$

where $\hat{\mathbf{k}}$ is the unit vector in the direction of propagation, as shown in Figure 8.26. Note that $\hat{\mathbf{k}}$ is given by

$$\hat{\mathbf{k}} = \frac{\hat{\mathbf{x}}\beta_x + \hat{\mathbf{y}}\beta_y + \hat{\mathbf{z}}\beta_z}{[\beta_x^2 + \beta_y^2 + \beta_z^2]^{1/2}} = \frac{\hat{\mathbf{x}}\beta_x + \hat{\mathbf{y}}\beta_y + \hat{\mathbf{z}}\beta_z}{\omega\sqrt{\mu\epsilon}}$$

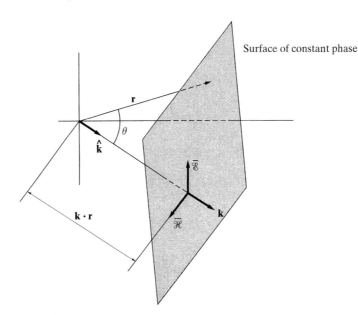

Figure 8.26   Constant-phase surface. The planar surface of constant phase, characteristic of a plane wave, and the orientation of the field vectors.

The vector $\hat{\mathbf{k}}\beta$ is the wavenumber vector, commonly referred in other texts as $\mathbf{k}$. For a
uniform plane wave propagating in the z direction as considered in previous sections,
the phase constant β represents the space rate of change of wave phase (in units of
rad-m^{-1}) along the z axis. In the general case of a uniform plane wave propagating
in an arbitrary direction, the component phase constants β_x, β_y, and β_z represent the
space rates of change of the wave phase as measured along the respective axes. These
component phase constants and the corresponding component wavelengths are further
discussed below in connection with Figure 8.27.

We note that $\hat{\mathbf{k}} \cdot \mathbf{r} =$ constant represents the equation for the planes over which
the wave phase is a constant, referred to as the planes of constant phase, or phase
fronts (Figure 8.26). By substituting (8.75) into (7.23b) for a source-free medium (i.e.,
$\nabla \cdot \mathbf{E} = 0$), we find

$$\nabla \cdot \mathbf{E}_0 e^{-j\beta\hat{\mathbf{k}}\cdot\mathbf{r}} = 0$$

$$\mathbf{E}_0 \cdot \nabla(e^{-j\beta\hat{\mathbf{k}}\cdot\mathbf{r}}) = 0$$

$$\mathbf{E}_0 \cdot \left(\hat{\mathbf{x}}\frac{\partial}{\partial x} + \hat{\mathbf{y}}\frac{\partial}{\partial y} + \hat{\mathbf{z}}\frac{\partial}{\partial z}\right) e^{-j(\beta_x x + \beta_y y + \beta_z z)} = 0$$

$$\mathbf{E}_0 \cdot [-j(\hat{\mathbf{x}}\beta_x + \hat{\mathbf{y}}\beta_y + \hat{\mathbf{z}}\beta_z)e^{-j(\beta_x x + \beta_y y + \beta_z z)}] = 0$$

$$\mathbf{E}_0 \cdot (-j\beta\hat{\mathbf{k}}e^{-j\beta\hat{\mathbf{k}}\cdot\mathbf{r}}) = 0$$

$$-j\beta(\mathbf{E}_0 \cdot \hat{\mathbf{k}})e^{-j\beta\hat{\mathbf{k}}\cdot\mathbf{r}} = 0$$

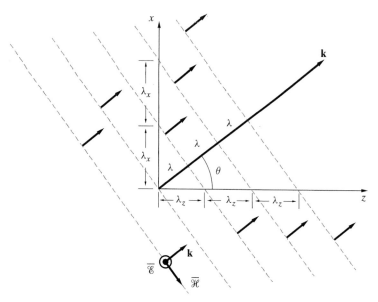

Figure 8.27   A uniform plane wave propagating at an angle θ to the z axis. Although the wavelength in the propagation direction is λ, the projections of the constant phase fronts on the z and x axes are separated by λ_z and λ_x, respectively.

which yields

$$\boxed{\hat{\mathbf{k}} \cdot \mathbf{E}_0 = 0}$$ (8.76)

Thus, $\mathbf{E}_0$ must be transverse to the propagation direction for (7.23b) ($\nabla \cdot \mathbf{E} = 0$) to be satisfied. (This result is the generalization of what we saw earlier for the simplest uniform plane wave propagating in the z direction, where we had to have $E_z = 0$ for the field to have zero divergence.)

To find the magnetic field $\mathbf{H}$ in terms of the electric field $\mathbf{E}$, we use Faraday's law (equation (7.23a)):

$$\mathbf{H}(\mathbf{r}) = \frac{1}{-j\omega\mu} \nabla \times \mathbf{E}(\mathbf{r})$$ (8.77)

We can further manipulate (8.77) as follows

$$\mathbf{H} = \frac{1}{-j\omega\mu} \nabla \times \mathbf{E}_0 e^{-j\beta\hat{\mathbf{k}}\cdot\mathbf{r}}$$

$$= \frac{-1}{-j\omega\mu} \mathbf{E}_0 \times \nabla e^{-j\beta\hat{\mathbf{k}}\cdot\mathbf{r}} \qquad \text{since } \mathbf{E}_0 \text{ is a constant vector}$$

$$= \frac{-1}{-j\omega\mu} \mathbf{E}_0 \times (-j\beta\hat{\mathbf{k}}) e^{-j\beta\hat{\mathbf{k}}\cdot\mathbf{r}} \qquad \text{see derivation of (8.76) above.}$$

$$= \frac{\beta}{\omega\mu}\hat{\mathbf{k}} \times \mathbf{E}_0 e^{-j\beta\hat{\mathbf{k}}\cdot\mathbf{r}}$$

$$= \frac{1}{\eta}\hat{\mathbf{k}} \times \underbrace{\mathbf{E}_0 e^{-j\beta\hat{\mathbf{k}}\cdot\mathbf{r}}}_{\mathbf{E}(\mathbf{r})}$$

or

$$\boxed{\mathbf{H}(\mathbf{r}) = \frac{1}{\eta}\hat{\mathbf{k}} \times \mathbf{E}(\mathbf{r})} \tag{8.78}$$

The magnetic field vector $\mathbf{H}$ is thus perpendicular to both the electric field vector $\mathbf{E}$ and $\hat{\mathbf{k}}$, as shown in Figure 8.26.

Equations (8.75) and (8.78) constitute general expressions for transverse electromagnetic (TEM) waves propagating in an arbitrary direction $\hat{\mathbf{k}}$. These expressions will be useful in the discussion of reflection and refraction of uniform plane waves obliquely incident on dielectric or conductor interfaces, which is the topic of the next chapter.

As an example, we consider a uniform plane wave propagating in the x-z plane (i.e., in a direction perpendicular to the y axis) but in a direction oriented at an angle θ with respect to the z axis, as shown in Figure 8.27. Assuming that the wave electric field is in the y direction, and noting that we have $\hat{\mathbf{k}} = \hat{\mathbf{x}}\sin\theta + \hat{\mathbf{z}}\cos\theta$, the expression for the wave electric field phasor can be written as

$$\mathbf{E}(\mathbf{r}) = \hat{\mathbf{y}}E_0 e^{-j\beta\hat{\mathbf{k}}\cdot\mathbf{r}} = \hat{\mathbf{y}}E_0 e^{-j\beta(x\sin\theta + z\cos\theta)}$$

whereas the wave magnetic field can be found using (8.78) as

$$\mathbf{H}(\mathbf{r}) = \frac{1}{\eta}\hat{\mathbf{k}} \times [\hat{\mathbf{y}}E_0 e^{-j\beta(x\sin\theta + z\cos\theta)}]$$

$$= \frac{1}{\eta}(\hat{\mathbf{x}}\sin\theta + \hat{\mathbf{z}}\cos\theta) \times [\hat{\mathbf{y}}E_0 e^{-j\beta(x\sin\theta + z\cos\theta)}]$$

$$= \frac{E_0}{\eta}(-\hat{\mathbf{x}}\cos\theta + \hat{\mathbf{z}}\sin\theta)e^{-j\beta(x\sin\theta + z\cos\theta)}$$

$$\mathbf{H}(\mathbf{r}) = \frac{E_0}{\eta}(-\hat{\mathbf{x}}\cos\theta + \hat{\mathbf{z}}\sin\theta)e^{-j(\beta_x x + \beta_z z)}$$

where $\beta_x = \beta\sin\theta$ and $\beta_z = \beta\cos\theta$. Note from Figure 8.27 that the expression for $\mathbf{H}(\mathbf{r})$ could also have been written by inspection, since $\mathbf{E} \times \mathbf{H}$ must be in the direction of $\hat{\mathbf{k}}$.

Further consideration of Figure 8.27 allows us to better understand the properties of a uniform plane wave. It was mentioned above that the component phase constants $\beta_{x,y,z}$ represent the space rate of change of the phase of the wave along the principal axes. Instead of considering the rate of change of wave phase, we can consider the distances between successive equiphase surfaces on which the wave phase differs by exactly 2π. These distances are *apparent wavelengths,* because the phase shift is linearly proportional to distance in any direction. Along the propagation direction, we have the conventional wavelength λ over which the wave phase changes by 2π. Measured along any other direction, the distances between equiphase surfaces are greater, as is apparent from Figure 8.27. The apparent wavelengths along the x and z axes are respectively λ_x and λ_z, which are both greater than λ. The apparent wavelengths λ_x and λ_z are related to λ by

$$\frac{1}{\lambda_x^2} + \frac{1}{\lambda_z^2} = \frac{1}{\lambda^2}$$

since

$$\beta^2 = \beta_x^2 + \beta_z^2$$

Similarly, because a particular planar phase front moves uniformly (i.e., with same speed at all points on the plane) with time along the propagation direction at the phase velocity $v_p = \omega/\beta$, its intercept along any other direction also moves uniformly in time but with a *greater* speed. In the time of one full cycle of the wave (i.e., $T_p = 1/f$), a phase front moves a distance along the $\hat{\mathbf{k}}$ direction of $\lambda = 2\pi/\beta$ and a distance along the x direction of $\lambda_x = \lambda/\sin\theta = 2\pi/(\beta\sin\theta) = 2\pi/\beta_x$. Thus, the various *apparent phase velocities* along the principal axes are in general given by

$$v_{x,y,\text{ or } z} = \frac{\omega}{\beta_{x,y,\text{ or } z}}$$

and, in the case shown in Figure 8.27, are related to one another as

$$\frac{1}{v_x^2} + \frac{1}{v_z^2} = \frac{1}{v_p^2} = \epsilon\mu$$

Component phase velocities greater than the speed of propagation in the medium can be observed in water waves approaching at an angle to a breakwater or a shoreline. Visualize water waves striking a beach obliquely. The distance between successive crests (i.e., maxima), for example, is large along the beach, especially if the waves are only slightly off normal incidence. To keep up with a given crest, one has to run much faster along the beach than the speed with which the waves progress in their own directions of propagation.

8.6.2 Nonuniform Plane Waves

Nonuniform plane waves are those electromagnetic waves for which the amplitudes of the electric and magnetic fields of the wave are not constant over the planes of constant phase. An example of a *nonuniform* plane wave is

$$\overline{\mathscr{H}}(y,z,t) = \hat{\mathbf{x}} \underbrace{C_1 \cos(\pi y)}_{H_0(y)} \cos(\omega t - \beta z)$$

which is a plane wave, because the constant phase surfaces are planes of $z = $ constant (i.e., planes parallel to the x-y plane), but it is nonuniform, because the amplitude of the field is a function of y and hence varies as a function of position along the planes of constant phase. We shall encounter nonuniform plane waves when we discuss reflection of uniform plane waves from obliquely oriented boundaries in Chapter 9, and also when we study guiding of electromagnetic waves by metallic and dielectric boundaries in Chapter 10.

8.7 NONPLANAR ELECTROMAGNETIC WAVES

We have seen that uniform plane waves are natural solutions of Maxwell's equations and the wave equations derived from them. The general character of these uniform plane waves is apparent in the functional form of any one of the electric or magnetic field components, namely

$$\overline{\mathcal{E}}(\mathbf{r}, t) = \mathcal{R}e\left\{\mathbf{E}_0 e^{-j\beta\hat{\mathbf{k}}\cdot\mathbf{r}} e^{j\omega t}\right\} = \mathcal{R}e\{\mathbf{E}_0\}\cos(\omega t - \beta\hat{\mathbf{k}}\cdot\mathbf{r}) - \mathcal{I}m\{\mathbf{E}_0\}\sin(\omega t - \beta\hat{\mathbf{k}}\cdot\mathbf{r})$$

where $\mathbf{E}_0$ is, in general, a complex vector given by $\mathbf{E}_0 = \mathcal{R}e\{\mathbf{E}_0\} + j\mathcal{I}m\{\mathbf{E}_0\}$. As discussed before, such waves[68] are *plane* waves because the surfaces of constant phase, that is, the surfaces over which $\overline{\mathcal{E}}(\mathbf{r}, t)$ is constant, are planes. These planar surfaces are defined by $\hat{\mathbf{k}} \cdot \mathbf{r} = $ constant. These waves are *uniform* because the field $\overline{\mathcal{E}}(\mathbf{r}, t)$ does not vary as a function of position over the surfaces of constant phase; in other words, E_0 is not a function of spatial coordinates.

Examples of *nonplanar* waves are those electromagnetic fields for which the variation of the field quantities exhibits a more complicated dependence on spatial coordinates than is implied by $(\omega t - \beta\hat{\mathbf{k}} \cdot \mathbf{r})$. In general, the electric field vector of a nonplanar time-harmonic wave can be expressed as

$$\overline{\mathcal{E}}(\mathbf{r}, t) = \overline{\mathcal{E}}_0 \cos[\omega t - g(\mathbf{r})]$$

where $g(\cdot)$ is some arbitrary function. The surfaces of constant phase are now given by

$$g(\mathbf{r}) = \text{constant}$$

which could be any arbitrary surface, depending on the function $g(\cdot)$. For example, if $g(\mathbf{r}) = g(x, y, z) = xy$, then the surfaces of constant phase are hyperbolic cylinders, with infinite extent in z (i.e., no variation in z) and with the functional form in planes parallel to the x-y plane given as $xy = $ constant. Many of the properties of plane waves that we may take for granted may not be applicable when the wave fronts are nonplanar. For example, the speed of wave propagation, or the phase velocity v_p, may not be constant in space and may instead be a function of position.[69]

---

[68]Note that the functional form of the quantity does not need to be a cosinusoid for it to be a uniform plane wave. In the general case, an electromagnetic field is considered to be a uniform plane wave if the field quantities vary with space and time as $f(\omega t - \beta\hat{\mathbf{k}} \cdot \mathbf{r})$, where $f(\cdot)$ is any function.

[69]See M. Born and E. Wolf, *Principles of Optics*, 5th ed., Section 1.3, Pergamon Press, Oxford, 1975.

8.8 SUMMARY

This chapter discussed the following topics:

- **Uniform plane waves.** The characteristics of electromagnetic waves in source-free and simple media are governed by the wave equation derived from Maxwell's equations:

$$\nabla^2 \overline{\mathscr{E}} - \mu\epsilon \frac{\partial^2 \overline{\mathscr{E}}}{\partial t^2} = 0$$

  Uniform plane waves are the simplest type of solution of the wave equation, with the electric and magnetic fields both lying in the direction transverse to the direction of propagation (i.e., z direction):

$$\mathscr{E}_x(z,t) = p_1(z - v_p t)$$

$$\mathscr{H}_y(z,t) = \frac{1}{\eta} p_1(z - v_p t)$$

  where $v_p = 1/\sqrt{\mu\epsilon}$ is the phase velocity, $\eta = \sqrt{\mu/\epsilon}$ is the intrinsic impedance of the medium, and p_1 is an arbitrary function.

- **Time-harmonic waves in a lossless medium.** The electric and magnetic fields of a time-harmonic uniform plane wave propagating in the z direction in a simple lossless medium are

$$\mathbf{E}(z) = \hat{\mathbf{x}} C_1 e^{-j\beta z} \qquad \overline{\mathscr{E}}(z,t) = \hat{\mathbf{x}} C_1 \cos(\omega t - \beta z)$$

  and

$$\mathbf{H}(z) = \hat{\mathbf{y}} \frac{C_1}{\eta} e^{-j\beta z} \qquad \overline{\mathscr{H}}(z,t) = \hat{\mathbf{y}} \frac{C_1}{\eta} \cos(\omega t - \beta z)$$

  where $\mathbf{E}$ and $\mathbf{H}$ are the phasor quantities, and $\beta = \omega\sqrt{\mu\epsilon}$ is the propagation constant (C_1 is assumed to be real).

- **Uniform plane waves in a lossy medium.** The electric and magnetic fields of a uniform plane wave propagating in the z direction in a lossy medium are

$$\mathbf{E}(z) = \hat{\mathbf{x}} C_1 e^{-\gamma z} \qquad \overline{\mathscr{E}}(z,t) = \hat{\mathbf{x}} C_1 e^{-\alpha z} \cos(\omega t - \beta z)$$

  and

$$\mathbf{H}(z) = \hat{\mathbf{y}} \frac{C_1}{\eta_c} e^{-\gamma z} \qquad \overline{\mathscr{H}}(z,t) = \hat{\mathbf{y}} \frac{C_1 e^{-\alpha z}}{|\eta_c|} \cos(\omega t - \beta z - \phi_\eta)$$

  where $\gamma = \alpha + j\beta$ is the complex propagation constant and $\eta_c = |\eta_c| e^{j\phi_\eta}$ is the complex intrinsic impedance of the medium.

  Tables 8.6, 8.7, and 8.8 summarize the equations for the (generally complex) propagation constants and intrinsic impedances considered in this chapter. Table 8.6

TABLE 8.6 SUMMARY OF PROPAGATION CONSTANT AND INTRINSIC IMPEDANCE EQUATIONS

Category	$\gamma = \alpha + j\beta$	η_c
General	$j\omega\sqrt{\mu_{\text{eff}}\epsilon_{\text{eff}}}$	$\sqrt{\dfrac{\mu_{\text{eff}}}{\epsilon_{\text{eff}}}}$
Lossless: $\epsilon_{\text{eff}} = \epsilon$ $\mu_{\text{eff}} = \mu$	$j\omega\sqrt{\mu\epsilon}$	$\sqrt{\dfrac{\mu}{\epsilon}}$
Lossy: $\epsilon_{\text{eff}} = \epsilon'(1 - j\tan\delta_c)$ $\mu_{\text{eff}} = \mu$	$j\omega\sqrt{\mu\epsilon'}\sqrt{1 - j\tan\delta_c}$	$\sqrt{\dfrac{\mu}{\epsilon'}}\dfrac{1}{\sqrt{1 - j\tan\delta_c}}$
Low-loss: $\tan\delta_c \ll 1$	$\simeq \omega\sqrt{\mu\epsilon'}\dfrac{\tan\delta_c}{2} + j\omega\sqrt{\mu\epsilon'}\left[1 + \dfrac{1}{8}(\tan\delta_c)^2\right]$	$\simeq \sqrt{\dfrac{\mu}{\epsilon'}}\left(1 + j\dfrac{\tan\delta_c}{2}\right)$
Good conductor: $\tan\delta_c \gg 1$	$\simeq \omega\sqrt{\dfrac{\mu\epsilon'\tan\delta_c}{2}} + j\omega\sqrt{\dfrac{\mu\epsilon'\tan\delta_c}{2}}$	$\simeq \sqrt{\dfrac{\mu}{\epsilon'\tan\delta_c}}e^{j45^o}$

TABLE 8.7 SUMMARY OF LOSS TANGENTS FOR MAGNETICALLY LOSSLESS MEDIA

Category	σ_{eff}	$\tan\delta_c = \dfrac{\sigma_{\text{eff}}}{\omega\epsilon'}$
Ohmic (σ) + Dielectric (ϵ'')	$\sigma + \omega\epsilon''$	$\dfrac{\sigma}{\omega\epsilon'} + \dfrac{\epsilon''}{\epsilon'}$
Dielectric (ϵ'')	$\omega\epsilon''$	$\dfrac{\epsilon''}{\epsilon'}$
Ohmic (σ)	σ	$\dfrac{\sigma}{\omega\epsilon'}$

shows γ and η_c for the general, lossless, and lossy cases, where the losses are strictly associated with a complex effective permittivity ($\mu_{\text{eff}} = \mu$ is assumed to be real). Approximate expressions with $\tan\delta_c \ll 1$ and $\tan\delta_c \gg 1$ are provided for the lossy case. Table 8.7 summarizes the effective conductivity and loss tangent equations for each permutation of losses due to Joule heating and dielectric losses. Finally, Table 8.8 shows γ and η_c for the low-loss dielectric and high-loss conductor special cases considered in Section 8.3. These equations are readily derived by substituting the relevant loss tangent values from Table 8.7 into the last two rows of Table 8.6.

TABLE 8.8 COMMON PROPAGATION CONSTANT AND INTRINSIC IMPEDANCE EQUATIONS
FOR MAGNETICALLY LOSSLESS MEDIA

Category	$\gamma = \alpha + j\beta$	η_c
Low-loss Dielectric: $\tan\delta_c = \dfrac{\epsilon''}{\epsilon'} \ll 1$	$\simeq \dfrac{\omega\epsilon''}{2}\sqrt{\dfrac{\mu}{\epsilon'}} + j\omega\sqrt{\mu\epsilon'}\left[1 + \dfrac{1}{8}\left(\dfrac{\epsilon''}{\epsilon'}\right)^2\right]$	$\simeq \sqrt{\dfrac{\mu}{\epsilon'}}\left(1 + j\dfrac{\epsilon''}{2\epsilon'}\right)$
Good conductor: $\tan\delta_c = \dfrac{\sigma}{\omega\epsilon} \gg 1$	$\simeq \sqrt{\dfrac{\mu\omega\sigma}{2}} + j\sqrt{\dfrac{\mu\omega\sigma}{2}}$	$\simeq \sqrt{\dfrac{\mu\omega}{\sigma}}e^{j45^o}$

Another important parameter for good conductors is the skin depth δ, given by

$$\delta = \frac{1}{\alpha} \simeq \frac{1}{\sqrt{\pi f \mu \sigma}}$$

The skin depth for metallic conductors is typically extremely small, being ~3.82 μm for copper at 300 MHz.

- **Electromagnetic power flow and the Poynting vector.** Poynting's theorem states that electromagnetic power flow entering into a given volume through the surface enclosing it equals the sum of the time rates of increase of the stored electric and magnetic energies and the ohmic power dissipated within the volume. The instantaneous power density of the electromagnetic wave is identified as

$$\overline{\mathscr{S}}(z,t) = \overline{\mathscr{E}}(z,t) \times \overline{\mathscr{H}}(z,t)$$

although in most cases the quantity of interest is the time-average power density, which can be found either from $\overline{\mathscr{S}}(z,t)$ or directly from the phasor fields $\mathbf{E}$ and $\mathbf{H}$ as

$$\mathbf{S}_{av}(z) = \frac{1}{T_p}\int_0^{T_p} \overline{\mathscr{S}}(z,t)\,dt \qquad \mathbf{S}_{av}(z) = \frac{1}{2}\mathscr{R}e\{\mathbf{E} \times \mathbf{H}^*\}$$

The time-average Poynting vector for a uniform plane wave propagating in the z direction in an unbounded medium is

$$\mathbf{S}_{av}(z) = \hat{\mathbf{z}}\frac{C_1^2}{2|\eta_c|}e^{-2\alpha z}\cos(\phi_\eta)$$

- **Wave polarization.** The polarization of an electromagnetic wave describes the behavior of its electric field vector as a function of time, at a fixed point in space. In the general case of an elliptically polarized wave, the two components of the electric field vector of a wave propagating in the z direction are given by

$$\mathscr{E}_x(z,t) = C_{1x}\cos(\omega t - \beta z)$$

$$\mathscr{E}_y(z,t) = C_{1y}\cos(\omega t - \beta z + \zeta)$$

A wave is linearly polarized if $\zeta = 0$ or π, and is circularly polarized if $C_{1x} = C_{1y}$ and $\zeta = \pm\pi/2$, with the negative sign corresponding to a right-hand circularly polarized wave. For any other value of ζ, the wave is elliptically polarized, in the left-hand sense for $0 < \zeta < \pi$, and right-hand sense for $\pi < \zeta < 2\pi$.

- **Uniform plane wave propagating in arbitrary direction.** The electric and magnetic field phasors of a uniform plane wave propagating in an arbitrary direction $\hat{\mathbf{k}}$ in a simple lossless medium are given as

$$\mathbf{E}(\mathbf{r}) = \mathbf{E}_0 e^{-j\beta\hat{\mathbf{k}}\cdot\mathbf{r}}$$

$$\mathbf{H}(\mathbf{r}) = \frac{1}{\eta}\hat{\mathbf{k}} \times \mathbf{E}(\mathbf{r})$$

PROBLEMS

8.1 **Uniform plane wave.** The electric field of a uniform plane wave in air is given by

$$\overline{\mathscr{E}}(z,t) = \hat{\mathbf{x}}3\cos(2\pi \times 10^{10}t - \beta z) \quad \text{V-m}^{-1}$$

(a) Find the phase constant β and the wavelength λ. (b) Sketch $\mathscr{E}_x(z,t)$ as a function of t at $z = 0$ and $z = \lambda/4$. (c) Sketch $\mathscr{E}_x(z,t)$ as a function of z at $t = 0$ and $t = \pi/\omega$.

8.2 **Uniform plane wave.** The electric field phasor of an 18 GHz uniform plane wave propagating in free space is given by

$$\mathbf{E}(y) = \hat{\mathbf{z}}15e^{-j\beta y} \quad \text{V-m}^{-1}$$

(a) Find the phase constant, β, and wavelength, λ. (b) Find the corresponding magnetic field phasor $\mathbf{H}(y)$.

8.3 **Uniform plane wave.** A uniform plane wave is traveling in the x direction in air with its magnetic field oriented in the z direction. At the instant $t = 0$, the wave magnetic field has two adjacent zero values, observed at locations $x = 2.5$ cm and $x = 7.5$ cm, with a maximum value of 70 mA-m^{-1} at $x = 5$ cm. (a) Find the wave magnetic field $\overline{\mathscr{H}}(x,t)$ and its phasor $\mathbf{H}(x)$. (b) Find the corresponding wave electric field $\overline{\mathscr{E}}(x,t)$ and its phasor $\mathbf{E}(x)$.

8.4 **Broadcast signal.** The electric field of a wireless communication signal travelling in air is given in phasor form as

$$\mathbf{E}(x) = 10e^{j60x}(\hat{\mathbf{y}} - j\hat{\mathbf{z}}) \quad \text{V-m}^{-1}$$

(a) Find the frequency f & wavelength λ. (b) Find the corresponding phasor form magnetic field $\mathbf{H}(x)$. (c) Find the total time-average power density carried by this wave. (d) Find the polarization of this wave.

8.5 **Uniform plane wave.** A NASA spacecraft orbiting around Mars receives a radio signal transmitted by the UHF antenna of Curiosity Rover, with an electric field given by

$$\overline{\mathscr{E}}(z,t) = \hat{\mathbf{x}}75\cos\left(\omega t + \frac{8\pi z}{3}\right) \quad \mu\text{V-m}^{-1}$$

(a) Determine the frequency f and the wavelength λ of this radio signal. (b) Find the corresponding magnetic field $\overline{\mathscr{H}}(z,t)$.

8.6   **Lossless nonmagnetic medium.** The magnetic field component of a uniform plane wave propagating in a lossless simple nonmagnetic medium ($\mu = \mu_0$) is given by

$$\overline{\mathcal{B}}(x, t) = \hat{z}\, 0.5 \sin[2\pi(10^8 t - 0.5x - 0.125)]\ \mu\text{T}$$

(a) Find the frequency, wavelength, and the phase velocity. (b) Find the relative permittivity, ϵ_r, and the intrinsic impedance, η, of the medium. (c) Find the corresponding $\overline{\mathcal{E}}$. (d) Find the time-average power density carried by this wave.

8.7   **A wireless communication signal.** The electric field of a wireless communication signal traveling in air is given in phasor form as

$$\mathbf{E}(x) = 10e^{j60x}(\hat{y} - j\hat{z})\ \text{V-m}^{-1}$$

(a) Find the frequency f and wavelength λ. (b) Find the corresponding phasor-form magnetic field $\mathbf{H}(x)$. (c) Find the total time-average power density carried by this wave.

8.8   **Uniform plane wave.** An 8 GHz uniform plane wave traveling in air is represented by a magnetic field vector given in phasor form as follows:

$$\mathbf{H}(y) = \hat{x}\, 0.015 e^{-j\beta y} + \hat{z}\, 0.03 e^{j(\pi - \beta y)}\ \text{mA-m}^{-1}$$

(a) Find β and frequency f. (b) Find the corresponding electric field vector in phasor form. (c) Find the total time-average power density carried by this wave.

8.9   **Cellular phones.** The electric field component of a uniform plane wave in air emitted by a mobile communication system is given by

$$\overline{\mathcal{E}}(x, z, t) = \hat{y}\, 50 \sin(\omega t - 7.8\pi x + 8.6\pi z + \theta)\ \text{mV-m}^{-1}$$

(a) Find the frequency f and wavelength λ. (b) Find θ if $\overline{\mathcal{E}}(x, z, t) \simeq -\hat{y}25$ mV-m^{-1} at $t = 0$ and at $x = 3$ m, $z = 2$ m. (c) Find the corresponding magnetic field $\overline{\mathcal{H}}(x, z, t)$. (d) Find the time-average power density carried by this wave.

8.10   **Superposition of two waves.** The sum of the electric fields of two time-harmonic (sinusoidal) electromagnetic waves propagating in opposite directions in air is given as

$$\overline{\mathcal{E}}(z, t) = \hat{x}\, 95 \sin(\beta z) \sin(21 \times 10^9 \pi t)\ \text{mV-m}^{-1}$$

(a) Find the constant β. (b) Find the corresponding $\overline{\mathcal{H}}$. (c) Assuming that this wave may be regarded as a sum of two uniform plane waves, determine the direction of propagation of the two component waves.

8.11   **Unknown material.** The intrinsic impedance and the wavelength of a uniform plane wave traveling in an unknown dielectric at 800 MHz are measured to be $\sim42\Omega$ and ~3.7 cm, respectively. Determine the constitutive parameters (i.e., ϵ_r and μ_r) of the material.

8.12   **Uniform plane wave.** A 10 GHz uniform plane wave with maximum electric field of 100 V-m^{-1} propagates in air in the direction of a unit vector given by $\hat{u} = 0.8\hat{x} - 0.6\hat{z}$. The wave magnetic field has only a y component, which is approximately equal to 40 mA-m^{-1} at $x = z = t = 0$. Find $\overline{\mathcal{E}}$ and $\overline{\mathcal{H}}$.

8.13 Standing waves. The electric field phasor of an electromagnetic wave in air is given by

$$\mathbf{E}(x) = \hat{\mathbf{y}}20 \sin(100\pi x) \text{ mV-m}^{-1}$$

(a) Find the wavelength, λ. (b) Find the corresponding magnetic field $\mathbf{H}(x)$. (c) Is this a traveling wave?

8.14 Propagation through wet versus dry earth. Assume the conductivities of wet and dry earth to be $\sigma_{\text{wet}} = 10^{-2}$ S-m^{-1} and $\sigma_{\text{dry}} = 10^{-4}$ S-m^{-1}, respectively, and the corresponding permittivities to be $\epsilon_{\text{wet}} = 10\epsilon_0$ and $\epsilon_{\text{dry}} = 3\epsilon_0$. Both media are known to be nonmagnetic (i.e., $\mu_r = 1$). Determine the attenuation constant, phase constant, wavelength, phase velocity, penetration depth, and intrinsic impedance for a uniform plane wave of 20 MHz propagating in (a) wet earth, (b) dry earth. Use approximate expressions whenever possible.

8.15 Propagation in lossy media. (a) Show that the penetration depth (i.e., the depth at which the field amplitude drops to $1/e$ of its value at the surface) in a lossy medium with $\mu = \mu_0$ is approximately given by

$$d \simeq \frac{0.225(c/f)}{\sqrt{\epsilon_r'}\sqrt{\sqrt{1 + \tan^2 \delta_c} - 1}}$$

where $\tan \delta_c$ is the loss tangent.[70] (b) For $\tan \delta_c \ll 1$, show that the above equation can be further approximated as $d \simeq [0.318(c/f)]/[\tan \delta_c \sqrt{\epsilon_r'}]$. (c) Assuming the properties of fat tissue at 2.45 GHz to be $\sigma = 0.12$ S-m^{-1}, $\epsilon_r = 5.5$, and $\mu_r = 1$, find the penetration depth of a 2.45 GHz plane wave in fat tissue using both expressions, and compare the results.

8.16 Glacier ice. A glacier is a huge mass of ice that, unlike sea ice, sits over land. Glaciers are formed in the cold polar regions and in high mountains. They spawn billions of tons of icebergs each year from tongues that reach the sea. The icebergs drift over an area of 70,000,000 km^2, which is more than 20% of the ocean area, and pose a serious threat to navigation and offshore activity in many areas of the world. Glacier ice is a low-loss dielectric material that permits significant microwave penetration.[71] The depth of penetration of an electromagnetic wave into glacier ice with loss tangent of $\tan \delta_c \simeq 0.001$ at X-band (assume 3 cm air wavelength) is found to be $\sim$5.41 m. (a) Calculate the dielectric constant and the effective conductivity of the glacier ice. (Note that $\mu_r = 1$.) (b) Calculate the attenuation of the signal measured in dB-m^{-1}.

8.17 Good dielectric. Alumina (Al$_2$O$_3$) is a low-loss ceramic material that is commonly used as a substrate for printed circuit boards. At 10 GHz, the relative permittivity and loss tangent of alumina are approximately equal to $\epsilon_r = 9.7$ and $\tan \delta_c = 2 \times 10^{-4}$. Assume $\mu_r = 1$. For a 10 GHz uniform plane wave propagating in a sufficiently large sample of alumina, determine the following: (a) attenuation constant, α, in np-m^{-1}; (b) penetration depth, d; (c) total attenuation in dB over thicknesses of 1 cm and 1 m.

8.18 Concrete wall. The effective complex dielectric constant of walls in buildings are investigated for wireless communication applications.[72] The relative dielectric constant of the

[70]J. M. Osepchuk, Sources and basic characteristics of microwave radiation, *Bull. N. Y. Acad. Med.,* 55(11), pp. 976–998, December 1979.

[71]E. O. Lewis, B. W. Currie, and S. Haykin, *Detection and Classification of Ice,* Chapter 3, John Wiley & Sons, New York, 1987.

[72]C. F. Yang, C. J. Ko, and B. C. Wu, A free space approach for extracting the equivalent dielectric constants of the walls of the buildings, *IEEE AP-S Int. Symp. URSI Radio Sci. Meet.,* pp. 1036–1039, Baltimore, MD, July 1996.

reinforced concrete wall of a building is found to be $\epsilon_r = 6.7 - j1.2$ at 900 MHz and $\epsilon_r = 6.2 - j0.69$ at 1.8 GHz, respectively. (a) Find the appropriate thickness of the concrete wall to cause a 10 dB attenuation in the field strength of the 900 MHz signal traveling over its thickness. Assume $\mu_r = 1$ and neglect the reflections from the surfaces of the wall. (b) Repeat the same calculations at 1.8 GHz.

8.19 **Unknown medium.** The magnetic field phasor of a 100 MHz uniform plane wave in a nonmagnetic medium is given by

$$\mathbf{H}(z) = \hat{\mathbf{x}}\, 3\, e^{j(16 - j10)z} \quad \text{A-m}^{-1}$$

(a) Find the conductivity σ and relative permittivity ϵ_r of the medium. (b) Find the corresponding time-domain electric field $\overline{\mathscr{E}}(z, t)$.

8.20 **Unknown biological tissue.** The electric field component of a uniform plane wave propagating in a biological tissue with relative dielectric constant $\epsilon_r = \epsilon_r' - j\epsilon_r''$ is given by

$$\overline{\mathscr{E}}(y, t) = \hat{\mathbf{x}}\, 0.5 e^{-39y} \cos(1.83\pi \times 10^9 t - 141y) \quad \text{V-m}^{-1}$$

(a) Find ϵ_r' and ϵ_r''. Assume $\sigma = 0$ and $\mu_r = 1$. (b) Write the corresponding expression for the wave magnetic field. (c) Write the mathematical expression for the time-average Poynting vector $\mathbf{S}_{av}$ and sketch its magnitude from $y = 0$ to 5 cm.

8.21 **Propagation in seawater.** Transmission of electromagnetic energy through the ocean is practically impossible at high frequencies because of the high attenuation rates encountered. For seawater, take $\sigma = 4$ S-m^{-1}, $\epsilon_r = 81$, and $\mu_r = 1$, respectively. (a) Show that seawater is a good conductor for frequencies much less than $\sim$890 MHz. (b) For frequencies less than 100 MHz, calculate, as a function of frequency (in Hz), the approximate distance over which the amplitude of the electric field is reduced by a factor of 10.

8.22 **Wavelength in seawater.** Find and sketch the wavelength in seawater as a function of frequency. Calculate λ_{sw} at the frequencies: 1 Hz, 1 kHz, 1 MHz & 1 GHz. Sketch $\log \lambda_{sw}$ vs. $\log f$. Use properties as: $\sigma = 4$ S-m^{-1}, $\epsilon_r = 81$ & $\mu_r = 1$.

8.23 **Communication in seawater.** ELF communication signals ($f \leq 3$ kHz) can more effectively penetrate seawater than VLF signals (3 kHz $\leq f \leq$ 30 kHz). In practice, an ELF signal used for communication can penetrate and be received at a depth of up to 80 m below the ocean surface.[73] (a) Find the ELF frequency at which the skin depth in seawater is equal to 80 m. For seawater, use $\sigma = 4$ S-m^{-1}, $\epsilon_r = 81$, and $\mu_r = 1$. (b) Find the ELF frequency at which the skin depth is equal to half of 80 m. (c) At 100 Hz, find the depth at which the peak value of the electric field propagating vertically downward in seawater is 40 dB less than its value immediately below the surface of the sea. (d) A surface vehicle-based transmitter operating at 1 kHz generates an electromagnetic signal of peak value 1 V-m^{-1} immediately below the sea surface. If the antenna and the receiver system of the submerged vehicle can measure a signal with a peak value of as low as 1 μV-m^{-1}, calculate the maximum depth beyond which the two vehicles cannot communicate.

8.24 **Submarine communication near a river delta.** A submarine submerged in the sea ($\sigma = 4$ S-m^{-1}, $\epsilon_r = 81$, $\mu_r = 1$) wants to receive the signal from a VLF transmitter operating at 25 kHz. (a) How close must the submarine be to the surface in order to receive 0.1% of the

---

[73]J. C. Kim and E. I. Muehldorf, *Naval Shipboard Communications Systems*, Prentice-Hall, Englewood Cliffs, NJ, 1995, p. 111.

signal amplitude immediately below the sea surface? (b) Repeat part (a) if the submarine is submerged near a river delta, where the average conductivity of seawater is ten times smaller.

8.25   **Human brain tissue.** Consider a 1.9 GHz electromagnetic wave produced by a wireless communication telephone inside a human brain tissue[74] ($\epsilon_r = 43.2$, $\mu_r = 1$, and $\sigma = 1.29$ S-m^{-1}) such that the peak electric field magnitude at the point of entry ($z = 0$) inside the tissue is about 100 V-m^{-1}. Assuming plane wave approximation, do the following: (a) Calculate the electric field magnitudes at points $z = 1$ cm, 2 cm, 3 cm, 4 cm, and 5 cm inside the brain tissue and sketch it with respect to z. (b) Calculate the time-average power density at the same points and sketch it with respect to z. (c) Calculate the time-average power absorbed in the first 1 cm thickness of a tissue sample having a cross-sectional area of 1 cm^2.

8.26   **Dispersion in sea water.** A uniform plane electromagnetic wave in free space propagates with the speed of light, namely, $c \simeq 3 \times 10^8$ m-s^{-1}. In a conducting medium, however, the velocity of propagation of a uniform plane wave depends on the signal frequency, leading to the "dispersion" of a signal consisting of a band of frequencies. (a) For sea water ($\sigma = 4$ S-m^{-1}, $\epsilon_r = 81$, and $\mu_r = 1$), show that for frequencies much less than ~890 MHz, the velocity of propagation is approximately given by $v_p \simeq k_1 \sqrt{f}$, where k_1 is a constant. What is the value of k_1? (b) Consider two different frequency components of a signal, one at 1 kHz, the other at 2 kHz. If these two signals propagate in the same direction in seawater and are in phase at $z = 0$, what is the phase delay (in degrees) between them (e.g., between their peak values) at a position 100 m away?

8.27   **Electromagnetic earthquake precursor.** A group of Stanford scientists measured[75] mysterious electromagnetic waves varying with ultralow frequencies in the range of 0.01–10 Hz during two different earthquakes which occurred in Santa Cruz, California, in 1989 and in Parkfield, California, in 1994. A member of the group speculates that these waves may result from a local disturbance in the earth's magnetic field caused by charged particles carried by water streams that flow along the fault lines deep in the earth's crust as a result of the shifts that led to the quake. These low-frequency waves can penetrate rock much more easily than those of higher frequencies but can still travel only about 15 km through the ground. Since this low-frequency electromagnetic activity was recorded close to a month before the quake and lasted about a month after, this phenomenon has a potential use as an earthquake predictor. Consider three plane waves of equal amplitudes with frequencies of 0.1 Hz, 1 Hz, and 10 Hz, all produced at a depth of 15 km below the earth's surface during an earthquake. Assuming each of these waves to be propagating vertically up toward the surface of the earth, (a) calculate the percentage time-average power of each wave that reaches the surface of the earth and (b) using the results of part (a), comment on which one of the three signals is more likely to be picked up by a receiver located on the earth's surface, based on their signal strengths. For simplicity, assume the earth's crust to be homogeneous, isotropic, and nonmagnetic with properties $\sigma = 10^{-3}$ S-m^{-1} and $\epsilon_r = 10$, respectively.

[74] O. P. Ghandi, G. Lazzi, and C. M. Furse, Electromagnetic absorption in the human head and neck for mobile telephones at 835 and 1900 MHz, *IEEE Trans. Microwave Theory Techniques*, 44(10), pp. 1884–1897, October 1996.

[75] A. C. Fraser-Smith, A. Bernardi, P. R. McGill, M. E. Ladd, R. A. Helliwell, and O. G. Villard, Jr., Low-frequency magnetic field measurements near the epicenter of the M_s 7.1 Loma Prieta earthquake, *Geophys. Res. Lett.*, 17, pp. 1465–1468, 1990; also see B. Holmes, Radio hum may herald quakes, *New Scientist*, p. 15, 23/30 December 1995.

8.28   **Phantom muscle tissue.** In order to develop radiofrequency (RF) heating techniques for treating tumors at various locations and depths in patients, it is necessary to carry out experiments to determine the energy absorbed by an object exposed to electromagnetic fields over a wide range of RF frequencies. An artificial muscle tissue ("muscle phantom") is designed to be used in these experiments to simulate actual muscle tissue for applications in the frequency range used for RF hyperthermia.[76] (a) Given the relative dielectric constant and the conductivity of the muscle phantom at 915 MHz and 22°C to be $\epsilon_r \simeq 51.1$ and $\sigma \simeq 1.27$ S-m^{-1}, calculate the depth of penetration in the phantom. Note that $\mu_r = 1$. (b) Repeat part (a) at 2.45 GHz when $\epsilon_r \simeq 47.4$ and $\sigma \simeq 2.17$ S-m^{-1}. Which frequency can penetrate deeper into the muscle phantom? (c) Calculate the total dB attenuation over a muscle phantom of 1.5 cm thickness at both frequencies.

8.29   **Unknown medium.** The skin depth and the loss tangent of a nonmagnetic conducting medium at 20 kHz are approximately equal to 1.72 m and 4.15×10^4, respectively. (a) Find the conductivity σ and the relative dielectric constant ϵ_r of the medium. What medium is this? (b) Write the mathematical expressions for the electric and magnetic field components of a 20 kHz uniform plane wave propagating in this medium, assuming the maximum peak value of the electric field to be 10 V-m^{-1}. (c) Repeat part (b) at 2.00 MHz. Assume the properties of the medium to be the same at both frequencies.

8.30   **Unknown medium.** A uniform plane wave propagates in the x direction in a certain type of material with unknown properties. At $t = 0$, the wave electric field is measured to vary with x as shown in Figure 8.28a. At $x = 40$ m, the temporal variation of the wave electric field is measured to be in the form shown in Figure 8.28b. Using the data in these two figures, find (a) σ and ϵ_r (assume nonmagnetic case), (b) the depth of penetration and the attenuation in dB-m^{-1}, and (c) the total dB attenuation and the phase shift over a distance of 100 m through this medium.

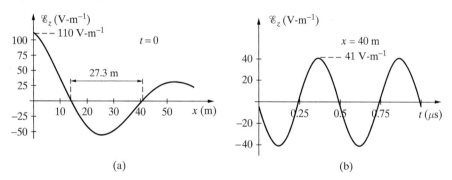

Figure 8.28   Unknown medium. Problem 8.30.

8.31   **Uniform plane waves.** The electric field of a 1 GHz uniform plane wave propagating in a low-loss dielectric is given by

$$\overline{\mathcal{E}}(x,y,t) = \hat{z}\, e^{-0.1\,(x+y)} \cos\left(2\pi\,10^9\,t - 30x - 30y\right) \qquad \text{mV-m}^{-1}$$

(a) Stating all assumptions, determine the conductivity σ and permittivity ϵ of this dielectric and calculate the depth of penetration d. (b) Write a complete (i.e., with all quantities

---

[76]C. K. Chou, G. W. Chen, A. W. Guy, and K. H. Luk, Formulas for preparing phantom muscle tissue at various radio frequencies, *Bioelectromagnetics*, 5, pp. 435–441, 1984.

specified in terms of numerical values) expression for the vector magnetic field intensity $\mathcal{H}(x,y,t)$ of this wave.

8.32 **Thickness of beef products.** Microwave heating is generally uniform over the entire body of the product being heated if the thickness of the product does not exceed about 1–1.5 times its penetration depth.[77] (a) Consider a beef product to be heated in a microwave oven operating at 2.45 GHz. The dielectric properties of raw beef at 2.45 GHz and 25°C are $\epsilon'_r = 52.4$, $\mu_r = 1$, and $\tan \delta_c = 0.33$.[78] What is the maximum thickness of this beef product for it to be heated uniformly? (b) Microwave ovens operating at 915 MHz are evidently more appropriate for cooking products with large cross sections and high dielectric loss factors. The dielectric properties of raw beef at 915 MHz and 25°C are $\epsilon'_r = 54.5$, $\mu_r = 1$, and $\tan \delta_c = 0.411$. Find the maximum thickness of the beef product at 915 MHz and compare it with the results of part (a).

8.33 **Beef versus bacon.** The dielectric properties of cooked beef and smoked bacon at 25°C are given by $\epsilon_r \simeq 31.1 - j10.3$ at 2.45 GHz and $\epsilon_r \simeq 2.5 - j0.125$ at 3 GHz, respectively (see the references in the preceding problem). Assuming $\mu_r = 1$, calculate the loss tangent and the penetration depth for each and explain the differences.

8.34 **Aluminum foil.** A sheet of aluminum foil of thickness ~ 25 μm is used to shield an electronic instrument at 100 MHz. Find the dB attenuation of a plane wave that travels from one side to the other side of the aluminum foil. (Neglect the effects from the boundaries.) For aluminum, $\sigma = 3.54 \times 10^7$ S-m^{-1} and $\epsilon_r = \mu_r = 1$.

8.35 **Unknown material.** Using the results of a reflection measurement technique, the intrinsic impedance of a material at 100 MHz is found to be approximately given by

$$\eta \simeq 22.5 e^{j37°}\ \Omega$$

Assuming that the material is nonmagnetic, determine its conductivity σ and the relative dielectric constant ϵ'_r.

8.36 **Poynting flux.** The electric and magnetic field expressions for a uniform plane wave propagating in a lossy medium are as follows:

$$\mathcal{E}_x(z,t) = 2e^{-4z}\cos(\omega t - \beta z)\quad \text{V-m}^{-1}$$

$$\mathcal{H}_y(z,t) = H_0 e^{-4z}\cos(\omega t - \beta z - \zeta)\quad \text{A-m}^{-1}$$

The frequency of operation is $f = 10^8$ Hz, and the electrical parameters of the medium are $\epsilon = 18.5\epsilon_0$, μ_0, and σ. (a) Find the time-average electromagnetic power density *entering* a rectangular box-shaped surface like that shown in Figure 8.29, assuming $a = d = 1$ m and $b = 0.5$ m. (b) Determine the power density *exiting* this region and compare with (a). (c) The difference between your results in (a) and (b) should represent electromagnetic power lost in the region enclosed by the square-box region. Can you calculate this dissipated power by any other method (i.e., without using the Poynting vector)? If yes, carry out this calculation. *Hint:* You may first need to find σ.

[77] J. Thuery, *Microwaves: Industrial, Scientific and Medical Applications,* Artech House, Boston-Norwood, Massachusetts, 1992.

[78] D. I. C. Wang and A. Goldblith, Dielectric properties of food, *Tech. Rep. TR-76-27-FEL,* MIT, Dept. of Nutrition and Food Science, Cambridge, Massachusetts, November, 1976.

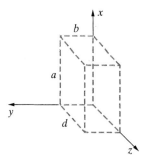

Figure 8.29 **Poynting flux.** Problem 8.36.

8.37 **Laser beams.** The electric field component of a laser beam propagating in the z direction is approximated by

$$\mathscr{E}_r = E_0 e^{-r^2/a^2} \cos(\omega t - \beta z)$$

where E_0 is the amplitude on the axis and a is the effective beam radius, where the electric field amplitude is a factor of e^{-1} lower than E_0. (a) Find the corresponding expression for the magnetic field $\overline{\mathscr{H}}$. (b) Show that the time-average power density at the center of the laser beam is given by

$$|\mathbf{S}_{av}| = \frac{E_0^2}{2\eta} e^{-2r^2/a^2} \text{W-m}^{-2}$$

where $\eta = 377\Omega$. (c) Find the total power of the laser beam. Consider a typical laboratory helium-neon laser with a total power of 5 mW and an effective radius of $a = 400$ μm. What is the power density at the center of the beam? (d) The power density of solar electromagnetic radiation at the Earth's surface is 1400 W-m^{-2}. At what distance from the Sun would its power density be equal to that for the helium-neon laser in part (c)? (e) One of the highest-power lasers built for fusion experiments operates at $\lambda = 1.6$ μm, produces 10.2 kJ for 0.2 ns and is designed for focusing on targets of 0.5 mm diameter. Estimate the electric field strength at the center of the beam. Is the field large enough to break down air? What is the radiation pressure of the laser beam? How much weight can be lifted with the pressure of this beam?

8.38 **Maxwell's equations.** Consider a parallel-plate transmission line with perfectly conducting plates of large extent, separated by a distance of d meters. As shown in Figure 8.30, an alternating surface current density J_s in the z direction flows on the conductor surface:

$$\overline{\mathscr{J}}_s(z,t) = \hat{z}J_0 \cos\left[\omega\left(t - \frac{z}{c}\right)\right] \text{ A-m}^{-1}$$

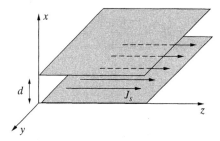

Figure 8.30 **Surface current.**
Problem 8.38.

(a) Find an expression for the electric field, and determine the voltage between the plates, for $d = 0.1$ m and $J_0 = 1$ A-m^{-1}. (b) Use the continuity equation to find an expression for the surface charge density $\rho_s(z, t)$.

8.39   Uniform plane wave. A uniform plane electromagnetic wave propagates in free space with electric and magnetic field components as shown in Figure 8.31:

$$\overline{\mathscr{E}} = \hat{\mathbf{x}} E_0 \cos(\omega t - \beta z)$$

$$\overline{\mathscr{H}} = \hat{\mathbf{y}} \frac{E_0}{\eta} \cos(\omega t - \beta z)$$

The wave frequency is 300 MHz and the electric field amplitude $E_0 = 1$ V-m^{-1}. A square loop antenna with side length $a = 10$ cm is placed at $z = 2$ m as shown. (a) Find the voltage $\mathscr{V}_{\text{ind}}(t)$ induced at the terminals of the loop. (b) Repeat (a) for the loop located at a distance of $d = 3$ m from the x axis instead of 2 m as shown. Compare your answers in (a) and (b).

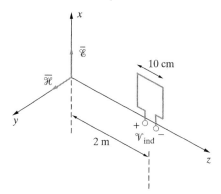

Figure 8.31   Uniform plane wave.
Problem 8.39.

8.40   FM radio. An FM radio station operating at 100 MHz radiates a circularly polarized plane wave with a total isotropically radiated power of 100 kW. The transmitter antenna is located on a tower 550 m above the ground. (a) Find the rms value of the electric field 1 km away from the base of the antenna tower. Neglect the effects of reflections from the ground and other boundaries. (b) If the primary coverage radius of this station is ~100 km, find the approximate time-average power density of the FM wave at this distance.

8.41   Mobile phones. Cellular phone antennas installed on cars have a maximum output power of 3 W, set by the Federal Communications Commission (FCC) standards. The incident electromagnetic energy to which the passengers in the car are exposed does not pose any health threats, both because they are some distance away from the antenna and also because the body of the car and glass window shield them from much of the radiation.[79] (a) For a car with a synthetic roof, the maximum localized power density in the passenger seat is about 0.3 mW-(cm)$^{-2}$. For cars with metal roofs, this value reduces to 0.02 mW-(cm)$^{-2}$ or less. Antennas mounted on the trunk or in the glass of the rear windshield deliver power densities of about 0.35–0.07 mW-(cm)$^{-2}$ to passengers in the back seat. Compare these values with the IEEE safety limit (IEEE Standard C95.1-1991) in the cellular phone frequency

---

[79]M. Fischetti, The cellular phone scare, *IEEE Spectr.*, pp. 43–47, June 1993.

range (which is typically 800–900 MHz) and comment on the safety of the passengers. Note that from 300 MHz to 15 GHz, IEEE safety limits[80] specify a maximum allowable power density that increases linearly with frequency as $|S_{av}|_{max} = f/1500$ in mW-(cm)$^{-2}$, where the frequency f is in MHz. For reference, the maximum allowable power density is 1 mW-(cm)$^{-2}$ at 1.5 GHz. (b) Calculate the maximum output power of a cellular phone antenna installed on the metal roof of a car such that the localized power density in the passenger compartment is equal to the IEEE safety limit at 850 MHz.

8.42 Radar aboard a Navy ship. Some shipboard personnel work daily in an environment where the radio frequency (RF) power density only a few feet above their heads may exceed safe levels. Some areas on the deck of the ship are not even allowed to personnel due to high power densities. Pilots of aircrafts routinely fly through the ship's radar beams during takeoff and landing operations. On one of the Navy aircraft carriers, the average power density along the axis of the main beam (where the field intensity is the greatest) 100 ft away from a 6 kW missile control radar operating in the C-band is measured[81] to be about 300 mW-(cm)$^{-2}$. (a) If it is assumed that human exposures in such environments are limited to less than 10 mW-(cm)$^{-2}$, calculate the approximate distance along the beam axis that can be considered as the hazardous zone. (b) Assuming the operation frequency of the C-band radar to be 5 GHz, recalculate the hazardous zone along the radar's main beam based on the IEEE safety limit[82] for the average power density given by $|S_{av}|_{max} = f/1500$ in mW-(cm)$^{-2}$ (valid over the frequency range 300 MHz–15 GHz), where f is in MHz.

8.43 Radio frequency exposure time. In 1965, the U.S. Army and Air Force amended their use of the prevailing 10 mW-(cm)$^{-2}$ exposure guideline to include a time limit for exposures, given by the formula

$$t_{max} = \frac{6000}{|S_{av}|^2}$$

where $|S_{av}|$ is the average power density [in mW-(cm)$^{-2}$] of exposure and t_{max} is the maximum recommended exposure duration, in minutes.[83] Consider a person standing on the deck of a Navy ship where the peak rms electric field strength due to a microwave radar transmitter is measured to be 140 V-m^{-1}. Using the above formula, calculate the maximum exposure time allowed (in hours) for this person to stay at that location.

8.44 Microwave cataracts in humans. Over 50 cases of human cataract induction have been attributed to microwave exposures, primarily encountered in occupational situations involving acute exposure to presumably relatively high-intensity fields.[84] The following are three reported incidents of cataracts caused by microwave radiation: (1) A 22-year-old technician exposed approximately five times to 3 GHz radiation at an estimated average power density of 300 mW-(cm)$^{-2}$ for 3 min/exposure developed bilateral cataracts. (2) A person was exposed to microwaves for durations of approximately 50 hour/month over a 4-year period at average power densities of less than 10 mW-(cm)$^{-2}$ in most instances, but with a period of 6 months or more during which the average power density was approximately

[80]American National Standards Institute—Safety levels with respect to human exposure to radio frequency electromagnetic fields, 3 kHz–300 GHz, ANSI/IEEE C95.1, 2005.

[81]Z. R. Glaser and G. M. Heimer, Determination and elimination of hazardous microwave fields aboard naval ships, *IEEE Trans. Microwave Theory Techn.*, 19(2), pp. 232–238, February 1971.

[82]American National Standards Institute—Safety levels with respect to exposure to radio frequency electromagnetic fields, 3 kHz to 300 GHz, ANSI/IEEE C95.1, 1992.

[83]W. F. Hammett, *Radio Frequency Radiation, Issues and Standards,* McGraw-Hill, New York, 1997.

[84]S. F. Cleary, Microwave cataractogenesis, *Proc. IEEE,* 68(1), pp. 49–55, January 1980.

1 W-(cm)$^{-2}$. (3) A 50-year-old woman was intermittently exposed to leakage radiation from a 2.45 GHz microwave oven of approximately 1 mW-(cm)$^{-2}$ during oven operation, with levels of up to 90 mW-(cm)$^{-2}$ when the oven door was open, presumably over a period of approximately 6 years prior to developing cataract. For each of these above cases, compare the power densities with the IEEE standards (see Problem 8.41) and comment.

8.45 **VHF TV signal.** The magnetic field component of a 10 μW-m^{-2}, 200 MHz TV signal in air is given by

$$\overline{\mathcal{H}}(x, y, t) = \hat{\mathbf{z}} H_0 \sin(\omega t - ax - ay + \pi/3)$$

(a) What are the values of H_0 and a? (b) Find the corresponding electric field $\overline{\mathcal{E}}(x, y, t)$. What is the polarization of the wave? (c) An observer at $z = 0$ is equipped with a wire antenna capable of detecting the component of the electric field along its length. Find the maximum value of the measured electric field if the antenna wire is oriented along the (i) x direction, (ii) y direction, (iii) 45° line between the x and y directions.

8.46 **FM polarization.** Find the type (linear, circular, elliptical) and sense (right- or left-handed) of the polarization of the FM broadcast signal given in Example 8.2.

8.47 **Unknown wave polarization.** The magnetic field component of a uniform plane wave in air is given by

$$\mathbf{H}(x) = 2e^{j20\pi x}[\hat{\mathbf{y}} e^{-j\pi/4} - \hat{\mathbf{z}} e^{ja\pi/4}] \quad \text{mA-m}^{-1}$$

where a is a real constant. (a) Find the wavelength λ and frequency f. (b) Find the total time-average power density carried by this wave. (c) Determine the type (linear, circular, elliptical) and sense (right- or left-handed) of the polarization of this wave when $a = 1$. (d) Repeat part (c) when $a = 3$.

8.48 **Linear and circularly polarized waves.** Two electromagnetic waves operating at the same frequency and propagating in the same direction (y direction) in air are such that one of them is linearly polarized in the x direction, whereas the other is left-hand circularly polarized (LHCP). However, the electric and magnetic field components of the two waves appear identical at one instant within every 50 ps time interval. The linearly polarized wave carries a time-average power density of 1.4 W-m^{-2}. An observer located at $y = 0$ uses a receiving antenna to measure the x component of the total electric field only and records a maximum field magnitude of about 65 V-m^{-1} over every time interval of 0.5 ns. (a) Write the mathematical expressions for the electric field components of each wave, using numerical values of various quantities whenever possible. (b) Find the ratio of the time-average power densities of the LHCP and the linearly polarized waves.

8.49 **Two circularly polarized waves.** Consider two circularly polarized waves traveling in the same direction transmitted by two different satellites operating at the same frequency given by

$$\mathbf{E}_1 = E_{01}(\hat{\mathbf{x}} - \hat{\mathbf{y}} e^{j\pi/2}) e^{-j\beta z}$$

$$\mathbf{E}_2 = E_{02}(\hat{\mathbf{x}} + \hat{\mathbf{y}} e^{j\pi/2}) e^{-j\beta z}$$

where E_{01} and E_{02} are real constants. (a) If the total time-average power densities of these two waves are equal, find the polarization of the total wave. (b) Repeat part (a) for the case when the total time-average power density of the first wave is four times the total time-average power density of the second wave.

8.50 **Wave polarization.** Consider the following complex phasor expression for a time-harmonic magnetic field in free space:

$$\mathbf{H}(x,z) = [40\hat{\mathbf{x}} + j50\hat{\mathbf{y}} + 30\hat{\mathbf{z}}]e^{j(6x-8z)\pi} \quad \mu\text{A-m}^{-1}$$

(a) Is this a uniform plane wave? What is its frequency? (b) What is the direction of propagation and the state of polarization (specify both the type and sense of polarization) of this electromagnetic field? (c) Find the associated electric field phasor and the total time-average power density in the direction of propagation.

8.51 **Wave polarization.** The electric field components of two electromagnetic waves at the same frequency & propogating in free space are represented by

$$\mathbf{E}_1 = 10(j\hat{\mathbf{x}} - \hat{\mathbf{y}})e^{j26\pi z} \quad \text{V m}^{-1}$$

$$\mathbf{E}_2 = \hat{\mathbf{y}} \cdot E_0 e^{-j26\pi z}$$

Find & sketch locus of the total electric field measured at the origin ($x = y = z = 0$) if E_0 is equal to (a) 10 V m^{-1} (b) 20 V m^{-1} & (c) 40 V m^{-1} respectively.

8.52 **Wave polarization.** A fellow engineer makes the following two measurements of the electric field vector of a uniform plane wave propagating in the $\hat{\mathbf{x}}$ direction in a simple, lossless and nonmagnetic ($\mu = \mu_0$) medium:

1st measurement at $\quad t = 0, x = 0$ $\qquad\qquad \overline{\mathscr{E}} = \dfrac{-5}{\sqrt{2}}\hat{\mathbf{y}} + \dfrac{5}{\sqrt{2}}\hat{\mathbf{z}} \quad \text{V-m}^{-1}$

2nd measurement at $\quad t = 0, x = 0.5\,\text{m}$ $\qquad \overline{\mathscr{E}} = \dfrac{5}{\sqrt{2}}\hat{\mathbf{y}} - \dfrac{5}{\sqrt{2}}\hat{\mathbf{z}} \quad \text{V-m}^{-1}$

(a) Are these two measurements enough to determine the polarization type and sense of the wave? What about the wavelength of the wave? (b) Now the engineer offers two other measurements made at the same time $t = 0$:

3rd measurement at $\quad t = 0, x = 0.25\,\text{m}$ $\qquad \overline{\mathscr{E}} = \dfrac{5}{\sqrt{2}}\hat{\mathbf{y}} + \dfrac{5}{\sqrt{2}}\hat{\mathbf{z}} \quad \text{V-m}^{-1}$

4th measurement at $\quad t = 0, x = 0.75\,\text{m}$ $\qquad \overline{\mathscr{E}} = \dfrac{-5}{\sqrt{2}}\hat{\mathbf{y}} - \dfrac{5}{\sqrt{2}}\hat{\mathbf{z}} \quad \text{V-m}^{-1}$

Furthermore, you are informed that the electric field amplitude at $t = 0$ does not exceed 5 V-m^{-1} at any point between $x = 0$ and $x = 0.75$ m. Given this information, determine the type and sense of polarization of the wave.

8.53 **Superposition of two waves.** The electric field components of two electromagnetic waves at the same frequency and propagating in free space are represented by

$$\mathbf{E}_1 = 10(j\hat{\mathbf{x}} - \hat{\mathbf{y}})e^{j28\pi z} \quad \text{V-m}^{-1}$$

$$\mathbf{E}_2 = \hat{\mathbf{y}}E_0 e^{-j28\pi x}$$

Find and sketch the locus of the total electric field measured at the origin ($x = y = z = 0$) if E_0 is equal to (a) 10 V-m^{-1}, (b) 20 V-m^{-1}, (c) 40 V-m^{-1}, respectively.

9

Reflection, Transmission, and Refraction of Waves at Planar Interfaces

Up to now, we have considered the relatively simple propagation of uniform plane waves in homogeneous and unbounded media. However, practical problems usually involve waves propagating in bounded regions, in which different media may be present, and require that we take account of the complicating effects at boundary surfaces. Typical boundary surfaces may lie between regions of different permittivity (e.g., glass and air) or between regions of different conductivity (e.g., air and copper). Boundary surfaces between regions of different permeability (e.g., air and iron) are also interesting but of less practical importance for electromagnetic wave phenomena.

When a wave encounters the boundary between two different homogeneous media, it is split into two waves: a *reflected* wave that propagates back to the first medium and a *transmitted* (or *refracted*) wave that proceeds into the second medium. Reflection of waves is very much a part of our everyday experience. When a guitar string is plucked, a wave is generated that runs back and forth between the ends of the string, reflecting repeatedly at each end. When we shout toward a cliff, we hear the reflection of the sound waves we generated as an echo. The basis for operation of radars, sonars, and lidars is the reflection of radio, acoustic, or light waves from various objects. When sound waves of sufficient intensity hit an object (such as an eardrum), they transmit energy into it by making it vibrate. Seismic waves transfer their energy quite efficiently through boundaries between different layers of the earth, as do ultrasound waves through tissues. Earthquake waves generated deep under the ocean floors transmit wave energy into the ocean, generating tsunamis.

When we look at the way ocean waves hit a beach or go around a ship standing still, we realize that the reflection of waves from boundaries can be very complicated indeed. The shape of a wave can be greatly altered on reflection, and a wave striking a rough (nonsmooth) surface can be reflected in all directions. We also know from our experiences with physical optics (e.g., refraction of light by lenses or reflection by mirrors) that the direction and magnitude of the waves reflected and refracted at a boundary depend on the angle of arrival of the waves at the boundary. In this book we confine our attention to planar waves of infinite extent incident on planar boundaries. When uniform plane waves encounter planar interfaces, the reflected and transmitted (refracted) waves are also planar, and expressions for their directions of propagation and their amplitudes and phases can be derived with relative ease. Restricting our attention to the reflection and refraction of uniform plane waves from planar boundaries is also appropriate because many practical situations can be modeled in this way to a high degree of accuracy. Optical beams usually have diameters much larger than their free-space wavelength, so their encounter with planar boundaries is quite accurately represented by treating them as uniform plane waves. Furthermore, nonuniform waves, such as those that propagate in planar waveguides, can be represented as a superposition of uniform plane waves reflecting from the conducting waveguide boundaries.

In this chapter, we discuss the simplest cases of propagation of electromagnetic waves in two or more different homogeneous media separated by plane boundaries of infinite extent. In general, when a uniform plane wave propagating in one medium is *incident* on a boundary, part of its energy reflects back into the first medium and is carried away from the boundary by a reflected wave, and the other part propagates into the second medium, carried by a transmitted wave. The existence of these two waves (reflected and transmitted) is a direct result of the boundary conditions imposed by the properties of the media, which in general cannot be satisfied without postulating the presence of both of these waves. Using Maxwell's equations with the appropriate boundary conditions, we can find mathematical expressions for the reflected and transmitted waves in terms of the properties of the incident wave and the two media. We limit our coverage to interfaces that are infinitely sharp, with the electromagnetic properties (e.g., dielectric constant) changing suddenly (within a distance very small compared with the wavelength) from that of one medium to that of the other. For example, for optical applications we require the medium properties near the interface to change from one medium to another over distances much smaller than the free-space wavelength of visible light (390–760 nm). This is why the surfaces of optical coatings have to be very smooth, since any thin layer of impurities will change the reflection properties.

The reflection and transmission of uniform plane waves from planar interfaces are in many ways analogous to the reflection and transmission of voltage and current waves at junctions between different transmission lines or at the load terminations of transmission lines that we studied in Chapters 2 and 3. When appropriate, we make references to this analogy as we study uniform plane wave reflection and transmission, and in some cases we utilize it for more efficient solution of interface problems.

9.1 NORMAL INCIDENCE ON A PERFECT CONDUCTOR

We first consider the case of a uniform plane wave propagating in the $+z$ direction in a simple lossless ($\sigma_1 = 0$) medium occupying the half-space $z < 0$ and normally incident from the left on a perfectly conducting ($\sigma_2 = \infty$) medium that occupies the $z > 0$ half-space, as shown in Figure 9.1. The interface between the two media is the entire x-y plane

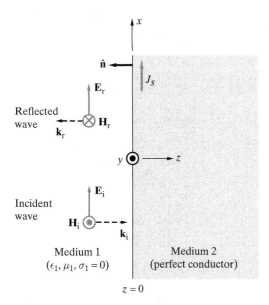

Figure 9.1 Uniform plane wave incident normally on a plane, perfectly conducting boundary. The unit vectors $\hat{\mathbf{k}}_i$ and $\hat{\mathbf{k}}_r$ represent the propagation directions of the incident and reflected waves, respectively.

(i.e., infinite in transverse extent). We arbitrarily[1] assume that the electric field of the incident wave is oriented in the x direction and that therefore its magnetic field is in the y direction. The phasor field expressions for the incident wave (presumably originating at $z = -\infty$) and the reflected wave are given as

Incident Wave:	**Reflected Wave:**
$\mathbf{E}_i(z) = \hat{\mathbf{x}} E_{i0} e^{-j\beta_1 z}$	$\mathbf{E}_r(z) = \hat{\mathbf{x}} E_{r0} e^{+j\beta_1 z}$
$\mathbf{H}_i(z) = \hat{\mathbf{y}} \dfrac{E_{i0}}{\eta_1} e^{-j\beta_1 z}$	$\mathbf{H}_r(z) = -\hat{\mathbf{y}} \dfrac{E_{r0}}{\eta_1} e^{+j\beta_1 z}$

where $\beta_1 = \omega\sqrt{\mu_1\epsilon_1}$ and $\eta_1 = \sqrt{\mu_1/\epsilon_1}$ are respectively, the propagation constant and the intrinsic impedance for medium 1. The presence of the boundary requires that a reflected wave propagating in the $-z$ direction exist in medium 1, with phasor field expressions as given above. Note that the reflected wave propagates in the $-z$ direction, so its magnetic field is oriented in the $-y$ direction for an x-directed electric field.

The boundary condition on the surface of the conductor requires the total tangential electric field to vanish for $z = 0$ for *all x and y,* since the electric field inside medium 2 (perfect conductor) must be zero. This condition requires that the electric field of the

---

[1]Note that for normal incidence on a planar boundary of infinite extent in two dimensions, we lose no generality by taking the direction in which the electric field of the incident wave vibrates to be along one of the principal axes.

reflected wave is also confined to the x direction. The total electric and magnetic fields in medium 1 are

$$\mathbf{E}_1(z) = \mathbf{E}_i(z) + \mathbf{E}_r(z) = \hat{\mathbf{x}}[E_{i0}e^{-j\beta_1 z} + E_{r0}e^{+j\beta_1 z}]$$

$$\mathbf{H}_1(z) = \mathbf{H}_i(z) + \mathbf{H}_r(z) = \hat{\mathbf{y}}\left[\frac{E_{i0}}{\eta_1}e^{-j\beta_1 z} - \frac{E_{r0}}{\eta_1}e^{+j\beta_1 z}\right]$$

Since $\mathbf{E}_1(z)$ is tangential to the boundary, application of the boundary condition at $z = 0$, namely that $\mathbf{E}_1(z = 0) = 0$, gives

$$\mathbf{E}_1(z = 0) = \hat{\mathbf{x}}[E_{i0}e^{-j\beta_1 z} + E_{r0}e^{+j\beta_1 z}]_{z=0} = \hat{\mathbf{x}}[E_{i0} + E_{r0}] = 0$$

from which

$$\boxed{E_{r0} = -E_{i0}}$$

so that

$$\mathbf{E}_1(z) = \hat{\mathbf{x}}E_{i0}\underbrace{[e^{-j\beta_1 z} - e^{+j\beta_1 z}]}_{-j2\sin(\beta_1 z)} = -\hat{\mathbf{x}}E_{i0}j2\sin(\beta_1 z)$$

With the constant E_{r0} determined in terms of E_{i0} (i.e., $E_{r0} = -E_{i0}$), we can also write the total magnetic field in medium 1 as

$$\mathbf{H}_1(z) = \mathbf{H}_i(z) + \mathbf{H}_r(z) = \hat{\mathbf{y}}\frac{1}{\eta_1}[E_{i0}e^{-j\beta_1 z} + E_{i0}e^{+j\beta_1 z}] = \hat{\mathbf{y}}\frac{2E_{i0}}{\eta_1}\cos(\beta_1 z)$$

The corresponding space–time functions $\overline{\mathscr{E}}_1(z, t)$ and $\overline{\mathscr{H}}_1(z, t)$ can be found from the phasors $\mathbf{E}_1(z)$ and $\mathbf{H}_1(z)$ as

$$\overline{\mathscr{E}}_1(z, t) = \mathcal{R}e\{\mathbf{E}_1(z)e^{j\omega t}\} = \mathcal{R}e\{\hat{\mathbf{x}}2E_{i0}\sin(\beta_1 z)e^{j\omega t}\underbrace{e^{-j\pi/2}}_{-j}\}$$

$$\overline{\mathscr{H}}_1(z, t) = \mathcal{R}e\{\mathbf{H}_1(z)e^{j\omega t}\} = \mathcal{R}e\left\{\hat{\mathbf{y}}\frac{2E_{i0}}{\eta_1}\cos(\beta_1 z)e^{j\omega t}\right\}$$

which gives

$$\boxed{\begin{aligned} \overline{\mathscr{E}}_1(z, t) &= \hat{\mathbf{x}}2E_{i0}\sin(\beta_1 z)\sin(\omega t) \\ \overline{\mathscr{H}}_1(z, t) &= \hat{\mathbf{y}}2\frac{E_{i0}}{\eta_1}\cos(\beta_1 z)\cos(\omega t) \end{aligned}} \tag{9.1}$$

assuming that E_{i0} is a real number.[2]

---

[2]If, instead, E_{i0} was a complex number, $E_{i0} = |E_{i0}|e^{j\zeta}$, the expressions for $\overline{\mathscr{E}}_1$ and $\overline{\mathscr{H}}_1$ would be

$$\overline{\mathscr{E}}_1(z, t) = \hat{\mathbf{x}}2|E_{i0}|\sin(\beta_1 z)\sin(\omega t + \zeta)$$

$$\overline{\mathscr{H}}_1(z, t) = \hat{\mathbf{y}}2\frac{|E_{i0}|}{\eta_1}\cos(\beta_1 z)\cos(\omega t + \zeta)$$

amounting to a simple shift of the time origin.

The fields described by (9.1) are plotted in Figure 9.2 as a function of distance z at different time instants. Note that the $\overline{\mathcal{E}}$ and $\overline{\mathcal{H}}$ fields do not represent a propagating wave, because as time advances, the peaks (or nulls) always occur at the same points in space. Such a wave is termed to be a *pure standing wave;* it consists of a superposition of two waves traveling in opposite directions. The maxima and minima *stand* at the same location as time advances. This standing wave, which is established when a uniform plane wave is incident on a perfect conductor, is analogous in all respects to the one that occurs in the case of sinusoidal excitation of a lossless transmission line terminated in a short circuit (see Section 3.2). This analogy is illustrated in Figure 9.3.

A pure standing wave such as the one just described does not carry electromagnetic energy, as expected on physical grounds, since the perfect conductor boundary reflects all of the incident energy. To verify this notion, we can consider the Poynting vector for the standing wave just described. Using the phasor forms of the fields and noting that

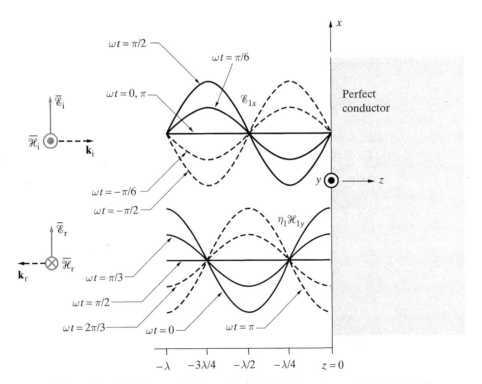

Figure 9.2 Instantaneous electric and magnetic field waveforms. Total electric and magnetic field waveforms in medium 1 are shown at selected instants of time. The "standing" nature of the waves is apparent as the peaks and nulls remain at the same point in space as time progresses. Note that medium 1 is a perfect dielectric, while medium 2 is a perfect conductor.

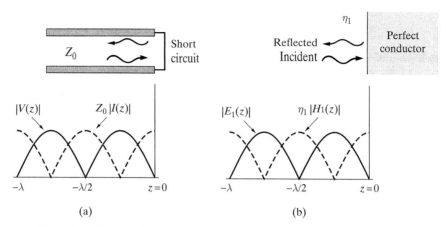

Figure 9.3   Transmission line analogy. (a) The voltage/current standing-wave patterns on a short-circuited lossless transmission line. Here $V(z)$ and $I(z)$ are the total voltage and current phasors along the lossless transmission line, respectively, and Z_0 is the characteristic impedance of the line, given by $Z_0 = \sqrt{L/C}$, with L and C, respectively, being the per-unit-length inductance and capacitance of the line. (b) The electric and magnetic fields for normal incidence of a uniform plane wave from a lossless medium on a perfect conductor. In this context, note that the intrinsic impedance η_1 is analogous to Z_0, while the electric field $\mathbf{E}_1$ and magnetic field $\mathbf{H}_1$ are analogous to the voltage $V(z)$ and current $I(z)$, respectively, although $\mathbf{E}_1$ and $\mathbf{H}_1$ are of course vector quantities while $V(z)$ and $I(z)$ are scalars.

E_{i0} is in general complex, we have

$$(\mathbf{S}_{av})_1 = \frac{1}{2}\mathcal{R}e\{\mathbf{E}_1 \times \mathbf{H}_1^*\} = \frac{1}{2}\mathcal{R}e\left\{\hat{\mathbf{x}}[-2jE_{i0}\sin(\beta_1 z)] \times \hat{\mathbf{y}}\left[\frac{2E_{i0}^*}{\eta_1}\cos(\beta_1 z)\right]\right\}$$

$$= \frac{1}{2}\mathcal{R}e\left\{\hat{\mathbf{z}}\left[-j\frac{2|E_{i0}|^2}{\eta_1}\sin(2\beta_1 z)\right]\right\} = 0$$

The instantaneous Poynting vector can also be obtained from $\overline{\mathscr{E}}_1$ and $\overline{\mathscr{H}}_1$ as

$$\overline{\mathscr{P}}_1(z,t) = \overline{\mathscr{E}}_1(z,t) \times \overline{\mathscr{H}}_1(z,t) = \hat{\mathbf{z}}\frac{4|E_{i0}|^2}{\eta_1}\sin(\beta_1 z)\cos(\beta_1 z)\sin(\omega t + \zeta)\cos(\omega t + \zeta)$$

$$= \hat{\mathbf{z}}\frac{|E_{i0}|^2}{\eta_1}\sin(2\beta_1 z)\sin[2(\omega t + \zeta)]$$

where ζ is the arbitrary phase angle of E_{i0}; that is, we assume that $E_{i0} = |E_{i0}|e^{j\zeta}$. The time-average value of $\overline{\mathscr{P}}_1$ is clearly zero. Although electromagnetic energy does not flow, it surges back and forth. In other words, at any given point z, the Poynting vector fluctuates between positive and negative values at a rate of 2ω.

With respect to the discussion of electromagnetic power flow in Section 8.4, we note from (8.44) that since there is no dissipation term (i.e., $\overline{\mathscr{J}}_1 = 0$, and thus $\overline{\mathscr{E}}_1 \cdot \overline{\mathscr{J}}_1 = 0$), the fluctuating Poynting vector $\mathcal{S}_1(z,t)$ represents the time variation of the stored electric and magnetic energy densities, w_e and w_m, which carry units J-m^{-3} and are given by

$$w_e(z,t) = \tfrac{1}{2}\epsilon_1|\overline{\mathscr{E}}_1(z,t)|^2 = 2\epsilon_1|E_{i0}|^2\sin^2(\beta_1 z)\sin^2(\omega t + \zeta)$$

$$w_m(z,t) = \tfrac{1}{2}\mu_1|\overline{\mathscr{H}}_1(z,t)|^2 = 2\epsilon_1|E_{i0}|^2\cos^2(\beta_1 z)\cos^2(\omega t + \zeta)$$

The variations of w_m and w_e with z at different times are shown in Figure 9.4. We can see that the stored energy at point z alternates between fully magnetic energy ($\omega t + \zeta = 0$) and fully electric energy ($\omega t + \zeta = \pi/2$), much like the fluctuation of stored energy between the inductance and capacitance in a resonant LC circuit.

Note that the total wave magnetic field phasor exhibits a maximum at the conductor surface ($z = 0$). This magnetic field is supported by a surface current whose density can be obtained from the boundary condition

$$\mathbf{J}_s = \hat{\mathbf{n}} \times \mathbf{H}_1(0) = (-\hat{\mathbf{z}} \times \hat{\mathbf{y}})\frac{2E_{i0}}{\eta_1} = \hat{\mathbf{x}}\frac{2E_{i0}}{\eta_1}$$

where $\hat{\mathbf{n}}$ is the outward normal unit vector to the first medium as shown in Figure 9.1. Note that the direction of the surface current $\mathbf{J}_s$ is related to that of $\mathbf{H}_1$ by Ampère's law, with its sense determined by the right-hand rule. If the conductor is not perfect, this current flow leads to dissipation of electromagnetic power, as discussed in Section 9.4.

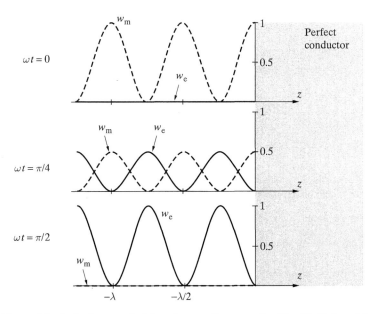

Figure 9.4 Instantaneous electric and magnetic energy densities. Oscillation of the total wave energy densities between electric and magnetic fields in a standing wave. For the purpose of these plots, we have taken E_{i0} to be real and equal to $1/\sqrt{2\epsilon_1}$, so that $\zeta = 0$.

Example 9.1: Linearly polarized wave incident on an air–conductor interface. A z-polarized uniform plane wave having a electric field with peak value 10 V-m^{-1} and operating at 1.5 GHz is normally incident from air on a perfectly conducting surface located at $y = 0$, as shown in Figure 9.5. (a) Write the phasor expressions for the total electric and magnetic fields in air. (b) Determine the nearest location to the reflecting surface in air where the total electric field is zero at all times. (c) Determine the nearest location to the reflecting surface in air where the total magnetic field is zero at all times.

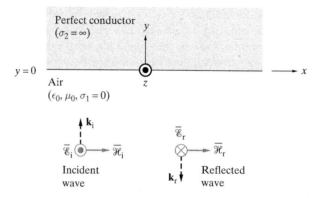

Figure 9.5   Air–perfect conductor interface. A uniform plane wave incident from air on a perfect conductor.

Solution:

(a) The phasor electric and magnetic fields of the incident wave are

$$\mathbf{E}_i(y) = \hat{\mathbf{z}} 10 e^{-j\beta_1 y} \quad \text{V-m}^{-1}$$

$$\mathbf{H}_i(y) = \hat{\mathbf{x}} \frac{10}{\eta_1} e^{-j\beta_1 y} \quad \text{A-m}^{-1}$$

where $\beta_1 \simeq [2\pi(1.5 \times 10^9) \text{ rad-s}^{-1}]/(3 \times 10^8 \text{ m-s}^{-1}) = 10\pi$ rad-m^{-1} and $\eta_1 \simeq 377\Omega$. Since the tangential component of the total electric field at the surface is zero, the electric and magnetic fields of the reflected wave are

$$\mathbf{E}_r(y) = -\hat{\mathbf{z}} 10 e^{+j\beta_1 y} \quad \text{V-m}^{-1}$$

$$\mathbf{H}_r(y) = \hat{\mathbf{x}} \frac{10}{\eta_1} e^{+j\beta_1 y} \quad \text{A-m}^{-1}$$

The total fields in air are

$$\mathbf{E}_1(y) = \mathbf{E}_i(y) + \mathbf{E}_r(y) \simeq -\hat{\mathbf{z}} j 20 \sin(10\pi y) \quad \text{V-m}^{-1}$$

$$\mathbf{H}_1(y) = \mathbf{H}_i(y) + \mathbf{H}_r(y) \simeq \hat{\mathbf{x}} \frac{20}{377} \cos(10\pi y) \simeq \hat{\mathbf{x}}(0.0531) \cos(10\pi y) \quad \text{A-m}^{-1}$$

The corresponding time-domain expressions for the total fields in air are

$$\overline{\mathscr{E}}_1(y, t) \simeq \hat{z}20 \sin(10\pi y) \sin(3\pi \times 10^9 t) \quad \text{V-m}^{-1}$$

$$\overline{\mathscr{H}}_1(y, t) \simeq \hat{x}53.1 \cos(10\pi y) \cos(3\pi \times 10^9 t) \quad \text{mA-m}^{-1}$$

(b) Therefore, the nearest location to $y = 0$ boundary in air where the total electric field is zero at all times can be found from

$$10\pi y = -\pi \quad \longrightarrow \quad y = -0.1 \text{ m} = -10 \text{ cm}$$

(c) Similarly, the nearest location to $y = 0$ where the total magnetic field is zero at all times can be found from

$$10\pi y = -\pi/2 \quad \longrightarrow \quad y = -5 \text{ cm}$$

Note that the two answers correspond to 10 cm $= \lambda/2$ and 5 cm $= \lambda/4$, as expected on the basis of Figure 9.2.

Example 9.2: Circularly polarized wave normally incident on an air–conductor interface. Show that circularly polarized light normally incident on the plane surface of a perfectly conducting mirror changes its sense of polarization (i.e., left-handed becomes right-handed and vice versa) upon reflection from the surface.

Solution: Assume the surface of the mirror to be the $z = 0$ plane. The phasor electric field of a circularly polarized wave propagating in air (half-space $z < 0$) can be written as

$$\mathbf{E}_i(z) = C_1(\hat{x} \pm \hat{y}j)e^{-j\beta_1 z}$$

where the plus sign represents an LHCP wave and the minus sign represents a RHCP wave. The corresponding time-domain expression is

$$\overline{\mathscr{E}}_i(z, t) = \hat{x}C_1 \cos(\omega t - \beta_1 z) \pm \hat{y}C_1 \underbrace{\cos(\omega t - \beta_1 z + \pi/2)}_{-\sin(\omega t - \beta_1 z)}$$

Since the tangential component of the total electric field at the surface is zero, the phasor reflected wave can be written as

$$\mathbf{E}_r(z) = -C_1(\hat{x} \pm \hat{y}j)e^{+j\beta_1 z}$$

The corresponding time-domain expression $\overline{\mathscr{E}}_r(z, t)$, written together with $\overline{\mathscr{E}}_i(z, t)$ for easy comparison, is

$$\overline{\mathscr{E}}_i(z, t) = \hat{x}C_1 \cos(\omega t - \beta_1 z) \pm \hat{y}C_1 \cos(\omega t - \beta_1 z + \pi/2)$$

$$\overline{\mathscr{E}}_r(z, t) = -\hat{x}C_1 \cos(\omega t + \beta_1 z) \mp \hat{y}C_1 \cos(\omega t + \beta_1 z + \pi/2)$$

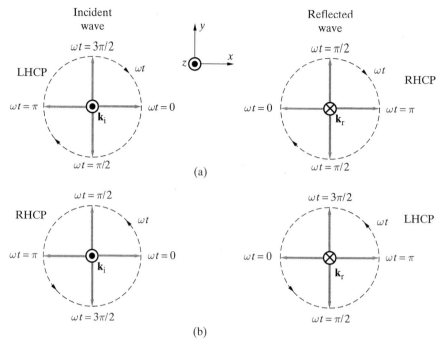

Figure 9.6   Circularly polarized wave incident on an air-conductor interface. An LHCP wave normally incident on a perfect conductor boundary results in a reflected wave which is a RHCP wave as shown in (a) and vice versa as shown in (b). In examining the motion of the electric field vectors as a function of time, it is useful to remember to use the right-hand rule for defining the sense of polarization as discussed in Chapter 8. With the thumb oriented *in the direction of propagation*, the electric field vector should rotate in the same (opposite) direction as the fingers of the right hand for a RHCP (LHCP) wave.

The orientation of the electric field vectors of the incident and reflected waves as obtained from the above expressions are shown in Figure 9.6 at $z = 0$ and at different time instants. Noting the directions of propagation of the waves $(\hat{\mathbf{k}}_i, \hat{\mathbf{k}}_r)$ as indicated, it is clear from these that the reflected wave has the opposite sense of polarization with respect to the incident wave.

9.1.1 More on the Reflection Process and Radiation Pressure

Consideration of the current flow on the conductor surface provides an alternative way of thinking about the reflection of the incident wave at the boundary. Allowing for the fact that the conductor may not be perfect, the current can be thought to result from the electromotive force induced in the conductor by the time-varying magnetic field (via Faraday's law), which, due to the very low resistivity (i.e., high conductivity), causes a large current flow. This current reradiates[3] a field of its own. In the ideal case of zero resistivity (perfect conductor), the reradiated field must exactly cancel the incident field within the conductor, since any finite field would produce an infinite current. Therefore, the polarity of the electric and magnetic fields of the reradiated field propagating in the $+z$ direction within the conductor is opposite to those of the incident wave, as shown in Figure 9.7. Thus, the electric field of the reradiated wave propagating in the $-z$ direction in medium 1 is $\overline{\mathscr{E}}_{rad}(z,t) = -\hat{x}E_{i0}\cos(\omega t + \beta_1 z)$. The magnetic field of the reradiated wave can be found from $\overline{\mathscr{J}}_s$ using the right-hand rule[4] and is in the y direction in medium 1 for an x-directed current. It should be noted that thinking about the reflection process in terms of a reradiated field is simply an alternative method that produces the same result previously arrived at via the resolution of boundary conditions.

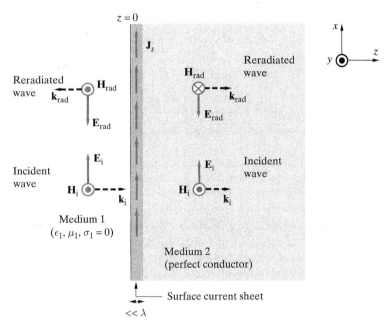

Figure 9.7   A reflected wave can be thought to be reradiated by the surface current in the perfect conductor. Note that the direction of the reradiated wave magnetic field is out of the page (into the page) on the left side (right side) of the current sheet, consistent with the right-hand rule as required by Ampère's law. The wave electric field is pointed in the $-x$ direction, as required for the Poynting vector of the two reradiated waves to be, respectively, in the $-z$ and $+z$ directions.

---

[3]An infinite planar sheet of alternating surface current $\overline{\mathscr{J}}_s = \hat{x}J_0\cos(\omega t)$ located at $z = 0$ in free space radiates uniform plane waves propagating away from the sheet in the $+z$ and $-z$ directions, with electric fields of $\overline{\mathscr{E}}(z,t) = \pm\hat{x}(1/2)\mu_0 cJ_0\cos(\omega t \mp \beta z)$. The derivation of this relationship involves a full treatment of radiation from current sources, which is beyond the scope of this book. For further information, see Section 10.7 of W. T. Scott, *Physics of Electricity and Magnetism*, 2nd ed., Wiley, New York, 1966.

[4]Which, by the way, is in essence is what the boundary condition $\overline{\mathscr{J}}_s = \hat{n} \times \mathscr{H}$ represents.

Both ways of looking at this problem are based on the steady-state (time-harmonic) solution of Maxwell's equations. Note also that, in considering the reradiated field, it is not appropriate to think in terms of a time-sequence of events (e.g., incident wave sets up the current, which *then* reradiates), since our discussion above is limited to sinusoidal steady-state and that therefore the surface current and the fields are different components of the same time-harmonic solution that have to coexist in order for Maxwell's equations to be satisfied.

Based on our discussions in Section 8.4.4, the uniform plane wave incident at the boundary carries a momentum per unit volume of $(\mu_1 \epsilon_1)(\overline{\mathscr{E}}_1 \times \overline{\mathscr{H}}_1)$. Since the wave travels at a speed of $(\mu_1 \epsilon_1)^{-1/2}$, a time-average momentum per unit area and per unit time (i.e., radiation pressure) of $(\mu_1 \epsilon_1)^{1/2} E_{i0}^2 / (2\eta_1)$ is delivered to the surface. The reradiated wave carries away the same amount of momentum in the opposite direction, thus recoiling the conductor (via Newton's third law), so the total radiation pressure on the perfectly reflecting surface is

$$\mathbf{P}_{av} = \frac{2\mathbf{S}_{av}}{v_{p1}} = \hat{\mathbf{z}} \frac{(\mu_1 \epsilon_1)^{1/2} E_{i0}^2}{\eta_1} = \hat{\mathbf{z}} \epsilon_1 E_{i0}^2$$

The magnitude of this pressure is of course very small, as discussed with specific examples in Section 8.4.4. Nevertheless, the device called Crookes's radiometer operates[5] by taking advantage of the fact that $\mathbf{P}_{av}$ on a perfectly reflecting surface is twice that on a perfectly absorbing surface. As mentioned in Section 8.4.4, this device consists of four vanes delicately mounted on a vertical axis, one side of each of the vanes being silvered, the other blackened. When light falls on the vanes, they start rotating, the sense of rotation depending on whether the system is in vacuum or whether there is residual gas left in the enclosure. In a near-perfect vacuum, the radiation pressure on the silvered (i.e., reflecting) surfaces is twice that on the absorbing (i.e., black) surfaces, moving the vanes in one way. However, if some residual gas is left in the enclosure, the black surfaces absorb more radiant energy and become warmer than the silvered surfaces. As a result, molecules that collide with the blackened surface rebound with higher velocities than those that strike the silvered surface, thus imparting more momentum to the blackened surface, causing the device to rotate in the opposite direction.

9.2 NORMAL INCIDENCE ON A LOSSLESS DIELECTRIC

In the previous section we considered the special case of normal incidence on a plane boundary in which the second medium was a perfect conductor, within which no electromagnetic field can exist. We now consider the more general case in which the second medium is a lossless dielectric, within which a nonzero electromagnetic field can propagate. When a uniform plane wave propagating in medium 1 is normally incident on an

---

[5]See Section 6.1 of J. R. Meyer-Arendt, *Introduction to Classical and Modern Optics*, Prentice-Hall, New Jersey, 1972.

interface with a second medium with a different ϵ and μ as shown in Figure 9.8, some of the incident wave energy is transmitted into medium 2 and continues to propagate to the right ($+z$ direction). In the following discussion, we assume both media to be lossless dielectrics (i.e., $\sigma_1 = \sigma_2 = 0$). Once again we assume, with no loss of generality under conditions of normal incidence on a planar boundary, that the incident electric field is oriented in the x direction. We also assume that the amplitude E_{i0} of the incident wave is real—once again, with no loss of generality, since this basically amounts to the choice of the time origin, as shown in the previous section. The phasor fields for the incident, reflected, and transmitted waves are given as

$$\mathbf{E}_i(z) = \hat{\mathbf{x}} E_{i0} e^{-j\beta_1 z}$$

Incident wave

$$\mathbf{H}_i(z) = \hat{\mathbf{y}} \frac{E_{i0}}{\eta_1} e^{-j\beta_1 z}$$

$$\mathbf{E}_r(z) = \hat{\mathbf{x}} E_{r0} e^{+j\beta_1 z}$$

Reflected wave

$$\mathbf{H}_r(z) = -\hat{\mathbf{y}} \frac{E_{r0}}{\eta_1} e^{+j\beta_1 z}$$

$$\mathbf{E}_t(z) = \hat{\mathbf{x}} E_{t0} e^{-j\beta_2 z}$$

Transmitted wave

$$\mathbf{H}_t(z) = \hat{\mathbf{y}} \frac{E_{t0}}{\eta_2} e^{-j\beta_2 z}$$

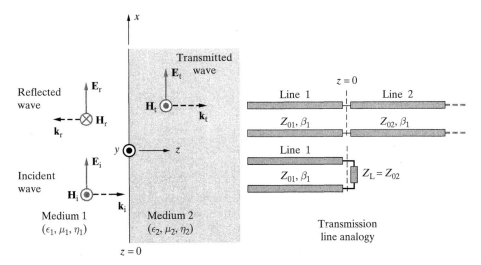

**Figure 9.8 Uniform plane wave normally incident on a lossless dielectric bound-
ary.** Also shown on the right is the transmission line equivalent. Note that the second
medium can either be thought of as an infinitely long line with characteristic impedance
Z_{02} or as a load with load impedance $Z_L = Z_{02}$.

where $\beta_2 = \omega\sqrt{\mu_2\epsilon_2}$ and $\eta_2 = \sqrt{\mu_2/\epsilon_2}$ are, respectively, the phase constant and the intrinsic impedance for medium 2. Note that E_{t0} is the amplitude (yet to be determined) of the transmitted wave. In Figure 9.8 we have defined the polarities of $\mathbf{E}_i$ and $\mathbf{E}_r$ to be the same, and taken (as before in Figure 9.1) $\mathbf{H}_r$ to be in the $-y$ direction so that $\mathbf{E}_r \times \mathbf{H}_r$ is in the $-z$ direction. Note that, at this point, the selected orientations of $\mathbf{E}$ and $\mathbf{H}$ for the different waves (incident, reflected, transmitted) are simply convenient choices.[6] The boundary conditions will determine whether the phasor fields at the boundary are positive or negative according to these assumed conventions.

9.2.1 Reflection and Transmission Coefficients

We now proceed by taking the incident wave as given and determine the properties of the reflected and transmitted waves so that the fundamental boundary conditions for electromagnetic fields are satisfied at the interface, where all three waves can be related to one another. We have two unknown quantities E_{r0} and E_{t0} to be determined in terms of the incident field amplitude E_{i0}. We will use two boundary conditions to determine them. These two conditions are the continuity of the tangential components of both the electric and magnetic fields across the interface. We thus have

$$\mathbf{E}_i(z = 0) + \mathbf{E}_r(z = 0) = \mathbf{E}_t(z = 0) \quad \longrightarrow \quad E_{i0} + E_{r0} = E_{t0}$$

$$\mathbf{H}_i(z = 0) + \mathbf{H}_r(z = 0) = \mathbf{H}_t(z = 0) \quad \longrightarrow \quad \left(\frac{E_{i0}}{\eta_1} - \frac{E_{r0}}{\eta_1}\right) = \frac{E_{t0}}{\eta_2}$$

The solution of these two equations yields

$$E_{r0} = \frac{\eta_2 - \eta_1}{\eta_2 + \eta_1}E_{i0} \qquad E_{t0} = \frac{2\eta_2}{\eta_2 + \eta_1}E_{i0}$$

We now define[7] the reflection and transmission coefficients as follows:

$$\boxed{\Gamma = \frac{E_{r0}}{E_{i0}} = \frac{\eta_2 - \eta_1}{\eta_2 + \eta_1}} \quad \text{Reflection coefficient} \tag{9.2}$$

$$\boxed{\mathcal{T} = \frac{E_{t0}}{E_{i0}} = \frac{2\eta_2}{\eta_2 + \eta_1}} \quad \text{Transmission coefficient} \tag{9.3}$$

where we note that $(1 + \Gamma) = \mathcal{T}$. Although we limited our discussions here to lossless dielectrics, the above relations are fully applicable when the media are dissipative, as

---

[6]The fact that selected orientations are simply convenient choices is important, especially since a different convention may be adopted in other texts.

[7]Note that definition of Γ as the ratio of the reflected to the incident *electric* fields (rather than magnetic fields) is simply a matter of convention.

long as the proper complex values of η_1 and η_2 are used. Note that, physically, the above coefficients are derived from application of the boundary conditions, which are valid for all media[8] in general. Complex reflection and transmission coefficients may result when η_2 and/or η_1 are complex (i.e., one or both of the media are lossy), meaning that in addition to the differences in amplitudes, phase shifts are also introduced between the incident, reflected, and transmitted fields at the interface.

The reflection coefficient expression also could have been obtained on the basis of the transmission line analogy depicted in Figure 9.8. For a transmission line of characteristic impedance Z_{01} terminated in a load impedance Z_L, the load reflection coefficient Γ_L is given by (3.19) which, upon substitution of $Z_L \to \eta_2$ and $Z_{01} \to \eta_1$, is identical to (9.2). Note that the load could just as well be an infinitely long transmission line with characteristic impedance $Z_{02} \neq Z_{01}$, presenting an impedance $Z_L = Z_{02}$ at the junction, as depicted in Figure 9.8. The analogy of the junction between two transmission lines of infinite extent and the interface between two different dielectric media of infinite extent is thus quite clear.

For most dielectrics and insulators, the magnetic permeability does not differ appreciably from its free-space value, so the expressions derived for Γ and $\mathcal{T}$ can also be simply rewritten in terms of ϵ_1 and ϵ_2. So, when $\mu_1 = \mu_2 = \mu_0$, we have

$$\Gamma = \frac{\sqrt{\epsilon_1} - \sqrt{\epsilon_2}}{\sqrt{\epsilon_1} + \sqrt{\epsilon_2}} \qquad \mathcal{T} = \frac{2\sqrt{\epsilon_1}}{\sqrt{\epsilon_1} + \sqrt{\epsilon_2}} \qquad (9.4)$$

In most optical applications, the electromagnetic properties of the dielectric materials are expressed in terms of the *refractive index,* defined as $n \equiv c/v_p = \beta c/\omega = \sqrt{\mu_r \epsilon_r}$, where v_p is the wave phase velocity. For the nonmagnetic case ($\mu_1 = \mu_2 = \mu_0$), the reflection and transmission coefficients can also be expressed in terms of the refractive indices as

$$\Gamma = \frac{n_1 - n_2}{n_1 + n_2} \qquad \mathcal{T} = \frac{2n_1}{n_1 + n_2} \qquad (9.5)$$

Example 9.3: Air–water interface. Calculate the reflection and transmission coefficients for a radio-frequency uniform plane wave traveling in air incident normally upon a calm lake. Assume the water in the lake to be lossless with a relative dielectric constant of $\epsilon_r \simeq 81$. Note that for water, we have $\mu_r = 1$.

Solution: Using (9.4), the reflection and transmission coefficients are given by

$$\Gamma = \frac{\sqrt{\epsilon_{1r}} - \sqrt{\epsilon_{2r}}}{\sqrt{\epsilon_{1r}} + \sqrt{\epsilon_{2r}}} = \frac{1 - \sqrt{81}}{1 + \sqrt{81}} = -0.8, \quad \mathcal{T} = \frac{2\sqrt{\epsilon_{1r}}}{\sqrt{\epsilon_{1r}} + \sqrt{\epsilon_{2r}}} = \frac{2 \times 1}{1 + \sqrt{81}} = 0.2$$

both of which satisfy the relationship $1 + \Gamma = \mathcal{T}$.

---

[8]Note that when medium 2 is a perfect conductor, for which we have $\eta_2 = 0$, (9.2) and (9.3) reduce to $\Gamma = -1$ and $\mathcal{T} = 0$, as expected.

Example 9.4: Air–germanium interface. Germanium (Ge) is a popular material used in infrared optical system designs in either the 3- to 5-μm or 8- to 12-μm spectral bands. Germanium has an index of refraction of approximately 4.0 at these wavelengths. Calculate the reflection and transmission coefficients for an uncoated air–germanium surface.

Solution: Using (9.5), the reflection coefficient is given by

$$\Gamma = \frac{n_1 - n_2}{n_1 + n_2} = \frac{1 - 4}{1 + 4} = -0.6$$

and the transmission coefficient is

$$\mathcal{T} = \frac{2n_1}{n_1 + n_2} = \frac{2 \times 1}{1 + 4} = 0.4$$

Once again, the relationship $1 + \Gamma = \mathcal{T}$ is satisfied.

9.2.2 Propagating and Standing Waves

In Section 9.1 we determined that for a uniform plane wave normally incident on a perfect conductor, the total electric field in medium 1 consisted of a purely standing wave. In the case of normal incidence on a lossless dielectric, we expect at least a portion of the total wave in medium 1 to be propagating in the z direction in order to supply the electromagnetic power taken away from the interface by the transmitted wave in medium 2. To determine the nature of the wave in medium 1, we now examine the total electric field in medium 1. We have

$$\mathbf{E}_1(z) = \mathbf{E}_i(z) + \mathbf{E}_r(z) = \hat{\mathbf{x}}E_{i0}(e^{-j\beta_1 z} + \Gamma e^{+j\beta_1 z})$$

$$= \hat{\mathbf{x}}E_{i0}[(1 + \Gamma)e^{-j\beta_1 z} - \Gamma e^{-j\beta_1 z} + \Gamma e^{+j\beta_1 z}]$$

$$\mathbf{E}_1(z) = \hat{\mathbf{x}}E_{i0}[(1 + \Gamma)e^{-j\beta_1 z} + \Gamma j2\sin(\beta_1 z)]$$

The corresponding space–time field is

$$\overline{\mathcal{E}}_1(z,t) = \hat{\mathbf{x}}E_{i0}[\ \underbrace{\mathcal{T}\cos(\omega t - \beta_1 z)}_{\text{Propagating wave}} + \underbrace{(-2\Gamma)\sin(\beta_1 z)\sin(\omega t)}_{\text{Standing wave}}\ ]$$

where we note once again that E_{i0} was assumed to be real. Note that the propagating wave in medium 1 sustains the transmitted wave in medium 2, whereas the standing wave is produced by the sum of the reflected wave and a portion of the incident wave. We can also express the total electric field phasor in medium 1 as

$$\mathbf{E}_1(z) = \hat{\mathbf{x}}E_{i0}e^{-j\beta_1 z}(1 + \Gamma e^{j2\beta_1 z}) \tag{9.6}$$

The associated total magnetic field phasor in medium 1 is

$$\mathbf{H}_1(z) = \hat{\mathbf{y}} \frac{E_{i0}}{\eta_1} (e^{-j\beta_1 z} - \Gamma e^{j\beta_1 z})$$

$$= \hat{\mathbf{y}} \frac{E_{i0}}{\eta_1} [(1 + \Gamma)e^{-j\beta_1 z} - \Gamma e^{-j\beta_1 z} - \Gamma e^{+j\beta_1 z}]$$

$$= \hat{\mathbf{y}} \frac{E_{i0}}{\eta_1} [(1 + \Gamma)e^{-j\beta_1 z} - 2\Gamma \cos(\beta_1 z)]$$

The corresponding space–time function is

$$\overline{\mathcal{H}}_1(z,t) = \hat{\mathbf{y}} \frac{E_{i0}}{\eta_1} [\ \underbrace{\mathcal{T} \cos(\omega t - \beta_1 z)}_{\text{Propagating wave}} + \underbrace{(-2\Gamma) \cos(\beta_1 z) \cos(\omega t)}_{\text{Standing wave}}\ ]$$

The total magnetic field phasor in medium 1 can also be expressed in a compact form as

$$\mathbf{H}_1(z) = \hat{\mathbf{y}} \frac{E_{i0}}{\eta_1} e^{-j\beta_1 z} (1 - \Gamma e^{j2\beta_1 z})$$

In medium 2, we have only one wave propagating in the $+z$ direction, represented as

$$\mathbf{E}_2(z) = \mathbf{E}_t(z) = \hat{\mathbf{x}} \mathcal{T} E_{i0} e^{-j\beta_2 z}$$

$$\mathbf{H}_2(z) = \mathbf{H}_t(z) = \hat{\mathbf{y}} \frac{\mathcal{T}}{\eta_2} E_{i0} e^{-j\beta_2 z}$$

with the corresponding space–time field expressions of

$$\overline{\mathcal{E}}_t(z,t) = \hat{\mathbf{z}} \mathcal{T} E_{i0} \cos(\omega t - \beta_2 z)$$
$$\overline{\mathcal{H}}_t(z,t) = \hat{\mathbf{y}} \frac{\mathcal{T}}{\eta_2} E_{i0} \cos(\omega t - \beta_2 z)$$

To understand better the relationships between the incident, reflected, and transmitted waves, and the matching of the fields at the boundary, we can examine the instantaneous electric field waveforms shown in Figure 9.9a. These displays are similar to the waveforms in Figure 9.2, except that the incident and reflected waves in medium 1 are shown separately and only at one instant of time, namely, at $\omega t = 0$. The waveforms are shown for the two different cases of $\epsilon_1 < \epsilon_2$ and $\epsilon_1 > \epsilon_2$ assuming nonmagnetic media or $\mu_1 = \mu_2 = \mu_0$. The difference in wavelength $(\lambda = 2\pi/\beta)$ in the two media is readily apparent; for the cases shown, ϵ_1 and ϵ_2 differ by a factor of 4, which means that wavelengths differ by a factor of 2.

For $\epsilon_1 < \epsilon_2$, once again assuming $\mu_1 = \mu_2$, we note that based on (9.4), the amplitude of the transmitted wave is smaller than that of the incident wave (i.e., $E_{t0} < E_{i0}$, or $\mathcal{T} < 1$) and the reflection coefficient Γ is negative. On the other hand, for $\epsilon_1 > \epsilon_2$, the

Instantaneous ($t = 0$) waveforms

(a)

Standing-wave patterns

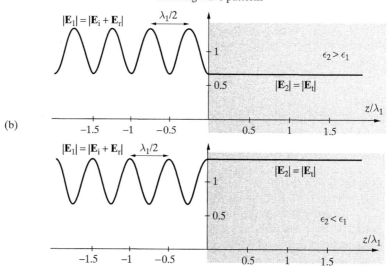

(b)

Figure 9.9 Electric field waveforms and standing-wave patterns on both sides of the interface. (a) Electric field waveforms are given separately for $\epsilon_1 < \epsilon_2$ (in the example shown, $\epsilon_{2r} = 4\epsilon_{1r}$) and for $\epsilon_1 > \epsilon_2$ ($\epsilon_{1r} = 4\epsilon_{2r}$). The electric field waveforms are shown as a function of position (z) at time instant $\omega t = 0$, with E_{i0} taken to be unity. (b) Also shown are the standing-wave patterns for the same two cases, $\epsilon_{2r} = 4\epsilon_{1r}$ and $\epsilon_{1r} = 4\epsilon_{2r}$. Note that the standing-wave patterns are identical in form to those for voltage waves on a transmission line.

transmitted wave amplitude is actually larger than that of the incident wave ($E_{t0} > E_{i0}$), and the reflected wave electric field is in the same direction as that of the incident wave (i.e., $\Gamma > 0$) as expected from (9.4). Note, however, that $E_{t0} > E_{i0}$ does not pose any problems with the conservation of wave energy, since the Poynting vector of the transmitted wave is given by $E_{t0}^2/(2\eta_2)$, and power conservation holds true, as we shall see in Section 9.2.3.

Also shown, in Figure 9.9b, are the standing-wave patterns: the magnitude of $\mathbf{E}_1(z)$ and $\mathbf{E}_2(z)$ for the two different cases of $\epsilon_1 < \epsilon_2$ and $\epsilon_1 > \epsilon_2$. Note that $\mathbf{E}_1(z)$ is given by (9.6), so its magnitude varies between a maximum value of $|\mathbf{E}_1(z)|_{\max} = |E_{i0}|(1 + |\Gamma|)$ and a minimum value of $|\mathbf{E}_1(z)|_{\min} = |E_{i0}|(1 - |\Gamma|)$, while $|\mathbf{E}_2(z)| = \mathcal{T}|E_{i0}|$. These patterns are entirely analogous to standing-wave patterns on transmission lines.[9] In medium 2, there is only one wave, and the standing-wave ratio is thus unity. In medium 1, however, the interference between the reflected and incident waves produces a standing-wave ratio of 2, since we have chosen $\epsilon_{2r} = 4\epsilon_{1r}$ or $\epsilon_{1r} = 4\epsilon_{2r}$ for the two cases shown. Note that in this context, the standing-wave ratio S_1 is defined as

$$S_1 = \frac{|\mathbf{E}_1(z)|_{\max}}{|\mathbf{E}_1(z)|_{\min}} \equiv \frac{1 + |\Gamma|}{1 - |\Gamma|} = \frac{1 + 1/3}{1 - 1/3} = 2$$

9.2.3 Electromagnetic Power Flow

We can now examine the electromagnetic power flow in both medium 1 and medium 2. Noting that using the complex Poynting vector and the phasor expressions for the fields is most convenient, using (8.58) we have in medium 2

$$(\mathbf{S}_{av})_2 = \frac{1}{2}\mathcal{R}e\{\mathbf{E}_t(z) \times \mathbf{H}_t^*(z)\} = \hat{\mathbf{z}}\frac{E_{i0}^2}{2\eta_2}\mathcal{T}^2$$

which follows from Section 8.4 assuming E_{i0} is real. In medium 1 we have

$$(\mathbf{S}_{av})_1 = \tfrac{1}{2}\mathcal{R}e\{\mathbf{E}_1 \times \mathbf{H}_1^*\}$$

$$(\mathbf{S}_{av})_1 = \hat{\mathbf{z}}\frac{E_{i0}^2}{2\eta_1}\mathcal{R}e\{e^{-j\beta_1 z}(1 + \Gamma e^{j2\beta_1 z})e^{+j\beta_1 z}(1 - \Gamma e^{-j2\beta_1 z})\}$$

$$= \hat{\mathbf{z}}\frac{E_{i0}^2}{2\eta_1}\mathcal{R}e\{(1 - \Gamma^2) + \Gamma(e^{j2\beta_1 z} - e^{-j2\beta_1 z})\}$$

$$= \hat{\mathbf{z}}\frac{E_{i0}^2}{2\eta_1}\mathcal{R}e\{(1 - \Gamma^2) + j2\Gamma \sin(2\beta_1 z)\}$$

$$(\mathbf{S}_{av})_1 = \hat{\mathbf{z}}\frac{E_{i0}^2}{2\eta_1}(1 - \Gamma^2)$$

[9]For transmission lines, standing-wave patterns are important in practice because, although the rapid temporal variations of the line voltages and currents are not easily accessible, the locations of the voltage minima and maxima and the ratio of the voltage maxima to minima are often readily measurable. A key parameter used to describe a standing-wave pattern is the standing-wave ratio, or S, defined as $S = |V(z)|_{\max}/|V(z)|_{\min}$ (see equation (3.23)).

Note that the time-average power in medium 1 propagates in the $+z$ direction and basically supplies the power carried by the transmitted wave in medium 2. The average power flows in both media must be equal, since no energy is stored or dissipated at the interface between the two media. We thus have

$$(\mathbf{S}_{\text{av}})_1 = (\mathbf{S}_{\text{av}})_2 \quad \longrightarrow \quad \boxed{1 - \Gamma^2 = \frac{\eta_1}{\eta_2} \mathcal{T}^2}$$

which can also be independently verified using the expressions (9.2) and (9.3) for Γ and $\mathcal{T}$.

Example 9.5: Reflectance of glass. A beam of light is incident normally from air on a BK-7 glass interface. Calculate the reflection coefficient and percent of incident energy reflected if the BK-7 glass has an index of refraction of $n = 1.52$.

Solution: Assuming $\mu_1 = \mu_2$, the reflection coefficient is given by

$$\Gamma = \frac{n_1 - n_2}{n_1 + n_2} = \frac{1 - 1.52}{1 + 1.52} \simeq -0.206$$

The power density of the incident wave is given by $|(\mathbf{S}_{\text{av}})_i| = \frac{1}{2} C_1^2 / \eta_1$, where C_1 is the magnitude of the electric field of the incident wave. The power density of the reflected wave is $|(\mathbf{S}_{\text{av}})_r| = \frac{1}{2} (\Gamma C_1)^2 / \eta_1 = \Gamma^2 |(\mathbf{S}_{\text{av}})_i|$. Therefore, the percent of incident energy reflected back is given by

$$\frac{|(\mathbf{S}_{\text{av}})_r|}{|(\mathbf{S}_{\text{av}})_i|} \times 100 \simeq \Gamma^2 \times 100 \simeq (0.206)^2 \times 100 \simeq 4.26\%$$

Thus, only ~4% of the incident power is reflected by the glass interface. In some applications, this loss may be considered as a significant loss. For example, a camera lens often consists of three or more separate lenses, representing six or more air–glass interfaces. If ~4% of the incoming energy reflects back every time light passes through one of these interfaces, up to ~26% of the original energy is lost during each traverse of light through the lens. It is possible to reduce these losses significantly by introducing antireflection coating on the glass surface, as discussed in the next section.

9.3 MULTIPLE DIELECTRIC INTERFACES

Many practical applications involve the reflection and refraction of electromagnetic waves from dielectric or metallic surfaces that are coated with another dielectric material to reduce reflections and to improve the coupling of the wave energy. The underlying principle in such cases is identical to that in the case of quarter-wave transformer matching of transmission lines discussed in Section 3.5. Some of the applications include antireflection coatings to improve light transmission of a lens, thin-film coatings on optical components to reduce losses selectively over narrow wavelength ranges, metal mirror coatings, flexible metal foils to provide EMI shielding, radar domes (radomes), and stealth technology in which aircraft are specially coated to reduce radar reflectivity.[10]

We formulate the multiple dielectric interface problem as shown in Figure 9.10, in terms of three different dielectric media characterized by intrinsic impedances η_1, η_2, and η_3 and separated by infinite extent planar boundaries located at $z = -d$ and $z = 0$, so that medium 2 is a layer of thickness d.

Figure 9.10   Normal incidence at multiple dielectric interfaces. Normal incidence of a uniform plane on a multiple dielectric interface is directly analogous to a transmission line with characteristic impedance Z_{01} (medium 1) connected to a load Z_L (or a third transmission line with characteristic impedance Z_{03}) via another transmission line (characteristic impedance Z_{02}) segment of length d, as shown above.

---

[10] An example of a coating material that reduces reflections is ferrite-titanate, for which $\mu_r \simeq \epsilon_r \simeq 60(2 - j1)$ at 100 MHz, so that its characteristic impedance, $\eta = \sqrt{\mu/\epsilon}$, is approximately equal to that of free space (377Ω). See Sections 12–16 of J. D. Kraus, *Electromagnetics,* 4th ed., McGraw-Hill, New York, 1992. For tabulated properties of a commercially used ferrite material as a function of frequency, see R. R. Delyser, C. L. Holloway, R. T. Johnk, A. R. Ondrejka, and M. Kauda, Figure of merit for low frequency analysis chambers based on absorber reflection coefficients, *IEEE Trans. Electromagn. Compatibility*, 38(4), pp. 576–584, November 1996. To provide radar invisibility over a range of frequencies, multilayered coatings are used. See J. A. Adam, How to design an "invisible" aircraft, *IEEE Spectr.,* p. 30, April 1988.

We assume a uniform plane wave propagating in medium 1 to be normally incident at the boundary located at $z = -d$. In attempting to determine the amount of wave energy reflected, we might at first be tempted to treat the problem in terms of a series of reflections in which a portion of the energy of the incident wave is transmitted into medium 2, which then propagates to the right and encounters the boundary at $z = 0$, where a portion of it is reflected; this reflected energy propagates to the left in medium 2, encountering the boundary at $z = -d$ from the right, where some of it is reflected back into medium 2 and some transmitted into medium 1, and so on. Even though it might lead to the "correct" answer in some cases, such thinking, which implies a "sequence of events" type of scenario, is inconsistent with our initial steady-state (or time-harmonic) assumption.

9.3.1 Application of Boundary Conditions

A fundamentally based determination of the reflected and transmitted waves requires the use of the general solutions of the wave equation with two undetermined constants (the amplitudes of the component waves propagating in the $+z$ and $-z$ directions) in each medium, and application of the boundary conditions to determine these unknown constants. An alternative method is to rely on a transmission line analogy as implied in Figure 9.10, which reduces the problem to one of impedance transformation. The latter approach is valid because *within* each medium the wave propagation is governed by equations identical to the transmission line equations (given by (2.5) and (2.6)) and because the boundary conditions of the continuity of the tangential electric and magnetic fields are precisely equivalent to the boundary conditions for the continuity of voltage and current at a line termination. We proceed with the general solution of the wave equation, but we also take advantage of the transmission line analogy in interpreting and generalizing our results. The impedance transformation method is discussed in the next subsection.

The expressions for the total wave fields in the three media shown in Figure 9.10 can be written in their most general form as

$$\mathbf{E}_1(z) = \mathbf{E}_{1i} + \mathbf{E}_{1r} = \hat{\mathbf{x}} E_{i0} [e^{-j\beta_1(z+d)} + \Gamma_{\text{eff}} e^{+j\beta_1(z+d)}]$$

$$\mathbf{H}_1(z) = \mathbf{H}_{1i} + \mathbf{H}_{1r} = \hat{\mathbf{y}} \frac{E_{i0}}{\eta_1} [e^{-j\beta_1(z+d)} - \Gamma_{\text{eff}} e^{+j\beta_1(z+d)}] \qquad z < -d \quad \text{Medium 1}$$

$$\mathbf{E}_2(z) = \mathbf{E}_{2f} + \mathbf{E}_{2r} = \hat{\mathbf{x}} E_{20} [e^{-j\beta_2 z} + \Gamma_{23} e^{+j\beta_2 z}]$$

$$\mathbf{H}_2(z) = \mathbf{H}_{2f} + \mathbf{H}_{2r} = \hat{\mathbf{y}} \frac{E_{20}}{\eta_2} [e^{-j\beta_2 z} - \Gamma_{23} e^{+j\beta_2 z}] \qquad -d < z < 0 \quad \text{Medium 2}$$

$$\mathbf{E}_3(z) = \mathbf{E}_{3f} = \hat{\mathbf{x}} \mathcal{T}_{\text{eff}} E_{i0} e^{-j\beta_3 z}$$

$$\mathbf{H}_3(z) = \mathbf{H}_{3f} = \hat{\mathbf{y}} \frac{\mathcal{T}_{\text{eff}} E_{i0}}{\eta_3} e^{-j\beta_3 z} \qquad z > 0 \quad \text{Medium 3}$$

where the wave components $\mathbf{E}_{2f}$ and $\mathbf{E}_{2r}$ in medium 2 are, respectively, the forward- and reverse-propagating waves that constitute the general solution of the wave equation, and where we recognize that there is no reverse-propagating wave in medium 3. Note that Γ_{23} is the reflection coefficient at the interface $z = 0$, whereas Γ_{eff} is an "effective" reflection coefficient[11] at the interface $z = -d$ that accounts for the effects of the presence of the third medium (and the second interface at $z = 0$). (Similarly, $\mathcal{T}_{\text{eff}}$ is an "effective" transmission coefficient.) However, due to the presence of the second boundary, Γ_{eff} is *not* simply the single-layer reflection coefficient $(\eta_2 - \eta_1)/(\eta_2 + \eta_1)$. The solution written above for medium 1 simply recognizes that the amplitudes of the wave components traveling in the $+z$ and $-z$ directions are represented by E_{i0} and $\Gamma_{\text{eff}}E_{i0}$, respectively. However, viewing Γ_{eff} as a reflection coefficient is useful, since in practice (e.g., dielectric coating of glass) the purpose of the second layer is typically to make $\Gamma_{\text{eff}} = 0$ in order to eliminate reflections from the entire multiple-dielectric interface system. Note that although the single-boundary reflection coefficient Γ was necessarily real (i.e., $\Gamma = \rho e^{j(0 \text{ or } \pi)}$) for lossless media, the effective reflection coefficient $\Gamma_{\text{eff}} = \rho_{\text{eff}}e^{j\phi_{\Gamma}}$ is in general complex even in the lossless case. That this should be so can be easily understood by considering the transmission line analogy shown in Figure 9.10; the dielectric layer of thickness d simply acts as an impedance transformer, which takes the intrinsic impedance η_3 of medium 3 and presents it at the $z = -d$ interface as another impedance, which in general is complex. Similarly, the effective transmission coefficient $\mathcal{T}_{\text{eff}} = |\mathcal{T}_{\text{eff}}|e^{j\phi_{\mathcal{T}}}$ is in general also complex.

We can now apply the electromagnetic boundary conditions, namely, the requirement that the tangential components of the electric and magnetic fields be continuous at the two interfaces. At $z = -d$, we find

$$E_{i0}(1 + \Gamma_{\text{eff}}) = E_{20}(e^{+j\beta_2 d} + \Gamma_{23}e^{-j\beta_2 d}) \tag{9.7a}$$

$$\frac{E_{i0}}{\eta_1}(1 - \Gamma_{\text{eff}}) = \frac{E_{20}}{\eta_2}(e^{+j\beta_2 d} - \Gamma_{23}e^{-j\beta_2 d}) \tag{9.7b}$$

Similarly, at $z = 0$ we find

$$E_{20}(1 + \Gamma_{23}) = \mathcal{T}_{\text{eff}}E_{i0} \tag{9.8a}$$

$$\frac{E_{20}}{\eta_2}(1 - \Gamma_{23}) = \frac{\mathcal{T}_{\text{eff}}E_{i0}}{\eta_3} \tag{9.8b}$$

Note that we have four simultaneous equations in terms of the four unknowns E_{20}, $\mathcal{T}_{\text{eff}}$, Γ_{eff}, and Γ_{23}, which can be determined in terms of the incident field amplitude E_{i0} and the parameters of the media, η_1, η_2, and η_3.

To solve, we start by multiplying (9.8b) by η_3 and subtracting from (9.8a) to find

$$E_{20}\left(1 - \frac{\eta_3}{\eta_2}\right) + \Gamma_{23}E_{20}\left(1 + \frac{\eta_3}{\eta_2}\right) = 0$$

---

[11]In this context, by an "effective" reflection coefficient we mean one that is defined as the ratio of the reflected to the incident electric field phasors at the boundary at $z = -d$, so that Γ_{eff} determines the amplitude and phase of the reflected wave in medium 1 in terms of that of the incident wave. Note, however, that $\Gamma_{\text{eff}} \neq (\eta_2 - \eta_1)/(\eta_2 + \eta_1)$, since Γ_{eff} must also include the effect of the second boundary at $z = 0$.

or

$$\Gamma_{23} = \frac{\dfrac{\eta_3}{\eta_2} - 1}{\dfrac{\eta_3}{\eta_2} + 1} = \frac{\eta_3 - \eta_2}{\eta_3 + \eta_2}$$

which is as expected, since there is only one wave in medium 3, and therefore the matching of the boundary conditions at the $z = 0$ interface is identical to that of a two-medium, single-interface problem as studied in the previous section. Accordingly, the expression for Γ_{23} is simply (9.2) applied to the interface between media 2 and 3.

We can now rewrite equations (9.7a) and (9.7b) using the above expression for Γ_{23}, divide them by one another, and manipulate further to find

$$\Gamma_{\text{eff}} = \frac{(\eta_2 - \eta_1)(\eta_3 + \eta_2) + (\eta_2 + \eta_1)(\eta_3 - \eta_2)e^{-2j\beta_2 d}}{(\eta_2 + \eta_1)(\eta_3 + \eta_2) + (\eta_2 - \eta_1)(\eta_3 - \eta_2)e^{-2j\beta_2 d}} \qquad (9.9)$$

The effective transmission coefficient $\mathcal{T}_{\text{eff}}$ can be also determined by similar manipulations of equations (9.7a) and (9.7b); we find

$$\mathcal{T}_{\text{eff}} = \frac{4\eta_2\eta_3 e^{-j\beta_2 d}}{(\eta_2 + \eta_1)(\eta_3 + \eta_2) + (\eta_2 - \eta_1)(\eta_3 - \eta_2)e^{-j2\beta_2 d}} \qquad (9.10)$$

Note that the foregoing expressions for Γ_{eff} and $\mathcal{T}_{\text{eff}}$ are completely general, since they were derived by applying the fundamental electromagnetic boundary conditions. Although our treatment in this section is limited to the cases of interfaces involving lossless dielectrics, (9.9) and (9.10) are also fully applicable when one or more of the three media is lossy, in which case η_1, η_2, and η_3 may be complex and $j\beta_2$ must be replaced by $\gamma_2 = \alpha_2 + j\beta_2$. We discuss multiple interfaces involving lossy media in Section 9.4, using the generalized forms of (9.9) and (9.10).

Based on power arguments similar to those used in Section 9.2.3 for single interfaces, the net time-average electromagnetic power propagating in the $+z$ direction in medium 1 must be equal to that carried in the same direction by the transmitted wave in medium 3, because medium 2 is lossless and therefore cannot dissipate power. Accordingly, the following relation between Γ_{eff} and $\mathcal{T}_{\text{eff}}$ must hold true:

$$1 - |\Gamma_{\text{eff}}|^2 = \frac{\eta_1}{\eta_3}|\mathcal{T}_{\text{eff}}|^2$$

To illustrate the dependence of the reflection coefficient on the thickness d of the dielectric slab, we study the practically useful case of nonmagnetic media with $\epsilon_1 < \epsilon_2 < \epsilon_3$ with $\epsilon_1 = \epsilon_0$, and for two different example media 3, namely, water ($\epsilon_3 = 81\epsilon_0$) and glass ($\epsilon_3 = 2.25\epsilon_0$). Plots of the magnitudes and phases of Γ_{eff} for the two cases are shown in Figure 9.11. For each case, we show results for three different values of ϵ_2.

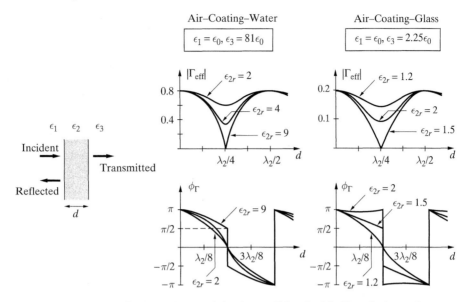

Figure 9.11   Reflection and transmission from a dielectric slab. The reflection coefficients of nonmagnetic dielectric slabs are shown as functions of the slab thickness d (given in terms of the wavelength λ_2 in the slab) and dielectric constant ϵ_{2r}. Results are given for two examples, namely, the air–slab–water and air–slab–glass interfaces.

We note from Figure 9.11 that Γ_{eff} varies significantly as a function of d. In each case, the minimum value of $|\Gamma_{\text{eff}}|$ (and hence the maximum value of $|\mathcal{T}_{\text{eff}}|$) occurs at $d = \lambda_2/4$, and further, $|\Gamma_{\text{eff}}| = 0$ when $\epsilon_2 = \sqrt{\epsilon_1 \epsilon_3}$. Under these conditions, all the energy is transmitted from medium 1 into medium 3. We note that with $\epsilon_2 = \sqrt{\epsilon_1 \epsilon_3}$ (or $\eta_2^2 = \eta_1 \eta_3$) and $d = \lambda_2/4$ we have from (9.10)

$$\mathcal{T}_{\text{eff}} = \sqrt{\frac{\eta_3}{\eta_1}} e^{-j\pi/2}$$

so that the condition

$$1 - |\Gamma_{\text{eff}}|^2 = \frac{\eta_1}{\eta_3} |\mathcal{T}_{\text{eff}}|^2$$

is satisfied with $\Gamma_{\text{eff}} = 0$ for $d = \lambda_2/4$. The $e^{-j\pi/2}$ factor simply represents the phase difference over a distance $d = \lambda_2/4$, or the fact that, for example, at $\omega t = 0$, the peak of the incident wave is at $z = -d = -\lambda_2/4$, whereas the peak of the transmitted wave at $z = 0$ as measured occurs at $\omega t = \pi/2$ rather that at $\omega t = 0$.

When the slab thickness is $d = \lambda_2/2$, we have from (9.9)

$$\Gamma_{\text{eff}} = \frac{\eta_3 - \eta_1}{\eta_3 + \eta_1}$$

so that the system behaves as if the slab were not present. In the air–slab–water example we have $\Gamma_{13} = (1 - \sqrt{81})/(1 + \sqrt{81}) = -0.8 = 0.8e^{j\pi}$, as can be also seen from Figure 9.11. These results are reminiscent of the quarter-wavelength or half-wavelength transmission line impedance transformers and in fact can be arrived at by using the transmission line analogy implied in Figure 9.10, as we show in the next subsection.

9.3.2 Impedance Transformation and Transmission Line Analogy

The transmission line analogy implied in Figure 9.10 provides a useful and system- atic method of solution of the problem of multiple dielectric interfaces, especially those involving more than three dielectric regions. Such a solution method, based on suc- cessive impedance transformations on transmission lines, can follow along lines very similar to methods used in transmission line analysis, specifically in determining the input impedance of cascaded transmission lines (see Sections 3.4 and 3.5). For this pur- pose, we define the concept of *wave impedance* of the total electromagnetic field at any position z in a given medium as

$$Z(z) \equiv \frac{[E_x(z)]_{\text{total}}}{[H_y(z)]_{\text{total}}}$$

where we implicitly assume a wave with an x-directed electric field propagating in the z direction. Note that, in this context, $E_x(z)$ and $H_y(z)$ are scalar quantities entirely analo- gous to the total line voltage $V(z)$ and the total line current $I(z)$ on a transmission line. To appreciate this analogy better, consider the expressions for the total fields $E_{2x}(z)$ and $H_{2y}(z)$ in medium 2 and the total line voltage and current expressions on the analogous line 2 (see Figure 9.10):

$$E_{2x}(z) = E_{20}[e^{-j\beta_2 z} + \Gamma_{23}e^{+j\beta_2 z}] \qquad V(z) = V_{20}[e^{-j\beta_2 z} + \Gamma_{\text{L}}e^{+j\beta_2 z}]$$

$$H_{2y}(z) = \frac{E_{20}}{\eta_2}[e^{-j\beta_2 z} - \Gamma_{23}e^{+j\beta_2 z}] \qquad I(z) = \frac{V_{20}}{Z_{02}}[e^{-j\beta_2 z} - \Gamma_{\text{L}}e^{+j\beta_2 z}]$$

where V_{20} is a constant and $\Gamma_{\text{L}} = (Z_{\text{L}} - Z_0)/(Z_{\text{L}} + Z_0)$. Note that the line impedance $Z(z)$ (analogous to the wave impedance) at any point on line 2 is similarly defined as $Z(z) = V(z)/I(z)$.

Referring back to the three-media problem depicted in Figure 9.10, the wave impedance of the total field in medium 3 is

$$Z_3(z) = \frac{E_{3x}(z)}{H_{3y}(z)} = \eta_3$$

Note that since the field in medium 3 consists only of a single wave traveling in the $+z$ direction, $Z_3(z) = \eta_3$ is independent of position z. This result is similar to the line impedance on an infinitely long or match-terminated transmission line, that is, $Z(z) = Z_{03}$, and η_3 is thus entirely analogous to the characteristic impedance of this line.

In media 2 and 1, on the other hand, the total fields are constituted by two waves traveling in the $+z$ and $-z$ directions, so the wave impedances are not simply η_2 or η_1. In medium 2, for example, the wave impedance has to take into account medium 3, which lies beyond the $z = 0$ interface and acts as a "load" (of impedance $Z_L = \eta_3$) on the second "transmission line." To find the ratio of E_{2x} and H_{2y}, we need to know the relative electric field amplitudes of the two traveling waves within the layer; that is, Γ_{23}. We already know that the reverse-propagating wave in medium 2 (i.e., $E_{20}\Gamma_{23}e^{+j\beta_2 z}$) is related to the forward-propagating wave in medium 3. In other words, the "reflection coefficient" that we "see" looking toward the load from a line with characteristic impedance Z_{02} at the $z = 0$ interface is

$$\Gamma_{23} = \frac{Z_L - Z_{02}}{Z_L + Z_{02}} \quad \rightarrow \quad \Gamma_{23} = \frac{\eta_3 - \eta_2}{\eta_3 + \eta_2}$$

This result is also expected because the field structure at the last interface is exactly the three-wave problem we studied in the case of a single dielectric interface, with an incident wave within the layer (here of unknown amplitude E_{20} rather than E_{i0}) and a reflected wave (of amplitude $\Gamma_{23}E_{20}$ rather than ΓE_{i0}). The above result for Γ_{23} then follows directly from expression (9.2) for Γ.

Using (3.29) and transmission line analogy the wave impedance in medium 2 is then given by

$$Z_2(z) = \frac{E_{2x}(z)}{H_{2y}(z)} = \eta_2 \frac{e^{-j\beta_2 z} + \Gamma_{23}e^{+j\beta_2 z}}{e^{-j\beta_2 z} - \Gamma_{23}e^{+j\beta_2 z}}$$

which in general results in a complex impedance. At $z = 0$ this expression reduces to

$$Z_2(z = 0) = \eta_2 \frac{1 + \Gamma_{23}}{1 - \Gamma_{23}} = \eta_2 \frac{1 + \dfrac{\eta_3 - \eta_2}{\eta_3 + \eta_2}}{1 - \dfrac{\eta_3 - \eta_2}{\eta_3 + \eta_2}} = \eta_2 \frac{2\eta_3}{2\eta_2} = \eta_3$$

which is as expected. The wave impedance in medium 2 at $z = -d$ is given by

$$\boxed{Z_2(-d) = \eta_2 \frac{e^{+j\beta_2 d} + \Gamma_{23}e^{-j\beta_2 d}}{e^{+j\beta_2 d} - \Gamma_{23}e^{-j\beta_2 d}} = \eta_2 \frac{\eta_3 + j\eta_2 \tan(\beta_2 d)}{\eta_2 + j\eta_3 \tan(\beta_2 d)} = \eta_{23}} \qquad (9.11)$$

Similarly, using the expressions for the total electric and magnetic fields in medium 1, we can write the wave impedance in medium 1 as

$$Z_1(z) = \frac{E_{1x}(z)}{H_{1y}(z)} = \eta_1 \frac{e^{-j\beta_1(z+d)} + \Gamma_{\text{eff}}e^{+j\beta_1(z+d)}}{e^{-j\beta_1(z+d)} - \Gamma_{\text{eff}}e^{+j\beta_1(z+d)}}$$

which at $z = -d$ reduces to

$$Z_1(z = -d) = \eta_1 \frac{1 + \Gamma_{\text{eff}}}{1 - \Gamma_{\text{eff}}}$$

Since the electromagnetic boundary conditions require the continuity of the tangential electric (E_x) and magnetic (H_y) field components, and since wave impedance is defined as the ratio of these two quantities, the wave impedances on both sides of any of the interfaces must be equal. In other words, we must have

$$Z_1(z = -d) = \eta_1 \frac{1 + \Gamma_{\text{eff}}}{1 - \Gamma_{\text{eff}}} = Z_2(z = -d) \quad \rightarrow \quad \Gamma_{\text{eff}} = \frac{Z_2(-d) - \eta_1}{Z_2(-d) + \eta_1} = \frac{\eta_{23} - \eta_1}{\eta_{23} + \eta_1}$$

where $Z_2(-d) = \eta_{23}$ is given by (9.11).

In summary, a dielectric slab of thickness d (i.e., medium 2) inserted between two media (i.e., media 1 and 3) essentially works like a transmission line impedance transformer, transforming the impedance of medium 3 from its intrinsic value of η_3 to η_{23}. The impedance η_{23} is what is "seen" looking toward the "load" at the $z = -d$ interface. Without the slab, the reflection coefficient at the interface of media 1 and 3 is given by

$$\Gamma_{13} = \frac{\eta_3 - \eta_1}{\eta_3 + \eta_1}$$

With the slab present, the effective reflection coefficient Γ_{eff} can be found by simply substituting η_{23} for η_3, namely,

$$\boxed{\Gamma_{\text{eff}} = \rho_{\text{eff}} e^{j\phi_\Gamma} = \frac{Z_2(-d) - \eta_1}{Z_2(-d) + \eta_1} = \frac{\eta_{23} - \eta_1}{\eta_{23} + \eta_1}} \qquad (9.12)$$

As an example, consider the case when medium 1 and medium 3 are the same (i.e., $\eta_1 = \eta_3$) and the thickness of medium 2 is an integer multiple of a half-wavelength in that medium. Since $\beta_2 d = (2\pi d)/\lambda_2$, where $d = (n\lambda_2)/2$, $n = 1, 2, \ldots$, we have $\tan(\beta_2 d) = 0$, yielding

$$Z_2(-d) = \eta_{23} = \eta_3$$

In addition, since $\eta_1 = \eta_3$, we find $\Gamma_{\text{eff}} = 0$, resulting in total transmission from medium 1 into medium 3. Note that this is true for any dielectric slab (i.e., a dielectric material with any ϵ_2) as long as its thickness is an integer multiple of a half-wavelength in that dielectric at the frequency of operation.

Another case of interest is that in which the intrinsic impedance of medium 2 is the geometric mean of the intrinsic impedances of media 1 and 3 (i.e., $\eta_2 = \sqrt{\eta_1 \eta_3}$), and the thickness of medium 2 is an odd integer multiple of a quarter-wavelength in that medium. Since $\beta_2 d = (2\pi d)/\lambda_2$, where $d = [(2n + 1)\lambda_2]/4$, $n = 0, 1, 2, \ldots$, we have $\tan(\beta_2 d) = \pm\infty$, yielding

$$\lim_{\tan(\beta_2 d) \to \pm\infty} [\eta_{23}] = \lim_{\tan(\beta_2 d) \to \pm\infty} \left[\eta_2 \frac{\eta_3 + j\eta_2 \tan(\beta_2 d)}{\eta_2 + j\eta_3 \tan(\beta_2 d)} \right] = \frac{\eta_2^2}{\eta_3}$$

Because $\eta_2^2 = \eta_1\eta_3$, however, we also have $\eta_{23} = \eta_1$, resulting in total transmission from medium 1 into medium 3.

The results displayed in Figure 9.11 can now be better understood as simply a manifestation of the quarter-wave type of transmission line matching. The variation of $|\Gamma_{\text{eff}}|$ with normalized slab thickness d/λ_2 provides a measure of the frequency bandwidth over which the matching is effective, since for a given value of d the curves show the variation of $|\Gamma_{\text{eff}}|$ with wavelength or frequency.

The transmission line analogy can easily be extended to problems involving more than two dielectric boundaries, which can be modeled in the same manner by using a series of cascaded transmission lines. Multiple-layer coatings to reduce reflections may be utilized where single-layer coating materials with the desired refractive index (i.e., $n_2 = \sqrt{\epsilon_{2r}}$) are not convenient to use (e.g., they are not structurally self-supporting at the thicknesses desired) or are not available. Also, multiple-layer coatings are generally more desirable in order to achieve the "matching" (i.e., no reflection) condition over a broader range of wavelengths,[12] in a manner quite analogous to multiple-stage quarter-wavelength transmission line systems utilized for wideband matching (see Example 3.19).

Example 9.6: Radome design. A radar dome, or *radome,* is a protective dielectric enclosure for a microwave antenna. A ground-based C-band microwave landing system used to help airplanes to land is to be protected from weather by enclosing it in a radome. The center frequency of the operating frequency band is 5 GHz. Thermoplastic PEI material (assume lossless, nonmagnetic, with $\epsilon_{2r} \simeq 3$) is chosen for the design. (a) Assuming a flat planar radome as shown in Figure 9.12a, determine the minimum thickness of the radome that will give no reflections at 5 GHz. (b) If the frequency is changed to 4 GHz and the thickness of the radome remains as in part (a), what percentage of the incident power is reflected? (c) Repeat part (b) for 6 GHz.

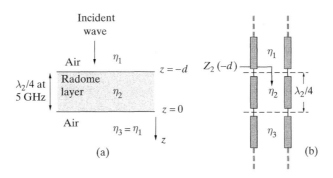

Figure 9.12   Radome design. (a) Geometry of the radome layer. (b) Transmission line analog.

---

[12]For extensive discussion of multiple-layer coatings for optical applications see J. D. Rancourt, *Optical Thin Films User's Handbook,* McGraw-Hill, New York, 1987.

Solution:

(a) At 5 GHz, the wavelength in thermoplastic PEI material is

$$\lambda_2 = \frac{\lambda_1}{\sqrt{\epsilon_{2r}}} = \frac{c}{f\sqrt{\epsilon_{2r}}} \simeq \frac{3 \times 10^8 \text{ m-s}^{-1}}{(5 \times 10^9 \text{ Hz})\sqrt{3}} \simeq 3.46 \times 10^{-2} \text{ m} = 3.46 \text{ cm}$$

In order not to affect the operation of the microwave landing system, the thickness d of the radome layer should be an integer multiple of $\lambda_2/2$, where $\lambda_2/2 \simeq 1.73$ cm, which is also the minimum thickness required.

(b) Using the transmission line method (see Figure 9.12b), we start by evaluating $Z_2(z = -d) = \eta_{23}$. Assuming that the radome is designed with the minimum thickness required (i.e., $d = d_{\min} \simeq 1.73$ cm) and noting that at $f = 4$ GHz, $\lambda_1 = (3 \times 10^8)/(4 \times 10^9) = 0.075$ m or 7.5 cm, $\lambda_2 = \lambda_1/\sqrt{\epsilon_{2r}} = 7.5/\sqrt{3} \simeq 4.33$ cm, $\tan(\beta_2 d) = \tan[(2\pi/\lambda_2)d] \simeq \tan[(2\pi/4.33)1.73] \simeq -0.727$, and $\eta_2 = \eta_1/\sqrt{\epsilon_{2r}} \simeq 377/\sqrt{3} \simeq 218\Omega$, we have from (9.11)

$$Z_2(z = -d) = \eta_{23} = \eta_2 \frac{\eta_3 + j\eta_2 \tan(\beta_2 d)}{\eta_2 + j\eta_3 \tan(\beta_2 d)}$$

$$\simeq 218\frac{377 + j218(-0.727)}{218 + j377(-0.727)} \simeq 218\frac{409e^{-j22.8°}}{350e^{-j51.5°}} \simeq 254e^{j28.8°} \, \Omega$$

Therefore, using (9.12), the effective reflection coefficient is

$$\Gamma_{\text{eff}} = \frac{\eta_{23} - \eta_1}{\eta_{23} + \eta_1} \simeq \frac{254e^{j28.8°} - 377}{254e^{j28.8°} + 377}$$

$$\simeq \frac{-154 + j122}{600 + j122} \simeq \frac{197e^{j142°}}{612e^{j11.5°}} \simeq 0.321e^{j130°}$$

Hence, the percentage of the incident power that reflects back can be calculated as

$$\frac{|(S_{av})_r|}{|(S_{av})_i|} \times 100 = |\Gamma_{\text{eff}}|^2 \times 100 \simeq (0.321)^2 \times 100 \simeq 10.3\%$$

(c) Following a similar procedure, at $f = 6$ GHz we have $\lambda_1 = 5$ cm, $\lambda_2 \simeq 2.89$ cm, $\tan(\beta_2 d) = \tan(216°) \simeq 0.727$, $\eta_2 \simeq 218\Omega$, and (9.11) yields

$$Z_2(z = -d) = \eta_{23} \simeq 218\frac{377 + j218(0.727)}{218 + j377(0.727)} \simeq 218\frac{409e^{j22.8°}}{350e^{j51.5°}} \simeq 254e^{-j28.8°} \, \Omega$$

Using (9.12), the effective reflection coefficient is

$$\Gamma_{\text{eff}} = \frac{254e^{-j28.8°} - 377}{254e^{-j28.8°} + 377} \simeq \frac{-154 - j122}{600 - j122} \simeq 0.321e^{-j130°}$$

Hence, the percentage of the incident power that reflects back is $(0.321)^2 \times 100 \simeq 10.3\%$. Thus, we see that the effective reflection coefficient varies quite symmetrically around the design frequency of 5 GHz, being down by the same amount in magnitude at 4 GHz as at 6 GHz.

Example 9.7: Coated glass surface. Consider Example 9.5 on the reflectance of glass. Determine the refractive index and minimum thickness of a film to be deposited on the glass surface ($n_3 = 1.52$) such that no normally incident visible light of free-space wavelength 550 nm (i.e., ~545 THz) is reflected.

Solution: Since the permittivity of air is equal to ϵ_0, the wavelength in air (medium 1) is $\lambda_1 = 550$ nm. Since we have nonmagnetic media, the requirement of $\eta_2 = \sqrt{\eta_1 \eta_3}$ is equivalent to $n_2 = \sqrt{n_1 n_3}$, and we can find the refractive index of the film as $n_2 = \sqrt{1 \times 1.52} \simeq 1.23$. The minimum thickness of the film is

$$d = d_{\min} = \frac{\lambda_2}{4} = \frac{\lambda_1}{4n_2} = \frac{550 \times 10^{-9}}{4 \times 1.23} \simeq 0.112 \ \mu\text{m}$$

Note that once this film is deposited on glass, it will eliminate reflections completely only at 550 nm. Although there would still be reflections at other wavelengths in the vicinity of 550 nm, the percentage of the reflected power will be less than the ~4% found in Example 9.5.

In practice, it is typically not possible to manufacture antireflection coating materials that have precisely the desired refractive index. A practical coating material that is commonly used is magnesium fluoride (MgF$_2$, $n_2 = 1.38$; see Figure 9.13a). For visible light with free-space wavelength of 550 nm, the wavelength in MgF$_2$ is $\lambda_2 = 550/1.38 \simeq 399$ nm, so the thickness of the quarter-wavelength MgF$_2$ layer is $d = \lambda_2/4 \simeq 99.6$ nm.

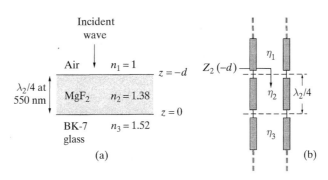

Figure 9.13   Air–glass interface with single MgF$_2$ coating layer. (a) Geometry of the antireflection coating layer. (b) Transmission line analog.

Because we know the matching is not perfect (i.e., $n_{\text{MgF}_2} \neq 1.23$), we can calculate the reflection coefficient for $\lambda_1 = 550$ nm. For this purpose, we use the transmission line method (Figure 9.13b), which can also be applied to multiple coating layers. We start by evaluating $Z_2(z = -d) = \eta_{23}$. Noting that $d = \lambda_2/4$, so that we have $\tan(\beta_2 d) = \tan[(2\pi/\lambda_2)(\lambda_2/4)] = \infty$, and that $\eta_1 \simeq 377\Omega$, $\eta_2 = \eta_1/1.38$, and $\eta_3 = \eta_1/1.52$, we have from (9.11)

$$Z_2(z = -d) = \eta_{23} = \eta_2 \frac{\eta_3 + j\eta_2 \tan(\beta_2 d)}{\eta_2 + j\eta_3 \tan(\beta_2 d)} = \frac{\eta_2^2}{\eta_3} \simeq \frac{(377)(1.52)}{(1.38)^2} \simeq 301\Omega$$

Therefore, using (9.12) we have

$$\Gamma_{\text{eff}} = \frac{Z_2(-d) - \eta_1}{Z_2(-d) + \eta_1} \simeq \frac{301 - 377}{301 + 377} \simeq -0.112$$

so the fraction of the incident power reflected at 550 nm is $\rho_{\text{eff}}^2 \simeq (0.112)^2 \simeq 0.0126$, or 1.26%; substantially less than the ~4% reflection that occurs without any coating. To assess the effectiveness of the MgF$_2$ coating layer in reducing reflections over the visible spectrum, we can calculate the reflection coefficient at other wavelengths. For example, at 400 nm (violet light) we have

$$\tan(\beta_2 d) = \tan\left(\frac{2\pi}{(400\,\text{nm}/1.38)}(99.6\,\text{nm})\right) = \tan(123.75°) \simeq -1.50$$

so using (9.11), we have

$$Z_2(z = -d) = \eta_{23} \simeq \left(\frac{377}{1.38}\right)\frac{(377/1.52) + j(377/1.38)(-1.50)}{(377/1.38) + j(377/1.52)(-1.50)}$$

$$= \left(\frac{377}{1.38}\right)\frac{1.38 - j(1.50)(1.52)}{1.52 - j(1.50)(1.38)} \simeq 283.45e^{-j5.11°}\,\Omega$$

and therefore from (9.12)

$$\Gamma_{\text{eff}} = \frac{Z_2(-d) - \eta_1}{Z_2(-d) + \eta_1} \simeq \frac{283e^{-j5.11°} - 377}{283e^{-j5.11°} + 377} \simeq \frac{-94.67 - j25.24}{659.3 - j25.24} \simeq 0.1485e^{-j163°}$$

so the percentage of reflected power in terms of the incident power is

$$\Gamma_{\text{eff}}^2 \times 100 = (0.149)^2 \times 100 \simeq 2.21\%$$

At the other end of the visible light spectrum, for red light at 750 nm, we have

$$\tan(\beta_2 d) = \tan\left(\frac{2\pi}{(750\,\text{nm}/1.38)}(99.6\,\text{nm})\right) \simeq 2.246$$

and using (9.11), we have

$$\dot{Z}_2(-d) = \eta_{23} \simeq \left(\frac{377}{1.38}\right)\frac{(1.52)^{-1} + j(1.38)^{-1}(2.246)}{(1.38)^{-1} + j(1.52)^{-1}(2.246)} \simeq 290.65 + j20.9\,\Omega$$

which from (9.12) gives

$$\Gamma_{\text{eff}} = \frac{Z_2(-d) - \eta_1}{Z_2(-d) + \eta_1} \simeq 0.133e^{j164.6°}$$

so the percentage of incident power that reflects back is ~1.77%, still substantially lower than the ~4% we have without any coating. The foregoing results for MgF$_2$ coating are consistent with the plot of $|\Gamma_{\text{eff}}|$ versus d given in Figure 9.11 for $\epsilon_{3r} = 2.25$ (glass with $n_3 = 1.5$) and $\epsilon_{2r} = 2$ (close to $\epsilon_{\text{MgF}_2} \simeq 1.9$ or $n_{\text{MgF}_2} \simeq 1.38$).

9.4 NORMAL INCIDENCE ON A LOSSY MEDIUM

Up to now, we have studied normal-incidence reflection and transmission of uniform plane waves at interfaces between different lossless media or between a lossless dielectric and a perfect conductor. In practice, all media have some loss, leading to absorption of the transmitted energy as it propagates through the lossy medium. In some cases such attenuation may be undesirable but unavoidable; in other cases the heat produced by the attenuation of the wave in the lossy material may constitute the primary application. In this section we consider two important cases involving reflection from an imperfect conducting plane and multiple lossy interfaces.

9.4.1 Reflection from an Imperfect Conducting Plane

We now consider the special case of incidence of a uniform plane wave on a "good" conductor with finite conductivity σ. We will show that the total current flowing within the conducting material is essentially independent of the conductivity. As the conductivity approaches infinity, the total current is squeezed into a narrower and narrower layer, until in the limit a true surface current (as discussed in Section 9.1 for the case of a perfectly conducting boundary) is obtained. We will further show that the conductor can be characterized as a boundary exhibiting a surface impedance of $Z_s = (\sigma\delta)^{-1}(1+j)$, where δ is the skin depth for the conductor, being $\delta = \sqrt{2/(\omega\mu\sigma)}$, as given in (8.41).

Let a uniform plane wave traveling in a lossless medium be normally incident on a conducting interface located at $z = 0$ (i.e., the half-space $z \geq 0$ is filled with a conducting medium), as shown in Figure 9.14. The phasor fields for the incident, reflected, and transmitted waves are

$$\mathbf{E}_i(z) = \hat{\mathbf{x}} E_{i0} e^{-j\beta_1 z}$$

$$\mathbf{H}_i(z) = \hat{\mathbf{y}} \frac{E_{i0}}{\eta_1} e^{-j\beta_1 z}$$

$$\mathbf{E}_r(z) = \hat{\mathbf{x}} E_{r0} e^{+j\beta_1 z}$$

$$\mathbf{H}_r(z) = -\hat{\mathbf{y}} \frac{E_{r0}}{\eta_1} e^{+j\beta_1 z}$$

$$\mathbf{E}_t(z) = \hat{\mathbf{x}} E_{t0} e^{-\gamma_2 z}$$

$$\mathbf{H}_t(z) = \hat{\mathbf{y}} \frac{E_{t0}}{\eta_c} e^{-\gamma_2 z}$$

where

$$\beta_1 = \omega\sqrt{\mu_1\epsilon_1} \qquad\qquad \eta_1 = \sqrt{\mu_1/\epsilon_1}$$

$$\gamma_2 = \sqrt{j\omega\mu_2(\sigma_2 + j\omega\epsilon_2)} \qquad\qquad \eta_c = \sqrt{j\omega\mu_2/(\sigma_2 + j\omega\epsilon_2)}$$

In a good conducting medium (i.e., $\sigma \gg \omega\epsilon$) we have $\nabla \times \mathbf{H} = (j\omega\epsilon + \sigma)\mathbf{E} \simeq \sigma\mathbf{E}$, since the conduction current is much greater than the displacement current. Rewriting this

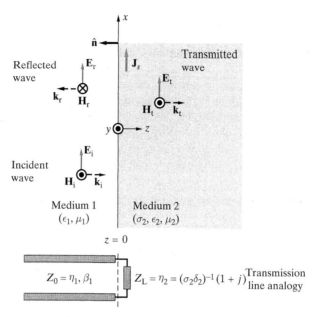

Figure 9.14   Uniform plane wave normally incident on the boundary between a dielectric and an imperfect conductor.

equation as $\nabla \times \mathbf{H} \simeq j\omega[\sigma/(j\omega)]\mathbf{E}$, we recall from Section 8.3 that $\epsilon_{\text{eff}} \simeq \sigma/(j\omega)$ may be considered as the permittivity ϵ in Maxwell's equations in a lossless medium. Thus, propagation constant and intrinsic impedance of the lossy medium can be obtained from that of a lossless medium by substituting ϵ with ϵ_{eff}. We thus have the propagation constant for medium 2 given by

$$\gamma_2 = j\omega\sqrt{\mu_2\epsilon_{\text{eff}}} \simeq j\omega\left(\frac{\mu_2\sigma_2}{j\omega}\right)^{1/2} = (j\omega\mu_2\sigma_2)^{1/2} = \delta_2^{-1}(1+j) \qquad (9.13)$$

and the complex intrinsic impedance given by

$$\eta_c \equiv Z_s = R_s + jX_s = \sqrt{\frac{\mu_2}{\epsilon_{\text{eff}}}} \simeq \left(\frac{j\omega\mu_2}{\sigma_2}\right)^{1/2} = \frac{\gamma_2}{\sigma_2} = (\sigma_2\delta_2)^{-1}(1+j) \qquad (9.14)$$

where $\delta_2 = (\pi f \mu_2\sigma_2)^{-1/2}$ is the skin depth for medium 2. Note that for medium 1 we have the propagation constant and intrinsic impedance of a lossless medium, namely, $\beta_1 = \omega\sqrt{\mu_1\epsilon_1}$ and $\eta_1 = \sqrt{\mu_1/\epsilon_1}$.

The conductor presents an impedance $Z_s = \eta_c$ to the electromagnetic wave with equal resistive and inductive parts, defined above as R_s and jX_s, respectively. The resistance part is simply the resistance of a sheet of metal of 1 meter square and of thickness δ_2, as illustrated in Figure 9.15; actually, the resistance is independent of the area (l^2) of

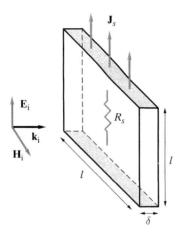

Figure 9.15 Surface resistance concept. The resistance between the shaded faces is independent of the linear dimension l and hence can be thought of as a surface resistance.

the square plate. Thus, the resistance between the two shaded faces (perpendicular to the current flow) is given by

$$R_s = \frac{l}{l\delta_2\sigma_2} = \frac{1}{\delta_2\sigma_2} \tag{9.15}$$

Since the resistance is independent of the linear dimension l, it is called a *surface resistance,* and the complex intrinsic impedance η_c can be thought of as the *surface impedance* of the conductor.

To find the reflection and transmission coefficients, we follow the same procedure as before and apply the boundary conditions:

$$E_{i0} + E_{r0} = E_{t0} \quad \text{and} \quad \frac{1}{\eta_1}(E_{i0} - E_{r0}) = \frac{E_{t0}}{Z_s}$$

Solving for E_{r0}/E_{i0} and E_{t0}/E_{i0}, we find

$$\Gamma = \frac{E_{r0}}{E_{i0}} = \frac{Z_s - \eta_1}{Z_s + \eta_1} = \rho e^{j\phi_\Gamma} \qquad \mathcal{T} = \frac{E_{t0}}{E_{i0}} = \frac{2Z_s}{Z_s + \eta_1} = \tau e^{j\phi_\mathcal{T}}$$

Note that we use the more general notation $\Gamma = \rho e^{j\phi_\Gamma}$ and $\mathcal{T} = \tau e^{j\phi_\mathcal{T}}$, since the reflection and transmission coefficients are in general complex.

For any reasonably good conductor, Z_s is very small compared with η_1 for free space (i.e., 377Ω), as was mentioned in Section 8.3. For example, for copper ($\sigma_2 = 5.8\times 10^7$ S-m^{-1}) at 1 MHz, $\delta_2 \simeq 66.1$ μm, and $R_s = 2.61 \times 10^{-4}\Omega$. The reflection and transmission coefficients, respectively, are $\Gamma \simeq 0.9999986e^{j\,179.999921°}$ and $\mathcal{T} \simeq 2 \times 10^{-6}e^{j45°}$. We thus note that for all practical purposes the field in front of the conductor ($z < 0$) is the same as exists for a perfect conductor, since $\rho = |\Gamma|$ is very close to unity. For the same reason, only a very small amount of power is transmitted into the conductor; that is, $\tau = |\mathcal{T}|$ is small. Nevertheless, this small amount of power that penetrates into the walls can still cause significant attenuation of waves propagating in waveguides and coaxial lines, especially when these types of transmission lines are relatively long. Next we discuss a method for calculating the finite amount of power dissipated in the conductor.

In the case when medium 2 is a good conductor (i.e., $\sigma_2 \gg \omega \epsilon_2$), we can obtain a useful approximate expression for Γ by using a Taylor series expansion. Noting that $\eta_2 \simeq \sqrt{j \mu_2 \omega / \sigma_2}$, and assuming $\mu_2 = \mu_1$, we have

$$\Gamma \simeq \frac{\sqrt{j \mu_2 \omega / \sigma_2} - \sqrt{\mu_1 / \epsilon_1}}{\sqrt{j \mu_2 \omega / \sigma_2} + \sqrt{\mu_1 / \epsilon_1}} = \frac{\sqrt{j \omega \epsilon_1 / \sigma_2} - 1}{\sqrt{j \omega \epsilon_1 / \sigma_2} + 1}$$

Since $|\sqrt{j \omega \epsilon_1 / \sigma_2}| \ll 1$, we can use the Taylor expansion

$$\frac{x - 1}{x + 1} \simeq -1 + 2x$$

with $x = \sqrt{j \omega \epsilon_1 / \sigma_2}$. Noting that $\sqrt{j} = (1 + j)/\sqrt{2}$, we obtain

$$\Gamma \simeq \left(-1 + \sqrt{\frac{2 \omega \epsilon_1}{\sigma_2}} \right) + j \sqrt{\frac{2 \omega \epsilon_1}{\sigma_2}}$$

Based on this expression, the fraction of the incident power reflected is approximately given by

$$|\Gamma|^2 \simeq 1 - 4 \sqrt{\frac{\omega \epsilon_1}{2 \sigma_2}}$$

and the fraction transmitted into the conductor is

$$1 - |\Gamma|^2 \simeq 4 \sqrt{\frac{\omega \epsilon_1}{2 \sigma_2}}$$

The current density $\mathbf{J}$ in the good conductor is $\mathbf{J} = \sigma \mathbf{E}_t = \hat{\mathbf{x}} \sigma_2 \mathcal{T} E_{i0} e^{-\gamma_2 z}$. The total current per unit width of the conductor is

$$J_s = \sigma_2 \mathcal{T} E_{i0} \int_0^\infty e^{-\gamma_2 z} \, dz = \frac{\sigma_2 \mathcal{T} E_{i0}}{\gamma_2} \tag{9.16}$$

Note that while $\mathbf{J}$ is in units of A-m^{-2}, the current per unit width is a surface current in units of A-m^{-1}. We can relate this surface current to the magnetic field at the surface of the conductor, namely, the total magnetic field $\mathbf{H}_1$ in medium 1. Note that the continuity of the tangential magnetic field at the interface requires that $\mathbf{H}_1(z = 0) = \mathbf{H}_t(z = 0)$. Since the magnetic field of the transmitted wave is $\mathbf{H}_t(z) = \hat{\mathbf{y}}(E_{t0}/\eta_c) e^{-\gamma_2 z}$, then substituting $\eta_c = Z_s$ and $\gamma_2 = \delta_2^{-1}(1 + j)$, we have

$$\mathbf{H}_1(0) = \mathbf{H}_t(0) \quad \rightarrow \quad \mathbf{H}_1(0) = \hat{\mathbf{y}} \frac{\mathcal{T} E_{i0}}{Z_s} \quad \rightarrow \quad \mathcal{T} E_{i0} = Z_s H_{1y}(0)$$

Substituting in (9.16) and using (9.13) and (9.14) we have

$$J_s = \frac{\sigma_2 Z_s H_{1y}(0)}{\gamma_2} = \frac{\sigma_2(\sigma_2\delta_2)^{-1}(1+j)}{\delta_2^{-1}(1+j)} H_{1y}(0) = H_{1y}(0)$$

It is thus apparent that if we let $\sigma_2 \to \infty$, we have $\delta_2 \to 0$, $\rho \to -1$, but the total current J_s does not vanish; it remains equal to H_{1y}. More generally, we have $\mathbf{J}_s = \hat{\mathbf{n}} \times \mathbf{H}_1(0)$, as was found in Section 9.1 for the case of normal incidence of a uniform plane wave on a perfect conductor. However, since $\delta_2 \to 0$ as $\sigma_2 \to \infty$, the current is squeezed into a narrower and narrower layer and in the limit becomes a true surface current. When σ_2 is finite, the current density $\mathbf{J}$ inside the conductor varies with z in the same manner as $\mathbf{E}_t(z)$ and is confined to a region of thickness δ_2, with the total current per unit width of the conductor being $\mathbf{J}_s$.

The time-average power loss per unit area on the x-y plane can be evaluated from the complex Poynting vector at the surface. We have

$$|\mathbf{S}_{\mathrm{av}}| = \left| \frac{1}{2}\mathcal{R}e\{\mathbf{E} \times \mathbf{H}^*\} \right| = \frac{1}{2}\mathcal{R}e\{E_{1x}H_{1y}^*\} = \frac{1}{2}\mathcal{R}e\left\{ \mathcal{T}E_{i0}\left(\frac{\mathcal{T}^*E_{i0}^*}{Z_s^*} \right) \right\}$$

$$= \frac{1}{2}|\mathcal{T}E_{i0}|^2 \mathcal{R}e\left\{ \frac{1}{Z_s^*} \right\} = \frac{1}{2}|\mathcal{T}E_{i0}|^2 \mathcal{R}e\left\{ \frac{1}{R_s - jX_s}\left(\frac{R_s + jX_s}{R_s + jX_s} \right) \right\}$$

$$= \frac{1}{2}|\mathcal{T}E_{i0}|^2 \mathcal{R}e\left\{ \frac{R_s + jX_s}{R_s^2 + X_s^2} \right\} = \frac{1}{2}|\mathcal{T}E_{i0}|^2 \frac{R_s}{R_s^2 + X_s^2} = \frac{1}{4}|\mathcal{T}E_{i0}|^2 \sigma_2\delta_2$$

where we have used the fact that $R_s = X_s = (\sigma_2\delta_2)^{-1}$. This Poynting flux represents the total time-average power per unit area entering the conductor. All of this power must be dissipated in the conductor due to $\mathbf{E} \cdot \mathbf{J}$ losses, which can be evaluated by integrating over a volume with 1 m^2 cross section in the x-y plane. In other words,

$$P_{\mathrm{loss}} = \frac{1}{2}\int_V \mathbf{E} \cdot \mathbf{J}^* dv = \frac{1}{2}\int_0^1 \int_0^1 \int_0^\infty (E_x)(\sigma_2 E_x^*)\, dx\, dy\, dz$$

$$= \frac{\sigma_2}{2}|\mathcal{T}E_{i0}|^2 \int_0^\infty e^{-2z/\delta_2} dz$$

$$= \frac{1}{4}|\mathcal{T}E_{i0}|^2 \sigma_2\delta_2$$

Thus, we find that, as expected, the total power entering the conductor through a unit area on its surface is equal to the total power dissipated in the volume of the conductor behind the unit area.

We can express P_{loss} in a more useful form by replacing $|\mathcal{T}E_{i0}|$ by $|J_s\gamma_2/\sigma_2|$ using (9.16). We find

$$P_{\mathrm{loss}} = \frac{1}{4}|J_s\gamma_2|^2\frac{\delta_2}{\sigma_2} = \frac{1}{2}|J_s|^2 R_s \tag{9.17}$$

which underscores the term *surface resistance* for R_s.

In practice, an approximate method is generally used to evaluate power loss per unit area in conducting walls of waveguides and coaxial lines. The electric and magnetic field configurations are found using the assumption that the conductors are perfect (i.e., $\sigma_2 = \infty$). The surface current density is then determined from the boundary condition (6.47)

$$\mathbf{J}_s = \hat{\mathbf{n}} \times \mathbf{H} \quad \rightarrow \quad J_s = H_t$$

where $\hat{\mathbf{n}}$ is the outward normal to the conductor surface and H_t is the tangential field at the surface, evaluated for $\sigma_2 = \infty$. Once $\mathbf{J}_s$ is found, we can use the surface resistance $R_s = (\sigma_2 \delta_2)^{-1}$ to calculate the power loss per unit area using (9.17).

Note that the value of H_t calculated assuming $\sigma_2 = \infty$ is a very good approximation of the actual field, since $\eta_1 \gg |Z_s|$. In the case of infinite conductivity, the tangential electric field at the conductor surface is zero. However, for finite σ_2 there has to be a finite value of tangential electric field to support a component of the Poynting vector directed into the conductor. It can be shown that this tangential electric field at the surface is $\mathbf{E}_t = \mathbf{J}_s Z_s$ and is thus normally quite small, since Z_s is small.

Example 9.8: Air–copper interface. Consider a uniform plane wave propagating in air incident normally on a large copper block. Find the percentage time-average power absorbed by the copper block at 1, 10, and 100 MHz and at 1 GHz.

Solution: The amount of time-average power absorbed by the copper block is the power transmitted into medium 2, which is equal to the Poynting flux moving in the $+z$ direction in medium 1:

$$(\mathbf{S}_{\text{av}})_1 = \hat{\mathbf{z}} \frac{E_{i0}^2}{2\eta_1}(1 - |\Gamma|^2)$$

as derived in Section 9.2.3. Since the power of the incident wave is $E_{i0}^2/(2\eta_1)$, the fraction of the incident power transmitted into medium 2 is $(1 - |\Gamma|^2)$. Thus, we need to determine the reflection coefficient Γ. The intrinsic impedance of copper is given by

$$\eta_c = \sqrt{\frac{j\omega\mu_0}{\sigma_2}} = \sqrt{\frac{\omega\mu_0}{\sigma_2}}e^{j45°} = \frac{2\pi \times 10^{-7}\sqrt{f}}{\sqrt{5.8}}(1+j)$$

$$\simeq 2.61 \times 10^{-7}\sqrt{f}(1+j)\,\Omega$$

where we have used $\mu_2 = \mu_0$ and $\sigma_2 = 5.8 \times 10^7$ S-m^{-1} for copper and $\sqrt{j} = e^{j\pi/4}$. The reflection coefficient at the air–copper interface is

$$\Gamma = \frac{\eta_c - \eta_0}{\eta_c + \eta_0} \simeq \frac{2.61 \times 10^{-7}\sqrt{f}(1+j) - 377}{2.61 \times 10^{-7}\sqrt{f}(1+j) + 377}$$

At $f = 1$ MHz, we have

$$\Gamma \simeq \frac{2.61 \times 10^{-7} \times 10^3(1+j) - 377}{2.61 \times 10^{-7} \times 10^3(1+j) + 377} \simeq 0.9999986 e^{j179.99992°}$$

so that the percentage of incident power absorbed by the copper block is

$$P_{\text{Cu}} \simeq (1 - |\Gamma|^2) \times 100 \simeq 0.000276\%$$

Similar calculations give the following results:

$$10 \text{ MHz} \quad \Gamma \simeq 0.9999956 e^{j\,179.99975°} \quad P_{\text{Cu}} \simeq 0.000873\%$$

$$100 \text{ MHz} \quad \Gamma \simeq 0.9999862 e^{j\,179.99921°} \quad P_{\text{Cu}} \simeq 0.00277\%$$

$$1000 \text{ MHz} \quad \Gamma \simeq 0.9999562 e^{j\,179.99749°} \quad P_{\text{Cu}} \simeq 0.00875\%$$

We see that the percentage of the incident power absorbed by the copper block increases with the frequency of the incident wave. However, note that as the frequency is increased, more of the total absorbed power is dissipated in a narrower region near the surface (i.e., the skin depth in copper, proportional to $f^{-1/2}$).

Example 9.9: Transmission through a metal foil: RF shielding. Consider an x-polarized uniform plane radio-frequency (RF) wave propagating in air, incident normally on a metal foil of thickness d at $z = 0$, as shown in Figure 9.16. (a) Find a relationship between the magnitude of the electric field of the transmitted wave (i.e., $E_{t0} = |\mathbf{E}_t|$) and that of the incident wave (i.e., $E_{i0} = |\mathbf{E}_i|$). Assume the metal foil to be thick enough that multiple reflections can be neglected. (b) Consider an ordinary aluminum foil, which is approximately 0.025 mm thick. If a 100-MHz plane wave is normally incident from one side of the foil, find the percentage of the incident power transmitted to the other side. For aluminum, take $\sigma_2 = 3.54 \times 10^7$ S-m^{-1}, $\epsilon_2 = \epsilon_0$, and $\mu_2 = \mu_0$.

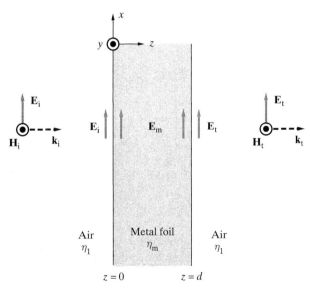

Figure 9.16   RF shielding. A metal foil of thickness d designed to shield RF energy. Note that typically we have $|\mathbf{E}_t| \ll |\mathbf{E}_i|$.

Solution:

(a) Neglecting multiple reflections allows us to treat each boundary of the metal foil as a single interface case. Thus, the amplitude of the wave transmitted into the metal foil due to the incident wave at the first boundary is given by

$$|\mathbf{E}_m(z = 0)| = \tau_1|\mathbf{E}_i(z = 0)|$$

where $\tau_1 = |\mathscr{T}_1|$ is the magnitude of the transmission coefficient at the air–metal interface. Note that from (9.3), we have

$$\mathscr{T}_1 = \frac{2\eta_m}{\eta_1 + \eta_m}$$

where $\eta_1 \simeq 377\Omega$ and $\eta_m \simeq \sqrt{j\omega\mu_0/\sigma_2} = \sqrt{\omega\mu_0/(2\sigma_2)}(1 + j)$. This transmitted wave attenuates exponentially as it propagates into the metal foil, so its amplitude at $z = d$ is

$$|\mathbf{E}_m(z = d)| = |\mathbf{E}_m(z = 0)|e^{-d/\delta}$$

where $\delta = (\pi f \mu_0\sigma_2)^{-1/2}$ is the skin depth in the metal. The attenuated transmitted wave incident on the metal–air boundary at $z = d$ transmits wave energy into medium 3 (air). The amplitude of the wave transmitted into medium 3 can be written as

$$|\mathbf{E}_t(z = d)| = \tau_2|\mathbf{E}_m(z = d)|$$

where τ_2 is the magnitude of the transmission coefficient $\mathscr{T}_2$ at the metal–air boundary, given by

$$\mathscr{T}_2 = \frac{2\eta_1}{\eta_m + \eta_1}$$

Therefore, the incident and transmitted electric field magnitudes are related by

$$|\mathbf{E}_t| = \tau_1 e^{-d/\delta}\tau_2|\mathbf{E}_i|$$

(b) We can first calculate the skin depth of aluminum at 100 MHz:

$$\delta = \frac{1}{\sqrt{\pi f \mu_0\sigma_2}} = \frac{1}{\sqrt{\pi(10^8)(4\pi \times 10^{-7})(3.54 \times 10^7)}}$$

$$\simeq 8.46 \times 10^{-6} \text{ m} = 8.46 \ \mu\text{m}$$

Noting that the thickness of the aluminum foil is $d = 0.025$ mm $\simeq 2.96\delta$, our assumption of neglecting multiple reflections is easily justified. In this connection, note that the reflected wave will be attenuated by an additional exponential factor of e^{-3} in the course of its propagation back through the metal to the first boundary. Thus, using the result found in (a) we have

$$\frac{|\mathbf{E}_t|}{|\mathbf{E}_i|} = \tau_1 e^{-d/\delta}\tau_2$$

where

$$\tau_1\tau_2 = \left|\frac{2\eta_m}{\eta_m + \eta_1}\right|\left|\frac{2\eta_1}{\eta_1 + \eta_m}\right| = \left|\frac{4\eta_m\eta_1}{(\eta_1 + \eta_m)^2}\right|$$

and $e^{-d/\delta} \simeq e^{-2.96} \simeq 0.0521$. Noting that $\eta_1 \simeq 377\Omega$ and

$$\eta_m = \sqrt{\frac{j\omega\mu_0}{\sigma_2}} \simeq \frac{2\pi \times 10^{-7}\sqrt{f}}{\sqrt{3.54}}(1+j) \simeq 3.34 \times 10^{-3}(1+j)\Omega$$

Since $\eta_1 \gg |\eta_m|$, we have

$$\tau_1\tau_2 \simeq \left|\frac{4\eta_1\eta_m}{\eta_1^2}\right| = \frac{4|\eta_m|}{\eta_1} \simeq \frac{4\sqrt{2}(3.34 \times 10^{-3})}{377} \simeq 5.01 \times 10^{-5}$$

so $|\mathbf{E_t}|/|\mathbf{E_i}| \simeq (0.0521)(5.01 \times 10^{-5}) \simeq 2.61 \times 10^{-6}$. Therefore, the percentage of the incident power that transmits to the other side of the foil is

$$\frac{|\mathbf{E_t}|^2/(2\eta_1)}{|\mathbf{E_i}|^2/(2\eta_1)} \times 100\% = \frac{|\mathbf{E_t}|^2}{|\mathbf{E_i}|^2} \times 100\% \simeq 6.8 \times 10^{-10}\%~!$$

Thus, the thin aluminum foil works quite well indeed as a shield for RF fields at 100 MHz.

9.4.2 Reflection from Multiple Lossy Interfaces

In the case of multiple interfaces involving conducting (lossy) media, the various expressions for the effective reflection and transmission coefficients that we derived in Section 9.3 apply if we substitute the complex propagation constant $\gamma_2 = \alpha_2 + j\beta_2$ instead of $j\beta_2$ and allow the various intrinsic impedances (η) to be complex. In other words, we have from (9.9) the effective reflection coefficient

$$\Gamma_{\text{eff}} = \rho e^{j\phi_\Gamma} = \frac{(\eta_2 - \eta_1)(\eta_3 + \eta_2) + (\eta_2 + \eta_1)(\eta_3 - \eta_2)e^{-2\gamma_2 d}}{(\eta_2 + \eta_1)(\eta_3 + \eta_2) + (\eta_2 - \eta_1)(\eta_3 - \eta_2)e^{-2\gamma_2 d}}$$

The effective transmission coefficient $\mathcal{T}_{\text{eff}}$ can also be determined in a similar manner from (9.10) as

$$\mathcal{T}_{\text{eff}} = \tau e^{j\phi_\mathcal{T}} = \frac{4\eta_2\eta_3 e^{-\gamma_2 d}}{(\eta_2 + \eta_1)(\eta_3 + \eta_2) + (\eta_2 - \eta_1)(\eta_3 - \eta_2)e^{-2\gamma_2 d}}$$

We note, however, that when the second medium is lossy, we have

$$1 - |\Gamma_{\text{eff}}|^2 \neq \frac{\eta_1}{|\eta_3|}|\mathcal{T}_{\text{eff}}|^2$$

since some of the incident power is absorbed in medium 2.

Reflection and penetration (transmission) of electromagnetic signals at biological tissue interfaces constitute an interesting and important application of multiple lossy interfaces and are briefly covered in Examples 9.10 through 9.12. In order to appreciate these applications better, it should be noted that the dielectric properties of different

TABLE 9.1 ϵ_r AND σ FOR BIOLOGICAL TISSUES

f (MHz)	Muscle, Skin, and Tissues with High Water Content		Fat, Bone, and Tissues with Low Water Content	
	$(\epsilon_m)_r$	σ_m (S-m^{-1})*	$(\epsilon_f)_r$	σ_f (mS-m^{-1})*
100	71.7	0.889	7.45	19.1–75.9
300	54	1.37	5.7	31.6–107
750	52	1.54	5.6	49.8–138
915	51	1.60	5.6	55.6–147
1,500	49	1.77	5.6	70.8–171
2,450	47	2.21	5.5	96.4–213
5,000	44	3.92	5.5	162–309
10,000	39.9	10.3	4.5	324–549

*Note that σ_m is in S-m^{-1}, whereas σ_f is in mS-m^{-1}.

biological tissues are different and have been studied over a very broad frequency range. The dielectric constant of tissues decreases gradually over many orders of magnitude as the frequency is varied from a few Hz to tens of GHz. The effective conductivity, on the other hand, rises with frequency, initially very slowly, and then more rapidly above 1 GHz. The permittivity of tissues depends on the tissue type and widely varies between different tissues. Tissues of higher water content, such as muscle, brain, kidney, heart, liver, and pancreas, have larger dielectric constant and conductivity than low-water-content tissues such as bone, fat, and lung. Table 9.1 lists the relative dielectric constant and conductivity of biological tissues with high water content versus those with low water content at discrete source frequencies over the radio-frequency spectrum.[13]

Example 9.10: Air–muscle interface. Consider a planar interface between air and muscle tissue. If a plane wave is normally incident at this boundary, find the percentage of incident power absorbed by the muscle tissue at (a) 100 MHz, (b) 300 MHz, (c) 915 MHz, and (d) 2.45 GHz. Assume $\mu_r = 1$.

Solution:

(a) The reflection coefficient is given by (9.2) as

$$\Gamma = \rho e^{j\phi} = \frac{\eta_m - \eta_1}{\eta_m + \eta_1}$$

where $\eta_1 \simeq 377\Omega$ and the intrinsic impedance of muscle tissue η_m is given by (8.26). So, at 100 MHz, the loss tangent of muscle is given by

$$\tan \delta_m = \frac{\sigma_m}{\omega \epsilon_m} \simeq \frac{0.889 \text{ S-m}^{-1}}{2\pi \times 10^8 \text{ rad-s}^{-1} \times 71.7 \times 8.85 \times 10^{-12} \text{ F-m}^{-1}} \simeq 2.23$$

---

[13]C. C. Johnson and A. W. Guy, Nonionizing electromagnetic wave effects in biological materials and systems, *Proc. IEEE,* 60(6), pp. 692–718, June 1972.

and the intrinsic impedance is

$$\eta_m \simeq \frac{377/\sqrt{71.7}}{[1 + (2.23)^2]^{1/4}} e^{j\,(1/2)\,\tan^{-1}(2.23)} \simeq 28.5 e^{j\,32.9°}\,\Omega$$

Substituting, we find

$$\Gamma \simeq \frac{28.5 e^{j\,32.9°} - 377}{28.5 e^{j\,32.9°} + 377} \simeq 0.881 e^{j\,175°}$$

So the percentage of the time-average incident power absorbed by the muscle tissue can be calculated as

$$\frac{|(\mathbf{S}_{av})_t|}{|(\mathbf{S}_{av})_i|} \times 100\% = (1 - \rho^2) \times 100\% \simeq [1 - (0.881)^2] \times 100\% \simeq 22.4\%$$

(b) Similarly, at 300 MHz, we have $\tan \delta_m \simeq 1.52$, $\eta_m \simeq 38 e^{j\,28.3°}\,\Omega$, and the reflection coefficient is

$$\Gamma \simeq \frac{38 e^{j\,28.3°} - 377}{38 e^{j\,28.3°} + 377} \simeq 0.837 e^{j\,174°}$$

so that the percentage absorbed power is $\sim[1 - (0.837)^2] \times 100 \simeq 29.9\%$.

(c) At 915 MHz, $\tan \delta_m \simeq 0.617$, $\eta_m \simeq 48.7 e^{j\,15.8°}\,\Omega$, and

$$\Gamma \simeq \frac{48.7 e^{j\,15.8°} - 377}{48.7 e^{j\,15.8°} + 377} \simeq 0.779 e^{j\,176°}$$

so that the percentage absorbed power is $\sim[1 - (0.779)^2] \times 100 \simeq 39.3\%$.

(d) At 2.45 GHz, $\tan \delta_m \simeq 0.345$, $\eta_m \simeq 53.5 e^{j\,9.52°}\,\Omega$, and

$$\Gamma \simeq \frac{53.5 e^{j\,9.52°} - 377}{53.5 e^{j\,9.52°} + 377} \simeq 0.755 e^{j\,177°}$$

So the percentage of incident power absorbed is

$$(1 - \rho^2) \times 100 \simeq [1 - (0.755)^2] \times 100 \simeq 43\%$$

Example 9.11: Muscle–fat interface. Consider a planar interface between muscle and fat tissues. If a 1 mW-(cm)$^{-2}$ plane wave in muscle is normally incident at this boundary at 2.45 GHz, find the time-average power density of the wave transmitted into the fat tissue. For fat tissue, assume $\mu_r = 1$ and take $\sigma_f = 155$ mS-m^{-1}.

Solution: At 2.45 GHz, the intrinsic impedances of muscle and fat tissues, η_m and η_f, are given by

$$\eta_m \simeq 53.5 e^{j\,9.52°}\,\Omega \text{ (from previous example)}$$

$$\eta_f \simeq \frac{377/\sqrt{5.5}}{[1 + (0.207)^2]^{1/4}} \exp\left[j\,\frac{1}{2}\tan^{-1}(0.207)\right] \simeq 159 e^{j\,5.84°}\,\Omega$$

The reflection coefficient can be calculated as

$$\Gamma = \frac{\eta_f - \eta_m}{\eta_f + \eta_m} \simeq \frac{159e^{j5.84°} - 53.5e^{j9.52°}}{159e^{j5.84°} + 53.5e^{j9.52°}}$$

$$\simeq \frac{158 + j16.2 - 52.7 - j8.84}{158 + j16.2 + 52.7 + j8.84} \simeq 0.498e^{-j2.78°}$$

Therefore, the time-average power density transmitted into the fat tissue is given by

$$|(\mathbf{S}_{av})_t| = (1 - \rho^2)|(\mathbf{S}_{av})_i| \simeq [1 - (0.498)^2](1 \text{ mW-(cm)}^{-2}) \simeq 0.752 \text{ mW-(cm)}^{-2}$$

Example 9.12: Microwave treatment of hypothermia in newborn piglets. Newly born piglets are very vulnerable to cold temperatures, and many of them die because of hypothermia. At the moment, hypothermia is treated by placing the piglets under infrared lamps, which are not very effective and are very costly. It has been proposed[14] that microwaves can be used to treat hypothermia. Compared with the infrared lamp, a microwave heater is more expensive to build but is more effective and consumes less power.

Consider a plane wave normally incident at the surface of the body of a pig (which to first order can be assumed to be a plane boundary). The body of the pig can be approximately modeled as a layer of fat tissue of a certain thickness followed by muscle tissue (which is assumed to be infinite in extent), as shown in Figure 9.17. For a newly born piglet, the fat layer is so thin (less than a mm) that it can be neglected, so the problem reduces to that of a single boundary. However, for a developed pig, the thickness of the fat layer can vary anywhere from 2 to 5 cm (taken below to be 4 cm for a mature pig), so it must be taken into account. Calculate the percentage of the incident microwave power reflected back into air, the percentage of power dissipated in the fat layer, the percentage of power transmitted into the muscle tissue, and the depth of penetration into the muscle tissue of the newly born piglet at (a) 915 MHz and (b) 2.45 GHz. (c) Repeat (a) and (b) for a mature pig. Use the same parameters for tissues as given in Table 9.1.

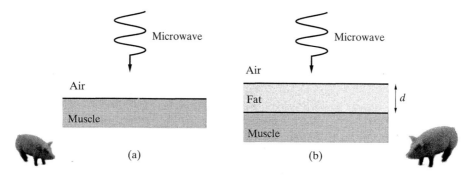

Figure 9.17   Microwave warming of pigs. (a) Air–piglet interface, modeled as a single air–muscle interface. (b) Air–adult pig interface, modeled as an air–fat–muscle interface.

[14]M. Allen, Pinky perks up when popped in the microwave, *New Scientist,* p. 19, January 23, 1993.

Solution:

(a) For the newly born piglet, using the results of Example 9.10, for 915 MHz, the percentage of incident power reflected is

$$\frac{|(S_{av})_r|}{|(S_{av})_i|} \times 100 = \rho^2 \times 100 \simeq (0.779)^2 \times 100 \simeq 60.7\%$$

and the percentage of incident power transmitted is

$$\frac{|(S_{av})_t|}{|(S_{av})_i|} \times 100 = (1 - \rho^2) \times 100 \simeq 39.3\%$$

No significant power is dissipated in the fat layer, since its thickness is negligible. The depth of penetration into the muscle layer at 915 MHz can be calculated from (8.34) as

$$d = \frac{c}{\omega} \left[\frac{2}{\epsilon_r(\sqrt{1 + \tan^2 \delta} - 1)} \right]^{1/2}$$

$$\simeq \frac{(3 \times 10^{10} \text{ cm-s}^{-1})}{2\pi(915 \times 10^6 \text{ Hz})} \left[\frac{2}{51(\sqrt{1 + (0.617)^2} - 1)} \right]^{1/2} \simeq 2.47 \text{ cm}$$

(b) Similarly, for 2.45 GHz, the percentage of the incident power reflected is $(0.755)^2 \times 100 \simeq 57\%$, and the percentage of incident power transmitted is $(1 - 0.755^2) \times 100 \simeq 43\%$. The depth of penetration at 2.45 GHz is

$$d \simeq \frac{(3 \times 10^{10} \text{ cm-s}^{-1})}{2\pi(2.45 \times 10^9 \text{ Hz})} \left[\frac{2}{47(\sqrt{1 + (0.345)^2} - 1)} \right]^{1/2} \simeq 1.67 \text{ cm}$$

As expected, the depth of penetration at 2.45 GHz is less than that at 915 MHz. Thus, to treat hypothermia it may be more desirable to use 915 MHz rather than the more typical commercial microwave oven frequency of 2.45 GHz. At the lower frequency the microwave energy can penetrate deeper into the piglet's body and thus provide more effective heating.

(c) For the mature pig we have a two-boundary problem. To calculate the percentages of incident power reflected back into air and transmitted into the muscle tissue, we need to use the effective reflection (Γ_{eff}) and transmission ($\mathcal{T}_{\text{eff}}$) coefficient expressions. First, we need to find the intrinsic impedances of fat and muscle tissues at 915 MHz and 2.45 GHz. At 915 MHz, for fat tissue (assume $\sigma_f = 0.1$ S-m^{-1}) we have

$$\tan \delta_f \simeq \frac{0.1 \text{ S-m}^{-1}}{(2\pi \times 915 \times 10^6 \text{ rad-s}^{-1})(5.6 \times 8.85 \times 10^{-12} \text{ F-m}^{-1})} \simeq 0.351$$

so that from (8.26), the intrinsic impedance is

$$\eta_f \simeq \frac{377/\sqrt{5.6}}{[1 + (0.351)^2]^{1/4}} \exp\left[j\frac{1}{2} \tan^{-1}(0.351) \right] \simeq 155 e^{j9.67°} \Omega$$

and the propagation constant is

$$\gamma_f = \alpha_f + j\beta_f$$

where from (8.21) and (8.22), we have

$$\begin{Bmatrix} \alpha_f \\ \beta_f \end{Bmatrix} \simeq \frac{2\pi \times 915 \times 10^6}{3 \times 10^8} \sqrt{\frac{5.6}{2}} [\sqrt{1 + (0.351)^2} \mp 1]^{1/2}$$

$$\simeq \begin{cases} 7.84 & \text{np-m}^{-1} \\ 46.0 & \text{rad-m}^{-1} \end{cases}$$

At 915 MHz, for muscle tissue, the depth of penetration is still 2.47 cm and we have from Example 9.10,

$$\eta_m \simeq 48.7 e^{j15.8°} \, \Omega$$

so the effective reflection and transmission coefficients are given by

$$\Gamma_{\text{eff}} = \frac{(\eta_f - \eta_1)(\eta_m + \eta_f) + (\eta_f + \eta_1)(\eta_m - \eta_f)e^{-2\gamma_f d}}{(\eta_f + \eta_1)(\eta_m + \eta_f) + (\eta_f - \eta_1)(\eta_m - \eta_f)e^{-2\gamma_f d}} \simeq 0.231 e^{-j159°}$$

and

$$\mathcal{T}_{\text{eff}} = \frac{4\eta_f \eta_m e^{-\gamma_f d}}{(\eta_f + \eta_1)(\eta_m + \eta_f) + (\eta_f - \eta_1)(\eta_m - \eta_f)e^{-2\gamma_f d}} \simeq 0.222 e^{-j99.2°}$$

Therefore, the percentages of the power reflected back to air and power transmitted into the muscle tissue are $\rho_{\text{eff}}^2 \times 100 \simeq (0.231)^2 \times 100 \simeq 5.33\%$ and $[\eta_1 \tau_{\text{eff}}^2 / |\eta_m|] \simeq [377(0.222)^2 / 48.7] \times 100 \simeq 38.2\%$, respectively, from which the percentage of power absorbed by the fat layer can be found as $\sim 56.4\%$. These results show that the mature pig absorbs, overall, a much higher percentage of the incident power than the newborn piglet does ($\sim 95\%$ versus $\sim 39\%$), because the fat layer acts as an impedance transformer and reduces the amount of reflections significantly. The calculations at 2.45 GHz are left as an exercise for the reader.

9.5 OBLIQUE INCIDENCE UPON A PERFECT CONDUCTOR

Up to now, we have exclusively considered reflection and transmission of uniform plane waves from interfaces upon which they are *normally* incident. Many interesting applications, especially in optics but also at microwave and radio frequencies, involve oblique incidence of uniform plane waves upon planar interfaces. The optical mirror arrangements in a camera involve reflections from slanted surfaces, and radio wave propagation at distances beyond line of sight involve oblique reflections from the ionized region (see Section 11.1) of the Earth's upper atmosphere known as the ionosphere. We have all observed the reflection of ocean waves from a beachfront upon which they are obliquely incident; careful observation of a particular isolated incident wavefront under relatively calm conditions clearly reveals the production of a reflected wave, propagating away at an angle from the normal that is the mirror image of that of the incident wave, and a "surface" wave, propagating rapidly along the beach. The reflection of electromagnetic waves occurs in a very similar manner, except that such waves are somewhat more complicated than water waves, because they are characterized by two vectors[15] (**E** and **H**) perpendicular to the direction of propagation.

We consider a uniform plane wave propagating in a lossless dielectric, obliquely incident upon a perfect conductor occupying the $z > 0$ half-space, with the interface between the two media being the entire x-y plane. As can be seen from Figure 9.18, it is necessary to distinguish between two distinctly different cases:

1. *Perpendicular polarization,*[16] in which the wave electric field is perpendicular to the plane of incidence, defined by the normal to the interface and the $\mathbf{k} = \beta\hat{\mathbf{k}}$ vector in the direction of the propagation (i.e., the x-z plane in Figure 9.18). The general geometry of oblique incidence and the identification of the plane of incidence and the interface plane are shown in Figure 9.18.

2. *Parallel polarization,* in which the electric field of the incident wave is parallel to the plane of incidence.

That the two cases lead to different results is evident when one considers the physical boundary conditions. For example, the continuity of the electric field at the boundary affects the entire electric field vector in the case of perpendicular polarization, whereas only one component of the wave electric field (tangential component) is affected in the parallel polarization case. In terms of the boundary condition on the tangential magnetic field, the surface currents[17] flowing on the conductor surface are in the x or y directions, respectively, depending on whether the incident wave is perpendicularly or parallel polarized. We first consider the perpendicular polarization case.

9.5.1 Perpendicular Polarization

We first consider the perpendicular polarization case as depicted in Figure 9.18a. In the context of the coordinate system shown in Figure 9.19, the incident plane wave can be described in phasor form as

$$\mathbf{E}_i(\mathbf{r}) = \hat{\mathbf{y}} E_{i0} e^{-j\beta_1 \hat{\mathbf{k}}_i \cdot \mathbf{r}}$$

---

[15]Rather than a single vector (i.e., velocity) in the direction of propagation and a scalar (i.e., density or pressure), as is the case for acoustic waves in a fluid.

[16]This terminology may be somewhat confusing until one realizes that the "perpendicular" or "parallel" labels are with respect to the plane of incidence and not the interface plane (Figure 9.18). Other ways of distinguishing between the two different cases can be encountered in various texts, including TE versus TM, horizontal versus vertical, E- versus H-, or s- versus p-waves. We believe that the perpendicular versus parallel polarization terminology is quite appropriate, since it actually refers to the polarization of the wave by identifying the orientation of the wave electric field, in accordance with the conventional understanding of wave polarization as discussed in Chapter 8. In this context, and taking, for example, the plane of incidence as the horizontal ground, perpendicular polarization is akin to a vertically polarized wave, whereas parallel polarization corresponds to a horizontally polarized one.

[17]Note that $\mathbf{J}_s = \hat{\mathbf{n}} \times \mathbf{H}$, so that the surface current must be orthogonal to the tangential component of $\mathbf{H}$ (i.e., H_t) *and* the outward normal to the interface, that is, $\hat{\mathbf{n}}$.

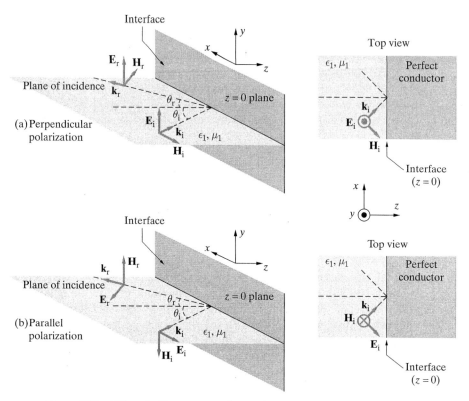

Figure 9.18   Oblique incidence on a perfect conductor. (a) Plane wave impinging on a perfect conductor with the electric field vector perpendicular to the plane of incidence (perpendicular polarization). (b) Electric field vector parallel to the plane of incidence (parallel polarization).

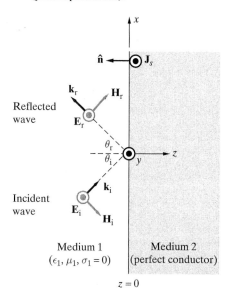

Figure 9.19   Perpendicularly polarized wave at a perfect conductor boundary. Uniform plane wave having perpendicular polarization obliquely incident on a perfectly conducting boundary.

where $\hat{\mathbf{k}}_i = \hat{\mathbf{x}} \sin \theta_i + \hat{\mathbf{z}} \cos \theta_i$ is the unit vector in the propagation direction of the incident wave, as shown in Figure 9.19, and $\mathbf{r}$ is the position vector $\mathbf{r} = \hat{\mathbf{x}}x + \hat{\mathbf{y}}y + \hat{\mathbf{z}}z$. Thus we have

$$\mathbf{E}_i(x, z) = \hat{\mathbf{y}} E_{i0} e^{-j\beta_1 (x \sin \theta_i + z \cos \theta_i)} \tag{9.18}$$

and the wave magnetic field is

$$\mathbf{H}_i(x, z) = \frac{1}{\eta_1}[\hat{\mathbf{k}}_i \times \mathbf{E}_i] = \frac{1}{\eta_1}[\hat{\mathbf{x}} \sin \theta_i + \hat{\mathbf{z}} \cos \theta_i] \times [\hat{\mathbf{y}} E_{i0} e^{-j\beta_1 (\hat{\mathbf{x}} \sin \theta_i + \hat{\mathbf{z}} \cos \theta_i) \cdot (\hat{\mathbf{x}}x + \hat{\mathbf{y}}y + \hat{\mathbf{z}}z)}]$$

$$= \frac{E_{i0}}{\eta_1} \underbrace{(-\hat{\mathbf{x}} \cos \theta_i + \hat{\mathbf{z}} \sin \theta_i)}_{\hat{\mathbf{h}}_i} e^{-j\beta_1 (x \sin \theta_i + z \cos \theta_i)} \tag{9.19}$$

where $\hat{\mathbf{h}}_i = -\hat{\mathbf{x}} \cos \theta_i + \hat{\mathbf{z}} \sin \theta_i$ is the unit vector in the direction of the magnetic field vector of the incident wave. Note that the orientation of $\mathbf{H}_i$ makes sense in that $\mathbf{E}_i \times \mathbf{H}_i$ is in the $\hat{\mathbf{k}}_i$ direction. To see this consider

$$\hat{\mathbf{e}}_i \times \hat{\mathbf{h}}_i = \hat{\mathbf{y}} \times (-\hat{\mathbf{x}} \cos \theta_i + \hat{\mathbf{z}} \sin \theta_i) = \hat{\mathbf{z}} \cos \theta_i + \hat{\mathbf{x}} \sin \theta_i = \hat{\mathbf{k}}_i$$

where $\hat{\mathbf{e}}_i$ is the unit vector in the direction of the electric field vector of the incident wave.

The reflected wave can be similarly described, except that the propagation direction vector $\hat{\mathbf{k}}_r$ is given[18] by

$$\hat{\mathbf{k}}_r = \hat{\mathbf{x}} \sin \theta_r - \hat{\mathbf{z}} \cos \theta_r$$

Thus we have

$$\mathbf{E}_r(x, z) = \hat{\mathbf{y}} E_{r0} e^{-j\beta_1 \hat{\mathbf{k}}_r \cdot \mathbf{r}} = \hat{\mathbf{y}} E_{r0} e^{-j\beta_1 (x \sin \theta_r - z \cos \theta_r)} \tag{9.20}$$

and

$$\mathbf{H}_r(x, z) = \frac{1}{\eta_1}(\hat{\mathbf{k}}_r \times \mathbf{E}_r) = \frac{E_{r0}}{\eta_1}(+\hat{\mathbf{x}} \cos \theta_r + \hat{\mathbf{z}} \sin \theta_r) e^{-j\beta_1 (x \sin \theta_r - z \cos \theta_r)} \tag{9.21}$$

Note that the direction of the magnetic field vector $\hat{\mathbf{h}}_r = \hat{\mathbf{x}} \cos \theta_r + \hat{\mathbf{z}} \sin \theta_r$ in (9.21) can be deduced by inspection since $\mathbf{E}_r \times \mathbf{H}_r$ must be along $\hat{\mathbf{k}}_r$. In expressions (9.18) through (9.21), we have two unknowns E_{r0} and θ_r, which we wish to find in terms of E_{i0} and θ_i. At the boundary surface, $z = 0$, the total electric field must vanish, since the total

---

[18] Note that the propagation direction of the reflected wave must lie in the x-z plane, because the presence of a y component of $\hat{\mathbf{k}}_r$ would require (since $\hat{\mathbf{k}}_r \perp \mathbf{E}_r$ for a uniform plane wave) the electric field of the reflected wave to have a component at an angle to the plane of incidence, in which case the boundary condition at the interface cannot be satisfied (since the incident wave electric field is entirely perpendicular to the plane of the interface).

electric field is tangential to the conductor and the electric field inside the conductor is zero. We thus have

$$\mathbf{E}_1(x,0) = \mathbf{E}_i(x,0) + \mathbf{E}_r(x,0) = 0 \tag{9.22}$$

where $\mathbf{E}_1$ is the total electric field in medium 1. Substituting (9.18) and (9.20) in (9.22) we find

$$\mathbf{E}_1(x,0) = \hat{\mathbf{y}}[E_{i0}e^{-j\beta_1 x \sin\theta_i} + E_{r0}e^{-j\beta_1 x \sin\theta_r}] = 0 \tag{9.23}$$

which must hold true at *all* values of x along the interface, which implies that

$$E_{r0} = -E_{i0}; \qquad \theta_i = \theta_r \tag{9.24}$$

The equality of the incident and reflected angles is known as Snell's law of reflection. Using (9.24), we can write the total electric field phasor in medium 1 as

$$\mathbf{E}_1(x,z) = \hat{\mathbf{y}}E_{i0}(e^{-j\beta_1 z \cos\theta_i} - e^{+j\beta_1 z \cos\theta_i})e^{-j\beta_1 x \sin\theta_i}$$

$$= -\hat{\mathbf{y}}j2E_{i0}\sin(\beta_1 z \cos\theta_i)e^{-j\beta_1 x \sin\theta_i}$$

Similarly by substituting (9.24) in (9.19) and (9.21) we can find an expression for the total magnetic field phasor $\mathbf{H}_1(x,z)$ in medium 1. We have

$$\mathbf{H}_1(x,z) = -\frac{2E_{i0}}{\eta_1}[\hat{\mathbf{x}}\cos\theta_i\cos(\beta_1 z \cos\theta_i) + \hat{\mathbf{z}}j\sin\theta_i\sin(\beta_1 z \cos\theta_i)]e^{-j\beta_1 x \sin\theta_i}$$

Assuming E_{i0} to be real, the time-averaged Poynting vector associated with the total fields is given by

$$\mathbf{S}_{av} = \frac{1}{2}\mathscr{R}e\{\mathbf{E}_1(x,z) \times \mathbf{H}_1^*(x,z)\}$$

$$= \frac{1}{2}\mathscr{R}e\left\{-\hat{\mathbf{z}}j\frac{2E_{i0}^2}{\eta_1}\cos\theta_i\sin(2\beta_1 z \cos\theta_i) + \hat{\mathbf{x}}\frac{4E_{i0}^2}{\eta_1}\sin\theta_i\sin^2(\beta_1 z \cos\theta_i)\right\}$$

$$= \hat{\mathbf{x}}\frac{2E_{i0}^2}{\eta_1}\sin\theta_i\sin^2(\beta_1 z \cos\theta_i)$$

Note that the z component of the Poynting vector does not have a real part, because $E_{1y}(x,z)$ and $H_{1x}(x,z)$ are 90° out of phase in time. In other words, since $E_{1y}(x,z)$ is multiplied by j while $H_{1x}(x,z)$ is not, the $j = e^{j\pi/2}$ term appears as a $\pi/2$ phase factor in the expression for $\mathscr{E}_{1y}(x,z,t)$. This result is to be expected because no average power is absorbed by the perfect conductor. In the z direction, $E_{1y}(x,z)$ and $H_{1x}(x,z)$ thus represent standing-wave patterns described by $\sin(\beta_{1z} z \cos\theta_i)$ and $\cos(\beta_{1z} z \cos\theta_i)$, respectively, where $\beta_{1z} = \beta_1 \cos\theta_i$ is the projection on the z axis of the wave phase constant, as was discussed in Section 8.6. On the other hand, the x component of the Poynting vector is real and positive, indicating nonzero average power flow in the x direction, parallel to the boundary. Note that the magnitude of $\mathbf{S}_{av}$ varies as a function of z.

We also observe that E_{1y} and H_{1z} are *in phase* in both space and time and represent a propagating wave with a velocity[19]

$$v_{1x} = \frac{\omega}{\beta_{1x}} = \frac{\omega}{\beta_1 \sin \theta_i} = \frac{v_{p1}}{\sin \theta_i}$$

where $v_{p1} = \omega/\beta_1$ is the phase velocity of a uniform plane wave in medium 1 (i.e., the phase velocity of the incident wave along its own propagation direction). The x component of the Poynting vector (resulting from $\frac{1}{2}\mathfrak{R}e\{E_y H_z^*\}$) is real and positive, indicating nonzero time-average power flow in the x direction, parallel to the interface. The wavelength of the wave propagating in the x direction is

$$\lambda_{1x} = \frac{2\pi}{\beta_{1x}} = \frac{\lambda_1}{\sin \theta_i}$$

where $\lambda_1 = 2\pi/\beta_1$ is the wavelength of the incident wave. We note that the wave propagating in the x direction is a *nonuniform* plane wave, since its amplitude varies[20] with z.

Physically, we can think of the wave field in the vicinity of the conductor as a pure standing wave extending along $-z$ that "slides" bodily (as a whole) along x. That there should be power flow in the x direction is intuitively obvious, because the incident wave carries power in both the x and z directions.

We further note that the total electric field in medium 1 is identically zero at a discrete set of planes parallel to the x-y plane, namely at

$$z = -\frac{m\lambda_1}{2 \cos \theta_i} \qquad m = 1, 2, 3, \ldots$$

It is thus possible to place a thin conducting sheet at any one of these positions at which the electric field is nulled without affecting[21] the field pattern to the left of the

---

[19]Note that since $v_{p1} = c$ when medium 1 is free space, v_{1x} can be larger than c. As discussed in Section 8.6, this circumstance does not pose any particular dilemma, because no energy or information travels at a speed of v_{1x}.

[20]Note that, based on our discussions in Section 8.7, the wave is still a *plane* wave because the phase fronts are planar, that is, $x \sin \theta_i = $ constant, describes planes parallel to the y-z plane. However, the wave is *nonuniform*, because the wave electric field varies as a function of position (namely z) over any given plane of constant phase.

[21]The only requirement such a thin conducting sheet imposes is to have the tangential electric field to be zero at its two surfaces. Since the electric field is in the y direction and is thus tangential to the sheet, and since it is already zero because of the interference between the incident and reflected waves, the conducting sheet can be placed without influencing the field. Note, however, that the tangential component of the wave magnetic field is nonzero on the conducting sheet, so surface currents flow on the sheet. To the degree that the thin conducting sheet consists of perfectly conducting material, these surface currents flow in an infinitesimally thin (macroscopically speaking) layer and do not lead to any power loss. If, however, the conductor is imperfect, there are losses and the current is confined to a layer of thickness of the order of one skin depth, as discussed in Section 9.4.

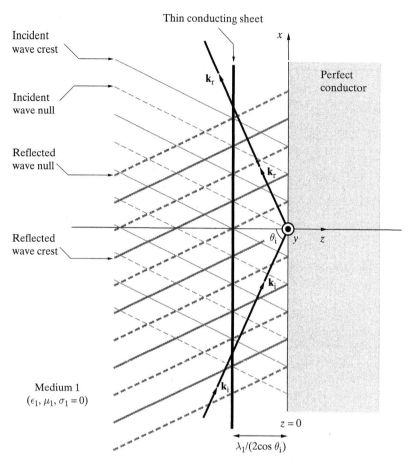

Figure 9.20 **Thin conducting sheet placed at** $z = -\lambda_1/(2\cos\theta_i)$. The interference between the incident and reflected wave fronts generates a null at $z = -\lambda_1/(2\cos\theta_i)$ and its other integer multiples. The crests (nulls) of the incident wave are shown with light solid (dashed) lines. The thicker solid (dashed) lines represent the crests (nulls) of the reflected wave.

sheet, as shown in Figure 9.20. Note that when such a thin conducting sheet is placed at a distance of $\lambda_1/(2\cos\theta_i)$ from the interface, the region between the sheet and the planar conducting boundary constitutes a parallel-plate waveguide. It appears, therefore, that the propagation of electromagnetic waves in such a guide may be interpreted as a superposition of multiple plane waves reflecting back and forth between the waveguide walls. We shall study parallel-plate waveguides extensively in Chapter 10, and refer back to Figure 9.20 when appropriate.

Current and charge induced in the conductor. As was the case for normal incidence of a uniform plane wave on a perfectly conducting boundary, the nonzero tangential component of the magnetic field (i.e., H_x) at $z = 0$ means that a current has to flow at the surface of the conductor to support this magnetic field. Furthermore, if the normal component of the wave electric field is nonzero at the surface of the conductor, surface charge has to be induced in accordance with the boundary condition $\rho_s = \hat{\mathbf{n}} \cdot (\epsilon \mathbf{E})$. The induced current can be estimated from the boundary condition $\mathbf{J}_s = \hat{\mathbf{n}} \times \mathbf{H}$, where $\hat{\mathbf{n}} = -\hat{\mathbf{z}}$ is the outward normal to the conductor surface. At $z = 0$ we have

$$\mathbf{H}_1(x, 0) = -\hat{\mathbf{x}}\frac{2E_{i0}}{\eta_1} \cos \theta_i e^{-j\beta_1 x \sin \theta_i}$$

so that

$$\mathbf{J}_s(x) = \hat{\mathbf{n}} \times \mathbf{H}_1(x, 0) = [(-\hat{\mathbf{z}}) \times (-\hat{\mathbf{x}})]\frac{2E_{i0}}{\eta_1} \cos \theta_i e^{-j\beta_1 x \sin \theta_i} = \hat{\mathbf{y}}\frac{2E_{i0}}{\eta_1} \cos \theta_i e^{-j\beta_1 x \sin \theta_i}$$

Note that $J_s(x)$ is a surface current density in units of A-m^{-1}. For this case of perpendicular polarization, the induced current flows in the direction perpendicular to the plane of incidence. The corresponding space–time expression for this surface current is

$$\overline{\mathcal{J}}_s(x, t) = \hat{\mathbf{y}}\frac{2E_{i0}}{\eta_1} \cos \theta_i \cos(\omega t - \beta_1 x \sin \theta_i) \tag{9.25}$$

where E_{i0} is assumed to be real. It was mentioned earlier that this induced current, which flows on the surface of the conductor, is necessary to support the wave; alternatively, we can think of the reflected wave as having been reradiated by this planar surface current, as was discussed in connection with the case of normal incidence on a perfect conductor. As we discussed in Section 9.4, if the conductor is imperfect, power is dissipated in the conductor by this current flow. We will see in Chapter 10 that these losses are an important consideration in the design of parallel-plate metallic waveguides.

    Because the wave electric field is entirely tangential to the conductor boundary (i.e., $E_z = 0$), no surface charge is induced in the conductor. This fact in turn means that the surface current must be divergence free,[22] which is true because $\overline{\mathcal{J}}$ has only a y component and does not depend on y.

9.5.2 Parallel Polarization

Now consider the case where the electric field vector $\mathbf{E}_i$ of the incident wave lies in the plane of incidence as shown in Figure 9.21 and in Figure 9.18b. The electric field of the incident wave is given by

$$\mathbf{E}_i(\mathbf{r}) = E_{i0}[\hat{\mathbf{x}} \cos \theta_i - \hat{\mathbf{z}} \sin \theta_i]e^{-j\beta_1 \hat{\mathbf{k}}_i \cdot \mathbf{r}}$$

where, as before, $\hat{\mathbf{k}}_i = \hat{\mathbf{x}} \sin \theta_i + \hat{\mathbf{z}} \cos \theta_i$ is the unit vector in the propagation direction of the incident wave as shown and $\mathbf{r}$ is the position vector $\mathbf{r} = \hat{\mathbf{x}}x + \hat{\mathbf{y}}y + \hat{\mathbf{z}}z$.

---

[22]Since $\nabla \cdot \overline{\mathcal{J}} = -\partial \tilde{\rho}/\partial t$ implies that $\nabla \cdot \overline{\mathcal{J}}_s = -\partial \tilde{\rho}_s \partial t$, and with $\tilde{\rho}_s = 0$, we must have $\nabla \cdot \overline{\mathcal{J}}_s = 0$.

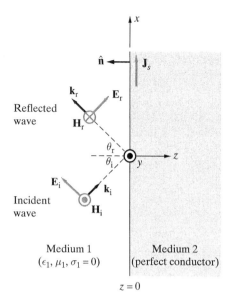

Reflected wave

Incident wave

Medium 1
$(\epsilon_1, \mu_1, \sigma_1 = 0)$

Medium 2
(perfect conductor)

$z = 0$

Figure 9.21   Parallel polarized wave at a perfectly conducting boundary. Uniform plane wave with parallel polarization obliquely incident on a perfectly conducting boundary.

We follow the same procedure as before; in this case $\mathbf{E}_i$ and $\mathbf{E}_r$ have components in the x and z directions, but $\mathbf{H}_i$ and $\mathbf{H}_r$ have only y components. Thus we have

$$\mathbf{E}_i(x, z) = E_{i0}[\hat{\mathbf{x}} \cos \theta_i - \hat{\mathbf{z}} \sin \theta_i] e^{-j\beta_1(x \sin \theta_i + z \cos \theta_i)}$$

$$\mathbf{H}_i(x, z) = \hat{\mathbf{y}} \frac{E_{i0}}{\eta_1} e^{-j\beta_1(x \sin \theta_i + z \cos \theta_i)}$$

Note that the orientations of $\mathbf{H}_i$ and $\mathbf{E}_i$ make sense in that $\mathbf{E}_i \times \mathbf{H}_i$ is in the $\hat{\mathbf{k}}_i$ direction. The reflected wave can be similarly described noting that

$$\hat{\mathbf{k}}_r = \hat{\mathbf{x}} \sin \theta_r - \hat{\mathbf{z}} \cos \theta_r$$

Thus we have

$$\mathbf{E}_r(x, z) = E_{r0}[\underbrace{\hat{\mathbf{x}} \cos \theta_r + \hat{\mathbf{z}} \sin \theta_r}_{\hat{\mathbf{e}}_r}] e^{-j\beta_1(x \sin \theta_r - z \cos \theta_r)}$$

$$\mathbf{H}_r(x, z) = -\hat{\mathbf{y}} \frac{E_{r0}}{\eta_1} e^{-j\beta_1(x \sin \theta_r - z \cos \theta_r)}$$

Note that the closer orientations for $\mathbf{H}_r$ and $\mathbf{E}_r$ make sense since $\hat{\mathbf{k}}_r \times \hat{\mathbf{e}}_r = -\hat{\mathbf{y}}$.

At the boundary surface, $z = 0$, the parallel component (i.e., x component) of the total electric field $\mathbf{E}_1(x, z)$ in medium 1 must vanish, since the field inside the conductor is zero. We thus have

$$E_{1x}(x, 0) = E_{ix}(x, 0) + E_{rx}(x, 0) = 0$$

$$(E_{i0} \cos \theta_i) e^{-j\beta_1 x \sin \theta_i} + (E_{r0} \cos \theta_r) e^{-j\beta_1 x \sin \theta_r} = 0$$

which requires $E_{r0} = -E_{i0}$ and $\theta_i = \theta_r$. The total electric field phasor in medium 1 can then be written as

$$
\begin{aligned}
\mathbf{E}_1(x,z) &= \mathbf{E}_i(x,z) + \mathbf{E}_r(x,z) \\
&= \hat{\mathbf{x}} E_{i0} \cos \theta_i (e^{-j\beta_1 z \cos \theta_i} - e^{+j\beta_1 z \cos \theta_i}) e^{-j\beta_1 x \sin \theta_i} \\
&\quad - \hat{\mathbf{z}} E_{i0} \sin \theta_i (e^{-j\beta_1 z \cos \theta_i} + e^{+j\beta_1 z \cos \theta_i}) e^{-j\beta_1 x \sin \theta_i} \\
&= -2E_{i0}[\hat{\mathbf{x}} j \cos \theta_i \sin(\beta_1 z \cos \theta_i) + \hat{\mathbf{z}} \sin \theta_i \cos(\beta_1 z \cos \theta_i)] e^{-j\beta_1 x \sin \theta_i}
\end{aligned}
$$

The associated total magnetic field in medium 1 is given by

$$
\begin{aligned}
\mathbf{H}_1(x,z) &= \mathbf{H}_i(x,z) + \mathbf{H}_r(x,z) \\
&= \hat{\mathbf{y}} \frac{2E_{i0}}{\eta_1} \cos(\beta_1 z \cos \theta_i) e^{-j\beta_1 x \sin \theta_i}
\end{aligned}
$$

We can make similar observations as for the case of perpendicular polarization. For example, a conducting plane can be placed at points $z = -m\lambda_1/(2\cos\theta_i)$ without affecting the fields to the left of the sheet. In the z direction, E_{1x} and H_{1y} maintain standing wave patterns with spacing of magnitude minima being $\lambda_{1z}/2$, where $\lambda_{1z} = \lambda_1/\cos\theta_i$. No time-average power is transmitted in the z direction, because E_{1x} and H_{1y} are $90°$ out of time phase. The time-average Poynting vector is given by (assuming E_{i0} is real)

$$
\mathbf{S}_{av} = \hat{\mathbf{x}} \frac{2E_{i0}^2}{\eta_1} \sin \theta_i \cos^2(\beta_1 z \cos \theta_i)
$$

In the direction along the boundary (i.e., x direction), E_{1z} and H_{1y} are in time and space phase and represent a propagating *nonuniform* plane wave with a phase velocity of $v_{1x} = v_{p1}/\sin\theta_i$.

Current and charge induced in the conductor.   As in the case of perpendicular polarization, we can find the current induced on the surface of the conductor using

$$
\mathbf{J}_s(x) = \hat{\mathbf{n}} \times \mathbf{H}_1(x,0)
$$

where $\hat{\mathbf{n}} = -\hat{\mathbf{z}}$ and

$$
\mathbf{H}_1(x,0) = \hat{\mathbf{y}} \frac{2E_{i0}}{\eta_1} e^{-j\beta_1 x \sin \theta_i}
$$

so that we find

$$
\mathbf{J}_s(x) = \hat{\mathbf{x}} \frac{2E_{i0}}{\eta_1} e^{-j\beta_1 x \sin \theta_i}
$$

which has dimensions of A·m^{-1}. In contrast to perpendicular polarization, the surface current for parallel polarization flows parallel to the plane of incidence. The space–time expression for the surface current is

$$\overline{\mathscr{J}}_s(x,t) = \hat{\mathbf{x}}\frac{2E_{i0}}{\eta_1}\cos(\omega t - \beta_1 x \sin\theta_i)$$

In the parallel polarization case, the wave electric field has a component that is oriented normally to the conductor surface (i.e., $E_z \neq 0$). Accordingly, through Gauss's law, or the boundary condition derived from it, we must have induced electric charge on the conductor surface given by

$$\rho_s(x) = \hat{\mathbf{n}} \cdot \mathbf{D} = \hat{\mathbf{n}} \cdot \epsilon_1 \mathbf{E} = -\epsilon_1 E_{1z}(x,0)$$

since $\hat{\mathbf{n}}$ is the outward normal from the conductor surface and is thus given by $\hat{\mathbf{n}} = -\hat{\mathbf{z}}$. We thus have

$$\tilde{\rho}_s(x,t) = 2\epsilon_1 E_{i0}\sin\theta_i \cos(\omega t - \beta_1 x \sin\theta_i)$$

We encourage the reader to check whether charge conservation is satisfied in the conductor, namely whether $\nabla \cdot \overline{\mathscr{J}}_s = -\partial\tilde{\rho}_s/\partial t$ holds.

Example 9.13: Unpolarized wave at a perfect conductor boundary. A 10 W-m^{-2}, 200 MHz uniform plane wave propagating in a nonmagnetic lossless medium with its electric field vector given by

$$\mathbf{E}_i(x,y) = (\hat{\mathbf{x}} - \hat{\mathbf{y}})\frac{E_0}{\sqrt{2}}e^{-j\sqrt{2}\pi(x+y)} + \hat{\mathbf{z}}jE_0 e^{-j\sqrt{2}\pi(x+y)} \quad \text{V-m}^{-1}$$

is obliquely incident upon a perfectly conducting surface located at the x-z plane, as shown in Figure 9.22a. Find (a) the angle of incidence θ_i and the relative dielectric constant of the lossless medium, (b) E_0, (c) the expression for $\mathbf{E}_r(x,y)$ of the reflected wave, and (d) the polarization of the incident and reflected waves.

Solution:

(a) From

$$\beta_1\hat{\mathbf{k}}_i \cdot \mathbf{r} = (\hat{\mathbf{x}}\beta_{1x} + \hat{\mathbf{y}}\beta_{1y} + \hat{\mathbf{z}}\beta_{1z}) \cdot (\hat{\mathbf{x}}x + \hat{\mathbf{y}}y + \hat{\mathbf{z}}z) = \sqrt{2}\pi(x+y)$$

we have

$$\beta_{1x} = \beta_1\sin\theta_i = \sqrt{2}\pi, \quad \beta_{1y} = \beta_1\cos\theta_i = \sqrt{2}\pi, \quad \beta_{1z} = 0$$

Taking the ratio β_{1x}/β_{1y} yields

$$\tan\theta_i = 1 \quad \rightarrow \quad \theta_i = 45°$$

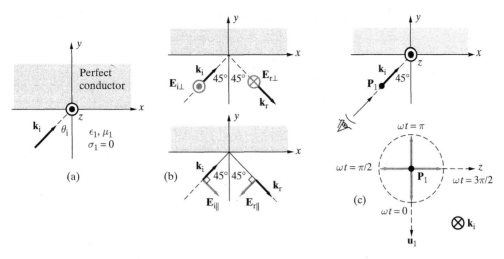

Figure 9.22   Unpolarized wave obliquely incident on a perfect conductor. (a) Incidence geometry. (b) Separation of the incident wave into parallel and perpendicularly polarized component waves. (c) The polarization of the field at point P_1 as viewed by an observer looking toward the propagation direction.

Substituting back,

$$\beta_1 \sin \theta_i = \beta_1 \sin 45° = \sqrt{2}\pi \to \beta_1 = \omega\sqrt{\mu_0\epsilon_1} = 2\pi$$

$$\frac{2\pi \times 2 \times 10^8}{3 \times 10^8}\sqrt{\epsilon_{1r}} \simeq 2\pi \to \epsilon_{1r} \simeq (1.5)^2 = 2.25$$

(b)

$$|\mathbf{S}_{av}| = \underbrace{\frac{(E_0/\sqrt{2})^2}{2\eta_1}}_{|E_x|^2/(2\eta_1)} + \underbrace{\frac{(E_0/\sqrt{2})^2}{2\eta_1}}_{|E_y|^2/(2\eta_1)} + \underbrace{\frac{E_0^2}{2\eta_1}}_{|E_z|^2/(2\eta_1)} = \frac{E_0^2}{\eta_1} = 10 \text{ W-m}^{-2}$$

$$E_0 \simeq \sqrt{\frac{10 \times 377}{\sqrt{2.25}}} \simeq 50.1 \text{ V-m}^{-1}$$

(c) Let us decompose the incident wave into its perpendicular and parallel components as

$$\mathbf{E}_{i\perp} = \hat{\mathbf{z}}jE_0e^{-j\sqrt{2}\pi(x+y)}$$

$$\mathbf{E}_{i\parallel} = (\hat{\mathbf{x}} - \hat{\mathbf{y}})\frac{E_0}{\sqrt{2}}e^{-j\sqrt{2}\pi(x+y)}$$

as shown in Figure 9.22b. Using the boundary condition (the tangential component of electric field at $y = 0$ must be continuous) we can write

$$\mathbf{E}_{r\perp} = -\hat{\mathbf{z}}jE_0e^{-j\sqrt{2}\pi(x-y)}$$

$$\mathbf{E}_{r\parallel} = (-\hat{\mathbf{x}} - \hat{\mathbf{y}})\frac{E_0}{\sqrt{2}}e^{-j\sqrt{2}\pi(x-y)}$$

or

$$\mathbf{E}_r(x,y) = \mathbf{E}_{r\perp} + \mathbf{E}_{r\parallel} = -(\hat{\mathbf{x}} + \hat{\mathbf{y}} + \hat{\mathbf{z}}j\sqrt{2})\frac{E_0}{\sqrt{2}}e^{-j\sqrt{2}\pi(x-y)}$$

(d) $\mathbf{E}_{i\perp}$ and $\mathbf{E}_{i\parallel}$ are both linearly polarized by themselves. $\overline{\mathscr{E}}_{i\perp}$ oscillates along the z direction as a function of time as

$$\overline{\mathscr{E}}_{i\perp}(x,y,z) = -\hat{\mathbf{z}}E_0 \sin[4\pi \times 10^8 t - \sqrt{2}\pi(x+y)]$$

$\overline{\mathscr{E}}_{i\parallel}$ oscillates along the $y = x$ direction (along a 45° line between the x and y axes) as a function of time as

$$\begin{aligned}
\overline{\mathscr{E}}_{i\parallel}(x,y,t) &= \hat{\mathbf{x}}\frac{E_0}{\sqrt{2}}\cos[4\pi \times 10^8 t - \sqrt{2}\pi(x+y)] \\
&\quad - \hat{\mathbf{y}}\frac{E_0}{\sqrt{2}}\cos[4\pi \times 10^8 t - \sqrt{2}\pi(x+y)] \\
&= \hat{\mathbf{u}}_1 E_0 \cos[4\pi \times 10^8 t - \sqrt{2}\pi(x+y)]
\end{aligned}$$

where $\hat{\mathbf{u}}_1$ is a unit vector given by

$$\hat{\mathbf{u}}_1 = \frac{1}{\sqrt{2}}(\hat{\mathbf{x}} - \hat{\mathbf{y}})$$

which is perpendicular to $\overline{\mathscr{E}}_{i\perp}(x,y,t)$. Since $\overline{\mathscr{E}}_{i\parallel}$ and $\overline{\mathscr{E}}_{i\perp}$ have the same magnitude but are out of phase by 90°, the total field $\overline{\mathscr{E}}_i = \overline{\mathscr{E}}_{i\parallel} + \overline{\mathscr{E}}_{i\perp}$ given by

$$\begin{aligned}
\overline{\mathscr{E}}_i(x,y,t) &= \hat{\mathbf{u}}_1 E_0 \cos[4\pi \times 10^8 t - \sqrt{2}\pi(x+y)] \\
&\quad - \hat{\mathbf{z}}E_0 \sin[4\pi \times 10^8 t - \sqrt{2}\pi(x+y)]
\end{aligned}$$

is circularly polarized. To find the sense of polarization, let us look at $\overline{\mathscr{E}}_i(x,y,t)$ at an arbitrary fixed point, such as $P_1(0,0,0)$ (i.e., the origin):

$$\overline{\mathscr{E}}_i(0,0,0) = \hat{\mathbf{u}}_1 E_0 \cos(4\pi \times 10^8 t) - \hat{\mathbf{z}}E_0 \sin(4\pi \times 10^8 t)$$

and examine its rotation as a function of time. The rotation of the electric field vector with respect to time is shown in Figure 9.22c as seen when viewed in the direction of propagation, which clearly indicates that this is a right-hand circularly polarized (RHCP) wave.

Similarly, the time-domain expression for the electric field of the reflected wave can be written as

$$\begin{aligned}
\overline{\mathscr{E}}_r(x,y,t) &= \hat{\mathbf{u}}_2 E_0 \cos[4\pi \times 10^8 t - \sqrt{2}\pi(x-y)] \\
&\quad + \hat{\mathbf{z}}E_0 \sin[4\pi \times 10^8 t - \sqrt{2}\pi(x-y)]
\end{aligned}$$

where $\hat{\mathbf{u}}_2 = \frac{1}{\sqrt{2}}(-\hat{\mathbf{x}} - \hat{\mathbf{y}})$. It can easily be shown that the reflected wave is left-hand circularly polarized (LHCP).

9.6 OBLIQUE INCIDENCE AT A DIELECTRIC BOUNDARY

We now analyze the more general case of oblique incidence of a uniform plane wave on an interface between two different lossless dielectric media. Many of our everyday optical experiences[23] involve refraction of light at dielectric interfaces, the most famous one being the rainbow. Reflection of waves from dielectric boundaries is also the basis of many useful applications, including polarizing filters for sunglasses, optical resonators, interferometers, and optical fibers.[24] Consider a uniform plane wave obliquely incident at the boundary between two lossless dielectrics, as shown in Figure 9.23. We rely on established concepts of uniform plane wave fronts (see Section 8.6) to derive the relationships between the incident, reflected, and refracted waves. It is useful to remember that all three waves considered are uniform plane waves, with phase surfaces of infinite

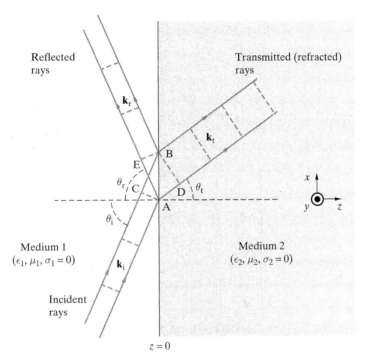

Figure 9.23 Oblique incidence at a dielectric boundary. Uniform plane wave obliquely incident at a dielectric boundary located at the $z = 0$ plane. Dashed lines AC, BD, and EB are the wavefronts (planar surfaces of constant phase).

---

[23]M. Minnaert, *The Nature of Light & Colour in the Open Air*, Dover Publications, Inc., New York, 1954.

[24]See, for example, H. A. Haus, *Waves and Fields in Optoelectronics*, Prentice-Hall, Inc., Englewood Cliffs, NJ, 1984.

extent in the direction orthogonal to their propagation direction. If, for example, the wave fronts shown in Figure 9.23 (i.e., the dashed lines AC, EB, and BD) are taken to represent the position of the wave crests, the successive crests are separated by the wavelength in each medium, being λ_1 in medium 1 and λ_2 in medium 2. On a purely geometric basis, the advancing wave fronts of the three waves must be "in step" with one another[25] along the interface. In other words, the phase velocities of the three waves as measured along the x axis must have the same magnitude and sign.

Starting with the incident wave front (AC), the incident ray travels a distance CB in medium 1, the transmitted ray travels a distance AD, and the reflected wave travels a distance of AE to their respective phase fronts. Noting that the phase velocities of uniform plane waves in media 1 and 2 are, respectively, v_{p1} and v_{p2}, in order for each wave to have a uniform phase we must have

$$\frac{\text{CB}}{\text{AD}} = \frac{v_{p1}}{v_{p2}}$$

We note, however, that $\text{CB} = \text{AB} \sin \theta_i$ and $\text{AD} = \text{AB} \sin \theta_t$, so

$$\boxed{\frac{\sin \theta_i}{\sin \theta_t} = \frac{v_{p1}}{v_{p2}} = \sqrt{\frac{\epsilon_2 \mu_2}{\epsilon_1 \mu_1}}} \qquad (9.26)$$

which is the well-known relationship between the angle of incidence and angle of refraction, commonly referred to as *Snell's law of refraction*. Also, since $\text{CB} = \text{AE}$, we have

$$\underbrace{\text{AB} \sin \theta_i}_{\text{CB}} = \underbrace{\text{AB} \sin \theta_r}_{\text{AE}}$$

which results in *Snell's law of reflection*, given by

$$\sin \theta_i = \sin \theta_r \quad \longrightarrow \quad \boxed{\theta_i = \theta_r}$$

Equation (9.26) is often expressed in terms of refractive indices of the media, defined as $n \equiv c/v_p = \sqrt{(\mu\epsilon)/(\mu_0\epsilon_0)} = \sqrt{\mu_r \epsilon_r}$

$$\frac{\sin \theta_i}{\sin \theta_t} = \frac{n_2}{n_1} \qquad \text{or} \qquad n_1 \sin \theta_i = n_2 \sin \theta_t$$

or, just in terms of the relative permittivities ϵ_{1r} and ϵ_{2r} for nonmagnetic media ($\mu_1 = \mu_2 = \mu_0$),

$$\frac{\sin \theta_i}{\sin \theta_t} = \sqrt{\frac{\epsilon_{2r}}{\epsilon_{1r}}} \qquad \text{or} \qquad \sqrt{\epsilon_{1r}} \sin \theta_i = \sqrt{\epsilon_{2r}} \sin \theta_t \qquad (9.27)$$

---

[25]The "in step" terminology is adopted from R. B. Adler, L. J. Chu, and R. M. Fano, *Electromagnetic Energy Transmission and Radiation*, Chapter 7, Wiley, New York, 1960.

It is important to note that Snell's laws are independent of the polarization of the incident wave, because they are purely based on the three waves being "in step" with one another in terms of their planar phase surfaces. This is further illustrated in Figure 9.24. Furthermore, we note that the functional forms of the fields have not entered into our derivation, so the wave fields of the uniform plane wave do not need to be time-harmonic (i.e., sinusoidal) for Snell's laws to be valid. If the incident wave fields involve any arbitrary functional form, for example $f(t - \beta\hat{\mathbf{k}} \cdot \mathbf{r}) = f(t - \beta_x x - \beta_z z)$, both the refracted and the reflected waves must have the same functional form at the boundary in order for the boundary conditions at $z = 0$ to be satisfied at all values of x and y (i.e., all along the boundary).

To derive the reflection and transmission coefficients, we can follow a procedure similar to that in previous sections, which involves matching boundary conditions and solving for the various unknown field amplitudes. However, for the case in hand, we can derive the reflection and transmission coefficients more easily by considering the conservation of energy flow. For each of the three waves (incident, reflected, and transmitted), the power transmitted per square meter is equal to $E^2/(2\eta)$ and is in the direction of propagation [i.e., $\mathbf{S}_{\mathrm{av}} = \hat{\mathbf{k}}E^2/(2\eta)$], where $\eta = \sqrt{\mu/\epsilon}$. Assuming E_{i0} to be real, we can

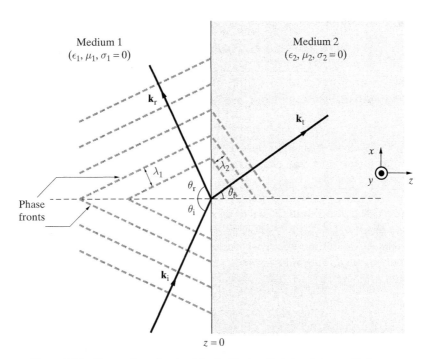

Figure 9.24 Planar phase fronts of the incident, reflected and transmitted waves. Illustration of the 'in step' condition for the planar phase fronts of the three waves. Note that Snell's law, derived from this condition, holds for any polarization of the incident wave.

consider the time-average power densities striking the surface AB as shown in Figure 9.23 carried by each of the waves as follows:

Incident wave: $\quad |\mathbf{S}_{\text{av}}|_i \cos \theta_i = \dfrac{1}{2\eta_1} E_{i0}^2 \cos \theta_i$

Reflected wave: $\quad |\mathbf{S}_{\text{av}}|_r \cos \theta_r = \dfrac{1}{2\eta_1} E_{r0}^2 \cos \theta_r$

Transmitted wave: $\quad |\mathbf{S}_{\text{av}}|_t \cos \theta_t = \dfrac{1}{2\eta_2} E_{t0}^2 \cos \theta_t$

where $\eta_1 = \sqrt{\mu_1/\epsilon_1}$ and $\eta_2 = \sqrt{\mu_2/\epsilon_2}$ are the intrinsic impedances of medium 1 and 2, respectively. Note that each of these can be obtained by taking the projection of the respective Poynting vectors on the area of the surface AB in Figure 9.23. For example, for the incident wave, we have

$$P_{\text{AB}} = (\mathbf{S}_{\text{av}})_i \cdot \hat{\mathbf{z}} = \frac{E_{i0}^2}{2\eta_1} \hat{\mathbf{k}}_i \cdot \hat{\mathbf{z}} = \frac{E_{i0}^2}{2\eta_1} \cos \theta_i$$

On the basis of conservation of power we must have[26]

$$|\mathbf{S}_{\text{av}}|_i \cos \theta_i = |\mathbf{S}_{\text{av}}|_r \cos \theta_r + |\mathbf{S}_{\text{av}}|_t \cos \theta_t$$

$$\rightarrow \quad \frac{1}{2\eta_1} E_{i0}^2 \cos \theta_i = \frac{1}{2\eta_1} E_{r0}^2 \cos \theta_r + \frac{1}{2\eta_2} E_{t0}^2 \cos \theta_t \tag{9.28}$$

or

$$\frac{E_{r0}^2}{E_{i0}^2} = 1 - \frac{\eta_1 E_{t0}^2 \cos \theta_t}{\eta_2 E_{i0}^2 \cos \theta_i} \tag{9.29}$$

We now separately consider the perpendicular and parallel polarizations. It is clear from Figure 9.25 that the boundary conditions will differently affect the electric and magnetic fields in the two cases.

9.6.1 Perpendicular Polarization

The expressions for the wave electric and magnetic field phasors of the incident, reflected, and refracted (transmitted) waves shown in Figure 9.26 can be expressed as

$$\mathbf{E}_i(x, z) = \hat{\mathbf{y}} E_{i0} e^{-j\beta_1 (x \sin \theta_i + z \cos \theta_i)}$$

$$\mathbf{H}_i(x, z) = \frac{E_{i0}}{\eta_1} (-\hat{\mathbf{x}} \cos \theta_i + \hat{\mathbf{z}} \sin \theta_i) e^{-j\beta_1 (x \sin \theta_i + z \cos \theta_i)}$$

$$\mathbf{E}_r(x, z) = \hat{\mathbf{y}} E_{r0} e^{-j\beta_1 (x \sin \theta_r - z \cos \theta_r)}$$

---

[26]Note that this condition is required simply so that there is no indefinite accumulation of electromagnetic energy at the interface, which would otherwise occur under the assumed steady-state conditions.

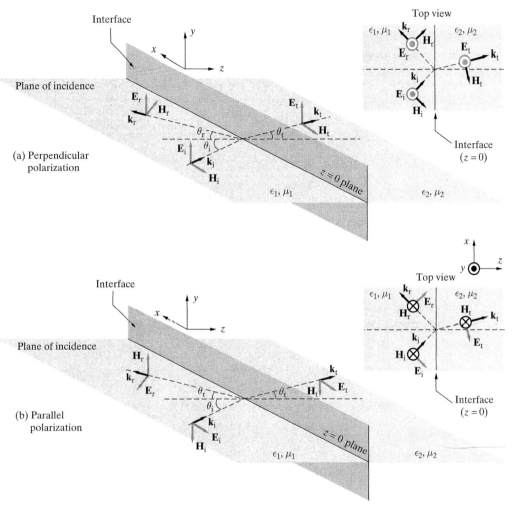

Figure 9.25 Oblique incidence on a dielectric boundary. (a) Plane wave impinging on a perfect dielectric with the wave electric field vector perpendicular to the plane of incidence (perpendicular polarization). (b) Wave electric field vector parallel to the plane of incidence (parallel polarization).

$$\mathbf{H}_r(x,z) = \frac{E_{r0}}{\eta_1}(\hat{\mathbf{x}}\cos\theta_r + \hat{\mathbf{z}}\sin\theta_r)e^{-j\beta_1(x\sin\theta_r - z\cos\theta_r)}$$

$$\mathbf{E}_t(x,z) = \hat{\mathbf{y}}E_{t0}e^{-j\beta_2(x\sin\theta_t + z\cos\theta_t)}$$

$$\mathbf{H}_t(x,z) = \frac{E_{t0}}{\eta_2}(-\hat{\mathbf{x}}\cos\theta_t + \hat{\mathbf{z}}\sin\theta_t)e^{-j\beta_2(x\sin\theta_t + z\cos\theta_t)}$$

To determine the amplitudes of the reflected and transmitted wave fields in terms of the incident field amplitude E_{i0}, we can apply the boundary condition concerning the

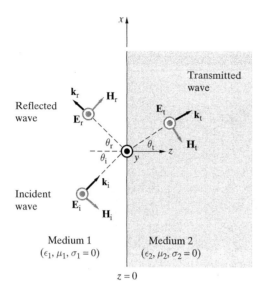

Figure 9.26 A perpendicularly polarized wave incident on a dielectric boundary. A perpendicularly polarized wave incident on a dielectric boundary at $z = 0$ at an incidence angle of θ_i.

continuity of the tangential component of the wave electric field across the interface. Considering the field orientations as defined in Figure 9.26, we have at $z = 0$,

$$E_{i0}e^{-j\beta_1 x \sin\theta_i} + E_{r0}e^{-j\beta_1 x \sin\theta_r} = E_{t0}e^{-j\beta_2 x \sin\theta_t}$$

Since this condition has to be satisfied at *all* values of x, all three exponents must be equal. Thus,

$$\beta_1 x \sin\theta_i = \beta_1 x \sin\theta_r = \beta_2 x \sin\theta_t$$

which leads to Snell's law,[27] as expressed in (9.26) and with $\theta_r = \theta_i$. We rewrite the boundary condition at any given value of x, say at $x = 0$:

$$E_{i0} + E_{r0} = E_{t0} \quad \longrightarrow \quad \frac{E_{t0}}{E_{i0}} = 1 + \frac{E_{r0}}{E_{i0}} \tag{9.30}$$

Substituting[28] (9.30) in (9.29) and manipulating to solve for E_{r0}/E_{i0} we find

$$\boxed{\Gamma_\perp \equiv \frac{E_{r0}}{E_{i0}} = \frac{\eta_2 \cos\theta_i - \eta_1 \cos\theta_t}{\eta_2 \cos\theta_i + \eta_1 \cos\theta_t}} \tag{9.31}$$

---

[27]Note that matching of the exponents is essentially equivalent to the phase-matching or "in-step" condition that we used on the basis of geometric arguments in deriving Snell's law using Figure 9.23.

[28]An alternative method would have been to use the second boundary condition, that is, the continuity of the tangential magnetic fields. However, since we have already derived (9.28) and (9.29) by matching the projections of the wave power densities at the boundary, we have in essence already utilized this second boundary condition. The same could be said about the boundary condition of the normal components of **D** and **B**; use of these conditions would not bring any additional information, because our analysis already yields complete information about the state of the reflected and refracted waves.

where $\Gamma_\perp$ is defined as the reflection coefficient for perpendicular polarization. Note that when the second medium is a perfect conductor, that is, $\eta_2 = 0$, (9.31) reduces to $\Gamma_\perp = -1$, consistent with (9.24). Note that although we assumed media 1 and 2 to be lossless dielectrics, the expression (9.31) is valid for lossy media also, as long as the proper complex values of η_1 and η_2 are used. Plots of the magnitude ($\rho_\perp = |\Gamma_\perp|$) of $\Gamma_\perp$ for selected cases are given in Figure 9.33 in Section 9.6.4.

For magnetically identical media ($\mu_1 = \mu_2$), and also using (9.27), we can find alternative expressions for $\Gamma_\perp$, such as

$$\Gamma_\perp \equiv \frac{E_{r0}}{E_{i0}} = \frac{\cos\theta_i - \sqrt{\epsilon_{2r}/\epsilon_{1r}}\cos\theta_t}{\cos\theta_i + \sqrt{\epsilon_{2r}/\epsilon_{1r}}\cos\theta_t} = \frac{\cos\theta_i - \sqrt{(\epsilon_{2r}/\epsilon_{1r}) - \sin^2\theta_i}}{\cos\theta_i + \sqrt{(\epsilon_{2r}/\epsilon_{1r}) - \sin^2\theta_i}}$$

that can be useful in some cases.

The transmission coefficient $\mathcal{T}_\perp$ can be found from (9.30) and (9.31) as

$$\boxed{\mathcal{T}_\perp \equiv \frac{E_{t0}}{E_{i0}} = \frac{2\eta_2\cos\theta_i}{\eta_2\cos\theta_i + \eta_1\cos\theta_t}} \quad \longrightarrow \quad \underbrace{\mathcal{T}_\perp = \frac{2\cos\theta_i}{\cos\theta_i + \sqrt{\epsilon_{2r}/\epsilon_{1r}}\cos\theta_t}}_{\mu_1 = \mu_2}$$

$$\longrightarrow \quad \mathcal{T}_\perp = \frac{2\cos\theta_i}{\cos\theta_i + \sqrt{(\epsilon_{2r}/\epsilon_{1r}) - \sin^2\theta_i}} \tag{9.32}$$

We note that

$$\boxed{1 + \Gamma_\perp = \mathcal{T}_\perp}$$

similar to the result obtained in Section 9.2.1 for normal incidence on a dielectric interface, where it was found that $1 + \Gamma = \mathcal{T}$.

Example 9.14: A perpendicularly polarized wave. A uniform plane wave traveling in air having an electric field given by

$$\mathbf{E}_i(y, z) = \hat{x}E_0 e^{-j\beta_1(y\cos\theta_i - z\sin\theta_i)}$$

is obliquely incident on the $y = 0$ interface between air and polystyrene, as shown in Figure 9.27. If the time-average power, the frequency, and the incidence angle of this wave are given to be 1.4 W-m^{-2}, 3 GHz, and $\sim 58°$, respectively, find (a) E_0 (assume real) and β_1, (b) $\mathbf{H}_i(y, z)$, (c) $\mathbf{E}_r(y, z)$ and $\mathbf{H}_r(y, z)$, and (d) the time-average power of the reflected and transmitted waves. For polystyrene, take $\sigma_2 = 0$, $\epsilon_{2r} \simeq 2.56$, $\mu_{2r} = 1$ at 3 GHz.

Solution:

(a) Using the time-average power of the incident wave, the amplitude of the electric field can be calculated as

$$|\mathbf{S}_{av}|_i = \frac{1}{2}\frac{E_0^2}{\eta_1} = 1.4 \text{ W-m}^{-2} \rightarrow E_0 \simeq \sqrt{2 \times 377 \times 1.4} \simeq 32.5 \text{ V-m}^{-1}$$

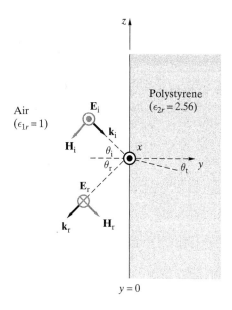

Figure 9.27 Air–polystyrene interface. Incident, reflected, and refracted waves for Example 9.14.

The phase constant is given by

$$\beta_1 = \omega\sqrt{\mu_1\epsilon_1} = \frac{\omega}{c} \simeq \frac{2\pi \times 3 \times 10^9}{3 \times 10^8} = 20\pi \quad \text{rad-m}^{-1}$$

(b) From Figure 9.27, the magnetic field of the incident wave can be found as

$$\mathbf{H}_i(y, z) = \frac{1}{\eta_1}\hat{\mathbf{k}}_i \times \mathbf{E}_i(y, z)$$

$$= \frac{1}{\eta_1}[\hat{\mathbf{y}}\cos\theta_i - \hat{\mathbf{z}}\sin\theta_i] \times [\hat{\mathbf{x}}E_0 e^{-j\beta_1(y\cos\theta_i - z\sin\theta_i)}]$$

$$= \frac{-E_0}{\eta_1}[\hat{\mathbf{z}}\cos\theta_i + \hat{\mathbf{y}}\sin\theta_i]\, e^{-j\beta_1(y\cos\theta_i - z\sin\theta_i)}$$

Substituting the numerical values, we have

$$\mathbf{H}_i(y, z) \simeq -86.2[\hat{\mathbf{y}}0.848 + \hat{\mathbf{z}}0.53]e^{-j20\pi(0.53y - 0.848z)} \quad \text{mA-m}^{-1}$$

(c) Using Snell's law of refraction as given by (9.27), we can find the transmission angle as

$$\sin 58° = \sqrt{2.56}\sin\theta_t \quad \rightarrow \quad \theta_t \simeq 32°$$

Since the incident wave is perpendicularly polarized, we can find the reflection coefficient using (9.31) as

$$\Gamma_\perp \simeq \frac{\cos 58° - \sqrt{2.56}\cos 32°}{\cos 58° + \sqrt{2.56}\cos 32°} \simeq -0.438$$

The electric field of the reflected wave can be written as

$$\mathbf{E}_r(y, z) = \hat{\mathbf{x}}\Gamma_\perp E_0 e^{j\beta_1(y\cos\theta_r + z\sin\theta_r)} \simeq -\hat{\mathbf{x}}14.2e^{j20\pi(0.53y + 0.848z)} \text{ V-m}^{-1}$$

The magnetic field of the reflected wave is given by

$$\mathbf{H}_r(y, z) = \frac{1}{\eta_1}\hat{\mathbf{k}}_r \times \mathbf{E}_r = \frac{\Gamma_\perp E_0}{\eta_1}[-\hat{\mathbf{y}}\sin\theta_r + \hat{\mathbf{z}}\cos\theta_r]e^{j\beta_1(y\cos\theta_r + z\sin\theta_r)}$$

$$\simeq -37.8[-\hat{\mathbf{y}}0.848 + \hat{\mathbf{z}}0.53]e^{j20\pi(0.53y + 0.848z)} \text{ mA-m}^{-1}$$

(d) The time-average power of the reflected wave is given by

$$|\mathbf{S}_{av}|_r = \frac{1}{2}\frac{(\Gamma_\perp E_0)^2}{\eta_1} \simeq \frac{1}{2}\frac{(14.2)^2}{377} \simeq 0.269 \text{ W-m}^{-2}$$

Using the principle of conservation of power (9.28), the time-average power of the transmitted wave can be found from

$$1.4\cos 58° \simeq 0.269\cos 58° + |\mathbf{S}_{av}|_t \cos 32°$$

yielding

$$|\mathbf{S}_{av}|_t \simeq 0.707 \text{ W-m}^{-2}$$

Note that one can also find this value using

$$|\mathbf{S}_{av}|_t = \frac{1}{2}\frac{(\mathcal{T}_\perp E_0)^2}{\eta_2}$$

where using (9.32),

$$\mathcal{T}_\perp \simeq 2\cos 58°/(\cos 58° + \sqrt{2.56}\cos 32°) \simeq 0.562$$

and $\eta_2 \simeq 377/\sqrt{2.56} \simeq 236\Omega$. Substituting,

$$|\mathbf{S}_{av}|_t \simeq \frac{1}{2}\frac{(0.562 \times 32.5)^2}{236} \simeq 0.707 \text{ W-m}^{-2}$$

Note that $|\mathbf{S}_{av}|_i \neq |\mathbf{S}_{av}|_r + |\mathbf{S}_{av}|_t$. Although this result may at first appear counterintuitive, there is in fact no reason to expect the sum of power densities of the reflected and transmitted waves to be equal to that of the incident wave. The Poynting flux of a wave is a directional quantity (i.e., a vector), and the summation of the magnitude of the vectors pointing in different directions has no physical significance. Projection of the three Poynting vectors on any area of the interface must satisfy the conservation of power, a fact used (9.28) in deriving the expressions for $\Gamma_\perp$ and $\Gamma_\parallel$.

9.6.2 Parallel Polarization

The expressions for the wave electric and magnetic field phasors of the incident, reflected, and refracted (transmitted) waves illustrated in Figure 9.28 are given by

$$\mathbf{E}_i(x, z) = E_{i0}(\hat{\mathbf{x}}\cos\theta_i - \hat{\mathbf{z}}\sin\theta_i)e^{-j\beta_1(x\sin\theta_i + z\cos\theta_i)}$$

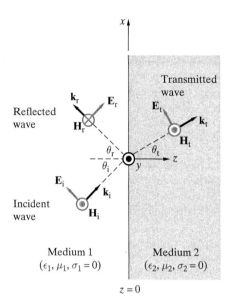

Figure 9.28 A parallel polarized wave at a dielectric boundary. A parallel polarized wave incident on a dielectric boundary at $z = 0$ at an incidence angle of θ_i.

$$\mathbf{H}_i(x, z) = \hat{\mathbf{y}}\frac{E_{i0}}{\eta_1}e^{-j\beta_1(x\sin\theta_i + z\cos\theta_i)}$$

$$\mathbf{E}_r(x, z) = E_{r0}(\hat{\mathbf{x}}\cos\theta_r + \hat{\mathbf{z}}\sin\theta_r)e^{-j\beta_1(x\sin\theta_r - z\cos\theta_r)}$$

$$\mathbf{H}_r(x, z) = -\hat{\mathbf{y}}\frac{E_{r0}}{\eta_1}e^{-j\beta_1(x\sin\theta_r - z\cos\theta_r)}$$

$$\mathbf{E}_t(x, z) = E_{t0}(\hat{\mathbf{x}}\cos\theta_t - \hat{\mathbf{z}}\sin\theta_t)e^{-j\beta_2(x\sin\theta_t + z\cos\theta_t)}$$

$$\mathbf{H}_t(x, z) = \hat{\mathbf{y}}\frac{E_{t0}}{\eta_2}e^{-j\beta_2(x\sin\theta_t + z\cos\theta_t)}$$

where $\eta_1 = \sqrt{\mu_1/\epsilon_1}$ and $\eta_2 = \sqrt{\mu_2/\epsilon_2}$ are, respectively, the intrinsic impedances of media 1 and 2.

We follow a procedure similar to that used for the perpendicular polarization case and set out to find the amplitudes of the reflected and transmitted waves in terms of E_{i0}. For this purpose, we apply the boundary condition concerning the continuity of the tangential component of the wave electric field across the interface. Therefore, at $z = 0$, we have

$$(E_{i0} + E_{r0})\cos\theta_i = E_{t0}\cos\theta_t$$

or

$$\frac{E_{t0}}{E_{i0}} = \left(1 + \frac{E_{r0}}{E_{i0}}\right)\frac{\cos\theta_i}{\cos\theta_t}$$

Substituting in (9.29) and manipulating to solve for E_{r0}/E_{i0} (by eliminating E_{t0}) we find

$$\Gamma_{\parallel} \equiv \frac{E_{r0}}{E_{i0}} = \frac{-\eta_1 \cos\theta_i + \eta_2 \cos\theta_t}{\eta_1 \cos\theta_i + \eta_2 \cos\theta_t} \quad \longrightarrow \quad \underbrace{\Gamma_{\parallel} = \frac{\cos\theta_t - \sqrt{\epsilon_2/\epsilon_1}\cos\theta_i}{\cos\theta_t + \sqrt{\epsilon_2/\epsilon_1}\cos\theta_i}}_{\mu_1=\mu_2}$$

$$\longrightarrow \quad \Gamma_{\parallel} = \frac{-\cos\theta_i + (\epsilon_{1r}/\epsilon_{2r})\sqrt{(\epsilon_{2r}/\epsilon_{1r}) - \sin^2\theta_i}}{\cos\theta_i + (\epsilon_{1r}/\epsilon_{2r})\sqrt{(\epsilon_{2r}/\epsilon_{1r}) - \sin^2\theta_i}} \tag{9.33}$$

as the reflection coefficient for parallel polarization and eliminating E_{r0} we find

$$\mathcal{T}_{\parallel} = \frac{E_{t0}}{E_{i0}} = \frac{2\eta_2\cos\theta_i}{\eta_1\cos\theta_i + \eta_2\cos\theta_t} \quad \longrightarrow \quad \underbrace{\mathcal{T}_{\parallel} = \frac{2\cos\theta_i}{\cos\theta_t + \sqrt{\epsilon_{2r}/\epsilon_{1r}}\cos\theta_i}}_{\mu_1=\mu_2}$$

$$\longrightarrow \quad \mathcal{T}_{\parallel} = \frac{2\sqrt{\epsilon_{1r}/\epsilon_{2r}}\cos\theta_i}{\cos\theta_i + \sqrt{\epsilon_{1r}/\epsilon_{2r}}\sqrt{1 - (\epsilon_{1r}/\epsilon_{2r})\sin^2\theta_i}} \tag{9.34}$$

as the transmission coefficient for parallel polarization. Plots of the magnitude $\rho_{\parallel}$ and the phase $\phi_{\parallel}$ of $\Gamma_{\parallel}$ (i.e., $\Gamma_{\parallel} = \rho_{\parallel}e^{j\phi_{\parallel}}$) are presented in Figure 9.33 in Section 9.6.4. From (9.33) and (9.34) we note that

$$1 + \Gamma_{\parallel} = \mathcal{T}_{\parallel}\left(\frac{\cos\theta_t}{\cos\theta_i}\right)$$

Example 9.15: Oblique incidence of light at air–glass interface. A light ray traveling in air is obliquely incident at an incidence angle of $40°$ on the air–glass ($n = 1.52$) plane boundary as shown in Figure 9.29. (a) Find the reflection and transmission coefficients for a perpendicularly polarized (s-polarized) light. (b) Find the reflection and transmission coefficients for a parallel polarized (p-polarized) light.

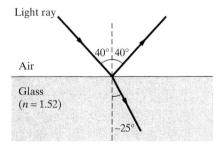

Figure 9.29   Air–glass boundary. Figure for Example 9.15.

Solution:

(a) For the perpendicularly polarized case, using (9.31) and (9.32) we have

$$\Gamma_\perp = \frac{\cos 40° - \sqrt{(1.52)^2 - \sin^2 40°}}{\cos 40° + \sqrt{(1.52)^2 - \sin^2 40°}} \simeq -0.285$$

$$\mathcal{T}_\perp = \frac{2\cos 40°}{\cos 40° + \sqrt{(1.52)^2 - \sin^2 40°}} \simeq 0.715$$

and $1 + \Gamma_\perp = \mathcal{T}_\perp$ as expected.

(b) For the parallel polarized case we have from (9.33) and (9.34),

$$\Gamma_\| = \frac{-\cos 40° + (1.52)^{-2}\sqrt{(1.52)^2 - \sin^2 40°}}{\cos 40° + (1.52)^{-2}\sqrt{(1.52)^2 - \sin^2 40°}} \simeq -0.125$$

$$\mathcal{T}_\| = \frac{2(1.52)^{-1}\cos 40°}{\cos 40° + (1.52)^{-1}\sqrt{1 - (1.52)^{-2}\sin^2 40°}} \simeq 0.74$$

From Snell's law of refraction given by (9.27), the transmission angle can be found as

$$\sin 40° = 1.52 \sin \theta_t \rightarrow \theta_t \simeq 25°$$

Substituting, we find that $1 + \Gamma_\| = \mathcal{T}_\|(\cos \theta_t / \cos \theta_i)$, as expected.

9.6.3 Brewster's Angle

Examination of (9.33) indicates the possibility that for some parameter values we may have $\Gamma_\| = 0$, in which case no reflection (i.e., total transmission) occurs. This condition is satisfied when

$$\Gamma_\| = 0 \longrightarrow -\eta_1 \cos \theta_i + \eta_2 \cos \theta_t = 0 \longrightarrow \cos \theta_i = \frac{\eta_2}{\eta_1} \cos \theta_t$$

From Snell's law of refraction, we have

$$\frac{\sin \theta_t}{\sin \theta_i} = \sqrt{\frac{\epsilon_{1r}\mu_{1r}}{\epsilon_{2r}\mu_{2r}}} \rightarrow \sin \theta_t = \sqrt{\frac{\epsilon_{1r}\mu_{1r}}{\epsilon_{2r}\mu_{2r}}} \sin \theta_i$$

and, using $\cos \theta_t = \sqrt{1 - \sin \theta_t^2}$, we can rewrite the special incidence angle θ_i at which $\Gamma_\| = 0$ as

$$\sin \theta_i = \sqrt{\frac{1 - \mu_{2r}\epsilon_{1r}/(\mu_{1r}\epsilon_{2r})}{1 - (\epsilon_{1r}/\epsilon_{2r})^2}}$$

This special incidence angle θ_i that satisfies the $\Gamma_\| = 0$ condition is known as the *Brewster angle,* denoted as θ_{iB}. For nonmagnetic materials $(\mu_1 = \mu_2 = \mu_0)$, the Brewster angle can be simplified as

$$\theta_{iB} = \sin^{-1}\left(1 + \frac{\epsilon_{1r}}{\epsilon_{2r}}\right)^{-1/2} = \tan^{-1}\left[\sqrt{\frac{\epsilon_{2r}}{\epsilon_{1r}}}\right] = \tan^{-1}\frac{n_2}{n_1} \qquad (9.35)$$

where n_1 and n_2 are the refractive indices of the two media. Note that if the the wave is incident at this angle, then there is no reflected wave when the incident wave is parallel polarized.

As a useful way of thinking about the Brewster condition we consider Figure 9.30. An interesting property of the Brewster condition is that the reflected and refracted rays are at 90° with respect to one another, or that $\theta_{iB} + \theta_t = \pi/2$. To see this, we can simply use the Snell's law of refraction and the Brewster condition:

Brewster angle: $\tan\theta_{iB} = \dfrac{\sin\theta_{iB}}{\cos\theta_{iB}} = \sqrt{\dfrac{\epsilon_{2r}}{\epsilon_{1r}}} \qquad \rightarrow \qquad \sin\theta_{iB} = \cos\theta_{iB}\sqrt{\dfrac{\epsilon_{2r}}{\epsilon_{1r}}}$

Snell's law: $\sin\theta_t = \sin\theta_i\sqrt{\dfrac{\epsilon_{1r}}{\epsilon_{2r}}} \qquad \rightarrow \qquad \sin\theta_t = \left[\cos\theta_{iB}\sqrt{\dfrac{\epsilon_{2r}}{\epsilon_{1r}}}\right]\sqrt{\dfrac{\epsilon_{1r}}{\epsilon_{2r}}}$

$$= \cos\theta_{iB}$$

and therefore θ_{iR} and its corresponding θ_t are complementary angles. Physically, the incident field gives rise to vibrations of electrons in the atoms of the second medium. These vibrations are in the direction of the electric vector of the transmitted wave.

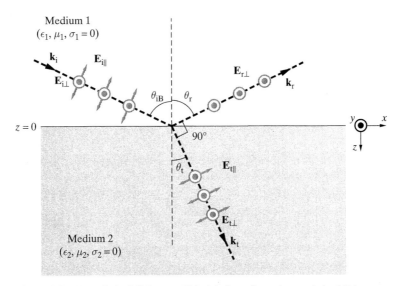

Figure 9.30   Unpolarized light at a dielectric boundary. An unpolarized light wave is incident at the dielectric boundary at the Brewster angle, that is, $\theta_i = \theta_{iB}$, produces a polarized reflected wave and a partially polarized transmitted wave. Note that for $\theta_i = \theta_{iB}$, the angle between the reflected and refracted rays is 90°.

The vibrating electrons essentially radiate a new wave in the direction transverse to their motion and give rise to the reflected wave, which propagates back into the first medium. When the reflected and transmitted rays are at right angles to each other, the reflected ray does not receive any energy for oscillations in the plane of incidence.[29]

It follows from Figure 9.30 that polarized light may be produced from unpolarized light. A randomly polarized wave incident on a dielectric at the Brewster angle becomes linearly polarized upon reflection.

Note that a Brewster angle may also be defined for perpendicular polarization. We find from (9.31) by setting $\Gamma_\perp = 0$ that

$$\sin^2 \theta_{iB\perp} = \frac{1 - \mu_{1r}\epsilon_{2r}/(\mu_{2r}\epsilon_{1r})}{1 - (\mu_{1r}/\mu_{2r})^2}$$

For nonmagnetic media ($\mu_1 = \mu_2 = \mu_0$), the right-hand side becomes infinite, so that $\theta_{iB\perp}$ does not exist. Thus, the case of a Brewster angle for perpendicular polarization is not of practical interest.

Example 9.16: Brewster angle at air–silicon interface. Find the Brewster angle for an air–silicon ($\epsilon_r = 11.7$, assume lossless and nonmagnetic) interface (Figure 9.31) for a parallel polarized wave incident (a) from air into silicon and (b) from silicon into air.

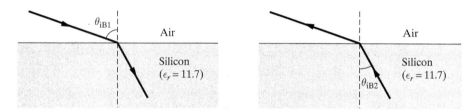

Figure 9.31   Example 9.16.

Solution:

(a) For the air–silicon interface, using (9.35), we have

$$\theta_{iB1} = \tan^{-1} \sqrt{\epsilon_r} = \tan^{-1} \sqrt{11.7} \simeq 73.7°$$

(b) For the silicon–air interface, we have

$$\theta_{iB2} = \tan^{-1} \left(\frac{1}{\sqrt{\epsilon_r}} \right) = \tan^{-1} \left(\frac{1}{\sqrt{11.7}} \right) \simeq 16.3°$$

(Note that θ_{iB1} and θ_{iB2} are complementary angles, as expected.)

[29]This particular interpretation of the lack of a reflected wave at $\theta_i = \theta_{iB}$ is often put forth and is undoubtedly useful. However, the argument is at the same time highly qualitative and somewhat superficial; for example, there is a Brewster angle when medium 2 is free space (i.e., no electrons) with permittivity ϵ_0. Also, at interfaces of magnetic media ($\mu_1 \neq \mu_2$), there can also exist a Brewster angle for perpendicular polarization [P. Lorrain, D. R. Corson, and F. Lorrain, *Electromagnetic Fields and Waves*, 3rd ed., Freeman, New York, p. 575, 1988].

Example 9.17: A parallel polarized wave. Consider the uniform plane wave incident on an air–polystyrene interface at an angle of $\sim 58°$ as considered in Example 9.14. Assuming that the wave is taken to be parallel polarized as shown in Figure 9.32, with its magnetic field represented by

$$\mathbf{H}_i(y, z) = \hat{\mathbf{x}} H_0 e^{-j\beta_1(\cos\theta_i y - \sin\theta_i z)}$$

and keeping the total time-average power, the frequency, and the incidence angle of the wave to be the same as in Example 9.14 (i.e., 1.4 W-m^{-2}, 3 GHz, and $\sim 58°$, respectively), find the following: (a) H_0 (assume real) and β_1, (b) $\mathbf{E}_i(y, z)$, (c) $\mathbf{E}_r(y, z)$ and $\mathbf{H}_r(y, z)$, and (d) $\mathbf{E}_t(y, z)$ and $\mathbf{H}_t(y, z)$.

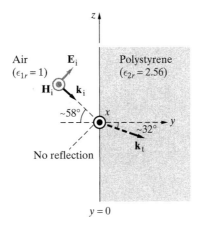

Figure 9.32   Air–polystyrene interface. Incident and refracted waves for Example 9.17.

Solution:

(a) We have

$$|\mathbf{S}_{av}|_i = \frac{1}{2}\eta_1 H_0^2 = 1.4 \text{ W-m}^{-2} \quad \rightarrow \quad H_0 \simeq \sqrt{2 \times 1.4/377} \simeq 0.0862 \text{ A-m}^{-1}$$

and

$$\beta_1 = \omega\sqrt{\mu_1\epsilon_1} = \frac{\omega}{c} \simeq \frac{2\pi \times 3 \times 10^9}{3 \times 10^8} = 20\pi \text{ rad-m}^{-1}$$

Substituting the numerical values, we have

$$\mathbf{H}_i(y, z) \simeq \hat{\mathbf{x}} 86.2 e^{-j20\pi(0.53y - 0.848z)} \text{ mA-m}^{-1}$$

(b) The electric field phasor of the incident wave can be written as

$$\mathbf{E}_i(y, z) = \eta_1 H_0(\hat{\mathbf{y}}\sin\theta_i + \hat{\mathbf{z}}\cos\theta_i)e^{-j\beta_1(\cos\theta_i y - \sin\theta_i z)}$$

$$\simeq 32.5(\hat{\mathbf{y}}0.848 + \hat{\mathbf{z}}0.53)e^{-j20\pi(0.53y - 0.848z)} \text{ V-m}^{-1}$$

(c) Since the incident wave is parallel-polarized, let us check the Brewster angle:

$$\theta_{iB} = \tan^{-1}\sqrt{\frac{\epsilon_{2r}}{\epsilon_{1r}}} = \tan^{-1}(1.6) \simeq 58°$$

So the incidence angle is the Brewster angle, which corresponds to the case of total transmission. As a result, we have

$$\Gamma_{\parallel} \simeq 0 \quad \rightarrow \quad \mathbf{E}_r(y,z) \simeq \mathbf{H}_r(y,z) \simeq 0$$

Note that if the incidence were not at the Brewster angle, we could have simply evaluated $\Gamma_{\parallel}$ from (9.33) to determine E_{r0}.

(d) From the conservation-of-power principle, we have

$$|\mathbf{S}_{av}|_i \cos\theta_i = |\mathbf{S}_{av}|_t \cos\theta_t \rightarrow \frac{1}{2}\frac{E_0^2}{\eta_1}\cos\theta_i = \frac{1}{2}\frac{E_{t0}^2}{\eta_2}\cos\theta_t$$

So

$$E_{t0} \simeq \sqrt{\frac{(32.5)^2}{1.6}\left[\frac{\cos 58°}{\cos 32°}\right]} \simeq 20.3 \text{ V-m}^{-1}$$

From Figure 9.32, the electric field phasor of the transmitted wave can be written as

$$\mathbf{E}_t(y,z) \simeq 20.3[\hat{\mathbf{y}}0.53 + \hat{\mathbf{z}}0.848]e^{-j32\pi(0.848y-0.53z)} \text{ V-m}^{-1}$$

The corresponding magnetic field phasor is

$$\mathbf{H}_t(y,z) \simeq \hat{\mathbf{x}}86.2e^{-j32\pi(0.848y-0.53z)} \text{ mA-m}^{-1}$$

9.6.4 Fresnel's Formulas

The reflection coefficient equations (9.31) and (9.33) are known as Fresnel's equations, or, respectively, Fresnel's sine law and Fresnel's tangent law,[30] after Augustin Fresnel, who put forth these and many other aspects of reflection, refraction, and diffraction of light waves during his short life.[31]

---

[30]The reason for this terminology is the fact that Fresnel's version of $\rho_{\perp}$ and $\rho_{\parallel}$ were

$$\Gamma_{\perp} = \frac{\sin(\theta_t - \theta_i)}{\sin(\theta_t + \theta_i)} \quad \text{and} \quad \Gamma_{\parallel} = \frac{\tan(\theta_t - \theta_i)}{\tan(\theta_t + \theta_i)}$$

It is clear from the above that when $\theta_t + \theta_i = \pi/2$, we have $\tan(\theta_t + \theta_i) = \infty$, which means $\Gamma_{\parallel} = 0$, that is, the Brewster condition. The reader is encouraged to determine that the above expressions for $\Gamma_{\perp}$ and $\Gamma_{\parallel}$ are indeed consistent with (9.31) and (9.33).

[31]For a very interesting discussion of Fresnel's scientific contributions, see Chapter IV of E. Whittaker, *History of the Theories of Aether and Electricity*, Thomas Nelson and Sons Lim., New York, 1951.

The behavior of the reflection coefficients can be investigated by plotting their magnitudes against incidence angle for different values of ϵ_1, ϵ_2. A plot of the magnitude and phases of $\Gamma_\perp$ and $\Gamma_\parallel$ for three different values of ϵ_2 (with $\epsilon_1 = \epsilon_0$) is given in Figure 9.33. Note that as θ_i is varied from 0 to 90°, $\Gamma_\parallel$ changes polarity at $\theta_i = \theta_{iB}$, where it is equal to zero. For the case of nonmagnetic media ($\mu_1 = \mu_2 = \mu_0$) and for $\epsilon_2 > \epsilon_1$, since $\theta_i > \theta_t$, both $\Gamma_\perp$ and $\Gamma_\parallel$ are negative for $\theta_i < \theta_{iB}$. For $\theta_i > \theta_{iB}$, $\Gamma_\perp$ remains negative, but $\Gamma_\parallel$ becomes positive. Note that the polarity of $\Gamma_{\perp,\parallel}$ is determined by our choice of the orientation of the incident and reflected electric field vectors in Figures 9.26 and 9.28.

Consider first the perpendicular polarization case. With the sense of orientation defined as the same for $\mathbf{E}_i$ and $\mathbf{E}_r$ (Figure 9.26), it is clear from the boundary condition that if the second medium is very dense, that is, $\epsilon_2 \gg \epsilon_1$ or $\eta_2 \ll \eta_1$, then most of the wave energy would be reflected and the magnitude of E_t would be very small. Accordingly, to satisfy the boundary condition, the electric field should switch polarity upon reflection, or $\Gamma_\perp \simeq -1$ for $\theta_i = 0$.

For parallel polarization, and with the electric field polarities as defined in Figure 9.28, $\Gamma_\parallel < 0$ at low values of θ_i simply means that the tangential component of the transmitted wave is smaller than that of the incident wave, so the electric field of the reflected wave has to be negative to cancel some of the incident wave field in medium 1. At $\theta_i = \theta_{iB}$, $\Gamma_\parallel$ changes sign, as can be seen from Figure 9.33. Thus, at large angles $\theta_i > \theta_{iB}$, the transmitted wave field is larger than the incident one, and the reflected wave needs to be positive to add to the incident wave field.

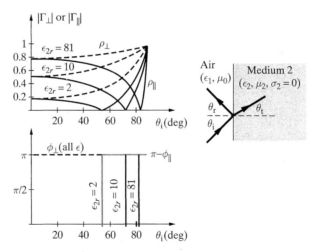

Figure 9.33   Reflection coefficient versus the angle of incidence. (a) Magnitude and phase of reflection coefficient for perpendicular ($\Gamma_\perp$) and parallel ($\Gamma_\parallel$) polarization versus angle of incidence θ_i for distilled water ($\epsilon_{2r} = 81$), flint glass ($\epsilon_{2r} = 10$), and paraffin ($\epsilon_{2r} = 2$), all assumed to be lossless. For clarity, the complement of the phase angle $\phi_\parallel$ (i.e., $\pi - \phi_\parallel$) is sketched, rather than $\phi_\parallel$ itself. In all cases, $\phi_\parallel = \pi$ for $\theta_i < \theta_{iB}$ and $\phi_\parallel = 0$ for $\theta_i > \theta_{iB}$, while $\phi_\perp = \pi$ for all θ_i.

Wave polarization upon reflection and refraction.   In all the above analyses we have implicitly assumed that the propagation directions of the reflected and refracted waves (i.e., $\hat{\mathbf{k}}_r$ and $\hat{\mathbf{k}}_t$) lie in the plane of incidence. For uniform plane waves, the incident fields are by definition independent of y, assuming the x-z plane to be the plane of incidence. Since the layout of the boundary is also independent of y, the reflected and refracted fields must also be independent of y. We further note that there is complete symmetry in the $+y$ versus $-y$ directions and that there is no preferred direction for the waves to bend upon reflection or refraction. Another argument against a y dependence of the reflected and refracted fields is the fact that any solution of Maxwell's equations satisfying a given set of boundary conditions must be unique; thus, because we can find a solution for the fields by assuming $\hat{\mathbf{k}}_r$ and $\hat{\mathbf{k}}_t$ to be in the plane of incidence, these solutions must be the only possible solutions.

In light of the above argument, an incident wave that is linearly polarized with $\mathbf{E}_i$ perpendicular to the plane of incidence is partially reflected, with the same polarization, and partially transmitted, again with the same polarization. When the incident wave is linearly polarized with $\mathbf{E}_i$ in the plane of incidence, the wave is again reflected and refracted with no change in polarization. This situation is analogous to dropping a uniform rigid rod on a smooth plane surface. If the rod is dropped with its axis either normal or parallel to the surface, then its orientation will not change as it bounces back from the surface (i.e., reflection). Otherwise, however, the rod rotates upon reflection.

For an incident wave that is linearly polarized, the polarizations of the reflected and refracted waves differs from that of the incident wave only if $\mathbf{E}_i$ is not entirely normal or parallel to the plane of incidence. That this is so is clear from Figure 9.33, which shows that, in general, the magnitudes of $\Gamma_\parallel$ and $\Gamma_\perp$ are different. If the incident wave is linearly polarized with its electric field $\mathbf{E}_i$ oriented at an angle to the plane of incidence, then we can decompose the field into two components, $\mathbf{E}_\perp$ and $\mathbf{E}_\parallel$, and separately calculate the reflection for each component, using $\Gamma_\perp$ and $\Gamma_\parallel$. Since $|\Gamma_\perp| \neq |\Gamma_\parallel|$, the reflected wave, while still linearly polarized, now has an electric field oriented at a different angle with respect to the plane of incidence. If the second medium is lossy, then the reflection coefficients are complex with different phase shifts for the parallel and perpendicular polarizations, in which case the reflected wave is elliptically polarized (Section 9.8 addresses oblique incidence on a lossy dielectric). A similar situation occurs under conditions of total reflection, as discussed in the next section.

When the incident wave is circularly polarized, the relative signs of the two reflection coefficients determine the polarization sense of the reflected wave, as illustrated in Figure 9.34. For $\theta_i < \theta_{iB}$, the reflected wave is elliptically polarized with the opposite sense of rotation to that of the incident wave. Both $\Gamma_\perp$ and $\Gamma_\parallel$ are negative in this case, so the sense of polarization changes (i.e., from RHEP to LHEP or vice versa), because the $\hat{\mathbf{k}}_r$ vector of the wave has a component in the $-z$ direction, while $\hat{\mathbf{k}}_i$ has a component in the $+z$ direction. At $\theta_i = \theta_{iB}$, there is no reflection for the parallel-polarized component, so the reflected wave would be linearly polarized, with electric field perpendicular to the plane of incidence. For $\theta_i > \theta_{iB}$, $\Gamma_\perp$ remains negative, while $\Gamma_\parallel$ is now positive, and the reflected wave is elliptically polarized with the same sense of rotation as that of the incident wave.

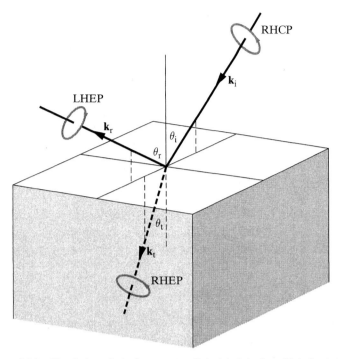

Figure 9.34   Circularly polarized wave at a dielectric interface. Right-hand circularly polarized (RHCP) wave is obliquely incident from air upon a dielectric. For $\theta_i < \theta_{iB}$, the reflected wave is in general left-hand elliptically polarized (LHEP) whereas the transmitted wave is also elliptically polarized but with the same sense of rotation as that of the incident wave (i.e., RHEP).

9.7 TOTAL INTERNAL REFLECTION

The phenomenon of *total internal reflection* has a broad range of applications, including reflections from prisms and the guiding of light waves in dielectric waveguides or optical fibers. Total internal reflection occurs when an electromagnetic wave is incident at a highly oblique angle from a dense medium onto a less dense medium.

We have so far implicitly assumed the incidence of plane waves from a less dense medium to a denser one (i.e., $\epsilon_1 < \epsilon_2$), as is evident from the various diagrams (e.g., Figures 9.26, 9.28, and 9.30) that illustrate the refracted wave to be "bent" toward the normal (i.e., $\theta_t < \theta_i$). Also, when we showed the magnitudes and phases of the Fresnel coefficients in Figure 9.33, we considered only cases with $\epsilon_2 > \epsilon_1$. In this section, we consider the incidence of plane waves from a denser medium into a less dense one, namely the cases in which $\eta_2 > \eta_1$ (or $\epsilon_2 < \epsilon_1$ when $\mu_1 = \mu_2$).

We start once again with Snell's law (Figure 9.23), which is valid regardless of the polarization of the incident wave. For nonmagnetic media ($\mu_1 = \mu_2 = \mu_0$), the sine of the angle of the transmitted (or refracted) ray is given by

$$\sin \theta_t = \sqrt{\frac{\epsilon_1}{\epsilon_2}} \sin \theta_i$$

and whenever $\epsilon_1 > \epsilon_2$, we have $\theta_t > \theta_i$. Thus, there exist large angles of incidence θ_i for which $\sin \theta_t$ exceeds unity, corresponding to total reflection of the incident wave (i.e., no transmitted wave), as illustrated in Figure 9.35. Specifically, $\sin \theta_t > 1$ for all angles $\theta_i > \theta_{ic}$, where the so-called critical angle θ_{ic}, corresponding to $\theta_t = 90°$, is given by

$$\sin \theta_{ic} = \sqrt{\frac{\epsilon_2}{\epsilon_1}} = \frac{n_2}{n_1} \qquad (9.36)$$

Thus, an observer near the interface in medium 2 is able to receive waves from medium 1 only over a range of incident angles of $0 \le \theta_i < \theta_{ic}$. In other words, there is no transmission into medium 2 for $\theta_i \ge \theta_{ic}$; that is, there is total internal reflection.

9.7.1 Reflection Coefficients

At low angles of incidence, $\theta_i < \theta_{ic}$, all of our previous discussion for $\epsilon_1 < \epsilon_2$ (assuming $\mu_1 = \mu_2$) is also valid[32] for $\epsilon_1 > \epsilon_2$. The Brewster condition is still realized at $\theta_{iB} = \tan^{-1}(n_2/n_1) < \theta_{ic} = \sin^{-1}(n_2/n_1)$ (see Figure 9.36).

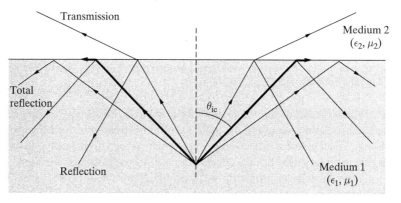

Figure 9.35   Critical angle of incidence and total internal reflection. Reflection and refraction characteristics of a plane wave passing from a relatively dense dielectric medium into a less dense medium (i.e., $\epsilon_1 > \epsilon_2$ for $\mu_1 = \mu_2$).

---

[32]Except that the phases of the reflected components (i.e., $\phi_\perp$ and $\phi_\parallel$) are reversed compared with the $\epsilon_1 < \epsilon_2$ case, as can be seen from (9.31) and (9.33). In other words, while for $\epsilon_2 > \epsilon_1$ we had $\Gamma_\perp$ and $\Gamma_\parallel$ negative for $\theta_i < \theta_{iB}$, for $\epsilon_1 > \epsilon_2$ the reflection coefficients $\Gamma_\perp$ and $\Gamma_\parallel$ are positive for $\theta_i < \theta_{iB}$.

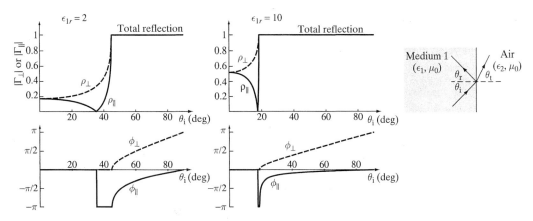

Figure 9.36   Reflection coefficient versus the angle of incidence. Magnitude and phase of the reflection coefficients for perpendicular ($\Gamma_\perp$) and parallel ($\Gamma_\parallel$) polarization versus angle of incidence for $\epsilon_1/\epsilon_2 = 2$ and $\epsilon_1/\epsilon_2 = 10$. The phase angle $\phi_\perp$ is shown as a dashed line, while $\phi_\parallel$ is plotted as a solid line.

At larger angles of incidence $\theta_i > \theta_{ic}$, we have total internal reflection and

$$\sin\theta_t = \sqrt{\frac{\epsilon_1}{\epsilon_2}}\sin\theta_i > 1 \qquad \longrightarrow \qquad \theta_t \text{ imaginary} \qquad (9.37)$$

Note that the condition $\sin\theta_t > 1$ cannot be satisfied if θ_t is a real angle. Although an imaginary angle does not pose any particular dilemma,[33] especially because we do not need to evaluate θ_t explicitly, since what is needed to determine the expressions for the fields is $\cos\theta_t$, which can be found from $\sin\theta_t$. For $\sin\theta_t > 1$, $\cos\theta_t$ is purely imaginary and can be found using (9.37), we can find $\cos\theta_t$ as

$$\cos\theta_t = \sqrt{1 - \sin^2\theta_t} = \pm j\sqrt{\frac{\epsilon_1}{\epsilon_2}\sin^2\theta_i - 1} \qquad (9.38)$$

where we choose the solution with the negative sign to ensure that the refracted electromagnetic field in medium 2 (see Section 9.7.2) attenuates with increasing distance

---

[33]For an arbitrary complex number $C = A + jB$, we can evaluate $\sin C$ as

$$\sin C = \frac{e^{jC} - e^{-jC}}{2j} = \frac{e^{jA}e^{-B} - e^{-jA}e^{+B}}{2j} = \sin A \cosh(B) + j\cos A \sinh(B)$$

In the case of $\theta_t = \mathcal{R}e\{\theta_t\} + j\mathcal{I}m\{\theta_t\}$, we note that $\sin\theta_t$ is a real number given by (9.37), so that we must have $\cos(\mathcal{R}e\{\theta_t\}) = 0$, or $\mathcal{R}e\{\theta_t\} = \pi/2$, so that

$$\theta_t = \frac{\pi}{2} + j\mathcal{I}m\{\theta_t\}$$

from the boundary. Substituting (9.38) in (9.31) and (9.33) shows that the reflection coefficients are now complex, given by

$$\Gamma_\perp = \frac{\cos\theta_i + j\sqrt{\sin^2\theta_i - \epsilon_{21}}}{\cos\theta_i - j\sqrt{\sin^2\theta_i - \epsilon_{21}}} = 1e^{j\phi_\perp} \tag{9.39}$$

$$\Gamma_\parallel = -\frac{\epsilon_{21}\cos\theta_i + j\sqrt{\sin^2\theta_i - \epsilon_{21}}}{\epsilon_{21}\cos\theta_i - j\sqrt{\sin^2\theta_i - \epsilon_{21}}} = 1e^{j\phi_\parallel} \tag{9.40}$$

where we have introduced the permittivity ratio parameter $\epsilon_{21} = \epsilon_2/\epsilon_1$ for convenience.

Plots of the magnitude and phase of $\Gamma_\parallel$ and $\Gamma_\perp$ for nonmagnetic media ($\mu_2 = \mu_1 = \mu_0$) and for $\epsilon_1/\epsilon_2 = 2$ and 10 are shown as a function of incidence angle θ_i in Figure 9.36. The Brewster condition, occurring at an incident angle somewhat below that of total reflection, is apparent for the parallel polarization case. Note that for $\theta_i > \theta_{ic}$, the reflection coefficients have unity magnitude but different phase angles, as is apparent from (9.39) and (9.40). Thus, upon total internal reflection an incident plane wave acquires a phase shift. By manipulating (9.39) and (9.40), explicit expressions for the phase shifts can be obtained. We find

$$\tan\frac{\phi_\perp}{2} = \frac{\sqrt{\sin^2\theta_i - \epsilon_{21}}}{\cos\theta_i}; \qquad \tan\frac{\phi_\parallel}{2} = -\frac{\epsilon_{21}\cos\theta_i}{\sqrt{\sin^2\theta_i - \epsilon_{21}}}$$

The phase change that takes place in total reflection may be used to produce a circularly or elliptically polarized wave from one that is linearly polarized. For this purpose, the incident wave must be polarized in a direction that is neither parallel nor normal to the plane of incidence. As an example, we consider the incident wave to be polarized at an angle of 45° with respect to the plane of incidence, so that it possesses two components of equal magnitude (i.e., $|E_\parallel| = |E_\perp|$). Upon total internal reflection, the two components acquire different phases, so the phase difference between them is given by

$$\Delta\phi = \phi_\perp - \phi_\parallel \;\rightarrow\; \tan\frac{\Delta\phi}{2} = \frac{\tan(\phi_\perp/2) - \tan(\phi_\parallel/2)}{1 + \tan(\phi_\perp/2)\tan(\phi_\parallel/2)} = \frac{\sin^2\theta_i}{\cos\theta_i\sqrt{\sin^2\theta_i - \epsilon_{21}}} \tag{9.41}$$

A plot of $\Delta\phi$ as a function of θ_i is given in Figure 9.37 for $\epsilon_1/\epsilon_2 = 2, 10, 81$. We note that the phase difference is zero at normal incidence, and π between the Brewster angle and $\theta_i = \theta_{ic}$. Between θ_{ic} and grazing incidence ($\theta_i = \pi/2$), the phase difference reaches a minimum. By differentiating the above expression for $\Delta\phi$ with respect to θ_i it can be shown that the incidence angle θ_{imin} at which the local minimum occurs and the local minimum value of the phase difference $(\Delta\phi)_{min}$ are given by

$$\sin^2\theta_{imin} = \frac{2\epsilon_{21}}{1 + \epsilon_{21}}; \qquad \tan\frac{(\Delta\phi)_{min}}{2} = \frac{2\sqrt{\epsilon_{21}}}{1 - \epsilon_{21}} \tag{9.42}$$

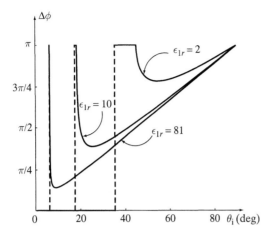

Figure 9.37 Phase difference between the two polarizations upon total internal reflection. Plots of $\Delta\phi = \phi_\perp - \phi_\parallel$ are shown for $\epsilon_{1r} = 2, 10, 81$, assuming $\epsilon_2 = \epsilon_0$. Note that $\Delta\phi = \pi$ when $\theta_i = \theta_{ic}$ or $\theta_i = 90°$. The critical angles θ_{ic} for the cases of $\epsilon_{1r} = 2, 10, 81$ (with $\epsilon_2 = \epsilon_0$) are, respectively, $\theta_{ic} = 45°, \sim 18°$, and $\sim 6.4°$.

It is evident from the foregoing discussion that circularly polarized light can be obtained from linearly polarized light by choosing the angle of incidence and the permittivity ratios in such a way that $\phi = 90°$. To obtain this in a single reflection, the refractive index of the first medium must be $n_1 \simeq 2.41$ (assuming the second medium is air, with $n_2 = 1$). In his first demonstration of this concept, Fresnel was not able to find any transparent material with such a high refractive index and instead made use of two total reflections on glass ($n_1 = 1.51$), as shown in Figure 9.38. For glass, we can show from (9.42) that the maximum relative phase difference of $(\Delta\phi)_{max} \simeq 45.93°$ occurs for $\theta_{imax} \simeq 51.33°$, and a $\Delta\phi = 45°$ phase difference can be obtained with either $\theta_i \simeq 48.62°$ or $\theta_i \simeq 54.62°$.

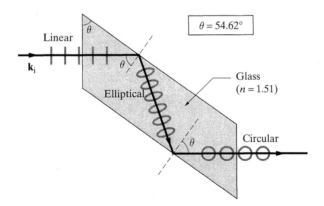

Figure 9.38 Fresnel's rhomb. Linearly polarized light incident on glass (let $n = 1.51$) with its electric field at 45° to the plane of incidence acquires a phase difference between its two components ($E_\perp$ and $E_\parallel$) of 45° at each total internal reflection. As a result, the wave exiting the rhomb is circularly polarized.

Example 9.18: Water–air interface. Consider a linearly polarized electromagnetic wave incident from water ($\epsilon_{1r} = 81$) on a water–air interface. The electric field vector of the wave is at an angle of $45°$ with respect to the plane of incidence. Find the incidence angle θ_i for which the reflected wave is circularly polarized. Neglect losses.

> **Solution:**   We have $\epsilon_1/\epsilon_2 = 81$ or $n_1/n_2 = 9$. From (9.42), we find $(\Delta\phi)_{min} \simeq 25.4°$, which occurs for $\theta_{imin} \simeq 8.98°$. By trial and error, or by using a graphical solution (see Figure 9.37), we find that $\phi = 90°$ occurs for $\theta_i \simeq 6.43°$ and $\theta_i \simeq 44.6°$. The first solution is very close to the critical angle (see Figure 9.37), which for the water–air interface is $\theta_{ic} = \sin^{-1}(n_2/n_1) \simeq 6.38°$. Furthermore the phase angle ϕ in the vicinity of $6.43°$ is highly sensitive to the incidence angle θ_i. The second solution is comfortably in the total internal reflection region and can be realized with relative ease.

9.7.2 The Electromagnetic Field in Medium 2: The Refracted "Wave"

Although the incident wave is *totally* reflected, there does exist a finite electromagnetic field in medium 2, and this field exhibits a number of surprisingly complicated properties. That there should be a field in the second medium is evident on the basis of the fact that the boundary conditions, as worked out in previous sections, must still be satisfied, and that $\mathbf{E}_i + \mathbf{E}_r = \mathbf{E}_i + \Gamma\mathbf{E}_i \neq 0$. Regardless of the fact that $\sin\theta_t > 1$, all of the field expressions that were derived earlier using the physical boundary conditions are still valid. Accordingly, the electric field of the wave in medium 2 for the perpendicular polarization[34] case is given by

$$\mathbf{E}_t(x, z) = \hat{\mathbf{y}}E_{t0}e^{-j\beta_2\hat{\mathbf{k}}_2\cdot\mathbf{r}} = \hat{\mathbf{y}}E_{t0}e^{-j\beta_2(x\sin\theta_t + z\cos\theta_t)} \tag{9.43}$$

where $\beta_2 = \omega\sqrt{\epsilon_2\mu_0}$, assuming nonmagnetic media. Using expression (9.38) for $\cos\theta_t$ in (9.43) we find

$$\mathbf{E}_t(x, z) = \hat{\mathbf{y}}E_{t0}e^{-\alpha_t z}e^{-j\beta_t x}$$

where

$$\alpha_t = \pm\beta_2\sqrt{\frac{\epsilon_1}{\epsilon_2}\sin^2\theta_i - 1} = \frac{2\pi}{\lambda_1}\sqrt{\sin^2\theta_i - \epsilon_{21}} \tag{9.44}$$

$$\beta_t = \beta_2\sqrt{\frac{\epsilon_1}{\epsilon_2}}\sin\theta_i = \underbrace{\omega\sqrt{\mu_0\epsilon_1}}_{\beta_1}\sin\theta_i = \frac{2\pi}{\lambda_1}\sin\theta_i \tag{9.45}$$

---

[34]Note that although (9.43) is written for perpendicular polarization, the spatial variation of both components of the electric field for the parallel polarization case is also proportional to $e^{-j\beta_2\hat{\mathbf{k}}_t\cdot\mathbf{r}}$. Thus, the properties of the refracted wave as discussed here are also exhibited by the refracted wave in the parallel polarization case.

Note that $\alpha_t = j\beta_2 \cos\theta_t$ must be positive in order for the refracted wave amplitude to decay (rather than grow) with distance into the second medium, meaning that $\cos\theta_t$ must be imaginary and negative; that is, we must choose the negative sign in (9.38). Thus, for $\theta_i > \theta_{ic}$ a wave exists along the interface, propagating in the x direction (i.e., parallel to the surface) with a phase velocity $v_{pt} = (\omega/\beta_t) = \omega/(\beta_1 \sin\theta_i) = v_{p1}/\sin\theta_i$. The wavelength of this propagation is $\lambda_t = 2\pi/\beta_t = \lambda_1/\sin\theta_i > \lambda_1$, as is clear from Figure 9.39. However, this wave is exponentially attenuated very rapidly in the z direction.[35] Because it is typically tightly bound to the surface,[36] such a wave is known as a *surface wave*. This nonuniform wave in medium 2 is also a *slow wave*, since its phase velocity $v_{pt} = \omega/\beta_t$

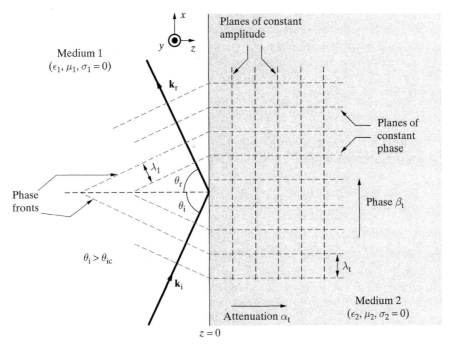

Figure 9.39   Total internal reflection: Constant amplitude and phase fronts. Oblique incidence beyond the critical angle $\theta_i > \theta_{ic}$ on a dielectric interface with $\mu_1 = \mu_2 = \mu_0$ and $\epsilon_1 > \epsilon_2$. The planes of constant amplitude are parallel to the interface, whereas the planes of constant phase are normal to it. This picture is valid for both perpendicular and parallel polarization.

[35] Naturally only the positive sign in α_t in (9.44) corresponds to a physical situation.

[36] The value of α_t is typically such that the depth of penetration d, defined as the distance $d = 1/\alpha_t$ at which the amplitude drops to e^{-1} of its value at the interface, is of order λ_1 or smaller. For example, for a glass–air interface with $\epsilon_1/\epsilon_2 = 2.31$, we have $\alpha_t \simeq 2.47/\lambda_1$ (i.e., $d \simeq 0.406\lambda_1$) at $\theta_i = 50°$, less than 10 degrees above the critical angle $\theta_{ic} \simeq 41°$. At this rate of decay, the amplitude is reduced to $e^{-2.47} \simeq 0.085$ or $\sim8.5\%$ of its value at the interface over a distance of λ_1. Note, however, that α_t becomes smaller for as $\theta_i \to \theta_{ic}$, becoming zero at $\theta_i = \theta_{ic}$, in which case the attenuation in medium 2 disappears and the transmitted wave travels as a uniform plane surface wave parallel to the boundary.

is less than the velocity of light in medium 2 (i.e., $v_{p2} = \omega/\beta_2 = 1/\sqrt{\mu_0\epsilon_2}$). To see this, we can use expression (9.45) for β_t and the fact under the total reflection condition we have $\sin\theta_t = \sqrt{\epsilon_1/\epsilon_2}\sin\theta_i > 1$. We have

$$v_{pt} = \frac{\omega}{\beta_t} = \frac{\omega}{\beta_2\sqrt{\dfrac{\epsilon_1}{\epsilon_2}}\sin\theta_i} = \frac{\omega}{\beta_2\sin\theta_t} < \frac{\omega}{\beta_2}$$

Figure 9.39 illustrates the structure of the phase fronts for the wave in medium 2. Since the planes of constant amplitude are parallel to the interface, whereas the planes of constant phase are orthogonal to it, the refracted wave is an inhomogeneous (or nonuniform) plane wave. In other words, the amplitude of the wave varies as a function of position (z) on the planes of constant phase (planes defined by $x =$ constant).

We can find the magnitude E_{t0} of the refracted field by substituting $\cos\theta_t$ as given by (9.38) into the transmission coefficient expression (9.32) of Section 9.6. We have

$$\mathcal{T}_\perp = \frac{E_{t0}}{E_{i0}} = \frac{2\cos\theta_i}{\cos\theta_i + j\sqrt{\sin^2\theta_i - \epsilon_{21}}} = \left(\frac{2\cos\theta_i}{\sqrt{1 - \epsilon_{21}}}\right)e^{-j(\phi_\perp/2)}$$

where $\phi_\perp$ is the phase of the reflection coefficient $\Gamma_\perp = 1e^{j\phi_\perp}$, plotted for two example cases in Figure 9.36.

The magnetic field of the refracted wave can be obtained from the electric field given in (9.43) (written below in terms of α_t and β_t) and Maxwell's equation (7.23a). We have

$$\nabla \times \mathbf{E}_t = -j\omega\mu_0\mathbf{H}_t$$
$$\mathbf{E}_t(x, z) = \hat{\mathbf{y}}E_{t0}e^{-\alpha_t z}e^{-j\beta_t x}$$

$$\rightarrow$$

$$-\frac{\partial E_{ty}}{\partial z} = -j\omega\mu_0 H_{tx}$$
$$\frac{\partial E_{ty}}{\partial x} = -j\omega\mu_0 H_{tz}$$

which upon manipulation gives

$$\mathbf{H}_t(x, z) = \frac{E_{t0}}{\omega\mu_0}(+\hat{\mathbf{x}}j\alpha_t + \hat{\mathbf{z}}\beta_t)e^{-\alpha_t z}e^{-j\beta_t x}$$

The spatial variation of the magnetic field is identical to that of the electric field in that the field components propagate in the x direction with a phase constant β_t and attenuate in the z direction with an attenuation constant α_t. Furthermore, the two components of the magnetic field are 90° out of phase with one another (i.e., the amplitude of one is multiplied by j, while the other is not), which means that the wave magnetic field vector, which is confined to the plane of incidence, rotates in time [as observed at a fixed point (x, z)] somewhat similar to an elliptically polarized wave.[37] The refracted

---

[37]Note, however, that the plane of this rotation is not transverse to the propagation direction, so the similarity with an elliptically polarized wave does not go beyond superficial.

wave that exists in the second medium to support the total internal reflection is indeed rather complex.

Although there do exist electric and magnetic fields in medium 2, it can be shown that no energy flows across the boundary. More precisely, although the component of the Poynting vector in the direction normal to the boundary is in general finite, its time-average vanishes; this implies that the energy flows to and fro, but that there is no lasting energy flow into the second medium. To see this, it suffices to note that the electric field (in the y direction) and the x component of the magnetic field are 90° out of phase, so that the z component of $\mathbf{E} \times \mathbf{H}^*$ is purely imaginary. Such a wave that does not carry power and attenuates with distance is sometimes referred to as an *evanescent* wave.

The time-average of the component of the Poynting vector in the x direction is finite in medium 2, indicating that real electromagnetic power flows parallel to the interface. This is clear from the expressions for $\mathbf{E}_t(x,z)$ and $\mathbf{H}_t(x,z)$, from which we see that E_{ty} and H_{tz} are in phase, indicating that the time-average power in the x direction is nonzero. We can find an expression for $\mathbf{S}_{av}$ after straightforward manipulation; for the perpendicular polarization case we have

$$(\mathbf{S}_{av})_{t\perp} = \frac{1}{2}\mathscr{R}e\{\mathbf{E}_t \times \mathbf{H}_t^*\} = \hat{\mathbf{x}}\frac{\beta_t}{\omega\mu_0}\frac{E_{i0}^2 2\cos^2\theta_i}{1 - \epsilon_{21}}e^{-2\alpha_t z}$$

This result, namely that real power flows unattenuated parallel to the interface, appears problematic when we consider an incident wave of finite cross section (e.g., a beam of light). Does the wave power continue to flow and extend beyond the illuminated region? This clearly cannot be the case, since the refracted wave exists simply to satisfy the boundary conditions. However, our existing formulation cannot resolve this dilemma, since it is based on using uniform plane waves of infinite transverse extent.[38]

The very existence of an electromagnetic field in the second medium seems somewhat paradoxical when one considers the fact that all of the incoming wave energy is reflected. Since the electric and magnetic fields of the refracted wave are clearly different from zero, the average energy densities $w_e = (1/2)\epsilon E^2$ and $w_m = (1/2)\mu H^2$ must be different from zero, and it seems strange that there can be energy in the second medium when all of the incoming energy is reflected. In other words, how is the energy transmitted into medium 2? The explanation lies in the fact that our analysis has been based on assumptions of steady-state, time-harmonic solutions. At the time the incident wave first strikes the surface, a small amount of energy penetrates into the second medium and establishes the field. Once established, this energy cannot escape the second medium, since the steady-state solution does not allow any energy transfer between the two media. More generally, the field in the second medium can be thought to arise from the fact that the incident field is bounded in both time and space.

The existence of an electric field in medium 2 under total internal reflection conditions is experimentally verified in optics by the fact that, if a prism is causing an

---

[38] A simplified treatment of the total reflection of a collimated beam of light is given in Section 3.6 of D. L. Lee, *Electromagnetic Principles of Integrated Optics,* Wiley, New York, 1986.

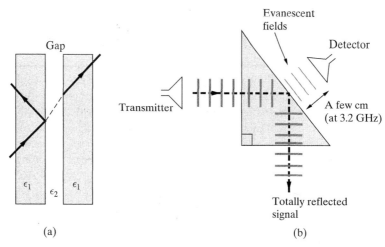

Figure 9.40   Coupling through an air gap.

internal total reflection, and another one is pressed closely against one of the sides at which the total reflection is occurring (see Figure 9.40a), most of the light goes into the second prism instead of being totally reflected. This means that there must have been some field *just* outside the original prism surface; otherwise, how could bringing up the second prism cause this change? Such coupling through the evanescent wave is illustrated in Figure 9.40b; however, note that for effective coupling, the second prism has to be within a few wavelengths, which means that at optical frequencies ($\lambda \simeq 10^{-5}$ cm), it must be physically pressed on to the first one. The refracted electromagnetic field that occurs under total internal reflection has also been measured in several experiments at microwave frequencies.[39] One should note, however, that any arrangement used to measure the refracted wave disturbs the boundary conditions, so that the experimental verifications noted can be considered only approximate confirmation of the existence of the wave in medium 2.

An interesting use of the refracted wave that exists during total internal reflection is as a primary diagnostic tool for evaluating the propagation characteristics of dielectric thin-film waveguides using a *prism coupler*. As depicted in Figure 9.41, the coupler consists of a high-refractive-index prism placed in close proximity to a slab dielectric waveguide. When an optical beam passing through the prism is incident upon its bottom at angle $\theta_i > \theta_{ic}$, the evanescent fields that extend below the prism base penetrate into the

---

[39]One set of experiments was carried out by G. F. Hull, Jr., Experiments with UHF Wave Guides, *Am. J. Phys.*, 13(6), pp. 384–389, December 1945. In Appendix J of J. Strong, *Concepts of Physical Optics,* Freeman, 1958, G. W. Hull describes his experiments carried out at wavelength of 3.2 cm using sizable prisms constructed out of paraffin ($\epsilon_r = 2.1$). The field behind the paraffin prism was found to decrease at a rate of $\sim$5 dB-(cm)$^{-1}$ (comparing well with the calculated value of α_t, which for $\lambda = 3.2$ cm and $\epsilon_{2r} = 2.1$ is $\sim$5.2 dB-(cm)$^{-1}$), being measurable out to a distance of $\sim$7 cm. In another experiment, 1.25-cm waves were used with a metal plate behind the totally reflecting boundary; see W. Culshaw and D. S. Jones, Effect of Metal Plate on Total Reflection, *Proc. Phys. Soc.,* B, 66, 859, 1954.

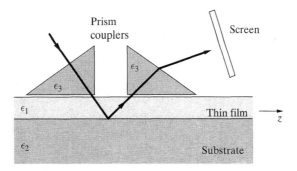

Figure 9.41   The prism coupler. The high-refractive-index prisms are in fact tightly clamped on the film, with dust particles serving as the air gap.

thin film waveguide and couple electromagnetic power into a single chosen (as determined by θ_i; see Section 10.2.3) waveguide mode. A second prism is used to decouple the energy at some distance away. The measurement of the output beam can be used to determine the refractive index and thickness of the thin film slab.[40]

Total internal reflection is widely utilized in a host of optical applications, including optical waveguides and binocular optics. Examples 9.19 and 9.20 illustrate the application of total internal reflection to the refraction of light by prisms.

Example 9.19: Refraction of light by prisms. Consider a light ray normally incident on one side of each of three different right angled glass ($n = 1.52$) prisms, as shown in Figure 9.42. Determine the side of the prism at which the beam of light first exits the prism and the exit transmission angle θ_{te} (measured with respect to the normal to that side) in all three cases.

Solution:

(a) Note that the critical angle for a glass–air interface is $\theta_{ic} = \sin^{-1}[(1.52)^{-1}] \simeq 41.14°$. Thus, there is no total internal reflection in this case, and, based on the ray geometry shown, light exits the prism at point A. Using Snell's law of refraction, we have

$$(1.52)\sin 30° = \sin \theta_t \quad \rightarrow \quad \theta_t = \theta_{te} \simeq 49.5°$$

(b) In this case, light is incident on the hypotenuse of the prism at an incidence angle $\theta_i = 45° > \theta_{ic}$. Accordingly, total internal reflection occurs at point A, and the light ray first exits the prism at point B. Since the incidence to the glass–air interface at B is normal to this side of the prism, the exit angle is $\theta_{te} = 0$.

(c) In this case, once again, total internal reflection occurs at point A. The light ray first exits the prism at point B, at an angle of $\theta_{te} \simeq 49.5°$. The incidence on the glass–air interface at point B in this case is identical to the incidence at point A in case (a).

---

[40]For a detailed discussion of the principles and operation of the prism coupler, see Chapter 6 of D. L. Lee, *Electromagnetic Principles of Integrated Optics,* Wiley, New York, 1986.

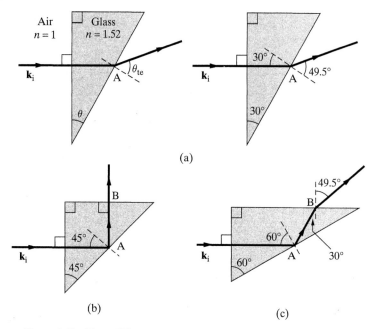

Figure 9.42   Three different right-angle prisms. Figure for Example 9.19.

Example 9.20: Effect of manufacturing errors. Consider the 45°–90°–45° right-angled prism analyzed in part (b) of Example 9.19. Assume that due to manufacturing error, one of the angles of the prism is $45 + \zeta$, where $\zeta < 1°$, and the other is the complementary angle $45 - \zeta$, as shown in Figure 9.43. Find the effect of this error on the exit transmission angle at point B.

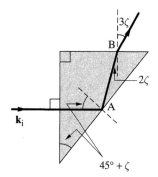

Figure 9.43   A 45° right-angled prism with a manufacturing error. Prism for Example 9.20.

Solution: Considering the worst-case error ($\zeta = 1°$), we have at point A

$$1.52 \sin 46° = \sin \theta_{tA} \quad \rightarrow \quad \sin \theta_{tA} \simeq 1.09 > 1$$

so total reflection occurs at point A. The ray is now incident on the glass–air interface at point B at an angle 2ζ. We apply Snell's law of refraction at point B at an incidence angle of 2ζ:

$$1.52\sin(2\zeta) \simeq (1.52)2\zeta = \sin\theta_{tB} \quad \rightarrow \quad \theta_{tB} = \sin^{-1}(3.04\zeta) \simeq 3.04\zeta$$

We note that small manufacturing errors can lead to relatively large deviations of the ray at its eventual exit point.[41]

9.8 OBLIQUE INCIDENCE ON A LOSSY MEDIUM

Reflection and refraction of uniform plane waves obliquely incident on a lossy medium is a topic of significant practical importance. In the context of HF, VHF, and UHF radio wave propagation, reflections from imperfectly conducting ground interfere with the direct signal between a transmitter and receiver (see Figure 9.44a), causing constructive or destructive interference at the receiver.[42] Accurate modeling of the propagation channel requires the knowledge of the reflection coefficient, which in general is complex, depending in a more complicated manner on the incidence angle θ_i. The rapid expansion of mobile radio, cellular telephone, and other wireless communications systems has brought renewed attention to the topic, because the propagation channel, which may be

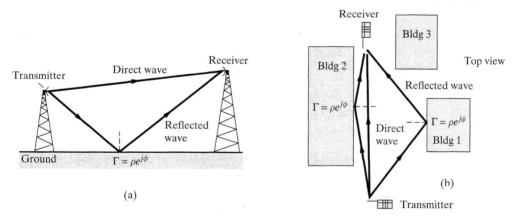

Figure 9.44   Illustration of direct and reflected waves. (a) Radio wave (e.g., HF, VHF, or UHF) or microwave link between a transmitter and receiver, with the ground (or seawater or a lake surface) behaving as a lossy reflector. (b) A mobile radio environment (shown in top view) where surrounding buildings or structures act as lossy reflecting surfaces.

[41] See Section 4.17 of W. J. Smith, *Modern Optical Engineering: The Design of Optical Systems,* McGraw-Hill, New York, 1966.

[42] See Chapter 6 of R. E. Collin, *Antennas and Radiowave Propagation,* McGraw-Hill, New York, 1985.

severely obstructed by buildings, mountains, and foliage, places fundamental limitations on the performance of such systems.[43] The problem of radio wave reflection from ground has been under extensive study for some time,[44] but characterization of the mobile radio environment requires new measurements and modeling[45] of reflection from boundaries such as buildings and concrete structures (see Figure 9.44b) at frequencies of a few GHz and higher. Given the enormous growth and demand for personal communication systems, and the anticipated expansion of these systems to nearly every aspect of our daily lives, a thorough understanding of the electromagnetic wave propagation in the presence of multiple lossy reflecting surfaces is of essential importance. Better understanding of reflecting properties of lossy surfaces is also important in other recently emerging consumer applications, such as millimeter-wave (30–100 GHz) systems used in car collision avoidance radar systems.[46]

The general treatment of this seemingly simple problem of a plane wave incident on a lossy plane surface is amazingly complex, because the refracted wave in the second medium is highly nonuniform and propagates in a direction determined by an effective refractive index whose magnitude is dependent on the incidence angle θ_i. While numerical results can be obtained for any general case, the physical nature of the reflected and refracted waves can be understood only in special cases (e.g., medium 2 being a good conductor, so that $\sigma_2 \gg \omega\epsilon_2$).

Before we proceed, it is useful to review the basis for Snell's law for the lossless case. The governing Maxwell's equations for a source-free and lossless medium are repeated here for convenience:

$$\nabla \times \mathbf{E} = -j\omega\mu\mathbf{H} \tag{9.46a}$$

$$\nabla \times \mathbf{H} = j\omega\epsilon\mathbf{E} \tag{9.46b}$$

Equations (9.46) are the basis for uniform plane wave propagation in each of the dielectric media with propagation constant $\beta = \omega\sqrt{\mu\epsilon}$, leading to wave solutions of the type

$$(\cdots)\,e^{-j\beta\,\hat{\mathbf{k}}\cdot\mathbf{r}}$$

and ultimately (on application of the boundary conditions) to Snell's law (equation (9.26)), repeated here as

$$\frac{\sin\theta_t}{\sin\theta_i} = \left(\frac{\mu_1\epsilon_1}{\mu_2\epsilon_2}\right)^{1/2} = \frac{\beta_1}{\beta_2} \tag{9.47}$$

---

[43] For a relatively simple discussion, see T. S. Rappaport, *Wireless Communications, Principles and Practices*, Prentice-Hall, Englewood Cliffs, New Jersey, 1996. For a recent discussion of current problems, see the Special Issue on Wireless Communications of *IEEE Trans. Antennas Propagation*, 46(6), June 1998.

[44] See Chapter 16 of E. C. Jordan and K. G. Balmain, *Electromagnetic Waves and Radiating Systems*, 2nd ed., Prentice-Hall, Englewood Cliffs, New Jersey, 1968.

[45] O. Landron, M. J. Feuerstein, and T. S. Rappaport, A comparison of theoretical and empirical reflection coefficients for typical exterior wall surfaces in a mobile radio environment, *IEEE Trans. Antennas Propagation*, 44(3), March 1996.

[46] K. Sato, T. Manabe, J. Polivka, T. Ihara, Y. Kasashima, and K. Yamaki, Measurement of the complex refractive index of concrete at 57.5 GHz, *IEEE Trans. Antennas Propagation*, 44(1), January 1996.

Now consider a plane wave obliquely incident on a lossy dielectric material, characterized by the parameters σ_2, ϵ_2, and μ_2. The governing equation corresponding to (9.46b) for the transmitted wave in medium 2 is now the complete Maxwell's equation (7.23c):

$$\nabla \times \mathbf{H}_2 = \sigma_2 \mathbf{E}_2 + j\omega\epsilon_2 \mathbf{E}_2 = j\omega\left(\epsilon_2 - \frac{j\sigma_2}{\omega}\right)\mathbf{E}_2 = j\omega\epsilon_{\text{eff}}\mathbf{E}_2 \qquad (9.48)$$

where $\epsilon_{\text{eff}} = [\epsilon_2 - j\sigma_2/\omega]$. The corresponding wave solutions can be written down by analogy with the lossless case and have the form

$$(\cdots)\,e^{-\gamma_2\hat{\mathbf{k}}\cdot\mathbf{r}} \qquad (9.49)$$

where the propagation constant γ_2 in the lossy medium 2 is given by

$$\gamma_2 \equiv \alpha_2 + j\beta_2 = j\omega\sqrt{\mu_2\,\epsilon_{\text{eff}}} = j\omega\left[\mu_2\left(\epsilon_2 - \frac{j\sigma_2}{\omega}\right)\right]^{1/2} \qquad (9.50)$$

Again by analogy with (9.47), we can write Snell's law for this case as

$$\frac{\sin\theta_t}{\sin\theta_i} = \frac{j\beta_1}{\gamma_2} = \frac{j\beta_1}{\alpha_2 + j\beta_2} = \left[\frac{\mu_1\epsilon_1}{\mu_2\left(\epsilon_2 - \dfrac{j\sigma_2}{\omega}\right)}\right]^{1/2} \qquad (9.51)$$

from which we realize that θ_t must in general be a complex number. Note that the sine of a complex number does not pose a mathematical problem, since for a general complex number $\theta_t = \theta_{tr} + j\theta_{ti}$ we can evaluate $\sin(\theta_t)$ as

$$\sin\theta_t = \frac{e^{j\theta_t} - e^{-j\theta_t}}{2j} = \sin(\theta_{tr})\cosh(\theta_{ti}) + j\cos(\theta_{tr})\sinh(\theta_{ti}) \qquad (9.52)$$

so that $\sin\theta_t$ is simply another complex number. It is possible to proceed at this point by expressing θ_t in terms of its real and imaginary parts and rewriting Snell's law in terms of two separate equations (obtained, respectively, from the real and imaginary parts of (9.52)).[47] However, better physical insight into the nature of the refracted wave in medium 2 is obtained in Section 9.8.1 by expanding[48] (into its real and imaginary parts) the quantity $\gamma_2\cos\theta_t$. We proceed by considering the refracted electromagnetic field in medium 2.

---

[47] For a compact treatment along such lines, see Section 14.4 of R. W. P. King and S. Prasad, *Fundamental Electromagnetic Theory and Applications*, Prentice-Hall, Englewood Cliffs, New Jersey, 1986.

[48] An expanded version of the brief discussion provided here can be found in Sections 9.8 and 9.10 of J. A. Stratton, *Electromagnetic Theory*, McGraw-Hill, New York, 1941.

9.8.1 The Refracted Electromagnetic Field in Medium 2

With reference to Figure 9.45, the electric field phasor of the refracted wave in medium 2, for the case of perpendicular polarization[49] is given by

$$\mathbf{E}_t(x, z) = \hat{\mathbf{y}} E_{t0}\, e^{-\gamma_2\, \hat{\mathbf{k}}_t \cdot \mathbf{r}} = \hat{\mathbf{y}} E_{t0}\, e^{-\gamma_2\, (x\, \sin\theta_t + z\, \cos\theta_t)} \tag{9.53}$$

which is identical to the expression provided in Section 9.6.1 for $\mathbf{E}_t(x, z)$ except for the fact that $j\beta_2$ is now replaced by γ_2. We now have to rewrite this expression by taking into account of the fact that γ_2, $\sin\theta_t$, and $\cos\theta_t$ are all complex numbers.

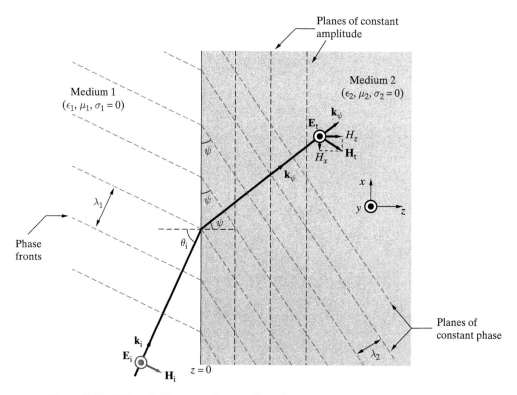

Figure 9.45   Oblique incidence on a lossy medium: Constant amplitude and phase fronts. The planes of constant amplitude are parallel to the interface, while the planes of constant phase are perpendicular to $\mathbf{k}_\psi$, which is at an angle ψ from the normal. This "true" angle ψ is given by (9.61) and is also equal to the real part of the complex angle θ_t defined by (9.57).

---

[49]Note that although (9.53) is written for perpendicular polarization, the spatial variation of both components of the electric field for the parallel polarization case is also proportional to $e^{-\gamma_2 \hat{\mathbf{k}}_t \cdot \mathbf{r}}$. Thus, the properties of the refracted wave as discussed here are also exhibited by the refracted wave in the parallel polarization case.

Noting that the total electric field phasor in medium 1 consists of the sum of the incident and reflected fields, we have

$$\mathbf{E}_1(x, z) = \hat{\mathbf{y}} E_{i0} \, e^{-j\beta_1 (x \, \sin\theta_i + z \, \cos\theta_i)} + \hat{\mathbf{y}} E_{r0} \, e^{-j\beta_1 (x \, \sin\theta_r - z \, \cos\theta_r)} \qquad (9.54)$$

The boundary condition on the continuity of the tangential electric field at the interface ($z = 0$) requires that $\mathbf{E}_1(x, 0) = \mathbf{E}_t(x, 0)$ or

$$\hat{\mathbf{y}} E_{i0} \, e^{-j\beta_1 x \, \sin\theta_i} + \hat{\mathbf{y}} E_{r0} \, e^{-j\beta_1 x \, \sin\theta_r} = \hat{\mathbf{y}} E_{t0} \, e^{-\gamma_2 x \, \sin\theta_t} \qquad (9.55)$$

For (9.55) to hold at all values of x, we must have three conditions:

$$\theta_i = \theta_r \quad \text{and} \quad E_{i0} + E_{r0} = E_{t0} \qquad (9.56)$$

and

$$\boxed{\gamma_2 \, \sin\theta_t = j\beta_1 \, \sin\theta_i} \qquad (9.57)$$

where (9.57) is the generalized form of Snell's law of refraction derived in Section 9.6 for an interface between two lossless dielectrics and also written in (9.51) by analogy with (9.47). We thus see that although γ_2 and $\sin\theta_t$ are both complex, their product is purely imaginary, as is evident from (9.57). On the other hand, the product $\gamma_2 \cos\theta_t$ is a complex number and we define its real and imaginary parts as

$$\gamma_2 \cos\theta_t = \gamma_2\sqrt{1 - \sin^2\theta_t} = \sqrt{\gamma_2^2 + \beta_1^2 \sin^2\theta_i} \equiv p + jq \qquad (9.58a)$$

$$p \equiv \mathfrak{Re}\left\{\sqrt{\gamma_2^2 + \beta_1^2 \sin^2\theta_i}\right\} \quad \text{and} \quad q \equiv \mathfrak{Im}\left\{\sqrt{\gamma_2^2 + \beta_1^2 \sin^2\theta_i}\right\} \qquad (9.58b)$$

By substituting (9.57) and (9.58a) into (9.53), we can rewrite the refracted electric field in medium 2 as

$$\boxed{\mathbf{E}_t(x, z) = \hat{\mathbf{y}} E_{t0} \, e^{-p\,z} e^{-j\,(x\,\beta_1 \, \sin\theta_i + q\,z)}} \qquad (9.59)$$

Examination of (9.59) indicates that the surfaces of constant amplitude (i.e., $pz = \text{constant}$) of the refracted wave are parallel to the interface. The planes of constant phase are given by

$$qz + \beta_1 x \sin\theta_i = \text{constant}$$

$$\mathbf{k}_\psi \cdot \mathbf{r} = \text{constant}$$

$$|\mathbf{k}_\psi| \, (\hat{\mathbf{k}}_\psi \cdot \mathbf{r}) = \text{constant} \qquad (9.60)$$

where $\mathbf{k}_\psi = \hat{\mathbf{z}} q + \hat{\mathbf{x}} \beta_1 \sin\theta_i$ and $\hat{\mathbf{k}}_\psi = \hat{\mathbf{x}} \sin\psi + \hat{\mathbf{z}} \cos\psi$ is the unit vector in the direction of $\mathbf{k}_\psi$. Note that $|\mathbf{k}_\psi| = \sqrt{q^2 + \beta_1^2 \sin^2\theta_i}$. We see from (9.60) that the refracted wave propagates at the "true" angle of refraction[50] ψ, which is defined by

$$\tan \psi = \frac{\beta_1 \sin \theta_i}{q} \tag{9.61}$$

It is clear that, as in the case of total internal reflection, the refracted wave in medium 2 is a nonuniform wave. The planes of constant amplitude are parallel to the boundary, as in the case of total internal reflection. However, in contrast to the case of total internal reflection, the planes of constant phase are not perpendicular to the boundary but are instead inclined at an angle ψ from the normal to the interface. This behavior is depicted in Figure 9.45.

Note that we have

$$\sin \psi = \frac{\mathbf{k}_\psi \cdot \hat{\mathbf{x}}}{|\mathbf{k}_\psi|} = \frac{\beta_1 \sin \theta_i}{\sqrt{q^2 + \beta_1^2 \sin^2 \theta_i}} \tag{9.62}$$

so that a geometrical refractive index for the lossy material can be defined which relates this true angle ψ to the incident angle θ_i via a modified Snell's law of refraction in terms of real angles as

$$\frac{\sin \theta_i}{\sin \psi} = \frac{\sqrt{q^2 + \beta_1^2 \sin^2 \theta_i}}{\beta_1} \equiv n(\theta_i) \tag{9.63}$$

Note that this geometrical index of refraction describing the angle ψ at which the refracted wave propagates is a function of the incident angle θ_i. The phase velocity in the lossy material, defined as the velocity of propagation of the planes of constant phase, is

$$v_{p2}(\theta_i) = \frac{\omega}{|\mathbf{k}_\psi|} = \frac{\omega}{\sqrt{q^2 + \beta_1^2 \sin^2 \theta_i}} = \frac{\omega}{\beta_1 n(\theta_i)} = \frac{v_{p1}}{n(\theta_i)} \tag{9.64}$$

---

[50]This true angle of refraction is in fact the real part θ_{tr} of the imaginary angle θ_t. By expanding $\sin \theta_t$ using (9.52), Snell's law (9.57) can be written as two equations

$$\beta_2 \sin(\theta_{tr})\cosh(\theta_{ti}) - \alpha_2 \cos(\theta_{tr})\sinh(\theta_{ti}) = \beta_1 \sin(\theta_i)$$

$$\alpha_2 \sin(\theta_{tr})\cosh(\theta_{ti}) + \beta_2 \cos(\theta_{tr})\sinh(\theta_{ti}) = 0$$

Solving these equations we find

$$\tanh(\theta_{ti}) = -\frac{\alpha_2}{\beta_2} \tan(\theta_{tr}) \qquad \text{and} \qquad \sin(\theta_{tr}) \left\{ \frac{\beta_2^2 + \alpha_2^2}{[\beta_2^2 + \alpha_2^2 \tan^2(\theta_{tr})]^{1/2}} \right\} = \beta_1 \sin(\theta_i)$$

the latter of which is Snell's law in terms of real angles, essentially equivalent to (9.63). For further details of an analysis in terms of θ_{tr} and θ_{ti} see Section 14.4 of R. W. P. King and S. Prasad, *Fundamental Electromagnetic Theory and Applications,* Prentice-Hall, Englewood Cliffs, New Jersey, 1986.

Note that the phase velocity depends on the incident angle and can exceed the velocity of light in medium 2. The wavelength in medium 2 can similarly be defined in terms of the refractive index $n(\theta_i)$ as

$$\lambda_2 = \frac{\lambda_1}{n(\theta_i)} = \frac{2\pi}{\sqrt{q^2 + \beta_1^2 \sin^2 \theta_i}} \tag{9.65}$$

The quantities $n(\theta_i)$, p, and q can be written out explicitly in terms of the phase constant β_2 and the attenuation constant α_2 of medium 2. Useful relations between these quantities are

$$\beta_1^2 \, n(\theta_i) = \beta_2^2 - \alpha_2^2; \qquad p(\theta_i) \, q(\theta_i) = \alpha_2 \beta_2; \qquad \beta_1 \, p(\theta_i) \, n(\theta_i) = \frac{\alpha_2 \beta_2}{\cos \psi} \tag{9.66}$$

where we note that p and q are also functions of the incidence angle θ_i. In principle, the incidence angle-dependent refractive index $n(\theta_i)$ can be optically measured (assuming that medium 2 is sufficiently transparent) and (9.66) can be used to determine the properties of medium 2 (i.e., α_2 and β_2).

As in the case of the refracted wave for total internal reflection (see Section 9.7.2), the refracted wave described by (9.59) has field components in its propagation direction $\hat{k}_\psi$. To see this, we note that with the electric field taken to be in the y direction, that is, $\mathbf{E}_t(x, z) = \hat{y} E_{ty}(x, z)$, we have

$$\nabla \times \mathbf{E}_t = -j\omega\mu_0 \mathbf{H}_t \qquad\qquad -\frac{\partial E_{ty}}{\partial z} = -j\omega\mu_0 H_{tx}$$

$$\mathbf{E}_t(x, z) = \hat{y} E_{t0} e^{-pz} e^{-j(x\beta_1 \sin\theta_i + qz)} \qquad \rightarrow \qquad \frac{\partial E_{ty}}{\partial x} = -j\omega\mu_0 H_{tz}$$

which upon manipulation yields

$$\mathbf{H}_t(x, z) = \frac{E_{t0}}{\omega\mu_0} \left[\hat{x} j \, (p + jq) + \hat{z} \, \beta_1 \sin \theta_i \right] e^{-pz} e^{-j(x\beta_1 \sin\theta_i + qz)}$$

Since in general $\mathbf{k}_\psi \cdot \mathbf{H}_t \neq 0$, we see that $\mathbf{H}_t$ in general has a component along the propagation direction $\mathbf{k}_\psi$. As in the case of total internal reflection, the two components of this magnetic field are not in phase; consequently, the magnetic field in medium 2 rotates in an elliptical path about the electric field $\mathbf{E}_t$ (i.e., in the xz plane).

Refraction into a good conductor.   As an important limiting case, we now consider oblique incidence of uniform plane waves on a good conductor, for which $\sigma_2 \gg \omega\epsilon_2$. In this case, we note from Section 8.3.2 that

$$\alpha_2 \simeq \beta_2 \simeq \sqrt{\frac{\omega\mu_2\sigma_2}{2}} \tag{9.67}$$

so that we have from (9.51) (also assuming $\mu_1 = \mu_2$)

$$\frac{\sin \theta_t}{\sin \theta_i} = \frac{j\beta_1}{\alpha_2 + j\beta_2} = \frac{j\omega\sqrt{\mu_1 \epsilon_1}}{\sqrt{\frac{\omega \mu_2 \sigma_2}{2}} + j\sqrt{\frac{\omega \mu_2 \sigma_2}{2}}} = \sqrt{\frac{\omega \epsilon_1}{\sigma_2}}\, e^{j\pi/4} \qquad (9.68)$$

indicating that $\sin \theta_t \to 0$ for large σ_2, which in turn means that $\cos \theta_t \to 1$, so that from (9.58a) we have

$$\gamma_2 \cos \theta_t \simeq \gamma_2 = \alpha_2 + j\beta_2 \simeq p + jq$$

$$\sqrt{\frac{\omega \mu_2 \sigma_2}{2}} + j\sqrt{\frac{\omega \mu_2 \sigma_2}{2}} \simeq p + jq$$

$$\to \quad p \simeq \alpha_2 \simeq \sqrt{\frac{\omega \mu_2 \sigma_2}{2}} \quad \text{and} \quad q \simeq \beta_2 \simeq \sqrt{\frac{\omega \mu_2 \sigma_2}{2}} \qquad (9.69)$$

and therefore from (9.61) we find

$$\psi = \tan^{-1}\left[\frac{\beta_1 \sin \theta_i}{q}\right] \simeq \tan^{-1}\left[\frac{\omega\sqrt{\mu_1 \epsilon_1}\, \sin \theta_i}{\sqrt{\frac{\omega \mu_2 \sigma_2}{2}}}\right] \to 0 \quad \text{for large } \sigma_2 \qquad (9.70)$$

Thus, we see that when medium 2 is a good conductor with $\sigma_2 \gg \omega \epsilon_2$, the true angle of refraction ψ tends to zero, and the planes of constant phase are oriented parallel to the reflecting plane boundary and to the planes of constant amplitude (which, based on (9.59), are generally parallel to the interface for any σ_2), as shown in Figure 9.46. The fact that the angle ψ is indeed extremely small in practice can be seen by assuming nonmagnetic materials ($\mu_1 = \mu_0$ and $\mu_2 = \mu_0$) and evaluating (9.70) at the highest end of microwave frequencies (e.g., 100 GHz) for a typical conductor (e.g., copper, so that $\sigma_2 \simeq 5.8 \times 10^7$ S-m^{-1} and $\epsilon_1 \simeq \epsilon_0$) and assuming the angle of incidence to be as large as possible (i.e., $\sin \theta_i \simeq 1$). We find using (9.70)

$$\psi \simeq \tan^{-1}\left[\frac{\omega\sqrt{\mu_0 \epsilon_1}\, \sin \theta_i}{\sqrt{\frac{\omega \mu_0 \sigma_2}{2}}}\right] \simeq \tan^{-1}\left[\sqrt{\frac{2\omega \epsilon_1}{\sigma_2}}\right] \simeq 0.025°$$

which means that $\mathbf{k}_\psi$ is indistinguishably close to being perpendicular to the interface. It is important to note that the angle ψ is very close to zero even for materials having much lower conductivity than metals. As another example, for seawater[51] at 1 MHz ($\sigma_2 \simeq 4$ S-m^{-1} and $\epsilon_2 \simeq 81\epsilon_0$), for $\sin \theta_i \simeq 1$ we find $\psi < 1°$. Using the approximate

---

[51]The fact that the transmitted wave penetrates the ocean vertically, independent of θ_i, is important for communication with deeply submerged submarines using extremely low-frequency (ELF) signals guided by the Earth–ionosphere waveguide (see Example 10.2).

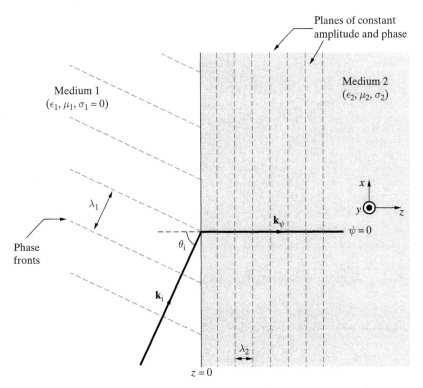

Figure 9.46   Oblique incidence on a good conductor ($\sigma_2 \gg \omega\epsilon_2$). The planes of constant amplitude and constant phase are both parallel to the interface, the former being exactly parallel as in the general case, while the planes of constant phase are very nearly so, with ψ differing from zero by an imperceptible amount.

expressions (9.69) for p and q in (9.59), we can write the electric field in medium 2 as

$$\mathbf{E}_t(x, z) \simeq \hat{\mathbf{y}} E_{t0} e^{-\alpha_2 z} e^{-j (x \beta_1 \sin \theta_i + \beta_2 z)} \qquad (9.71)$$

which indicates that the skin depth as defined in Section 8.3.2, namely $\delta \simeq \alpha_2^{-1} = \sqrt{2/(\omega\mu_2\sigma_2)}$, is indeed a valid measure of effective penetration of an electromagnetic wave into a good conductor, regardless of the angle θ_i at which the wave is incident on the medium. The concept of skin depth thus applies to an electromagnetic wave incident at any angle on a good conductor. Independent of the angle of incidence θ_i, the wave transmitted (or refracted) into medium 2 can be considered to be a plane wave propagating normal to the surface, which is heavily attenuated as it propagates, as expected for an electromagnetic wave in a good conductor.

The fact that the wave transmitted into a good conductor propagates nearly normal to the surface regardless of the incident angle θ_i greatly broadens the applicability of the surface resistance method of characterizing power loss in a conductor, discussed in Section 9.4. Thus, the procedure described in Section 9.4 for the specific case of normal incidence is also valid for oblique incidence. In most cases, the calculation of the power loss using $\frac{1}{2}|H_t|^2 R_s$ (equation (9.17)) is entirely valid for arbitrary conducting surfaces having any incident electromagnetic fields.[52]

Note that the approximate expression (9.69) for q when $\sigma_2 \gg \omega\epsilon_2$ can be used in (9.63) to write an expression for the refractive index $n(\theta_i)$. We find

$$n(\theta_i) = \frac{\sqrt{q^2 + \beta_1^2 \sin^2\theta_i}}{\beta_1} = \sqrt{\frac{q^2}{\beta_1^2} + \sin^2\theta_i} \simeq \sqrt{\frac{\mu_2\sigma_2}{2\omega\mu_1\epsilon_1} + \sin^2\theta_i} \simeq \sqrt{\frac{\mu_2\sigma_2}{2\omega\mu_1\epsilon_1}} \quad (9.72)$$

since typically $\sigma_2 \gg \omega\epsilon_1$ and $\mu_1 = \mu_2 = \mu_0$, while $\sin\theta_i < 1$. We thus see that the refractive index for a good conductor is independent of the incident angle θ_i. Substituting numerical values for the previously discussed cases of copper at 100 GHz and seawater at 1 MHz yield very large refractive index values of, respectively, $n \simeq 2280$ and $n \simeq 190$, which in turn indicate from (9.65) that the wavelength λ_2 in a good conducting medium 2 is typically much smaller than it is in medium 1.

9.8.2 Oblique Incidence Reflection from a Lossy Medium

We have so far focused our attention on the refracted wave in the second medium. More important in practice—especially in applications involving reflection of radio waves from the ground and seawater or reflection of cellular phone signals from buildings and walls—are the characteristics of the reflected signals. The problem of reflection in oblique incidence from a lossy medium is generally quite involved, because the reflection coefficients are both complex and also dependent on the polarization of the incident wave.[53] Fortunately, however, the numerical evaluation of the reflection coefficients is quite straightforward, since the reflection coefficients given by equations (9.31) (for perpendicular polarization) and (9.33) (for parallel polarization) are fully valid as long as we use the complex values of $\cos\theta_t$ and η_2. Using $\sin\theta_t$ as given in (9.57), adopting $\epsilon_{eff} = [\epsilon_2 - j\sigma_2/\omega]$ as defined in (9.48) to be the effective permittivity of medium 2, and noting that $\gamma_2 = j\omega\sqrt{\mu_2\epsilon_{eff}}$ as given in (9.50), we can write

$$\cos\theta_t = \sqrt{1 - \sin^2\theta_t} = \sqrt{1 + \frac{\beta_1^2 \sin^2\theta_i}{\gamma_2^2}}$$

$$= \sqrt{1 - \frac{\mu_1\epsilon_1 \sin^2\theta_i}{\mu_2\epsilon_{eff}}} = \sqrt{1 - \frac{j\omega\mu_1\epsilon_1 \sin^2\theta_i}{\mu_2(\sigma_2 + j\omega\epsilon_2)}} \quad (9.73a)$$

---

[52]This method is not applicable if the conductor surface is curved with a radius of curvature comparable to the skin depth δ. Given the extremely small skin depths for most good conductors (see Section 8.3.2), conductors with such small radii of curvature are not encountered in practice.

[53]The rich variety of resulting polarization effects have been well utilized to determine the optical constants of metals from measurements of reflected light. Note that while it is in general not possible to observe the refracted wave in a highly absorbing (i.e., lossy) medium, the reflected wave can be well measured and the dielectric properties of the reflecting medium can be deduced from polarization properties of the reflected wave. See Section 13.2 of M. Born and E. Wolf, *Principles of Optics*, 5th ed., Pergamon Press, Oxford, 1975. Also see Section II-20 of A. R. Von Hippel, *Dielectrics and Waves*, Wiley, New York, 1954.

$$\eta_2 = \sqrt{\frac{\mu_2}{\epsilon_{\text{eff}}}} = \sqrt{\frac{j\omega\mu_2}{\sigma_2 + j\omega\epsilon_2}} \tag{9.73b}$$

With $\cos\theta_t$ and η_2 as given in (9.73), the reflection coefficients for perpendicular and parallel polarization cases are given by (9.31) and (9.33) as

$$\Gamma_\perp = \frac{\eta_2\cos\theta_i - \eta_1\cos\theta_t}{\eta_2\cos\theta_i + \eta_1\cos\theta_t} = \rho_\perp e^{j\phi_\perp} \tag{9.74a}$$

$$\Gamma_\parallel = \frac{-\eta_1\cos\theta_i + \eta_2\cos\theta_t}{\eta_1\cos\theta_i + \eta_2\cos\theta_t} = \rho_\parallel e^{j\phi_\parallel} \tag{9.74b}$$

It is clear from (9.74) that both $\Gamma_\perp$ and $\Gamma_\parallel$ are in general complex, and that furthermore, in general we have $\phi_\perp \neq \phi_\parallel$. This in turn means that a linearly polarized incident wave, polarized such that its electric field is neither parallel nor perpendicular to the interface, in general becomes elliptically polarized upon reflection by the lossy medium.[54] However, under certain circumstances (e.g., when the incident wave is polarized at an angle of 45° with respect to the plane of incidence, $\Delta\phi = \phi_\perp - \phi_\parallel = 90°$, $\rho_\perp = \rho_\parallel$), the reflected wave is circularly polarized.

Numerical evaluation of the reflection coefficients as given in (9.74) is straightforward, using $\cos\theta_t$ and η_2 as given in (9.73a) and (9.73b), respectively, for any given set of medium parameter values ϵ_1, μ_1, ϵ_2, μ_2, and σ_2. If medium 2 is a lossy dielectric, we may have $\sigma_2 = 0$ but instead have a complex permittivity $\epsilon_{2c} = \epsilon_2' - j\epsilon_2''$. In such a case, (9.73) can still be used by making the substitution $\epsilon_{\text{eff}} \leftrightarrow \epsilon_{2c}$.

9.9 SUMMARY

This chapter discussed the following topics:

- **Normal incidence on a perfect conductor.** When a uniform plane wave is normally incident from a dielectric (medium 1) onto a perfect conductor, a pure standing wave pattern is produced with the total electric and magnetic fields in medium 1 given by

$$\overline{\mathscr{E}}_1(z,t) = \hat{\mathbf{x}} 2E_{i0}\sin(\beta_1 z)\sin(\omega t)$$

$$\overline{\mathscr{H}}_1(z,t) = \hat{\mathbf{y}} 2\frac{E_{i0}}{\eta_1}\cos(\beta_1 z)\cos(\omega t)$$

The standing wave is entirely analogous to the standing wave that is produced on a short-circuited lossless uniform transmission line and does not carry electromagnetic energy, that is, the time-average power is zero.

---

[54]Note that this aspect of the problem of oblique reflection from a lossy medium is similar to the case of total internal reflection discussed in Section 9.7, except for the fact that in the present case $\rho_\perp$ and $\rho_\parallel$ are not unity.

- **Normal incidence on a lossless dielectric.** The reflection and transmission coefficients for the case of normal incidence of a uniform plane wave on a lossless dielectric are, respectively, given by

$$\Gamma = \frac{E_{r0}}{E_{i0}} = \frac{\eta_2 - \eta_1}{\eta_2 + \eta_1} \qquad \mathcal{T} = \frac{E_{t0}}{E_{i0}} = \frac{2\eta_2}{\eta_2 + \eta_1}$$

where η_1 and η_2 are the intrinsic impedances of the two dielectric media. Note that $(1 + \Gamma) = \mathcal{T}$. The total electric and magnetic fields in medium 1 each consist of a component propagating in the z direction and another that is a standing wave, and can be expressed in phasor form as

$$\mathbf{E}_1(z) = \hat{\mathbf{x}} E_{i0} e^{-j\beta_1 z}(1 + \Gamma e^{j2\beta_1 z})$$

$$\mathbf{H}_1(z) = \hat{\mathbf{y}} \frac{E_{i0}}{\eta_1} e^{-j\beta_1 z}(1 - \Gamma e^{j2\beta_1 z})$$

The time-average Poynting flux in medium 1 is equal to that in medium 2 and is given as

$$(\mathbf{S}_{av})_1 = \hat{\mathbf{z}} \frac{E_{i0}^2}{2\eta_1}(1 - \Gamma^2)$$

so the fraction of the incident power transmitted into medium 2 is determined by $(1 - \Gamma^2)$.

- **Multiple dielectric interfaces.** The effective reflection and transmission coefficients for a uniform plane wave incident on an interface consisting of two dielectrics are given as

$$\Gamma_{\text{eff}} = \frac{(\eta_2 - \eta_1)(\eta_3 + \eta_2) + (\eta_2 + \eta_1)(\eta_3 - \eta_2)e^{-j2\beta_2 d}}{(\eta_2 + \eta_1)(\eta_3 + \eta_2) + (\eta_2 - \eta_1)(\eta_3 - \eta_2)e^{-j2\beta_2 d}}$$

$$\mathcal{T}_{\text{eff}} = \frac{4\eta_2\eta_3 e^{-j\beta_2 d}}{(\eta_2 + \eta_1)(\eta_3 + \eta_2) + (\eta_2 - \eta_1)(\eta_3 - \eta_2)e^{-j2\beta_2 d}}$$

These expressions are valid even for lossy media, as long as the proper complex intrinsic impedances are used and $j\beta_2$ is replaced by $\gamma_2 = \alpha_2 + j\beta_2$. In general, interfaces involving multiple dielectrics can be analyzed by relying on the transmission line analogy and treating the problem as one involving impedance transformation by a set of cascaded transmission lines.

- **Normal incidence on a lossy medium.** When a uniform plane wave is normally incident on an imperfect conductor, the surface current that flows within the conductor leads to power dissipation, although the magnitude of the reflection coefficient is very nearly unity. The amount of power lost in the conductor is given by

$$P_{\text{loss}} = \tfrac{1}{2}|J_s|^2 R_s$$

where the surface resistance $R_s = (\delta\sigma)^{-1}$, with δ being the skin depth of the imperfect conductor, and $|J_s| = |\mathbf{H}_1(0)|$ with $\mathbf{H}_1(0)$ being the total tangential magnetic field on the conductor surface.

- **Oblique incidence on a perfect conductor.** When a uniform plane wave is obliquely incident at an angle θ_i on a perfect conductor, a reflected wave propagating away from the conductor at an angle $\theta_r = \theta_i$ is produced. The combination of the incident and reflected waves forms a standing wave pattern in the direction orthogonal to the boundary, with the electric field being identically zero at a discrete set of planes defined by

$$z = -\frac{m\lambda_1}{2\cos\theta_i} \qquad m = 1, 2, 3, \ldots$$

The orientation of the total electric and magnetic fields, and the associated surface currents that flow within the conductor, differ depending on whether the incident wave is parallel or perpendicularly polarized.

- **Oblique incidence at a dielectric boundary.** The reflection and refraction of a uniform plane wave obliquely incident at an angle θ_i to an interface between two dielectrics occurs in accordance with Snell's law, namely:

$$\frac{\sin\theta_i}{\sin\theta_t} = \frac{v_{p1}}{v_{p2}} = \sqrt{\frac{\epsilon_{2r}\mu_{2r}}{\epsilon_{1r}\mu_{1r}}} \quad \text{and} \quad \theta_r = \theta_i$$

where θ_r and θ_t are the angles from the vertical of, respectively the reflected and refracted (transmitted) rays. The reflection and transmission coefficients for perpendicular polarization are

$$\Gamma_\perp = \frac{\eta_2\cos\theta_i - \eta_1\cos\theta_t}{\eta_2\cos\theta_i + \eta_1\cos\theta_t}$$

$$\mathcal{T}_\perp = \frac{2\eta_2\cos\theta_i}{\eta_2\cos\theta_i + \eta_1\cos\theta_t}$$

with $(1 + \Gamma_\perp) = \mathcal{T}_\perp$. For parallel polarization we have

$$\Gamma_\| = \frac{-\eta_1\cos\theta_i + \eta_2\cos\theta_t}{\eta_1\cos\theta_i + \eta_2\cos\theta_t}$$

$$\mathcal{T}_\| = \frac{2\eta_2\cos\theta_i}{\eta_1\cos\theta_i + \eta_2\cos\theta_t}$$

with $(1 + \Gamma_\|) = \mathcal{T}_\|(\cos\theta_t/\cos\theta_i)$. An interesting case with many applications occurs for the case of parallel polarization when in the nonmagnetic case, the angle of incidence θ_i is equal to the Brewster angle θ_{iB}, given by

$$\theta_{iB} = \tan^{-1}\left[\sqrt{\frac{\epsilon_{2r}}{\epsilon_{1r}}}\right]$$

in which case $\Gamma_\| = 0$ and there is no reflected wave.

- **Total internal reflection.** When a uniform plane wave is incident at an angle θ_i from a denser medium into a less dense medium (i.e., $\epsilon_1 > \epsilon_2$), total internal reflection occurs for $\theta_i \geq \theta_{ic}$, where θ_{ic} satisfies

$$\sin \theta_{ic} = \sqrt{\frac{\epsilon_{2r}}{\epsilon_{1r}}}$$

Under conditions of total internal reflection, the reflection coefficients for parallel and perpendicular polarizations are both complex, with unity magnitude but, in general, different phase angles. When an incident wave is totally reflected, a nonuniform plane electromagnetic wave does exist in medium 2, but is heavily attenuated with distance into the medium, being confined to a spatial region of size comparable to the wavelength.

- **Oblique incidence on a lossy medium.** When a uniform plane wave is obliquely incident on a lossy material having a propagation constant $\gamma_2 = \alpha_2 + j\beta_2$, the generalized form of Snell's law is

$$\gamma_2 \sin \theta_t = j\beta_1 \sin \theta_i$$

indicating that the transmitted angle θ_t is a complex number. A modified version of Snell's law of refraction can be derived as follows

$$\frac{\sin \theta_i}{\sin \psi} = \frac{\sqrt{q^2 + \beta_1^2 \sin^2 \theta_i}}{\beta_1} \qquad \begin{aligned} q &\equiv \mathcal{I}m\{\gamma_2 \cos \theta_t\} \\ &= \mathcal{I}m\left\{\sqrt{\gamma_2^2 + \omega^2 \mu_1 \epsilon_1 \sin^2 \theta_i}\right\} \end{aligned}$$

where ψ is the "true" angle of refraction, $\beta_1 = \omega\sqrt{\mu_1 \epsilon_1}$, and q is the imaginary part of $\gamma_2 \cos \theta_t$. The refracted wave is a nonuniform plane wave, with planes of constant amplitude parallel to the interface, whereas the planes of constant phase are at an angle ψ with respect to the normal at the interface. When medium 2 is a good conductor, the true angle $\psi \simeq 0$, so that the planes of constant phase are nearly parallel to the interface, regardless of the incident angle θ_i. The reflection coefficients for oblique incidence on a lossy medium can be simply calculated using the general expressions (9.74) and the complex values of $\cos \theta_t$ and η_2.

PROBLEMS

9.1   **Air–perfect conductor interface.** A uniform plane electromagnetic wave traveling in air with its electric field given by

$$\overline{\mathscr{E}}_i(x,t) = \hat{\mathbf{y}}15 \cos(44 \times 10^9 t - \beta x) \quad \text{V-m}^{-1}$$

is normally incident on a perfect conductor boundary located at $x = 0$. (a) Find the phase constant β. (b) Find the corresponding magnetic field $\overline{\mathscr{H}}_i(x,t)$. (c) Find the electric and magnetic fields $\overline{\mathscr{E}}_r(x,t)$ and $\overline{\mathscr{H}}_r(x,t)$ of the reflected wave. (d) Find the nearest two positions in air away from the boundary where the total electric field is always zero.

9.2   Air–perfect conductor interface. A uniform plane wave of time-average power density 75 mW-cm^{-2} in air is normally incident on the surface of a perfect conductor located at $y = 0$, as shown in Figure 9.47. The total magnetic field phasor in air is given by

$$\mathbf{H}_1(y) = \hat{\mathbf{x}} H_0 \cos(100\pi y)$$

(a) What is H_0? (b) What is the frequency, f? (c) Find the total electric field $\overline{\mathcal{E}}_1(y,t)$ at $y = -3.5$ cm.

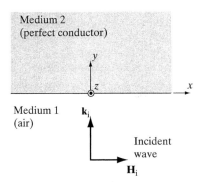

Figure 9.47   Normal incidence on a perfect conductor. Problem 9.2.

9.3   Air–perfect conductor interface. A uniform plane wave propagating in air given by

$$\mathbf{E}_i(z) = 40e^{-j30\pi z}(\hat{\mathbf{x}} - j\hat{\mathbf{y}}) \quad \text{V-m}^{-1}$$

is normally incident on a perfectly conducting plane located at $z = 0$. (a) Find the frequency and the wavelength of this wave. (b) Find the corresponding magnetic field $\mathbf{H}_i(z)$. (c) Find the electric and magnetic field vectors of the reflected wave [i.e., $\mathbf{E}_r(z)$ and $\mathbf{H}_r(z)$]. (d) Find the total electric field in air [i.e., $\mathbf{E}_1(z)$], and plot the magnitude of each of its components as a function of z.

9.4   Air–perfect conductor interface. A uniform plane wave of frequency 12 GHz traveling in free space having a magnetic field given by

$$\overline{\mathcal{H}}_i(y,t) = \hat{\mathbf{x}}60\cos(\omega t - \beta y) - \hat{\mathbf{z}}30\sin(\omega t - \beta y) \quad \text{mA-m}^{-1}$$

is normally incident on a perfect conductor boundary located at $y = 0$. (a) Find the real-time expression for the reflected wave, $\overline{\mathcal{H}}_r(y,t)$. (b) Compare the polarizations of the incident and the reflected waves. Is there any difference? (c) Find the maximum value of the total magnetic field at $y = 0$, -1.25 cm, -2.5 cm, -3.75 cm, and -5 cm, respectively.

9.5   Air-perfect conductor interface. A 10 mW-m^{-2}, 3 GHz uniform plane wave traveling in air is normally incident on a perfect conductor boundary located at the $x = 0$ plane, as shown in Figure 9.48. (a) Find the phasor-form electric and magnetic field vectors of the incident and reflected waves: $\mathbf{E}_i(x)$, $\mathbf{H}_i(x)$, $\mathbf{E}_r(x)$, and $\mathbf{H}_r(x)$. (b) Find the two closest positions to the boundary in air where the total magnetic field is always zero.

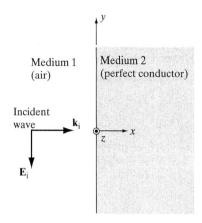

Figure 9.48   Normal incidence on a perfect conductor. Problem 9.5.

9.6   **Air-dielectric interface.** A 65 mW-m^{-2}, 3 GHz uniform plane wave traveling in the z direction in air is normally incident onto a planar, lossless, nonmagnetic dielectric interface with $\epsilon_r = 6$ located at $z = 0$, as shown in Figure 9.49. (a) Find the phasor-form electric and magnetic fields of the incident, reflected, and transmitted waves. (b) Find the two nearest positions in air with respect to the boundary where the total magnetic field is at a local minimum.

Figure 9.49   Normal incidence on a dielectric. Problem 9.6.

9.7   **Air–GaAs interface.** A uniform plane wave having a magnetic field given by

$$\mathbf{H}_i(y) = \hat{\mathbf{z}}20e^{-j200\pi y} \text{ mA-m}^{-1}$$

is normally incident from air onto a plane air–gallium arsenide (GaAs) interface located at $y = 0$. Assume GaAs to be a perfect dielectric with $\epsilon_r \simeq 12$ and $\mu_r = 1$. (a) Find the reflected ($\mathbf{H}_r(y)$) and the transmitted ($\mathbf{H}_t(y)$) magnetic fields. (b) Calculate the power density of the incident, reflected, and transmitted waves independently and verify the conservation of energy principle.

9.8   **Air–salty lake interface.** Consider a uniform plane wave traveling in air with its electric field given by

$$\overline{\mathscr{E}}_i(y, t) = \hat{\mathbf{y}}100 \sin(2\pi \times 10^9 t - \beta z) \text{ V-m}^{-1}$$

normally incident on the surface ($z = 0$) of a salty lake ($\sigma = 0.88$ S-m^{-1}, $\epsilon_r = 78.8$, and $\mu_r = 1$ at 1 GHz). (a) Assuming the lake to be perfectly flat and lossless (i.e., assume $\sigma = 0$), find the electric fields of the reflected and transmitted waves (i.e., $\overline{\mathscr{E}}_r$ and $\overline{\mathscr{E}}_t$). (b) Modify the $\overline{\mathscr{E}}_t$ expression found in part (a) by introducing in it the exponential attenuation term due to the nonzero conductivity of lake water, given in part (a), and justify this approximation. (c) Using this expression, find the thickness of the lake over which 90% of the power of the transmitted wave is dissipated. What percentage of the power of the incident wave corresponds to this amount?

9.9   **Aircraft–submarine communication.** A submarine submerged in the ocean is trying to communicate with a Navy airplane, equipped with a VLF transmitter operating at 20 kHz, approximately 10 km immediately overhead from the location of the submarine. If the output power of the VLF transmitter is 200 kW and the receiver sensitivity of the submarine is 1 μV-m^{-1}, calculate the maximum depth of the submarine from the surface of the ocean for it to be able to communicate with the transmitter. Assume the transmitter is radiating its power isotropically and assume normal incidence at the air–ocean boundary. Use $\sigma = 4$ S-m^{-1}, $\epsilon_r = 81$, and $\mu_r = 1$ for the properties of the ocean.

9.10   **Vertical incidence on plasma.** An ionized gas (plasma) consists of free electrons and ions, the motion of which can be represented in terms of a frequency-dependent dielectric constant given by

$$\epsilon_p = \epsilon_0 \left[1 - \left(\frac{f_p}{f} \right)^2 \right]$$

where f_p is known as the plasma frequency. Consider a plasma with $f_p = 1$ MHz onto which a uniform plane wave at frequency $f = 500$ kHz is normally ($\theta_i = 0$) incident from air ($\epsilon_1 = \epsilon_0$, $\mu_1 = \mu_0$). In this case, the dielectric constant for the second medium is $\epsilon_2 = 3\epsilon_0$. (a) Determine the reflection coefficient Γ and the complete time-domain expression for the electric field reflected wave. What fraction of the power carried by the incident wave is reflected? (b) Determine the complete time-domain expression for the electric field of the transmitted wave. (c) If the incident wave is left hand circularly polarized (LHCP), what is the polarization of the reflected and transmitted waves? Specify both the type (linear, circular, elliptical) and sense (RH or LH) of the polarization.

9.11   **Air–fat interface.** Consider a planar interface between air and fat tissue (assume it to be of semi-infinite extent). If a plane wave is normally incident from air at this boundary, find the percentage of the power absorbed by the fat tissue at (a) 100 MHz, (b) 300 MHz, (c) 915 MHz, and (d) 2.45 GHz, and compare your results with the results of Example 9.10. Use Table 9.1 for the parameters of the fat tissue and assume $\mu_r = 1$.

9.12   **Air–concrete interface.** A uniform plane wave operating at 1 GHz is normally incident from air onto the air–concrete interface. At 1 GHz, the complex relative dielectric constants of wet and dry concrete are measured as $\epsilon_{wr} \simeq 14.8 - j1.73$ and $\epsilon_{dr} \simeq 4.5 - j0.03$, respectively.[55] For each case, calculate (a) the percentage of the incident power reflected and (b) the penetration depth in the concrete. Assume concrete to be semi-infinite in extent with $\mu_r = 1$.

9.13   **Shielding with a copper foil.** A 1-GHz, 1-kW-m^{-2} microwave beam is incident upon a sheet of copper foil of 10 μm thickness (see Example 9.8 for the electromagnetic properties of copper). Consider neglecting multiple reflections, if justified. (a) Find the power density

[55]H. C. Rhim and O. Buyukozturk, Electromagnetic properties of concrete at microwave frequency range, *ACI Mater. J.*, 95(3), pp. 262–271, May–June 1998.

of the reflected wave. (b) Find the power density transmitted into the foil. (c) Find the power density of the wave that emerges from the other side of the foil. Comment on the shielding effectiveness of this thin copper foil.

9.14 **Absorbing material.** Consider a commercial absorber slab[56] made of EHP-48 material of 1 m thickness backed by a perfectly conducting metal plate. A 100-MHz uniform plane wave is normally incident from the air side at the air–absorber–metal interface. Find the percentage of incident power lost in the absorber material. For EHP-48, use $\epsilon_r = 6.93 - j8.29$ and $\mu_r = 1$ at 100 MHz.

9.15 **Radome design.** A common material in dielectric radomes for aeronautical applications is fiberglass. For L-band (1–2 GHz), fiberglass has a typical relative dielectric constant of approximately $\epsilon_r \simeq 4.6$. (a) Assuming a flat-plane radome, determine the minimum thickness of fiberglass that causes no reflections at the center of the L-band. (b) Using the thickness found in part (a), find the percentage of the incident power which transmits to the other side of the radome at each end of the L-band (i.e., 1 GHz and 2 GHz). Assume nonmagnetic case.

9.16 **Radome design.** A radome is to be designed for the nose of an aircraft to protect an X-band weather radar operating between 8.5 and 10.3 GHz. A new type of foam material with $\epsilon_r = 2.2$ (assume lossless) is chosen for the design. (a) Assuming a flat planar radome, determine the minimum thickness of the foam that will give no reflections at the center frequency of the band. Assume $\mu_r = 1$. What percent of the incident power is reflected at the center frequency?

9.17 **Transmission through a multilayered dielectric.** (a) Find the three lowest frequencies at which all of the incident power would be transmitted through the three-layer structure shown in Figure 9.50. The permeability of all three media is μ_0. (b) If complete transmission is required for *any* thickness of the center medium, what is the lowest usable frequency? (c) Find the bandwidth of the transmission, defined as the range between the two lowest percentage values adjacent to and on either side of the frequency found in (b). Also find the lowest percentage values of transmission. (d) Why does the reflection from multiply coated optical lenses tend to be purple in color?

Figure 9.50   Multilayered dielectric. Problem 9.17.

9.18 **Glass slab.** Consider a 0.5-cm thick slab of crown glass, with index of refraction $n = 1.48$. (a) If a beam of visible light at 490 nm is normally incident from one side of the

[56]C. L. Holloway and E. F. Kuester, A low frequency model for wedge or pyramid absorber arrays—II: Computed and measured results, *IEEE Trans. Electromagn. Compatibility*, 36(4), pp. 307–313, November 1994.

slab, what percentage of the incident power transmits to the other side? (b) Repeat for 600 nm.

9.19 **Refractive index of a liquid.** To measure the refractive index of a liquid, a container is designed as shown in Figure 9.51 to hold the liquid sample.[57] Consider a container made of Teflon ($\epsilon_r \simeq 2.08$ and $\mu_r = 1$ at 10 GHz) with wall thickness of $L_1 \simeq 1.04$ cm on each side. A liquid with refractive index n is poured inside the container's compartment with thickness $L_2 \simeq 1.49$ cm. When a 10-GHz plane wave is normally incident from one side of the container, the effective reflection coefficient at that side is measured to be $\Gamma_{\text{eff}} \simeq -0.39$. (a) Find the refractive index of the liquid (assume lossless case). (b) Recalculate Γ_{eff} at 20 GHz (assume the same material properties apply). (c) Repeat part (b) at 5 GHz.

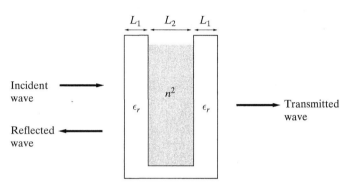

Figure 9.51   Refractive index of a liquid. Liquid with unknown index of refraction n in a container of known dielectric constant ϵ_r. Problem 9.19.

9.20 **Antireflection (AR) coating on a glass slab.** A beam of light is normally incident on one side of a 1-cm thick slab of flint glass (assume $n = 1.86$) at 550 nm. (a) What percentage of the incident power reflects back? (b) To minimize reflections, the glass is coated with a thin layer of antireflection coating material on both sides. The material chosen is magnesium fluoride (MgF_2), which has a refractive index around 1.38 at 550 nm. Find the approximate thickness of each coating layer of MgF_2 needed. (c) If a beam of light at 400 nm is normally incident on the coated glass with the thickness of the coating layers found in part (b), what percentage of the incident power reflects back? (d) Repeat part (c) for a light beam at 700 nm.

9.21 **AR coating on a glass slab.** A 1-cm thick slab of flint glass ($n = 1.86$) is to be coated only on one side so that when a beam of light is incident on the uncoated side, a sample of the light beam that reflects back from that side can be used to monitor the power of the incident beam. Assuming the light beam to be normally incident at 550 nm and the coating material used on the other side to be MgF_2, calculate the percentage of the incident power that reflects back.

9.22 **Infrared antireflection coating.** To minimize reflections at the air–germanium interface in the infrared frequency spectrum, a coating material with an index of refraction of 2.04 is introduced as shown in Figure 9.52. (a) If the thickness of the coating material is adjusted to be a quarter-wavelength in the coating material for a free-space wavelength of 4 μm,

[57]D. Kralj and L. Carin, Wideband dispersion measurements of water in reflection and transmission, *IEEE Trans. Microwave Theory Techn.*, 42(4), pp. 553–557, April 1994.

find the effective reflection coefficient at free-space wavelengths of 4 and 8 μm. (b) Repeat part (a) if the thickness of the coating material is adjusted to be a quarter-wavelength in the coating material at 8 μm.

Figure 9.52 **Infrared antireflection coat-ing.** Problem 9.22.

9.23 Reflection from ferrite-titanate slab. A 30-GHz uniform plane wave is normally incident from air onto an interface consisting of a second section (medium 2) of electrical length d having a complex permittivity of $\epsilon_c = \epsilon_0(3 - j4)$ and a complex permeability of $\mu_c = \mu_0(3 - j4)$ followed by a third section (medium 3) consisting of air, as shown in Figure 9.53. (a) Calculate the values of the real and imaginary parts of the propagation constant γ_2 (i.e., α_2 and β_2) in medium 2 and determine all of the value(s) of d for which the effective reflection coefficient Γ_{eff} is zero. (b) For $d = 0.25$ cm, determine the fraction of the incident wave power that is transmitted into medium 3 (air).

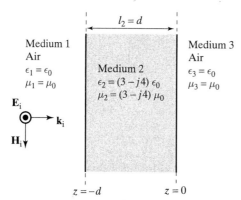

Figure 9.53 **Reflection from ferrite-titanate slab.** Problem 9.23.

9.24 Superwide infrared antireflection coating. A wideband antireflection coating system as shown in Figure 9.54 is designed to be used between air and germanium at infrared

Figure 9.54 **Infrared antireflection coat-ing.** Problem 9.24.

frequencies. If the thicknesses of the coating layers are each quarter wavelength for operation at $\lambda_1 = 3.1$ μm, find and sketch the magnitude of the effective reflection coefficient over the range from 3 to 12 μm.

9.25 **A snow-covered glacier.** Glaciers are huge masses of ice formed in the cold polar regions and in high mountains. Most glaciers range in thickness from about 100 m to 3000 m. In Antarctica, the deepest ice on the polar plateau is 4.7 km. Consider a large glacier in Alaska covered with a layer of snow of 1 m thickness during late winter. A radar signal operating at 56 MHz is normally incident from air onto the air–snow interface. Assume both the snow and the ice to be lossless and nonmagnetic; for ice, $\epsilon_r = 3.2$, and for snow, ϵ_r can vary between 1.2 and 1.8. Assuming both the snow and the ice to be homogeneous and the ice to be semi-infinite in extent, calculate the reflection coefficient at the air–snow interface for three different permittivity values of snow: $\epsilon_r = 1.2$, 1.5, and 1.8. For which case is the snow layer most transparent (invisible) to the radar signal? Why?

9.26 **Minimum ice thickness.** Consider a 500-MHz uniform plane wave radiated by an aircraft radar normally incident on a freshwater ($\epsilon_r = 80$, $\mu_r = 1$) lake covered with a layer of ice ($\epsilon_r = 3$, $\mu_r = 1$), as shown in Figure 9.55. (a) Find the minimum thickness of the ice such that the reflected wave has maximum strength. Assume the lake water to be very deep. (b) What is the ratio of the amplitudes of the reflected and incident electric fields?

Figure 9.55 Aircraft radar signal incident on an icy lake surface. Problem 9.26.

9.27 **A snow–ice-covered lake.** An interior lake in Alaska can be 30–100 m deep and is covered with ice and snow on the top, each of which can be about a meter deep in late winter.[58] Consider a 5-GHz C-band radar signal normally incident from air onto the surface of a lake that is 50 m deep, covered with a layer of snow (assume $\epsilon_r = 1.5$) of 60 cm thickness over a layer of ice ($\epsilon_r = 3.2$) of 1.35 m. Assume both the snow and the ice to be lossless and nonmagnetic. Also assume the lake water, with $\epsilon_{cr} = 68 - j35$ at 5 GHz at 0°C, to be slightly brackish (salty), with an approximate conductivity of $\sigma = 0.01$ S-m^{-1}, and the bottom of the lake to consist of thick silt with $\epsilon_r \simeq 50$. (a) Calculate the reflection coefficient at the air interface with and without the snow layer. (b) Repeat part (a) at an X-band radar frequency of 10 GHz. Assume all the other parameters to be the same except for the lake water, $\epsilon_{cr} = 42 - j41$ at 10 GHz and 0°C. Use any approximations possible, with the condition that sufficient justifications are provided.

9.28 **Incidence on an air gap between two dielectrics.** A 1 GHz uniform plane wave having an incident electric field amplitude of

$$\mathscr{E}_{i0}(z,t) = 100 \cos(2\pi \times 10^9 \, t - \beta_1 z) \qquad \text{mV-m}^{-1}$$

[58]S. A. Arcone, N. E. Yankielun, and E. F. Chacho, Jr., Reflection profiling of Arctic lake ice using microwave FM-CW radar, *IEEE Trans. Geosci. Remote Sensing,* 35(2), pp. 436–443, March 1997.

and propagating in a lossless dielectric medium ($\epsilon_1 = 4\epsilon_0$, $\mu_1 = \mu_0$) is normally incident on an airgap (i.e., $\epsilon_2 = \epsilon_0$, $\mu_2 = \mu_0$) of length d between itself and an identical dielectric ($\epsilon_3 = 4\epsilon_0$, $\mu_3 = \mu_0$). (a) Determine the magnitude of the reflected wave and the complete time-domain expression for the electric field $\mathscr{E}_r(z,t)$ of the reflected wave for $d = \lambda_2/8$ and $d = \lambda_2/4$. (b) Determine the time-average electromagnetic power density (in W-m^{-2}) transmitted through the airgap and the complete time-domain expression for the transmitted electric field $\mathscr{E}_3(z,t)$ for $d = \lambda_2/8$ and $d = \lambda_2/4$. (c) Determine the the complete time-domain expression for the electric field of the electromagnetic wave within the airgap (i.e., $\mathscr{E}_2(z,t)$) for $d = \lambda_2/8$ and $d = \lambda_2/4$.

9.29 **Air–water–air interface.** A uniform plane wave operating at 2.45 GHz in air is normally incident onto a planar water boundary at 20°C ($\epsilon_r = \epsilon_r' - j\epsilon_r'' = 79 - j11$).[59] (a) Calculate the percentage of the incident power that is transmitted into the water (assume the water region to be nonmagnetic and semi-infinite in extent). (b) What percentage of the incident power is absorbed in the first 1-cm thick layer of water? (c) If the water layer has a finite thickness of 1 cm with air on the other side (i.e., air–water–air interface), calculate the percentage of the incident power that is absorbed in the water layer and compare it with the result of part (b).

9.30 **Incidence on a coated perfect conductor.** A 1.8 GHz uniform plane wave having an incident electric field amplitude of $E_{i0} = 1e^{j0}$ V-m^{-1} and propagating in a lossless dielectric medium (ϵ_1, μ_0) is incident on a perfect conductor coated with a lossy dielectric as shown in Figure 9.56. (a) Determine the total amount of power absorbed in the lossy dielectric. (b) Is there a nonzero surface current induced on the surface of the perfect conductor (at $z = 0$)? If so, determine its magnitude, phase, and direction and write the real time dependent expression, that is, $\overline{\mathscr{J}}_s(z,t)$.

$d = 0.5$ cm

Lossless dielectric
$\epsilon_1 = 9\epsilon_0$
$\mu_1 = \mu_0$

Lossy dielectric
$\epsilon_2 = 20\epsilon_0$
$\mu_2 = \mu_0$
$\sigma_2 = 2$ S-m^{-1}

Perfect Conductor

$\mathbf{E}_i$ ⊙ → $\mathbf{k}_i$

$\mathbf{H}_i$ ↓

$z = -d$ $z = 0$

Figure 9.56   Normal incidence on a coated perfect conductor. Problem 9.30.

9.31 **Air–concrete wall–air.** A 900-MHz wireless communication signal is normally incident from one side onto a reinforced concrete wall of thickness d having air on both sides. (a) Find the percentages of the incident power that is reflected back and that is transmitted to the other side of the wall for three different wall thicknesses: 10, 20, and 30 cm. (See Problem 8.18 for data on the properties of reinforced concrete wall.) (b) Repeat part (a) at 1.8 GHz.

[59]W. Fu and A. Metaxas, A mathematical derivation of power penetration depth for thin lossy materials, *J. Microwave Power Electromagn. Energy*, 27(4), pp. 217–222, 1992.

9.32 **Reflection from multiple interfaces.** A uniform plane wave is normally incident on a multiple dielectric interface consisting of two sections (mediums 2 and 3) of the same electrical length ($l_1 = d$ and $l_2 = 2d$) as shown in Figure 9.57. Determine the value of d (if any) such that the all of the incident wave power is transmitted to medium 4 (i.e., such that the reflection coefficient in medium 1 is zero).

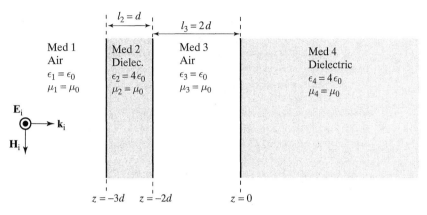

Figure 9.57   Multiple dielectric interfaces. Problem 9.32.

9.33 **Oblique incidence on a perfect conductor.** A 30 W-m^{-2} uniform plane wave in air is obliquely incident on a perfect conductor boundary located at the $y = 0$ plane. The electric field of the incident wave is given by

$$\mathbf{E}_i(x, y) = \hat{\mathbf{z}} E_0 e^{-j6.4\pi x} e^{-j4.8\pi y}$$

(a) Find $E_0, f,$ and θ_i. (b) Find $\mathbf{E}_r$. (c) Find the total electric field $\mathbf{E}_1$ and the nearest positions (with respect to the conductor surface) of the minima and maxima of its magnitude.

9.34 **Oblique incidence on a perfect conductor.** A parallel-polarized (with respect to the plane of incidence) 100 μW-(cm)$^{-2}$, 4-GHz wireless communication signal in air is incident on a perfect conductor surface located at $y = 0$ at an incidence angle of $\theta_i = 30°$ as shown in Figure 9.58. The signal can be approximated as a uniform plane wave. (a) Write the instantaneous expressions for $\overline{\mathscr{E}}_i(y, z, t)$ and $\overline{\mathscr{H}}_i(y, z, t)$. (b) Find $\mathbf{E}_r(y, z)$ and $\mathbf{H}_r(y, z)$ of the reflected wave. (c) Find the magnitude of the total magnetic field phasor $\mathbf{H}_1(y, z)$ and sketch it as a function of y.

Figure 9.58   Oblique incidence on a perfect conductor. Problem 9.34.

9.35   **Plane wave at 45° angle.** A plane wave in air is incident at 45° upon a perfectly conducting surface located at $x = 0$. The plane wave consists of two components as follows:

$$\mathbf{E}_{i\perp} = \hat{\mathbf{y}} E_0 e^{jk(x-z)/\sqrt{3}}$$

$$\mathbf{E}_{i\parallel} = (\hat{\mathbf{x}} + \hat{\mathbf{z}}) \frac{jE_0}{\sqrt{2}} e^{jk(x-z)/\sqrt{3}}$$

(a) Write the perpendicular and parallel polarization components of the electric fields of the reflected wave, and show that the tangential electric fields satisfy the boundary conditions.
(b) What are the polarizations of the incident and the reflected waves?

9.36   **Oblique incidence on a dielectric medium.** Consider oblique incidence of a uniform plane wave on the interface between two nonmagnetic dielectric media with permittivities ϵ_A and ϵ_B. When a perpendicularly polarized plane wave is incident from medium A (i.e., $\epsilon_1 = \epsilon_A$) onto medium B (i.e., $\epsilon_2 = \epsilon_B$) at an incidence angle $\theta_i = \theta_1$, the transmitted angle is $\theta_t = \theta_2$, and the reflection coefficient is $\left[\Gamma_\perp\right]_{A \to B}^{\text{at } \theta_i = \theta_1} = -1/2$. If the propagation direction is reversed, that is, if a new incident wave with the same polarization is now incident from medium B (i.e., $\epsilon_1 = \epsilon_B$) onto medium A (i.e., $\epsilon_2 = \epsilon_A$) at an incidence angle $\theta_i = \theta_2$, what is the numerical value of the reflection coefficient $\left[\Gamma_\perp\right]_{B \to A}^{\text{at } \theta_i = \theta_2}$?

9.37   **Oblique incidence.** A uniform plane wave is obliquely incident at an angle θ_i at the interface between two nonmagnetic ($\mu_1 = \mu_2 = \mu_0$) dielectric media as shown in Figure 9.59. The relative permittivity of the second medium is known to be $\epsilon_{2r} = 3$, and the electric field of the incident wave is given by

$$\overline{\mathscr{E}}_i(x,z,t) = \hat{\mathbf{y}} E_0 \cos[12 \times 10^9 t - 40\sqrt{3}(x + z)]$$

(a) Calculate the relative dielectric constant ϵ_{1r} and the angle of incidence θ_i. (b) Write the corresponding expression for the magnetic field of the incident wave (i.e., $\overline{\mathscr{H}}_i(x,z,t)$). (c) Determine the percentage of the incident power that will be transmitted across the interface.

Figure 9.59   Oblique incidence. Problem 9.37.

9.38   **Reflection from ground.** A 8-GHz and 200-μW-m^{-2} microwave communication signal in air is obliquely incident at $\theta_i = 60°$ onto the ground (assume lossless and $\epsilon_r = 14$) located at $z = 0$. (a) If the incident wave is perpendicularly polarized, write the complete expressions for $\mathbf{E}_i$, $\mathbf{E}_r$, and $\mathbf{E}_t$. (b) Repeat part (a) for a parallel-polarized wave.

9.39   **Air–dielectric interface.** A uniform plane wave propagating in air has an electric field given by

$$\mathbf{E}_i(x,y) = E_0(0.5\hat{\mathbf{x}} + 0.5\sqrt{3}\hat{\mathbf{y}} - e^{j\pi/2}\hat{\mathbf{z}}) e^{-j2\sqrt{3}\pi x + j2\pi y}$$

where E_0 is a real constant. The wave is incident on the planar interface (located at $y = 0$) of a dielectric with $\mu_r = 1$, $\epsilon_r = 3$, as shown in Figure 9.60. (a) What are the values of the wave frequency and the angle of incidence? (b) What is the polarization of the incident wave (i.e., linear, circular, elliptical, right-handed or left-handed)? (c) Write the complete expression for the electric field of the reflected wave in a simplified form. (d) What is the polarization of the reflected wave?

Figure 9.60   Air–dielectric interface.
Figure for Problem 9.39.

9.40   Oblique incidence on a dielectric. An elliptically polarized uniform plane wave is incident from free space (ϵ_0, μ_0) onto on a lossless nonmagnetic dielectric (ϵ_2, μ_0) at an incidence angle $\theta_i = 70°$. Assuming the coordinate system and the placement of the interface to be as in Section 9.6 (e.g., Figure 9.23), the electric field vector of the incident wave can be written as:

$$\mathbf{E}_i(x,z) = \underbrace{\hat{\mathbf{y}}\, e^{-j\beta_1(x\sin\theta_i + z\cos\theta_i)}}_{\mathbf{E}_\perp} + \underbrace{j\, 2\, [\hat{\mathbf{x}}\cos\theta_i - \hat{\mathbf{z}}\sin\theta_i]\, e^{-j\beta_1(x\sin\theta_i + z\cos\theta_i)}}_{\mathbf{E}_\parallel}\ \ \text{mV-m}^{-1}$$

(a) Determine the dielectric constant ϵ_2 for which the reflected wave is right-hand (RH) circularly polarized. (b) For the value of ϵ_2 found in part (a), determine the fraction of the incident wave power that is transmitted into the second medium. (c) Specify the polarization (the type, i.e., linear/circular/elliptical, and the sense, i.e., RH or LH) of the wave transmitted into the second medium.

9.41   Brewster angle at the air–water interface. A perpendicularly polarized uniform plane wave is obliquely incident from air onto the surface of a smooth freshwater lake (assume lossless and nonmagnetic with $\epsilon_r \simeq 84$) at the Brewster angle (i.e., $\theta_i = \theta_{iB}$). Calculate the reflection and transmission coefficients.

9.42   Air–ice interface. A 1-W-m^{-2}, 1-GHz radar signal is obliquely incident at an angle $\theta_i = 30°$ from air onto an air–ice interface. (a) Assuming the ice to be lossless, nonmagnetic and semi-infinite in extent, with $\epsilon_r \simeq 3.0$, calculate the reflection and the transmission coefficients and the average power densities of the reflected and the transmitted waves if the incident wave is perpendicularly polarized. (b) Repeat part (a) for an incident wave that is parallel-polarized.

9.43   Communication over a lake. Consider a ground-to-air communication system as shown in Figure 9.61. The receiver antenna is on an aircraft over a huge lake circling at a horizontal distance of $\sim$10 km from the transmitter antenna as it waits for a landing time. The transmitter antenna is located right at the shore mounted on top of a 100-m tower above the lake surface overlooking the lake and transmits a parallel polarized (with respect to the plane of incidence) signal. The transmitter operates in the VHF band. The pilot of the aircraft experiences noise (sometimes called *ghosting* effect) in his receiver due to the destructive interference between the direct wave and the ground-reflected wave and needs to adjust his altitude to minimize this interference. Assuming the lake to be flat and lossless with $\epsilon_r \simeq 80$, calculate the critical height of the aircraft in order to achieve clear transmission between the transmitter and the receiver.

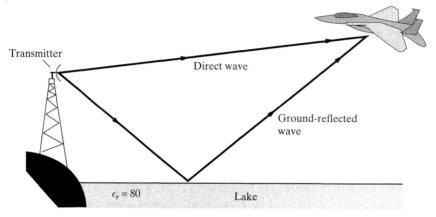

Figure 9.61   Communication over a lake. Problem 9.43.

9.44   Air-to-air communication. Consider two helicopters flying in air separated by a horizontal distance of 2 km over a flat terrain as shown in Figure 9.62. The pilot of one of the helicopters, located at an altitude of 100 m, transmits a parallel-polarized (with respect to the plane of incidence) VHF-band signal (assume 200 MHz) to communicate with the other helicopter. The pilot of the other helicopter needs to adjust her altitude to eliminate the noise on her receiver due to the interference of the ground-reflected wave with the direct wave. (a) Assuming the ground to be homogeneous, nonmagnetic and lossless with $\epsilon_r \simeq 16$, find the critical altitude of the receiver helicopter in order to minimize this interference. (b) Consider another scenario when both helicopters are at 250 m altitude. In this case, what

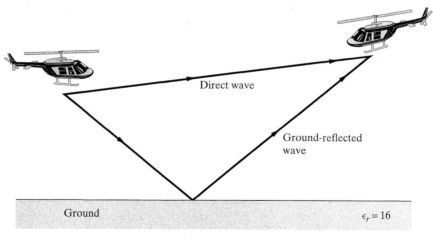

Figure 9.62   Air-to-air communication. Problem 9.44.

should be the horizontal separation distance between the helicopters in order to achieve clear signal transmission? (c) Repeat both (a) and (b) for the case in which the helicopters try to land at a remote site in Alaska where the ground is covered with permafrost (assume $\epsilon_r \simeq 4$).

9.45 **Signal to Interference Ratio (SIR).** You have just launched your start-up company, with your headquarters located at a distance of $d = 140$ m from a cellular tower of height $h_1 = 50$ m. As the CEO you choose to have your office on the top floor of your building, at a height of $h_2 = 20$ m, as shown in Figure 9.63. (a) Determine the signal-to-interference ratio (SIR):

$$SIR \equiv \frac{|\mathbf{S}_{av}|_{direct}}{|\mathbf{S}_{av}|_{reflected}}$$

assuming that the permittivity of the ground is $\epsilon = 5.55\epsilon_0$. Provide separate results for both parallel and perpendicular polarization. (b) After a few dropped calls, you feel that you are not satisfied with the SIR you have and decide that you can move your office to a lower floor to maximize SIR. Calculate the height h_2 that results in maximum SIR and the value of maximum SIR. Provide separate results for both parallel and perpendicular polarization.

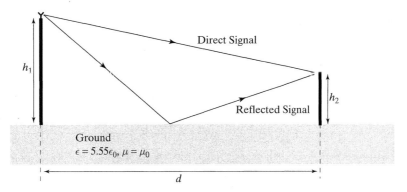

Figure 9.63   Cell phone reception. Problem 9.45.

9.46 **Dry asphalt roads acting as mayfly traps?** Adult mayflies have only a few hours in which to find a mate and reproduce. Direct sunlight reflected off the surface of water is strongly polarized in the horizontal plane (i.e., parallel polarized), and many water-dwelling insects, including mayflies, use this reflected polarized light to identify open stretches of water where they can lay their eggs during their brief mating period. However, researchers discovered that light reflected from dry asphalt roads is also horizontally polarized and visually deceives mayflies into laying their eggs on roads instead of in rivers.[60] The higher the degree of polarization of the reflected light, the more attractive it is for mayflies. Hence, mayflies swarming, mating, and egg-laying on asphalt roads are predominantly deceived by and attracted to the asphalt surface because the largely horizontally polarized reflected light imitates a water surface. Note that although sunlight has mixed polarization and is incident

[60]G. Kriska, G. Horvath, and S. Andrikovics, Why do mayflies lay their eggs *en masse* on dry asphalt roads? Water-imitating polarized light reflected from asphalt attracts Ephemeroptera, *The J. Exp. Biol.,* 201, pp. 2273–2286, 1998; J. Copley, Tar babies: Swarms of mayflies are laying their eggs on roads rather than rivers, *New Scientist,* p. 16, July 25, 1998.

on the asphalt over a range of angles, the polarization of the reflected light is predominantly horizontal because of the deep minimum for $\rho_\parallel$ in the vicinity of the Brewster angle (see Figure 9.33). (a) If the reflected light from asphalt is almost 100% horizontally polarized when the light is incident on the asphalt surface at an incidence angle of about $57.5°$, calculate the effective refractive index of asphalt. (b) Assuming the refractive index of water to be $n_w \simeq 1.33$, find the angle at which the reflected light from the water surface is 100% horizontally polarized.

9.47 Total internal reflection. A uniform plane wave with a magnetic field given by

$$\overline{\mathcal{H}}(y,z,t) = \hat{\mathbf{x}}H_0 \cos[9\pi \times 10^9 t - 45\pi(y + \sqrt{3}z)]$$

is obliquely incident at an interface at $z = 0$ separating two nonmagnetic lossless media as shown in Figure 9.64. (a) Calculate the relative dielectric constant ϵ_{1r} of medium 1 and the angle of incidence θ_i. (b) Find the maximum value of the relative dielectric constant ϵ_{2r} of medium 2 for total internal reflection to occur. (c) Is it possible to achieve total transmission by adjusting the incidence angle? If yes, use the maximum value of ϵ_{2r} found in part (b) to determine the incidence angle at which total transmission would occur.

Figure 9.64 Total internal reflection. Problem 9.47.

9.48 Reflection from prisms. Consider the various right-angled prisms shown in Figure 9.65. (a) What is the minimum index of refraction n_1 of the prism necessary in each case for there to be no time-average power transmitted across the hypotenuse when the prisms are (i) in free space, (ii) in water (assume $n_2 \simeq 1.33$). (b) At these refractive index values (found in (a)), what are the exit angles θ_{te}?

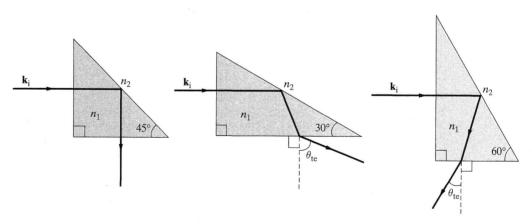

Figure 9.65 Reflection from prisms. Problem 9.48.

9.49 **MgF$_2$ prism.** A 30°–90°–30° prism is constructed from MgF$_2$ ($n = 1.4$) to be used to turn a light beam around by 90° by internal reflection at its hypotenuse. Does the light beam exit at the hypotenuse, and if so, what is its exit angle?

9.50 **Refractive index of a prism.** An experiment is designed to measure the refractive index of a prism using the principle of total internal reflection.[61] In this experiment, a plane-polarized, collimated, monochromatic beam of light is obliquely incident from one side of the prism at an incidence angle of θ_i as shown in Figure 9.66. The incidence angle θ_i of the beam is adjusted to a critical value ψ_c such that the incidence angle on the other side of the prism is equal to the critical angle of incidence θ_{ic}. Thus, by measuring the refracting angle A of the prism and the critical incidence angle ψ_c, the refractive index n_p of the prism can be calculated. (a) Show that

$$n_p^2 = 1 + \left(\frac{\sin \psi_c + \cos A}{\sin A} \right)^2$$

(b) For a prism under test, the refracting angle of the prism and the critical incidence angle adjusted are measured to be $A = 60°$ and $\psi_c = 42°$, respectively. Calculate the refractive index n_p of this prism.

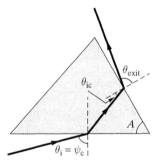

Figure 9.66   Refractive index of a prism.
Problem 9.50.

9.51 **Right-angle prism.** A new technique is proposed to measure small angles in optical systems using right-angle prisms.[62] Consider a 45°–90°–45° prism as shown in Figure 9.67 made of glass having a refractive index of $n \simeq 1.515$ to be used in an experiment. A light beam that is obliquely incident at an incidence angle θ_1 on the entrance face undergoes reflection and refraction at the entrance face, the hypotenuse face, and the exit face of the prism. (a) For an incidence angle of $\theta_1 = 30°$, find the exit angles θ_4 and θ_6. (b) Find the critical incidence angle θ_1 of the incident beam on the entrance face that results in no transmission at the hypotenuse face. (c) What happens to the critical angle found in part (b) when the hypotenuse face of the prism is coated with an antireflection coating (single or multiple layers)? (d) Repeat parts (a) and (b) if the prism is made of a different type of glass with $n \simeq 1.845$.

[61] S. P. Talim, Measurement of the refractive index of a prism by a critical angle method, *Opt. Acta,* 25(2), pp. 157–165, 1978.

[62] P. S. Huang and J. Ni, Angle measurement based on the internal-reflection effect and the use of right-angle prisms, *Appl. Optics,* 34(22), pp. 4976–4981, August 1995.

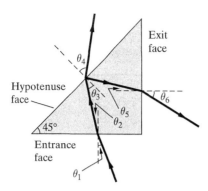

Figure 9.67   Right-angle prism.
Problem 9.51.

9.52   A V-shaped prism. A V-shaped right-angled prism is designed to measure accurately the refractive index of liquids[63] as shown in Figure 9.68. A laser beam normally incident from one side is refracted at the inclined faces, between which the liquid sample under test is placed, and leaves the prism on the other side at a deflection angle ψ. (a) A liquid of known refractive index $n \simeq 1.512$ is placed in the top compartment of the prism with $n_p \simeq 1.628$. Find the deflection angle ψ_d. (b) Using the same prism, the deflection angle for a different liquid with unknown refractive index is approximately measured to be $\psi \simeq 55.5°$. Calculate the refractive index n of this liquid.

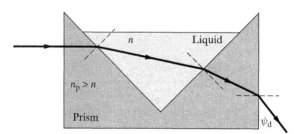

Figure 9.68   A V-shaped prism. Problem 9.52.

9.53   Turning a perfect corner. An interesting optical phenomenon is the invariance of the angle between the incoming and outgoing light rays passing through a right-angle isosceles prism with a silver-coated hypotenuse.[64] For the isosceles right-angle prism shown in Figure 9.69, the incident ray enters at side AB and exits at side AC after being reflected twice and refracted twice. Show that the total deviation angle ψ_d between the incident and the exiting rays is exactly $90°$.

[63]E. Moreels, C. de Greef, and R. Finsy, Laser light refractometer, *Appl. Optics,* 23(17), pp. 3010–3013, September 1984.

[64]S. Taylor and E. Hafner, Turning a perfect corner, *Am. J. Phys.,* 47(1), pp. 113–114, January 1979.

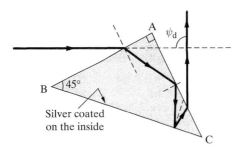

Figure 9.69   Turning a perfect corner.
Problem 9.53.

9.54   Air–oil–water. Consider a layer of oil (assume $n = 1.3$) about 5 mm thick floating over a body of water ($n \simeq 1.21$). (a) If a ray is obliquely incident from air onto the air surface, find range of incidence angles that result in total internal reflection at oil–water interface. (b) If a light ray is obliquely incident from water onto oil surface, find the range of incidence angle that results TIR at oil–air interface.

9.55   An in-line Brewster angle prism. Brewster angle prisms are optical elements that use light at the polarizing angle to obtain perfect transmission of parallel polarized light. An in-line Brewster angle polarizing prism is designed[65] as shown in Figure 9.70, to polarize a light beam without changing its direction. Consider an unpolarized light beam incident on this prism at point A at an incidence angle of $\theta_i = 31.639°$. Given the prism angles to be $\alpha = 63.278°$ and $\beta = 103.939°$ and the prism refractive index to be $n = 1.623$, (a) determine whether light exits the prism at points B, C, and/or D, (b) find the exit angles θ_B, θ_C, and/or θ_D at these exit points, (c) specify the polarization of each exiting beam, and (d) find the angle between each exiting beam and the incident beam. Discuss your results in terms of the stated purpose of this particular prism design.

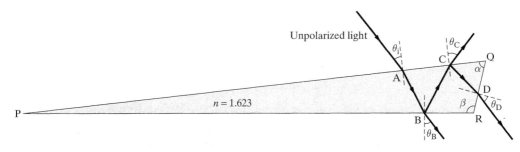

Figure 9.70   In-line polarizing prism. Problem 9.55.

9.56   V-shaped prism polarizer. A symmetrical three-reflection silicon (Si) polarizer prism is designed as shown in Figure 9.71a, based on the Brewster-angle internal reflection that

---

[65]R. D. Tewari, A. M. Ghodgaonkar, and V. Bhattacharyya, Modified polarizing prism, *Optics and Laser Technology,* 30, pp. 63–70, 1998.

occurs at the base of the prism.[66] (a) In Figure 9.71a, the prism angle A is adjusted for use at 1.3 μm light wave communication wavelength (the refractive index of silicon at 1.30 μm is $n_{Si} \simeq 3.5053$), to $A \simeq 52.9613°$. Assuming the incident light beam entering the prism on the left side to be unpolarized, find the polarization of the beam exiting on the right side. (b) Find the new value of A for the prism to be used as a polarizer at 1.55-μm wavelength (the refractive index of silicon at 1.55 μm is $n_{Si} \simeq 3.4777$). (c) Another interesting design is to coat the base of the silicon prism with silicon dioxide (SiO_2) as shown in Figure 9.71b. At 1.3 μm, the prism angle is adjusted to $A \simeq 56.217°$. Find the refractive index of SiO_2 at 1.3 μm. (d) Repeat part (c) at 1.55 μm when the prism angle is adjusted to $A \simeq 56.277°$.

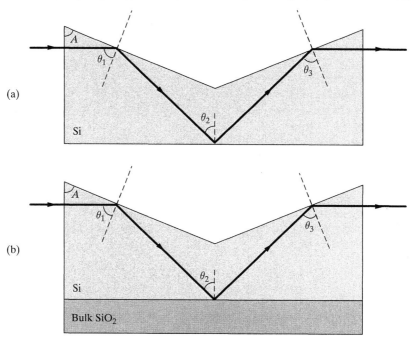

Figure 9.71   V-shaped prism polarizer. Problem 9.56.

9.57   Limestone wall versus brick wall. Consider two buildings, one with limestone ($\epsilon_r = 7.51$, $\sigma = 0.03$ S-m^{-1}) exterior walls and the other with brick ($\epsilon_r = 4.44$, $\sigma = 0.01$ S-m^{-1}) exterior walls, with the material properties cited measured[67] at 4 GHz. These walls represent some of the typical building surfaces that affect the propagation of mobile radio signals. Assuming the walls to be lossless, nonmagnetic, semi-infinite in extent, and neglecting the roughness of their surfaces, calculate the reflection coefficients at the surface of each building for both perpendicular and parallel polarizations at three different angles of incidence of $\theta_i = 30°$, $45°$, and $60°$ and compare the results.

[66]R. M. A. Azzam and M. M. K. Howlader, Silicon-based polarization optics for the 1.30 and 1.55 μm communication wavelengths, *J. Lightwave Technol.,* 14(5), pp. 873–878, May 1996.

[67]O. Landron, M. Feuerstein, T. S. Rappaport, A comparison of theoretical and empirical reflection coefficients for typical exterior wall surfaces in a mobile radio environment, *IEEE Trans. Antennas and Propagation,* 44(3), pp. 341–351, March 1996.

9.58   **Refractive index of concrete.** Knowledge of the dielectric properties of construction material is important because the reflection and transmission characteristics of buildings and rooms are governed by these properties. The complex refractive index of a plain concrete plate mixed from Portland cement, gravel, sand, and water was measured at 57.5 GHz for use in designing and testing millimeter-wave communication systems.[68] The measured refractive index of the concrete 14 months after concreting was $n = 2.55 - j0.084$. (a) Using the measured values for concrete, calculate and sketch the magnitude of the reflection coefficient at the air–concrete interface at 57.5 GHz as a function of the incidence angle varying between $0°$ and $90°$ for both perpendicular and parallel polarization cases. Assume the concrete to be semi-infinite in extent. (b) For a 5-cm thick concrete wall having air on both sides, calculate the magnitude of the normal incidence reflection coefficient at 57.5 GHz and compare it with the result of part (a). (c) Repeat part (b) for a thickness of 10 cm.

9.59   **Oblique incidence on a multiple dielectric interface.** Suppose that a parallel polarized uniform plane wave is incident, in air, on a nonmagnetic glass slab (assume $\epsilon_r = 2$) of thickness d, as shown in Figure 9.72. The angle of incidence θ_i is chosen to be the air-glass Brewster angle so that there is no reflection at the first interface ($z = 0$). Find the complete expression for the electric field phasor of the wave transmitted into air ($z > d$), that is, $\mathbf{E}_t(x, z)$. *Hint:* First find the reflection coefficient at the $z = d$ interface.

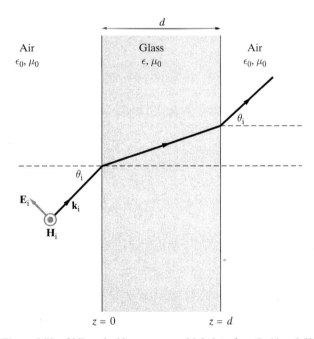

Figure 9.72   Oblique incidence on a multiple interface. Problem 9.59.

[68]K. Sato, T. Manabe, J. Polivka, T. Ihara, Y. Kasashima, and K. Yamaki, Measurement of complex refractive index of concrete at 57.5 GHz, *IEEE Trans. Antennas Propagation*, 44(1), pp. 35–40, January 1996.

9.60   **Oblique incidence on a multiple dielectric interface.** A perpendicularly polarized uniform plane wave propagating in air is incident obliquely (at an angle θ_i) on a structure consisting of two lossless and nonmagnetic dielectrics: a coating layer with permittivity ϵ_{1r} and thickness d coated on another dielectric with permittivity ϵ_{2r} and of infinite thickness, as shown in Figure 9.73. Derive expressions for the effective reflection and transmission coefficients (Γ_{eff} and $\mathcal{T}_{\text{eff}}$) in terms of parameters of the media ($\epsilon_{1r}, \epsilon_{2r}$), the angle of incidence θ_i, and the slab thickness d.

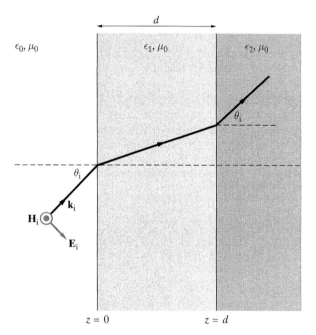

Figure 9.73   Oblique incidence on a multiple interface. Problem 9.60.

10

Parallel-Plate and Dielectric Slab Waveguides

When an electromagnetic wave propagates out from a point source through empty space, it does so as a spherical wave, traveling with equal speed in all directions. At large distances ($r \gg \lambda$) from its source, a spherical electromagnetic wave is well approximated as a uniform plane wave; in Chapter 8, we considered the propagation of uniform plane waves in unbounded media, and in Chapter 9 their reflections from planar interfaces between different simple media. We now consider the guiding of electromagnetic waves from one point to another by means of metallic or dielectric boundaries.

In a wide range of electromagnetic applications it is necessary to convey electromagnetic waves efficiently from one point to another so as to transmit energy or information (signals). The energy or information can be conveyed by means of *unguided* electromagnetic waves propagating in free space. In unguided transmission of electromagnetic energy, the characteristics of the transmitting antennas determine the intensity of the waves radiated in different directions. Some applications, such as line-of-sight microwave (radio, television, and satellite) links, optical links (bar-code readers, infrared smoke detectors, infrared remote controllers), high-power lasers (in materials processing, laser printers, and surgery), and radar require directional transmission of waves, whereas others, such as navigation, radio communication, and radio and TV broadcast use omnidirectional transmissions. In addition to the spherical spreading of the available energy, broadcast applications do not represent efficient use of the electromagnetic spectrum, since only one transmitter can operate at a given frequency in a given region.

In many other applications it is necessary to transmit electromagnetic energy or signals by using waves that are *guided* by metallic or dielectric structures to minimize radiation losses and unnecessary spreading of the energy. Guided transmission of waves also facilitates the use of one frequency band to convey multiple signals simultaneously, because most waveguides can be laid close to one another without significant coupling between them. Electromagnetic waves are generally guided by confining them in two dimensions and allowing them to propagate freely in the third dimension. The simplest example is a coaxial line, which completely confines the energy to the region between the inner and outer conductors. The principle of operation of the two-wire transmission line is identical; although the electric and magnetic fields are not completely confined in an enclosure, their amplitudes decrease rapidly with transverse distance away from the conductors, so most of the electromagnetic energy is confined to the region in the immediate vicinity of the conductors. The surfaces of the conducting wires provide boundaries on which the electric field lines can terminate, so that the wave can propagate as a plane wave following the conductor from end to end. A power transmission line and a telephone wire are examples of such two-wire lines.

Wave-guiding structures of all types can be called waveguides, although the term *waveguide* is sometimes used specifically for the particular guiding structure consisting of a metallic tube with a two-dimensional cross-section. The two guiding systems explicitly considered in this chapter are the parallel-plate waveguide and the dielectric slab waveguide. Both of these are planar structures, guiding electromagnetic waves in one direction (typically taken to be the z direction) by confining the wave energy in one of the transverse directions (e.g., the x direction) while being of infinite extent[1] in the other (e.g., the y direction). We confine our attention in this chapter to these two types of waveguides, using them as examples by means of which we introduce the fundamental principles of guiding of waves by metallic (parallel-plate waveguide) and dielectric (slab waveguide) boundaries. Primarily because of their infinite extent in one of the two transverse dimensions, these waveguides can be analyzed in a relatively straightforward manner, at a level appropriate for this book. These two waveguides are examples of a class of planar waveguide structures that can be manufactured relatively easily (hence at low cost) using planar integrated circuit technologies. Other planar waveguides, such as the stripline and the microstrip line, may in fact be more commonly utilized in practice than the parallel-plate and dielectric slab waveguides studied here. However, the analysis of these structures is considerably more involved and typically requires approximate or numerical treatments whose description is not appropriate for our coverage of fundamental principles and is thus beyond the scope of this book.[2]

The parallel-plate waveguide is arguably the simplest example of transmission systems that utilize two separate conductors for guiding electromagnetic waves. All of the methodologies used in its analysis are directly applicable to other transmission systems that use two separate conductors symmetrically arranged across the transverse plane (e.g., the coaxial line, two-wire line, twisted-pair, striplines), which can support the propagation of transverse electromagnetic (TEM) waves. At the same time, the well-developed techniques for analysis of voltage and current waves on transmission lines that we discussed in Chapters 2 and 3 are directly applicable to the parallel-plate waveguide

---

[1] Practical parallel-plate and dielectric slab waveguides necessarily also have a finite extent in the y direction; however, typically the extent of the waveguide in the y direction is much larger than the thickness of the waveguide in the x direction, so the fringing fields can be neglected, and the field structure, propagation characteristics, and attenuation rates can be evaluated by assuming the waveguide to be of infinite extent.

[2] For an excellent and brief treatment, see Section 8.6 of S. Ramo, J. R. Whinnery, and T. Van Duzer, *Fields and Waves in Communications Electronics*, 3rd ed., Wiley, New York, 1994.

or any other two-conductor structures, as long as the waveguide voltage and current waveforms are properly defined (see Section 10.1.5) in terms of the propagating electric and magnetic fields. In most cases, TEM waves guided by two-conductor systems are not *uniform* plane waves, because the amplitudes of the wave field components vary over the planar constant-phase fronts. For example, the amplitudes decrease with distance away from the conductors for the two-wire line; hence it is the confinement of the energy to the vicinity of the conductors, which "guide" the wave.

In addition to supporting TEM waves, the parallel-plate waveguide also allows the propagation of more complicated electromagnetic waves, for which the wave magnetic or the wave electric field has a component in the direction of propagation. The presence of magnetic or electric field components along the propagation direction requires[3] that the wave field components vary with position along the planar phase fronts. Such waves are, respectively, called transverse electric (TE) or transverse magnetic (TM) waves. The different types of possible wave solutions are referred to as *modes*, analogous to the different ways (modes) in which a string with both ends held fixed may vibrate.

Dielectric slab waveguides of one form or another are the workhorses of integrated optics,[4] primarily concerned with constructing optical devices and networks on substrates. The field of integrated optics (sometimes also called integrated photonics) combines optics and electronics on a single substrate, often with integrated components having dimensions of the order of the wavelength of light. The dielectric slab waveguide is important in view of its common use as the means by which light is transferred between components of an integrated optics network. In addition, propagation of light waves in a dielectric slab waveguide is fundamentally similar to the guiding of light in cylindrical dielectric waveguide structures (optical fibers), where the electromagnetic energy is confined in both of the transverse dimensions. Thus, our analyses of the dielectric slab waveguide also serve to illustrate and introduce the basic principles of guiding of light waves within dielectric boundaries, with much broader application to the field of fiber optics, which is beyond the scope of this book but well covered elsewhere.[5]

The mathematical foundation of guided electromagnetic waves and modern transmission line theory was laid down by O. Heaviside, who wrote a series of 47 papers between 1885 and 1887, introducing the vector notation currently in use today and many other concepts that have since proven to be so useful in electrical engineering. Most of the early experimental work on guided waves was carried out by H. Hertz; both Hertz and Heaviside exclusively considered two-conductor transmission lines, with Heaviside even declaring that two conductors were definitely necessary for transmission of

---

[3]Otherwise we cannot have $\nabla \cdot \mathbf{B} = 0$ or $\nabla \cdot \mathbf{D} = 0$.

[4]R. G. Hunsberger, *Integrated Optics: Theory and Technology*, 4th ed., Springer-Verlag, New York, 1995.

[5]For an up-to-date coverage of technology and applications see J. C. Palais, *Fiber Optic Communications*, 5th ed., Prentice-Hall, Upper Saddle River, New Jersey, 2004; for coverage of fundamentals of guiding of light waves in different modes, see A. W. Snyder and J. D. Love, *Optical Waveguide Theory*, Chapman and Hall, London, 1983, and D. B. Keck, in *Fundamentals of Optical Fiber Communications* (M. L. Barnoski, Ed.), Academic Press, San Diego, California, 1976.

electromagnetic energy.[6] (Heaviside was wrong in this prophecy, since electromagnetic waves propagate very effectively in hollow single metallic cylindrical guides.) Nevertheless, the series of papers written by Heaviside and his three-volume work published in 1893 stand out as the mathematical foundations of guided electromagnetic waves.

Our purpose in this chapter is to provide the fundamental principles of the guiding of waves, using parallel-plate and dielectric slab planar waveguides as examples. The techniques for deriving propagating waveguide solutions, and the practical considerations such as power loss, discrete propagation modes, and waveguide dispersion, have close analogues in axial waveguides that are confined in two dimensions. A treatment at the appropriate level of rectangular waveguides, circular waveguides, coaxial lines, and optical fibers is given in the on-line addendum (see the Preface for a link). Interested readers may also consult the texts listed in the General Bibliography listed after the appendices.

This chapter also constitutes an important step in our coverage of electromagnetic wave propagation that we started in Chapter 8, where we constrained our coverage to waves in unbounded, simple, and source-free media. Chapter 9 was our first step in removing the first constraint, where we considered the reflection and refraction of waves from planar boundaries. We investigate the propagation of waves in regions bounded by conducting or dielectric structures, while still assuming that the medium within which the waves propagate is simple and source-free. An important aspect of the guided propagation of electromagnetic waves is that the phase velocity of the propagating modes becomes frequency-dependent, as opposed to the frequency-independent character of uniform plane waves in a simple, unbounded, and lossless medium. This property is important for the guided propagation of an information-carrying signal consisting of a band of frequencies (which occurs at the so-called group velocity, introduced and discussed in Section 10.3). The specific conditions on the propagation constant for guided waves considered in this chapter arise from the presence of the metallic or dielectric boundaries. In Chapter 11, on the other hand, we revert back to considering uniform plane waves in an unbounded medium once again as in Chapter 8, but we now allow the medium not to be simple, so that the medium characteristics (ϵ, μ, and σ) can be functions of frequency. In such cases, special conditions on the propagation constant, wavelength, and phase velocity arise as a result of the material properties, rather than the presence of boundaries.

10.1 WAVES BETWEEN PARALLEL METAL PLATES

The simplest type of waveguide consists of two perfectly conducting plates, between which electromagnetic waves of various types can be guided. Accordingly, we start our considerations of guided waves by analyzing the so-called *parallel-plate waveguide,* as

---

[6]On p. 399 of his three-volume work *Electromagnetic Theory,* Benn, London, 1893, or on p. 100 of the reprinted edition by Dover, 1950, Heaviside stated that guided waves needed "two leads as a pair of parallel wires; or if but one be used, there is the earth, or something equivalent, to make another." For an interesting discussion of the early history, see J. H. Bryant, Coaxial transmission lines, related two-conductor transmission lines, connectors, and components: A U.S. historical perspective, *IEEE Trans. Microwave Theory Techn.,* 32(9), pp. 970–983, September 1984.

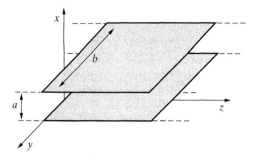

Figure 10.1 Parallel-plate waveguide. In an ideal parallel-plate waveguide, the plates are assumed to be infinite in extent in the y and z directions. In practice, the width b of the guide in the y direction is finite but is typically much larger than a. The material between the plates is assumed to be a lossless dielectric.

shown in Figure 10.1. The plates have an infinite extent in the z direction, in which the waves are to be guided. In practice the conductor plates would have a finite extent b in the y direction, but it is typically much larger than the plate separation a, so the effects of fringing fields at the edges are negligible. For our purposes here, we assume that the conductor plates have infinite extent ($b \gg a$) in the y direction, so the electric and magnetic fields do not vary with y.

Just as in the case of waves propagating in unbounded media, the configuration and propagation of the electromagnetic fields in any guiding structure are governed by Maxwell's equations and the wave equations derived from them. However, in the case of metallic waveguides, the solutions of the wave equations are now subject to the following boundary conditions at the conductor surfaces:

$$E_{\text{tangential}} = 0 \qquad H_{\text{normal}} = 0$$

Assuming that the medium between the perfectly conducting plates is source-free, simple (i.e., ϵ and μ are simple constants), and lossless[7] (i.e., $\sigma = 0$, μ and ϵ are real), the governing equations are the time-harmonic phasor form of Maxwell's equations, namely

$$\nabla \times \mathbf{H} = j\omega\epsilon\mathbf{E} \tag{10.1a}$$

$$\nabla \cdot \mathbf{E} = 0 \tag{10.1b}$$

$$\nabla \times \mathbf{E} = -j\omega\mu\mathbf{H} \tag{10.1c}$$

$$\nabla \cdot \mathbf{H} = 0 \tag{10.1d}$$

from which the wave equations for $\mathbf{E}$ and $\mathbf{H}$ can be derived in a straightforward manner following the same procedure as in Section 8.2. We have

$$\nabla^2\mathbf{E} - (j\beta)^2\mathbf{E} = 0 \tag{10.2a}$$

$$\nabla^2\mathbf{H} - (j\beta)^2\mathbf{H} = 0 \tag{10.2b}$$

where $\beta = \omega\sqrt{\mu\epsilon}$.

---

[7]We shall later relax this requirement and calculate the losses in the dielectric medium using a perturbation solution.

Equations (10.1a) and (10.1c) can be written in component form for rectangular coordinates (note that the conductor configuration is rectangular; otherwise, for example for circularly shaped conductors, it is more appropriate to write the equations in terms of cylindrical coordinates r, ϕ, and z) for the region between plates (assumed to be a lossless dielectric, with both σ and ϵ'' equal to zero) as

$$\underbrace{\begin{aligned}\frac{\partial H_z}{\partial y} - \frac{\partial H_y}{\partial z} &= j\omega\epsilon E_x \\[1ex] \frac{\partial H_x}{\partial z} - \frac{\partial H_z}{\partial x} &= j\omega\epsilon E_y \\[1ex] \frac{\partial H_y}{\partial x} - \frac{\partial H_x}{\partial y} &= j\omega\epsilon E_z\end{aligned}}_{\nabla\times\mathbf{H}=j\omega\epsilon\mathbf{E}} \qquad \underbrace{\begin{aligned}\frac{\partial E_z}{\partial y} - \frac{\partial E_y}{\partial z} &= -j\omega\mu H_x \\[1ex] \frac{\partial E_x}{\partial z} - \frac{\partial E_z}{\partial x} &= -j\omega\mu H_y \\[1ex] \frac{\partial E_y}{\partial x} - \frac{\partial E_x}{\partial y} &= -j\omega\mu H_z\end{aligned}}_{\nabla\times\mathbf{E}=-j\omega\mu\mathbf{H}} \qquad (10.3)$$

Note that since the extent of the conductor plates in the y direction is infinite, none of the field quantities vary in the y direction, and as a result, the $\partial(\cdots)/\partial y$ terms in (10.3) are all zero.

We assume propagation in the $+z$ direction, so all field components vary as $e^{-\bar{\gamma}z}$, where $\bar{\gamma} = \bar{\alpha} + j\bar{\beta}$, with $\bar{\alpha}$ and $\bar{\beta}$ being real.[8] In the rest of this chapter, we consider the two special cases for which $\bar{\beta} = 0$ or for which $\bar{\alpha} = 0$. When the phasor fields vary in this manner, the real fields vary as

$$\mathscr{R}e\{e^{-\bar{\gamma}z}e^{j\omega t}\} = \begin{cases} e^{-\bar{\alpha}z}\cos(\omega t) & \bar{\gamma} = \bar{\alpha} & \text{Evanescent wave} \\ \cos(\omega t - \bar{\beta}z) & \bar{\gamma} = j\bar{\beta} & \text{Propagating wave} \end{cases}$$

An *evanescent*[9] wave is one that does not propagate but instead exponentially decays with distance.

The wave equation(s) (10.2) can be rewritten as

$$\nabla^2\mathbf{E} - (j\beta)^2\mathbf{E} = 0 \qquad \rightarrow \qquad \frac{\partial^2\mathbf{E}}{\partial x^2} + \frac{\partial^2\mathbf{E}}{\partial y^2} + \frac{\partial^2\mathbf{E}}{\partial z^2} - (j\beta)^2\mathbf{E} = 0$$

$$\nabla^2\mathbf{H} - (j\beta)^2\mathbf{H} = 0 \qquad \rightarrow \qquad \frac{\partial^2\mathbf{H}}{\partial x^2} + \frac{\partial^2\mathbf{H}}{\partial y^2} + \frac{\partial^2\mathbf{H}}{\partial z^2} - (j\beta)^2\mathbf{H} = 0$$

---

[8]Note here that $\bar{\gamma}$ is different from the complex propagation constant of the medium $\gamma = \alpha + j\beta$ used earlier for uniform plane waves. Although γ, α, and β are dependent only on the wave frequency and the medium parameters (ϵ, μ, σ), the waveguide propagation constant $\bar{\gamma}$ depends in general also on the dimensions of the guiding structure. Even though the propagation medium may be nondissipative, $\bar{\gamma}$ may be either imaginary ($\bar{\gamma} = j\bar{\beta}$) or real ($\bar{\gamma} = \bar{\alpha}$), depending on the frequency of operation, as we shall see later. Note also that the quantities $\bar{\gamma}$, $\bar{\alpha}$, and $\bar{\beta}$ are not vectors; the bar in this case is simply used to distinguish them from corresponding quantities for waves propagating in simple, unbounded media. The $\bar{\gamma}$, $\bar{\alpha}$, and $\bar{\beta}$ notation used here is adapted from E. C. Jordan and K. G. Balmain, *Electromagnetic Waves and Radiating Systems,* Prentice-Hall, Englewood Cliffs, New Jersey, 1968.

[9]The dictionary meaning of the word *evanescent* is "tending to vanish like vapor."

Once again, we assume that none of the field quantities vary in the y direction. In other words, we have

$$\frac{\partial^2}{\partial y^2}(\cdots) = 0$$

The field variations in the x direction cannot be similarly limited, because the perfect conductors impose boundary conditions in the x direction at $x = 0$ and $x = a$. Note that for any of the field components, for example, for H_y, we can explicitly show the variations in the z direction by expressing $H_y(x, z)$ as a product of two functions, one having x and the other z, respectively, as their independent variables:

$$H_y(x, z) = H_y^0(x) e^{-\overline{\gamma} z}$$

where $H_y^0(x)$ is a function only of x. We then have

$$\frac{\partial}{\partial z} H_y(x, z) = -\overline{\gamma} H_y^0(x) e^{-\overline{\gamma} z} = -\overline{\gamma} H_y(x, z) \quad \longrightarrow \quad \frac{\partial}{\partial z} \quad \rightarrow \quad -\overline{\gamma}$$

The wave equations (10.2) can thus be rewritten as

$$\frac{\partial^2 \mathbf{E}}{\partial x^2} + \overline{\gamma}^2 \mathbf{E} - (j\beta)^2 \mathbf{E} = 0 \tag{10.4}$$

$$\frac{\partial^2 \mathbf{H}}{\partial x^2} + \overline{\gamma}^2 \mathbf{H} - (j\beta)^2 \mathbf{H} = 0 \tag{10.5}$$

Substituting $\partial/\partial z \to -\overline{\gamma}$ and $\partial/\partial y = 0$ in (10.3), we find

$$
\begin{aligned}
\overline{\gamma} E_y &= -j\omega\mu H_x & \overline{\gamma} H_y &= j\omega\epsilon E_x \\
-\overline{\gamma} E_x - \frac{\partial E_z}{\partial x} &= -j\omega\mu H_y & -\overline{\gamma} H_x - \frac{\partial H_z}{\partial x} &= j\omega\epsilon E_y \\
\frac{\partial E_y}{\partial x} &= -j\omega\mu H_z & \frac{\partial H_y}{\partial x} &= j\omega\epsilon E_z
\end{aligned}
\tag{10.6}
$$

The foregoing relationships between the field components are valid in general for any time-harmonic electromagnetic wave solution for which (i) there are no variations in the y direction (i.e., $\partial(\cdot)/\partial y = 0$) and (ii) the field components vary in the z direction as $e^{-\overline{\gamma} z}$. In general, if a solution of the wave equation subject to the boundary conditions is obtained for any one of the field components, the other components can be found from (10.6).

In some cases, it is convenient to express the various field components explicitly in terms of the axial components (i.e., E_z and H_z). Introducing

$$h^2 = \overline{\gamma}^2 - (j\beta)^2 \tag{10.7}$$

we can rewrite equations (10.6) in the form

$$H_x = -\frac{\overline{\gamma}}{h^2}\frac{\partial H_z}{\partial x} \tag{10.8a}$$

$$H_y = -\frac{j\omega\epsilon}{h^2}\frac{\partial E_z}{\partial x} \tag{10.8b}$$

$$E_x = -\frac{\overline{\gamma}}{h^2}\frac{\partial E_z}{\partial x} \tag{10.8c}$$

$$E_y = +\frac{j\omega\mu}{h^2}\frac{\partial H_z}{\partial x} \tag{10.8d}$$

These expressions are useful because the different possible solutions of the wave equation are categorized in terms of the axial components (i.e., z components), as discussed in the next section.

10.1.1 Field Solutions for TE and TM Waves

In general, all of the field components, including both E_z and H_z, may need to be nonzero to satisfy the boundary conditions imposed by an arbitrary conductor or dielectric structure. However, it is convenient and customary to divide the solutions of guided-wave equations into three categories:

$$\left.\begin{array}{l} E_z = 0 \\ H_z \neq 0 \end{array}\right\} = \text{Transverse electric (TE) waves}$$

$$\left.\begin{array}{l} E_z \neq 0 \\ H_z = 0 \end{array}\right\} = \text{Transverse magnetic (TM) waves}$$

$$E_z, H_z = 0 \quad \longrightarrow \quad \text{Transverse electromagnetic (TEM) waves}$$

where z is the axial direction of the guide along which the wave propagation takes place. In the following, we separately examine the field solutions for TE, TM, and TEM waves.

    Transverse electric (TE) waves. We first consider transverse electric (TE) waves, for which we have $E_z = 0$, $H_z \neq 0$. With $E_z = 0$, the electric field vector of a TE wave is always and everywhere transverse to the direction of propagation. We can solve the wave equation for any of the field components; for the parallel-plate case, it is most convenient[10] to solve for E_y, from which all the other field components can be

---

[10] Since $E_z \equiv 0$ for a TE wave, and since there is no variation with y, it is clear that there can be a y component of the electric field, as long as E_y, which is tangential to the conductors, varies with x such that it is zero at the perfectly conducting walls.

found using equations (10.6). Rewriting the wave equation (10.4) for the y component, we have

$$\frac{\partial^2 E_y}{\partial x^2} + \overline{\gamma}^2 E_y - (j\beta)^2 E_y = 0 \tag{10.9}$$

or

$$\frac{\partial^2 E_y}{\partial x^2} + h^2 E_y = 0 \tag{10.10}$$

where $h^2 = \overline{\gamma}^2 - (j\beta)^2 = \overline{\gamma}^2 + \omega^2\mu\epsilon$. We now express $E_y(x,z)$ as a product of two functions, one varying with x and another varying with z as

$$E_y(x,z) = E_y^0(x)e^{-\overline{\gamma}z}$$

Note that z variation of all the field components has the form $e^{-\overline{\gamma}z}$.

We can now rewrite (10.10) as an ordinary differential equation in terms of $E_y^0(x)$:

$$\frac{d^2 E_y^0(x)}{dx^2} + h^2 E_y^0(x) = 0 \tag{10.11}$$

Note that (10.11) is a second-order differential equation whose general solution is

$$E_y^0(x) = C_1 \sin(hx) + C_2 \cos(hx) \tag{10.12}$$

where C_1, C_2, and h are constants to be determined by the boundary conditions. Note that the complete phasor for the y component of the electric field is

$$E_y(x,z) = [C_1 \sin(hx) + C_2 \cos(hx)]e^{-\overline{\gamma}z}$$

We now consider the boundary condition that requires the tangential electric field to be zero at the perfectly conducting plates. In other words,

$$\left.\begin{array}{l} E_y = 0 \text{ at } x = 0 \\ E_y = 0 \text{ at } x = a \end{array}\right\} \quad \text{for all } y \text{ and } z$$

Thus, $C_2 = 0$, and

$$E_y(x,z) = C_1 \sin(hx)e^{-\overline{\gamma}z}$$

results from the first boundary condition. The second boundary condition (i.e., $E_y = 0$ at $x = a$) imposes a restriction on h; that is,

$$\boxed{h = \frac{m\pi}{a} \qquad m = 1, 2, 3, \ldots} \tag{10.13}$$

This result illustrates the particular importance of the constant h. Waveguide field solutions such as (10.12) can satisfy the necessary boundary conditions only for a discrete set of values of h. These values of h for which the field solutions of the wave equation can satisfy the boundary conditions are known as the *characteristic values* or *eigenvalues*.[11]

The other wave field components can all be obtained from E_y using (10.6). We find $E_x = H_y = 0$, and

Parallel-plate TE_m, $m = \pm 1, \pm 2, \ldots$

$$
\begin{aligned}
E_y &= C_1 \sin\left(\frac{m\pi}{a}x\right) e^{-\bar{\gamma}z} \\[2mm]
H_z &= -\frac{1}{j\omega\mu}\frac{\partial E_y}{\partial x} = -\frac{m\pi}{j\omega\mu a}C_1 \cos\left(\frac{m\pi}{a}x\right) e^{-\bar{\gamma}z} \\[2mm]
H_x &= -\frac{\bar{\gamma}}{j\omega\mu}E_y = -\frac{\bar{\gamma}}{j\omega\mu}C_1 \sin\left(\frac{m\pi}{a}x\right) e^{-\bar{\gamma}z}
\end{aligned}
\tag{10.14}
$$

Note that $m = 0$ makes all fields vanish. Each integer value of m specifies a given field configuration, or *mode,* referred to in this case as the TE_m mode. The configurations of fields for the cases of $m = 1$ and 2 are shown in Figure 10.2.

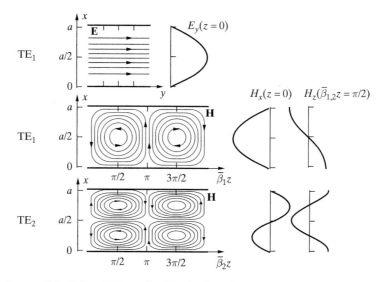

Figure 10.2 TE_1 and TE_2 modes. The electric and magnetic field distributions for the TE_1 and the magnetic field distribution for the TE_2 modes in a parallel-plate waveguide. Careful examination of the field structure for the TE_1 mode indicates that the magnetic field lines encircle the electric field lines (i.e., displacement current) in accordance with (10.1a) and the right-hand rule. The same is true for the TE_2 mode, although the electric field distribution for this mode is not shown.

[11] *Eigen* is from the German word for "characteristic" or "one's own."

Transverse magnetic (TM) waves.   For TM waves the magnetic field vector is always and everywhere transverse to the direction of propagation (i.e., $H_z = 0$), whereas the axial component of the electric field is nonzero (i.e., $E_z \neq 0$). We proceed with the solution in a similar manner as was done for TE waves, except that it is more convenient to use the wave equation for **H**, namely (10.5). The y component of this equation for H_y is in the form of (10.9) for E_y. Following the same procedure as for the TE case for E_y, the general solution for H_y is

$$H_y = [C_3 \sin(hx) + C_4 \cos(hx)]e^{-\bar{\gamma}z}$$

Since the boundary condition for the magnetic field is in terms of its component normal to the conductor boundary, it cannot be applied to H_y directly. However, using (10.6), we can write the tangential component of the electric field E_z in terms of H_y as

$$E_z = \frac{1}{j\omega\epsilon}\frac{\partial H_y}{\partial x} = \frac{h}{j\omega\epsilon}[C_3 \cos(hx) - C_4 \sin(hx)]e^{-\bar{\gamma}z}$$

We now note that for all y and z:

$$E_z = 0 \quad \text{at } x = 0 \quad \rightarrow \quad C_3 = 0$$

and

$$E_z = 0 \quad \text{at } x = a \quad \rightarrow \quad \boxed{h = \frac{m\pi}{a} \qquad m = 0, 1, 2, 3, \ldots} \qquad (10.15)$$

The only other nonzero field component is E_x, which can be simply found from H_y or E_z. The nonzero field components for this mode, denoted as the TM_m mode, are then

Parallel-plate TM_m,
$m = 0, \pm 1, \pm 2, \ldots$

$$\boxed{\begin{aligned} H_y &= C_4 \cos\left(\frac{m\pi}{a}x\right)e^{-\bar{\gamma}z} \\ E_x &= \frac{\bar{\gamma}}{j\omega\epsilon}H_y = \frac{\bar{\gamma}}{j\omega\epsilon}C_4 \cos\left(\frac{m\pi}{a}x\right)e^{-\bar{\gamma}z} \\ E_z &= \frac{jm\pi}{\omega\epsilon a}C_4 \sin\left(\frac{m\pi}{a}x\right)e^{-\bar{\gamma}z} \end{aligned}} \qquad (10.16)$$

The configurations of the fields for $m = 1$ and 2 (i.e., for the TM_1 and TM_2 modes) are shown in Figure 10.3.

Transverse electromagnetic (TEM) waves.   Note that contrary to the TE case, the TM solutions do not all vanish for $m = 0$. Since E_z is zero for $m = 0$, the TM_0

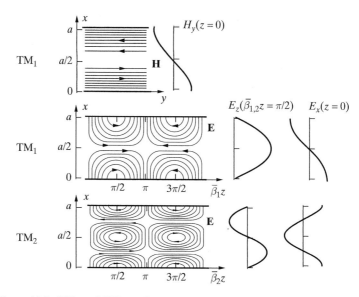

Figure 10.3   TM$_1$ and TM$_2$ modes. The electric and magnetic field distributions for the TM$_1$ and TM$_2$ modes in a parallel-plate waveguide. Note that for TM$_1$ the electric field lines encircle the magnetic field lines (Faraday's law) in accordance with (10.1c). The same is true for TM$_2$, although the magnetic field distribution for TM$_2$ is not shown.

mode is actually a transverse electromagnetic (or TEM) wave. In this case, we have

$$\text{Parallel-plate TEM} \qquad \boxed{\begin{aligned} H_y &= C_4 e^{-\overline{\gamma} z} \\ E_x &= \frac{\overline{\gamma}}{j\omega\epsilon} C_4 e^{-\overline{\gamma} z} \\ E_z &= 0 \end{aligned}} \qquad (10.17)$$

For this case, we have $h = 0$, so $\overline{\gamma} = j\overline{\beta} = j\beta$, where $\beta = \omega\sqrt{\mu\epsilon}$ is the phase constant for a uniform plane wave in the unbounded lossless medium. The field configuration for the TM$_0$ or TEM wave is shown in Figure 10.4.

10.1.2 Cutoff Frequency, Phase Velocity, and Wavelength

Close examination of the solutions for TE and TM waves reveal a number of common characteristics: (i) **E** and **H** have sinusoidal standing-wave distributions in the x direction, (ii) any x-y plane is an equiphase plane, or, in other words, the TE and TM waves are plane waves with surfaces of constant phase given by $z = \text{const.}$, and (iii) these equiphase surfaces progress (propagate) along the waveguide with a velocity $\overline{v}_p = \omega/\overline{\beta}$. To see this, consider any of the field components—say, E_y, for TE waves. For a propagating wave,

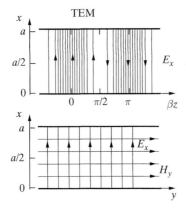

Figure 10.4 TEM mode. Electric and magnetic fields between parallel planes for the TEM (TM_0) mode. Only electric field lines are shown in the top panel; the magnetic field lines are out of (or into) the page.

with $\overline{\gamma} = j\overline{\beta}$, and assuming C_1 to be real, from (10.14), we have

$$\mathscr{E}_y(x,z,t) = C_1 \sin\left(\frac{m\pi}{a}x\right)\cos(\omega t - \overline{\beta}z)$$

To determine the conditions under which we can have a propagating wave (i.e., $\overline{\gamma}$ purely imaginary, that is, $\overline{\gamma} = j\overline{\beta}$), consider the definition of the eigenvalue parameter h:

$$h^2 = \overline{\gamma}^2 - (j\beta)^2 \quad \rightarrow \quad \overline{\gamma} = \sqrt{h^2 + (j\beta)^2} = \sqrt{\left(\frac{m\pi}{a}\right)^2 - \omega^2\mu\epsilon}$$

Note that this expression for h is valid for both TE and TM modes. Thus, the expressions given below for propagation constant, phase velocity, and wavelength are also equally valid for both TE and TM modes.

From the preceding definition of h it is apparent that for each mode (TE or TM) identified by mode index m, there exists a critical or *cutoff* frequency f_{c_m} (or wavelength λ_{c_m}) such that

$$\overline{\gamma} = 0 \quad \rightarrow \quad f_{c_m} = \frac{m v_p}{2a} = \frac{m}{2a\sqrt{\mu\epsilon}} \quad \text{or} \quad \lambda_{c_m} = \frac{v_p}{f_{c_m}} = \frac{2a}{m} \tag{10.18}$$

where $v_p = 1/\sqrt{\mu\epsilon}$ is the *intrinsic* phase velocity for uniform plane waves in the unbounded lossless medium. Note that $v_p = (\mu\epsilon)^{-1/2}$ depends only on ϵ and μ and is thus indeed an intrinsic property of the medium. For $f > f_{c_m}$, the propagation constant $\overline{\gamma}$ is purely imaginary and is given by

$$\overline{\gamma} = j\overline{\beta}_m = j\sqrt{\omega^2\mu\epsilon - \left(\frac{m\pi}{a}\right)^2} = j\beta\sqrt{1 - \left(\frac{f_{c_m}}{f}\right)^2} \qquad f > f_{c_m} \tag{10.19}$$

where we now start to use $\overline{\beta}_m$ instead of $\overline{\beta}$ to underscore the dependence of the phase constant on the mode index m. Note that $\overline{\beta}_m$ is the phase constant (sometimes referred to as the longitudinal phase constant or the propagation constant) for mode m, and $\beta = \omega\sqrt{\mu\epsilon}$ is the intrinsic phase constant, being that for uniform plane waves in the unbounded lossless medium. For $f < f_{c_m}$, $\overline{\gamma}$ is a real number, given by

$$\overline{\gamma} = \overline{\alpha}_m = \sqrt{\left(\frac{m\pi}{a}\right)^2 - \omega^2\mu\epsilon} = \beta\sqrt{\left(\frac{f_{c_m}}{f}\right)^2 - 1} \qquad f < f_{c_m} \qquad (10.20)$$

where $\overline{\alpha}_m$ is the attenuation constant, in which case the fields attenuate exponentially in the z direction, without any wave motion. For example, the expression for the $\mathscr{E}_y$ component for TE waves with C_1 real for the case when $f < f_{c_m}$ is

$$\mathscr{E}_y(x,z,t) = C_1 e^{-\overline{\alpha}z}\sin\left(\frac{m\pi}{a}x\right)\cos(\omega t)$$

Such a nonpropagating mode is called an evanescent wave.[12] For $f > f_{c_m}$, the propagating wave has a wavelength $\overline{\lambda}_m$ (sometimes referred to as the guide wavelength) and phase velocity $\overline{v}_{p_m}$ given as

$$\begin{aligned}
\overline{\lambda}_m &= \frac{2\pi}{\overline{\beta}_m} = \frac{\lambda}{\sqrt{1-(f_{c_m}/f)^2}} \\
\overline{v}_{p_m} &= \frac{\omega}{\overline{\beta}_m} = \frac{v_p}{\sqrt{1-(f_{c_m}/f)^2}}
\end{aligned} \qquad (10.21)$$

where $\lambda = 2\pi/\beta = v_p/f$ is the intrinsic wavelength, being the wavelength for uniform plane waves in the unbounded lossless medium, and thus depending only on μ, ϵ, and frequency f. Note that for the TM$_0$ mode (which is in fact a TEM wave, as discussed earlier), $\overline{v}_p$ is equal to the intrinsic phase velocity $v_p = (\sqrt{\mu\epsilon})^{-1}$ in the unbounded loss-less medium and is thus independent of frequency. Furthermore, the phase constant $\overline{\beta}_0$ and wavelength $\overline{\lambda}_0$ are also equal to their values for a uniform plane wave in an unbounded lossless medium. There is no cutoff frequency for this mode, since the propagating field solutions are valid at any frequency.

For TE and TM waves in general, it is clear from this discussion that the wavelength $\overline{\lambda}_m$, as observed along the guide, is longer than the corresponding wavelength in an unbounded lossless medium by the factor $[1 - (f_{c_m}/f)^2]^{-1/2}$. Also, the velocity of phase progression inside the guide is likewise greater than the corresponding intrinsic phase velocity. Note, however, that $\overline{v}_p$ is not the velocity with which energy or information propagates,[13] so $\overline{v}_p > v_p$ does not pose any particular dilemma.

---

[12]It is important to note that the attenuation of the field in the z direction is not due to any energy losses; this condition results simply from the fact that the boundary conditions cannot be satisfied by a TE or TM wave at frequency ω. If a TE or TM field configuration is somehow excited (e.g., by an appropriate configuration of source currents or charges) at a given point z, the fields attenuate exponentially with distance away from the excitation point, with no energy being carried away.

[13]The velocity at which information (e.g., the envelope of a modulated signal) travels is the so-called *group velocity* v_g, given by $v_g = d\omega/d\beta$, which for metallic waveguides is $v_g = c^2/\overline{v}_p$. The concept of group velocity is discussed extensively in Section 10.3.

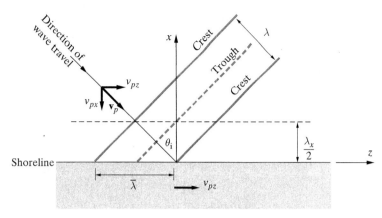

Figure 10.5   Water wave. Water wave approaching a shoreline or a breakwater.

The difference between $\bar{v}_p$ and v_p can be further understood using an analogy with water waves. Consider water waves approaching a shoreline or a breakwater at an angle θ_i from the line perpendicular to it as shown in Figure 10.5. The velocity of the wave in its direction of propagation can be found from the distance between successive crests (λ) and the frequency (f) with which crests pass a given point, in other words, $v_p = \lambda f$. However, we all remember from our own experiences watching waves hit a shoreline that the velocity of water flow along the shoreline can appear to be much faster. To understand this effect, we can observe from Figure 10.5 that the velocity of flow along the shoreline is given by $v_{pz} = \bar{\lambda} f$, where $\bar{\lambda}$ is the distance between successive crests along the shoreline. We thus have $\bar{\lambda} = \lambda / \sin\theta_i$ or $v_{pz} = v_p / \sin\theta_i$, so that in general v_{pz} is indeed larger then v_p.

Example 10.1: Parallel-plate waveguide modes. An air-filled parallel-plate waveguide has a plate separation of 1.25 cm. Find (a) the cutoff frequencies of the TE_0, TM_0, TE_1, TM_1, and TM_2 modes, (b) the phase velocities of these modes at 15 GHz, (c) the lowest-order TE and TM mode that cannot propagate in this waveguide at 25 GHz.

Solution:

(a) As we have seen in earlier sections, the TE_0 mode does not exist in a parallel-plate waveguide. TM_0 is equivalent to the TEM mode, and this mode can propagate at all frequencies (i.e., $f_{c_0} = 0$). The cutoff frequencies of the other modes can be calculated using (10.18) as

$$f_{c_m} = \frac{m}{2a\sqrt{\mu_0\epsilon_0}} = \frac{m(3\times10^{10}\ \text{cm-s}^{-1})}{2(1.25\ \text{cm})} = 1.2\times10^{10}m\ \text{Hz} = 12m\ \text{GHz}$$

The results are summarized in the following table:

Mode	TM_0	TE_1	TM_1	TE_2	TM_2
f_{c_m} (GHz)	0	12	12	24	24

(b) For the TM_0 mode, $\bar{v}_p = v_p = c$. For the TE_1 and TM_1 modes, using (10.21), we have

$$\bar{v}_p = \frac{c}{\sqrt{1 - (f_{c_1}/f)^2}} = \frac{c}{\sqrt{1 - (12/15)^2}} = \frac{5c}{3}$$

Note that the TE_2 and TM_2 modes do not propagate at 15 GHz; that is, their cutoff frequency is above 15 GHz.

(c) The lowest-order modes that cannot propagate in this waveguide at 25 GHz are TE_3 and TM_3, since

$$f_{c_3} = 12 \times 3 = 36 \text{ GHz} > 25 \text{ GHz}$$

Example 10.2: ELF propagation in the Earth–ionosphere waveguide. Extremely low frequencies (ELF) are ideal for communicating with deeply submerged submarines, because below 1 kHz, electromagnetic waves penetrate into seawater.[14] Propagation at these frequencies takes place in the "waveguide" formed between the earth and the ionosphere (Figure 10.6); low propagation losses allow nearly worldwide communication from a single ELF transmitter.

In J. R. Wait's simple model,[15] the surface of the earth and the bottom of the ionosphere form the boundaries of a terrestrial "parallel"-plate waveguide with lossy walls. The ionosphere is approximated by an isotropic layer beginning at a given altitude and extending to infinity with no horizontal variations allowed. Energy is lost through the "walls" either into the finitely conducting

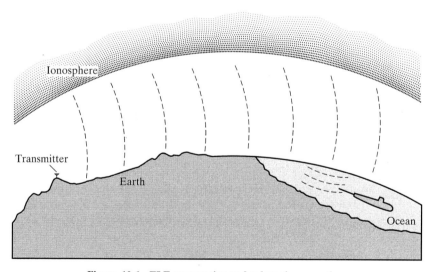

Figure 10.6   ELF propagation and submarine reception.

[14]S. L. Bernstein, et al., Long-range communications at extremely low frequencies, *Proc. IEEE,* 62(3), pp. 292–312, March 1974.

[15]J. R. Wait, Earth-ionosphere cavity resonances and the propagation of ELF radio waves, *Radio Sci.,* 69D, pp. 1057–1070, August 1965.

ionosphere or into the ground, with the former loss being dominant. The important feature of propagation below 1 kHz is that there is a single propagating mode, a so-called quasi-TEM mode. All the other modes are evanescent and are almost undetectable at distances in excess of 1000 km. In the far field, the wave consists of a vertical electric field and a horizontal magnetic field transverse to the direction of propagation (Figure 10.6). The leakage of energy from this wave into the ocean (see Section 9.8) gives rise to a plane wave propagating vertically downward, and it is this signal that the submarine receiver detects.

Consider an idealized earth–ionosphere waveguide where both the ionosphere and the earth are assumed to be perfect conductors. In addition, neglect the curvature of the waveguide and assume it to be flat. The height of the terrestrial waveguide can vary anywhere from 70 km to 90 km depending on conditions; for our purposes, assume it to be 80 km. Find all the propagating modes at an operating frequency of (a) 100 Hz, (b) 1 kHz, and (c) 10 kHz.

Solution: As we already know, the TEM mode (or TM_0 mode) exists in all cases. The cutoff frequencies of other modes can be found using (10.18) as

$$f_{c_m} = \frac{m(3 \times 10^8 \text{ m-s}^{-1})}{2 \times 80 \times 10^3 \text{ m}} = 1875m \text{ Hz}$$

So the cutoff frequencies of some of the lower-order modes are

Mode	TM_0	TE_1	TM_1	TE_2	TM_2	TE_3	TM_3
f_{c_m} (Hz)	0	1875	1875	3750	3750	5625	5625

Therefore, (a) at 100 Hz and (b) at 1 kHz, only one mode (the TEM mode) can propagate, whereas (c) at 10 kHz, eleven different modes (i.e., TEM, TE_1, TM_1, TE_2, TM_2, TE_3, TM_3, TE_4, TM_4, TE_5, and TM_5) can propagate in the earth–ionosphere waveguide.

Example 10.3: Waveguide propagation constant versus frequency. Sketch the magnitude of the waveguide propagation constant $\overline{\gamma}$ as a function of the operating frequency f for TM_0, TE_1, TM_1, TE_2, and TM_2 modes in a parallel-plate waveguide of plate separation a and filled with a lossless dielectric with properties ϵ and μ.

Solution: For TM_0 or TEM mode, $|\overline{\gamma}| = \overline{\beta} = \omega\sqrt{\mu\epsilon} = 2\pi f \sqrt{\mu\epsilon}$ for all operating frequencies. For higher-order TE_m and TM_m modes, we have from (10.19) and (10.20)

$$\left. \begin{array}{l} \overline{\gamma} = \overline{\alpha} = 2\pi f \sqrt{\mu\epsilon}\sqrt{(f_{c_m}/f)^2 - 1} \\[2mm] \overline{\beta} = 0 \text{ (no propagation)} \end{array} \right\} \quad \text{for } f < f_{c_m}$$

and

$$\left. \begin{array}{l} \overline{\alpha} = 0 \text{ (no attenuation)} \\[2mm] |\overline{\gamma}| = \overline{\beta} = 2\pi f \sqrt{\mu\epsilon}\sqrt{1 - (f_{c_m}/f)^2} \end{array} \right\} \quad \text{for } f > f_{c_m}$$

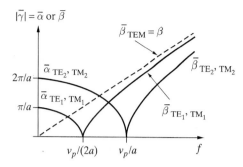

Figure 10.7   Waveguide propagation constant versus frequency. Magnitude of the propagation constant $|\overline{\gamma}|$ versus frequency f for the five lowest-order modes in a parallel-plate waveguide. Note that $v_p = 1/\sqrt{\mu\epsilon}$.

where f_{c_m} is the cutoff frequency of the mth-order mode, given by

$$f_{c_m} = \frac{m}{2a\sqrt{\mu\epsilon}}$$

Using these expressions, we can sketch the magnitudes of $\overline{\gamma}$, $\overline{\alpha}$, and $\overline{\beta}$ as a function of the operating frequency f for TM_0, TE_1, TM_1, TE_2, and TM_2 modes, as shown in Figure 10.7.

10.1.3 TE and TM Modes as Superpositions of TEM Waves

It is often instructive to think of the different TE or TM waveguide modes as a superposition of two or more TEM waves. This picture is helpful, as it provides a better physical understanding of the mode field structures as well as concepts such as guide wavelength, cutoff frequency, and losses in the waveguide walls.

To demonstrate the decomposition of the mode fields analytically, consider, for example, the phasor electric field component of the propagating TE_m mode, which can be written from (10.14) as

$$E_y(x,z) = C_1 \sin\left(\frac{m\pi}{a}x\right)e^{-j\overline{\beta}z} = \frac{C_1}{2j}\left[e^{j(m\pi x/a)} - e^{-j(m\pi x/a)}\right]e^{-j\overline{\beta}z}$$

$$= \frac{C_1}{2}\left[e^{j(m\pi x/a - \overline{\beta}z - \pi/2)} - e^{-j(m\pi x/a + \overline{\beta}z + \pi/2)}\right]$$

Using the notation introduced in Section 8.6 and Figure 8.27, the first term in the above expression represents a perpendicularly polarized (with respect to the x-z plane) TEM wave (denoted as TEM$_1$) propagating in the direction $\hat{\mathbf{k}}_1 = \beta^{-1}(-\beta_x\hat{\mathbf{x}} + \beta_z\hat{\mathbf{z}}) = -\hat{\mathbf{x}}\cos\theta_{i_m} + \hat{\mathbf{z}}\sin\theta_{i_m}$, where θ_{i_m} is the angle between $\hat{\mathbf{k}}_1$ and the x axis (see Figure 10.8).[16]

---

[16]In comparison with Figure 9.20 in Chapter 9, it should be noted that the angle θ_i defined in Figure 9.20 is the angle between $\hat{\mathbf{k}}$ and the z axis, or the complement (i.e., $90° - \theta_i$) of the angle θ_{i_m} defined here (Figure 10.8).

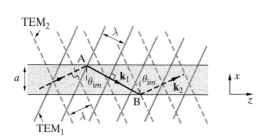

Figure 10.8 Two TEM waves forming a TE$_m$ wave in a parallel-plate waveguide. Waveguide modes represented as a superposition of two TEM waves. TEM$_1$ propagates in the direction $\hat{\mathbf{k}}_1 = \mathbf{k}_1/|\mathbf{k}_1|$ with surfaces of constant phase (i.e., phase fronts) indicated by solid lines. TEM$_2$ propagates in the direction $\hat{\mathbf{k}}_2 = \mathbf{k}_2/|\mathbf{k}_2|$, with phase fronts shown in dashed lines. The same type of decomposition, and thus the same picture shown here, is also valid for the TM$_m$ modes.

The phase constants β_x and β_z are given as

$$\beta_x = \beta\cos\theta_{i_m} = \frac{m\pi}{a} = \omega_{c_m}\sqrt{\mu\epsilon}$$

$$\beta_z = \beta\sin\theta_{i_m} = \sqrt{\beta^2 - \left(\frac{m\pi}{a}\right)^2} = \bar{\beta} = \omega\sqrt{\mu\epsilon}\sqrt{1 - \left(\frac{\omega_{c_m}}{\omega}\right)^2}$$

where $\beta = \omega\sqrt{\mu\epsilon}$. Similarly, the second term in the above expression for $E_y(x,z)$ represents another perpendicularly polarized TEM plane wave (denoted as TEM$_2$) propagating in the direction $\hat{\mathbf{k}}_2 = \beta^{-1}(\beta_x\hat{\mathbf{x}} + \beta_z\hat{\mathbf{z}}) = \hat{\mathbf{x}}\cos\theta_{i_m} + \hat{\mathbf{z}}\sin\theta_{i_m}$ with the same phase constants β_x and β_z. These component TEM waves are shown in Figure 10.8. Note that this picture of the superposition of two TEM waves is identical to that for TEM waves obliquely incident on a conducting sheet, which was described in Chapter 9 in connection with Figure 9.20. The field solutions in front of the conducting sheet also consisted of a superposition of two TEM waves, propagating in the $\hat{\mathbf{k}}_i$ and $\hat{\mathbf{k}}_r$ directions, at an incidence angle θ_{i_m} with respect to the line perpendicular to the conductor boundary.

The angle θ_{i_m} can be found from the above relationship for β_x as

$$\theta_{i_m} = \cos^{-1}\left[\frac{\omega_{c_m}}{\omega}\right] = \cos^{-1}\left[\frac{f_{c_m}}{f}\right] = \cos^{-1}\left[\frac{m}{2af\sqrt{\mu\epsilon}}\right]$$

$$= \cos^{-1}\left[\frac{m\lambda}{2a}\right] = \cos^{-1}\left[\frac{\lambda}{\lambda_{c_m}}\right] \qquad m = 0, 1, 2, 3, \ldots \qquad (10.22)$$

where we have used the facts that $f_{c_m} = m/(2a\sqrt{\mu\epsilon}) = (\lambda_{c_m}\sqrt{\mu\epsilon})^{-1}$ and $\lambda f = (\sqrt{\mu\epsilon})^{-1}$. Note that each mode (i.e., each integer value of m) corresponds to a discrete value of θ_{i_m}. Note also that a real value for θ_{i_m} exists only when $f > f_{c_m}$ (or $\lambda < \lambda_{c_m}$), which corresponds to a propagating mode. In the special case when $f = f_{c_m}$, we have $\cos\theta_{i_m} = 1$, and $\theta_{i_m} = 0$, in which case the two component TEM waves that constitute the TE$_m$ wave propagate, respectively, in the x and $-x$ directions, with no propagation of wave energy along the waveguide in the z direction.

Although we have illustrated the decomposition for the case for one field component of the TE$_m$ mode, a similar decomposition can be done for each of the other field components of the TE_m mode or for the field components of TM_m modes. The only difference for the case of TM_m modes is that the component TEM waves are parallel polarized with respect to the x-z plane, and the electric field vectors lie in the x-z plane.

The decomposition of TE or TM mode fields into component TEM waves, and the fact that this is possible for only a discrete set of values of θ_{i_m}, can also be understood graphically by considering Figure 10.9. The combination of the two component waves (TEM_1 and TEM_2) can be thought in terms of a "ray" reflecting back and forth between the walls, with the wave energy regarded as being transported along the ray path.[17] Consider the ray path ABC shown in the top panel of Figure 10.9. The solid (dashed) lines perpendicular to the ray paths represent the constant phase fronts corresponding to the crests (nulls) of the TEM wave during its travel from A to B (the constant phase fronts corresponding to the ray path from B to C are not shown). Before the ray undergoes reflection at point B, the line B–B′ constitutes a constant phase front. In order for the ray to continue down the guide as a uniform plane wave, the line C–C′ must also be a constant phase front. Thus, the amount of wave phase change along paths B–C and B′–C′ must be an integer multiple of 2π, that is,

$$\beta(\overline{BC} - \overline{B'C'}) = m2\pi \qquad m = 0, 1, 2, 3, \ldots$$

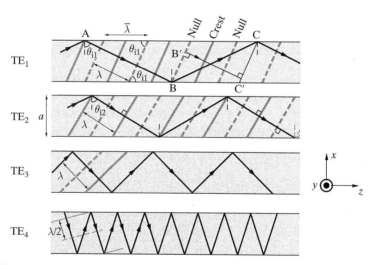

Figure 10.9   Representation of propagating waveguide modes as "rays" reflecting between the parallel plates. The null and crest lines represent the phase fronts, and the solid lines represent the ray paths.

[17]It is sometimes convenient to represent an electromagnetic wave by *rays* rather than by wave fronts (or the surfaces of constant phase). This picture is most appropriate for light rays, and rays were used to describe light long before its wave nature was firmly established. In the context of the wave nature of light, and in an isotropic medium, a ray is an imaginary line along the direction of propagation of the wave and perpendicular to the wave fronts. In the context of the particle theory of light, rays can be thought of as paths of the photons.

where $\beta = \omega\sqrt{\mu\epsilon}$ is the propagation constant in the unbounded lossless medium. From geometrical considerations we have

$$\overline{BC} = \frac{a}{\cos\theta_{i_m}} \qquad \overline{B'C'} = \overline{BC'}\sin\theta_{i_m}$$

where

$$\overline{BC'} = a\tan\theta_{i_m} - \frac{a}{\tan\theta_{i_m}}$$

Substituting in the above and performing algebraic manipulation, we find

$$2\beta a\cos\theta_{i_m} = m2\pi \qquad m = 0, 1, 2, 3, \dots$$

Noting that $\beta = 2\pi/\lambda$, we can write the angle θ_{i_m} as

$$\cos\theta_{i_m} = \frac{m\lambda}{2a} \quad \rightarrow \quad \theta_{i_m} = \cos^{-1}\left[\frac{m\lambda}{2a}\right] = \cos^{-1}\left[\frac{\lambda}{\lambda_{c_m}}\right] \quad m = 0, 1, 2, 3, \dots \quad (10.23)$$

which is an expression identical to that determined above on the basis of an analytical decomposition of the mode fields into component TEM waves.[18] We find once again that a uniform plane wave (represented by a ray) can propagate down the waveguide by undergoing multiple reflections, as long as the angle θ_{i_m} between its ray path and the normal to the waveguide walls is equal to one of the set of discrete values of θ_{i_m} given by (10.23). Note that each one of these discrete values θ_{i_m} corresponds to a particular waveguide mode, TE$_m$ or TM$_m$. The ray path representation is therefore equally valid for both TM and TE waves; the difference between the two mode types is the orientation of the wave electric field with respect to the x-z plane. For TE modes the ray represents a perpendicularly polarized wave, with the electric field perpendicular to the page (i.e., the y direction) in Figure 10.9. For TM modes, the uniform plane wave represented by the ray is parallel polarized, with the electric field lying in the plane of Figure 10.9 (i.e., on the x-z plane).

As defined earlier, the wavelength $\lambda_{c_m} = 2a/m$ is the cutoff wavelength. At this wavelength, $\theta_{i_m} = 0$, and, as a result, there is no energy propagation along the z axis, because the ray path is perpendicular to the z direction and the ray simply reflects up and down between the waveguide walls. From the condition on θ_{i_m}, namely $\cos\theta_{i_m} = m\lambda/(2a)$, it is clear that for a given value of a the ray paths for waves with very

---

[18]Note that we could have arrived at the same condition for θ_{i_m} by simply noting that for the picture of the ray paths and phase fronts of Figure 10.9 to be valid, the separation between the waveguide walls (i.e., a) must be an integral multiple of half-wavelengths in the x direction, or

$$\frac{m\lambda_x}{2} = a \qquad \rightarrow \qquad \cos\theta_{i_m} = \frac{m\lambda}{2a} = \frac{\lambda}{\lambda_{c_m}}$$

since $\lambda_x = \lambda/\cos\theta_{i_m}$.

short wavelengths $\lambda \ll a$ (very high frequencies) are almost parallel to the z axis. The wavelength parallel to the walls (or along the z direction), which is the guide wavelength $\overline{\lambda}_m$ introduced in the previous section, and the velocity $\overline{v}_{p_m}$ are given by

$$\overline{\lambda}_m = \frac{\lambda}{\sin \theta_{i_m}} = \frac{\lambda}{\sqrt{1 - (m\lambda/2a)^2}} = \frac{\lambda}{\sqrt{1 - (\lambda/\lambda_{c_m})^2}}$$

$$\overline{v}_{p_m} = \frac{v_p}{\sin \theta_{i_m}} = \frac{v_p}{\sqrt{1 - (m\lambda/2a)^2}} = \frac{v_p}{\sqrt{1 - (\lambda/\lambda_{c_m})^2}}$$

Because of the zigzag path that is followed by the ray, the velocity with which the energy carried by the ray propagates along the guide has a smaller magnitude than that of the free space velocity in the unbounded lossless medium (i.e., v_p). We shall see in Section 10.3 that this velocity is known as the *group velocity* v_g, and is the projection of the ray velocity v_p along the guide given by

$$v_g = v_p \sin \theta_{i_m} = v_p \sqrt{1 - \left(\frac{m\lambda}{2a}\right)^2} = v_p \sqrt{1 - (\lambda/\lambda_{c_m})^2}$$

Example 10.4: Incidence angle of TEM waves. Consider an air-filled parallel-plate waveguide having a plate separation of 3 cm. At an operating frequency of 20 GHz, find the oblique incidence angle θ_{i_m} and sketch the ray paths corresponding to the TE$_1$, TE$_2$, TE$_3$, and TE$_4$ modes and determine the mode which propagates with the highest phase velocity.

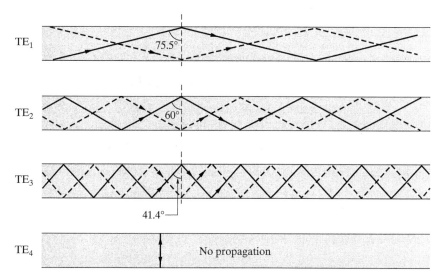

Figure 10.10　Incidence angle of component TEM waves. Different TE$_m$ modes represented by ray paths propagating inside the parallel-plate waveguide at different angles of incidence θ_{i_m}.

Solution: Using (10.22) with $f_{c_m} = m/(2a\sqrt{\mu_0\epsilon_0}) \simeq 5 \times 10^9 m$ Hz $= 5m$ GHz, we find $\theta_{i_1} = \cos^{-1}(5/20) \simeq 75.5°$ (TE$_1$), $\theta_{i_2} = 60°$ (TE$_2$), $\theta_{i_3} \simeq 41.4°$ (TE$_3$), and $\theta_{i_4} = 0$ (TE$_4$). Figure 10.10 shows a graphical representation of each TE$_m$ mode in terms of the superposition of two TEM waves. Note that the component TEM waves corresponding to the TE$_4$ mode bounce back and forth between the waveguide walls without any propagation along the guide length, because its cutoff frequency is equal to the operating frequency.

10.1.4 Attenuation in Parallel-Plate Waveguides

Up to now, we have considered the field structures and other characteristics of propagating modes, assuming the waveguide walls to be perfectly conducting and the dielectric between the plates to be lossless, and thus implicitly neglecting losses. In this section we consider attenuation of the fields due to conduction or dielectric losses.[19]

Attenuation due to conduction losses. In all of the TE and TM mode solutions we have discussed, the wave magnetic field has a nonzero component parallel to the waveguide walls. As a result of the boundary condition $\hat{\mathbf{n}} \times \mathbf{H} = \mathbf{J}_s$, this parallel magnetic field component causes surface currents to flow in the metallic waveguide walls. If the walls are made of perfectly conducting materials, this current flow does not have any effect on the wave fields. However, in actual waveguides, the fact that the metallic walls have finite conductivity results in losses, and the fields attenuate with distance.[20]

Determining attenuation in a waveguide in the general case, when the waveguide walls consist of an arbitrary lossy material, is a rather difficult problem. When the walls are not perfect conductors, the basic boundary conditions, namely that $E_{\text{tan}} = 0$ and $H_{\text{nor}} = 0$, no longer hold, so the field configurations (e.g., those given by (10.14) for TE$_m$ waves) derived using these conditions are not valid either. In other words, although the losses can in principle be calculated (e.g., using (9.17) with the surface currents determined from $\mathbf{J}_s = \hat{\mathbf{n}} \times \mathbf{H}$) if the field configurations were known, the latter in fact depends on the losses and cannot be calculated without accounting for the losses. We resolve this dilemma by resorting to a so-called "perturbation" solution, which is often used in many engineering problems, and which gives highly accurate results in the case of metallic waveguides. The basic premise of our solution is that the losses are small enough that they have a negligible effect on the field distribution within the waveguide. This assumption makes sense, since the waveguides we work with are presumably designed for and used in applications for which the losses are actually small—otherwise they would not be used for that application. In the context of this basic assumption, we can use the field distributions derived for the lossless case to determine the tangential magnetic field on the conductor surfaces, from which we can in turn find the surface

---

[19]Attenuation in waveguides can occur because the walls are not perfectly conducting and also if the dielectric material between the plates is lossy (i.e., $\epsilon'' \neq 0$ or $\sigma \neq 0$).

[20]Note that this attenuation is in general quite small, unlike the very rapid attenuation of an evanescent wave. In other words, typically the value of $\bar{\alpha}$ for a nonpropagating mode is much larger than the attenuation constant for conducting losses; that is, α_c.

currents and use (9.17) to compute the wall losses. If deemed necessary, a second and improved estimate of losses can then be made, using a field distribution corrected (to first order) to account for the calculated losses.[21] However, since practical waveguides are constructed of high-conductivity metals such as copper or brass or are coated with silver, the original assumption of determining the field configuration as if the walls were perfect conductors yields very accurate results, and a second assumption is rarely necessary.

We now proceed by adopting the common practice of assuming that the attenuation is exponential in form (i.e., the fields vary as $e^{-\alpha_c z}$) and determine the attenuation constant α_c using a perturbation solution. Since the fields vary as $e^{-\alpha_c z}$, the time-average wave power P_{av} transmitted along the waveguide (i.e., in the z direction) varies as $e^{-2\alpha_c z}$. The rate of decrease of transmitted power along the waveguide is then

$$-\frac{\partial P_{av}}{\partial z} = +2\alpha_c P_{av}$$

The reduction in power per unit length must be equal to the power lost or dissipated per unit length. Therefore we have

$$\frac{\text{Power lost per unit length}}{\text{Power transmitted}} = \frac{2\alpha_c P_{av}}{P_{av}} = 2\alpha_c$$

or

$$\alpha_c = \frac{\text{Power lost per unit length}}{2 \times \text{Power transmitted}} = \frac{P_{loss}}{2P_{av}} \tag{10.24}$$

We now use this definition of α_c for the parallel-plate waveguide.

We first consider attenuation due to conductor losses for a TEM wave. Using the solutions (10.17) obtained in the previous subsection, we have

$$H_y = C_4 e^{-j\bar{\beta}z} \qquad E_x = \frac{\bar{\beta}}{\omega\epsilon}C_4 e^{-j\bar{\beta}z}$$

The surface current density on each of the conducting plates is

$$\mathbf{J}_s = \hat{\mathbf{n}} \times \mathbf{H}$$

---

[21] As an example of the manner in which such a correction may be introduced, consider the TEM mode, for which the magnetic field at the conductor surface is in the y direction, so the surface current is in the $+z$ direction on the lower wall and in the $-z$ direction on the upper wall. Since the conductor is not perfect, the presence of a current $\mathbf{J}_s = \hat{\mathbf{z}}J_z$ implies that there is a nonzero z component of the electric field (i.e., E_z) within the conductor. The continuity of the tangential component of an electric field across any boundary between two materials (i.e., equation (7.27)) in turn requires that there must be a nonzero electric field component E_z immediately outside the conductor and inside the parallel-plate waveguide. Our original assumption of a TEM mode, with both electric and magnetic fields completely transverse, clearly cannot be precisely valid. Assuming that the magnitude of E_z is small, we may approximate its variation within the waveguide as $E_z = K(1 - 2x/a)$, where K is a constant, dependent on the surface impedance $Z_s = (1 + j)R_s$ of the conductor. Note that we then have a linear variation of E_z from $E_z = +K$ at the lower wall to $E_z = -K$ at the upper wall. We can now take the total electric field to be $\mathbf{E} = \hat{\mathbf{x}}E_x + \hat{\mathbf{z}}E_z$ and use Maxwell's equations to determine the associated magnetic field, to arrive at the first-order corrected field distribution.

The surface current density vector in each plate is $\mathbf{J}_s = \mp\hat{\mathbf{z}}\, C_4 e^{-j\bar{\beta}z}$, so $|J_{sz}| = C_4$, assuming that C_4 is positive and real.

The total time-average power loss (sum of that in the upper and lower plates) for a length of 1 meter and a width of b meters of the guide is

$$P_{\text{loss}} = 2\int_0^1 \int_0^b \tfrac{1}{2}|J_{sz}|^2 R_s \; dy \; dz = C_4^2 R_s b$$

where R_s is the resistive part of the surface impedance discussed in Section 9.4.1. From (9.14), the surface impedance of a conductor is

$$\eta_c = Z_s = R_s + jX_s = (\sigma\delta)^{-1}(1+j)$$

where $\delta = \sqrt{2/(\omega\mu_0\sigma)}$ is the skin depth, assuming nonmagnetic, conducting walls with permeability $\mu = \mu_0$. Another expression for R_s is $R_s = \sqrt{\omega\mu_0/(2\sigma)}$.

The time-average power density transmitted in the z direction along the guide per unit cross-sectional area is

$$\mathbf{S}_{\text{av}} = \tfrac{1}{2}\Re e\{\mathbf{E} \times \mathbf{H}^*\}$$

Since $|E_x| = \eta|H_y|$, with C_4 assumed to be real, we have $|\mathbf{S}_{\text{av}}| = \tfrac{1}{2}\eta C_4^2$. For a spacing of a meters and a width of b meters, the cross-sectional area is ba, thus, the total time-average power transmitted through such a cross-sectional area of a TEM wave is

$$P_{\text{av}} = \tfrac{1}{2}\eta C_4^2 ba \tag{10.25}$$

Using the definition of α_c in (10.24), we then have

$$
\begin{array}{c}
\text{Parallel-plate} \\
\text{TEM (TM}_0\text{)}
\end{array}
\quad
\boxed{\;\alpha_{c\text{TEM}} = \frac{C_4^2 R_s b}{2(\tfrac{1}{2}\eta C_4^2 ba)} = \frac{R_s}{\eta a} = \frac{1}{\eta a}\sqrt{\frac{\omega\mu_0}{2\sigma}}\;}
\tag{10.26}
$$

Next we consider conductor losses for TE waves. Note that the nonzero field components for TE waves in a parallel-plate waveguide are as given in (10.14). The magnitude of the surface current density is determined by the tangential $\mathbf{H}$ (i.e., H_z) at $x = 0$ and $x = a$ to be (for C_1 real and positive)

$$|J_{sy}| = |H_z| = \frac{m\pi C_1}{\omega\mu a}$$

It is interesting to note that J_{sy} is the only nonzero surface current component and does not flow in the propagation direction (i.e., z direction). The total time-average power loss (sum of the losses in the two plates) for a length of 1 meter and a width of b meters of the guide is

$$P_{\text{loss}} = 2\int_0^1 \int_0^b \left[\frac{1}{2}|J_{sy}|^2 R_s\right] dy \; dz = \frac{bm^2\pi^2 C_1^2 \sqrt{\omega\mu_0/(2\sigma)}}{\omega^2\mu^2 a^2}$$

The time-average power density transmitted in the z direction per unit cross-sectional area is

$$|\mathbf{S}_{av}| = \frac{1}{2}\mathscr{R}e\{\mathbf{E} \times \mathbf{H}^*\} \cdot \hat{\mathbf{z}} = -\frac{1}{2}E_yH_x^* = \frac{\overline{\beta}C_1^2}{2\omega\mu}\sin^2\left(\frac{m\pi}{a}x\right)$$

The total power through a guide with cross-sectional area of width b and a height of a meters can be determined by integrating $|\mathbf{S}_{av}|$ in x and y:

$$P_{av} = \int_0^b \int_0^a \frac{\overline{\beta}C_1^2}{2\omega\mu}\sin^2\left(\frac{m\pi}{a}x\right)\,dx\,dy = \frac{\overline{\beta}C_1^2ab}{4\omega\mu}$$

Based on (10.24), the attenuation constant α_c is then given by

$$\text{Parallel-plate TE}_m \quad \boxed{\alpha_{c\text{TE}_m} = \frac{2m^2\pi^2\sqrt{\omega\mu_0/(2\sigma)}}{\overline{\beta}\omega\mu a^3} = \frac{2R_s\,(f_{c_m}/f)^2}{\eta a\sqrt{1-(f_{c_m}/f)^2}}} \quad (10.27)$$

where we have used $\overline{\beta} = \beta\sqrt{1-(f_{c_m}/f)^2}$.

The derivation of a corresponding expression for α_c for TM waves proceeds in the same manner as for TE waves.

$$\text{Parallel-plate TM}_m \quad \boxed{\alpha_{c\text{TM}_m} = \frac{2R_s}{\eta a\sqrt{1-(f_{c_m}/f)^2}}} \quad (10.28)$$

Attenuation constants due to conduction losses for TEM, TE, and TM waves are plotted as a function of frequency in Figure 10.11. It is interesting to note that the losses are generally higher for TM than for TE waves. This is because the surface current $\mathbf{J}_s$ flows in

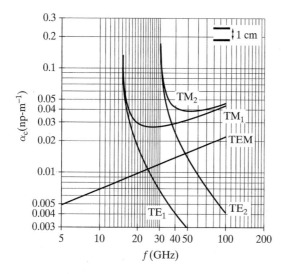

Figure 10.11 Attenuation versus frequency for parallel-plate waveguide. Attenuation-versus-frequency characteristics of waves guided by parallel plates.

support of the tangential component of **H**, and for TM waves the transverse component of the magnetic field (H_y) is tangential. For TE waves, the currents are related to H_z, whose magnitude approaches zero as frequency increases, as can be seen by considering the fact that, for a given value of a, the component TEM waves (see Section 10.1.3) of any given mode have increasingly higher values of incident angle θ_i for higher frequencies. The fact that the losses are very large near the cutoff frequency can also be understood in view of the uniform plane wave decomposition of the waveguide modes (Figure 10.9). Near cutoff, the component plane waves have to reflect many more times (per distance z traveled) from the walls, losing more power in each reflection.

Example 10.5: Parallel-copper-plate waveguide. Consider an air-filled parallel-copper-plate waveguide with an inner plate separation distance of 1 cm. Calculate the attenuation constant due to conductor losses, α_c, in decibels per meter (dB-m^{-1}), for TEM, TE$_1$, and TM$_1$ modes at operating frequencies of (a) 20 GHz and (b) 30 GHz.

Solution: For copper, the surface resistance R_s is given by

$$R_s = \sqrt{\frac{\omega\mu_0}{2\sigma}} = \sqrt{\frac{2\pi f \times 4\pi \times 10^{-7}}{2 \times 5.8 \times 10^7}} \simeq 2.61 \times 10^{-7}\sqrt{f} \ \Omega$$

Thus, at 20 GHz we have $R_s \simeq 3.69 \times 10^{-2}\Omega$, whereas at 30 GHz $R_s \simeq 4.52 \times 10^{-2}\Omega$. From (10.26), for the TEM mode, the attenuation constant is given by

$$\alpha_{c\text{TEM}} = \frac{R_s}{\eta a}$$

So at 20 GHz, we have $\alpha_{c\text{TEM}} \simeq 9.79 \times 10^{-3}$ np-m^{-1} (or[22] $\sim 8.50 \times 10^{-2}$ dB-m^{-1}). At 30 GHz, we have $\alpha_{c\text{TEM}} \simeq 0.104$ dB-m^{-1}.

The cutoff frequencies for the TE$_1$ and TM$_1$ modes are $f_{c_1} = (3 \times 10^8)/0.02 = 15$ GHz < 20 GHz. Thus, both TE$_1$ and TM$_1$ modes propagate at 20 GHz and 30 GHz. At 20 GHz for the TE$_1$ mode, using (10.27), we have

$$\alpha_{c\text{TE}_1} = \frac{2R_s(f_{c_1}/f)^2}{\eta a\sqrt{1 - (f_{c_1}/f)^2}} \simeq \frac{2 \times 3.69 \times 10^{-2}(15/20)^2}{377 \times 10^{-2} \times \sqrt{1 - (15/20)^2}} \simeq 1.66 \times 10^{-2} \text{ np-m}^{-1}$$

corresponding to an attenuation rate of ~ 0.145 dB-m^{-1}. Similarly, at 30 GHz the attenuation rate is $\sim 6.01 \times 10^{-2}$ dB-m^{-1}.

For the TM$_1$ mode at 20 GHz, we have from (10.28)

$$\alpha_{c\text{TM}_1} = \frac{2R_s}{\eta a\sqrt{1 - (f_{c_1}/f)^2}} \simeq \frac{2 \times 3.69 \times 10^{-2}}{377 \times 10^{-2} \times \sqrt{1 - (15/20)^2}} \simeq 2.96 \times 10^{-2} \text{ np-m}^{-1}$$

or an attenuation rate of ~ 0.257 dB-m^{-1}. At 30 GHz, the attenuation rate is ~ 0.240 dB-m^{-1}.

---

[22]As discussed in Section 3.7.1, conversion from np-m^{-1} to dB-m^{-1} simply requires multiplication by 20 $\log_{10} e \simeq 8.686$.

Attenuation due to dielectric losses. We have so far considered losses due to imperfectly conducting parallel plate walls. In practice, losses also occur because of imperfections of the dielectric material between the conductors. Such losses can typically be accounted for by taking the permittivity of the dielectric to be complex, or

$$\epsilon_c = \epsilon' - j\epsilon''$$

For the TEM mode of propagation, simple substitution of $\epsilon_c = \epsilon' - j\epsilon''$ for ϵ in all field expressions gives an attenuation constant in the same manner as we obtained the fields for a uniform TEM plane wave in a lossy dielectric (Section 8.3) by making the substitution $\sigma \to \omega\epsilon''$.

For TE or TM modes, the analysis of the field structures (i.e., mode configurations) was developed by assuming $\overline{\gamma}^2$ to be real ($\overline{\gamma}^2 = \overline{\alpha}^2$ or $(j\overline{\beta})^2$). If the dielectric losses are substantial, the field configurations may well have to be different; however, in most cases, the losses are small enough that we can assume that the character of the fields remains the same, and that the modes and their cutoff frequencies can be calculated assuming no losses. Consider the expression (10.19) for the propagation constant

$$\overline{\gamma} = j[\mu\epsilon(\omega^2 - \omega_{c_m}^2)]^{1/2} = j\left[\omega^2\mu\epsilon - \left(\frac{m\pi}{a}\right)^2\right]^{1/2}$$

where we have used that $\omega_{c_m} = \pi m/(a\sqrt{\mu\epsilon})$ to show the dependence of $\overline{\gamma}$ on ϵ explicitly. Now substituting for $\epsilon \to \epsilon' - j\epsilon''$, we have

$$\overline{\gamma} = j\left[\omega^2\mu(\epsilon' - j\epsilon'') - \left(\frac{m\pi}{a}\right)^2\right]^{1/2}$$

$$= j\left[\omega^2\mu\epsilon' - \left(\frac{m\pi}{a}\right)^2\right]^{1/2}\left[1 - \frac{j\,\omega^2\mu\epsilon''}{\omega^2\mu\epsilon' - \left(\frac{m\pi}{a}\right)^2}\right]^{1/2}$$

Assuming $\tan\delta = \epsilon''/\epsilon' \ll 1$ and expanding the second square root term using the binomial expansion given in footnote 31 of Chapter 8, we can approximate the complex waveguide propagation constant as

$$\overline{\gamma} \simeq \frac{\omega^2\mu\epsilon''}{2\left[\omega^2\mu\epsilon' - \left(\frac{m\pi}{a}\right)^2\right]^{1/2}} + j\left[\omega^2\mu\epsilon' - \left(\frac{m\pi}{a}\right)^2\right]^{1/2}$$

For a dielectric with complex permittivity, the cutoff frequency ω_{c_m} can naturally be defined in terms of the real part of ϵ_c, namely $\omega_{c_m} = \pi m/(a\sqrt{\mu\epsilon'})$, in which case we can rewrite $\overline{\gamma}$ as

$$\overline{\gamma} \simeq \alpha_d + j\overline{\beta}_m \simeq \frac{\omega\sqrt{\mu\epsilon'}\,\epsilon''/\epsilon'}{2\sqrt{1 - (\omega_{c_m}/\omega)^2}} + j\sqrt{(\omega^2 - \omega_{c_m}^2)\mu\epsilon'} \qquad (10.29)$$

Thus, we see that the phase constant (i.e., the imaginary part of $\overline{\gamma}$) is

$$\boxed{\overline{\beta}_m = \sqrt{(\omega^2 - \omega_{c_m}^2)\mu\epsilon'}} \qquad (10.30)$$

which is simply the waveguide propagation constant for a lossless dielectric with permittivity ϵ'. The attenuation constant due to dielectric losses is the real part of $\overline{\gamma}$ in (10.29):

$$\boxed{\alpha_d \simeq \frac{\omega\sqrt{\mu\epsilon'}\,\epsilon''/\epsilon'}{2\sqrt{1 - (\omega_{c_m}/\omega)^2}}} \qquad (10.31)$$

Note that in most cases the dielectric materials used are nonmagnetic, so $\mu = \mu_0$. Expressions (10.30) and (10.31) for $\overline{\beta}_m$ and α_d are valid for dielectrics with small losses such that $\epsilon'' \ll \epsilon'$. Note that equation (10.31) for α_d is valid for all TE or TM modes, and in fact also for all cross-sectional shapes of the waveguide, as long as f_{c_m} is the cutoff frequency for the particular mode under consideration. The attenuation rate due to dielectric losses for the TEM mode can be found from (10.31) by substituting $\omega_{c_m} = 0$, which gives $\alpha_d \simeq \omega\epsilon''\sqrt{\mu_0/\epsilon'}/2$.

10.1.5 Voltage, Current, and Impedance in a Parallel-Plate Waveguide

The TEM wave on a parallel-plate line is the propagation mode that corresponds to the voltage and current waves on two-conductor transmission lines that are the subject of most transmission line analyses and applications, including those discussed in Chapters 2 and 3. The voltage between the two conductor plates with separation a and the total current flowing through a width b of the upper plate are related to the electric and magnetic fields of the TEM wave given by (10.17) as

$$V_{\text{TEM}}(z) = -\int_{x=0}^{a} \mathbf{E} \cdot d\mathbf{l} = \int_{a}^{0} E_x(z)\,dx = \frac{\beta}{\omega\epsilon}C_4 e^{-j\beta z}\int_{a}^{0} dx = V^+ e^{-j\beta z}$$

$$I_{\text{TEM}}(z) = \mathbf{J}_s \cdot \hat{\mathbf{z}}b = (\hat{\mathbf{n}} \times [\mathbf{H}]_{x=a}) \cdot \hat{\mathbf{z}}b = (-\hat{\mathbf{x}} \times \hat{\mathbf{y}}[H_y]_{x=a}) \cdot \hat{\mathbf{z}}b = -C_4 b e^{-j\beta z}$$

$$= \frac{bV^+}{a\eta}e^{-j\beta z} = \frac{b}{a}\frac{V^+}{Z_{\text{TEM}}}e^{-j\beta z}$$

where we have noted that for the TEM mode $\overline{\beta} = \beta = \omega\sqrt{\mu\epsilon}$, we have defined $V^+ = -\beta C_4 a/(\omega\epsilon) = -\eta a C_4$ as the peak voltage, and we have introduced the quantity Z_{TEM}, called the wave impedance, also discussed in Section 9.3.2. The wave impedance is defined as the ratio of the transverse components of the electric and magnetic fields:

$$Z_{\text{TEM}} \equiv \frac{E_x}{H_y} = \frac{[\beta/(\omega\epsilon)]C_4 e^{-j\beta z}}{C_4 e^{-j\beta z}} = \sqrt{\frac{\mu}{\epsilon}} = \eta$$

If we identify $V(z) = V_{TEM}$, $I(z) = I_{TEM}$, and $Z_0 = Z_{TEM}a/b$, respectively, as the line voltage, line current, and characteristic line impedance, all of the standard transmission line analysis techniques covered in Chapters 2 and 3 are applicable to parallel-plate waveguides operating in the TEM mode. As an example, the time-average power flowing down the transmission line through a cross-sectional area ab, given by $\frac{1}{2}|V^+|^2/Z_0$, is equal to the time-average transmitted power P_{av} evaluated in the previous section using the phasor expressions for the electric and magnetic fields (equation (10.25)). We note that the phase velocity $\overline{v}_p = \omega/\overline{\beta} = 1/\sqrt{\mu\epsilon}$ and the wave impedance Z_{TEM} for TEM waves are both independent of frequency.

In general, for TE or TM modes, the definitions of equivalent voltage and current become ambiguous. To see this, consider the propagating TE_m mode, the fields for which are

$$E_y = C_1 \sin\left(\frac{m\pi}{a}x\right)e^{-j\overline{\beta}z}$$

$$H_z = -\frac{m\pi}{j\omega\mu a}C_1 \cos\left(\frac{m\pi}{a}x\right)e^{-j\overline{\beta}z}$$

$$H_x = -\frac{\overline{\beta}}{\omega\mu}C_1 \sin\left(\frac{m\pi}{a}x\right)e^{-j\overline{\beta}z}$$

where $\overline{\beta} = \sqrt{\omega^2\mu\epsilon - (m\pi/a)^2}$. We note that the wave electric field is in the y direction, so the quantity that we would normally define as voltage—that is, the line integral of the electric field from one plate to another—is zero. In other words,

$$-\int_0^a \mathbf{E} \cdot d\mathbf{l} = -\int_0^a \hat{\mathbf{y}}E_y \cdot \hat{\mathbf{x}}\, dx = 0$$

Similarly, the line current, which we may normally think of as a z-directed current in the plates, actually flows in the y direction. To see this, note that the only nonzero magnetic field at the surfaces of the conducting plates is H_z, which requires a surface current density of

$$\mathbf{J}_s = \hat{\mathbf{n}} \times \hat{\mathbf{z}}[H_z]_{x=0} = \hat{\mathbf{x}} \times \hat{\mathbf{z}}\left(\frac{-m\pi}{j\omega\mu a}\right)e^{-j\overline{\beta}z} = \hat{\mathbf{y}}\left(\frac{m\pi}{j\omega\mu a}\right)e^{-j\overline{\beta}z}$$

along the $x = 0$ plate. In practice, the lack of unique definitions of voltage and current for TE and TM modes does not preclude the use of transmission line techniques in analyzing the behavior of waveguides or other microwave circuit components. In most cases, useful definitions suitable for the particular problems at hand can be put forth.[23]

It is common practice to define voltage and current in such a way that (i) the line voltage is proportional to the transverse component of the wave electric field, (ii) the line current is taken to be proportional to the transverse component of the wave magnetic field, and (iii) the product of the line voltage and current, $\frac{1}{2}\Re e\{VI^*\}$, is equal to the time-average power transmitted P_{av} through a given cross-sectional area. The ratio of the voltage to current for a forward propagating wave is defined as the characteristic

[23] See R. E. Collin, *Foundations of Microwave Engineering,* 2nd ed., McGraw-Hill, New York, 1992.

impedance of the transmission system. This impedance is usually taken to be the wave impedance, defined as the ratio of the transverse components of the electric and magnetic fields:

$$Z_{\text{TM}} \equiv \frac{E_x}{H_y} \qquad Z_{\text{TE}} \equiv -\frac{E_y}{H_x} \tag{10.32}$$

For propagating TE_m modes (i.e., $f > f_{c_m}$ so that $\overline{\gamma} = j\overline{\beta}$), using (10.14) and (10.19) we have

$$Z_{\text{TE}_m} = -\frac{E_y}{H_x} = -\frac{C_1 \sin\left(\frac{m\pi}{a}x\right) e^{-j\overline{\beta}z}}{-\frac{\overline{\beta}}{\omega\mu} C_1 \sin\left(\frac{m\pi}{a}x\right) e^{-j\overline{\beta}z}} = \frac{\omega\mu}{\overline{\beta}} = \frac{\eta}{\sqrt{1 - (f_{c_m}/f)^2}} \tag{10.33}$$

where $\eta = \sqrt{\mu/\epsilon}$ is the intrinsic impedance of the dielectric inside the waveguide. We thus see that the wave impedance of propagating TE_m modes in a parallel-plate waveguide with a lossless dielectric is purely resistive and is always *larger* than the intrinsic impedance of the dielectric medium. The higher the operating frequency, the closer Z_{TE_m} is to η, meaning that the wave is more like a TEM wave. This behavior can be understood in terms of the component TEM wave representation of the TE and TM modes discussed in Section 10.1.3. The higher the operating frequency, the smaller is the wavelength, which in turn means that more half-wavelengths "fit" into the waveguide dimension a and that the angle θ_{i_m} that the component TEM wave ray path makes with the normal to the plates is larger. In other words, the higher the frequency, the closer are the ray paths of the component TEM waves to being parallel to the waveguide walls along the axial direction. The frequency dependence of Z_{TE_m} is illustrated in Figure 10.12. At frequencies below cutoff, $f < f_{c_m}$, when we have $\overline{\gamma} = \overline{\alpha}$ and thus an attenuating (evanescent) wave, it is clear from (10.33) that the wave impedance becomes purely imaginary. A purely reactive wave impedance is consistent with the fact that there is no electromagnetic power flow associated with evanescent waves.

The following general relation between the electric and magnetic fields holds for TE_m waves:

$$\mathbf{E} = -Z_{\text{TE}_m}(\hat{\mathbf{z}} \times \mathbf{H}) \tag{10.34}$$

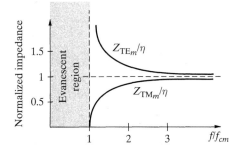

Figure 10.12 TE_m and TM_m wave impedances. The frequency dependence of normalized wave impedances (normalized to η) for TE_m and TM_m modes. Note that $\eta = \sqrt{\mu/\epsilon}$ and $f_{c_m} = m/(2a\sqrt{\mu\epsilon})$.

Equation (10.34) is valid for TE waves in any waveguide, including rectangular and circular waveguides.

For TM_m waves, the wave impedance is defined in the same manner, namely, as the ratio of the transverse components of the wave electric and magnetic fields as given in (10.32). Using the field expressions derived earlier for propagating TM_m modes given by (10.16), we have

$$Z_{\text{TM}_m} = \frac{E_x}{H_y} = \frac{[\overline{\beta}/(\omega\epsilon)]C_4 \cos\left(\dfrac{m\pi}{a}x\right) e^{-j\overline{\beta}z}}{C_4 \cos\left(\dfrac{m\pi}{a}x\right) e^{-j\overline{\beta}z}} = \frac{\overline{\beta}}{\omega\epsilon} = \eta\sqrt{1 - \left(\frac{f_{c_m}}{f}\right)^2} \qquad (10.35)$$

Equation (10.35) indicates that the wave impedance of propagating TM_m modes in a parallel-plate waveguide with a lossless dielectric is purely resistive and is always *less* than the intrinsic impedance of the dielectric medium. The frequency dependence of Z_{TM_m} is illustrated in Figure 10.12. At frequencies below cutoff ($f \leq f_{c_m}$) the wave impedance Z_{TM_m} is purely imaginary, as expected for an evanescent wave. At frequencies much larger than the cutoff frequency ($f \gg f_{c_m}$), the wave impedance approaches the intrinsic impedance of the medium, as the TM_m wave becomes more and more like a TEM wave.

The following general relation between the electric and magnetic fields holds for TM_m waves:

$$\mathbf{H} = \frac{1}{Z_{\text{TM}_m}} (\hat{\mathbf{z}} \times \mathbf{E}) \qquad (10.36)$$

As for the case of TE_m waves given by (10.34), equation (10.36) is valid for TM_m waves in any waveguide, including rectangular and circular waveguides.

10.1.6 Electric and Magnetic Field Distributions

The mode field distributions shown in Figures 10.2 and 10.3 are useful tools for visualizing the behavior of the particular modes shown. Using the fact that the electric and magnetic fields are locally tangent to the corresponding lines shown in such a plot, it is possible to derive mathematical expressions for the electric and magnetic field lines. Considering first the TE modes, we note that the time-domain field expressions for the propagating TE_m mode can be obtained from the phasor fields found earlier as

$$\mathscr{E}_y(x, z, t) = \mathscr{R}e\{E_y^0(x)e^{-j\overline{\beta}z}e^{j\omega t}\} = C_1 \sin\left(\frac{m\pi}{a}x\right)\cos(\omega t - \overline{\beta}z)$$

$$\mathscr{H}_x(x, z, t) = -\frac{\overline{\beta}}{\omega\mu}C_1 \sin\left(\frac{m\pi}{a}x\right)\cos(\omega t - \overline{\beta}z)$$

$$\mathscr{H}_z(x, z, t) = -\frac{m\pi}{\omega\mu a}C_1 \cos\left(\frac{m\pi}{a}x\right)\sin(\omega t - \overline{\beta}z)$$

Similarly, for the propagating TM$_m$ mode, we have

$$\mathcal{E}_x(x,z,t) = \frac{\overline{\beta}}{\omega\epsilon}C_4 \cos\left(\frac{m\pi}{a}x\right)\cos(\omega t - \overline{\beta}z)$$

$$\mathcal{E}_z(x,z,t) = -\frac{m\pi}{\omega\epsilon a}C_4 \sin\left(\frac{m\pi}{a}x\right)\sin(\omega t - \overline{\beta}z)$$

$$\mathcal{H}_y(x,z,t) = C_4 \cos\left(\frac{m\pi}{a}x\right)\cos(\omega t - \overline{\beta}z)$$

For the TE$_m$ modes, the magnetic field consists of two components, $\mathcal{H}_x(x,z,t)$ and $\mathcal{H}_z(x,z,t)$. To find a parametric equation to trace the locus of magnetic field lines, we first note that the tangent to the vector field line at any point is in the direction of the magnetic field vector. Thus the equations for the magnetic field lines satisfy the differential equation

$$\frac{dx}{dz} = \frac{\mathcal{H}_x(x,z,t)}{\mathcal{H}_z(x,z,t)} = \left(\frac{\overline{\beta}a}{m\pi}\right)\frac{\sin\left(\frac{m\pi}{a}x\right)\cos(\omega t - \overline{\beta}z)}{\cos\left(\frac{m\pi}{a}x\right)\sin(\omega t - \overline{\beta}z)}$$

Rearranging terms, we can write

$$\frac{\left(\frac{m\pi}{a}\right)\cos\left(\frac{m\pi}{a}x\right)dx}{\sin\left(\frac{m\pi}{a}x\right)} = \frac{\overline{\beta}\cos(\omega t - \overline{\beta}z)\,dz}{\sin(\omega t - \overline{\beta}z)}$$

Substituting

$$\left(\frac{m\pi}{a}\right)\cos\left(\frac{m\pi}{a}x\right)dx = d\left[\sin\left(\frac{m\pi}{a}x\right)\right]$$

and

$$\overline{\beta}\cos(\omega t - \overline{\beta}z)\,dz = -d[\sin(\omega t - \overline{\beta}z)]$$

we can rewrite the preceding expression as

$$\frac{d\left[\sin\left(\frac{m\pi}{a}x\right)\right]}{\sin\left(\frac{m\pi}{a}x\right)} = -\frac{d[\sin(\omega t - \overline{\beta}z)]}{\sin(\omega t - \overline{\beta}z)}$$

Integrating both sides yields

$$\ln\left[\sin\left(\frac{m\pi}{a}x\right)\right] = -\ln[\sin(\omega t - \overline{\beta}z)] + \text{const.}$$

or

$$\sin\left(\frac{m\pi}{a}x\right)\sin(\omega t - \overline{\beta}z) = \text{const.}$$

which represents the equation describing the locus of the magnetic field lines in the x-z plane. Plots of the field lines for TE$_1$ and TE$_2$ were shown in Figure 10.2.

Using a similar procedure, we can find the locus in the x-z plane of the electric field lines for TM$_m$ modes. We find

$$\cos\left(\frac{m\pi}{a}x\right)\sin(\omega t - \overline{\beta}z) = \text{const.}$$

The resultant field distributions were shown in Figure 10.3.

10.2 DIELECTRIC WAVEGUIDES

We have seen that the parallel-plate waveguide with metallic boundaries can effectively guide electromagnetic energy in the axial (i.e., z) direction. The metallic waveguides are very effective for microwave frequencies, at which the conductivity of the walls is high and the reflectivity is correspondingly good. However, at optical frequencies, the dimensions of metallic waveguides become too small and wall losses become unacceptably large. In addition, it is difficult to fabricate such waveguides with the required precision. Because light has a characteristic wavelength on the order of 1 μm, waveguides must have a cross-sectional dimension of around the same order.

Although research work to extend the wavelength range of metallic waveguides to the μm range continues,[24] a low-cost and potentially much more powerful alternative guiding structure that works extremely well at optical frequencies is the optical fiber. Although detailed coverage of optical fiber transmission systems is well beyond the scope of this book,[25] we consider in this section the so-called dielectric waveguide, which consists of a slab of dielectric surrounded by another dielectric. These structures lend themselves to complete analytical solution and also encompass the basic principles of wave propagation in optical fibers. In addition, waveguiding structures consisting of rectangular dielectric slabs or dielectric slabs on a conducting surface are used in microwave integrated circuits, in integrated optics, and in optoelectronics applications. There are no fundamental reasons why dielectric waveguides should not be used at lower microwave frequencies, except that metal waveguides are more practical, provide better shielding from interference, and allow operation at higher power levels.

In a dielectric waveguide, there are no conducting surfaces, and the electromagnetic energy is guided through the structure not by reflections from conducting metal surfaces but by means of total internal reflections from dielectric boundaries. The potential guiding of electromagnetic waves by dielectric structures was recognized in the late nineteenth century,[26] and a theoretical solution was obtained around the turn of the century,

---

[24]Two papers were cited in the beginning of this chapter; another example is Y. Matsuura, A. Hongo, and M. Miyagi, Dielectric-coated metallic hollow waveguide for 3-μm Er:YAG, 5-μm CO, and 10.6-μm CO$_2$ laser light transmission, *Appl. Optics*, 29(15), pp. 2213–2217, May 1990.

[25]For a reference, see D. J. H. Maclean, *Optical Line Systems*, Wiley, New York, 1996.

[26]The first demonstration of the guiding of light by a thin stream of water was achieved by John Tyndall in 1870.

in 1910.[27] Early experimental efforts considered water to be a good dielectric candidate and involved attempts to transmit waves along water-filled metallic tubes. Initial results were interpreted as being due to guiding by the dielectric, but it was soon realized that the water was unnecessary and that the wave was actually supported by the metal tube.[28] Although technological demonstrations of transmission of light through optical fibers were first realized in the 1930s, the practical use of fiber technology became feasible only after the invention of the laser in 1960 and the development of low-loss fibers in the late 1970s. At present, most optical fibers operate in the 1.3 μm and the 1.5 μm optical windows. The former has the advantage of being the wavelength of minimum distortion, and the latter provides lower loss (0.2 dB-(km)$^{-1}$ as opposed to 0.4 dB-(km)$^{-1}$ for 1.3 μm).

In this section, we illustrate how dielectric regions with boundaries in the transverse direction can support electromagnetic fields propagating in the axial direction. The guiding action of dielectric waveguides is more complicated than for metal waveguides, and it is often not possible to obtain analytical solutions. For simplicity, we confine our attention in this section to structures with planar geometry, namely, the dielectric slab waveguide. The dielectric slab waveguide is extensively used in integrated optics systems to transfer light between components. The basic principles and techniques of guiding light waves by dielectric boundaries are well illustrated by our coverage of this planar structure. These techniques can be readily extended to cylindrical and other geometries.

10.2.1 Dielectric Slab Waveguide: Mode Theory

In general, analysis of dielectric waveguides can be based on "mode theory" or "ray theory." Here, we first discuss mode theory by considering wave propagation along a dielectric slab waveguide as shown in Figure 10.13, consisting of a rectangular dielectric

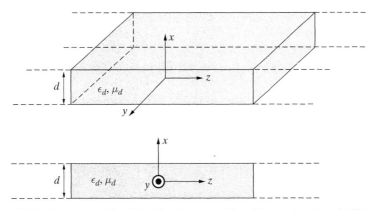

Figure 10.13   Geometry of the dielectric slab waveguide. (a) Perspective view. (b) Side view.

[27]D. Hondros and P. Debye, Elektromagnetische Wellen an dielektrischen Drähten, *Ann. d. Phys.,* 49(4), p. 465, 1910.

[28]See K. S. Packard, The origin of waveguides: A case of multiple rediscovery, *IEEE Trans. Microwave Theory Techn.,* 32(9), pp. 961–969, September 1984, and references therein.

slab sandwiched by free space. The procedure we follow to arrive at the field solutions is similar to that used for parallel-plate metallic waveguides. We take advantage of the fact that the structure has infinite extent in the y direction and look for solutions of the type $e^{-\overline{\gamma}z}$ in terms of the z variation. We also separately investigate the case of TM or TE modes. We first consider TM waves for which equations are written in terms of $E_z(x,z)$.

Unlike the case of metallic waveguides, dielectric waveguides in general have nonzero fields not only inside but also outside the dielectric slab. In the free space region, waves must decay exponentially in the transverse direction, in order for the energy to be in any sense "guided" by the dielectric slab.

Transverse magnetic (TM) waves.   Noting that there is no y dependence (i.e., $\partial(\cdot)/\partial y = 0$) for any of the field components, we have

$$E_z(x,z) = E_z^0(x)e^{-\overline{\gamma}z}$$

where $E_z^0(x)$ is that portion of $E_z(x,z)$ that depends on x and is governed by a wave equation similar to (10.11), which was used in Section 10.1 to solve for the E_y component of TE modes in a parallel-plate waveguide. We have

$$\frac{d^2 E_z^0(x)}{dx^2} + h^2 E_z^0(x) = 0 \tag{10.37}$$

where $h^2 = \overline{\gamma}^2 - (j\beta)^2 = \overline{\gamma}^2 + \omega^2\mu\epsilon$. Note that, unlike the case of the parallel-plate waveguide, the wave equation (10.37) must be satisfied both within and outside the dielectric slab, since the fields are in general nonzero both inside and outside. Also, as in the case of (10.11), we have in general $h^2 = \overline{\gamma}^2 + \omega^2\mu\epsilon$, so that $h^2 = h_d^2 = \overline{\gamma}^2 + \omega^2\mu_d\epsilon_d$ in the dielectric region and $h^2 = h_0^2 = \overline{\gamma}^2 + \omega^2\mu_0\epsilon_0$ in the surrounding free space.[29] As before, h are the eigenvalues to be determined by the boundary conditions, and $\overline{\gamma}$ can be either purely real ($\overline{\gamma} = \overline{\alpha}$, for evanescent, nonpropagating waves) or purely imaginary ($\overline{\gamma} = j\overline{\beta}$, for propagating waves). We look for forms of solutions appropriate for inside and outside the dielectric slab and then match them via boundary conditions at the interfaces. Inside the slab we have

$$E_z^0(x) = C_o \sin(\beta_x x) + C_e \cos(\beta_x x) \qquad |x| \leq \frac{d}{2} \tag{10.38}$$

where β_x is the transverse phase constant and is given by

$$\beta_x^2 = \omega^2\mu_d\epsilon_d - \overline{\beta}^2 = h_d^2 \tag{10.39}$$

assuming propagating solutions with $\overline{\gamma} = j\overline{\beta}$. The quantities μ_d and ϵ_d are the permeability and permittivity for the dielectric, which is assumed to be lossless. The subscripts

---

[29] In practice, the medium outside the dielectric waveguide is actually another dielectric with $\epsilon < \epsilon_d$. In such a case, all of the following solutions apply if we replace ϵ_0 with ϵ.

on the constants C_o and C_e are the respective coefficients of the "odd" [i.e., $\sin(\beta_x x)$ is an odd function of x] and "even" [i.e., $\cos(\beta_x x)$ is an even function of x] parts of $E_z^0(x)$, which is the basis for classification of the different modes.

In order for the dielectric slab to be "guiding" the wave energy, the wave fields must be confined to the vicinity of the slab. Thus, outside the slab the fields of the propagating wave must decay exponentially with distance from the slab. In other words, we must have

$$E_z^0(x) = \begin{cases} C_a e^{-\alpha_x(x-d/2)} & x \ge \dfrac{d}{2} \\[2mm] C_b e^{\alpha_x(x+d/2)} & x \le -\dfrac{d}{2} \end{cases} \tag{10.40}$$

where the subscripts on the constants C_a and C_b are, respectively, associated with the solutions "above" and "below" the slab, and where α_x is the transverse attenuation constant, given by

$$\alpha_x^2 = \bar{\beta}^2 - \omega^2 \mu_0 \epsilon_0 = -h_0^2 \quad \text{or} \quad \alpha_x = \sqrt{\bar{\beta}^2 - \omega^2 \mu_0 \epsilon_0} > 0 \tag{10.41}$$

with μ_0 and ϵ_0 being the permeability and permittivity of free space. Note that we must have $\alpha_x > 0$ for physically realizable solutions.

To obtain the full solutions, the relationships between C_o, C_e, C_a, C_b need to be determined and the values of β_x and α_x need to be found. In general, the odd and even mode fields have x dependencies as shown in Figure 10.14.

The other TM mode field components can be obtained from E_z using (10.8b,c) and noting that $E_z(x,z)$ varies as $E_z^0(x)e^{-j\bar{\beta}z}$. Because E_z depends only on x and z and because we have assumed $H_z = 0$ (i.e., TM mode), from (10.8b), $H_y(x,z)$ is the only

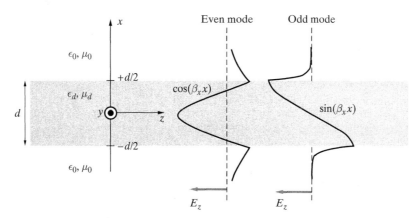

Figure 10.14 Field distributions of even and odd modes. Even and odd mode field distributions in a dielectric slab. The case shown corresponds to 30 GHz operation with $d = 0.75$ cm and $\epsilon_d = 2\epsilon_0$.

nonzero magnetic field component. In addition because E_z varies with z, and because there is no variation of any field quantities with y, the electric field must also have an x component in order that $\nabla \cdot \mathbf{E} = 0$.

We now examine the *odd* and *even* TM modes separately.[30]

Odd TM modes: Odd TM modes are the solutions for which we have $C_e = 0$ in (10.38), and so

$$E_z^0(x) = C_o \sin(\beta_x x) \qquad\qquad |x| \leq \frac{d}{2} \qquad (10.42)$$

We can use the fact that E_z is continuous across the boundary at $x = \pm d/2$ to determine C_a and C_b in terms of C_o. Using (10.8b,c) and (10.41), the field components for odd TM modes are then found to be

Free space
above the slab
$(x \geq d/2)$

$$E_z^0(x) = \left[C_o \sin\left(\frac{\beta_x d}{2}\right) \right] e^{-\alpha_x(x-d/2)}$$

$$E_x^0(x) = -\frac{j\overline{\beta}}{\alpha_x}\left[C_o \sin\left(\frac{\beta_x d}{2}\right) \right] e^{-\alpha_x(x-d/2)}$$

$$H_y^0(x) = -\frac{j\omega\epsilon_0}{\alpha_x}\left[C_o \sin\left(\frac{\beta_x d}{2}\right) \right] e^{-\alpha_x(x-d/2)}$$

Dielectric region
$(|x| \leq d/2)$

$$E_z^0(x) = C_o \sin(\beta_x x)$$

$$E_x^0(x) = -\frac{j\overline{\beta}}{\beta_x} C_o \cos(\beta_x x) \qquad (10.43)$$

$$H_y^0(x) = -\frac{j\omega\epsilon_d}{\beta_x} C_o \cos(\beta_x x)$$

Free space
below the slab
$(x \leq -d/2)$

$$E_z^0(x) = -\left[C_o \sin\left(\frac{\beta_x d}{2}\right) \right] e^{\alpha_x(x+d/2)}$$

$$E_x^0(x) = -\frac{j\overline{\beta}}{\alpha_x}\left[C_o \sin\left(\frac{\beta_x d}{2}\right) \right] e^{\alpha_x(x+d/2)}$$

$$H_y^0(x) = -\frac{j\omega\epsilon_0}{\alpha_x}\left[C_o \sin\left(\frac{\beta_x d}{2}\right) \right] e^{\alpha_x(x+d/2)}$$

---

[30]We choose to define the odd and even designations with respect to the $\sin(\beta_x x)$ and $\cos(\beta_x x)$ functional form of the transverse variation of the axial field component (E_z for TM and H_z for TE). In general, the odd/even classification can also be made with respect to the functional forms of the other components, such as H_y for TM modes or E_y for TE modes. The modes so identified would in general be different from those identified on the basis of the axial components; in other words, for example, the odd TM_3 mode defined on the basis of the oddness/evenness of the H_y component would correspond to a different distribution of fields than the odd TM_3 defined on the basis of the functional form of the E_z component. The relative arbitrariness of this nomenclature is important to note here since different definitions are used in different texts.

We observe from (10.43) that the time-average electromagnetic power flow is in the z direction; that is, H_y and E_x are in time phase. Also, since E_z and H_y are 90° out of phase, no real time-average power flows in the x direction.

We can now apply the boundary conditions to determine β_x and α_x. Note that the continuity of E_z was used to find C_a and C_b in terms of C_0. The continuity of H_y at the dielectric surface, that is, at $x = \pm d/2$, requires that

$$\boxed{\text{Odd TM modes:} \qquad \frac{\alpha_x}{\beta_x} = \frac{\epsilon_0}{\epsilon_d} \tan\left(\frac{\beta_x d}{2}\right)} \qquad (10.44)$$

From expressions (10.39) and (10.41) for β_x and α_x, we also have

$$\alpha_x^2 + \beta_x^2 = \omega^2(\mu_d \epsilon_d - \mu_0 \epsilon_0) \qquad (10.45)$$

or

$$\boxed{\alpha_x = [\omega^2(\mu_d \epsilon_d - \mu_0 \epsilon_0) - \beta_x^2]^{1/2}} \qquad (10.46)$$

which is valid for both odd and even modes.

For given values of ω, μ_d, ϵ_d, and d, (10.44) and (10.46) can be plotted as α_x versus β_x and a graphical solution for α_x and β_x can be obtained. An example for the case of $\epsilon_d = 2\epsilon_0$ and $d = 1.25\lambda$, where λ is the free space wavelength ($\lambda = c/f$), is illustrated in Figure 10.15. For the case shown, there are three solutions possible, denoted as S1, S2, and S3, which are, respectively, the odd TM_1, even TM_2, and odd TM_3 modes.

Even TM modes: Even TM modes (see Figure 10.14) are the solutions for which we have $C_0 = 0$ in (10.38), yielding

$$E_z^0(x) = C_e \cos(\beta_x x) \qquad |x| \leq \frac{d}{2} \qquad (10.47)$$

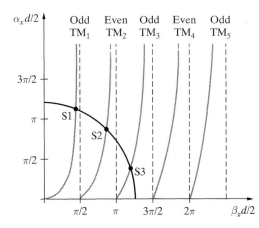

Figure 10.15   Graphical solution for TM modes. Graphical solution representing transverse attenuation and phase constants for a dielectric slab waveguide. Both odd and even TM modes are shown. The quarter-circle is the contour corresponding to (10.46); the curves marked "odd" correspond to (10.44) and the curves marked "even" correspond to (10.48). The parameters chosen for this example are $\epsilon_d = 2\epsilon_0$ and $d = 1.25\lambda$, where λ is the free space wavelength.

Much of the analysis is the same as above; instead of (10.44), we have

$$\boxed{\text{Even TM modes:} \qquad \frac{\alpha_x}{\beta_x} = -\frac{\epsilon_0}{\epsilon_d} \cot\left(\frac{\beta_x d}{2}\right)} \tag{10.48}$$

which correspond to the curves marked "even TM" in Figure 10.15.

Propagation and cutoff characteristics. Note that the propagation constant $\bar{\beta}$ has values between the intrinsic propagation constant of free space and that of the dielectric, that is,

$$\omega\sqrt{\mu_0\epsilon_0} < \bar{\beta} < \omega\sqrt{\mu_d\epsilon_d}$$

As $\bar{\beta}$ approaches $\omega\sqrt{\mu_0\epsilon_0}$, we note that α_x approaches zero, indicating that there is no attenuation with distance away from the dielectric slab, and so the wave energy is not "guided" by the slab and the wave fields are not confined to the vicinity of the slab. The condition $\alpha_x = 0$ thus defines the cutoff condition, and the cutoff frequencies at which this occurs can be determined from (10.46) by setting $\alpha_x = 0$:

$$\beta_x = \omega_c\sqrt{\mu_d\epsilon_d - \mu_0\epsilon_0}$$

In other words, guided propagation occurs only for frequencies $\omega > \omega_c$, for which α_x is real and positive.

The cutoff frequencies for a given mode can be determined by substituting into the respective equations for that mode. Using (10.44) and (10.48), we find

$$\tan\left(\frac{\beta_x d}{2}\right) = 0 \qquad \text{for odd TM modes}$$

$$\cot\left(\frac{\beta_x d}{2}\right) = 0 \qquad \text{for even TM modes}$$

evaluating at $\beta_x = \omega_c\sqrt{\mu_d\epsilon_d - \mu_0\epsilon_0}$, we find

$$\boxed{f_{c\text{TM}_m} = \frac{(m-1)}{2d\sqrt{\mu_d\epsilon_d - \mu_0\epsilon_0}} \qquad \begin{array}{l} m = 1,3,5,\ldots \\ m = 2,4,6,\ldots \end{array} \qquad \begin{array}{l} \text{Odd TM}_m \\ \text{Even TM}_m \end{array}} \tag{10.49}$$

as the cutoff frequencies.[31]

For the odd TM$_1$ mode, $f_{c\text{TM}_1} = 0$, and so this lowest-order odd TM mode can propagate for any slab thickness (analogous to the TEM mode on a parallel-plate waveguide).

---

[31]Note once again that the "odd" and "even" designations simply refer, respectively, to the $\sin(\beta_x x)$ and $\cos(\beta_x x)$ functional forms of the transverse variation of the axial field (E_z for TM and H_z for TE) component. In addition, however, the dependence of f_{c_m} on the index m is written such that the odd/even values of m correspond respectively to odd/even modes.

Example 10.6: Propagating modes in a dielectric slab waveguide. Consider a dielectric slab waveguide with $d = 0.75$ cm, $\mu_r = 1, \epsilon_r = 2$ surrounded by air, operating at 30 GHz. (a) Find all TM modes and cutoff frequencies for propagating modes. (b) Calculate $\beta_x, \alpha_x, \overline{\beta}$, and $(\overline{\beta}/\beta_0)$ for propagating modes.

Solution: The only modes with cutoff frequencies smaller than 30 GHz are TM_1 (odd) and TM_2 (even). We have

$$f_{c_{TM_1}} = 0 \qquad f_{c_{TM_2}} = \frac{1}{(2 \times 0.75)\sqrt{\mu_r \mu_0 \epsilon_r \epsilon_0 - \epsilon_0 \mu_0}} \simeq 20 \text{ GHz}$$

To find the corresponding phase constants we can either carry out a graphical solution or solve a transcendental equation. Combining (10.44) and (10.46), we have for odd TM modes

$$\alpha_x = \frac{\beta_x \epsilon_0}{\epsilon_d} \tan\left(\frac{\beta_x d}{2}\right) = [\omega^2(\mu_d \epsilon_d - \mu_0 \epsilon_0) - \beta_x^2]^{1/2}$$

and similarly, combining (10.46) and (10.48), we have for even TM modes

$$\alpha_x = -\frac{\beta_x \epsilon_0}{\epsilon_d} \cot\left(\frac{\beta_x d}{2}\right) = [\omega^2(\mu_d \epsilon_d - \mu_0 \epsilon_0) - \beta_x^2]^{1/2}$$

Solutions of each of these two equations give the following roots:

$$TM_1 \text{ odd: } \beta_x \simeq 3.37 \text{ rad-(cm)}^{-1} \quad \text{and} \quad TM_2 \text{ even: } \beta_x \simeq 5.89 \text{ rad-(cm)}^{-1}$$

The axial electric field distributions for the odd TM_1 and even TM_2 modes are shown in Figure 10.16.

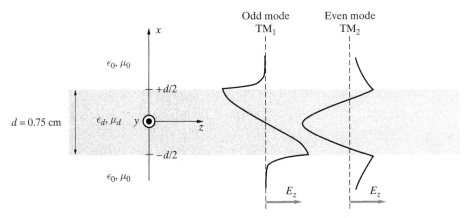

Figure 10.16   The propagating modes for Example 10.6. The plots show the variation of the axial component E_z over the vertical cross section of the dielectric slab. Parameters are $f = 30$ GHz, $d = 0.75$ cm, and $\epsilon_d = 2\epsilon_0$.

From β_x, we can directly calculate α_x using (10.46). We find

$$
\begin{array}{cc}
\text{Odd TM}_1 & \text{Even TM}_2 \\
\beta_x \simeq 3.37 \text{ rad-(cm)}^{-1} & \beta_x \simeq 5.89 \text{ rad-(cm)}^{-1} \\
\alpha_x \simeq 5.31 \text{ np-(cm)}^{-1} & \alpha_x \simeq 2.18 \text{ np-(cm)}^{-1} \\
\simeq 46.16 \text{ dB-(cm)}^{-1} & \simeq 18.92 \text{ dB-(cm)}^{-1}
\end{array}
$$

The propagation constants in the axial direction can be obtained from (10.39), which is

$$\beta_x^2 = \omega^2 \mu_d \epsilon_d - \overline{\beta}_m^2$$

We find

$$\overline{\beta}_m \simeq 8.23 \ \text{rad-(cm)}^{-1} \quad \rightarrow \quad \frac{\overline{\beta}_1}{\beta_0} \simeq 1.31 \qquad \text{for TM}_1 \ \text{(odd)}$$

$$\overline{\beta}_m \simeq 6.66 \ \text{rad-(cm)}^{-1} \quad \rightarrow \quad \frac{\overline{\beta}_2}{\beta_0} \simeq 1.06 \qquad \text{for TM}_2 \ \text{(even)}$$

The dependencies of the mode structures on the operating frequency and slab thickness are illustrated in the lower two panels of Figure 10.17, for the same parameter values as in Example 10.6. As seen from the sketches, increasing the operating frequency increases the number of propagating modes in the guide. Also, for a fixed operating frequency, using a thicker slab increases the number of propagating modes.

Figures 10.16 and 10.17 show the transverse variation of the axial electric field for different TM modes. Figure 10.18 shows the complete electric and magnetic field distributions for the odd TM$_1$ mode, in a manner similar to that given in Figure 10.3 for TM modes in a parallel-plate waveguide. In this plot, the intensity of the electric field at any point is indicated by the density of the field lines at that location. We see that the electric field intensity falls off (i.e., the density of lines decreases) with distance away from the slab, consistent with the $e^{-\alpha_x(x \pm d/2)}$ variation of the fields as given in (10.43).

Transverse electric (TE) modes. The solutions for TE modes can be found in a completely analogous manner, for which we start with $H_z^0(x)$ and write similar solutions for inside and outside the slab. Note that we have

$$H_z^0(x) = C_o \sin(\beta_x x) + C_e \cos(\beta_x x) \tag{10.50}$$

and that the odd or even modes are now defined with respect to the functional form of $H_z^0(x)$ being, respectively, $\sin(\beta_x x)$ or $\cos(\beta_x x)$. The nonzero field components are now H_z, H_x, and E_y. We find the following relations:

$$\boxed{\frac{\alpha_x}{\beta_x} = \frac{\mu_0}{\mu_d} \tan\left(\frac{\beta_x d}{2}\right) \qquad \text{Odd TE modes}} \tag{10.51}$$

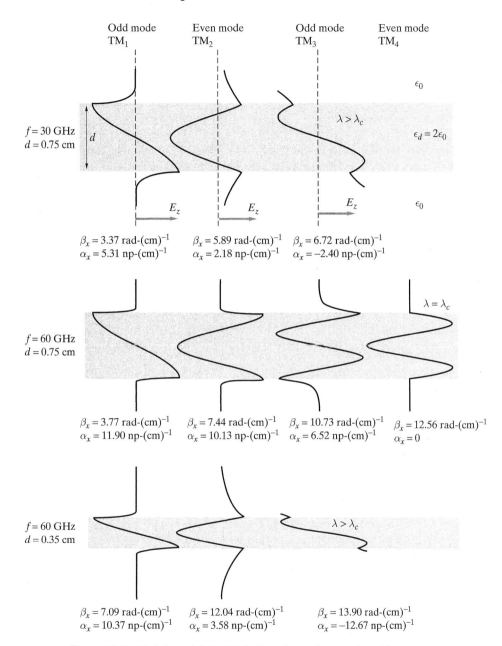

Figure 10.17   Variation of field distributions for various modes with respect to thickness and frequency. Dependence of the dielectric waveguide mode structure on frequency and slab thickness. The plots show the variation of the axial field component E_z over the vertical cross section of the dielectric slab. For all cases, $\epsilon_d = 2\epsilon_0$. Note that the numerical values are the same as those for Example 10.6.

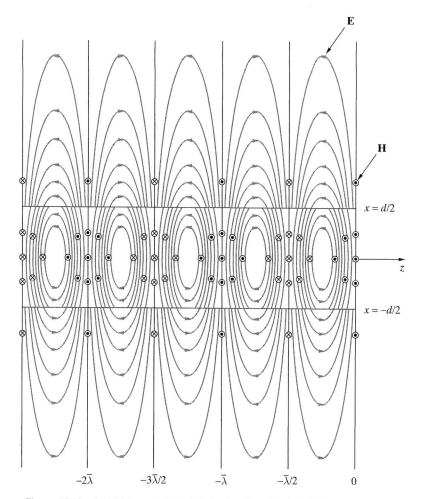

Figure 10.18 Odd TM₁ mode field distribution. The electric field lines are shown as solid lines, while the magnetic field lines are orthogonal to the page and are indicated alternately with circles or crosses. Figure taken (with permission) from H. A. Haus, *Waves and Fields in Optoelectronics*, Prentice-Hall, Englewood Cliffs, New Jersey, 1984.

and

$$\boxed{\frac{\alpha_x}{\beta_x} = -\frac{\mu_0}{\mu_d}\cot\left(\frac{\beta_x d}{2}\right) \qquad \text{Even TE modes}} \qquad (10.52)$$

Equations (10.51) and (10.52), as well as (10.46), which is valid for both odd and even modes, must be solved simultaneously to determine the values of β_x and α_x for the different modes.

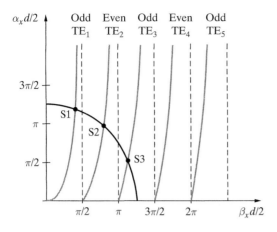

Figure 10.19 Graphical solution for TE modes. Graphical solution representing attenuation and phase constants for TE modes in a dielectric slab waveguide. Both odd and even TE lower-order modes are shown. The parameters chosen for this example are $\epsilon_d = 2\epsilon_0$ and $d = 1.25\lambda$, where $\lambda = c/f$ is the free space wavelength.

Solutions can once again be obtained either graphically or by numerical solution of the transcendental equations. The cutoff frequencies are determined by the $\alpha_x = 0$ condition and are the same as for TM modes. An illustration of a graphical solution for TE modes is given in Figure 10.19. The three possible solutions correspond to the intersection points S1, S2, and S3 as marked. The case treated here is the same as that for which the TM mode solutions are shown in Figure 10.15, for which $\epsilon_d = 2\epsilon_0$ and $d = 1.25\lambda$, where λ is the free space wavelength.

Note that the cutoff frequencies for the TE_m modes are given by the same expression as that for TM_m modes, namely,

$$f_{cTE_m} = \frac{(m-1)}{2d\sqrt{\mu_d \epsilon_d - \mu_0 \epsilon_0}} \qquad \begin{array}{ll} m = 1, 3, 5, \ldots & \text{Odd } TE_m \\ m = 2, 4, 6, \ldots & \text{Even } TE_m \end{array} \qquad (10.53)$$

For example, the lowest-order odd TE_1 mode has no cutoff frequency (similar to the odd TM_1 mode). The next higher mode (even TE_2 mode) has a cutoff frequency given by

$$f_{cTE_2} = \frac{1}{2d\sqrt{\mu_d \epsilon_d - \mu_0 \epsilon_0}}$$

10.2.2 Dielectric Covered Ground Plane

In the previous section, we considered wave propagation in planar dielectric slabs surrounded by another dielectric. In a wide range of applications, especially those involving miniature microwave circuits, guiding structures with dielectrics on a conducting ground plane or sandwiched between two conductors (microstrips) are used. In this subsection, we apply our previous analyses to a simple planar geometry as shown in Figure 10.20.

Note here that the boundary conditions on the metal end require both $E_y = 0$ and $E_z = 0$ at $x = 0$. We can obtain the possible solutions for the fields by examining the solutions obtained earlier for the dielectric slab. The electric field components of the

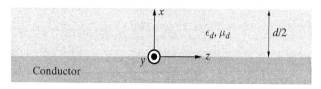

Figure 10.20 Dielectric slab with a ground plane. Geometry for a dielectric covered ground plane.

propagating TM modes within the slab were

$$
\left.
\begin{aligned}
E_y &= 0 \\[4pt]
E_x &= \frac{j\bar{\beta}}{\beta_x} C_e \sin(\beta_x x) e^{-j\bar{\beta}z} \\[4pt]
E_z &= C_e \cos(\beta_x x) e^{-j\bar{\beta}z}
\end{aligned}
\right\} \quad \text{Even TM} \qquad 0 \le x \le \frac{d}{2}
$$

$$
\left.
\begin{aligned}
E_y &= 0 \\[4pt]
E_x &= -\frac{j\bar{\beta}}{\beta_x} C_o \cos(\beta_x x) e^{-j\bar{\beta}z} \\[4pt]
E_z &= C_o \sin(\beta_x x) e^{-j\bar{\beta}z}
\end{aligned}
\right\} \quad \text{Odd TM} \qquad 0 \le x \le \frac{d}{2}
$$

Note that even TM modes *do not* satisfy boundary conditions at $x = 0$ (i.e., $E_z \neq 0$). For TE modes, we have

$$
\left.
\begin{aligned}
E_y &= \frac{-j\omega\mu}{\beta_x} C_e \sin(\beta_x x) e^{-j\bar{\beta}z} \\[4pt]
E_x &= 0 \\[4pt]
E_z &= 0
\end{aligned}
\right\} \quad \text{Even TE} \qquad 0 \le x \le \frac{d}{2}
$$

$$
\left.
\begin{aligned}
E_y &= \frac{j\omega\mu}{\beta_x} C_o \cos(\beta_x x) e^{-j\bar{\beta}z} \\[4pt]
E_x &= 0 \\[4pt]
E_z &= 0
\end{aligned}
\right\} \quad \text{Odd TE} \qquad 0 \le x \le \frac{d}{2}
$$

Note that odd TE modes also *do not* satisfy the boundary conditions at $x = 0$, since $E_y(x = 0, z) \neq 0$.

It thus appears that the geometry of the dielectric covered ground plane supports only TM odd and TE even modes. The governing equations for these modes are

$$
\frac{\epsilon_0}{\epsilon_d} \beta_x \tan\left(\frac{\beta_x d}{2}\right) = \alpha_x \qquad \text{Odd TM}
$$

$$
-\frac{\mu_0}{\mu_d} \beta_x \cot\left(\frac{\beta_x d}{2}\right) = \alpha_x \qquad \text{Even TE}
$$

and for all modes, we have

$$\left.\begin{array}{l} \beta_x^2 + \overline{\beta}^2 = \omega^2 \mu_d \epsilon_d \\ -\alpha_x^2 + \overline{\beta}^2 = \omega^2 \mu_0 \epsilon_0 \end{array}\right\} \quad \rightarrow \quad \alpha_x^2 + \beta_x^2 = \omega^2(\mu_d \epsilon_d - \mu_0 \epsilon_0)$$

The cutoff frequencies are

$$f_{c\text{TM or TE}} = \frac{(m-1)}{2d\sqrt{\mu_d \epsilon_d - \mu_0 \epsilon_0}} \qquad \begin{array}{l} m = 1, 3, 5, \ldots \quad \text{Odd TM}_m \\ m = 2, 4, 6, \ldots \quad \text{Even TE}_m \end{array} \qquad (10.54)$$

The dominant mode is the odd TM$_1$ with cutoff frequency of zero. All of the modes are called surface wave modes, and solutions are obtained exactly in the same way as for the dielectric slab discussed before. The only difference between the dielectric slab and the dielectric above a ground plane is that the latter structure cannot support the propagation of the TM even and TE odd modes.

Note that the transverse attenuation rate α_x for $f > f_c$ is given as

$$\alpha_x = \sqrt{\overline{\beta}^2 - \omega^2 \mu_0 \epsilon_0}$$

For very thick dielectrics (d large compared with the transverse wavelength $\lambda_x = 2\pi/\beta_x$), $\overline{\beta}$ approaches $\beta_d = \omega\sqrt{\mu_d \epsilon_d}$ and, as a result, we have

$$\alpha_x \simeq \sqrt{\omega^2(\mu_d \epsilon_d - \mu_0 \epsilon_0)} = \beta\sqrt{\frac{\mu_d \epsilon_d}{\mu_0 \epsilon_0} - 1} \qquad (10.55)$$

which at microwave and higher frequencies is usually very large, indicating that the fields decrease rapidly with distance from the dielectric, or that the guiding provided by the slab is effective. For very thin dielectrics ($d \ll \lambda_x$), the propagation constant $\overline{\beta}$ approaches that of uniform plane waves in free space, that is, $\beta = \omega\sqrt{\mu_0 \epsilon_0}$, and we have

$$\beta_x^2 = \omega^2 \mu_d \epsilon_d - \overline{\beta}^2 \simeq \omega^2(\mu_d \epsilon_d - \mu_0 \epsilon_0)$$

In the limit $(d/\lambda_x) \rightarrow 0$ we can also write

$$\alpha_x = \frac{\epsilon_0}{\epsilon_d} \beta_x \tan\left(\frac{\beta_x d}{2}\right) \simeq \frac{\epsilon_0}{\epsilon_d} \beta_x \left(\beta_x \frac{d}{2}\right) + \cdots \simeq \frac{d}{2} \frac{\epsilon_0}{\epsilon_d} \beta_x^2$$

Substituting in the preceding expression for β_x, we find

$$\alpha_x \simeq \frac{d}{2} \frac{\epsilon_0}{\epsilon_d} \left[\omega^2 \mu_0 \epsilon_0 \left(\frac{\epsilon_d \mu_d}{\epsilon_0 \mu_0} - 1\right)\right] = 2\pi\beta \left[\frac{\mu_d}{\mu_0} - \frac{\epsilon_0}{\epsilon_d}\right] \frac{d}{2\lambda} \qquad (10.56)$$

where β and λ are, respectively, the phase constant and wavelength in free space. The attenuation constant as given above for thin dielectrics (d small compared with λ_x) is usually quite small, indicating that the wave is not well guided (i.e., the fields are loosely confined to the dielectric slab).

Example 10.7: Polystyrene waveguide with ground plane. A ground plane is covered with polystyrene ($\epsilon_r = 2.56$, $\mu_r = 1$) of thickness ($d/2$). Determine the distance above the dielectric-air interface over which the fields decrease to e^{-1} of their value at the interface, for the cases in which (i) d is very large ($d \gg \lambda$) and (ii) d is very small (assume $d \simeq 2 \times 10^{-3}\lambda$, where λ is the free space wavelength).

> **Solution:** Note that the distance at which the fields decrease to e^{-1} of their value at the surface is simply $(\alpha_x)^{-1}$. Thus, using (10.55) and (10.56), we have
>
> $$d \gg \lambda \qquad (\alpha_x)^{-1} = \frac{1}{\beta\sqrt{\dfrac{\mu_d\epsilon_d}{\mu_0\epsilon_0} - 1}} \simeq 0.127\lambda \qquad \text{Tightly bound wave}$$
>
> $$d \simeq 2 \times 10^{-3}\lambda \qquad (\alpha_x)^{-1} = \frac{\lambda}{\pi\beta d\left(\dfrac{\mu_d}{\mu_0} - \dfrac{\epsilon_0}{\epsilon_d}\right)} \simeq 41.6\lambda \qquad \text{Loosely bound wave}$$

The lowest-order modes. In both the case of the dielectric slab waveguide and the dielectric above a conducting ground plane, the existence of a propagating TM_1 mode that persists down to zero frequency (i.e., $f_c = 0$) is quite remarkable. For the dielectric slab in free space, a TE_1 mode also exists with $f_c = 0$. It thus appears that very low-frequency waves can in principle be guided by a dielectric sheet. However, further consideration indicates that as the frequency is reduced, the transverse decay rate α_x decreases and the resultant extension of the fields far beyond the slab requires large amounts of power to be supplied in order to sustain such a mode.

First note that for any TM or TE mode at very high frequencies ($\omega \to \infty$), we have from (10.39)

$$\overline{\beta} = \sqrt{\omega^2\mu_d\epsilon_d - \beta_x^2} \simeq \omega\sqrt{\mu_d\epsilon_d} \qquad (10.57)$$

so that $\overline{\beta} \to \infty$ as $\omega \to \infty$. Thus, at high frequencies, the phase constant (and thus the phase velocity $\overline{v}_p = \omega/\overline{\beta}$) is the same as that for a uniform plane wave in the unbounded dielectric, so that most of the real axial power flow is confined within the slab. Consistent with this observation is the fact that for $\omega \to \infty$, the transverse attenuation constant α_x is given from (10.46) by

$$\alpha_x = \sqrt{\omega^2(\mu_d\epsilon_d - \mu_0\epsilon_0) - \beta_x^2} \simeq \omega\sqrt{\mu_d\epsilon_d - \mu_0\epsilon_0} \qquad (10.58)$$

It thus appears that the transverse attenuation constant α_x increases without limit proportionally with frequency, indicating that the depth of penetration (given by α_x^{-1} in view of the $e^{-\alpha_x x}$ type of variation of the fields) into the free space surrounding the slab approaches zero.

On the other hand, we see that at the cutoff frequency $\omega = \omega_c$, where

$$\omega_c = \frac{(m-1)\pi}{(d\sqrt{\mu_d \epsilon_d - \mu_0 \epsilon_0})}$$

we have

$$\overline{\beta} = \omega\sqrt{\mu_0\epsilon_0} = \beta \qquad \text{and} \qquad \alpha_x = 0$$

so that the fields extend to infinity outside the slab. In other words, an infinite amount of real power is carried in the axial direction *outside* the slab! The propagation constant is dominated by free space. The trend, therefore, is for the fields to spread to larger distances from the slab as frequency decreases toward cutoff, even in the case where the cutoff frequency is zero. Thus, although the lowest order mode can in principle be excited at very low frequencies, the extension of the fields far beyond the dielectric slab requires unreasonably large amounts of power to be supplied by the excitation.

For the case of the dielectric above a ground plane, the TM_1 solution is maintained, so that it appears that this geometry can guide waves of arbitrarily low frequency. However, we note once again that the decay rate α_x decreases as the frequency is reduced, so that increasingly large amounts of power must be supplied in order to sustain the TM_1 mode, subject to the same power limitation discussed previously. However, note that for the TE case, there is no mode that propagates at very low frequencies for the dielectric above a ground plane.

Example 10.8: An InP–Ga$_x$In$_{1-x}$As–InP optical waveguide. Consider a Ga$_x$In$_{1-x}$As planar waveguide with a thickness d sandwiched between two InP regions, as shown in Figure 10.21. (a) Find the maximum thickness d_{max} (in μm) so that only the TE$_1$ mode is guided at $\lambda = 1.65$ μm. (b) If the thickness of the waveguide is $d = 0.3$ μm, find the attenuation constant α_x in the InP region, and the phase constants β_x and $\overline{\beta}$ in the waveguide, all in units of $(\mu m)^{-1}$. Assume all materials to be nonmagnetic.

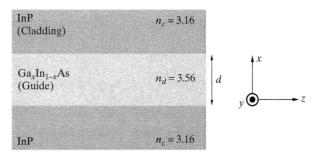

Figure 10.21   An InP–Ga$_x$In$_{1-x}$As–InP planar waveguide.

Solution:

(a) As we know, the TE$_1$ mode can propagate regardless of the thickness of the guide. The next higher mode (even TE$_2$) has a cutoff frequency given by (equation (10.53))

$$f_{c_2} = \left[\frac{(m-1)c}{2d\sqrt{n_d^2 - n_c^2}} \right]_{m=2} \simeq \frac{3 \times 10^8}{2d\sqrt{(3.56)^2 - (3.16)^2}} \geq \frac{3 \times 10^8 \text{ m-s}^{-1}}{1.65 \ \mu\text{m}}$$

which yields $d_{\max} \simeq 0.503 \ \mu$m.

(b) Noting that the TE$_1$ mode is an odd mode, we need to solve the following equations in order to find α_x and β_x:

$$\alpha_x = \beta_x \tan\left(\frac{\beta_x d}{2}\right)$$

$$\alpha_x^2 + \beta_x^2 = \beta^2(n_d^2 - n_c^2)$$

Using $d = 0.3 \ \mu$m, $n_d = 3.56$, and $n_c = 3.16$, we find (by either graphical or numerical solution) $\beta_x \simeq 4.73$ rad-$(\mu\text{m})^{-1}$ and $\alpha_x \simeq 4.07$ np-$(\mu\text{m})^{-1}$. To find $\overline{\beta}$, we can use

$$\beta_x^2 = \omega^2 \mu_d \epsilon_d - \overline{\beta}^2 \quad \text{or} \quad \alpha_x^2 = \overline{\beta}^2 - \omega^2 \mu_c \epsilon_c$$

both of which give $\overline{\beta} \simeq 12.70$ rad-$(\mu\text{m})^{-1}$, which indicates the wavelength in the guide is $\overline{\lambda} = 2\pi/\overline{\beta} \simeq 0.49 \ \mu$m.

10.2.3 Dielectric Slab Waveguide: Ray Theory

Our previous analysis of the dielectric slab waveguide was based on waveguide mode theory. It is also possible to analyze the wave propagation in a dielectric slab using ray theory, in a manner similar to the decomposition of waveguide modes into two uniform plane waves for the case of parallel-plate or hollow metallic waveguides. In some ways, the ray approach provides better physical insight into the guiding properties of dielectric waveguides, since dielectric guiding is a manifestation of the total internal reflection phenomena.

Consider wave propagation in the dielectric slab represented by a ray with a zigzag path, as shown in Figure 10.22, due to a TEM wave launched from the left.[32] Two possible cases are illustrated in Figure 10.22a. When $\theta_i < \theta_{ic}$, where θ_{ic} is the critical angle of incidence given by $\theta_{ic} = \sin^{-1}\sqrt{\epsilon_0/\epsilon_d}$, the wave refracts out of the dielectric slab and into the surrounding medium, loses a fraction of its power at each reflection, and thus eventually vanishes. In other words, the dielectric does not 'guide' the wave or confine the wave energy in any useful sense; to sustain such a wave at steady state, an infinite amount of input wave power would be required, since the waves outside the slab have infinite extent. When $\theta_i > \theta_{ic}$, on the other hand, the waves undergo total reflection at the boundaries and are guided or confined inside the dielectric slab, traveling

[32]Actual launching of the wave may be realized using a prism coupler as shown in Figure 9.41.

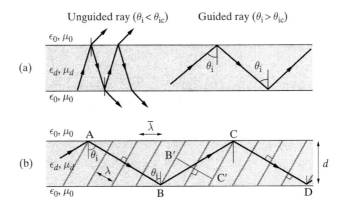

Figure 10.22   Illustration of dielectric waveguide operation using a ray approach.
(a) The wave is guided if the incidence angle θ_i is bigger than the critical angle θ_{ic}.
(b) Constructive interference condition and geometry for rays in a dielectric waveguide
for permitted angles of reflection. The dotted lines represent the wave phase fronts for
the ray segments AB and CD.

in the z direction by bouncing back and forth between the slab boundaries without loss
of power. This circumstance is very similar to propagation of uniform plane waves
between two perfectly conducting parallel plates, with the important difference that the
tangential component of the wave electric field does not need to be zero at the dielectric–
dielectric interface. Note that, as discussed in Chapter 9 in connection with total internal
reflection, the fields outside the dielectric slab decrease rapidly with transverse distance
from the slab.

At first it might appear that any wave for which $\theta_i > \theta_{ic}$ propagates in the dielectric
slab. However, due to the nature (constructive or destructive) of the interference between
different rays, only waves at certain discrete angles of incidence can actually propagate,
for a given slab thickness d and operating frequency f. Consider the typical ray path
ABCD shown in Figure 10.22b. The dotted lines represent the constant phase fronts of
the uniform plane wave during its travel from A to B and from C to D (the constant
phase fronts corresponding to path B–C are not shown). Before the ray goes through
total internal reflection at B, the line B–B$'$ constitutes a constant phase front. For constructive interference, we impose a self-consistency condition that after two successive
reflections, at B and C, the wave front must remain in phase. For example, waves starting
at two different points on the constant phase front B–B$'$ must go through the same phase
difference as one moves directly from B$'$ to C$'$ while the other reflects at B, moves from
B to C, and reflects at C. This means that any phase difference between these two paths
may amount only to an integer multiple of 2π, that is,

$$\beta_d(\overline{\text{BC}} - \overline{\text{B}'\text{C}'}) - 2\phi_r = (m - 1)2\pi \qquad m = 1, 2, 3, \ldots \qquad (10.59)$$

where $\beta_d = 2\pi/\lambda_d = (2\pi/\lambda)\sqrt{\epsilon_d/\epsilon_0}$ and ϕ_r is the phase shift that the wave acquires
during total internal reflection at point B or C from either of the boundaries of the slab.

From geometrical considerations, we have

$$\overline{BC} = \frac{d}{\cos\theta_i} \quad\text{and}\quad \overline{B'C'} = \overline{BC'}\sin\theta_i$$

where

$$\overline{BC'} = d\tan\theta_i - \frac{d}{\tan\theta_i}$$

Substituting in (10.59) and using algebraic manipulation, we find

$$2\beta_d d\cos\theta_i - 2\phi_r = (m-1)2\pi \qquad m = 1,2,3,\ldots$$

Now we restrict our attention to waves with **E** perpendicular to the plane of incidence (i.e., **E** out of the page). In that case, we know from Chapter 9 that the reflection coefficient is given as

$$\Gamma_\perp = \frac{\cos\theta_i + j\sqrt{\sin\theta_i^2 - (\epsilon_0/\epsilon_d)}}{\cos\theta_i - j\sqrt{\sin\theta_i^2 - (\epsilon_0/\epsilon_d)}} = 1e^{j\phi_\perp} \tag{10.60}$$

where $\phi_r = \phi_\perp = +2\tan^{-1}(\sqrt{\sin^2\theta_i - (\epsilon_0/\epsilon_d)}/\cos\theta_i)$. In other words, we must have

$$2\beta_d d\cos\theta_i - 4\tan^{-1}\left[\frac{\sqrt{\sin^2\theta_i - (\epsilon_0/\epsilon_d)}}{\cos\theta_i}\right] = (m-1)2\pi \qquad m = 1,2,3,\ldots$$

or

$$\tan\left(\frac{\beta_d d\cos\theta_i}{2} - \frac{(m-1)\pi}{2}\right) = \frac{\sqrt{\sin^2\theta_i - \epsilon_0/\epsilon_d}}{\cos\theta_i} \qquad m = 1,2,3,\ldots \tag{10.61}$$

Example 10.9: Glass slab. Consider the nonmagnetic dielectric slab waveguide shown in Figure 10.21, which consists of glass ($n_d = 1.5$ or $\epsilon_d = 2.25\epsilon_0$) of thickness $d = 1$ cm surrounded by air. For an operating frequency of 30 GHz, find the propagating TE$_m$ modes and their corresponding angles of incidence θ_i.

Solution: Using (9.36), the critical angle of incidence for the glass–air interface is

$$\theta_{ic} = \sin^{-1}\sqrt{\frac{\epsilon_0}{\epsilon_d}} = \sin^{-1}\left(\frac{1}{1.5}\right) \simeq 41.8°$$

Thus, for total internal reflection we must have $\theta_i \geq\sim 41.8°$. Using the transcendental equation (10.61), we have

$$\tan\left(\frac{2\pi(1.5)\cos\theta_i}{2} - \frac{(m-1)\pi}{2}\right) = \frac{\sqrt{\sin^2\theta_i - (2.25)^{-1}}}{\cos\theta_i} \qquad m = 1,2,3,\ldots$$

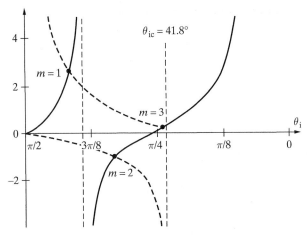

Figure 10.23 The solid line is a plot of $\tan(1.5\pi \cos\theta_i)$, and the dashed lines above and below the horizontal axis, respectively, correspond to $\sqrt{\sin^2\theta_i - (2.25)^{-1}}/\cos\theta_i$ and $-\cos\theta_i/\sqrt{\sin^2\theta_i - (2.25)^{-1}}$. The θ_i values corresponding to the three solutions $m = 1, 2, 3$ are, respectively, $\theta_i \simeq 75.03°$, $59.47°$, and $43.86°$.

or

$$\tan(1.5\pi \cos\theta_i) = \frac{\sqrt{\sin^2\theta_i - (2.25)^{-1}}}{\cos\theta_i} \qquad m = 1, 3, 5, \ldots$$

$$\tan(1.5\pi \cos\theta_i) = -\frac{\cos\theta_i}{\sqrt{\sin^2\theta_i - (2.25)^{-1}}} \qquad m = 2, 4, 6, \ldots$$

The right- and left-hand sides of these equations are plotted in Figure 10.23. The values of θ_i corresponding to the different modes can be either read from the graph or numerically evaluated from the equations. We find $\theta_i \simeq 75.03°$ (TE_1), $\theta_i \simeq 59.47°$ (TE_2), and $\theta_i \simeq 43.86°$ (TE_3). The three θ_i values at which the curves intersect are the three solutions, or eigenvalues, for which (10.61) is satisfied. These three angles correspond to the three transverse electric modes TE_1, TE_2, and TE_3. Note that all of these modes may exist simultaneously, but if the thickness d is decreased or the wavelength λ increased (or both), fewer solutions (eigenvalues) or modes are possible. However, we note that one solution, the TE_1 mode, will always exist, so that, at least in theory, waves at zero frequency can be guided by a dielectric slab. Note, however, the caveat in the previous section, namely, that an increasingly large amount of power is required to support such a mode as the frequency is decreased.

Relation between ray and mode theory parameters. The mode theory treatment of the dielectric slab waveguide discussed in Section 10.2.1 led to transcendental equations that needed to be solved in order to determine the transverse attenuation rate α_x and the transverse wave number β_x, from which we can find the axial phase constant $\overline{\beta}$ using (10.39). The ray theory treatment discussed in the present section leads to a different type of transcendental equation (10.61), which needs to be solved in order to determine the θ_i values corresponding to the different modes. The relationship between these ray and mode theory parameters can be seen by considering the decomposition of the mode solutions into component TEM waves, in the same manner as was discussed in Section 10.1.3. Consider the E_y component of the odd TE mode:

$$E_y = \frac{j\omega\mu}{\beta_x} C_0 \cos(\beta_x x) e^{-j\overline{\beta}z}$$

$$= \frac{j\omega\mu}{\beta_x} \frac{C_0}{2} \left[e^{j\beta_x x} + e^{-j\beta_x x}\right] e^{-j\overline{\beta}z}$$

$$= \frac{j\omega\mu}{\beta_x} \frac{C_0}{2} \left[e^{j(\beta_x x - \overline{\beta}z)} + e^{-j(\beta_x x + \overline{\beta}z)}\right]$$

$$= \frac{j\omega\mu}{\beta_x} \frac{C_0}{2} \left[e^{j\beta_d(x\cos\theta_i - z\sin\theta_i)} + e^{-j\beta_d(x\cos\theta_i + z\sin\theta_i)}\right]$$

where $\beta_d^2 = \beta_x^2 + \overline{\beta}^2 = \omega^2\mu_d\epsilon_d$. It is thus clear that the component uniform plane (or TEM) waves propagate in the x and z directions, respectively, with phase constants β_x and $\overline{\beta}$, so the tangent of the angle they make with the line that is vertical to the boundary of the dielectric is given by

$$\tan\theta_i = \frac{\overline{\beta}}{\beta_x}$$

Thus, once we know the values of β_x and $\overline{\beta}$ for a particular dielectric waveguide mode, the corresponding incidence angle θ_i of the ray model representation of that mode can be readily determined without having to solve an additional transcendental equation such as (10.61). In practice, one uses either the ray or mode theory description to analyze or design a dielectric waveguide. The values of β_x and $\overline{\beta}$ are generally not available when we attempt to use a ray analysis, in which case (10.61) can be used to find the θ_i values.

10.3 WAVE VELOCITIES AND WAVEGUIDE DISPERSION

In our study of guided waves, we have exclusively considered a special class of electromagnetic fields that vary harmonically in time t and where the phasors of the field quantities vary with axial distance z as $e^{-\overline{\gamma}z}$, so that for propagating waves ($\overline{\gamma} = j\overline{\beta}$), the field components (e.g., $\mathscr{E}_z$) in the time domain vary with axial distance and time as $\mathscr{E}_z(z,t) = \cos(\omega t \pm \overline{\beta}z)$. The axial direction is the direction in which we want to convey (or guide) the wave energy. Accordingly, all of the wave-guiding structures considered in this chapter have cross-sectional shapes that are identical at different points along the z axis. For such applications, it is natural to define the phase velocity of the propagating wave as that speed with which one has to move along the waveguide in order to keep up with points at which the fields have a specific instantaneous phase. For example, we may want to consider a particular electric field component (e.g., $\mathscr{E}_z$) and want to stay with the crest (or maximum) of the field as it travels along the guide. A specific phase of the wave corresponds to a constant value of $\mathcal{E}_z^+(z,t)$, so that we have

$$\mathcal{E}_z^+(z,t) = \text{const.} \quad \rightarrow \quad \omega t - \bar{\beta} z = \text{const.}$$

$$\xrightarrow{d/dt} \quad \omega - \bar{\beta}\,\frac{dz}{dt} = 0$$

$$\rightarrow \quad \bar{v}_p = \frac{dz}{dt} = \frac{\omega}{\bar{\beta}}$$

where we exclusively consider propagation in the $+z$ direction. The different wave-guiding systems considered in this chapter have different functional forms for the phase constant $\bar{\beta}$; for example, for parallel-plate metal waveguides, $\bar{\beta} = \omega\sqrt{\mu\epsilon}\sqrt{1 - (f_c/f)^2}$, so that the phase velocity is given as

$$\bar{v}_p = \frac{1}{\sqrt{\mu\epsilon}\sqrt{1 - (f_c/f)^2}} \tag{10.62}$$

where μ and ϵ are, respectively, the permeability and permittivity of the lossless dielectric material within the waveguide walls. We noted earlier that the phase velocity $\bar{v}_p$ in a waveguide is always greater than the phase velocity $v_p = \omega/\beta = 1/\sqrt{\mu\epsilon}$ of uniform plane waves in the unbounded lossless dielectric filling, and it is clear from (10.62) and other expressions for $\bar{v}_p$ in earlier sections that $\bar{v}_p$ increases toward infinity as the operating frequency f approaches the cutoff frequency (i.e., $f \rightarrow f_c$). The fact that $\bar{v}_p$ is larger than v_p does not pose any dilemma or violate any fundamental principle, since no energy or information can be transmitted at $\bar{v}_p$.

The most important aspect of equation (10.62) is that the phase velocity $\bar{v}_p$ depends on frequency, since the phase constant $\bar{\beta}$ is a nonlinear function of frequency. This result is to be contrasted with the phase velocity of uniform plane waves propagating (with phase constant $\beta = \omega\sqrt{\mu\epsilon}$) in a lossless unbounded dielectric medium, which is a constant independent of frequency,[33] assuming that the dielectric medium between the guide walls is simple (i.e., ϵ and μ are simple constants) and lossless ($\sigma = 0$, $\epsilon'' = 0$, and $\mu'' = 0$). The frequency dependence of $\bar{v}_p$ in (10.62) is not due to the nature of the material filling but is rather a consequence of the configuration of the waveguide as it influences the behavior of the fields within it. Note that for lossy material media, ($\sigma \neq 0$, $\epsilon'' \neq 0$, or $\mu'' \neq 0$), the propagation constant is complex and the phase velocity and attenuation rate are both functions of frequency, as discussed in Section 8.3.

The phase velocity $\bar{v}_p$ as given in (10.62) is for continuous, steady-state waves of a single frequency. In order to convey any information, it is necessary to "modulate" the signal by changing its amplitude, phase, or frequency, or by interrupting it in accordance

---

[33]In many material media, either or both ϵ and μ may be functions of frequency. The phase velocity of uniform plane waves in such media is a function of frequency even if the medium is unbounded and lossless ($\sigma = 0$, $\epsilon'' = 0$, and $\mu'' = 0$), and is given by

$$v_p = \frac{\omega}{\beta} = \frac{\omega}{\omega\sqrt{\mu(f)\epsilon(f)}} = \frac{1}{\sqrt{\mu(f)\epsilon(f)}}$$

with the information to be transmitted. Any information-carrying signal consists of a band of Fourier components with different amplitudes and phases such that the signal at the input to the waveguide is constituted by a superposition of these components. If the phase velocity $\bar{v}_p$ is the same for different frequency components and there is no attenuation, the spectral components add in proper phase at all locations in the waveguide, to precisely reproduce the original signal shape. However, when the phase velocity $\bar{v}_p$ is a function of frequency, the different frequency components acquire different phase delays as the signal travels through the waveguide, so that the superposition of the individual Fourier components at some point down the waveguide is no longer a precise replica of the source transmitted signal. The signal intended for transmission (e.g., a rectangular pulse riding on a carrier) smears out as a result, losing its sharp edges and broadening as it propagates along the guide. In other words, the fact that $\bar{v}_p$ is a function of frequency leads to the *distortion* of the signal, by *dispersing* its frequency components. This phenomenon is referred to as *dispersion*. This type of dispersion, which is a result of the propagation characteristics imposed by the waveguide structure, is sometimes referred to as *waveguide dispersion*, to distinguish it from *material dispersion*, which occurs because the material media is either lossy or exhibits properties such that ϵ and μ depend on frequency.

When the propagation of an electromagnetic signal is subject to either waveguide or material dispersion, an important question arises concerning the velocity of a wave packet containing a band of frequency components. The "information" carried by a modulated signal is represented by the "shape" or the envelope of the wave. Although this shape itself changes somewhat with distance, it is nevertheless possible to obtain a useful measure of its "speed" of propagation. This measure of the speed of propagation of a *group* of frequencies that constitute a wave packet is called the *group velocity*, defined in the next section. In many (but not all) cases, the group velocity also represents the velocity of travel of electromagnetic energy. Further discussion of different types of wave velocities and related references can be found elsewhere.[34]

10.3.1 Group Velocity

In introducing the concept of group velocity, we recognize that, in general, its utility is not confined to the subject of waveguides and that the concept is equally applicable in cases of material dispersion, for example, in the case of propagation of uniform plane waves in a lossy dielectric or in cases of more complicated media where the permittivity ϵ and/or permeability μ may be functions of frequency (Section 11.2). To understand dispersion and group velocity, we consider propagating wave solutions for any one of the field components (say, $\mathscr{E}_x$) with the general form

$$\mathscr{E}_x(z,t) = \mathscr{R}e\{Ce^{j(\omega t - \bar{\beta}z)}\} \tag{10.63}$$

Note that because any electromagnetic field solutions must obey the wave equation derived from Maxwell's equations, the frequency ω and the phase constant $\bar{\beta}$ are related.

---

[34] See Section 5.17 of J. A. Stratton, *Electromagnetic Theory*, McGraw-Hill, New York, 1941.

The phase constant $\bar{\beta}$ can in general be a complicated function of frequency. Note that for uniform plane waves in a lossless unbounded dielectric medium, we have $\bar{\beta} = \beta = \omega\sqrt{\mu\epsilon}$, so that $\bar{\beta}$ is a linear function of frequency, as long as μ and ϵ are both independent of frequency. The phase velocity $\bar{v}_p = \omega/\bar{\beta}$ is then simply $\bar{v}_p = v_p = (\sqrt{\mu\epsilon})^{-1}$, which is independent of frequency. Thus, there is no signal dispersion in cases where the phase constant $\bar{\beta}$ is linearly proportional to frequency. Fundamentally, dispersion occurs when the phase constant $\bar{\beta}$ is a nonlinear function of frequency, that is, $\bar{\beta}(\omega) \neq K\omega$, where K is a constant.

In general, an information-carrying signal is constituted by a superposition of many Fourier components, each of which is given by (10.63), with both $\bar{\beta}$ and C being functions of frequency. Before we consider a general wave packet, we consider the simplest case of the superposition of two harmonic waves of equal amplitudes but slightly different frequencies and phase constants, represented by

$$\mathscr{E}_{1x}(z,t) = C\cos(\omega t - \bar{\beta}z)$$

$$\mathscr{E}_{2x}(z,t) = C\cos[(\omega + \Delta\omega)t - (\bar{\beta} + \Delta\bar{\beta})z]$$

where $\Delta\omega \ll \omega$ and $\Delta\bar{\beta} \ll \bar{\beta}$, and C is a constant. Using the trigonometric identity $\cos\zeta + \cos\xi = 2\cos[(\zeta + \xi)/2]\cos[(\zeta - \xi)/2]$, the sum of these two waves can be written as

$$\mathscr{E}_x(z,t) = \mathscr{E}_{1x} + \mathscr{E}_{2x} = 2C\cos\left[\frac{1}{2}\left(\Delta\omega t - \Delta\bar{\beta}z\right)\right]\cos\left[\left(\omega + \frac{\Delta\omega}{2}\right)t - \left(\bar{\beta} + \frac{\Delta\bar{\beta}}{2}\right)z\right]$$

$$(10.64)$$

Examination of (10.64) indicates that the total field structure is similar to a uniform plane wave oscillating at a frequency of $\omega + \Delta\omega/2$, which is negligibly different from ω for $\Delta\omega \ll \omega$. The amplitude of the wave, however, is not a constant, but varies slowly with time and position between $-2C$ and $2C$, giving rise to the well-known phenomenon of *beats*, as illustrated in Figure 10.24. The amplitude envelope is designated $A_{\text{env}}(z,t)$ and is given by

$$A_{\text{env}}(z,t) = 2C\cos\left[\frac{1}{2}\left(\Delta\omega t - \Delta\bar{\beta}z\right)\right]$$

The successive maxima of the slowly varying amplitude function $A_{\text{env}}(z,t)$ at a fixed position occurs every $4\pi/\Delta\omega$ seconds and at a fixed time occurs at distance intervals of $4\pi/\Delta\bar{\beta}$. The velocity at which the amplitude envelope propagates can be found by equating the argument of the cosine term to a constant as

$$\Delta\omega t - \Delta\bar{\beta}z = \text{const.}$$

from which it follows that the envelope propagates with the special velocity

$$v_g = \frac{dz}{dt} = \frac{\Delta\omega}{\Delta\bar{\beta}}$$

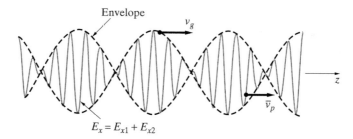

$$E_x = E_{x1} + E_{x2}$$

Figure 10.24 Two-wave beat pattern. The superposition of two harmonic waves of equal amplitude at slightly different frequencies. The envelope shows a beat pattern that propagates at the group velocity v_g.

or, in the limit of $\Delta\omega \to 0$,

$$v_g \equiv \frac{d\omega}{d\overline{\beta}} \tag{10.65}$$

which constitutes the definition of the *group velocity*. If the medium is nondispersive, with, for example, $\overline{\beta} = \beta = \omega\sqrt{\mu\epsilon}$, we have

$$v_g = \frac{d\omega}{d\overline{\beta}} = \frac{1}{\dfrac{d\overline{\beta}}{d\omega}} = (\sqrt{\mu\epsilon})^{-1} = v_p$$

so that the group velocity is equal to the phase velocity (i.e., $v_g = v_p$). In a dispersive medium, however, where $\overline{\beta}$ is a nonlinear function of ω, the phase and group velocities are different.

In general, when the phase constant $\overline{\beta}$ is a nonlinear function of frequency, the phase velocity $\overline{v}_p = \omega/\overline{\beta}(\omega)$ is also a function of frequency. In this case, we can relate v_g to $\overline{v}_p$ as

$$v_g = \frac{d\omega}{d\overline{\beta}} = \frac{1}{\dfrac{d\overline{\beta}}{d\omega}} = \frac{1}{\dfrac{d}{d\omega}\left(\dfrac{\omega}{\overline{v}_p}\right)} = \frac{\overline{v}_p}{1 - \dfrac{\omega}{\overline{v}_p}\dfrac{d\overline{v}_p}{d\omega}} \tag{10.66}$$

It is clear from equation (10.66) that for $d\overline{v}_p/d\omega = 0$ the medium is not dispersive and $v_g = \overline{v}_p$. It is also clear from (10.66) that whether the group velocity v_g is greater or smaller than the phase velocity $\overline{v}_p$ depends on the sign of $d\overline{v}_p/d\omega$. In cases where the phase velocity $\overline{v}_p$ decreases with increasing frequency (i.e., $d\overline{v}_p/d\omega < 0$), as in a metallic parallel-plate waveguide, we have *normal dispersion*, with $v_g < \overline{v}_p$. The case in which the dependence of $\overline{v}_p$ on frequency is such that $d\overline{v}_p/d\omega > 0$ is called *anomalous dispersion*, with $v_g > \overline{v}_p$.

In the preceding discussion we have considered the superposition of two harmonic waves of equal amplitude as the simplest "group" of waves. A single group or pulse of any desired shape may be constructed using Fourier superposition by noting that the amplitudes (and, in general, also the phases) of the component harmonic waves are functions of frequency. For the general case of nonperiodic signals, the Fourier summation must extend continuously over the entire range of frequencies. Thus, we can represent a wave packet as

$$\mathscr{E}_x(z,t) = \mathscr{R}e \left\{ \int_{-\infty}^{\infty} A(\omega) e^{j(\omega t - \overline{\beta} z)} \, d\omega \right\} \tag{10.67}$$

Since ω and $\overline{\beta}$ are one-to-one related as discussed above, we can equally take $\overline{\beta}$ as the independent variable in the Fourier decomposition of the wave packet:

$$\mathscr{E}_x(z,t) = \mathscr{R}e \left\{ \int_{-\infty}^{\infty} A(\overline{\beta}) e^{j(\omega(\overline{\beta})t - \overline{\beta} z)} \, d\overline{\beta} \right\} \tag{10.68}$$

where $A(\overline{\beta})$ is obviously a different function from $A(\omega)$. In general, the concept of group velocity applies only to groups of waves with frequency components that are confined to a narrowband spectrum centered around a frequency ω_0 corresponding to a wave number $\overline{\beta}_0$. In such cases, the amplitude function $A(\overline{\beta})$ has negligible magnitude outside the frequency and phase constant ranges

$$\omega_0 \quad \Delta\omega \le \omega \le \omega_0 + \Delta\omega \quad \text{or} \quad \overline{\beta}_0 - \Delta\overline{\beta} \le \overline{\beta} \le \overline{\beta}_0 + \Delta\overline{\beta}$$

where $\Delta\omega \ll \omega_0$ (or $\Delta\overline{\beta} \ll \overline{\beta}_0$). We can thus rewrite the superposition integral (10.68) as

$$\mathscr{E}_x(z,t) \simeq \mathscr{R}e \left\{ \int_{\overline{\beta}_0 - \Delta\overline{\beta}}^{\overline{\beta}_0 + \Delta\overline{\beta}} A(\overline{\beta}) e^{j[\omega(\overline{\beta})t - \overline{\beta}(\omega)z]} \, d\overline{\beta} \right\} \tag{10.69}$$

which represents a *wave packet*. Since the range $\Delta\overline{\beta}$ is small compared with $\overline{\beta}$ (i.e., $\Delta\overline{\beta} \ll \overline{\beta}_0$), $\omega(\overline{\beta})$ is only slightly different from its value at $\overline{\beta} = \overline{\beta}_0$. Consider a Taylor series representation of $\omega(\overline{\beta})$:

$$\omega(\overline{\beta}) = \omega(\overline{\beta}_0) + \left(\frac{d\omega}{d\overline{\beta}} \right)_{\overline{\beta} = \overline{\beta}_0} (\overline{\beta} - \overline{\beta}_0) + \cdots \tag{10.70}$$

The higher-order terms in (10.70) can be neglected since $\Delta\omega \ll \omega_0$. Using (10.70) for $\omega(\overline{\beta})$, we can rewrite $[\omega t - \overline{\beta} z]$ as

$$[\omega t - \overline{\beta} z] = \omega_0 t - \overline{\beta}_0 z + (\overline{\beta} - \overline{\beta}_0) \left[t \left(\frac{d\omega}{d\overline{\beta}} \right)_{\overline{\beta} = \overline{\beta}_0} - z \right] + \cdots \tag{10.71}$$

where $\omega_0 = \omega(\overline{\beta}_0)$. The wave packet of (10.69) can thus be represented as

$$\mathscr{E}_x(z,t) = \mathscr{R}e \left\{ A_{\text{env}}(z,t) e^{j(\omega_0 t - \overline{\beta}_0 z)} \right\}$$

where $A_{\text{env}}(z,t)$ is the slowly varying amplitude given approximately by

$$A_{\text{env}}(z,t) \simeq \int_{\overline{\beta}_0 - \Delta\overline{\beta}}^{\overline{\beta}_0 + \Delta\overline{\beta}} A(\overline{\beta}) e^{j(\overline{\beta} - \overline{\beta}_0)[t(d\omega/d\overline{\beta})_{\overline{\beta}=\overline{\beta}_0} - z]} \, d\overline{\beta}$$

We note that the amplitude envelope $A_{\text{env}}(z,t)$ is constant over surfaces defined by

$$t \left(\frac{d\omega}{d\overline{\beta}} \right)_{\overline{\beta}=\overline{\beta}_0} - z = \text{const.}$$

from which it is clear that the wave packet (or the "shape" of the envelope of the wave packet, as discussed earlier) propagates at the special velocity called the group velocity, given by

$$v_g = \frac{dz}{dt} = \left(\frac{d\omega}{d\overline{\beta}} \right)_{\overline{\beta}=\overline{\beta}_0}$$

The concepts of phase and group velocities can be further illustrated by considering the superposition of two harmonic waves (e.g., two sidebands) at a certain instant of time as shown in Figure 10.25. If the component waves a and b have the same velocity, the two corresponding crests a_1 and b_1 move along together, and the maximum of the modulation envelope (i.e., beat pattern) moves along at the same velocity. Under these circumstances, phase and group velocities are the same. If, however, the lower-frequency (i.e., longer-wavelength) wave b has a velocity slightly greater (normal dispersion) than that of wave a, the crests a_1 and b_1 move apart in time, while the crests a_2 and b_2 move closer together. Therefore, at some later instant the maximum of the beat pattern occurs at the point where a_2 and b_2 coincide in time, and at a still later time where a_3 and b_3 coincide. It is evident from Figure 10.25 that the envelope (i.e., amplitude of the sum of the two waves) slips backward with respect to the component waves; in other words, it moves forward with a different velocity, that is, the group velocity v_g, which in this case is less than the phase velocity of either of the two component waves.

    If the shorter-wavelength (i.e., higher-frequency) wave (in this case, wave a) has a phase velocity greater than that of wave b (anomalous dispersion), the relative motion of

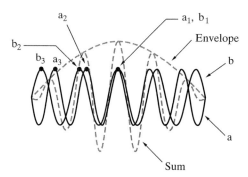

Figure 10.25 Physical interpretation of phase versus group velocity. Two waves a and b of slightly different frequencies form an amplitude-modulated "beat" wave pattern. Wave a has a slightly higher frequency than wave b.

the crests of the two waves is reversed, and the beat pattern slips forward with respect to the component waves. The group velocity is then greater than the phase velocity of the component waves.[35]

10.3.2 Dispersion ($\bar{\beta}$–ω) Diagrams

A useful way of visualizing and analyzing dispersion properties for different types of waves is to plot the phase constant $\bar{\beta}$ as a function of frequency ω, as shown for the case of TE or TM modes in a parallel-plate metal waveguide in Figure 10.26. For these modes, the *dispersion relation* (i.e., the relation between the phase constant $\bar{\beta}$ and frequency ω) has the form

$$\bar{\beta} = \omega\sqrt{\mu\epsilon}\sqrt{1 - \left(\frac{\omega_c}{\omega}\right)^2} \quad \rightarrow \quad \omega^2 = \omega_c^2 + \bar{\beta}^2 v_p^2 \tag{10.72}$$

where ω_c is the cutoff frequency and $v_p = (\mu\epsilon)^{-1/2}$ is the phase velocity in the unbounded dielectric medium. Equation (10.72) is a rather common form of dispersion relation that holds also for rectangular and circular waveguides as well as for other propagation media.[36] The dispersion characteristics of such systems can be understood by examining the dispersion curve, or a $\bar{\beta}$–ω diagram, as shown in Figure 10.26.

The phase velocity in a waveguide at any given frequency is given by (10.21),

$$\bar{v}_p = \bar{v}_p(\omega) = \frac{\omega}{\bar{\beta}} = \frac{v_p}{\sqrt{1 - (\omega_c/\omega)^2}}$$

whereas the group velocity can be found using (10.72) as

$$v_g = v_g(\omega) = \frac{d\omega}{d\bar{\beta}} = \frac{1}{(d\bar{\beta}/d\omega)} = v_p\sqrt{1 - \left(\frac{\omega_c}{\omega}\right)^2} \tag{10.73}$$

---

[35]Note that the fact that v_g can be greater than the phase velocity does not violate special relativity by implying that energy or information can travel faster than the speed of light. In the case of anomalous dispersion, group velocity as defined in (10.65) is simply not a useful concept. The fact that $d\bar{\beta}/d\omega$ is large means that the approximations made in connection with (10.70), where we dropped the higher-order terms in characterizing the envelope, do not hold. In other words, the envelope of the signal behaves in a more complicated manner than that described by the first-order term in (10.70).

[36]For example, the dispersion relation for an isotropic plasma is identical. A plasma is an ionized medium in which free electrons exist and are able to move under the influence of the electric and magnetic fields of the wave. This motion of charged particles constitutes a current $\mathbf{J}$ that in turn modifies the fields via $\nabla \times \mathbf{H} = \mathbf{J} + j\omega\mathbf{D}$. It turns out (see Section 11.1) that the physical effects of these charged particle motions can be accounted for by using an effective permittivity $\epsilon_{\text{eff}} = \epsilon_0(1 - \omega_p^2/\omega^2)$, where ω_p is a characteristic frequency of the plasma medium that is a function of the density of free electrons. The phase constant for uniform plane waves in an unbounded isotropic plasma is thus given as

$$\bar{\beta} = \omega\sqrt{\mu_0\epsilon_0}\sqrt{1 - \left(\frac{\omega_p^2}{\omega^2}\right)} \quad \rightarrow \quad \omega^2 = \omega_p^2 + \bar{\beta}^2 c^2$$

Thus, the plasma medium allows the propagation of waves above a cutoff frequency $\omega_c = \omega_p$.

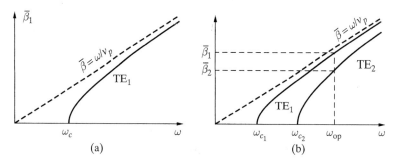

Figure 10.26 Dispersion relation for TE or TM modes in a parallel-plate wave-guide. $\overline{\beta} - \omega$ diagram for a parallel-plate waveguide. (a) TE$_1$ (or TM$_1$) mode. (b) TE$_1$ and TE$_2$ modes.

or can be evaluated from (10.66). Note that for this case, the phase and group velocities are related by

$$\overline{v}_p v_g = v_p^2$$

where $v_p = 1/\sqrt{\mu\epsilon}$. This interesting relationship between group and phase velocity is true only for propagation media for which the $\overline{\beta}-\omega$ relationship is described by (10.72). Note from Figure 10.26b that each waveguide mode has its own dispersion curve. If care is not taken in the choice of the operational frequency ω_{op}, different modes with different phase constants $\overline{\beta}$ can coexist. For example, if the operational frequency ω_{op} is chosen to be above the cutoff frequencies of both the TE$_1$ and TE$_2$ modes, as indicated in Figure 10.26b, the dispersion relations of these two modes both allow real values for $\overline{\beta}$ (i.e., propagating solutions). On the other hand, if ω_{op} is chosen such that $\omega_{c_1} < \omega_{op} < \omega_{c_2}$, a real solution for the phase constant $\overline{\beta}$ exists only for the TE$_1$ mode.

The variation with frequency of the phase and group velocities for a waveguide is shown in Figure 10.27. We note that the group velocity is zero at the cutoff frequency, corresponding to the situation where the component uniform plane waves constituting the waveguide mode reflect back and forth between the upper and lower plates with no propagation along the guide. For waveguides, and for all media for which (10.72) holds, we have normal dispersion since $d\overline{\beta}/d\omega$ is always larger than $\overline{\beta}/\omega$, so that $\overline{v}_p$ is always

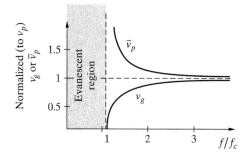

Figure 10.27 Phase and group veloci-ties. Phase and group velocities in a waveguide plotted as a function of frequency. Note that the quantities plotted are normalized by v_p, the phase velocity of uniform plane waves in the unbounded dielectric medium.

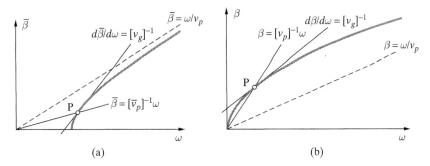

Figure 10.28 $\bar{\beta}$–ω **diagrams.** (a) Dispersion diagram for metallic waveguides ($v_g < \bar{v}_p$). (b) Dispersion diagram for uniform plane waves in an unbounded lossy dielectric ($v_g > v_p$).

greater than v_g. This fact can be seen graphically from Figure 10.28a, since the slope of the line from the origin to an arbitrary point P (which is equal to the inverse of $\bar{v}_p$) is smaller than the slope of the tangent line at the same point (which is the inverse of v_g). It is also interesting to note that $\bar{v}_p$ and v_g both approach v_p as $\omega \to \infty$.

Under other circumstances the phase velocity can be less than the group velocity. This case of anomalous dispersion ($v_g > v_p$) is illustrated in Figure 10.28b, which shows the β–ω diagram for uniform plane waves propagating in an unbounded lossy dielectric. For this case, we recall from Section 8.3 that

$$\beta = \omega \sqrt{\frac{\mu\epsilon}{2}} \left[\sqrt{1 + \left(\frac{\sigma}{\omega\epsilon}\right)^2} + 1 \right]^{1/2}$$

so that the phase constant is clearly a function of frequency, and the medium is thus dispersive. We can see from Figure 10.28b that $d\beta/d\omega$ for this case is always less than β/ω, so that the group velocity v_g is always greater than the phase velocity v_p. As will be noted in footnote 37, group velocity as defined in (10.65) is not a useful concept in such cases. The fact that $d\omega/d\beta$ is large means that the approximations made in connection with (10.70), namely, dropping the higher-order terms in characterizing the envelope, do not hold.

10.3.3 Group Velocity as Energy Velocity

We have seen that the envelope of a group of frequency components representing any given field quantity travels at the group velocity. If the magnitude of the quantity involved (e.g., wave electric field) is associated with the energy density of the electromagnetic wave, we can expect that the transport of energy carried by an electromagnetic wave also occurs at the group velocity. In this section, we demonstrate this interpretation of group velocity. Consider the propagating TM_m mode in a parallel-plate waveguide, for

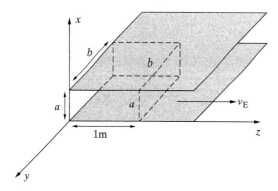

Figure 10.29 A parallel-plate waveguide. A segment of a parallel-plate waveguide of height a, width b, and 1 m axial length.

which the field components are given by (10.16) as

$$H_y = C \cos\left(\frac{m\pi}{a}x\right) e^{-j\overline{\beta}_m z}$$

$$E_x = \frac{\overline{\beta}}{\omega\epsilon} C \cos\left(\frac{m\pi}{a}x\right) e^{-j\overline{\beta}_m z}$$

$$E_z = \frac{jm\pi}{\omega\epsilon a} C \sin\left(\frac{m\pi}{a}x\right) e^{-j\overline{\beta}_m z}$$

The velocity of energy flow represented by this wave can be expressed in terms of the time-average power flowing through a cross-sectional area of width b (see Figure 10.29) divided by the energy stored per unit length of the guide of the same width. In other words, we can write the velocity of energy flow as

$$v_E = \frac{P_{av}}{\overline{W}_{str}} \qquad \frac{J\text{-}s^{-1}}{J\text{-}m^{-1}} = m\text{-}s^{-1}$$

where $\overline{W}_{str}$ is the energy stored per unit length of waveguide of width b. Note that since the power flow is in the z direction, the energy transport velocity v_E is also in the z direction.

The total time-average power flow through the cross-sectional area of height a and width b is then

$$P_{av} = b \int_0^a \frac{1}{2} [\mathfrak{Re}\{\mathbf{E} \times \mathbf{H}^*\}] \cdot \hat{\mathbf{z}} \, dx = b \int_0^a \frac{1}{2} \mathfrak{Re}(E_x H_y^*) \, dx$$

$$= \frac{b}{2} C^2 \frac{\overline{\beta}_m}{\omega\epsilon} \int_0^a \cos^2\left(\frac{m\pi}{a}x\right) dx = \frac{bC^2\overline{\beta}_m}{2\omega\epsilon} \frac{a}{2} = \frac{baC^2\overline{\beta}_m}{4\omega\epsilon}$$

The time-average stored energy per unit length of the same waveguide of height a and width b, including both electric and magnetic fields, is[37]

$$\overline{W}_{str} = \frac{1}{4} \int_0^a \int_0^b \int_0^1 [\epsilon \mathbf{E} \cdot \mathbf{E}^* + \mu \mathbf{H} \cdot \mathbf{H}^*] \, dz \, dy \, dx$$

$$= b \int_0^a \left\{ \frac{\epsilon}{4} \left[|E_x|^2 + |E_z^2| \right] + \frac{\mu}{4} |H_y|^2 \right\} dx$$

$$= \frac{bC^2}{4} \int_0^a \left\{ \epsilon \left[\frac{\overline{\beta}_m^2}{\omega^2 \epsilon^2} \cos^2 \left(\frac{m\pi}{a} x \right) + \frac{m^2 \pi^2}{\omega^2 \epsilon^2 a^2} \sin^2 \left(\frac{m\pi}{a} x \right) \right] + \mu \cos^2 \left(\frac{m\pi}{a} x \right) \right\} dx$$

$$= \frac{bC^2}{4} \frac{a}{2} \left[\frac{\overline{\beta}_m^2}{\omega^2 \epsilon} + \frac{m^2 \pi^2}{\omega^2 \epsilon a^2} + \frac{\omega^2 \epsilon \mu}{\omega^2 \epsilon} \right] = \frac{baC^2}{8\omega^2 \epsilon} \left[\overline{\beta}_m^2 + \frac{m^2 \pi^2}{a^2} + \omega^2 \mu \epsilon \right]$$

$$\overline{W}_{str} = \frac{baC^2 \mu}{4}$$

where we have recognized that for TM_m modes,

$$\overline{\beta}_m^2 + (m^2 \pi^2 / a^2) = \omega^2 \mu \epsilon$$

Using $\overline{W}_{str}$ and P_{av} in the expression for v_E, we have

$$v_E = \frac{P_{av}}{\overline{W}_{str}} = \frac{baC^2 \overline{\beta}_m / 4\omega\epsilon}{baC^2 \mu / 4} = \frac{\overline{\beta}_m}{\omega\mu\epsilon}$$

Noting that the waveguide phase velocity $\overline{v}_p = \omega / \overline{\beta}_m$, we can write the energy velocity as

$$v_E = \frac{\dfrac{1}{(\mu\epsilon)}}{\overline{v}_p} = \frac{v_p^2}{\overline{v}_p} \quad \rightarrow \quad v_E \overline{v}_p = v_p^2 \quad \rightarrow \quad v_E = v_g$$

[37]Note that the instantaneous stored energy in an electric (or magnetic) field is given by $\frac{1}{2}\epsilon |\overline{\mathscr{E}}|^2$ (or $\frac{1}{2}\mu |\overline{\mathscr{H}}|^2$), where $\overline{\mathscr{E}}(z,t) = \hat{x} E_0 \cos(\omega t - \overline{\beta} z + \phi)$ is the "real" electric field, related to the phasor $\mathbf{E}(z) = \hat{x} E_0 \, e^{-j\overline{\beta} z} \, e^{j\phi}$ through

$$\overline{\mathscr{E}}(z,t) = \Re e\{\mathbf{E}(z) e^{j\omega t}\}$$

The time-average stored energy can be expressed as

$$\frac{1}{T_p} \int_0^{T_p} \frac{1}{2} \epsilon E_0^2 \cos^2(\omega t - \beta z + \phi) \, dt = \frac{1}{4} \epsilon E_0^2$$

where $T_p = 2\pi / \omega$ is the period of the assumed sinusoidal variation.

where we have used the fact that for waveguides $v_g \bar{v}_p = v_p^2$. Also note that since from (10.19) we have $\bar{\beta}_m = \sqrt{\omega^2 \mu \epsilon - (m^2 \pi^2 / a^2)} = \beta \sqrt{1 - (f_c/f)^2}$, the energy velocity can be expressed as

$$v_{\mathrm{E}} = v_g = \frac{\beta \sqrt{1 - (f_c/f)^2}}{\omega \mu \epsilon}$$

$$= v_p \sqrt{1 - \left(\frac{m\lambda}{2a}\right)^2}$$

Note that a similar result can be obtained for any of the other TE or TM modes for the parallel-plate waveguide. The fact that the group velocity is the velocity of energy transport for TE or TM modes in waveguides is consistent with the decomposition of such modes into component TEM waves, as discussed in Section 10.1.3. In this context, we can think of the energy of the electromagnetic wave as being associated with the component plane waves that move along the waveguide by bouncing back and forth between the waveguide walls, being incident on the walls at an angle θ_{i_m} from the vertical. Although these component TEM waves propagate along at velocity v_p, the component of their velocity in the z direction (i.e., the velocity at which energy travels along the waveguide) is $v_{p_z} = v_p \sin \theta_{i_m} = v_p \sqrt{1 - (\omega_{c_m}/\omega)^2}$, which is identical to the group velocity v_g.

It should be noted that, in general, both the velocity of energy transport and the velocity at which information travels (sometimes called the signal velocity) are difficult to define.[38] The concept of group velocity is particularly useful when a signal consists of component frequencies concentrated in a narrow band around a carrier frequency. In such cases, the approximation made in connection with (10.70) is valid, and the group velocity is a very good approximation of both the signal velocity and the velocity of energy transport. However, in other cases, for example, in the case of a step excitation, the frequency content of the signal is not restricted to a narrow range of frequencies, and the concept of a group velocity is not useful.

10.4 SUMMARY

This chapter discussed the following topics:

- **General relationships for waves guided by metallic conductors.** The configuration and propagation of electromagnetic waves guided by metallic conductors are governed by the solutions of the wave equations subject to the boundary conditions

$$E_{\mathrm{tangential}} = 0 \qquad H_{\mathrm{normal}} = 0$$

[38] A complete discussion of the underlying issues and of phase, group, and signal velocities is given by L. Brillouin, *Wave Propagation and Group Velocity*, Academic Press, New York, 1960. Although the level of mathematical analysis is quite high, clear discussions of the underlying physics are provided. Also see Section 5.18 of J. A. Stratton, *Electromagnetic Theory*, McGraw-Hill, New York, 1941. For an excellent discussion at a simpler level, see the questions and answers sections of *Am. J. Phys.*, 64(11), p. 1353, November 1996 and 66(8), pp. 656–661, August 1998.

The basic method of solution assumes propagation in the z direction, so that all field components vary as $e^{-\overline{\gamma}z}$, where $\overline{\gamma}$ is equal to either $\overline{\alpha}$ or $j\overline{\beta}$. The z variations of real fields, then, have the form

$$\overline{\gamma} = \overline{\alpha} \qquad\qquad e^{-\overline{\alpha}z}\cos(\omega t) \quad \text{Evanescent wave}$$

$$\overline{\gamma} = j\overline{\beta} \qquad\qquad \cos(\omega t - \overline{\beta}z) \quad \text{Propagating wave}$$

The different type of possible solutions are classified as TE, TM, or TEM waves:

$$E_z = 0, \; H_z \neq 0 \qquad \text{Transverse electric (TE) waves}$$

$$E_z \neq 0, \; H_z = 0 \qquad \text{Transverse magnetic (TM) waves}$$

$$E_z, \; H_z = 0 \qquad \text{Transverse electromagnetic (TEM) waves}$$

Assuming that the wave-guiding structures extend in the z direction, the transverse field components can be derived from the axial (z direction) components, so that the z components act as a generating function for the transverse field components.

- **Parallel-plate waveguide.** The forms of the z components of the propagating TE_m and TM_m modes between parallel plates are

$$\text{TE}_m: \quad H_z = H_0 \cos\left(\frac{m\pi}{a}x\right)e^{-j\overline{\beta}_m z}$$

$$\text{TM}_m: \quad E_z = E_0 \sin\left(\frac{m\pi}{a}x\right)e^{-j\overline{\beta}_m z}$$

The cutoff frequency f_c and cutoff wavelength λ_c for the TE_m or TM_m modes are

$$f_{c_m} = \frac{m}{2a\sqrt{\mu\epsilon}} \qquad \lambda_{c_m} = \frac{2a}{m}$$

Note that f_{c_m} determines all the other quantities, such as $\overline{\beta}_m$, $\overline{\lambda}_m$, and $\overline{v}_{p_m}$, via (10.19) and (10.21). The TEM mode electric and magnetic field phasors for the parallel-plate waveguide are given by

$$\mathbf{E} = \hat{\mathbf{x}}E_0 e^{-j\beta z} \qquad \text{and} \qquad \mathbf{H} = \hat{\mathbf{y}}\frac{E_0}{\eta}e^{-j\beta z}$$

where $\beta = \omega\sqrt{\mu\epsilon}$. TEM modes do not exhibit any cutoff frequency and can in principle propagate at any frequency.

The attenuation rate due to conductor losses typically increases with frequency for TEM and TM_m modes, exhibiting a minimum as a function of frequency for TM_m modes. For TE_m modes, the attenuation rate due to conductor losses decreases with increasing frequency. For both TE_m and TM_m modes, the attenuation increases rapidly as the frequency approaches the cutoff frequency of the particular modes. The attenuation constants due to conduction losses for parallel-plate waveguide modes can be summarized as follows:

$$\text{TEM:}\quad \alpha_{c\text{TEM}} = \frac{1}{\eta a}\sqrt{\frac{\omega\mu}{2\sigma}}$$

$$\text{TM}_m:\quad \alpha_{c\text{TM}_m} = \frac{2R_s}{\eta a\sqrt{1 - (f_{c_m}/f)^2}}$$

$$\text{TE}_m:\quad \alpha_{c\text{TE}_m} = \frac{2R_s\,(f_{c_m}/f)^2}{\eta a\sqrt{1 - (f_{c_m}/f)^2}}$$

where f_{c_m} is the cutoff frequency for the particular mode. The attenuation coefficients for small dielectric losses in a parallel-plate waveguide are

$$\text{TEM:}\quad \alpha_d = \frac{\omega\epsilon''\sqrt{\mu/\epsilon'}}{2} \qquad \text{TM}_m \text{ or TE}_m:\quad \alpha_d \simeq \frac{\pi f\sqrt{\mu_0\epsilon'}\,\epsilon''/\epsilon'}{\sqrt{1 - (f_{c_m}/f)^2}}$$

- **Dielectric waveguides.** Electromagnetic waves guided by a dielectric slab of thickness d surrounded by free space must have electric and magnetic fields that decay exponentially (at a rate of α_x) with transverse distance outside the slab, while having variations of the form $\sin(\beta_x d/2)$ or $\cos(\beta_x d/2)$ inside the slab. The modes are classified as odd or even depending on whether the axial field component (E_z for TM and H_z for TE modes) varies in the form of $\sin(\beta_x d/2)$ or $\cos(\beta_x d/2)$, respectively. The determination of the values of α_x and β_x requires the simultaneous solution of the equation

$$\alpha_x = [\omega^2(\mu_d\epsilon_d - \mu_0\epsilon_0) - \beta_x^2]^{1/2}$$

together with one of the following equations, depending on the mode under consideration:

$$\text{Odd TM:}\quad \frac{\alpha_x}{\beta_x} = \frac{\epsilon_0}{\epsilon_d}\tan\left(\frac{\beta_x d}{2}\right) \qquad \text{Odd TE:}\quad \frac{\alpha_x}{\beta_x} = \frac{\mu_0}{\mu_d}\tan\left(\frac{\beta_x d}{2}\right)$$

$$\text{Even TM:}\quad \frac{\alpha_x}{\beta_x} = -\frac{\epsilon_0}{\epsilon_d}\cot\left(\frac{\beta_x d}{2}\right) \qquad \text{Even TE:}\quad \frac{\alpha_x}{\beta_x} = -\frac{\mu_0}{\mu_d}\cot\left(\frac{\beta_x d}{2}\right)$$

The propagation constant $\overline{\beta}$ is related to β_x as

$$\overline{\beta} = \sqrt{\omega^2\mu_d\epsilon_d - \beta_x^2}$$

and, as usual, determines the phase velocity through $\overline{v}_p = \omega/\overline{\beta}$ and wavelength via $\overline{\lambda} = 2\pi/\overline{\beta}$. The cutoff frequencies of the modes are given by

$$f_c = \frac{(m-1)}{2d\sqrt{\mu_d\epsilon_d - \mu_0\epsilon_0}} \qquad \begin{array}{l} m = 1, 3, 5, \ldots \\ m = 2, 4, 6, \ldots \end{array} \qquad \begin{array}{l} \text{Odd TM}_m \text{ or TE}_m \\ \text{Even TM}_m \text{ or TE}_m \end{array}$$

• **Group velocity.** When the propagation constant $\overline{\beta}$ is not linearly proportional to frequency, the phase velocity $\overline{v}_p$ is a function of frequency, and the different frequency components of a wave packet travel along a waveguide at different speeds. In such cases, and for electromagnetic wave packets consisting of frequency components closely clustered around a high-frequency carrier such that $\Delta\omega \ll \omega_0$, the signal envelope travels at the group velocity, which for the parallel-plate waveguide is always less than the phase velocity and is given by

$$v_g \equiv \frac{d\omega}{d\overline{\beta}} = v_p \sqrt{1 - \left(\frac{f_c}{f}\right)^2}$$

In the parallel-plate waveguide, the group velocity is also the velocity at which the electromagnetic energy travels along the waveguide.

PROBLEMS

10.1 **Parallel-plate waveguide modes.** A parallel-plate waveguide consists of two perfectly conducting infinite plates spaced 1 cm apart in air. Find the propagation constant $\overline{\gamma}$ for the TM_0, TE_1, TM_1, TE_2, and TM_2 modes at an operating frequency of (a) 5 GHz, (b) 10 GHz, and (c) 20 GHz.

10.2 **Parallel-plate waveguide modes.** A parallel-plate air waveguide has a plate separation of 6 mm and width of 10 cm. (a) List the cutoff frequencies of the seven lowest-order modes (TE_m and TM_m) that can propagate in this guide. (b) Find all the propagating modes (TE_m and TM_m) at 40 GHz. (c) Find all the propagating modes at 60 GHz. (d) Repeat part (c) if the waveguide is filled with polyethylene (assume it is lossless, with $\epsilon_r \simeq 2.25$, $\mu_r = 1$).

10.3 **Parallel-plate waveguide modes.** An air-filled parallel-plate waveguide with a plate separation of 2 cm is to be used to connect a 12-GHz microwave transmitter to an antenna. (a) Find all the propagating modes. (b) Repeat part (a) if the waveguide is filled with polyethylene (assume it is lossless, with $\epsilon_r \simeq 2.25$, $\mu_r = 1$).

10.4 **VLF propagation in the Earth–ionosphere waveguide.** The height of the terrestrial earth–ionosphere waveguide considered in Example 10.2 varies for VLF (3–30 kHz) from 70 km during the day to about 90 km during the night.[39] (a) Find the total number of propagating modes during the day at 12 kHz. (Assume both the ionosphere and the earth to be flat and perfect conductors.) (b) Repeat part (a) for during the night. (c) Find the total number of propagation modes during the day at 24 kHz. (d) Repeat part (c) for during the night. (e) Does the TM_{17} mode propagate during the day at 30 kHz? (f) Does the TM_{17} mode propagate during the night at 30 kHz?

10.5 **Waveguide in the earth's crust.** It has been proposed[40] that radio waves may propagate in a waveguide deep in the earth's crust, where the basement rock has very low conductivity and is sandwiched between the conductive layers near the surface and the high-temperature

---

[39]J. Galejs, *Terrestrial Propagation of Long Electromagnetic Waves,* Pergamon Press, New York, 1972.

[40]H. A. Wheeler, Radio-wave propagation in the Earth's crust, *J. Res. NBS,* 65D(2), pp. 189–191, March–April 1961. Also see H. Mott and A. W. Biggs, Very-low-frequency propagation below the bottom of the sea, *IEEE Trans. Antennas Propagation,* AP-11, pp. 323–329, May 1963.

conductive layer far below the surface. The upper boundary of this waveguide is on the order of 1 km to several kilometers below the earth's surface. The depth of the dielectric layer of the waveguide (the basement rock) can vary anywhere from 2 to 20 km, with a conductivity of 10^{-6} to 10^{-11} S-m^{-1} and a relative dielectric constant of ~6. This waveguide may be used for communication from a shore sending station to an underwater receiving station. Consider such a waveguide and assume it to be an ideal nonmagnetic parallel-plate waveguide. (a) If the depth of the dielectric layer of the guide is 2 km, find all the propagating modes of this guide below an operating frequency of 2 kHz. (b) Repeat part (a) (same depth) below 5 kHz. (c) Repeat parts (a) and (b) for a dielectric layer depth of 20 km.

10.6 **Single-mode propagation.** Consider a parallel-plate air waveguide with plate separation a. (a) Find the maximum plate separation a_{max} that results in single-mode propagation along the guide at 20 GHz. (b) Repeat part (a) for the waveguide filled with a dielectric with $\epsilon_r \simeq 2.54$, $\mu_r = 1$.

10.7 **Evanescent wave attenuator design.** A section of a parallel-plate air waveguide with a plate separation of 7.11 mm is constructed to be used at 15 GHz as an evanescent wave attenuator to attenuate all the modes except the TEM mode along the guide. Find the minimum length of this attenuator needed to attenuate each mode by at least 100 dB. Assume perfect conductor plates.

10.8 **Evanescent wave filter design.** Consider a parallel-plate air waveguide to be designed such that no other mode but the TEM mode will propagate along the guide at 6 GHz. If it is required by the design constraints that the lowest-order nonpropagating mode should face a minimum attenuation of 20 dB-(cm)$^{-1}$ along the guide, (a) find the maximum plate separation a_{max} of the guide needed to satisfy this criterion. (b) Using the maximum value of a found in part (a), find the dB-(cm)$^{-1}$ attenuation experienced by the TE$_2$ and the TM$_2$ modes.

10.9 **Cutoff and waveguide wavelengths.** Consider a parallel-plate waveguide in air with a plate separation of 3 cm to be used at 8 GHz. (a) Determine the cutoff wavelengths of the three lowest-order nonpropagating TM$_m$ modes. (b) Determine the guide wavelengths of all the propagating TM$_m$ modes.

10.10 **Guide wavelength of an unknown mode.** The waveguide wavelength of a propagating mode along an air-filled parallel-plate waveguide at 30 GHz is found to be $\bar{\lambda} = 5$ cm. (a) Find the cutoff frequency of this mode. (b) Recalculate the guide wavelength of this same mode for a waveguide filled with polyethylene (lossless, with $\epsilon_r \simeq 2.25$, $\mu_r = 1$).

10.11 **Guide wavelength, phase velocity, and wave impedance.** Consider a parallel-plate air waveguide having a plate separation of 10 cm. Find the following: (a) The cutoff frequencies of the TM$_0$, TM$_1$, and TM$_2$ modes. (b) The phase velocity $\bar{v}_{p_m}$, waveguide wavelength $\bar{\lambda}_m$ and wave impedance Z_{TM_m} of the above modes at 16 GHz. (c) The highest-order TM$_m$ mode that can propagate in this guide at 10 GHz.

10.12 **Parallel-plate waveguide design.** Design a parallel-plate air waveguide to operate at 5 GHz such that the cutoff frequency of the TE$_1$ mode is at least 25% less than 5 GHz, the cutoff frequency of the TE$_2$ mode is at least 25% greater than 5 GHz, and the power-carrying capability of the guide is maximized.

10.13 **TEM decomposition of TE$_m$ modes.** For an air-filled parallel-plate waveguide with 4 cm plate separation, find the oblique incidence angle θ_i and sketch TE$_1$, TE$_2$, TE$_3$, and TE$_4$ modes in terms of two TEM waves at an operating frequency of 20 GHz.

10.14 **Unknown waveguide mode.** The electric field of a particular mode in a parallel-plate air waveguide with a plate separation of 1 cm is given by

$$E_x(y, z) = 5e^{-60\pi y} \sin(100\pi z) \text{ kV-m}^{-1}$$

(a) What is this mode? Is it a propagating or a nonpropagating mode? (b) What is the operating frequency? (c) What is the similar highest-order mode (TE_m or TM_m) that can propagate in this waveguide?

10.15 Unknown waveguide mode. The magnetic field of a particular mode in a parallel-plate air waveguide with a plate separation of 2.5 cm is given by

$$H_z(x,y) = C_1 e^{-j640\pi x/3} \cos(160\pi y)$$

where x and y are both in meters. (a) What is this mode? Is it a propagating or a nonpropagating mode? (b) What is the operating frequency? (c) Find the corresponding electric field $\mathbf{E}(x,y)$. (d) Find the lowest-order similar mode (TE_m or TM_m) that does not propagate in this waveguide at the same operating frequency.

10.16 Maximum power capacity. (a) In Problem 10.15, determine the value of constant C_1 (assumed to be real) which maximizes the power carried by all the propagating modes without causing any dielectric breakdown (use 15 kV-m^{-1} for maximum allowable electric field in air, which is half of the breakdown electric field in air at sea level). (b) Using the value of C_1 found in part (a), find the maximum time-average power per unit width carried by the mode found in part (a) of Problem 10.15.

10.17 Power-handling capacity of a parallel-plate waveguide. A parallel-plate air waveguide with a plate separation of 1.5 cm is operated at a frequency of 15 GHz. Determine the maximum time-average power per unit guide width in units of kW-(cm)$^{-1}$ that can be carried by the TE_1 mode in this guide, using a breakdown strength of air of 15 kV (cm)$^{-1}$ (safety factor of approximately 2 to 1) at sea level.

10.18 Power capacity of a parallel-plate waveguide. Show that the maximum power-handling capability of a TM_m mode propagating in a parallel-plate waveguide without dielectric breakdown is determined only by the longitudinal component of the electric field for $f_{c_m} < f < \sqrt{2}f_{c_m}$ and by the transverse component of the electric field for $f > \sqrt{2}f_{c_m}$.

10.19 Power capacity of a parallel-plate waveguide. For a parallel-plate waveguide formed of two perfectly conducting plates separated by air at an operating frequency of $f = 1.5f_{c_m}$, find the maximum time-average power per unit area of the waveguide that can be carried without dielectric breakdown [use 15 kV-(cm)$^{-1}$ for maximum allowable electric field in air, which is half of the breakdown electric field in air at sea level] for the following modes: (a) TEM, (b) TE_1, and (c) TM_1.

10.20 Attenuation in a parallel-plate waveguide. Consider a parallel-plate waveguide with plate separation a having a lossless dielectric medium with properties ϵ and μ. (a) Find the frequency in terms of the cutoff frequency (i.e., find f/f_{c_m}) such that the attenuation constants α_c due to conductor losses of the TEM and the TE_m modes are equal. (b) Find the attenuation α_c for the TEM, TE_m, and TM_m modes at that frequency.

10.21 Attenuation in a parallel-plate waveguide. For a TM_m mode propagating in a parallel-plate waveguide, do the following: (a) Show that the attenuation constant α_c due to conductor losses for the propagating TM_m mode is given by

$$\alpha_c = \frac{2\omega\epsilon R_s}{\bar{\beta}a}$$

(b) Find the frequency in terms of f_{c_m} (i.e., find f/f_{c_m}) such that the attenuation constant α_c found in part (a) is minimum. (c) Find the minimum for α_c. (d) For an air-filled waveguide made of copper plates 2.5 cm apart, find α_c for TEM, TE_1, and TM_1 modes at the frequency found in part (b).

10.22 Parallel-plate waveguide: phase velocity, wavelength, and attenuation. A parallel-plate waveguide is formed by two parallel brass plates ($\sigma = 2.56 \times 10^7$ S-m^{-1}, $\epsilon_r = 1$, $\mu_r = 1$) separated by a 2-cm thick polyethylene slab ($\epsilon_r' \simeq 2.25$, $\tan \delta \simeq 4 \times 10^{-4}$, $\mu_r = 1$) to operate at a frequency of 20 GHz. For TEM, TE$_1$, and TM$_1$ modes, find (a) the phase velocity $\bar{v}_p$ and the waveguide wavelength $\bar{\lambda}$ and (b) the attenuation constants α_c due to conductor losses and α_d due to dielectric losses.

10.23 Losses in a parallel-plate waveguide. Consider TEM wave propagation in a parallel-plate waveguide. Although ideally the only nonzero wave components for this mode are E_x and H_y (as given by equation (10.17)), it is clear that there should be a small E_z component due to the finite ($\mathbf{J}_s = \hat{\mathbf{n}} \times \mathbf{H}$) current that flows in the conductors, as mentioned in Footnote 21 on page 856. Since J_z is equal and oppositely directed at the top ($x = a$) and the bottom ($x = 0$) conductors, we expect E_z to also have the same magnitude and the opposite sign at $x = a$ and $x = 0$. Thus, it is reasonable to assume a linear variation between the two values of E_z as:

$$E_z(x, z) = K \left(1 - \frac{2x}{a}\right) e^{-j\bar{\beta}z}$$

where K is a constant. Since we must have $\nabla \cdot \mathbf{E} = 0$ within the waveguide, we expect that the fields E_x and H_y may have to be modified to be consistent with this nonzero E_z. Find the modified expressions for E_x and H_y and determine all components of the time-average Poynting flux $\mathbf{S}_{av}$ at $x = a/4$ inside the waveguide.

10.24 Semi-infinite parallel-plate waveguide. Two perfectly conducting and infinitesimally thin sheets in air form a semi-infinite parallel-plate waveguide, with mouth in the plane $z = 0$ and sides parallel to the y-z plane as shown in Figure 10.30. Two perpendicularly polarized (i.e., electric field in the y direction) uniform plane waves 1 and 2 of equal strength are incident upon the mouth of the guide at angles θ_i as shown. The two waves are in phase, so their separate surfaces of constant phase intersect in lines lying in the y-z plane. The peak electric field strength for each wave is known to be 1 V-m^{-1}. (a) Find an expression (in terms of θ_i) for the time-average power flowing down the inside of the guide, per unit meter of the guide in the y direction. (b) If the wavelength λ of the incident waves is such that $\sin \theta_i = \lambda/(2a) = \sqrt{3}/2$ and $a = 1$ cm, find the numerical value of the time-average power transmitted, per unit guide width in the y direction.

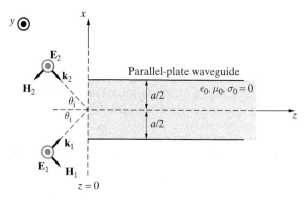

Figure 10.30 Semi-infinite parallel-plate waveguide. Problem 10.24.

10.25 Reflection and transmission in a parallel plate waveguide. The region $z > 0$ in a parallel-plate waveguide is filled with nonmagnetic dielectric material with permittivity $\epsilon_2 = 3\epsilon_0$, as shown in Figure 10.31. (a) Assuming that the TM$_1$ wave is incident from the left with an incident magnetic field intensity given by

$$\mathbf{H}_i(x,z) = \hat{y}H_{i0}\cos\left(\frac{\pi x}{a}\right)e^{-j\bar{\beta}_1 z}$$

determine the magnitude of the magnetic fields of the reflected and transmitted TM_1 waves.
(b) A fellow engineer claims that the TM_1 wave can be completely transmitted (i.e., without any reflection) across the interface, as long as ϵ_2 has a specific value. Comment on whether the engineer is correct and if so, determine the value of ϵ_2 for which there is no reflection.

Figure 10.31 Parallel-plate waveguide with a dielectric. Problem 10.25.

10.26 Propagating modes in a dielectric slab waveguide. Consider a dielectric slab waveguide with thickness d and refractive indices of 1.5 (for the guide) and 1.48 (for the cladding). (a) Find all the propagating modes at air wavelength $\lambda = 2\,\mu$m if $d = 5\,\mu$m. (b) Repeat part (a) if $d = 15\,\mu$m. (c) Repeat part (a) if the refractive index of the cladding is 1.49.

10.27 Single TE_1 mode dielectric slab design. Consider a dielectric slab waveguide with guide thickness d and refractive index 1.55 sandwiched between two cladding regions each with refractive index 1.5. (a) Design the guide such that only the TE_1 mode is guided at air wavelength $\lambda = 2\,\mu$m. (b) Repeat part (a) for a cladding region with a refractive index of 1.53.

10.28 Millimeter-wave dielectric slab waveguide. Consider a dielectric slab waveguide with $\epsilon_d = 4\epsilon_0$ and $\mu_d = \mu_0$ surrounded by air to be used to guide millimeter waves. Find the guide thickness d such that only the TE_1 mode can propagate at frequencies up to 300 GHz.

10.29 TE_2 mode in a dielectric slab waveguide. Find the cutoff frequency for the TE_2 mode in a dielectric slab waveguide with $\epsilon = 7\epsilon_0$ and thickness 4 cm, which is embedded in another dielectric with $\epsilon = 3\epsilon_0$. Assume nonmagnetic case.

10.30 Dielectric slab waveguide modes. A dielectric slab waveguide in air is used to guide electromagnetic energy along its axis. Assume that the slab is 2 cm in thickness, with $\epsilon = 5\epsilon_0$ and $\mu = \mu_0$. (a) Find all of the propagating modes for an operating frequency of 4 GHz and specify their cutoff frequencies. (b) Find α_x (in np-m^{-1}) and β_x (in rad-m^{-1}) at 8 GHz for each of the propagating modes. (c) Considering a ray theory analysis of the propagating modes, find the incidence angles θ_i of the component TEM waves within the slab for all of the propagating modes at 8 GHz.

10.31 Dielectric slab waveguide modes. Consider a dielectric slab waveguide surrounded by air made of a dielectric core material of $d = 10\,\mu$m thickness with a refractive index of $n_d = 1.5$ covered with a cladding material of refractive index $n_c = 1.45$, which is assumed to be of infinite extent. (a) Find all the propagating modes at 1 μm air wavelength. (b) Determine the shortest wavelength allowed in the single-mode transmission.

10.32 Dielectric waveguide. Consider a nonmagnetic dielectric slab waveguide consisting of a slab with relative permittivity $\epsilon_{1r} = 2.19$ surrounded by another dielectric with relative permittivity $\epsilon_{2r} = 2.13$. The frequency of operation is 50 GHz. Determine the thickness of the dielectric slab if the lowest order TE mode propagates at a ray angle of $\theta_i = 85°$.

10.33 Dielectric slab waveguide thickness. A dielectric slab waveguide is made of a dielectric core and cladding materials with refractive indices of $n_d = 1.5$ and $n_c = 1$, respectively.

If the number of propagating modes is 100 at a free-space wavelength of 500 nm, calculate the thickness of the core material.

10.34 **TEM decomposition of TM$_m$ modes.** A dielectric slab waveguide is designed using a core dielectric material of refractive index $n_d = 3$ and thickness 5 μm covered by a cladding dielectric material of refractive index $n_c = 2.5$. Find all the propagating TM$_m$ modes and corresponding angles of incidence with which they are bouncing back and forth between the two boundaries at 3 μm air wavelength.

10.35 **Dielectric above a ground plane.** A planar perfect conductor of infinite dimensions is coated with a dielectric material ($\epsilon_r = 5$, $\mu_r = 1$) of thickness 5.625 cm. (a) Find the cutoff frequencies of the first four TE and/or TM modes, and specify whether they are odd or even. (b) For an operating frequency of 1 GHz, find all of the propagating TE modes. (c) For each of the TE modes found in part (b), find the corresponding propagation constant $\overline{\beta}$. Assume the medium above the coating to be free space.

10.36 **Group velocity.** Derive the expression (equation (10.73)) for the group velocity for TE$_m$ or TM$_m$ modes in a parallel-plate waveguide.

10.37 **Group velocity.** Consider the propagation in seawater ($\sigma = 4$ S/m, $\epsilon = 81\epsilon_0$, $\mu = \mu_0$) of a uniform plane wave signal consisting of the superposition two frequency components at 19 and 21 kHz. (a) Stating all assumptions, determine the phase and group velocities at 20 kHz. (b) Assuming that the two signals are in phase at position $z_0 = 0$ and at time $t_0 = 0$, determine the minimum distance z_1 (and the time t_1) at which the two signals are once again in phase, that is, have a phase difference of a multiple of 2π. Is the group velocity v_g at 20 kHz equal to z_1/t_1? If not, why not?

10.38 **Dielectric waveguide.** Consider the TE$_m$ mode in a nonmagnetic dielectric slab waveguide consisting of a slab with thickness d and permittivity ϵ_d surrounded by air. (a) Show that the dispersion relation (i.e., $\overline{\beta}$-ω relation) is given by

$$\tan^2\left[\left(\frac{d}{2}\right)\left(\omega^2\mu_0\epsilon_d - \overline{\beta}^2\right)^{\frac{1}{2}} - \frac{(m-1)\pi}{2}\right] = \frac{\overline{\beta}^2 - \omega^2\mu_0\epsilon_0}{\omega^2\mu_0\epsilon_d - \overline{\beta}^2}$$

(b) For $\epsilon_d = 4\epsilon_0$ and for the TE$_1$ mode, find the value of d such that the waveguide phase velocity $\overline{v}_p = \omega/\overline{\beta}$ is equal to the geometric mean of $c = (\mu_0\epsilon_0)^{-1/2}$ and $v_{pd} = (\mu_d\epsilon_d)^{-1/2}$.

10.39 **Dispersion in seawater.** Reconsider Problem 8.26 in view of your knowledge of group velocity. Assuming that the frequency dependence of the phase velocity is as given in Problem 8.26, derive an expression for the group velocity in seawater, and plot both v_p and v_g as a function of frequency between 0.5 and 2.5 kHz.

10.40 **Group velocity in a plasma.** A cold ionized gas consisting of equal numbers of electrons and protons behaves (see Section 11.1) as a medium with an effective permittivity $\epsilon_{\text{eff}} = \epsilon_0(1 - \omega_p^2/\omega^2)$, where $\omega_p = \sqrt{Nq_e^2/(m_e\epsilon_0)}$ is known as the *plasma frequency*, with N being the volume density of free electrons, $q_e \simeq -1.6 \times 10^{-19}$ C the electronic charge, and $m_e \simeq 9.11 \times 10^{-31}$ kg the electronic mass. (a) Derive the expression for the group velocity v_g in this medium. (b) Evaluate the group velocity (and express it as a fraction of the speed of light in free space) for a 1 MHz radio signal propagating through the earth's ionosphere, where $N \simeq 10^{11}$ m^{-3}. (c) Repeat part (b) for 100 kHz.

10.41 **Group velocity in a dielectric slab waveguide.** Derive an expression for the group velocity v_g for the odd TM$_m$ modes in a dielectric slab waveguide of slab thickness d and permittivity ϵ_d.

11

Field–Matter Interactions
and Metamaterials

In Chapter 8 we considered electromagnetic wave propagation in an *unbounded, simple,* and *source-free* medium. In Chapters 9 and 10, we essentially removed the first of these three conditions, considering how waves reflect and refract in the presence of boundaries, and how they can be guided by conducting or dielectric structures. In particular, we saw in Chapter 10 that the presence of boundaries necessitates that the wave phase velocity be a function of frequency, leading to waveguide dispersion. In this chapter, we revert to considering the case of an *unbounded* medium (i.e., no boundaries) but one that is not *simple,* and illustrate that material properties alone can also lead to dispersion effects.

Electromagnetic fields interact with matter via their influence on *charged particles* (electrons and ions), whether these particles are free to move about (e.g., electrons in a conductor or electrons and ions in a low-density gas) or are tightly bound (e.g., atoms or molecules of a dielectric). In Section 11.1, we consider the interaction between *freely mobile charged particles* and electromagnetic fields. To illustrate the basic physical principles in the context of an interesting and important example, we discuss the case of wave propagation in a relatively dilute ionized gas in which particles can move freely. It should be noted, however, that the dynamics of charged particles studied in Section 11.1 are also equally applicable to current carriers in solids, such as electrons and holes in semiconductors. We shall see that the behavior of the freely mobile particles under the influence of the wave fields can be represented in terms of an effective permittivity ϵ_{eff}, which is in general a function of frequency, so an ionized gas is a good example of a medium that exhibits *material dispersion.*

Interactions of electromagnetic waves with solid, liquid, or densely gaseous media are in general more difficult to treat on the microscopic level (i.e., by describing the motions of individual particles), because material properties play a significant role. At one level, we have already accounted for the interaction of such media with electromagnetic fields by describing their behavior in terms of *macroscopic* parameters ϵ, σ, and μ. These macroscopic parameters represent the response of the free or bound electrons within the material to the fields on a *microscopic* scale. To the degree that these parameters can be considered to be simple (real or complex) constants, the formulations used in previous chapters are adequate. However, in many important applications, the microscopic response of the particles to the fields becomes complicated, so the macroscopic parameters ϵ, μ, and σ need to be formulated as functions of frequency, intensity, and orientation of the electromagnetic fields. Although a general study of such effects are well beyond the scope of this book, we devote Section 11.2 to the discussion of material dispersion in dielectrics and conductors (for which ϵ and σ depend on frequency).

It is also possible to engineer a material to have a desired set of electromagnetic properties. While Sections 11.1 and 11.2 deal with naturally occurring media, in Section 11.3 we consider wave propagation in an artificial medium that has been designed to have simultaneously negative permittivity and permeability. The most striking property of these novel materials is so-called backward wave propagation, where the phase fronts travel in the opposite direction of the energy flow.

In Sections 11.1 and 11.2, we derive effective relative permittivity expressions by considering the interaction of the electric field in a propagating electromagnetic wave with the charges in a medium. Our starting point in this discussion is the Lorentz force equation (see (6.51)) describing the electrical and magnetic forces acting on a charged particle of rest mass m_0 and charge q:

$$\overline{\mathfrak{F}}_{\text{elec}} + \overline{\mathfrak{F}}_{\text{mag}} = q\overline{\mathscr{E}} + q\tilde{\mathbf{v}} \times \overline{\mathscr{B}} = q[\overline{\mathscr{E}} + \tilde{\mathbf{v}} \times \overline{\mathscr{B}}] \tag{11.1}$$

where $\overline{\mathfrak{F}}_{\text{elec}}$ and $\overline{\mathfrak{F}}_{\text{mag}}$ are, respectively, the electric and magnetic Lorentz forces. For freely mobile elementary particles such as an electron, proton, or even heavier ions (e.g., O^+), the electromagnetic forces on these particles are generally much stronger than other forces (e.g., gravity) for reasonable field intensities (e.g., an electric field of >0.1 nV-m^{-1} for electrons). Thus, the acceleration of the free charge is dominated by the electromagnetic force, so its motion is described by Newton's second law as

$$\overline{\mathfrak{F}} = \frac{d(m\tilde{\mathbf{v}})}{dt} = m\frac{d^2\mathbf{r}}{dt^2}$$

where $\tilde{\mathbf{v}}$ and $\mathbf{r}$ are the velocity and position vectors of the charged particle, respectively, and $m = m_0(1 - \tilde{v}^2/c^2)^{-1/2}$ is the mass of the charged particle to allow for a relativistic correction, with m_0 being the rest mass and $\tilde{v} = |\tilde{\mathbf{v}}|$. The notation $\tilde{\mathbf{v}}$ is used for the velocity to distinguish it from the time-harmonic velocity $\mathbf{v}$, although we do not always make such a distinction for position variables $\mathbf{r}$, x, y, or z.

In this chapter, we exclusively consider[1] nonrelativistic motion ($\tilde{v} \ll c$), so $m \simeq m_0$ is a constant, and we thus have

$$q[\overline{\mathscr{E}} + \tilde{\mathbf{v}} \times \overline{\mathscr{B}}] = m_0\frac{d\tilde{\mathbf{v}}}{dt} = m_0\frac{d^2\mathbf{r}}{dt^2} \tag{11.2}$$

---

[1] By the same token, we neglect any radiation produced by the acceleration of charged particles. At nonrelativistic velocities, such radiation is quite negligible; the radiated electric field at a distance R from the particle is proportional to $q^2a^2/(c^2R^2)$, where q is the charge of the particle, c is the speed of light in free space, and a is the acceleration. For a discussion at an appropriate level, see Chapter 7 of J. B. Marion, *Classical Electromagnetic Radiation,* Academic Press, New York, 1965.

as the equation of motion for a charged particle under the influence of external electric and magnetic fields. Accordingly, we drop the subscript distinguishing mass from rest mass and simply denote the particle mass as m.

Equation (11.2) is applicable in a wide variety of situations. In the simplest case, $\overline{\mathscr{E}}$ and $\overline{\mathscr{B}}$ are externally applied, and the number of charged particles is small enough that their collective motion does not alter the fields (by, e.g., creating currents $\overline{\mathscr{J}}$, which in turn generate magnetic fields through $\nabla \times \overline{\mathscr{H}} = \overline{\mathscr{J}}$, which in turn generate electric fields and further modify the particle motion). In such a case, $\widetilde{\mathscr{E}}$ is a specified function and $\tilde{\mathbf{v}}(t)$ and $\mathbf{r}(t)$ can be obtained by successive integration of (11.2). In general, when the charged particle density is sufficiently high, the moving charges constitute electric current, which in turn influences the electromagnetic field, so the equations of motion must be solved simultaneously with Maxwell's equations. Proper treatment of the electromagnetic wave-charged particle interaction is the subject of *plasma physics* and is well beyond the scope of this book. Nevertheless, we provide an introduction to the fundamental principles involved, in the context of the propagation of electromagnetic waves in an ionized medium discussed in Section 11.1.

11.1 WAVE PROPAGATION IN IONIZED GASES (PLASMAS)

In our everyday environment we observe matter in solid, liquid, or gaseous form. However, these three states of matter, which are found on the surfaces of Earth and some other planets, are not typical in the universe at large. More than 99% of the matter in the universe exists as the "fourth state of matter," a *plasma*. Plasmas are a special class of gases that include a large number of electrons and ionized atoms and molecules as well as the neutral atoms and molecules that are present in a normal gas. Although one might think of an ionized collection of positive and negatively charged particles as a gas, the bulk behavior of plasmas more closely resembles that of electrically conducting solids (metals) and can often be effectively described using formulations developed for fluids. The most important distinction between a plasma and a normal gas is the fact that mutual Coulomb interactions between charged particles are important in the dynamics of a plasma and cannot be disregarded. When a neutral gas is raised to a sufficiently high temperature, or when it is subjected to electric fields of sufficient intensity, the atoms and molecules of the gas may become ionized as electrons are stripped off by collisions as a result of the heightened thermal agitation of the particles. Ionization in gases can also be produced by illumination with ultraviolet light or X rays, by bombarding the substance with energetic electrons and ions, and in other ways. When a gas is ionized, even to a rather small degree, its dynamical behavior is typically dominated by the electromagnetic forces acting on the free ions and electrons, and it begins to conduct electricity. The charged particles in such an ionized gas interact with electromagnetic fields, and the organized motion of these charge carriers (i.e., electric currents, fluctuations in charge density) can in turn produce electromagnetic fields.

During the 1920s, I. Langmuir and colleagues first showed that characteristic electrical oscillations of very high frequency can exist in an ionized gas that is neutral or quasi-neutral, and they introduced the terms *plasma*[2] and *plasma oscillations* in recognition of the fact that these oscillations resembled those of jellylike substances. When subjected to a static electric field, the charge carriers in an ionized gas rapidly redistribute themselves in such a way that most of the gas is shielded from the field, in a manner quite similar to the redistribution of charge that occurs within a metallic conductor placed in an electric field, resulting in zero electric field everywhere inside.

The plasma medium is often referred to as the fourth state of matter because it has properties profoundly different from those of the gaseous, liquid, and solid states. All states of matter represent different degrees of organization, corresponding to certain values of binding energy. In the solid state, the important quantity is the binding energy of molecules in a crystal. If the average kinetic energy of a molecule exceeds the binding energy (typically a small fraction of an electron volt), the crystal structure breaks up, either into a liquid or directly into a gas (e.g., iodine). Similarly, a certain minimum kinetic energy is required in order to break the bonds of the van der Waals forces to change a liquid into a gas. For matter to pass transition into its fourth state and exist as a plasma, the kinetic energy per plasma particle must exceed the ionizing potential of atoms (typically a few electron volts). Thus, the state of matter is basically determined by the average kinetic energy per particle. Using water as a convenient example, we note that at low temperatures the bond between the H_2O molecules holds them tightly together against the low energy of molecular motion, so the matter is in the solid state (ice). At room temperature, the increased molecular energy permits more widespread movements and currents of molecular motion, and we have the liquid state (water). Since the particle motions are random, not all particles have the same energy, and the more energetic ones escape from the liquid surface to form a vapor above it. As the temperature of the water is further increased, a larger fraction of molecules escapes, until the whole substance is in the gaseous phase (steam). If steam is subjected to further heating, illumination by ultraviolet or X rays, or bombardment by energetic particles, it becomes ionized (plasma).

Although we live in a bubble of essentially un-ionized gas in the midst of an otherwise ionized environment, examples of partially ionized gases or plasmas have long been part of our natural environment, including flaming fire, aurora borealis, and lightning. The immediate environment of our planet Earth, including the radiation belts and the vast upper atmospheric regions referred to as the *ionosphere* and the *magnetosphere,* are also in a plasma state.[3] Early natural philosophers held that the material Universe is built of four "roots": earth, water, air, and fire, curiously resembling our modern terminology of solid, liquid, gas, and plasma states of matter. A transient plasma exists in Earth's atmosphere every time a lightning stroke occurs, but it is clearly not very much at home

[2]I. Langmuir, *Phys. Rev.,* 33, p. 954, 1929; L. Tonks and I. Langmuir, *Phys. Rev.,* 33, p. 195, 990, 1929. The word *plasma* first appeared as a scientific term in 1839, when the Czech biologist J. Purkynie coined the term *protoplasma* to describe the jellylike medium, containing a large number of floating particles, that makes up the interior of the cells. The word *plasma* thus means a mold or form and is also used for the liquid part of blood in which corpuscles are suspended.

[3]J. K. Hargreaves, *The Solar-Terrestrial Environment,* Cambridge University Press, 1992. Also see M. C. Kelley, *The Earth's Ionosphere,* Academic Press, Cambridge, 1989.

and is short-lived. Early work on electrical discharges included generation of electric sparks by rubbing a large rotating sphere of sulfur against a cloth [O. von Guericke, 1672], production of sparks by harnessing atmospheric electricity in rather hazardous experiments [B. Franklin, 1751], and studies of dust patterns left by a spark discharge passing through a surface of an insulator [G. C. Lichtenberg, 1777]. However, only when electrical and vacuum techniques were developed to the point that long-lived and relatively stable electrical discharges were available did the physics of ionized gases emerge as a field of study. In 1879, W. Crookes published the results of his investigations of discharges at low pressure and remarked: "The phenomena in these exhausted tubes reveal to physical science a new world, a world where matter may exist in a fourth state..." A rich period of discoveries followed, leading to Langmuir's coining the word *plasma* in 1929, and continuing on as a most fascinating branch of physics until now.

Although plasma is often considered to be the fourth state of matter, it has many properties in common with the gaseous state. However, at the same time, the plasma is an ionized gas in which the long range of Coulomb forces gives rise to collective interaction effects, resembling a fluid with a density higher than that of a gas. In its most general sense, a plasma is any state of matter that contains enough number of free charged particles for its dynamical behavior to be dominated by electromagnetic forces.[4] Most applications of plasma physics are concerned with ionized gases. It turns out that a very low degree of ionization is sufficient for a gas to exhibit electromagnetic properties and behave as a plasma; a gas achieves an electrical conductivity of about half of its possible maximum at about 0.1% ionization and has a conductivity nearly equal to that of a fully ionized gas at 1% ionization. The degree of ionization of a gas can be defined as the ratio $N_e/(N_e + N_0)$, where N_e is the free electron density and N_0 is the density of neutral molecules. As an example, the degree of ionization in a fluorescent tube is $\sim 10^{-5}$, with $N_0 \simeq 10^{16}$ (cm)$^{-3}$ and $N_e \simeq 10^{11}$ (cm)$^{-3}$. Typically, a gas is considered to be a weakly (strongly) ionized gas if the degree of ionization is less than (greater than) 10^{-4}.

The sun and the stars are hot enough to be almost completely ionized with enormous densities ($N_e \simeq 10^{27}$ (cm)$^{-3}$), and the interstellar gas is sparse enough to be almost completely ionized by stellar radiation. Starting at about 60 km altitude, the sun bathes our atmosphere in a variety of radiations, and the energy in the ultraviolet part of the spectrum is absorbed by the atmospheric gas. In the process, a significant number of air molecules and atoms receive enough energy to become ionized. The resulting free electrons and positive ions constitute the *ionosphere*. Maximum ionization density occurs in the ionosphere at about 350 km altitude, where $N_e = 10^6$ (cm)$^{-3}$. The atmospheric density at 350 km altitude is $N_0 \simeq 3.3 \times 10^8$ (cm)$^{-3}$, so the degree of ionization is $\sim 3 \times 10^{-3}$. At even higher altitudes, air is thin enough (i.e., density of air is low enough) that it is almost completely ionized, and the motions of charged particles are dominated by the earth's magnetic field in a region known as the *magnetosphere.*

---

[4]Plasma physics therefore encompasses the solid state, since electrons in metals and semiconductors fall into this category. However, the redistribution of charge and the screening of the inner regions of a metal occurs extremely quickly (typically $\sim 10^{-19}$ s), in view of the very high density of free charges.

In the last few decades, an important objective of plasma physics research has been to reproduce thermonuclear fusion under controlled conditions on earth, for possible use as an environmentally "clean" source of energy.[5] In terms of practical engineering applications, plasma etching is now widely used in integrated circuit manufacturing and other materials processing.[6]

11.1.1 Plasma Oscillations and the Electron Plasma Frequency

The most fundamental and elementary collective behavior of plasmas consists of the so-called *plasma oscillations.* If a plasma is disturbed, powerful electrical restoring forces are set up, leading to oscillatory motion of the particles around their equilibrium positions at a characteristic frequency, referred to as the plasma frequency. To describe this behavior quantitatively, we consider a plasma that consists of an equal number of electrons, each with charge $q_e \simeq -1.6 \times 10^{-19}$ C, and ions (atomic nuclei) with charge $q_i = -q_e$. We assume that these particles are initially uniformly distributed, so that the plasma is electrically neutral over any macroscopic spatial scale. We further assume that the plasma is "cold"—that is, the thermal motion of the electrons and ions is negligible, so they are initially at rest.

We now perturb this system by transferring a group of electrons (assumed for simplicity to be in a one-dimensional slab) from a given region of space to a nearby region, leaving net positive charge behind (i.e., the ions), as shown in Figure 11.1. This local charge separation gives rise to an electric field $\overline{\mathscr{E}}$, which exerts a force on the electrons and ions. Since the electrons are much lighter[7] than the ions, they respond more rapidly to the electric field $\overline{\mathscr{E}}$, so the motion of the ions can be neglected. The electric field $\overline{\mathscr{E}}$ acts to reduce the charge separation by pulling the electrons back to their initial locations. The electrons are thus accelerated by the $\overline{\mathscr{E}}$ field toward their initial positions. However, as they acquire kinetic energy in this process, their inertia carries them past their neutral positions. The plasma once again becomes nonneutral, and again an electric field is set up (now pointing in the direction opposite to that shown in Figure 11.1) to retard their motion. Now the electrons accelerate toward the right and go past their equilibrium positions due to their inertia, and once again the charge displacement depicted in Figure 11.1 is set up. In the absence of any damping (e.g, due to collisions of the electrons with ions or other electrons), this oscillatory motion continues forever. In relatively tenuous plasmas, collisional damping can be neglected, so any slight disturbance of the system leads to the oscillatory process just described.

We now consider the frequency of this oscillation, or the period of time in which the electrons move from one extreme to the other and back. Intuitively, we expect that the restoring electric field force depends on the amount of the charge displaced, and hence

---

[5]See Chapter 9 of F. F. Chen, *Introduction to Plasma Physics,* 1st ed., Plenum Press, New York, 1977.

[6]M. A. Lieberman and A. J. Lichtenberg, *Principles of Plasma Discharges and Materials Processing,* John Wiley & Sons., Inc., New York, 1995.

[7]Note that even if the ions consisted simply of the hydrogen ions (H^+), or protons, the mass of the proton is $m_{H^+} \simeq 1831 m_e$.

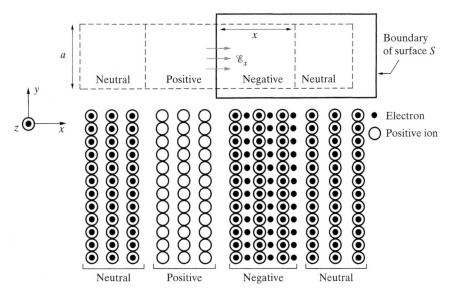

Figure 11.1 A one-dimensional plasma slab. The positively charged region consists of atoms which lost one electron, while the negatively charged region has excess electrons.

the electronic charge q_e and the density or number of electrons per unit volume. Since the inertia the particles exhibit depends on their mass, the oscillation frequency should also depend on the electron mass m_e. The frequency with which displaced electrons oscillate in a plasma is known as the *plasma frequency* and is denoted ω_p. We can determine ω_p using the equation of motion (11.2) for a single electron in the presence of only an electric field $\overline{\mathscr{E}} = \hat{\mathbf{x}}\mathscr{E}_x$ given by

$$m_e \frac{d^2x}{dt^2} = q_e \mathscr{E}_x \qquad (11.3)$$

where x is the direction parallel to the electric field as shown in Figure 11.1. Consider Gauss's law, applied to a closed rectangular box-shaped surface as indicated in Figure 11.1, noting that only the boundary of the surface in the x-y plane is shown. We have

$$\oint_S \overline{\mathscr{E}} \cdot d\mathbf{s} = \frac{Q}{\epsilon_0}$$

where Q is the total charge contained within the closed surface S. If the equilibrium density of electrons is N_e, we must have $Q = AxN_eq_e$, where A is the cross-sectional area, and x denotes the displacement of the electrons as shown in Figure 11.1. Assuming that a is the dimension of the rectangular-box surface in the y direction and that the depth of the box is Δz, we have

$$\oint_S \overline{\mathscr{E}} \cdot d\mathbf{s} = -a\Delta z \mathscr{E}_x$$

$$= \frac{Q}{\epsilon_0} = \frac{ax\Delta z N_e q_e}{\epsilon_0} \quad \rightarrow \quad \mathscr{E}_x = -\frac{xN_e q_e}{\epsilon_0}$$

Substituting back into the equation of motion (11.3), we have

$$m_e \frac{d^2x}{dt^2} + \frac{N_e q_e^2}{\epsilon_0}x = 0 \quad \rightarrow \quad \frac{d^2x}{dt^2} + \omega_p^2 x = 0 \tag{11.4}$$

where

$$\omega_p \equiv \sqrt{N_e q_e^2/(m_e \epsilon_0)} \tag{11.5}$$

is the oscillation frequency (the plasma frequency).

Equation (11.4) is a second-order differential equation that we have encountered many times before. Its general solution is

$$x = C_1 \sin(\omega_p t) + C_2 \cos(\omega_p t) \tag{11.6}$$

where C_1 and C_2 are to be determined by initial conditions. Equation (11.6) describes the displacement for free oscillations of the plasma slab. This is the natural frequency of oscillation of the plasma. Oscillations at other frequencies can only occur if the plasma slab is driven by an external field.

As expected, the oscillation frequency depends on the total amount of displaced electrons, having a total charge of $N_e q_e$, which determines the magnitude of the restoring electric field. The dependence on mass m_e is also expected, since if the electron had no mass, the electrostatic energy could not be transformed into kinetic energy, and the particle would have zero inertia. The fact that the mass of the electrons is small accounts for their rapid response to the electric field. In typical laboratory plasmas, the value of ω_p is in the microwave range. For example, for a plasma with an electron density of $N_e = 10^{12}$ cm^{-3}, $f_p = \omega_p/(2\pi) \simeq 9$ GHz. In the earth's ionosphere, where the typical maximum density of free electrons is $\sim 10^{10}$ to $\sim 10^{12}$ m^{-3}, the corresponding plasma frequency f_p is $\sim$1–10 MHz. Note that since q_e, m_e, and ϵ_0 are fundamental constants, the plasma frequency is essentially a direct measure of the density of electrons N_e. An often useful approximate expression for the plasma frequency is $f_p = \omega_p/(2\pi) \simeq 9\sqrt{N_e}$.

11.1.2 Electromagnetic Wave Propagation in a Plasma

To describe the propagation of electromagnetic waves in a plasma medium, we use Maxwell's equations together with the equations of motion and determine which oscillations are possible, assuming, as usual, sinusoidal steady-state or time-harmonic solutions with $e^{j\omega t}$ type of time dependence. We continue to assume the plasma to be made up solely of electrons and ions, with the ions being relatively immobile (because of their heavier mass). The motion of the electrons under the influence of the wave electric and magnetic fields constitute a current, which must be accounted for in Maxwell's equations via the $\overline{\mathscr{J}}$ term, and their displacement creates localized regions of finite charge, represented by the volume charge density term $\tilde{\rho}$. In general, we have

$$\overline{\mathscr{J}} = \mathscr{N}_e q_e \tilde{\mathbf{v}} \qquad \text{and} \qquad \tilde{\rho} = \mathscr{N}_e q_e$$

where $\mathscr{N}_e$ is the total density of electrons given by

$$\mathscr{N}_e = N_e + \mathscr{R}e\{n_e e^{j\omega t}\}$$

with N_e being the steady or ambient density of electrons, and n_e being the "phasor" electron density representing the time-harmonic variations. The electron velocity $\tilde{\mathbf{v}}$ is assumed to vary harmonically with no steady (or drift) components so that we have

$$\tilde{\mathbf{v}} = \mathscr{R}e\{\mathbf{v}e^{j\omega t}\}$$

where $\mathbf{v}$ is the velocity phasor. Note that the quantities n_e, $\mathbf{J}$, and $\mathbf{v}$ are phasors, while $\tilde{\mathbf{v}}$, $\overline{\mathscr{J}}$, $\tilde{\rho}$, and $\mathscr{N}_e$ are time-varying quantities. Substituting in the expression for the current density, we find

$$\overline{\mathscr{J}} = [N_e + \mathscr{R}e\{n_e e^{j\omega t}\}]q_e[\mathscr{R}e\{\mathbf{v}e^{j\omega t}\}]$$

$$= N_e q_e \mathscr{R}e\{\mathbf{v}e^{j\omega t}\} + q_e[\mathscr{R}e\{n_e e^{j\omega t}\}][\mathscr{R}e\{\mathbf{v}e^{j\omega t}\}]$$

Here, we make the assumption that we consider the case of small oscillations, so $|n_e| \ll N_e$. As a result, the second term above can then be neglected compared to the first, and we have

$$\overline{\mathscr{J}} \simeq N_e q_e \mathscr{R}e\{\mathbf{v}e^{j\omega t}\} \tag{11.7}$$

Since $\overline{\mathscr{J}} = \mathscr{R}e\{\mathbf{J}e^{j\omega t}\}$, in phasor form we have

$$\boxed{\mathbf{J} \simeq N_e q_e \mathbf{v}} \tag{11.8}$$

This is the current density resulting from the motions of the electrons, which we now have to consider in Maxwell's equations. Note that similar relations can also be written for the current due to the motion of the ions. However, the current density term due to the ions is generally small and negligible because they are heavy and relatively immobile. It can be shown that the charge density term due to ions cancels out the steady part of the charge density (N_e) due to the electrons, so that there are no steady (dc) charge density or currents and electric fields. In other words, $\tilde{\rho} \simeq \mathscr{R}e\{n_e e^{j\omega t}\}$, or $\rho \simeq n_e$.

We can now write the time-harmonic (or phasor) form of Maxwell's equations (7.23) and the equation of motion (11.2) as

$$\nabla \times \mathbf{H} \simeq j\omega\epsilon_0\mathbf{E} + N_e q_e \mathbf{v} \tag{11.9a}$$

$$\nabla \times \mathbf{E} = -j\omega\mu_0\mathbf{H} \tag{11.9b}$$

$$\nabla \cdot \mathbf{E} \simeq \frac{n_e q_e}{\epsilon_0} \qquad (11.9c)$$

$$\nabla \cdot \mathbf{H} = 0 \qquad (11.9d)$$

$$q_e \mathbf{E} \simeq j\omega m_e \mathbf{v} \qquad (11.9e)$$

whereas the continuity of equation (contained within the above) is

$$\nabla \cdot (N_e \mathbf{v}) \simeq -j\omega n_e \qquad (11.10)$$

Note that equation (11.9e) is the phasor form of the equation of motion, the complete form of which is

$$m_e \frac{d\tilde{\mathbf{v}}}{dt} = m_e \left[\frac{\partial \tilde{\mathbf{v}}}{\partial t} + (\tilde{\mathbf{v}} \cdot \nabla)\tilde{\mathbf{v}} \right] \simeq m_e \frac{\partial \tilde{\mathbf{v}}}{\partial t} = q_e[\overline{\mathscr{E}} + \tilde{\mathbf{v}} \times \overline{\mathscr{B}}] \qquad (11.11)$$

where the $(\tilde{\mathbf{v}} \cdot \nabla)\tilde{\mathbf{v}}$ term represents the change in velocity that an observer moving with the particle would observe (since in general the velocity would be varying with space coordinates). Note that the $\overline{\mathscr{B}}$ in (11.11) is the magnetic field of the electromagnetic wave that must necessarily accompany a time-harmonic electric field $\overline{\mathscr{E}}$. However, for the small-oscillation case considered here, the $\tilde{\mathbf{v}} \times \overline{\mathscr{B}}$ term is negligible, since it represents the cross product of two small quantities. For the time-harmonic case, (11.11) can be written in terms of phasor quantities as

$$j\omega m_e \mathbf{v} \simeq q_e[\mathbf{E} + \mathbf{v} \times \mathbf{B}]$$

where the $\mathbf{v} \times \mathbf{B}$ term also represents the cross product of two small terms (both $\mathbf{v}$ and $\mathbf{B}$ are sinusoidal oscillations each with a small amplitude), so that it can also be neglected,[8] resulting in (11.9e).

We now consider the solutions of equations (11.9) for propagating waves. By eliminating $\mathbf{v}$ from equations (11.9e) and (11.9a) we find

$$\nabla \times \mathbf{H} = j\omega\epsilon_0 \left(1 - \frac{N_e q_e^2}{\omega^2 m_e \epsilon_0} \right) \mathbf{E} \qquad (11.12)$$

The similarity of (11.12) with the fundamental Maxwell's equation (7.23c), that is, $\nabla \times \mathbf{H} = j\omega\epsilon\mathbf{E}$, suggests that the plasma can be represented by an *effective dielectric permittivity* given by

$$\boxed{\epsilon_{\text{eff}} = \epsilon_0 \left(1 - \frac{\omega_p^2}{\omega^2} \right)} \qquad (11.13)$$

[8]This term cannot be neglected when a steady magnetic field $\mathbf{B}_0$ (such as the magnetic field of Earth) is present, since $\mathbf{v} \times \mathbf{B}_0$ is no longer a cross product of two small quantities.

where $\omega_p = \sqrt{N_e q_e^2/(m_e \epsilon_0)}$ is the plasma frequency introduced in the previous subsection. Note that equation (11.9c) can also be rewritten[9] using this effective permittivity as

$$\nabla \cdot (\epsilon_{eff}\mathbf{E}) = 0$$

With the effects of the plasma accounted for in terms of an effective permittivity, *all* of our previous solutions for wave propagation in a dielectric with permittivity ϵ can be used. For example, from (8.10), the phasor form of the wave equation for uniform plane waves in an unbounded medium for a single component of $\mathbf{E}$ is

$$\frac{d^2 E_x}{dz^2} + \omega^2 \mu_0 \epsilon_{eff} E_x = 0$$

and has the general solution (8.11)

$$E_x(z) = C_1 e^{-j\beta z} + C_2 e^{+j\beta z}$$

where $\beta = \omega\sqrt{\mu_0\epsilon_{eff}} = \omega\sqrt{\mu_0\epsilon_0(1 - \omega_p^2/\omega^2)}$. Introducing the notation $X = \omega_p^2/\omega^2$, we have

$$E_x(z) = C_1 e^{-jz\omega\sqrt{\mu_0\epsilon_0(1-X)}} + C_2 e^{+jz\omega\sqrt{\mu_0\epsilon_0(1-X)}} \tag{11.14}$$

We see that the physical properties of the plasma medium—more specifically, the fact that electrons move under the influence of the wave fields, constituting current, which in turn influences the wave fields through (11.9a)—require that uniform plane waves can propagate in the plasma only if they have a phase constant of $\beta = \omega\sqrt{\mu_0\epsilon_0(1 - X)}$. For $X < 1$, or $\omega > \omega_p$, the propagation constant β is real, and wave propagation in a plasma is like that in an ordinary dielectric. At $X = 1$, $\beta = 0$, and plane waves cease to propagate. By analogy to a waveguide, ω_p is the *cutoff* frequency of the plasma medium. It is interesting to note that in the case of the plasma, the cutoff behavior (i.e., β is a function of frequency) is brought about by the physical properties of the medium, rather than by the presence of the boundaries as in the case of waveguides. For $X > 1$, or $\omega < \omega_p$, β is imaginary, and plane waves do not propagate. In that case, we can write (11.14) as

$$E_x(z) = C_1 e^{-z\omega\sqrt{\mu_0\epsilon_0(X-1)}} + C_2 e^{+z\omega\sqrt{\mu_0\epsilon_0(X-1)}}$$

so the waves are evanescent and attenuate with distance.

In a similar manner, the intrinsic impedance of the plasma medium is

$$\eta_p = \sqrt{\frac{\mu_0}{\epsilon_{eff}}} = \frac{\sqrt{\mu_0/\epsilon_0}}{\sqrt{1 - \omega_p^2/\omega^2}} \tag{11.15}$$

[9]Using (11.9e), (11.9c), and the continuity equation (11.10).

and the wave magnetic field corresponding to the electric field given in (11.14) is

$$H_y(z) = \frac{C_1}{\eta_p} e^{-jz\omega\sqrt{\mu_0\epsilon_0(1-X)}} + \frac{C_2}{\eta_p} e^{+jz\omega\sqrt{\mu_0\epsilon_0(1-X)}} \qquad (11.16)$$

Note that for $\omega > \omega_p$, the impedance η_p is purely real, so the electric and magnetic fields are in phase, and their product represents real time-average power flow. However, for $\omega < \omega_p$, the waves are evanescent, and we note from (11.15) that η_p is imaginary. Thus, no power is carried by evanescent waves, since $\mathbf{E}$ and $\mathbf{H}$ are out of phase by $90°$. Note also that the evanescent attenuation of the wave for $\omega < \omega_p$ does not represent any absorption of power and conversion into heat, since any such damping (e.g., due to collisions) was neglected.

Considering only a wave propagating in the $+z$ direction, and $\omega > \omega_p$, the real electric and magnetic fields are given by

$$\overline{\mathscr{E}}(z,t) = \hat{\mathbf{x}} C_1 \cos[\omega t - z\omega\sqrt{\mu_0\epsilon_0(1-X)}] \qquad (11.17a)$$

$$\overline{\mathscr{H}}(z,t) = \hat{\mathbf{y}} \frac{C_1\sqrt{1-X}}{\sqrt{\mu_0/\epsilon_0}} \cos[\omega t - z\omega\sqrt{\mu_0\epsilon_0(1-X)}] \qquad (11.17b)$$

where we have assumed C_1 to be real. It is interesting to note that for a constant peak electric field amplitude given as $C_1 = E_0$, the amplitude of the magnetic field is frequency-dependent. The impedance of the plasma medium is high for frequencies in the vicinity of the cutoff frequency ω_p. In other words, for $\omega = \omega_p$ the plasma behaves like an open circuit. For frequencies much higher than ω_p, that is, $\omega \gg \omega_p$, we have $\eta_p \simeq \sqrt{\mu_0/\epsilon_0}$, so that the plasma behaves much like free space. Note that the time-average power flow for uniform plane waves in a unmagnetized plasma (i.e., a plasma without a steady magnetic field $\mathbf{B}_0$) is given by

$$\mathbf{S}_{av} = \hat{\mathbf{z}} \frac{E_0^2}{2\eta_p} = \frac{E_0^2}{2\sqrt{\mu_0/\epsilon_0}} \sqrt{1 - \omega_p^2/\omega^2} \qquad (11.18)$$

The phase velocity for uniform plane waves in an ionized medium is given by

$$v_p = \frac{\omega}{\beta} = \frac{\omega}{\sqrt{\omega^2\mu_0\epsilon_0(1-\omega_p^2/\omega^2)}} = \frac{c}{\sqrt{1-\omega_p^2/\omega^2}} \qquad (11.19)$$

Thus, we see that vibration of electrons in the ionized region results in a wave phase velocity greater than that of the speed of light in free space. The group velocity can be found in the usual manner, namely

$$v_g = \frac{d\omega}{d\beta} = c\sqrt{1 - \frac{\omega_p^2}{\omega^2}} \qquad (11.20)$$

We note that $v_p v_g = c^2$, as in the case of waveguides. This result is simply a consequence of the fact that the functional frequency dependence of the propagation constant β for an ionized gas is similar to that for $\overline{\beta}$ in a waveguide.

Total reflection from free space–plasma interface.

An interesting and important consequence of the fact that uniform plane waves can propagate in a plasma only at frequencies $\omega > \omega_p$ involves the total reflection of a uniform plane wave at the interface between a dielectric and a plasma medium or an ionized gas when $\omega < \omega_p$. A natural example of such an interface is that between free space and Earth's ionosphere (although the electron density in the ionosphere increases relatively gradually with height, so the interface is not a single sharp interface between two media like those considered in Chapter 9). It is this type of reflection that makes radio waves "bounce" off the ionosphere, making long-range radio communication possible, with reflection occurring when ω is less than ω_p. For the ionosphere, the peak value of f_p is approximately 10 MHz; thus, AM radio broadcast frequencies are reflected from the ionospheric conducting layer. Microwave, television, and FM radio signals are typically above 40 MHz and are thus easily transmitted through the conducting ionospheric layer with negligible reflection. To illustrate the basic concept of total reflection at such an interface, we consider a sharp single interface between free space and an ionized region (characterized with plasma frequency ω_p) as illustrated in Figure 11.2. Assuming that the incident wave in Figure 11.2 is at a frequency ω, that the first medium is free space, and that the ionized region behaves as a dielectric with dielectric constant $\epsilon_{\text{eff}} = \epsilon_0 \sqrt{1 - \omega_p^2/\omega^2}$, the reflection and transmission coefficients for the case of normal incidence are given by the expressions

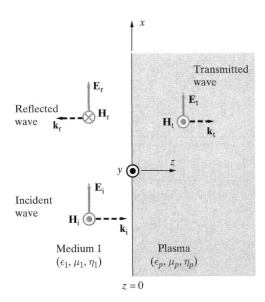

Figure 11.2   Reflection from free space–plasma interface. Normal incidence at a sharp interface between free space and an ionized medium (plasma) characterized by a plasma frequency of ω_p.

similar to those that were given in Chapter 9 for the case of reflection from the interface between two ordinary dielectrics, namely

$$\Gamma = \frac{\sqrt{\epsilon_1} - \sqrt{\epsilon_2}}{\sqrt{\epsilon_1} + \sqrt{\epsilon_2}} = \frac{\sqrt{\epsilon_0} - \sqrt{\epsilon_0(1 - \omega_p^2/\omega^2)}}{\sqrt{\epsilon_0} + \sqrt{\epsilon_0(1 - \omega_p^2/\omega^2)}} = \frac{\omega - \sqrt{\omega^2 - \omega_p^2}}{\omega + \sqrt{\omega^2 - \omega_p^2}} \qquad (11.21\text{a})$$

$$\mathcal{T} = \frac{2\sqrt{\epsilon_1}}{\sqrt{\epsilon_1} + \sqrt{\epsilon_2}} = \frac{2\omega}{\omega + \sqrt{\omega^2 - \omega_p^2}} \qquad (11.21\text{b})$$

For $\omega > \omega_p$, we note from the above that both Γ and $\mathcal{T}$ are real, and portions of the incident wave energy are reflected and transmitted accordingly. However, for $\omega < \omega_p$, Γ becomes imaginary, in which case we can write it as

$$\Gamma = \frac{\omega - j\sqrt{(\omega_p^2 - \omega^2)}}{\omega + j\sqrt{(\omega_p^2 - \omega^2)}} = 1e^{j\phi_\Gamma} \qquad (11.22)$$

We note that (11.22) is similar to the form of the reflection coefficients $\Gamma_{\parallel}$ and $\Gamma_{\perp}$ derived in Chapter 9 for total internal reflection upon oblique incidence at a dielectric interface (i.e., (9.39) and (9.40)). This similarity indicates that an electromagnetic wave passing from free space into a plasma with $\omega < \omega_p$ is thus totally reflected, even at normal incidence. The amplitude of the reflected wave is equal to that of the incident wave, but the wave acquires phase upon reflection, so the phase of the reflected wave is different from that of the incident one by an amount ϕ_Γ, which depends on frequency as given in (11.22). Note that—as with the refracted wave in the case of total internal reflection at oblique incidence discussed in Section 9.7.2—the transmission coefficient $\mathcal{T}$ also becomes complex, so the transmitted wave is also out of phase with the incident wave. More importantly, the transmitted wave is evanescent, decays rapidly with distance in the z direction, and carries no real power, since for $\omega < \omega_p$ the intrinsic impedance of the plasma η_p is purely imaginary. As a result, the electric and magnetic fields of the wave in the ionized medium are 90° out of phase.

Attenuation due to collisions.   In practice, some electromagnetic power is always lost in a plasma, because the electrons frequently collide with gas molecules, ions, and even other electrons. These collisions cause electromagnetic power to be transformed into heat. For $\omega > \omega_p$, the collisions cause the wave to be attenuated with distance. Also, for $\omega < \omega_p$, the losses due to collisions lead to partial reflection instead of the total reflection just discussed.

Collisional effects may be taken into account by including in the equations of motion a frictional force term in (11.9e) such as

$$q_e \mathbf{E} = j\omega m_e \mathbf{v} + m_e \nu \mathbf{v} = j\omega m_e \left(1 - j\frac{\nu}{\omega}\right) \mathbf{v} \qquad (11.23)$$

where ν is the effective collision frequency of electrons with other particles, in units of collisions per second or simply s^{-1}.

With the collision term included, we can once again eliminate **v** from the equation (11.9a) to obtain

$$\nabla \times \mathbf{H} = j\omega\epsilon_0 \mathbf{E} + \frac{N_e q_e^2 \mathbf{E}}{j\omega m_e \left(1 - j\dfrac{\nu}{\omega}\right)}$$

or

$$\nabla \times \mathbf{H} = j\omega\epsilon_0 \left(1 - \frac{X}{1 - jZ}\right)\mathbf{E}$$

where $X = \omega_p^2/\omega^2$ and $Z = \nu/\omega$. Note that X and Z are dimensionless quantities. The effective permittivity of the plasma with collisions included is thus

$$\epsilon_{\text{eff}} = \epsilon_0 \left(1 - \frac{X}{1 - jZ}\right) = \epsilon'_{\text{eff}} - j\epsilon''_{\text{eff}}$$

As was shown in Chapter 8 for lossy dielectrics, the imaginary part ϵ''_{eff} represents power loss and results in attenuation of the wave. The expressions for uniform plane waves in a collisional plasma can be obtained by using the general form of uniform plane waves in a lossy medium represented by σ, ϵ, and μ, as was discussed in Section 8.3, by replacing ϵ with ϵ'_{eff} and σ with $\omega\epsilon''_{\text{eff}}$.

11.2 FREQUENCY RESPONSE OF DIELECTRICS AND CONDUCTORS

When an external electric field is applied to a dielectric, it polarizes the material by displacing particles. The simplest type of polarization, referred to as electronic polarization, occurs when electrons are displaced with respect to their nuclei. The electric permittivity ϵ of a material is a macroscopic parameter that represents the combined effect of microscopic electric dipoles produced by such displacement.[10] Since electrons or other displaced particles have finite mass, it is clear that the amount of polarization must be a function of the frequency of the applied field. At a qualitative level, we expect that if the electric field alternates rapidly (i.e., high frequencies), inertia of the electrons will tend to prevent them from following the rapid oscillations. At low enough frequencies, the electron can vibrate in phase with the applied field; however, at very high frequencies, the displacements of the electrons are out of phase with the applied field, because of their slow inertial response. In between the two regimes, a resonance phenomenon occurs, leading to significant absorption of electromagnetic energy. These microscopic frequency-dependent behaviors of the materials are generally represented in terms of a complex permittivity, as discussed in Section 8.3. In this section, we provide a first-order

---

[10]Section 4.10 contains a relatively simple discussion of atomic polarization and the permittivity concept. For further discussion on ionic, orientational, and other polarizations, see Chapter 13 of C. Kittel, *Introduction to Solid State Physics,* 5th ed., Wiley, New York, 1976.

description of the frequency dependence of electronic polarization and note that other types of polarization exhibit generally similar behavior.

Electrical conductivities of materials also vary with frequency, primarily because of the inertia of the electrons, whose average drift motion results in electric current flow. In most metals, the frequency dependence of conductivity becomes significant only at optical frequencies. In the following, we provide a very brief discussion of frequency response of conducting materials.

A good example of a material that exhibits frequency-dependent polarization and conductivity is pure water.[11] At low frequencies, distilled water is essentially nonconducting, and its permittivity is $\epsilon \simeq 81\epsilon_0$. At microwave frequencies, however, the permittivity of water ranges from $10\epsilon_0$ to $20\epsilon_0$, and furthermore it becomes quite lossy. Similarly, a salt solution is a good conductor at low frequencies but behaves as an insulator at optical frequencies (i.e., such a solution is transparent to visible light). Thin films of silver are good conductors up through optical frequencies but become transparent in the ultraviolet region. At X ray frequencies, metals and insulators react quite similarly to electromagnetic radiation.

11.2.1 Frequency Response of Dielectric Materials

To analyze the frequency dependence of polarization quantitatively, we start by considering a nonconducting ($\sigma = 0$) isotropic gas. At the simplest level, gases and solids are made up of molecules, each of which can be considered a classical oscillator. In gases, the separation of the oscillators is large enough so that they do not interact with each other. To approximate the response to an applied electromagnetic field of a real gas, we consider the motion of an electron of a typical gas atom. Since the nucleus is relatively heavy, we consider it to be fixed in space and to be surrounded by an electron cloud of charge q_e and mass m_e that oscillates back and forth in response to the applied electromagnetic field. Consider the Lorentz force acting on an electron in a plane electromagnetic wave in a lossless medium:

$$\overline{\tilde{\mathfrak{F}}} = \overline{\tilde{\mathfrak{F}}}_{\text{elec}} + \overline{\tilde{\mathfrak{F}}}_{\text{mag}} = q_e\overline{\tilde{\mathscr{E}}} + q_e\tilde{\mathbf{v}} \times \overline{\mathscr{B}}$$

where $\overline{\mathscr{E}}$ and $\overline{\mathscr{B}}$ are, respectively, the electric and magnetic fields of an electromagnetic wave and $\tilde{\mathbf{v}}$ is the velocity of the electron. We can recall from Chapter 8 that for a uniform plane wave we have $|\overline{\mathscr{B}}| = \mu|\overline{\mathscr{H}}| = \mu|\overline{\mathscr{E}}|/\eta = (v_p)^{-1}|\overline{\mathscr{E}}|$, so that unless the electrons have a velocity approaching that of the phase velocity v_p of the electromagnetic wave (i.e., the speed of light, in the gaseous medium under consideration), the force due to the magnetic field can be neglected.[12] The electron cloud is thus primarily driven by the electric field

---

[11] An excellent brief discussion of the variation of the refractive index (and hence ϵ) of water as a function of frequency is given in Section 7.5 of J. D. Jackson, *Classical Electrodynamics,* 2nd ed., Wiley, New York, 1975.

[12] That is, if $|\tilde{\mathbf{v}}| \ll v_p$, then

$$|\tilde{\mathbf{v}} \times \overline{\mathscr{B}}| = |\tilde{\mathbf{v}}||\overline{\mathscr{B}}| \sin\theta = (v_p)^{-1}|\tilde{\mathbf{v}}||\overline{\mathscr{E}}| \sin\theta \ll |\overline{\mathscr{E}}|$$

and therefore $\overline{\tilde{\mathfrak{F}}} = q_e\overline{\mathscr{E}} + q_e\tilde{\mathbf{v}} \times \overline{\mathscr{B}} \simeq q_e\overline{\mathscr{E}}$.

of the applied electromagnetic wave. For the rest of this discussion, we assume the wave to be polarized in the x direction, so that $\overline{\mathscr{E}} = \hat{x}\mathscr{E}_x$, in which case the electron cloud is displaced in the x direction. Once the electron cloud is displaced with respect to the nucleus, it is acted upon by a restoring force due to the Coulomb attraction of the positive nucleus. This restoring force is proportional to the displacement $\tilde{x}$ and is given[13] by $-k\tilde{x}$, where the constant k is $k = q_e^2/(4\pi\epsilon_0 a^3)$, with a being the radius of the electron cloud. In effect, k is an elastic constant, and the electron can be considered to be elastically bound to its nucleus. The energy that is transferred by the electron to its surroundings during its rapid motion can be represented by a viscous (or frictional) damping force proportional to the mass and to the velocity of the electron, that is, $-m_e\kappa(d\tilde{x}/dt)$, where κ is a positive constant. What is described here is naturally a rather crude classical model; the damping term is often due to the reradiation of energy, and this system should actually be described using quantum mechanics. Nevertheless, more detailed considerations lead to a model very similar to the classical model described here, the validity of which is based on its success in describing the propagation of electromagnetic waves through gases.

The equation of motion for the electron, under the influence of the three forces described above, is thus given by

$$m_e \frac{d^2\tilde{x}}{\partial t^2} = -k\tilde{x} - m_e\kappa\frac{d\tilde{x}}{\partial t} + q_e\mathscr{E}_x \tag{11.24}$$

We can simplify this equation by considering sinusoidal oscillations and rewriting it in terms of a phasor displacement x such that $\tilde{x} = \mathscr{R}e\{xe^{j\omega t}\}$. We have

$$-m_e\omega^2 x = -kx - j\omega m_e\kappa x + q_e E_x \tag{11.25}$$

where E_x is the electric field phasor; that is, $\overline{\mathscr{E}} = \mathscr{R}e\{\hat{x}E_x e^{j\omega t}\}$. Equation (11.25) can be readily solved for the displacement x to find

$$x = \frac{(q_e/m_e)E_x}{\omega_0^2 - \omega^2 + j\omega\kappa} \tag{11.26}$$

where $\omega_0 = \sqrt{k/m_e}$ is the characteristic angular frequency of the elastically bound electron. The constants ω_0 and κ depend on the atomic and molecular structure of the particular material and must in general be determined from experimental data.

The displacement of a single electron by an amount x creates an electric dipole moment of $\mathbf{p} = \hat{x}q_e x$. If there are N_e electrons of this type per unit volume, the dipole moment per unit volume, or the polarization, is $\mathbf{P} = \hat{x}Nq_e x = \hat{x}P_x$ where P_x is given by

$$P_x = \frac{N_e(q_e^2/m_e)E_x}{\omega_0^2 - \omega^2 + j\omega\kappa} \tag{11.27}$$

---

[13]The restoring force can be simply calculated by considering the charge q_e of the electron to be distributed in a negatively charged spherical cloud of atomic radius a, with charge density $\rho = 3q_e/(4\pi a^3)$ and assuming that the center of this cloud is displaced by an amount $\tilde{x}$ with respect to a positive charge (nucleus) of $+|q_e|$. Gauss's law can then be used to determine the electric field E_x at a distance $\tilde{x}$ from the center of the cloud that exerts the restoring force of $+|q_e|E_x$ on the positive charge.

Recalling (see Section 4.10) that the susceptibility χ_e is defined by the relation $P_x = \epsilon_0 \chi_e E_x$, and that the permittivity is given by $\epsilon = \epsilon_0(1 + \chi_e)$, we have

$$\chi_e = \frac{N_e q_e^2}{m_e \epsilon_0}\left[\frac{1}{\omega_0^2 - \omega^2 + j\omega\kappa}\right]$$

and thus the relative complex permittivity follows as

$$\epsilon_r = \frac{\epsilon}{\epsilon_0} = 1 + \chi_e = 1 + \frac{N_e q_e^2}{m_e \epsilon_0}\left[\frac{1}{\omega_0^2 - \omega^2 + j\omega\kappa}\right] \tag{11.28a}$$

$$= 1 + \frac{N_e q_e^2}{m_e \epsilon_0}\left[\frac{\omega_0^2 - \omega^2}{(\omega_0^2 - \omega^2)^2 + \omega^2\kappa^2} - j\frac{\omega\kappa}{(\omega_0^2 - \omega^2)^2 + \omega^2\kappa^2}\right] \tag{11.28b}$$

$$= \epsilon_r' - j\epsilon_r'' \tag{11.28c}$$

We thus see that the underlying physical cause of the complex permittivity concept that was introduced in Chapter 8 is the elastically bound nature of the electrons to their nuclei.[14] The variations of $\epsilon' = \epsilon_r'\epsilon_0$ and $\epsilon'' = \epsilon_r''\epsilon_0$ as a function of frequency are illustrated in Figure 11.3. The frequency behavior of the real and imaginary parts of the

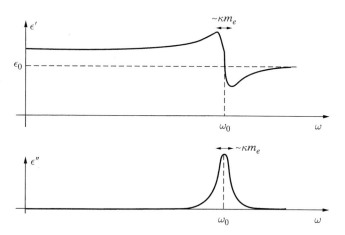

Figure 11.3 Resonance effect in electronic polarization. At low frequencies, the permittivity is different from ϵ_0 by a constant multiplier. In the vicinity of the resonance, ϵ'' goes through a pronounced peak.

[14] Actually, complex permittivity comes about not only from the displacement of electrons but also from the displacement of charged ions or from the change in orientation of existing permanent dipole moments in response to the applied field. Although we have chosen to discuss only electronic polarization, the frequency response of other polarizations is quite similar, as indicated in Figure 11.4.

permittivity are not independent of one another; instead, they are related to one another through the so-called Kramers–Kronig relations,[15] which are similar to the relations between the real and imaginary parts of the complex impedance of an electrical circuit. At low frequencies, ϵ is essentially real, but in the vicinity of $\omega \simeq \omega_0$ a resonance takes place and ϵ becomes complex. The width of the resonance is determined by the damping coefficient κ. From earlier discussions in Section 8.3.1, we know that the imaginary part of the permittivity is associated with power loss or dissipation of electromagnetic energy within the dielectric. At any given frequency, ϵ'' produces the same macroscopic effect as the conductivity σ. Since the values of ϵ'' are often determined by measurement, losses due to nonzero conductivity σ and the dielectric losses discussed here are often indistinguishable. In practice, losses are represented by σ for metals and by ϵ'' for dielectrics.

For simplicity, we have only considered electronic (sometimes also called atomic) polarization, for which the resonance frequency ω_0 is typically quite high, usually in the ultraviolet range. However, total polarizability of a material under the influence of an electric field can be separated into three parts: electronic, ionic, and dipolar (or orientational) polarization. The electronic polarization arises from the displacement of the electron cloud with respect to the nucleus, as discussed above. The ionic polarization contribution comes from the displacement of a charged ion with respect to other ions. In some materials, two different atoms may join together as a molecule by forming a chemical bond. We can think of such molecules as consisting of positively and negatively charged ions, with the Coulomb forces between them serving as the binding force. Examples of such materials are HCl, CO_2, and H_2O. Depending on whether the electrons are transferred from one atom to the other or shared between atoms, the bond can be *ionic* or *covalent*. In either case the material may possess a permanent dipole moment, and furthermore, the application of an electric field to any such molecule displaces the positive ions with respect to negative ones and thereby induces a dipole moment. The dipolar or orientational polarizability is due to molecules with a permanent electric dipole that change orientation in response to the applied electromagnetic field. Some polyatomic molecules, such as H_2O, may be at least partially ionic and may consist of *polar* molecules, which carry a permanent dipole moment. With no electric field, the individual dipole moments point in random directions, so that the net dipole moment is zero. When an electric field is applied, such materials exhibit the electronic and ionic polarization effects discussed above. In addition, the electric field tends to line up the individual dipoles to produce an additional net dipole moment per unit volume. If all the dipoles in a material were to line up, the polarization would be very large; however, at ordinary temperatures and relatively small electric field levels, the collisions of the molecules in their thermal motion allow only a small fraction of the dipoles to line up with the field.

The frequency response of the dipolar and ionic polarizations is qualitatively similar to that of electronic polarization, but they exhibit much lower resonance frequencies.

---

[15]See p. 311 of J. D. Jackson, *Classical Electrodynamics,* 2nd ed., John Wiley & Sons, New York, 1975.

The reason for this behavior is the fact that the effective masses of the microscopic bodies that are displaced are substantially larger, consistent with the fact that for electronic polarization, the resonant frequency increases with decreasing particle mass (we have $\omega_0 = \sqrt{k/m_e}$). Although the elastic constant k is different for the different types of polarizations, the resonant frequency is nevertheless inversely proportional to the square root of the particle mass. The frequency response of a hypothetical dielectric is shown in Figure 11.4, with the contributions from the different types of polarizations identified in terms of the frequency ranges in which they significantly contribute.

The frequency response of dielectric materials and the various resonances illustrated in Figure 11.4 have many important practical implications. The broad dipolar resonance in the microwave region is responsible for microwave heating applications. A resonance in water vapor molecules at 1.25-cm wavelength places a limit on the frequency of long-range radar. Another strong resonance at 0.5 cm from oxygen, together with other higher-frequency absorptions, prevents propagation of signals above 0.5 cm wavelength. Very strong resonances, due mostly to ozone and nitrogen at high altitudes, keep most of the ultraviolet energy of the sun from penetrating to lower altitudes. Another example of an important resonance occurs in sodium vapor at a wavelength of about 589 nm. Different gases exhibit a large variety of such resonances that have their origins in the basic energy level structure of the atoms and molecules.[16] In most gases, there exist many

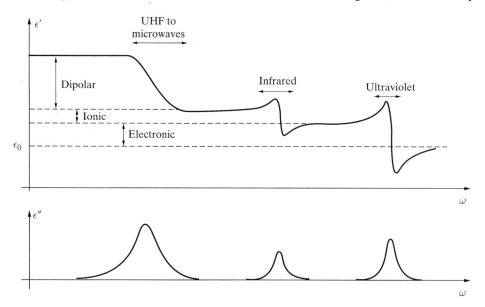

Figure 11.4 Dielectric constant as a function of frequency. The behavior of the dipolar and ionic polarization effects as a function of frequency are similar to that of electronic polarization. However, the resonances for these polarizations occur at lower frequencies.

[16]As an example of the kind of complexity that is possible, see an interesting discussion of the properties of liquid water as a function of frequency in Section 7.5 of J. D. Jackson, *Classical Electrodynamics,* 2nd ed., John Wiley & Sons, New York, 1975.

resonant frequencies for each type of polarization, each exhibiting different damping coefficients κ.

The above discussion is applicable for gases, for which the separation of the different elastically bound oscillators is large enough so that they do not interact with one another. Similar phenomena take place in solids, except that the oscillators are packed much more closely together, so that the effect of neighboring dipoles must be included. Relatively simple analysis of this problem[17] indicates that the effective local field acting to polarize the material is $E_{\text{loc}} = E + P/(3\epsilon_0)$, which in turn leads to the so-called *Clausius–Mossotti* relation,[18] defining the frequency dependence of the complex permittivity ϵ:

$$\frac{\epsilon - 1}{\epsilon + 2} = \frac{1}{3}\left[\frac{Nq_e^2}{m_e \epsilon_0}\right]\left[\frac{1}{\omega_0^2 - \omega^2 + j\omega\kappa}\right]$$

11.2.2 Frequency Response of Metals

We now consider conducting materials and imagine a cloud of freely circulating electrons under statistical equilibrium, which slowly drift in the direction of an applied field.[19] The drift velocity v_d is the solution of an equation of motion with a damping term, namely

$$m_e \frac{dv_d}{dt} + \frac{m_e}{t_c}v_d = q_e \mathscr{E}_x \tag{11.29}$$

where we have once again assumed that the electric field is in the x direction, so that the drift velocity is $\mathbf{v}_d = \hat{\mathbf{x}}v_d$. The free electrons are acted upon by an external force $q_e\mathscr{E}_x$, and their motion is opposed by a force $-(m_e/t_c)v_d$ proportional to their velocity, which empirically accounts for the dissipative effect of collisions, where t_c is the mean free time between collisions and is typically of order 10^{-14} s. Note that since $v_d = dx/dt$, equation (11.29) is similar to (11.24), except that the electrons in a metal are not elastically bound so that $k = 0$, and thus $\omega_0 = 0$. For sinusoidal oscillations, the solution for the displacement is similar to (11.25), that is,

$$x = \frac{(q_e/m_e)E_x}{-\omega^2 + j(\omega/t_c)} \tag{11.30}$$

If there are n_c conduction electrons per unit volume, the current density $\mathscr{J}_x$ is then given by

$$\mathscr{J}_x = n_c q_e v_d = n_c q_e \frac{d\tilde{x}}{dt} \quad \rightarrow \quad J_x = n_c q_e (j\omega)x = \frac{(j\omega n_c q_e^2/m_e)E_x}{-\omega^2 + j(\omega/t_c)}$$

[17]See Chapter 13 of C. Kittel, *Introduction to Solid State Physics,* 5th ed., Wiley, New York, 1971.

[18]See Section I-B of *Dielectric Materials and Applications,* A. von Hippel (Ed.), Artech House, Boston, 1995.

[19]See Section 5.1. For further discussion, see Section 8.5 of R. L. Sproull, *Modern Physics: The Quantum Physics of Atoms, Solids, and Nuclei,* 3rd ed., John Wiley & Sons, 1980.

where $\mathcal{J}_x = \mathcal{R}e\{J_x e^{j\omega t}\}$, so J_x is the current density phasor. Simplifying further, and by analogy with the static conductivity defined by $\mathbf{J} = o\,\mathbf{E}$, we have

$$\sigma = \frac{n_c q_e^2}{(m_e/t_c) + jm_e \omega} \tag{11.31}$$

Note that for $\omega \to 0$ we have $\sigma = n_c q_e^2 t_c/m_e$, which is the classic expression for the the static conductivity of a material. In view of the fact that t_c is of order 10^{-14} s, the second term in the denominator of (11.31) is negligible compared with the first term at frequencies well into the infrared region. It has been experimentally verified[20] that conductivity values measured under static conditions can be used for wavelengths greater than ~ 25 μm. For wavelengths ranging from ~ 25 μm to the visible region, however, strong frequency dependence is observed, and the current and electric field are no longer in phase. In the far ultraviolet and X ray regions, the conductivity σ approaches zero, as can be seen from (11.31).

To determine the effects of frequency dependence of the conductivity of metals on the propagation constant β and attenuation constant α of a uniform plane wave in a metal, we recall from Section 8.3 that the propagation constant γ in a conducting medium with $\epsilon = \epsilon_0$ and $\mu = \mu_0$ is given by (see (8.20))

$$\gamma = j\omega\sqrt{\mu_0 \epsilon_0}\sqrt{1 - j\frac{\sigma}{\omega\epsilon_0}}$$

Assuming the frequency dependence of ϵ_r to be of the form (11.28a) and σ of the form (11.31), we have

$$\gamma = j\omega\sqrt{\mu_0\epsilon_0}\sqrt{1 + \frac{Nq_e^2}{m_e\epsilon_0}\left[\frac{1}{\omega_0^2 - \omega^2 + j\omega\kappa}\right] + \frac{n_c q_e^2}{m_e\epsilon_0}\left[\frac{1}{-\omega^2 + j(\omega/t_c)}\right]} \tag{11.32}$$

The second term on the right of (11.32) is due to the bound electrons, whose resonance frequencies for metallic atoms lie well in the violet or ultraviolet frequency range. Thus, in the visible, red, infrared, microwave, and lower frequencies, metals can be treated as if their dielectric constant is ϵ_0. The third term in (11.32) is that due to the free conduction electrons, whose frequency dependence is significant in the low infrared through visible range. As frequency is further increased into the far ultraviolet and X ray regions, the third term in (11.32) approaches zero, so the propagation constant for uniform plane waves in most metals becomes similar to that in dielectrics.

11.3 METAMATERIALS

From (11.13), we see that in a lossless cold plasma, the effective permittivity is *negative* when the wave frequency ω is less than the plasma frequency ω_p. Similarly, the real part

---

[20]Extensive measurements of reflectivity of a wide range of metals were conducted by Hagen and Rubens in 1903 for frequencies in the range 10^{13} to 10^{14} Hz. See Chap. IX, p. 508 of *Ann. Physik,* 11, p. 873, 1903.

of the permittivity in dielectrics, given by (11.28), may become negative near the resonance frequency (see Problem 11.7). When $\epsilon_{\text{eff}} < 0$, the solutions to the wave equation are evanescent and attenuate with distance. Media with negative permeability also exist, but are less common in nature since most materials only weakly interact with magnetic fields. One exception is ferrimagnetic materials, which have sufficiently strong magnetic interactions and sufficiently low losses to produce resonant frequency bands with negative magnetic permeability. As with low-loss plasmas, wave propagation through such a medium is evanescent, since $\beta = \omega\sqrt{\mu\epsilon}$ becomes strictly imaginary.

An interesting case arises when both the permittivity *and* the permeability of a material are negative, which opens up the possibility for wave propagation to exist. Materials with simultaneously negative permittivity and permeability do not occur in nature. Rather, they are carefully manufactured materials that are designed to achieve specific macroscopic constitutive parameters. Materials with these engineered properties are a subclass of *electromagnetic metamaterials* (MTMs), which are a broad class of artificial structures with unusual electromagnetic properties that are not readily available in nature. In this section, we use Maxwell's equations to derive the resulting wave propagation parameters for a general class of media with constitutive parameters satisfying both $\Re e\{\epsilon\} < 0$ and $\Re e\{\mu\} < 0$.

Figure 11.5 summarizes the four sign permutations of ϵ and μ. Quadrant I, with $\epsilon > 0$ and $\mu > 0$, corresponds to simple media that we presumed before this chapter. For reasons that become clear in the following section, simple media with $\epsilon > 0$ and $\mu > 0$ are called right-handed (RH) media. In quadrants II and IV, ϵ and μ have opposite signs. In such cases, $\beta = \pm j\omega\sqrt{|\mu||\epsilon|}$ is purely imaginary, and so the solutions to the wave equation are evanescent waves. The final class of materials, with simultaneously negative permittivity and permeability (quadrant III), belongs to a new class of so-called left-handed (LH) materials.

As we saw in Sections 11.1 and 11.2, the interaction dynamics of the electric field in an electromagnetic wave with the free (bound) electrons in a plasma (dielectric)

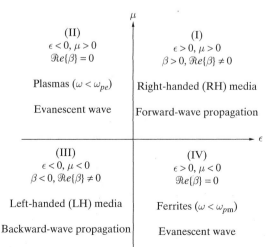

Figure 11.5 Permittivity-permeability diagram. The angular frequencies ω_{pe} and ω_{pm} represent the electric and magnetic plasma frequencies, respectively. In lossy media, this diagram should be modified by replacing ϵ with $\Re e\{\epsilon\}$ and μ with $\Re e\{\mu\}$.

may be captured by a frequency-dependent effective permittivity. As a result of this frequency dependence, unbounded waves propagating in these media experience material dispersion, not due to bounding structures, but rather due to properties of the material itself. In Section 11.3.3, we show that materials with both $\epsilon < 0$ and $\mu < 0$ must also be dispersive. However, in addition to having different group and phase velocities, the group and phase velocities in materials with both $\epsilon < 0$ and $\mu < 0$ also carry a different *sign*.

The constitutive parameters are macroscopic descriptions of a material that result from averaging the effects from the interaction between the electric and magnetic fields of the wave with unit structures that make up the material. For example, in a simple dielectric, these unit structures may be composed of individual atoms or molecules. In order for a macroscopic description to be applicable, the feature size of each basic unit must be small enough so that electromagnetic waves respond to the average effect from a large collection of units. Thus, these novel materials must be constructed using periodic arrays of structures that have feature sizes much smaller than a wavelength. If the average unit size is L, then the resulting structure is effectively homogeneous when $L < \lambda/5$, where λ is the wavelength in the medium. Because of this restriction, early work in MTMs focused on microwave applications, where the wavelength is suitably long such that the atomic elements of the material may be constructed without novel manufacturing techniques.[21] However, due to the wavelengths in visible light, MTMs manufactured for the optical spectrum must use nanoscale fabrication technology. The current standard manufacturing method uses *electron-beam lithography* (EBL), a process that serially scans a 2D surface with an electron beam in the desired pattern. Bulk 3D metamaterials in the optical range may be constructed by stacking a sequence of etched 2D layers.[22]

The general solution to the time-harmonic vector wave equation for a lossless medium is (see (8.75))

$$\mathbf{E(r)} = \mathbf{E}_0 e^{-j\beta \hat{\mathbf{k}} \cdot \mathbf{r}} \tag{11.33}$$

where $\hat{\mathbf{k}}$ is the unit vector in the direction of propagation of the wavefronts, and the wavenumber β satisfies

$$\beta^2 = \omega^2 \mu\epsilon \tag{11.34}$$

Mathematically, there are two solutions to (11.34): $\beta = \pm\omega\sqrt{\mu\epsilon}$. On physical grounds, in simple media with $\epsilon > 0$ and $\mu > 0$, we ascribed the $\beta > 0$ solution to waves traveling away from the source. In this section, we will see that when $\epsilon < 0$ and $\mu < 0$, the phase fronts travel *toward* the source, and so β assumes the negative sign ($\beta < 0$).

Several texts use the index of refraction in place of the wavenumber β, defined as

$$n = \beta\frac{c}{\omega} = \pm c\sqrt{\mu\epsilon} = \pm\sqrt{\mu_r\epsilon_r} \tag{11.35}$$

---

[21] One of the first experimental verifications of wave propagation in media with $\epsilon < 0$ and $\mu < 0$ was reported by R. Shelby, D. Smith, and S. Schulz, Experimental verification of a negative index of refraction, *Science*, 292:77–79, 2001.

[22] For a review of modern MTM manufacturing techniques, see M. A. Noginov and V. A. Podolskiy (Eds.), *Tutorials in Metamaterials*, CRC Press, 2012, and references therein.

Since both the speed of light and frequency are always positive, the sign of n follows the sign of β. Hence, for media with $\epsilon < 0$ and $\mu < 0$, since $\beta < 0$, the index of refraction also takes the negative sign. For this reason, media with simultaneously negative permittivity and permeability are often referred to as *negative refractive index* (NRI) media.

The electromagnetic properties of materials with simultaneously negative electric permittivity and magnetic permeability were first derived in the 1960s by the Russian physicist Viktor Veselago.[23] In his pioneering paper, he predicted several phenomena that have since been verified experimentally using MTMs. These properties include the necessary material dispersion in such a medium, the reversal of Snell's law, and the resulting negative refraction at the interface between a RH medium and a LH medium. He discussed potential candidates from natural materials that might have simultaneously negative permittivity and permeability, but conceded that no such substances were known at that time. It would take more than 30 years before the phenomena he predicted were experimentally demonstrated, not with a natural substance, but rather with an artificially manufactured MTM.

This section introduces some of the characteristics of electromagnetic waves propagating in materials with simultaneously negative permittivity and permeability that were first predicted in Veselago's paper. We do not concern ourselves with how to construct these materials. Instead, we assume the macro material properties of $\mathcal{R}e\{\epsilon\} < 0$ and $\mathcal{R}e\{\mu\} < 0$, and use Maxwell's equations to derive the effects of this condition on wave propagation in such a medium.

11.3.1 Left-Handedness

Consider a lossless medium with no sources. The time-harmonic Maxwell's equations governing wave propagation are

$$\nabla \times \mathbf{E} = -j\omega\mu\mathbf{H} \tag{11.36a}$$

$$\nabla \times \mathbf{H} = j\omega\epsilon\mathbf{E} \tag{11.36b}$$

The general solution for a wave propagating in the direction[24] $\mathbf{k}$ is $e^{-j\mathbf{k}\cdot\mathbf{r}}$. Inserting this solution into (11.36) gives

$$\mathbf{k} \times \mathbf{E} = \omega\mu\mathbf{H} \tag{11.37a}$$

$$\mathbf{k} \times \mathbf{H} = -\omega\epsilon\mathbf{E} \tag{11.37b}$$

For simple media with $\epsilon > 0$ and $\mu > 0$, the triad $\{\mathbf{E}, \mathbf{H}, \mathbf{k}\}$ conform to the right-hand rule, as illustrated in Figure 11.6a. Hence materials with $\epsilon > 0$ and $\mu > 0$ are referred to as *right-handed (RH) media.*

---

[23] V. Veselago, The electrodynamics of substances with simultaneously negative values of ϵ and μ, *Soviet Physics Uspekhi*, 10(4), pp. 509–514, January–February 1968.

[24] In order to avoid confusion, in this section we begin with $\mathbf{k}$ for the wavenumber vector instead of $\beta\hat{\mathbf{k}}$. Since β is negative for LH materials, we have the cumbersome result where $\mathbf{k}$ and $\hat{\mathbf{k}}$ are antiparallel. By using $\mathbf{k}$ instead, we are free to focus on the direction of the wavenumber vector, which describes the direction of propagation of the wavefronts.

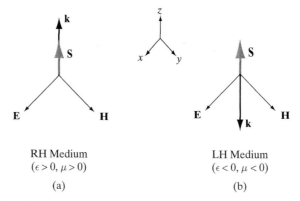

Figure 11.6 Relationship between fields, wavenumber vector, and Poynting vector in RH and LH media. (a) In a RH medium, {**E**, **H**, **k**} form a right-handed triad. (b) In a LH medium, {**E**, **H**, **k**} form a left-handed triad. In both cases, {**E**, **H**, **S**} form a right-handed triad, since **S** only depends on **E** and **H**.

Let us now consider a medium with $\epsilon < 0$ and $\mu < 0$. Under these conditions, we may write (11.37) as

$$\mathbf{k} \times \mathbf{E} = -\omega|\mu|\mathbf{H} \tag{11.38a}$$

$$\mathbf{k} \times \mathbf{H} = \omega|\epsilon|\mathbf{E} \tag{11.38b}$$

We see that under these conditions, {**E**, **H**, **k**} form a left-handed triad, as shown in Figure 11.6b. This left-handed relationship is the reason negative refractive index materials are often called *left-handed (LH) media*.

Unlike the wavenumber vector, the complex Poynting vector for a plane wave, which describes the direction and magnitude of power flow, only depends on the field quantities **E** and **H**: $\mathbf{S} = \mathbf{E} \times \mathbf{H}^*$. Assuming a plane wave of the form $\mathbf{E} = \mathbf{E}_0 e^{-j\mathbf{k} \cdot \mathbf{r}}$ and $|\mathbf{E}_0| = E_0$, and assuming **k** and μ are purely real, we have, using the triple cross product vector identity and noting that $\mathbf{E} \cdot \mathbf{k} = 0$,

$$\mathbf{S} = \mathbf{E} \times \mathbf{H}^* \tag{11.39a}$$

$$= \mathbf{E} \times \left(\frac{\mathbf{k} \times \mathbf{E}^*}{\omega\mu} \right) \tag{11.39b}$$

$$= \left(\mathbf{E} \cdot \mathbf{E}^* \right) \frac{\mathbf{k}}{\omega\mu} \tag{11.39c}$$

$$= \frac{E_0^2}{\omega\mu} \mathbf{k} \tag{11.39d}$$

The relationship between **E**, **H**, **k**, and **S** for plane waves propagating in RH and LH media are summarized in Figure 11.6. In waves propagating through RH media, **k** and **S** are parallel; in LH media where $\mu < 0$, **k** and **S** are antiparallel (colinear with opposite directions). Thus, in LH media, the time-average energy (described by **S**) and constant phase fronts (described by **k**) propagate in opposite directions.

From (11.38), we see that in LH media, the wavenumber vector **k** is in the opposite direction with respect to the Poynting vector **S** compared to propagation in RH media. Since β is positive in RH media, this reversal in wavenumber vector direction is captured by reversing the sign of β. Hence, for LH media we need to choose the negative root of (11.34):

$$\text{LH Medium:} \rightarrow \beta_{\text{LH}} = -\omega\sqrt{\mu\epsilon} < 0 \tag{11.40}$$

Since the frequency is always positive, this finding supports our earlier claim that the index of refraction is also negative in LH media: $n_{\text{LH}} = \beta c/\omega < 0$.

While β reverses sign in LH media, the intrinsic impedance $\eta = \sqrt{\mu/\epsilon}$ is always positive for passive media, both RH and LH. If we write $\mathbf{k} = \beta\hat{\mathbf{k}}$, (11.37a) can be rewritten as

$$\mathbf{H} = \frac{\beta}{\omega\mu}\hat{\mathbf{k}} \times \mathbf{E} \tag{11.41a}$$

$$= \frac{1}{\eta}\hat{\mathbf{k}} \times \mathbf{E} \tag{11.41b}$$

This result is exactly the same as that of a plane wave propagating in a RH medium (see (8.78)). The difference here is that $\hat{\mathbf{k}}$ is antiparallel to **k** since $\beta < 0$. Similarly, we may write (11.37b) as

$$\mathbf{E} = -\frac{\beta}{\omega\epsilon}\hat{\mathbf{k}} \times \mathbf{H} \tag{11.42a}$$

$$= -\eta\hat{\mathbf{k}} \times \mathbf{H} \tag{11.42b}$$

Since $\beta < 0$ in LH media, the constant phase fronts travel toward the source. At first, a phase velocity directed toward the source does not seem to make physical sense. However, we note that the phase fronts are only perturbations in the wave; they do not describe energy flow. The Poynting vector, which does describe the direction of power flow (and therefore information transfer), points away from the source in both RH and LH media.

11.3.2 Lossy Media

In the previous section, we considered electromagnetic wave propagation in a lossless LH medium. We now let ϵ and μ be complex quantities, and consider their effects on the propagation constant β. Consider a region filled with a lossy medium, either LH or RH, that is free from any sources. Since there are no sources, the time-average power flow out of a closed surface S must be negative:

$$\frac{1}{2}\mathcal{R}e\left\{\oint_S \mathbf{S} \cdot d\mathbf{s}\right\} = \frac{1}{2}\mathcal{R}e\left\{\oint_S (\mathbf{E} \times \mathbf{H}^*) \cdot d\mathbf{s}\right\} < 0 \tag{11.43}$$

If we associate all the losses in a medium with the imaginary components of its complex permittivity and permeability, the complex Poynting theorem (equation (8.53) with $\mathbf{J}^* = 0$ since we assume $\sigma = 0$) becomes

$$\nabla \cdot (\mathbf{E} \times \mathbf{H}^*) = -j\omega \left(\mathbf{B} \cdot \mathbf{H}^* - \mathbf{E} \cdot \mathbf{D}^*\right) = -j\omega \left(\mu |\mathbf{H}|^2 - \epsilon^* |\mathbf{E}|^2\right) \qquad (11.44)$$

Integrating (11.44) over a volume V enclosed by the surface S and using the divergence theorem, we have

$$\oint_S (\mathbf{E} \times \mathbf{H}^*) \cdot d\mathbf{s} = -j\omega \int_V \left(\mu |\mathbf{H}|^2 - \epsilon^* |\mathbf{E}|^2\right) dv \qquad (11.45)$$

and so[25]

$$\mathcal{R}e \left\{ \oint_S (\mathbf{E} \times \mathbf{H}^*) \cdot d\mathbf{s} \right\} = \omega \mathcal{I}m \left\{ \int_V \left(\mu |\mathbf{H}|^2 - \epsilon^* |\mathbf{E}|^2\right) dv \right\} < 0 \qquad (11.46)$$

Noting that $\mathcal{I}m\{\epsilon^*\} = -\mathcal{I}m\{\epsilon\}$, (11.46) is equivalent to

$$\omega \int_V \left(\mathcal{I}m\{\mu\} |\mathbf{H}|^2 + \mathcal{I}m\{\epsilon\} |\mathbf{E}|^2\right) dv < 0 \qquad (11.47)$$

Since this relation must hold for any volume V, we have the general result

$$\mathcal{I}m\{\epsilon\} < 0 \quad \text{and} \quad \mathcal{I}m\{\mu\} < 0 \qquad (11.48)$$

This restriction on the sign of the imaginary components of ϵ and μ is the reason why the complex permittivity was defined as $\epsilon' - j\epsilon''$ and the complex permeability as $\mu' - j\mu''$ in Section 8.3.

We now restrict our consideration to a LH medium. Using (11.48) and noting that for LH medium $\mathcal{R}e\{\epsilon\} < 0$, $\mathcal{R}e\{\mu\} < 0$, it can easily be shown (see Problem 11.5) that $\mathcal{I}m\{\beta^2\} = \mathcal{I}m\{\omega^2 \mu\epsilon\} > 0$. This condition also implies that $\mathcal{R}e\{\beta\}\,\mathcal{I}m\{\beta\} > 0$ (see Problem 11.6), and so the real and imaginary component of the wavevector are either both positive or both negative:

$$\left\{ \begin{array}{l} \mathcal{R}e\{\beta\} > 0 \\ \mathcal{I}m\{\beta\} > 0 \end{array} \right. \quad \text{or} \quad \left\{ \begin{array}{l} \mathcal{R}e\{\beta\} < 0 \\ \mathcal{I}m\{\beta\} < 0 \end{array} \right. \qquad (11.49)$$

To reiterate this result, we derive β for propagation in a low-loss LH material. In LH media, since $\mathcal{R}e\{\epsilon\} < 0$, we can write $\epsilon = -|\epsilon'| - j\epsilon''$. In media with small dielectric losses, so that $\epsilon'' \ll \epsilon'$, we may write

$$\sqrt{\epsilon} = \pm\sqrt{-|\epsilon'| - j\epsilon''}$$

$$= \pm j\sqrt{|\epsilon'|}\sqrt{1 + j\frac{\epsilon''}{|\epsilon'|}}$$

---

[25] Note that for a complex number z, $\mathcal{R}e\{z\} = \mathcal{I}m\{jz\}$.

$$\approx \pm j \sqrt{|\epsilon'|} \left(1 + j \frac{\epsilon''}{2|\epsilon'|} \right)$$

$$= \pm j \left(\sqrt{|\epsilon'|} + j \frac{\epsilon''}{2\sqrt{|\epsilon'|}} \right)$$

where we used the Taylor series expansion $\sqrt{1+x} = 1 + x/2 + \ldots$. If the magnetic losses are also small, an analogous derivation gives

$$\sqrt{\mu} \approx \pm j \left(\sqrt{|\mu'|} + j \frac{\mu''}{2\sqrt{|\mu'|}} \right) \tag{11.50}$$

So, for a LH media with small losses, the wavevector number is approximately

$$\beta = \beta_r + j\beta_i = \omega\sqrt{\epsilon}\sqrt{\mu} \approx \mp\omega\sqrt{|\epsilon'||\mu'|} \left[1 + \frac{j}{2} \left(\frac{\epsilon''}{|\epsilon'|} + \frac{\mu''}{|\mu'|} \right) \right] \tag{11.51}$$

where we have dropped the term $-\epsilon''\mu''/(4|\epsilon'||\mu'|) \ll 1$. We see that since $\epsilon'' > 0$ and $\mu'' > 0$, the real and imaginary components of β have the same sign, that is, they are either both positive or both negative.

As an example, consider a plane wave propagating in the z direction. All field components have a z-dependence of

$$e^{-j\beta z} = e^{-j(\beta_r + j\beta_i)z} = e^{-j\beta_r z} e^{\beta_i z} \tag{11.52}$$

For waves with constant phase fronts travelling in the $+z$ direction, the amplitude grows exponentially. Physically, power flows in the direction of amplitude attenuation in a lossy material, so the wavefront direction is antiparallel to the Poynting vector direction, which is consistent with the notion of backwards propagation (wavefronts traveling toward the source). In a RH medium, the real and imaginary components of β have opposite sign, and so the waves grow in the opposite direction of wavefront propagation.

11.3.3 Material Dispersion

In this section, we show that LH media are always frequency dispersive (i.e., the propagation constant, β, is a nonlinear function of frequency). If β is a nonlinear function of frequency, it follows that β/ω is frequency dependent, and so based on (11.40), either ϵ and/or μ must be frequency dependent. Physically, this condition corresponds to a medium where the relationship between $\overline{\mathcal{E}}$ and $\overline{\mathcal{D}}$ (if ϵ is frequency dependent) or between $\overline{\mathcal{H}}$ and $\overline{\mathcal{B}}$ (if μ is frequency dependent) is not instantaneous but instead also depends on field values in the past. In Section 11.2, we showed that the interaction between the electric field of an electromagnetic wave with the bound electrons in a material yields a frequency-dependent permittivity. In this case, the electric field $\overline{\mathcal{E}}$ created the polarization vector $\overline{\mathcal{P}}$ and hence the flux density $\overline{\mathcal{D}} = \epsilon_0\overline{\mathcal{E}} + \overline{\mathcal{P}}$ by acting on the bound electrons of the atoms in the medium. If the bound electrons redistributed instantaneously, then the flux density $\overline{\mathscr{D}}$ would be related to $\overline{\mathscr{E}}$ at each point in space by a scalar proportionality constant. However, if there is a delay between the applied electric field and the polarization, then the flux density $\overline{\mathscr{D}}$ would depend not only on the current value of $\overline{\mathscr{E}}$ but also on values of $\overline{\mathscr{E}}$ in the past. Assuming linear and time invariant conditions, this relationship is captured using a convolution integral:

$$\overline{\mathscr{D}}(\mathbf{r},t) = \epsilon(t) * \overline{\mathscr{E}}(\mathbf{r},t) = \int_{-\infty}^{t} \epsilon(t-t')\overline{\mathscr{E}}(\mathbf{r},t')dt' \tag{11.53a}$$

$$\overline{\mathscr{B}}(\mathbf{r},t) = \mu(t) * \overline{\mathscr{H}}(\mathbf{r},t) = \int_{-\infty}^{t} \mu(t-t')\overline{\mathscr{H}}(\mathbf{r},t')dt' \tag{11.53b}$$

The upper limit of the convolution integral in (11.53) reflects the fact that $\overline{\mathscr{D}}(\mathbf{r},t)$ and $\overline{\mathscr{B}}(\mathbf{r},t)$ depend on the superposition of the effects of $\overline{\mathscr{E}}(\mathbf{r},t)$ and $\overline{\mathscr{H}}(\mathbf{r},t)$, respectively, over the time interval $t' \leq t$.

It is well known that a convolution in the time domain corresponds to multiplication in the frequency domain. Using the Fourier transform pair

$$\widehat{x}(\omega) = \int_{-\infty}^{\infty} x(t)e^{-j\omega t}dt \tag{11.54a}$$

$$x(t) = \frac{1}{2\pi}\int_{-\infty}^{\infty} \widehat{x}(\omega)e^{j\omega t}d\omega \tag{11.54b}$$

we have

$$\widehat{\overline{\mathscr{D}}}(\mathbf{r},\omega) = \widehat{\epsilon}(\omega)\widehat{\overline{\mathscr{E}}}(\mathbf{r},\omega) \tag{11.55a}$$

$$\widehat{\overline{\mathscr{B}}}(\mathbf{r},\omega) = \widehat{\mu}(\omega)\widehat{\overline{\mathscr{H}}}(\mathbf{r},\omega) \tag{11.55b}$$

Written in terms of phasors, and dropping the $\widehat{\cdot}$ marker on the permittivity and permeability values with an understanding that each is now written as a time-harmonic entity, the time-harmonic constituent relations are

$$\mathbf{D}(\mathbf{r},\omega) = \epsilon(\omega)\mathbf{E}(\mathbf{r},\omega) \tag{11.56a}$$

$$\mathbf{B}(\mathbf{r},\omega) = \mu(\omega)\mathbf{H}(\mathbf{r},\omega) \tag{11.56b}$$

Thus a frequency-dependent permittivity and/or permeability corresponds to a medium with "memory," where the physical medium has a delay property in response to the applied electric and magnetic fields.

With the constitutive relationships given by (11.53), the derivation of the Poynting vector following (8.43) is no longer valid. More specifically, the instantaneous electric and magnetic energy densities are no longer given by $\frac{1}{2}\epsilon|\overline{\mathscr{E}}|^2$ and $\frac{1}{2}\mu|\overline{\mathscr{H}}|^2$, respectively. For band-limited wavepackets traveling in a dispersive media, if the bandwidth is sufficiently limited around the center frequency ω_0 such that the field amplitude varies slowly in comparison to $e^{j\omega_0 t}$, it can be shown[26] that the time-average energy density per unit volume can be approximated as

$$\overline{W} = \frac{1}{4}\left[\frac{\partial(\omega\epsilon)}{\partial\omega}|\mathbf{E}|^2 + \frac{\partial(\omega\mu)}{\partial\omega}|\mathbf{H}|^2\right] \tag{11.57}$$

Since $\overline{W} > 0$, we must have

$$\frac{\partial(\omega\epsilon)}{\partial\omega} = \epsilon + \omega\frac{\partial\epsilon}{\partial\omega} > 0 \tag{11.58a}$$

$$\frac{\partial(\omega\mu)}{\partial\omega} = \mu + \omega\frac{\partial\mu}{\partial\omega} > 0 \tag{11.58b}$$

and therefore we have the following conditions on the amplitudes of the frequency derivatives of the permittivity and permeability for LH media (noting that $\epsilon < 0$, $\mu < 0$):

$$\frac{\partial\epsilon}{\partial\omega} > \frac{-\epsilon}{\omega} \tag{11.59a}$$

$$\frac{\partial\mu}{\partial\omega} > \frac{-\mu}{\omega} \tag{11.59b}$$

Thus, LH media must be highly dispersive, that is, ϵ and μ, and thus the phase velocity, vary with frequency.

We note that if ϵ and μ are not frequency dependent, (11.57) reduces to

$$\overline{W} = \frac{1}{4}\left[\epsilon|\mathbf{E}|^2 + \mu|\mathbf{H}|^2\right] \tag{11.60}$$

which would have led to the nonphysical result of $\overline{W} < 0$ for LH media. We also note that, for dispersive media, and therefore all LH media, the interpretation of the imaginary component of the complex Poynting theorem (equation (8.57)), which relates the flux of the reactive power through a closed surface with the difference between the electric and magnetic energies inside the surface, is no longer valid.

The conditions (11.58), together with a negative permittivity and permeability, constitute another derivation of backward propagating waves. The inequalities given in (11.58), together with negative ϵ and μ, yield the following inequality on the derivative of $\beta^2 = \omega^2\mu\epsilon$ with respect to angular frequency:

$$\frac{\partial\beta^2}{\partial\omega} = \omega\epsilon\frac{\partial(\omega\mu)}{\partial\omega} + \omega\mu\frac{\partial(\omega\epsilon)}{\partial\omega} < 0 \tag{11.61}$$

Also,

$$\frac{\partial\beta^2}{\partial\omega} = 2\beta\frac{\partial\beta}{\partial\omega} = 2\frac{\omega}{v_p v_g} \tag{11.62}$$

---

[26]For a sample derivation, see Section 2.2 of C. Caloz and T. Itoh, *Electromagnetic Metamaterials: Transmission Line Theory and Microwave Applications,* Wiley IEEE Press, New York, 2005.

where $v_p = \omega/\beta$ is the phase velocity and $v_g = \partial\omega/\partial\beta$ is the group velocity. Hence, from (11.62) and (11.61), for LH media we have

$$v_p v_g < 0 \tag{11.63}$$

That is, wavefronts travel in the opposite direction from wavepackets, which is again consistent with the physical interpretation of backward wave propagation.

11.3.4 Planar Interfaces

A large number of emerging practical applications leverage the unique properties of LH MTMs, particularly backward wave propagation and their highly dispersive properties. These applications range from using a segment of LH material for phase compensation in a transmission line to using slabs of LH materials in novel optical lens designs.[27] These and other applications need to carefully consider the reflection and transmission properties of electromagnetic waves at the interface between different materials, where at least one material is a LH medium. In this section, we revisit the treatment of reflection and refraction at a planar boundary, but this time we assume at least one side of the interface be a LH medium. We begin by summarizing the electromagnetic boundary conditions at the boundary of a LH and RH medium. We then revisit Snell's law and the reflection and refraction at an interface for both perpendicular and parallel polarizations.

Boundary conditions.   The relationships among the normal and tangential components of the electric and magnetic field vectors on either side of an interface between two materials were summarized in Section 7.5. In the absence of a surface charge density and surface current density, the electromagnetic boundary conditions may be summarized as:

$$\mathscr{E}_{1t} = \mathscr{E}_{2t} \tag{11.64a}$$

$$\mathscr{H}_{1t} = \mathscr{H}_{2t} \tag{11.64b}$$

$$\mathscr{D}_{1n} = \mathscr{D}_{2n} \tag{11.64c}$$

$$\mathscr{B}_{1n} = \mathscr{B}_{2n} \tag{11.64d}$$

Thus, assuming no surface currents, both the tangential electric and magnetic ($\mathscr{H}$) field are continuous across the boundary. The normal components of the electric and magnetic field experience a discontinuity according to

$$\mathscr{E}_{1n} = \frac{\epsilon_2}{\epsilon_1}\mathscr{E}_{2n} \tag{11.65a}$$

$$\mathscr{H}_{1n} = \frac{\mu_2}{\mu_1}\mathscr{H}_{2n} \tag{11.65b}$$

[27]L. Billings, Exotic optics: Metamaterial world, *Nature*, 500.7461:138, 2013.

At an interface between a LH and a RH medium, the normal components of $\overline{\mathscr{E}}$ and $\overline{\mathscr{H}}$ change sign across the boundary since $\epsilon_2/\epsilon_1 < 0$ and $\mu_2/\mu_1 < 0$. The relationships between the tangential field components are the same as with two RH media since they do not involve ϵ and μ. Thus, at a RH/LH interface, boundary conditions for the normal and tangential electric and magnetic field vectors are

$$\mathscr{E}_{1t} = \mathscr{E}_{2t} \tag{11.66a}$$

$$\mathscr{H}_{1t} = \mathscr{H}_{2t} \tag{11.66b}$$

$$\mathscr{E}_{1n} = -\frac{|\epsilon_2|}{|\epsilon_1|}\mathscr{E}_{2n} \tag{11.66c}$$

$$\mathscr{H}_{1n} = -\frac{|\mu_2|}{|\mu_1|}\mathscr{H}_{2n} \tag{11.66d}$$

Snell's law.   In the process of deriving the perpendicular and parallel reflection coefficients in Section 9.6, we showed that Snell's law was a consequence of phase matching between the tangential fields at the boundary. We now revisit this result for a RH and LH media interface.

Consider an incident wave with a wavenumber vector $\mathbf{k}_i = \beta_1 (\sin\theta_i\hat{\mathbf{x}} + \cos\theta_i\hat{\mathbf{z}})$ in the $x-z$ plane, as shown in Figure 11.7. We assume the incident wave to have an arbitrary polarization (i.e., it is a linear combination of perpendicular and parallel polarizations):

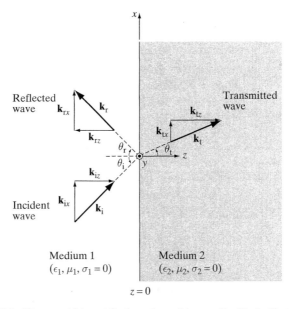

Figure 11.7   Phase matching at the boundary of two media. The incident, reflected, and transmitted wavenumber vectors are shown along with their x- and z-axis projections. The angle θ_t is defined to be positive in rotational directions counterclockwise from the z axis.

$E_i(x,z) = E_{i0}e^{-j\beta_1(x\sin\theta_i + z\cos\theta_i)}$, where $E_{i0} = E_{i0x}\hat{x} + E_{i0y}\hat{y} + E_{i0z}\hat{z}$. The incident, reflected, and transmitted waves may be expressed as

$$E_i(x,z) = \left(E_{i0x}\hat{x} + E_{i0y}\hat{y} + E_{i0z}\hat{z}\right) e^{-j\beta_1(x\sin\theta_i + z\cos\theta_i)} \tag{11.67a}$$

$$E_r(x,z) = \left(E_{r0x}\hat{x} + E_{r0y}\hat{y} + E_{r0z}\hat{z}\right) e^{-j\beta_1(x\sin\theta_r - z\cos\theta_r)} \tag{11.67b}$$

$$E_t(x,z) = \left(E_{t0x}\hat{x} + E_{t0y}\hat{y} + E_{t0z}\hat{z}\right) e^{-j\beta_2(x\sin\theta_t + z\cos\theta_t)} \tag{11.67c}$$

As discussed in the previous section, the total tangential electric field is continuous across the boundary between the two materials. Let us consider the $\hat{y}$ component of the electric field (this component represents the perpendicular polarization; however, we would arrive at the same result if we instead considered the $\hat{x}$ component). At the boundary ($z = 0$), the $\hat{y}$ component of the incident, reflected, and transmitted electric fields are related by

$$E_{i0y}e^{-j\beta_1 x\sin\theta_i} + E_{r0y}e^{-j\beta_1 x\sin\theta_r} = E_{t0y}e^{-j\beta_2 x\sin\theta_t} \tag{11.68}$$

At the origin ($x = 0$), (11.68) reduces to $E_{i0y} + E_{r0y} = E_{t0y}$. Additionally, since (11.68) holds for all x, we must have

$$\beta_1 \sin\theta_i = \beta_1 \sin\theta_r = \beta_2 \sin\theta_t \tag{11.69}$$

Equation (11.69) states that the tangential component of the wavenumber vector is continuous across a boundary between two disparate media. This result is a direct consequence of the continuity of the tangential components of E at the boundary (note that we could have derived the same result by enforcing continuity with the tangential components of H), and so it holds for both RH and LH medium. The left-side equality of (11.69) recovers Snell's law of reflection ($\theta_i = \theta_r$), since the incident and reflected waves are travelling in the same medium.

We now consider specifically an interface between a RH medium and a LH medium. Since based on (11.40), $\beta_2 < 0$, from (11.69), we also have $\sin\theta_t < 0$ and therefore, $\theta_t < 0$. Furthermore, the $\hat{z}$ component of k_t is negative, since $\beta_2\cos\theta_t < 0$. Thus, the transmitted wavenumber vector points toward the boundary, and as a result, the normal components of the incident and transmitted wavenumber vectors are antiparallel. The resulting geometry of a reflected and refracted wave at the interface between a RH and LH medium is depicted in Figure 11.8. A wave incident on a RH:RH or LH:LH medium boundary is said to undergo positive refraction, since the refraction angle θ_t is positive. However, a wave incident on a RH:LH or LH:RH medium boundary undergoes *negative refraction*, since the refraction angle θ_t is negative.

Since the wavenumber vector and the Poynting vector are antiparallel in LH media, one consequence of negative refraction is that the tangential component of the Poynting vector changes direction at the interface between two media of different handedness. This remarkable consequence enables a large collection of potential applications in optics and microwave engineering. For example, a LH material slab may be used to create a flat lens without an optical axis (see Example 11.1).

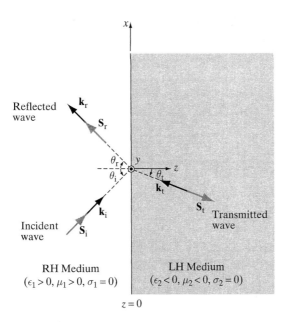

Figure 11.8 Negative refraction. The refracted wave in a LH medium is described by a negative angle of refraction.

We now revisit the reflection and transmission coefficients for oblique incidence on a dielectric interface where the transmission medium is a LH material. We first consider the perpendicular polarization case, followed by the parallel polarization case.

Perpendicular polarization.   As we did in Section 9.6.1, we begin by writing down the equations for the incident, reflected, and transmitted electric and magnetic fields. Using the same coordinate system of Figure 9.26, we have

$$\mathbf{E}_i(x,z) = \hat{\mathbf{y}} E_{i0} e^{-j\beta_1(x\sin\theta_i + z\cos\theta_i)} \tag{11.70a}$$

$$\mathbf{H}_i(x,z) = \frac{1}{\eta_1}\hat{\mathbf{k}}_i \times \mathbf{E}_i$$

$$= \frac{E_{i0}}{\eta_1}(-\hat{\mathbf{x}}\cos\theta_i + \hat{\mathbf{z}}\sin\theta_i)\, e^{-j\beta_1(x\sin\theta_i + z\cos\theta_i)} \tag{11.70b}$$

$$\mathbf{E}_r(x,z) = \hat{\mathbf{y}} E_{r0} e^{-j\beta_1(x\sin\theta_r - z\cos\theta_r)} \tag{11.70c}$$

$$\mathbf{H}_r(x,z) = \frac{E_{r0}}{\eta_1}(\hat{\mathbf{x}}\cos\theta_r + \hat{\mathbf{z}}\sin\theta_r)\, e^{-j\beta_1(x\sin\theta_r - z\cos\theta_r)} \tag{11.70d}$$

$$\mathbf{E}_t(x,z) = \hat{\mathbf{y}} E_{t0} e^{-j\beta_2(x\sin\theta_t + z\cos\theta_t)} \tag{11.70e}$$

$$\mathbf{H}_t(x,z) = \frac{E_{t0}}{\eta_2}(-\hat{\mathbf{x}}\cos\theta_t + \hat{\mathbf{z}}\sin\theta_t)\, e^{-j\beta_2(x\sin\theta_t + z\cos\theta_t)} \tag{11.70f}$$

These field equations are identical to the expressions written in Section 9.6.1. If medium 2 is a LH material with $\mu_2 < 0$, $\epsilon_2 < 0$, and therefore $\beta_2 < 0$, then, based on Figure 11.6, the $\{\mathbf{E}, \mathbf{H}, \mathbf{k}\}$ triad conforms to the left-handed rule and, from the previous section, we also have $\theta_t < 0$. The resulting orientations of the field and wavenumber vectors when medium 2 is LH are shown in Figure 11.9. Enforcing the continuity of the tangential electric field component at the boundary $z = 0$ yields

$$E_{i0}e^{-j\beta_1 x \sin\theta_i} + E_{r0}e^{-j\beta_1 x \sin\theta_r} = E_{t0}e^{-j\beta_2 x \sin\theta_t} \tag{11.71}$$

Since this relation must hold for all x, we again recover the phase matching condition:

$$\beta_1 \sin\theta_i = \beta_1 \sin\theta_r = \beta_2 \sin\theta_t \tag{11.72}$$

Defining $\Gamma_\perp \equiv E_{r0}/E_{i0}$ and $\mathcal{T}_\perp \equiv E_{t0}/E_{i0}$, we can also write

$$1 + \Gamma_\perp = \mathcal{T}_\perp \tag{11.73}$$

The derivation of the transmission and reflection coefficients at a boundary between two RH media in Chapter 9 used conservation of power as a second boundary condition. We follow here a slightly different approach, using instead the continuity of the tangential magnetic field as a second boundary condition that must be satisfied at the $z = 0$ interface. From the magnetic field components in (11.70) and the definitions of

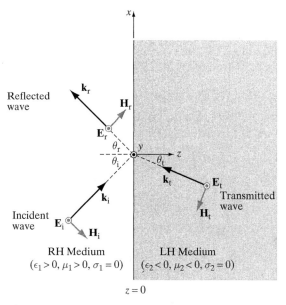

Figure 11.9 Perpendicular polarization. Negative refraction with a perpendicularly polarized incident wave.

$\Gamma_\perp$ and $\mathcal{T}_\perp$, matching the $\hat{\mathbf{x}}$ component of **H** at either side of the boundary yields

$$-\frac{1}{\eta_1}\cos\theta_i + \frac{\Gamma_\perp}{\eta_1}\cos\theta_r = -\frac{\mathcal{T}_\perp}{\eta_2}\cos\theta_t \tag{11.74}$$

Noting that $\theta_i = \theta_r$, (11.74) reduces to

$$\mathcal{T}_\perp = \frac{\eta_2\cos\theta_i}{\eta_1\cos\theta_t}(1 - \Gamma_\perp) \tag{11.75}$$

Solving (11.73) and (11.75) for $\Gamma_\perp$ and $\mathcal{T}_\perp$ gives

$$\Gamma_\perp = \frac{\eta_2\cos\theta_i - \eta_1\cos\theta_t}{\eta_2\cos\theta_i + \eta_1\cos\theta_t} \tag{11.76a}$$

$$\mathcal{T}_\perp = \frac{2\eta_2\cos\theta_i}{\eta_2\cos\theta_i + \eta_1\cos\theta_t} \tag{11.76b}$$

Thus, the magnitude and the phase of the perpendicular reflection and transmission coefficients are the same at the interface of a RH and LH medium as in the case of two RH media. This result is not surprising, since the derivation of the reflection and transmission coefficients depends on the continuity of the tangential electric and magnetic fields, a property that is always valid in the absence of surface currents. Furthermore, the transmission and reflection coefficient expressions involve intrinsic impedances, which are positive for both RH and LH materials, and cosines of the incident and reflected angles, which are invariant with respect to the change in the sign of the angles themselves.

Parallel polarization.   As with perpendicular polarization, the continuity of the tangential fields at the boundary again determine the reflection and transmission coefficients for a parallel polarized wave incident on a dielectric boundary. The field equations of the incident, reflected, and transmitted waves are (see Section 9.6.2)

$$\mathbf{E}_i(x,z) = E_{i0}\,(\hat{\mathbf{x}}\cos\theta_i - \hat{\mathbf{z}}\sin\theta_i)\,e^{-j\beta_1(x\sin\theta_i + z\cos\theta_i)} \tag{11.77a}$$

$$\mathbf{H}_i(x,z) = \frac{1}{\eta_1}\hat{\mathbf{k}}_i \times \mathbf{E}_i$$

$$= \hat{\mathbf{y}}\frac{E_{i0}}{\eta_1}e^{-j\beta_1(x\sin\theta_i + z\cos\theta_i)} \tag{11.77b}$$

$$\mathbf{E}_r(x,z) = E_{r0}\,(\hat{\mathbf{x}}\cos\theta_r + \hat{\mathbf{z}}\sin\theta_r)\,e^{-j\beta_1(x\sin\theta_r - z\cos\theta_r)} \tag{11.77c}$$

$$\mathbf{H}_r(x,z) = -\hat{\mathbf{y}}\frac{E_{r0}}{\eta_1}e^{-j\beta_1(x\sin\theta_r - z\cos\theta_r)} \tag{11.77d}$$

$$\mathbf{E}_t(x,z) = E_{t0}\,(\hat{\mathbf{x}}\cos\theta_t - \hat{\mathbf{z}}\sin\theta_t)\,e^{-j\beta_2(x\sin\theta_t + z\cos\theta_t)} \tag{11.77e}$$

$$\mathbf{H}_t(x,z) = \hat{\mathbf{y}}\frac{E_{t0}}{\eta_2}e^{-j\beta_2(x\sin\theta_t + z\cos\theta_t)} \tag{11.77f}$$

The resulting orientations of the field and wavenumber vectors when medium 2 is LH are shown in Figure 11.10. For the parallel polarization case, we begin by using the continuity of the tangential magnetic field (H_y) at $z = 0$:

$$\frac{E_{i0}}{\eta_1}e^{-j\beta_1 x \sin\theta_i} - \frac{E_{r0}}{\eta_1}e^{-j\beta_1 x \sin\theta_r} = \frac{E_{t0}}{\eta_2}e^{-j\beta_2 x \sin\theta_t} \qquad (11.78)$$

Since (11.78) must be satisfied for all x, we once again recover the phase matching criteria at the boundary:

$$\beta_1 \sin\theta_i = \beta_1 \sin\theta_r = \beta_2 \sin\theta_t \qquad (11.79)$$

As before we define the reflection and transmission coefficients in terms of the electric field magnitudes, so that $\Gamma_\parallel = E_{r0}/E_{i0}$ and $\mathcal{T}_\parallel = E_{t0}/E_{i0}$. With these definitions, (11.78) can be rewritten as

$$\mathcal{T}_\parallel = \frac{\eta_2}{\eta_1}\left(1 - \Gamma_\parallel\right) \qquad (11.80)$$

Using the continuity of the tangential electric field (E_x) as a second boundary condition, we have

$$\cos\theta_i + \Gamma_\parallel \cos\theta_r = \mathcal{T}_\parallel \cos\theta_t \qquad (11.81)$$

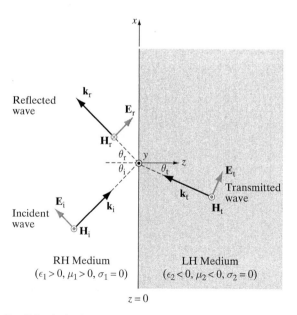

Figure 11.10 Parallel polarization. Negative refraction with a parallel polarized incident wave.

Using $\theta_r = \theta_i$, we have

$$\mathcal{T}_\parallel = \frac{\cos \theta_i}{\cos \theta_t} \left(1 + \Gamma_\parallel \right) \tag{11.82}$$

Solving (11.80) and (11.82) simultaneously to obtain the reflection and transmission coefficients for parallel polarization, we have

$$\Gamma_\parallel = \frac{\eta_2 \cos \theta_t - \eta_1 \cos \theta_i}{\eta_2 \cos \theta_t + \eta_1 \cos \theta_i} \tag{11.83a}$$

$$\mathcal{T}_\parallel = \frac{2\eta_2 \cos \theta_i}{\eta_2 \cos \theta_t + \eta_1 \cos \theta_i} \tag{11.83b}$$

As with the perpendicular polarization, the reflection and transmission coefficients have the same magnitude and sign for a RH:LH medium boundary compared to a RH:RH medium boundary.

Example 11.1: Flat Lens. Consider a point source in a RH medium radiating from a distance x to a LH slab of thickness d, as shown in Figure 11.11. (a) Determine the distance l_1 from the left face of the slab to the focal point P_1 and the distance l_2 from the right face of the slab to the focal point P_2 in terms of x and θ_{RH}. (b) Determine the conditions on ϵ_2 and μ_2 to focus incident plane waves of arbitrary angles θ_{RH} to the same points P_1 and P_2. (c) Determine the conditions on ϵ_2 and μ_2 to have a "perfect" focusing lens, with no loss between the source and point P_2.

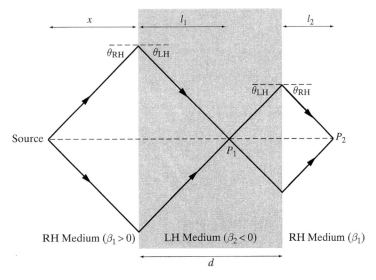

Figure 11.11   LH slab "flat" lens. Double focusing from a LH slab.

Solution:

(a) The vertical distance between the line connecting the source point and focal point P_1 and the reflection point between the RH medium and the LH slab is

$$x \tan \theta_{RH} = l_1 \tan |\theta_{LH}| \tag{11.84}$$

from which we obtain

$$l_1 = x \frac{\tan \theta_{RH}}{\tan |\theta_{LH}|} \tag{11.85}$$

Using an equivalent derivation, one can obtain

$$l_2 = d \frac{x}{s_1} - x \tag{11.86}$$

By Snell's law, $\beta_1 \sin \theta_{RH} = \beta_2 \sin \theta_{LH}$, so

$$\theta_{LH} = \sin^{-1} \left(\frac{\beta_1}{\beta_2} \sin \theta_{RH} \right) \tag{11.87}$$

Solving (11.87) along with (11.85) and (11.86) produces the desired expressions.

(b) From (11.87) and (11.85), we see that if $\beta_1 = -\beta_2$, then $l_1 = x$ and is independent of the incident angle θ_{RH}. Under this condition, $l_2 = d - x$ is also independent of θ_{RH}. If $\beta_1 \neq -\beta_2$, then the distances l_1 and l_2 are dependent on the angle θ_{RH}. If we consider a point source emitting spherical waves, then each component plane wave would have a different focal point based on its incidence angle. Thus, to focus a point source, we must have $\beta_1/\beta_2 = -1$, or

$$\frac{\mu_1 \epsilon_1}{|\mu_2||\epsilon_2|} = 1$$

(c) In order to have no losses, we need $\theta_{RH} = \theta_{LH}$ and $\eta_1 = \eta_2$ (in which case $\Gamma_\perp = \Gamma_\| = 0$ and $\mathcal{T}_\perp = \mathcal{T}_\| = 1$). Hence, we must also have

$$\frac{|\mu_2|\epsilon_1}{|\epsilon_2|\mu_1} = 1$$

That is, both the ratios and the products of the permittivity and the permeability must be same. The only way to satisfy this condition is to have

$$\epsilon_1 = -\epsilon_2$$

$$\mu_1 = -\mu_2$$

11.4 SUMMARY

This chapter discussed the following topics:

- **Wave propagation in a plasma.** In terms of propagation of uniform plane electromagnetic waves, an ionized medium with electron density N_e behaves as if it has an effective permittivity

$$\epsilon_{\text{eff}} = \epsilon_0 \left(1 - \frac{f_p^2}{f^2} \right)$$

where $f_p = (2\pi)^{-1}\sqrt{N_e q_e^2/(m_e \epsilon_0)} \simeq 9\sqrt{N_e}$ is the plasma frequency. The cutoff frequency of uniform plane waves in a plasma is $f_c = f_p$, with propagation being possible only for frequencies $f > f_p$.

- **Frequency response of dielectrics and metals.** The microscopic response of a dielectric to an applied electric field is highly frequency-dependent, with resonances occurring due to electronic, ionic, and dipolar polarizations. In the vicinity of such resonances, the imaginary part of the complex permittivity peaks, leading to losses and absorption of the energy of an electromagnetic wave. Similar frequency-dependent microscopic behavior in metals makes the conductivity of metals frequency dependent, typically at infrared and higher frequencies. At far ultraviolet and higher frequencies, metals respond to uniform plane waves in a similar manner as dielectrics.

- **Metamaterials.** MTMs are bulk materials manufactured to have specific constitutive parameters over a specified frequency range. A class of MTMs includes media with simultaneously negative permittivity and permeability. Electromagnetic waves propagating in media with both $\epsilon < 0$ and $\mu < 0$ have antiparallel wavenumber and Poynting vectors. That is, the group and phase velocities have opposite signs:

$$v_p v_g < 0$$

This property is captured by assigning a negative wavenumber β to solutions of the wave equation. Since the relationship between the electric field, magnetic field, and wavenumber vectors satisfy the left-hand rule, these materials are often called LH media.

It can be shown that LH media are intrinsically dispersive, that is, the propagation constant β is a nonlinear function of frequency. The lower bound for the derivatives of the permittivity and permeability versus frequency are given by

$$\frac{\partial \epsilon}{\partial \omega} > \frac{-\epsilon}{\omega}$$

$$\frac{\partial \mu}{\partial \omega} > \frac{-\mu}{\omega}$$

At a plane boundary between a simple RH medium and LH material, incident waves undergo negative refraction. Snell's law takes the usual form, given by

$$\beta_1 \sin\theta_i = \beta_2 \sin\theta_t$$

If the wavenumber β_2 is negative, then the angle of refraction θ_t must also be negative. This striking property invites a plethora of novel engineering applications, including a flat LH slab capable of acting as a focusing lens.

PROBLEMS

11.1   **GPS signal transmission through the ionosphere.** The Global Positioning System (GPS) uses frequencies of 1.228 and 1.575 GHz. The highest electron density in the earth's ionosphere occurs near the so-called F-region (at altitudes of $\sim$300 km) of the ionosphere, reaching values of $N_e = 10^7$ cm^{-3} during periods of high solar activity. By modeling the ionosphere as a single layer of 50 km thickness with uniform electron density N_e, assess the effect of the ionosphere on the GPS signal. What fraction of a GPS signal vertically incident from space onto the ionospheric layer is reflected?

11.2   **Space vehicle reentry.** The intense friction around a space vehicle reentering the atmosphere generates a plasma sheath which is 10 meter thick and is characterized by an electron density of $N_e = 10^{13}$ cm^{-3}, and a collision frequency of $v = 10^{11}$ s^{-1}. What frequency is required in order to transmit plane waves through the plasma sheath with a minimum power loss of 10 dB?

11.3   **Reflection from the ionosphere.** Consider the propagation of waves by reflection from the ionosphere as shown in Figure 11.12. (a) On a particular date and time of day, and above a particular geographic location, the highest electron density occurs at an altitude of ~500 km, and the plasma frequency at this altitude is $f_p \simeq 5$ MHz. Neglecting any ionization below this densest layer (i.e., assume electron density to be zero below 500 km altitude), determine the greatest possible angle of incidence θ_i at which an electromagnetic wave originating at a ground-based transmitter could possibly strike the layer. (b) How far away from the transmitter will the reflected radiation return to earth? (c) What is the highest frequency at which this obliquely incident radiation will be totally reflected?

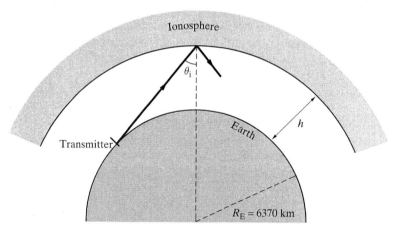

Figure 11.12   Reflection from the ionosphere. Problem 11.3. The sketch shown is not to scale, since $h = 400$ km while the radius of the Earth is ~6370 km as indicated.

11.4   **Attenuation rate in the ionosphere.** Show that the effective permittivity of a plasma with collisions can be expressed as

$$\epsilon_{\text{eff}}(\omega) = \epsilon_0 \left[1 - \frac{\omega_p^2}{\omega^2 + v^2} + j \frac{v \omega_p}{\omega(\omega^2 + v^2)} \right]$$

where v is the collision frequency and ω_p is the plasma frequency. Using this expression to find a simple approximation for the attenuation rate (in dB-m^{-1}) for a 30 MHz wave passing through the lower ionosphere (assume $v \simeq 10^7$ s^{-1}, $N_e \simeq 10^8$ m^{-3}).

11.5 **Lossy LH Media.** (a) In a lossy LH medium with $\Re e\{\epsilon\} < 0$ and $\Re e\{\mu\} < 0$, show that $\Im m\{\beta^2\} = \Im m\{\omega^2\mu\epsilon\} > 0$. (b) In a lossy RH medium with $\Re e\{\epsilon\} > 0$ and $\Re e\{\mu\} > 0$, show that $\Im m\{\beta^2\} = \Im m\{\omega^2\mu\epsilon\} < 0$.

11.6 **Lossy LH Media.** In a lossy LH medium with $\Re e\{\epsilon\} < 0$ and $\Re e\{\mu\} < 0$, use the result from part(a) of Problem 11.5 to show that $\Re e\{\beta\}\Im m\{\beta\} > 0$.

11.7 **Lorentz model of a metamaterial.** In Section 11.2.1, we developed a model for the relative permittivity by considering the dynamics of a bound electron in the presence of an external electric field. The formula for ϵ_r given by (11.28) is called the Lorentz model. Using (11.5), we can rewrite (11.28a) as

$$\epsilon_r = 1 + \frac{\omega_{pe}^2}{\omega_{0e}^2 - \omega^2 + j\omega\kappa_e}$$

where we added the subscript e to associate the characteristic frequencies with the bound electron model. A similar model may be used to approximate the permeability of artificially constructed metamaterials:[28]

$$\mu_r = 1 + \frac{\omega_{pm}^2}{\omega_{0m}^2 - \omega^2 + j\omega\kappa_m}$$

(a) First consider the lossless case, where $\kappa_e = \kappa_m = 0$. Using $\omega_{pe} = 2\pi \times 8.0 \times 10^9$ rad-s^{-1}, $\omega_{pm} = 2\pi \times 7.0 \times 10^9$ rad-s^{-1}, $\omega_{0e} = 2\pi \times 2.8 \times 10^9$ rad-s^{-1}, and $\omega_{0m} = 2\pi \times 2.5 \times 10^9$ rad-s^{-1}: (i) Plot the real and imaginary components of ϵ_r and μ_r in the frequency range $4 < f < 11$ GHz. (ii) Plot the real and imaginary components of the index of refraction n in the same frequency range. Use the convention that $\Re e\{\beta\} < 0$ for backward propagation. (iii) In the frequency range $4 < f < 11$ GHz, which frequencies allow for wave propagation? What is the frequency range of backward wave propagation? At what frequency is $\Re e\{n\} = -1$? (iv) Plot βc versus f over the domain $4 < f < 11$ GHz, and comment on the sign of the phase and group velocities over frequencies where propagating solutions exist.

(b) Repeat part (a) using $\kappa_e = \kappa_m = 0.05\omega_{pe}$.

---

[28] See, for example, G. Lubkowski, R. Schuhmann, and T. Weiland, Extraction of effective metamaterial parameters by parameter fitting of dispersive models, *Microwave Opt. Technol. Lett.* 49(2), pp. 285–288, 2007.

A

Vector Analysis

Vector analysis is used extensively in electromagnetics and in other fields of engineering and physics in which the physical quantities involved have both direction and magnitude; that is, they are *vectors*. Examples of physical quantities that are vectors include velocity, momentum, force, displacement, electric field, current, and magnetic field. Vector quantities are inherently different from *scalars,* which are physical quantities that are specified entirely by one value, such as the number of coins in your pocket, time, mass, the temperature in a room, electric charge density, electric potential and energy. In this appendix, we review some basic rules and techniques of vector analysis that are particularly relevant to this book. More general topics of vector algebra or vector calculus are covered elsewhere[1] and are not discussed here.

Many of the essential aspects of vector manipulations and definitions needed to work with the material in this book were covered either in the body of the text or in footnotes in those sections where they first appeared. This appendix simply collects these vector rules and techniques in one handy place. The appendix also covers some of the most basic definitions that were taken for granted, such as the definition of vectors, position vectors, unit vectors, and coordinate systems. Some of the more important concepts of vector calculus, including gradient, divergence, curl, and the associated theorems (the divergence theorem and Stokes's theorem), are covered extensively in the text and are not included here.

Throughout this text, vector quantities are often written either using boldface symbols (e.g., $\mathbf{G}$) or with a bar above the symbol (e.g., $\overline{\mathscr{G}}$). Starting with Section 7.4, the latter notation is used to represent real physical quantities and the former for their corresponding complex phasors, as needed for sinusoidal (or time-harmonic) applications. In this appendix, we represent vectors using boldface symbols.

---

[1] Simpler treatments are available in most textbooks; for a complete treatment of the subject, see Chapter 1 of G. B. Arfken and H. J. Weber, *Mathematical Methods for Physicists,* Academic Press, 1995.

To accurately describe vectors requires that we specify their direction as well as their magnitude. To do this, we nccd a framework to orient the vector in three-dimensional space and show its different projections or components. The most common means of doing this is to use the so-called rectangular or Cartesian coordinate system, which we use in Sections A.1 and A.2 to describe the basic aspects of vector addition and multiplication. We then briefly review, in Section A.3, two other commonly used coordinate systems, namely the cylindrical and spherical coordinate systems. Section A.4 presents some commonly encountered vector identities.

A.1 VECTOR COMPONENTS, UNIT VECTORS, AND VECTOR ADDITION

The Cartesian or rectangular coordinate system is illustrated in Figure A.1. To describe a vector $\mathbf{A}$ in the rectangular coordinate system, we can represent it as extending outward

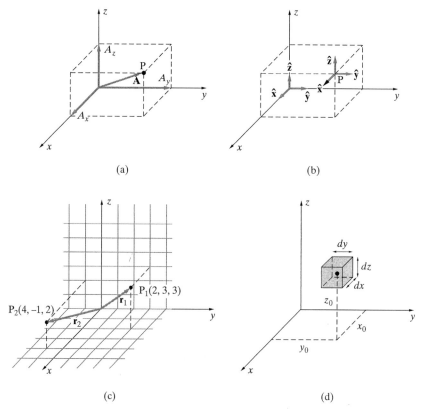

Figure A.1   Rectangular coordinate system. (a) Vector $\mathbf{A}$ pointing from the origin to point $P(A_x, A_y, A_z)$. (b) The three rectangular coordinate unit vectors. (c) Position vectors $\mathbf{r}_1 = \hat{\mathbf{x}}2 + \hat{\mathbf{y}}3 + \hat{\mathbf{z}}3$ and $\mathbf{r}_2 = \hat{\mathbf{x}}4 - \hat{\mathbf{y}} + \hat{\mathbf{z}}2$. (d) The differential volume element in a rectangular coordinate system.

from the origin in a direction determined by the magnitudes of each of its three components, A_x, A_y, and A_z, as shown in Figure A.1a. It is customary to write the vector $\mathbf{A}$ as

$$\mathbf{A} = \hat{\mathbf{x}}A_x + \hat{\mathbf{y}}A_y + \hat{\mathbf{z}}A_z$$

where $\hat{\mathbf{x}}$, $\hat{\mathbf{y}}$, and $\hat{\mathbf{z}}$ are the rectangular coordinate unit vectors, as shown in Figure A.1b.
In this text, we represent unit vectors with the "hat," or circumflex, notation.

The magnitude of vector $\mathbf{A}$ is denoted $|\mathbf{A}|$ (or sometimes simply A) and is given
by

$$A = |\mathbf{A}| = \sqrt{A_x^2 + A_y^2 + A_z^2}$$

Sometimes it is convenient to define a unit vector that is in the direction of a given
vector $\mathbf{A}$. The unit vector in the direction of vector $\mathbf{A}$ is denoted $\hat{\mathbf{A}}$ and is defined as

$$\hat{\mathbf{A}} = \frac{\mathbf{A}}{|\mathbf{A}|} = \frac{\mathbf{A}}{\sqrt{A_x^2 + A_y^2 + A_z^2}}$$

Note that $|\hat{\mathbf{A}}| = 1$. An alternative way of expressing vector $\mathbf{A}$ is $\mathbf{A} = \hat{\mathbf{A}}|\mathbf{A}| = \hat{\mathbf{A}}A$.

Two vectors $\mathbf{A}$ and $\mathbf{B}$ can be added together to produce another vector $\mathbf{C} = \mathbf{A} + \mathbf{B}$.
In rectangular coordinates, vector addition can be carried out component by component.
In other words,

$$\mathbf{C} = \mathbf{A} + \mathbf{B} = \hat{\mathbf{x}}(A_x + B_x) + \hat{\mathbf{y}}(A_y + B_y) + \hat{\mathbf{z}}(A_z + B_z) = \hat{\mathbf{x}}C_x + \hat{\mathbf{y}}C_y + \hat{\mathbf{z}}C_z$$

Geometrically, the orientation and magnitude of the sum vector $\mathbf{C}$ can be determined
using the parallelogram method depicted in Figure A.2a. Subtraction of two vectors is

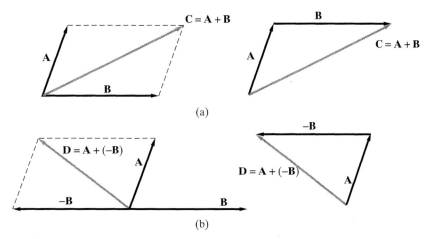

Figure A.2   **Addition and subtraction of vectors.**

done similarly; subtraction of vector **B** from vector **A** gives the same result as the addition of vectors **A** and −**B**, as depicted in Figure A.2b.

A point P in a rectangular coordinate system is represented by its coordinates $P(x, y, z)$. The position vector **r** of a point P is defined as the directed distance from the origin to point P. For example, the position vector of point P_1 in Figure A.1c is $\mathbf{r}_1 = \hat{\mathbf{x}}2 + \hat{\mathbf{y}}3 + \hat{\mathbf{z}}3$. The position vector of point P_2 in Figure A.1c is $\mathbf{r}_2 = \hat{\mathbf{x}}4 - \hat{\mathbf{y}} + \hat{\mathbf{z}}2$. The vector pointing from point P_1 to point P_2 is known as the distance vector (or separation vector) and is given by

$$\mathbf{R} = \mathbf{r}_2 - \mathbf{r}_1 = \hat{\mathbf{x}}4 - \hat{\mathbf{y}} + \hat{\mathbf{z}}2 - (\hat{\mathbf{x}}2 + \hat{\mathbf{y}}3 + \hat{\mathbf{z}}3)$$

$$= \hat{\mathbf{x}}2 - \hat{\mathbf{y}}4 - \hat{\mathbf{z}}$$

Note that the magnitude of **R** is $R = |\mathbf{R}| = \sqrt{2^2 + 4^2 + 1^2} = \sqrt{21}$. The unit vector in the direction of **R** is thus

$$\hat{\mathbf{R}} = \frac{\mathbf{R}}{R} = \hat{\mathbf{x}}\frac{2}{\sqrt{21}} - \hat{\mathbf{y}}\frac{4}{\sqrt{21}} - \hat{\mathbf{z}}\frac{1}{\sqrt{21}}$$

A.2 VECTOR MULTIPLICATION

A.2.1 The Dot Product

The *dot product* of two vectors **A** and **B** is a scalar denoted by **A** · **B**. It is equal to the product of the magnitudes |**A**| and |**B**| of vectors **A** and **B** and the cosine of the angle ψ_{AB} between vectors **A** and **B**. Namely,

$$\mathbf{A} \cdot \mathbf{B} \equiv |\mathbf{A}||\mathbf{B}| \cos \psi_{AB}$$

(The dot product is sometimes referred to as the *scalar product* since the result is a scalar quantity.) Noting that in rectangular coordinates we have $\mathbf{A} = \hat{\mathbf{x}}A_x + \hat{\mathbf{y}}A_y + \hat{\mathbf{z}}A_z$ and $\mathbf{B} = \hat{\mathbf{x}}B_x + \hat{\mathbf{y}}B_y + \hat{\mathbf{z}}B_z$, an alternative expression for the dot product is

$$\mathbf{A} \cdot \mathbf{B} = (\hat{\mathbf{x}}A_x + \hat{\mathbf{y}}A_y + \hat{\mathbf{z}}A_z) \cdot (\hat{\mathbf{x}}B_x + \hat{\mathbf{y}}B_y + \hat{\mathbf{z}}B_z)$$

$$\mathbf{A} \cdot \mathbf{B} = A_x B_x + A_y B_y + A_z B_z$$

where we have used the fact that the dot product of one rectangular unit vector with a different rectangular unit vector is zero, whereas the dot product of any rectangular unit vector with itself is unity. For example, $\hat{\mathbf{x}} \cdot \hat{\mathbf{y}} = 0$ and $\hat{\mathbf{x}} \cdot \hat{\mathbf{z}} = 0$, but $\hat{\mathbf{x}} \cdot \hat{\mathbf{x}} = 1$. This feature of the dot product facilitates determining the component of a vector in a given direction, that is, determining the *projection* of the vector along a given direction. For example, the projection along the y axis of any arbitrary vector $\mathbf{B} = \hat{\mathbf{x}}B_x + \hat{\mathbf{y}}B_y + \hat{\mathbf{z}}B_z$ is simply given by

$$\mathbf{B} \cdot \hat{\mathbf{y}} = (\hat{\mathbf{x}}B_x + \hat{\mathbf{y}}B_y + \hat{\mathbf{z}}B_z) \cdot \hat{\mathbf{y}} = B_y$$

Figure A.3   The dot product. The dot product of **B** with any arbitrary unit vector **â** gives the projection of **B** along **â**, which is negative if $|\psi_{B\hat{a}}| > \pi/2$ or positive if $|\psi_{B\hat{a}}| < \pi/2$.

Similarly, the projection of **B** along any other arbitrary unit vector **â** is represented by **B** · **â**, with the resulting scalar being equal to the length of the vector **B** projected on vector **â**. Note that, depending on whether the angle $\psi_{B\hat{a}}$ is acute or obtuse, the projection could be positive or negative, as illustrated in Figure A.3.

The dot product is both commutative and distributive; in other words,

$$\mathbf{A} \cdot \mathbf{B} = \mathbf{B} \cdot \mathbf{A}$$

$$\mathbf{A} \cdot (\mathbf{B} + \mathbf{C}) = \mathbf{A} \cdot \mathbf{B} + \mathbf{A} \cdot \mathbf{C}$$

A.2.2 The Cross Product

The *cross product* of two vectors **A** and **B** is a vector **C** denoted by **C** = (**A** × **B**). The magnitude of **C** is equal to the product of the magnitudes of vectors **A** and **B** and the sine of the angle ψ_{AB} between vectors **A** and **B**. The direction of **C**, according to the right-hand rule, follows that of the thumb of the right hand when the fingers rotate from **A** to **B** through the angle ψ_{AB}, as depicted in Figure A.4. In other words,

$$\mathbf{A} \times \mathbf{B} \equiv \hat{\mathbf{n}}|\mathbf{A}||\mathbf{B}|\sin\psi_{AB}$$

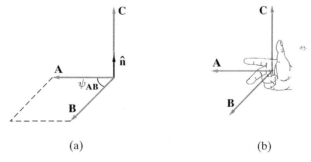

(a)   (b)

Figure A.4   The cross product. The cross product of two vectors **A** and **B** is another vector **C** = **A** × **B**, which is perpendicular to both **A** and **B** and is related to them via the right-hand rule.

where $\hat{\mathbf{n}}$ is the unit vector normal to both $\mathbf{A}$ and $\mathbf{B}$ and points in the direction in which a right-handed screw advances as $\mathbf{A}$ is turned toward $\mathbf{B}$. (The cross product is sometimes referred to as the *vector* product since the result is a vector quantity.)

Noting that $\hat{\mathbf{x}} \times \hat{\mathbf{y}} = \hat{\mathbf{z}}$, $\hat{\mathbf{y}} \times \hat{\mathbf{z}} = \hat{\mathbf{x}}$, and $\hat{\mathbf{z}} \times \hat{\mathbf{x}} = \hat{\mathbf{y}}$, in rectangular coordinates we can use the distributive property of the cross product to write

$$\begin{aligned}
\mathbf{C} = \mathbf{A} \times \mathbf{B} &= (\hat{\mathbf{x}}A_x + \hat{\mathbf{y}}A_y + \hat{\mathbf{z}}A_z) \times (\hat{\mathbf{x}}B_x + \hat{\mathbf{y}}B_y + \hat{\mathbf{z}}B_z) \\
&= \hat{\mathbf{x}}(A_y B_z - A_z B_y) + \hat{\mathbf{y}}(A_z B_x - A_x B_z) + \hat{\mathbf{z}}(A_x B_y - A_y B_x) \\
&= \hat{\mathbf{x}}C_x + \hat{\mathbf{y}}C_y + \hat{\mathbf{z}}C_z
\end{aligned}$$

It is often convenient to write the cross product in determinant form:

$$\mathbf{A} \times \mathbf{B} = \begin{vmatrix} \hat{\mathbf{x}} & \hat{\mathbf{y}} & \hat{\mathbf{z}} \\ A_x & A_y & A_z \\ B_x & B_y & B_z \end{vmatrix}$$

Using the definition of the cross product and following the right-hand rule, it is clear that

$$\mathbf{B} \times \mathbf{A} = -\mathbf{A} \times \mathbf{B}$$

so that the cross product is not commutative. The cross product is also not associative since we have

$$\mathbf{A} \times (\mathbf{B} \times \mathbf{C}) \neq (\mathbf{A} \times \mathbf{B}) \times \mathbf{C}$$

It is clear, however that the cross product is distributive; in other words,

$$\mathbf{A} \times (\mathbf{B} + \mathbf{C}) = \mathbf{A} \times \mathbf{B} + \mathbf{A} \times \mathbf{C}$$

A.2.3 Triple Products

Given three vectors $\mathbf{A}$, $\mathbf{B}$, and $\mathbf{C}$, there are in general two different ways we can form a triple product. The *scalar triple product* is defined as

$$\mathbf{A} \cdot (\mathbf{B} \times \mathbf{C}) = \mathbf{B} \cdot (\mathbf{C} \times \mathbf{A}) = \mathbf{C} \cdot (\mathbf{A} \times \mathbf{B})$$

Note that the result of the scalar triple product is a scalar.

The *vector triple product* is defined as $\mathbf{A} \times (\mathbf{B} \times \mathbf{C})$ and can be expressed as the difference of two simpler vectors as follows:

$$\mathbf{A} \times (\mathbf{B} \times \mathbf{C}) = \mathbf{B}(\mathbf{A} \cdot \mathbf{C}) - \mathbf{C}(\mathbf{A} \cdot \mathbf{B})$$

A.3 CYLINDRICAL AND SPHERICAL COORDINATE SYSTEMS

The physical properties of electromagnetic fields and waves as studied in this book do not depend on the particular coordinate system used to describe the vector quantities. An electric field exists, and may or may not have nonzero divergence or curl, depending on its physical properties with no regard to any system of coordinates. Scalar products, cross products, and other vector operations are all independent of the mathematical frame of reference used to describe them. In practice, however, description, manipulation, and computation of vector quantities in a given problem may well be easier in coordinate systems other than the rectangular system, depending on the symmetries involved. In this section, we describe the two most commonly used additional coordinate systems: namely, the circular cylindrical and the spherical coordinate systems. These two systems are examples of orthogonal curvilinear coordinate systems; other examples not covered here include the elliptic cylindrical, parabolic cylindrical, conical, and prolate spheroidal systems.

A.3.1 Cylindrical Coordinates

The circular cylindrical system is an extension to three dimensions of the polar coordinate system of analytical geometry. In view of its common use, it is often referred to simply as the cylindrical coordinate system, although cylindrical systems with other cross-sectional shapes, such as elliptic or hyperbolic, are also used in special applications. The three cylindrical coordinates are r, ϕ, and z. The radial coordinate r of any point P is the closest distance from the z axis to that point. A given value of r specifies a circular cylindrical surface on which the point P resides, with the particular position of P on the sphere further specified by ϕ and z. The convention is to measure ϕ from the x axis, in the right-handed sense. In other words, with the thumb pointed in the z direction, the fingers trace the direction of increasing ϕ. The angle ϕ is commonly called the azimuthal angle, and it varies from 0 to 2π.

The cylindrical coordinate unit vectors $\hat{\mathbf{r}}$, $\hat{\boldsymbol{\phi}}$, and $\hat{\mathbf{z}}$ are shown in Figure A.5b. Note that the unit vectors are mutually orthogonal and that $\hat{\mathbf{r}} \times \hat{\boldsymbol{\phi}} = \hat{\mathbf{z}}$. A good way to remember the orientation of the unit vectors is to note that they obey the right-hand rule, with the thumb, forefinger, and middle finger representing the directions of $\hat{\mathbf{r}}$, $\hat{\boldsymbol{\phi}}$, and $\hat{\mathbf{z}}$, respectively.

A vector $\mathbf{A}$ in cylindrical coordinates is specified by means of the values of its components A_r, A_ϕ, and A_z and is typically written as

$$\mathbf{A} = \hat{\mathbf{r}}A_r + \hat{\boldsymbol{\phi}}A_\phi + \hat{\mathbf{z}}A_z$$

The magnitude of $\mathbf{A}$ is given by $|\mathbf{A}| = \sqrt{\mathbf{A} \cdot \mathbf{A}} = \sqrt{A_r^2 + A_\phi^2 + A_z^2}$.

The expressions for the scalar and vector products of vectors for cylindrical coordinates are very similar to those for rectangular coordinates. We have

$$\mathbf{A} \cdot \mathbf{B} = A_r B_r + A_\phi B_\phi + A_z B_z$$

$$\mathbf{A} \times \mathbf{B} = \begin{vmatrix} \hat{\mathbf{r}} & \hat{\boldsymbol{\phi}} & \hat{\mathbf{z}} \\ A_r & A_\phi & A_z \\ B_r & B_\phi & B_z \end{vmatrix}$$

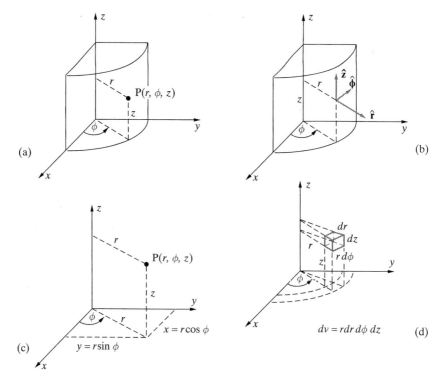

Figure A.5 The cylindrical coordinate system. (a) The three cylindrical coordinates. (b) The three cylindrical coordinate unit vectors. (c) The relationships between rectangular and cylindrical coordinates. (d) The differential volume element in a cylindrical coordinate system.

Cylindrical-to-rectangular transformations.   A vector specified in cylindrical coordinates can be transformed into rectangular coordinates and vice versa. The relationships between the (r, ϕ, z) and (x, y, z) are illustrated in Figure A.5c. Consider a vector

$$\mathbf{A} = \hat{\mathbf{r}}A_r + \hat{\boldsymbol{\phi}}A_\phi + \hat{\mathbf{z}}A_z$$

where in general each of the component values A_r, A_ϕ, and A_z may themselves be functions of spatial coordinates r, ϕ, and z. The rules for transforming from cylindrical to rectangular coordinates are given as follows, where the set of expressions on the left are for transforming scalar quantities (such as A_r), and those on the right are for transforming the vectors.

$$r = \sqrt{x^2 + y^2} \qquad\qquad \hat{\mathbf{r}} = \hat{\mathbf{x}}\cos\phi + \hat{\mathbf{y}}\sin\phi$$

$$\phi = \tan^{-1}\left[\frac{y}{x}\right] \qquad\qquad \hat{\boldsymbol{\phi}} = -\hat{\mathbf{x}}\sin\phi + \hat{\mathbf{y}}\cos\phi$$

$$z = z \qquad\qquad\qquad \hat{\mathbf{z}} = \hat{\mathbf{z}}$$

TABLE A.1   DOT PRODUCTS OF UNIT VECTORS

	Cylindrical Coordinates			Spherical Coordinates		
	$\hat{\mathbf{r}}$	$\hat{\boldsymbol{\phi}}$	$\hat{\mathbf{z}}$	$\hat{\mathbf{r}}$	$\hat{\boldsymbol{\theta}}$	$\hat{\boldsymbol{\phi}}$
$\hat{\mathbf{x}}\cdot$	$\cos\phi$	$-\sin\phi$	0	$\sin\theta\cos\phi$	$\cos\theta\cos\phi$	$-\sin\phi$
$\hat{\mathbf{y}}\cdot$	$\sin\phi$	$\cos\phi$	0	$\sin\theta\sin\phi$	$\cos\theta\sin\phi$	$\cos\phi$
$\hat{\mathbf{z}}\cdot$	0	0	1	$\cos\theta$	$-\sin\theta$	0

Note that the preceding relationships can be deduced from careful examination of Figure A.5b (unit vectors) and Figure A.5c (coordinates). A similar set of rules can be given for transformation from rectangular to cylindrical coordinates:

$$x = r\cos\phi \qquad\qquad \hat{\mathbf{x}} = \hat{\mathbf{r}}\cos\phi - \hat{\boldsymbol{\phi}}\sin\phi$$

$$y = r\sin\phi \qquad\qquad \hat{\mathbf{y}} = \hat{\mathbf{r}}\sin\phi + \hat{\boldsymbol{\phi}}\cos\phi$$

$$z = z \qquad\qquad \hat{\mathbf{z}} = \hat{\mathbf{z}}$$

The transformation of a vector given in the cylindrical (rectangular) coordinate system can be accomplished by direct substitution of the preceding unit vector and coordinate variable expressions. However, doing so in fact amounts to taking the dot product of the vector as written in cylindrical (rectangular) coordinates with the rectangular (cylindrical) coordinate unit vectors and substituting coordinate variables, as illustrated in Example A.1. The dot products of the cylindrical coordinate unit vectors with their rectangular coordinate counterparts are provided in Table A.1.

Example A.1: Cylindrical-to-rectangular and rectangular-to-cylindrical transformations.

(a) Transform a vector

$$\mathbf{A} = \hat{\mathbf{r}}\left(\frac{1}{r}\right) + \hat{\boldsymbol{\phi}}\left(\frac{2}{r}\right)$$

given in cylindrical coordinates to rectangular coordinates.

(b) Transform a vector $\mathbf{B} = \hat{\mathbf{x}}x^2 + \hat{\mathbf{y}}xy$ given in rectangular coordinates to cylindrical coordinates.

Solution:

(a) We set out to find A_x, A_y, and A_z. We have

$$A_x = \mathbf{A}\cdot\hat{\mathbf{x}} = \left[\hat{\mathbf{r}}\left(\frac{1}{r}\right) + \hat{\boldsymbol{\phi}}\left(\frac{2}{r}\right)\right]\cdot\hat{\mathbf{x}} = \frac{\cos\phi}{r} - \frac{2\sin\phi}{r} = \frac{x - 2y}{x^2 + y^2}$$

where we have used the unit vector dot products as given in Table A.1 and substituted $\cos\phi = x/r$, $\sin\phi = y/r$, and $r = \sqrt{x^2 + y^2}$. Similarly, we have

$$A_y = \mathbf{A}\cdot\hat{\mathbf{y}} = \left[\hat{\mathbf{r}}\left(\frac{1}{r}\right) + \hat{\boldsymbol{\phi}}\left(\frac{2}{r}\right)\right]\cdot\hat{\mathbf{y}} = \frac{\sin\phi}{r} + \frac{2\cos\phi}{r} = \frac{y + 2x}{x^2 + y^2}$$

Note that $A_z = 0$, since $\mathbf{A} \cdot \hat{\mathbf{z}} = 0$. Thus we have

$$\mathbf{A} = \hat{\mathbf{x}}\frac{x - 2y}{x^2 + y^2} + \hat{\mathbf{y}}\frac{y + 2x}{x^2 + y^2}$$

(b) We set out to find B_r, B_ϕ, and B_z. We have

$$B_r = \mathbf{B} \cdot \hat{\mathbf{r}} = (\hat{\mathbf{x}}x^2 + \hat{\mathbf{y}}xy) \cdot \hat{\mathbf{r}} = x^2 \cos\phi + xy \sin\phi$$

$$= [r^2 \cos^2\phi \cos\phi + (r\cos\phi)(r\sin\phi)\sin\phi] = r^2 \cos\phi[\cos^2\phi + \sin^2\phi] = r^2 \cos\phi$$

where we have used the unit vector dot products as given in Table A.1 and substituted $x = r\cos\phi$ and $y = r\sin\phi$. Similarly, we have

$$B_\phi = \mathbf{B} \cdot \hat{\boldsymbol{\phi}} = (\hat{\mathbf{x}}x^2 + \hat{\mathbf{y}}xy) \cdot \hat{\boldsymbol{\phi}} = -x^2 \sin\phi + xy \cos\phi$$

$$= [-r^2 \cos^2\phi \sin\phi + (r\cos\phi)(r\sin\phi)\cos\phi] = 0$$

Note that $B_z = 0$ since $\mathbf{B} \cdot \hat{\mathbf{z}} = 0$. Thus, we have

$$\mathbf{B} = \hat{\mathbf{r}}r^2 \cos\phi$$

Length, surface, and volume elements. The general expression for differential length $d\mathbf{l}$ in cylindrical coordinates is

$$d\mathbf{l} = \hat{\mathbf{r}}\,dr + \hat{\boldsymbol{\phi}}r\,d\phi + \hat{\mathbf{z}}\,dz$$

as is evident from Figure A.5d. By inspection of the volume element sketched in Figure A.5d, we can see that the cylindrical coordinate surface elements in the three coordinate directions are

$$(ds)_r = r\,d\phi\,dz$$

$$(ds)_\phi = dr\,dz$$

$$(ds)_z = r\,dr\,d\phi$$

and the volume element in cylindrical coordinates is:

$$dv = r\,dr\,d\phi\,dz$$

A.3.2 Spherical Coordinates

The three spherical coordinates are r, θ, and ϕ. The radial coordinate r of any point P is simply the distance from the origin to that point. A given value of r specifies a sphere on which the point P resides, with the particular position of P on the sphere further specified by θ and ϕ. The best way to think of the spherical coordinate system and especially the coordinates θ and ϕ is in terms of the latitude/longitude system of identifying a point on the earth's surface. The earth is a sphere with radius $r \simeq 6370$ km. With respect to

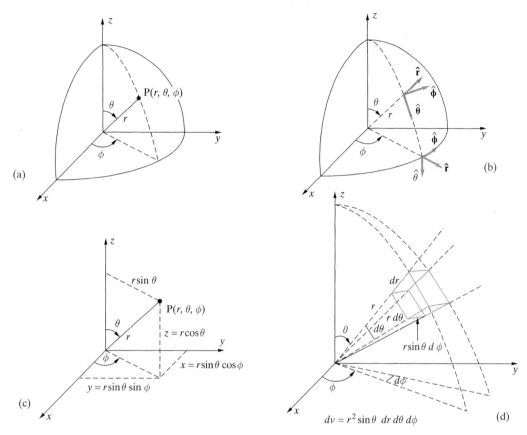

Figure A.6 The spherical coordinate system. (a) The three spherical coordinates. (b) The three spherical coordinate unit vectors. (c) The relationships between rectangular and spherical coordinates. (d) The differential volume element in a spherical coordinate system.

Figure A.6a, latitude corresponds to θ, while longitude corresponds to ϕ. The convention is that the positive z axis points toward north, and ϕ is measured from the x axis, in the right-handed sense. In other words, with the thumb pointed in the z direction, the fingers trace the direction of increasing ϕ. The Greenwich meridian, from which positive (east) longitude is measured, coincides with the positive x axis. The angle ϕ, commonly called the azimuthal angle, varies from 0 to 2π. Unlike geographic latitude, which is commonly measured with respect to the equatorial plane and which thus varies in the range from $-\pi/2$ to $+\pi/2$, the spherical coordinate θ is measured from the positive z axis (see Figure A.6) and thus varies between $\theta = 0$ (north pole) and $\theta = \pi$ (south pole).

The spherical coordinate unit vectors $\hat{\mathbf{r}}$, $\hat{\boldsymbol{\phi}}$, and $\hat{\boldsymbol{\theta}}$ are shown in Figure A.6b. Note that $\hat{\mathbf{r}}$ points radially outward, $\hat{\boldsymbol{\phi}}$ points "east," and $\hat{\boldsymbol{\theta}}$ points "south." Note that the unit vectors are mutually orthogonal and that $\hat{\mathbf{r}} \times \hat{\boldsymbol{\theta}} = \hat{\boldsymbol{\phi}}$. A good way to remember the orientation of the unit vectors is to note that they obey the right-hand rule, with the thumb, forefinger, and middle finger representing the directions of $\hat{\mathbf{r}}$, $\hat{\boldsymbol{\theta}}$, and $\hat{\boldsymbol{\phi}}$, respectively.

A vector $\mathbf{A}$ in spherical coordinates is specified by means of the values of its components A_r, A_θ, and A_ϕ and is typically written as

$$\mathbf{A} = \hat{\mathbf{r}}A_r + \hat{\boldsymbol{\theta}}A_\theta + \hat{\boldsymbol{\phi}}A_\phi$$

The magnitude of $\mathbf{A}$ is given by $|\mathbf{A}| = \sqrt{\mathbf{A} \cdot \mathbf{A}} = \sqrt{A_r^2 + A_\theta^2 + A_\phi^2}$.

The expressions for the scalar and vector products of vectors for spherical coordinates are very similar to those for rectangular coordinates. We have

$$\mathbf{A} \cdot \mathbf{B} = A_r B_r + A_\theta B_\theta + A_\phi B_\phi$$

$$\mathbf{A} \times \mathbf{B} = \begin{vmatrix} \hat{\mathbf{r}} & \hat{\boldsymbol{\theta}} & \hat{\boldsymbol{\phi}} \\ A_r & A_\theta & A_\phi \\ B_r & B_\theta & B_\phi \end{vmatrix}$$

Spherical to rectangular transformations.   A vector specified in spherical coordinates can be transformed into rectangular coordinates and vice versa. The relationship between (r, θ, ϕ) and (x, y, z) is illustrated in Figure A.6c. Consider a vector

$$\mathbf{A} = \hat{\mathbf{r}}A_r + \hat{\boldsymbol{\theta}}A_\theta + \hat{\boldsymbol{\phi}}A_\phi$$

where in general each of the component values A_r, A_θ, and A_ϕ may themselves be functions of spatial coordinates r, θ, and ϕ. The rules for transformation from spherical coordinates to rectangular coordinates are given as follows, where the set of expressions on the left-hand side are for transforming scalar quantities (such as A_r), and those on the right are for transforming vectors.

Coordinate Variables:	Unit Vectors:
$r = \sqrt{x^2 + y^2 + z^2}$	$\hat{\mathbf{r}} = \hat{\mathbf{x}}\sin\theta\cos\phi + \hat{\mathbf{y}}\sin\theta\sin\phi + \hat{\mathbf{z}}\cos\theta$
$\theta = \cos^{-1}\left[\dfrac{z}{\sqrt{x^2 + y^2 + z^2}}\right]$	$\hat{\boldsymbol{\theta}} = \hat{\mathbf{x}}\cos\theta\cos\phi + \hat{\mathbf{y}}\cos\theta\sin\phi - \hat{\mathbf{z}}\sin\theta$
$\phi = \tan^{-1}\left[\dfrac{y}{x}\right]$	$\hat{\boldsymbol{\phi}} = -\hat{\mathbf{x}}\sin\phi + \hat{\mathbf{y}}\cos\phi$

Note that the preceding relationships can be deduced from careful examination of Figure A.6b (unit vectors) and FigureA.6c (quantities). A similar set of rules can be given for transformation from rectangular to spherical coordinates:

Coordinate Variables:	Unit Vectors:
$x = r\sin\theta\cos\phi$	$\hat{\mathbf{x}} = \hat{\mathbf{r}}\sin\theta\cos\phi + \hat{\boldsymbol{\theta}}\cos\theta\cos\phi - \hat{\boldsymbol{\phi}}\sin\phi$
$y = r\sin\theta\sin\phi$	$\hat{\mathbf{y}} = \hat{\mathbf{r}}\sin\theta\sin\phi + \hat{\boldsymbol{\theta}}\cos\theta\sin\phi + \hat{\boldsymbol{\phi}}\cos\phi$
$z = r\cos\theta$	$\hat{\mathbf{z}} = \hat{\mathbf{r}}\cos\theta - \hat{\boldsymbol{\theta}}\sin\theta$

The transformation of a vector given in the spherical (rectangular) coordinate system can be accomplished by direct substitution of the unit vector and the preceding coordinate variable expressions. However, doing so in fact amounts to taking the dot products of the vector, as written in spherical (rectangular) coordinates, with the rectangular (spherical) coordinate unit vectors and substituting coordinate variables, as illustrated in Example A.2. The dot products of the spherical coordinate unit vectors with their rectangular coordinate counterparts are provided in Table A.1.

Example A.2: Spherical-to-rectangular and rectangular-to-spherical transformations.
  (a) Transform the vector

$$\mathbf{E} = \hat{\mathbf{r}}\frac{2\cos\theta}{r^3} + \hat{\boldsymbol{\theta}}\sin\theta$$

  given in spherical coordinates to rectangular coordinates.

  (b) Transform a vector $\mathbf{B} = \hat{\mathbf{x}}x^2 + \hat{\mathbf{y}}xy$ given in rectangular coordinates to spherical coordinates.

Solution:
  (a) Note that this vector $\mathbf{E}$ has the same form as the distant electric field of an electric dipole (equation (4.28)) or the distant magnetic field of a magnetic dipole (equation (6.22)). We set out to determine E_x, E_y, and E_z. We have

$$E_x = \mathbf{E} \cdot \hat{\mathbf{x}} = \frac{2\cos\theta}{r^3}\sin\theta\cos\phi + \sin\theta\cos\theta\cos\phi = \left(\frac{2}{r^3} + 1\right)\sin\theta\cos\theta\cos\phi$$

$$= \left(\frac{2 + r^3}{r^3}\right)\left(\frac{x}{r}\right)\left(\frac{z}{r}\right) = \frac{[2 + (x^2 + y^2 + z^2)^{3/2}]xz}{(x^2 + y^2 + z^2)^{5/2}}$$

where we have used the unit vector dot products as given in Table A.1 and substituted $x = r\sin\theta\cos\phi$ and $z = r\cos\theta$. Similarly we have

$$E_y = \mathbf{E} \cdot \hat{\mathbf{y}} = \frac{2\cos\theta}{r^3}\sin\theta\sin\phi + \sin\theta\cos\theta\sin\phi = \left(\frac{2}{r^3} + 1\right)\cos\theta\sin\theta\cos\phi$$

which is identical to E_x. The z component is given by

$$E_z = \mathbf{E} \cdot \hat{\mathbf{z}} = \frac{2\cos\theta}{r^3}\cos\theta - \sin\theta\sin\theta = \frac{2\cos^2\theta}{r^3} - \sin^2\theta = \frac{2\cos^2\theta}{r^3} - (1 - \cos^2\theta)$$

$$= \frac{2z^2}{r^5} - 1 + \frac{z^2}{r^2} = \frac{2z^2}{(x^2 + y^2 + z^2)^{5/2}} + \frac{z^2}{x^2 + y^2 + z^2} - 1$$

Thus, we have

$$\mathbf{E} = [\hat{\mathbf{x}} + \hat{\mathbf{y}}]\left\{\frac{[2 + (x^2 + y^2 + z^2)^{3/2}]xz}{(x^2 + y^2 + z^2)^{3/2}}\right\} + \hat{\mathbf{z}}\left[\frac{2z^2}{(x^2 + y^2 + z^2)^{5/2}} + \frac{z^2}{x^2 + y^2 + z^2} - 1\right]$$

The rather complicated nature of the rectangular form of vector $\mathbf{E}$ underscores the usefulness of spherical coordinates for problems involving spherical symmetry, such as the electric and magnetic dipoles.

(b) Note that the vector **B** is the same as that considered in Example A.1. We set out to find B_r, B_θ, and B_ϕ. We have

$$B_r = \mathbf{B} \cdot \hat{\mathbf{r}} = x^2 \sin\theta \cos\phi + xy \sin\theta \sin\phi$$

$$= r^2 \sin^2\theta \cos^2\phi \sin\theta \cos\phi + (r\sin\theta\cos\phi)(r\sin\theta\sin\phi)\sin\theta\sin\phi$$

$$= r^2 \sin^3\theta \cos\phi(\cos^2\phi + \sin^2\phi) = r^2 \sin^3\theta \cos\phi$$

where we have used the unit vector dot products as given in Table A.1 and substituted $x = r\sin\theta\cos\phi$ and $y = r\sin\theta\sin\phi$. Similarly, we have

$$B_\theta = \mathbf{B} \cdot \hat{\boldsymbol{\theta}} = x^2 \cos\theta\cos\phi + xy\cos\theta\sin\phi$$

$$= r^2 \sin^2\theta \cos^2\phi \cos\theta\cos\phi + (r\sin\theta\cos\phi)(r\sin\theta\sin\phi)\cos\theta\sin\phi$$

$$= r^2 \sin^2\theta \cos\theta\cos\phi(\cos^2\phi + \sin^2\phi)$$

$$= r^2 \sin^2\theta \cos\theta\cos\phi$$

And the ϕ component is given by

$$B_\phi = \mathbf{B} \cdot \hat{\boldsymbol{\phi}} = x^2(-\sin\phi) + xy\cos\phi$$

$$= -r^2 \sin^2\theta\cos^2\phi\sin\phi + (r\sin\theta\cos\phi)(r\sin\theta\sin\phi)\cos\phi = 0$$

Thus, we have

$$\mathbf{B} = \hat{\mathbf{r}}r^2\sin^3\theta\cos\phi + \hat{\boldsymbol{\theta}}r^2\sin^2\theta\cos\theta\cos\phi$$

Length, surface, and volume elements.   The general expression for differential length $d\mathbf{l}$ in spherical coordinates is

$$d\mathbf{l} = \hat{\mathbf{r}}\,dr + \hat{\boldsymbol{\theta}}r\,d\theta + \hat{\boldsymbol{\phi}}r\sin\theta\,d\phi$$

as is evident from Figure A.6d. By inspection of the volume element sketched in Figure A.6d, we can see that the spherical coordinate surface elements in the three coordinate directions are

$$(ds)_r = r^2 \sin\theta\,d\theta\,d\phi$$

$$(ds)_\theta = r\sin\theta\,dr\,d\phi$$

$$(ds)_\phi = r\,dr\,d\theta$$

and the volume element in spherical coordinates is:

$$dv = r^2 \sin\theta\,dr\,d\theta\,d\phi$$

A.4 VECTOR IDENTITIES

We start with some general vector relations involving dot and cross products of arbitrary vectors **A**, **B**, **C**, and **D**:

$$\mathbf{A} \cdot \mathbf{B} \times \mathbf{C} = \mathbf{A} \times \mathbf{B} \cdot \mathbf{C} = \mathbf{B} \cdot \mathbf{C} \times \mathbf{A}$$

$$\mathbf{A} \times (\mathbf{B} \times \mathbf{C}) = (\mathbf{A} \cdot \mathbf{C})\mathbf{B} - (\mathbf{A} \cdot \mathbf{B})\mathbf{C}$$

$$\mathbf{A} \times (\mathbf{B} \times \mathbf{C}) + \mathbf{B} \times (\mathbf{C} \times \mathbf{A}) + \mathbf{C} \times (\mathbf{A} \times \mathbf{B}) = 0$$

$$(\mathbf{A} \times \mathbf{B}) \cdot (\mathbf{C} \times \mathbf{D}) = \mathbf{A} \cdot [\mathbf{B} \times (\mathbf{C} \times \mathbf{D})] = (\mathbf{A} \cdot \mathbf{C})(\mathbf{B} \cdot \mathbf{D}) - (\mathbf{A} \cdot \mathbf{D})(\mathbf{B} \cdot \mathbf{C})$$

$$(\mathbf{A} \times \mathbf{B}) \times (\mathbf{C} \times \mathbf{D}) = (\mathbf{A} \times \mathbf{B} \cdot \mathbf{D})\mathbf{C} - (\mathbf{A} \times \mathbf{B} \cdot \mathbf{C})\mathbf{D}$$

A.4.1 Identities Involving ∇

The nabla (del or grad) operator is defined in rectangular coordinates as

$$\nabla \equiv \hat{\mathbf{x}}\frac{\partial}{\partial x} + \hat{\mathbf{y}}\frac{\partial}{\partial y} + \hat{\mathbf{z}}\frac{\partial}{\partial z}$$

and has been used extensively in this book. A number of vector identities involving the nabla operator are quite useful and are given in this section. Each of these identities can be verified by direct reduction of both sides of the equation. Arbitrary vectors are represented as **A** and **B**, whereas Φ and Ψ denote scalars.

Note that the preceding definition of the nabla operator is valid and useful only for a rectangular coordinate system. In cylindrical or spherical coordinate systems, the orientation of the unit vectors depends on their position, so that the quantity $\nabla \times \mathbf{A}$ is not simply the cross product of a corresponding nabla operator and the vector **A**. If any of the following vector identities were used in a cylindrical or spherical coordinate system, the quantities $\nabla \cdot \mathbf{A}$ and $\nabla \times \mathbf{A}$ should be taken to be symbolic representations of the cylindrical- or spherical-coordinate divergence and curl expressions, given in Sections 4.6 and 6.4. For example, in spherical coordinates, the quantity $\nabla \cdot \mathbf{A}$ is the quantity div **A** given by equation (4.38).

$$\nabla(\Phi + \Psi) \equiv \nabla\Phi + \nabla\Psi$$

$$\nabla(\Phi\Psi) \equiv \Phi\nabla\Psi + \Psi\nabla\Phi$$

$$\nabla \cdot (\Phi\mathbf{A}) \equiv \mathbf{A} \cdot \nabla\Phi + \Phi\nabla \cdot \mathbf{A}$$

$$\nabla \cdot (\mathbf{A} \times \mathbf{B}) \equiv \mathbf{B} \cdot (\nabla \times \mathbf{A}) - \mathbf{A} \cdot (\nabla \times \mathbf{B})$$

$$\nabla \times (\Phi\mathbf{A}) \equiv \nabla\Phi \times \mathbf{A} + \Phi\nabla \times \mathbf{A}$$

$$\nabla \times (\mathbf{A} \times \mathbf{B}) \equiv \mathbf{A}(\nabla \cdot \mathbf{B}) - \mathbf{B}(\nabla \cdot \mathbf{A}) + (\mathbf{B} \cdot \nabla)\mathbf{A} - (\mathbf{A} \cdot \nabla)\mathbf{B}$$

$$\nabla(\mathbf{A} \cdot \mathbf{B}) \equiv (\mathbf{A} \cdot \nabla)\mathbf{B} + (\mathbf{B} \cdot \nabla)\mathbf{A} + \mathbf{A} \times (\nabla \times \mathbf{B}) + \mathbf{B} \times (\nabla \times \mathbf{A})$$

$$\nabla \cdot \nabla \Phi \equiv \nabla^2 \Phi$$

$$\nabla \cdot \nabla \times \mathbf{A} \equiv 0$$

$$\nabla \times \nabla \Phi \equiv 0$$

$$\nabla \times \nabla \times \mathbf{A} \equiv \nabla(\nabla \cdot \mathbf{A}) - \nabla^2 \mathbf{A}$$

The operator formed by the dot product of a vector $\mathbf{A}$ and the del operator forms a new operator that is a scalar operation of the form

$$\mathbf{A} \cdot \nabla = A_x \frac{\partial}{\partial x} + A_y \frac{\partial}{\partial y} + A_z \frac{\partial}{\partial z}$$

which, when applied to a vector $\mathbf{B}$, gives

$$(\mathbf{A} \cdot \nabla)\mathbf{B} = \hat{\mathbf{x}}\left[A_x \frac{\partial B_x}{\partial x} + A_y \frac{\partial B_x}{\partial y} + A_z \frac{\partial B_x}{\partial z} \right] + \hat{\mathbf{y}}\left[A_x \frac{\partial B_y}{\partial x} + A_y \frac{\partial B_y}{\partial y} + A_z \frac{\partial B_y}{\partial z} \right]$$

$$+ \hat{\mathbf{z}}\left[A_x \frac{\partial B_z}{\partial x} + A_y \frac{\partial B_z}{\partial y} + A_z \frac{\partial B_z}{\partial z} \right]$$

Special caution is needed in interpreting the Laplacian $\nabla^2 \mathbf{A}$, which in rectangular coordinates can be simply expanded as $\nabla^2 \mathbf{A} = \hat{\mathbf{x}}\nabla^2 A_x + \hat{\mathbf{y}}\nabla^2 A_y + \hat{\mathbf{z}}\nabla^2 A_z$. In spherical and cylindrical coordinates, the last identity given here, namely $\nabla^2 \mathbf{A} \equiv \nabla(\nabla \cdot \mathbf{A}) - \nabla \times \nabla \times \mathbf{A} = \mathrm{grad}(\mathrm{div}\ \mathbf{A}) - \mathrm{curl}(\mathrm{curl}\ \mathbf{A})$ is in fact the defining relation for the quantity $\nabla^2 \mathbf{A}$.

Special relations involving r.   Some special relations involving the position vector $\mathbf{r} = \hat{\mathbf{x}}x + \hat{\mathbf{y}}y + \hat{\mathbf{z}}z$, with $r = |\mathbf{r}|$, and any arbitrary constant vector $\mathbf{K}$ are

$$\nabla \cdot \mathbf{r} = 3; \qquad \nabla \times \mathbf{r} = 0; \qquad \nabla r = \frac{\mathbf{r}}{r}$$

$$\nabla\left(\frac{1}{r}\right) = \frac{-\mathbf{r}}{r^3}; \qquad \nabla\left(\frac{\mathbf{r}}{r^3}\right) = -\nabla^2\left(\frac{1}{r}\right) = 0 \qquad \text{if } r \neq 0$$

$$\nabla \cdot \left(\frac{\mathbf{K}}{r}\right) = \mathbf{K} \cdot \left[\nabla\left(\frac{1}{r}\right)\right] = \frac{-(\mathbf{K} \cdot \mathbf{r})}{r^3}$$

$$\nabla^2\left(\frac{\mathbf{K}}{r}\right) = \mathbf{K}\nabla^2\left(\frac{1}{r}\right) = 0 \qquad \text{if } r \neq 0$$

$$\nabla \times (\mathbf{K} \times \mathbf{B}) = \mathbf{K}(\nabla \cdot \mathbf{B}) + \mathbf{K} \times (\nabla \times \mathbf{B}) - \nabla(\mathbf{K} \cdot \mathbf{B})$$

Integral relations.   For any arbitrary vector field $\mathbf{G}$ and a scalar field Φ we have

$$\oint_S \mathbf{G} \cdot d\mathbf{s} = \int_V (\nabla \cdot \mathbf{G})\, dv \qquad\qquad \text{Divergence theorem}$$

$$\oint_S \Phi\, d\mathbf{s} = \int_V (\nabla \Phi)\, dv$$

$$\oint_S \mathbf{G} \times d\mathbf{s} = -\int_V (\nabla \times \mathbf{G})\, dv$$

where V is a volume bounded by a closed surface S, with the surface element $d\mathbf{s}$ defined outward from the enclosed volume.

If S' is an open surface bounded by contour C, the line element of which is $d\mathbf{l}$, we have

$$\oint_C \Phi\, d\mathbf{l} = -\int_{S'} \nabla \Phi \times d\mathbf{s}$$

$$\oint_C \mathbf{G} \cdot d\mathbf{s} = \int_{S'} (\nabla \times \mathbf{G}) \cdot d\mathbf{l} \qquad\qquad \text{Stokes' theorem}$$

where the defined direction of $d\mathbf{s}$ and the sense of the contour integration (i.e., direction of $d\mathbf{l}$) are related by the right-hand rule.

Uniqueness Theorem

It is of practical interest to ask what boundary conditions we need to specify in order to solve Poisson's equation (equation (4.44)). This question is addressed by the *uniqueness theorem*, which specifies the boundary conditions that we must know in order for the solution of the electric potential Φ that satisfies (4.44) to be uniquely determined.

To derive the uniqueness theorem, we begin by substituting $\mathbf{G} = \nabla\Phi$ into the vector identity $\nabla \cdot (\Phi\mathbf{G}) \equiv \mathbf{G} \cdot \nabla\Phi + \Phi\nabla \cdot \mathbf{G}$. If Φ is a solution to (4.45), then

$$\nabla \cdot (\Phi\nabla\Phi) = \nabla\Phi \cdot \nabla\Phi + \Phi\nabla \cdot \nabla\Phi$$
$$= |\nabla\Phi|^2 + \Phi\nabla^2\Phi$$
$$= |\nabla\Phi|^2 \tag{B.1}$$

since $\nabla \cdot \nabla\Phi = \nabla^2\Phi = 0$. Integrating both sides over a volume V enclosed by a surface S gives

$$\int_V \nabla \cdot (\Phi\nabla\Phi)dv = \oint_S \Phi\nabla\Phi \cdot d\mathbf{s}$$
$$= \oint_S \Phi\,(\nabla\Phi \cdot \hat{\mathbf{n}})\,ds$$
$$= \oint_S \Phi\frac{\partial\Phi}{\partial n}ds = \int_V |\nabla\Phi|^2 dv \tag{B.2}$$

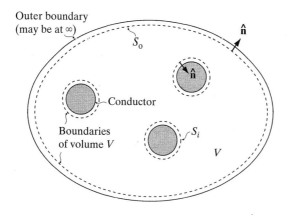

Figure B.1   Uniqueness Theorem.
The dotted lines indicate the boundary surface S enclosing the volume V.

where $\partial\Phi/\partial n$ is the partial derivative of Φ in the direction of the normal[1] to the surface $d\mathbf{s}$.

Let us now consider the geometry depicted in Figure B.1. A given number of conductors are interspersed in this region. In general, we may have a nonzero charge density ρ specified throughout the volume V, which excludes the space occupied by the conductors. The surface S enclosing the volume V includes all surfaces S_i that enclose each conductor, as well as the outer boundary S_0, which may in general be at infinity. The normal unit vector $\hat{\mathbf{n}}$ on each surface S_i is antiparallel (in the opposite direction) to the outward surface vector of the corresponding enclosed conductor.

Suppose we have found two solutions, Φ_1 and Φ_2, that each satisfy Poisson's equation (4.44). Define $\Phi_3 = \Phi_1 - \Phi_2$ as the difference between these two solutions.

---

[1]$\partial\Phi/\partial n$ is a *directional derivative*, which evaluates the partial derivative of the multivariable scalar function Φ in a specific direction. In general, the directional derivative of a scalar function $\Phi(\mathbf{r})$ along the unit vector $\hat{\mathbf{v}}$ is defined as

$$\nabla_{\hat{\mathbf{v}}}\Phi(\mathbf{r}) = \lim_{h \to 0} \frac{\Phi(\mathbf{r} + h\hat{\mathbf{v}}) - \Phi(\mathbf{r})}{h}$$

If Φ is differentiable at $\mathbf{r}$, then

$$\nabla_{\hat{\mathbf{v}}}\Phi(\mathbf{r}) = \nabla\Phi(\mathbf{r}) \cdot \hat{\mathbf{v}} = \frac{\partial\Phi}{\partial x}v_x + \frac{\partial\Phi}{\partial y}v_y + \frac{\partial\Phi}{\partial z}v_z$$

where $\hat{\mathbf{v}} = v_x\hat{\mathbf{x}} + v_y\hat{\mathbf{y}} + v_z\hat{\mathbf{z}}$. It is customary to denote the directional derivative in the direction $\hat{\mathbf{v}}$ as

$$\nabla\Phi(\mathbf{r}) \cdot \hat{\mathbf{v}} \sim \frac{\partial\Phi(\mathbf{r})}{\partial v}$$

Since the gradient of a scalar function is the steepest slope of that function at each point, the directional derivative is maximized when

$$\hat{\mathbf{v}} = \frac{\nabla\Phi}{|\nabla\Phi|}$$

The quantity Φ_3 satisfies Laplace's equation, since

$$\nabla^2\Phi_3 = \nabla^2\Phi_1 - \nabla^2\Phi_2 = -\frac{\rho}{\epsilon_0} + \frac{\rho}{\epsilon_0} = 0$$

Thus, we can substitute Φ_3 into (B.2) to get

$$\oint_S (\Phi_1 - \Phi_2)\frac{\partial(\Phi_1 - \Phi_2)}{\partial n}ds = \int_V |\nabla(\Phi_1 - \Phi_2)|^2 dv \qquad (B.3)$$

The closed surface of the surface integral on the left side of (B.3) consists of each surface S_i and the outer surface S_o enclosing the entire solution space:

$$\oint_S (\Phi_1 - \Phi_2)\frac{\partial(\Phi_1 - \Phi_2)}{\partial n}ds = \oint_{S_o} (\Phi_1 - \Phi_2)\frac{\partial(\Phi_1 - \Phi_2)}{\partial n}ds$$
$$+ \sum_i \oint_{S_i} (\Phi_1 - \Phi_2)\frac{\partial(\Phi_1 - \Phi_2)}{\partial n}ds$$

If we let the radius of the outer surface S_o go to infinity, the first surface integral over S_o vanishes since the electric potential Φ decreases at least as fast as r^{-1}, $\nabla\Phi$ decreases at least as fast as r^{-2}, and the surface area increases as r^2.

The remaining terms in the summation all correspond to the surface integrals of the product of the electric potentials and their derivatives at the surface S_i of each conductor. Since the potential at the surface S_i of each conductor is a constant (the boundary of each conductor is an equipotential surface), we may rewrite (B.3) as

$$\sum_i (\Phi_{1i} - \Phi_{2i}) \oint_{S_i} \frac{\partial(\Phi_1 - \Phi_2)}{\partial n}ds = \int_V |\nabla(\Phi_1 - \Phi_2)|^2 dv \qquad (B.4)$$

Suppose we specify the potential of each conductor. In this case, since we assumed Φ_1 and Φ_2 to be valid solutions of (4.44) under the specified boundary conditions, Φ_1 and Φ_2 must be equal on the surface of each conductor and so (B.4) reduces to

$$\int_V |\nabla(\Phi_1 - \Phi_2)|^2 dv = 0 \qquad (B.5)$$

Since the integrand is always positive, we must have

$$\nabla(\Phi_1 - \Phi_2) = 0 \qquad (B.6)$$

everywhere, and hence Φ_1 and Φ_2 differ at most by a constant. However, since $\Phi_1 = \Phi_2$ on each conductor, this constant must be 0, and so the two solutions must be equal. Hence, we have the result that if we specify Φ everywhere on the closed surface S, that is, we specify the potential of each conductor (and let the potential at infinity be zero), then there exists a unique solution to (4.44). This result has practical implications. If we know the potential on each of the conductors, and we find a solution Φ that satisfies (4.44) and the potential boundary conditions on the conductor surfaces, then we know that this solution is the only possible solution. So, regardless of how we found this solution (even if by guessing), this is indeed the correct unique solution for the electric potential Φ.

If, instead of specifying the potential on each conductor, we specify the normal derivative of the potential at each surface, then the $\partial(\Phi_1 - \Phi_2)/\partial n$ terms in (B.4) would be zero and (B.4) would again reduce to (B.6). In this case, since we only specified the derivative, potentials Φ_1 and Φ_2 may differ by a constant. However, the electric fields, which are recovered by taking the gradient of the potential functions, are necessarily equal.

With the normal of the integration surfaces S_i surrounding each conductor pointing inward toward the conductor surfaces, we have

$$\frac{\partial \Phi}{\partial n} = \frac{\rho_s}{\epsilon} \tag{B.7}$$

where we have used (4.24) and (4.42), and noted that the normal derivative points inward toward the conductor surface (otherwise there would be a minus sign on the right-hand side of (B.7)). Substituting (B.7) into (B.4) gives

$$\sum_i (\Phi_{1i} - \Phi_{2i}) \oint_{S_i} \frac{(\rho_{s1} - \rho_{s2})}{\epsilon_0} ds = \sum_i (\Phi_{1i} - \Phi_{2i}) \frac{(Q_{1i} - Q_{2i})}{\epsilon_0}$$

$$= \int_V |\nabla(\Phi_1 - \Phi_2)|^2 dv \tag{B.8}$$

where Q_i represents the total charge residing on the surface of conductor i. Hence, if we specify the total charge on each conductor so that $Q_{1i} = Q_{2i}$, then the potential Φ is everywhere determined within a constant.

We can combine the two boundary conditions specified above. In general, if we have N conductors, with the potential specified on some and the total charge specified on the rest, then the potential (and thus the electric field) in the region surrounding these conductors is uniquely determined. In the limiting case where we specify the charge Q_i on each conductor, the potential is determined within a constant, but the electric field is still uniquely determined. If we specify the potential on at least one conductor, this condition removes the ambiguity of the additive constant to the potential Φ.

Derivation of Ampère's Circuital Law from the Biot–Savart Law

In Section 6.3, we introduced Ampère's circuital law, expressed in (6.9), as a mathematical consequence of the Biot–Savart law, which is given by expression (6.6). In this appendix, we mathematically derive (6.11) from (6.6). Since (6.9) can be derived from (6.11) simply by using Stokes's theorem, the analysis that follows[1] constitutes a derivation of Ampère's circuital law.

Our starting point is thus the experimentally established Biot–Savart law:

$$\mathbf{B}(\mathbf{r}) = \frac{\mu_0}{4\pi} \int_{V'} \frac{\mathbf{J}(\mathbf{r'}) \times \hat{\mathbf{R}}}{R^2}\, dv' \qquad (6.6)$$

which, as we recall from Section 6.5, can be rewritten as (6.13), or

$$\mathbf{B}(\mathbf{r}) = \nabla \times \frac{\mu_0}{4\pi} \int_{V'} \frac{\mathbf{J}(\mathbf{r'})}{R}\, dv' \qquad (6.13)$$

Since our goal is to show that (6.13) (and thus (6.6)) leads to (6.11), we start by taking the curl of both sides of (6.13):

$$\nabla \times \mathbf{B}(\mathbf{r}) = \nabla \times \nabla \times \frac{\mu_0}{4\pi} \int_{V'} \frac{\mathbf{J}(\mathbf{r'})}{R}\, dv' \qquad (\text{C.1})$$

We now use the same vector identity used in deriving the wave equation in Section 8.1, namely equation (8.2), to expand the curl–curl operator and to rewrite (C.1) as

$$\nabla \times \mathbf{B}(\mathbf{r}) = \frac{\mu_0}{4\pi} \int_{V'} \left\{ \nabla \left[\nabla \cdot \left(\frac{\mathbf{J}(\mathbf{r'})}{R} \right) \right] - \mathbf{J}(\mathbf{r'}) \nabla^2 \left(\frac{1}{R} \right) \right\} dv' \qquad (\text{C.2})$$

[1]The derivation presented here was adapted from that given in Section 6.6 of R. Plonsey and R. E. Collin, *Principles and Applications of Electromagnetic Fields,* McGraw-Hill, New York, 1961.

First we consider the first term in (C.2), which can be rewritten as

$$\int_{V'} \nabla \left[\nabla \cdot \left(\frac{\mathbf{J}(\mathbf{r}')}{R} \right) \right] dv' = \nabla \int_{V'} \left[\nabla \cdot \left(\frac{\mathbf{J}(\mathbf{r}')}{R} \right) \right] dv' \qquad (C.3)$$

where one of the del operators is brought outside the integral. We can now rewrite the integrand as

$$\nabla \cdot \left(\frac{\mathbf{J}(\mathbf{r}')}{R} \right) = \mathbf{J} \cdot \nabla \left(\frac{1}{R} \right) = -\mathbf{J} \cdot \nabla' \left(\frac{1}{R} \right) = -\nabla' \cdot \left(\frac{\mathbf{J}}{R} \right)$$

where ∇' is the del operator applied to the source coordinates ($\mathbf{r}'$), and where we have used $\nabla' \cdot (\mathbf{J}/R) = (1/R)\nabla' \cdot \mathbf{J} + \mathbf{J} \cdot \nabla'(1/R)$ and noted that $\nabla' \cdot \mathbf{J} = 0$ for stationary currents, as required by the equation of continuity. Also note that since $R = |\mathbf{r} - \mathbf{r}'| = [(x - x')^2 + (y - y')^2 + (z - z')^2]^{1/2}$, we have $\nabla(1/R) = -\nabla'(1/R)$. Equation (C.3) can now be written as

$$\nabla \int_{V'} \left[\nabla \cdot \left(\frac{\mathbf{J}(\mathbf{r}')}{R} \right) \right] dv' = \nabla \int_{V'} \left[\nabla' \cdot \left(\frac{\mathbf{J}(\mathbf{r}')}{R} \right) \right] dv' = -\nabla \oint_{S'} \left(\frac{\mathbf{J}}{R} \right) \cdot d\mathbf{s}' \qquad (C.4)$$

where we have invoked the divergence theorem. Since $\mathbf{J}$ is a stationary current and is confined to a finite region of space, the surface S' can always be chosen to be large enough to include all currents within it, so that $\mathbf{J} \cdot d\mathbf{s}' = 0$ on it. Thus, (C.4) vanishes, and thus (C.3) is zero, and (C.2) is left with only the second term on the right-hand side. We have

$$\nabla \times \mathbf{B}(\mathbf{r}) = -\frac{\mu_0}{4\pi} \int_{V'} \mathbf{J}(\mathbf{r}') \nabla^2 \left(\frac{1}{R} \right) dv' \qquad (C.5)$$

Since our goal is to arrive at (6.11), we need to show that the right-hand side of (C.5) reduces to $\mu_0 \mathbf{J}(\mathbf{r})$. Direct differentiation indicates that $\nabla^2(1/R) = 0$ for all finite values of R. Thus, if the observation point $\mathbf{r}$ is outside a source region of finite extent, then $R = |\mathbf{r} - \mathbf{r}'|$ is never zero, and we have $\nabla \times \mathbf{B} = 0$. However, if the observation point $\mathbf{r}$ is within the region where the sources are located, then in the course of the integration of (C.5), R can be zero at certain points. We must thus be able to show that the singularity of $\nabla^2(1/R)$ at $R = 0$ is integrable and yields a finite result.

Consider a point at which $R = 0$, where $\mathbf{r} = \mathbf{r}'$ or $x = x'$, $y = y'$, and $z = z'$. In the immediate neighborhood of this point, we can assume that the current density does not vary much from its value at $\mathbf{r} = \mathbf{r}'$. Since the integrand of (C.5) is zero everywhere except at $R = 0$, we can integrate (C.5) over a sphere centered at $\mathbf{r}$ that is small enough so that the current density everywhere within it is the same (i.e., $\mathbf{J}(\mathbf{r}') = \mathbf{J}(\mathbf{r})$). Thus, (C.5) becomes

$$\nabla \times \mathbf{B}(\mathbf{r}) = -\frac{\mu_0}{4\pi} \mathbf{J}(\mathbf{r}) \int_{V'_s} \nabla^2 \left(\frac{1}{R} \right) dv' \qquad (C.6)$$

where V_s' is the small spherical volume surrounding the point at which $R = 0$. As noted previously in connection with (C.3), we have $\nabla(1/R) = -\nabla'(1/R)$ and $\nabla^2(1/R) = (\nabla')^2(1/R) = \nabla' \cdot \nabla'(1/R)$. Thus, we can write (C.6) as

$$\nabla \times \mathbf{B}(\mathbf{r}) = -\frac{\mu_0}{4\pi}\mathbf{J}(\mathbf{r}) \int_{V_s'} \nabla' \cdot \nabla' \left(\frac{1}{R}\right) dv' = -\frac{\mu_0}{4\pi}\mathbf{J}(\mathbf{r}) \oint_{S_s'} \nabla' \left(\frac{1}{R}\right) \cdot d\mathbf{s}' \quad \text{(C.7)}$$

We have invoked the divergence theorem, where S_s' is the spherical surface that encloses the spherical volume V_s' centered at $\mathbf{r}$. As was shown in footnote 40 in Section 6.5, we have $\nabla(1/R) = -\hat{\mathbf{R}}/R^2$, and thus $\nabla'(1/R) = -\nabla(1/R) = \hat{\mathbf{R}}/R^2$. Note that since we took the volume V_s' to be centered at $\mathbf{r}$, the vector $\mathbf{R} = \mathbf{r} - \mathbf{r}'$ points inward toward the center point $\mathbf{r}$. Thus, we have $\hat{\mathbf{R}} = -\hat{\mathbf{r}}$ so that $\nabla'(1/R) = -\hat{\mathbf{r}}/R^2$. Noting that the surface element for spherical coordinates is $\hat{\mathbf{r}}ds' = \hat{\mathbf{r}}R^2 \sin\theta' \, d\phi' \, d\theta'$, we can write (C.7) as

$$\nabla \times \mathbf{B}(\mathbf{r}) = -\frac{\mu_0}{4\pi}\mathbf{J}(\mathbf{r}) \int_0^\pi \int_0^{2\pi} \left(\frac{-\hat{\mathbf{r}}}{R^2}\right) \cdot \hat{\mathbf{r}}R^2 \sin\theta' \, d\phi' \, d\theta'$$

$$= \frac{\mu_0}{4\pi}\mathbf{J}(\mathbf{r}) \int_0^\pi \int_0^{2\pi} \sin\theta' \, d\phi' \, d\theta' = \mu_0\mathbf{J}(\mathbf{r})$$

which is (6.11). We have thus shown that (6.11) is a mathematical consequence of the Biot–Savart law, or (6.6). Since Ampère's circuital law, or (6.9), can be obtained from (6.6) simply by using Stokes's theorem, we have also shown that Ampère's circuital law is a direct requirement of the Biot–Savart law, which is an expression of one of the three experimental pillars of our formulation of electromagnetics.

Symbols and Units
for Basic Quantities

Symbol	Quantity	SI Unit	Comments
$\mathcal{A}$	Magnetic vector potential	webers (Wb)-m^{-1}	
A	Magnetic vector potential	Wb-m^{-1}	Phasor in Section 7.4– Chapter 11
B	Susceptance	siemens (S)	Chapter 3 only
B	Magnetic field (**B** field)	Wb-m^{-2} or tesla (T)	Phasor in Section 7.4– Chapter 11
$\overline{\mathcal{B}}$	Magnetic field (**B** field)	Wb-m^{-2}	
C	Capacitance	farads (F)	Per unit length (F-m^{-1}) in Chapter 2 & 3
C_u	Capacitance per unit length	F-m^{-1}	
D	Electric flux density	C-m^{-2}	Phasor in Section 7.4– Chapter 11
$\overline{\mathcal{D}}$	Electric flux density	C-m^{-2}	
E	Electric field intensity	V-m^{-1}	Phasor in Section 7.4– Chapter 11
$\overline{\mathcal{E}}$	Electric field intensity	V-m^{-1}	
F	Force	Newtons (N)	Phasor in Section 7.4– Chapter 11
$\overline{\mathcal{F}}$	Force	N	
f	Frequency	hertz (Hz)	
f_p	Plasma frequency	Hz	

Symbol	Quantity	SI Unit	Comments
G	Conductance	S	Per unit length (S-m^{-1}) in Chapter 2 & 3
G_{u}	Conductance per unit length	S-m^{-1}	
$\mathcal{G}$	Electromagnetic momentum	N-s	
$\overline{g}$	Momentum density	N-s-m^{-3}	
H	Magnetic field intensity	A-m^{-1}	Phasor in Section 7.4– Chapter 11
$\overline{\mathcal{H}}$	Magnetic field intensity	A-m^{-1}	
I	Current	amperes (A)	Phasor in Chapter 3, Section 7.4– Chapter 11
$\mathcal{I}$	Current	A	
J	Current density	A-m^{-2}	Phasor in Section 7.4– Chapter 11
$\overline{\mathcal{J}}$	Current density	A-m^{-2}	
J$_s$	Surface current density	A-m^{-1}	Phasor in Section 7.4– Chaptcr 11
$\overline{\mathcal{J}}_s$	Surface current density	A-m^{-1}	
k	Mechanical spring constant	N-m^{-1}	
k	Wavenumber vector	radians(rad)-m^{-1}	
L	Inductance	henrys (H)	Per unit length (H-m^{-1}) in Chapter 2 & 3
L_{u}	Inductance per unit length	H-m^{-1}	
$\overline{\mathcal{L}}$	Angular momentum	kg-m^2-s^{-1}	
l	Length	m	
M	Magnetization vector	A-m^{-1}	Phasor in Section 7.4– Chapter 11
$\overline{\mathcal{M}}$	Magnetization vector	A-m^{-1}	
m	Mass	kg	
m	Magnetic dipole moment	A-m^{-2}	
n	Index of refraction		
P	Power	watts (W)	
$\overline{\mathcal{P}}$	Electric polarization vector	C-m^{-2}	
P	Electric polarization vector	C-m^{-2}	Phasor in Section 7.4– Chapter 11
P$_{\mathrm{av}}$	Time-average pressure	N-m^{-2}	
p	Electric dipole moment	C-m	

Symbol	Quantity	SI Unit	Comments
Q, q	Electric charge	coulombs (C)	
R	Resistance	Ω	Per unit length (Ω-m^{-1}) in Chapter 2 & 3
R_s	Surface resistance	Ω	
S	Standing wave ratio		
$\overline{\mathcal{S}}$	Poynting vector	W-m^{-2}	
$\mathbf{S}_{av}$	Poynting vector	W-m^{-2}	Time-average quantity
$\mathbf{S}$	Complex Poynting vector	W-m^{-2}	
T	Temperature	$^{\circ}$C	Chapter 4 and 5 only
$\overline{\mathcal{T}}$	Torque	N-m	
$\mathbf{T}$	Torque	N-m	Phasor in Section 7.4– Chapter 11
$\mathcal{T}$	Transmission coefficient		$\mathcal{T} = \tau e^{j\phi_{\mathcal{T}}}$
t	Time	s	
t_d	One-way travel time	s	
t_r	Rise time	s	
$\mathcal{N}_e$	Electron density	m^{-3}	
N_e	Electron density	m^{-3}	
$\tilde{\mathbf{v}}$	Velocity vector	m-s^{-1}	
$\mathbf{v}$	Velocity vector	m-s^{-1}	Phasor in Section 7.4– Chapter 11
v_p	Phase velocity	m-s^{-1}	
$\overline{v}_p$	Phase velocity in waveguide	m-s^{-1}	
V	Voltage	volts (V)	Phasor
$\mathcal{V}$	Voltage	V	
$\mathcal{V}_{emf}$	Electromotive force	V	
$\mathcal{V}_{ind}$	Induced emf	V	
W	Work (energy)	joules (J)	
w	Energy density	J-m^{-3}	
X	Reactance	Ω	
Y	Admittance	S	
Z	Impedance	Ω	
Z_0	Characteristic impedance	Ω	$Z_0 = \sqrt{L/C}$ for lossless line

Symbol	Quantity	SI Unit	Comments
α	Attenuation constant	nepers (np)-m^{-1}	Real part of γ
$\overline{\alpha}$	Attenuation constant in waveguide	np-m^{-1}	Real part of $\overline{\gamma}$
α_c	Attenuation constant	np-m^{-1}	Due to conduction losses
α_d	Attenuation constant	np-m^{-1}	Due to dielectric losses
$\alpha_e, \alpha_i, \alpha_T$	Polarizability	m^{-3}	Section 4.10 only
β	Propagation constant	radians (rad)-m^{-1}	$\beta = \omega\sqrt{\mu\epsilon}$
β	Phase constant	rad-m^{-1}	Imaginary part of γ
$\overline{\beta}$	Phase constant in waveguide	rad-m^{-1}	Imaginary part of $\overline{\gamma}$
Γ	Reflection coefficient		$\Gamma = \rho e^{j\phi_\Gamma}$
γ	Propagation constant	m^{-1}	$\gamma = \alpha + j\beta$
$\overline{\gamma}$	Propagation constant in waveguide	m^{-1}	$\overline{\gamma} = \overline{\alpha}$ or $\overline{\gamma} = j\overline{\beta}$
δ	Skin depth	m	$\delta = (\pi f \mu \sigma)^{-1/2}$
$\tan\delta_c$	Loss tangent		
ϵ, ϵ_0	Permittivity	F-m^{-1}	
ϵ_r	Relative permittivity		
ϵ_c	Complex permittivity	F-m^{-1}	$\epsilon_c = \epsilon' - j\epsilon''$
$\theta_i, \theta_r, \theta_t$	Incident, reflected, transmitted angles	rad	
θ_{iB}	Brewster angle	rad	
θ_{ic}	Critical angle	rad	
η	Intrinsic impedance	Ω	$\eta = \sqrt{\mu/\epsilon}$
η_c	Impedance of a lossy medium	Ω	$\eta_c = |\eta_c|e^{j\phi_\eta}$
λ	Wavelength	m	$\lambda = 2\pi/\beta$
$\overline{\lambda}$	Wavelength in waveguide	m	$\overline{\lambda} = 2\pi/\overline{\beta}$
μ, μ_0	Permeability	H-m^{-1}	
μ_r	Relative permeability		
μ_c	Complex permeability	H-m^{-1}	$\mu_c = \mu' - j\mu''$
ρ	Magnitude of Γ		
ρ	Volume charge density	C-m^{-3}	Phasor in Section 7.4– Chapter 11
$\tilde{\rho}$	Volume charge density	C-m^{-3}	
ρ_s	Surface charge density	C-m^{-2}	
ρ_l	Linear charge density	C-m^{-1}	

Symbol	Quantity	SI Unit	Comments
σ	Conductivity	S	
τ	Magnitude of $\mathcal{T}$		
τ	Time constant	s	
τ_r	Relaxation time	s	$\tau_\mathrm{r} = \epsilon/\sigma$
Φ	Electrostatic potential	volts (V)	Phasor in Section 7.4–Chapter 11
$\tilde{\Phi}$	Electrostatic potential	volts (V)	
χ_e	Electrical susceptibility		
χ_m	Magnetic susceptibility		
ψ	Phase of reflection coefficient		
ψ	True angle of refraction	rad	
Ψ	Magnetic flux	Wb	
ω	Angular frequency	$\mathrm{rad\text{-}s}^{-1}$	
ω_p	Plasma frequency	$\mathrm{rad\text{-}s}^{-1}$	
ω_0	Characteristic frequency of bound electrons	$\mathrm{rad\text{-}s}^{-1}$	Section 11.2

General Bibliography

In addition to the specific references to numerous books and articles provided in the footnotes, the following texts on electromagnetic fields and waves were found to be generally useful in developing this book.

Adler, R. B., L. J. Chu, and R. M. Fano, *Electromagnetic Energy Transmission and Radiation*, Wiley, 1960.

Atwater, H. A., *Introduction to Microwave Theory*, McGraw-Hill, Inc., 1962.

Caloz, C. and T. Itoh, *Electromagnetic Metamaterials: Transmission Line Theory and Microwave Applications*, John Wiley & Sons, 2005.

Cheng, D. K., *Field and Wave Electromagnetics,* Addison-Wesley, 2nd ed., 1989.

Cheston, W. B., *Elementary Theory of Electric and Magnetic Fields,* Wiley, 1964.

Elliott, R. S., *Electromagnetics: History, Theory, and Applications,* IEEE Press, 1993.

Feynman, R. P., R. B. Leighton, and M. Sands, *The Feynman Lectures on Physics,* Addison-Wesley, 1964.

Griffiths, D. J., *Introduction to Electrodynamics*, 3rd ed., Prentice-Hall, 1999.

Johnk, C. T. A., *Engineering Electromagnetic Fields and Waves*, 2nd ed., John Wiley & Sons, Inc., 1988.

Johnson, C. C., *Field and Wave Electrodynamics*, McGraw-Hill, Inc., 1965.

Jordan, E. C. and K. G. Balmain, *Electromagnetic Waves and Radiating Systems,* 2nd ed., Prentice-Hall, 1968.

Korvink, J. and O. Paul. *MEMS: A Practical Guide to Design, Analysis and Applications*, William Andrew, 2005.

Kraus, J. D., *Electromagnetics,* 4th ed., McGraw-Hill, 1992.

Liu, C., *Foundations of MEMS*, 2nd ed., Pearson, 2012.

Marqués, R., F. Martín, and M. Sorolla, *Metamaterials with Negative Parameters: Theory, Design and Microwave Applications*, John Wiley & Sons, 2008.

Moore, R. K., *Travelling-Wave Engineering,* McGraw-Hill, 1960.

Paris, D. T. and F. K. Hurd, *Basic Electromagnetic Theory*, McGraw-Hill, Inc., 1969.

Plonsey, R. and R. E. Collin, *Principles and Applications of Electromagnetic Fields,* McGraw-Hill, 1961.

Plonus, M. A., *Applied Electromagnetics,* McGraw-Hill, 1978.

Ramo, S., J. R. Whinnery, and T. Van Duzer, *Fields and Waves in Communication Electronics,* 3rd ed., John Wiley & Sons, 1994.

Scott, W. T., *The Physics of Electricity and Magnetism,* 2nd ed., Wiley, 1966.

Skilling, H. H., *Fundamentals of Electric Waves,* 2nd ed., Wiley, 1948; reprinted by Robert E. Krieger Publishing Co., 1974.

Answers to Odd-Numbered Problems

Chapter 1

1.1. $t \simeq 0.1334$ s.

1.3. $c \simeq 2.847 \times 10^5$ km-s^{-1}.

1.5. $d = 510$ m.

1.7. $d \simeq 2.775$ km.

1.9. $d = 1.7$ km.

1.11. $l \simeq 0.3$ m.

1.13. $d \simeq 4.225$ ly.

1.15. $v_{\text{average}} \simeq 223$ m-s^{-1}.

1.17. $t \simeq 58.8$ ms.

1.19. $l_{\text{max}} \simeq 0.417$ cm.

1.21. $t_d \simeq 0.529$ ns.

1.23. Not appropriate.

1.25. $l_{\text{max}} \simeq 0.67$ mm.

Chapter 2

2.1. (a) $Z_0 = 50\Omega$, $t_d = 0.4$ ns.

(b) Not appropriate.

2.3. $\mathcal{V}_{\text{L}} = \{0, t < t_d;\ 2$ V, $t_d < t < 3t_d;\ 0, 3t_d < t < 5t_d\}$ and periodic for $t > t_d$ with $T_p = 4t_d$.

2.5. (a) $\mathcal{V}_L = \{0, t < 0.5$ ns; 2 V, 0.5 ns $< t <$ 1.5 ns; 2.67 V, 1.5 ns $< t <$ 2.5 ns; 2.89 V, 2.5 ns $< t < 3$ ns$\}$.
      (b) $\mathcal{V}_L = \{0, t < 0.5$ ns; 3 V, 0.5 ns $< t < 3$ ns$\}$.
      (c) $\mathcal{V}_L = \{0, t < 0.5$ ns; 4 V, 0.5 ns $< t <$ 1.5 ns; 2.67 V, 1.5 ns $< t <$ 2.5 ns; 3.11 V, 2.5 ns $< t < 3$ ns$\}$.

2.7. (a) $\mathcal{V}_L = \{3$ V, $t < 0.5$ ns; -1 V, 0.5 ns $< t <$ 1.5 ns; 0.33 V, 1.5 ns $< t <$ 2.5 ns; -0.11 V, 2.5 ns $< t < 3$ ns$\}$.
      (b) $\mathcal{V}_L = \{3$ V, $t < 0.5$ ns; $0, 0.5$ ns $< t < 3$ ns$\}$.
      (c) $\mathcal{V}_L = \{3$ V, $t < 0.5$ ns; 1 V, 0.5 ns $< t <$ 1.5 ns; 0.33 V, 1.5 ns $< t <$ 2.5 ns; 0.11 V, 2.5 ns $< t < 3$ ns$\}$.

2.9. (a) $\mathcal{V}_s = \{0.75$ V, $t < 1$ ns; 0.375 V, 1 ns $< t < 2$ ns; 0.1875 V, 2 ns $< t < 3$ ns; $\ldots\}$. (b) $\mathcal{V}_s = \{0.75$ V, $t < 1$ ns; -0.375 V, 1 ns $< t < 2$ ns; -0.1875 V, 2 ns $< t < 3$ ns; $\ldots\}$.

2.11. (a) $Z_0 = 100\Omega$, $R_L = 400\Omega$. (b) $\mathcal{V}_{ctr} = \{0, t < t_d/2$; 0.5 V, $t_d/2 < t <$ $3t_d/2$; 0.8 V, $3t_d/2 < t < 2t_d$; 0.3 V, $2t_d < t < 3t_d$; 0, $t > 3t_d\}$.

2.13. $Z_{01} = 50\Omega$, $l_1 = 10$ cm, $R_1 = 50\Omega$.

2.15. $Z_{01} = 75\Omega$, $l = 37.5$ cm, $Z_{02} = 125\Omega$, $R_L = \infty$.

2.17. (a) $\mathcal{V}_{L1} = \mathcal{V}_{L2} = \{0, t < 3$ ns; 0.5 V, $t > 3$ ns$\}$.
      (b) $\mathcal{V}_{L1} = \{0, t < 3$ ns; 0.5 V, 3 ns $< t < 6$ ns; 0.75 V, 6 ns $< t < 9$ ns; $\ldots\}$, $\mathcal{V}_{L2} = \{0, t < 3$ ns; 1 V, 3 ns $< t < 6$ ns; 0.5 V, 6 ns $< t < 9$ ns; $\ldots\}$, $V_{ss} \simeq 0.667$ V.

2.19. $\mathcal{V}_1 = \{4$ V, $t < 4$ ns; 3.68 V, 4 ns $< t < 8$ ns; $\sim$3.95 V, 8 ns $< t <$ 12 ns; $\ldots\}$, $\mathcal{V}_2 = \{0, t < 4$ ns; 3.84 V, 4 ns $< t < 8$ ns; $\sim$4.45 V, 8 ns $< t < 12$ ns; $\ldots\}$.

2.21. $\mathcal{V}_s = \{4.5$ V, $t < 5.4$ ns; 4.2 V, 5.4 ns $< t < 16.2$ ns; $\sim$4.319 V, 16.2 ns $<$ $t\}$, $\mathcal{V}_L = \{0, t < 5.4$ ns; 3.3 V, 5.4 ns $< t < 10.8$ ns; 4.29 V, 10.8 ns $< t <$ 16.2 ns; $\sim$4.319 V, 16.2 ns $< t\}$.

2.23. $\mathcal{V}_L = \{0, t < 2t_d$; 0.8 V, $2t_d < t < 3t_d$; 0.533 V, $3t_d < t < 5t_d$; 0.64 V, $5t_d <$ $t < 6t_d\}$.

2.25. (a) Series termination: $R_T = 36\Omega$, parallel: $R_T = 50\Omega$. (b) No steady-state power dissipation for the series case.

2.27. Gate 1 turns ON permanently at $t = 22$ ns; Gate 2 turns ON permanently at $t = 19$ ns.

2.29. $\mathcal{V}_1(t) = 1.25u(t - 2$ ns$) + 1.25u(t - 6$ ns$)[1 - 2e^{-(t-6\ \text{ns})/(0.25\ \text{ns})}]$ V.

2.31. Capacitor; $C = 4A(V_0 Z_0)^{-1}$.

2.33. Resistor (25Ω) and Inductor (15 nH).

2.35. $\mathcal{V}_s(t) = 3u(t) + u(t - 4$ ns$) + \frac{1}{2}e^{-(t-4\ \text{ns})/0.8\ \text{ns}}u(t - 4$ ns$)$ V; $\mathcal{V}_L(t) = 4u(t - 4$ ns$) + \frac{1}{2}e^{-(t-4\ \text{ns})/0.8\ \text{ns}}u(t - 4$ ns$)$ V.

2.37. $\mathcal{V}_s(t) = 4u(t - 3$ ns$) - 8e^{-(t-3\ \text{ns})/0.3\ \text{ns}}u(t - 3$ ns$)$ V.

2.39. $\mathcal{V}_L(t) = 2.5u(t - 10 \text{ ns})\left[1 + \frac{1}{3}e^{-(t-10 \text{ ns})/(0.4 \text{ ns})}\right]$ V.

2.41. (a) $\mathcal{V}_L = \{0,\, t < 3t_d;\, 2.5 \text{ V},\, 3t_d < t < 5t_d;\, 0,\, 5t_d < t < 7t_d;\, -1.25 \text{ V},\, 7t_d < t < 11t_d;\, \ldots\}$. (b) $\mathcal{V}_L = \{0,\, t < 3t_d;\, 2.5 \text{ V},\, 3t_d < t < 5t_d;\, 0,\, 5t_d < t < 7t_d;\, -3.75 \text{ V},\, 7t_d < t < 9t_d;\, -1.25 \text{ V},\, 9t_d < t < 11t_d;\, \ldots\}$.

2.43. $\mathcal{V}_{s1} = \{V_0,\, t < 2t_d;\, V_0/2,\, t > 2t_d\}$, $\mathcal{V}_{s2} = \{0,\, t < 2t_d;\, V_0/2,\, t > 2t_d\}$.

2.45. $L \simeq 1.42\mu\text{H-m}^{-1}$, $C \simeq 7.82$ pF-m^{-1}, $R \simeq 1.96\Omega$-m^{-1}, $G = 0$, and $Z_0 \simeq 426.5\Omega$.

Chapter 3

3.1. (a) $l_{\min} \simeq 0.75$ cm. (b) $l_{\min} \simeq 1.50$ cm. (c) Open-circuited line.

3.3. (a) $(l/\lambda)_{\min} = 0.125$. (b) $(l/\lambda)_{\min} \simeq 0.301$. (c) $(l/\lambda)_{\min} = 0.25$. (d) $(l/\lambda)_{\min} = 0.5$ (or 0).

3.5. (a) $Z_{\text{in}} = 50\Omega$. (b) $Z_{\text{in}} = 12.5\Omega$. (c) $Z_{\text{in}} = 20 + j15\Omega$.

3.7. (a) $\Gamma_L \simeq 0.669e^{-j48.01^o}$. (b) $S \simeq 5.042$. (c) $z_{\min} \simeq -55.0$ cm. (d) $z_{\max} \simeq -130$ cm.

3.9. $(l/\lambda)_{\min} \simeq 0.1592$; $R_{\text{in}} \simeq 461.7\Omega$.

3.11. (a) $S_1 \simeq 2.618$; $S_2 \simeq 4.2656$. (b) $P_L = 0.05$ W.

3.13. $Z_{\text{in}}|_{f_2} = 160\Omega$.

3.15. (a) $Z_{\text{in}} = \infty$. (b) $Z_{\text{in}} = Z_0/2$. (c) $Z_{\text{in}} = Z_0$.

3.17. (a) $Z_0/2$. (b) 0. (c) $Z_0/4$.

3.19. $Z_{01}/Z_{02} = 1.5$

3.21. (a) $f \simeq 750$ MHz. (b) $Z_L \simeq 16.67\Omega$.

3.23. $Z_L = Z_0[S - j\tan(\beta l_{\max})]/[1 - jS\tan(\beta l_{\max})]$

3.25. (a) $\Gamma_L = 0.1857e^{j21.8°}$; $S \simeq 1.456$.
(b) $z_{\max}$ (m) : $-0.1817, -3.1817, -6.1817, -9.1817$;
$z_{\min}$ (m) : $-1.6817, -4.6817, -7.6817$.
(c) $z_{\max}$ (m) : $-1.6817, -4.6817, -7.6817$;
$z_{\min}$ (m) : $-0.1817, -3.1817, -6.1817, -9.1817$. (d) $Z_{\text{in}}^{\max} \simeq 72.8\Omega$;
$Z_{\text{in}}^{\min} \simeq 34.34\Omega$. (e) $Z_{\text{in}} = 35 - j5\Omega$. (f) $V_s \simeq 4.1523e^{-j4.7636°}$ V;
$V^+ = 5e^{j90°}$ V; $V^- \simeq 0.9285e^{j111.8°}$ V; $V_L \simeq 5.872e^{j93.37°}$ V.
(g) $V_{\max} \simeq 5.9285$ V; $V_{\min} \simeq 4.0715$ V. (h) $I_{\max} \simeq 0.1186$ A;
$I_{\min} \simeq 0.08143$ A. (i) $P^+ = 0.25$ W; $P^- \simeq -8.621$ mW; $P_{R_s} \simeq 0.3448$ W;
$P_L \simeq 0.2414$ W; $P_{\text{source}} \simeq 0.5862$ W; 3.448%. (j) $\Gamma_L = 0.6705e^{-j106.7°}$;
$S \simeq 5.0694$; $z_{\min}$ (m) : $-0.6108, -3.6108, -6.6108, -9.6108$;
$z_{\max}$ (m) : $-2.1108, -5.1108, -8.1108$;
$z_{\min}$ (m) : $-2.1108, -5.1108, -8.1108$;
$z_{\max}$ (m) : $-0.6108, -3.6108, -6.6108, -9.6108$; $Z_{\text{in}}^{\max} \simeq 253.5\Omega$;
$Z_{\text{in}}^{\min} \simeq 9.863\Omega$; $Z_{\text{in}} \simeq 25.86 + j60.34\Omega$; $V_s \simeq 6.773e^{j28.3°}$ V, $V^+ \simeq 5e^{j90°}$ V,
$V^- \simeq 3.352e^{-j163.3°}$ V, $V_L \simeq 5.158e^{j51.5°}$ V; $V_{\max} \simeq 8.352$ V,

$V_{min} \simeq 1.648$ V; $I_{max} \simeq 0.167$ A, $I_{min} \simeq 0.03295$ A; $P^+ = 0.25$ W,
$P^- \simeq -0.1122$ W, $P_{R_s} \simeq 0.2661$ W, $P_L \simeq 0.1376$ W, $P_{source} \simeq 0.4037$ W,
44.95%.

3.27. (a) $P_L = 0.5$ W. (b) $P_L \simeq 0.222$ W.

3.29. (a) $S_1 \simeq 6.1713$; $S_2 = \infty$. (b) $P_{Z_1} = 30$ mW; $P_{Z_2} = 0$.

3.31. $P_L = 0.25$ W.

3.33. $P_{source} \simeq 111$ mW; $P_2 \simeq 24.73$ mW; $P_3 \simeq 24.73$ mW; $P_4 \simeq 12.37$ mW.

3.35. (a) $l_{min} \simeq 0.0669\lambda$. (b) Capacitor; $C \simeq 5.69$ pF. (c) $l_{min} \simeq 5.49$ cm; Capacitor; $C \simeq 4.45$ nF.

3.37. (a) $l_s = 0.375\lambda$. (b) $S = \infty$. (c) $S \simeq 2.36$.

3.39. (a) $Z_Q \simeq 83.7\Omega$. (b) $S \simeq 1.374$. (c) $S \simeq 1.30$.

3.41. (a) $\Gamma_L \simeq 0.351$; $S \simeq 2.08$. (b) Use two cascaded quarter-wave coaxial transformers: 75Ω, 52Ω.

3.43. 23.2 pF $\leq C \leq$ 36.4 pF.

3.45. (a) $C \simeq 0.107$ pF, $L \simeq 4.27$ nH. (b) $l_{s1} \simeq 1.66$ cm, $l_{s2} \simeq 1.16$ cm.

3.47. $l/\lambda \simeq 0.3087$; $Z_Q \simeq 89.64\Omega$.

3.49. $Z_L \simeq (85 - j72.1)\Omega$.

3.51. $R \simeq 241$ Ω-(cm)$^{-1}$, $L \simeq 9.10$ nH-(cm)$^{-1}$, $C \simeq 0.869$ pF-(cm)$^{-1}$, and $G \simeq 7.3 \times 10^{-5}$ S-(cm)$^{-1}$.

Chapter 4

4.1. $\sim 10.5\mu$C.

4.3. $\sim 29.9°$.

4.5. $\sqrt{3}kQ^2/a^2$.

4.7. $\mathbf{E}_1 \simeq (kQ/a^2)(\hat{\mathbf{y}}\,0.032 + \hat{\mathbf{z}}\,0.0871)$, $\mathbf{E}_2 \simeq (kQ/a^2)(\hat{\mathbf{y}}\,0.032 - \hat{\mathbf{z}}\,0.0871)$.

4.9. (a) $(0,\sim 2,0)$. (b) $(0,\sim 0,0)$. (c) Yes: $(0,\sim -1,0)$.

4.11. (a) $\Phi \simeq 22.5$ kV. (b) $W \simeq -5.62 \times 10^{-4}$ J.

4.13. $\Phi \simeq 17.1$ kV.

4.15. (a) $Q_2 = -2Q_1/3$. (b) $Q_2 = 8Q_1/(5\pi)$.

4.17. (a) $\Phi_P = Q/(4\pi\epsilon_0)[(a+z)/(az) - \sqrt{a^2+z^2}/(az)]$.
   (b) $E(z) = -[1/(az) - (a+z)/(az^2) - 1/(a\sqrt{a^2+z^2}) + \sqrt{a^2+z^2}/(az^2)]$.
   (c) $\Phi_P = 3Q/(4\pi\epsilon a^3)[z^2/3 + a^2(2a+3z)/(6z) - (a^2+z^2)^{3/2}/(3z)]$,
   $E(z) = 3Q/(4\pi\epsilon_0 a^3)[-2z/3 - a^2/(2z) + a^2(2a+3z)/(6z^2) + \sqrt{a^2+z^2} - (a^2+z^2)^{3/2}/(3z^2)]$.

4.19. (a) $Q = K(2\pi)(2)[2/b^3 - e^{-ba}(a^2/b + 2a/b^2 + 2/b^3)]$.
   (b) $E_r = K/(\epsilon_0 r^2)[2/b^3 - e^{-br}(r^2/b + 2r/b^2 + 2/b^3)]$, $r \leq a$,
   $E_r = K/(\epsilon_0 r^2)[2/b^3 - e^{-ar}(a^2/b + 2a/b^2 + 2/b^3)]$, $r > a$.

(c) $\Phi(r) = K/(\epsilon_0 r)(2/b^3) + Ke^{-br}/(\epsilon_0 b^2) + 2K/(\epsilon_0 b) \int_\infty^r (e^{-br}/r)dr + 2K/(\epsilon_0 b^3) \int_\infty^r (e^{-br}/r^2)dr$, $r \le a$.
$\Phi(r) = K/(\epsilon_0 r)[2/b^3 - e^{-ba}(a^2/b + 2a/b^2 + 2/b^3)]$, $r > a$.

4.21. (a) $4\pi K(b-a)$.
    (b) $E_r(r) = 0, r < a; K(r-a)/(\epsilon_0 r^2), a \le r \le b; K(b-a)/(\epsilon_0 r^2), r > b$.
    (c) $\Phi(r) = K(b-a)/(\epsilon_0 b) - K/(\epsilon_0)\ln(a/b) + Ka/(\epsilon_0)(1/b - 1/a), r < a; K(b-a)/(\epsilon_0 b) - K/(\epsilon_0)\ln(r/b) + Ka/(\epsilon_0)(1/b - 1/r), a \le r \le b; K(b-a)/(\epsilon_0 r), r > b$.

4.23. (a) $Q = 8\pi\rho_0 a^3/15$. (b) $E_r = 2\rho_0 r/(15\epsilon_0), r \le a; E_r = 2\rho_0 a^3/(15\epsilon_0 r^2), r > a$. (c) $\Phi = [\rho_0/(5\epsilon_0)][a^2 - r^2/3], r \le a; \Phi = 2\rho_0 a^3/(15\epsilon_0 r), r \ge a$.

4.25. $\mathbf{E}_c = 0$ and $\Phi_c \simeq 90$ kV.

4.27. $A \simeq 1.13 \times 10^8$ m^2.

4.29. $b/a \simeq 4.90$.

4.31. $\Phi_{max} = 6.25$ kV.

4.33. (a) $C_{max} = 14.4$ pF. (b) $C_{max} = 451$ pF.

4.35. (a) $C = [(2\pi - 2\phi)\epsilon_0 + 2\phi\epsilon]/\ln(b/a)$. (b) $\phi \simeq 3.27°$. (c) $a = b/e$.

4.37. $C_u = 20\pi\epsilon_0 a/(b-a)$; $C_u < C_{u,air}, b > 36.6a$; $C_u > C_{u,air}, b < 36.6a$.

4.39. (a) $\mathbf{E}_1 \simeq 13.56\hat{\mathbf{x}} + 91\hat{\mathbf{y}}$ V-m^{-1}, $\mathbf{E}_2 \simeq -13.56\hat{\mathbf{x}} + 91\hat{\mathbf{y}}$ V-m^{-1}.
    (b) $\Phi_{12} \simeq -13.77$ V.

4.41. (a) $C_u = \pi/\{\epsilon_1^{-1}\ln[b(a-d)/(a(b-d))] + \epsilon_0^{-1}\ln[(d-b)/b]\}$.
    (b) $\Phi_{max} \simeq 324$ kV.

4.43. (a) $W_e = Q^2/(8\pi\epsilon_0)[1/a + 1/b - 1/(d-b) - 1/(d-a)]$.
    (b) $W_e = Q^2/(8\pi\epsilon_0)[1/a + 4/b + 2/(d-a) + 2/(d-b)]$.

4.45. $\Delta C \simeq 81$ fF.

Chapter 5

5.1. $\sim 6.24 \times 10^{18}$ electrons.

5.3.   (a) $E \simeq 26.2$ mV-m^{-1}. (b) $\Phi \simeq 26.2$ mV.

5.5.   (a) $\mu_e \simeq 42.8$ cm^2-V^{-1}s^{-1}. (b) $v_d \simeq 9.41 \times 10^{-5}$ m-s^{-1}.

5.7.   (a) $R \simeq 0.0185\Omega$-m^{-1}. (b) $R \simeq 0.022\Omega$-m^{-1}. (c) $R \simeq 0.0254\Omega$-m^{-1}.

5.9.   (a) $N_i \simeq 6.13 \times 10^9$ (cm)$^{-3}$, $\sigma \simeq 1.87 \times 10^{-4}$ S-m^{-1}.
    (b) $N_i \simeq 2.25 \times 10^{13}$ (cm)$^{-3}$, $\sigma \simeq 2.09$ S-m^{-1}.

5.11. $R \simeq 85.4\Omega$, $R_{pure\ Si} \simeq 5.36 \times 10^8\Omega$.

5.13. (c) $R = 1.5$ kΩ.

5.15. $R_{sq} \simeq 347\Omega$-sq^{-1}, $l \simeq 17.3$ μm.

5.17. ~ 142 μm.

5.19. (a) 4, $\sim$6.67, $\sim$8.33, 10, $\sim$11.9 fS-(mm)$^{-1}$. (b) 4, $\sim$5.29, $\sim$7.68, 10, $\sim$11.5 fS-(mm)$^{-1}$.

5.21. $R \simeq 12.7$ kΩ.

5.23. $R \simeq 11$ kΩ-km.

5.25. (a) $\sigma_1 = \sigma$, $\sigma_2 = 3\sigma$;
$\ln\{[(2b)/(a+b)]^{1/3}(a+b)/(2a)\} / \ln\{[(a+b)/(2a)]^{1/3}(2b)/(a+b)\}$.
(b) $w_1 = \sqrt{ab} - a$, $w_2 = b - \sqrt{ab}$.

5.29. (a) $Q_{inner} = (4\pi\epsilon_0 V_0 a^3)/(b^2 - ab)$, $Q_{outer} = (-4\pi\epsilon_0 V_0 b^3)/(b^2 - ab)e^{(b-a)/b}$,
$\rho(r) = [V_0\epsilon_0(3b+r)]/[(b^2 - ab)b]e^{(r-a)/b}$.
(b) $Q_{space} = (4\pi\epsilon_0 V_0)/(b^2 - ab)[b^3 e^{(b-a)/b} - a^3]$.

5.31. $R \simeq \ln(2h/a)/(2\pi\sigma)$.

Chapter 6

6.1. $I \simeq 1118$ A.

6.3. $F = 4000$ N-m^{-1}.

6.5. Expand.

6.7. (a) $\mathbf{F}_1 = -\mathbf{F}_3 = \hat{\mathbf{z}}15$ N, $\mathbf{F}_2 = \mathbf{F}_4 = 0$. (b) $\mathbf{F}_1 = -\mathbf{F}_3 = \hat{\mathbf{x}}15$ N,
$\mathbf{F}_2 = -\mathbf{F}_4 = \hat{\mathbf{y}}15$ N.

6.9. $\mathbf{B} = \hat{\mathbf{z}}\mu_0 I/(4\pi a)$. $\mathbf{B} = \hat{\mathbf{z}}\mu_0 I/(4\pi a)$. $\mathbf{B} = (\hat{\mathbf{x}} + \hat{\mathbf{y}})\mu_0 I/(4\pi a)$.

6.11. $\mathbf{B} = \hat{\mathbf{z}}\mu_0 I(4\pi)^{-1} \sin\phi_0 (1/b - 1/a)$.

6.13. $\mathbf{B}_P = -\hat{\mathbf{z}}\frac{2}{3}\left(2\sqrt{2} + \sqrt{10}\right)\mu_0 I(\pi a)^{-1}$.

6.15. $\mathbf{B}_P = -\mu_0(4\pi)^{-1}\left[\hat{\mathbf{x}} + \sqrt{2}\hat{\mathbf{y}} + (\sqrt{2} - 1)\hat{\mathbf{z}}\right]$ T.

6.17. (a) $\mathbf{B}_P = \hat{\mathbf{z}}\frac{1}{2}\mu_0 NIa^2\{(a^2 + z^2)^{-3/2} + [a^2 + (d-z)^2]^{-3/2}\}$.
(e) $B_z \simeq 0.8505 \times 10^{-6}NI/a$.

6.19. $B_z = \mu_0 NI(2a)^{-1}\{[1 + (z/a)^2]^{-3/2} - [1 + (1 - z/a)^2]^{-3/2}\}$.

6.21. $B_z(z) = 2\mu_0 NI(2a)^2\left\{\pi[4z^2 + (2a)^2]\sqrt{4z^2 + 5(2a)^2}\right\}^{-1} + \frac{1}{2}\mu_0 NIa^2[a^2 + (d-z)^2]^{-3/2}$.

6.23. $\mathbf{B} = \hat{\mathbf{z}}\left[(a^3 + 3az^2)/(2\sqrt{a^2 + z^2}) + 3z^2 \ln(z) - \frac{3}{2}z^2 \ln(a + \sqrt{a^2 + z^2})\right]$.

6.25. (a) $\sim$4 turns-(cm)$^{-1}$. (b) $\sim$1194 turns-(cm)$^{-1}$.

6.27. $\mathbf{B} \simeq \hat{\boldsymbol{\phi}}2 \times 10^{-7} - \hat{\mathbf{z}}7.85 \times 10^{-8}$ T.

6.29. (a) $\Psi = \mu_0 I \tan\phi_0(2\pi)^{-1}\left[b \ln(b/a) - (b - a)\right]$. (b) $\Psi \simeq 5.09 \times 10^{-6}$ Wb.

6.31. $\mathbf{A} = -\mu_0 Ir^2(4\pi a^2)^{-1}\hat{\mathbf{z}}$, $r \le a$; $-\mu_0 I(2\pi)^{-1}\ln(r)\hat{\mathbf{z}}$, $r > a$ (answer not unique).

6.33. (a) $A_z = [\mu_0 I/(4\pi)]\{\ln[(z-a) - \sqrt{r^2 + (z-a)^2}] - \ln[(z+a) - \sqrt{r^2 + (z+a)^2}]\}$.

6.35. $\mathbf{B}_1 = \mathbf{B}_2 = \hat{\mathbf{z}}\,\mu_0 J_0 d/2$.

6.37. (a) $L \simeq 41.8$ mH. (b) $L \simeq 38$ mH.

6.39. (a) $N \simeq 141$. (b) $N \simeq 100$.

6.43. (a) $L \simeq 207$ nH. (b) $L \simeq 676$ nH.

6.45. $L_{12} = \mu_0(d - \sqrt{d^2 - a^2})$.

6.47. (a) $L \simeq 0.329$ mH. (b) $W \simeq 16.4$ mJ.

6.49. (a) ~ 5 circular turns. (b) ~ 169 circular turns.

6.51. 10^4 N-m^{-1}. This huge repulsion force causes the container of the two wires to explode.

Chapter 7

7.1. (a) $\mathcal{V}_{\text{ind}} = B_0 A e^{-\alpha t}(\alpha t - 1)$. (b) $\mathcal{V}_{\text{ind}} = B_0 A e^{-\alpha t}[\alpha \sin(\omega t) - \omega \cos(\omega t)]$.

7.3. (a) $\mathcal{V}_{\text{ind2}} \simeq 7.72$ mV. (b) $\mathcal{V}_{\text{ind2}} \simeq 77.2$ mV.

7.5. $I_0 \simeq 6.96$ A.

7.7. $\mathcal{V}_{\text{ind}} \simeq 13\sin(377t)$ V.

7.9. (a) $\mathbf{a} = \hat{\mathbf{y}}1.5$ m-s^{-2}, $\mathbf{v}_{\text{final}} \simeq \hat{\mathbf{y}}13.3$ m-s^{-1}.

7.11. $\mathcal{V} = 5.0$ mV, $0 < x < 5$ cm; -5.0 mV, $10 < x < 15$ cm; 0 otherwise.

7.13. $I_{\text{ind}} = B_0 \pi a^2 \omega (2R)^{-1} \sin\omega t$.

7.15. $\mathcal{V}_{\text{ind}} = -0.12\cos(\omega t) + 0.36\sin^2(\omega t)\cos(\omega t)$ V.

7.17. (a) $\mathcal{V}_{\text{ind}} = \mu_0(2\pi)^{-1}\{b\omega \ln[(d+a)/d]\sin(\omega t) + a\omega \ln[(d+b)/d]\cos(\omega t)\}$.
(b) $\theta_{\text{max}} = \tan^{-1}[(a+d)/(b+d)]$;
$(\mathcal{V}_{\text{ind}})_{\text{max}} = v\mu_0 I a b (2\pi d)^{-1}\left[\sin\theta_{\text{max}}/(b+d) + \cos\theta_{\text{max}}/(a+d)\right]$.

7.19. (a) Pulse in Figure 7.35 shifted to the right by $d - l/2$, d, $d + l/2$, and $d + 3l$.
(b) $\mathcal{E}_z = E_0/3, E_0,$ and 0, respectively.

7.21. $\mathcal{E}_{Bx} = \{-E_0 z/l + E_0[1 + d/(2l)], d/2 < z \le (d+l)/2;$
$E_0 z/l - E_0 d/(2l), (d+l)/2 < z < d/2 + l;$ zero otherwise$\}$.

7.23. (a) $\mathbf{E}(z) = \hat{\mathbf{y}}\,e^{-jz}$ V-m^{-1}. (b) $\mathbf{H}(x) = 0.1[\hat{\mathbf{y}}\,e^{-j0.3x} + \hat{\mathbf{z}}0.5e^{j(0.3x - \pi/2)}]$
mA-m^{-1}. (c) $\mathbf{B}(y,z) = \hat{\mathbf{x}}\,40e^{j(0.8y - 0.6z - \pi/4)}$ μT. (d) $\mathbf{E}(x,y) = \hat{\mathbf{z}}\,E_0\sin(ax)e^{jby}$.

7.25. (a) $I_d \simeq -3.34\sin(2\pi \times 10^4 t)$ μA. (b) $I_d \simeq -0.334\sin(2\pi \times 10^6 t)$ mA.

7.27. (a) $\sim 8.88 \times 10^4$, ~ 888, ~ 8.88, and $\sim 8.88 \times 10^{-2}$, respectively.
(b) $f \simeq 888$ MHz.

7.29. $\mathcal{H} = \hat{\mathbf{y}}\,(\mu v_p)^{-1}[p_1(z - v_p t) - p_2(z + v_p t)]$.

7.31. $\beta \simeq 0.025$ rad-m^{-1}, $\eta \simeq 377\Omega$.

7.33. (a) $\beta \simeq 70\pi$ rad-m^{-1}. (b) $\mathcal{H} \simeq \hat{\mathbf{y}}0.252\cos(21 \times 10^9 \pi t)\cos(70\pi z)$ μA-m^{-1}.

7.35. (a) $\mathbf{H} = \hat{\boldsymbol{\phi}}\,E_0 e^{-j\beta z}/(r\sqrt{\mu/\epsilon})$. (b) $\mathcal{E} = \hat{\mathbf{r}}\,E_0\cos(\omega t - \beta z)/r$ and
$\mathcal{H} = \hat{\boldsymbol{\phi}}\,E_0\cos(\omega t - \beta z)/(r\sqrt{\mu/\epsilon})$.

Chapter 8

8.1.   (a) $\beta \simeq 20.9$ rad-m^{-1}, $\lambda \simeq 30$ cm. (b) $\mathscr{E}_x(0,t) = 3\cos(2\pi \times 10^9 t)$ V-m^{-1},
       $\mathscr{E}_x(\lambda/4,t) = 3\sin(2\pi \times 10^9 t)$ V-m^{-1}. (c) $\mathscr{E}_x(z,0) = 3\cos(20\pi z/3)$ V-m^{-1},
       $\mathscr{E}_x(z,\pi/\omega) \simeq -3\cos(20\pi z/3)$ V-m^{-1}.

8.3.   (a) $\overline{\mathscr{H}}(x,t) = \hat{\mathbf{z}}\,70\cos(6\pi \times 10^9 t - 20\pi x + \pi)$ mA-m^{-1},
       $\mathbf{H}(x) = \hat{\mathbf{z}}\,70e^{-j(20\pi x - \pi)}$ mA-m^{-1}. (b) $\mathbf{E}(x) \simeq -\hat{\mathbf{y}}\,26.4e^{-j(20\pi x - \pi)}$ V-m^{-1},
       $\overline{\mathscr{E}}(x,t) \simeq -\hat{\mathbf{y}}\,26.4\cos(6\pi \times 10^9 t - 20\pi x + \pi)$ V-m^{-1}.

8.5.   (a) 400 MHz and 0.75 m.
       (b) $\overline{\mathscr{H}}(z,t) \simeq -\hat{\mathbf{y}}0.199\cos(8\pi \times 10^8 t + 8\pi z/3)$ μA-m^{-1}.

8.7.   (a) $f \simeq 2.39 \times 10^9$ Hz, $\lambda \simeq 0.126$ m.
       (b) $\mathbf{H}(x) \simeq 2.65 \times 10^{-2}e^{j50x}\left[-j\hat{\mathbf{y}} - \hat{\mathbf{z}}\right]$ A-m^{-1}. (c) $|\mathbf{S}_{av}| \simeq 0.265$ W-m^{-2}.

8.9.   (a) $f = 1.8$ GHz, $\lambda \simeq 16.7$ cm. (b) $\theta \simeq 0.733$ rad. (c) $\overline{\mathscr{H}}(x,z,t) \simeq$
       $[\hat{\mathbf{x}}0.106 + \hat{\mathbf{z}}0.0796]\sin(3.6\pi \times 10^9 t - 7.2\pi x + 9.6\pi z + 0.733)$ mA-m^{-1}.
       (d) $|\mathbf{S}_{av}| \simeq 3.32$ μW-m^{-2}.

8.11. $\epsilon_r \simeq 80.9$, $\mu_r \simeq 1$.

8.13. (a) $\lambda = 20$ cm. (b) $\mathbf{H}(x) \simeq \hat{\mathbf{z}}j53.1\cos(100\pi x)$ μA-m^{-1}. (c) No: standing
       wave.

8.15. $d \simeq 10.4$ cm for both expressions.

8.17. (a) $\alpha \simeq 0.0653$ np-m^{-1}. (b) $d \simeq 15.3$ m. (c) $\sim$$-0.00567$ dB and $\sim$$-0.567$ dB.

8.19. (a) $\sigma \simeq 0.431$ S-m^{-1}, $\epsilon_r \simeq 43.1$.
       (b) $\overline{\mathscr{E}}(z,t) \simeq \hat{\mathbf{y}}\,120e^{10z}\cos(2\pi \times 10^8 t + 17z + 30.5°)$ V-m^{-1}.

8.21. (b) $\sim \frac{579.4}{\sqrt{f\,(\mathrm{Hz})}}$ m.

8.23. (a) $f \simeq 9.89$ Hz. (b) $f \simeq 39.6$ Hz. (c) $\sim$116 m. (d) $\sim$110 m.

8.25. (a) $\sim$69.3 V-m^{-1}, $\sim$48.1 V-m^{-1}, $\sim$33.3 V-m^{-1}, $\sim$23.1 V-m^{-1}, and $\sim$16.0
       V-m^{-1}. (b) $\sim$42.3 W-m^{-2}, $\sim$20.4 W-m^{-2}, $\sim$9.79 W-m^{-2}, $\sim$4.71 W-m^{-2}, and
       $\sim$2.26 W-m^{-2}. (c) $\sim$4.57 mW.

8.27. (a) $\sim$55%, 15.2%, and 0.258%. (b) The 0.1 Hz signal.

8.29. (a) $\sigma \simeq 4$ S-m^{-1}, $\epsilon_r \simeq 81$ (sea water).
       (b) $\overline{\mathscr{E}}(z,t) \simeq \hat{\mathbf{x}}\,10e^{-0.581z}\cos(2\pi(21.4 \times 10^3)t - 0.581z)$ V-m^{-1},
       $\overline{\mathscr{H}}(z,t) \simeq \hat{\mathbf{y}}\,48.7e^{-0.581z}\cos(2\pi(21.4 \times 10^3)t - 0.581z - 45°)$ A-m^{-1}.
       (c) $\overline{\mathscr{E}}(z,t) \simeq \hat{\mathbf{x}}\,10e^{-5.81z}\cos(2\pi(2.14 \times 10^6)t - 5.81z)$ V-m^{-1},
       $\overline{\mathscr{H}}(z,t) \simeq \hat{\mathbf{y}}\,4.87e^{-5.81z}\cos(2\pi(2.14 \times 10^6)t - 5.81z - 45°)$ A-m^{-1}.

8.31. (a) $\sigma \simeq 1.52 \times 10^{-3}$ np-m^{-1}, $\epsilon \simeq 3.62 \times 10^{-11}$ F-m^{-1}, $d \simeq 7.07$ m.
       (b) $\overline{\mathscr{H}} \simeq \frac{1}{263}(\hat{\mathbf{x}} - \hat{\mathbf{y}})\,e^{-0.1(x+y)}\cos(2\pi 10^9 t - 30x - 30y)$ mA-m^{-1}

8.33. For cooked beef, $\tan\delta_c \simeq 0.331$ and $d \simeq 2.14$ cm. For smoked bacon,
       $\tan\delta_c = 0.05$ and $d \simeq 40.2$ cm. Therefore, cooked beef has a much higher
       microwave absorption rate.

8.35. $\sigma \simeq 3$ S-m^{-1}, $\epsilon_r' \simeq 77.4$.

8.37. (a) $\overline{\mathcal{H}}_\phi = (E_0/r)e^{-r^2/\omega^2}\cos(\omega t - \beta z)$. (c) $P_{\text{total}} = E_0^2\pi\omega^2/(4\eta)$.
$|\mathbf{S}_{\text{av}}(r=0)| \simeq 20$ kW-m^{-2}. (d) $\sim 0.265 R_{\text{SE}}$ where $R_{\text{SE}} \simeq 1.5 \times 10^8$ km.
(e) $E_0 \simeq 625$ GV-m$^{-1} \gg 3$ MV-m^{-1}, yes, $\sim 1.73 \times 10^{12}$ N-m^{-2}, $m \simeq 34.7$ kg.

8.39. (a) $\mathcal{V}_{\text{in}} \simeq 61.8\sin(\omega t)$ mV. (b) Same as (a).

8.41. (a) All safe. (b) $P = 85$ W.

8.43. $t_{\text{max}} \simeq 3.7$ hrs.

8.45. (a) $H_0 \simeq 0.23$ mA-m^{-1}, $a \simeq 2.96$ rad-m^{-1}.
(b) $\overline{\mathcal{E}}(x,y,t) \simeq 61.3[-\hat{\mathbf{x}} + \hat{\mathbf{y}}]\sin(\omega t - 2.96x - 2.96y + \pi/3)$ mV-m^{-1};
linearly polarized at $-45°$. (c) (i) $\mathcal{E}_{\text{max}} \simeq 61.3$ mV-m^{-1}, (ii)
$\mathcal{E}_{\text{max}} \simeq 61.3$ mV-m^{-1}, (iii) $\mathcal{E}_{\text{max}} = 0$.

8.47. (a) $\lambda = 20$ cm, $f \simeq 1.5$ GHz. (b) $|\mathbf{S}_{\text{av}}| \simeq 1.51$ mW-m^{-2}. (c) LHCP. (d) Linearly polarized.

8.49. (a) Linearly polarized. (b) RHEP.

8.51. (a) $\mathbf{H}(z) \simeq [(1-j)\hat{\mathbf{y}} - \hat{\mathbf{x}}]\left(\frac{10}{377}\right)e^{-j80\pi z}$ A-m^{-1}. (b) $|\mathbf{S}_{\text{av}}| \simeq 0.398$ W-m^{-2}.
(c) LHEP.

8.53. (a) Straight line (along the x axis). (b) Circle (on the x-y plane). (c) Ellipse (on the x-y plane).

Chapter 9

9.1. (a) $\beta_1 \simeq 147$ rad-m^{-1}. (b) $\overline{\mathcal{H}}_i(x,t) \simeq \hat{\mathbf{z}}\,39.8\cos(44 \times 10^9 t - 147x)$ mA-m^{-1}.
(c) $\overline{\mathcal{E}}_r(x,t) \simeq -\hat{\mathbf{y}}\,15\cos(44 \times 10^9 + 147x)$ V-m^{-1},
$\overline{\mathcal{H}}_r(x,t) \simeq \hat{\mathbf{z}}\,39.8\cos(44 \times 10^9 + 147x)$ mA-m^{-1}. (d) ~ -2.14 cm,
~ -4.28 cm.

9.3. (a) $f \simeq 9$ GHz, $\lambda \simeq 3.33$ cm. (b) $\mathbf{H}_i(z) \simeq [\hat{\mathbf{x}}j0.106 + \hat{\mathbf{y}}0.106]e^{-j60\pi z}$ A-m^{-1}.
(c) $\mathbf{E}_r(z) = 40e^{j60\pi z}\left[-\hat{\mathbf{x}} + j\hat{\mathbf{y}}\right]$ V-m^{-1}, $\mathbf{H}_r \simeq 0.106e^{j60\pi z}[\hat{\mathbf{y}} + j\hat{\mathbf{x}}]$ A-m^{-1}.
(d) $\mathbf{E}_1(x) = (-j\hat{\mathbf{x}} - \hat{\mathbf{y}})80\sin(60\pi z)$ V-m^{-1}, so
$|E_{1x}(z)| = |E_{1y}(z)| = 80|\sin(60\pi z)|$.

9.5. (a) $\mathbf{E}_i(x) \simeq -\hat{\mathbf{y}}\,2.75e^{-j20\pi x}$ V-m^{-1}, $\mathbf{H}_i(x) \simeq -\hat{\mathbf{z}}\,7.28 \times 10^{-3}e^{-j20\pi x}$ A-m^{-1},
$\mathbf{E}_r(x) \simeq \hat{\mathbf{y}}\,2.75e^{j20\pi x}$ V-m^{-1}, $\mathbf{H}_r(x) \simeq -\hat{\mathbf{z}}\,7.28 \times 10^{-3}e^{j20\pi x}$ A-m^{-1}.
(b) ~ -2.5 cm, ~ -7.5 cm.

9.7. (a) $\mathbf{H}_r(y) \simeq \hat{\mathbf{z}}\,11.3e^{j100\pi y}$ mA-m^{-1}, $\mathbf{H}_t(y) \simeq \hat{\mathbf{z}}\,31.3e^{-j361\pi y}$ mA-m^{-1}.
(b) $|(\mathbf{S}_{\text{av}})_i| \simeq 75.4$ mW-m^{-2}, $|(\mathbf{S}_{\text{av}})_r| \simeq 24.1$ mW-m^{-2}, and $|(\mathbf{S}_{\text{av}})_t| \simeq 51.3$ mW-m^{-2}. (c) $\mathbf{H}_1(y) \simeq \hat{\mathbf{z}}\,[20e^{-j100\pi y} + 11.3e^{j100\pi y}]$ mA-m^{-1}.

9.9. $d_{\text{max}} \simeq 10.5$ m.

9.11. (a) $\sim 66.4\%$. (b) $\sim 77.1\%$. (c) $\sim 81.8\%$. (d) $\sim 83.2\%$. The muscle tissue is more reflective than the fat tissue.

9.13. (a) $|(\mathbf{S}_{\text{av}})_r| \simeq 999.912$ W-m^{-2}. (b) $|(\mathbf{S}_{\text{av}})_t| \simeq 0.0876$ W-m^{-2}.
(c) $|(\mathbf{S}_{\text{av}})_{\text{emerging}}| \simeq 6.2$ μW-m^{-2}.

9.15. (a) $d_{min} \simeq 4.66$ cm. (b) ~65.43% at 1 GHz and ~65.43% at 2 GHz.

9.17. (a) $f_1 \simeq 2.04$ GHz, $f_2 \simeq 4.06$ GHz, $f_3 \simeq 5.30$ GHz. (b) $f \simeq 5.30$ GHz.
(c) Bandwidth $\simeq 10.13$ GHz. (d) Γ_{eff} rises faster on the high frequency end of the visible light spectrum (i.e., reflections increase as λ approaches 400 nm) so that higher frequencies (blue, violet) are reflected more effectively.

9.19. (a) $n \simeq 1.51$ (benzene). (b) $\Gamma_{eff} \simeq 0$. (c) $\Gamma_{eff} \simeq 0.224e^{-j136.5°}$.

9.21. ~8.63%.

9.23. (a) $\alpha_2 \simeq 2513$ np-m^{-1}, $\beta_2 \simeq 1885$ rad-m^{-1}, any d. (b) ~3.5×10^{-6}.

9.25. $\Gamma_{eff} \simeq 0.2055e^{j40.1°}$, $\simeq 0.0954e^{32.3°}$, $\simeq 0.00321e^{-j165°}$, $\epsilon_r = 1.8$ most transparent.

9.27. (a) $\Gamma_{eff, \text{ with snow}} \simeq 0.632e^{-j129°}$, $\Gamma_{eff, \text{ without snow}} \simeq 0.57e^{j72.3°}$.
(b) $\Gamma_{eff, \text{ with snow}} \simeq 0.705e^{-j125°}$, $\Gamma_{eff, \text{ without snow}} \simeq 0.707e^{-j129°}$.

9.29. (a) ~36.2%. (b) ~17%. (c) ~15%.

9.31. (a) ~34%, ~9.8%; ~16.2%, ~2.1%; ~21.3%, ~0.34%. (b) ~8.83%, ~9.47%; ~14.8%, ~1.1%; ~17.1%, ~0.13%.

9.33. (a) $E_0 \simeq 150.4$ V-m^{-1}, $f \simeq 1.2$ GHz, $\theta_i \simeq 53.1°$.
(b) $\mathbf{E}_r \simeq -\hat{\mathbf{z}}150.4e^{-j6.4\pi x}e^{+j4.8\pi y}$ V-m^{-1}.
(c) $\mathbf{E}_1 \simeq -\hat{\mathbf{z}}300.8je^{-j6.4\pi x}\sin(4.8\pi y)$ V-m^{-1}. $y_{min1} \simeq -20.8$ cm, $y_{min2} \simeq -41.7$ cm, etc.; $y_{max1} \simeq -10.4$ cm, $y_{max2} \simeq -31.3$ cm, etc..

9.35. (a) $\mathbf{E}_{r\perp} = -\hat{\mathbf{y}}E_0e^{-jk(x+z)/\sqrt{2}}$, $\mathbf{E}_{r\parallel} = (\hat{\mathbf{x}} - \hat{\mathbf{z}})(jE_0/\sqrt{2})e^{-jk(x+z)/\sqrt{2}}$. (b) $\mathbf{E}_i$ is RHCP, $\mathbf{E}_r$ is LHCP.

9.37. (a) $\epsilon_{1r} \simeq 6$, $\theta_i = 45°$. (b) $\overline{\mathcal{H}}_i(x,z,t) \simeq \frac{E_0\sqrt{6}}{377}\cos[12 \times 10^9 t - 40\sqrt{3}(x+z)]$.
(c) 0.

9.39. (a) $\theta_i = 60°$, $f \simeq 600$ MHz. (b) RHCP. (c) $\mathbf{E}_r(x,y) = j\frac{E_0}{2}\hat{\mathbf{z}}e^{-j2\pi\left(\sqrt{3}x-y\right)}$.
(d) Linear.

9.41. $\Gamma_\perp \simeq -0.976$, $\mathcal{T}_\perp \simeq 0.0244$.

9.43. $h \simeq 1.018$ km.

9.45. (a) SIR$_\perp \simeq 2810$, SIR$_\parallel \simeq 290000$. (b) $h'_{2,\perp} = 20$ m, SIR$_\perp \simeq 2810$; $h'_{2,\parallel} \simeq 9.45$ m, SIR$_\parallel = \infty$.

9.47. (a) $\epsilon_{1r} \simeq 9$, $\theta_i = 30°$. (b) $[\epsilon_{2r}]_{max} = 2.25$. (c) Yes; $\theta_{iB} \simeq 26.57°$.

9.49. Yes; $\sim 77.4°$.

9.51. (a) $\theta_4 \simeq 41.12°$, $\theta_6 \simeq 30°$. (b) $\theta_{1c} \simeq 5.6°$. (c) Critical angle changes.
(d) $\theta_4 \simeq 64.45°$, $\theta_6 \simeq 30°$, $\theta_{1c} \simeq 22.9°$.

9.55. (a) Light exits at B and D only. (b) $\theta_B \simeq 58.364°$, $\theta_D \simeq 31.635°$.
(c) B: unpolarized; D: perpendicularly polarized; (d) Polarized beam exiting D is in the same direction as the incident beam.

9.57. $[\Gamma_\perp]_{lm} \simeq 0.526e^{j179.7°}$ ($\theta_i = 30°$), $0.6e^{j179.7°}$ ($\theta_i = 45°$), $0.704e^{j179.8°}$ ($\theta_i = 60°$),

$\left[\Gamma_\parallel\right]_{lm} \simeq 0.4e^{j179.4°}$ $(\theta_i = 30°)$, $0.305e^{j179.2°}$ $(\theta_i = 45°)$, $0.133e^{j177.9°}$
$(\theta_i = 60°)$.
$\left[\Gamma_\perp\right]_{br} \simeq 0.428e^{j179.7°}$ $(\theta_i = 30°)$, $0.517e^{j179.7°}$ $(\theta_i = 45°)$, $0.64e^{j179.9°}$
$(\theta_i = 60°)$,
$\left[\Gamma_\parallel\right]_{br} \simeq 0.279e^{j179.5°}$ $(\theta_i = 30°)$, $0.171e^{j179.1°}$ $(\theta_i = 45°)$, $0.013e^{j12.67°}$
$(\theta_i = 60°)$.

9.59. $\mathbf{E}_t(x, z) = E_0 \left(0.578\,\hat{\mathbf{x}} - 0.816\,\hat{\mathbf{z}}\right) e^{-j\omega\sqrt{\epsilon_0\mu_0}\,(0.816x + 0.578z)}$.

Chapter 10

10.1. (a) $\overline{\gamma}_0 \simeq j100\pi/3$ rad-m^{-1}, $\overline{\gamma}_1 \simeq 117$ np-m^{-1}, $\overline{\gamma}_2 \simeq 296$ np-m^{-1}.
(b) $\overline{\gamma}_0 \simeq j200\pi/3$ rad-m^{-1}, $\overline{\gamma}_1 \simeq j44.1\pi$ rad-m^{-1}, $\overline{\gamma}_2 \simeq 234$ np-m^{-1}.
(c) $\overline{\gamma}_0 \simeq j400\pi/3$ rad-m^{-1}, $\overline{\gamma}_1 \simeq j124\pi$ rad-m^{-1}, $\overline{\gamma}_2 \simeq j277$ rad-m^{-1}.

10.3. (a) TM$_0$, TE$_1$, TM$_1$. (b) TM$_0$, TE$_1$, TM$_1$, TE$_2$, TM$_2$.

10.5. (a) TEM. (b) TEM. (c) TEM, TE$_1$, TM$_1$.

10.7. $l_{min} \simeq 3.71$ cm.

10.9. (a) $\lambda_{c_2} = 3$ cm, $\lambda_{c_3} = 2$ cm, $\lambda_{c_4} = 1.5$ cm. (b) $\overline{\lambda}_0 \simeq 3.75$ cm, $\overline{\lambda}_1 \simeq 4.8$ cm.

10.11. (a) $f_{c_0} = 0$, $f_{c_1} \simeq 3$ GHz, $f_{c_2} \simeq 6$ GHz. (b) $\overline{v}_{p_0} = c$, $\overline{v}_{p_1} \simeq 1.08c$, $\overline{v}_{p_2} \simeq 1.51c$,
$\overline{\lambda}_0 \simeq 3.75$ cm, $\overline{\lambda}_1 \simeq 4.05$ cm, $\overline{\lambda}_2 \simeq 5.67$ cm, $Z_{TM_0} \simeq 377\Omega$, $Z_{TM_1} \simeq 349.5\Omega$,
$Z_{TM_2} \simeq 249\Omega$. (c) TM$_6$.

10.13. $\theta_{i_1} \simeq 79.19°$, $\theta_{i_2} \simeq 67.98°$, $\theta_{i_3} \simeq 55.77°$, $\theta_{i_4} \simeq 41.41°$.

10.15. (a) Propagating TM$_4$ mode. (b) $f \simeq 40$ GHz.
(c) $E_x \simeq j226C_1 \sin(160\pi y)e^{-j640\pi x/3}$, $E_y \simeq 301.3C_1 \cos(160\pi y)e^{-j640\pi x/3}$.
(d) TM$_7$.

10.17. $P_{av} \simeq 167$ kW-(cm)$^{-1}$.

10.19. $P_{av}^{TEM} \simeq 2.98$ GW-m^{-2}, $P_{av}^{TE_1} \simeq 1.12$ GW-m^{-2}, $P_{av}^{TM_1} \simeq 1.99$ GW-m^{-2}.

10.21. (b) $f = \sqrt{3}f_{cm}$. (c) $\alpha_{c_{min}} = [2(3^{0.75})/(a\eta)]\sqrt{\pi\mu_0 f_{cm}/(2\sigma)}$.
(d) $\alpha_{c_{TEM}} \simeq 2.82 \times 10^{-3}$ np-m^{-1}, $\alpha_{c_{TE_1}} \simeq 2.3 \times 10^{-3}$ np-m^{-1},
$\alpha_{c_{TM_1}} \simeq 6.91 \times 10^{-3}$ np-m^{-1}.

10.23. $E_x = \left[j\overline{\beta}K\left(x - x^2/a\right) + \overline{\beta}C_4/(\omega\epsilon)\right]e^{-j\overline{\beta}z}$,
$H_y = j(\omega\mu)^{-1}\left[\overline{\beta}^2 K\left(x - x^2/a\right) - j\overline{\beta}^2 C_4/(\omega\epsilon) + 2K/a\right]e^{-j\overline{\beta}z}$, $S_{av}(x = \frac{a}{4}) = \frac{1}{2}\left\{\overline{\beta}^3 C_4^2(\omega^3\epsilon^2\mu)^{-1} - \overline{\beta}K^2(\omega\mu)^{-1}(3a/16)\left[\overline{\beta}(3a/16) + 2/a\right]\right\}\hat{\mathbf{z}} - \frac{1}{2}\left[K\overline{\beta}^2 C_4(2\omega^2\mu\epsilon)^{-1}\right]\hat{\mathbf{x}}$.

10.25. (a) $H_{r0} = H_{i0}(1 - x)/(1 + x)$, $H_{t0} = 2H_{i0}/(1 + x)$, where $x = \epsilon_0\overline{\beta}_2/(\epsilon_2\overline{\beta})$.
(b) Yes: $\epsilon_2 = \epsilon_0$ and $\epsilon_2 = (\omega^2\mu_0 a^2/\pi^2 - 1/\epsilon_0)^{-1}$.

10.27. (a) $d_{max} \simeq 1.28$ μm. (b) $d_{max} \simeq 2.014$ μm.

10.29. $f_{c_2} \simeq 3.75$ GHz.

10.31. (a) Odd TE_1 and TM_1, Even TE_2 and TM_2, Odd TE_3 and TM_3, Even TE_4 and TM_4, Odd TE_5 and TM_5, Even TE_6 and TM_6, Odd TE_7 and TM_7, and Even TE_8 and TM_8. (b) ~ 7.681 μm.

10.33. $d \simeq 11.8$ μm.

10.35. (a) $f_{c_1} = 0$ (Odd TM_1), $f_{c_2} \simeq 667$ MHz (Even TE_2), $f_{c_3} \simeq 1.33$ GHz (Odd TM_3), and $f_{c_4} = 2$ GHz (Even TE_4). (b) Even TE_2. (c) $\bar{\beta} \simeq 29.01$ rad-m^{-1}.

10.37. (a) $v_p \simeq 223.6$ km-s^{-1}, $v_g \simeq 447.2$ km-s^{-1}. (b) $z_1 \simeq 224$ m, $t_1 = 500$ μs.

10.39. $v_g \simeq 10^3 \sqrt{10f} = 2v_p$.

10.41. $v_g = (\bar{\beta}/\omega) \left[1 + (\alpha_x/\beta_x)(d\alpha/d\beta_x) \right] / \left[(\alpha_x/\beta_x)(d\alpha_x/d\beta_x)(\mu_d \epsilon_d) + \mu_0 \epsilon_0 \right]$, where $(d\alpha_x/d\beta_x) =$
$(\beta_x/\alpha_x)(\epsilon_0/\epsilon_d)^2 \tan(\beta_x d/2)[\tan(\beta_x d/2) + (\beta_x d/2) \sec^2(\beta_x d/2)]$.

Chapter 11

11.1. $\Gamma \simeq 1.34 \times 10^{-4}$ ($f = 1.228$ GHz), $\Gamma \simeq 8.16 \times 10^{-5}$ ($f = 1.575$ GHz).

11.3. (a) $\theta_i \simeq 70.2°$. (b) ~ 4401 km. (c) $f_{\max} \simeq 14.8$ MHz.

11.7. (a) (iii) 4 GHz $< f <\sim 7.43$ GHz: LH propagation; ~ 8.47 GHz $< f < 11$ GHz: RH propagation. $\Re e\{n(f \simeq 5.89$ GHz)$\} = -1$. (iv) Phase velocity negative over 4 GHz $< f <\sim 7.43$ GHz, positive over ~ 8.47 GHz $< f < 11$ GHz. Group velocity always nonnegative. (b) (iii) 4 GHz $< f <\sim 7.9$ GHz: LH propagation; ~ 7.9 GHz $< f < 11$ GHz: RH propagation. $\Re e\{n(f \simeq 5.87$ GHz)$\} = -1$. (iv) Phase velocity negative over 4 GHz $< f <\sim 7.9$ GHz, positive over ~ 7.9 GHz$< f < 11$ GHz. Group velocity always positive.

Index

Fundamental Physical Constants

Quantity	Symbol	Value	Comments
Speed of light	c	2.99792458×10^8 m-s^{-1}	
Charge of an electron	q_e	$-1.602177 \times 10^{-19}$ C	
Gravitational constant	G	6.67259×10^{-11} N-m^2-kg^{-2}	
Gravitational acceleration	g	9.78049 m-s^{-2}	Sea level, at equator
Planck's constant	h	$6.6260755 \times 10^{-34}$ J-s	
Boltzmann's constant	k_B	1.38066×10^{-23} J-(K)$^{-1}$	
Avogadro's number	N_A	6.022×10^{23} (mole)$^{-1}$	Molecules per mole
Permittivity of free space	ϵ_0	8.854×10^{-12} F-m^{-1}	$\sim(36\pi)^{-1} \times 10^{-9}$ F-m^{-1}
Permeability of free space	μ_0	$4\pi \times 10^{-7}$ H-m^{-1}	
Rest mass of an electron	m_e	9.10939×10^{-31} kg	
Rest mass of a proton	m_p	1.67262×10^{-27} kg	$(m_p/m_e) \simeq 1836$
Rest mass of a neutron	m_n	1.67493×10^{-27} kg	
Bohr radius	a	0.529177×10^{-10} m	Radius of hydrogen atom
1 electron-volt	eV	1.60217×10^{-19} J	
Frequency of 1-eV photon	f	2.41796×10^{14} Hz	

Powers of Ten

Power	Symbol	Prefix
10^{18}	E	Exa
10^{15}	P	Peta
10^{12}	T	Tera
10^{9}	G	Giga
10^{6}	M	Mega
10^{3}	k	kilo
10^{-3}	m	milli
10^{-6}	μ	micro
10^{-9}	n	nano
10^{-12}	p	pico
10^{-15}	f	femto
10^{-18}	a	atto

Gradient, Divergence, and Curl Operators

Rectangular coordinates:

$$\nabla\Phi = \hat{\mathbf{x}}\frac{\partial\Phi}{\partial x} + \hat{\mathbf{y}}\frac{\partial\Phi}{\partial y} + \hat{\mathbf{z}}\frac{\partial\Phi}{\partial z}$$

$$\text{div }\mathbf{A} = \nabla\cdot\mathbf{A} = \frac{\partial A_x}{\partial x} + \frac{\partial A_y}{\partial y} + \frac{\partial A_z}{\partial z}$$

$$\text{curl }\mathbf{A} = \nabla\times\mathbf{A} = \begin{vmatrix} \hat{\mathbf{x}} & \hat{\mathbf{y}} & \hat{\mathbf{z}} \\ \frac{\partial}{\partial x} & \frac{\partial}{\partial y} & \frac{\partial}{\partial z} \\ A_x & A_y & A_z \end{vmatrix}$$

$$= \hat{\mathbf{x}}\left(\frac{\partial A_z}{\partial y} - \frac{\partial A_y}{\partial z}\right) + \hat{\mathbf{y}}\left(\frac{\partial A_x}{\partial z} - \frac{\partial A_z}{\partial x}\right) + \hat{\mathbf{z}}\left(\frac{\partial A_y}{\partial x} - \frac{\partial A_x}{\partial y}\right)$$

Cylindrical coordinates:

$$\nabla\Phi = \left[\hat{\mathbf{r}}\frac{\partial\Phi}{\partial r} + \hat{\boldsymbol{\phi}}\frac{1}{r}\frac{\partial\Phi}{\partial\phi} + \hat{\mathbf{z}}\frac{\partial\Phi}{\partial z}\right]$$

$$\text{div }\mathbf{A} = \nabla\cdot\mathbf{A} = \frac{1}{r}\frac{\partial}{\partial r}(rA_r) + \frac{1}{r}\frac{\partial A_\phi}{\partial\phi} + \frac{\partial A_z}{\partial z}$$

$$\text{curl }\mathbf{A} = \nabla\times\mathbf{A} = \hat{\mathbf{r}}\left[\frac{1}{r}\frac{\partial A_z}{\partial\phi} - \frac{\partial A_\phi}{\partial z}\right] + \hat{\boldsymbol{\phi}}\left[\frac{\partial A_r}{\partial z} - \frac{\partial A_z}{\partial r}\right] + \hat{\mathbf{z}}\frac{1}{r}\left[\frac{\partial}{\partial r}(rA_\phi) - \frac{\partial A_r}{\partial\phi}\right]$$

Spherical coordinates:

$$\nabla\Phi = \left[\hat{\mathbf{r}}\frac{\partial\Phi}{\partial r} + \hat{\boldsymbol{\theta}}\frac{1}{r}\frac{\partial\Phi}{\partial\theta} + \hat{\boldsymbol{\phi}}\frac{1}{r\sin\theta}\frac{\partial\Phi}{\partial\phi}\right]$$

$$\text{div }\mathbf{A} = \nabla\cdot\mathbf{A} = \frac{1}{r^2}\frac{\partial}{\partial r}(r^2 A_r) + \frac{1}{r\sin\theta}\frac{\partial}{\partial\theta}(\sin\theta A_\theta) + \frac{1}{r\sin\theta}\frac{\partial A_\phi}{\partial\phi}$$

$$\text{curl }\mathbf{A} = \nabla\times\mathbf{A} = \hat{\mathbf{r}}\frac{1}{r\sin\theta}\left[\frac{\partial}{\partial\theta}(A_\phi\sin\theta) - \frac{\partial A_\theta}{\partial\phi}\right] + \hat{\boldsymbol{\theta}}\frac{1}{r}\left[\frac{1}{\sin\theta}\frac{\partial A_r}{\partial\phi} - \frac{\partial}{\partial r}(rA_\phi)\right]$$

$$+ \hat{\boldsymbol{\theta}}\frac{1}{r}\left[\frac{\partial}{\partial r}(rA_\theta) - \frac{\partial A_r}{\partial\theta}\right]$$